GOD CREATED
THE INTEGERS

GOD CREATED
THE INTEGERS

The Mathematical Breakthroughs that Changed History

Edited, with commentary, by

STEPHEN
HAWKING

RUNNING PRESS

9   8   7   6   5   4   3
Digit on the right indicates the number of this printing

Library of Congress Control Number: 2005924493
ISBN-13:   978-0-7624-1922-7
ISBN-10:   0-7624-1922-9

Cover design by Bill Jones
Interior design by Bill Jones and Techbooks
Typography: Adobe Garamond
Editor: Diana von Glahn

This book may be ordered by mail from the publisher.
Please include $2.50 for postage and handling.
But try your bookstore first!

Running Press Book Publishers
2300 Chestnut Street Suite 200
Philadelphia, PA 19103-4371

Visit us on the web!
www.runningpress.com

A NOTE ON THE TEXTS

The texts in this book are based on translations of the original printed editions. We have made no attempt to modernize the authors' own distinct usage, spelling or punctuation, or to make the texts consistent with each other in this regard.

The Editors

TEXT AND PICTURE CREDITS

Text Permissions:

Selections from Euclid's *Elements* courtesy of Dover Publications.

Selections from *The Works of Archimedes* courtesy of Dover Publications.

Selections from *Diophantus of Alexandria, A Study in the History of Greek Algebra* reprinted with permission of Cambridge University Press.

The Geometry of Rene Descartes courtesy of Dover Publications.

Isaac Newton's *Principia* courtesy of New York: Daniel Adee, © 1848.

Pierre-Simon Laplace's *A Philosophical Essay on Probabilities* courtesy of Dover Publications.

Selection from Jean Baptiste Joseph Fourier's *The Analytical Theory of Heat* courtesy of Dover Publications.

Selections from Carl Friederich Gauss's *Disquisitiones Arithmeticae* courtesy of Yale University Press.

Selections from *Oeuvres complètes d'Augustin Cauchy* reprinted from 1882 version published by Gauthier-Villars, Paris.

George Boole's *An Investigation of the Laws of Thought* courtesy of Dover Publications.

Georg Friederich Bernhard Riemann's *Ueber di Anzahl of Primzahlen unter eine gegeben Grosse, Ueber die Darstellbarkeit einer Function durch einer trigonometrische Reihe*, and *On the Hypotheses which lie at the Bases of Geometry* courtesy of Dover Publications.

Karl Weierstrass reprinted from R. Siegmund-Schultze (Ed.), Teubner, Leipzig, 1886 (Springer, Berlin).

Richard Dedekind's *Essays on the Theory of Numbers* courtesy of Dover Publications.

Selections from Georg Cantor's *Contributions to the Founding of the Theory of Transfinite Numbers* courtesy of Dover Publications.

Selections from Henri Lebesgue's *Integrale, Longeur, Aire* reprinted from *Annali di Matematica, Pura ed Applicata*, 1902, Ser. 3, vol. 7, pp. 231–359.

Kurt Godel's *On Formally Undecidable Propositions of Principia Mathematica and Related Systems* courtesy of Dover Publications.

Alan Turing's *On computable numbers with an application to the Entscheidungsproblem, Proceedings of the London Mathematical Society* courtesy of the London Mathematical Society.

Picture Credits:

Euclid: Getty Images.

Archimedes: Getty Images.

Diophantus: Title page of *Diophanti Alexandrini Arthimeticorum libri sex. . . . , 1621*: Library of Congress, call number QA31.D5, Rare Book/Special Collections Reading Room, (Jefferson LJ239).

Rene Descartes: Getty Images.

Isaac Newton: Time Life Pictures/Getty Images.

Pierre-Simon Laplace: Getty Images.

Jean Baptiste Joseph Fourier: Science and Society Picture Library, London.

Carl Friederich Gauss: Getty Images. ·

Augustin Louis Cauchy: © Bettmann/CORBIS.

George Boole: © CORBIS.

Georg Friederich Bernhard Riemann: © Bettmann/CORBIS.

Karl Weierstrass: © Bettmann/CORBIS.

Richard Dedekind: Frontispiece from *Richard Dedekind Gesammelte mathematische Werke*. Reprint by Chelsea Publishing Company, Bronx, NY, 1969, of the edition published in Brunswick by F. Vieweg, 1930–32. Reprinted by permission of the Chelsea Publishing Company. Photograph provided by the Library of Congress, call number QA3.D42.

Georg Cantor: © CORBIS.

Henri Lebesgue: Frontispiece from *Henri Lebesgue Oeuvres Scientifiques, volume I*. Reproduced by permission of L'Enseignement Mathématique, Universite De Geneve, Switzerland. Photograph provided by the Library of Congress, call number QA3.L27, vol. 1, copy 1.

Kurt Godel: Time Life Pictures/Getty Images.

Alan Turing: Photo provided by King's College Archive Centre, Cambridge, UK, AMT/K/7/9. Copyright holder is unknown, please contact the publisher if you have copyright information.

CONTENTS

CONTENTS

GEORGE BOOLE
(1815–1864)

GEORG FRIEDRICH BERNHARD RIEMANN
(1826–1866)

KARL WEIERSTRASS
(1815–1897)

RICHARD JULIUS WILHELM DEDEKIND
(1831–1916)

GEORG CANTOR
(1845–1918)

HENRI LEBESGUE
(1875–1941)

CONTENTS

KURT GÖDEL
(1906–1978)

ALAN MATHISON TURING
(1912–1954)

INTRODUCTION

WE ARE LUCKY TO LIVE IN AN AGE IN WHICH WE ARE STILL MAKING DISCOVERIES. IT IS LIKE THE DISCOVERY OF AMERICA–YOU ONLY DISCOVER IT ONCE. THE AGE IN WHICH WE LIVE IS THE AGE IN WHICH WE ARE DISCOVERING THE FUNDAMENTAL LAWS OF NATURE . . .

—AMERICAN PHYSICIST RICHARD FEYNMAN,

SPOKEN IN 1964

Showcasing excerpts from thirty-one of the most important works in the history of mathematics (four of which have been translated into English for the very first time), this book is a celebration of the mathematicians who helped move us forward in our understanding of the world and who paved the way for our current age of science and technology.

Over the centuries, the efforts of these mathematicians have helped the human race to achieve great insight into nature, such as the realization that the earth is round, that the same force that causes an apple to fall here on earth is also responsible for the motions of the heavenly bodies, that space is finite and not eternal, that time and space are intertwined and warped by matter and energy, and that the future can only be determined probabilistically. Such revolutions in the way we perceive the world have always gone hand in hand with revolutions in mathematical thought. Isaac Newton could never have formulated his laws without the analytic geometry of René Descartes and Newton's own invention of calculus. It is hard to imagine the development of either electrodynamics or quantum theory without the methods of Jean Baptiste Joseph Fourier or the work on calculus and the theory of complex functions pioneered by Carl Friedrich Gauss and Augustin Louis Cauchy—and it was Henri Lebesgue's work on the theory of measure that enabled John von Neumann to formulate the rigorous understanding of quantum theory that we have today. Albert Einstein could not have completed his general theory of relativity had it not been for the geometric ideas of Bernhard Riemann. And practically all of modern science would be far less potent (if it existed at all) without the concepts of probability and statistics pioneered by Pierre-Simon Laplace.

All through the ages, no intellectual endeavor has been more important to those studying physical science than has the field of mathematics. But mathematics is more than a tool and language for science. It is also an end in itself, and as such, it has, over the centuries, affected our worldview in its own right. Karl Weierstrass provided a new idea of what it means for a function to be continuous, and Georg Cantor's work revolutionized people's idea of infinity. George Boole's *Laws of Thought* revealed logic as a system of processes subject to laws identical to the laws of algebra, thus illuminating the very nature of thought and eventually enabling to some degree its mechanization, that is, modern digital computing. Alan Turing illuminated the power and the limits of digital computing, long before sophisticated computations were even possible. Kurt Gödel proved a theorem troubling to many philosophers, as well as anyone else believing in absolute truth: that in any sufficiently complex logical system (such as arithmetic) there must exist statements that can neither be proven nor disproven. And if that weren't bad enough, he also showed that the question of whether the system itself is logically consistent cannot be proven within the system.

This fascinating volume presents all these and other groundbreaking developments, the central ideas in twenty-five centuries of mathematics, employing the original texts to trace the evolution, and sometimes revolution, in mathematical thinking from its beginnings to today.

Though the first work presented here is that of Euclid, c.300 B.C., the Egyptians and Babylonians had developed an impressive ability to perform mathematical calculations as early as 3,500 B.C. The Egyptians employed this skill to build the great pyramids and to accomplish other impressive ends, but their computations lacked one quality considered essential to mathematics ever since: rigor. For example, the ancient Egyptians equated the area of a circle to the area of a square whose sides were 8/9 the diameter of the circle. This method amounts to employing a value of the mathematical constant pi that is equal to 256/81. In one sense this is impressive—it is only about one half of one percent off of the exact answer. But in another sense it is completely wrong. Why worry about an error of one half of one percent? Because the Egyptian approximation overlooks one of the deep and fundamental mathematical properties of the true number π: that it cannot be written as any fraction. That is a matter of principle, unrelated to any issue of mere quantitative accuracy. Though the irrationality of π wasn't proved until the late eighteenth century, the early Greeks did discover that numbers existed which could not be written as fractions, and this was both puzzling and shocking to them. This was the brilliance of the Greeks: to recognize the importance of principle plura in mathematics, and that in its essence mathematics is a subject in which one begins with a set of concepts and rules and then rigorously works out their precise consequences.

Euclid detailed the Greek understanding of geometry in his *Elements*, in Alexandria, around 300 B.C. In the ensuing centuries the Greeks made great strides in both algebra and geometry. Archimedes, the greatest mathematician of antiquity, studied the properties of geometric shapes and created ingenious methods of finding areas and volumes and new approximations for π. Another Alexandrian, Diophantus, looking over the clutter of words and numbers in algebraic problems, saw that an abstraction could be a great simplification. And so, Diophantus took the first step toward introducing symbolism into algebra. Over a millennium later, Frenchman René Descartes united the two fields: geometry and algebra, with his creation of analytic geometry. His work paved the way for Isaac Newton to invent calculus, and with it, a new way of doing science. Since Newton's day, the pace of mathematical innovation has been almost frenetic, as the fundamental mathematical fields of algebra, geometry, and calculus (or function theory) have fed on and in turn nourished one another, yielding insights into applications as diverse as probability, numbers, and the theory of heat. And as mathematics matured, so did the range of questions it addresses: Kurt Gödel and Alan Turing, the last two thinkers represented in this volume, address perhaps the deepest issue—the question of what is knowable. Like those of the past, future developments in mathematics are sure to affect, directly or indirectly, our ways of living and thinking. The wonders of the ancient world were physical, like the pyramids in Egypt. As this volume illustrates, the greatest wonder of the modern world is our own understanding.

Euclid

(c.325B.C.–c.265B.C.)

HIS LIFE AND WORK

With the possible exception of Isaac Newton, Euclid is the best known mathematician of all time. Until sometime in the twentieth century, his one and only surviving work, *The Elements*, was the second best selling book of all time, surpassed only by the Bible. Yet there is no reason to believe that Euclid was anything other than a compiler of the mathematics known in his times—akin to Noah Webster, the great nineteenth-century lexicographer who gave his name to the leading American dictionary.

There is very little known about Euclid. He taught at an academy in Alexandria, the Hellenic city that Alexander the Great founded at the mouth of the Nile in Egypt. Because of his work as a compiler, Euclid was familiar with all of the Greek mathematics that had preceded him, and especially familiar with the first crisis in mathematics: the crisis of the irrational.

Pythagoras (died c.475 B.C.) is a mysterious figure from early Greek mathematics. If we know little of Euclid, we know even less of Pythagoras. We do, however, know something about the Pythagorean School. The Pythagoreans thought that the whole of the cosmos could be described in terms of the whole numbers: 1, 2, 3, etc. As Aristotle wrote, "The Pythagoreans . . . having been brought up in the study of

1

mathematics, thought that things are numbers . . . and that the whole cosmos is a scale and a number."

The Pythagorean theorem (presented in this chapter) demonstrated this. Small whole numbers, like 3, 4, and 5, could be found that not only described the lengths of a right triangle but also had the property that when the areas of the squares erected on the two smaller sides were added together they equaled the area of the square erected on the longest side, the hypotenuse, the side opposite the right angle. Notice how the ancient Greeks phrased the Pythagorean theorem in terms of geometric objects and not in terms of numbers!

Then someone asked an interesting question. If there were a square with each side a unit in length, and a second square with double the area of the first square, how would the side of the second square compare to the side of the first square? This was how the question of the square root of 2 was originally proposed.

The ancient Egyptians found a good approximation to the answer. The side of the second square was to the side of the first square almost as 7 is to 5. This should certainly be no surprise to us since we know $\frac{7}{5}$ can also be expressed as 1.4, which is very close to what to we know to be the decimal expansion of $\sqrt{2}$. But close wasn't good enough for the Pythagoreans. After all, the Pythagorean theorem didn't assert that the areas of the squares were *close* to being equal. It asserted that they were, in fact, equal.

Then, someone (whose name we don't know) hit upon a deep insight. Suppose that the square root of 2 can be expressed as the ratio of two whole numbers and that these two whole numbers share no common divisors except 1, their common unit. Call these whole numbers p and q with the property that the square P erected on a side of length p is exactly twice the area of a square Q erected on a side of length q. Now if P contains twice the number of units than Q, then P must contain an even number of units! The Pythagoreans already knew that if a square contains an even number of units, the number of units the square contains must be a multiple of 4 units of area and its sides must contain an even number of whole units of length.

Still, everyone knew how, given one square, you could find another with one fourth the area: Just erect a square on a side equal to half of that side. In this case, erect a square T whose side t is one half of the side p. Because p contains an even number of whole units of length, the side t must still contain some number of whole units of length. But then, if the square T has one fourth the area of the square P, the square Q must contain twice the number of units of area as the square T. So because the square T contains an even number of units of area, then, just like the square P, the square Q must also contain a multiple of 4 units of area. So, its side q had to contain an even number of units of length. So far, this mathematical argument has been like a tennis match—going back and forth between the players.

Eventually, the argument reaches its climax. It began by supposing that the sides p and q had no divisor in common except 1 and ended up with a contradiction: that they shared the divisor 2! Try as they might, the Pythagoreans could not find a flaw with the argument. Knowing that no one had in fact ever found a way to express the square root of 2 as the ratio of two whole numbers, the Pythagoreans faced up to the reality that they had proven that the square root of 2 could not be expressed as the ratio of two whole numbers.

Thus were born the irrational numbers, mathematical objects that could not be expressed by means of whole numbers, at least not for two millennia: the first of what Kronecker called the works of man.

The Pythagoreans carefully guarded this great discovery because it created a crisis that reached to the very roots of their cosmology. When the Pythagoreans learned that one of their members had divulged the secret to someone outside their circle, they quickly made plans to throw the betrayer overboard and drown him while on the high seas. Whoever he was, this man was the first martyr for mathematics!

The crisis of the irrational also instructed the ancient Greeks that they could not look to arithmetic to form the foundation for the rest of mathematics and with it an explanation for the structure of the cosmos. They had to look elsewhere. They turned to geometry.

Euclid's *Elements* is most remembered for its geometry, and especially remembered for its treatment of parallel lines in the definition of parallel lines:

> *Parallel straight lines are straight lines which, being in the same plane and produced indefinitely in both directions, do not meet one another in either direction.*

And in the fifth postulate, the parallel postulate:

> *That, if a straight line falling on two straight lines makes the interior angles on the same side less than two right angles, the straight lines, if produced indefinitely, meet on that side on which are the angles less than the two right angles.*

This is very different from the form in which it is often presented:

> *Given a line and a point not on the line, it is possible to draw exactly one line through the given point parallel to the line.*

This is an equivalent, but different form given by the Scottish mathematician John Playfair in 1795.

3

During the height of the Newtonian era, philosophers, such as Immanuel Kant, never doubted the truth of Euclid's parallel postulate. They merely inquired about the nature of its truth. Was the parallel postulate necessarily true of the cosmos or only contingently true? Of course, since the advent of the Einsteinian revolution, we know that the parallel postulate isn't true at all about the cosmos. The Einsteinian space-time cosmos that we inhabit is curved. Euclidean geometry and Newtonian physics are only approximations.

So, we may ask, what did the Greeks think about the nature of the parallel postulate? I believe that a brief consideration of the ancient Greek conception of the world will explain that they too viewed the parallel postulate as a useful fiction rather than as a true description of the physical world. After all, the Greeks believed that we inhabited what the historian of science Alexandre Koyré has called a "closed world," a spherical cosmos in which there were no straight lines actually extending to infinity. Below the moon's orbits, bodies moved in straight lines either towards the center of the earth or away from it. Above the moon's orbit, bodies orbited in perfect circles around the center of the earth. In this cosmos, there weren't actually any straight lines at all.

But the Greeks had a problem. They needed to find a foundation for their mathematics. The Pythagoreans had pursued arithmetic as a foundation and reached a crisis. Needing to find an alternative, another school descending from Thales (died c.547 B.C.) sought to base mathematics on geometry. This school found that they could achieve very little without the parallel postulate! They couldn't prove the Pythagorean theorem, for example. In fact, they couldn't prove much geometry at all. This should come as no surprise to us moderns who have the benefit of 2,500 years of hindsight and know that the Pythagorean theorem is false in non-Euclidean geometries. I believe that the ancient Greeks knew that the parallel postulate was only a useful, no, let me say a very useful, approximation.

If the proof of the irrationality of the square root gave us the first crisis in mathematics, it also gave us the first example of the form of argument known since ancient times as *reductio ad absurdum*, reduction to absurdity. A second example of this form of argument can be seen in Euclid's proof the infinitude of prime numbers, another proof that certainly originated with someone else.

A prime number is a positive integer, such as 3 or 23, whose only positive integer divisors are 1 and itself. Proving the infinitude of prime numbers is stunningly simple. Suppose there is a largest prime number P. Multiply together all of the prime numbers up to an including P. Now add 1. The result is not divisible by P, nor is it divisible by any of the prime numbers less than P, because P and all of the prime numbers less than it evenly divides into their product before 1 is added to it. Supposing that there is a largest prime number leads to a contradiction. *Reductio ad absurdum*!

The Greeks noticed that many prime numbers come together in pairs: 11 and 13, 17 and 19, 29 and 31, for example. These are called twin primes. The Greeks speculated that there were not only an infinitude of prime numbers but also an infinitude of twin primes. But they weren't able to prove that, nor has any mathematician since.

Neither has any mathematician been able to disprove the existence of an *odd perfect* number. A *perfect* number: That sounds quite odd indeed! What is a perfect number? A perfect number is the sum of its integer divisors greater than or equal to 1 but less than itself, what are called its *aliquot* divisors. The ancient Greeks found all of the even perfect numbers as follows:

Notice that the sum of the powers of 2 from 1, which is 2^0, to 2^{n-1} is equal to $2^n - 1$. In a very simple case, for n = 3, $1 + 2 + 4 = 7 = 8 - 1$. Now let's do some simple arithmetic.

$$7 \quad = 1 + 2 + 4 \quad = 2^0 + 2^1 + 2^2$$
$$7 \quad = 8 - 1 \quad = 2^3 - 2^0$$
$$14 \quad = 16 - 2 \quad = 2^4 - 2^1$$

These columns all add up to 28. Writing that sum in yet another we see that

$$28 = 1 + 2 + 4 + 7 + 14 = 2^2 + 2^3 + 2^4 = 2^2 \times (2^0 + 2^1 + 2^2) = 2^2 \times (2^3 - 1)$$

28 is the sum of its divisors. Notice how these divisors are first all the powers of 2 up to a given exponent, then 1 less than the next power of 2, call this the turning point divisor, and then the turning point divisor times all of the powers of 2 up to that given exponent. And notice that if 7 were not a prime number, then 28 would not be equal to the sum of its aliquot divisors. If the turning divisor had just one prime aliquot divisor that would cause the sum of all the aliquot divisors to overflow.

With these observations the Greeks had proven that

If $(2^n - 1)$ is a prime number then $2^{n-1} \times (2^{n-1} - 1)$ is a perfect number and that even perfect numbers must have this form.

More than two millennia later, no one has ever discovered an odd perfect number. No mathematician believes that an odd perfect number exists. But none has been able to prove that no odd perfect number exists!

The Pythagoreans tried and failed to found all of mathematics on arithmetic. Re-founding mathematics on geometry meant founding arithmetic on geometry.

Quick, which is larger: $\frac{7}{5}$ or $\frac{10}{7}$? That may be too easy for you. Try this one instead, without a calculator: which is larger: $\frac{19}{12}$ or $\frac{30}{19}$? Try doing it just using multiplication and not division. The Eudoxian theory of proportion, presented in Book V of *Euclid's Elements* provides the tools, namely, multiplication, for arriving at an answer.

Following Eudoxus (died c.355 B.C.), Euclid posed the problem as follows: consider 4 magnitudes of length—*a*, *b*, *c*, and *d*. How can one determine whether the ratio of *a* to *b* is greater than, less than, or equal to the ratio of *c* to *d*? Eudoxus began by asserting that "magnitudes are said to have a ratio to one another which is capable, when a multiple of either may exceed the other." He recognized that if the ratio of *a* to *b* is greater than the ratio of *c* to *d*, all multiples of the ratio of *a* to *b* are greater than the same number of multiples of the ratio of *c* to *d*. Recognizing this fact, Eudoxus realized that all he had to do was to find one multiple to use and he could solve the problem. The multiple he chose was the product of *b* and *d*. Multiplying the ratio of *a* to *b* by the product of *b* and *d*, gave the ratio of the product of *a*, *b* and *d* to *b*, or the area of the rectangle with sides of length *a* and *d*. Similarly, multiplying the ratio of *c* to *d* by the product of *b* and *d* gave the ratio of the product of *c*, *b*, and *d* to *d*, or the area of a rectangle with sides of *c* and *b*.

Thus, the area of the rectangle with side *a* and *d* is greater than the area of the rectangle with sides of length *c* and *b*, if and only if the ratio of *a* to *b* is greater than the ratio of *c* to *d*. Whereas the Pythagoreans had tried to arithmetize geometry and failed, when Eudoxus tried to geometrize arithmetic he succeeded! By the way, since 19×19 is greater than 30×12, $\frac{19}{12}$ is greater than $\frac{30}{19}$!

Euclid is the greatest mathematical encyclopediast of all times. Today, when mathematicians from different specialties have difficulty understanding work at the frontiers of each others' specialty, no one mathematician could hope to edit a compendium of all known mathematics. Yet it remains an ideal in the mathematical community. In the second half of the twentieth century, the French mathematical community offered a pale imitation of Euclid in the person of Nicholas Bourbaki. In fact, Bourbaki wasn't even a person. He was the fictitious *nom de plume* of a collective of more than twenty French mathematicians working in diverse branches of mathematics! Euclid remains, to this day, our model for mathematical texts.

Selections from Euclid's *Elements*
Book I

BASIC GEOMETRY—DEFINITIONS, POSTULATES, COMMON NOTIONS AND PROPOSITION 47 (LEADING UP TO THE PYTHAGOREAN THEOREM)

Definitions

1. A **point** is that which has no part.

2. A **line** is breadthless length.

3. The extremities of a line are points.

4. A **straight line** is a line which lies evenly with the points on itself.

5. A **surface** is that which has length and breadth only.

6. The extremities of a surface are lines.

7. A **plane surface** is a surface which lies evenly with the straight lines on itself.

8. A **plane angle** is the inclination to one another of two lines in a plane which meet one another and do not lie in a straight line.

9. And when the lines containing the angle are straight, the angle is called **rectilineal**.

10. When a straight line set up on a straight line makes the adjacent angles equal to one another, each of the equal angles is **right**, and the straight line standing on the other is called a **perpendicular** to that on which it stands.

11. An **obtuse angle** is an angle greater than a right angle.

12. An **acute angle** is an angle less than a right angle.

13. A **boundary** is that which is an extremity of anything.

14. A **figure** is that which is contained by any boundary or boundaries.

15. A **circle** is a plane figure contained by one line such that all the straight lines falling upon it from one point among those lying within the figure are equal to one another;

16. And the point is called the **centre** of the circle.

17. A **diameter** of the circle is any straight line drawn through the centre and terminated in both directions by the circumference of the circle, and such a straight line also bisects the circle.

18. A **semicircle** is the figure contained by the diameter and the circumference cut off by it. And the centre of the semicircle is the same as that of the circle.

7

19. **Rectilineal figures** are those which are contained by straight lines, **trilateral** figures being those contained by three, **quadrilateral** those contained by four, and **multilateral** those contained by more than four straight lines.

20. Of trilateral figures, an **equilateral triangle** is that which has its three sides equal, an **isosceles triangle** that which has two of its sides alone equal, and a **scalene triangle** that which has its three sides unequal.

21. Further, of trilateral figures, a **right-angled triangle** is that which has a right angle, an **obtuse-angled triangle** that which has an obtuse angle, and an **acute-angled triangle** that which has its three angles acute.

22. Of quadrilateral figures, a **square** is that which is both equilateral and right-angled; an **oblong** that which is right-angled but not equilateral; a **rhombus** that which is equilateral but not right-angled; and a **rhomboid** that which has its opposite sides and angles equal to one another but is neither equilateral nor right-angled. And let quadrilaterals other than these be called **trapezia**.

23. **Parallel** straight lines are straight lines which, being in the same plane and being produced indefinitely in both directions, do not meet one another in either direction.

POSTULATES

Let the following be postulated:

1. To draw a straight line from any point to any point.

2. To produce a finite straight line continuously in a straight line.

3. To describe a circle with any centre and distance.

4. That all right angles are equal to one another.

5. That, if a straight line falling on two straight lines make the interior angles on the same side less than two right angles, the two straight lines, if produced indefinitely, meet on that side on which are the angles less than the two right angles.

COMMON NOTIONS

1. Things which are equal to the same thing are also equal to one another.

2. If equals be added to equals, the wholes are equal.

3. If equals be subtracted from equals, the remainders are equal.

[7] 4. Things which coincide with one another are equal to one another.

[8] 5. The whole is greater than the part.

PROPOSITION 47

In right-angled triangles the square on the side subtending the right angle is equal to the squares on the sides containing the right angle.

Let *ABC* be a right-angled triangle having the angle *BAC* right;

I say that the square on *BC* is equal to the squares on *BA*, *AC*.

5   For let there be described on *BC* the square *BDEC*,

8

and on *BA*, *AC* the squares *GB*, *HC*;   [i. 46]
through *A* let *AL* be drawn parallel to either *BD*
or *CE*, and let *AD*, *FC* be joined.

Then, since each of the angles *BAC*, *BAG* is
right, it follows that with a straight line *BA*, and
at the point *A* on it, the two straight lines *AC*, *AG*
not lying on the same side make the adjacent
angles equal to two right angles;

therefore *CA* is in a straight line with
AG.   [i. 14]

For the same reason

BA is also in a straight line with *AH*.

And, since the angle *DBC* is equal to the angle *FBA*: for each is right:
let the angle *ABC* be added to each;

therefore the whole angle *DBA* is equal to the whole angle *FBC*.   [C. N. 2]

And, since *DB* is equal to *BC*, and *FB* to *BA*,
the two sides *AB*, *BD* are equal to the two sides *FB*, *BC* respectively,

and the angle *ABD* is equal to the angle *FBC*;

therefore the base *AD* is equal to the base *FC*,
and the triangle *ABD* is equal to the triangle *FBC*.   [i. 4]

Now the parallelogram *BL* is double of the triangle *ABD*, for they have the same
base *BD* and are in the same parallels *BD*, *AL*.   [i. 41]

And the square *GB* is double of the triangle *FBC*,
for they again have the same base *FB* and are in the same parallels *FB*, *GC*.   [i. 41]

[But the doubles of equals are equal to one another.]

Therefore the parallelogram *BL* is also equal to the square *GB*.

Similarly, if *AE*, *BK* be joined,
the parallelogram *CL* can also be proved equal to the square *HC*;

therefore the whole square *BDEC* is equal to the two squares *GB*, *HC*.

[C. N. 2]

And the square *BDEC* is described on *BC*,
and the squares *GB*, *HC* on *BA*, *AC*.

Therefore the square on the side *BC* is equal to the squares on the sides *BA*, *AC*.

Therefore etc.   Q.E.D.

1. **the square on,** τὸ ἀπὸ . . . τετράγωνον, the word ἀναγραφέν or ἀναγεγραμμένον being understood.

subtending the right angle. Here ὑποτεινούσης, "subtending," is used with the simple accusative (τὴν ὀρθὴν γωνίαν) instead of being followed by ὑπό and the accusative, which seems to be the original and more orthodox construction.

22. **the two sides AB, BD. . . .** Euclid actually writes "*DB, BA*," and therefore the equal sides in the two triangles are not mentioned in corresponding order, though he adheres to the words ἑκατέρα ἑκατέρα "respectively." Here *DB* is equal to *BC* and *BA* to *FB*.

30. **[But the doubles of equals are equal to one another.]** Heiberg brackets these words as an interpolation, since it quotes a *Common Notion* which is itself interpolated.

"If we listen," says Proclus, "to those who wish to recount ancient history, we may find some of them referring this theorem to Pythagoras and saying that he sacrificed an ox in honour of his discovery. But for my part, while I admire those who first observed the truth of this theorem, I marvel more at the writer of the Elements, not only because he made it fast (κατεδήσατο) by a most lucid demonstration, but because he compelled assent to the still more general theorem by the irrefragable arguments of science in the sixth Book. For in that Book he proves generally that, in right-angled triangles, the figure on the side subtending the right angle is equal to the similar and similarly situated figures described on the sides about the right angle."

In addition, Plutarch, Diogenes Laertius (viii. 12) and Athenaeus (x. 13) agree in attributing this proposition to Pythagoras. It is easy to point out, as does G. Junge ("Wann haben die Griechen das Irrationale entdeckt?" in *Novae Symbolae Joachimicae*, Halle a. S., 1907), that these are late witnesses, and that the Greek literature which we possess belonging to the first five centuries after Pythagoras contains no statement specifying this or any other particular great geometrical discovery as due to him. Yet the distich of Apollodorus the "calculator," whose date (though it cannot be fixed) is at least earlier than that of Plutarch and presumably of Cicero, is quite definite as to the existence of *one* "famous proposition" discovered by Pythagoras, whatever it was. Nor does Cicero, in commenting apparently on the verses (*De nat. deor.* iii. c. 36, § 88), seem to dispute the fact of the geometrical discovery, but only the story of the sacrifice. Junge naturally emphasises the apparent uncertainty in the statements of Plutarch and Proclus. But, as I read the passages of Plutarch, I see nothing in them inconsistent with the supposition that Plutarch unhesitatingly accepted as discoveries of Pythagoras *both* the theorem of the square of the hypotenuse and the problem of the application of an area, and the only doubt he felt was as to which of the two discoveries was the more appropriate occasion for the supposed sacrifice. There is also other evidence not without bearing on the question. The theorem is closely connected with the whole of the matter of Eucl. Book II., in which one of the most prominent features is the use of the *gnomon*. Now the gnomon was a well-understood term with the Pythagoreans. Aristotle also (*Physics* iii. 4, 203 a 10–15) clearly attributes to the Pythagoreans the placing of odd numbers as *gnomons* round successive squares beginning with 1, thereby forming new squares, while in another place (*Categ.* 14, 15 a 30) the word *gnomon* occurs in the same (obviously familiar) sense: "e.g. a square, when a gnomon is placed round it, is increased in size but is not altered in form." The inference must therefore be that practically the whole doctrine of Book II. is Pythagorean. Again Heron (? 3rd cent. A.D.), like Proclus, credits Pythagoras with a general rule for forming right-angled triangles with rational whole numbers for sides. Lastly, the "summary" of Proclus appears to credit Pythagoras with the discovery of the theory, or study, of irrationals (τὴν τῶν ἀλόγων πραγματείαν). But it is now more or less agreed that the reading here should be, not τῶν ἀλόγων, but τῶν ἀναλόγων, or rather τῶν ἀνὰ λόγον ("of proportionals"), and that the author intended to attribute to Pythagoras a theory of *proportion*, i.e. the (arithmetical) theory of proportion applicable only to commensurable magnitudes, as distinct from the theory of Eucl. Book V., which was due to Eudoxus. It is not however disputed that the *Pythagoreans* discovered the irrational (cf. the scholium No. 1 to Book X.). Now everything goes to show that this discovery of the irrational was made with reference to √2, the ratio of the diagonal of a square to its side. It is clear that this presupposes the knowledge that i. 47 is true of an isosceles right-angled triangle; and the fact that some triangles of which it had been discovered to be true were *rational* right-angled triangles was doubtless what suggested the inquiry whether the ratio between the lengths of the diagonal and the side of a square could also be expressed in whole numbers. On the whole, therefore, I see no sufficient reason to question the tradition that, *so far as Greek geometry is concerned* (the possible priority of the discovery of the same proposition in India will be considered later), Pythagoras was the first to introduce the theorem of i. 47 and to give a general proof of it.

On this assumption, how was Pythagoras led to this discovery? It has been suggested and commonly assumed that the Egyptians were aware that a triangle with its sides in the ratio 3, 4, 5

10

was right-angled. Cantor inferred this from the fact that this was precisely the triangle with which Pythagoras began, if we may accept the testimony of Vitruvius (ix. 2) that Pythagoras taught how to make a right angle by means of three lengths measured by the numbers 3, 4, 5. If then he took from the Egyptians the triangle 3, 4, 5, he presumably learnt its property from them also. Now the Egyptians must certainly be credited from a period at least as far back as 2000 B.C. with the knowledge that $4^2 + 3^2 = 5^2$. Cantor finds proof of this in a fragment of papyrus belonging to the time of the 12th Dynasty newly discovered at Kahun. In this papyrus we have extractions of square roots: e.g. that of 16 is 4, that of $1\frac{9}{16}$ is $1\frac{1}{4}$, that of $6\frac{1}{4}$ is $2\frac{1}{2}$, and the following equations can be traced:

$$1^2 + (\tfrac{3}{4})^2 = (1\tfrac{1}{4})^2$$
$$8^2 + 6^2 = 10^2$$
$$2^2 + (1\tfrac{1}{2})^2 = (2\tfrac{1}{2})^2$$
$$16^2 + 12^2 = 20^2.$$

It will be seen that $4^2 + 3^2 = 5^2$ can be derived from each of these by multiplying, or dividing out, by one and the same factor. We may therefore admit that the Egyptians knew that $3^2 + 4^2 = 5^2$. But there seems to be no evidence that they knew that the triangle (3, 4, 5) is *right-angled*; indeed, according to the latest authority (T. Eric Peet, *The Rhind Mathematical Papyrus*, 1923), nothing in Egyptian mathematics suggests that the Egyptians were acquainted with this or any special cases of the Pythagorean theorem.

How then did Pythagoras discover the general theorem? Observing that 3, 4, 5 was a right-angled triangle, while $3^2 + 4^2 = 5^2$, he was probably led to consider whether a similar relation

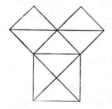

was true of the sides of right-angled triangles other than the particular one. The simplest case (geometrically) to investigate was that of the *isosceles* right-angled triangle; and the truth of the theorem in this particular case would easily appear from the mere construction of a figure. Cantor and Allman (*Greek Geometry from Thales to Euclid*) illustrate by a figure in which the squares are drawn outwards, as in i. 47, and divided by diagonals into equal triangles; but I think that the truth was more likely to be first observed from a figure of the kind suggested by Bürk (*Das Āpastamba-Śulba-Sūtra* in *Zeitschrift der deuts hen morgenländ. Gesellschaft*, LV., 1901) to explain how the Indians arrived at the same thing. The two figures are as shown above. When the geometrical consideration of the figure had shown that the isosceles right-angled triangle had the property in question, the investigation of the same fact from the arithmetical point of view would ultimately lead to the other momentous discovery of the irrationality of the length of the diagonal of a square expressed in terms of its side.

The *irrational* will come up for discussion later; and our next question is: Assuming that Pythagoras had observed the geometrical truth of the theorem in the case of the two particular triangles, and doubtless of other rational right-angled triangles, how did he establish it generally?

There is no positive evidence on this point. Two possible lines are however marked out. (1) Tannery says (*La Géométrie grecque*) that the geometry of Pythagoras was sufficiently advanced to make it possible for him to prove the theorem by *similar triangles*. He does not say in what particular manner similar triangles would be used, but their use must apparently have involved the use of *proportions*, and, in order that the proof should be conclusive, of the theory of proportions in its complete form applicable to incommensurable as well as commensurable magnitudes. Now Eudoxus was the first to make the theory of proportion independent of the hypothesis of commensurability; and as, before Eudoxus' time, this had not been done, any proof of the general theorem by means of proportions given by Pythagoras must at least have been inconclusive. But this does not constitute any objection to the supposition that the truth of the general theorem may have been discovered in such a manner; on the contrary, the supposition that Pythagoras proved it by means of an imperfect theory of proportions would better than anything else account for the fact that Euclid had to devise an entirely new proof, as Proclus says he did in i. 47. This proof had to be independent of the theory of proportion even in its rigorous form, because the plan of the *Elements* postponed that theory to Books V. and VI., while the Pythagorean theorem was required as early as Book II. On the other hand, if the Pythagorean proof had been based on the doctrine of Books I. and II. only, it would scarcely have been necessary for Euclid to supply a new proof.

The possible proofs by means of proportion would seem to be practically limited to two.

(*a*) One method is to prove, from the similarity of the triangles *ABC, DBA*, that the rectangle *CB, BD* is equal to the square on *BA*, and, from the similarity of the triangles *ABC, DAC*, that the rectangle *BC, CD* is equal to the square on *CA*; whence the result follows by addition.

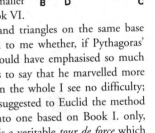

It will be observed that this proof is *in substance* identical with that of Euclid, the only difference being that the equality of the two smaller squares to the respective rectangles is inferred by the method of Book VI. instead of from the relation between the areas of parallelograms and triangles on the same base and between the same parallels established in Book I. It occurred to me whether, if Pythagoras' proof had come, even in substance, so near to Euclid's, Proclus would have emphasised so much as he does the originality of Euclid's, or would have gone so far as to say that he marvelled more at that proof than at the original discovery of the theorem. But on the whole I see no difficulty; for there can be little doubt that the proof by proportion is what suggested to Euclid the method of i. 47, and the transformation of the method of proportions into one based on Book I. only, effected by a construction and proof so extraordinarily ingenious, is a veritable *tour de force* which compels admiration, notwithstanding the ignorant strictures of Schopenhauer, who wanted something as obvious as the second figure in the case of the isosceles right-angled triangle, and accordingly (*Sämmtliche. Werke*, iii. § 39 and i. § 15) calls Euclid's proof "a mouse-trap proof" and "a proof walking on stilts, nay, a mean, underhand, proof" ("Des Eukleides stelzbeiniger, ja, hinterlistiger Beweis").

(*b*) The other possible method is this. As it would be seen that the triangles into which the original triangle is divided by the perpendicular from the right angle on the hypotenuse are similar to one another and to the whole triangle, while in these three triangles the two sides about the right angle in the original triangle, and the hypotenuse of the original triangle, are corresponding sides, and that the sum of the two former similar triangles is identically equal to the similar triangle on the hypotenuse, it might be inferred that the same would also be true of *squares* described on the corresponding three sides respectively, because squares as well as similar triangles are to one another in the duplicate ratio of corresponding sides. But the same thing is equally true of any similar rectilineal figures, so that this proof would practically establish the extended theorem of Eucl. vi. 31, which theorem, however, Proclus appears to regard as being entirely Euclid's discovery.

On the whole, the most probable supposition seems to me to be that Pythagoras used the first method (*a*) of proof by means of the theory of proportion as he knew it, i.e. in the defective form which was in use up to the date of Eudoxus.

(2) I have pointed out the difficulty in the way of the supposition that Pythagoras' proof depended upon the principles of Eucl. Books I. and II. only. Were it not for this difficulty, the

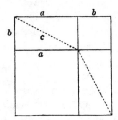 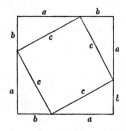

conjecture of Bretschneider, followed by Hankel, would be the most tempting hypothesis. According to this suggestion, we are to suppose a figure like that of Eucl. II. 4 in which a, b are the sides of the two inner squares respectively, and $a + b$ is the side of the complete square. Then, if the two complements, which are equal, are divided by their two diagonals into four equal triangles of sides a, b, c, we can place these triangles round another square of the same size as the whole square, in the manner shown in the second figure, so that the sides a, b of *successive* triangles make up one of the sides of the square and are arranged in cyclic order. It readily follows that the remainder of the square when the four triangles are deducted is, in the one case, a square whose side is c, and in the other the sum of two squares whose sides are a, b respectively. Therefore the square on c is equal to the sum of the squares on a, b. All that can be said against this conjectural proof is that it has no specifically Greek colouring but rather recalls the Indian method. Thus Bhāskara (born 1114 A.D.) simply draws four right-angled triangles equal to the original one inwards, one on each side of the square on the hypotenuse, and says "see!", without even adding that inspection shows that

$$c^2 = 4\frac{ab}{2} + (a - b)^2 = a^2 + b^2.$$

Though, for the reason given, there is difficulty in supposing that Pythagoras used a general proof of this kind, which applies of course to right-angled triangles with sides incommensurable as well as commensurable, there is no objection, I think, to supposing that the truth of the proposition in the case of the first *rational* right-angled triangles discovered, e.g. 3, 4, 5, was proved by a method of this sort. Where the sides are commensurable in this way, the squares can be divided up into small (unit) squares, which would much facilitate the comparison between them. That this subdivision was in fact resorted to in adding and subtracting squares is made probable by Aristotle's allusion to odd numbers as *gnomons* placed round unity to form successive squares in *Physics* III. 4; this must mean that the squares were represented by dots arranged in the form of a square and a gnomon formed of dots put round, or that (if the given square was drawn in the usual way) the gnomon was divided up into unit squares. Zeuthen has shown (*"Théorème de Pythagore," Origine de la Géométrie scientifique* in *Comptes rendus du IIme Congrès international de Philosophie*, Genève, 1904), how easily the proposition could be proved by a method of this kind for the triangle 3, 4, 5. To admit of the two smaller squares being shown side by side, take a square on a line containing 7 units of length (4 + 3), and divide it up into 49 small squares. It would be obvious that the whole square could be exhibited as containing four rectangles of sides 4, 3 cyclically arranged round the figure with one unit square in the middle. (This same figure is given by Cantor to illustrate the method given in the Chinese "Chóu-peï".) It would be seen that

(I) the whole square (7^2) is made up of two squares 3^2 and 4^2, and two rectangles 3, 4;

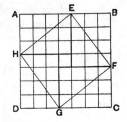

(II) the same square is made up of the square *EFGH* and the halves of four of the same rectangles 3, 4, whence the square *EFGH*, being equal to the sum of the squares 3^2 and 4^2, must contain 25 unit squares and its side, or the diagonal of one of the rectangles, must contain 5 units of length.

Or the result might equally be seen by observing that

(I) the square *EFGH* on the diagonal of one of the rectangles is made up of the halves of four rectangles and the unit square in the middle, while

(II) the squares 3^2 and 4^2 placed at adjacent corners of the large square make up two rectangles 3, 4 with the unit square in the middle.

The procedure would be equally easy for any *rational* right-angled triangle, and would be a natural method of trying to *prove* the property when it had once been *empirically* observed that triangles like 3, 4, 5 did in fact contain a right angle.

Zeuthen has, in the same paper, shown in a most ingenious way how the property of the triangle 3, 4, 5 could be verified by a sort of combination of the second possible method by similar triangles, (*b*) on p. 13 above, with subdivision of *rectangles* into similar small *rectangles*. I give the method on account of its interest, although it is no doubt too advanced to have been used by those who first proved the property of the particular triangle.

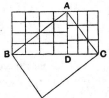

Let *ABC* be a triangle right-angled at *A*, and such that the lengths of the sides *AB*, *AC* are 4 and 3 units respectively.

Draw the perpendicular *AD*, divide up *AB*, *AC* into unit lengths, complete the rectangle on *BC* as base and with *AD* as altitude, and subdivide this rectangle into small rectangles by drawing parallels to *BC*, *AD* through the points of division of *AB*, *AC*.

Now, since the diagonals of the small rectangles are all equal, each being of unit length, it follows by similar triangles that the small rectangles are all equal. And the rectangle with *AB* for diagonal contains 16 of the small rectangles, while the rectangle with diagonal *AC* contains 9 of them.

But the sum of the triangles *ABD*, *ADC* is equal to the triangle *ABC*.

Hence the rectangle with *BC* as diagonal contains 9 + 16 or 25 of the small rectangles; and therefore *BC* = 5.

Rational right-angled triangles from the arithmetical standpoint.

Pythagoras investigated the *arithmetical* problem of finding rational numbers which could be made the sides of right-angled triangles, or of finding square numbers which are the sum of two squares; and herein we find the beginning of the *indeterminate analysis* which reached so high a stage of development in Diophantus. Fortunately Proclus has preserved Pythagoras' method of solution in the following passage. "Certain methods for the discovery of triangles of this kind are handed down, one of which they refer to Plato, and another to Pythagoras. [The latter] starts from odd numbers. For it makes the odd number the smaller of the sides about the right angle; then it takes the square of it, subtracts unity, and makes half the difference the greater of the sides about the right angle; lastly it adds unity to this and so forms the remaining side, the hypotenuse. For example, taking 3, squaring it, and subtracting unity from the 9, the method takes half of the 8, namely 4; then, adding unity to it again, it makes 5, and a right-angled triangle has been found with one side 3, another 4 and another 5. But the method of Plato argues from even numbers. For it takes the given even number and makes it one of the sides about the right angle; then, bisecting this number and squaring the half, it adds' unity to the square to form the hypotenuse, and subtracts unity from the square to form the other side about the right angle. For example, taking 4, the method squares half of this, or 2, and so makes 4; then, subtracting unity; it produces 3, and adding unity it produces 5. Thus it has formed the same triangle as that which was obtained by the other method."

The formula of Pythagoras amounts, if m be an odd number, to

$$m^2 + \left(\frac{m^2 - 1}{2}\right)^2 = \left(\frac{m^2 + 1}{2}\right)^2,$$

the sides of the right-angled triangle being m, $\frac{m^2 - 1}{2}$, $\frac{m^2 + 1}{2}$. Cantor, taking up an idea of Röth (*Geschichte der abendländischen Philosophie*, II. 527), gives the following as a possible explanation of the way in which Pythagoras arrived at his formula. If $c^2 = a^2 + b^2$, it follows that

$$a^2 = c^2 - b^2 = (c + b)(c - b).$$

Numbers can be found satisfying the first equation if (1) $c + b$ and $c - b$ are either both even or both odd, and if further (2) $c + b$ and $c - b$ are such numbers as, when multiplied together, produce a square number. The first condition is necessary because, in order that c and b may both be whole numbers, the sum and difference of $c + b$ and $c - b$ must both be even. The second condition is satisfied if $c + b$ and $c - b$ are what were called *similar numbers* (ὅμοιοι ἀριθμοί); and that such numbers were most probably known in the time before Plato may be inferred from their appearing in Theon of Smyrna (*Expositio rerum mathematicarum ad legendum Platonem utilium*, ed. Hiller), who says that similar plane numbers are, first, all square numbers and, secondly, such oblong numbers as have the sides which contain them proportional. Thus 6 is an oblong number with length 3 and breadth 2; 24 is another with length 6 and breadth 4. Since therefore 6 is to 3 as 4 is to 2, the numbers 6 and 24 are similar.

Now the simplest case of two similar numbers is that of 1 and a^2, and, since 1 is odd, the condition (1) requires that a^2, and therefore a, is also odd. That is, we may take 1 and $(2n + 1)^2$ and equate them respectively to $c - b$ and $c + b$, whence we have

$$b = \frac{(2n + 1)^2 - 1}{2},$$

$$c = \frac{(2n + 1)^2 - 1}{2} + 1,$$

while

$$a = 2n + 1.$$

As Cantor remarks, the form in which c and b appear correspond sufficiently closely to the description in the text of Proclus.

Another obvious possibility would be, instead of equating $c - b$ to unity, to put $c - b = 2$, in which case the similar number $c + b$ must be equated to double of some square, i.e. to a number of the form $2n^2$, or to the half of an even square number, say $\frac{(2n)^2}{2}$. This would give

$$a = 2n,$$
$$b = n^2 - 1,$$
$$c = n^2 + 1,$$

which is Plato's solution, as given by Proclus.

The two solutions supplement each other. It is interesting to observe that the method suggested by Röth and Cantor is very like that of Eucl. x. (Lemma 1 following Prop. 28). We shall come to this later, but it may be mentioned here that the problem is *to find two square numbers such that their sum is also a square*. Euclid there uses the property of ii. 6 to the effect that, if AB is bisected at C and produced to D,

$$AD \cdot DB + BC^2 = CD^2.$$

We may write this

$$uv = c^2 - b^2,$$

where

$$u = c + b, \quad v = c - b.$$

In order that uv may be a square, Euclid points out that u and v must be similar numbers, and further that u and v must be either both odd or both even in order that b may be a whole number. We may then put for the similar numbers, say, $\alpha\beta^2$ and $\alpha\gamma^2$, whence (if $\alpha\beta^2$, $\alpha\gamma^2$ are either both odd or both even) we obtain the solution

$$\alpha\beta^2 \cdot \alpha\gamma^2 + \left(\frac{\alpha\beta^2 - \alpha\gamma^2}{2}\right)^2 = \left(\frac{\alpha\beta^2 + \alpha\gamma^2}{2}\right)^2.$$

But I think a serious, and even fatal, objection to the conjecture of Cantor and Röth is the very fact that the method enables both the Pythagorean and the Platonic series of triangles to be deduced with equal ease. If this had been the case with the method used by Pythagoras, it would not, I think, have been left to Plato to discover the second series of such triangles. It seems to me therefore that Pythagoras must have used some method which would produce his rule *only*; and further it would be some less recondite method, suggested by direct *observation* rather than by argument from general principles.

One solution satisfying these conditions is that of Bretschneider, who suggests the following simple method. Pythagoras was certainly aware that the successive odd numbers are *gnomons*, or the differences between successive square numbers. It was then a simple matter to write down in three rows (*a*) the natural numbers, (*b*) their squares, (*c*) the successive odd numbers constituting the differences between the successive squares in (*b*), thus:

1	2	3	4	5	6	7	8	9	10	11	12	13	14
1	4	9	16	25	36	49	64	81	100	121	144	169	196
1	3	5	7	9	11	13	15	17	19	21	23	25	27

Pythagoras had then only to pick out the numbers in the third row which are squares, and his rule would be obtained by finding the formula connecting the square in the third line with the two adjacent squares in the second line. But even this would require some little argument; and I think a still better suggestion, because making pure observation play a greater part, is that of P. Treutlein (*Zeitschrift für Mathematik und Physik*, XXVIII., 1883, Hist.-litt. Abtheilung).

We have the best evidence (e.g. in Theon of Smyrna) of the practice of representing square numbers and other figured numbers, e.g. oblong, triangular, hexagonal, by dots or signs arranged in the shape of the particular figure. (Cf. Aristotle, *Metaph.* 1092 b 12). Thus, says Treutlein, it would be easily seen that any square number can be turned into the next higher square by putting a single row of dots round two adjacent sides, in the form of a gnomon (see figures on next page).

If a is the side of a particular square, the gnomon round it is shown by simple inspection to contain $2a + 1$ dots or units. Now, in order that $2a + 1$ may itself be a square, let us suppose

$$2a + 1 = n^2,$$

whence

$$a = \tfrac{1}{2}(n^2 - 1),$$

and

$$a + 1 = \tfrac{1}{2}(n^2 + 1).$$

In order that a and $a + 1$ may be integral, n must be odd, and we have at once the Pythagorean formula

$$n^2 + \left(\frac{n^2 - 1}{2}\right)^2 = \left(\frac{n^2 + 1}{2}\right)^2.$$

I think Treutlein's hypothesis is shown to be the correct one by the passage in Aristotle's *Physics* already quoted, where the reference is undoubtedly to the Pythagoreans, and odd numbers are clearly identified with *gnomons* "placed round 1." But the ancient commentaries on the passage make the matter clearer still. Philoponus says: "As a proof . . . the Pythagoreans refer to what

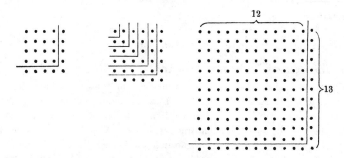

happens with the addition of numbers; for when the odd numbers are successively added to a square number they keep it square and equilateral. . . . Odd numbers are accordingly called *gnomons* because, when added to what are already squares, they preserve the square form. . . . Alexander has excellently said in explanation that the phrase 'when gnomons are placed round' means *making a figure* with the odd numbers (τὴν κατὰ τοὺς περιττοὺς ἀριθμοὺς σχηματογραφίαν) . . . for it is the practice with the Pythagoreans to *represent things in figures* (σχηματογραφεῖν)."

The next question is: assuming this explanation of the Pythagorean formula, what are we to say of the origin of Plato's? It could of course be obtained as a particular case of the general formula of Eucl. x. already referred to; but there are two simple alternative explanations in this case also. (1) Bretschneider observes that, to obtain Plato's formula, we have only to double the sides of the squares in the Pythagorean formula,

for
$$(2n)^2 + (n^2 - 1)^2 = (n^2 + 1)^2,$$

where however n is not necessarily odd.

(2) Treutlein would explain by means of an extension of the gnomon idea. As, he says, the Pythagorean formula was obtained by placing a gnomon consisting of a single row of dots round two adjacent sides of a square, it would be natural to try whether another solution could not be found by placing round the square a gnomon consisting of a *double* row of dots. Such a gnomon would equally turn the square into a larger square; and the question would be whether the double-row gnomon itself could be a square. If the side of the origi- nal square was a, it would easily be seen that the number of units in the double-row gnomon would be $4a + 4$, and we have only to put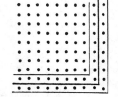

$$4a + 4 = 4n^2,$$
whence
$$a = n^2 - 1,$$
$$a + 2 = n^2 + 1,$$

and we have the Platonic formula

$$(2n)^2 + (n^2 - 1)^2 = (n^2 + 1)^2.$$

I think this is, in substance, the right explanation, but, in form, not quite correct. The Greeks would not, I think, have treated the *double* row as a gnomon. Their com- parison would have been between (1) a certain square *plus* a single-row gnomon and (2) the same square *minus* a single-row gnomon. As the appli- cation of Eucl. ii. 4 to the case where the segments of the side of the square are a, 1 enables the Pythagorean formula to be obtained as Treutlein obtains it, so I think that Eucl. ii. 8 confirms the idea that the Platonic formula was obtained by comparing a square *plus* a gnomon with the same square *minus* a gnomon. For ii. 8 proves that

$$4ab + (a - b)^2 = (a + b)^2,$$

whence, substituting 1 for b, we have
$$4a + (a - 1)^2 = (a + 1)^2,$$
and we have only to put $a = n^2$ to obtain Plato's formula.

The "theorem of Pythagoras" in India.

This question has been discussed anew in the last few years as the result of the publication of two important papers by Albert Bürk on *Das Āpastamba-Sulba-Sūtra* in the *Zeitschrift der deutschen morgenländischen Gesellschaft* (lv., 1901, and lvi., 1902). The first of the two papers contains the introduction and the text, the second the translation with notes. A selection of the most important parts of the material was made and issued by G. Thibaut in the *Journal of the Asiatic Society of Bengal*, XLIV., 1875, Part I. (reprinted also at Calcutta, 1875, as *The Śulvasūtras*, by G. Thibaut). Thibaut in this work gave a most valuable comparison of extracts from the three Śulvasūtras by Bāudhāyana, Āpastamba and Kātyāyana respectively, with a running commentary and an estimate of the date and originality of the geometry of the Indians. Bürk has however done good service by making the Āpastamba-Ś.-S. accessible in its entirety and investigating the whole subject afresh. With the natural enthusiasm of an editor for the work he is editing, he roundly maintains, not only that the Pythagorean theorem was known and proved in all its generality by the Indians long before the date of Pythagoras (about 580–500 B.C.), but that they had also discovered the irrational; and further that, so far from Indian geometry being indebted to the Greek, the much-travelled Pythagoras probably obtained his theory from India. Three important notices and criticisms of Bürk's work have followed, by H. G. Zeuthen (*"Théorème de Pythagore," Origine de la Géométrie scientifique*, 1904, already quoted), by Moritz Cantor (*Über die älteste indische Mathematik* in the *Archiv der Mathematik und Physik*, VIII., 1905) and by Heinrich Vogt (*Haben die alten Inder den Pythagoreischen Lehrsatz und das Irrationale gekannt ?* in the *Bibliotheca Mathematica*, VII$_3$, 1906. See also Cantor's *Geschichte der Mathematik*, I$_3$.

The general effect of the criticisms is, I think, to show the necessity for the greatest caution, to say the least, in accepting Bürk's conclusions.

I proceed to give a short summary of the portions of the contents of the Āpastamba-Ś.-S. which are important in the present connexion. It may be premised that the general object of the book is to show how to construct altars of certain shapes, and to vary the dimensions of altars without altering the form. It is a collection of *rules* for carrying out certain constructions. There are no proofs, the nearest approach to a proof being in the rule for obtaining the area of an isosceles trapezium, which is done by drawing a perpendicular from one extremity of the smaller of the two parallel sides to the greater, and then taking away the triangle so cut off and placing it, the other side up, adjacent to the other equal side of the trapezium, thereby transforming the trapezium into a rectangle. It should also be observed that Āpastamba does not speak of *right-angled triangles*, but of two adjacent sides and the diagonal of a *rectangle*. For brevity, I shall use the expression "rational rectangle" to denote a rectangle the two sides and the diagonal of which can be expressed in terms of rational numbers. The references in brackets are to the chapters and numbers of Āpastamba's work.

(1) Constructions of right angles by means of cords of the following relative lengths respectively:

$$\begin{cases} 3, & 4, & 5 & \text{(i. 3, v. 3)} \\ 12, & 16, & 20 & \text{(v. 3)} \\ 15, & 20, & 25 & \text{(v. 3)} \end{cases}$$

$$\begin{cases} 5, & 12, & 13 & \text{(v. 4)} \\ 15, & 36, & 39 & \text{(i. 2, v. 2, 4)} \end{cases}$$

$$8, \quad 15, \quad 17 \quad \text{(v. 5)}$$
$$12, \quad 35, \quad 37 \quad \text{(v. 5)}$$

(2) A general enunciation of the Pythagorean theorem thus: "The diagonal of a rectangle produces [i.e. the square on the diagonal is equal to] the sum of what the longer and shorter sides separately produce [i.e. the squares on the two sides]."   (i. 4)

(3) The application of the Pythagorean theorem to a *square* instead of a rectangle [i.e. to an *isosceles* right-angled triangle]: "The diagonal of a square produces an area double [of the original square]." (i. 5)

(4) An approximation to the value of $\sqrt{2}$; the diagonal of a square is $\left(1 + \frac{1}{3} + \frac{1}{3 \cdot 4} - \frac{1}{3 \cdot 4 \cdot 34}\right)$ times the side. (i. 6)

(5) Application of this approximate value to the construction of a square with side of any length. (ii. 1)

(6) The construction of $a\sqrt{3}$, by means of the Pythagorean theorem, as the diagonal of a rectangle with sides a and $a\sqrt{2}$. (ii. 2)

(7) Remarks equivalent to the following:

(a) $a\sqrt{\frac{1}{3}}$ is the side of $\frac{1}{9}(a\sqrt{3})^2$, or $a\sqrt{\frac{1}{3}} = \frac{1}{3}a\sqrt{3}$. (ii. 3)

(b) A square on length of 1 unit gives 1 unit square (iii. 4)

		2 units gives 4 unit squares	(iii. 6)
"	"	3 " 9 "	(iii. 6)
"	"	$1\frac{1}{2}$ " $2\frac{1}{4}$ "	(iii. 8)
"	"	$2\frac{1}{2}$ " $6\frac{1}{4}$ "	(iii. 8)
"	"	$\frac{1}{2}$ " $\frac{1}{4}$ "	(iii. 10)
"	"	$\frac{1}{3}$ " $\frac{1}{9}$ "	(iii. 10)

(c) Generally, the square on any length contains as many rows (of small, unit, squares) as the length contains units. (iii. 7)

(8) Constructions, by means of the Pythagorean theorem, of

(a) the *sum* of two squares as one square, (ii. 4)

(b) the *difference* of two squares as one square. (ii. 5)

(9) A transformation of a rectangle into a square. (ii. 7)

[This is not directly done as by Euclid in II. 14, but the rectangle is first transformed into a gnomon, i.e. into the difference between two squares, which difference is then transformed into one square by the preceding rule. If *ABCD* be the given rectangle of which *BC* is the longer side, cut off the square *ABEF*, bisect the rectangle *DE* left over by *HG* parallel to *FE*, move the upper half *DG* and place it on *AF* as base in the position *AK*. Then the rectangle *ABCD* is equal to the gnomon which is the difference between the square *LB* and the square *LF*. In other words, Āpastamba transforms the rectangle *ab* into the difference between the squares $\left(\frac{a+b}{2}\right)^2$ and $\left(\frac{a-b}{2}\right)^2$.]

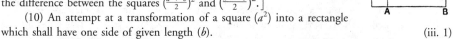

(10) An attempt at a transformation of a square (a^2) into a rectangle which shall have one side of given length (b). (iii. 1)

[This shows no sign of such a procedure as that of Eucl. I. 44, and indeed does no more than say that we must subtract *ab* from a^2 and then adapt the remainder $a^2 - ab$ so that it may "fit on" to the rectangle *ab*. The problem is therefore only reduced to another of the same kind, and presumably it was only solved *arithmetically* in the case where *a*, *b* are given numerically. The Indian was therefore far from the general, geometrical, solution.]

(11) Increase of a given square into a larger square. (iii. 9)

[This amounts to saying that you must add two rectangles (a, b) and another square (b^2) in order to transform a square a^2 into a square $(a + b)^2$. The formula is therefore that of Eucl. ii. 4, $a^2 + 2ab + b^2 = (a + b)^2$.]

The first important question in relation to the above is that of date. Bürk assigns to the *Āpastamba-Śulba-Sūtra* a date at least as early as the 5th or 4th century B.C. He observes however (what is likely enough) that the matter of it must have been much older than the book itself.

19

Further, as regards one of the constructions for right angles, that by means of cords of lengths 15, 36, 39, he shows that it was known at the time of the *Tāittirīya-Samhitā* and the *Satapatha-Brā hmana*, still older works belonging to the 8th century B.C. at latest. It may be that (as Bürk maintains) the discovery that triangles with sides (a, b, c) in rational numbers such that $a^2 + b^2 = c^2$ are right-angled was nowhere made so early as in India. We find however in two ancient Chinese treatises (1) a statement that the diagonal of the rectangle (3, 4) is 5 and (2) a rule for finding the hypotenuse of a "right triangle" from the sides, while tradition connects both works with the name of Chou Kung who died 1105 B.C. (D. E. Smith, *History of Mathematics*).

As regards the various "rational rectangles" used by Apastamba, it is to be observed that two of the seven, viz. 8, 15, 17 and 12, 35, 37, do not belong to the Pythagorean series, the others consist of two which belong to it, viz. 3, 4, 5 and 5, 12, 13, and multiples of these. It is true, as remarked by Zeuthen, that the rules of ii. 7 and iii. 9, numbered (9) and (11) above respectively, would furnish the means of finding any number of "rational rectangles." But it would not appear that the Indians had been able to formulate any general rule; otherwise their list of such rectangles would hardly have been so meagre. Āpastamba mentions seven only, really reducible to four (though one other, 7, 24, 25, appears in the Bāudhāyanaś.-S., supposed to be older than Āpastamba). These are all that Āpastamba knew of, for he adds (v. 6): "So many *recognisable* (erkennbare) constructions are there," implying that he knew of no other "rational rectangles" that could be employed. But the words also imply that the theorem of the square on the diagonal is also true of other rectangles not of the "recognisable" kind, i.e. rectangles in which the sides and the diagonal are not in the ratio of integers; this is indeed implied by the constructions for $\sqrt{2}$, $\sqrt{3}$ etc. up to $\sqrt{6}$ (cf. ii. 2, viii. 5). This is all that can be said. The theorem is, it is true, enunciated as a general proposition, but there is no sign of anything like a general proof; there is nothing to show that the assumption of its universal truth was founded on anything better than an imperfect induction from a certain number of cases, discovered empirically, of triangles with sides in the ratio of whole numbers in which the property (1) that the square on the longest side is equal to the sum of the squares on the other two sides was found to be always accompanied by the property (2) that the latter two sides include a right angle.

It remains to consider Bürk's claim that the Indians had discovered the *irrational.* This is based upon the approximate value of $\sqrt{2}$ given by Āpastamba in his rule i. 6 numbered (4) above. There is nothing to show how this was arrived at, but Thibaut's suggestion certainly seems the best and most natural. The Indians may have observed that $17^2 = 289$ is nearly double of $12^2 = 144$. If so, the next question which would naturally occur to them would be, by how much the side 17 must be diminished in order that the square on it may be 288 *exactly.* If, in accordance with the Indian fashion, a gnomon with unit area were to be subtracted from a square with 17 as side, this would approximately be secured by giving the gnomon the breadth $\frac{1}{34}$, for $2 \times 17 \times \frac{1}{34} = 1$. The side of the smaller square thus arrived at would be $17 - \frac{1}{34} = 12 + 4 + 1 - \frac{1}{34}$, whence, dividing out by 12, we have

$$\sqrt{2} = 1 + \frac{1}{3} + \frac{1}{3 \cdot 4} - \frac{1}{3 \cdot 4 \cdot 34}, \text{ approximately.}$$

But it is a far cry from this calculation of an approximate value to the discovery of the *irrational.* First, we ask, is there any sign that this value was known to be inexact? It comes directly after the statement (i. 6) that the square on the diagonal of a square is double of that square, and the rule is quite boldly stated without any qualification: "lengthen the unit by one-third and the latter by one-quarter of itself less one-thirty-fourth of this part." Further, the approximate value is actually used for the purpose of constructing a square when the side is given (ii. 1). So familiar was the formula that it was apparently made the basis of a sub-division of measures of length. Thibaut observes (*Journal of the Asiatic Society of Bengal*, XLIX.) that, according to Bāudhāyana, the unit of length was divided into 12 *finger-breadths*, and that one of two divisions of the *fingerbreadth* was into 34 *sesame-corns*, and he adds that he has no doubt that this division, which he has not elsewhere met, owes its origin to the formula for $\sqrt{2}$. The result of using this subdivision would be that, in a square with side equal to 12 *fingerbreadths*, the diagonal would be 17 *fingerbreadths* less 1 *sesame-corn*. Is it conceivable that a sub-division of a measure of length would be based on an evaluation known to be inexact? No doubt the first discoverer would be aware that the area of a gnomon with breadth $\frac{1}{34}$ and outer side 17 is not exactly equal to 1 but less than it by the square of $\frac{1}{34}$ or by $\frac{1}{1156}$, and therefore that, in taking that gnomon as the proper area to be subtracted from 17^2, he was leaving out of account the small fraction $\frac{1}{1156}$; as, however, the object of the whole proceeding was purely practical, he would, without hesitation, ignore this as being of no practical importance, and, thereafter, the formula would be handed down and taken as a matter of course without arousing suspicion as to its accuracy. This supposition is confirmed by reference to the sort of rules which the Indians allowed themselves to regard as accurate. Thus Āpastamba himself gives a construction for a circle equal in area to a given square, which is equivalent to taking $\pi = 3.09$, and yet observes that it gives the required circle "*exactly*" (iii. 2), while his construction of a square equal to a circle, which he equally calls "exact," makes the side of the square equal to $\frac{13}{15}$ths of the diameter of the circle (iii. 3), and is equivalent to taking $\pi = 3.00\overset{.}{4}$. But, even if some who used the approximation for $\sqrt{2}$ were conscious that it was not quite accurate (of which there is no evidence), there is an immeasurable difference between arrival at this consciousness and the discovery of the irrational. As Vogt says, three stages had to be passed through before the irrationality of the diagonal of a square was discovered in any real sense. (1) All values found by direct measurement or calculations based thereon have to be recognised as being inaccurate. Next (2) must supervene the conviction that it is *impossible* to arrive at an accurate arithmetical expression of the value. And lastly (3) the impossibility must be proved. Now there is no real evidence that the Indians, at the date in question, had even reached the first stage, still less the second or third.

The net results then of Bürk's papers and of the criticisms to which they have given rise appear to be these. (1) It must be admitted that Indian geometry had reached the stage at which we find it in Āpastamba quite independently of Greek influence. But (2) the old Indian geometry was purely empirical and practical, far removed from abstractions such as the irrational. The Indians had indeed, by-trial in particular cases, persuaded themselves of the truth of the Pythagorean theorem and enunciated it in all its generality; but they had not established it by scientific proof.

Alternative proofs.

I. The well-known proof of i. 47 obtained by putting two squares side by side, with their bases continuous, and cutting off right-angled triangles which can then be put on again in different positions, is attributed by an-Nairīzī to Thābit b. Qurra (826–901 A.D.).

His actual construction proceeds thus.

Let *ABC* be the given triangle right-angled at *A*.

Construct on *AB* the square *AD*;

produce *AC* to *F* so that *EF* may be equal to *AC*.

Construct on *EF* the square *EG*, and produce *DH* to *K* so that *DK* may be equal to *AC*.

It is then proved that, in the triangles *BAC*, *CFG*, *KHG*, *BDK*, the sides *BA*, *CF*, *KH*, *BD* are all equal, and the sides *AC*, *FG*, *HG*, *DK* are all equal.

The angles included by the equal sides are all right angles; hence the four triangles are equal in all respects.　　　　[i. 4]

Hence *BC*, *CG*, *GK*, *KB* are all equal.

Further the angles *DBK*, *ABC* are equal; hence, if we add to each the angle *DBC*, the angle *KBC* is equal to the angle *ABD* and is therefore a right angle.

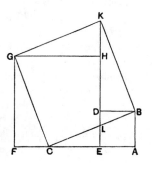

In the same way the angle *CGK* is right;
therefore *BCGK* is a square, i.e. the square on *BC*.

Now the sum of the quadrilateral *GCLH* and the triangle *LDB* together with two of the equal triangles make the squares on *AB*, *AC*, and together with the other two make the square on *BC*.

Therefore etc.

II. Another proof is easily arrived at by taking the particular case of Pappus' more general proposition given below in which the given triangle is right-angled and the parallelograms on the sides containing the right angles are squares. If the figure is drawn, it will be seen that, with no more than one additional line inserted, it contains Thābit's figure, so that Thābit's proof may have been practically derived from that of Pappus.

III. The most interesting of the remaining proofs seems to be that shown in the accompanying figure. It is given by J. W. Müller, *Systematische Zusammenstellung der wichtigsten bisher bekannten Beweise des Pythag. Lehrsatzes* (Nürnberg, 1819), and in the second edition (Mainz, 1821) of Ign. Hoffmann, *Der Pythag. Lehrsatz mit 32 theils bekannten theils neuen Beweisen* [3 more in second edition]. It appears to come from one of the scientific papers of Lionardo da Vinci (1452–1519).

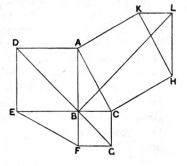

The triangle *HKL* is constructed on the base *KH* with the side *KL* equal to *BC* and the side *LH* equal to *AB*.

Then the triangle *HLK* is equal in all respects to the triangle *ABC*, and to the triangle *EBF*.

Now *DB*, *BG*, which bisect the angles *ABE*, *CBF* respectively, are in a straight line. Join *BL*.

It is easily proved that the four quadrilaterals *ADGC*, *EDGF*, *ABLK*, *HLBC* are all equal.

Hence the hexagons *ADEFGC*, *ABCHLK* are equal.

Subtracting from the former the two triangles *ABC*, *EBF*, and from the latter the two equal triangles *ABC*, *HLK*, we prove that the square *CK* is equal to the sum of the squares *AE*, *CF*.

Pappus' extension of i. 47.

In this elegant extension the triangle may be *any* triangle (not necessarily right-angled), and *any* parallelograms take the place of squares on two of the sides.

Pappus enunciates the theorem as follows:

If ABC *be a triangle, and any parallelograms whatever* ABED, BCFG *be described on* AB, BC, *and if* DE, FG *be produced to* H, *and* HB *be joined, the parallelograms* ABED, BCFG *are equal to the parallelogram contained by* AC, HB *in an angle which is equal to the sum of the angles* BAC, DHB.

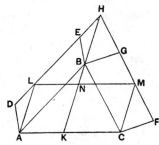

Produce *HB* to *K*; through *A*, *C* draw *AL*, *CM* parallel to *HK*, and join *LM*.

Then, since *ALHB* is a parallelogram, *AL*, *HB* are equal and parallel.

Similarly *MC*, *HB* are equal and parallel.

Therefore *AL*, *MC* are equal and parallel;
whence *LM*, *AC* are also equal and parallel,

and *ALMC* is a parallelogram.

Further, the angle *LAC* of this parallelogram is equal to the sum of the angles *BAC*, *DHB*, since the angle *DHB* is equal to the angle *LAB*.

Now, since the parallelogram *DABE* is equal to the parallelogram *LABH* (for they are on the same base *AB* and in the same parallels *AB*, *DH*),
and likewise *LABH* is equal to *LAKN* (for they are on the same base *LA* and in the same parallels *LA*, *HK*),

the parallelogram *DABE* is equal to the parallelogram *LAKN*.

For the same reason,

the parallelogram *BGFC* is equal to the parallelogram *NKCM*.

Therefore the sum of the parallelograms *DABE*, *BGFC* is equal to the parallelogram *LACM*, that is, to the parallelogram which is contained by *AC*, *HB* in an angle *LAC* which is equal to the sum of the angles *BAC*, *BHD*.

"And this is far more general than what is proved in the Elements about squares in the case of right-angled (triangles)."

Heron's proof that AL, BK, CF in Euclid's figure meet in a point.

The final words of Proclus' note on i. 47 are historically interesting. He says: "The demonstration by the writer of the Elements being clear, I consider that it is unnecessary to add anything further, and that we may be satisfied with what has been written, since in fact those who have added anything more, like Pappus and Heron, were obliged to draw upon what is proved in the sixth Book, for no really useful object." These words cannot of course refer to the extension of i. 47 given by Pappus; but the key to them, so far as Heron is concerned, is to be found in the commentary of an-Nairīzī on i. 47, wherein he gives Heron's proof that the lines *AL*, *FC*, *BK* in Euclid's figure meet in a point. Heron proved this by means of three lemmas which would most naturally be proved from the principle of similitude as laid down in Book VI., but which Heron, as a *tour de force*, proved on the principles of Book I. only. The *first* lemma is to the following effect.

If, in a triangle ABC, DE *be drawn parallel to the base* BC, *and if* AF *be drawn from the vertex* A *to the middle point* F *of* BC, *then* AF *will also bisect* DE.

This is proved by drawing *HK* through *A* parallel to *DE* or *BC*, and *HDL*, *KEM* through *D*, *E* respectively parallel to *AGF*, and lastly joining *DF*, *EF*.

Then the triangles *ABF*, *AFC* are equal (being on equal bases), and the triangles *DBF*, *EFC* are also equal (being on equal bases and between the same parallels).

Therefore, by subtraction, the triangles *ADF*, *AEF* are equal, and hence the parallelograms *AL*, *AM* are equal.

These parallelograms are between the same parallels *LM*, *HK*; therefore *LF*, *FM* are equal, whence *DG*, *GE* are also equal.

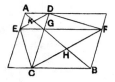

The *second* lemma is an extension of this to the case where *DE* meets *BA*, *CA* produced beyond *A*.

The *third* lemma proves the converse of Euclid i. 43, that, *If a parallelogram* AB *is cut into four others* ADGE, DF, FGCB, CE, *so that* DF, CE *are equal, the common vertex* G *will be on the diagonal* AB.

Heron produces *AG* till it meets *CF* in *H*. Then, if we join *HB*, we have to prove that *AHB* is one straight line. The proof is as follows. Since the areas *DF*, *EC* are equal, the triangles *DGF*, *ECG* are equal.

If we add to each the triangle *GCF*,

the triangles *ECF*, *DCF* are equal;

therefore *ED*, *CF* are parallel.

Now it follows from i. 34, 29 and 26 that the triangles *AKE*, *GKD* are equal in all respects;

therefore *EK* is equal to *KD*.

Hence, by the second lemma,

CH is equal to *HF*.

Therefore, in the triangles *FHB, CHG,*
the two sides *BF, FH* are equal to the two sides *GC, CH,*

and the angle *BFH* is equal to the angle *GCH*;
hence the triangles are equal in all respects,
and the angle *BHF* is equal to the angle *GHC*.

Adding to each the angle *GHF*, we find that the angles *BHF, FHG* are equal to the angles *CHG, GHF,*

and therefore to two right angles.

Therefore *AHB* is a straight line.

Heron now proceeds to prove the proposition that, in the accompanying figure, if *AKL* perpendicular to *BC* meet
EC in *M*, and if *BM, MG* be joined,

BM, MG are in one straight line.

Parallelograms are completed as shown in the figure, and
the diagonals *OA, FH* of the parallelogram *FH* are drawn.

Then the triangles *FAH, BAC* are clearly equal in all
respects;

therefore the angle *HFA* is equal to the angle *ABC*, and
therefore to the angle *CAK* (since *AK* is perpendicular to *BC*).

But, the diagonals of the rectangle *FH* cutting one another
in *Y*,

FY is equal to *YA,*
and the angle *HFA* is equal to the
angle *OAF*.

Therefore the angles *OAF, CAK* are equal, and accordingly

OA, AK are in a straight line.

Hence *OM* is the diagonal of *SQ*;

therefore *AS* is equal to *AQ*,
and, if we add *AM* to each,

FM is equal to *MH*.

But, since *EC* is the diagonal of the parallelogram *FN*,

FM is equal to *MN*.

Therefore *MH* is equal to *MN*;
and, by the third lemma, *BM, MG* are in a straight line.

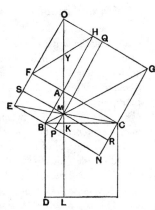

BOOK V

THE EUDOXIAN THEORY OF PROPORTION—
DEFINITIONS AND PROPOSITIONS

DEFINITIONS

1. A magnitude is a **part** of a magnitude, the less of the greater, when it measures the greater.

2. The greater is a **multiple** of the less when it is measured by the less.

3. A **ratio** is a sort of relation in respect of size between two magnitudes of the same kind.

4. Magnitudes are said to **have a ratio** to one another which are capable, when multiplied, of exceeding one another.

5. Magnitudes are said to **be in the same ratio**, the first to the second and the third to the fourth, when, if any equimultiples whatever be taken of the first and third, and any equimultiples whatever of the second and fourth, the former equimultiples alike exceed, are alike equal to, or alike fall short of, the latter equimultiples respectively taken in corresponding order.

6. Let magnitudes which have the same ratio be called **proportional**.

7. When, of the equimultiples, the multiple of the first magnitude exceeds the multiple of the second, but the multiple of the third does not exceed the multiple of the fourth, then the first is said to **have a greater ratio** to the second than the third has to the fourth.

8. A proportion in three terms is the least possible.

9. When three magnitudes are proportional, the first is said to have to the third the **duplicate ratio** of that which it has to the second.

10. When four magnitudes are <continuously> proportional, the first is said to have to the fourth the **triplicate ratio** of that which it has to the second, and so on continually, whatever be the proportion.

11. The term **corresponding magnitudes** is used of antecedents in relation to antecedents, and of consequents in relation to consequents.

12. **Alternate ratio** means taking the antecedent in relation to the antecedent and the consequent in relation to the consequent.

13. **Inverse ratio** means taking the consequent as antecedent in relation to the antecedent as consequent.

14. **Composition of a ratio** means taking the antecedent together with the consequent as one in relation to the consequent by itself.

15. **Separation of a ratio** means taking the excess by which the antecedent exceeds the consequent in relation to the consequent by itself.

16. **Conversion of a ratio** means taking the antecedent in relation to the excess by which the antecedent exceeds the consequent.

17. A ratio **ex aequali** arises when, there being several magnitudes and another set equal to them in multitude which taken two and two are in the same proportion, as the first is to the last among the first magnitudes, so is the first to the last among the second magnitudes;

Or, in other words, it means taking the extreme terms by virtue of the removal of the intermediate terms.

18. A **perturbed proportion** arises when, there being three magnitudes and another set equal to them in multitude, as antecedent is to consequent among the first magnitudes, so is antecedent to consequent among the second magnitudes, while, as the consequent is to a third among the first magnitudes, so is a third to the antecedent among the second magnitudes.

Proposition 1

If there be any number of magnitudes whatever which are, respectively, equimultiples of any magnitudes equal in multitude, then, whatever multiple one of the magnitudes is of one, that multiple also will all be of all.

Let any number of magnitudes whatever *AB*, *CD* be respectively equimultiples of any magnitudes *E*, *F* equal in multitude;

I say that, whatever multiple *AB* is of *E*, that multiple will *AB*, *CD* also be of *E*, *F*.

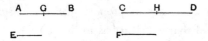

For, since *AB* is the same multiple of *E* that *CD* is of *F*, as many magnitudes as there are in *AB* equal to *E*, so many also are there in *CD* equal to *F*.

Let *AB* be divided into the magnitudes *AG*, *GB* equal to *E*,
and *CD* into *CH*, *HD* equal to *F*;
then the multitude of the magnitudes *AG*, *GB* will be equal to the multitude of the magnitudes *CH*, *HD*.

Now, since *AG* is equal to *E*, and *CH* to *F*,
therefore *AG* is equal to *E*, and *AG*, *CH* to *E*, *F*.

For the same reason
GB is equal to *E*, and *GB*, *HD* to *E*, *F*;

therefore, as many magnitudes as there are in *AB* equal to *E*,
so many also are there in *AB*, *CD* equal to *E*, *F*;
therefore, whatever multiple *AB* is of *E*, that multiple will *AB*, *CD* also be of *E*, *F*.
Therefore etc.

Q.E.D.

De Morgan remarks of v. 1–6 *that they are "simple propositions of concrete arithmetic, covered in language which makes them unintelligible to modern ears. The first, for instance, states no more than that* ten *acres and* ten *roods make* ten *times as much as one acre and one rood." One aim therefore of notes on these as well as the other propositions of Book V. should be to make their purport clearer to the learner by setting them side by side with the same truths expressed in the much shorter and more familiar modern (algebraical) notation. In doing so, we shall express magnitudes by the first letters of the alphabet,* a, b, c *etc., adopting small instead of capital letters so as to avoid confusion with Euclid's lettering; and we shall use the small letters* m, n, p *etc. to represent integral numbers. Thus* ma *will always mean* m *times* a *or the* mth *multiple of* a *(counting* 1. a *as the first,* 2. a *as the second multiple, and so on).*

Prop. 1 then asserts that, if ma, mb, mc *etc. be any equimultiples of* a, b, c *etc., then*

$$ma + mb + mc + \ldots = m(a + b + c + \ldots).$$

Proposition 2

If a first magnitude be the same multiple of a second that a third is of a fourth, and a fifth also be the same multiple of the second that a sixth is of the fourth, the sum of the first and fifth will also be the same multiple of the second that the sum of the third and sixth is of the fourth.

Let a first magnitude, *AB*, be the same multiple of a second, *C*, that a third, *DE*, is of a fourth, *F*, and let a fifth, *BG*, also be the same multiple of the second, *C*, that a sixth, *EH*, is of the fourth *F*;

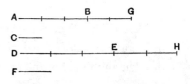

I say that the sum of the first and fifth, *AG*, will be the same multiple of the second, *C*, that the sum of the third and sixth, *DH*, is of the fourth, *F*.

For, since *AB* is the same multiple of *C* that *DE* is of *F*, therefore, as many magnitudes as there are in *AB* equal to *C*, so many also are there in *DE* equal to *F*.

For the same reason also,
as many as there are in *BG* equal to *C*, so many are there also in *EH* equal to *F*;
therefore, as many as there are in the whole *AG* equal to *C*, so many also are there in the whole *DH* equal to *F*.

Therefore, whatever multiple *AG* is of *C*, that multiple also is *DH* of *F*.

Therefore the sum of the first and fifth, *AG*, is the same multiple of the second, *C*, that the sum of the third and sixth, *DH*, is of the fourth, *F*.

Therefore etc.

Q.E.D.

To find the corresponding formula for the result of this proposition, we may suppose a *to be the "second" magnitude and* b *the "fourth." If now the "first" magnitude is* ma, *the "third" is, by hypothesis,* mb; *and, if the "fifth" magnitude is* na, *the "sixth" is* nb. *The proposition then asserts that* ma + na *is the same multiple of* a *that* mb + nb *is of* b.

More generally, if pa, qa . . . *and* pb, qb . . . *be any further equimultiples of* a, b *respectively,* ma + na + pa + qa + . . . *is the same multiple of* a *that* mb + nb + pb + qb + . . . *is of* b. *This extension is stated in Simson's corollary to* v. 2 *thus:*

"From this it is plain that, if any number of magnitudes AB, BG, GH *be multiples of another* C; *and as many* DE, EK, KL *be the same multiples of* F, *each of each; the whole of the first, viz.* AH, *is the same multiple of* C *that the whole of the last, viz.* DL, *is of* F."

The course of the proof, which separates m *into its units and also* n *into its units, practically tells us that the multiple of* a *arrived at by adding the* two *multiples is the* (m + n)th *multiple; or practically we are shown that*

$$ma + na = (m + n)a,$$

or, more generally, that

$$ma + na + pa + \ldots = (m + n + p + \ldots)a."$$

Proposition 3

If a first magnitude be the same multiple of a second that a third is of a fourth, and if equimultiples be taken of the first and third, then also ex aequali the magnitudes taken will be equimultiples respectively, the one of the second and the other of the fourth.

Let a first magnitude *A* be the same multiple of a second *B* that a third *C* is of a fourth *D*, and let equimultiples *EF*, *GH* be taken of *A*, *C*;
I say that *EF* is the same multiple of *B* that *GH* is of *D*.

For, since *EF* is the same multiple of *A* that *GH* is of *C*, therefore, as many magnitudes as there are in *EF* equal to *A*, so many also are there in *GH* equal to *C*.

Let *EF* be divided into the magnitudes *EK*, *KF* equal to *A*, and *GH* into the magnitudes *GL*, *LH* equal to *C*;
then the multitude of the magnitudes *EK*, *KF* will be equal to the multitude of the magnitudes *GL*, *LH*.

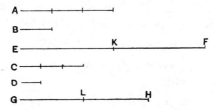

And, since *A* is the same multiple of *B* that *C* is of *D*, while *EK* is equal to *A*, and *GL* to *C*,
therefore *EK* is the same multiple of *B* that *GL* is of *D*.

For the same reason
KF is the same multiple of *B* that *LH* is of *D*.

Since, then, a first magnitude *EK* is the same multiple of a second *B* that a third *GL* is of a fourth *D*,

28

and a fifth *KF* is also the same multiple of the second *B* that a sixth *LH* is of the fourth *D*,

therefore the sum of the first and fifth, *EF*, is also the same multiple of the second *B* that the sum of the third and sixth, *GH*, is of the fourth *D*. [v. 2]

Therefore etc.

Q.E.D.

Heiberg remarks of the use of ex aequali *in the enunciation of this proposition that, strictly speaking, it has no reference to the definition (17) of a* ratio ex aequali. *But the uses of the expression here and in the definition are, I think, sufficiently parallel, as may be seen thus. The proposition asserts that, if*

na, nb *are equimultiples of* a, b,

and if m · na, m · nb *are equimultiples of* na, nb,

then m · na *is the same multiple of* a *that* m · nb *is of* b. *Clearly the proposition can be extended by taking further equimultiples of the last equimultiples and so on; and we can prove that*

p · q ... m · na *is the same multiple of* a *that* p · q ... m · nb *is of* b,

where the series of numbers p · q ... m · n *is exactly the same in both expressions; and* ex aequali (δι ἴσου) *expresses the fact that the equimultiples are at the same distance from* a, b *in the series* na, m · na ... *and* nb, m · nb ... *respectively.*

Here again the proof breaks m *into its units, and then breaks* n *into its units; and we are practically shown that the multiple of* a *arrived at, viz.* m · na, *is the multiple denoted by the product of the numbers* m, n, *i.e. the* (mn)*th multiple, or in other words that*

m · na = mn · a.

Proposition 4

If a first magnitude have to a second the same ratio as a third to a fourth, any equimultiples whatever of the first and third will also have the same ratio to any equimultiples whatever of the second and fourth respectively, taken in corresponding order.

For let a first magnitude *A* have to a second *B* the same ratio as a third *C* to a fourth *D*; and let equimultiples *E, F* be taken of *A, C*, and *G, H* other, chance, equimultiples of *B, D*;

I say that, as *E* is to *G*, so is *F* to *H*.

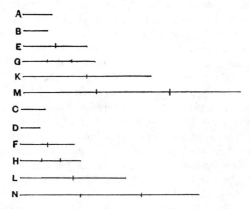

For let equimultiples *K, L* be taken of *E, F*, and other, chance, equimultiples *M, N* of *G, H*.

29

Since E is the same multiple of A that F is of C,
and equimultiples K, L of E, F have been taken,
therefore K is the same multiple of A that L is of C.   [v. 3]

For the same reason

M is the same multiple of B that N is of D.

And, since, as A is to B, so is C to D,
and of A, C equimultiples K, L have been taken,
and of B, D other, chance, equimultiples M, N,
therefore, if K is in excess of M, L also is in excess of N,
if it is equal, equal, and if less, less.   [v. Def. 5]

And K, L are equimultiples of E, F,
and M, N other, chance, equimultiples of G, H;
therefore, as E is to G, so is F to H.   [v. Def. 5]

Therefore etc.

Q.E.D.

This proposition shows that, if a, b, c, d *are proportionals, then*

ma *is to* nb *as* mc *is to* nd;

and the proof is as follows:

Take pma, pmc *any equimultiples of* ma, mc, *and* qnb, qnd *any equimultiples of* nb, nd.
Since a : b = c : d, *it follows* [v. Def. 5] *that,*

according as pma > = < qnb, pmc > = < qnd.

But the p- *and* q-*equimultiples are* any *equimultiples;*
therefore [v. Def. 5]

ma : nb = mc : nd.

It will be observed that Euclid's phrase for taking *any* equimultiples of A, C and *any* other equimultiples of B, D is "let there be taken equimultiples E, F of A, C, and G, H other, *chance,* equimultiples of B, D," E, F being called ἰσάκις πολλαπλάσια simply, and G, H ἄλλα, ἅ ἔτυχεν, ἰσάκιςπολ λαπλάσια. And similarly, when *any* equimultiples (K, L) of E, F come to be taken, and *any* other equimultiples (M, N) of G, H. But later on Euclid uses the same phrases about the *new* equimultiples with reference to the original magnitudes, reciting that "there have been taken, of A, C, equimultiples K, L and of B, D, other, *chance,* equimultiples M, N"; whereas M, N are not *any equimultiples whatever* of B, D, but are *any* equimultiples of the *particular* multiples (G, H) which have been taken of B, D respectively, though *these latter* have been taken at random. Simson would, in the first place, add ἅ ἔτυχεν in the passages where *any* equimultiples E, F are taken of A, C and *any* equimultiples K, L are taken of E, F, because the words are "wholly necessary" and, in the second place, would leave them out where M, N are called ἄλλα, ἅ ἔτυχεν, ἰσάκις πολλαπλάσια of B, D, because it is not true that of B, D have been taken "any equimultiples whatever (ἅετυχε), M, N." Simson adds: "And it is strange that neither Mr. Briggs, who did right to leave out these words in one place of Prop. 13 of this book, nor Dr. Gregory, who changed them into the word 'some' in three places, and left them out in a fourth of that same Prop. 13, did not also leave them out in this place of Prop. 4 and in the second of the two places where they occur in Prop. 17 of this book, in neither of which they can stand consistent with truth: And in none of all these places, even in those which they corrected in their Latin translation, have they cancelled the words ἅ ἔτυχε in the Greek text, as they ought to have done. The same words α ετυχε are found in four places of Prop. 11 of this book, in the first and last of which they are necessary, but in the second and third, though they are true, they are quite super-fluous; as they likewise are in the second of the two places in which they are found in the 12th prop. and in the like places of Prop. 22, 23 of this book; but are wanting in the last place of Prop. 23, as also in Prop. 25, Book XI."

As will be seen, Simson's emendations amount to alterations of the text so considerable as to suggest doubt whether we should be justified in making them in the absence of MS. authority. The phrase "equimultiples of *A, C* and other, chance, equimultiples of *B, D*" recurs so constantly as to suggest that it was for Euclid a quasi-stereotyped phrase, and that it is equally genuine wher-ever it occurs. Is it then absolutely necessary to insert ἃ ἔτυχε in places where it does not occur, and to leave it out in the places where Simson holds it to be wrong? I think the text can be defended as it stands. In the first place to say "take equimultiples of *A, C*" is a fair enough way of saying take *any* equimultiples *whatever* of *A, C.* The other difficulty is greater, but may, I think, be only due to the adoption of *any whatever* as the translation of ἃ ἔτυχε. As a matter of fact, the words only mean *chance* equimultiples, equimultiples which are the result of random selection. Is it not justifiable to describe the product of two *chance* numbers, numbers selected at random, as being a "chance number," since it is the result of two random selections? I think so, and I have translated ἃ ἔτυχε accordingly as implying, in the case in question, "other equimultiples whatever they may happen to be."

To this proposition Theon added the following:

"Since then it was proved that, if *K* is in excess of *M, L* is also in excess of *N*, if it is equal, (the other is) equal, and if less, less,

it is clear also that,

if *M* is in excess of *K, N* is also in excess of *L*, if it is equal, (the other is) equal, and if less, less;

and for this reason,

as *G* is to *E*, so also is *H* to *F.*

PORISM. From this it is manifest that, if four magnitudes be proportional, they will also be proportional inversely."

Simson rightly pointed out that the demonstration of what Theon intended to prove, viz. that, if *E, G, F, H* be proportionals, they are proportional inversely, i.e. *G* is to *E* as *H* is to *F*, does not in the least depend upon this 4th proposition or the proof of it; for, when it is said that, "if *K* exceeds *M, L* also exceeds *N* etc.," this is not proved from the fact that *E, G, F, H* are pro-portionals (which is the conclusion of Prop. 4), but from the fact that *A, B, C, D* are proportionals.

The proposition that, if *A, B, C, D* are proportionals, they are also proportionals inversely is not given by Euclid, but Simson supplies the proof in his Prop. B. The fact is really obvious at once from the 5th definition of Book V., and Euclid probably omitted the proposition as unnecessary.

Simson added, in place of Theon's corollary, the following:

"Likewise, if the first has the same ratio to the second which the third has to the fourth, then also any equimultiples whatever of the first and third have the same ratio to the second and fourth : And, in like manner, the first and the third have the same ratio to any equimultiples what-ever of the second and fourth."

The proof, of course, follows exactly the method of Euclid's proposition itself, with the only difference that, instead of one of the two pairs of equimultiples, the magnitudes themselves are taken. In other words, the conclusion that

ma is to nb as mc is to nd

is equally true when either m or n is equal to unity.

As De Morgan says, Simson's corollary is only necessary to those who will not admit *M* into the list *M, 2M, 3M* etc.; the exclusion is grammatical and nothing else. The same may be said of Simson's Prop. A to the effect that, "If the first of four magnitudes has to the second the same ratio which the third has to the fourth: then, if the first be greater than the second, the third is also greater than the fourth; and if equal, equal; if less, less." This is needless to those who believe *once* A to be a proper component of the list of multiples, in spite of *multus* signifying many.

Proposition 5

If a magnitude be the same multiple of a magnitude that a part subtracted is of a part subtracted, the remainder will also be the same multiple of the remainder that the whole is of the whole.

For let the magnitude *AB* be the same multiple of the magnitude *CD* that the part *AE* subtracted is of the part *CF* subtracted;

I say that the remainder *EB* is also the same multiple of the remainder *FD* that the whole *AB* is of the whole *CD*.

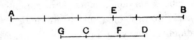

For, whatever multiple *AE* is of *CF*, let *EB* be made that multiple of *CG*.

Then, since *AE* is the same multiple of *CF* that *EB* is of *GC*,

therefore *AE* is the same multiple of *CF* that *AB* is of *GF*.   [v. 1]

But, by the assumption, *AE* is the same multiple of *CF* that *AB* is of *CD*.

Therefore *AB* is the same multiple of each of the magnitudes *GF*, *CD*;

therefore *GF* is equal to *CD*.

Let *CF* be subtracted from each;

therefore the remainder *GC* is equal to the remainder *FD*.

And, since *AE* is the same multiple of *CF* that *EB* is of *GC*,

and *GC* is equal to *DF*,

therefore *AE* is the same multiple of *CF* that *EB* is of *FD*.

But, by hypothesis,

AE is the same multiple of *CF* that *AB* is of *CD*;

therefore *EB* is the same multiple of *FD* that *AB* is of *CD*.

That is, the remainder *EB* will be the same multiple of the remainder *FD* that the whole *AB* is of the whole *CD*.

Therefore etc.

Q.E.D.

8. **let EB be made that multiple of CG**, τοσαυτάπλάσιον γεγονέτω καὶ τὸ ΕΒ τοῦ ΓΗ. From this way of stating the construction one might suppose that *CG* was given and *EB* had to be found equal to a certain multiple of it. But in fact *EB* is what is given and *CG* has to be found, i.e. *CG* has to be constructed as a certain *sub*multiple of *EB*.

This proposition corresponds to v. 1, *with subtraction taking the place of addition. It proves the formula*

$$ma - mb = m(a - b).$$

Euclid's construction assumes that, if AE *is any multiple of* CF, *and* EB *is any other magnitude, a fourth straight line can be found such that* EB *is the same multiple of it that* AE *is of* CF, *or in other words that, given any magnitude, we can divide it into any number of equal parts. This is however not proved, even of straight lines, much less other magnitudes, until* vi. 9. *Peletarius had already seen this objection to the construction. The difficulty is not got over by regarding it merely as a* hypothetical *construction; for hypothetical constructions are not in Euclid's manner. The remedy is to substitute the alternative construction given by Simson, after Peletarius and Campanus' translation from the Arabic, which only requires us to add a magnitude to itself a certain number of times. The demonstration follows Euclid's line exactly.*

"*Take* AG *the same multiple of* FD *that* AE *is of* CF;

therefore AE *is the same multiple of* CF *that* EG *is of* CD.   [v. 1]*

But AE, *by hypothesis, is the same multiple of* CF *that* AB *is of* CD; *therefore* EG *is the same multiple of* CD *that* AB *is of* CD;

wherefore EG *is equal to* AB.

Take from them the common magnitude AE; *the remainder* AG *is equal to the remainder* EB.

Wherefore, since AE *is the same multiple of* CF *that* AG *is of* FD, *and since* AG *is equal to* EB, *therefore* AE *is the same multiple of* CF *that* EB *is of* FD.

But AE *is the same multiple of* CF *that* AB *is of* CD; *therefore* EB *is the same multiple of* FD *that* AB *is of* CD."

Q.E.D.

Euclid's proof amounts to this.

Suppose a magnitude x *taken such that*

$$ma - mb = mx, say.$$

Add mb *to each side, whence (by* v. 1*)*

$$ma = m(x + b).$$

Therefore $a = x + b, or x = a - b,$

so that $ma - mb = m(a - b).$

Simson's proof, on the other hand, argues thus.

Take x = m(a − b), *the same multiple of* (a − b) *that* mb *is of* b.

Then, by addition of mb *to both sides, we have [v. 1]*

$$x + mb = ma,$$

or $x = ma - mb.$

That is, $ma - mb = m(a - b).$

Proposition 6

If two magnitudes be equimultiples of two magnitudes, and any magnitudes subtracted from them be equimultiples of the same, the remainders also are either equal to the same or equimultiples of them.

For let two magnitudes *AB, CD* be equimultiples of two magnitudes *E, F,* and let *AG, CH* subtracted from them be equimultiples of the same two *E, F*; I say that the remainders also, *GB, HD,* are either equal to *E, F* or equimultiples of them.

For, first, let *GB* be equal to *E*; I say that *HD* is also equal to *F*.

For let *CK* be made equal to *F*.

Since *AG* is the same multiple of *E* that *CH* is of *F*, while *GB* is equal to *E* and *KC* to *F*, therefore *AB* is the same multiple of *E* that *KH* is of *F*.   [v. 2]

But, by hypothesis, *AB* is the same multiple of *E* that *CD* is of *F*; therefore *KH* is the same multiple of *F* that *CD* is of *F*.

Since then each of the magnitudes *KH, CD* is the same multiple of *F*, therefore *KH* is equal to *CD*.

Let *CH* be subtracted from each;
therefore the remainder *KC* is equal to the remainder *HD*.

But *F* is equal to *KC*;
therefore *HD* is also equal to *F*.

Hence, if *GB* is equal to *E*, *HD* is also equal to *F*.

Similarly we can prove that, even if *GB* be a multiple of *E*, *HD* is also the same multiple of *F*.

Therefore etc.

Q.E.D.

This proposition corresponds to v. 2, with subtraction taking the place of addition. It asserts namely that, if n *is less than* m, ma − na *is the same multiple of* a *that* mb − nb *is of* b. *The enunciation distinguishes the cases in which* m − n *is equal to 1 and greater than 1 respectively.*

Simson observes that, while only the first case (the simpler one) is proved in the Greek, both are given in the Latin translation from the Arabic; and he supplies accordingly the proof of the second case, which Euclid leaves to the reader. The fact is that it is exactly the same as the other except that, in the construction, CK *is made the same multiple of* F *that* GB *is of* E, *and at the end, when it has been proved that* KC *is equal to* HD, *instead of concluding that* HD *is equal to* F, *we have to say "Because* GB *is the same multiple of* E *that* KC *is of* F, *and* KC *is equal to* HD, *therefore* HD *is the same multiple of* F *that* GB *is of* E."*

Proposition 7

Equal magnitudes have to the same the same ratio, as also has the same to equal magnitudes.

Let *A*, *B* be equal magnitudes and *C* any other, chance, magnitude;
I say that each of the magnitudes *A*, *B* has the same ratio to *C*, and *C* has the same ratio to each of the magnitudes *A*, *B*.

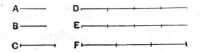

For let equimultiples *D*, *E* of *A*, *B* be taken, and of *C* another, chance, multiple *F*.

Then, since *D* is the same multiple of *A* that *E* is of *B*,
while *A* is equal to *B*,
therefore *D* is equal to *E*.

But *F* is another, chance, magnitude.

If therefore *D* is in excess of *F*, *E* is also in excess of *F*, if equal to it, equal; and, if less, less.

And *D*, *E* are equimultiples of *A*, *B*,
while *F* is another, chance, multiple of *C*;
therefore, as *A* is to *C*, so is *B* to *C*. [v. Def. 5]

I say next that *C* also has the same ratio to each of the magnitudes *A*, *B*.

34

For, with the same construction, we can prove similarly that D is equal to E; and F is some other magnitude.

If therefore F is in excess of D, it is also in excess of E, if equal, equal; and, if less, less.

And F is a multiple of C, while D, E are other, chance, equimultiples of A, B; therefore, as C is to A, so is C to B. [v. Def. 5]

Therefore etc.

PORISM. From this it is manifest that, if any magnitudes are proportional, they will also be proportional inversely.

Q.E.D.

In this proposition there is a similar use of ὃ ἔτυχεν to that which has been discussed under Prop. 4. Any multiple F of C is taken and then, four lines lower down, we are told that "F is another, chance, magnitude." It is of course not any magnitude whatever, and Simson leaves out the sentence, but this time without calling attention to it.

Of the Porism to this proposition Heiberg says that it is properly put here in the best MS.; for, as August had already observed, if it was in its right place where Theon put it (at the end of v. 4), the second part of the proof of this proposition would be unnecessary. But the truth is that the Porism is no more in place here. The most that the proposition proves is that, if A, B are equal, and C any other magnitude, then two conclusions are simultaneously established, (1) that A is to C as B is to C and (2) that C is to A as C is to B. The second conclusion is not established from the first conclusion (as it ought to be in order to justify the inference in the Porism), but from a hypothesis on which the first conclusion itself depends; and moreover it is not a proportion in its general form, i.e. between four magnitudes, that is in question, but only the particular case in which the consequents are equal.

Aristotle tacitly assumes inversion (combined with the solution of the problem of Eucl. vi. 11) in Meteorologica iii. 5, 376 a 14–16.

Proposition 8

Of unequal magnitudes, the greater has to the same a greater ratio than the less has; and the same has to the less a greater ratio than it has to the greater.

Let AB, C be unequal magnitudes, and let AB be greater; let D be another, chance, magnitude; I say that AB has to D a greater ratio than C has to D, and D has to C a greater ratio than it has to AB.

For, since AB is greater than C, let BE be made equal to C;

then the less of the magnitudes AE, EB, if multiplied, will sometime be greater than D. [v. Def. 4]

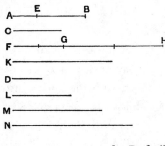

[*Case 1.*]

First, let AE be less than EB;

let AE be multiplied, and let FG be a multiple of it which is greater than D; then, whatever multiple FG is of AE, let GH be made the same multiple of EB and K of C;

35

and let *L* be taken double of *D*, *M* triple of it, and successive multiples increasing by one, until what is taken is a multiple of *D* and the first that is greater than *K*. Let it be taken, and let it be *N* which is quadruple of *D* and the first multiple of it that is greater than *K*.

Then, since *K* is less than *N* first,

therefore *K* is not less than *M*.

And, since *FG* is the same multiple of *AE* that *GH* is of *EB*,

therefore *FG* is the same multiple of *AE* that *FH* is of *AB*.   [v. 1]

But *FG* is the same multiple of *AE* that *K* is of *C*;

therefore *FH* is the same multiple of *AB* that *K* is of *C*;

    therefore *FH*, *K* are equimultiples of *AB*, *C*.

Again, since *GH* is the same multiple of *EB* that *K* is of *C*,

and *EB* is equal to *C*,

    therefore *GH* is equal to *K*.

But *K* is not less than *M*;

therefore neither is *GH* less than *M*.

And *FG* is greater than *D*;

therefore the whole *FH* is greater than *D*, *M* together.

But *D*, *M* together are equal to *N*, inasmuch as *M* is triple of *D*, and *M*, *D* together are quadruple of *D*, while *N* is also quadruple of *D*; whence *M*, *D* together are equal to *N*.

But *FH* is greater than *M*, *D*;

therefore *FH* is in excess of *N*,

while *K* is not in excess of *N*.

And *FH*, *K* are equimultiples of *AB*, *C*, while *N* is another, chance, multiple of *D*;   [v. Def. 7]

therefore *AB* has to *D* a greater ratio than *C* has to *D*.

I say next, that *D* also has to *C* a greater ratio than *D* has to *AB*.

For, with the same construction, we can prove similarly that *N* is in excess of *K*, while *N* is not in excess of *FH*.

And *N* is a multiple of *D*,

while *FH*, *K* are other, chance, equimultiples of *AB*, *C*;

    therefore *D* has to *C* a greater ratio than *D* has to *AB*.   [v. Def. 7]

[*Case* 2.]

Again, let *AE* be greater than *EB*.

Then the less, *EB*, if multiplied, will sometime be greater than *D*.   [v. Def. 4]

Let it be multiplied, and let *GH* be a multiple of *EB* and greater than *D*;

and, whatever multiple *GH* is of *EB*, let *FG* be made the same multiple of *AE*, and *K* of *C*.

Then we can prove similarly that *FH*, *K* are equimultiples of *AB*, *C*;

and, similarly, let *N* be taken a multiple of *D* but the first that is greater than *FG*,

so that *FG* is again not less than *M*.

But *GH* is greater than *D*;

therefore the whole *FH* is in excess of *D*, *M*, that is, of *N*.

Now *K* is not in excess of *N*, inasmuch as *FG* also, which is greater than *GH*, that is, than *K*, is not in excess of *N*.

And in the same manner, by following the above argument, we complete the demonstration.

Therefore etc.

Q.E.D.

The two separate cases found in the Greek text of the demonstration can practically be compressed into one. Also the expositor of the two cases makes them differ more than they need. It is necessary in each case to select the smaller *of the two segments AE, EB of AB with a view to taking a multiple of it which is greater than D; in the first case therefore AE is taken, in the second EB. But, while in the first case successive multiples of D are taken in order to find the first multiple that is greater than GH (or K), in the second case the multiple is taken which is the first that is greater than FG. This difference is not necessary; the first multiple of D that is greater than GH would equally serve in the second case. Lastly, the use of the magnitude K might have been dispensed with in both cases; it is of no practical use and only lengthens the proofs. For these reasons Simson considers that Theon, or some other unskilful editor, has vitiated the proposition. This however seems an unsafe assumption; for, while it was not the habit of the great Greek geometers to discuss separately a number of different cases (e.g. in* i. 7 *and* i. 35 *Euclid proves one case and leaves the others to the reader), there are many exceptions to prove the rule, e.g. Eucl.* iii. 25 *and* 33; *and we know that many fundamental propositions, afterwards proved generally, were first discovered in relation to particular cases and then generalised, so that Book V., presenting a comparatively new theory, might fairly be expected to exhibit more instances than the earlier books do of unnecessary subdivision. The use of the K is no more conclusive against the genuineness of the proof.*

Nevertheless Simson's version of the proof is certainly snorter, and moreover it takes account of the case in which AE is equal to EB, and of the case in which AE, EB are both greater than D (though these cases are scarcely worth separate mention).

"If the magnitude which is not the greater of the two AE, EB be (1) not less than D, take FG, GH the doubles of AE, EB.

But if that which is not the greater of the two AE, EB be (2) less than D, this magnitude can be multiplied so as to become greater than D whether it be AE or EB.

Let it be multiplied until it becomes greater than D, and let the other be multiplied as often; let FG be the multiple thus taken of AE and GH the same multiple of EB,

therefore FG and GH are each of them greater than D.

And, in every one of the cases, take L the double of D, M its triple and so on, till the multiple of D be that which first becomes greater than GH.

Let N be that multiple of D which is first greater than GH, and M the multiple of D which is next less than N.

Then, because N is the multiple of D which is the first that becomes greater than GH, the next preceding multiple is not greater than GH;

that is, GH is not less than M.

And, since FG is the same multiple of AE that GH is of EB,

GH is the same multiple of EB that FH is of AB; [v. 1]

wherefore FH, GH are equimultiples of AB, EB.

And it was shown that GH *was not less than* M;
and, by the construction, FG *is greater than* D;
therefore the whole FH *is greater than* M, D *together.*
But M, D *together are equal to* N;
therefore FH *is greater than* N.
But GH *is not greater than* N;
and FH, GH *are equimultiples of* AB, BE,
and N *is a multiple of* D;
therefore AB *has to* D *a greater ratio than* BE *(or* C*) has to* D.   [v. Def. 7]
Also D *has to* BE *a greater ratio than it has to* AB.
For, *having made the same construction, it may be shown, in like manner, that* N *is greater than* GH *but that it is not greater than* FH;
and N *is a multiple of* D,
and GH, FH *are equimultiples of* EB, AB;
Therefore D *has to* EB *a greater ratio than it has to* AB."   [v. Def. 7]
The proof may perhaps be more readily grasped in the more symbolical form thus.
Take the m*th equimultiples of* C, *and of the excess of* AB *over* C *(that is, of* AE *), such that each is greater than* D;
and, of the multiples of D, *let* pD *be the first that is greater than* mC, *and* nD *the next less multiple of* D.
Then, since mC *is not less than* nD,
and, by the construction, m(AE *) is greater than* D,
the sum of mC *and* m(AE*) is greater than the sum of* nD *and* D.
That is, m(AB*) is greater than* pD.
And, by the construction, mC *is less than* pD.
Therefore [v. Def. 7] AB *has to* D *a greater ratio than* C *has to* D.
Again, since pD *is less than* m(AB*),*
and pD *is greater than* mC,
D *has to* C *a greater ratio than* D *has to* AB.

Proposition 9

Magnitudes which have the same ratio to the same are equal to one another; and magnitudes to which the same has the same ratio are equal.

For let each of the magnitudes *A, B* have the same ratio to *C*;
I say that *A* is equal to *B*.

For, otherwise, each of the magnitudes *A, B* would not have had the same ratio to *C*; but it has;   [v. 8]

therefore *A* is equal to *B*.
Again, let *C* have the same ratio to each of the magnitudes *A, B*;
I say that *A* is equal to *B*.

For, otherwise, *C* would not have had the same ratio to each of the magnitudes *A, B*;   [v. 8]
but it has;

therefore *A* is equal to *B*.
Therefore etc.

Q.E.D.

If A *is to* C *as* B *is to* C,
or if C *is to* A *as* C *is to* B, *then* A *is equal to* B.

Simson gives a more explicit proof of this proposition which has the advantage of referring back to the fundamental 5th and 7th definitions, instead of quoting the results of previous propositions, which, as will be seen from the next note, may be, in the circumstances, unsafe.

"*Let* A, B *have each of them the same ratio to* C;
A *is equal to* B.
For, if they are not equal, one of them is greater than the other;
let A *be the greater.*

Then, by what was shown in the preceding proposition, there are some equimultiples of A *and* B, *and some multiple of* C, *such that the multiple of* A *is greater than the multiple of* C, *but the multiple of* B *is not greater than that of* C.

Let such multiples be taken, and let D, E *be the equimultiples of* A, B, *and* F *the multiple of* C, *so that* D *may be greater than* F, *and* E *not greater than* F.

But, because A *is to* C *as* B *is to* C,
and of A, B *are taken equimultiples* D, E, *and of* C *is taken a multiple* F, *and* D *is greater than* F,
E must also be greater than F. [v. Def. 5]

But E *is not greater than* F: *which is impossible.*

Next, let C *have the same ratio to each of the magnitudes* A *and* B;
A *is equal to* B.
For, if not, one of them is greater than the other;
let A *be the greater.*

Therefore, as was shown in Prop. 8, there is some multiple F *of* C, *and some equimultiples* E *and* D *of* B *and* A, *such that* F *is greater than* E *and not greater than* D.

But, because C *is to* B *as* C *is to* A,
and F *the multiple of the first is greater than* E *the multiple of the second,*
F the multiple of the third is greater than D the multiple of the fourth. [v. Def. 5]

But F *is not greater than* D: *which is impossible.*

Therefore A *is equal to* B."

Proposition 10

Of magnitudes which have a ratio to the same, that which has a greater ratio is greater; and that to which the same has a greater ratio is less.

For let A have to C a greater ratio than B has to C;
I say that A is greater than B.

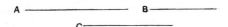

For, if not, A is either equal to B or less.

Now A is not equal to B;
for in that case each of the magnitudes A, B would have had the same ratio to C;
[v. 7]
but they have not;

therefore A is not equal to B.

Nor again is A less than B;
for in that case A would have had to C a less ratio than B has to C;   [v. 8]
but it has not;

therefore A is not less than B.

But it was proved not to be equal either;

therefore *A* is greater than *B*.

Again, let *C* have to *B* a greater ratio than *C* has to *A*;
I say that *B* is less than *A*.

For, if not, it is either equal or greater.

Now *B* is not equal to *A*;
for in that case *C* would have had the same ratio to each of the magnitudes *A*, *B*;
[v. 7]

but it has not;

therefore *A* is not equal to *B*.

Nor again is *B* greater than *A*;
for in that case *C* would have had to *B* a less ratio than it has to *A*;   [v. 8]
but it has not;

therefore *B* is not greater than *A*.

But it was proved that it is not equal either;

therefore *B* is less than *A*.

Therefore etc.

Q.E.D.

No better example can, I think, be found of the acuteness which Simson brought to bear in his critical examination of the Elements, *and of his great services to the study of Euclid, than is furnished by the admirable note on this proposition where he points out a serious flaw in the proof as given in the text.*

For the first time Euclid is arguing about greater *and* less *ratios, and it will be found by an examination of the steps of the proof that he assumes more with regard to the meaning of the terms than he is entitled to assume, having regard to the fact that the definition of greater ratio (Def. 7) is all that, as yet, he has to go upon. That we cannot argue, at present, about* greater *and* less *as applied to ratios in the same way as about the same terms in relation to* magnitudes *is indeed sufficiently indicated by the fact that Euclid does not assume for ratios what is in Book I. an axiom, viz. that things which are equal to the same thing are equal to one another; on the contrary, he proves, in Prop. 11, that ratios which are the same with the same ratio are the same with one another.*

Let us now examine the steps of the proof in the text. First we are told that
    "A *is greater than* B.
    For, if not, it is either equal to B *or less than it.*
    Now A *is not equal to* B;
    for in that case each of the two magnitudes A, B *would have had the*
        same ratio to C: [v. 7]
    but they have not:
                        therefore A *is not equal to* B."
As Simson remarks, the force of this reasoning is as follows.
If A *has to* C *the same ratio as* B *has to* C,
then—supposing any equimultiples of A, B *to be taken and any multiple of* C—
by Def. 5, if the multiple of A *be greater than the multiple of* C, *the multiple of* B *is also greater than that of* C.
But it follows from the hypothesis (that A *has a greater ratio to* C *than* B *has to* C) *that,*
by Def. 7, there must be some equimultiples of A, B *and some multiple of* C *such that the multiple of* A *is greater than the multiple of* C, *but the multiple of* B *is* not *greater than the same multiple of* C.
And this directly contradicts the preceding deduction from the supposition that A *has to* C *the same ratio as* B *has to* C;
                        therefore that supposition is impossible.

The proof now goes on thus:
    "Nor again is A *less than* B;
    for, in that case, A *would have had to* C *a less ratio than* B *has to* C;   [v. 8]
    but it has not;
                        therefore A *is not less than* B."

It is here that the difficulty arises. As before, we must use Def. 7. *"A would have had to* C *a less ratio than* B *has to* C,*" or the equivalent statement that* B *would have had to* C *a greater ratio than* A *has to* C, *means that there would have been some equimultiples of* B, A *and some multiple of* C *such that*

  (1) *the multiple of* B *is greater than the multiple of* C, *but*
  (2) *the multiple of* A *is* not *greater than the multiple of* C,

and it ought to have been proved that this can never happen if the hypothesis of the proposition is true, viz. that A *has to* C *a greater ratio than* B *has to* C: *that is, it should have been proved that, in the latter case, the multiple of* A *is* always *greater than the multiple of* C *whenever the multiple of* B *is greater than the multiple of* C *(for, when this is demonstrated, it will be evident that* B *cannot have a greater ratio to* C *than* A *has to* C *). But this is not proved (cf. the remark of De Morgan quoted in the note on v. Def. 7, p. 130), and hence it is not proved that the above inference from the supposition that* A *is less than* B *is inconsistent with the hypothesis in the enunciation. The proof therefore fails.*

Simson suggests that the proof is not Euclid's, but the work of some one who apparently "has been deceived in applying what is manifest, when understood of magnitudes, unto ratios, viz. that a magnitude cannot be both greater and less than another."

The proof substituted by Simson is satisfactory and simple.

"Let A *have to* C *a greater ratio than* B *has to* C;
A *is greater than* B.

For, because A *has a greater ratio to* C *than* B *has to* C, *there are some equimultiples of* A, B *and some multiple of* C *such that*

  the multiple of A *is greater than the multiple of* C, *but the multiple of* B *is* not *greater than it.*   [v. Def. 7]
  Let them be taken, and let D, E *be equimultiples of* A, B, *and* F *a multiple of* C, *such that*
                            D *is greater than* F,
  but   E *is not greater than* F.
Therefore D *is greater than* E.
And, because D *and* E *are equimultiples of* A *and* B, *and* D *is greater than* E,
                  therefore A *is greater than* B.   [Simson's 4th Ax.]
  Next, let C *have a greater ratio to* B *than it has to* A;
B *is less than* A.

For there is some multiple F *of* C *and some equimultiples* E *and* D *of* B *and* A *such that*
F *is greater than* E *but not greater than* D.   [v. Def. 7]
Therefore E *is less than* D;
and, because E *and* D *are equimultiples of* B *and* A,
                  therefore B *is less than* A."

Proposition 11

Ratios which are the same with the same ratio are also the same with one another.

For, as *A* is to *B*, so let *C* be to *D*,
and, as *C* is to *D*, so let *E* be to *F*;
I say that, as *A* is to *B*, so is *E* to *F.*

For of *A*, *C*, *E* let equimultiples *G*, *H*, *K* be taken, and of *B*, *D*, *F* other, chance, equimultiples *L*, *M*, *N*.

Then since, as *A* is to *B*, so is *C* to *D*,
and of *A*, *C* equimultiples *G*, *H* have been taken,
and of *B*, *D* other, chance, equimultiples *L*, *M*,
therefore, if *G* is in excess of *L*, *H* is also in excess of *M*,
if equal, equal,

and if less, less.

Again, since, as *C* is to *D*, so is *E* to *F*,
and of *C*, *E* equimultiples *H*, *K* have been taken,
and of *D*, *F* other, chance, equimultiples *M*, *N*,
therefore, if *H* is in excess of *M*, *K* is also in excess of *N*,
if equal, equal,
and if less, less.

But we saw that, if *H* was in excess of *M*, *G* was also in excess of *L*; if equal, equal; and if less, less;
so that, in addition, if *G* is in excess of *L*, *K* is also in excess of *N*,
if equal, equal,
and if less, less.

And *G*, *K* are equimultiples of *A*, *E*,
while *L*, *N* are other, chance, equimultiples of *B*, *F*;
therefore, as *A* is to *B*, so is *E* to *F*.
Therefore etc.

Q.E.D.

Algebraically, if   a : b = c : d,
and   c : d = e : f,
then   a : b = e : f.

The idiomatic use of the imperfect in quoting a result previously obtained is noteworthy. Instead of saying "But it was proved that, if H *is in excess of* M, G *is also in excess of* L," *the Greek text has "But if* H *was in excess of* M, G *was also in excess of* L," ἀλλὰ εἰ ὑπερεῖχε τὸ Θ τοῦ M, ὑπερεῖχε καὶ τὸ H τοῦ Λ.*

This proposition is tacitly used in combination with v. 16 *and* v. 24 *in the geometrical passage in Aristotle,* Meteorologica iii. 5, 376 a 22–26.

Proposition 12

If any number of magnitudes be proportional, as one of the antecedents is to one of the consequents, so will all the antecedents be to all the consequents.

Let any number of magnitudes *A*, *B*, *C*, *D*, *E*, *F* be proportional, so that, as *A* is to *B*, so is *C* to *D* and *E* to *F*;
I say that, as *A* is to *B*, so are *A*, *C*, *E* to *B*, *D*, *F*.

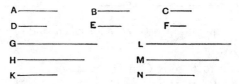

For of *A*, *C*, *E* let equimultiples *G*, *H*, *K* be taken,
and of *B*, *D*, *F* other, chance, equimultiples *L*, *M*, *N*.
Then since, as *A* is to *B*, so is *C* to *D*, and *E* to *F*,

and of *A*, *C*, *E* equimultiples *G*, *H*, *K* have been taken,
and of *B*, *D*, *F* other, chance, equimultiples *L*, *M*, *N*,
therefore, if *G* is in excess of *L*, *H* is also in excess of *M*,
and *K* of *N*,
if equal, equal,
and if less, less;
so that, in addition,
if *G* is in excess of *L*, then *G*, *H*, *K* are in excess of *L*, *M*, *N*,
if equal, equal,
and if less, less.

Now *G* and *G*, *H*, *K* are equimultiples of *A* and *A*, *C*, *E*, since, if any number of magnitudes whatever are respectively equimultiples of any magnitudes equal in multitude, whatever multiple one of the magnitudes is of one, that multiple also will all be of all.

[v. 1]

For the same reason
L and *L*, *M*, *N* are also equimultiples of *B* and *B*, *D*, *F*;
therefore, as *A* is to *B*, so are *A*, *C*, *E* to *B*, *D*, *F*. [v. Def. 5]
Therefore etc.

Q.E.D.

Algebraically, if $a:a' = b:b' = c:c'$ *etc., each ratio is equal to the ratio* $(a + b + c + ...):(a' + b' + c' + ...)$. *This theorem is quoted by Aristotle, Eth. Nic. v. 7, 1131 b 14, in the shortened form "the whole is to the whole what each part is to each part (respectively)."*

Proposition 13

If a first magnitude have to a second the same ratio as a third to a fourth, and the third have to the fourth a greater ratio than a fifth has to a sixth, the first will also have to the second a greater ratio than the fifth to the sixth.

For let a first magnitude *A* have to a second *B* the same ratio as a third *C* has to a fourth *D*,
and let the third *C* have to the fourth *D* a greater ratio than a fifth *E* has to a sixth *F*;
I say that the first *A* will also have to the second *B* a greater ratio than the fifth *E* to the sixth *F*.

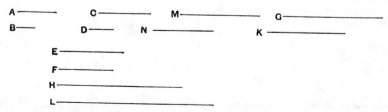

For, since there are some equimultiples of *C*, *E*,

and of D, F other, chance, equimultiples, such that the multiple of C is in excess of the multiple of D,

while the multiple of E is not in excess of the multiple of F,　　　　[v. Def. 7]

let them be taken,

and let G, H be equimultiples of C, E,

and K, L other, chance, equimultiples of D, F,

so that G is in excess of K, but H is not in excess of L;

and, whatever multiple G is of C, let M be also that multiple of A,

and, whatever multiple K is of D, let N be also that multiple of B.

Now, since, as A is to B, so is C to D,

and of A, C equimultiples M, G have been taken,

and of B, D other, chance, equimultiples N, K,

therefore, if M is in excess of N, G is also in excess of K,

if equal, equal,

and if less, less.　　　　[v. Def. 5]

But G is in excess of K;

therefore M is also in excess of N.

But H is not in excess of L;

and M, H are equimultiples of A, E,

and N, L other, chance, equimultiples of B, F;

therefore A has to B a greater ratio than E has to F.　　　　[v. Def. 7]

Therefore etc.

Q.E.D.

Algebraically, if　　　　　$a:b = c:d,$
and　　　　　　　　　　$c:d > e:f,$
then　　　　　　　　　$a:b > e:f.$
After the words "for, since" in the first line of the proof, Theon added "C has to D a greater ratio than E has to F,"
so that "there are some equimultiples" began, with him, the principal sentence.
　　The Greek text has, after "of D, F other, chance, equimultiples," "and the multiple of C is in excess of the multiple of D. . . ." The meaning being "such that," I have substituted this for "and," after Simson.
　　The following will show the method of Euclid's proof.
Since　　　　　　　　$c:d > e:f,$
there will be some equimultiples mc, me of c, e, and some equimultiples nd, nf of d, f, such that
　　　　　　　　　mc > nd, *while* me $\not>$ nf.
But, since　　　　　　$a:b = c:d,$
therefore, according as ma $> = <$ nb,　mc $> = <$ nd.
And mc > nd;
　　　　　　therefore ma $>$ nb, *while (from above)* me $\not>$ nf.
　　Therefore　　　　　$a:b > e:f.$
Simson adds as a corollary the following:
　　"If the first have a greater ratio to the second than the third has to the fourth, but the third the same ratio to the fourth which the fifth has to the sixth, it may be demonstrated in like manner that the first has a greater ratio to the second than the fifth has to the sixth."
　　This however scarcely seems to be worth separate statement, since it only amounts to changing the order of the two parts of the hypothesis.

Proposition 14

If a first magnitude have to a second the same ratio as a third has to a fourth, and the first be greater than the third, the second will also be greater than the fourth; if equal, equal; and if less, less.

For let a first magnitude A have the same ratio to a second B as a third C has to a fourth D; and let A be greater than C;
I say that B is also greater than D.

For, since A is greater than C,
and B is another, chance, magnitude,
therefore A has to B a greater ratio than C has to B.   [v. 8]
    But, as A is to B, so is C to D;
        therefore C has also to D a greater ratio than C has to B.   [v. 13]
    But that to which the same has a greater ratio is less;   [v. 10]
        therefore D is less than B;
    so that B is greater than D.

Similarly we can prove that, if A be equal to C, B will also be equal to D;
and, if A be less than C, B will also be less than D.
    Therefore etc.

                                                                    Q.E.D.

Algebraically, if   a : b = c : d,
                            then, according as a $>=<$ c,   b $>=<$ d.
    Simson adds the specific proof of the second and third parts of this proposition, which Euclid dismisses with "Similarly we can prove. . . ."
    "Secondly, if A *be equal to* C, B *is equal to* D; *for* A *is to* B *as* C, *that is* A, *is to* D;
        therefore B *is equal to* D.
    Thirdly, if A *be less than* C, B *shall be less than* D.   [v. 9]
    For C *is greater than* A;
and, because C *is to* D *as* A *is to* B,
                            D *is greater than* B, *by the first case.*
    Wherefore B *is less than* D."
    Aristotle, Meteorol. iii. 5, 376 a 11–14, quotes the equivalent proposition that, if a $>$ b, c $>$ d.

Proposition 15

Parts have the same ratio as the same multiples of them taken in corresponding order.

For let AB be the same multiple of C that DE is of F; I say that, as C is to F, so is AB to DE.

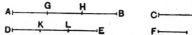

For, since *AB* is the same multiple of *C* that *DE* is of *F*, as many magnitudes as there are in *AB* equal to *C*, so many are there also in *DE* equal to *F*.

Let *AB* be divided into the magnitudes *AG*, *GH*, *HB* equal to *C*, and *DE* into the magnitudes *DK*, *KL*, *LE* equal to *F*; then the multitude of the magnitudes *AG*, *GH*, *HB* will be equal to the multitude of the magnitudes *DK*, *KL*, *LE*.

And, since *AG*, *GH*, *HB* are equal to one another, and *DK*, *KL*, *LE* are also equal to one another, therefore, as *AG* is to *DK*, so is *GH* to *KL*, and *HB* to *LE*.   [v. 7]

Therefore, as one of the antecedents is to one of the consequents, so will all the antecedents be to all the consequents;   [v. 12]

therefore, as *AG* is to *DK*, so is *AB* to *DE*.

But *AG* is equal to *C* and *DK* to *F*;

therefore, as *C* is to *F*, so is *AB* to *DE*.

Therefore etc.

Q.E.D.

Algebraically,   a : b = ma : mb.

Proposition 16

If four magnitudes be proportional, they will also be proportional alternately.

Let *A*, *B*, *C*, *D* be four proportional magnitudes, so that, as *A* is to *B*, so is *C* to *D*;
I say that they will also be so alternately, that is, as *A* is to *C*, so is *B* to *D*.

For of *A*, *B* let equimultiples *E*, *F* be taken, and of *C*, *D* other, chance, equimultiples *G*, *H*.

Then, since *E* is the same multiple of *A* that *F* is of *B*, and parts have the same ratio as the same multiples of them,   [v. 15]

therefore, as *A* is to *B*, so is *E* to *F*.

But as *A* is to *B*, so is *C* to *D*;

therefore also, as *C* is to *D*, so is *E* to *F*.   [v. 11]

Again, since *G*, *H* are equimultiples of *C*, *D*,

therefore, as *C* is to *D*, so is *G* to *H*.   [v. 15]

But, as *C* is to *D*, so is *E* to *F*;

therefore also, as *E* is to *F*, so is *G* to *H*.   [v. 11]

But, if four magnitudes be proportional, and the first be greater than the third,
the second will also be greater than the fourth;
if equal, equal;
and if less, less. [v. 14]
Therefore, if E is in excess of G, F is also in excess of H,
if equal, equal,
and if less, less.
Now E, F are equimultiples of A, B,
and G, H other, chance, equimultiples of C, D;
therefore, as A is to C, so is B to D. [v. Def. 5]
Therefore etc.

Q.E.D.

2. "Let A, B, C, D be four proportional magnitudes, so that, as A is to B, so is C to D." In a number of expressions like this it is absolutely necessary, when translating into English, to interpolate words which are not in the Greek. Thus the Greek here is: "Ἔστω τέσσαρα μεγέθη ἀνάλογον τὰ
A, B, Γ, Δ, ὡς τὸ A πρὸς τὸ B, οὕτως τὸ Γ πρὸς τὸ Δ, literally "Let A, B, C, D be four proportional magnitudes, as A to B, so C to D." The same
remark applies to the corresponding expressions in the next propositions, v. 17, 18, and to other forms of expression in v. 20–23 and later propositions:
e.g. in v. 20 we have a phrase meaning literally "Let there be magnitudes . . . which taken two and two are in the same ratio, as A to B, so D to E," etc.:
in v. 21 "(magnitudes) . . . which taken two and two are in the same ratio, and let the proportion of them be perturbed, as A to B, so E to F," etc. In all
such cases (where the Greek is so terse as to be almost-ungrammatical) I shall insert the words necessary in English, without further remark.

Algebraically, if $\qquad$ a : b = c : d,
then $\qquad$ a : c = b : d.
Taking equimultiples ma, mb *of* a, b, *and equimultiples* nc, nd *of* c, d, *we have, by* v. 15,
$$a : b = ma : mb,$$
$$c : d = nc : nd.$$
And, since $\qquad$ a : b = c : d,
we have [v. 11] $\qquad$ ma : mb = nc : nd.
Therefore [v. 14], *according as* ma $> = <$ nc, $\quad$ mb $> = <$ nd,
so that $\qquad$ a : c = b : d.
Aristotle tacitly uses the theorem in Meteorologica iii. 5, *376 a 22–24.*
The four magnitudes in this proposition must all be of the same kind, *and Simson inserts* "of the same kind" *in the
enunciation.*
This is the first of the propositions of Eucl. v. *which Smith and Bryant* (Euclid's Elements of Geometry, 1901)
prove by means of vi. 1 *so far as the only geometrical magnitudes in question are* straight lines *or* rectilineal areas; *and
certainly the proofs are more easy to follow than Euclid's. The proof of this proposition is as follows.*
To prove that, If four magnitudes of the same kind *[straight lines or rectilineal areas]* be proportionals, they will
be proportionals when taken alternately.
Let P, Q, R, S *be the four magnitudes of the same kind such that*
$$P : Q = R : S;$$
then it is required to prove that
$$P : R = Q : S.$$
First, *let all the magnitudes be areas.*
Construct a rectangle abcd *equal to the area* P, *and to* bc *apply the rectangle* bcef *equal to* Q.
Also to ab, bf *apply rectangles* ag, bk *equal to* R, S *respectively.*
Then, since the rectangles ac, be *have the same height, they are to one another as
their bases.* [vi. 1]

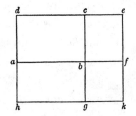

Hence $\qquad$ P : Q = ab : bf.
But $\qquad$ P : Q = R : S.
Therefore $\qquad$ R : S = ab : bf, $\qquad$ [v. 11]
i.e. $\qquad$ rect. ag : rect. bk = ab : bf.
Hence (by the converse of vi. 1) *the rectangles* ag, bk *have the same height, so that*
k *is on the line* hg.
Hence the rectangles ac, ag *have the same height, namely* ab; *also* be, bk *have the
same height, namely* bf.

Therefore	rect. ac : rect. ag = bc : bg,	
and	rect. be : rect. bk = bc : bg.	[vi. 1]
Therefore	rect. ac : rect. ag = rect. be : rect. bk.	[v. 11]
That is,	P : R = Q : S.	

Secondly, *let the magnitudes be straight lines* AB, BC, CD, DE.
Construct the rectangles Ab, Bc, Cd, De *with the same height.*

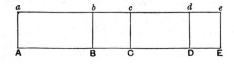

Then	Ab : Bc = AB : BC,	
and	Cd : De = CD : DE.	[vi. 1]
But	AB : BC = CD : DE.	
Therefore	Ab : Bc = Cd : De.	[v. 11]

Hence, by the first case,

$$Ab : Cd = Bc : De,$$

and, since these rectangles have the same height,

$$AB : CD = BC : DE.$$

Proposition 17

If magnitudes be proportional componendo, *they will also be proportional* separando.

Let *AB, BE, CD, DF* be magnitudes proportional *componendo*, so that, as *AB* is to *BE*, so is *CD* to *DF*;

I say that they will also be proportional *separando*, that is, as *AE* is to *EB*, so is *CF* to *DF*.

For of *AE, EB, CF, FD* let equimultiples *GH, HK, LM, MN* be taken, and of *EB, FD* other, chance, equimultiples, *KO, NP*.

Then, since *GH* is the same multiple of *AE* that *HK* is of *EB*, therefore *GH* is the same multiple of *AE* that *GK* is of *AB*.   [v. 1]

But *GH* is the same multiple of *AE* that *LM* is of *CF*;
therefore *GK* is the same multiple of *AB* that *LM* is of *CF*.

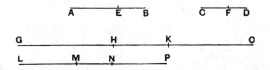

Again, since *LM* is the same multiple of *CF* that *MN* is of *FD*, therefore *LM* is the same multiple of *CF* that *LN* is of *CD*.   [v. 1]

But *LM* was the same multiple of *CF* that *GK* is of *AB*;
therefore *GK* is the same multiple of *AB* that *LN* is of *CD*.

Therefore *GK, LN* are equimultiples of *AB, CD*.

Again, since *HK* is the same multiple of *EB* that *MN* is of *FD*, and *KO* is also the same multiple of *EB* that *NP* is of *FD*,

therefore the sum *HO* is also the same multiple of *EB* that *MP* is of *FD*.　　　[v. 2]

And, since, as *AB* is to *BE*, so is *CD* to *DF*,

and of *AB*, *CD* equimultiples *GK*, *LN* have been taken,

and of *EB*, *FD* equimultiples *HO*, *MP*,

therefore, if *GK* is in excess of *HO*, *LN* is also in excess of *MP*,

if equal, equal,

and if less, less.

Let *GK* be in excess of *HO*;

then, if *HK* be subtracted from each,

　　　　　　　　GH is also in excess of *KO*.

But we saw that, if *GK* was in excess of *HO*, *LN* was also in excess of *MP*;

　　　therefore *LN* is also in excess of *MP*,

and, if *MN* be subtracted from each,

　　　　　　　　LM is also in excess of *NP*;

so that, if *GH* is in excess of *KO*, *LM* is also in excess of *NP*.

Similarly we can prove that,

if *GH* be equal to *KO*, *LM* will also be equal to *NP*,

and if less, less.

And *GH*, *LM* are equimultiples of *AE*, *CF*,

while *KO*, *NP* are other, chance, equimultiples of *EB*, *FD*;

　　　therefore, as *AE* is to *EB*, so is *CF* to *FD*.

Therefore etc.

　　　　　　　　　　　　　　　　　　　　　　Q.E.D.

Algebraically, if　　　　　　　　　　$a : b = c : d,$
　　then　　　　　　　　　　　$(a - b) : b = (c - d) : d.$

I have already noted the somewhat strange use of the participles of συγκεῖσθαι *and* διαιρεῖσθαι *to convey the sense of the technical* σύνθεσις *and* διαίρεσις λόγου, *or what we denote by* componendo *and* separando. ἐὰν συγκείμενα μεγέθη ἀνάλογν ἦ, καὶ διαιρεθέντα ἀνάλογον ἔσται *is, literally, "if magnitudes compounded be proportional, they will also be proportional separated," by which is meant "if one magnitude made up of two parts is to one of its parts as another magnitude made up of two parts is to one of its parts, the remainder of the first whole is to the part of it first taken as the remainder of the second whole is to the part of it first taken." In the algebraical formula above* a, c *are the wholes and* b, a − b *and* d, c − d *are the parts and remainders respectively. The formula might also be stated thus:*
　If　　　　　　　　　　$a + b : b = c + d : d,$
then　　　　　　　　　　　$a : b = c : d,$
in which case a + b, c + d *are the wholes and* b, a *and* d, c *the parts and remainders respectively. Looking at the last formula, we observe that "separated,"* διαιρεθέντα, *is used with reference not to the magnitudes* a, b, c, d *but to the compounded magnitudes* a + b, b, c + d, d.

As the proof is somewhat long, it will be useful to give a conspectus of it in the more symbolical form. To avoid minuses, we will take for the hypothesis the form
　　　　　　　　　　a + b *is to* b *as* c + d *is to* d.
Take any equimultiples of the four magnitudes a, b, c, d, *viz.*
　　　　　　　　　　ma,　mb,　mc,　md,
and any other equimultiples of the consequents, viz.
　　　　　　　　　　nb *and* nd.

49

Then, by v. 1, m (a + b), m (c + d) *are equimultiples of* a + b, c + d,
and, by v. 2, (m + n) b, (m + n) d *are equimultiples of* b, d.
Therefore, by Def. 5, *since* a + b *is to* b *as* c + d *is to* d,
  according as m (a + b) > = < (m + n) b,   m (c + d) > = < (m + n) d.
Subtract from m (a + b), (m + n) b *the common part* mb, *and from* m (c + d), (m + n) d *the common part*
md; *and we have,*
          according as ma > = < nb,   mc > = < nd.
But ma, mc *are any equimultiples of* a, c, *and* nb, nd *any equimultiples of* b, d,
therefore, by v. Def. 5,
              a *is to* b *as* c *is to* d.
Smith and Bryant's proof follows, mutatis mutandis, *their alternative proof of the next proposition.*

Proposition 18

If magnitudes be proportional separando, *they will also be proportional* componendo.

Let *AE, EB, CF, FD* be magnitudes proportional *separando,* so that, as *AE* is to
EB, so is *CF* to *FD*;
I say that they will also be proportional *componendo,*
that is, as *AB* is to *BE,* so is *CD* to *FD.*

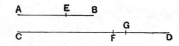

For, if *CD* be not to *DF* as *AB* to *BE,*
then, as *AB* is to *BE,* so will *CD* be either to some
magnitude less than *DF* or to a greater.

First, let it be in that ratio to a less magnitude *DG.*
Then, since, as *AB* is to *BE,* so is *CD* to *DG,*
they are magnitudes proportional *componendo*;
  so that they will also be proportional *separando.*   [v. 17]
Therefore, as *AE* is to *EB,* so is *CG* to *GD.*
But also, by hypothesis,
        as *AE* is to *EB,* so is *CF* to *FD.*
Therefore also, as *CG* is to *GD,* so is *CF* to *FD.*   [v. 11]
But the first *CG* is greater than the third *CF*;
  therefore the second *GD* is also greater than the fourth *FD.*   [v. 14]
But it is also less: which is impossible.
Therefore, as *AB* is to *BE,* so is not *CD* to a less magnitude than *FD.*
Similarly we can prove that neither is it in that ratio to a greater;
it is therefore in that ratio to *FD* itself.
Therefore etc.

Q.E.D.

Algebraically, if   a : b = c : d,
then   (a + b) : b = (c + d) : d.
   In the enunciation of this proposition there is the same special use of διηρημένα *and* συντεθέντα *as there was of*
συγκείμενα *and* διαιρεθέντα *in the last enunciation. Practically, as the algebraical form shows,* διηρημε'να *might
have been left out.*
   The following is the method of proof employed by Euclid.

Given that $\quad\quad\quad$ a : b = c : d,

suppose, if possible, that

$$(a + b) : b = (c + d) : (d \pm x).$$

Therefore, separando [v. 17],

$$a : b = (c \mp x) : (d \pm x),$$

whence, by v. 11, $\quad\quad$ (c $\mp$ x) : (d $\pm$ x) = c : d.

But $\quad\quad\quad\quad$ (c $-$ x) $<$ c, *while* (d $+$ x) $>$ d.

and $\quad\quad\quad\quad$ (c $+$ x) $>$ c, *while* (d $-$ x) $<$ d,

which relations respectively contradict v. 14.

Simson pointed out (as Saccheri before him saw) that Euclid's demonstration is not legitimate, because it assumes without proof that to any three magnitudes, two of which, at least, are of the same kind, there exists a fourth proportional. Clavius and, according to him, other editors made this an axiom. But it is far from axiomatic; it is not till vi. 12 *that Euclid shows, by construction, that it is true even in the particular case where the three given magnitudes are all straight lines.*

In order to remove the defect it is necessary either (1) to prove beforehand the proposition thus assumed by Euclid or (2) to prove v. 18 *independently of it.*

Saccheri ingeniously proposed that the assumed proposition should be proved, for areas and straight lines, by means of Euclid vi. 1, 2 *and* 12. *As he says, there was nothing to prevent Euclid from interposing these propositions immediately after* v. 17 *and then proving* v. 18 *by means of them.* vi. 12 *enables us to construct the fourth proportional when the three given magnitudes are straight lines; and* vi. 12 *depends only on* vi. 1 *and* 2. *"Now," says Saccheri, "when we have once found the means of constructing a straight line which is a fourth proportional to three given straight lines, we obviously have the solution of the general problem 'To construct a straight line which shall have to a given straight line the same ratio which two polygons have (to one another).'" For it is sufficient to transform the polygons into two triangles of equal height and then to construct a straight line which shall be a fourth proportional to the bases of the triangles and the given straight line.*

The method of Saccheri is, as will be seen, similar to that adopted by Smith and Bryant in proving the theorems of Euclid v. 16, 17, 18, 22, *so far as straight lines and rectilineal areas are concerned, by means of* vi. 1.

De Morgan gives a sketch of a general proof of the assumed proposition that, B being any magnitude, and P and Q two magnitudes of the same kind, there does exist a magnitude A which is to B in the same ratio as P to Q.

"The right to reason upon any aliquot part of any magnitude is assumed; though, in truth, aliquot parts obtained by continual bisection would suffice: and it is taken as previously proved that the tests of greater and of less ratio are never both presented in any one scale of relation as compared with another."

"(1) If M be to B in a greater ratio than P to Q, so is every magnitude greater than M, and so are some less magnitudes; and if M be to B in a less ratio than P to Q, so is every magnitude less than M, and so are some greater magnitudes. Part of this is in every system: the rest is proved thus. If M be to B in a greater ratio than P to Q, say, for instance, we find that 15M lies between 22B and 23B, while 15P lies before 22Q. Let 15M exceed 22B by Z; then, if N be less than M by anything less than the 15th part of Z, 15N is between 22B and 23B: or N, less than M, is in a greater ratio to B than P to Q. And similarly for the other case.

(2) M can certainly be taken so small as to be in a less ratio to B than P to Q, and so large as to be in a greater; and since we can never pass from the greater ratio back again to the smaller by increasing M, it follows that, while we pass from the first designated value to the second, we come upon an intermediate magnitude A such that every smaller is in a less ratio to B than P to Q, and every greater in a greater ratio. Now A cannot be in a less ratio to B than P to Q, for then some greater magnitudes would also be in a less ratio; nor in a greater ratio, for then some less magnitudes would be in a greater ratio; therefore A is in the same ratio to B as P to Q. The previously proved proposition above mentioned shows the three alternatives to be the only ones."

Alternative proofs of V. 18.

Simson bases his alternative on v. 5, 6. *As the 18th proposition is the converse of the 17th, and the latter is proved by means of* v. 1 *and* 2, *of which* v. 5 *and* 6 *are converses, the proof of* v. 18 *by* v. 5 *and* 6 *would be natural; and Simson holds that Euclid must have proved* v. 18 *in this way because "the 5th and 6th do not enter into the demonstration of any proposition in this book as we have it, nor can they be of any use in any proposition of the Elements," and "the 5th and 6th had undoubtedly been put into the 5th book for the sake of some propositions in it, as all the other propositions about equimultiples have been."*

Simson's proof is however, as it seems to me, intolerably long and difficult to follow unless it be put in the symbolical form as follows.

Suppose that a *is to* b *as* c *is to* d;

it is required to prove that a $+$ b *is to* b *as* c $+$ d *is to* d.

Take any equimultiples of the last four magnitudes, say

$$m(a + b), \quad mb, \quad m(c + d), \quad md,$$

and any equimultiples of b, d, *as*

$$nb, \quad nd.$$

Clearly, if   nb *is greater than* mb,

                    nd *is greater than* md;

    if equal, equal; and if less, less.

    I. *Suppose* nb *not greater than* mb, *so that* nd *is also not greater than* md.

       Now   m (a + b) *is greater than* mb:

       therefore   m (a + b) *is greater than* nb.

Similarly   m (c + d) *is greater than* nd.

    II. *Suppose* nb *greater than* mb.

    Since m (a + b), mb, m (c + d), md *are equimultiples of* (a + b), b, (c + d), d,

                 ma *is the same multiple of* a *that* m (a + b) *is of* (a + b),

and   mc *is the same multiple of* c *that* m (c + d) *is of* (c + d),

so that ma, mc *are equimultiples of* a, c.   [v. 5]

    Again nb, nd *are equimultiples of* b, d,

       and so are mb, md;

therefore (n − m) b, (n − m) d *are equimultiples of* b, d *and, whether* n − m *is equal to unity or to any other integer* [v. 6], *it follows, by* Def. 5, *that, since* a, b, c, d *are proportionals,*

if   ma *is greater than* (n − m) b,

then   mc *is greater than* (n − m) d;

    if equal, equal; and if less, less.

    (1) *If now* m (a + b) *is greater than* nb, *subtracting* mb *from each, we have*

                ma *is greater than* (n − m) b;

therefore   mc *is greater than* (n − m) d,

and, if we add md *to each,*

              m (c + d) *is greater than* nd.

    (2) *Similarly it may be proved that,*

if   m (a + b) *is equal to* nb,

then   m (c + d) *is equal to* nd,

and (3) *that, if*   m (a + b) *is less than* nb,

then   m (c + d) *is less than* nd.

    But (under 1. above) it was proved that, in the case where nb *is not greater than* mb,

            m (a + b) *is always greater than* nb,

and   m (c + d) *is always greater than* nd.

    Hence, whatever be the values of m *and* n, m (c + d) *is always greater than, equal to, or less than* nd *according as* m (a + b) *is greater than, equal to, or less than* nb.

    Therefore, by Def. 5,

            a + b *is to* b *as* c + d *is to* d.

    Todhunter gives the following short demonstration from Austin (Examination of the first six books of Euclid's Elements).

    "*Let* AE *be to* EB *as* CF *is to* FD:

            AB *shall be to* BE *as* CD *is to* DF.

    For, because AE *is to* EB *as* CF *is to* FD,

therefore, alternately,

            AE *is to* CF *as* EB *is to* FD.   [v. 16]

    And, as one of the antecedents is to its consequent, so is the sum of the antecedents to the sum of the consequents: [v. 12]

       therefore, as EB *is to* FD, *so are* AE, EB *together to* CF, FD *together;*

       that is,   AB *is to* CD *as* EB *is to* FD.

       Therefore, alternately,

            AB *is to* BE *as* CD *is to* FD. "

    The objection to this proof is that it is only valid in the case where the proposition v. 16 *used in it is valid, i.e. where all four magnitudes are of the same kind.*

    Smith and Bryant's proof avails where all four magnitudes are straight lines, where all four magnitudes are rectilineal areas, or where one antecedent and its consequent are straight lines and the other antecedent and its consequent rectilineal areas.

    Suppose that   A : B = C : D.

    First, *let all the magnitudes be areas.*

    Construct a rectangle abcd *equal to* A, *and to* bc *apply the rectangle* bcef *equal to* B.

Also to ab, bf *apply the rectangles* ag, bk *equal to* C, D *respectively.*
Then, since the rectangles ac, be *have equal heights* bc, *they are to one another
as their bases.*   [vi. 1]
     Hence   ab : bf = *rect.* ac : *rect.* be
                          = A : B
                          = C : D
                          = *rect.* ag : *rect.* bk.
*Therefore [*vi. 1, *converse] the rectangles* ag, bk *have the same height, so that* k *is on the straight line* hg.
     Hence   A + B : B = *rect.* ae : *rect.* be
                          = af : bf
                          = *rect.* ak : *rect.* bk
                          = C + D : D.
Secondly, let the magnitudes A, B *be straight lines and the magnitudes* C, D *areas.*
Let ab, bf *be equal to the straight lines* A, B, *and to* ab, bf *apply the rectangles* ag, bk *equal to* C, D *respectively.*
Then, as before, the rectangles ag, bk *have the same height.*
     Now   A + B : B = af : bf
                          = *rect.* ak : *rect.* bk
                          = C + D : D.
Thirdly, *let all the magnitudes be straight lines.*
Apply to the straight lines C, D *rectangles* P, Q *having the same height.*
     Then   P : Q = C : D.   [vi. 1]
Hence, by the second case,
                    A + B : B = P + Q : Q.
Also   P + Q : Q = C + D : D.
Therefore   A + B : B = C + D : D.

Proposition 19

If, as a whole is to a whole, so is a part subtracted to a part subtracted, the remainder will also be to the remainder as whole to whole.

For, as the whole *AB* is to the whole *CD*, so let the part *AE* subtracted be to the part *CF* subtracted;

I say that the remainder *EB* will also be to the remainder *FD* as the whole *AB* to the whole *CD*.

For since, as *AB* is to *CD*, so is *AE* to *CF*,

alternately also, as *BA* is to *AE*, so is *DC* to *CF*.   [v. 16]

And, since the magnitudes are proportional *componendo*, they will also be proportional *separando*,   [v. 17]

that is, as *BE* is to *EA*, so is *DF* to *CF*,

and, alternately,

             as *BE* is to *DF*, so is *EA* to *FC*.   [v. 16]

But, as *AE* is to *CF*, so by hypothesis is the whole *AB* to the whole *CD*.

Therefore also the remainder *EB* will be to the remainder *FD* as the whole *AB* is to the whole *CD*.   [v. 11]

Therefore etc.

[PORISM. From this it is manifest that, if magnitudes be proportional *componendo*, they will also be proportional *convertendo*.]

                                       Q.E.D.

Algebraically, if a : b = c : d *(where* c < a *and* d < b), *then*

$$(a - c):(b - d) = a:b.$$

The *"Porism" at the end of this proposition is led up to by a few lines which Heiberg brackets because it is not Euclid's habit to explain a Porism, and indeed a Porism, from its very nature, should not need any explanation, being a sort of by-product appearing without effort or trouble, ἀπραγματεύτως (Proclus). But Heiberg thinks that Simson does wrong in finding fault with the argument leading to the "Porism," and that it does contain the true demonstration of* conversion *of a ratio. In this it appears to me that Heiberg is clearly mistaken, the supposed proof on the basis of Prop. 19 being no more correct than the similar attempt to prove the* inversion *of a ratio from Prop. 4. The words are: "And since it was proved that, as* AB *is to* CD,

alternately also, as AB *is to* BE, *so is* CD *to* FD:

therefore magnitudes when compounded are proportional.

But it was proved that, as BA *is to* AE, *so is* DC *to* CF, *and this is* convertendo."

It will be seen that this amounts to proving from the hypothesis a : b = c : d *that the following transformations are simultaneously true, viz.:*

$$a:a - c = b:b - d,$$

and $$a:c = b:d.$$

The *former is not proved from the latter as it ought to be if it were intended to prove* conversion.

The inevitable conclusion is that both the "Porism" and the argument leading up to it are interpolations, though no doubt made, as Heiberg says, before Theon's time.

The conversion *of ratios does not depend upon* v. 19 *at all but, as Simson shows in his Proposition E (containing a proof already given by Clavius), on Props. 17 and 18. Prop. E is as follows.*

If four magnitudes be proportionals, they are also proportionals by conversion, that is, the first is to its excess above the second as the third is to its excess above the fourth.

Let AB *be to* BE *as* CD *to* DF:
then BA *is to* AE *as* DC *to* CF.

Because AB *is to* BE *as* CD *to* DF,
by division [separando],

AE *is to* EB *as* CF *to* FD, [v. 17]

and, by inversion,

BE *is to* EA *as* DF *to* FC.
[Simson's Prop. B directly obtained from v. Def. 5]

Wherefore, by composition [componendo],

BA *is to* AE *as* DC *to* CF. [v. 18]

Proposition 20

If there be three magnitudes, and others equal to them in multitude, which taken two and two are in the same ratio, and if ex aequali *the first be greater than the third, the fourth will also be greater than the sixth; if equal, equal; and, if less, less.*

Let there be three magnitudes *A*, *B*, *C*, and others *D*, *E*, *F* equal to them in multitude, which taken two and two are in the same ratio, so that,

as *A* is to *B*, so is *D* to *E*,

and as *B* is to *C*, so is *E* to *F*;

and let *A* be greater than *C* ex aequali;

I say that *D* will also be greater than *F*; if *A* is equal to *C*, equal; and, if less, less.

For, since *A* is greater than *C*,
and *B* is some other magnitude,

and the greater has to the same a greater ratio than the less has,   [v. 8]
therefore A has to B a greater ratio than C has to B.

But, as A is to B, so is D to E,
and, as C is to B, inversely, so is F to E;
therefore D has also to E a greater ratio than F has to E.   [v. 13]

But, of magnitudes which have a ratio to the same, that which has a greater ratio
is greater;   [v. 10]

therefore D is greater than F.

Similarly we can prove that, if A be equal to C, D will also be equal to F; and
if less, less.

Therefore etc.

Q.E.D.

Though, as already remarked, Euclid has not yet given us any definition of compounded ratios, *Props. 20–23 contain an important part of the theory of such ratios. The term "compounded ratio" is not used, but the propositions connect themselves with the definitions of* ex aequali *in its two forms, the ordinary form defined in Def. 17 and that called* perturbed proportion *in Def. 18. The compounded ratios dealt with in these propositions are those compounded of successive ratios in which the consequent of one is the antecedent of the next, or the antecedent of one is the consequent of the next.*

Prop. 22 states the fundamental proposition about the ratio ex aequali *in its ordinary form, to the effect that,*

if	a *is to* b *as* d *is to* e,
and	b *is to* c *as* e *is to* f,
then	a *is to* c *as* d *is to* f,

with the extension to any number of such ratios; Prop. 23 gives the corresponding theorem for the case of perturbed proportion, *namely that,*

if	a *is to* b *as* e *is to* f,
and	b *is to* c *as* d *is to* e,
then	a *is to* c *as* d *is to* f.

Each depends on a preliminary proposition, Prop. 22 on Prop. 20 and Prop. 23 on Prop. 21. The course of the proof will be made most clear by using the algebraic notation.

The preliminary Prop. 20 asserts that,

if	$a : b = d : e$,
and	$b : c = e : f$,

$$\text{then, according as } a > = < c, \quad d > = < f.$$

For, according as a *is greater than, equal to, or less than* c,
the ratio a : b *is greater than, equal to, or less than the ratio* c : b,   [v. 8 or v. 7]

or (since	$d : e = a : b$,
and	$c : b = f : e$)

the ratio d : e *is greater than, equal to, or less than the ratio* f : e,

    [by aid of v. 13 and v. 11]

and therefore d *is greater than, equal to, or less than* f.   [v. 10 or v. 9]

It is next proved in Prop. 22 that, by v. 4, *the given proportions can be transformed into*

$$ma : nb = md : ne,$$

and $$nb : pc = ne : pf,$$

whence, by v. 20,

according as $$ma \text{ is greater than, equal to, or less than } pc,$$
$$md \text{ is greater than, equal to, or less than } pf,$$

so that, by Def. 5,

$$a : c = d : f.$$

Prop. 23 depends on Prop. 21 in the same way as Prop. 22 on Prop. 20, but the transformation of the ratios in Prop. 23 is to the following:

(1) $$ma : mb = ne : nf$$

(by a double application of v. 15 *and by* v. 11*),*

(2)   mb : nc = md : ne

      (by v. 4, *or equivalent steps),*
and *Prop.* 21 *is then used.*

    Simson makes the proof of Prop. 20 *slightly more explicit, but the main difference from the text is in the addition of the two other cases which Euclid dismisses with "Similarly we can prove." These cases are:*

    "Secondly, *let* A *be equal to* C; *then shall* D *be equal to* F.

    Because A *and* C *are equal to one another,*

	A *is to* B *as* C *is to* B.	[v. 7]
But	A *is to* B *as* D *is to* E,	
and	C *is to* B *as* F *is to* E,	
wherefore	D *is to* E *as* F *to* E;	[v. 11]
	and therefore D *is equal to* F.	[v. 9]

    Next, *let* A *be less than* C; *then shall* D *be less than* F.

    For C *is greater than* A,
and, *as was shown in the first case,*

           C *is to* B *as* F *to* E,
and, *in like manner,*

           B *is to* A *as* E *to* D;
therefore F *is greater than* D, *by the first case; and therefore* D *is less than* F."

Proposition 21

If there be three magnitudes, and others equal to them in multitude, which taken two and two together are in the same ratio, and the proportion of them be perturbed, then, if ex aequali *the first magnitude is greater than the third, the fourth will also be greater than the sixth; if equal, equal; and if less, less.*

Let there be three magnitudes *A*, *B*, *C*, and others *D*, *E*, *F* equal to them in multitude, which taken two and two are in the same ratio, and let the proportion of them be perturbed, so that,

           as *A* is to *B*, so is *E* to *F*,
and,   as *B* is to *C*, so is *D* to *E*,

and let *A* be greater than *C* ex aequali;
I say that *D* will also be greater than *F*; if *A* is equal to *C*, equal; and if less, less.

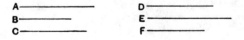

For, since *A* is greater than *C*,
and *B* is some other magnitude,
therefore *A* has to *B* a greater ratio than *C* has to *B*.   [v. 8]

    But, as *A* is to *B*, so is *E* to *F*,
and, as *C* is to *B*, inversely, so is *E* to *D*.
Therefore also *E* has to *F* a greater ratio than *E* has to *D*.   [v. 13]

    But that to which the same has a greater ratio is less;   [v. 10]

    therefore *F* is less than *D*;

    therefore *D* is greater than *F*.

Similarly we can prove that,

if *A* be equal to *C*, *D* will also be equal to *F*;

and if less, less.

Therefore etc.

Q.E.D.

Algebraically, if   a : b = e : f,
     and   b : c = d : e,
        then, according as a $> = <$ c,   d $> = <$ f.
Simson's alterations correspond to those which he makes in Prop. 20. After the first case he proceeds thus.
    "*Secondly, let* A *be equal to* C; *then shall* D *be equal to* F.
Because A and C are equal,
                A *is to* B *as* C *is to* B.   [v. 7]
   But   A *is to* B *as* E *is to* F,
and
wherefore   C *is to* B *as* E *is to* D:
                E *is to* F *as* E *to* D,   [v. 11]
          and therefore D *is equal to* F.   [v. 9]
    Next, let A *be less than* C; *then shall* D *be less than* F.
    For C *is greater than* A,
and, as was shown,
                C *is to* B *as* E, *to* D,
and, in like manner,
                B *is to* A *as* F *to* E;
    therefore F *is greater than* D, *by the first case,*
and therefore D *is less than* F."
    The proof may be shown thus.
           According as a $> = <$ c,   a : b $> = <$ c : b.
But a : b = e : f, *and, by inversion,* c : b = e : d.
       Therefore, according as a $> = <$ c,   e : f $> = <$ e : d,
            and therefore d $> = <$ f.

Proposition 22

If there be any number of magnitudes whatever, and others equal to them in mul-
titude, which taken two and two together are in the same ratio, they will also be in the
same ratio ex aequali.

Let there be any number of magnitudes *A*, *B*, *C*, and others *D*, *E*, *F* equal to
them in multitude, which taken two and two together are in the same ratio, so that,

              as *A* is to *B*, so is *D* to *E*,

and,   as *B* is to *C*, so is *E* to *F*;

I say that they will also be in the same ratio *ex aequali*,

       $<$ that is, as *A* is to *C*, so is *D* to *F* $>$.

For of *A*, *D* let equimultiples *G*, *H* be taken,

and of *B*, *E* other, chance, equimultiples *K*, *L*;

and, further, of *C*, *F* other, chance, equimultiples *M*, *N*.

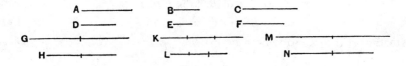

Then, since, as A is to B, so is D to E,

and of A, D equimultiples G, H have been taken,

and of B, E other, chance, equimultiples K, L,

therefore, as G is to K, so is H to L.   [v. 4]

For the same reason also,

$$\text{as } K \text{ is to } M, \text{ so is } L \text{ to } N.$$

Since, then, there are three magnitudes G, K, M, and others H, L, N equal to them in multitude, which taken two and two together are in the same ratio, therefore, *ex aequali*, if G is in excess of M, H is also in excess of N; if equal, equal; and if less, less.   [v. 20]

And G, H are equimultiples of A, D,

and M, N other, chance, equimultiples of C, F.

Therefore, as A is to C, so is D to F.   [v. Def. 5]

Therefore etc.

Q.E.D.

Euclid enunciates this proposition as true of any number of magnitudes whatever forming two sets connected in the manner described, but his proof is confined to the case where each set consists of three magnitudes only. The extension to any number of magnitudes is, however, easy, as shown by Simson.

"Next let there be four magnitudes A, B, C, D, *and other four* E, F, G, H, *which two and two have the same ratio, viz.:*

$$\text{as A is to B, so is E to F,}$$

and $$\text{as B is to C, so is F to G,}$$

and $$\text{as C is to D, so is G to H;}$$

$$\text{A shall be to D as E to H.}$$

A	B	C	D
E	F	G	H

Because A, B, C *are three magnitudes, and* E, F, G *other three, which taken two and two have the same ratio,*

by the foregoing case,

$$\text{A is to C as E to G.}$$

But C *is to* D *as* G *is to* H;

wherefore again, by the first case,

$$\text{A is to D as E to H.}$$

And so on, whatever be the number of magnitudes."

Proposition 23

If there be three magnitudes, and others equal to them in multitude, which taken two and two together are in the same ratio, and the proportion of them be perturbed, they will also be in the same ratio ex aequali.

Let there be three magnitudes A, B, C, and others equal to them in multitude, which, taken two and two together, are in the same proportion, namely D, E, F; and let the proportion of them be perturbed, so that,

$$\text{as } A \text{ is to } B, \text{ so is } E \text{ to } F,$$

and, $$\text{as } B \text{ is to } C, \text{ so is } D \text{ to } E;$$

I say that, as A is to C, so is D to F.

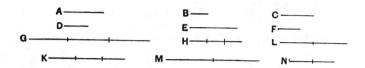

Of *A*, *B*, *D* let equimultiples *G*, *H*, *K* be taken,
and of *C*, *E*, *F* other, chance, equimultiples *L*, *M*, *N*.

Then, since *G*, *H* are equimultiples of *A*, *B*,
and parts have the same ratio as the same multiples of them,   [v. 15]
therefore, as *A* is to *B*, so is *G* to *H*.

For the same reason also,

as *E* is to *F*, so is *M* to *N*.

And, as *A* is to *B*, so is *E* to *F*;
therefore also, as *G* is to *H*, so is *M* to *N*.   [v. 11]

Next, since, as *B* is to *C*, so is *D* to *E*,
alternately, also, as *B* is to *D*, so is *C* to *E*.   [v. 16]

And, since *H*, *K* are equimultiples of *B*, *D*,
and parts have the same ratio as their equimultiples,
therefore, as *B* is to *D*, so is *H* to *K*.   [v. 15]

But, as *B* is to *D*, so is *C* to *E*;
therefore also, as *H* is to *K*, so is *C* to *E*.   [v. 11]

Again, since *L*, *M* are equimultiples of *C*, *E*,
therefore, as *C* is to *E*, so is *L* to *M*.   [v. 15]

But, as *C* is to *E*, so is *H* to *K*;
therefore also, as *H* is to *K*, so is *L* to *M*,   [v. 11]
and, alternately, as *H* is to *L*, so is *K* to *M*.   [v. 16]

But it was also proved that,

as *G* is to *H*, so is *M* to *N*.

Since, then, there are three magnitudes *G*, *H*, *L*, and others equal to them in
multitude *K*, *M*, *N*, which taken two and two together are in the same ratio,
and the proportion of them is perturbed,
therefore, *ex aequali*, if *G* is in excess of *L*, *K* is also in excess of *N*;
if equal, equal; and if less, less.   [v. 21]

And *G*, *K* are equimultiples of *A*, *D*,
and *L*, *N* of *C*, *F*.

Therefore, as *A* is to *C*, so is *D* to *F*.
Therefore etc.

                                          Q.E.D.

There is an important difference between the version given by Simson of one part of the proof of this proposition and that found in the Greek text of Heiberg. Peyrard's MS. *has the version given by Heiberg, but Simson's version has the authority of other* MSS. *The Basel* editio princeps *gives both versions (Simson's being the first). After it has been proved by means of* v. 15 *and* v. 11 *that,*

as G *is to* H, *so is* M *to* N,

or, with the notation used in the note on Prop. 20,

ma : mb = ne : nf,

it has to be proved further that,

as H *is to* L, *so is* K *to* M,

or   mb : nc = md : ne,

and it is clear that the latter result may be directly inferred from v. 4. *The reading translated by Simson makes this inference:*

"And because, as B *is to* C, *so is* D *to* E,
and H, K *are equimultiples of* B, D,
and L, M *of* C, E,

therefore, as H *is to* L, *so is* K *to* M."   [v. 4]

The version in Heiberg's text is not only much longer (it adopts the roundabout method of using each of three Propositions v. 11, 15, 16 *twice over), but it is open to the objection that it uses* v. 16 *which is only applicable if the four magnitudes are of the same kind; whereas* v. 23, *the proposition now in question, is not subject to this restriction.*

*Simson rightly observes that in the last step of the proof it should be stated that "*G, K *are any equimultiples whatever of* A, D, *and* L, N *any whatever of* C, F."

He also gives the extension of the proposition to any number of magnitudes, enunciating it thus:

"If there be any number of magnitudes, and as many others, which, taken two and two, in a cross order, have the same ratio; the first shall have to the last of the first magnitudes the same ratio which the first of the others has to the last"; and adding to the proof as follows:

"Next, let there be four magnitudes A, B, C, D, *and other four* E, F, G, H, *which, taken two and two in a cross order, have the same ratio, viz.:*

A *to* B *as* G *to* H,
B *to* C *as* F *to* G,
and   C *to* D *as* E *to* F;

then A *is to* D *as* E *to* H.

Because A, B, C *are three magnitudes, and* F, G, H *other three which, taken two and two in a cross order, have the same ratio,*

by the first case,   A *is to* C *as* F *to* H.
     But   C *is to* D *as* E *is to* F;
wherefore again, by the first case,

A *is to* D *as* E *to* H.

And so on, whatever be the number of magnitudes."

Proposition 24

If a first magnitude have to a second the same ratio as a third has to a fourth, and also a fifth have to the second the same ratio as a sixth to the fourth, the first and fifth added together will have to the second the same ratio as the third and sixth have to the fourth.

Let a first magnitude *AB* have to a second *C* the same ratio as a third *DE* has to a fourth *F*;

and let also a fifth *BG* have to the second *C* the same ratio as a sixth *EH* has to the fourth *F*;

I say that the first and fifth added together, *AG*, will have to the second *C* the same ratio as the third and sixth, *DH*, has to the fourth *F*.

For since, as *BG* is to *C*, so is *EH* to *F*,
inversely, as *C* is to *BG*, so is *F* to *EH*.

Since, then, as *AB* is to *C*, so is *DE* to *F*,

and, as *C* is to *BG*, so is *F* to *EH*,
therefore, *ex aequali*, as *AB* is to *BG*, so is *DE* to *EH*.   [v. 22]

And, since the magnitudes are proportional *separando*, they will also be proportional *componendo*;   [v. 18]

therefore, as *AG* is to *GB*, so is *DH* to *HE*.

But also, as *BG* is to *C*, so is *EH* to *F*;
therefore, *ex aequali*, as *AG* is to *C*, so is *DH* to *F*.   [v. 22]

Therefore etc.

Q.E.D.

Algebraically, if   a : c = d : f,
and   b : c = e : f,
then   (a + b) : c = (d + e) : f.
This proposition is of the same character as those which precede the propositions relating to compounded ratios; but
it could not be placed earlier than it is because v. 22 is used in the proof of it.
Inverting the second proportion to

c : b = f : e,
it follows, by v. 22, *that*   a : b = d : e,
whence, by v. 18,   (a + b) : b = (d + e) : e,
and from this and the second of the two given proportions we obtain, by a fresh application of v. 22,
(a + b) ; c = (d + e) : f.
The first use of v. 22 is important as showing that the opposite process to compounding ratios, or what we should now
call division of one ratio by another, does not require any new and separate propositions.
Aristotle tacitly uses v. 24 in combination with v. 11 and v. 16, Meteorologica iii. 5, 376 a 22–26.
Simson adds two corollaries, one of which (Cor. 2) notes the extension to any number of magnitudes.
"The proposition holds true of two ranks of magnitudes whatever be their number, of which each of the first rank has to
the second magnitude the same ratio that the corresponding one of the second rank has to a fourth magnitude; as is manifest."
Simson's Cor. 1 states the corresponding proposition to the above with separando taking the place of componendo,
viz., that corresponding to the algebraical form
(a − b) : c = (d − e) : f.
"COR. 1. If the same hypothesis be made as in the proposition, the excess of the first and fifth shall be to the second
as the excess of the third and sixth to the fourth. The demonstration of this is the same with that of the proposition if
division be used instead of composition." That is, we use v. 17 instead of v. 18, and conclude that
(a − b) : b = (d − e) : e.

Proposition 25

*If four magnitudes be proportional, the greatest and the least are greater than the
remaining two.*

Let the four magnitudes *AB*, *CD*, *E*, *F* be proportional so that, as *AB* is to *CD*,
so is *E* to *F*, and let *AB* be the greatest of them and *F* the
least;

I say that *AB*, *F* are greater than *CD*, *E*.

For let *AG* be made equal to *E*, and *CH* equal to *F*.

Since, as *AB* is to *CD*, so is *E* to *F*,

and E is equal to AG, and F to CH,

therefore, as AB is to CD, so is AG to CH.

And since, as the whole AB is to the whole CD, so is the part AG subtracted to the part CH subtracted,

the remainder GB will also be to the remainder HD as the whole AB is to the whole CD.   [v. 19]

But AB is greater than CD;

therefore GB is also greater than HD.

And, since AG is equal to E, and CH to F, therefore AG, F are equal to CH, E.

And if, GB, HD being unequal, and GB greater, AG, F be added to GB and CH, E be added to HD,

it follows that AB, F are greater than CD, E.

Therefore etc.

                                                                                                    Q.E.D.

Algebraically, if $a : b = c : d$,
and a is the greatest of the four magnitudes and d the least,
$$a + d > b + c.$$
    Simson is right in inserting a word in the setting-out, "let AB *be the greatest of them and* <consequently> F *the least." This follows from the particular case, really included in Def. 5, which Simson makes the subject of his proposition A, the case namely where the equimultiples taken are* once *the several magnitudes.*
    The proof is as follows.
        Since $a : b = c : d$,
$$a - c : b - d = a : b.$$ [v. 19]
But a $>$ b ; *therefore* (a $-$ c) $>$ (b $-$ d).   [v. 16 and 14]
    Add to each (c $+$ d);
therefore $(a + d) > (b + c)$.
    There is an important particular case of this proposition, which is, however, not mentioned here, viz. the case where b $=$ c. *The result shows, in this case, that the arithmetic mean between two magnitudes is greater than their geometric mean. The truth of this is proved for straight lines in* vi. 27 *by "geometrical algebra,s" and the theorem forms the* διορισμός *for equations of the second degree.*
    Simson adds at the end of Book V. four propositions, F, G, H, K, which, however, do not seem to be of sufficient practical use to justify their inclusion here. But he adds at the end of his notes to the Book the following paragraph which deserves quotation word for word.
    "The 5th book being thus corrected, I most readily agree to what the learned Dr. Barrow says, 'that there is nothing in the whole body of the elements of a more subtile invention, nothing more solidly established, and more accurately handled than the doctrine of proportionals.' And there is some ground to hope that geometers will think that this could not have been said with as good reason, since Theon's time till the present."
    Simson's claim herein will readily be admitted by all readers who are competent to form a judgment upon his criticisms and elucidations of Book V.

BOOK VII

ELEMENTARY NUMBER THEORY— DEFINITIONS AND PROPOSITIONS

DEFINITIONS

1. An **unit** is that by virtue of which each of the things that exist is called one.

2. A **number** is a multitude composed of units.

3. A number is **a part** of a number, the less of the greater, when it measures the greater;

4. but **parts** when it does not measure it.

5. The greater number is a **multiple** of the less when it is measured by the less.

6. An **even number** is that which is divisible into two equal parts.

7. An **odd number** is that which is not divisible into two equal parts, or that which differs by an unit from an even number.

8. An **even-times even number** is that which is measured by an even number according to an even number.

9. An **even-times odd number** is that which is measured by an even number according to an odd number.

10. An **odd-times odd number** is that which is measured by an odd number according to an odd number.

11. A **prime number** is that which is measured by an unit alone.

12. Numbers **prime to one another** are those which are measured by an unit alone as a common measure.

13. A **composite number** is that which is measured by some number.

14. Numbers **composite to one another** are those which are measured by some number as a common measure.

15. A number is said to **multiply** a number when that which is multiplied is added to itself as many times as there are units in the other, and thus some number is produced.

16. And, when two numbers having multiplied one another make some number, the number so produced is called **plane**, and its **sides** are the numbers which have multiplied one another.

17. And, when three numbers having multiplied one another make some number, the number so produced is **solid**, and its **sides** are the numbers which have multiplied one another.

18. A **square number** is equal multiplied by equal, or a number which is contained by two equal numbers.

19. And a **cube** is equal multiplied by equal and again by equal, or a number which is contained by three equal numbers.

20. Numbers are **proportional** when the first is the same multiple, or the same part, or the same parts, of the second that the third is of the fourth.

21. **Similar plane** and **solid** numbers are those which have their sides proportional.

22. A **perfect number** is that which is equal to its own parts.

Proposition 1

Two unequal numbers being set out, and the less being continually subtracted in turn from the greater, if the number which is left never measures the one before it until an unit is left, the original numbers will be prime to one another.

For, the less of two unequal numbers *AB*, *CD* being continually subtracted from the greater, let the number which is left never measure the one before it until an unit is left;

I say that *AB*, *CD* are prime to one another, that is, that an unit alone measures *AB*, *CD*.

For, if *AB*, *CD* are not prime to one another, some number will measure them.

Let a number measure them, and let it be *E*; let *CD*, measuring *BF*, leave *FA* less than itself,

let *AF*, measuring *DG*, leave *GC* less than itself,

and let *GC*, measuring *FH*, leave an unit *HA*.

Since, then, *E* measures *CD*, and *CD* measures *BF*,

therefore *E* also measures *BF*.

But it also measures the whole *BA*;

therefore it will also measure the remainder *AF*.

But *AF* measures *DG*;

therefore *E* also measures *DG*.

But it also measures the whole *DC*;

therefore it will also measure the remainder *CG*.

But *CG* measures *FH*;

therefore *E* also measures *FH*.

But it also measures the whole *FA*;

therefore it will also measure the remainder, the unit *AH*, though it is a number: which is impossible.

Therefore no number will measure the numbers *AB, CD*; therefore *AB, CD* are
prime to one another. [vii. Def. 12]

Q.E.D.

It is proper to remark here that the representation in Books VII. to IX. of numbers by straight lines is adopted by Heiberg from the MSS. *The method of those editors who substitute* points *for lines is open to objection because it practically necessitates, in many cases, the use of specific numbers, which is contrary to Euclid's manner.*

"Let CD, measuring BF, *leave* FA *less than itself." This is a neat abbreviation for saying, measure along* BA *successive lengths equal to* CD *until a point* F *is reached such that the length* FA *remaining is less than* CD; *in other words, let* BF *be the largest exact multiple of* CD *contained in* BA.

Euclid's method in this proposition is an application to the particular case of prime numbers of the method of finding the greatest common measure of two numbers not prime to one another, which we shall find in the next proposition. With our notation, the method may be shown thus. Supposing the two numbers to be a, b, *we have, say,*

$$
\begin{array}{r}
b)\,a\,(p \\
\underline{pb} \\
c)\,b\,(q \\
\underline{qc} \\
d)\,c\,(r \\
\underline{rd} \\
1
\end{array}
$$

If now a, b *are not prime to one another, they must have a common measure* e, *where* e *is some integer, not unity. And since* e *measures* a, b, *it measures* a − pb, *i.e.* c.

Again, since e *measures* b, c, *it measures* b − qc, *i.e.* d,
and lastly, since e *measures* c, d, *it measures* c − rd, *i.e.* 1:
which is impossible.

Therefore there is no integer, except unity, that measures a, b, *which are accordingly prime to one another.*

Observe that Euclid assumes as an axiom that, if a, b *are both divisible by* c, *so is* a − pb. *In the next proposition he assumes as an axiom that* c *will in the case supposed divide* a + pb.

Proposition 2

Given two numbers not prime to one another, to find their greatest common measure.

Let *AB, CD* be the two given numbers not prime to one another.

Thus it is required to find the greatest common measure of *AB, CD*.

If now *CD* measures *AB*—and it also measures itself—*CD* is a
common measure of *CD, AB*.

And it is manifest that it is also the greatest; for no greater number than *CD* will measure *CD*.

But, if *CD* does not measure *AB*, then, the less of the numbers
AB, CD being continually subtracted from the greater, some number will be left
which will measure the one before it.

For an unit will not be left; otherwise *AB, CD* will be prime to one another
[vii. 1], which is contrary to the hypothesis.

Therefore some number will be left which will measure the one before it.

Now let *CD*, measuring *BE*, leave *EA* less than itself,
let *EA*, measuring *DF*, leave *FC* less than itself,
and let *CF* measure *AE*.

Since then, *CF* measures *AE*, and *AE* measures *DF*, therefore *CF* will also measure *DF*.

But it also measures itself; therefore it will also measure the whole *CD*.

But *CD* measures *BE*; therefore *CF* also measures *BE*.

But it also measures *EA*; therefore it will also measure the whole *BA*.

But it also measures *CD*; therefore *CF* measures *AB*, *CD*.

Therefore *CF* is a common measure of *AB*, *CD*.

I say next that it is also the greatest.

For, if *CF* is not the greatest common measure of *AB*, *CD*, some number which is greater than *CF* will measure the numbers *AB*, *CD*.

Let such a number measure them, and let it be *G*.

Now, since *G* measures *CD*, while *CD* measures *BE*, *G* also measures *BE*.

But it also measures the whole *BA*; therefore it will also measure the remainder *AE*.

But *AE* measures *DF*; therefore *G* will also measure *DF*.

But it also measures the whole *DC*; therefore it will also measure the remainder *CF*, that is, the greater will measure the less: which is impossible.

Therefore no number which is greater than *CF* will measure the numbers *AB*, *CD*; therefore *CF* is the greatest common measure of *AB*, *CD*.

PORISM. From this it is manifest that, if a number measure two numbers, it will also measure their greatest common measure.

Q.E.D.

Here we have the exact method of finding the greatest common measure given in the text-books of algebra, including the reductio ad absurdum *proof that the number arrived at is not only a common measure but the greatest common measure. The process of finding the greatest common measure is simply shown thus:*

$$b)\, a\, (p$$
$$\underline{pb}$$
$$c)\, b\, (q$$
$$\underline{qc}$$
$$d)\, c\, (r$$
$$\underline{rd}$$

We shall arrive, says Euclid, at some number, say d, *which measures the one before it, i.e. such that* c = rd. *Otherwise the process would go on until we arrived at unity. This is impossible because in that case* a, b *would be prime to one another, which is contrary to the hypothesis.*

Next, like the text-books of algebra, he goes on to show that d *will be* some *common measure of* a, b. *For* d *measures* c; *therefore it measures* qc + d, *that is,* b, *and hence it measures* pb + c, *that is,* a.

Lastly, he proves that d *is the* greatest *common measure of* a, b *as follows.*

Suppose that e *is a common measure greater than* d.

Then e, *measuring* a, b, *must measure* a − pb, *or* c.

Similarly e *must measure* b − qc, *that is,* d: *which is impossible, since* e *is by hypothesis greater than* d. *Therefore etc.*

Euclid's proposition is thus identical *with the algebraical proposition as generally given, e.g. in Todhunter's algebra, except that of course Euclid's numbers are integers.*

Nicomachus gives the same rule (though without proving it) when he shews how to determine whether two given odd *numbers are prime or not prime to one another, and, if they are not prime to one another, what is their common measure. We are, he says, to compare the numbers in turn by continually taking the less from the greater as many times as possible, then taking the remainder as many times as possible from the less of the original numbers, and so on; this process "will finish either at an unit or at some one and the same number," by which it is implied that the division of a greater number by a less is done by* separate subtractions *of the less. Thus, with regard to 21 and 49, Nicomachus says, "I subtract the less from the greater; 28 is left; then again I subtract from this the same 21 (for this is possible); 7 is left; I subtract this from 21, 14 is left; from which I again subtract 7 (for this is possible); 7 will be left, but 7 cannot be subtracted from 7." The last phrase is curious, but the meaning of it is obvious enough, as also the meaning of the phrase about ending "at one and the same number."*

The proof of the Porism is of course contained in that part of the proposition which proves that G, *a common measure different from* CF, *must measure* CF. *The supposition, thereby proved to be false, that* G *is greater than* CF *does not affect the validity of the proof that* G *measures* CF *in any case.*

Proposition 3

Given three numbers not prime to one another, to find their greatest common measure.

Let A, B, C be the three given numbers not prime to one another; thus it is required to find the greatest common measure of A, B, C.

For let the greatest common measure, D, of the two numbers A, B be taken;   [vii. 2]

then D either measures, or does not measure, C.

First, let it measure it.

But it measures A, B also;

therefore D measures A, B, C;

therefore D is a common measure of A, B, C.

I say that it is also the greatest.

For, if D is not the greatest common measure of A, B, C, some number which is greater than D will measure the numbers A, B, C.

Let such a number measure them, and let it be E.

Since then E measures A, B, C,

it will also measure A, B;

therefore it will also measure the greatest common measure of A, B.   [vii. 2, Por.]

But the greatest common measure of A, B is D;

therefore E measures D, the greater the less: which is impossible.

Therefore no number which is greater than D will measure the numbers A, B, C;

therefore D is the greatest common measure of A, B, C.

Next, let D not measure C;

I say first that C, D are not prime to one another.

For, since A, B, C are not prime to one another, some number will measure them.

Now that which measures A, B, C will also measure A, B, and will measure D, the greatest common measure of A, B.   [vii. 2, Por.]

But it measures C also;

therefore some number will measure the numbers D, C;

therefore D, C are not prime to one another.

Let then their greatest common measure E be taken.   [vii. 2]

Then, since E measures D,

and D measures A, B,

therefore E also measures A, B.

But it measures C also;

therefore E measures A, B, C;

therefore E is a common measure of A, B, C.

I say next that it is also the greatest.

For, if E is not the greatest common measure of A, B, C, some number which is greater than E will measure the numbers A, B, C.

Let such a number measure them, and let it be F.

Now, since F measures A, B, C,

it also measures A, B;

therefore it will also measure the greatest common measure of A, B.   [vii. 2, Por.]

But the greatest common measure of A, B is D;

therefore F measures D.

And it measures C also;

therefore F measures D, C;

therefore it will also measure the greatest common measure of D, C.   [vii. 2, Por.]

But the greatest common measure of D, C is E;

therefore F measures E, the greater the less: which is impossible.

Therefore no number which is greater than E will measure the numbers A, B, C;

therefore E is the greatest common measure of A, B, C.

Q.E.D.

Euclid's proof is here longer than we should make it because he distinguishes two cases, the simpler of which is really included in the other.

Having taken the greatest common measure, say d, *of* a, b, *two of the three given numbers* a, b, c, *he distinguishes the cases*

(1) in which d *measures* c,

(2) in which d *does not measure* c.

In the first case the greatest common measure of d, c *is* d *itself; in the second case it has to be found by a repetition of the process of* vii. 2. *In either case the greatest common measure of* a, b, c *is the greatest common measure of* d, c.

But, after disposing of the simpler case, Euclid thinks it necessary to prove that, if d *does not measure* c, d *and* c *must necessarily have a greatest common measure. This he does by means of the original hypothesis that* a, b, c *are not prime to one another. Since they are not prime to one another, they must have a common measure; any common measure of* a, b *is a measure of* d, *and therefore any common measure of* a, b, c *is a common measure of* d, c; *hence* d, c *must have a common measure, and are therefore not prime to one another.*

The proofs of cases (1) and (2) repeat exactly the same argument as we saw in vii. 2, *and it is proved separately for* d *in case (1) and* e *in case (2), where* e *is the greatest common measure of* d, c,

(α) *that it is a common measure of* a, b, c,

(β) *that it is the* greatest *common measure.*

Heron remarks (an-Nairīzī, ed. Curtze) that the method does not only enable us to find the greatest common measure of three numbers; it can be used to find the greatest common measure of as many numbers as we please. This is because any number measuring two numbers also measures their greatest common measure; and hence we can find the G.C.M. of pairs, then the G.C.M. of pairs of these, and so on, until only two numbers are left and we find the G.C.M. of these. Euclid tacitly assumes this extension in vii. 33, *where he takes the greatest common measure of as many numbers as we please.*

Proposition 4

Any number is either a part or parts of any number, the less of the greater.

Let *A, BC* be two numbers, and let *BC* be the less;
I say that *BC* is either a part, or parts, of *A*.

For *A, BC* are either prime to one another or not.

First, let *A, BC* be prime to one another.

Then, if *BC* be divided into the units in it, each unit of those in *BC* will be some part of *A*; so that *BC* is parts of *A*.

Next let *A, BC* not be prime to one another; then *BC* either measures, or does not measure, *A*.

If now *BC* measures *A, BC* is a part of *A*.

But, if not, let the greatest common measure *D* of *A, BC* be taken;   [vii. 2]
and let *BC* be divided into the numbers equal to *D*, namely *BE, EF, FC*.

Now, since *D* measures *A, D* is a part of *A*.

But *D* is equal to each of the numbers *BE, EF, FC*;
therefore each of the numbers *BE, EF, FC* is also a part of *A*; so that *BC* is parts of *A*.

Therefore etc.

Q.E.D.

The meaning of the enunciation is of course that, if a, b *be two numbers of which* b *is the less, then* b *is either a sub-multiple or some proper fraction of* a.

(1) If a, b *are prime to one another, divide each into its units; then* b *contains* b *of the same parts of which* a *contains* a. *Therefore* b *is "parts" or a proper fraction of* a.

(2) If a, b *be not prime to one another, either* b *measures* a, *in which case* b *is a submultiple or "part" of* a, *or, if* g *be the greatest common measure of* a, b, *we may put* a = mg *and* b = ng, *and* b *will contain* n *of the same parts* (g) *of which* a *contains* m, *so that* b *is again "parts," or a proper fraction, of* a.

Proposition 5

If a number be a part of a number, and another be the same part of another, the sum will also be the same part of the sum that the one is of the one.

For let the number *A* be a part of *BC*,
and another, *D*, the same part of another *EF* that *A* is of *BC*;
I say that the sum of *A*, *D* is also the same part of the sum of *BC*,
EF that *A* is of *BC*.

For since, whatever part *A* is of *BC*, *D* is also the same part of *EF*,
therefore, as many numbers as there are in *BC* equal to *A*, so many
numbers are there also in *EF* equal to *D*.

Let *BC* be divided into the numbers equal to *A*, namely *BG*, *GC*,
and *EF* into the numbers equal to *D*, namely *EH*, *HF*;
then the multitude of *BG*, *GC* will be equal to the multitude of *EH*, *HF*.

And, since *BG* is equal to *A*, and *EH* to *D*,
therefore *BG*, *EH* are also equal to *A*, *D*.

For the same reason
GC, *HF* are also equal to *A*, *D*.

Therefore, as many numbers as there are in *BC* equal to *A*, so many are there
also in *BC*, *EF* equal to *A*, *D*.

Therefore, whatever multiple *BC* is of *A*, the same multiple also is the sum of
BC, *EF* of the sum of *A*, *D*.

Therefore, whatever part *A* is of *BC*, the same part also is the sum of *A*, *D* of
the sum of *BC*, *EF*.

Q.E.D.

If
$$a = \frac{1}{n}b, \ and \ c = \frac{1}{n}d, \ then$$
$$a + c = \frac{1}{n}(b + d).$$

*The proposition is of course true for any quantity of pairs of numbers similarly related, as is the next proposition also;
and both propositions are used in the extended form in* vii. 9, 10.

Proposition 6

*If a number be parts of a number, and another be the same parts of another, the
sum will also be the same parts of the sum that the one is of the one.*

For let the number *AB* be parts of the number *C*, and another, *DE*, the same
parts of another, *F*, that *AB* is of *C*;
I say that the sum of *AB*, *DE* is also the same parts of the sum of
C, *F* that *AB* is of *C*.

For since, whatever parts *AB* is of *C*, *DE* is also the same parts
of *F*,
therefore, as many parts of *C* as there are in *AB*, so many parts of
F are there also in *DE*.

Let *AB* be divided into the parts of *C*, namely *AG*, *GB*, and *DE* into the parts of *F*, namely *DH*, *HE*;

thus the multitude of *AG*, *GB* will be equal to the multitude of *DH*, *HE*.

And since, whatever part *AG* is of *C*, the same part is *DH* of *F* also, therefore, whatever part *AG* is of *C*, the same part also is the sum of *AG*, *DH* of the sum of *C*, *F*.   [vii. 5]

For the same reason,

whatever part *GB* is of *C*, the same part also is the sum of *GB*, *HE* of the sum of *C*, *F*.

Therefore, whatever parts *AB* is of *C*, the same parts also is the sum of *AB*, *DE* of the sum of *C*, *F*.

Q.E.D.

If
$$a = \frac{m}{n}b, \; and \; c = \frac{m}{n}d,$$

then
$$a + c = \frac{m}{n}(b + d).$$

More generally, if
$$a = \frac{m}{n}b, \; c = \frac{m}{n}d, \; e = \frac{m}{n}f, \ldots$$

then
$$(a + c + e + g + \ldots) = \frac{m}{n}(b + d + f + h + \ldots).$$

In Euclid's proposition m < n, *but the generality of the result is of course not affected. This proposition and the last are complementary to* v. 1, *which proves the corresponding result with* multiple *substituted for "part" or "parts."*

Proposition 7

If a number be that part of a number, which a number subtracted is of a number subtracted, the remainder will also be the same part of the remainder that the whole is of the whole.

For let the number *AB* be that part of the number *CD* which *AE* subtracted is of *CF* subtracted;

I say that the remainder *EB* is also the same part of the remainder *FD* that the whole *AB* is of the whole *CD*.

For, whatever part *AE* is of *CF*, the same part also let *EB* be of *CG*.

Now since, whatever part *AE* is of *CF*, the same part also is *EB* of *CG*, therefore, whatever part *AE* is of *CF*, the same part also is *AB* of *GF*.   [vii. 5]

But, whatever part *AE* is of *CF*, the same part also, by hypothesis, is *AB* of *CD*; therefore, whatever part *AB* is of *GF*, the same part is it of *CD* also;

therefore *GF* is equal to *CD*.

Let *CF* be subtracted from each;

therefore the remainder *GC* is equal to the remainder *FD*.

Now since, whatever part *AE* is of *CF*, the same part also is *EB* of *GC*, while *GC* is equal to *FD*,

therefore, whatever part *AE* is of *CF*, the same part also is *EB* of *FD*.

But, whatever part *AE* is of *CF*, the same part also is *AB* of *CD*;

therefore also the remainder *EB* is the same part of the remainder *FD* that the whole *AB* is of the whole *CD*.

Q.E.D.

If a $= \frac{1}{n}$ b *and* c $= \frac{1}{n}$ d, *we are to prove that*

$$a - c = \frac{1}{n}(b - d),$$

a result differing from that of vii. 5 *in that* minus *is substituted for* plus. *Euclid's method is as follows.*
  Suppose that e *is taken such that*

$$a - c = \frac{1}{n}e. \quad \dots\dots\dots\dots\dots\dots\dots\dots\dots\dots\dots\dots\dots\dots\dots\dots(1)$$

Now
$$c = \frac{1}{n}d.$$

Therefore
$$a = \frac{1}{n}(d + e), \qquad\qquad\qquad [\text{vii. 5}]$$

whence, from the hypothesis, $d + e = b,$
so that $e = b - d,$
and, substituting this value of e *in (1), we have*

$$a - c = \frac{1}{n}(b - d).$$

Proposition 8

If a number be the same parts of a number that a number subtracted is of a number subtracted, the remainder will also be the same parts of the remainder that the whole is of the whole.

For let the number *AB* be the same parts of the number *CD* that *AE* subtracted is of *CF* subtracted;

I say that the remainder EB is also the same parts of the remainder FD that the whole AB is of the whole CD.

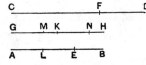

For let *GH* be made equal to *AB*.

Therefore, whatever parts *GH* is of *CD*, the same parts also is *AE* of *CF*.

Let *GH* be divided into the parts of *CD*, namely *GK*, *KH*, and *AE* into the parts of *CF*, namely *AL*, *LE*;

72

thus the multitude of *GK*, *KH* will be equal to the multitude of *AL*, *LE*.

Now since, whatever part *GK* is of *CD*, the same part also is *AL* of *CF*, while *CD* is greater than *CF*,

therefore *GK* is also greater than *AL*.

Let *GM* be made equal to *AL*.

Therefore, whatever part *GK* is of *CD*, the same part also is *GM* of *CF*; therefore also the remainder *MK* is the same part of the remainder *FD* that the whole *GK* is of the whole *CD*. [vii. 7]

Again, since, whatever part *KH* is of *CD*, the same part also is *EL* of *CF*, while *CD* is greater than *CF*,

therefore *HK* is also greater than *EL*.

Let *KN* be made equal to *EL*.

Therefore, whatever part *KH* is of *CD*, the same part also is *KN* of *CF*; therefore also the remainder *NH* is the same part of the remainder *FD* that the whole *KH* is of the whole *CD*. [vii. 7]

But the remainder *MK* was also proved to be the same part of the remainder *FD* that the whole *GK* is of the whole *CD*;

therefore also the sum of *MK*, *NH* is the same parts of *DF* that the whole *HG* is of the whole *CD*.

But the sum of *MK*, *NH* is equal to *EB*,

and *HG* is equal to *BA*;

therefore the remainder *EB* is the same parts of the remainder *FD* that the whole *AB* is of the whole *CD*.

Q.E.D.

If
$$a = \frac{m}{n}b \ and \ c = \frac{m}{n}d, \quad (m < n)$$

then
$$a - c = \frac{m}{n}(b - d).$$

Euclid's proof amounts to the following.
Take e equal to $\frac{1}{n}$b, and f equal to $\frac{1}{n}$d.
Then since, by hypothesis, b > d,

$$e > f,$$

and, by vii. 7,
$$e - f = \frac{1}{n}(b - d).$$

Repeat this for all the parts equal to e and f that there are in a, b respectively, and we have, by addition (a, b containing m of such parts respectively),

$$m(e - f) = \frac{m}{n}(b - d).$$

But
$$m(e - f) = a - c.$$

Therefore
$$a - c = \frac{m}{n}(b - d).$$

The propositions vii. 7, 8 *are complementary to* v. 5 *which gives the corresponding result with* multiple *in the place of "part" or "parts."*

Proposition 9

If a number be a part of a number, and another be the same part of another, alternately also, whatever part or parts the first is of the third, the same part, or the same parts, will the second also be of the fourth.

For let the number A be a part of the number BC, and another, D, the same part of another, EF, that A is of BC;

I say that, alternately also, whatever part or parts A is of D, the same part or parts is BC of EF also.

For since, whatever part A is of BC, the same part also is D of EF, therefore, as many numbers as there are in BC equal to A, so many also are there in EF equal to D.

Let BC be divided into the numbers equal to A, namely BG, GC, and EF into those equal to D, namely EH, HF;

thus the multitude of BG, GC will be equal to the multitude of EH, HF.

Now, since the numbers BG, GC are equal to one another, and the numbers EH, HF are also equal to one another,

while the multitude of BG, GC is equal to the multitude of EH, HF,

therefore, whatever part or parts BG is of EH, the same part or the same parts is GC of HF also;

so that, in addition, whatever part or parts BG is of EH, the same part also, or the same parts, is the sum BC of the sum EF.   [vii. 5, 6]

But BG is equal to A, and EH to D;

therefore, whatever part or parts A is of D, the same part or the same parts is BC of EF also.

Q.E.D.

If $a = \frac{1}{n}b$ *and* $c = \frac{1}{n}d$, *then, whatever fraction ("part" or "parts")* a *is of* c, *the same fraction will* b *be of* d.

Dividing b *into each of its parts equal to* a, *and* d *into each of its parts equal to* c, *it is clear that, whatever fraction one of the parts* a *is of one of the parts* c, *the same fraction is any other of the parts* a *of any other of the parts* c.

And the number of the parts a *is equal to the number of the parts* c, viz. n.

Therefore, by vii. 5, 6, na *is the same fraction of* nc *that* a *is of* c, *i.e.* b *is the same fraction of* d *that* a *is of* c.

Proposition 10

If a number be parts of a number, and another be the same parts of another, alternately also, whatever parts or part the first is of the third, the same parts or the same part will the second also be of the fourth.

For let the number AB be parts of the number C, and another, DE, the same parts of another, F;

I say that, alternately also, whatever parts or part AB is of DE, the same parts or the same part is C of F also.

For since, whatever parts *AB* is of *C*, the same parts also is *DE* of *F*, therefore, as many parts of *C* as there are in *AB*, so many parts also of *F* are there in *DE*.

Let *AB* be divided into the parts of *C*, namely *AG*, *GB*, and *DE* into the parts of *F*, namely *DH*, *HE*;

thus the multitude of *AG*, *GB* will be equal to the multitude of *DH*, *HE*.

Now since, whatever part *AG* is of *C*, the same part also is *DH* of *F*, alternately also, whatever part or parts *AG* is of *DH*,

the same part or the same parts is *C* of *F* also.   [vii. 9]

For the same reason also,

whatever part or parts *GB* is of *HE*, the same part or the same parts is *C* of *F* also; so that, in addition, whatever parts or part *AB* is of *DE*, the same parts also, or the same part, is *C* of *F*.   [vii. 5, 6]

Q.E.D.

If $a = \frac{m}{n} b$ *and* $c = \frac{m}{n} d$, *then, whatever fraction* a *is of* c, *the same fraction is* b *of* d.

To prove this, a *is divided into its* m *parts equal to* b/n, *and* c *into its* m *parts equal to* d/n.

Then, by vii. 9, *whatever fraction one of the* m *parts of* a *is of one of the* m *parts of* c, *the same fraction is* b *of* d.

And, by vii. 5, 6, *whatever fraction one of the* m *parts of* a *is of one of the* m *parts of* c, *the same fraction is the sum of the parts of* a *(that is,* a) *of the sum of the parts of* c *(that is,* c).

Whence the result follows.

In the Greek text, after the words "so that, in addition" in the last line but one, is an additional explanation making the reference to vii. 5, 6 *clearer, as follows: "whatever part or parts* AG *is of* DH, *the same part or the same parts is* GB *of* HE *also;*

therefore also, whatever part or parts AG *is of* DH, *the same part or the same parts is* AB *of* DE *also.*   [vii. 5, 6]

But it was proved that, whatever part or parts AG *is of* DH, *the same part or the same parts is* C *of* F *also; therefore also" etc. as in the last two lines of the text.*

Heiberg concludes, on the authority of P, *which only has the words in the margin in a later hand, that they may be attributed to Theon.*

Proposition 11

If, as whole is to whole, so is a number subtracted to a number subtracted, the remainder will also be to the remainder as whole to whole.

As the whole *AB* is to the whole *CD*, so let *AE* subtracted be to *CF* subtracted; I say that the remainder *EB* is also to the remainder *FD* as the whole *AB* to the whole *CD*.

Since, as *AB* is to *CD*, so is *AE* to *CF*,

whatever part or parts *AB* is of *CD*, the same part or the same parts is *AE* of *CF* also;   [vii. Def. 20]

Therefore also the remainder *EB* is the same part or parts of *FD* that *AB* is of *CD*.   [vii. 7, 8]

Therefore, as *EB* is to *FD*, so is *AB* to *CD*.   [vii. Def. 20]

Q.E.D.

It will be observed that, in dealing with the proportions in Props. 11–13, Euclid only contemplates the case where the first number is "a part" or "parts" of the second, while in Prop. 13 he assumes the first to be "a part" or "parts" of the third also; that is, the first number is in all three propositions assumed to be less than the second, and in Prop. 13 less than the third also. Yet the figures in Props. 11 and 13 are inconsistent with these assumptions. If the facts are taken to correspond to the figures in these propositions, it is necessary to take account of the other possibilities involved in the definition of proportion (vii. Def. 20), that the first number may also be a multiple, or a multiple plus "a part" or "parts" (including once as a multiple in this case), of each number with which it is compared. Thus a number of different cases would have to be considered. The remedy is to make the ratio which is in the lower terms the first ratio, and to invert the ratios, if necessary, in order to make "a part" or "parts" literally apply.

If $\qquad\qquad\qquad\qquad$ $a : b = c : d,$ $\qquad\qquad\qquad$ $(a > c, b > d)$
then $\qquad\qquad\qquad$ $(a - c) : (b - d) = a : b.$

This proposition for numbers corresponds to v. 19 for magnitudes. The enunciation is the same except that the masculine (agreeing with ἀριθμός) takes the place of the neuter (agreeing with μέγεθος).

The proof is no more than a combination of the arithmetical definition of proportion (vii. Def. 20) with the results of vii. 7, 8. The language of proportions is turned into the language of fractions by Def. 20; the results of vii. 7, 8 are then used and the language retransformed by Def. 20 into the language of proportions.

Proposition 12

If there be as many numbers as we please in proportion, then, as one of the antecedents is to one of the consequents, so are all the antecedents to all the consequents.

Let A, B, C, D be as many numbers as we please in proportion, so that,

as A is to B, so is C to D;

I say that, as A is to B, so are A, C to B, D.

For since, as A is to B, so is C to D,

whatever part or parts A is of B, the same part or parts is C of D also.

[vii. Def. 20]

Therefore also the sum of A, C is the same part or the same parts of the sum of B, D that A is of B. $\qquad$ [vii. 5, 6]

Therefore, as A is to B, so are A, C to B, D. $\qquad$ [vii. Def. 20]

If $\qquad\qquad\qquad$ $a : a' = b : b' = c : c' = \ldots,$
then each ratio is equal to $(a + b + c + \ldots) : (a' + b' + c' + \ldots).$

The proposition corresponds to v. 12, and the enunciation is word for word the same with that of v. 12 except that ἀριθμός takes the place of μέγεθος.

Again the proof merely connects the arithmetical definition of proportion (vii. def. 20) with the results of vii. 5, 6, which are quoted as true for any number of numbers, and not merely for two numbers as in the enunciations of vii. 5, 6.

Proposition 13

If four numbers be proportional, they will also be proportional alternately.

Let the four numbers A, B, C, D be proportional, so that,

as A is to B, so is C to D;

I say that they will also be proportional alternately, so that,

as A is to C, so will B be to D.

For since, as A is to B, so is C to D,

therefore, whatever part or parts A is of B,

the same part or the same parts is C of D also. $\qquad$ [vii. Def. 20]

76

Therefore, alternately, whatever part or parts A is of C, the same part or the same parts is B of D also. [vii. 10]

Therefore, as A is to C, so is B to D. [vii. Def. 20]

Q.E.D.

If $a : b = c : d,$
then, alternately, $a : c = b : d.$
The proposition corresponds to v. 16 *for magnitudes, and the proof consists in connecting* vii. Def. 20 *with the result of* vii. 10.

Proposition 14

If there be as many numbers as we please, and others equal to them in multitude, which taken two and two are in the same ratio, they will also be in the same ratio ex aequali.

Let there be as many numbers as we please A, B, C, and others equal to them in multitude D, E, F, which taken two and two are in the same ratio, so that,

as A is to B, so is D to E,

and,   as B is to C, so is E to F;

I say that, *ex aequali*,

as A is to C, so also is D to F.

For, since, as A is to B, so is D to E,
therefore, alternately,

as A is to D, so is B to E. [vii. 13]

Again, since, as B is to C, so is E to F,
therefore, alternately,

as B is to E, so is C to F. [vii. 13]

But, as B is to E, so is A to D;
therefore also, as A is to D, so is C to F.

Therefore, alternately,

as A is to C, so is D to F. [id.]

If $a : b = d : e,$
and $b : c = e : f,$
    then, ex aequali, $a : c = d : f;$
and the same is true however many successive numbers are so related.
    The proof is simplicity itself.
    By vii. 13, *alternately,* $a : b = b : e,$
        and $b : e = c : f,$
    Therefore $a : d = c : f,$
        and, again alternately, $a : c = d : f.$

Observe that this simple method cannot be used to prove the corresponding proposition for magnitudes, v. 22, although v. 22 has been preceded by the two propositions in that Book corresponding to the propositions used here, viz. v. 16 and v.11. The reason of this is that this method would only prove v. 22 for six magnitudes all of the same kind, whereas the magnitudes in v. 22 are not subject to this limitation.

Heiberg remarks in a note on vii.19 that, while Euclid has proved several propositions of Book V. over again, by a separate proof, for numbers, he has neglected to do so in certain cases; e.g., he often uses v. 11 in these propositions of Book VII., v. 9 in vii. 19, v. 7 in the same proposition, and so on. Thus Heiberg would apparently suppose Euclid to use v. 11 in the last step of the present proof (Ratios which are the same with the same ratio are also the same with one another). I think it preferable to suppose that Euclid regarded the last step as axiomatic; since, by the definition of proportion, the first number is the same multiple or the same part or the same parts of the second that the third is of the fourth: the assumption is no more than an assumption that the numbers or proper fractions which are respectively equal to the same number or proper fraction are equal to one another.

Though the proposition is only proved of six numbers, the extension to as many as we please (as expressed in the enunciation) is obvious.

Proposition 15

If an unit measure any number, and another number measure any other number the same number of times, alternately also, the unit will measure the third number the same number of times that the second measures the fourth.

For let the unit A measure any number BC, and let another number D measure any other number EF the same number of times;

I say that, alternately also, the unit A measures the number D the same number of times that BC measures EF.

For, since the unit A measures the number BC the same number of times that D measures EF,

therefore, as many units as there are in BC, so many numbers equal to D are there in EF also.

Let BC be divided into the units in it, BG, GH, HC,

and EF into the numbers EK, KL, LF equal to D.

Thus the multitude of BG, GH, HC will be equal to the multitude of EK, KL, LF.

And, since the units BG, GH, HC are equal to one another, and the numbers EK, KL, LF are also equal to one another, while the multitude of the units BG, GH, HC is equal to the multitude of the numbers EK, KL, LF,

therefore, as the unit BG is to the number EK, so will the unit GH be to the number KL, and the unit HC to the number LF.

Therefore also, as one of the antecedents is to one of the consequents, so will all the antecedents be to all the consequents;   [vii. 12]

therefore, as the unit BG is to the number EK, so is BC to EF.

But the unit BG is equal to the unit A,

and the number EK to the number D.

Therefore, as the unit A is to the number D, so is BC to EF.

Therefore the unit A measures the number D the same number of times that BC measures EF.

Q.E.D.

If there be four numbers 1, m, a, ma *(such that* 1 *measures* m *the same number of times that* a *measures* ma*),* 1 *measures* a *the same number of times that* m *measures* ma.

Except that the first number is unity and the numbers are said to measure *instead of being a* part *of others, this proposition and its proof do not differ from* vii. 9; *in fact this proposition is a particular case of the other.*

Proposition 16

If two numbers by multiplying one another make certain numbers, the numbers so produced will be equal to one another.

Let A, B be two numbers, and let A by multiplying B make C, and B by multiplying A make D;

I say that C is equal to D.

For, since A by multiplying B has made C, therefore B measures C according to the units in A.

But the unit E also measures the number A according to the units in it;

therefore the unit E measures A the same number of times that B measures C.

Therefore, alternately, the unit E measures the number B the same number of times that A measures C.   [vii. 15]

Again, since B by multiplying A has made D, therefore A measures D according to the units in B.

But the unit E also measures B according to the units in it;

therefore the unit E measures the number B the same number of times that A measures D.

But the unit E measured the number B the same number of times that A measures C;

therefore A measures each of the numbers C, D the same number of times.

Therefore C is equal to D.

Q.E.D.

1–2. **The numbers so produced.** The Greek has οἱ γενόμενοι ἐξ αὐτῶν, "the (numbers) produced *from them.*" By "from them" Euclid means "from the original numbers," though this is not very clear even in the Greek. I think ambiguity is best avoided by leaving out the words.

This proposition proves that, if any numbers be multiplied together, the order of multiplication is indifferent, or ab = ba.

It is important to get a clear understanding of what Euclid means when he speaks of one number multiplying another. vii. Def. 15 *states that the effect of* "a *multiplying* b" *is taking* a *times* b. *We shall always represent* "a *times* b" *by* ab *and* "b *times* a" *by* ba. *This being premised, the proof that* ab = ba *may be represented as follows in the language of proportions.*

By vii. Def. 20,	1 : a = b : ab.	
Therefore, alternately,	1 : b = a : ab.	[vii. 13]
Again, by vii. Def. 20,	1 : b = a : ba.	

Therefore a : ab = a : ba,
or ab = ba.
Euclid does not use the language of proportions but that of fractions or their equivalent measures, quoting vii. 15, *a particular case of* vii. 13 *differently expressed, instead of* vii. 13 *itself.*

Proposition 17

If a number by multiplying two numbers make certain numbers, the numbers so produced will have the same ratio as the numbers multiplied.

For let the number A by multiplying the two numbers B, C make D, E;
I say that, as B is to C, so is D to E.

For, since A by multiplying B has made D,
therefore B measures D according to the units in A.

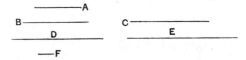

But the unit F also measures the number A according to the units in it;
therefore the unit F measures the number A the same number of times that B measures D.

Therefore, as the unit F is to the number A, so is B to D.　　　[vii. Def. 20]
For the same reason,
as the unit F is to the number A, so also is C to E;
therefore also, as B is to D, so is C to E.

Therefore, alternately, as B is to C, so is D to E.　　　[vii. 13]

Q.E.D.

b : c = ab : ac.
In this case Euclid translates the language of measures into that of proportions, and the proof is exactly like that set out in the last note.

By vii. Def. 20,	1 : a = b : ab,
and	1 : a = c : ac.
Therefore	b : ab = c : ac,
and, alternately,	b : c = ab : ac.　　　[vii. 13]

Proposition 18

If two numbers by multiplying any number make certain numbers, the numbers so produced will have the same ratio as the multipliers.

For let two numbers A, B by multiplying any number C make D, E;
I say that, as A is to B, so is D to E.

For, since A by multiplying C has made D,
therefore also C by multiplying A has made D.

[vii. 16]

For the same reason also

C by multiplying B has made E.

Therefore the number C by multiplying the two numbers A, B has made D, E.

Therefore, as A is to B, so is D to E.   [vii. 17]

It is here proved that	$a : b = ac : bc.$	
The argument is as follows.		
	$ac = ca.$	[vii. 16]
Similarly	$bc = cb.$	
And	$a : b = ca : cb;$	[vii. 17]
therefore	$a : b = ac : bc.$	

Proposition 19

If four numbers be proportional, the number produced from the first and fourth will be equal to the number produced from the second and third; and, if the number produced from the first and fourth be equal to that produced from the second and third, the four numbers will be proportional.

Let A, B, C, D be four numbers in proportion, so that,

as A is to B, so is C to D;

and let A by multiplying D make E, and let B by multiplying C make F;

I say that E is equal to F.

For let A by multiplying C make G.

Since, then, A by multiplying C has made G, and by multiplying D has made E, the number A by multiplying the two numbers C, D has made G, E.

Therefore, as C is to D, so is G to E.   [vii. 17]

But, as C is to D, so is A to B;

therefore also, as A is to B, so is G to E.

Again, since A by multiplying C has made G,

but, further, B has also by multiplying C made F,

the two numbers A, B by multiplying a certain number C

have made G, F.

Therefore, as A is to B, so is G to F.   [vii. 18]

But further, as A is to B, so is G to E also;

therefore also, as G is to E, so is G to F.

Therefore G has to each of the numbers E, F the same ratio;

therefore E is equal to F.   [cf. v. 9]

Again, let E be equal to F;

I say that, as A is to B, so is C to D.

For, with the same construction,

since E is equal to F,

therefore, as *G* is to *E*, so is *G* to *F*.   [cf. v. 7]

But, as *G* is to *E*, so is *C* to *D*,   [vii. 17]

and, as *G* is to *F*, so is *A* to *B*.   [vii. 18]

Therefore also, as *A* is to *B*, so is *C* to *D*.   Q.E.D.

If $a : b = c : d,$

then $ad = bc$; *and conversely.*

The proof is equivalent to the following.

(1)	$ac : ad = c : d$	[vii. 17]
	$= a : b.$	
But	$a : b = ac : bc.$	[vii. 18]
Therefore	$ac : ad = ac : bc,$	
or	$ad = bc.$	
(2) Since	$ad = bc,$	
	$ac : ad = ac : bc.$	
But	$ac : ad = c : d,$	[vii. 17]
and	$ac : bc = a : b.$	[vii. 18]
Therefore	$a : b = c : d.$	

As indicated in the note on vii. 14 *above, Heiberg regards Euclid as basing the inferences contained in the last step of part (1) of this proof and in the first step of part (2) on the propositions* v. 9 *and* v. 7 *respectively, since he has not proved those propositions separately for numbers in this Book. I prefer to suppose that he regarded the inferences as obvious and not needing proof, in view of the definition of numbers which are in proportion. E.g., if* ac *is the same fraction ("part" or "parts") of* ad *that* ac *is of* bc, *it is obvious that* ad *must be equal to* bc.

Heiberg *omits from his text here, and relegates to an Appendix, a proposition appearing in the manuscripts* v, p, φ *to the effect that, if three numbers be proportional, the product of the extremes is equal to the square of the mean, and conversely. It does not appear in* P *in the first hand,* B *has it in the margin only, and Campanus omits it, remarking that Euclid does not give the proposition about* three *proportionals as he does in* vi. 17, *since it is easily proved by the proposition just given. Moreover an-Nairīzī quotes the proposition about three proportionals as an observation on* vii. 19 *probably due to Heron (who is mentioned by name in the preceding paragraph).*

Proposition 20

The least numbers of those which have the same ratio with them measure those which have the same ratio the same number of times, the greater the greater and the less the less.

For let *CD*, *EF* be the least numbers of those which have the same ratio with *A*, *B*;
I say that *CD* measures *A* the same number of times that *EF* measures *B*.

Now *CD* is not parts of *A*.

For, if possible, let it be so;

therefore *EF* is also the same parts of *B* that *CD* is of *A*.

[vii. 13 and Def. 20]

Therefore, as many parts of *A* as there are in *CD*, so many parts of *B* are there also in *EF*.

Let *CD* be divided into the parts of *A*, namely *CG*, *GD*, and *EF* into the parts of *B*, namely *EH*, *HF*;

thus the multitude of *CG*, *GD* will be equal to the multitude of *EH*, *HF*.

Now, since the numbers *CG*, *GD* are equal to one another, and the numbers *EH*, *HF* are also equal to one another,

while the multitude of *CG*, *GD* is equal to the multitude of *EH*, *HF*, therefore, as *CG* is to *EH*, so is *GD* to *HF*.

Therefore also, as one of the antecedents is to one of the consequents, so will all the antecedents be to all the consequents. [vii. 12]

Therefore, as *CG* is to *EH*, so is *CD* to *EF*.

Therefore *CG*, *EH* are in the same ratio with *CD*, *EF*, being less than they: which is impossible, for by hypothesis *CD*, *EF* are the least numbers of those which have the same ratio with them.

Therefore *CD* is not parts of *A*;

therefore it is a part of it. [vii. 4]

And *EF* is the same part of *B* that *CD* is of *A*; [vii. 13 and Def. 20]

therefore *CD* measures *A* the same number of times that *EF* measures *B*.

Q.E.D.

If a, b *are the least numbers among those which have the same ratio (i.e. if* a/b *is a fraction in its lowest terms), and* c, d *are any others in the same ratio, i.e. if*

$$a : b = c : d,$$

then $a = \frac{1}{n}c$, $b = \frac{1}{n}d$ *where* n *is some integer.*

The proof is by reductio ad absurdum, *thus.*

[Since a<c, a *is some proper fraction ("part" or "parts") of* c, *by* vii. 4.]

Now a *cannot be equal to* $\frac{m}{n}c$, *where* m *is an integer less than* n *but greater than 1.*

For, if $a = \frac{m}{n}c$, $b = \frac{m}{n}d$ *also.* [vii. 13 and Def. 20]

Take each of the m *parts of* a *with each of the* m *parts of* b, *two and two; the ratio of the members of all pairs is the same ratio* $\frac{1}{m}a : \frac{1}{m}b$.

Therefore

$$\frac{1}{m}a : \frac{1}{m}b = a:b.$$ [vii. 12]

But $\frac{1}{m}$ a *and* $\frac{1}{m}$ b *are respectively less than* a, b *and they are in the same ratio: which contradicts the hypothesis.* H. E. 11.

Hence a *can only be "a part" of* c, *or*

$$a \text{ is of the form } \frac{1}{n}c,$$

and therefore

$$b \text{ is of the form } \frac{1}{n}d.$$

Here also Heiberg omits a proposition which was no doubt interpolated by Theon (B, V, p, φ have it as vii. 22, *but P only has it in the margin and in a later hand; Campanus also omits it) proving for numbers the* ex aequali *proposition when "the proportion is perturbed," i.e. (cf. enunciation of* v. 22) *if*

	a : b = e : f, ...(1)
and	b : c = d : e, ...(2)
then	a : c = d : f.

The proof (see Heiberg's Appendix) depends on vii. 19.

From (1) we have	af = be,	
and from (2)	be = cd.	[vii. 19]
Therefore	af = cd,	
and accordingly	a : c = d : f.	[vii. 19]

Proposition 21

Numbers prime to one another are the least of those which have the same ratio with them.

Let *A*, *B* be numbers prime to one another;

I say that A, B are the least of those which have the same ratio with them.

For, if not, there will be some numbers less than A, B which are in the same ratio with A, B.

Let them be C, D.

Since, then, the least numbers of those which have the same ratio measure those which have the same ratio the same number of times, the greater the greater and the less the less, that is, the antecedent the antecedent and the consequent the consequent, [vii. 20]

therefore C measures A the same number of times that D measures B.

Now, as many times as C measures A, so many units let there be in E.

Therefore D also measures B according to the units in E.

And, since C measures A according to the units in E, therefore E also measures A according to the units in C. [vii. 16]

For the same reason

E also measures B according to the units in D. [vii. 16]

Therefore E measures A, B which are prime to one another: which is impossible. [vii. Def. 12]

Therefore there will be no numbers less than A, B which are in the same ratio with A, B.

Therefore A, B are the least of those which have the same ratio with them.

Q.E.D.

In other words, if a, b *are prime to one another, the ratio* a : b *is "in its lowest terms."*
The proof is equivalent to the following.
If not, suppose that c, d *are the* least *numbers for which*

$$a : b = c : d.$$

[Euclid only supposes some numbers c, d *in the ratio of* a *to* b *such that* c < a, *and (consequently)* d < b. *It is however necessary to suppose that* c, d *are the* least *numbers in that ratio in order to enable* vii. 20 *to be used in the proof.]*
Then [vii. 20] a = mc, *and* b = md, *where* m *is some integer.*
Therefore $\qquad\qquad$ a = cm, $\quad$ b = dm, $\qquad\qquad\qquad\qquad$ [vii. 16]
and m *is a common measure of* a, b, *though these are prime to one another which is impossible.* [vii. Def. 12]
Thus the least numbers in the ratio of a *to* b *cannot be less than* a, b *themselves.*
Where I have quoted vii. 16 *Heiberg regards the reference as being to* vii. 15. *I think the phraseology of the text combined with that of* Def. 15 *suggests the former rather than the latter.*

Proposition 22

The least numbers of those which have the same ratio with them are prime to one another.

Let A, B be the least numbers of those which have the same ratio with them;

I say that A, B are prime to one another.

For, if they are not prime to one another, some number will measure them.

84

Let some number measure them, and let it be *C*.

And, as many times as *C* measures *A*, so many units let there be in *D*, and, as many times as *C* measures *B*, so many units let there be in *E*

Since *C* measures *A* according to the units in *D*,

therefore *C* by multiplying *D* has made *A*.   [vii. Def. 15]

For the same reason also

C by multiplying *E* has made *B*.

Thus the number *C* by multiplying the two numbers *D*, *E* has made *A*, *B*;

therefore, as *D* is to *E*, so is *A* to *B*;   [vii. 17]

therefore *D*, *E* are in the same ratio with *A*, *B*, being less than they: which is impossible.

Therefore no number will measure the numbers *A*, *B*.

Therefore *A*, *B* are prime to one another.

Q.E.D.

If a : b *is "in its lowest terms,"* a, b *are prime to one another.*
Again the proof is indirect.
If a, b *are not prime to one another, they have some common measure* c,
and

$$a = mc, \quad b = nc.$$

Therefore $m : n = a : b.$ [vii. 17 or 18]

But m, n *are less than* a, b *respectively, so that* a : b *is not in its lowest terms: which is contrary to the hypothesis.*
Therefore etc.

Proposition 23

If two numbers be prime to one another, the number which measures the one of them will be prime to the remaining number.

Let *A*, *B* be two numbers prime to one another, and let any number *C* measure *A*; I say that *C*, *B* are also prime to one another.

For, if *C*, *B* are not prime to one another, some number will measure *C*, *B*.

Let a number measure them, and let it be *D*.

Since *D* measures *C*, and *C* measures *A*,

therefore *D* also measures *A*.

But it also measures *B*;

therefore *D* measures *A*, *B* which are prime to one another: which is impossible.

[vii. Def. 12]

Therefore no number will measure the numbers *C*, *B*.

Therefore *C*, *B* are prime to one another.

Q.E.D.

If a, mb *are prime to one another,* b *is prime to* a. *For, if not, some number* d *will measure both* a *and* b, *and therefore both* a *and* mb: *which is contrary to the hypothesis.*
Therefore etc.

Proposition 24

If two numbers be prime to any number, their product also will be prime to the same.

For let the two numbers A, B be prime to any number C, and let A by multiplying B make D;

I say that C, D are prime to one another.

For, if C, D are not prime to one another, some number will measure C, D.

Let a number measure them, and let it be E.

Now, since C, A are prime to one another,

and a certain number E measures C,

therefore A, E are prime to one another. [vii. 23]

As many times, then, as E measures D, so many units let there be in F;

therefore F also measures D according to the units in E. [vii. 16]

Therefore E by multiplying F has made D. [vii. Def. 15]

But, further, A by multiplying B has also made D;

therefore the product of E, F is equal to the product of A, B.

But, if the product of the extremes be equal to that of the means, the four numbers are proportional; [vii. 19]

therefore, as E is to A, so is B to F.

But A, E are prime to one another,

numbers which are prime to one another are also the least of those which have the same ratio, [vii. 21]

and the least numbers of those which have the same ratio with them measure those which have the same ratio the same number of times, the greater the greater and the less the less, that is, the antecedent the antecedent and the consequent the consequent; [vii. 20]

therefore E measures B.

But it also measures C;

therefore E measures B, C which are prime to one another:

which is impossible. [vii. Def. 12]

Therefore no number will measure the numbers C, D.

Therefore C, D are prime to one another.

Q.E.D.

---

1. **their product.** ὁ ἐξ αὐτῶν γενόμενος, literally "the (number) produced from them," will henceforth be translated as "their product."

If a, b *are both prime to* c, *then* ab, c *are prime to one another.*
The proof is again by reductio ad absurdum.
If ab, c *are not prime to one another, let them be measured by* a *and be equal to* md, nd, *say, respectively.*
Now, since a, c *are prime to one another and* d *measures* c,

a, d *are prime to one another.*   [vii. 23]

But, since   ab = md,

d : a = b : m.   [vii. 19]

Therefore [vii. 20]   d *measures* b,

or   b = pd, *say.*

But   c = nd.

Therefore d *measures both* b *and* c, *which are therefore not prime to one another: which is impossible.*
Therefore etc.

Proposition 25

If two numbers be prime to one another, the product of one of them into itself will be prime to the remaining one.

Let A, B be two numbers prime to one another,
and let A by multiplying itself make C:
I say that B, C are prime to one another.

For let D be made equal to A.

Since A, B are prime to one another, and A is equal to D,
therefore D, B are also prime to one another.

Therefore each of the two numbers D, A is prime to B;
therefore the product of D, A will also be prime to B.   [vii. 24]

But the number which is the product of D, A is C.

Therefore C, B are prime to one another.

Q.E.D.

1. **the product of one of them into itself.** The Greek, ὁ ἐκ τοῦ ἑνὸς αὐτῶν γενόμενος, literally "the number produced from the one of them," leaves "multiplied into itself" to be understood.

If a, b *are prime to one another,*

a^2 *is prime to* b.

Euclid takes d *equal to* a, *so that* d, a *are both prime to* b.

Hence, by vii. 24, da, *i.e.* a^2, *is prime to* b.

The proposition is a particular case of the preceding proposition; and the method of proof is by substitution of different numbers in the result of that proposition.

Proposition 26

If two numbers be prime to two numbers, both to each, their products also will be prime to one another.

For let the two numbers A, B be prime to the two numbers C, D; both to each,
and let A by multiplying B make E, and let C by
multiplying D make F;
I say that E, F are prime to one another.

For, since each of the numbers A, B is prime to
C, therefore the product of A, B will also be prime to C.   [vii. 24]

But the product of A, B is E;
therefore E, C are prime to one another.

For the same reason

E, D are also prime to one another.

Therefore each of the numbers *C, D* is prime to *E*.

Therefore the product of *C, D* will also be prime to *E*.   [vii. 24]

But the product of *C, D* is *F*.

Therefore *E, F* are prime to one another.

Q.E.D.

If both a *and* b *are prime to each of two numbers* c, d, *then* ab, cd *will be prime to one another.*
Since a, b *are both prime to* c,

	ab, c *are prime to one another.*	[vii. 24]
Similarly	ab, d *are prime to one another.*	
Therefore	c, d *are both prime to* ab,	
and so therefore is cd.		[vii. 24]

Proposition 27

If two numbers be prime to one another, and each by multiplying itself make a certain number, the products will be prime to one another; and, if the original numbers by multiplying the products make certain numbers, the latter will also be prime to one another [and this is always the case with the extremes].

Let *A, B* be two numbers prime to one another, let *A* by multiplying itself make *C*, and by multiplying *C* make *D*,

and let *B* by multiplying itself make *E*, and by multiplying *E* make *F*;

I say that both *C, E* and *D, F* are prime to one another.

For, since *A, B* are prime to one another, and *A* by multiplying itself has made *C*,

therefore *C, B* are prime to one another.   [vii. 25]

Since then *C, B* are prime to one another, and *B* by multiplying itself has made *E*,

therefore *C, E* are prime to one another.   [id.]

Again, since *A, B* are prime to one another,

and *B* by multiplying itself has made *E*,

therefore *A, E* are prime to one another.   [id.]

Since then the two numbers *A, C* are prime to the two numbers *B, E*, both to each,

therefore also the product of *A, C* is prime to the product of *B, E*.   [vii. 26]

And the product of *A, C* is *D*, and the product of *B, E* is *F*.

Therefore *D, F* are prime to one another.

Q.E.D.

If a, b *are prime to one another, so are* a^2, b^2 *and so are* a^3, b^3; *and, generally,* a^n, b^n *are prime to one another.*
The words in the enunciation which assert the truth of the proposition for any powers are suspected and bracketed by Heiberg because (1) in περὶ τοὺς ἄκρους *the use of* ἄκροι *is peculiar, for it can only mean "the last products," and*

(2) the words have nothing corresponding to them in the proof, much less is the generalisation proved. Campanus omits the words in the enunciation, though he adds to the proof a remark that the proposition is true of any, the same or different, powers of a, b. *Heiberg concludes that the words are an interpolation of date earlier than Theon.*

Euclid's proof amounts to this.

Since a, b *are prime to one another, so are* a^2, b [vii. 25], *and therefore also* a^2, b^2.   [vii. 25]

Similarly [vii. 25] a, b^2 *are prime to one another.*

Therefore a, a^2 *and* b, b^2 *satisfy the description in the enunciation of* vii. 26.

Hence a^3, b^3 *are prime to one another.*

Proposition 28

If two numbers be prime to one another, the sum will also be prime to each of them; and, if the sum of two numbers be prime to any one of them, the original numbers will also be prime to one another.

For let two numbers *AB*, *BC* prime to one another be added; I say that the sum *AC* is also prime to each of the numbers *AB*, *BC*.

For, if *CA*, *AB* are not prime to one another, some number will measure *CA*, *AB*.

Let a number measure them, and let it be *D*.

Since then *D* measures *CA*, *AB*,

therefore it will also measure the remainder *BC*.

But it also measures *BA*;

therefore *D* measures *AB*, *BC* which are prime to one another: which is impossible.   [vii. Def. 12]

Therefore no number will measure the numbers *CA*, *AB*; therefore *CA*, *AB* are prime to one another.

For the same reason

AC, *CB* are also prime to one another.

Therefore *CA* is prime to each of the numbers *AB*, *BC*.

Again, let *CA*, *AB* be prime to one another;

I say that *AB*, *BC* are also prime to one another.

For, if *AB*, *BC* are not prime to one another, some number will measure *AB*, *BC*.

Let a number measure them, and let it be *D*.

Now, since *D* measures each of the numbers *AB*, *BC*, it will also measure the whole *CA*.

But it also measures *AB*;

therefore *D* measures *CA*, *AB* which are prime to one another: which is impossible.   [vii. Def. 12]

Therefore no number will measure the numbers *AB*, *BC*.

Therefore *AB, BC* are prime to one another.

Q.E.D.

If a, b *are prime to one another,* a + b *will be prime to both* a *and* b; *and conversely.*
For suppose (a + b), a *are not prime to one another. They must then have some common measure* d.
Therefore d *also divides the difference* (a + b) − a, *or* b, *as well as* a; *and therefore* a, b *are not prime to one another: which is contrary to the hypothesis.*
Therefore   a + b *is prime to* a.
Similarly   a + b *is prime to* b.
The converse is proved in the same way.
    Heiberg remarks on Euclid's assumption that, if c *measures both* a *and* b, *it also measures* a ± b. *But it has already* (vii. 1, 2) *been assumed, more generally, as an axiom that, in the case supposed,* c *measures* a ± pb.

Proposition 29

Any prime number is prime to any number which it does not measure.

Let *A* be a prime number, and let it not measure *B*;
I say that *B, A* are prime to one another.

For, if *B, A* are not prime to one another,
some number will measure them.

Let *C* measure them.

Since *C* measures *B*,
and *A* does not measure *B*,
therefore *C* is not the same with *A*.

Now, since *C* measures *B, A*,
therefore it also measures *A* which is prime, though it is not the same with it:
which is impossible.

Therefore no number will measure *B, A*.

Therefore *A, B* are prime to one another.

Q.E.D.

If a *is prime and does not measure* b, *then* a, b *are prime to one another. The proof is self-evident.*

Proposition 30

If two numbers by multiplying one another make some number, and any prime number measure the product, it will also measure one of the original numbers.

For let the two numbers *A, B* by multiplying one another make *C*, and let any prime number *D* measure *C*;
I say that *D* measures one of the numbers *A, B*.

For let it not measure *A*.

Now *D* is prime;
therefore *A, D* are prime to one another.   [vii. 29]

And, as many times as *D* measures *C*, so many units let there be in *E*.

Since then *D* measures *C* according to the units in *E*,

therefore D by multiplying E has made C.   [vii. Def. 15]

Further, A by multiplying B has also made C;

therefore the product of D, E is equal to the product of A, B.

Therefore, as D is to A, so is B to E.   [vii. 19]

But D, A are prime to one another,

primes are also least,   [vii. 21]

and the least measure the numbers which have the same ratio the same number of times, the greater the greater and the less the less, that is, the antecedent the antecedent and the consequent the consequent;   [vii. 20]

therefore D measures B.

Similarly we can also show that, if D do not measure B, it will measure A.

Therefore D measures one of the numbers A, B.

        Q.E.D.

If c, *a prime number, measure* ab, c *will measure either* a *or* b.
Suppose c *does not measure* a.

    Therefore c, a *are prime to one another.*   [vii. 29]

    Suppose $ab = mc.$

        Therefore $c : a = b : m.$ [vii. 19]

    Hence [vii. 20, 21]   c *measures* b.

    Similarly, if c *does not measure* b, *it measures* c.

    Therefore it measures one or other of the two numbers a, b.

Proposition 31

Any composite number is measured by some prime number.

Let A be a composite number;

I say that A is measured by some prime number.

For, since A is composite, some number will measure it.

Let a number measure it, and let it be B.

Now, if B is prime, what was enjoined will have been done.

But if it is composite, some number will measure it.

Let a number measure it, and let it be C.

Then, since C measures B,

and B measures A,

therefore C also measures A.

And, if C is prime, what was enjoined will have been done.

But if it is composite, some number will measure it.

Thus, if the investigation be continued in this way, some prime number will be found which will measure the number before it, which will also measure A.

For, if it is not found, an infinite series of numbers will measure the number A, each of which is less than the other: which is impossible in numbers.

Therefore some prime number will be found which will measure the one before it, which will also measure *A*.

Therefore any composite number is measured by some prime number.

6. **if B is prime, what was enjoined will have been done,** i.e. the implied *problem* of finding a prime number which measures *A*.

14. **some prime number will be found which will measure.** In the Greek the sentence stops here, but it is necessary to add the words "the number before it, which will also measure *A*," which are found a few lines further down. It is possible that the words may have fallen out of P here by a simple mistake due to ὁμοιοτέλευτον (Heiberg).

Heiberg relegates to the Appendix an alternative proof of this proposition, to the following effect. Since A is composite, some number will measure it. Let B be the least such number. I say that B is prime. For, if not, B is composite, and some number will measure it, say C; so that C is less than B. But, since C measures B, and B measures A, C must measure A. And C is less than B: which is contrary to the hypothesis.

Proposition 32

Any number either is prime or is measured by some prime number.

Let *A* be a number;

I say that *A* either is prime or is measured by some prime number.

If now *A* is prime, that which was enjoined will have been done.

A ————————

But if it is composite, some prime number will measure it. [vii. 31]

Therefore any number either is prime or is measured by some prime number.

Q.E.D.

Proposition 33

Given as many numbers as we please, to find the least of those which have the same ratio with them.

Let *A*, *B*, *C* be the given numbers, as many as we please; thus it is required to find the least of those which have the same ratio with *A*, *B*, *C*.

A, *B*, *C* are either prime to one another or not.

Now, if *A*, *B*, *C* are prime to one another, they are the least of those which have the same ratio with them. [vii. 21]

But, if not, let *D* the greatest common measure of *A*, *B*, *C* be taken, [vii. 3]

and, as many times as *D* measures the numbers *A*, *B*, *C* respectively, so many units let there be in the numbers *E*, *F*, *G* respectively.

Therefore the numbers *E*, *F*, *G* measure the numbers *A*, *B*, *C* respectively according to the units in *D*. [vii. 16]

Therefore *E*, *F*, *G* measure *A*, *B*, *C* the same number of times;

therefore *E*, *F*, *G* are in the same ratio with *A*, *B*, *C*. [vii. Def. 20]

I say next that they are the least that are in that ratio.

For, if *E*, *F*, *G* are not the least of those which have the same ratio with *A*, *B*, *C*, there will be numbers less than *E*, *F*, *G* which are in the same ratio with *A*, *B*, *C*.

Let them be *H*, *K*, *L*;

therefore *H* measures *A* the same number of times that the numbers *K*, *L* measure the numbers *B*, *C* respectively.

Now, as many times as *H* measures *A*, so many units let there be in *M*;

therefore the numbers *K*, *L* also measure the numbers *B*, *C* respectively according to the units in *M*.

And, since *H* measures *A* according to the units in *M*,

therefore *M* also measures *A* according to the units in *H*.   [vii. 16]

For the same reason

M also measures the numbers *B*, *C* according to the units in the numbers *K*, *L* respectively;

Therefore *M* measures *A*, *B*, *C*.

Now, since *H* measures *A* according to the units in *M*,

therefore *H* by multiplying *M* has made *A*.   [vii. Def. 15]

For the same reason also

E by multiplying *D* has made *A*.

Therefore the product of *E*, *D* is equal to the product of *H*, *M*.

Therefore, as *E* is to *H*, so is *M* to *D*.   [vii. 19]

But *E* is greater than *H*;

therefore *M* is also greater than *D*.

And it measures *A*, *B*, *C*:

which is impossible, for by hypothesis *D* is the greatest common measure of *A*, *B*, *C*.

Therefore there cannot be any numbers less than *E*, *F*, *G* which are in the same ratio with *A*, *B*, *C*.

Therefore *E*, *F*, *G* are the least of those which have the same ratio with *A*, *B*, *C*.

Q.E.D.

13. **the numbers E, F, G measure the numbers A, B, C respectively,** literally (as usual) "each of the numbers *E*, *F*, *G* measures each of the numbers *A*, *B*, *C*."

Given any numbers a, b, c, . . . , *to find the least numbers that are in the same ratio.*

Euclid's method is the obvious one, and the result is verified by reductio ad absurdum.

We will, like Euclid, take three numbers only, a, b, c.

Let g, *their greatest common measure, be found* [vii. 3], *and suppose that*

$$a = mg, \textit{i.e. } gm,$$ [vii. 16]
$$b = ng, \textit{i.e. } gn,$$
$$c = pg, \textit{i.e. } gp.$$

It follows, by vii. Def. 20, *that*

$$m : n : p = a : b : c.$$

m, n, p *shall be the numbers required.*

For, if not, let x, y, z *be the least numbers in the same ratio as* a, b, c, *being less than* m, n, p.

Therefore
$$a = kx \ (or \ xk, \ vii. \ 16),$$
$$b = ky \ (or \ yk),$$
$$c = kz \ (or \ zk),$$
where k *is some integer.*   [vii. 20]

Thus
$$mg = a = xk.$$
Therefore
$$m : x = k : g.$$ [vii. 19]

And m > x; *therefore* k > g.

Since then k *measures* a, b, c, *it follows that* g *is not the greatest common measure: which contradicts the hypothesis. Therefore etc.*

It is to be observed that Euclid merely supposes that x, y, z *are smaller numbers than* m, n, p *in the ratio of* a, b, c; *but, in order to justify the next inference, which apparently can only depend on* vii. 20, x, y, z *must also be assumed to be the* least *numbers in the ratio of* a, b, c.

The inference from the last proportion that, since m > x, k > g *is supposed by Heiberg to depend upon* vii. 13 *and* v. 14 *together. I prefer to regard Euclid as making the inference quite independently of Book V. E.g., the proportion could just as well be written*
$$x : m = g : k,$$
when the definition of proportion in Book VII. (Def. 20) gives all that we want, since, whatever proper fraction x *is of* m, *the same proper fraction is* g *of* k.

Proposition 34

Given two numbers, to find the least number which they measure.

Let *A, B* be the two given numbers;

thus it is required to find the least number which they measure.

Now *A, B* are either prime to one another or not.

First, let *A, B* be prime to one another, and let *A* by multiplying *B* make *C*;

therefore also *B* by multiplying *A* has made *C*.   [vii. 16]

Therefore *A, B* measure *C*

I say next that it is also the least number they measure.

For, if not, *A, B* will measure some number which is less than *C*.

Let them measure *D*.

Then, as many times as *A* measures *D*, so many units let there be in *E*,

and, as many times as *B* measures *D*, so many units let there be in *F*;

therefore *A* by multiplying *E* has made *D*,

and *B* by multiplying *F* has made *D*;   [vii. Def. 15]

therefore the product of *A, E* is equal to the product of *B, F*.

Therefore, as *A* is to *B*, so is *F* to *E*.   [vii. 19]

But *A, B* are prime,

primes are also least,   [vii. 21]

and the least measure the numbers which have the same ratio the same number of times, the greater the greater and the less the less;   [vii. 20]

therefore *B* measures *E*, as consequent consequent.

And, since *A* by multiplying *B, E* has made *C, D*,

therefore, as *B* is to *E*, so is *C* to *D*.   [vii. 17]

But *B* measures *E*;

therefore *C* also measures *D*, the greater the less:

which is impossible.

Therefore *A*, *B* do not measure any number less than *C*; therefore *C* is the least that is measured by *A*, *B*.

Next, let *A*, *B* not be prime to one another,

and let *F*, *E*, the least numbers of those which have the same ratio with *A*, *B*, be taken;   [vii. 33]

therefore the product of *A*, *E* is equal to the product of *B*, *F*.   [vii. 19]

And let *A* by multiplying *E* make *C*;

therefore also *B* by multiplying *F* has made *C*;

therefore *A*, *B* measure *C*.

I say next that it is also the least number that they

measure.

For, if not, *A*, *B* will measure some number which is

less than *C*.

Let them measure *D*.

And, as many times as *A* measures *D*, so many units let there be in *G*,

and, as many times as *B* measures *D*, so many units let there be in *H*.

Therefore *A* by multiplying *G* has made *D*,

and *B* by multiplying *H* has made *D*.

Therefore the product of *A*, *G* is equal to the product of *B*, *H*;

therefore, as *A* is to *B*, so is *H* to *G*.   [vii. 19]

But, as *A* is to *B*, so is *F* to *E*,

Therefore also, as *F* is to *E*, so is *H* to *G*.

But *F*, *E* are least,

and the least measure the numbers which have the same ratio the same number of times, the greater the greater and the less the less;   [vii. 20]

therefore *E* measures *G*.

And, since *A* by multiplying *E*, *G* has made *C*, *D*,

therefore, as *E* is to *G*, so is *C* to *D*.   [vii. 17]

But *E* measures *G*;

therefore *C* also measures *D*, the greater the less:

which is impossible.

H. E. 11.

Therefore *A*, *B* will not measure any number which is less than *C*.

Therefore *C* is the least that is measured by *A*, *B*.

Q.E.D.

This is the problem of finding the least common multiple *of two numbers, as* a, b.
I. If a, b *be prime to one another, the* L.C.M. *is* ab.
For, if not, let it be d, *some number less than* ab.
Then $\qquad$ d = ma = nb, *where* m, n *are integers.*
Therefore $\qquad$ a : b = n : m, $\qquad\qquad$ [vii. 19]
and hence, a, b *being prime to one another,*

$\qquad\qquad$ b *measures* m. $\qquad\qquad$ [vii. 20, 21]
But $\qquad\qquad$ b : m = ab : am $\qquad\qquad$ [vii. 17]
$\qquad\qquad$ = ab : d.
Therefore ab *measures* d: *which is impossible.*
II. If a, b *be not prime to one another, find the numbers which are the least of those having the ratio of* a *to* b, *say*
m, n; $\qquad\qquad\qquad\qquad\qquad\qquad$ [vii. 33]
then $\qquad\qquad$ a : b = m : n,
and $\qquad\qquad$ an = bm (= c, *say*); $\qquad\qquad$ [vii. 19]
c *is then the* L.C.M.
For, if not, let it be d (< c), *so that*
$\qquad\qquad$ ap = bq = d, *where* p, q *are integers.*
Then $\qquad\qquad$ a : b = q : p, $\qquad\qquad$ [vii. 19]
whence $\qquad\qquad$ m : n = q : p,
so that $\qquad\qquad$ n *measures* p. $\qquad\qquad$ [vii. 20, 21]
And $\qquad\qquad$ n : p = an : ap = c : d,
so that $\qquad\qquad$ c *measures* d,
which is impossible.
Therefore *etc.*

By vii. 33, $\qquad$ $\left. \begin{array}{l} m = \dfrac{a}{g} \\ n = \dfrac{b}{g} \end{array} \right\}$, *where* g *is the* G.C.M. *of* a, b.

Hence the L.C.M. *is* $\dfrac{ab}{g}$.

Proposition 35

If two numbers measure any number, the least number measured by them will also measure the same.

For let the two numbers *A*, *B* measure any number *CD*, and let *E* be the least that they measure;

I say that *E* also measures *CD*.

For, if *E* does not measure *CD*, let *E*, measuring *DF*, leave *CF* less than itself.

Now, since *A*, *B* measure *E*,
and *E* measures *DF*,
therefore *A*, *B* will also measure *DF*.

But they also measure the whole *CD*;
therefore they will also measure the remainder *CF* which is less than *E*:
which is impossible.

Therefore *E* cannot fail to measure *CD*; therefore it measures it.

$\qquad\qquad\qquad\qquad\qquad\qquad\qquad\qquad\qquad$ Q.E.D.

The least common multiple *of any two numbers must measure any other common multiple.*
The proof is obvious, depending on the fact that, if any number divides a *and* b, *it also divides* a − pb.

Proposition 36

Given three numbers, to find the least number which they measure.

Let *A, B, C* be the three given numbers;

thus it is required to find the least number which they measure.

Let *D*, the least number measured by the two numbers *A, B*, be taken. [vii. 34]

Then *C* either measures, or does not measure, *D*.

First, let it measure it.

But *A, B* also measure *D*;

therefore *A, B, C* measure *D*.

I say next that it is also the least that they measure.

For, if not, *A, B, C* will measure some number which is less than *D*.

Let them measure *E*.

Since *A, B, C* measure *E*,

therefore also *A, B* measure *E*.

Therefore the least number measured by *A, B* will also measure *E*. [vii. 35]

But *D* is the least number measured by *A, B*;

therefore *D* will measure *E*, the greater the less:

which is impossible.

Therefore *A, B, C* will not measure any number which is less than *D*;

therefore *D* is the least that *A, B, C* measure.

Again, let *C* not measure *D*,

and let *E*, the least number measured by *C, D*, be taken. [vii. 34]

Since *A, B* measure *D*,

and *D* measures *E*,

therefore also *A, B* measure *E*.

But *C* also measures *E*;

therefore also *A, B, C* measure *E*.

I say next that it is also the least that they measure.

For, if not, *A, B, C* will measure some number which is less than *E*.

Let them measure *F*.

Since *A, B, C* measure *F*,

therefore also *A, B* measure *F*;

therefore the least number measured by *A, B* will also measure *F*. [vii. 35]

But *D* is the least number measured by *A, B*;

therefore *D* measures *F*.

But *C* also measures *F*;

therefore *D, C* measure *F*,

so that the least number measured by D, C will also measure F.

But E is the least number measured by C, D;

therefore E measures F, the greater the less:

which is impossible.

Therefore A, B, C will not measure any number which is less than E.

Therefore E is the least that is measured by A, B, C.

Q. E. D.

Euclid's rule for finding the L.C.M. of three numbers a, b, c is the rule with which we are familiar. The L.C.M. of a, b is first found, say d, and then the L.C.M. of d and c is found.

Euclid distinguishes the cases (1) in which c measures d, (2) in which c does not measure d. We need only reproduce the proof of the general case (2). The method is that of reductio ad absurdum.

Let e be the L.C.M. of d, c.

Since a, b both measure d, and d measures e,

a, b both measure e.

So does c.

Therefore e is some common multiple of a, b, c.

If it is not the least, let f be the L.C.M.

Now a, b both measure f;

therefore d, their L.C.M., also measures f.　　　　　　　　　　　　　　　　　　　　　[vii. 35]

Thus d, c both measure f;

therefore e, their L.C.M., measures f:

which is impossible, since f < e.　　　　　　　　　　　　　　　　　　　　　　　　[vii. 35]

Therefore etc.

The process can be continued ad libitum, so that we can find the L.C.M., not only of three, but of as many numbers as we please.

Proposition 37

If a number be measured by any number, the number which is measured will have a part called by the same name as the measuring number.

For let the number A be measured by any number B;

I say that A has a part called by the same name as B.

For, as many times as B measures A, so many units let there be in C.

Since B measures A according to the units in C,

and the unit D also measures the number C according to the units in it,

therefore the unit D measures the number C the same number of times as B measures A.

Therefore, alternately, the unit D measures the number B the same number of times as C measures A;　　　　　　　　　　　　　　　　　　　　　　　　[vii. 15]

therefore, whatever part the unit D is of the number B, the same part is C of A also.

But the unit D is a part of the number B called by the same name as it;

therefore C is also a part of A called by the same name as B, so that A has a part C which is called by the same name as B.

Q.E.D.

If b *measures* a, *then* $\frac{1}{b}$th *of* a *is a whole number.*

Let
$$a = m \cdot b.$$
Now
$$m = m \cdot 1.$$
Thus *1*, m, b, a, *satisfy the enunciation of* vii. 15;
therefore m *measures* a *the same number of times that 1 measures* b.

But
$$1 \text{ } is \text{ } \frac{1}{b}\text{th part of b;}$$

therefore
$$m \text{ } is \text{ } \frac{1}{b}\text{th part of a.}$$

Proposition 38

If a number have any part whatever, it will be measured by a number called by the same name as the part.

For let the number *A* have any part whatever, *B*, and let *C* be a number called by the same name as the part *B*; I say that *C* measures *A*.

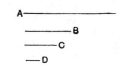

For, since *B* is a part of *A* called by the same name as *C*, and the unit *D* is also a part of *C* called by the same name as it, therefore, whatever part the unit *D* is of the number *C*, the same part is *B* of *A* also; therefore the unit *D* measures the number *C* the same number of times that *B* measures *A*.

Therefore, alternately, the unit *D* measures the number *B* the same number of times that *C* measures *A*.   [vii. 15]

Therefore *C* measures *A*.

Q.E.D.

This proposition is practically a restatement of the preceding proposition. It asserts that, if b *is* $\frac{1}{m}$ th part of a,

i.e., if
$$b = \frac{1}{m}a,$$

then
$$m \text{ measures a.}$$

We have
$$b = \frac{1}{m}a,$$

and
$$1 = \frac{1}{m}m.$$

Therefore 1, m, b, a, *satisfy the enunciation of* vii. 15, *and therefore* m *measures* a *the same number of times as 1 measures* b, *or*

$$m = \frac{1}{b}a.$$

Proposition 39

To find the number which is the least that will have given parts.

Let *A*, *B*, *C* be the given parts;

thus it is required to find the number which is the least that will have the parts *A*, *B*, *C*.

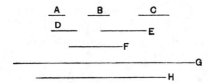

Let *D*, *E*, *F* be numbers called by the same name as the parts *A*, *B*, *C*,

and let *G*, the least number measured by *D*, *E*, *F*, be taken. [vii. 36]

Therefore *G* has parts called by the same name as *D*, *E*, *F*. [vii. 37]

But *A*, *B*, *C* are parts called by the same name as *D*, *E*, *F*; therefore *G* has the parts *A*, *B*, *C*.

I say next that it is also the least number that has.

For, if not, there will be some number less than *G* which will have the parts *A*, *B*, *C*.

Let it be *H*.

Since *H* has the parts *A*, *B*, *C*,

therefore *H* will be measured by numbers called by the same name as the parts *A*, *B*, *C*. [vii. 38]

But *D*, *E*, *F* are numbers called by the same name as the parts *A*, *B*, *C*; therefore *H* is measured by *D*, *E*, *F*.

And it is less than *G*: which is impossible.

Therefore there will be no number less than *G* that will have the parts *A*, *B*, *C*.

Q.E.D.

This again is practically a restatement in another form of the problem of finding the L.C.M.
To find a number which has $\frac{1}{a}$th, $\frac{1}{b}$th and $\frac{1}{c}$th parts.
Let d *be the L.C.M. of* a, b, c.
Thus d *has $\frac{1}{a}$th, $\frac{1}{b}$th and $\frac{1}{c}$th parts.* [vii. 37]

If it is not the least number which has, let the least such number be e.
Then, since e *has those parts,*
e *is measured by* a, b, c; *and* e < d :
which is impossible.

BOOK IX

THE INFINITUDE OF PRIME NUMBERS
PROPOSITION 20

Prime numbers are more than any assigned multitude of prime numbers.

Let *A*, *B*, *C* be the assigned prime numbers;
I say that there are more prime numbers than
A, *B*, *C*.

A—
B—   G———
C—
E————————————D
                      +F

For let the least number measured by *A*, *B*, *C*
be taken,
and let it be *DE*;
let the unit *DF* be added to *DE*.

Then *EF* is either prime or not.

First, let it be prime;

then the prime numbers *A*, *B*, *C*, *EF* have been found which are more than *A*, *B*, *C*.

Next, let *EF* not be prime;

therefore it is measured by some prime number.   [vii. 31]

Let it be measured by the prime number *G*.

I say that *G* is not the same with any of the numbers *A*, *B*, *C*.

For, if possible, let it be so.

Now *A*, *B*, *C* measure *DE*;

therefore *G* also will measure *DE*.

But it also measures *EF*.

Therefore *G*, being a number, will measure the remainder, the unit *DF*:
which is absurd.

Therefore *G* is not the same with any one of the numbers *A*, *B*, *C*.

And by hypothesis it is prime.

Therefore the prime numbers *A*, *B*, *C*, *G* have been found which are more than
the assigned multitude of *A*, *B*, *C*.

Q.E.D.

BOOK IX

EVEN PERFECT NUMBERS
PROPOSITION 36

If as many numbers as we please beginning from an unit be set out continuously in double proportion, until the sum of all becomes prime, and if the sum multiplied into the last make some number, the product will be perfect.

For let as many numbers as we please, *A, B, C, D*, beginning from an unit be set out in double proportion, until the sum of all becomes prime,

let *E* be equal to the sum, and let *E* by multiplying *D* make *FG*;

I say that *FG* is perfect.

For, however many *A, B, C, D* are in multitude, let so many *E, HK, L, M* be taken in double proportion beginning from *E*;

therefore, *ex aequali*, as *A* is to *D*, so is *E* to *M*.   [vii. 14]

Therefore the product of *E, D* is equal to the product of *A, M*.   [vii. 19]

And the product of *E, D* is *FG*;

therefore the product of *A, M* is also *FG*.

Therefore *A* by multiplying *M* has made *FG*;

therefore *M* measures *FG* according to the units in *A*.

And *A* is a dyad;

therefore *FG* is double of *M*.

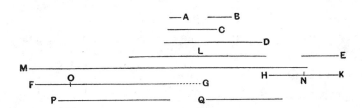

But *M, L, HK, E* are continuously double of each other; therefore *E, HK, L, M, FG* are continuously proportional in double proportion.

Now let there be subtracted from the second *HK* and the last *FG* the numbers *HN, FO*, each equal to the first *E*;

therefore, as the excess of the second is to the first, so is the excess of the last to all those before it.   [ix. 35]

Therefore, as *NK* is to *E*, so is *OG* to *M, L, KH, E*.

And *NK* is equal to *E*;

therefore OG is also equal to M, L, HK, E.

But FO is also equal to E,

and E is equal to A, B, C, D and the unit.

Therefore the whole FG is equal to E, HK, L, M and A, B, C, D and the unit; and it is measured by them.

I say also that FG will not be measured by any other number except A, B, C, D, E, HK, L, M and the unit.

For, if possible, let some number P measure FG, and let P not be the same with any of the numbers A, B, C, D, E, HK, L, M.

And, as many times as P measures FG, so many units let there be in Q; therefore Q by multiplying P has made FG.

But, further, E has also by multiplying D made FG;

therefore, as E is to Q, so is P to D.   [vii. 19]

And, since A, B, C, D are continuously proportional beginning from an unit, therefore D will not be measured by any other number except A, B, C.   [ix. 13]

And, by hypothesis, P is not the same with any of the numbers A, B, C; therefore P will not measure D.

But, as P is to D, so is E to Q;

therefore neither does E measure Q.   [vii. Def. 20]

And E is prime;

and any prime number is prime to any number which it does not measure. [vii. 29]

Therefore E, Q are prime to one another.

But primes are also least,   [vii. 21]

and the least numbers measure those which have the same ratio the same number of times, the antecedent the antecedent and the consequent the consequent;

[vii. 20]

and, as E is to Q, so is P to D;

therefore E measures P the same number of times that Q measures D.

But D is not measured by any other number except A, B, C;

therefore Q is the same with one of the numbers A, B, C.

Let it be the same with B.

And, however many B, C, D are in multitude, let so many E, HK, L be taken beginning from E.

Now E, HK, L are in the same ratio with B, C, D;

therefore, *ex aequali*, as B is to D, so is E to L.   [vii. 14]

Therefore the product of B, L is equal to the product of D, E.   [vii. 19]

But the product of D, E is equal to the product of Q, P; therefore the product of Q, P is also equal to the product of B, L.

Therefore, as Q is to B, so is L to P. [vii. 19]

And Q is the same with B;

therefore L is also the same with P:

which is impossible, for by hypothesis P is not the same with any of the numbers set out.

Therefore no number will measure FG except A, B, C, D, E, HK, L, M and the unit.

And FG was proved equal to A, B, C, D, E, HK, L, M and the unit;

and a perfect number is that which is equal to its own parts; [vii. Def. 22]

therefore FG is perfect.

Q.E.D.

BOOK X

COMMENSURABLE AND INCOMMENSURABLE MAGNITUDES

DEFINITIONS

1. Those magnitudes are said to be **commensurable** which are measured by the same measure, and those **incommensurable** which cannot have any common measure.

2. Straight lines are **commensurable in square** when the squares on them are measured by the same area, and **incommensurable in square** when the squares on them cannot possibly have any area as a common measure.

3. With these hypotheses, it is proved that there exist straight lines infinite in multitude which are commensurable and incommensurable respectively, some in length only, and others in square also, with an assigned straight line. Let then the assigned straight line be called **rational**, and those straight lines which are commensurable with it, whether in length and in square or in square only, **rational**, but those which are incommensurable with it **irrational**.

4. And let the square on the assigned straight line be called **rational** and those areas which are commensurable with it **rational**, but those which are incommensurable with it **irrational**, and the straight lines which produce them **irrational**, that is, in case the areas are squares, the sides themselves, but in case they are any other rectilineal figures, the straight lines on which are described squares equal to them.

Proposition 1

Two unequal magnitudes being set out, if from the greater there be subtracted a magnitude greater than its half, and from that which is left a magnitude greater than its half, and if this process be repeated continually, there will be left some magnitude which will be less than the lesser magnitude set out.

Let *AB*, *C* be two unequal magnitudes of which *AB* is the greater:
I say that, if from *AB* there be subtracted a magnitude greater than its half, and from that which is left a magnitude greater than its half, and if this process

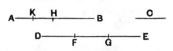

be repeated continually, there will be left some magnitude which will be less than the magnitude *C*.

For *C* if multiplied will sometime be greater than *AB*.   [cf. v. Def. 4]

Let it be multiplied, and let *DE* be a multiple of *C*, and greater than *AB*;
let *DE* be divided into the parts *DF*, *FG*, *GE* equal to *C*,
from *AB* let there be subtracted *BH* greater than its half,
and, from *AH*, *HK* greater than its half,
and let this process be repeated continually until the divisions in *AB* are equal in multitude with the divisions in *DE*.

Let, then, *AK*, *KH*, *HB* be divisions which are equal in multitude with *DF*, *FG*, *GE*.

Now, since *DE* is greater than *AB*,
and from *DE* there has been subtracted *EG* less than its half,
and, from *AB*, *BH* greater than its half,
therefore the remainder *GD* is greater than the remainder *HA*.

And, since *GD* is greater than *HA*,
and there has been subtracted, from *GD*, the half *GF*,
and, from *HA*, *HK* greater than its half,
therefore the remainder *DF* is greater than the remainder *AK*.

But *DF* is equal to *C*;
therefore *C* is also greater than *AK*.

Therefore *AK* is less than *C*.

Therefore there is left of the magnitude *AB* the magnitude *AK* which is less than the lesser magnitude set out, namely *C*.

Q.E.D.

And the theorem can be similarly proved even if the parts subtracted be halves.

This proposition will be remembered because it is the lemma required in Euclid's proof of xii. 2 *to the effect that circles are to one another as the squares on their diameters. Some writers appear to be under the impression that* xii. 2 *and the other propositions in Book XII. in which the method of exhaustion is used are the only places where Euclid makes use of* x. 1; *and it is commonly remarked that* x. 1 *might just as well have been deferred till the beginning of Book XII. Even Cantor (Gesch. d. Math.) remarks that "Euclid draws no inference from it* [x. 1], *not even that which we should more*

than anything else expect, namely that, if two magnitudes are incommensurable, we can always form a magnitude com-
mensurable with the first which shall differ from the second magnitude by as little as we please." But, so far from making
no use of x. 1 *before* xii. 2, *Euclid actually uses it in the very next proposition,* x. 2. *This being so, as the next note will*
show, it follows that, since x. 2 *gives the criterion for the incommensurability of two magnitudes (a very necessary prelim-*
inary to the study of incommensurables), x. 1 *comes exactly where it should be.*

    Euclid uses x. 1 *to prove not only* xii. 2 *but* xii. 5 *(that pyramids with the same height and triangular bases are to*
one another as their bases), by means of which he proves (xii. 7 *and Por.) that any pyramid is a third part of the prism*
which has the same base and equal height, and xii. 10 *(that any cone is a third part of the cylinder which has the same*
base and equal height), besides other similar propositions. Now xii. 7 *Por. and* xii. 10 *are theorems specifically attributed*
to Eudoxus by Archimedes (On the Sphere and Cylinder, Preface), who says in another place (Quadrature of the
Parabola, Preface) that the first of the two, and the theorem that circles are to one another as the squares on their diameters,
were proved by means of a certain lemma which he states as follows: "Of unequal lines, unequal surfaces, or unequal
solids, the greater exceeds the less by such a magnitude as is capable, if added [continually] to itself, of exceeding any
magnitude of those which are comparable with one another," i.e. of magnitudes of the same kind as the original magni-
tudes. Archimedes also says (loc. cit.) *that the second of the two theorems which he attributes to Eudoxus (Eucl. xii. 10)*
was proved by means of "a lemma similar to the aforesaid." The lemma stated thus by Archimedes is decidedly different
from x. 1, *which, however, Archimedes himself uses several times, while he refers to the use of it in* xii. 2 *(On the Sphere*
and Cylinder, i. 6). As I have before suggested (The Works of Archimedes), the apparent difficulty caused by the men-
tion of two lemmas in connexion with the theorem of Eucl. xii. 2 *may be explained by the proof of* x. 1.
Euclid there takes the lesser magnitude and says that it is possible, by multiplying it, to make it some time exceed the
greater, and this statement he clearly bases on the 4th definition of Book V., to the effect that "magnitudes are said to bear
a ratio to one another which can, if multiplied, exceed one another." Since then the smaller magnitude in x. 1 *may be*
regarded as the difference between some two unequal magnitudes, it is clear that the lemma stated by Archimedes is in
substance used to prove the lemma in x. 1, *which appears to play so much larger a part in the investigations of quadra-*
ture and cubature which have come down to us.

    Besides being employed in Eucl. x. 1, *the "Axiom of Archimedes" appears in Aristotle, who also practically quotes the re-*
sult of x. 1 *itself. Thus he says,* Physics viii. 10, 266 b 2, *"By continually adding to a finite (magnitude) I shall exceed any*
definite (magnitude), and similarly by continually subtracting from it I shall arrive at something less than it," and ibid. iii.
7, 207 b 10 *"For bisections of a magnitude are endless." It is thus somewhat misleading to use the term "Archimedes' Axiom"*
for the "lemma" quoted by him, since he makes no claim to be the discoverer of it, and it was obviously much earlier.

    Stolz (see G. Vitali in Questioni riguardanti le matematiche elementari, *1.) showed how to prove the so-called*
Axiom or Postulate of Archimedes by means of the Postulate of Dedekind, thus. Suppose the two magnitudes to be straight
lines. It is required to prove that, given two straight lines, there always exists a multiple of the smaller which is greater
than the other.

    Let the straight lines be so placed that they have a common extremity and the smaller lies along the other on the same
side of the common extremity.

    If AC *be the greater and* AB *the smaller, we have to prove that there exists an integral number* n *such that*
n · AB > AC.

    Suppose that this is not true but that there are some points, like B, *not coincident with the extremity* A, *and such*
that, n *being any integer however great,* n · AB < AC; *and we have to prove that this assumption leads to an absurdity.*

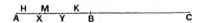

    The points of AC *may be regarded as distributed into two "parts," namely*
(1) points H *for which there exists no integer* n *such that* n · AH > AC,
(2) points K *for which an integer* n *does exist such that* n · AK > AC.

    This division into parts satisfies the conditions for the application of Dedekind's Postulate, and therefore there exists
a point M *such that the points of* AM *belong to the first part and those of* MC *to the second part.*

    Take now a point Y *on* MC *such that* MY < AM. *The middle point* (X) *of* AY *will fall between* A *and* M *and*
will therefore belong to the first part; but, since there exists an integer n *such that* n · AY > AC, *it follows that*
2n · AX > AC: *which is contrary to the hypothesis.*

Proposition 2

If, when the less of two unequal magnitudes is continually subtracted in turn from
the greater, that which is left never measures the one before it, the magnitudes will be
incommensurable.

For, there being two unequal magnitudes *AB*, *CD*, and *AB* being the less, when the less is continually subtracted in turn from the greater, let that which is left over never measure the one before it;

I say that the magnitudes *AB*, *CD* are incommensurable.

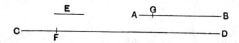

For, if they are commensurable, some magnitude will measure them.

Let a magnitude measure them, if possible, and let it be *E*;

let *AB*, measuring *FD*, leave *CF* less than itself,

let *CF* measuring *BG*, leave *AG* less than itself,

and let this process be repeated continually, until there is left some magnitude which is less than *E*.

Suppose this done, and let there be left *AG* less than *E*.

Then, since *E* measures *AB*,

while *AB* measures *DF*,

therefore *E* will also measure *FD*.

But it measures the whole *CD* also;

therefore it will also measure the remainder *CF*.

But *CF* measures *BG*;

therefore *E* also measures *BG*.

But it measures the whole *AB* also;

therefore it will also measure the remainder *AG*, the greater the less: which is impossible.

Therefore no magnitude will measure the magnitudes *AB*, *CD*;

therefore the magnitudes *AB*, *CD* are incommensurable.   [x. Def. 1]

Therefore etc.

Q.E.D.

This proposition states the test for incommensurable magnitudes, founded on the usual operation for finding the greatest common measure. The sign of the incommensurability of two magnitudes is that this operation never comes to an end, while the successive remainders become smaller and smaller until they are less than any assigned magnitude.

Observe that Euclid says "let this process be repeated continually until there is left some magnitude which is less than E." Here he evidently assumes that the process will some time produce a remainder less than any assigned magnitude E. Now this is by no means self-evident, and yet Heiberg (though so careful to supply references) and Lorenz do not refer to the basis of the assumption, which is in reality x. 1, as Billingsley and Williamson were shrewd enough to see. The fact is that, if we set off a smaller magnitude once or oftener along a greater which it does not exactly measure, until the remainder is less than the smaller magnitude, we take away from the greater more than its half. Thus, in the figure, FD is more than the half of CD, and BG more than the half of AB. If we continued the process, AG marked off along CF as many times as possible would cut off more than its half; next, more than half AG would be cut off, and so on. Hence along CD, AB alternately the process would cut off more than half, then more than half the remainder and so on, so that on both lines we should ultimately arrive at a remainder less than any assigned length.

The method of finding the greatest common measure exhibited in this proposition and the next is of course again the same as that which we use and which may be shown thus:

$$b)\ a\ (p$$
$$\underline{pb}$$
$$c)\ b\ (q$$
$$\underline{qc}$$
$$d)\ c\ (r$$
$$\underline{rd}$$
$$e$$

The proof too is the same as ours, taking just the same form, as shown in the notes to the similar propositions vii. 1, 2 *above. In the present case the hypothesis is that the process never stops, and it is required to prove that* a, b *cannot in that case have any common measure, as* f. *For suppose that* f *is a common measure, and suppose the process to be continued until the remainder* e, *say, is less than* f.

Then, since f *measures* a, b, *it measures* a − pb, *or* c.

Since f *measures* b, c, *it measures* b − qc, *or* d; *and, since* f *measures* c, d, *it measures* c − rd, *or* e: *which is impossible, since* e < f.

Euclid assumes as axiomatic that, if f *measures* a, b, *it measures* ma ± nb.

In practice, of course, it is often unnecessary to carry the process far in order to see that it will never stop, and consequently that the magnitudes are incommensurable. A good instance is pointed out by Allman (Greek Geometry from Thales to Euclid). *Euclid proves in* xiii. 5 *that, if* AB *be cut in extreme and mean ratio at* C, *and if* DA *equal to* AC *be added, then* DB *is also cut in extreme and mean ratio at* A. *This is indeed obvious from the proof of* ii. 11. *It follows conversely that, if* BD *is cut into extreme and mean ratio at* A, *and* AC, *equal to the lesser segment* AD, *be subtracted from the greater* AB, AB *is similarly divided at* C. *We can then mark off from* AC *a portion equal to* CB, *and* AC *will then be similarly divided, and so on. Now the greater segment in a line thus divided is greater than half the line, but it follows from* xiii. 3 *that it is less than twice the lesser segment, i.e. the lesser segment can never be marked off more than once from the greater. Our process of marking off the lesser segment from the greater continually is thus exactly that of finding the greatest common measure. If, therefore, the segments were commensurable, the process would stop. But it clearly does not; therefore the segments are incommensurable.*

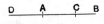

Allman expresses the opinion that it was rather in connexion with the line cut in extreme and mean ratio than with reference to the diagonal and side of a square that the Pythagoreans discovered the incommensurable. But the evidence seems to put it beyond doubt that the Pythagoreans did discover the incommensurability of √2 *and devoted much attention to this particular case. The view of Allman does not therefore commend itself to me, though it is likely enough that the Pythagoreans were aware of the incommensurability of the segments of a line cut in extreme and mean ratio. At all events the Pythagoreans could hardly have carried their investigations into the incommensurability of the segments of this line very far, since Theaetetus is said to have made the first classification of irrationals, and to him is also, with reasonable probability, attributed the substance of the first part of Eucl.* xiii., *in the sixth proposition of which occurs the proof that the segments of a rational straight line cut in extreme and mean ratio are apotomes.*

Again, the incommensurability of √2 *can be proved by a method practically equivalent to that of* x. 2, *and without carrying the process very far. This method is given in Chrystal's* Textbook of Algebra. *Let* d, a *be the diagonal and side respectively of a square* ABCD. *Mark off* AF *along* AC *equal to* a. *Draw* FE *at right angles to* AC *meeting* BC *in* E.

It is easily proved that

$$BE = EF = FC,$$
$$CF = AC - AB = d - a \ldots\ldots\ldots (1).$$
$$CE = CB - CF = a - (d - a)$$
$$= 2a - d \ldots\ldots\ldots (2).$$

Suppose, if possible, that d, a *are commensurable. If* d, a *are both commensurably expressible in terms of any finite unit, each must be an integral multiple of a certain finite unit.*

But from (1) *it follows that* CF, *and from* (2) *it follows that* CE, *is an integral multiple of the same unit.*

And CF, CE *are the side and diagonal of a square* CFEG, *the side of which is less than half the side of the original square. If* a_1, d_1 *are the side and diagonal of this square,*

$$\left.\begin{array}{l}a_1 = d - a \\ d_1 = 2a - d\end{array}\right\}.$$

Similarly we can form a square with side a_2 *and diagonal* d_2 *which are less than half* a_1, d_1 *respectively, and* a_2, d_2 *must be integral multiples of the same unit, where*

$$a_2 = d_1 - a_1,$$
$$d_2 = 2a_1 - d_1;$$

and this process may be continued indefinitely until (x. 1) *we have a square as small as we please, the side and diagonal of which are integral multiples of a finite unit: which is absurd.*

Therefore a, d *are incommensurable.*

It will be observed that this method is the opposite of that shown in the Pythagorean series of side- and diagonal-numbers, the squares being successively smaller instead of larger.

Proposition 3

Given two commensurable magnitudes, to find their greatest common measure.

Let the two given commensurable magnitudes be *AB*, *CD* of which *AB* is the less; thus it is required to find the greatest common measure of *AB*, *CD*.

Now the magnitude *AB* either measures *CD* or it does not.

If then it measures it—and it measures itself also—*AB* is a common measure of *AB*, *CD*.

And it is manifest that it is also the greatest;

for a greater magnitude than the magnitude *AB* will not measure *AB*.

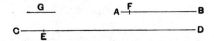

Next, let *AB* not measure *CD*.

Then, if the less be continually subtracted in turn from the greater, that which is left over will sometime measure the one before it, because *AB*, *CD* are not incommensurable;   [cf. x. 2]

let *AB*, measuring *ED*, leave *EC* less than itself,

let *EC*, measuring *FB*, leave *AF* less than itself,

and let *AF* measure *CE*.

Since, then, *AF* measures *CE*,

while *CE* measures *FB*,

therefore *AF* will also measure *FB*.

But it measures itself also;

therefore *AF* will also measure the whole *AB*.

But *AB* measures *DE*;

therefore *AF* will also measure *ED*.

But it measures *CE* also;

therefore it also measures the whole *CD*.

Therefore *AF* is a common measure of *AB*, *CD*.

I say next that it is also the greatest.

For, if not, there will be some magnitude greater than *AF* which will measure *AB*, *CD*.

Let it be *G*.

Since then *G* measures *AB*,
while *AB* measures *ED*,
therefore *G* will also measure *ED*.

But it measures the whole *CD* also;
therefore *G* will also measure the remainder *CE*.

But *CE* measures *FB*;
therefore *G* will also measure *FB*.

But it measures the whole *AB* also,
and it will therefore measure the remainder *AF*, the greater the less:
which is impossible.

Therefore no magnitude greater than *AF* will measure *AB*, *CD*;
therefore *AF* is the greatest common measure of *AB*, *CD*.

Therefore the greatest common measure of the two given commensurable magnitudes *AB*, *CD* has been found.

Q.E.D.

PORISM. From this it is manifest that, if a magnitude measure two magnitudes, it will also measure their greatest common measure.

This proposition for two commensurable magnitudes *is,* mutatis mutandis, *exactly the same as* vii. 2 *for numbers. We have the process*

$$\begin{array}{l} b)\ a\ (p \\ \underline{pb} \\ \quad c)\ b\ (q \\ \quad \underline{qc} \\ \quad\quad d)\ c\ (r \\ \quad\quad \underline{rd} \end{array}$$

where c *is equal to* rd *and therefore there is no remainder.*

It is then proved that d *is a common measure of* a, b; *and next, by a* reductio ad absurdum, *that it is the greatest common measure, since any common measure must measure* d, *and no magnitude greater than* d *can measure* d. *The* reductio ad absurdum *is of course one of form only.*

The Porism corresponds exactly to the Porism to vii. 2.

The process of finding the greatest common measure is probably given in this Book, not only for the sake of completeness, but because in x. 5 *a common measure of two magnitudes* A, B *is assumed and used, and therefore it is important to show that such a measure can be* found *if not already known.*

Proposition 4

Given three commensurable magnitudes, to find their greatest common measure.

Let *A*, *B*, *C* be the three given commensurable magnitudes;
thus it is required to find the greatest common measure of *A*, *B*, *C*.

Let the greatest common measure of the two magnitudes *A*, *B* be taken, and let it be *D*;   [x. 3]
then *D* either measures *C*, or does not measure it.

First, let it measure it.

A ————————
B ——————
C ————
D —— E— F

Since then D measures C,
while it also measures A, B,
therefore D is a common measure of A, B, C.

And it is manifest that it is also the greatest;
for a greater magnitude than the magnitude D does not measure A, B.

Next, let D not measure C.

I say first that C, D are commensurable.

For, since A, B, C are commensurable,
some magnitude will measure them,
and this will of course measure A, B also;
so that it will also measure the greatest common measure of A, B, namely D.

[x. 3, Por.]

But it also measures C;
so that the said magnitude will measure C, D;
therefore C, D are commensurable.

Now let their greatest common measure be taken, and let it be E.   [x. 3]

Since then E measures D,
while D measures A, B,
therefore E will also measure A, B.

But it measures C also;
therefore E measures A, B, C;
therefore E is a common measure of A, B, C.

I say next that it is also the greatest.

For, if possible, let there be some magnitude F greater than E, and let it measure A, B, C.

Now, since F measures A, B, C,
it will also measure A, B,
and will measure the greatest common measure of A, B.   [x. 3, Por.]

But the greatest common measure of A, B is D;
therefore F measures D.

But it measures C also;
therefore F measures C, D;
therefore F will also measure the greatest common measure of C, D.   [x. 3, Por.]

But that is E;
therefore F will measure E, the greater the less:
which is impossible.

Therefore no magnitude greater than the magnitude E will measure A, B, C;
therefore E is the greatest common measure of A, B, C if D do not measure C,

and, if it measure it, *D* is itself the greatest common measure.

Therefore the greatest common measure of the three given commensurable magnitudes has been found.

PORISM. From this it is manifest that, if a magnitude measure three magnitudes, it will also measure their greatest common measure.

Similarly too, with more magnitudes, the greatest common measure can be found, and the porism can be extended.

Q.E.D.

This proposition again corresponds exactly to vii. 3 for numbers. As there Euclid thinks it necessary to prove that, a, b, c not being prime to one another, d and c are also not prime to one another, so here he thinks it necessary to prove that d, c are commensurable, as they must be since any common measure of a, b must be a measure of their greatest common measure d (x. 3, Por.).

The argument in the proof that e, the greatest common measure of d, c, is the greatest common measure of a, b, c, is the same as that in vii. 3 and x. 3.

The Porism contains the extension of the process to the case of four or more magnitudes, corresponding to Heron's remark with regard to the similar extension of vii. 3 to the case of four or more numbers.

Proposition 5

Commensurable magnitudes have to one another the ratio which a number has to a number.

Let *A*, *B* be commensurable magnitudes;
I say that *A* has to *B* the ratio which a number has to a number.

For, since *A*, *B* are commensurable, some magnitude will measure them.
Let it measure them, and let it be *C*.

And, as many times as *C* measures *A*, so many units let there be in *D*;
and, as many times as *C* measures *B*, so many units let there be in *E*.

Since then *C* measures *A* according to the units in *D*,
while the unit also measures *D* according to the units in it,
therefore the unit measures the number *D* the same number of times as the magnitude *C* measures *A*;
therefore, as *C* is to *A*, so is the unit to *D*;   [vii. Def. 20]
therefore, inversely, as *A* is to *C*, so is *D* to the unit.   [cf. v. 7, Por.]

Again, since *C* measures *B* according to the units in *E*,
while the unit also measures *E* according to the units in it,
therefore the unit measures *E* the same number of times as *C* measures *B*;
therefore, as *C* is to *B*, so is the unit to *E*.

But it was also proved that,

as A is to C, so is D to the unit;

therefore, *ex aequali*,

as A is to B, so is the number D to E.   [v. 22]

Therefore the commensurable magnitudes A, B have to one another the ratio which the number D has to the number E.

Q.E.D.

The argument is as follows. If a, b *be commensurable magnitudes, they have some common measure* c, *and*

$$a = mc,$$
$$b = nc,$$

where m, n *are integers.*

It follows that $c : a = 1 : m$ ....(1),

or, inversely, $a : c = m : 1;$

and also that $c : b = 1 : n,$

so that, ex aequali, $a : b = m : n.$

It will be observed that, in stating the proportion (1), Euclid is merely expressing the fact that a *is the same multiple of* c *that* m *is of 1. In other words, he rests the statement on the definition of proportion in* vii. Def. 20. *This, however, is applicable only to four* numbers, *and* c, a *are not numbers but magnitudes. Hence the statement of the proportion is not legitimate unless it is proved that it is true in the sense of* v. Def. 5 *with regard to magnitudes in general, the numbers* 1, m *being magnitudes. Similarly with regard to the other proportions in the proposition.*

There is, therefore, a hiatus. Euclid ought to have proved that magnitudes which are proportional in the sense of vii. Def. 20 *are also proportional in the sense of* v. Def. 5, *or that the proportion of numbers is included in the proportion of magnitudes as a particular case. Simson has proved this in his Proposition C inserted in Book V. The portion of that proposition which is required here is the proof that,*

if $a = mb$
          $c = md$,

then $a : b = c : d,$ *in the sense of* v. Def. 5.

Take any equimultiples pa, pc *of* a, c *and any equimultiples* qb, qd *of* b, d.

Now $pa = pmb$
          $pc = pmd$.

But, according as pmb $> = <$ qb, pmd $> = <$ qd.

Therefore, according as pa $> = <$ qb, pa $> = <$ qd.

And pa, pc *are any equimultiples of* a, c, *and* qb, qd *any equimultiples of* b, d.

Therefore $a : b = c : d.$ [v. Def. 5.]

Proposition 6

If two magnitudes have to one another the ratio which a number has to a number, the magnitudes will be commensurable.

For let the two magnitudes A, B have to one another the ratio which the number D has to the number E;

I say that the magnitudes A, B are commensurable.

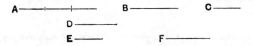

For let A be divided into as many equal parts as there are units in D,

and let C be equal to one of them;

and let F be made up of as many magnitudes equal to C as there are units in E.

Since then there are in A as many magnitudes equal to C as there are units in D, whatever part the unit is of D, the same part is C of A also; therefore, as C is to A, so is the unit to D.   [vii. Def. 20]

But the unit measures the number D;

therefore C also measures A.

And since, as C is to A, so is the unit to D,

therefore, inversely, as A is to C, so is the number D to the unit.   [cf. v. 7, Por.]

Again, since there are in F as many magnitudes equal to C as there are units in E, therefore, as C is to F, so is the unit to E.   [vii. Def. 20]

But it was also proved that,

as A is to C, so is D to the unit;

therefore, *ex aequali*, as A is to F, so is D to E.   [v. 22]

But, as D is to E, so is A to B;

therefore also, as A is to B, so is it to F also.   [v. 11]

Therefore A has the same ratio to each of the magnitudes B, F;

therefore B is equal to F.   [v. 9]

But C measures F;

therefore it measures B also.

Further it measures A also;

therefore C measures A, B.

Therefore A is commensurable with B.

Therefore etc.

PORISM. From this it is manifest that, if there be two numbers, as D, E, and a straight line, as A, it is possible to make a straight line $[F]$ such that the given straight line is to it as the number D is to the number E.

And, if a mean proportional be also taken between A, F, as B,

as A is to F, so will the square on A be to the square on B, that is, as the first is to the third, so is the figure on the first to that which is similar and similarly described on the second.   [vi. 19, Por.]

But, as A is to F, so is the number D to the number E; therefore it has been contrived that, as the number D is to the number E, so also is the figure on the straight line A to the figure on the straight line B.

Q.E.D.

12. **But the unit measures the number D; therefore C also measures A.** These words are redundant, though they are apparently found in all the MSS.

The same link to connect the proportion of numbers with the proportion of magnitudes as was necessary in the last proposition is necessary here. This being premised, the argument is as follows.
   Suppose $a : b = m : n$,
where m, n *are (integral) numbers.*
   Divide a *into* m *parts, each equal to* c, *say,*

so that
  Now take d *such that*
   Therefore we have
and
so that, ex aequali,

$$a = mc.$$
$$d = nc.$$
$$a : c = m : 1,$$
$$c : d = 1 : n,$$
$$a : d = m : n$$
$$= a : b, \text{ } by \text{ } hypothesis.$$

  Therefore b = d = nc,
so that c *measures* b n *times, and* a, b *are commensurable.*
  The Porism is often used in the later propositions. It follows (1) that, if a *be a given straight line, and* m, n *any numbers, a straight line* x *can be found such that*

$$a : x = m : n.$$

(2) We can find a straight line y *such that*

$$a^2 : y^2 = m : n.$$

For we have only to take y, *a mean proportional between* a *and* x, *as previously found, in which case* a, y, x *are in continued proportion and*   [v. Def. 9]

$$a^2 : y^2 = a : x$$
$$= m : n.$$

Proposition 7

Incommensurable magnitudes have not to one another the ratio which a number has to a number.

Let A, B be incommensurable magnitudes;
I say that A has not to B the ratio which a number has to a number.

For, if A has to B the ratio which a number has to a number, A will be commensurable with B.   [x. 6]

But it is not;
therefore A has not to B the ratio which a number has to a number.

                          A

                          B

Therefore etc.

Proposition 8

If two magnitudes have not to one another the ratio which a number has to a number, the magnitudes will be incommensurable.

For let the two magnitudes A, B not have to one another the ratio which a number has to a number;
I say that the magnitudes A, B are incommensurable.

                          A

                          B

For, if they are commensurable, A will have to B the ratio which a number has to a number.   [x. 5]

But it has not;
therefore the magnitudes A, B are incommensurable.

Therefore etc.

Proposition 9

The squares on straight lines commensurable in length have to one another the ratio which a square number has to a square number; and squares which have to one another the ratio which a square number has to a square number will also have their sides commensurable in length. But the squares on straight lines incommensurable in length have not to one another the ratio which a square number has to a square number; and squares which have not to one another the ratio which a square number has to a square number will not have their sides commensurable in length either.

For let *A, B* be commensurable in length;
I say that the square on *A* has to the square on
B the ratio which a square number has to a square
number.

For, since *A* is commensurable in length with *B*,
therefore *A* has to *B* the ratio which a number has to a number. [x. 5]

Let it have to it the ratio which *C* has to *D*.

Since then, as *A* is to *B*, so is *C* to *D*,
while the ratio of the square on *A* to the square on *B* is duplicate of the ratio of *A*
to *B*,
for similar figures are in the duplicate ratio of their corresponding sides; [vi. 20, Por.]
and the ratio of the square on *C* to the square on *D* is duplicate of the ratio of *C*
to *D*,
for between two square numbers there is one mean proportional number, and the
square number has to the square number the ratio duplicate of that which the side
has to the side; [viii. 11]
therefore also, as the square on *A* is to the square on *B*, so is the square on *C* to the
square on *D*.

Next, as the square on *A* is to the square on *B*, so let the square on *C* be to the
square on *D*;
I say that *A* is commensurable in length with *B*.

For since, as the square on *A* is to the square on *B*, so is the square on *C* to the
square on *D*,
while the ratio of the square on *A* to the square on *B* is duplicate of the ratio of
A to *B*,
and the ratio of the square on *C* to the square on *D* is duplicate of the ratio of
C to *D*,
therefore also, as *A* is to *B*, so is *C* to *D*.

Therefore *A* has to *B* the ratio which the number *C* has to the number *D*;
therefore *A* is commensurable in length with *B*. [x. 6]

Next, let *A* be incommensurable in length with *B*;
I say that the square on *A* has not to the square on *B* the ratio which a square number has to a square number.

116

For, if the square on *A* has to the square on *B* the ratio which a square number has to a square number, *A* will be commensurable with *B*.

But it is not;

therefore the square on *A* has not to the square on *B* the ratio which a square number has to a square number.

Again, let the square on *A* not have to the square on *B* the ratio which a square number has to a square number;

I say that *A* is incommensurable in length with *B*.

For, if *A* is commensurable with *B*, the square on *A* will have to the square on *B* the ratio which a square number has to a square number.

But it has not;

therefore *A* is not commensurable in length with *B*.

Therefore etc.

PORISM. And it is manifest from what has been proved that straight lines commensurable in length are always commensurable in square also, but those commensurable in square are not always commensurable in length also.

[LEMMA. It has been proved in the arithmetical books that similar plane numbers have to one another the ratio which a square number has to a square number,

[viii. 26]

and that, if two numbers have to one another the ratio which a square number has to a square number, they are similar plane numbers.

[Converse of viii. 26]

And it is manifest from these propositions that numbers which are not similar plane numbers, that is, those which have not their sides proportional, have not to one another the ratio which a square number has to a square number.

For, if they have, they will be similar plane numbers: which is contrary to the hypothesis.

Therefore numbers which are not similar plane numbers have not to one another the ratio which a square number has to a square number.]

A scholium to this proposition (Schol. X. No. 62) says categorically that the theorem proved in it was the discovery of Theaetetus.

If a, b *be straight lines, and*
where m, n are numbers,
then
and conversely.

$$a : b = m : n,$$
$$a^2 : b^2 = m^2 : n^2;$$

Archimedes

(287–212 B.C.)

HIS LIFE AND WORK

Archimedes is best remembered for jumping out of his bath and running naked through the streets shouting, "Eureka! Eureka!" ("I have found it!"), when he discovered how to distinguish a genuine gold crown from a counterfeit crown. Less well known is that he devised a test to distinguish the greatest mathematician of antiquity, himself, from impostors. In the Greek world of mathematics, it was common for mathematicians to send out announcements of their newly discovered mathematical theorems without accompanying proofs. When Archimedes suspected others of claiming his results as their own, he inserted into his own announcements two or three propositions that required all of his mathematical talent to demonstrate to be false. When the impostors claimed the false statements as their own newly discovered truths, he exposed them by sending counterexamples.

We know little of Archimedes' life. The Roman general Marcellus wrote that one of his soldiers killed Archimedes in 212 B.C. during the Second Punic War. Tradition says that he worked at geometry until the very end of his life in his seventy-fifth year, which would place his birth in 287 B.C. Archimedes' father was an astronomer named Phidias who lived and worked in the Greek city of Syracuse on the island of Sicily. Their family may have been related to the royal house of Syracuse. Archimedes was on intimate terms with King Hieron II.

Although, like Euclid, we know very little of Archimedes' life, the comparison ends there. Although Euclid was a compiler who may have achieved little if any new mathematical results on his own, Archimedes was a pioneer, many centuries ahead of his time in both mathematics and engineering. In fact, Archimedes was best known in antiquity for his engineering accomplishments on behalf of the royal house of Syracuse.

When King Hieron challenged him to move a great weight with a small force, Archimedes conceived the idea of the compound pulley and showed how he could easily pull in to shore a three-masted ship that 100 men could only pull in with much difficulty. According to the ancient Roman biographer Plutarch, it is in connection with this story that Archimedes uttered his famous remark, "Give me a place to stand on, and I will move the earth."

Plutarch and other ancient commentators such as Polybius and Livy mention the fantastic ballistic war engines Archimedes devised for the defense of Syracuse against the Roman army led by the general Marcellus. Plutarch wrote:

> . . . when Archimedes began to ply his engines, he at once shot against the land forces all sorts of missile weapons, and immense masses of stone that came down with incredible noise and violence; against which no man could stand; for they knocked down those upon whom they fell in heaps, breaking all their ranks and files. In the meantime huge poles thrust out from the walls over the ships and sunk some by great weights which they let down from on high upon them; others they lifted up into the air by an iron hand or beak like a crane's beak and, when they had drawn them up by the prow, and set them on end upon the poop, they plunged them to the bottom of the sea; or else the ships, drawn by engines within, and whirled about, were dashed against steep rocks that stood jutting out under the walls, with great destruction of the soldiers that were aboard them. A ship was frequently lifted up to a great height in the air (a dreadful thing to behold), and was rolled to and fro, and kept swinging, until the mariners were all thrown out, when at length it was dashed against the rocks, or let fall.

Archimedes could completely immerse himself in a problem and stay unaware of his surroundings. As Plutarch wrote:

> Often times Archimedes' servants got him against his will to the baths, to wash and anoint him, and yet being there, he would ever be drawing

*out of the geometrical figures, even in the very embers of the chimney.
And while they were anointing of him with oils and sweet savors, with
his fingers he drew lines upon his naked body, so far was he taken from
himself, and brought into ecstasy or trance, with the delight he had in
the study of geometry.*

His habit of ignoring his surroundings would end up costing him his life.
Archimedes' engineering accomplishments building war engines made him a prime
target of the Roman army that invaded Sicily in 287 B.C., during the Second Punic
War. Legend records that the Roman soldier found Archimedes drawing figures in
the sand. The soldier commanded Archimedes to stop what he was doing and leave
immediately. Archimedes's asked for more time to work out a problem in the sand.
Enraged, the soldier ruined Archimedes's figures in the sand and ran him through
with his sword!

Archimedes's mathematical works fall into three groups:

1. Those that prove theorems concerning areas and solids bounded by curves
   and surfaces. These include *On the Sphere and the and Circle, On the Measurement of the Circle,* and *The Method.*
2. Works that geometrically analyze problems in statics and hydrostatics.
3. Miscellaneous works, especially ones that emphasize counting, such as *The
   Sand Reckoner.*

This volume includes *The Sand Reckoner, On the Sphere and the Circle, On the
Measurement of the Circle,* and *The Method.*

The version of *On the Measurement of the Circle* known to us may be a far cry
from the work written by Archimedes. It contains just three propositions, and the second one in order, which compares the area of a circle to a square on its diameter,
depends on the third proposition which states that the ratio of a circle's circumference
to its diameter is greater than $3\frac{1}{7}$ but less than $3\frac{10}{71}$, a fairly good approximation of π.

It is the first proposition that holds the most interest for us. It states that the
area of a circle is equal to the area of a right triangle in which one of the sides about
the right angle is equal to the radius and the other is equal to the circumference.
Notice the interesting way in which Archimedes' statement of the theorem equates
an area bounded by a curve, the circle, with an area bounded by straight lines, the
right triangle. (We moderns express the area of the circle as πr^2. In modern notation Archimedes expresses it as $\frac{1}{2}(2\pi r)r$.)

When proving theorems about the area or volume of a figure bounded by curves
or surfaces, Archimedes employs the "method of exhaustion," sometimes called "indirect passage to the limit" along with the eponymous Archimedes' lemma. The lemma

states, "given two unequal lines, surfaces, or solids the greater exceeds the lesser by an amount such that, when added to itself, it may exceed any assigned magnitude of the type of magnitudes compared with one another." Archimedes generally employs these tools in a *reductio ad absurdum* proof.

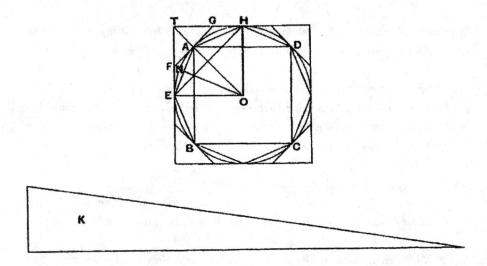

Archimedes begins his proof by recognizing that if the circle *ABCD* is not equal to the right triangle *K*, then it must be larger or smaller than *K*. The first half of the proof supposes the area of the circle to be greater than the area of the right triangle *K*. He begins his proof by inscribing the square *ABCD* inside the circle. He continues by bisecting the arcs *AB*, *BC*, *CD*, and *DA* and then bisecting these halves again and again until the sides of the inscribed polygon whose vertices are the points of continued bisection are so close to the circle that the area remaining between the circle and the inscribed polygon is less than the supposed difference between the areas of the circle and *K*. Therefore, he reasons, the area of the inscribed polygon must be greater than the area of *K*.

He continues by considering the perpendicular *ON* from the center of the circle *O* to the side *AE*. Because *AE* is part of a polygon inscribed inside of the circle, the segment *ON* must be less than the circle's radius, and the perimeter of the polygon must be less than the circle's circumference, the other side of the right triangle *K*. Therefore, the area of the inscribed polygon must be less than the area of right triangle *K*, contrary to what has just been proven. The other half of the proof demonstrates that the area of the circle cannot be less than the area of *K*.

Many of the theorems Archimedes proved had been stated and proved inadequately prior to him. However, he also discovered many remarkable new results.

The Method gives us a glimpse at how Archimedes discovered new theorems to be proven. In his preface to *On the Sphere and Cylinder*, Archimedes wrote:

> *Since then certain theorems not hitherto demonstrated have occurred to me, and I have worked out the proofs of them. They are these: first, that the surface of any sphere is four times its greatest circle. [$4\pi r^2$ in modern terms]. . . . Now, these properties were all along naturally inherent in the figures referred to . . . but remained unknown to those who were before my time engaged in the study of geometry. Having, however, now discovered that the properties are true of these figures, I cannot feel any hesitation in setting them side by side both with my former investigations and with those theorems of Eudoxus on solids which are held to be most irrefragably established, namely, that any pyramid is one third part of the cylinder which has the same base with the cone and equal height. For, though these properties also were naturally inherent in the figures all along, yet they were in fact unknown to all the many able geometers who lived before Eudoxus, and had not been observed by anyone. Now, however, it will be open to those who possess the requisite ability to examine these discoveries of mine.*

The Method has the most interesting history of all of Archimedes known works. For many centuries, it was only known by an obscure reference made by the tenth-century encylopediast Suidas, who mentioned a commentary written by Theodosius of Bithynia about a century after Archimedes' death. Mathematicians were tantalized by the prospect of finding a universal method for finding their results. Descartes, in fact, suspected Archimedes of suppressing *The Method* so that no one else would be able to benefit from it.

In 1899, the Greek scholar Papadapulos Kerameus reported the discovery of a mathematical palimpsest he had found in a library in Istanbul, Turkey. A palimpsest is an ancient document that has had its original contents washed off so that new contents could be written on it. Upon reading the few lines of the manuscript published by Kerameus, the Danish classics scholar Johan Ludvig Heiberg recognized characteristic Archimedean traits. He suspected that the underlying manuscript must be a work of Archimedes. Heiberg must have been amazed when he examined the palimpsest firsthand. Kerameus had found the long lost treatise, *The Method*, which begins, "Archimedes to Eratosthenes greeting." The presence of other Archimedean works in the palimpsest only confirmed its authorship.

The Kerameus–Heiberg palimpsest was originally written in the tenth century. In the thirteenth century, a monk had washed away the original ink so that he could write a book of devotional prayers. The monk must have had no idea of what he had washed away. Nor could he have imagined the palimpsest's future value. In 1998, Christie's auction house sold it for two million dollars!

Archimedes' technique in *The Method* supposes that solids are composed of planar elements of uniform density. The volume of a given solid **X** can be discovered by placing it on an axis with one or two other figures **B** and **C**. Archimedes supposes that all of the figures are cut by parallel planes, all perpendicular to the axis. He has chosen the figures **B** and **C** so that

1. Their centers of gravity and the centers of gravity of their planar elements all lie on the chosen axis.
2. Their volumes are known.
3. The planar elements of **B** can be compared to the planar elements of **X** (possibly together with **C**).

The last requirement requires the investigator to discover figures **B** and **C** appropriate for computing the volume of **X**. Once Archimedes establishes the relationship between the corresponding planar elements, he produces the axis to a point τ so that the length $\tau\phi$ is appropriate to the problem, another demand on the investigator's ingenuity, and it is taken to act as a lever about the fulcrum ϕ.

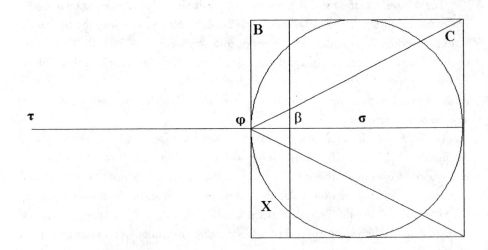

For each planar element **x** of **X** (and possibly **c** of **C** if **C** is needed for the problem) he places a corresponding element **y** of equal area so that its center of gravity is at τ, so that **y** acting at τ balances the corresponding planar element b acting where it is (β), that is, $\mathbf{y}{:}\mathbf{b} = \phi\beta : \phi\tau$. For his next step, Archimedes takes the composition

Y of the planar elements **y**. Its volume is then equal to the volume of **X** (possibly together with **C**), and when **Y** is taken to be a solid of uniform density, Archimedes concludes that its center of gravity is τ. Because all the parts acting at τ balance all of the parts of **B** acting at the fulcrum β, the whole of the composition **Y** acting at τ balances all the whole of **B** acting at its center of gravity σ. However, **B** has been chosen so that both its volume (and its weight) and its center of gravity (on the axis) are known. Consequently, Archimedes can determine the volume of the composition **Y** from the equation **Y:B** $= \phi\sigma : \phi\tau$, from which he concludes that **X = Y** (or **X = Y − C**).

The full title of this work, *The Method of Archimedes Treating of Mechanical Problems*, underscores Archimedes' view that the proofs contained in the works did not qualify as *mathematical* proofs. The proofs in *The Method* depend on physical assumptions about the figures involved and the principle of the lever, a *mechanical* principle!

Given the right lever, Archimedes could not only move the earth, he could also discover new mathematical truths!

ON THE SPHERE AND CYLINDER

BOOK I

"Archimedes to Dositheus greeting.

On a former occasion I sent you the investigations which I had up to that time completed, including the proofs, showing that any segment bounded by a straight line and a section of a right-angled cone [a parabola] is four-thirds of the triangle which has the same base with the segment and equal height. Since then certain theorems not hitherto demonstrated ($\dot{\alpha}\nu\epsilon\lambda\acute{\epsilon}\gamma\kappa\tau\omega\nu$) have occurred to me, and I have worked out the proofs of them. They are these: first, that the surface of any sphere is four times its greatest circle ($\tau o\hat{v}\,\mu\epsilon\gamma\acute{\iota}\sigma\tau o\upsilon\,\kappa\acute{\upsilon}\kappa\lambda o\upsilon$); next, that the surface of any segment of a sphere is equal to a circle whose radius ($\dot{\eta}\,\dot{\epsilon}\kappa\,\tau o\hat{v}\,\kappa\acute{\epsilon}\nu\tau\rho o\upsilon$) is equal to the straight line drawn from the vertex ($\kappa o\rho\upsilon\phi\acute{\eta}$) of the segment to the circumference of the circle which is the base of the segment; and, further, that any cylinder having its base equal to the greatest circle of those in the sphere, and height equal to the diameter of the sphere, is itself [*i.e.* in content] half as large again as the sphere, and its surface also [including its bases] is half as large again as the surface of the sphere. Now these properties were all along naturally inherent in the figures referred to ($\alpha\dot{\upsilon}\tau\hat{\eta}\,\tau\hat{\eta}\,\phi\acute{\upsilon}\sigma\epsilon\iota\,\pi\rho o\upsilon\pi\hat{\eta}\rho\chi\epsilon\nu\,\pi\epsilon\rho\grave{\iota}\,\tau\grave{\alpha}\,\epsilon\dot{\iota}\rho\eta\mu\acute{\epsilon}\nu\alpha\,\sigma\chi\acute{\eta}\mu\alpha\tau\alpha$), but remained unknown to those who were before my time engaged in the study of geometry. Having, however, now discovered that the properties are true of these figures, I cannot feel any hesitation in setting them side by side both with my former investigations and with those of the theorems of Eudoxus on solids which are held to be most irrefragably established, namely, that any pyramid is one third part of the prism which has the same base with the pyramid and equal height, and that any cone is one third part of the cylinder which has the same base with the cone and equal height. For, though these properties also were naturally inherent in the figures all along, yet they were in fact unknown to all the many able geometers who lived before Eudoxus, and had not been observed by any one. Now, however, it will be open to those who possess the requisite ability to examine these discoveries of mine. They ought to have been published while Conon was still alive, for I should conceive that he would best have been able to grasp them and to pronounce upon them the appropriate verdict; but, as I judge it well to communicate them to those who are conversant with mathematics, I send them to you with the proofs written out, which it will be open to mathematicians to examine. Farewell.

I first set out the axioms[1] and the assumptions which I have used for the proofs of my propositions.

DEFINITIONS

1. There are in a plane certain terminated bent lines (καμπύλαι γραμμαὶ πεπερασμέναι)[2], which either lie wholly on the same side of the straight lines joining their extremities, or have no part of them on the other side.

2. I apply the term **concave in the same direction** to a line such that, if any two points on it are taken, either all the straight lines connecting the points fall on the same side of the line, or some fall on one and the same side while others fall on the line itself, but none on the other side.

3. Similarly also there are certain terminated surfaces, not themselves being in a plane but having their extremities in a plane, and such that they will either be wholly on the same side of the plane containing their extremities, or have no part of them on the other side.

4. I apply the term **concave in the same direction** to surfaces such that, if any two points on them are taken, the straight lines connecting the points either all fall on the same side of the surface, or some fall on one and the same side of it while some fall upon it, but none on the other side.

5. I use the term **solid sector**, when a cone cuts a sphere, and has its apex at the centre of the sphere, to denote the figure comprehended by the surface of the cone and the surface of the sphere included within the cone.

6. I apply the term **solid rhombus**, when two cones with the same base have their apices on opposite sides of the plane of the base in such a position that their axes lie in a straight line, to denote the solid figure made up of both the cones.

ASSUMPTIONS

1. *Of all lines which have the same extremities the straight line is the least.*[3]

2. Of other lines in a plane and having the same extremities, [any two] such are unequal whenever both are concave in the same direction and one of them is either

---

[1]Though the word used is ἀξιώματα, the "axioms" are more of the nature of definitions; and in fact Eutocius in his notes speaks of them as such (ὅροι).

[2]Under the term *bent line* Archimedes includes not only curved lines of continuous curvature, but lines made up of any number of lines which may be either straight or curved.

[3]This well-known Archimedean assumption is scarcely, as it stands, a *definition* of a straight line, though Proclus says [ed. Friedlein] "Archimedes defined (ὡρίσατο) the straight line as the least of those [lines] which have the same extremities. For because, as Euclid's definition says, ἐξ ἴσου κεῖται τοῖς ἐφ᾽ἑαυτῆς σημείοις, it is in consequence the least of those which have the same extremities." Proclus had just before explained Euclid's definition, which, as will be seen, is different from the ordinary version given in our textbooks; a straight line is not "that which lies evenly between its extreme points," but "that which ἐξ ἴσου τοῖς ἐφ᾽ἑαυτῆς σημείοις κεῖται." The words of Proclus are, "He [Euclid] shows by means of this that the straight line alone [of all lines] occupies a distance (κατέχειν διάστημα) equal to that between the points on it. For, as far as one of its points is removed from another, so great is the length (μέγεθος) of the straight line of which the points are the extremities; and this is the meaning of τὸ ἐξ ἴσου κεῖσθαι τοῖς ἐφ᾽ἑαυτῆς σημείοις. But, if you take two points on a circumference or any other line, the distance cut off between them along the line is greater than the interval separating them; and this is the case with every line except the straight line." It appears then from this that Euclid's definition should be understood in a sense very like that of Archimedes' assumption, and we might perhaps translate as follows, "A straight line is that which extends equally (ἐξ ἴσου κεῖται) with the points on it," or, to follow Proclus' interpretation more closely, "A straight line is that which represents equal extension with [the distances separating] the points on it."

wholly included between the other and the straight line which has the same extremities with it, or is partly included by, and is partly common with, the other; and that [line] which is included is the lesser [of the two].

3. Similarly, of surfaces which have the same extremities, if those extremities are in a plane, the plane is the least [in area].

4. Of other surfaces with the same extremities, the extremities being in a plane, [any two] such are unequal whenever both are concave in the same direction and one surface is either wholly included between the other and the plane which has the same extremities with it, or is partly included by, and partly common with, the other; and that [surface] which is included is the lesser [of the two in area].

5. Further, of unequal lines, unequal surfaces, and unequal solids, the greater exceeds the less by such a magnitude as, when added to itself, can be made to exceed any assigned magnitude among those which are comparable with [it and with] one another.

These things being premised, *if a polygon be inscribed in a circle, it is plain that the perimeter of the inscribed polygon is less than the circumference of the circle;* for each of the sides of the polygon is less than that part of the circumference of the circle which is cut off by it."

Proposition 1

If a polygon be circumscribed about a circle, the perimeter of the circumscribed polygon is greater than the perimeter of the circle.

Let any two adjacent sides, meeting in A, touch the circle at P, Q respectively.

Then [*Assumptions*, 2]

$$PA + AQ > (\text{arc } PQ).$$

A similar inequality holds for each angle of the polygon; and, by addition, the required result follows.

Proposition 2

Given two unequal magnitudes, it is possible to find two unequal straight lines such that the greater straight line has to the less a ratio less than the greater magnitude has to the less.

Let AB, D represent the two unequal magnitudes, AB being the greater.

Suppose BC measured along BA equal to D, and let GH be any straight line.

Then, if *CA* be added to itself a sufficient number of times, the sum will exceed *D*. Let *AF* be this sum, and take *E* on *GH* produced such that *GH* is the same multiple of *HE* that *AF* is of *AC*.

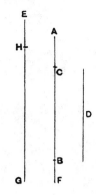

Thus $EH : HG = AC : AF$.

But, since $AF > D$ (or CB),

$$AC : AF < AC : CB.$$

Therefore, *componendo*,

$$EG : GH < AB : D.$$

Hence *EG*, *GH* are two lines satisfying the given condition.

Proposition 3

Given two unequal magnitudes and a circle, it is possible to inscribe a polygon in the circle and to describe another about it so that the side of the circumscribed polygon may have to the side of the inscribed polygon a ratio less than that of the greater magnitude to the less.

Let *A*, *B* represent the given magnitudes, *A* being the greater.

Find [Prop. 2] two straight lines *F*, *KL*, of which *F* is the greater, such that

$$F : KL < A : B \dots\dots\dots\dots\dots\dots\dots\dots\dots\dots(1).$$

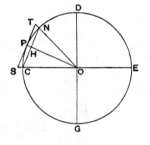
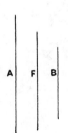

Draw *LM* perpendicular to *LK* and of such length that $KM = F$.

In the given circle let *CE*, *DG* be two diameters at right angles. Then, bisecting the angle *DOC*, bisecting the half again, and so on, we shall arrive ultimately at an angle (as *NOC*) less than twice the angle *LKM*.

Join *NC*, which (by the construction) will be the side of a regular polygon inscribed in the circle. Let *OP* be the radius of the circle bisecting the angle *NOC* (and therefore bisecting *NC* at right angles, in *H*, say), and let the tangent at *P* meet *OC*, *ON* produced in *S*, *T* respectively.

Now, since $\angle CON < 2 \angle LKM,$

$$\angle HOC < \angle LKM,$$

and the angles at H, L are right;

$$\text{therefore } MK : LK > OC : OH$$
$$> OP : OH.$$

Hence $$ST : CN < MK : LK$$
$$< F : LK;$$

therefore, *a fortiori*, by (1),

$$ST : CN < A : B.$$

Thus two polygons are found satisfying the given condition.

Proposition 4

Again, given two unequal magnitudes and a sector, it is possible to describe a polygon about the sector and to inscribe another in it so that the side of the circumscribed polygon may have to the side of the inscribed polygon a ratio less than the greater magnitude has to the less.

[The "inscribed polygon" found in this proposition is one which has for two sides the two radii bounding the sector, while the remaining sides (the number of which is, by construction, some power of 2) subtend equal parts of the arc of the sector; the "circumscribed polygon" is formed by the tangents parallel to the sides of the inscribed polygon and by the two bounding radii produced.]

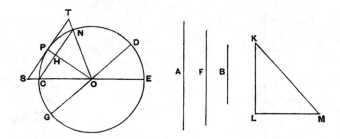

In this case we make the same construction as in the last proposition except that we bisect the angle COD of the sector, instead of the right angle between two diameters, then bisect the half again, and so on. The proof is exactly similar to the preceding one.

Proposition 5

Given a circle and two unequal magnitudes, to describe a polygon about the circle and inscribe another in it, so that the circumscribed polygon may have to the inscribed a ratio less than the greater magnitude has to the·less.

Let A be the given circle and B, C the given magnitudes, B being the greater.

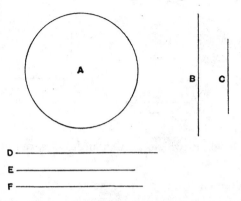

Take two unequal straight lines D, E, of which D is the greater, such that $D : E < B : C$ [Prop. 2], and let F be a mean proportional between D, E, so that D is also greater than F.

Describe (in the manner of Prop. 3) one polygon about the circle, and inscribe another in it, so that the side of the former has to the side of the latter a ratio less than the ratio $D : F$.

Thus the duplicate ratio of the side of the former polygon to the side of the latter is less than the ratio $D^2 : F^2$.

But the said duplicate ratio of the sides is equal to the ratio of the areas of the polygons, since they are similar;

therefore the area of the circumscribed polygon has to the area of the inscribed polygon a ratio less than the ratio $D^2 : F^2$, or $D : E$, and *a fortiori* less than the ratio $B : C$.

Proposition 6

"Similarly we can show that, *given two unequal magnitudes and a sector, it is possible to circumscribe a polygon about the sector and inscribe in it another similar one so that the circumscribed may have to the inscribed a ratio less than the greater magnitude has to the less.*

And it is likewise clear that, *if a circle or a sector, as well as a certain area, be given, it is possible, by inscribing regular polygons in the circle or sector, and by continually inscribing such in the remaining segments, to leave segments of the circle or sector which are [together] less than the given area.* For this is proved in the *Elements* [Eucl. xii. 2].

But it is yet to be proved that, *given a circle or sector and an area, it is possible to describe a polygon about the circle or sector, such that the area remaining between the circumference and the circumscribed figure is less than the given area.*"

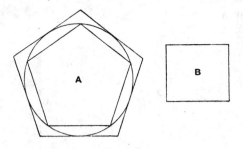

.The proof for the circle (which, as Archimedes says, can be equally applied to a sector) is as follows.

Let A be the given circle and B the given area.

Now, there being two unequal magnitudes $A + B$ and A, let a polygon (C) be circumscribed about the circle and a polygon (I) inscribed in it [as in Prop. 5], so that

$$C : I < A + B : A \quad..(1).$$

The circumscribed polygon (C) shall be that required.

For the circle (A) is greater than the inscribed polygon (I).

Therefore, from (1), *a fortiori*,

$$C : A < A + B : A,$$

whence

$$C < A + B,$$

or

$$C - A < B.$$

Proposition 7

If in an isosceles cone [i.e. *a right circular cone*] *a pyramid be inscribed having an equilateral base, the surface of the pyramid excluding the base is equal to a triangle having its base equal to the perimeter of the base of the pyramid and its height equal to the perpendicular drawn from the apex on one side of the base.*

Since the sides of the base of the pyramid are equal, it follows that the perpendiculars from the apex to all the sides of the base are equal; and the proof of the proposition is obvious.

Proposition 8

If a pyramid be circumscribed about an isosceles cone, the surface of the pyramid excluding its base is equal to a triangle having its base equal to the perimeter of the base of the pyramid and its height equal to the side [i.e. *a generator*] *of the cone.*

The base of the pyramid is a polygon circumscribed about the circular base of the cone, and the line joining the apex of the cone or pyramid to the point of contact of any side of the polygon is perpendicular to that side. Also all these perpendiculars, being generators of the cone, are equal; whence the proposition follows immediately.

Proposition 9

If in the circular base of an isosceles cone a chord be placed, and from its extremities straight lines be drawn to the apex of the cone, the triangle so formed will be less than the portion of the surface of the cone intercepted between the lines drawn to the apex.

Let *ABC* be the circular base of the cone, and *O* its apex.

Draw a chord *AB* in the circle, and join *OA*, *OB*. Bisect the arc *ACB* in *C*, and join *AC*, *BC*, *OC*.

Then $\triangle OAC + \triangle OBC > \triangle OAB.$

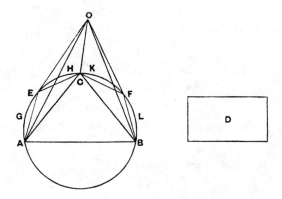

Let the excess of the sum of the first two triangles over the third be equal to the area *D*.

Then *D* is either less than the sum of the segments *AEC, CFB,* or not less.

I. Let *D* be not less than the sum of the segments referred to.

We have now two surfaces

(1) that consisting of the portion *OAEC* of the surface of the cone together with the segment *AEC*, and

(2) the triangle *OAC*;

and, since the two surfaces have the same extremities (the perimeter of the triangle *OAC*), the former surface is greater than the latter, which is *included* by it [*Assumptions,* 3 or 4].

Hence (surface *OAEC*) + (segment *AEC*) > △ *OAC*.

Similarly (surface *OCFB*) + (segment *CFB*) > △ *OBC*.

Therefore, since *D* is not less than the sum of the segments, we have, by addition,

(surface *OAECFB*) + *D* > △ *OAC* + △ *OBC*

> △ *OAB* + *D*, by hypothesis.

Taking away the common part *D*, we have the required result.

II. Let *D* be less than the sum of the segments *AEC, CFB.*

If now we bisect the arcs *AC*, *CB*, then bisect the halves, and so on, we shall ultimately leave segments which are together less than *D*.   [Prop. 6]

Let *AGE*, *EHC*, *CKF*, *FLB* be those segments, and join *OE*, *OF*.

Then, as before,

$$(\text{surface } OAGE) + (\text{segment } AGE) > \triangle OAE$$

and $$(\text{surface } OEHC) + (\text{segment } EHC) > \triangle OEC.$$

Therefore $$(\text{surface } OAGHC) + (\text{segments } AGE, EHC)$$
$$> \triangle OAE + \triangle OEC$$
$$> \triangle OAC, \text{ a fortiori.}$$

Similarly for the part of the surface of the cone bounded by *OC*, *OB* and the arc *CFB*.

Hence, by addition,

$$(\text{surface } OAGEHCKFLB) + (\text{segments } AGE, EHC, CKF, FLB)$$
$$> \triangle OAC + \triangle OBC$$
$$> \triangle OAB + D, \text{ by hypothesis.}$$

But the sum of the segments is less than *D*, and the required result follows.

Proposition 10

If in the plane of the circular base of an isosceles cone two tangents be drawn to the circle meeting in a point, and the points of contact and the point of concourse of the tangents be respectively joined to the apex of the cone, the sum of the two triangles formed by the joining lines and the two tangents are together greater than the included portion of the surface of the cone.

Let *ABC* be the circular base of the cone, *O* its apex, *AD*, *BD* the two tangents to the circle meeting in *D*. Join *OA*, *OB*, *OD*.

Let *ECF* be drawn touching the circle at *C*, the middle point of the arc *ACB*, and therefore parallel to *AB*. Join *OE*, *OF*.

Then $$ED + DF > EF,$$

and, adding *AE* + *FB* to each side,

$$AD + DB > AE + EF + FB.$$

Now *OA*, *OC*, *OB*, being generators of the cone, are equal, and they are respectively perpendicular to the tangents at *A*, *C*, *B*.

It follows that

$$\triangle OAD + \triangle ODB > \triangle OAE + \triangle OEF + \triangle OFB.$$

Let the area *G* be equal to the excess of the first sum over the second.

G is then either less, or not less, than the sum of the spaces *EAHC*, *FCKB* remaining between the circle and the tangents, which sum we will call *L*.

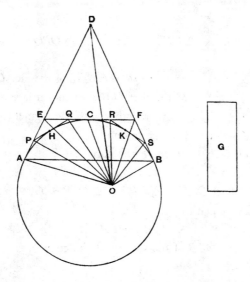

I. Let G be not less than L.

We have now two surfaces

(1) that of the pyramid with apex O and base $AEFB$, excluding the face OAB,

(2) that consisting of the part $OACB$ of the surface of the cone together with the segment ACB.

These two surfaces have the same extremities, viz. the perimeter of the triangle OAB, and, since the former *includes* the latter, the former is the greater [*Assumptions*, 4].

That is, the surface of the pyramid exclusive of the face OAB is greater than the sum of the surface $OACB$ and the segment ACB.

Taking away the segment from each sum, we have

$$\triangle OAE + \triangle OEF + \triangle OFB + L > \text{the surface } OAHCKB.$$

And G is not less than L.

It follows that

$$\triangle OAE + \triangle OEF + \triangle OFB + G,$$

which is by hypothesis equal to $\triangle OAD + \triangle ODB$, is greater than the same surface.

II. Let G be less than L.

If we bisect the arcs AC, CB and draw tangents at their middle points, then bisect the halves and draw tangents, and so on, we shall lastly arrive at a polygon such that the sum of the parts remaining between the sides of the polygon and the circumference of the segment is less than G.

Let the remainders be those between the segment and the polygon $APQRSB$, and let their sum be M. Join OP, OQ, etc.

135

Then, as before,

$$\triangle\, OAE + \triangle\, OEF + \triangle\, OFB > \triangle\, OAP + \triangle\, OPQ + \ldots + \triangle\, OSB.$$

Also, as before,

> (surface of pyramid *OAPQRSB* excluding the face *OAB*)
> $>$ the part *OACB* of the surface of the
> cone together with the segment *ACB*.

Taking away the segment from each sum,

$$\triangle\, OAP + \triangle\, OPQ + \ldots + M > \text{the part } OACB \text{ of the surface of the cone.}$$

Hence, *a fortiori*,

$$\triangle\, OAE + \triangle\, OEF + \triangle\, OFB + G,$$

which is by hypothesis equal to

$$\triangle\, OAD + \triangle\, ODB,$$

is greater than the part *OACB* of the surface of the cone.

Proposition 11

If a plane parallel to the axis of a right cylinder cut the cylinder, the part of the surface of the cylinder cut off by the plane is greater than the area of the parallelogram in which the plane cuts it.

Proposition 12

If at the extremities of two generators of any right cylinder tangents be drawn to the circular bases in the planes of those bases respectively, and if the pairs of tangents meet, the parallelograms formed by each generator and the two corresponding tangents respectively are together greater than the included portion of the surface of the cylinder between the two generators.

[The proofs of these two propositions follow exactly the methods of Props. 9, 10 respectively, and it is therefore unnecessary to reproduce them.]

"From the properties thus proved it is clear (1) that, *if a pyramid be inscribed in an isosceles cone, the surface of the pyramid excluding the base is less than the surface of the cone [excluding the base]*, and (2) that, *if a pyramid be circumscribed about an isosceles cone, the surface of the pyramid excluding the base is greater than the surface of the cone excluding the base.*

"It is also clear from what has been proved both (1) that, *if a prism be inscribed in a right cylinder, the surface of the prism made up of its parallelograms [i.e. excluding its bases] is less than the surface of the cylinder excluding its bases*, and (2) that, *if a prism be circumscribed about a right cylinder, the surface of the prism made up of its parallelograms is greater than the surface of the cylinder excluding its bases.*"

Proposition 13

The surface of any right cylinder excluding the bases is equal to a circle whose radius is a mean proportional between the side [i.e. a generator] of the cylinder and the diameter of its base.

Let the base of the cylinder be the circle A, and make CD equal to the diameter of this circle, and EF equal to the height of the cylinder.

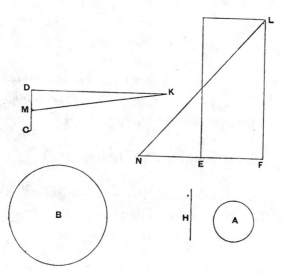

Let H be a mean proportional between CD, EF, and B a circle with radius equal to H.

Then the circle B shall be equal to the surface of the cylinder (excluding the bases), which we will call S.

For, if not, B must be either greater or less than S.

I. Suppose $B < S$.

Then it is possible to circumscribe a regular polygon about B, and to inscribe another in it, such that the ratio of the former to the latter is less than the ratio $S : B$.

Suppose this done, and circumscribe about A a polygon similar to that described about B; then erect on the polygon about A a prism of the same height as the cylinder. The prism will therefore be circumscribed to the cylinder.

Let KD, perpendicular to CD, and FL, perpendicular to EF, be each equal to the perimeter of the polygon about A. Bisect CD in M, and join MK.

Then $\triangle KDM$ = the polygon about A.

Also $\square EL$ = surface of prism (excluding bases).

Produce FE to N so that $FE = EN$, and join NL.

Now the polygons about A, B, being similar, are in the duplicate ratio of the radii of A, B.

Thus

$$\triangle KDM : (\text{polygon about } B) = MD^2 : H^2$$
$$= MD^2 : CD \cdot EF$$
$$= MD : NF$$
$$= \triangle KDM : \triangle LFN$$
$$\text{(since } DK = FL\text{)}.$$

Therefore $(\text{polygon about } B) = \triangle LFN$
$$= \square EL$$
$$= (\text{surface of prism about } A),$$
$$\text{from above.}$$

But $\qquad (\text{polygon about } B) : (\text{polygon in } B) < S : B.$

Therefore

$$(\text{surface of prism about } A) : (\text{polygon in } B) < S : B,$$

and, alternately,

$$(\text{surface of prism about } A) : S < (\text{polygon in } B) : B;$$

which is impossible, since the surface of the prism is greater than S, while the polygon inscribed in B is less than B.

Therefore $\qquad\qquad B \not< S.$

II. Suppose $B > S$.

Let a regular polygon be circumscribed about B and another inscribed in it so that
$$(\text{polygon about } B) : (\text{polygon in } B) < B : S.$$

Inscribe in A a polygon similar to that inscribed in B, and erect a prism on the polygon inscribed in A of the same height as the cylinder.

Again, let DK, FL, drawn as before, be each equal to the perimeter of the polygon inscribed in A.

Then, in this case,

$$\triangle KDM > (\text{polygon inscribed in } A)$$

(since the perpendicular from the centre on a side of the polygon is less than the radius of A).

Also $\triangle LFN = \square EL = $ surface of prism (excluding bases).

Now

$$(\text{polygon in } A) : (\text{polygon in } B) = MD^2 : H^2,$$
$$= \triangle KDM : \triangle LFN, \text{ as before.}$$

And $\qquad\qquad \triangle KDM > (\text{polygon in } A).$

Therefore

$$\triangle LFN, \text{ or (surface of prism)} > (\text{polygon in } B).$$

138

But this is impossible, because

$$\text{(polygon about } B) : \text{(polygon in } B) < B : S,$$
$$< \text{(polygon about } B) : S, \textit{a fortiori},$$

so that

$$\text{(polygon in } B) > S,$$
$$> \text{(surface of prism)}, \textit{a fortiori}.$$

Hence B is neither greater nor less than S, and therefore

$$B = S.$$

Proposition 14

The surface of any isosceles cone excluding the base is equal to a circle whose radius is a mean proportional between the side of the cone [a generator] and the radius of the circle which is the base of the cone.

Let the circle A be the base of the cone; draw C equal to the radius of the circle, and D equal to the side of the cone, and let E be a mean proportional between C, D.

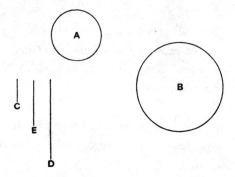

Draw a circle B with radius equal to E.

Then shall B be equal to the surface of the cone (excluding the base), which we will call S.

If not, B must be either greater or less than S.

I. Suppose $B < S$.

Let a regular polygon be described about B and a similar one inscribed in it such that the former has to the latter a ratio less than the ratio $S : B$.

Describe about A another similar polygon, and on it set up a pyramid with apex the same as that of the cone.

Then
$$\text{(polygon about } A) : \text{(polygon about } B)$$
$$= C^2 : E^2$$
$$= C : D$$
$$= \text{(polygon about } A) : \text{(surface of pyramid excluding base)}.$$

Therefore

(surface of pyramid) = (polygon about B).

Now   (polygon about B) : (polygon in B) < $S : B$.

Therefore

(surface of pyramid) : (polygon in B) < $S : B$,

which is impossible, (because the surface of the pyramid is greater than S, while the polygon in B is less than B).

Hence $B \nless S$.

II. Suppose $B > S$.

Take regular polygons circumscribed and inscribed to B such that the ratio of the former to the latter is less than the ratio $B : S$.

Inscribe in A a similar polygon to that inscribed in B, and erect a pyramid on the polygon inscribed in A with apex the same as that of the cone.

In this case

$$\text{(polygon in } A) : \text{(polygon in } B) = C^2 : E^2$$
$$= C : D$$

> (polygon in A) : (surface of pyramid excluding base).

This is clear because the ratio of C to D is greater than the ratio of the perpendicular from the centre of A on a side of the polygon to the perpendicular from the apex of the cone on the same side.[4]

Therefore

(surface of pyramid) > (polygon in B).

But   (polygon about B) : (polygon in B) < $B : S$.

Therefore, *a fortiori*,

(polygon about B) : (surface of pyramid) < $B : S$;

which is impossible.

Since therefore B is neither greater nor less than S,

$$B = S.$$

Proposition 15

The surface of any isosceles cone has the same ratio to its base as the side of the cone has to the radius of the base.

By Prop. 14, the surface of the cone is equal to a circle whose radius is a mean proportional between the side of the cone and the radius of the base.

Hence, since circles are to one another as the squares of their radii, the proposition follows.

---

[4] This is of course the geometrical equivalent of saying that, if α, β be two angles each less than a right angle, and $\alpha > \beta$, then $\sin\alpha > \sin\beta$.

Proposition 16

If an isosceles cone be cut by a plane parallel to the base, the portion of the surface of the cone between the parallel planes is equal to a circle whose radius is a mean proportional between (1) the portion of the side of the cone intercepted by the parallel planes and (2) the line which is equal to the sum of the radii of the circles in the parallel planes.

Let *OAB* be a triangle through the axis of a cone, *DE* its intersection with the plane cutting off the frustum, and *OFC* the axis of the cone.

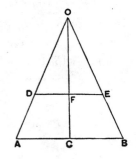

Then the surface of the cone *OAB* is equal to a circle whose radius is equal to $\sqrt{OA \cdot AC}$.　　　[Prop. 14.]

Similarly the surface of the cone *ODE* is equal to a circle whose radius is equal to $\sqrt{OD \cdot DF}$.

And the surface of the frustum is equal to the difference between the two circles.

Now
$$OA \cdot AC - OD \cdot DF = DA \cdot AC + OD \cdot AC - OD \cdot DF.$$

But　　　　　　　　　$$OD \cdot AC = OA \cdot DF,$$

since　　　　　　　　$$OA : AC = OD : DF.$$

Hence $OA \cdot AC - OD \cdot DF = DA \cdot AC + DA \cdot DF$
$$= DA \cdot (AC + DF).$$

And, since circles are to one another as the squares of their radii, it follows that the difference between the circles whose radii are $\sqrt{OA \cdot AC}$, $\sqrt{OD \cdot DF}$ respectively is equal to a circle whose radius is $\sqrt{DA \cdot (AC + DF)}$.

Therefore the surface of the frustum is equal to this circle.

Lemmas

"1. *Cones having equal height have the same ratio as their bases; and those having equal bases have the same ratio as their heights.*[5]

2. *If a cylinder be cut by a plane parallel to the base, then, as the cylinder is to the cylinder, so is the axis to the axis.*[6]

3. *The cones which have the same bases as the cylinders [and equal height] are in the same ratio as the cylinders.*

4. *Also the bases of equal cones are reciprocally proportional to their heights; and those cones whose bases are reciprocally proportional to their heights are equal.*[7]

[5]Euclid xii. 11. "Cones and cylinders of equal height are to one another as their bases."
Euclid xii. 14. "Cones and cylinders on equal bases are to one another as their heights."
[6]Euclid xii. 13. "If a cylinder be cut by a plane parallel to the opposite planes [the bases], then, as the cylinder is to the cylinder, so will the axis be to the axis."
[7]Euclid xii. 15. "The bases of equal cones and cylinders are reciprocally proportional to their heights; and those cones and cylinders whose bases are reciprocally proportional to their heights are equal."

5. *Also the cones, the diameters of whose bases have the same ratio as their axes, are to one another in the triplicate ratio of the diameters of the bases.*[8]

And all these propositions have been proved by earlier geometers."

Proposition 17

If there be two isosceles cones, and the surface of one cone be equal to the base of the other, while the perpendicular from the centre of the base [of the first cone] on the side of that cone is equal to the height [of the second], the cones will be equal.

Let *OAB, DEF* be triangles through the axes of two cones respectively, *C, G* the centres of the respective bases, *GH* the perpendicular from *G* on *FD*; and suppose

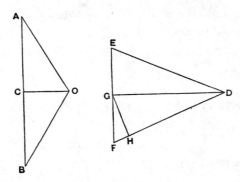

that the base of the cone *OAB* is equal to the surface of the cone *DEF*, and that *OC = GH*.

Then, since the base of *OAB* is equal to the surface of *DEF*,

(base of cone *OAB*) : (base of cone *DEF*)
= (surface of *DEF*) : (base of *DEF*)
= *DF : FG*   [Prop. 15]
= *DG : GH*, by similar triangles,
= *DG : OC*.

Therefore the bases of the cones are reciprocally proportional to their heights; whence the cones are equal. [*Lemma 4*.]

Proposition 18

Any solid rhombus consisting of isosceles cones is equal to the cone which has its base equal to the surface of one of the cones composing the rhombus and its height equal to the perpendicular drawn from the apex of the second cone to one side of the first cone.

---

[8]Euclid xii. 12. "Similar cones and cylinders are to one another in the triplicate ratio of the diameters of their bases."

Let the rhombus be *OABD* consisting of two cones with apices *O, D* and with a common base (the circle about *AB* as diameter).

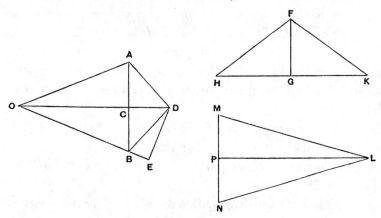

Let *FHK* be another cone with base equal to the surface of the cone *OAB* and height *FG* equal to *DE*, the perpendicular from *D* on *OB*.

Then shall the cone *FHK* be equal to the rhombus.

Construct a third cone *LMN* with base (the circle about *MN*) equal to the base of *OAB* and height *LP* equal to *OD*.

Then, since $\qquad LP = OD$,

$$LP : CD = OD : CD.$$

But [*Lemma* 1] $OD : CD = $ (rhombus *OADB*) : (cone *DAB*),

and $\qquad LP : CD = $ (cone *LMN*) : (cone *DAB*).

It follows that

(rhombus *OADB*) = (cone *LMN*) ....(1).

Again, since $AB = MN$, and

$$(\text{surface of } OAB) = (\text{base of } FHK),$$

(base of *FHK*) : (base of *LMN*)

$$= (\text{surface of } OAB) : (\text{base of } OAB)$$

$$= OB : BC \qquad\qquad\qquad [\text{Prop. 15}]$$

$$= OD : DE, \text{ by similar triangles,}$$

$$= LP : FG, \text{ by hypothesis.}$$

Thus, in the cones *FHK, LMN*, the bases are reciprocally proportional to the heights.

Therefore the cones *FHK, LMN* are equal,

and hence, by (1), the cone *FHK* is equal to the given solid rhombus.

Proposition 19

If an isosceles cone be cut by a plane parallel to the base, and on the resulting circular section a cone be described having as its apex the centre of the base [of the first cone], and if the rhombus so formed be taken away from the whole cone, the part remaining will be equal to the cone with base equal to the surface of the portion of the first cone between the parallel planes and with height equal to the perpendicular drawn from the centre of the base of the first cone on one side of that cone.

Let the cone *OAB* be cut by a plane parallel to the base in the circle on *DE* as diameter. Let *C* be the centre of the base of the cone, and with *C* as apex and the circle about *DE* as base describe a cone, making with the cone *ODE* the rhombus *ODCE*.

Take a cone *FGH* with base equal to the surface of the frustum *DABE* and height equal to the perpendicular (*CK*) from *C* on *AO*.

Then shall the cone *FGH* be equal to the difference between the cone *OAB* and the rhombus *ODCE*.

Take (1) a cone *LMN* with base equal to the surface of the cone *OAB*, and height equal to *CK*,

(2) a cone *PQR* with base equal to the surface of the cone *ODE* and height equal to *CK*.

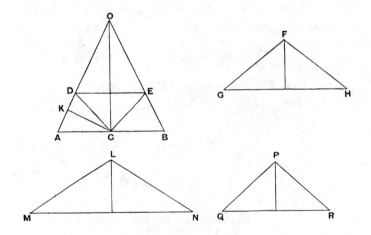

Now, since the surface of the cone *OAB* is equal to the surface of the cone *ODE* together with that of the frustum *DABE*, we have, by the construction,

(base of *LMN*) = (base of *FGH*) + (base of *PQR*)

and, since the heights of the three cones are equal,

(cone *LMN*) = (cone *FGH*) + (cone *PQR*).

But the cone *LMN* is equal to the cone *OAB* [Prop. 17], and the cone *PQR* is equal to the rhombus *ODCE* [Prop. 18].

Therefore (cone *OAB*) = (cone *FGH*) + (rhombus *ODCE*), and the proposition is proved.

Proposition 20

If one of the two isosceles cones forming a rhombus be cut by a plane parallel to the base and on the resulting circular section a cone be described having the same apex as the second cone, and if the resulting rhombus be taken from the whole rhombus, the remainder will be equal to the cone with base equal to the surface of the portion of the cone between the parallel planes and with height equal to the perpendicular drawn from the apex of the second[9] cone to the side of the first cone.

Let the rhombus be *OACB*, and let the cone *OAB* be cut by a plane parallel to its base in the circle about *DE* as diameter. With this circle as base and *C* as apex describe a cone, which therefore with *ODE* forms the rhombus *ODCE*.

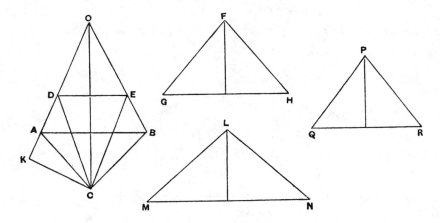

Take a cone *FGH* with base equal to the surface of the frustum *DABE* and height equal to the perpendicular (*CK*) from *C* on *OA*.

The cone *FGH* shall be equal to the difference between the rhombi *OACB*, *ODCE*.

For take (1) a cone *LMN* with base equal to the surface of *OAB* and height equal to *CK*,

(2) a cone *PQR*, with base equal to the surface of *ODE*, and height equal to *CK*.

Then, since the surface of *OAB* is equal to the surface of *ODE* together with that of the frustum *DABE*, we have, by construction,

(base of *LMN*) = (base of *PQR*) + (base of *FGH*),

and the three cones are of equal height;

therefore   (cone *LMN*) = (cone *PQR*) + (cone *FGH*).

[9]There is a slight error in Heiberg's translation "prioris coni". The perpendicular is not drawn from the apex of the cone which is cut by the plane but from the apex of the other.

But the cone *LMN* is equal to the rhombus *OACB*, and the cone *PQR* is equal to the rhombus *ODCE* [Prop. 18].

Hence the cone *FGH* is equal to the difference between the two rhombi *OACB*, *ODCE*.

Proposition 21

A regular polygon of an even number of sides being inscribed in a circle, as ABC...A'...C'B'A, so that AA' is a diameter, if two angular points next but one to each other, as B, B', be joined, and the other lines parallel to BB' and joining pairs of angular points be drawn, as CC', DD' ..., then

$$(BB' + CC' + ...) : AA' = A'B : BA.$$

Let *BB'*, *CC'*, *DD'*, . . . meet *AA'* in *F*, *G*, *H*, . . .; and let *CB'*, *DC'*, . . . be joined meeting *AA'* in *K*, *L*, . . . respectively.

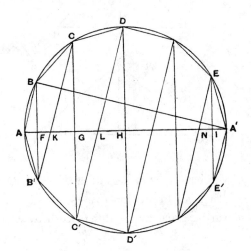

Then clearly *CB'*, *DC'*, ... are parallel to one another and to *AB*.

Hence, by similar triangles,

$$BF : FA = B'F : FK$$
$$= CG : GK$$
$$= C'G : GL$$
$$\ldots\ldots\ldots$$
$$= E'I : IA';$$

and, summing the antecedents and consequents respectively, we have

$$(BB' + CC' + ...) : AA' = BF : FA$$
$$= A'B : BA.$$

Proposition 22

If a polygon be inscribed in a segment of a circle LAL' so that all its sides excluding the base are equal and their number even, as $LK \ldots A \ldots K'L'$, A being the middle point of the segment, and if the lines BB', CC', ... parallel to the base LL' and joining pairs of angular points be drawn, then

$$(BB' + CC' + \ldots + LM) : AM = A'B : BA,$$

where M is the middle point of LL' and AA' is the diameter through M.

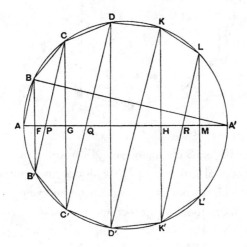

Joining CB', DC', ... LK', as in the last proposition, and supposing that they meet AM in P, Q, ... R, while BB', CC', ..., KK' meet AM in F, G, ... H, we have, by similar triangles,

$$BF : FA = B'F : FP$$
$$= CG : PG$$
$$= C'G : GQ$$
$$\ldots\ldots\ldots\ldots\ldots$$
$$= LM : RM;$$

and, summing the antecedents and consequents, we obtain

$$(BB' + CC' + \ldots + LM) : AM = BF : FA$$
$$= A'B : BA.$$

Proposition 23

Take a great circle $ABC \ldots$ of a sphere, and inscribe in it a regular polygon whose sides are a multiple of four in number. Let AA', MM' be diameters at right angles and joining opposite angular points of the polygon.

Then, if the polygon and great circle revolve together about the diameter AA', the angular points of the polygon, except A, A', will describe circles on the surface

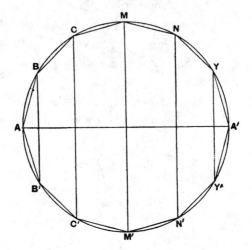

of the sphere at right angles to the diameter AA'. Also the sides of the polygon will describe portions of conical surfaces, e.g. BC will describe a surface forming part of a cone whose base is a circle about CC' as diameter and whose apex is the point in which CB, $C'B'$ produced meet each other and the diameter AA'.

Comparing the hemisphere MAM' and that half of the figure described by the revolution of the polygon which is included in the hemisphere, we see that the surface of the hemisphere and the surface of the inscribed figure have the same boundaries in one plane (viz. the circle on MM' as diameter), the former surface entirely includes the latter, and they are both concave in the same direction.

Therefore [*Assumptions,* 4] the surface of the hemisphere is greater than that of the inscribed figure; and the same is true of the other halves of the figures.

Hence *the surface of the sphere is greater than the surface described by the revolution of the polygon inscribed in the great circle about the diameter of the great circle.*

Proposition 24

If a regular polygon $AB \ldots A' \ldots B'A$, the number of whose sides is a multiple of four, be inscribed in a great circle of a sphere, and if BB' subtending two sides be joined, and all the other lines parallel to BB' and joining pairs of angular points be drawn, then the surface of the figure inscribed in the sphere by the revolution of the polygon about the diameter AA' is equal to a circle the square of whose radius is equal to the rectangle

$$BA(BB' + CC' + \ldots).$$

The surface of the figure is made up of the surfaces of parts of different cones.

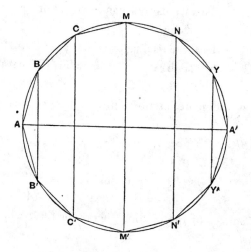

Now the surface of the cone ABB' is equal to a circle whose radius is
$\sqrt{BA \cdot \frac{1}{2}BB'}$.

[Prop. 14]

The surface of the frustum $BB'C'C$ is equal to a circle of radius
$\sqrt{BC \cdot \frac{1}{2}(BB' + CC')}$,

[Prop. 16]

and so on.

It follows, since $BA = BC = \ldots$, that the whole surface is equal to a circle whose radius is equal to

$$\sqrt{BA\,(BB' + CC' + \ldots + MM' + \ldots + YY')}.$$

Proposition 25

The surface of the figure inscribed in a sphere as in the last propositions, consisting of portions of conical surfaces, is less than four times the greatest circle in the sphere.

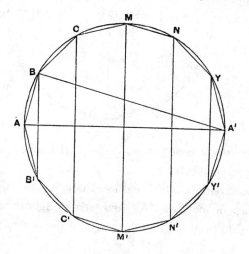

Let $AB \ldots A' \ldots B'A$ be a regular polygon inscribed in a great circle, the number of its sides being a multiple of four.

As before, let BB' be drawn subtending two sides, and $CC', \ldots YY'$ parallel to BB'.

Let R be a circle such that the square of its radius is equal to

$$AB(BB' + CC' + \ldots + YY'),$$

so that the surface of the figure inscribed in the sphere is equal to R.   [Prop. 24]

Now

$$(BB' + CC' + \ldots + YY') : AA' = A'B : AB, \qquad \text{[Prop. 21]}$$

whence
$$AB(BB' + CC' + \ldots + YY') = AA' \cdot A'B.$$

Hence
$$(\text{radius of } R)^2 = AA' \cdot A'B$$
$$< AA'^2.$$

Therefore the surface of the inscribed figure, or the circle R, is less than four times the circle $AMA'M'$.

Proposition 26

The figure inscribed as above in a sphere is equal [in volume] to a cone whose base is a circle equal to the surface of the figure inscribed in the sphere and whose height is equal to the perpendicular drawn from the centre of the sphere to one side of the polygon.

Suppose, as before, that $AB \ldots A' \ldots B'A$ is the regular polygon inscribed in a great circle, and let BB', $CC', \ldots$ be joined.

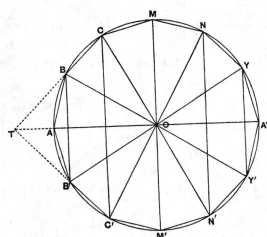

With apex O construct cones whose bases are the circles on BB', $CC', \ldots$ as diameters in planes perpendicular to AA'.

Then $OBAB'$ is a solid rhombus, and its volume is equal to a cone whose base is equal to the surface of the cone ABB' and whose height is equal to the perpendicular from O on AB [Prop. 18]. Let the length of the perpendicular be p.

Again, if CB, $C'B'$ produced meet in T, the portion of the solid figure which is described by the revolution of the triangle BOC about AA' is equal to the difference between the rhombi $OCTC'$ and $OBTB'$, i.e. to a cone whose base is equal to the surface of the frustum$BB'C'C$ and whose height is p [Prop. 20].

Proceeding in this manner, and adding, we prove that, since cones of equal height are to one another as their bases, the volume of the solid of revolution is equal to a cone with height p and base equal to the sum of the surfaces of the cone BAB', the frustum $BB'C'C$, etc., i.e. a cone with height p and base equal to the surface of the solid.

Proposition 27

The figure inscribed in the sphere as before is less than four times the cone whose base is equal to a great circle of the sphere and whose height is equal to the radius of the sphere.

By Prop. 26 the volume of the solid figure is equal to a cone whose base is equal to the surface of the solid and whose height is p, the perpendicular from O on any side of the polygon. Let R be such a cone.

Take also a cone S with base equal to the great circle, and height equal to the radius, of the sphere.

Now, since the surface of the inscribed solid is less than four times the great circle [Prop. 25], the base of the cone R is less than four times the base of the cone S.

Also the height (p) of R is less than the height of S.

Therefore the volume of R is less than four times that of S; and the proposition is proved.

Proposition 28

Let a regular polygon, whose sides are a multiple of four in number, be circumscribed about a great circle of a given sphere, as $AB \ldots A' \ldots B'A$; and about

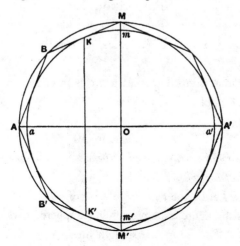

the polygon describe another circle, which will therefore have the same centre as the great circle of the sphere. Let AA' bisect the polygon and cut the sphere in a, a'.

If the great circle and the circumscribed polygon revolve together about AA', the great circle will describe the surface of a sphere, the angular points of the polygon except A, A' will move round the surface of a larger sphere, the points of contact of the sides of the polygon with the great circle of the inner sphere will describe circles on that sphere in planes perpendicular to AA', and the sides of the polygon themselves will describe portions of conical surfaces. *The circumscribed figure will thus be greater than the sphere itself.*

Let any side, as BM, touch the inner circle in K, and let K' be the point of contact of the circle with $B'M'$.

Then the circle described by the revolution of KK' about AA' is the boundary in one plane of two surfaces

(1) the surface formed by the revolution of the circular segment KaK', and

(2) the surface formed by the revolution of the part $KB \ldots A \ldots B'K'$ of the polygon.

Now the second surface entirely includes the first, and they are both concave in the same direction;

therefore [*Assumptions*, 4] the second surface is greater than the first.

The same is true of the portion of the surface on the opposite side of the circle on KK' as diameter.

Hence, adding, we see that *the surface of the figure circumscribed to the given sphere is greater than that of the sphere itself.*

Proposition 29

In a figure circumscribed to a sphere in the manner shown in the previous proposition the surface is equal to a circle the square on whose radius is equal to $AB(BB' + CC' + \ldots)$.

For the figure circumscribed to the sphere is inscribed in a larger sphere, and the proof of Prop. 24 applies.

Proposition 30

The surface of a figure circumscribed as before about a sphere is greater than four times the great circle of the sphere.

Let $AB \ldots A' \ldots B'A$ be the regular polygon of $4n$ sides which by its revolution about AA' describes the figure circumscribing the sphere of which $ama'm'$ is a great circle. Suppose aa', AA' to be in one straight line.

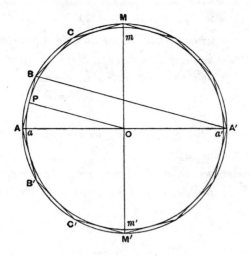

Let R be a circle equal to the surface of the circumscribed solid.

Now $\qquad (BB' + CC' + \ldots) : AA' = A'B : BA,$ [as in Prop. 21]

so that $\qquad AB(BB' + CC' + \ldots) = AA' \cdot A'B.$

Hence $\qquad$ (radius of R) $= \sqrt{AA' \cdot A'B}$ [Prop. 29]

$$> A'B.$$

But $A'B = 20P$, where P is the point in which AB touches the circle $ama'm'$.

Therefore (radius of R) $>$ (diameter of circle $ama'm'$);

whence R, and therefore the surface of the circumscribed solid, is greater than four times the great circle of the given sphere.

Proposition 31

The solid of revolution circumscribed as before about a sphere is equal to a cone whose base is equal to the surface of the solid and whose height is equal to the radius of the sphere.

The solid is, as before, a solid inscribed in a larger sphere; and, since the perpendicular on any side of the revolving polygon is equal to the radius of the inner sphere, the proposition is identical with Prop. 26.

COR. *The solid circumscribed about the smaller sphere is greater than four times the cone whose base is a great circle of the sphere and whose height is equal to the radius of the sphere.*

For, since the surface of the solid is greater than four times the great circle of the inner sphere [Prop. 30], the cone whose base is equal to the surface of the solid and whose height is the radius of the sphere is greater than four times the cone of the same height which has the great circle for base. [*Lemma* 1.]

Hence, by the proposition, the volume of the solid is greater than four times the latter cone.

Proposition 32

If a regular polygon with 4n sides be inscribed in a great circle of a sphere, as ab ... a' ... b' a, and a similar polygon AB ... A' ... B' A be described about the great circle, and if the polygons revolve with the great circle about the diameters aa', AA' respectively, so that they describe the surfaces of solid figures inscribed in and circumscribed to the sphere respectively, then

(1) *the surfaces of the circumscribed and inscribed figures are to one another in the duplicate ratio of their sides, and*

(2) *the figures themselves [i.e. their volumes] are in the triplicate ratio of their sides.*

(1) Let AA', aa' be in the same straight line, and let $MmOm'M'$ be a diameter at right angles to them.

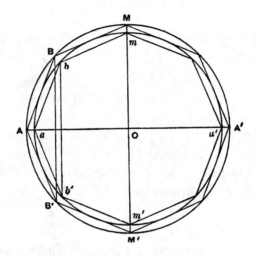

Join BB', CC', ... and bb', cc', ... which will all be parallel to one another and MM'.

Suppose R, S to be circles such that

$$R = \text{(surface of circumscribed solid)},$$
$$S = \text{(surface of inscribed solid)}.$$

Then   (radius of R)^2 = $AB(BB' + CC' + ...)$ [Prop. 29]

          (radius of S)^2 = $ab(bb' + cc' + ...)$.   [Prop. 24]

And, since the polygons are similar, the rectangles in these two equations are similar, and are therefore in the ratio of

$$AB^2 : ab^2.$$

Hence

(surface of circumscribed solid) : (surface of inscribed solid)
$$= AB^2 : ab^2.$$

(2) Take a cone V whose base is the circle R and whose height is equal to Oa, and a cone W whose base is the circle S and whose height is equal to the perpendicular from O on ab, which we will call p.

Then, V, W are respectively equal to the volumes of the circumscribed and inscribed figures. [Props. 31, 26]

Now, since the polygons are similar,

$$AB : ab = Oa : p$$
$$= (\text{height of cone } V) : (\text{height of cone } W);$$

and, as shown above, the bases of the cones (the circles R, S) are in the ratio of AB^2 to ab^2.

Therefore $$V : W = AB^3 : ab^3.$$

Proposition 33

The surface of any sphere is equal to four times the greatest circle in it.

Let C be a circle equal to four times the great circle.

Then, if C is not equal to the surface of the sphere, it must either be less or greater.

I. Suppose C less than the surface of the sphere.

It is then possible to find two lines β, γ, of which β is the greater, such that

$$\beta : \gamma < (\text{surface of sphere}) : C. \qquad [\text{Prop. 2}]$$

Take such lines, and let δ be a mean proportional between them.

Suppose similar regular polygons with $4n$ sides circumscribed about and inscribed in a great circle such that the ratio of their sides is less than the ratio $\beta : \delta$.

[Prop. 3]

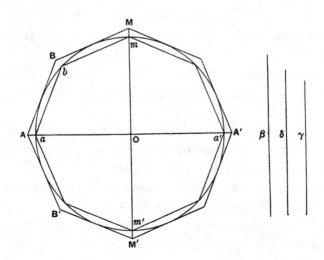

Let the polygons with the circle revolve together about a diameter common to all, describing solids of revolution as before.

Then (surface of outer solid) : (surface of inner solid)

$$= \text{(side of outer)}^2 : \text{(side of inner)}^2 \qquad \text{[Prop. 32]}$$
$$< \beta^2 : \delta^2, \text{ or } \beta : \gamma$$
$$< \text{(surface of sphere)} : C, \text{ a fortiori.}$$

But this is impossible, since the surface of the circumscribed solid is greater than that of the sphere [Prop. 28], while the surface of the inscribed solid is less than C [Prop. 25].

Therefore C is not less than the surface of the sphere.

II. Suppose C greater than the surface of the sphere.

Take lines β, γ, of which β is the greater, such that

$$\beta : \gamma < C : \text{(surface of sphere)}.$$

Circumscribe and inscribe to the great circle similar regular polygons, as before, such that their sides are in a ratio less than that of β to δ, and suppose solids of revolution generated in the usual manner.

Then, in this case,

$$\text{(surface of circumscribed solid)} : \text{(surface of inscribed solid)}$$
$$< C : \text{(surface of sphere)}.$$

But this is impossible, because the surface of the circumscribed solid is greater than C [Prop. 30], while the surface of the inscribed solid is less than that of the sphere [Prop. 23].

Thus C is not greater than the surface of the sphere.

Therefore, since it is neither greater nor less, C is equal to the surface of the sphere.

Proposition 34

Any sphere is equal to four times the cone which has its base equal to the greatest circle in the sphere and its height equal to the radius of the sphere.

Let the sphere be that of which $ama'm'$ is a great circle.

If now the sphere is not equal to four times the cone described, it is either greater or less.

I. If possible, let the sphere be greater than four times the cone.

Suppose V to be a cone whose base is equal to four times the great circle and whose height is equal to the radius of the sphere.

Then, by hypothesis, the sphere is greater than V; and two lines β, γ can be found (of which β is the greater) such that

$$\beta : \gamma < \text{(volume of sphere)} : V.$$

Between β and γ place two arithmetic means δ, ϵ.

As before, let similar regular polygons with sides $4n$ in number be circumscribed about and inscribed in the great circle, such that their sides are in a ratio less than $\beta : \delta$.

Imagine the diameter aa' of the circle to be in the same straight line with a diameter of both polygons, and imagine the latter to revolve with the circle about aa', describing the surfaces of two solids of revolution. The volumes of these solids are therefore in the triplicate ratio of their sides. [Prop. 32]

Thus (vol. of outer solid) : (vol. of inscribed solid)

$< \beta^3 : \delta^3$, by hypothesis,

$< \beta : \gamma$, *a fortiori* (since $\beta : \gamma > \beta^3 : \delta^3$),[10]

$<$ (volume of sphere) : V, *a fortiori.*

But this is impossible, since the volume of the circumscribed solid is greater than

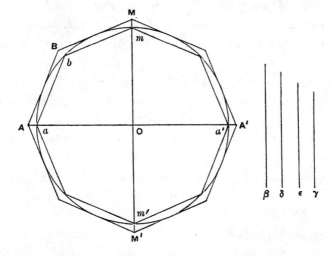

that of the sphere [Prop. 28], while the volume of the inscribed solid is less than V [Prop. 27].

Hence the sphere is not greater than V, or four times the cone described in the enunciation.

[10]That $\beta : \gamma > \beta^3 : \delta^3$ is assumed by Archimedes. Eutocius proves the property in his commentary as follows.

Take x such that $\qquad\qquad\qquad\qquad \beta : \delta = \delta : x.$

Thus $\qquad\qquad\qquad\qquad\qquad \beta - \delta : \beta = \delta - x : \delta$

and, since $\beta > \delta$, $\beta - \delta > \delta - x$.

But, by hypothesis, $\qquad\qquad\qquad\quad \beta - \delta = \delta - \epsilon.$

Therefore $\qquad\qquad\qquad\qquad\qquad \delta - \epsilon > \delta - x,$

or $\qquad\qquad\qquad\qquad\qquad\qquad\qquad x > \epsilon.$

Again, suppose $\qquad\qquad\qquad\qquad\quad \delta : x = x : y,$

and, as before, we have $\qquad\qquad\qquad \delta - x > x - y,$

so that, *a fortiori,* $\qquad\qquad\qquad\qquad \delta - \epsilon > x - y.$

Therefore $\qquad\qquad\qquad\qquad\qquad \epsilon - \gamma > x - y;$

and, since $x > \epsilon$, $y > \gamma$.

Now, by hypothesis, β, δ, x, y are in continued proportion ;

therefore $\qquad\qquad\qquad\qquad\qquad \beta^3 : \delta^3 = \beta : y$

$\qquad\qquad\qquad\qquad\qquad\qquad\qquad\quad < \beta : \gamma.$

II. If possible, let the sphere be less than *V.*

In this case we take β, γ (β being the greater) such that

$$\beta : \gamma < V : (\text{volume of sphere}).$$

The rest of the construction and proof proceeding as before, we have finally

(volume of outer solid) : (volume of inscribed solid)

$$< V : (\text{volume of sphere}).$$

But this is impossible, because the volume of the outer solid is greater than *V* [Prop. 31, Cor.], and the volume of the inscribed solid is less than the volume of the sphere.

Hence the sphere is not less than *V.*

Since then the sphere is neither less nor greater than *V*, it is equal to *V*, or to four times the cone described in the enunciation.

COR. From what has been proved it follows that *every cylinder whose base is the greatest circle in a sphere and whose height is equal to the diameter of the sphere is $\frac{3}{2}$ of the sphere, and its surface together with its bases is $\frac{3}{2}$ of the surface of the sphere.*

For the cylinder is three times the cone with the same base and height [Eucl. xii. 10], i.e. six times the cone with the same base and with height equal to the radius of the sphere.

But the sphere is four times the latter cone [Prop. 34]. Therefore the cylinder is $\frac{3}{2}$ of the sphere.

Again, the surface of a cylinder (excluding the bases) is equal to a circle whose radius is a mean proportional between the height of the cylinder and the diameter of its base [Prop. 13].

In this case the height is equal to the diameter of the base and therefore the circle is that whose radius is the diameter of the sphere, or a circle equal to four times the great circle of the sphere.

Therefore the surface of the cylinder with the bases is equal to six times the great circle.

And the surface of the sphere is four times the great circle [Prop. 33]; whence

$$(\text{surface of cylinder with bases}) = \tfrac{3}{2} \cdot (\text{surface of sphere}).$$

Proposition 35

If in a segment of a circle LAL' (where A is the middle point of the arc) a polygon LK...A...K'L' be inscribed of which LL' is one side, while the other sides are 2n in number and all equal, and if the polygon revolve with the segment about the diameter AM, generating a solid figure inscribed in a segment of a sphere, then

158

the surface of the inscribed solid is equal to a circle the square on whose radius is equal to the rectangle

$$AB\left(BB' + CC' + \dots + KK' + \frac{LL'}{2}\right).$$

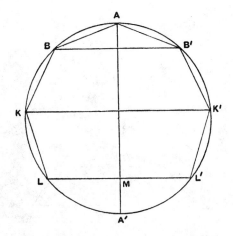

The surface of the inscribed figure is made up of portions of surfaces of cones.

If we take these successively, the surface of the cone BAB' is equal to a circle whose radius is

$$\sqrt{AB \cdot \tfrac{1}{2}BB'}. \qquad \text{[Prop. 14]}$$

The surface of the frustum of a cone $BCC'B'$ is equal to a circle whose radius is

$$\sqrt{AB \cdot \frac{BB' + CC'}{2}}; \qquad \text{[Prop. 16]}$$

and so on.

Proceeding in this way and adding, we find, since circles are to one another as the squares of their radii, that the surface of the inscribed figure is equal to a circle whose radius is

$$\sqrt{AB\left(BB' + CC' + \dots + KK' + \frac{LL'}{2}\right)}.$$

Proposition 36

The surface of the figure inscribed as before in the segment of a sphere is less than that of the segment of the sphere.

This is clear, because the circular base of the segment is a common boundary of each of two surfaces, of which one, the segment, includes the other, the solid, while both are concave in the same direction [*Assumptions*, 4].

Proposition 37

The surface of the solid figure inscribed in the segment of the sphere by the revolution of LK...A...K'L' about AM is less than a circle with radius equal to AL.

Let the diameter AM meet the circle of which LAL' is a segment again in A'. Join $A'B$.

As in Prop. 35, the surface of the inscribed solid is equal to a circle the square on whose radius is

$$AB(BB' + CC' + \ldots + KK' + LM).$$

But this rectangle $= A'B \cdot AM$ [Prop. 22]

$$< A'A \cdot AM$$
$$< AL^2.$$

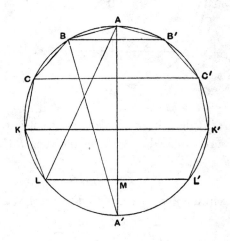

Hence the surface of the inscribed solid is less than the circle whose radius is AL.

Proposition 38

The solid figure described as before in a segment of a sphere less than a hemisphere, together with the cone whose base is the base of the segment and whose apex is the centre of the sphere, is equal to a cone whose base is equal to the surface of the inscribed solid and whose height is equal to the perpendicular from the centre of the sphere on any side of the polygon.

Let O be the centre of the sphere, and p the length of the perpendicular from O on AB.

Suppose cones described with O as apex, and with the circles on BB', CC', ... as diameters as bases.

Then the rhombus $OBAB'$ is equal to a cone whose base is equal to the surface of the cone BAB', and whose height is p.   [Prop. 18]

Again, if CB, $C'B'$ meet in T, the solid described by the triangle BOC as the polygon revolves about AO is the difference between the rhombi $OCTC'$ and $OBTB'$, and is therefore equal to a cone whose base is equal to the surface of the frustum $BCC'B'$ and whose height is p.   [Prop. 20]

Similarly for the part of the solid described by the triangle COD as the polygon revolves; and so on.

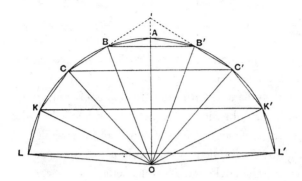

Hence, by addition, the solid figure inscribed in the segment together with the cone OLL' is equal to a cone whose base is the surface of the inscribed solid and whose height is p.

COR. *The cone whose base is a circle with radius equal to AL and whose height is equal to the radius of the sphere is greater than the sum of the inscribed solid and the cone OLL'.*

For, by the proposition, the inscribed solid together with the cone OLL' is equal to a cone with base equal to the surface of the solid and with height p.

This latter cone is less than a cone with height equal to OA and with base equal to the circle whose radius is AL, because the height p is less than OA, while the surface of the solid is less than a circle with radius AL.   [Prop. 37]

Proposition 39

Let lal' be a segment of a great circle of a sphere, being less than a semicircle. Let O be the centre of the sphere, and join Ol, Ol'. Suppose a polygon circumscribed about the sector $Olal'$ such that its sides, excluding the two radii, are $2n$ in number and all equal, as LK, ... BA, AB', ... $K'L'$; and let OA be that radius of the great circle which bisects the segment lal'.

The circle circumscribing the polygon will then have the same centre O as the given great circle.

161

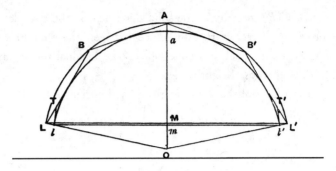

Now suppose the polygon and the two circles to revolve together about *OA*. The two circles will describe spheres, the angular points except *A* will describe circles on the outer sphere, with diameters *BB'* etc., the points of contact of the sides with the inner segment will describe circles on the inner sphere, the sides themselves will describe the surfaces of cones or frusta of cones, and the whole figure circumscribed to the segment of the inner sphere by the revolution of the equal sides of the polygon will have for its base the circle on *LL'* as diameter.

The surface of the solid figure so circumscribed about the sector of the sphere [excluding its base] will be greater than that of the segment of the sphere whose base is the circle on ll' as diameter.

For draw the tangents *lT*, *l'T'* to the inner segment at *l*, *l'*. These with the sides of the polygon will describe by their revolution a solid whose surface is greater than that of the segment [*Assumptions*, 4].

But the surface described by the revolution of *lT* is less than that described by the revolution of *LT*, since the angle *TlL* is a right angle, and therefore *LT* > *lT*.

Hence, *a fortiori*, the surface described by *LK*...*A*...*K'L'* is greater than that of the segment.

COR. *The surface of the figure so described about the sector of the sphere is equal to a circle the square on whose radius is equal to the rectangle*

$$AB(BB' + CC' + \ldots + KK' + \tfrac{1}{2}LL').$$

For the circumscribed figure is inscribed in the outer sphere, and the proof of Prop. 35 therefore applies.

Proposition 40

The surface of the figure circumscribed to the sector as before is greater than a circle whose radius is equal to al.

Let the diameter *AaO* meet the great circle and the circle circumscribing the revolving polygon again in *a'*, *A'*. Join *A'B*, and let *ON* be drawn to *N*, the point of contact of *AB* with the inner circle.

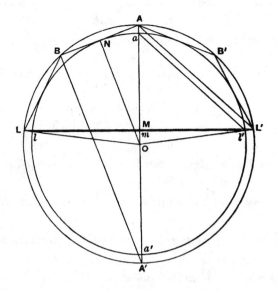

Now, by Prop. 39, Cor., the surface of the solid figure circumscribed to the sector $OlAl'$ is equal to a circle the square on whose radius is equal to the rectangle

$$AB\left(BB' + CC' + \ldots + KK' + \frac{LL'}{2} \right).$$

But this rectangle is equal to $A'B \cdot AM$ [as in Prop. 22].

Next, since AL', al' are parallel, the triangles AML', aml' are similar. And $AL' > al'$; therefore $AM > am$.

Also $\qquad\qquad\qquad\qquad\qquad A'B = 2ON = aa'.$

Therefore $\qquad\qquad\qquad\quad A'B \cdot AM > am \cdot aa'$
$$> al'^2.$$

Hence the surface of the solid figure circumscribed to the sector is greater than a circle whose radius is equal to al', or al.

COR. 1. *The volume of the figure circumscribed about the sector together with the cone whose apex is O and base the circle on LL' as diameter, is equal to the volume of a cone whose base is equal to the surface of the circumscribed figure and whose height is ON.*

For the figure is inscribed in the outer sphere which has the same centre as the inner. Hence the proof of Prop. 38 applies.

COR. 2. *The volume of the circumscribed figure with the cone OLL' is greater than the cone whose base is a circle with radius equal to al and whose height is equal to the radius (Oa) of the inner sphere.*

For the volume of the figure with the cone OLL' is equal to a cone whose base is equal to the surface of the figure and whose height is equal to ON.

And the surface of the figure is greater than a circle with radius equal to *al* [Prop. 40], while the heights *Oa*, *ON* are equal.

Proposition 41

Let *lal'* be a segment of a great circle of a sphere which is less than a semicircle.

Suppose a polygon inscribed in the sector *Olal'* such that the sides *lk*, . . . *ba*, *ab'*, ... *k'l'* are 2*n* in number and all equal. Let a similar polygon be circumscribed about the sector so that its sides are parallel to those of the first polygon; and draw the circle circumscribing the outer polygon.

Now let the polygons and circles revolve together about *OaA*, the radius bisecting the segment *lal'*.

Then (1) *the surfaces of the outer and inner solids of revolution so described are in the ratio of AB² to ab²*, and (2) *their volumes together with the corresponding cones with the same base and with apex O in each case are as AB³ to ab³.*

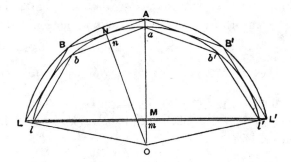

(1) For the surfaces are equal to circles the squares on whose radii are equal respectively to

$$AB\left(BB' + CC' + \ldots + KK' + \frac{LL'}{2} \right),$$ [Prop. 39, Cor.]

and $$ab\left(bb' + cc' + \ldots + kk' + \frac{ll'}{2} \right).$$ [Prop. 35]

But these rectangles are in the ratio of *AB²* to *ab²*. Therefore so are the surfaces.

(2) Let *OnN* be drawn perpendicular to *ab* and *AB*; and suppose the circles which are equal to the surfaces of the outer and inner solids of revolution to be denoted by *S*, *s* respectively.

Now the volume of the circumscribed solid together with the cone *OLL'* is equal to a cone whose base is *S* and whose height is *ON* [Prop. 40, Cor. 1].

And the volume of the inscribed figure with the cone Oll' is equal to a cone with base s and height On [Prop. 38].

But $$S : s = AB^2 : ab^2,$$
and $$ON : On = AB : ab.$$

Therefore the volume of the circumscribed solid together with the cone OLL' is to the volume of the inscribed solid together with the cone Oll' as AB^3 is to ab^3 [*Lemma 5*].

Proposition 42

If lal' be a segment of a sphere less than a hemisphere and Oa the radius perpendicular to the base of the segment, the surface of the segment is equal to a circle whose radius is equal to al.

Let R be a circle whose radius is equal to al. Then the surface of the segment, which we will call S, must, if it be not equal to R, be either greater or less than R.

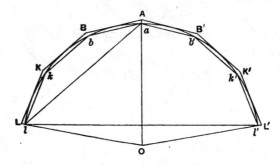

I. Suppose, if possible, $S > R$.

Let lal' be a segment of a great circle which is less than a semicircle. Join Ol, Ol', and let similar polygons with $2n$ equal sides be circumscribed and inscribed to the sector, as in the previous propositions, but such that

(circumscribed polygon) : (inscribed polygon) $< S : R$.   [Prop. 6]

Let the polygons now revolve with the segment about OaA, generating solids of revolution circumscribed and inscribed to the segment of the sphere.

Then

(surface of outer solid) : (surface of inner solid)
$$= AB^2 : ab^2$$ [Prop. 41]
$$= \text{(circumscribed polygon)} : \text{(inscribed polygon)}$$
$$< S : R, \text{ by hypothesis.}$$

But the surface of the outer solid is greater than S [Prop. 39].

Therefore the surface of the inner solid is greater than R; which is impossible, by Prop. 37.

II. Suppose, if possible, $S < R$.

In this case we circumscribe and inscribe polygons such that their ratio is less than $R : S$; and we arrive at the result that

(surface of outer solid) : (surface of inner solid)
$$< R : S.$$

But the surface of the outer solid is greater than R [Prop. 40]. Therefore the surface of the inner solid is greater than S : which is impossible [Prop. 36].

Hence, since S is neither greater nor less than R,
$$S = R.$$

Proposition 43

Even if the segment of the sphere is greater than a hemisphere, its surface is still equal to a circle whose radius is equal to al.

For let $lal'a'$ be a great circle of the sphere, aa' being the diameter perpendicular to ll'; and let $la'l'$ be a segment less than a semicircle.

Then, by Prop. 42, the surface of the segment $la'l'$ of the sphere is equal to a circle with radius equal to $a'l$.

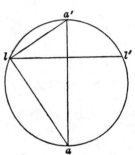

Also the surface of the whole sphere is equal to a circle with radius equal to aa' [Prop. 33].

But $aa'^2 - a'l^2 = al^2$, and circles are to one another as the squares on their radii.

Therefore the surface of the segment lal', being the difference between the surfaces of the sphere and of $la'l'$, is equal to a circle with radius equal to al.

Proposition 44

The volume of any sector of a sphere is equal to a cone whose base is equal to the surface of the segment of the sphere included in the sector, and whose height is equal to the radius of the sphere.

Let R be a cone whose base is equal to the surface of the segment lal' of a sphere and whose height is equal to the radius of the sphere; and let S be the volume of the sector $Olal'$.

Then, if S is not equal to R, it must be either greater or less.

I. Suppose, if possible, that $S > R$.

Find two straight lines β, γ, of which β is the greater, such that
$$\beta : \gamma < S : R;$$
and let δ, ϵ be two arithmetic means between β, γ.

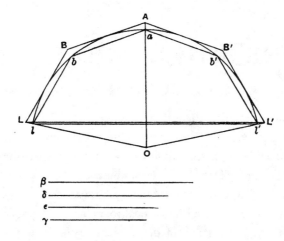

Let lal' be a segment of a great circle of the sphere. Join Ol, Ol', and let similar polygons with $2n$ equal sides be circumscribed and inscribed to the sector of the circle as before, but such that their sides are in a ratio less than $\beta : \delta$. [Prop. 4].

Then let the two polygons revolve with the segment about OaA, generating two solids of revolution.

Denoting the volumes of these solids by V, v respectively, we have

$$(V + \text{cone } OLL') : (v + \text{cone } Oll') = AB^3 : ab^3 \qquad \text{[Prop. 41]}$$
$$< \beta^3 : \delta^3$$
$$< \beta : \gamma, \textit{a fortiori}, [11]$$
$$< S : R, \text{by hypothesis.}$$

Now $\qquad\qquad (V + \text{cone } OLL') > S.$

Therefore also $\qquad (v + \text{cone } Oll') > R.$

But this is impossible, by Prop. 38, Cor. combined with Props. 42, 43.

Hence $\qquad\qquad\qquad S \not> R.$

II. Suppose, if possible, that $S < R$.

In this case we take β, γ such that

$$\beta : \gamma < R : S,$$

and the rest of the construction proceeds as before.

We thus obtain the relation

$$(V + \text{cone } OLL') : (v + \text{cone } Oll') < R : S.$$

Now $\qquad\qquad (v + \text{cone } Oll') < S.$

Therefore $\qquad (V + \text{cone } OLL') < R;$

which is impossible, by Prop. 40, Cor. 2 combined with Props. 42, 43.

Since then S is neither greater nor less than R,

$$S = R.$$

[11] Cf. note on Prop. 34, p. 157.

ON THE SPHERE AND CYLINDER
BOOK II

"ARCHIMEDES to Dositheus greeting.

On a former occasion you asked me to write out the proofs of the problems the enunciations of which I had myself sent to Conon. In point of fact they depend for the most part on the theorems of which I have already sent you the demonstrations, namely (1) that the surface of any sphere is four times the greatest circle in the sphere, (2) that the surface of any segment of a sphere is equal to a circle whose radius is equal to the straight line drawn from the vertex of the segment to the circumference of its base, (3) that the cylinder whose base is the greatest circle in any sphere and whose height is equal to the diameter of the sphere is itself in magnitude half as large again as the sphere, while its surface [including the two bases] is half as large again as the surface of the sphere, and (4) that any solid sector is equal to a cone whose base is the circle which is equal to the surface of the segment of the sphere included in the sector, and whose height is equal to the radius of the sphere. Such then of the theorems and problems as depend on these theorems I have written out in the book which I send herewith; those which are discovered by means of a different sort of investigation, those namely which relate to spirals and the conoids, I will endeavour to send you soon.

The first of the problems was as follows: *Given a sphere, to find a plane area equal to the surface of the sphere.*

The solution of this is obvious from the theorems aforesaid. For four times the greatest circle in the sphere is both a plane area and equal to the surface of the sphere.

The second problem was the following."

Proposition 1 (Problem)

Given a cone or a cylinder, to find a sphere equal to the cone or to the cylinder.

If V be the given cone or cylinder, we can make a cylinder equal to $\frac{3}{2}V$. Let this cylinder be the cylinder whose base is the circle on AB as diameter and whose height is OD.

Now, if we could make another cylinder, equal to the cylinder (OD) but such that its height is equal to the diameter of its base, the problem would be solved, because this latter cylinder would be equal to $\frac{3}{2}V$, and the sphere whose diameter is equal to the height (or to the diameter of the base) of the same cylinder would then be the sphere required [i. 34, Cor.].

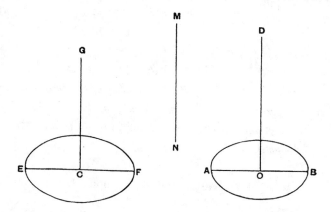

Suppose the problem solved, and let the cylinder (CG) be equal to the cylinder (OD), while EF, the diameter of the base, is equal to the height CG.

Then, since in equal cylinders the heights and bases are reciprocally proportional,

$$AB^2 : EF^2 = CG : OD$$
$$= EF : OD \dotfill (1).$$

Suppose MN to be such a line that

$$EF^2 = AB \cdot MN \dotfill (2).$$

Hence $\qquad AB : EF = EF : MN,$

and, combining (1) and (2), we have

$$AB : MN = EF : OD,$$

or $\qquad\qquad AB : EF = MN : OD.$

Therefore $AB : EF = EF : MN = MN : OD,$

and EF, MN *are two mean proportionals between* AB, OD.

The synthesis of the problem is therefore as follows. Take two mean proportionals EF, MN between AB and OD, and describe a cylinder whose base is a circle on EF as diameter and whose height CG is equal to EF.

Then, since

$$AB : EF = EF : MN = MN : OD,$$
$$EF^2 = AB \cdot MN,$$

and therefore $\qquad AB^2 : EF^2 = AB : MN$
$$= EF : OD$$
$$= CG : OD;$$

whence the bases of the two cylinders (OD), (CG) are reciprocally proportional to their heights.

Therefore the cylinders are equal, and it follows that

$$\text{cylinder } (CG) = \tfrac{3}{2}V.$$

169

The sphere on *EF* as diameter is therefore the sphere required, being equal to *V.*

Proposition 2

If BAB' be a segment of a sphere, BB' a diameter of the base of the segment, and O the centre of the sphere, and if AA' be the diameter of the sphere bisecting BB' in M, then the volume of the segment is equal to that of a cone whose base is the same as that of the segment and whose height is h, where

$$h : AM = OA' + A'M : A'M.$$

Measure *MH* along *MA* equal to *h*, and *MH'* along *MA'* equal to *h'*, where

$$h' : A'M = OA + AM : AM.$$

Suppose the three cones constructed which have *O*, *H H'* for their apices and the base (*BB'*) of the segment for their common base. Join *AB*, *A'B*.

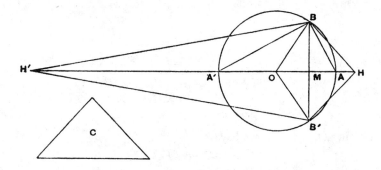

Let *C* be a cone whose base is equal to the surface of the segment *BAB'* of the sphere, i.e. to a circle with radius equal to *AB* [i. 42], and whose height is equal to *OA*.

Then the cone *C* is equal to the solid sector *OBAB'* [i. 44].

Now, since $\qquad HM : MA = OA' + A'M : A'M,$

dividendo, $\qquad HA : AM = OA : A'M,$

and, alternately, $\qquad HA : AO = AM : MA',$

so that

$$HO : OA = AA' : A'M$$
$$= AB^2 : BM^2$$
$$= (\text{base of cone } C) : (\text{circle on } BB' \text{ as diameter}).$$

But *OA* is equal to the height of the cone *C*; therefore, since cones are equal if their bases and heights are reciprocally proportional, it follows that the cone *C* (or the solid sector *OBAB'*) is equal to a cone whose base is the circle on *BB'* as diameter and whose height is equal to *OH.*

And this latter cone is equal to the sum of two others having the same base and with heights OM, MH, i.e. to the solid rhombus $OBHB'$.

Hence the sector $OBAB'$ is equal to the rhombus $OBHB'$.

Taking away the common part, the cone OBB',

$$\text{the segment } BAB' = \text{the cone } HBB'.$$

Similarly, by the same method, we can prove that

$$\text{the segment } BA'B' = \text{the cone } H'BB'.$$

Alternative proof of the latter property.

Suppose D to be a cone whose base is equal to the surface of the whole sphere and whose height is equal to OA.

Thus D is equal to the volume of the sphere.   [i. 33, 34]

Now, since $OA' + A'M : A'M = HM : MA$,

dividendo and *alternando*, as before,

$$OA : AH = A'M : MA.$$

Again, since $\quad\quad\quad H'M : MA' = OA + AM : AM$,

$$H'A' : OA = A'M : MA$$
$$= OA : AH, \text{ from above.}$$

Componendo, $\quad\quad\quad H'O : OA = OH : HA \ldots\ldots\ldots(1).$

Alternately, $\quad\quad\quad H'O : OH = OA : AH \ldots\ldots\ldots(2),$

and, *componendo*, $\quad\quad HH' : HO = OH : HA$,

$$= H'O : OA, \text{ from (1)},$$

whence $\quad\quad\quad HH' \cdot OA = H'O \cdot OH \ldots\ldots\ldots(3).$

Next, since $\quad\quad\quad H'O : OH = OA : AH$, by (2),

$$= A'M : MA,$$

$$(H'O + OH)^2 : H'O \cdot OH = (A'M + MA)^2 : A'M \cdot MA,$$

whence, by means of (3),

$$HH'^2 : HH' \cdot OA = AA'^2 : A'M \cdot MA,$$

or $\quad\quad\quad HH' : OA = AA'^2 : BM^2.$

Now the cone D, which is equal to the sphere, has for its base a circle whose radius is equal to AA', and for its height a line equal to OA.

Hence this cone D is equal to a cone whose base is the circle on BB' as diameter and whose height is equal to HH';

therefore $\quad\quad$ the cone $D = $ the rhombus $HBH'B'$,

or $\quad\quad\quad$ the rhombus $HBH'B' = $ the sphere.

But $\quad\quad\quad$ the segment $BAB' = $ the cone HBB';

therefore the remaining segment $BA'B' = $ the cone $H'BB'$.

COR. *The segment BAB' is to a cone with the same base and equal height in the ratio of $OA' + A'M$ to $A'M$.*

Proposition 3 (Problem)

To cut a given sphere by a plane so that the surfaces of the segments may have to one another a given ratio.

Suppose the problem solved. Let AA' be a diameter of a great circle of the sphere, and suppose that a plane perpendicular to AA' cuts the plane of the great

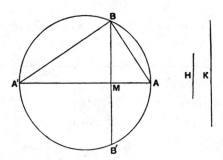

circle in the straight line BB', and AA' in M, and that it divides the sphere so that the surface of the segment BAB' has to the surface of the segment $BA'B'$ the given ratio.

Now these surfaces are respectively equal to circles with radii equal to AB, $A'B$ [i. 42, 43].

Hence the ratio $AB^2 : A'B^2$ is equal to the given ratio, i.e. AM is to MA' in the given ratio.

Accordingly the synthesis proceeds as follows.

If $H : K$ be the given ratio, divide AA' in M so that
$$AM : MA' = H : K.$$
Then $AM : MA' = AB^2 : A'B^2$
$$= \text{(circle with radius } AB) : \text{(circle with radius } A'B)$$
$$= \text{(surface of segment } BAB') : \text{(surface of segment } BA'B').$$
Thus the ratio of the surfaces of the segments is equal to the ratio $H : K$.

Proposition 4 (Problem)

To cut a given sphere by a plane so that the volumes of the segments are to one another in a given ratio.

Suppose the problem solved, and let the required plane cut the great circle ABA' at right angles in the line BB'. Let AA' be that diameter of the great circle which bisects BB' at right angles (in M), and let O be the centre of the sphere.

Take H on OA produced, and H' on OA' produced, such that
$$OA' + A'M : A'M = HM : MA, \dots\dots\dots\dots\dots\dots\dots(1),$$
and
$$OA + AM : AM = H'M : MA' \dots\dots\dots\dots\dots\dots(2).$$

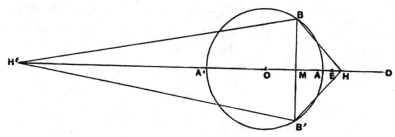

Join BH, $B'H$, BH', $B'H'$.

Then the cones HBB', $H'BB'$ are respectively equal to the segments BAB', $BA'B'$ of the sphere [Prop. 2].

Hence the ratio of the cones, and therefore of their altitudes, is given, i.e.

$$HM : H'M = \text{the given ratio} \quad\dots\dots(3).$$

We have now three equations (1), (2), (3), in which there appear three as yet undetermined points M, H, H'; and it is first necessary to find, by means of them, another equation in which only one of these points (M) appears, i.e. we have, so to speak, to *eliminate* H, H'.

Now, from (3), it is clear that $HH' : H'M$ is also a given ratio; and Archimedes' method of elimination is, *first*, to find values for each of the ratios $A'H' : H'M$ and $HH' : H'A$ which are alike independent of H, H', and then, *secondly*, to equate the ratio compounded of these two ratios to the known value of the ratio $HH' : H'M$.

(*a*) To find such a value for $A'H' : H'M$.

It is at once clear from equation (2) above that

$$A'H' : H'M = OA : OA + AM \quad\dots\dots(4).$$

(*b*) To find such a value for $HH' : A'H'$.

From (1) we derive

$$A'M : MA = OA' + A'M : HM$$
$$= OA' : AH \quad\dots\dots(5);$$

and, from (2),

$$A'M : MA = H'M : OA + AM$$
$$= A'H' : OA \quad\dots\dots(6).$$

Thus

$$HA : AO = OA' : A'H',$$

whence

$$OH : OA' = OH' : A'H',$$

or

$$OH : OH' = OA' : A'H'.$$

It follows that

$$HH' : OH' = OH' : A'H',$$

or

$$HH' \cdot H'A' = OH'^2.$$

Therefore

$$HH' : H'A' = OH'^2 : H'A'^2$$
$$= AA'^2 : A'M^2, \text{ by means of (6)}$$

(c) To express the ratios $A'H' : H'M$ and $HH' : H'M$ more simply we make the following construction. Produce OA to D so that $OA = AD$. (D will lie beyond H, for $A'M > MA$, and therefore, by (5), $OA > AH$.)

Then
$$A'H' : H'M = OA : OA + AM$$
$$= AD : DM \dots\dots\dots\dots\dots\dots\dots\dots\dots\dots\dots\dots\dots(7).$$

Now divide AD at E so that
$$HH' : H'M = AD : DE \dots\dots\dots\dots\dots\dots\dots\dots\dots\dots\dots(8).$$

Thus, using equations (8), (7) and the value of $HH' : H'A'$ above found, we have
$$AD : DE = HH' : H'M$$
$$= (HH' : H'A') \cdot (A'H' : H'M)$$
$$= (AA'^2 : A'M^2) \cdot (AD : DM).$$

But
$$AD : DE = (DM : DE) \cdot (AD : DM).$$

Therefore
$$MD : DE = AA'^2 : A'M^2 \dots\dots\dots\dots\dots\dots\dots\dots\dots\dots(9).$$

And D is given, since $AD = OA$. Also $AD : DE$ (being equal to $HH' : H'M$) is a given ratio. Therefore DE is given.

Hence the problem reduces itself to the problem of dividing $A'D$ into two parts at M so that
$$MD : (\text{a given length}) = (\text{a given area}) : A'M^2.$$

Archimedes adds: "If the problem is propounded in this general form, it requires a διορισμός [i.e. it is necessary to investigate the limits of possibility], but, if there be added the conditions subsisting in the present case, it does not require a διορισμός."

In the present case the problem is:

Given a straight line $A'A$ produced to D so that $A'A = 2AD$, and given a point E on AD, to cut AA' in a point M so that
$$AA'^2 : A'M^2 = MD : DE.$$

"And the analysis and synthesis of both problems will be given at the end."[12]

The synthesis of the main problem will be as follows. Let $R : S$ be the given ratio, R being less than S. AA' being a diameter of a great circle, and O the centre, produce OA to D so that $OA = AD$, and divide AD in E so that
$$AE : ED = R : S.$$

Then cut AA' in M so that
$$MD : DE = AA'^2 : A'M^2.$$

Through M erect a plane perpendicular to AA'; this plane will then divide the sphere into segments which will be to one another as R to S.

---

[12]See the note following this proposition.

Take H on $A'A$ produced, and H' on AA' produced, so that

$$OA' + A'M : A'M = HM : MA, \dots\dots\dots\dots\dots\dots\dots(1),$$
$$OA + AM : AM = H'M : MA' \dots\dots\dots\dots\dots\dots(2).$$

We have then to show that

$$HM : MH' = R : S, \text{ or } AE : ED.$$

(a) We first find the value of $HH' : H'A'$ as follows.

As was shown in the analysis (b),

$$HH' \cdot H'A' = OH'^2,$$

or
$$HH' : H'A' = OH'^2 : H'A'^2$$
$$= AA'^2 : A'M^2$$
$$= MD : DE, \text{ by construction.}$$

(β) Next we have

$$H'A' : H'M = OA : OA + AM$$
$$= AD : DM.$$

Therefore
$$HH' : H'M = (HH' : H'A') \cdot (H'A' : H'M)$$
$$= (MD : DE) \cdot (AD : DM)$$
$$= AD : DE,$$

whence
$$HM : MH' = AE : ED$$
$$= R : S.$$

Q.E.D.

Note. The solution of the subsidiary problem to which the original problem of Prop. 4 is reduced, and of which Archimedes promises a discussion, is given in a highly interesting and important note by Eutocius, who introduces the subject with the following explanation.

"He [Archimedes] promised to give a solution of this problem at the end, but we do not find the promise kept in any of the copies. Hence we find that Dionysodorus too failed to light upon the promised discussion and, being unable to grapple with the omitted lemma, approached the original problem in a different way, which I shall describe later. Diocles also expressed in his work περὶ πυρίων the opinion that Archimedes made the promise but did not perform it, and tried to supply the omission himself. His attempt I shall also give in its order. It will however be seen to have no relation to the omitted discussion but to give, like Dionysodorus, a construction arrived at by a different method of proof. On the other hand, as the result of unremitting and extensive research, I found in a certain old book some theorems discussed which, although the reverse of clear owing to errors and in many ways faulty as regards the figures, nevertheless gave the substance of what I sought, and moreover to some extent kept to the Doric dialect affected by Archimedes, while they retained the names familiar in old usage, the parabola being called a section of a right-angled cone, and the hyperbola a section of an obtuse-angled cone; whence I was led to consider whether these theorems might not in fact be what he promised he would give at the end. For this reason I paid them the closer attention, and, after finding great difficulty with the actual text owing to the multitude of the mistakes above referred to, I made out the sense gradually and now proceed to set it out, as well as I can, in more familiar and clearer language. And first the theorem will be treated generally, in order that what Archimedes says about the limits of possibility may be made clear; after which there will follow the special application to the conditions stated in his analysis of the problem."

The investigation which follows may be thus reproduced. The general problem is: *Given two straight lines AB, AC and an area D, to divide AB at M so that*

$$AM : AC = D : MB^2.$$

Analysis.

Suppose M found, and suppose AC placed at right angles to AB. Join CM and produce it. Draw EBN through B parallel to AC meeting CM in N, and through C draw CHE parallel to AB meeting EBN in E. Complete the parallelogram $CENF$, and through M draw PMH parallel to AC meeting FN in P.

Measure EL along EN so that $CE \cdot EL$ (or $AB \cdot EL$) = D.

Then, by hypothesis,

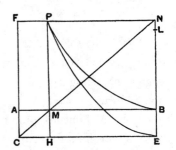

$$AM : AC = CE \cdot EL : MB^2.$$

And $AM : AC = CE : EN,$

by similar triangles,

$$= CE \cdot EL : EL \cdot EN.$$

It follows that $PN^2 = MB^2 = EL \cdot EN.$
Hence, if a parabola be described with vertex E, axis EN, and parameter equal to EL, it will pass through P; and it will be given in position, since EL is given.

Therefore P lies on a given parabola.

Next, since the rectangles FH, AE are equal,

$$FP \cdot PH = AB \cdot BE.$$

Hence, if a rectangular hyperbola be described with CE, CF as asymptotes and passing through B, it will pass through P. And the hyperbola is given in position.

Therefore P lies on a given hyperbola.

Thus P is determined as the intersection of the parabola and hyperbola. And since P is thus given, M is also given.

διορισμός.

Now, since
$$AM : AC = D : MB^2,$$
$$AM \cdot MB^2 = AC \cdot D.$$

But $AC \cdot D$ is given, and *it will be proved later that the maximum value of $AM \cdot MB^2$ is that which it assumes when $BM = 2AM$.*

Hence *it is a necessary condition of the possibility of a solution that $AC \cdot D$ must not be greater than* $\frac{1}{3}AB \cdot (\frac{2}{3}AB)^2$, *or* $\frac{4}{27}AB^2$.

Synthesis.

If O be such a point on AB that $BO = 2AO$, we have seen that, in order that the solution may be possible,
$$AC \cdot D \not> AO \cdot OB^2.$$
Thus $AC \cdot D$ is either equal to, or less than, $AO \cdot OB^2$.

(1) if $AC \cdot D = AO \cdot OB^2$, then the point O itself solves the problem.

(2) Let $AC \cdot D$ be less than $AO \cdot OB^2$.

Place AC at right angles to AB. Join CO, and produce it to R. Draw EBR through B parallel to AC meeting CO in R, and through C draw CE parallel to AB meeting EBR in E. Complete the parallelogram $CERF$, and through O draw QOK parallel to AC meeting FR in Q and CE in K.

Then, since
$$AC \cdot D < AO \cdot OB^2,$$
measure RQ' along RQ so that
$$AC \cdot D = AO \cdot Q'R^2,$$
or $\qquad\qquad AO : AC = D : Q'R^2.$

Measure EL along ER so that
$$D = CE \cdot EL \text{ (or } AB \cdot EL).$$
Now, since $\qquad AO : AC = D : Q'R^2$, by hypothesis,
$$= CE \cdot EL : Q'R^2,$$
and $\qquad\qquad AO : AC = CE : ER$, by similar triangles,
$$= CE \cdot EL : EL \cdot ER,$$
it follows that
$$Q'R^2 = EL \cdot ER.$$
Describe a parabola with vertex E, axis ER, and parameter equal to EL. This parabola will then pass through Q'.

Again, $\qquad\qquad$ rect. $FK = $ rect. AE,
or $\qquad\qquad FQ \cdot QK = AB \cdot BE$;
and, if we describe a rectangular hyperbola with asymptotes CE, CF and passing through B, it will also pass through Q.

Let the parabola and hyperbola intersect at P, and through P draw PMH parallel to AC meeting AB in M and CE in H, and GPN parallel to AB meeting CF in G and ER in N.

Then shall M be the required point of division.

Since
$$PG \cdot PH = AB \cdot BE,$$
$$\text{rect. } GM = \text{rect. } ME,$$
and therefore CMN is a straight line.

Thus
$$AB \cdot BE = PG \cdot PH = AM \cdot EN \quad\text{.................................(1)}.$$

Again, by the property of the parabola,
$$PN^2 = EL \cdot EN,$$
or
$$MB^2 = EL \cdot EN \quad\text{...(2)}.$$

From (1) and (2)
$$AM : EL = AB \cdot BE : MB^2,$$
or
$$AM \cdot AB : AB \cdot EL = AB \cdot AC : MB^2.$$

Alternately,
$$AM \cdot AB : AB \cdot AC = AB \cdot EL : MB^2,$$
or
$$AM : AC = D : MB^2.$$

Proof of διορισμός.

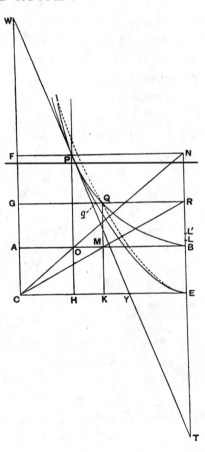

It remains to be proved that, *if AB be divided at O so that $BO = 2AO$, then $AO \cdot OB^2$ is the maximum value of $AM \cdot MB^2$,*
or
$$AO \cdot OB^2 > AM \cdot MB^2,$$
where M is any point on AB other than O.

Suppose that
$$AO : AC = CE \cdot EL' : OB^2,$$
so that $AO \cdot OB^2 = CE \cdot EL' \cdot AC$.
Join CO, and produce it to N; draw EBN through B parallel to AC, and complete the parallelogram $CENF$.

Through O draw POH parallel to AC meeting FN in P and CE in H.

With vertex E, axis EN, and parameter EL', describe a parabola. This will pass through P, as shown in the analysis above, and beyond P will meet the diameter CF of the parabola in some point.

Next draw a rectangular hyperbola with asymptotes CE, CF and passing through B. This hyperbola will also pass through P, as shown in the analysis.

Produce NE to T so that $TE = EN$. Join TP meeting CE in Y, and produce it to meet CF in W. Thus TP will touch the parabola at P.

Then, since $\quad BO = 2AO$,

$$TP = 2PW.$$

And $\qquad TP = 2PY.$

Therefore $\qquad PW = PY.$

Since, then, WY between the asymptotes is bisected at P, the point where it meets the hyperbola,

$$WY \text{ is a tangent to the hyperbola.}$$

Hence the hyperbola and parabola, having a common tangent at P, touch one another at P.

Now take any point M on AB, and through M draw QMK parallel to AC meeting the hyperbola in Q and CE in K. Lastly, draw $GqQR$ through Q parallel to AB meeting CF in G, the parabola in q, and EN in R.

Then, since, by the property of the hyperbola, the rectangles GK, AE are equal, CMR is a straight line.

By the property of the parabola,

$$qR^2 = EL' \cdot ER,$$

so that $\qquad QR^2 < EL' \cdot ER.$

Suppose $\qquad QR^2 = EL \cdot ER,$

and we have $\qquad AM : AC = CE : ER$

$$= CE \cdot EL : EL \cdot ER$$
$$= CE \cdot EL : QR^2$$
$$= CE \cdot EL : MB^2,$$

or $\qquad AM \cdot MB^2 = CE \cdot EL \cdot AC.$

Therefore $\qquad AM \cdot MB^2 < CE \cdot EL' \cdot AC$

$$< AO \cdot OB^2.$$

If $AC \cdot D < AO \cdot OB^2$, there are two solutions because there will be two points of intersection between the parabola and the hyperbola.

For, if we draw with vertex E and axis EN a parabola whose parameter is equal to EL, the parabola will pass through the point Q (see the last figure); and, since the parabola meets the diameter CF beyond Q, it must meet the hyperbola again (which has CF for its asymptote).

[If we put $AB = a$, $BM = x$, $AC = c$, and $D = b^2$, the proportion

$$AM : AC = D : MB^2$$

is seen to be equivalent to the equation

$$x^2(a - x) = b^2c,$$

being a *cubic equation* with the term containing x omitted.

Now suppose EN, EC to be axes of coordinates, EN being the axis of y.

Then the parabola used in the above solution is the parabola

$$x^2 = \frac{b^2}{a} \cdot y,$$

and the rectangular hyperbola is

$$y(a - x) = ac.$$

Thus the solution of the cubic equation and the conditions under which there are no positive solutions, or one, or two positive solutions are obtained by the use of the two conics.]

For the sake of completeness, and for their intrinsic interest, the solutions of the original problem in Prop. 4 given by Dionysodorus and Diocles are here appended.

Dionysodorus' solution.

Let AA' be a diameter of the given sphere. It is required to find a plane cutting AA' at right angles (in a point M, suppose) so that the segments into which the sphere is divided are in a given ratio, as $CD : DE$.

Produce $A'A$ to F so that $AF = OA$, where O is the centre of the sphere.

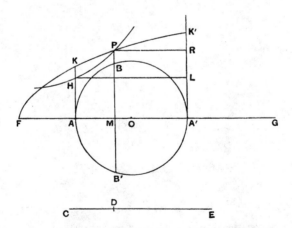

Draw AH perpendicular to AA' and of such length that

$$FA : AH = CE : ED,$$

and produce AH to K so that

$$AK^2 = FA \cdot AH \quad\text{................................}(\alpha).$$

With vertex F, axis FA, and parameter equal to AH describe a parabola. This will pass through K, by the equation (α).

Draw $A'K'$ parallel to AK and meeting the parabola in K'; and with $A'F$, $A'K'$ as asymptotes describe a rectangular hyperbola passing through H. This hyperbola will meet the parabola at some point, as P, between K and K'.

Draw PM perpendicular to AA' meeting the great circle in B, B', and from H, P draw HL, PR both parallel to AA' and meeting $A'K'$ in L, R respectively.

Then, by the property of the hyperbola,

$$PR \cdot PM = AH \cdot HL,$$

i.e. $$PM \cdot MA' = HA \cdot AA',$$

or $$PM : AH = AA' : A'M,$$

and $$PM^2 : AH^2 = AA'^2 : A'M^2.$$

Also, by the property of the parabola,

$$PM^2 = FM \cdot AH,$$

i.e. $$FM : PM = PM : AH,$$

or $$FM : AH = PM^2 : AH^2$$
$$= AA'^2 : A'M^2, \text{ from above.}$$

Thus, since circles are to one another as the squares of their radii, the cone whose base is the circle with $A'M$ as radius and whose height is equal to FM, and the cone whose base is the circle with AA' as radius and whose height is equal to AH, have their bases and heights reciprocally proportional.

Hence the cones are equal; i.e., if we denote the first cone by the symbol $c(A'M)$, FM, and so on,

$$c(A'M), FM = c(AA'), AH.$$

Now $$c(AA'), FA : c(AA'), AH = FA : AH$$
$$= CE : ED, \text{ by construction.}$$

Therefore

$$c(AA'), FA : c(A'M), FM = CE : ED \dots\dots\dots\dots\dots(\beta).$$

But (1) $$c(AA'), FA = \text{the sphere.} \qquad \text{[i. 34]}$$

(2) $c(A'M)$, FM can be proved equal to the segment of the sphere whose vertex is A' and height $A'M$.

For take G on AA' produced such that

$$GM : MA' = FM : MA$$
$$= OA + AM : AM.$$

Then the cone GBB' is equal to the segment $A'BB'$ [Prop. 2].

And $$FM : MG = AM : MA', \text{ by hypothesis,}$$
$$= BM^2 : A'M^2.$$

Therefore

$$(\text{circle with rad. } BM) : (\text{circle with rad. } A'M)$$
$$= FM : MG,$$

so that $\qquad c(A'M), FM = c(BM), MG$
$$= \text{the segment } A'BB'.$$

We have therefore, from the equation (β) above,

$$(\text{the sphere}) : (\text{segmt. } A'BB') = CE : ED,$$

whence $(\text{segmt. } ABB') : (\text{segmt. } A'BB') = CD : DE.$

Diocles' solution.

Diocles starts, like Archimedes, from the property, proved in Prop. 2, that, if the plane of section cut a diameter AA' of the sphere at right angles in M, and if H, H' be taken on OA, OA' produced respectively so that

$$OA' + A'M : A'M = HM : MA,$$
$$OA + AM : AM = H'M : MA',$$

then the cones HBB', $H'BB'$ are respectively equal to the segments ABB', $A'BB'$.

Then, drawing the inference that

$$HA : AM = OA' : A'M,$$
$$H'A' : A'M = OA : AM,$$

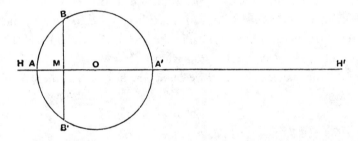

he proceeds to state the problem in the following form, slightly generalising it by the substitution of *any* given straight line for OA or OA':

Given a straight line AA', its extremities A, A', a ratio $C : D$, and another straight line as AK, to divide AA' at M and to find two points H, H' on $A'A$ and AA' produced respectively so that the following relations may hold simultaneously,

$$C : D = HM : MH' \qquad \dots\dots\dots\dots\dots\dots\dots(\alpha),$$
$$HA : AM = AK : A'M \qquad \dots\dots\dots\dots\dots\dots\dots(\beta),$$
$$H'A' : A'M = AK : AM \qquad \dots\dots\dots\dots\dots\dots\dots(\gamma).$$

Analysis.

Suppose the problem solved and the points M, H, H' all found.

Place AK at right angles to AA', and draw $A'K'$ parallel and equal to AK. Join KM, $K'M$, and produce them to meet $K'A'$, KA respectively in E, F. Join KK', draw EG through E parallel to $A'A$ meeting KF in G, and through M draw QMN parallel to AK meeting EG in Q and KK' in N.

Now $\qquad HA : AM = A'K' : A'M$, by (β),
$$= FA : AM, \text{ by similar triangles,}$$
whence $\qquad HA = FA.$
Similarly $\qquad H'A' = A'E.$
Next,
$$FA + AM : A'K' + A'M = AM : A'M$$
$$= AK + AM : EA' + A'M, \text{ by similar triangles.}$$
Therefore
$$(FA + AM) \cdot (EA' + A'M) = (KA + AM) \cdot (K'A' + A'M).$$
Take AR along AH and $A'R'$ along $A'H'$ such that
$$AR = A'R' = AK.$$
Then, since $FA + AM = HM$, $EA' + A'M = MH'$, we have
$$HM \cdot MH' = RM \cdot MR' \quad \dots\dots\dots\dots\dots\dots\dots\dots\dots\dots(\delta).$$
(Thus, if R falls between A and H, R' falls on the side of H' remote from A', and *vice versa*.)

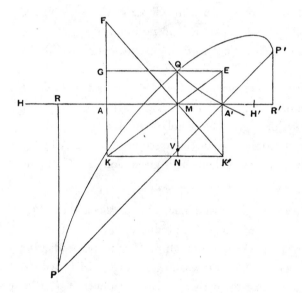

Now $\qquad C : D = HM : MH'$, by hypothesis,
$$= HM \cdot MH' : MH'^2$$
$$= RM \cdot MR' : MH'^2, \text{ by } (\delta).$$
Measure MV along MN so that $MV = A'M$. Join $A'V$ and produce it both ways. Draw RP, $R'P'$ perpendicular to RR' meeting $A'V$ produced in P, P' respectively. Then, the angle $MA'V$ being half a right angle, PP' is given in position, and, since R, R' are given, so are P, P'.

And, by parallels,
$$P'V : PV = R'M : MR.$$
Therefore $\quad\quad PV \cdot P'V : PV^2 = RM \cdot MR' : RM^2.$

But $\quad\quad\quad\quad\quad PV^2 = 2RM^2.$

Therefore $\quad\quad PV \cdot P'V = 2RM \cdot MR'.$

And it was shown that
$$RM \cdot MR' : MH'^2 = C : D.$$
Hence $\quad\quad\quad PV \cdot P'V : MH'^2 = 2C : D.$

But $\quad\quad MH' = A'M + A'E = VM + MQ = QV.$

Therefore $\quad\quad QV^2 : PV \cdot P'V = D : 2C,$ a given ratio.

Thus, if we take a line p such that
$$D : 2C = p : PP', [13]$$
and if we describe an ellipse with PP' as a diameter and p as the corresponding parameter [$= DD'^2/PP'$ in the ordinary notation of geometrical conics], and such that the ordinates to PP' are inclined to it at an angle equal to half a right angle, i.e. are parallel to QV or AK, then the ellipse will pass through Q.

Hence Q lies on an ellipse given in position.

Again, since EK is a diagonal of the parallelogram GK',
$$GQ \cdot QN = AA' \cdot A'K'.$$

If therefore a rectangular hyperbola be described with KG, KK' as asymptotes and passing through A', it will also pass through Q.

Hence Q lies on a given rectangular hyperbola.

Thus Q is determined as the intersection of a given ellipse and a given hyperbola, and is therefore given. Thus M is given, and H, H' can at once be found.

Synthesis.

Place AA', AK at right angles, draw $A'K'$ parallel and equal to AK, and join KK'.

Make AR (measured along $A'A$ produced) and $A'R'$ (measured along AA' produced) each equal to AK, and through R, R' draw perpendiculars to RR'.

Then through A' draw PP' making an angle $(AA'P)$ with AA' equal to half a right angle and meeting the perpendiculars just drawn in P, P' respectively.

Take a length p such that
$$D : 2C = p : PP', [14]$$

[13] There is a mistake in the Greek text here which seems to have escaped the notice of all the editors up to the present. The words are ἐὰν ἄρα ποιήσωμεν, ὡς τὴν Δ πρὸς τὴν διπλασίαν τῆς Γ, οὕτως τὴν ΤΥ πρὸς ἄλλην τινὰ ὡς τὴν Φ, i.e. (with the lettering above) "If we take a length p such that $D : 2C = PP' : p$." This cannot be right, because we should then have
$$QV^2 : PV \cdot P'V = PP' : p,$$
whereas the two latter terms should be reversed, the correct property of the ellipse being
$$QV^2 : PV \cdot P'V = p : PP'. \quad\quad\quad\quad\quad\quad\quad\text{[Apollonius i. 21]}$$
The mistake would appear to have originated as far back as Eutocius, but I think that Eutocius is more likely to have made the slip than Diocles himself, because any intelligent mathematician would be more likely to make such a slip in writing out another man's work than to overlook it if made by another.

[14] Here too the Greek text repeats the same error as that noted above.

and with PP' as diameter and p as the corresponding parameter describe an ellipse such that the ordinates to PP' are inclined to it at an angle equal to $AA'P$, i.e. are parallel to AK.

With asymptotes KA, KK' draw a rectangular hyperbola passing through A'.

Let the hyperbola and ellipse meet in Q, and from Q draw $QMVN$ perpendicular to AA' meeting AA' in M, PP' in V and KK' in N. Also draw GQE parallel to AA' meeting AK, $A'K'$ respectively in G, E.

Produce KA, $K'M$ to meet in F.

Then, from the property of the hyperbola,

$$GQ \cdot QN = AA' \cdot A'K',$$

and, since these rectangles are equal, KME is a straight line.

Measure AH along AR equal to AF, and $A'H'$ along $A'R'$ equal to $A'E$.

From the property of the ellipse,

$$QV^2 : PV \cdot P'V = p : PP'$$
$$= D : 2C.$$

And, by parallels,

$$PV : P'V = RM : R'M,$$

or
$$PV \cdot P'V : P'V^2 = RM \cdot MR' : R'M^2,$$

while $P'V^2 = 2R'M^2$, since the angle $RA'P$ is half a right angle.

Therefore
$$PV \cdot P'V = 2RM \cdot MR',$$

whence
$$QV^2 : 2RM \cdot MR' = D : 2C.$$

But
$$QV = EA' + A'M = MH'.$$

Therefore
$$RM \cdot MR' : MH'^2 = C : D.$$

Again, by similar triangles,

$$FA + AM : K'A' + A'M = AM : A'M$$
$$= KA + AM : EA' + A'M.$$

Therefore

$$(FA + AM) \cdot (EA' + A'M) = (KA + AM) \cdot (K'A' + A'M)$$

or
$$HM \cdot MH' = RM \cdot MR'.$$

It follows that

$$HM \cdot MH' : MH'^2 = C : D,$$

or
$$HM : MH' = C : D \dots\dots\dots\dots\dots\dots\dots\dots\dots (\alpha).$$

Also
$$HA : AM = FA : AM,$$
$$= A'K' : A'M, \text{ by similar}$$
$$\text{triangles} \dots (\beta),$$

and
$$H'A : A'M = EA' : A'M$$
$$= AK : AM \dots\dots\dots\dots\dots\dots\dots\dots\dots (\gamma).$$

Hence the points M, H, H' satisfy the three given relations.

Proposition 5 (Problem)

To construct a segment of a sphere similar to one segment and equal in volume to another.

Let *ABB'* be one segment whose vertex is *A* and whose base is the circle on *BB'* as diameter; and let *DEF* be another segment whose vertex is *D* and whose base is the circle on *EF* as diameter. Let *AA'*, *DD'* be diameters of the great circles passing through *BB'*, *EF* respectively, and let *O*, *C* be the respective centres of the spheres.

Suppose it required to draw a segment similar to *DEF* and equal in volume to *ABB'*.

Analysis.

Suppose the problem solved, and let *def* be the required segment, *d* being the vertex and *ef* the diameter of the base. Let *dd'* be the diameter of the sphere which bisects *ef* at right angles, *c* the centre of the sphere.

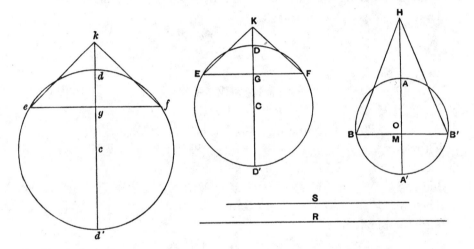

Let *M*, *G*, *g* be the points where *BB'*, *EF*, *ef* are bisected at right angles by *AA'*, *DD'*, *dd'* respectively, and produce *OA*, *CD*, *cd* respectively to *H*, *K*, *k*, so that

$$\left. \begin{array}{c} OA' + A'M : A'M = HM : MA \\ CD' + D'G : D'G = KG : GD \\ cd' + d'g : d'g = kg : gd \end{array} \right\} ,$$

and suppose cones formed with vertices *H*, *K*, *k* and with the same bases as the respective segments. The cones will then be equal to the segments respectively [Prop. 2].

Therefore, by hypothesis,

the cone *HBB'* = the cone *kef*.

Hence

(circle on diameter *BB'*) : (circle on diameter *ef*) = *kg* : *HM*,

so that $BB'^2 : ef^2 = kg : HM$ .... (1).

But, since the segments *DEF*, *def* are similar, so are the cones *KEF*, *kef*.

186

Therefore $\qquad\qquad\qquad KG : EF = kg : ef.$

And the ratio $KG : EF$ is given. Therefore the ratio $kg : ef$ is given.

Suppose a length R taken such that

$$kg : ef = HM : R \qquad\qquad\qquad\qquad (2).$$

Thus R is given.

Again, since $kg : HM = BB'^2 : ef^2 = ef : R$, by (1) and (2), suppose a length S taken such that

$$ef^2 = BB' \cdot S,$$

or $\qquad\qquad\qquad BB'^2 : ef^2 = BB' : S.$

Thus $\qquad\qquad\qquad BB' : ef = ef : S = S : R,$

and *ef, S are two mean proportionals in continued proportion between BB', R.*

Synthesis.

Let ABB', DEF be great circles, AA', DD' the diameters bisecting BB', EF at right angles in M, G respectively, and O, C the centres.

Take H, K in the same way as before, and construct the cones HBB', KEF, which are therefore equal to the respective segments ABB', DEF.

Let R be a straight line such that

$$KG : EF = HM : R,$$

and between BB', R take two mean proportionals ef, S.

On ef as base describe a segment of a circle with vertex d and similar to the segment of a circle DEF. Complete the circle, and let dd' be the diameter through d, and c the centre. Conceive a sphere constructed of which def is a great circle, and through ef draw a plane at right angles to dd'.

Then shall def be the required segment of a sphere.

For the segments DEF, def of the spheres are similar, like the circular segments DEF, def.

Produce cd to k so that

$$cd' + d'g : d'g = kg : gd.$$

The cones KEF, kef are then similar.

Therefore $\qquad\qquad kg : ef = KG : EF = HM : R,$

whence $\qquad\qquad\qquad kg : HM = ef : R.$

But, since BB', ef, S, R are in continued proportion,

$$BB'^2 : ef^2 = BB' : S$$
$$= ef : R$$
$$= kg : HM.$$

Thus the bases of the cones HBB', kef are reciprocally proportional to their heights. The cones are therefore equal, and def is the segment required, being equal in volume to the cone kef. $\qquad\qquad\qquad\qquad$ [Prop. 2]

Proposition 6 (Problem)

Given two segments of spheres, to find a third segment of a sphere similar to one of the given segments and having its surface equal to that of the other.

Let *ABB'* be the segment to whose surface the surface of the required segment is to be equal, *ABA'B'* the great circle whose plane cuts the plane of the base of the segment *ABB'* at right angles in *BB'*. Let *AA'* be the diameter which bisects *BB'* at right angles.

Let *DEF* be the segment to which the required segment is to be similar, *DED'F* the great circle cutting the base of the segment at right angles in *EF*. Let *DD'* be the diameter bisecting *EF* at right angles in *G*.

Suppose the problem solved, *def* being a segment similar to *DEF* and having its surface equal to that of *ABB'*; and complete the figure for *def* as for *DEF*, corresponding points being denoted by small and capital letters respectively.

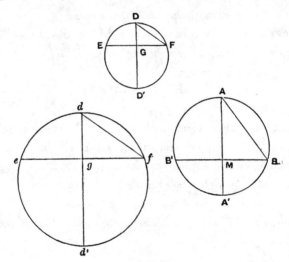

Join *AB*, *DF*, *df*.

Now, since the surfaces of the segments *def*, *ABB'* are equal, so are the circles on *df*, *AB* as diameters;   [i. 42, 43]

that is, $df = AB$.

From the similarity of the segments *DEF*, *def* we obtain

$$d'd : dg = D'D : DG,$$

and $$dg : df = DG : DF;$$

whence $$d'd : df = D'D : DF,$$

or $$d'd : AB = D'D : DF.$$

But *AB*, *D'D*, *DF* are all given;

therefore *d'd* is given.

Accordingly the synthesis is as follows.

Take $d'd$ such that

$$d'd : AB = D'D : DF \quad\text{..} \text{(1)}.$$

Describe a circle on $d'd$ as diameter, and conceive a sphere constructed of which this circle is a great circle.

Divide $d'd$ at g so that

$$d'g : gd = D'G : GD,$$

and draw through g a plane perpendicular to $d'd$ cutting off the segment def of the sphere and intersecting the plane of the great circle in ef. The segments def, DEF are thus similar,

and $$dg : df = DG : DF.$$

But from above, *componendo*,

$$d'd : dg = D'D : DG.$$

Therefore, *ex aequali*, $\quad d'd : df = D'D : DF$,

whence, by (1), $df = AB$.

Therefore the segment def has its surface equal to the surface of the segment ABB' [i. 42, 43], while it is also similar to the segment DEF.

Proposition 7 (Problem)

From a given sphere to cut off a segment by a plane so that the segment may have a given ratio to the cone which has the same base as the segment and equal height.

Let AA' be the diameter of a great circle of the sphere. It is required to draw a plane at right angles to AA' cutting off a segment, as ABB', such that the segment ABB' has to the cone ABB' a given ratio.

Analysis.

Suppose the problem solved, and let the plane of section cut the plane of the great circle in BB', and the diameter AA' in M. Let O be the centre of the sphere.

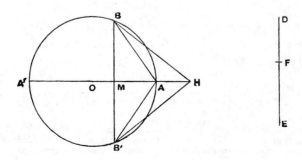

Produce OA to H so that

$$OA' + A'M : A'M = HM : MA \dots\dots\dots\dots\dots\dots\dots(1).$$

Thus the cone HBB' is equal to the segment ABB'. [Prop. 2]

Therefore the given ratio must be equal to the ratio of the cone HBB' to the cone ABB', *i.e.* to the ratio $HM : MA$.

Hence the ratio $OA' + A'M : A'M$ is given; and therefore $A'M$ is given.

διορισμός.

Now $\qquad\qquad OA' : A'M > OA' : A'A,$

so that $\qquad OA' + A'M : A'M > OA' + A'A : A'A$

$$> 3 : 2.$$

Thus, *in order that a solution may be possible, it is a necessary condition that the given ratio must be greater than* 3 : 2.

The **synthesis** proceeds thus.

Let AA' be a diameter of a great circle of the sphere, O the centre.

Take a line DE, and a point F on it, such that $DE : EF$ is equal to the given ratio, being greater than 3 : 2.

Now, since $\qquad OA' + A'A : A'A = 3 : 2,$

$$DE : EF > OA' + A'A : A'A,$$

so that $\qquad\qquad DF : FE > OA' : A'A.$

Hence a point M can be found on AA' such that

$$DF : FE = OA' : A'M. \dots\dots\dots\dots\dots\dots\dots\dots(2).$$

Through M draw a plane at right angles to AA' intersecting the plane of the great circle in BB', and cutting off from the sphere the segment ABB'.

As before, take H on OA produced such that

$$OA' + A'M : A'M = HM : MA.$$

Therefore $HM : MA = DE : EF$, by means of (2).

It follows that the cone HBB', or the segment ABB', is to the cone ABB' in the given ratio $DE : EF$.

Proposition 8

If a sphere be cut by a plane not passing through the centre into two segments $A'BB'$, ABB', of which $A'BB'$ is the greater, then the ratio

$$(segmt.\ A'BB') : (segmt.\ ABB')$$

$$< (surface\ of\ A'BB')^2 : (surface\ of\ ABB')^2$$

$$but > (surface\ of\ A'BB')^{\frac{3}{2}} : (surface\ of\ ABB')^{\frac{3}{2}}.[15]$$

[15] This is expressed in Archimedes' phrase by saying that the greater segment has to the lesser a ratio "less than the duplicate (διπλάσιον) of that which the surface of the greater segment has to the surface of the lesser, but greater than the sesquialterate (ἡμιόλιον) [of that ratio]."

Let the plane of section cut a great circle $A'BAB'$ at right angles in BB', and let AA' be the diameter bisecting BB' at right angles in M.

Let O be the centre of the sphere.

Join $A'B$, AB.

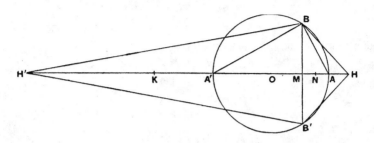

As usual, take H on OA produced, and H' on OA' produced, so that

$$OA' + A'M : A'M = HM : MA \dots\dots\dots\dots\dots\dots\dots\dots\dots(1),$$
$$OA + AM : AM = H'M : MA' \dots\dots\dots\dots\dots\dots\dots\dots(2),$$

and conceive cones drawn each with the same base as the two segments and with apices H, H' respectively. The cones are then respectively equal to the segments [Prop. 2], and they are in the ratio of their heights HM, $H'M$.

Also

$$(\text{surface of } A'BB') : (\text{surface of } ABB') = A'B^2 : AB^2 \qquad [\text{i. 42, 43}]$$
$$= A'M : AM.$$

We have therefore to prove

(a) that $\qquad\qquad H'M : MH < A'M^2 : MA^2,$

(b) that $\qquad\qquad H'M : MH > A'M^{\frac{3}{2}} : MA^{\frac{3}{2}}.$

(a) From (2) above,

$$A'M : AM = H'M : OA + AM$$
$$= H'A' : OA', \text{ since } OA = OA'.$$

Since $A'M > AM$, $H'A' > OA'$; therefore, if we take K on $H'A'$ so that $OA' = A'K$, K will fall between H' and A'.

And, by (1), $\qquad\qquad A'M : AM = KM : MH.$

Thus $\qquad KM : MH = H'A' : A'K, \text{ since } A'K = OA',$
$$> H'M : MK.$$

Therefore $\qquad\qquad H'M \cdot MH < KM^2.$

It follows that

$$H'M \cdot MH : MH^2 < KM^2 : MH^2,$$

or $\qquad\qquad H'M : MH < KM^2 : MH^2$
$$< A'M^2 : AM^2, \text{ by (1)}.$$

(b) Since $OA' = OA$,

$$A'M \cdot MA < A'O \cdot OA,$$

or
$$A'M : OA' < OA : AM$$
$$< H'A' : A'M, \text{ by means of (2).}$$

Therefore
$$A'M^2 < H'A' \cdot OA'$$
$$< H'A' \cdot A'K.$$

Take a point N on $A'A$ such that
$$A'N^2 = H'A' \cdot A'K.$$

Thus
$$H'A' : A'K = A'N^2 : A'K^2 \quad\text{.....................................(3).}$$

Also
$$H'A' : A'N = A'N : A'K,$$

and, *componendo*,

$$H'N' : A'N = NK : A'K,$$

whence
$$A'N^2 : A'K^2 = H'N^2 : NK^2.$$

Therefore, by (3),

$$H'A' : A'K = H'N^2 : NK^2.$$

Now
$$H'M : MK > H'N : NK.$$

Therefore
$$H'M^2 : MK^2 > H'A' : A'K$$
$$> H'A' : OA'$$
$$> A'M : MA, \text{ by (2), as above,}$$
$$> OA' + A'M : MH, \text{ by (1),}$$
$$> KM : MH.$$

Hence
$$H'M^2 : MH^2 = (H'M^2 : MK^2) \cdot (KM^2 : MH^2)$$
$$> (KM : MH) \cdot (KM^2 : MH^2).$$

It follows that

$$H'M : MH > KM^{\frac{3}{2}} : MH^{\frac{3}{2}}$$
$$> A'M^{\frac{3}{2}} : AM^{\frac{3}{2}}, \text{ by (1).}$$

[The text of Archimedes adds an alternative proof of this proposition, which is here omitted because it is in fact neither clearer nor shorter than the above.]

Proposition 9 (Problem)

Of all segments of spheres which have equal surfaces the hemisphere is the greatest in volume.

Let $ABA'B'$ be a great circle of a sphere, AA' being a diameter, and O the centre. Let the sphere be cut by a plane, not passing through O, perpendicular to AA' (at M), and intersecting the plane of the great circle in BB'. The segment ABB' may then be either less than a hemisphere as in Fig. 1, or greater than a hemisphere as in Fig. 2.

Let $DED'E'$ be a great circle of another sphere, DD' being a diameter and C the centre. Let the sphere be cut by a plane through C perpendicular to DD' and intersecting the plane of the great circle in the diameter EE'.

Suppose the surfaces of the segment ABB' and of the hemisphere DEE' to be equal.

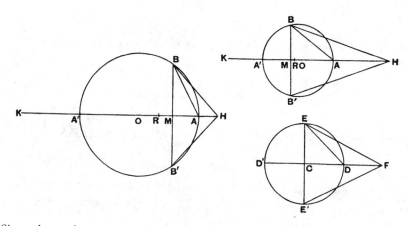

Since the surfaces are equal, $AB = DE$.   [i. 42, 43]

Now, in Fig. 1, $AB^2 > 2AM^2$ and $< 2AO^2$,

and, in Fig. 2, $AB^2 < 2AM^2$ and $> 2AO^2$.

Hence, if R be taken on AA' such that

$$AR^2 = \tfrac{1}{2}AB^2,$$

R will fall between O and M.

Also, since $AB^2 = DE^2$, $AR = CD$.

Produce OA' to K so that $OA' = A'K$, and produce $A'A$ to H so that

$$A'K : A'M = HA : AM,$$

or, *componendo*, $A'K + A'M : A'M = HM : MA$ ....(1).

Thus the cone HBB' is equal to the segment ABB'.

                                          [Prop. 2]

Again, produce CD to F so that $CD = DF$, and the cone FEE' will be equal to the hemisphere DEE'.

                                          [Prop. 2]

Now $AR \cdot RA' > AM \cdot MA'$,

and $AR^2 = \tfrac{1}{2}AB^2 = \tfrac{1}{2}AM \cdot AA' = AM \cdot A'K.$

Hence

$$AR \cdot RA' + RA^2 > AM \cdot MA' + AM \cdot A'K,$$

or $AA' \cdot AR > AM \cdot MK$

                        $> HM \cdot A'M$, by (1).

Therefore $AA' : A'M > HM : AR,$

or $AB^2 : BM^2 > HM : AR,$

i.e.
$$AR^2 : BM^2 > HM : 2AR, \text{ since } AB^2 = 2AR^2,$$
$$> HM : CF.$$

Thus, since $AR = CD$, or CE,

(circle on diam. EE') : (circle on diam. BB') $> HM : CF$.

It follows that
$$(\text{the cone } FEE') > (\text{the cone } HBB'),$$
and therefore the hemisphere DEE' is greater in volume than the segment ABB'.

MEASUREMENT OF A CIRCLE

Proposition 1

The area of any circle is equal to a right-angled triangle in which one of the sides about the right angle is equal to the radius, and the other to the circumference, of the circle.

Let $ABCD$ be the given circle, K the triangle described.

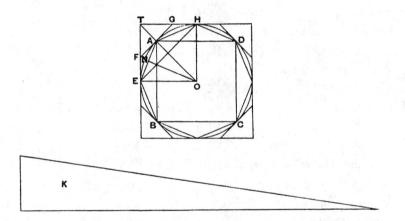

Then, if the circle is not equal to K, it must be either greater or less.

I. If possible, let the circle be greater than K.

Inscribe a square $ABCD$, bisect the arcs AB, BC, CD, DA, then bisect (if necessary) the halves, and so on, until the sides of the inscribed polygon whose angular points are the points of division subtend segments whose sum is less than the excess of the area of the circle over K.

Thus the area of the polygon is greater than K.

Let AE be any side of it, and ON the perpendicular on AE from the centre O.

Then *ON* is less than the radius of the circle and therefore less than one of the sides about the right angle in *K*. Also the perimeter of the polygon is less than the circumference of the circle, i.e. less than the other side about the right angle in *K*.

Therefore the area of the polygon is less than *K*; which is inconsistent with the hypothesis.

Thus the area of the circle is not greater than *K*.

II. If possible, let the circle be less than *K*.

Circumscribe a square, and let two adjacent sides, touching the circle in *E*, *H*, meet in *T*. Bisect the arcs between adjacent points of contact and draw the tangents at the points of bisection. Let *A* be the middle point of the arc *EH*, and *FAG* the tangent at *A*.

Then the angle *TAG* is a right angle.

Therefore
$$TG > GA$$
$$> GH.$$

It follows that the triangle *FTG* is greater than half the area *TEAH*.

Similarly, if the arc *AH* be bisected and the tangent at the point of bisection be drawn, it will cut off from the area *GAH* more than one-half.

Thus, by continuing the process, we shall ultimately arrive at a circumscribed polygon such that the spaces intercepted between it and the circle are together less than the excess of *K* over the area of the circle.

Thus the area of the polygon will be less than *K*.

Now, since the perpendicular from *O* on any side of the polygon is equal to the radius of the circle, while the perimeter of the polygon is greater than the circumference of the circle, it follows that the area of the polygon is greater than the triangle *K*; which is impossible.

Therefore the area of the circle is not less than *K*.

Since then the area of the circle is neither greater nor less than *K*, it is equal to it.

Proposition 2

The area of a circle is to the square on its diameter as 11 *to* 14.

[The text of this proposition is not satisfactory, and Archimedes cannot have placed it before Proposition 3, as the approximation depends upon the result of that proposition.]

Proposition 3

The ratio of the circumference of any circle to its diameter is less than $3\frac{1}{7}$ *but greater than* $3\frac{10}{71}$:

[In view of the interesting questions arising out of the arithmetical content of this proposition of Archimedes, it is necessary, in reproducing it, to distinguish carefully the actual steps set out in the text as we have it from the intermediate steps (mostly supplied by Eutocius) which it is convenient to put in for the purpose of making the proof easier to follow. Accordingly all the steps not actually appearing in the text have been enclosed in square brackets, in order that it may be clearly seen how far Archimedes omits actual calculations and only gives results. It will be observed that he gives two fractional approximations to $\sqrt{3}$ (one being less and the other greater than the real value) without any explanation as to how he arrived at them; and in like manner approximations to the square roots of several large numbers which are not complete squares are merely stated.]

I. Let AB be the diameter of any circle, O its centre, AC the tangent at A; and let the angle AOC be one-third of a right angle.

Then $OA : AC[= \sqrt{3} : 1] > 265 : 153$ ....(1),

and $OC : CA[= 2 : 1] = 306 : 153$ ....(2).

First, draw OD bisecting the angle AOC and meeting AC in D.

Now $CO : OA = CD : DA$, [Eucl. vi. 3]

so that $[CO + OA : OA = CA : DA,$ or$]$

$CO + OA : CA = OA : AD.$

Therefore [by (1) and (2)]

$OA : AD > 571 : 153$ ....(3).

Hence $OD^2 : AD^2[= (OA^2 + AD^2) : AD^2$

$> (571^2 + 153^2) : 153^2]$

$> 349450 : 23409,$

so that $OD : DA > 591\frac{1}{8} : 153$ ....(4).

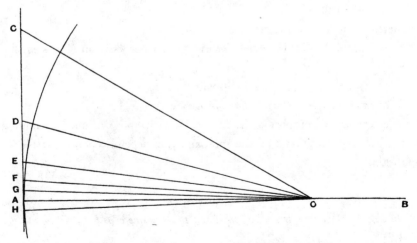

Secondly, let OE bisect the angle AOD, meeting AD in E.

[Then $DO : OA = DE : EA,$

196

so that $\qquad DO + OA : DA = OA : AE.]$

Therefore $\qquad OA : AE [> (591\frac{1}{8} + 571) : 153$, by (3) and (4)$]$

$$> 1162\frac{1}{8} : 153 \dots\dots\dots\dots\dots\dots(5).$$

[It follows that

$$OE^2 : EA^2 > \{(1162\frac{1}{8})^2 + 153^2\} : 153^2$$
$$> (1350534\tfrac{33}{64} + 23409) : 23409$$
$$> 1373943\tfrac{33}{64} : 23409.]$$

Thus $\qquad OE : EA > 1172\frac{1}{8} : 153 \dots\dots\dots\dots\dots\dots(6).$

Thirdly, let OF bisect the angle AOE and meet AE in F.

We thus obtain the result [corresponding to (3) and (5) above] that

$$OA : AF [> (1162\tfrac{1}{8} + 1172\tfrac{1}{8}) : 153]$$
$$> 2334\tfrac{1}{4} : 153 \dots\dots\dots\dots\dots\dots(7).$$

[Therefore $\qquad OF^2 : FA^2 > \{(2334\frac{1}{4})^2 + 153^2\} : 153^2$

$$> 5472132\tfrac{1}{16} : 23409.]$$

Thus $\qquad OF : FA > 2339\frac{1}{4} : 153 \dots\dots\dots\dots\dots\dots(8).$

Fourthly, let OG bisect the angle AOF, meeting AF in G.

We have then

$$OA : AG [> (2334\tfrac{1}{4} + 2339\tfrac{1}{4}) : 153, \text{ by means of (7) and (8)}]$$
$$> 4673\tfrac{1}{2} : 153.$$

Now the angle AOC, which is one-third of a right angle, has been bisected four times, and it follows that

$$\angle AOG = \tfrac{1}{48} \text{ (a right angle)}.$$

Make the angle AOH on the other side of OA equal to the angle AOG, and let GA produced meet OH in H.

Then $\qquad \angle GOH = \frac{1}{24}$ (a right angle).

Thus GH is one side of a regular polygon of 96 sides circumscribed to the given circle.

And, since $\qquad OA : AG > 4673\frac{1}{2} : 153,$

while $\qquad AB = 2OA, \quad GH = 2AG,$

it follows that

$$AB : \text{(perimeter of polygon of 96 sides)} [> 4673\tfrac{1}{2} : 153 \times 96]$$
$$> 4673\tfrac{1}{2} : 14688.$$

But $\qquad \dfrac{14688}{4673\frac{1}{2}} = 3 + \dfrac{667\frac{1}{2}}{4673\frac{1}{2}}$

$$\left[< 3 + \dfrac{667\frac{1}{2}}{4672\frac{1}{2}} \right]$$

$$< 3\tfrac{1}{7}.$$

Therefore the circumference of the circle (being less than the perimeter of the polygon) is *a fortiori* less than $3\frac{1}{7}$ times the diameter *AB*.

II. Next let *AB* be the diameter of a circle, and let *AC*, meeting the circle in *C*, make the angle *CAB* equal to one-third of a right angle. Join *BC*.

Then $\qquad\qquad AC : CB \left[= \sqrt{3} : 1 \right] < 1351 : 780.$

First, let *AD* bisect the angle *BAC* and meet *BC* in *d* and the circle in *D*. Join *BD*.

Then $\qquad\qquad\qquad \angle BAD = \angle dAC$
$$= \angle dBD,$$

and the angles at *D*, *C* are both right angles.

It follows that the triangles *ADB*, [*ACd*], *BDd* are similar.

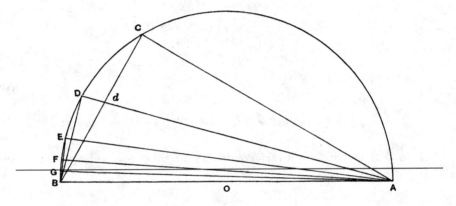

Therefore $\qquad\qquad AD : DB = BD : Dd$
$$\left[= AC : Cd \right]$$
$$= AB : Bd \qquad\qquad\qquad \text{[Eucl. vi.3]}$$
$$= AB + AC : Bd + Cd$$
$$= AB + AC : BC$$

or $\qquad\qquad\qquad BA + AC : BC = AD : DB.$

[But $\qquad\qquad AC : CB < 1351 : 780$, from above,

while $\qquad\qquad\qquad BA : BC = 2 : 1$
$$= 1560 : 780.]$$

Therefore $\qquad\qquad AD : DB < 2911 : 780$ ....(1).

[Hence $\qquad\qquad AB^2 : BD^2 < (2911^2 + 780^2) : 780^2$
$$< 9082321 : 608400.]$$

Thus $\qquad\qquad AB : BD < 3013\frac{3}{4} : 780$ ....(2).

Secondly, let *AE* bisect the angle *BAD*, meeting the circle in *E*; and let *BE* be joined.

Then we prove, in the same way as before, that

$$AE : EB [= BA + AD : BD$$
$$< (3013\tfrac{3}{4} + 2911) : 780, \text{ by (1) and (2)]}$$
$$< 5924\tfrac{3}{4} : 780$$
$$< 5924\tfrac{3}{4} \times \tfrac{4}{13} : 780 \times \tfrac{4}{13}$$
$$< 1823 : 240 \dots\dots\dots\dots\dots\dots\dots\dots\dots\dots\dots\dots\dots(3).$$

[Hence
$$AB^2 : BE^2 < (1823^2 + 240^2) : 240^2$$
$$< 3380929 : 57600.]$$

Therefore
$$AB : BE < 1838\tfrac{9}{11} : 240 \dots\dots\dots\dots\dots\dots\dots\dots\dots\dots\dots(4).$$

Thirdly, let AF bisect the angle BAE, meeting the circle in F.

Thus
$$AF : FB [= BA + AE : BE$$
$$< 3661\tfrac{9}{11} : 240, \text{ by (3) and (4)]}$$
$$< 3661\tfrac{9}{11} \times \tfrac{11}{40} : 240 \times \tfrac{11}{40}$$
$$< 1007 : 66 \dots\dots\dots\dots\dots\dots\dots\dots\dots\dots\dots\dots\dots\dots\dots(5).$$

[It follows that
$$AB^2 : BF^2 < (1007^2 + 66^2) : 66^2$$
$$< 1018405 : 4356.]$$

Therefore
$$AB : BF < 1009\tfrac{1}{6} : 66 \dots\dots\dots\dots\dots\dots\dots\dots\dots\dots\dots\dots(6).$$

Fourthly, let the angle BAF be bisected by AG meeting the circle in G.

Then
$$AG : GB [= BA + AF : BF]$$
$$< 2016\tfrac{1}{6} : 66, \text{ by (5) and (6).}$$

[And
$$AB^2 : BG^2 < \{(2016\tfrac{1}{6})^2 + 66^2\} : 66^2$$
$$< 4069284\tfrac{1}{36} : 4356.]$$

Therefore $AB : BC < 2017\tfrac{1}{4} : 66,$

whence
$$BG : AB > 66 : 2017\tfrac{1}{4} \dots\dots\dots\dots\dots\dots\dots\dots\dots\dots\dots\dots(7).$$

[Now the angle BAG which is the result of the fourth bisection of the angle BAC, or of one-third of a right angle, is equal to one-fortyeighth of a right angle.

Thus the angle subtended by BG at the centre is

$\tfrac{1}{24}$ (a right angle).]

Therefore BG is a side of a regular inscribed polygon of 96 sides.

It follows from (7) that

$$(\text{perimeter of polygon}): AB [> 96 \times 66 : 2017\tfrac{1}{4}]$$
$$> 6336 : 2017\tfrac{1}{4}.$$

And
$$\frac{6336}{2017\tfrac{1}{4}} > 3\tfrac{10}{71}.$$

Much more then is the circumference of the circle greater than $3\tfrac{10}{71}$ times the diameter.

Thus the ratio of the circumference to the diameter

$$< 3\tfrac{1}{7} \text{ but } > 3\tfrac{10}{71}.$$

THE SAND RECKONER

"THERE are some, king Gelon, who think that the number of the sand is infinite in multitude; and I mean by the sand not only that which exists about Syracuse and the rest of Sicily but also that which is found in every region whether inhabited or uninhabited. Again there are some who, without regarding it as infinite, yet think that no number has been named which is great enough to exceed its multitude. And it is clear that they who hold this view, if they imagined a mass made up of sand in other respects as large as the mass of the earth, including in it all the seas and the hollows of the earth filled up to a height equal to that of the highest of the mountains, would be many times further still from recognising that any number could be expressed which exceeded the multitude of the sand so taken. But I will try to show you by means of geometrical proofs, which you will be able to follow, that, of the numbers named by me and given in the work which I sent to Zeuxippus, some exceed not only the number of the mass of sand equal in magnitude to the earth filled up in the way described, but also that of a mass equal in magnitude to the universe. Now you are aware that 'universe' is the name given by most astronomers to the sphere whose centre is the centre of the earth and whose radius is equal to the straight line between the centre of the sun and the centre of the earth. This is the common account (τὰ γραφόμενα), as you have heard from astronomers. But Aristarchus of Samos brought out a book consisting of some hypotheses, in which the premisses lead to the result that the universe is many times greater than that now so called. His hypotheses are that the fixed stars and the sun remain unmoved, that the earth revolves about the sun in the circumference of a circle, the sun lying in the middle of the orbit, and that the sphere of the fixed stars, situated about the same centre as the sun, is so great that the circle in which he supposes the earth to revolve bears such a proportion to the distance of the fixed stars as the centre of the sphere bears to its surface. Now it is easy to see that this is impossible; for, since the centre of the sphere has no magnitude, we cannot conceive it to bear any ratio whatever to the surface of the sphere. We must however take Aristarchus to mean this: since we conceive the earth to be, as it were, the centre of the universe, the ratio which the earth bears to what we describe as the 'universe' is the same as the ratio which the sphere containing the circle in which he supposes the earth to revolve bears to the sphere of the fixed stars. For he adapts the proofs of his results to a hypothesis of this kind, and in particular he appears to suppose the magnitude of the sphere in which he represents the earth as moving to be equal to what we call the 'universe.'

I say then that, even if a sphere were made up of the sand, as great as Aristarchus supposes the sphere of the fixed stars to be, I shall still prove that, of the numbers named in the *Principles*,[1] some exceed in multitude the number of the sand which is equal in magnitude to the sphere referred to, provided that the following assumptions be made.

1. *The perimeter of the earth is about* 3,000,000 *stadia and not greater.*

It is true that some have tried, as you are of course aware, to prove that the said perimeter is about 300,000 stadia. But I go further and, putting the magnitude of the earth at ten times the size that my predecessors thought it, I suppose its perimeter to be about 3,000,000 stadia and not greater.

2. *The diameter of the earth is greater than the diameter of the moon, and the diameter of the sun is greater than the diameter of the earth.*

In this assumption I follow most of the earlier astronomers.

3. *The diameter of the sun is about* 30 *times the diameter of the moon and not greater.*

It is true that, of the earlier astronomers, Eudoxus declared it to be about nine times as great, and Pheidias my father[2] twelve times, while Aristarchus tried to prove that the diameter of the sun is greater than 18 times but less than 20 times the diameter of the moon. But I go even further than Aristarchus, in order that the truth of my proposition may be established beyond dispute, and I suppose the diameter of the sun to be about 30 times that of the moon and not greater.

4. *The diameter of the sun is greater than the side of the chiliagon inscribed in the greatest circle in the (sphere of the) universe.*

I make this assumption[3] because Aristarchus discovered that the sun appeared to be about $\frac{1}{720}$ th part of the circle of the zodiac, and I myself tried, by a method which I will now describe, to find experimentally ($\mathit{\grave{o}\rho\gamma\alpha\nu\iota\kappa\hat{\omega}s}$) the angle subtended by the sun and having its vertex at the eye ($\tau\grave{\alpha}\nu~\gamma\omega\nu\acute{\iota}\alpha\nu,~\epsilon\grave{\iota}s~\grave{\alpha}\nu~\acute{o}~\mathring{\alpha}\lambda\iota os~\acute{\epsilon}\nu\alpha\rho$ $\mu\acute{o}\zeta\epsilon\iota~\tau\grave{\alpha}\nu~\kappa o\rho\upsilon\phi\grave{\alpha}\nu~\acute{\epsilon}\chi o\upsilon\sigma\alpha\nu~\pi o\tau\grave{\iota}~\tau\^{\alpha}~\mathring{o}\psi\epsilon\iota$)."

[Up to this point the treatise has been literally translated because of the historical interest attaching to the *ipsissima verba* of Archimedes on such a subject. The rest of the work can now be more freely reproduced, and, before proceeding to the mathematical contents of it, it is only necessary to remark that Archimedes next describes how he arrived at a higher and a lower limit for the angle subtended by the sun. This he did by taking a long rod or ruler ($\kappa\alpha\nu\acute{o}\nu$), fastening on the end of it a small cylinder or disc, pointing the rod in the direction of the sun just after its rising (so that it was possible to look directly at it), then putting the cylinder at such a distance that it just concealed, and just failed to conceal, the sun, and lastly measuring the angles

---

[1]'Αρχαί was apparently the title of the work sent to Zeuxippus.
[2]τοῦ ἁμοῦ πατρὸς is the correction of Blass for τοῦ 'Ακούπατρο ς (*Jahrb. f. Philol.* CXXVII. 1883).
[3]This is not, strictly speaking, an assumption; it is a proposition proved later by means of the result of an experiment about to be described.

subtended by the cylinder. He explains also the correction which he thought it necessary to make because "the eye does not see from one point but from a certain area" (ἐπεὶ αἱ ὄψιες οὐκ ἀφ᾽ ἑνὸς σαμείου βλέποντι, ἀλλὰ ἀπό τινος μεγέθεος).]

The result of the experiment was to show that the angle subtended by the diameter of the sun was less than $\frac{1}{164}$th part, and greater than $\frac{1}{200}$th part, of a right angle.

To prove that (on this assumption) the diameter of the sun is greater than the side of a chiliagon, or figure with 1000 equal sides, inscribed in a great circle of the 'universe.'

Suppose the plane of the paper to be the plane passing through the centre of the sun, the centre of the earth and the eye, at the time when the sun has just risen above the horizon. Let the plane cut the earth in the circle *EHL* and the sun in the circle *FKG*, the centres of the earth and sun being *C*, *O* respectively, and *E* being the position of the eye.

Further, let the plane cut the sphere of the 'universe' (i.e. the sphere whose centre is *C* and radius *CO*) in the great circle *AOB*.

Draw from *E* two tangents to the circle *FKG* touching it at *P*, *Q*, and from *C* draw two other tangents to the same circle touching it in *F*, *G* respectively.

Let *CO* meet the sections of the earth and sun in *H*, *K* respectively; and let *CF*, *CG* produced meet the great circle *AOB* in *A*, *B*.

Join *EO*, *OF*, *OG*, *OP*, *OQ*, *AB*, and let *AB* meet *CO* in *M*.

Now *CO* > *EO*, since the sun is just above the horizon.
Therefore $\angle PEQ > \angle FCG.$

And $\angle PEQ > \frac{1}{200}R$ } where *R* represents a right angle.
but $< \frac{1}{164}R$ }

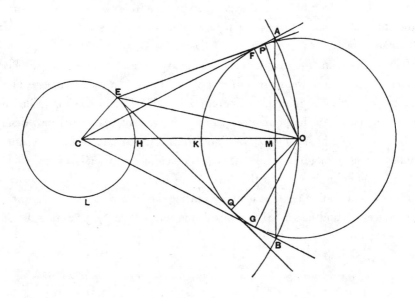

Thus $\angle FCG < \frac{1}{164}R$, *a fortiori*,

and the chord AB subtends an arc of the great circle which is less than $\frac{1}{656}$th of the circumference of that circle, i.e.

$AB <$ (side of 656-sided polygon inscribed in the circle).

Now the perimeter of any polygon inscribed in the great circle is less than $\frac{44}{7}CO$.

[Cf. *Measurement of a circle*, Prop. 3.]

Therefore $AB : CO < 11 : 1148,$

and, *a fortiori*, $AB < \frac{1}{100}CO$ ....(α).

Again, since $CA = CO$, and AM is perpendicular to CO, while OF is perpendicular to CA,

$$AM = OF.$$

Therefore $AB = 2AM =$ (diameter of sun).

Thus (diameter of sun) $< \frac{1}{100}CO$, by (α),

and, *a fortiori*,

(diameter of earth) $< \frac{1}{100}CO.$ [Assumption 2]

Hence $CH + OK < \frac{1}{100}CO,$

so that $HK > \frac{99}{100}CO,$

or $CO : HK < 100 : 99.$

And $CO > CF,$

while $HK < EQ.$

Therefore $CF : EQ < 100 : 99$..............................(β).

Now in the right-angled triangles CFO, EQO, of the sides about the right angles, $OF = OQ$, but $EQ < CF$ (since $EO < CO$).

Therefore $\angle OEQ : \angle OCF > CO : EO,$

but $< CF : EQ.$[4]

Doubling the angles,

$\angle PEQ : \angle ACB > CF : EQ$

$< 100 : 99,$ by (β) above.

But $\angle PEQ > \frac{1}{200}R$, by hypothesis.

Therefore $\angle ACB > \frac{99}{20000}R$

$> \frac{1}{203}R.$

It follows that the arc AB is greater than $\frac{1}{812}$th of the circumference of the great circle AOB.

---

[4] The proposition here assumed is of course equivalent to the trigonometrical formula which states that, if α, β are the circular measures of two angles, each less than a right angle, of which α is the greater, then

$$\frac{\tan \alpha}{\tan \beta} > \frac{\alpha}{\beta} > \frac{\sin \alpha}{\sin \beta}.$$

Hence, *a fortiori*,

$$AB > \text{(side of chiliagon inscribed in great circle)},$$

and AB is equal to the diameter of the sun, as proved above.

———

The following results can now be proved:

$$(\textit{diameter of 'universe'}) < 10,000 \ (\textit{diameter of earth}),$$

and $\qquad\qquad (\textit{diameter of 'universe'}) < 10,000,000,000 \ \textit{stadia}.$

(1) Suppose, for brevity, that d_u represents the diameter of the 'universe,' d_s that of the sun, d_e that of the earth, and d_m that of the moon.

By hypothesis, $\qquad\qquad d_s \not> 30d_m,$ [Assumption 3]

and $\qquad\qquad\qquad\qquad\quad d_e > d_m;$ [Assumption 2]

therefore $\qquad\qquad\qquad\quad d_s < 30d_e.$

Now, by the last proposition,

$$d_s > \text{(side of chiliagon inscribed in great circle)},$$

so that $\qquad\qquad$ (perimeter of chiliagon) $< 1000d_s$

$$< 30,000d_e.$$

But the perimeter of any regular polygon with more sides than 6 inscribed in a circle is greater than that of the inscribed regular hexagon, and therefore greater than three times the diameter. Hence

$$\text{(perimeter of chiliagon)} > 3d_u.$$

It follows that $\qquad\quad d_u < 10,000d_e.$

(2) $\qquad\qquad$ (Perimeter of earth) $\not> 3,000,000$ stadia. [Assumption 1]

and $\qquad\qquad$ (perimeter of earth) $> 3d_e.$

Therefore $\qquad\qquad\qquad\quad d_e < 1,000,000$ stadia,

whence $\qquad\qquad\qquad\quad d_u < 10,000,000,000$ stadia.

———

Assumption 5.

Suppose a quantity of sand taken not greater than a poppy-seed, and suppose that it contains not more than 10,000 grains.

Next suppose the diameter of the poppy-seed to be not less than $\frac{1}{40}$th of a finger-breadth.

Orders and periods of numbers.

I. We have traditional names for numbers up to a myriad (10,000); we can therefore express numbers up to a myriad myriads (100,000,000). Let these numbers be called numbers of the *first order*.

Suppose the 100,000,000 to be the unit of the *second order*, and let the *second order* consist of the numbers from that unit up to $(100,000,000)^2$.

Let this again be the unit of the *third order* of numbers ending with $(100,000,000)^3$; and so on, until we reach the 100,000,000*th order* of numbers ending with $(100,000,000)^{100,000,000}$, which we will call *P.*

II. Suppose the numbers from 1 to *P* just described to form the *first period.*

Let *P* be the unit of the *first order of the second period,* and let this consist of the numbers from *P* up to 100,000,000*P.*

Let the last number be the unit of the *second order of the second period,* and let this end with $(100,000,000)^2 P$.

We can go on in this way till we reach the 100,000,000*th order of the second period* ending with $(100,000,000)^{100,000,000} P$, or P^2.

III. Taking P^2 as the unit of the *first order of the third period,* we proceed in the same way till we reach the 100,000,000*th order of the third period* ending with P^3.

IV. Taking P^3 as the unit of the *first order of the fourth period,* we continue the same process until we arrive at the 100,000,000*th order of the* 100,000,000*th period* ending with $P^{100,000,000}$. This last number is expressed by Archimedes as "a myriad-myriad units of the myriad-myriad-th order of the myriad-myriad-th period (αἱ μυριακισμυριοστᾶς περιόδου μυριακισμυριοστῶν ἀριθμῶν μυρίαι μυριάδες)," which is easily seen to be 100,000,000 times the product of $(100,000,000)^{99,999,999}$ and $P^{99,999,999}$, i.e. $P^{100,000,000}$.

[The scheme of numbers thus described can be exhibited more clearly by means of *indices* as follows.

FIRST PERIOD.

First order.	Numbers from		1 to 10^8.
Second order.	"	"	10^8 to 10^{16}.
Third order.	"	"	10^{16} to 10^{24}.

$\vdots$

(10^8)*th order.*	"	"	$10^{8\cdot(10^8-1)}$ to $10^{8\cdot10^8}$ (*P*, say).

SECOND PERIOD.

First order.	"	"	$P \cdot 1$ to $P \cdot 10^8$.
Second order.	"	"	$P \cdot 10^8$ to $P \cdot 10^{16}$.

$\vdots$

(10^8)*th order.*	"	"	$P \cdot 10^{8\cdot(10^8-1)}$ to $P \cdot 10^{8\cdot10^8}$
			(or P^2).

$\vdots$

(10^8)TH PERIOD.

First order.	"	"	$P^{10^8-1} \cdot 1$ to $P^{10^8-1} \cdot 10^8$.

Second order. Numbers from $P^{10^8-1} \cdot 10^8$ to $P^{10^8-1} \cdot 10^{16}$.

⋮

$(10^8)th$ *order.* „ „ $P^{10^8-1} \cdot 10^{8 \cdot (10^8-1)}$ to

„ „ $P^{10^8-1} \cdot 10^{8 \cdot 10^8}$ (i.e. P^{10^8}).

The prodigious extent of this scheme will be appreciated when it is considered that the last number in the *first period* would be represented now by 1 followed by 800,000,000 ciphers, while the last number of the $(10^8)th$ *period* would require 100,000,000 times as many ciphers, i.e. 80,000 million millions of ciphers.]

Octads.

Consider the series of terms in continued proportion of which the first is 1 and the second 10 [i.e. the geometrical progression 1, 10^1, 10^2, 10^3, ...]. The *first octad* of these terms [*i.e.* 1, 10^1, 10^2, ... 10^7] fall accordingly under the *first order of the first period* above described, the *second octad* [i.e. 10^8, 10^9, ... 10^{15}] under the *second order of the first period*, the first term of the octad being the unit of the corresponding order in each case. Similarly for the *third octad*, and so on. We can, in the same way, place any number of octads.

Theorem.

If there be any number of terms of a series in continued proportion, say A_1, A_2, A_3, ... A_m, ... A_n, ... A_{m+n-1}, ... *of which* $A_1 = 1$, $A_2 = 10$ [so that the series forms the geometrical progression 1, 10^1, 10^2, ... 10^{m-1}, ... 10^{n-1}, ... 10^{m+n-2}, ...], *and if any two terms as* A_m, A_n *be taken and multiplied, the product* $A_m \cdot A_n$ *will be a term in the same series and will be as many terms distant from* A_n *as* A_m *is distant from* A_1; *also it will be distant from* A_1 *by a number of terms less by one than the sum of the numbers of terms by which* A_m *and* A_n *respectively are distant from* A_1.

Take the term which is distant from A_n by the same number of terms as A_m is distant from A_1. This number of terms is m (the first and last being both counted). Thus the term to be taken is m terms distant from A_n, and is therefore the term A_{m+n-1}.

We have therefore to prove that

$$A_m \cdot A_n = A_{m+n-1}.$$

Now terms equally distant from other terms in the continued proportion are proportional.

Thus

$$\frac{A_m}{A_1} = \frac{A_{m+n-1}}{A_n}.$$

But

$$A_m = A_m \cdot A_1, \text{ since } A_1 = 1.$$

Therefore $$A_{m+n-1} = A_m \cdot A_n \dots\dots\dots\dots\dots\dots\dots\dots\dots(1).$$

The second result is now obvious, since A_m is m terms distant from A_1, A_n is n terms distant from A_1, and A_{m+n-1} is $(m + n - 1)$ terms distant from A_1.

Application to the number of the sand.

By Assumption 5,

$$(\text{diam. of poppy-seed}) \not\lessgtr \tfrac{1}{40} \text{ (finger-breadth)};$$

and, since spheres are to one another in the triplicate ratio of their diameters, it follows that

$$(\text{sphere of diam. 1 finger-breadth}) \not> 64{,}000 \text{ poppy-seeds}$$

$$\not> 64{,}000 \times 10{,}000$$

$$\not> 640{,}000{,}000$$

$$\not> 6 \text{ units of } second$$

$$order + 40{,}000{,}000 \quad \left.\begin{array}{l}\text{grains}\\\text{of}\\\text{sand.}\end{array}\right.$$

$$\text{units of } first\ order$$

$$(a\ fortiori) < 10 \text{ units of } second$$

$$order \text{ of numbers.}$$

We now gradually increase the diameter of the supposed sphere, multiplying it by 100 each time. Thus, remembering that the sphere is thereby multiplied by 100^3 or 1,000,000, the number of grains of sand which would be contained in a sphere with each successive diameter may be arrived at as follows.

Diameter of sphere.	Corresponding number of grains of sand.	
(1) 100 finger-breadths	$<1{,}000{,}000 \times 10$ units of *second order*	
	$<$(7th term of series) $\times$ (10th term of series)	
	$<$16th term of series	[i.e. 10^{15}]
	$<[10^7$ or$]$ 10,000,000 units of the *second order.*	
(2) 10,000 finger-breadths	$<1{,}000{,}000 \times$ (last number)	
	$<$(7th term of series) $\times$ (16th term)	
	$<$22nd term of series	[i.e. 10^{21}]
	$<[10^5$ or$]$ 100,000 units of *third order.*	
(3) 1 stadium ($<$10,000 finger-breadths)	$<$100,000 units of *third order.*	
(4) 100 stadia	$<1{,}000{,}000 \times$ (last number)	
	$<$(7th term of series) $\times$ (22nd term)	
	$<$28th term of series	[10^{27}]
	$<[10^3$ or$]$ 1,000 units of *fourth order.*	
(5) 10,000 stadia	$<1{,}000{,}000 \times$ (last number)	
	$<$(7th term of series) $\times$ (28th term)	
	$<$34th term of series	[10^{33}]
	$<$10 units of *fifth order.*	
(6) 1,000,000 stadia	$<$(7th term of series) $\times$ (34th term)	
	$<$40th term	[10^{39}]
	$<[10^7$ or$]$ 10,000,000 units of *fifth order.*	

Diameter of sphere.	*Corresponding number of grains of sand.*
(7) 100,000,000 stadia	$<$(7th term of series) $\times$ (40th term)
	$<$ 46th term $\quad\quad\quad\quad\quad\quad\quad\quad$ [10^{45}]
	$<$[10^5 or] 100,000 units of *sixth order.*
(8) 10,000,000,000 stadia	$<$(7th term of series) $\times$ (46th term)
	$<$52nd term of series $\quad\quad\quad\quad\quad$ [10^{51}]
	$<$[10^3 or] 1,000 units of *seventh order.*

But, by the proposition above,

(diameter of 'universe') $<$ 10,000,000,000 stadia.

Hence *the number of grains of sand which could be contained in a sphere of the size of our 'universe' is less than* 1,000 *units of the seventh order of numbers* [or 10^{51}].

From this we can prove further that *a sphere of the size attributed by Aristarchus to the sphere of the fixed stars would contain a number of grains of sand less than* 10,000,000 *units of the eighth order of numbers* [or $10^{56+7} = 10^{63}$].

For, by hypothesis,

(earth) : ('universe') = ('universe') : (sphere of fixed stars).

And

(diameter of 'universe') $<$ 10,000 (diam. of earth);

whence

(diam. of sphere of fixed stars) $<$ 10,000 (diam. of 'universe').

Therefore

(sphere of fixed stars) $<$ $(10,000)^3 \cdot$ ('universe').

It follows that the number of grains of sand which would be contained in a sphere equal to the sphere of the fixed stars

$<$ $(10,000)^3 \times$ 1,000 units of *seventh order*

$<$ (13th term of series) $\times$ (52nd term of series)

$<$ 64th term of series $\quad\quad\quad\quad\quad\quad\quad\quad$ [i.e. 10^{63}]

$<$ [10^7 or] 10,000,000 units of *eighth order* of numbers.

Conclusion.

"I conceive that these things, king Gelon, will appear incredible to the great majority of people who have not studied mathematics, but that to those who are conversant therewith and have given thought to the question of the distances and sizes of the earth the sun and moon and the whole universe the proof will carry conviction. And it was for this reason that I thought the subject would be not inappropriate for your consideration."

THE METHOD OF ARCHIMEDES TREATING OF MECHANICAL PROBLEMS–TO ERATOSTHENES

"Archimedes to Eratosthenes greeting.

I sent you on a former occasion some of the theorems discovered by me, merely writing out the enunciations and inviting you to discover the proofs, which at the moment I did not give. The enunciations of the theorems which I sent were as follows.

1. If in a right prism with a parallelogrammic base a cylinder be inscribed which has its bases in the opposite parallelograms,[1] and its sides [i.e. four generators] on the remaining planes (faces) of the prism, and if through the centre of the circle which is the base of the cylinder and (through) one side of the square in the plane opposite to it a plane be drawn, the plane so drawn will cut off from the cylinder a segment which is bounded by two planes and the surface of the cylinder, one of the two planes being the plane which has been drawn and the other the plane in which the base of the cylinder is, and the surface being that which is between the said planes; and the segment cut off from the cylinder is one sixth part of the whole prism.

2. If in a cube a cylinder be inscribed which has its bases in the opposite parallelograms[2] and touches with its surface the remaining four planes (faces), and if there also be inscribed in the same cube another cylinder which has its bases in other parallelograms and touches with its surface the remaining four planes (faces), then the figure bounded by the surfaces of the cylinders, which is within both cylinders, is two-thirds of the whole cube.

Now these theorems differ in character from those communicated before; for we compared the figures then in question, conoids and spheroids and segments of them, in respect of size, with figures of cones and cylinders : but none of those figures have yet been found to be equal to a solid figure bounded by planes; whereas each of the present figures bounded by two planes and surfaces of cylinders is found to be equal to one of the solid figures which are bounded by planes. The proofs then of these theorems I have written in this book and now send to you. Seeing moreover in you, as I say, an earnest student, a man of considerable eminence in philosophy, and an admirer [of mathematical inquiry], I thought fit to write out for you and explain in

---

[1] The parallelograms are apparently *squares*.
[2] i.e. squares.

detail in the same book the peculiarity of a certain method, by which it will be possible for you to get a start to enable you to investigate some of the problems in mathematics by means of mechanics. This procedure is, I am persuaded, no less useful even for the proof of the theorems themselves; for certain things first became clear to me by a mechanical method, although they had to be demonstrated by geometry afterwards because their investigation by the said method did not furnish an actual demonstration. But it is of course easier, when we have previously acquired, by the method, some knowledge of the questions, to supply the proof than it is to find it without any previous knowledge. This is a reason why, in the case of the theorems the proof of which Eudoxus was the first to discover, namely that the cone is a third part of the cylinder, and the pyramid of the prism, having the same base and equal height, we should give no small share of the credit to Democritus who was the first to make the assertion with regard to the said figure[3] though he did not prove it. I am myself in the position of having first made the discovery of the theorem now to be published [by the method indicated], and I deem it necessary to expound the method partly because I have already spoken of it and I do not want to be thought to have uttered vain words, but equally because I am persuaded that it will be of no little service to mathematics; for I apprehend that some, either of my contemporaries or of my successors, will, by means of the method when once established, be able to discover other theorems in addition, which have not yet occurred to me.

First then I will set out the very first theorem which became known to me by means of mechanics, namely that

Any segment of a section of a right-angled cone (i.e. a parabola) is four-thirds of the triangle which has the same base and equal height,

and after this I will give each of the other theorems investigated by the same method. Then, at the end of the book, I will give the geometrical [proofs of the propositions] . . .

[I premise the following propositions which I shall use in the course of the work.]

1. If from [one magnitude another magnitude be subtracted which has not the same centre of gravity, the centre of gravity of the remainder is found by] producing [the straight line joining the centres of gravity of the whole magnitude and of the subtracted part in the direction of the centre of gravity of the whole] and cutting off from it a length which has to the distance between the said centres of gravity the ratio which the weight of the subtracted magnitude has to the weight of the remainder.

[*On the Equilibrium of Planes*, i. 8]

---

[3] περὶ τοῦ εἰρημένου σχήματος, in the singular. Possibly Archimedes may have thought of the case of the pyramid as being the more fundamental and as really involving that of the cone. Or perhaps "figure" may be intended for "type of figure."

2. If the centres of gravity of any number of magnitudes whatever be on the same straight line, the centre of gravity of the magnitude made up of all of them will be on the same straight line. [Cf. *Ibid.* i. 5]

3. The centre of gravity of any straight line is the point of bisection of the straight line. [Cf. *Ibid.* i. 4]

4. The centre of gravity of any triangle is the point in which the straight lines drawn from the angular points of the triangle to the middle points of the (opposite) sides cut one another. [*Ibid.* i. 13, 14]

5. The centre of gravity of any parallelogram is the point in which the diagonals meet. [*Ibid.* i. 10]

6. The centre of gravity of a circle is the point which is also the centre [of the circle].

7. The centre of gravity of any cylinder is the point of bisection of the axis.

8. The centre of gravity of any cone is [the point which divides its axis so that] the portion [adjacent to the vertex is] triple [of the portion adjacent to the base].

[All these propositions have already been] proved.[4] [Besides these I require also the following proposition, which is easily proved:

If in two series of magnitudes those of the first series are, in order, proportional to those of the second series and further] the magnitudes [of the first series], either all or some of them, are in any ratio whatever [to those of a third series], and if the magnitudes of the second series are in the same ratio to the corresponding magnitudes [of a fourth series], then the sum of the magnitudes of the first series has to the sum of the selected magnitudes of the third series the same ratio which the sum of the magnitudes of the second series has to the sum of the (correspondingly) selected magnitudes of the fourth series. [*On Conoids and Spheroids*, Prop. 1.]"

Proposition 1

Let *ABC* be a segment of a parabola bounded by the straight line *AC* and the parabola *ABC*, and let *D* be the middle point of *AC*. Draw the straight line *DBE* parallel to the axis of the parabola and join *AB*, *BC*.

Then shall the segment *ABC* be $\frac{4}{3}$ of the triangle *ABC*.

From *A* draw *AKF* parallel to *DE*, and let the tangent to the parabola at *C* meet *DBE* in *E* and *AKF* in *F*. Produce *CB* to meet *AF* in *K*, and again produce *CK* to *H*, making *KH* equal to *CK*.

---

[4]The problem of finding the centre of gravity of a cone is not solved in any extant work of Archimedes. It may have been solved either in a separate treatise, such as the περὶ ζυγῶν which is lost, or perhaps in a larger mechanical work of which the extant books *On the Equilibrium of Planes* formed only a part.

Consider *CH* as the bar of a balance, *K* being its middle point.

Let *MO* be any straight line parallel to *ED*, and let it meet *CF*, *CK*, *AC* in *M*, *N*, *O* and the curve in *P*.

Now, since *CE* is a tangent to the parabola and *CD* the semi-ordinate,

$$EB = BD;$$

"for this is proved in the Elements [of Conics]."[5]

Since *FA*, *MO* are parallel to *ED*, it follows that

$$FK = KA, \quad MN = NO.$$

Now, by the property of the parabola, "proved in a lemma,"

$$MO : OP = CA : AO \qquad \text{[Cf. } Quadrature \ of \ Parabola, \text{ Prop. 5]}$$
$$= CK : KN \qquad \text{[Eucl. vi. 2]}$$
$$= HK : KN.$$

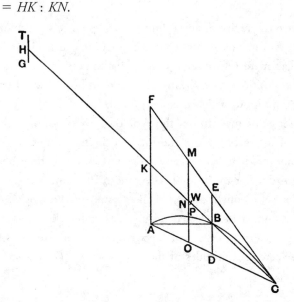

Take a straight line *TG* equal to *OP*, and place it with its centre of gravity at *H*, so that *TH* = *HG*; then, since *N* is the centre of gravity of the straight line *MO*, and

$$MO : TG = HK : KN,$$

it follows that *TG* at *H* and *MO* at *N* will be in equilibrium about *K*.

[*On the Equilibrium of Planes*, i. 6, 7]

Similarly, for all other straight lines parallel to *DE* and meeting the arc of the parabola, (1) the portion intercepted between *FC*, *AC* with its middle point on *KC* and (2) a length equal to the intercept between the curve and *AC* placed with its centre of gravity at *H* will be in equilibrium about *K*.

[5] i.e. the works on conics by Aristaeus and Euclid. Cf. the similar expression in *On Conoids and Spheroids*, Prop. 3, and *Quadrature of Parabola*, Prop. 3.

Therefore K is the centre of gravity of the whole system consisting (1) of all the straight lines as MO intercepted between FC, AC and placed as they actually are in the figure and (2) of all the straight lines placed at H equal to the straight lines as PO intercepted between the curve and AC.

And, since the triangle CFA is made up of all the parallel lines like MO, and the segment CBA is made up of all the straight lines like PO within the curve, it follows that the triangle, placed where it is in the figure, is in equilibrium about K with the segment CBA placed with its centre of gravity at H.

Divide KC at W so that $CK = 3KW$;

then W is the centre of gravity of the triangle ACF; "for this is proved in the books on equilibrium" ($\acute{\epsilon}\nu$ $\tau o\hat{\iota}\varsigma$ $\grave{\iota}\sigma o\rho\rho o\pi\iota\kappa o\hat{\iota}\varsigma$).

[Cf. *On the Equilibrium of Planes* i. 15]

Therefore $\triangle ACF$: (segment ABC) $= HK : KW$

$$= 3 : 1.$$

Therefore   segment $ABC = \frac{1}{3} \triangle ACF.$

But $\triangle ACF = 4 \triangle ABC.$

Therefore   segment $ABC = \frac{4}{3} \triangle ABC.$

"Now the fact here stated is not actually demonstrated by the argument used; but that argument has given a sort of indication that the conclusion is true. Seeing then that the theorem is not demonstrated, but at the same time suspecting that the conclusion is true, we shall have recourse to the geometrical demonstration which I myself discovered and have already published."[6]

Proposition 2

We can investigate by the same method the propositions that

(1) *Any sphere is (in respect of solid content) four times the cone with base equal to a great circle of the sphere and height equal to its radius; and*

(2) *the cylinder with base equal to a great circle of the sphere and height equal to the diameter is* $1\frac{1}{2}$ *times the sphere.*

(1) Let $ABCD$ be a great circle of a sphere, and AC, BD diameters at right angles to one another.

Let a circle be drawn about BD as diameter and in a plane perpendicular to AC, and on this circle as base let a cone be described with A as vertex. Let the surface

---

[6]The word governing $\tau\grave{\eta}\nu$ $\gamma\epsilon\omega\mu\epsilon\tau\rho o\upsilon\mu\acute{\epsilon}\nu\eta\nu$ $\grave{\alpha}\pi\acute{o}\delta\epsilon\iota\xi\iota\nu$ in the Greek text is $\tau\acute{\alpha}\xi o\mu\epsilon\nu$, a reading which seems to be doubtful and is certainly difficult to translate. Heiberg translates as if $\tau\acute{\alpha}\xi o\mu\epsilon\nu$ meant "we shall give lower down" or "later on," but I agree with Th. Reinach (*Revue ge'ne'rale des sciences pures et appliquées*, 30 November 1907) that it is questionable whether Archimedes would really have written out in full once more, as an appendix, a proof which, as he says, had already been published (i.e. presumably in the *Quadrature of a Parabola*). $\tau\acute{\alpha}\xi o\mu\epsilon\nu$, if correct, should apparently mean "we shall appoint," "prescribe" or "assign."

of this cone be produced and then cut by a plane through C parallel to its base; the section will be a circle on EF as diameter. On this circle as base let a cylinder be erected with height and axis AC, and produce CA to H, making AH equal to CA.

Let CH be regarded as the bar of a balance, A being its middle point.

Draw any straight line MN in the plane of the circle $ABCD$ and parallel to BD. Let MN meet the circle in O, P, the diameter AC in S, and the straight lines AE, AF in Q, R respectively. Join AO.

Through MN draw a plane at right angles to AC;
this plane will cut the cylinder in a circle with diameter MN, the sphere in a circle with diameter OP, and the cone in a circle with diameter QR.

Now, since $MS = AC$, and $QS = AS$,
$$MS \cdot SQ = CA \cdot AS$$
$$= AO^2$$
$$= OS^2 + SQ^2.$$

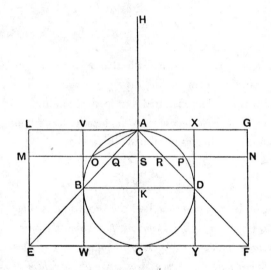

And, since $HA = AC$,
$$HA : AS = CA : AS$$
$$= MS : SQ$$
$$= MS^2 : MS \cdot SQ$$
$$= MS^2 : (OS^2 + SQ^2), \text{ from above,}$$
$$= MN^2 : (OP^2 + QR^2)$$
$$= (\text{circle, diam. } MN) : (\text{circle, diam. } OP) + \text{circle, diam. } QR).$$

That is,
$$HA : AS = (\text{circle in cylinder}) : (\text{circle in sphere} + \text{circle in cone}).$$

Therefore the circle in the cylinder, placed where it is, is in equilibrium, about A, with the circle in the sphere together with the circle in the cone, if both the latter circles are placed with their centres of gravity at H.

Similarly for the three corresponding sections made by a plane perpendicular to AC and passing through any other straight line in the parallelogram LF parallel to EF.

If we deal in the same way with all the sets of three circles in which planes perpendicular to AC cut the cylinder, the sphere and the cone, and which make up those solids respectively, it follows that the cylinder, in the place where it is, will be in equilibrium about A with the sphere and the cone together, when both are placed with their centres of gravity at H.

Therefore, since K is the centre of gravity of the cylinder,

$$HA : AK = \text{(cylinder)} : \text{(sphere} + \text{cone } AEF).$$

But $HA = 2AK$;

therefore $\qquad\qquad$ cylinder $= 2$ (sphere $+$ cone AEF).

Now $\qquad\qquad\qquad$ cylinder $= 3$ (cone AEF); $\qquad\qquad\qquad$ [Eucl. xii. 10]

therefore $\qquad\qquad$ cone $AEF = 2$ (sphere).

But, since $EF = 2BD$,

$$\text{cone } AEF = 8 \text{ (cone } ABD);$$

therefore $\qquad\qquad$ sphere $= 4$ (cone ABD).

(2) Through B, D draw VBW, XDY parallel to AC; and imagine a cylinder which has AC for axis and the circles on VX, WY as diameters for bases.

Then $\qquad\qquad$ cylinder $VY = 2$(cylinder VD)

$$= 6 \text{(cone } ABD) \qquad\qquad \text{[Eucl. xii. 10]}$$

$$= \tfrac{3}{2} \text{(sphere), from above.} \qquad\qquad \text{Q.E.D.}$$

"From this theorem, to the effect that a sphere is four times as great as the cone with a great circle of the sphere as base and with height equal to the radius of the sphere, I conceived the notion that the surface of any sphere is four times as great as a great circle in it; for, judging from the fact that any circle is equal to a triangle with base equal to the circumference and height equal to the radius of the circle, I apprehended that, in like manner, any sphere is equal to a cone with base equal to the surface of the sphere and height equal to the radius."[7]

Proposition 3

By this method we can also investigate the theorem that

A cylinder with base equal to the greatest circle in a spheroid and height equal to the axis of the spheroid is $1\frac{1}{2}$ times the spheroid;

[7]That is to say, Archimedes originally solved the problem of finding the solid content of a sphere before that of finding its surface, and he inferred the result of the latter problem from that of the former. Yet in *On the Sphere and Cylinder* i. the surface is independently found (Prop. 33) and *before* the volume, which is found in Prop. 34: another illustration of the fact that the order of propositions in the treatises of the Greek geometers as finally elaborated does not necessarily follow the order of discovery.

and, when this is established, it is plain that

If any spheroid be cut by a plane through the centre and at right angles to the axis, the half of the spheroid is double of the cone which has the same base and the same axis as the segment (i.e. the half of the spheroid).

Let a plane through the axis of a spheroid cut its surface in the ellipse *ABCD*, the diameters (i.e. axes) of which are *AC*, *BD*; and let *K* be the centre.

Draw a circle about *BD* as diameter and in a plane perpendicular to *AC*; imagine a cone with this circle as base and *A* as vertex produced and cut by a plane through *C* parallel to its base; the section will be a circle in a plane at right angles to *AC* and about *EF* as diameter.

Imagine a cylinder with the latter circle as base and axis *AC*; produce *CA* to *H*, making *AH* equal to *CA*.

Let *HC* be regarded as the bar of a balance, *A* being its middle point.

In the parallelogram *LF* draw any straight line *MN* parallel to *EF* meeting the ellipse in *O*, *P* and *AE*, *AF*, *AC* in *Q*, *R*, *S* respectively.

If now a plane be drawn through *MN* at right angles to *AC*, it will cut the cylinder in a circle with diameter *MN*, the spheroid in a circle with diameter *OP*, and the cone in a circle with diameter *QR*.

Since *HA = AC*,

$$HA : AS = CA : AS$$
$$= EA : AQ$$
$$= MS : SQ.$$

Therefore $\qquad HA : AS = MS^2 : MS \cdot SQ.$

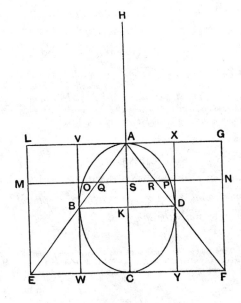

But, by the property of the ellipse,

$$AS \cdot SC : SO^2 = AK^2 : KB^2$$
$$= AS^2 : SQ^2;$$

therefore
$$SQ^2 : SO^2 = AS^2 : AS \cdot SC$$
$$= SQ^2 : SQ \cdot QM,$$

and accordingly
$$SO^2 = SQ \cdot QM.$$

Add SQ^2 to each side, and we have

$$SO^2 + SQ^2 = SQ \cdot SM.$$

Therefore, from above, we have

$$HA : AS = MS^2 : (SO^2 + SQ^2)$$
$$= MN^2 : (OP^2 + QR^2)$$
$$= \text{(circle, diam. } MN) : \text{(circle, diam. } OP + \text{circle, diam. } QR)$$

That is,

$$HA : AS = \text{(circle in cylinder)} : \text{(circle in spheroid} + \text{circle in cone)}.$$

Therefore the circle in the cylinder, in the place where it is, is in equilibrium, about A, with the circle in the spheroid and the circle in the cone together, if both the latter circles are placed with their centres of gravity at H.

Similarly for the three corresponding sections made by a plane perpendicular to AC and passing through any other straight line in the parallelogram LF parallel to EF.

If we deal in the same way with all the sets of three circles in which planes perpendicular to AC cut the cylinder, the spheroid and the cone, and which make up those figures respectively, it follows that the cylinder, in the place where it is, will be in equilibrium about A with the spheroid and the cone together, when both are placed with their centres of gravity at H.

Therefore, since K is the centre of gravity of the cylinder,

$$HA : AK = \text{(cylinder)} : \text{(spheroid} + \text{cone } AEF).$$

But $HA = 2AK$;

therefore
$$\text{cylinder} = 2 \text{ (spheroid} + \text{cone } AEF).$$

And
$$\text{cylinder} = 3 \text{ (cone } AEF);$$
[Eucl. xii. 10]

therefore
$$\text{cone } AEF = 2 \text{ (spheroid)}.$$

But, since $EF = 2BD$,

$$\text{cone } AEF = 8 \text{ (cone } ABD);$$

therefore
$$\text{spheroid} = 4 \text{ (cone } ABD),$$

and
$$\text{half the spheroid} = 2 \text{ (cone } ABD).$$

Through B, D draw VBW, XDY parallel to AC; and imagine a cylinder which has AC for axis and the circles on VX, WY as diameters for bases.

Then cylinder $VY = 2$(cylinder VD)

$= 6$(cone ABD)

$= \frac{3}{2}$ (spheroid), from above.   Q.E.D.

Proposition 4

Any segment of a right-angled conoid (i.e. a paraboloid of revolution) cut off by a plane at right angles to the axis is $1\frac{1}{2}$ times the cone which has the same base and the same axis as the segment.

This can be investigated by our method, as follows.

Let a paraboloid of revolution be cut by a plane through the axis in the parabola *BAC*;

and let it also be cut by another plane at right angles to the axis and intersecting the former plane in *BC*. Produce *DA*, the axis of the segment, to *H*, making *HA* equal to *AD*.

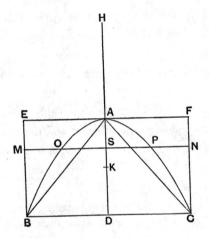

Imagine that *HD* is the bar of a balance, *A* being its middle point.

The base of the segment being the circle on *BC* as diameter and in a plane perpendicular to *AD*,

imagine (1) a cone drawn with the latter circle as base and *A* as vertex, and (2) a cylinder with the same circle as base and *AD* as axis.

In the parallelogram *EC* let any straight line *MN* be drawn parallel to *BC*, and through *MN* let a plane be drawn at right angles to *AD*; this plane will cut the cylinder in a circle with diameter *MN* and the paraboloid in a circle with diameter *OP*.

Now, *BAC* being a parabola and *BD*, *OS* ordinates,

$$DA : AS = BD^2 : OS^2,$$

or
$$HA : AS = MS^2 : SO^2.$$

Therefore
$$HA : AS = (\text{circle, rad. } MS) : (\text{circle, rad. } OS)$$
$$= (\text{circle in cylinder}) : (\text{circle in paraboloid}).$$

Therefore the circle in the cylinder, in the place where it is, will be in equilibrium, about A, with the circle in the paraboloid, if the latter is placed with its centre of gravity at H.

Similarly for the two corresponding circular sections made by a plane perpendicular to AD and passing through any other straight line in the parallelogram which is parallel to BC.

Therefore, as usual, if we take all the circles making up the whole cylinder and the whole segment and treat them in the same way, we find that the cylinder, in the place where it is, is in equilibrium about A with the segment placed with its centre of gravity at H.

If K is the middle point of AD, K is the centre of gravity of the cylinder;

therefore $\qquad HA : AK = (\text{cylinder}) : (\text{segment}).$

Therefore $\qquad \text{cylinder} = 2(\text{segment}).$

And $\qquad \text{cylinder} = 3(\text{cone } ABC); \qquad\qquad$ [Eucl. xii. 10]

therefore $\qquad \text{segment} = \frac{3}{2}(\text{cone } ABC).$

Proposition 5

The centre of gravity of a segment of a right-angled conoid (i.e. a paraboloid of revolution) cut off by a plane at right angles to the axis is on the straight line which is the axis of the segment, and divides the said straight line in such a way that the portion of it adjacent to the vertex is double of the remaining portion.

This can be investigated by the method, as follows.

Let a paraboloid of revolution be cut by a plane through the axis in the parabola BAC;

and let it also be cut by another plane at right angles to the axis and intersecting the former plane in BC.

Produce DA, the axis of the segment, to H, making HA equal to AD; and imagine DH to be the bar of a balance, its middle point being A.

The base of the segment being the circle on BC as diameter and in a plane perpendicular to AD,

imagine a cone with this circle as base and A as vertex, so that AB, AC are generators of the cone.

In the parabola let any double ordinate OP be drawn meeting AB, AD, AC in Q, S, R respectively.

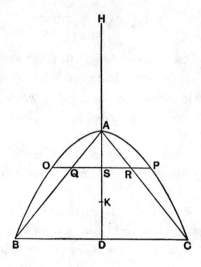

Now, from the property of the parabola,

$$BD^2 : OS^2 = DA : AS$$
$$= BD : QS$$
$$= BD^2 : BD \cdot QS.$$

Therefore $\qquad OS^2 = BD \cdot QS,$

or $\qquad BD : OS = OS : QS,$

whence $\qquad BD : QS = OS^2 : QS^2.$

But $\qquad BD : QS = AD : AS$
$$= HA : AS.$$

Therefore $\qquad HA : AS = OS^2 : QS^2$
$$= OP^2 : QR^2.$$

If now through OP a plane be drawn at right angles to AD, this plane cuts the paraboloid in a circle with diameter OP and the cone in a circle with diameter QR.

We see therefore that

$$HA : AS = (\text{circle, diam. } OP) : (\text{circle, diam. } QR)$$
$$= (\text{circle in paraboloid}) : (\text{circle in cone});$$

and the circle in the paraboloid, in the place where it is, is in equilibrium about A with the circle in the cone placed with its centre of gravity at H.

Similarly for the two corresponding circular sections made by a plane perpendicular to AD and passing through any other ordinate of the parabola.

Dealing therefore in the same way with all the circular sections which make up the whole of the segment of the paraboloid and the cone respectively, we see that the segment of the paraboloid, in the place where it is, is in equilibrium about A with the cone placed with its centre of gravity at H.

Now, since A is the centre of gravity of the whole system as placed, and the centre of gravity of part of it, namely the cone, as placed, is at H, the centre of gravity of the rest, namely the segment, is at a point K on HA produced such that

$$HA : AK = \text{(segment)} : \text{(cone)}.$$

But $\qquad\qquad\qquad$ segment $= \frac{3}{2}$(cone). $\qquad\qquad\qquad\qquad$ [Prop. 4]

Therefore $\qquad\qquad\qquad HA = \frac{3}{2}AK$;

that is, K divides AD in such a way that $AK = 2KD$.

PROPOSITION 6

The centre of gravity of any hemisphere [is on the straight line which] is its axis, and divides the said straight line in such a way that the portion of it adjacent to the surface of the hemisphere has to the remaining portion the ratio which 5 has to 3.

Let a sphere be cut by a plane through its centre in the circle $ABCD$; let AC, BD be perpendicular diameters of this circle, and through BD let a plane be drawn at right angles to AC.

The latter plane will cut the sphere in a circle on BD as diameter.

Imagine a cone with the latter circle as base and A as vertex.

Produce CA to H, making AH equal to CA, and let HC be regarded as the bar of a balance, A being its middle point.

In the semicircle BAD, let any straight line OP be drawn parallel to BD and cutting AC in E and the two generators AB, AD of the cone in Q, R respectively. Join AO.

Through OP let a plane be drawn at right angles to AC; this plane will cut the hemisphere in a circle with diameter OP and the cone in a circle with diameter QR.

Now

$$HA : AE = AC : AE$$
$$= AO^2 : AE^2$$
$$= (OE^2 + AE^2) : AE^2$$
$$= (OE^2 + QE^2) : QE^2$$
$$= \text{(circle, diam. } OP + \text{circle, diam. } QR) : \text{(circle, diam. } QR).$$

Therefore the circles with diameters OP, QR, in the places where they are, are in equilibrium about A with the circle with diameter QR if the latter is placed with its centre of gravity at H.

And, since the centre of gravity of the two circles with diameters OP, QR taken together, in the place where they are, is......

[There is a lacuna here; but the proof can easily be completed on the lines of the corresponding but more difficult case in Prop. 8.

221

We proceed thus from the point where the circles with diameters OP, QR, in the place where they are, balance, about A, the circle with diameter QR placed with its centre of gravity at H.

A similar relation holds for all the other sets of circular sections made by other planes passing through points on AG and at right angles to AG.

Taking then all the circles which fill up the hemisphere BAD and the cone ABD respectively, we find that the hemisphere BAD and the cone ABD, in the places where they are, together balance, about A, a cone equal to ABD placed with its centre of gravity at H.

Let the cylinder $M + N$ be equal to the cone ABD.

Then, since the cylinder $M + N$ placed with its centre of gravity at H balances the hemisphere BAD and the cone ABD in the places where they are,

suppose that the portion M of the cylinder, placed with its centre of gravity at H, balances the cone ABD (alone) in the place where it is; therefore the portion N of the cylinder placed with its centre of gravity at H balances the hemisphere (alone) in the place where it is.

Now the centre of gravity of the cone is at a point V such that $AG = 4GV$; therefore, since M at H is in equilibrium with the cone,

$$M : (\text{cone}) = \tfrac{3}{4}AG : HA = \tfrac{3}{8}AC : AC,$$

whence $\qquad\qquad\qquad M = \tfrac{3}{8} (\text{cone}).$

But $M + N = (\text{cone})$; therefore $N = \tfrac{5}{8} (\text{cone})$.

Now let the centre of gravity of the hemisphere be at W, which is somewhere on AG.

Then, since N at H balances the hemisphere alone,

$$(\text{hemisphere}) : N = HA : AW.$$

But the hemisphere $BAD = $ twice the cone ABD;

[On the Sphere and Cylinder i. 34 and Prop. 2 above]

and $N = \tfrac{5}{8} (\text{cone})$, from above.

Therefore $\qquad\qquad 2 : \tfrac{5}{8} = HA : AW$

$$= 2AG : AW,$$

whence $AW = \tfrac{5}{8}AG$, so that W divides AG in such a way that

$$AW : WG = 5 : 3.]$$

Proposition 7

We can also investigate by the same method the theorem that

[*Any segment of a sphere has*] *to the cone* [*with the same base and height the ratio which the sum of the radius of the sphere and the height of the complementary segment has to the height of the complementary segment.*]

[There is a lacuna here; but all that is missing is the construction, and the construction is easily understood by means of the figure. *BAD* is of course the segment

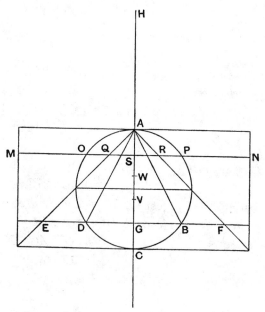

of the sphere the volume of which is to be compared with the volume of a cone with the same base and height.]

The plane drawn through *MN* and at right angles to *AC* will cut the cylinder in a circle with diameter *MN*, the segment of the sphere in a circle with diameter *OP*, and the cone on the base *EF* in a circle with diameter *QR*.

In the same way as before [cf. Prop. 2] we can prove that the circle with diameter *MN*, in the place where it is, is in equilibrium about *A* with the two circles with diameters *OP*, *QR* if these circles are both moved and placed with their centres of gravity at *H*.

The same thing can be proved of all sets of three circles in which the cylinder, the segment of the sphere, and the cone with the common height *AG* are all cut by any plane perpendicular to *AC*.

Since then the sets of circles make up the whole cylinder, the whole segment of the sphere and the whole cone respectively, it follows that the cylinder, in the place where it is, is in equilibrium about *A* with the sum of the segment of the sphere and the cone if both are placed with their centres of gravity at *H*.

Divide *AG* at *W*, *V* in such a way that

$$AW = WG, \quad AV = 3VG.$$

Therefore *W* will be the centre of gravity of the cylinder, and *V* will be the centre of gravity of the cone.

Since, now, the bodies are in equilibrium as described,

(cylinder) : (cone AEF + segment BAD of sphere)
$$= HA : AW.$$

....

[The rest of the proof is lost; but it can easily be supplied thus.
We have

$$(\text{cone } AEF + \text{segmt. } BAD) : (\text{cylinder}) = AW : AC$$
$$= AW \cdot AC : AC^2.$$

But $\qquad$ (cylinder) : (cone AEF) $= AC^2 : \tfrac{1}{3}EG^2$
$$= AC^2 : \tfrac{1}{3}AG^2.$$

Therefore, *ex aequali*,

$$(\text{cone } AEF + \text{segmt. } BAD) : (\text{cone } AEF) = AW \cdot AC : \tfrac{1}{3}AG^2$$
$$= \tfrac{1}{2}AC : \tfrac{1}{3}AG,$$

whence $\qquad$ (segmt. BAD) : (cone AEF) $= (\tfrac{1}{2}AC - \tfrac{1}{3}AG) : \tfrac{1}{3}AG.$

Again $\qquad$ (cone AEF) : (cone ABD) $= EG^2 : DG^2$
$$= AG^2 : AG \cdot GC$$
$$= AG : GC$$
$$= \tfrac{1}{3}AG : \tfrac{1}{3}GC.$$

Therefore, *ex aequali*,

$$(\text{segment } BAD) : (\text{cone } ABD) = (\tfrac{1}{2}AC - \tfrac{1}{3}AG) : \tfrac{1}{3}GC$$
$$= (\tfrac{3}{2}AC - AG) : GC$$
$$= (\tfrac{1}{2}AC + GC) : GC. \quad \text{Q.E.D.}]$$

Proposition 8

[The enunciation, the setting-out, and a few
words of the construction are missing.

The enunciation however can be supplied from
that of Prop. 9, with which it must be identical
except that it cannot refer to "*any* segment," and
the presumption therefore is that the proposition
was enunciated with reference to one kind of segment only, i.e. either a segment greater than a hemisphere or a segment less than a hemisphere.

Heiberg's figure corresponds to the case of a
segment greater than a hemisphere. The segment
investigated is of course the segment *BAD*. The
setting-out and construction are self-evident from
the figure.]

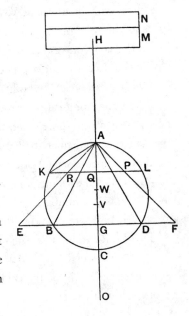

Produce AC to H, O, making HA equal to AC and CO equal to the radius of the sphere;

and let HC be regarded as the bar of a balance, the middle point being A.

In the plane cutting off the segment describe a circle with G as centre and radius (GE) equal to AG; and on this circle as base, and with A as vertex, let a cone be described. AE, AF are generators of this cone.

Draw KL, through any point Q on AG, parallel to EF and cutting the segment in K, L, and AE, AF in R, P respectively. Join AK.

Now $$HA : AQ = CA : AQ$$
$$= AK^2 : AQ^2$$
$$= (KQ^2 + QA^2) : QA^2$$
$$= (KQ^2 + PQ^2) : PQ^2$$
$$= (\text{circle, diam. } KL + \text{circle, diam. } PR)$$
$$: (\text{circle, diam. } PR).$$

Imagine a circle equal to the circle with diameter PR placed with its centre of gravity at H;

therefore the circles on diameters KL, PR, in the places where they are, are in equilibrium about A with the circle with diameter PR placed with its centre of gravity at H.

Similarly for the corresponding circular sections made by any other plane perpendicular to AG.

Therefore, taking all the circular sections which make up the segment ABD of the sphere and the cone AEF respectively, we find that the segment ABD of the sphere and the cone AEF, in the places where they are, are in equilibrium with the cone AEF assumed to be placed with its centre of gravity at H.

Let the cylinder $M + N$ be equal to the cone AEF which has A for vertex and the circle on EF as diameter for base.

Divide AG at V so that $AG = 4VG$;

therefore V is the centre of gravity of the cone AEF; "for this has been proved before."[8]

Let the cylinder $M + N$ be cut by a plane perpendicular to the axis in such a way that the cylinder M (alone), placed with its centre of gravity at H, is in equilibrium with the cone AEF.

Since $M + N$ suspended at H is in equilibrium with the segment ABD of the sphere and the cone AEF in the places where they are,

while M, also at H, is in equilibrium with the cone AEF in the place where it is, it follows that

[8] cf. note on p. 211 above.

N at *H* is in equilibrium with the segment *ABD* of the sphere in the place where it is.

$$\text{Now (segment } ABD \text{ of sphere)} : (\text{cone } ABD)$$
$$= OG : GC;$$

"for this is already proved" [Cf. *On the Sphere and Cylinder* ii. 2 Cor. as well as Prop. 7 *ante*].

And (cone *ABD*) : (cone *AEF*)
$$= (\text{circle, diam. } BD) : (\text{circle, diam. } EF)$$
$$= BD^2 : EF^2$$
$$= BG^2 : GE^2$$
$$= CG \cdot GA : GA^2$$
$$= CG : GA.$$

Therefore, *ex aequali*,
$$(\text{segment } ABD \text{ of sphere}) : (\text{cone } AEF)$$
$$= OG : GA.$$

Take a point *W* on *AG* such that
$$AW : WG = (GA + 4GC) : (GA + 2GC).$$

We have then, inversely,
$$GW : WA = (2GC + GA) : (4GC + GA),$$

and, *componendo*,
$$GA : AW = (6GC + 2GA) : (4GC + GA).$$

But $GO = \frac{1}{4}(6GC + 2GA)$, [for $GO - GC = \frac{1}{2}(CG + GA)$]

and $CV = \frac{1}{4}(4GC + GA)$;

therefore $GA : AW = OG : CV$,

and, alternately and inversely,
$$OG : GA = CV : WA.$$

It follows, from above, that
$$(\text{segment } ABD \text{ of sphere}) : (\text{cone } AEF) = CV : WA.$$

Now, since the cylinder *M* with its centre of gravity at *H* is in equilibrium about *A* with the cone *AEF* with its centre of gravity at *V*,
$$(\text{cone } AEF) : (\text{cylinder } M) = HA : AV$$
$$= CA : AV;$$

and, since the cone *AEF* = the cylinder *M* + *N*, we have, *dividendo* and *invertendo*,
$$(\text{cylinder } M) : (\text{cylinder } N) = AV : CV.$$

Hence, *componendo*,
$$(\text{cone } AEF) : (\text{cylinder } N) = CA : CV \text{[9]}$$
$$= HA : CV.$$

---

[9]Archimedes arrives at this result in a very roundabout way, seeing that it could have been obtained at once *convertendo*. Cf. Euclid x. 14.

But it was proved that

(segment *ABD* of sphere) : (cone *AEF*) = *CV* : *WA*;

therefore, *ex aequali*,

(segment *ABD* of sphere) : (cylinder *N*) = *HA* : *AW*.

And it was above proved that the cylinder *N* at *H* is in equilibrium about *A* with the segment *ABD*, in the place where it is;

therefore, since *H* is the centre of gravity of the cylinder *N*, *W* is the centre of gravity of the segment *ABD* of the sphere.

Proposition 9

In the same way we can investigate the theorem that

The centre of gravity of any segment of a sphere is on the straight line which is the axis of the segment, and divides this straight line in such a way that the part of it adjacent to the vertex of the segment has to the remaining part the ratio which the sum of the axis of the segment and four times the axis of the complementary segment has to the sum of the axis of the segment and double the axis of the complementary segment.

[As this theorem relates to "*any* segment" but states the same result as that proved in the preceding proposition, it follows that Prop. 8 must have related to one kind of segment, either a segment greater than a semicircle (as in Heiberg's figure of Prop. 8) or a segment less than a semicircle; and the present proposition completed the proof for both kinds of segments. It would only require a slight change in the figure, in any case.]

Proposition 10

By this method too we can investigate the theorem that

[*A segment of an obtuse-angled conoid (i.e. a hyperboloid of revolution) has to the cone which has*] the same base [*as the segment and equal height the same ratio as the sum of the axis of the segment and three times*] the "annex to the axis" (*i.e. half the transverse axis of the hyperbolic section through the axis of the hyperboloid or, in other words, the distance between the vertex of the segment and the vertex of the enveloping cone*) has to the sum of the axis of the segment and double of the "annex"[10] [this is the theorem proved in *On Conoids and Spheroids*, Prop. 25], "and also many other theorems, which, as the method has been made clear by means of the foregoing examples, I will omit, in order that I may now proceed to compass the proofs of the theorems mentioned above."

[10] The text has "triple" ($\tau\rho\iota\pi\lambda\alpha\sigma\iota\alpha\nu$) in the last line instead of "double." As there is a considerable lacuna before the last few lines, a theorem about the centre of gravity of a segment of a hyperboloid of revolution may have fallen out.

Proposition 11

If in a right prism with square bases a cylinder be inscribed having its bases in opposite square faces and touching with its surface the remaining four parallelogrammic faces, and if through the centre of the circle which is the base of the cylinder and one side of the opposite square face a plane be drawn, the figure cut off by the plane so drawn is one sixth part of the whole prism.

"This can be investigated by the method, and, when it is set out, I will go back to the proof of it by geometrical considerations."

[The investigation by the mechanical method is contained in the two Propositions, 11, 12. Prop. 13 gives another solution which, although it contains no mechanics, is still of the character which Archimedes regards as inconclusive, since it assumes that the solid is actually *made up* of parallel plane sections and that an auxiliary parabola is actually *made up* of parallel straight lines in it. Prop. 14 added the conclusive geometrical proof.]

Let there be a right prism with a cylinder inscribed as stated.

Let the prism be cut through the axis of the prism and cylinder by a plane perpendicular to the plane which cuts off the portion of the cylinder; let this plane make, as section, the parallelogram *AB*, and let it cut the plane cutting off the portion of the cylinder (which plane is perpendicular to *AB*) in the straight line *BC*.

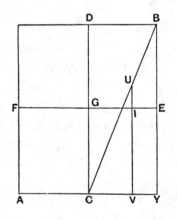

Let *CD* be the axis of the prism and cylinder, let *EF* bisect it at right angles, and through *EF* let a plane be drawn at right angles to *CD*; this plane will cut the prism in a square and the cylinder in a circle.

Let *MN* be the square and *OPQR* the circle, and let the circle touch the sides of the square in *O*, *P*, *Q*, *R* [*F*, *E* in the first figure are identical with *O*, *Q* respectively]. Let *H* be the centre of the circle.

Let *KL* be the intersection of the plane through *EF* perpendicular to the axis of the cylinder and the plane cutting off the portion of the cylinder; *KL* is bisected by *OHQ* [and passes through the middle point of *HQ*].

Let any chord of the circle, as *ST*, be drawn perpendicular to *HQ*, meeting *HQ* in *W*; and through *ST* let a plane be drawn at right angles to *OQ* and produced on both sides of the plane of the circle *OPQR*.

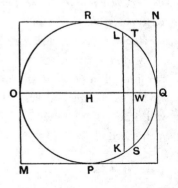

The plane so drawn will cut the half cylinder having the semicircle *PQR* for section and the axis of the prism for height in a parallelogram, one side of which is equal to *ST* and another is a generator of the cylinder; and it will also cut the portion of the cylinder cut off in a parallelogram, one side of which is equal to *ST* and the other is equal and parallel to *UV* (in the first figure).

UV will be parallel to *BY* and will cut off, along *EG* in the parallelogram *DE*, the segment *EI* equal to *QW.*

Now, since *EC* is a parallelogram, and *VI* is parallel to *GC,*

$$EG : GI = YC : CV$$
$$= BY : UV$$
$$= (\square \text{ in half cyl.}) : (\square \text{ in portion of cyl.}).$$

And *EG = HQ, GI = HW, QH = OH;*

therefore $\qquad OH : HW = (\square \text{ in half cyl.}) : (\square \text{ in portion}).$

Imagine that the parallelogram in the portion of the cylinder is moved and placed at *O* so that *O* is its centre of gravity, and that *OQ* is the bar of a balance, *H* being its middle point.

Then, since *W* is the centre of gravity of the parallelogram in the half cylinder, it follows from the above that the parallelogram in the half cylinder, in the place where it is, with its centre of gravity at *W*, is in equilibrium about *H* with the parallelogram in the portion of the cylinder when placed with its centre of gravity at *O*.

Similarly for the other parallelogrammic sections made by any plane perpendicular to *OQ* and passing through any other chord in the semicircle *PQR* perpendicular to *OQ*.

If then we take all the parallelograms making up the half cylinder and the portion of the cylinder respectively, it follows that the half cylinder, in the place where it is, is in equilibrium about *H* with the portion of the cylinder cut off when the latter is placed with its centre of gravity at *O*.

Proposition 12

Let the parallelogram (square) *MN* perpendicular to the axis, with the circle *OPQR* and its diameters *OQ, PR*, be drawn separately.

Join *HG, HM*, and through them draw planes at right angles to the plane of the circle, producing them on both sides of that plane.

This produces a prism with triangular section *GHM* and height equal to the axis of the cylinder; this prism is $\frac{1}{4}$ of the original prism circumscribing the cylinder.

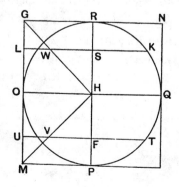

Let *LK, UT* be drawn parallel to *OQ* and equidistant from it, cutting the circle in *K, T, RP* in S, *F*, and *GH, HM* in *W, V* respectively.

Through *LK, UT* draw planes at right angles to *PR*, producing them on both sides of the plane of the circle; these planes produce as sections in the half cylinder *PQR* and in the prism *GHM* four parallelograms in which the heights are equal to the axis of the cylinder, and the other sides are equal to *KS, TF, LW, UV* respectively. . . . .

. . . .

[The rest of the proof is missing, but, as Zeuthen says,[11] the result obtained and the method of arriving at it are plainly indicated by the above.

Archimedes wishes to prove that the half cylinder *PQR*, in the place where it is, balances the prism *GHM*, in the place where it is, about *H* as fixed point.

He has first to prove that the elements (1) the parallelogram with side = *KS* and (2) the parallelogram with side = *LW*, in the places where they are, balance about S, or, in other words that the straight lines *SK, LW*, in the places where they are, balance about S.

Now $\qquad$ (radius of circle $OPQR)^2 = SK^2 + SH^2$,

or $$SL^2 = SK^2 + SW^2.$$

Therefore $\qquad LS^2 - SW^2 = SK^2$,

and accordingly $\qquad (LS + SW) \cdot LW = SK^2$,

whence $\qquad \frac{1}{2}(LS + SW) : \frac{1}{2}SK = SK : LW.$

And $\frac{1}{2}(LS + SW)$ is the distance of the centre of gravity of *LW* from S, while $\frac{1}{2}SK$ is the distance of the centre of gravity of *SK* from S.

Therefore *SK* and *LW*, in the places where they are, balance about S.

Similarly for the corresponding parallelograms.

Taking *all* the parallelogrammic elements in the half cylinder and prism respectively, we find that the half cylinder *PQR* and the prism *GHM*, in the places where they are respectively, balance about *H*.

From this result and that of Prop. 11 we can at once deduce the volume of the portion cut off from the cylinder. For in Prop. 11 the portion of the cylinder, placed with its centre of gravity at *O*, is shown to balance (about *H*) the half-cylinder in the place where it is. By Prop. 12 we may substitute for the half-cylinder in the place where it is the prism *GHM* of that proposition turned the opposite way relatively to *RP*. The centre of gravity of the prism as thus placed is at a point (say *Z*) on *HQ* such that $HZ = \frac{2}{3}HQ$.

---

[11]Zeuthen in *Bibliotheca Mathematica* VII, 1906–7.

Therefore, assuming the prism to be applied at its centre of gravity, we have

$$\text{(portion of cylinder)} : \text{(prism)} = \tfrac{2}{3}HQ : OH$$
$$= 2 : 3;$$

therefore
$$\text{(portion of cylinder)} = \tfrac{2}{3}(\text{prism } GHM)$$
$$= \tfrac{1}{6}(\text{original prism}).$$

Note. This proposition of course solves the problem of finding the centre of gravity of a half cylinder or, in other words, of a semicircle.

For the triangle *GHM* in the place where it is balances, about *H*, the semicircle *PQR* in the place where it is.

If then *X* is the point on *HQ* which is the centre of gravity of the semicircle,
$$\tfrac{2}{3}HO \cdot (\triangle GHM) = HX \cdot (\text{semicircle } PQR),$$
or
$$\tfrac{2}{3}HO \cdot HO^2 = HX \cdot \tfrac{1}{2}\pi \cdot HO^2;$$

that is,
$$HX = \frac{4}{3\pi} \cdot HQ.]$$

Proposition 13

Let there be a right prism with square bases, one of which is *ABCD*; in the prism let a cylinder be inscribed, the base of which is the circle *EFGH* touching the sides of the square *ABCD* in *E*, *F*, *G*, *H*.

Through the centre and through the side corresponding to *CD* in the square face *opposite* to *ABCD* let a plane be drawn; this will cut off a prism equal to $\frac{1}{4}$ of the original prism and formed by three parallelograms and two triangles, the triangles forming opposite faces.

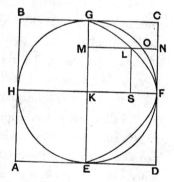

In the semicircle *EFG* describe the parabola which has *FK* for axis and passes through *E*, *G*; draw *MN* parallel to *KF* meeting *GE* in *M*, the parabola in *L*, the semicircle in *O* and *CD* in *N*.

Then
$$MN \cdot NL = NF^2;$$
"for this is clear."

[Cf. Apollonius, *Conics* i. 11]

[The parameter is of course equal to *GK* or *KF*.]

Therefore
$$MN : NL = GK^2 : LS^2.$$

Through *MN* draw a plane at right angles to *EG*;

this will produce as sections (1) in the prism cut off from the whole prism a right-angled triangle, the base of which is *MN*, while the perpendicular is perpendicular at *N* to the plane *ABCD* and equal to the axis of the cylinder, and the hypotenuse is in the plane cutting the cylinder, and (2) in the portion of the cylinder cut off a right-angled triangle the base of which is *MO*, while the perpendicular is the generator of the cylinder perpendicular at *O* to the plane *KN*, and the hypotenuse is . . . .

. . . .

[There is a lacuna here, to be supplied as follows.

Since
$$MN : NL = GK^2 : LS^2$$
$$= MN^2 : LS^2,$$

it follows that
$$MN : ML = MN^2 : (MN^2 - LS^2)$$
$$= MN^2 : (MN^2 - MK^2)$$
$$= MN^2 : MO^2.$$

But the triangle (1) in the prism is to the triangle (2) in the portion of the cylinder in the ratio of $MN^2 : MO^2$.

Therefore

($\triangle$ in prism) : ($\triangle$ in portion of cylinder)

$= MN : ML$

$=$ (straight line in rect. *DG*) : (straight line in parabola).

We now take all the corresponding elements in the prism, the portion of the cylinder, the rectangle *DG* and the parabola *EFG* respectively;]

and it will follow that

(all the $\triangle$s in prism) : (all the $\triangle$s in portion of cylinder)

$=$ (all the str. lines in $\square$ *DG*)

: (all the straight lines between parabola and *EG*).

But the prism is made up of the triangles in the prism, [the portion of the cylinder is made up of the triangles in it], the parallelogram *DG* of the straight lines in it parallel to *KF*, and the parabolic segment of the straight lines parallel to *KF* intercepted between its circumference and *EG*;

therefore
(prism) : (portion of cylinder)

$=$ ($\square$ *GD*) : (parabolic segment *EFG*).

But
$\square$ $GD = \frac{3}{2}$ (parabolic segment *EFG*);

"for this is proved in my earlier treatise."

[*Quadrature of Parabola*]

Therefore
prism $= \frac{3}{2}$ (portion of cylinder).

If then we denote the portion of the cylinder by 2, the prism is 3, and the original prism circumscribing the cylinder is 12 (being 4 times the other prism);

therefore the portion of the cylinder $= \frac{1}{6}$(original prism).

Q.E.D.

[The above proposition and the next are peculiarly interesting for the fact that the parabola is an auxiliary curve introduced for the sole purpose of analytically reducing the required cubature to the known quadrature of the parabola.]

Proposition 14

Let there be a right prism with square bases [and a cylinder inscribed therein having its base in the square *ABCD* and touching its sides at *E, F, G, H*; let the cylinder be cut by a plane through *EG* and the side corresponding to *CD* in the square face opposite to *ABCD*].

This plane cuts off from the prism a prism, and from the cylinder a portion of it.

It can be proved that the portion of the cylinder cut off by the plane is $\frac{1}{6}$ of the whole prism.

But we will first prove that it is possible to inscribe in the portion cut off from the cylinder, and to circumscribe about it, solid figures made up of prisms which have equal height and similar triangular bases, in such a way that the circumscribed figure exceeds the inscribed by less than any assigned magnitude. . . . .

. . . .

But it was proved that

(prism cut off by oblique plane)

$$< \tfrac{3}{2}(\text{figure inscribed in portion of cylinder}).$$

Now

(prism cut off) : (inscribed figure)

$$= \square \, DG : (\square\text{s inscribed in parabolic segment});$$

therefore $\qquad \square \, DG < \tfrac{3}{2}(\square\text{s in parabolic segment}):$

which is impossible, since "it has been proved elsewhere" that the parallelogram *DG* is $\frac{3}{2}$ of the parabolic segment.

Consequently. . . . .

. . . . not greater.

. . . .

And (all the prisms in prism cut off)

$\qquad\qquad\qquad$: (all prisms in circumscr. figure)

$$= (\text{all } \square\text{s in } \square \, DG)$$

$\qquad\qquad$: (all $\square$s in fig. circumscr. about parabolic segmt.);

therefore

(prism cut off) : (figure circumscr. about portion of cylinder)

$$= (\square \, DG) : (\text{figure circumscr. about parabolic segment}).$$

But the prism cut off by the oblique plane is $> \frac{3}{2}$ of the solid figure circumscribed about the portion of the cylinde. . . . .

. . . .

[There are large gaps in the exposition of this geometrical proof, but the way in which the method of exhaustion was applied, and the parallelism between this and other applications of it, are clear. The first fragment shows that solid figures made up of prisms were circumscribed and inscribed to the portion of the cylinder. The parallel triangular faces of these prisms were perpendicular to GE in the figure of Prop. 13; they divided GE into equal portions of the requisite smallness; each section of the portion of the cylinder by such a plane was a triangular face common to an inscribed and a circumscribed right prism. The planes also produced prisms in the prism cut off by the same oblique plane as cuts off the portion of the cylinder and standing on GD as base.

The number of parts into which the parallel planes divided GE was made great enough to secure that the circumscribed figure exceeded the inscribed figure by less than a small assigned magnitude.

The second part of the proof began with the assumption that the portion of the cylinder is $> \frac{2}{3}$ of the prism cut off; and this was proved to be impossible, by means of the use of the auxiliary parabola and the proportion

$$MN : ML = MN^2 : MO^2$$

which are employed in Prop. 13.

We may supply the missing proof as follows.[12]

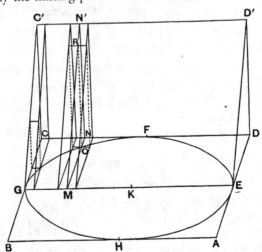

In the accompanying figure are represented (1) the first element-prism circumscribed to the portion of the cylinder, (2) two element-prisms adjacent to the ordinate

[12]It is right to mention that this has already been done by Th. Reinach in his version of the treatise ("Un Traité de Géométrie inédit d'Archimède" in *Revue générale des sciences pures et appliquées*, 30 Nov. and 15 Dec. 1907); but I prefer my own statement of the proof.

OM, of which that on the left is circumscribed and that on the right (equal to the other) inscribed, (3) the corresponding element-prisms forming part of the prism cut off (*CC'GEDD'*) which is $\frac{1}{4}$ of the original prism.

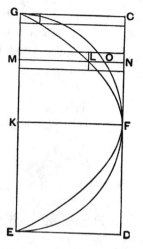

In the second figure are shown element-rectangles circumscribed and inscribed to the auxiliary parabola, which rectangles correspond exactly to the circumscribed and inscribed element-prisms represented in the first figure (the length of *GM* is the same in both figures, and the breadths of the element-rectangles are the same as the heights of the element-prisms); the corresponding element-rectangles forming part of the rectangle *GD* are similarly shown.

For convenience we suppose that *GE* is divided into an even number of equal parts, so that *GK* contains an integral number of these parts.

For the sake of brevity we will call each of the two element-prisms of which *OM* is an edge "el. prism (*O*)" and each of the element-prisms of which *MNN'* is a common face "el. prism (*N*)." Similarly we will use the corresponding abbreviations "el. rect. (*L*)" and "el. rect. (*N*)" for the corresponding elements in relation to the auxiliary parabola as shown in the second figure.

Now it is easy to see that the figure made up of all the inscribed prisms is less than the figure made up of the circumscribed prisms by twice the final circumscribed prism adjacent to *FK*, i.e. by twice "el. prism (*N*)"; and, as the height of this prism may be made as small as we please by dividing *GK* into sufficiently small parts, it follows that inscribed and circumscribed solid figures made up of element-prisms can be drawn differing by less than any assigned solid figure.

(1) Suppose, if possible, that

$$\text{(portion of cylinder)} > \tfrac{2}{3}\text{(prism cut off)},$$

or

$$\text{(prism cut off)} < \tfrac{3}{2}\text{(portion of cylinder)}.$$

Let (prism cut off) $= \tfrac{3}{2}$(portion of cylinder $- X$), say.

Construct circumscribed and inscribed figures made up of element-prisms, such that

$$\text{(circumscr. fig.)} - \text{(inscr. fig.)} < X.$$

Therefore

$$\text{(inscr. fig.)} > \text{(circumscr. fig.} - X),$$

and *a fortiori*

$$> \text{(portion of cyl.} - X).$$

It follows that

$$\text{(prism cut off)} < \tfrac{3}{2}\text{(inscribed figure)}.$$

Considering now the element-prisms in the prism cut off and those in the inscribed figure respectively, we have

$$\text{el. prism } (N) : \text{el. prism } (O) = MN^2 : MO^2$$
$$= MN : ML \qquad\qquad \text{[as in Prop. 13]}$$
$$= \text{el. rect. } (N) : \text{el. rect. } (L).$$

It follows that

$$\Sigma\{\text{el. prism } (N)\} : \Sigma\{\text{el. prism } (O)\}$$
$$= \Sigma \{\text{el. rect. } (N)\} : \Sigma \{\text{el. rect. } (L)\}.$$

(There are really two more prisms and rectangles in the first and third than there are in the second and fourth terms respectively; but this makes no difference because the first and third terms may be multiplied by a common factor as $n/(n-2)$ without affecting the truth of the proportion.)

Therefore

$$(\text{prism cut off}) : (\text{figure inscr. in portion of cyl.})$$
$$= (\text{rect. } GD) : (\text{fig. inscr. in parabola}).$$

But it was proved above that

$$(\text{prism cut off}) < \tfrac{3}{2} (\text{fig. inscr. in portion of cyl.});$$

therefore $\qquad\qquad (\text{rect. } GD) < \tfrac{3}{2} (\text{fig. inscr. in parabola}),$

and, *a fortiori*,

$$(\text{rect. } GD) < \tfrac{3}{2} (\text{parabolic segmt.}) :$$

which is impossible, since

$$(\text{rect. } GD) = \tfrac{3}{2} (\text{parabolic segmt.}).$$

Therefore

$$(\text{portion of cyl.}) \text{ is } not \text{ greater than } \tfrac{2}{3} (\text{prism cut off}).$$

(2) In the second lacuna must have come the beginning of the next *reductio ad absurdum* demolishing the other possible assumption that the portion of the cylinder is $< \tfrac{2}{3}$ of the prism cut off.

In this case our assumption is that

$$(\text{prism cut off}) > \tfrac{3}{2} (\text{portion of cylinder});$$

and we circumscribe and inscribe figures made up of element-prisms, such that

$$(\text{prism cut off}) > \tfrac{3}{2} (\text{fig. circumscr. about portion of cyl.}).$$

We now consider the element-prisms in the prism cut off and in the circumscribed figure respectively, and the same argument as above gives

$$(\text{prism cut off}) : (\text{fig. circumscr. about portion of cyl.})$$
$$= (\text{rect. } GD) : (\text{fig. circumscr. about parabola}),$$

whence it follows that

$$(\text{rect. } GD) > \tfrac{3}{2} (\text{fig. circumscribed about parabola}),$$

and, *a fortiori*,
$$(\text{rect. } GD) > \tfrac{3}{2}(\text{parabolic segment}) :$$
which is impossible, since
$$(\text{rect. } GD) = \tfrac{3}{2}(\text{parabolic segmt.}).$$

Therefore
$$(\text{portion of cyl.}) \text{ is } not \text{ less than } \tfrac{2}{3}(\text{prism cut off}).$$

But it was also proved that neither is it greater;

therefore
$$(\text{portion of cyl.}) = \tfrac{2}{3}(\text{prism cut off})$$
$$= \tfrac{1}{6}(\text{original prism}).]$$

Proposition 15

[This proposition, which is lost, would be the mechanical investigation of the second of the two special problems mentioned in the preface to the treatise, namely that of the cubature of the figure included between two cylinders, each of which is inscribed in one and the same cube so that its opposite bases are in two opposite faces of the cube and its surface touches the other four faces.

Zeuthen has shown how the mechanical method can be applied to this case.[13]

In the accompanying figure *VWYX* is a section of the cube by a plane (that of the paper) passing through the axis *BD* of one of the cylinders inscribed in the cube and parallel to two opposite faces.

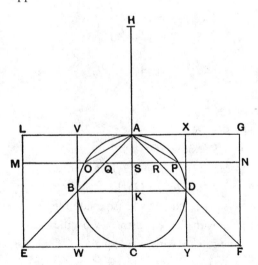

The same plane gives the circle *ABCD* as the section of the other inscribed cylinder with axis perpendicular to the plane of the paper and extending on each side of the plane to a distance equal to the radius of the circle or half the side of the cube.

---

[13]Zeuthen in *Bibliotheca Mathematica* VII₃, 1906–7.

AC is the diameter of the circle which is perpendicular to BD.

Join AB, AD and produce them to meet the tangent at C to the circle in E, F. Then $EC = CF = CA$.

Let LG be the tangent at A, and complete the rectangle $EFGL$.

Draw straight lines from A to the four corners of the section in which the plane through BD perpendicular to AK cuts the cube. These straight lines, if produced, will meet the plane of the face of the cube opposite to A in four points forming the four corners of a square in that plane with sides equal to EF or double of the side of the cube, and we thus have a pyramid with A for vertex and the latter square for base.

Complete the prism (parallelepiped) with the same base and height as the pyramid.

Draw in the parallelogram LF any straight line MN parallel to EF, and through MN draw a plane at right angles to AC.

This plane cuts–

(1) the solid included by the two cylinders in a square with side equal to OP,

(2) the prism in a square with side equal to MN, and

(3) the pyramid in a square with side equal to QR.

Produce CA to H, making HA equal to AC, and imagine HC to be the bar of a balance.

Now, as in Prop. 2, since $MS = AC$, $QS = AS$,
$$MS \cdot SQ = CA \cdot AS$$
$$= AO^2$$
$$= OS^2 + SQ^2.$$

Also
$$HA : AS = CA : AS$$
$$= MS : SQ$$
$$= MS^2 : MS \cdot SQ$$
$$= MS^2 : (OS^2 + SQ^2), \text{ from above,}$$
$$= MN^2 : (OP^2 + QR^2)$$
$$= (\text{square, side } MN) : (\text{sq., side } OP + \text{sq., side } QR).$$

Therefore the square with side equal to MN, in the place where it is, is in equilibrium about A with the squares with sides equal to OP, QR respectively placed with their centres of gravity at H.

Proceeding in the same way with the square sections produced by other planes perpendicular to AC, we finally prove that the prism, in the place where it is, is in equilibrium about A with the solid included by the two cylinders and the pyramid, both placed with their centres of gravity at H.

Now the centre of gravity of the prism is at K.

Therefore $\quad HA : AK = $ (prism) : (solid + pyramid)

or $\qquad\qquad 2 : 1 = $ (prism) : (solid + $\frac{1}{3}$ prism).

Therefore 2 (solid) + $\frac{2}{3}$(prism) = (prism).

It follows that

$$\text{(solid included by cylinders)} = \frac{1}{6}\text{(prism)}$$
$$= \frac{2}{3}\text{(cube)}.$$

Q.E.D.

There is no doubt that Archimedes proceeded to, and completed, the rigorous geometrical proof by the method of exhaustion.

As observed by Prof. C. Juel (Zeuthen l.c.), the solid in the present proposition is made up of 8 pieces of cylinders of the type of that treated in the preceding proposition. As however the two propositions are separately stated, there is no doubt that Archimedes' proofs of them were distinct.

In this case AC would be divided into a very large number of equal parts and planes would be drawn through the points of division perpendicular to AC. These planes cut the solid, and also the cube VY, in square sections. Thus we can inscribe and circumscribe to the solid the requisite solid figures made up of element-prisms and differing by less than any assigned solid magnitude; the prisms have square bases and their heights are the small segments of AC. The element-prism in the inscribed and circumscribed figures which has the square equal to OP^2 for base corresponds to an element-prism in the cube which has for base a square with side equal to that of the cube; and as the ratio of the element-prisms is the ratio $OS^2 : BK^2$, we can use the same auxiliary parabola, and work out the proof in exactly the same way, as in Prop. 14.]

Diophantus

(Third Century A.D.)

HIS LIFE AND WORK

Although historians can only date Diophantus to the third century A.D. with some uncertainty, we do know something of his life. We know this from a word problem left by a friend of his shortly after his death:

- His boyhood lasted $\frac{1}{6}$ of his life.
- He grew a beard after another $\frac{1}{12}$.
- After $\frac{1}{7}$ more he married
- And had a son 5 years later.
- The son lived to half the father's age
- And the father died 4 years later.

This is a typical example of Diophantus' great work, the *Arithmetica*, which is excerpted here. See if you can solve the riddle of Diophantus' life history. (The answer can be found at the end of this introduction.)

The best evidence for his dates comes from the works he cites and the works that cite him. In his book on polygonal numbers, Diophantus presents the definition given by the Greek mathematician Hypsicles in the middle of the second century B.C. At the other end of the range, the Alexandrian mathematician Theon mentions Diophantus in a work from the middle of the fourth century A.D. Historians date

Diophantus to the middle of the third century A.D. based on a letter of Michael Psellus, an eleventh-century Byzantine scholar and statesman, which states that Anatolius, Bishop of Laodicea around 270 had dedicated a treatise on ancient Egyptian computation to his friend Diophantus of Alexandria.

This is consistent with the dedication Diophantus makes at the opening of the *Arithmetica*:

> *Knowing, my most esteemed friend Dionysius, that you are anxious to learn how to investigate problems in numbers, I have tried, beginning from the foundations on which the science is built up, to set forth to you the nature and power subsiding in numbers.*

Most likely, Diophantus made this dedication to the Dionysius who was Bishop of Alexandria between 247 and 264. Alexandria had become the center of Hellenistic culture shortly after Alexander the Great founded it in 331 B.C. By the third century A.D., it was often the scene of conflict between the growing Christian community and the established Roman and Greek communities. By the early 260s, persecution of the Christian community was an everyday event in Alexandria. The great road through the city was usually clogged with the bodies of Christian martyrs. Famine and pestilence often beset the city. Given this environment, we can only wonder if the friendship between Diophantus and Dionysius meant that Diophantus was a Christian.

The *Arithmetica* is not a theoretical work in the tradition of the Pythagoreans. Euclid's proof that there is no largest prime number would have been completely out of place in the *Arithmetica*. Instead, the *Arithmetica* is a work of logistic or computational arithmetic. Its roots lie more in the mathematics of ancient Egypt, Babylonia, and India than in those of ancient Greece.

In his introduction, Diophantus promises a work comprised of thirteen books. However, only six books were known until very recently. The other books must have been separated very early. The first female mathematician known in the Western world, Hypatia, the daughter of Theon of Alexandria, wrote the earliest commentary on the *Arithmetica*. We know little of Hypatia beyond the fact that she was murdered by a mob of Christian fanatics in 415 as they burned Alexandria's great library to the ground, destroying the greatest repository of classical thought. According to the tenth-century encyclopediast Suidas, she wrote commentaries on the *Arithmetica*, Ptolemy's *Astronomy*, and the *Conics* of Apollonius.

In the Western world, the *Arithmetica* may have survived the destruction of the great library of Alexandria in the form of the manuscript in Constantinople seen by Michael Psellus. This one manuscript is the progenitor of all of the copies of the

Arithmetica known to the West until very recently. A little more than thirty years ago, scholars found an Arabic manuscript containing four previously unknown books from the *Arithmetica* at the shrine of Imam Rezā in the holy Islamic city of Mashad in northeastern Iran. It had been translated by Qusta in Luqa, a Christian of Greek origins, born in the city of Heliopolis in Syria.

In the introduction, Diophantus begins by defining the different species of numbers from the second through the sixth powers of a variable; that is:

- (x^2) A square, and its sign is a D with a Y superposed, thus D^Y.
- (x^3) A cube, and its sign is K^Y.
- (x^4) A square-square, and its sign is D^YD.
- (x^5) A square cube, and its sign is DK^Y.
- (x^6) A cube-cube, and its sign is K^YK.

He continues:

> But the number which has none of these characteristics, but merely has in it an indeterminate multitude of units is called arithmos, *and its sign is* ς.

This symbol, something like the Greek letter sigma as found at the end of words, is how Diophantus represents the unknown quantity, which he literally described as "the number." (We will use *x* in conformance with contemporary convention.)

Diophantus was the first mathematician to introduce a sign for subtraction. It looked something like this:

$$\bigwedge$$

perhaps representing an abbreviation for the Greek verb *leipein* "to want" which begins with the Greek letters lambda (Λ) and iota (I). Diophantus did not employ signs for either addition or multiplication. He represented addition simply by juxtaposing numbers or variables next to each other. He never needs a symbol for multiplication because his coefficients are always definite integers or fractions.

To Diophantus, positive terms represent a "forthcoming"; and negative terms, a "wanting." Positive terms always come before negative terms in his problems. The two general rules he gives concern multiplication:

- A "wanting" times a "wanting" yields a "forthcoming."
- A "forthcoming" times a "wanting" yields a "wanting."

For Diophantus, a negative quantity simply could not exist without some positive quantity from which it would be subtracted. He regarded as useless any equation leading to a negative solution, much less an irrational or imaginary solution.

Diophantus' commentators have all despaired at their inability to provide any real categorization of the problems in the *Arithmetica*. The first book consists of determinate algebraic equations. Books II through V mostly contain indeterminate problems consisting of expressions in two or more variables to the first or second power that are to be made equal to a square or a cube. Book VI focuses on right triangles considered arithmetically with some linear or quadratic function of their sides to be made into a square or a cube.

Because every problem requires its own special method, I will analyze one of the problems for you (Book V, problem 9). The problem is:

> *To divide unity [i.e., 1] into two parts such that if the same given number be added to either part, the result will be a square.*

The given number will be an integer and, tantalizingly, Diophantus tells us:

> *The given number must not be odd and the double of it +1 must not be divisible by any prime number, which, when increased by 1, is divisible by 4.*

This condition indicates that Diophantus knew that the square of an integer cannot leave a remainder of 3 when divided by 4. (You can easily prove this by squaring 0, 1, 2, and 3 and taking the remainders after dividing by 4.) Without any further conditions, Diophantus takes 6 as the given number.

> *Therefore 13 must be divided into two squares each of which >6. If we then divide 13 into 2 squares the difference of which is <1, we solve the problem.*
>
> *Take half of 13 or $6\frac{1}{2}$, and we have to add to $6\frac{1}{2}$ a small fraction which will make it a square, or multiplying by 4, we have to make $(\frac{1}{x^2}) + 26$ a square,*
>
> *i.e., $26x^2 + 1 = a\ square = (5x + 1)^2$, say whence x = 10.*

Multiplying by 2 would have eliminated the fraction; however, 2 is not a square, therefore, Diophantus multiplies by 4, a square number. The step changing $(\frac{1}{x^2}) + 26$ to $26x^2 + 1$ is ordinary, straightforward algebra. However, the step *re*-expressing $26x^2 + 1$ as the square $(5x + 1)^2$ is pure Diophantus, a step to get a term in x^2 near to $26x^2$, because he is solving for 2 numbers summing to 1. Having done this, he continues:

> *That is, in order to make 26 a square, we must add $\frac{1}{100}$, or to make $6\frac{1}{2}$ a square, we must add $\frac{1}{400}$, and*
> $$\frac{1}{400} + 6\frac{1}{2} = \left(\frac{51}{20}\right)^2.$$

Therefore we must divide 13 *into two squares such that their sides may be as nearly as possible equal to* $\left(\frac{51}{20}\right)$.

This becomes a bit more understandable if we rewrite this equality as

$$6\frac{1}{2} = \left(\frac{51}{20}\right)^2 - \left(\frac{1}{20}\right)^2 \quad \text{or} \quad 13 = 2 * \left\{\left(\frac{51}{20}\right)^2 - \left(\frac{1}{20}\right)^2\right\}$$

Diophantus continues:

Now 13 = $2^2 + 3^2$. *Therefore we seek two numbers such that* 3 *minus the first* = $\frac{51}{20}$, *so that the first* = $\frac{9}{20}$, *and* 2 *plus the second* = $\frac{51}{20}$, *so that the second* = $\frac{11}{20}$.

Diophantus next substitutes x for $\frac{1}{20}$. He continues:

We write accordingly $(11x + 2)^2, (3 - 9x)^2$ *for the required squares.*

Recall from basic algebra that these expand to $(121x^2 + 44x + 4)$ and $(9 - 54x + 81x^2)$, respectively, so

The sum = $202x^2 - 10x + 13 = 13$.
Therefore, $x = \left(\frac{5}{101}\right)$, and the sides are $\left(\frac{257}{101}\right)$ and $\left(\frac{258}{101}\right)$.
Subtracting 6 from the squares of each, we have, as the parts of unity, $\left(\frac{4843}{10201}\right), \left(\frac{5358}{10201}\right)$

Have you been able to solve the problem at the beginning of this sketch? In modern notation it resolves to solving the equation

$$\left(\frac{1}{6}\right)x + \left(\frac{1}{12}\right)x + \left(\frac{1}{7}\right)x + 5 + \left(\frac{1}{2}\right)x + 4 = x$$

whose solution is $x = 84$. Thus,

- Diophantus' boyhood lasted until he was 14.
- He grew a beard 7 years later at the age of 21.
- And married 12 years later at the age of 33.
- 5 years later he had a son when he was 38.
- His son died at the age of 42 when Diophantus was 80.
- 4 years before his own death at the age of 84.

SELECTIONS FROM *DIOPHANTUS* OF *ALEXANDRIA*, *A STUDY IN THE HISTORY OF GREEK ALGEBRA*

BOOK II

PROBLEMS 8–35

8. To divide a given square number into two squares[1].

Given square number 16.

x^2 one of the required squares. Therefore $16 - x^2$ must be equal to a square.

Take a square of the form[2] $(mx - 4)^2$, m being any integer and 4 the number which is the square root of 16, *e.g.* take $(2x - 4)^2$, and equate it to $16 - x^2$.

Therefore $4x^2 - 16x + 16 = 16 - x^2$,

or $5x^2 = 16x$, and $x = \frac{16}{5}$.

The required squares are therefore $\dfrac{256}{25}, \dfrac{144}{25}$.

9. To divide a given number which is the sum of two squares into two other squares[3].

Given number $13 = 2^2 + 3^2$.

As the roots of these squares are 2, 3, take $(x + 2)^2$ as the first square and $(mx - 3)^2$ as the second (where m is an integer), say $(2x - 3)^2$.

---

[1] It is to this proposition that Fermat appended his famous note in which he enunciates what is known as the "great theorem" of Fermat. The text of the note is as follows:

"On the other hand it is impossible to separate a cube into two cubes, or a biquadrate into two biquadrates, or generally *any power except a square into two powers with the same exponent*. I have discovered a truly marvellous proof of this, which however the margin is not large enough to contain."

Did Fermat really possess a proof of the general proposition that $x^m + y^m = z^m$ cannot be solved in rational numbers where m is any number > 2? As Wertheim says, one is tempted to doubt this, seeing that, in spite of the labours of Euler, Lejeune-Dirichlet, Kummer and others, a general proof has not even yet been discovered. Euler proved the theorem for $m = 3$ and $m = 4$, Dirichlet for $m = 5$, and Kummer, by means of the higher theory of numbers, produced a proof which only excludes certain particular values of m, which values are rare, at all events among the smaller values of m; thus there is no value of m below 100 for which Kummer's proof does not serve. (I take these facts from Weber and Wellstein's *Encyclopädie der Elementar-Mathematik*, I₂, where a proof of the formula for $m = 4$ is given.)

It appears that the Göttingen Academy of Sciences has recently awarded a prize to Dr A. Wieferich, of Münster, for a proof that the equation $x^p + y^p = z^p$ cannot be solved in terms of positive integers not multiples of p, if $2^p - 2$ is not divisible by p^2. "This surprisingly simple result represents the first advance, since the time of Kummer, in the proof of the last Fermat theorem" (*Bulletin of the American Mathematical Society*, February 1910).

Fermat says ("Relation des nouvelles découvertes en la science des nombres," August 1659, *Oeuvres*, II.) that he proved that *no cube is divisible into two cubes* by a variety of his method of *infinite diminution* (*descente infinie* or *indéfinie*) different from that which he employed for other negative or positive theorems; as to the other cases, see Supplement, sections I., II.

[2] Diophantus' words are: "I form the square from any number of ἀριθμοί *minus* as many units as there are in the side of 16." It is implied throughout that m must be so chosen that the result may be *rational* in Diophantus' sense, *i.e.* rational and positive.

[3] Diophantus' solution is substantially the same as Euler's (*Algebra*, tr. Hewlett, Part II. Art. 219), though the latter is expressed more generally. Required to find x, y such that

$$x^2 + y^2 = f^2 + g^2.$$

If $x \gtreqless f$, then $y \lesseqgtr g$.
Put therefore

$$x = f + pz, \quad y = g - qz:$$
$$2fpz + p^2z^2 - 2gqz + q^2z^2 = 0,$$

hence

$$z = \frac{2gq - 2fp}{p^2 + q^2},$$

and

so that

$$x = \frac{2gpq + f(q^2 - p^2)}{p^2 + q^2}, \quad y = \frac{2fpq + g(p^2 - q^2)}{p^2 + q^2},$$

in which we may substitute all possible numbers for p, q.

H. D.

Therefore $(x^2 + 4x + 4) + (4x^2 + 9 - 12x) = 13$,

or $\qquad\qquad\qquad 5x^2 + 13 - 8x = 13.$

Therefore $x = \frac{8}{5}$, and

$$\text{the required squares are } \frac{324}{25}, \frac{1}{25}.$$

10. To find two square numbers having a given difference.

Given difference 60.

Side of one number x, side of the other x *plus* any number the square of which is not greater than 60, say 3.

Therefore $\qquad\qquad\qquad (x + 3)^2 - x^2 = 60;$

$x = 8\frac{1}{2}$, and

$$\text{the required squares are } 72\frac{1}{4}, 132\frac{1}{4}.$$

11. To add the same (required) number to two given numbers so as to make each of them a square.

(1) Given numbers 2, 3; required number x.

Therefore $\begin{rcases} x + 2 \\ x + 3 \end{rcases}$ must both be squares.

This is called a double-equation ($\delta\iota\pi\lambda o\ddot{\iota}\sigma\acute{o}\tau\eta\varsigma$).

To solve it, *take the difference between the two expressions and resolve it into two factors*[4]; in this case let us say 4, $\frac{1}{4}$.

Then *take either*

(*a*) *the square of half the difference between these factors and equate it to the lesser expression,*

or (*b*) *the square of half the sum and equate it to the greater.*

In this case (*a*) the square of half the difference is $\frac{225}{64}$.

Therefore $x + 2 = \frac{225}{64}$, and $x = \frac{97}{64}$, the squares being $\frac{225}{64}, \frac{289}{64}$.

Taking (*b*) the square of half the sum, we have $x + 3 = \frac{289}{64}$, which gives the same result.

(2) To avoid a double-equation[5],

first find a number which when added to 2, or to 3, gives a square.

---

[4] Here, as always, the factors chosen must be suitable factors, *i.e.* such as will lead to a "rational" result, in Diophantus' sense.
[5] This is the same procedure as that of Euler, who does not use double-equations. Euler (*Algebra*, tr. Hewlett, Part II. Art. 214) solves the problem

$$\begin{rcases} x + 4 = u^2 \\ x + 7 = v^2 \end{rcases}.$$

Suppose $\qquad\qquad\qquad x + 4 = p^2$;

therefore $\qquad\qquad\qquad x = p^2 - 4$, and $x + 7 = p^2 + 3$.

Suppose that $\qquad\qquad\qquad p^2 + 3 = (p + q)^2$;

therefore $\qquad\qquad\qquad p = (3 - q^2)/2q$.

Thus $\qquad\qquad\qquad x = (9 - 22q^2 + q^4)/4q^2$,

or, if we take a fraction r/s instead of q,

$$x = (9s^4 - 22r^2s^2 + r^4)/4r^2s^2.$$

Take *e.g.* the number $x^2 - 2$, which when added to 2 gives a square.

Therefore, since this same number added to 3 gives a square,

$$x^2 + 1 = \text{a square} = (x - 4)^2, \text{ say,}$$

the number of units in the expression (in this case 4) being so taken that the solution may give $x^2 > 2$.

Therefore $x = \frac{15}{8}$, and

$$\text{the required number is } \frac{97}{64}, \text{ as before.}$$

12. To subtract the same (required) number from two given numbers so as to make both remainders squares.

Given numbers 9, 21.

Assuming $9 - x^2$ as the required number, we satisfy one condition, and the other requires that $12 + x^2$ shall be a square.

Assume as the side of this square x *minus* some number the square of which >12, say 4.

Therefore $\quad (x - 4)^2 = 12 + x^2$,

and $\quad x = \frac{1}{2}$.

$$\text{The required number is then } 8\frac{3}{4}.$$

[Diophantus does not reduce to lowest terms, but says $x = \frac{4}{8}$ and then subtracts $\frac{16}{64}$ from 9 or $\frac{576}{64}$.]

13. From the same (required) number to subtract two given numbers so as to make both remainders squares.

Given numbers 6, 7.

(1) Let x be the required number.

Therefore $\left.\begin{array}{c} x - 6 \\ x - 7 \end{array}\right\}$ are both squares.

The difference is 1, which is the product of, say, 2 and $\frac{1}{2}$; and, by the rule for solving a double equation,

$$x - 7 = \frac{9}{16}, \text{ and } x = \frac{121}{16}.$$

(2) To avoid a double-equation, seek a number which exceeds a square by 6, say $x^2 + 6$.

Therefore $x^2 - 1$ must also be a square $= (x - 2)^2$, say.

Therefore $x = \frac{5}{4}$, and

$$\text{the required number is } \frac{121}{16}.$$

14. To divide a given number into two parts and to find a square which when added to each of the two parts gives a square number.

Given number 20.

Take two numbers[6] such that the sum of their squares < 20, say 2, 3.

Add x to each and square.

We then have

$$\left.\begin{array}{c} x^2 + 4x + 4 \\ x^2 + 6x + 9 \end{array}\right\},$$

and, if $\left.\begin{array}{c} 4x + 4 \\ 6x + 9 \end{array}\right\}$ are respectively subtracted, the remainders are the same square.

Let then x^2 be the required square, and we have only to make $\left.\begin{array}{c} 4x + 4 \\ 6x + 9 \end{array}\right\}$ the required parts of 20.

Thus

$$10x + 13 = 20,$$

and

$$x = \tfrac{7}{10}.$$

The required parts are then $\left(\tfrac{68}{10}, \tfrac{132}{10}\right)$, and the required square is $\tfrac{49}{100}$.

15. To divide a given number into two parts and to find a square which, when each part is respectively subtracted from it, gives a square.

Given number 20.

Take $(x + m)^2$ for the required square[7], where m^2 is not greater than 20, *e.g.* take $(x + 2)^2$.

This leaves a square if either $\left.\begin{array}{c} 4x + 4 \\ \text{or} \quad 2x + 3 \end{array}\right\}$ is subtracted.

---

[6] Diophantus implies here that the two numbers chosen *must* be such that the sum of their squares < 20. Tannery pointed out (*Bibliotheca Mathematica*, 1887) that this is not so and that the condition actually necessary to ensure a real solution in Diophantus' sense is something different. We have to solve the equations

$$x + y = a, \quad z^2 + x = u^2, \quad z^2 + y = v^2.$$

We assume $u = z + m$, $v = z + n$, and, eliminating x, y, we obtain

$$z = \frac{a - (m^2 + n^2)}{2(m + n)}.$$

In order that z may be positive, we must have $m^2 + n^2 < a$; but z need not be positive in order to satisfy the above equations. What is really required is that x, y shall both be positive.

Now from the above we derive

$$x - y = (u^2 - v^2) = 2z(m - n) + m^2 - n^2$$
$$= \frac{(m - n)(a + 2mn)}{m + n}.$$

Solving for x, y, we have

$$x = \frac{m(a + mn - n^2)}{m + n}, \quad y = \frac{n(a + mn - m^2)}{m + n}.$$

If, of the two assumed numbers, $m > n$, the condition necessary to secure that x, y shall both be positive is $a + mn > m^2$.

[7] Here again the implied condition, namely that m^2 is not greater than 20, is not necessary; the condition necessary for a real solution is something different.

The equations to be solved are $x + y = a$, $z^2 - x = u^2$, $z^2 - y = v^2$.

Diophantus here puts $(\xi + m)^2$ for z^2, so that, if $x = 2m\xi + m^2$, the second equation is satisfied. Now $(\xi + m)^2 - y$ must also be a square, and if this square is equal to $(\xi + m - n)^2$, say, we must have

$$y = 2n\xi + 2mn - n^2.$$

Therefore, since $x + y = a$,

$$2(m + n)\xi + m^2 + 2mn - n^2 = a,$$

whence

$$\xi = \frac{a - m^2 + n^2 - 2mn}{2(m + n)},$$

and it follows that

$$x = \frac{m(a - mn + n^2)}{m + n}, \quad y = \frac{n(a - mn + m^2)}{m + n}.$$

If $m > n$, it is necessary, in order that x, y may both be positive, that $a + n^2 > mn$, which is the true condition for a real solution.

Let these then be the parts of 20.

Therefore $6x + 7 = 20$, and $x = \frac{13}{6}$.

The required parts are therefore $\left(\frac{76}{6}, \frac{44}{6}\right)$, and the required square is $\frac{625}{36}$.

16. To find two numbers in a given ratio and such that each when added to an assigned square gives a square.

Given square 9, given ratio 3:1.

If we take a square of side $(mx + 3)$ and subtract 9 from it, the remainder may be taken as one of the numbers required.

Take, *e.g.*, $(x + 3)^2 - 9$, or $x^2 + 6x$, for the lesser number.

Therefore $3x^2 + 18x$ is the greater number, and $3x^2 + 18x + 9$ must be made a square $= (2x - 3)^2$, say.

Therefore $x = 30$, and

the required numbers are 1080, 3240.

17. To find three numbers such that, if each give to the next following a given fraction of itself and a given number besides, the results after each has given and taken may be equal[8].

First gives to second $\frac{1}{5}$ of itself $+6$, second to third $\frac{1}{6}$ of itself $+7$, third to first $\frac{1}{7}$ of itself $+8$.

Let first and second be $5x$, $6x$ respectively.

When second has taken $x + 6$ from first it becomes $7x + 6$, and when it has given $x + 7$ to third it becomes $6x - 1$.

But first when it has given $x + 6$ to second becomes $4x - 6$; and this too when it has taken $\frac{1}{7}$ of third $+8$ must become $6x - 1$.

Therefore $\frac{1}{7}$ of third $+8 = 2x + 5$, and

third $= 14x - 21$.

Next, third after receiving $\frac{1}{6}$ of second $+7$ and giving $\frac{1}{7}$ of itself $+8$ must become $6x - 1$.

Therefore $13x - 19 = 6x - 1$, and $x = \frac{18}{7}$.

The required numbers are $\frac{90}{7}, \frac{108}{7}, \frac{105}{7}$.

18. To divide a given number into three parts satisfying the conditions of the preceding problem[9].

[8] Tannery is of opinion that the problems ii. 17 and 18 have crept into the text from an ancient commentary to Book I. to which they would more appropriately belong.

[9] Though the solution is not given in the text, it is easily obtained from the *general* solution of the preceding problem, which again, at least with our notation, is easy.

Let us assume, with Wertheim, that the numbers required in ii. 17 are $5x$, $6y$, $7z$.

Then by the conditions of the problem

$$4x - 6 + z + 8 = 5y - 7 + x + 6 = 6z - 8 + y + 7,$$

from which two equations we can find x, z in terms of y.

In fact

$$x = (26y - 18)/19 \text{ and } z = (17y - 3)/19,$$

and the general solution is

$$5(26y - 18)/19, \quad 6y, \quad 7(17y - 3)/19.$$

[In his solution Diophantus assumes $x = y$, whence $y = \frac{18}{7}$].

250

Given number 80.

Let first give to second $\frac{1}{5}$ of itself $+6$, second to third $\frac{1}{6}$ of itself $+7$, and third to first $\frac{1}{7}$ of itself $+8$.

[What follows in the text is not a solution of the problem but an alternative solution of the preceding. The first two numbers are assumed to be $5x$ and 12, and the numbers found are $\frac{170}{19}, \frac{228}{19}, \frac{217}{19}$.]

19. To find three squares such that the difference between the greatest and the middle has to the difference between the middle and the least a given ratio.

Given ratio 3 : 1.

Assume the least square $= x^2$, the middle $= x^2 + 2x + 1$.

Therefore the greatest $= x^2 + 8x + 4 =$ square $= (x + 3)^2$, say.

Thus $x = 2\frac{1}{2}$, and

$$\text{the squares are } 30\frac{1}{4},\ 12\frac{1}{4},\ 6\frac{1}{4}.$$

20. To find two numbers such that the square of either added to the other gives a square[10].

Assume for the numbers x, $2x + 1$, which by their form satisfy one condition.

The other condition gives

$$4x^2 + 5x + 1 = \text{square} = (2x - 2)^2, \text{ say.}$$

Therefore $x = \frac{3}{13}$, and

$$\text{the numbers are } \frac{3}{13},\ \frac{19}{13}.$$

21. To find two numbers such that the square of either *minus* the other number gives a square.

$x + 1$, $2x + 1$ are assumed, satisfying one condition.

The other condition gives

$$4x^2 + 3x = \text{square} = 9x^2, \text{ say.}$$

Now, to solve ii. 18, we have only to equate the sum of the three expressions to 80, and so find y. We have

$$y(5 \cdot 26 + 6 \cdot 19 + 7 \cdot 17) - 5 \cdot 18 - 7 \cdot 3 = 80 \cdot 19, \quad y = \frac{1631}{363};$$

and the required numbers are

$$\frac{9440}{363},\ \frac{9786}{363},\ \frac{9814}{363}.$$

[10] Euler (*Algebra*, Part II. Art. 239) solves this problem more generally thus.
Required to find x, y such that $x^2 + y$ and $y^2 + x$ are squares.
If we begin by supposing $x^2 + y = p^2$, so that $y = p^2 - x^2$, and then substitute the value of y in terms of x in the second expression, we must have
$$p^4 - 2p^2x^2 + x^4 + x = \text{square.}$$
But, as this is difficult to solve, let us suppose instead that
$$x^2 + y = (p - x)^2 = p^2 - 2px + x^2,$$
$$y^2 + x = (q - y)^2 = q^2 - 2qy + y^2.$$
and that
It follows that
$$y + 2px = p^2,$$
$$x + 2qy = q^2,$$
whence
$$x = \frac{2gp^2 - q^2}{4pq - 1}, \quad y = \frac{2pq^2 - p^2}{4pq - 1}.$$
Suppose, for example, $p = 2$, $q = 3$, and we have $x = \frac{15}{23}$, $y = \frac{32}{23}$; and so on. We must of course choose p, q such that x, y are both positive. Diophantus' solution is obtained by putting $p = -1$, $q = 3$.

Therefore $x = \frac{3}{5}$, and

$$\text{the numbers are } \frac{8}{5}, \frac{11}{5}.$$

22. To find two numbers such that the square of either added to the sum of both gives a square.

Assume x, $x + 1$ for the numbers. Thus one condition is satisfied.

It remains that

$$x^2 + 4x + 2 = \text{square} = (x - 2)^2, \text{ say.}$$

Therefore $x = \frac{1}{4}$, and

$$\text{the numbers are } \frac{1}{4}, \frac{5}{4}.$$

[Diophantus has $\frac{2}{8}$, $\frac{10}{8}$.]

23. To find two numbers such that the square of either *minus* the sum of both gives a square.

Assume x, $x + 1$ for the numbers, thus satisfying one condition.

Then $x^2 - 2x - 1 = \text{square} = (x - 3)^2$, say.

Therefore $x = 2\frac{1}{2}$, and

$$\text{the numbers are } 2\frac{1}{2}, 3\frac{1}{2}.$$

24. To find two numbers such that either added to the square of their sum gives a square.

Since $x^2 + 3x^2$, $x^2 + 8x^2$ are both squares, let the numbers be $3x^2$, $8x^2$ and their sum x.

Therefore $121x^4 = x^2$, whence $11x^2 = x$, and $x = \frac{1}{11}$.

$$\text{The numbers are therefore } \frac{3}{121}, \frac{8}{121}.$$

25. To find two numbers such that the square of their sum *minus* either number gives a square.

If we subtract 7 or 12 from 16, we get a square.

Assume then $12x^2$, $7x^2$ for the numbers, and $16x^2$ for the square of their sum.

Hence $19x^2 = 4x$, and $x = \frac{4}{19}$.

$$\text{The numbers are } \frac{192}{361}, \frac{112}{361}.$$

26. To find two numbers such that their product added to either gives a square, and the sides of the two squares added together produce a given number.

Let the given number be 6.

Since $x(4x - 1) + x$ is a square, let x, $4x - 1$ be the numbers.

Therefore $4x^2 + 3x - 1$ is a square, and the side of this square must be $6 - 2x$ [since $2x$ is the side of the first square and the sum of the sides of the square is 6].

Since $$4x^2 + 3x - 1 = (6 - 2x)^2,$$

we have $x = \frac{37}{27}$, and

$$\text{the numbers are } \frac{37}{27}, \frac{121}{27}.$$

27. To find two numbers such that their product *minus* either gives a square, and the sides of the two squares so arising when added together produce a given number.

Let the given number be 5.

Assume $4x + 1$, x for the numbers, so that one condition is satisfied.

Also $$4x^2 - 3x - 1 = (5 - 2x)^2.$$

Therefore $x = \frac{26}{17}$, and

$$\text{the numbers are } \frac{26}{17}, \frac{121}{17}.$$

28. To find two square numbers such that their product added to either gives a square.

Let the numbers[11] be x^2, y^2.

Therefore $\left.\begin{matrix} x^2y^2 + y^2 \\ x^2y^2 + x^2 \end{matrix}\right\}$ are both squares.

To make the first expression a square we make $x^2 + 1$ a square, putting

$$x^2 + 1 = (x - 2)^2, \text{ say.}$$

Therefore $x = \frac{3}{4}$, and $x^2 = \frac{9}{16}$.

We have now to make $\frac{9}{16}(y^2 + 1)$ a square [and y must be different from x].

Put $$9y^2 + 9 = (3y - 4)^2, \text{ say,}$$

and $y = \frac{7}{24}$.

$$\text{Therefore the numbers are } \frac{9}{16}, \frac{49}{576}.$$

29. To find two square numbers such that their product *minus* either gives a square.

Let x^2, y^2 be the numbers.

Then $\left.\begin{matrix} x^2y^2 - y^2 \\ x^2y^2 - x^2 \end{matrix}\right\}$ are both squares.

A solution of $x^2 - 1 = $ (a square) is $x^2 = \frac{25}{16}$.

We have now to solve

$$\frac{25}{16}y^2 - \frac{25}{16} = \text{ a square.}$$

[11] Diophantus does not use two unknowns, but assumes the numbers to be x^2 and 1 until he has found x. Then he uses the same unknown (x) to find what he had first taken to be unity. The same remark applies to the next problem.

Put $$y^2 - 1 = (y - 4)^2, \text{ say.}$$
Therefore $y = \frac{17}{8}$, and

$$\text{the numbers are } \frac{289}{64}, \frac{100}{64}.$$

30. To find two numbers such that their product ± their sum gives a square.

Now $m^2 + n^2 \pm 2mn$ is a square.

Put 2, 3, say, for m, n respectively, and of course
$$2^2 + 3^2 \pm 2 \cdot 2 \cdot 3 \text{ is a square.}$$
Assume then product of numbers $= (2^2 + 3^2) x^2$ or $13x^2$, and
sum $= 2 \cdot 2 \cdot 3x^2$ or $12x^2$.

The product being $13x^2$, let x, $13x$ be the numbers.

Therefore their sum $14x = 12x^2$, and $x = \frac{7}{6}$.

$$\text{The numbers are therefore } \frac{7}{6}, \frac{91}{6}.$$

31. To find two numbers such that their sum is a square and their product ± their sum gives a square.

$2 \cdot 2m \cdot m = $ a square, and $(2m)^2 + m^2 \pm 2 \cdot 2m \cdot m = $ a square.

If $m = 2$, $4^2 + 2^2 \pm 2 \cdot 4 \cdot 2 = 36$ or 4.

Let then the product of the numbers be $(4^2 + 2^2)x^2$ or $20x^2$, and their sum $2 \cdot 4 \cdot 2x^2$ or $16x^2$, and take $2x$, $10x$ for the numbers.

Then $\quad 12x = 16x^2$, and $x = \frac{3}{4}$.

$$\text{The numbers are } \frac{6}{4}, \frac{30}{4}.$$

32. To find three numbers such that the square of any one of them added to the next following gives a square.

Let the first be x, the second $2x + 1$, and the third $2(2x + 1) + 1$ or $4x + 3$, so that two conditions are satisfied.

The last condition gives $(4x + 3)^2 + x = $ square $= (4x - 4)^2$, say.

Therefore $x = \frac{7}{57}$, and

$$\text{the numbers are } \frac{7}{57}, \frac{71}{57}, \frac{199}{57}.$$

33. To find three numbers such that the square of any one of them *minus* the next following gives a square.

Assume $x + 1$, $2x + 1$, $4x + 1$ for the numbers, so that two conditions are satisfied.

Lastly, $16x^2 + 7x = $ square $= 25x^2$, say,

and $\quad x = \frac{7}{9}$.

$$\text{The numbers are } \frac{16}{9}, \frac{23}{9}, \frac{37}{9}.$$

34. To find three numbers such that the square of any one added to the sum of all three gives a square.

$\{\frac{1}{2}(m - n)\}^2 + mn$ is a square. Take a number separable into two factors (m, n) in three ways, say 12, which is the product of $(1, 12)$, $(2, 6)$ and $(3, 4)$.

The values then of $\frac{1}{2}(m - n)$ are $5\frac{1}{2}$, 2, $\frac{1}{2}$.

Take $5\frac{1}{2}x$, $2x$, $\frac{1}{2}x$ for the numbers, and for their sum $12x^2$.

Therefore $8x = 12x^2$, and $x = \frac{2}{3}$.

$$\text{The numbers are } \frac{11}{3}, \frac{4}{3}, \frac{1}{3}.$$

[Diophantus says $\frac{4}{6}$, and $\frac{22}{6}, \frac{8}{6}, \frac{2}{6}$.]

35. To find three numbers such that the square of any one *minus* the sum of all three gives a square.

$\{\frac{1}{2}(m + n)\}^2 - mn$ is a square. Take, as before, a number divisible into factors in three ways, as 12.

Let then $6\frac{1}{2}x$, $4x$, $3\frac{1}{2}x$ be the numbers, and their sum $12x^2$.

Therefore $14x = 12x^2$, and $x = \frac{7}{6}$.

$$\text{The numbers are } \frac{45\frac{1}{2}}{6}, \frac{28}{6}, \frac{24\frac{1}{2}}{6}.$$

BOOK III

PROBLEMS 5–21

5. To find three numbers such that their sum is a square and the sum of any pair exceeds the third by a square.

Let the sum of the three be $(x + 1)^2$; let first + second = third +1, so that third $= \frac{1}{2}x^2 + x$; let second + third = first $+x^2$, so that first $= x + \frac{1}{2}$.

Therefore second $= \frac{1}{2}x^2 + \frac{1}{2}$.

It remains that first + third = second + a square.

Therefore $2x = $ square $= 16$, say, and $x = 8$.

$$\text{The numbers are } 8\frac{1}{2}, 32\frac{1}{2}, 40.$$

Otherwise thus[1].

First find three squares such that their sum is a square.

Find *e.g.* what square number $+4$ $+9$ gives a square, that is, 36;

4, 36, 9 are therefore squares with the required property.

Next find three numbers such that the sum of each pair $=$ the third $+$ a given number; in this case suppose

$$\text{first} + \text{second} - \text{third} = 4,$$
$$\text{second} + \text{third} - \text{first} = 9,$$
$$\text{third} + \text{first} - \text{second} = 36.$$

This problem has already been solved [i. 18].

The numbers are respectively 20, $6\frac{1}{2}$, $22\frac{1}{2}$.

6. To find three numbers such that their sum is a square and the sum of any pair is a square.

Let the sum of all three be $x^2 + 2x + 1$, sum of first and second x^2, and therefore the third $2x + 1$; let sum of second and third be $(x - 1)^2$.

Therefore the first $= 4x$, and the second $= x^2 - 4x$.

But first $+$ third $=$ square,

that is, $6x + 1 =$ square $= 121$, say.

Therefore $x = 20$, and

the numbers are 80, 320, 41.

[An alternative solution, obviously interpolated, is practically identical with the above except that it takes the square 36 as the value of $6x + 1$, so that $x = \frac{35}{6}$, and the numbers are $\frac{140}{6} = \frac{840}{36}, \frac{385}{36}, \frac{456}{36}$.]

7. To find three numbers in A.P. such that the sum of any pair gives a square.

First find three square numbers in A.P. and such that half their sum is greater than any one of them. Let $x^2, (x + 1)^2$ be the first and second of these; therefore the third is $x^2 + 4x + 2 = (x - 8)^2$, say.

Therefore $x = \frac{62}{20}$ or $\frac{31}{10}$;

and we may take as the numbers 961, 1681, 2401.

We have now to find three numbers such that the sums of pairs are the numbers just found.

[1] We should naturally suppose that this alternative solution, like others, was interpolated. But we are reluctant to think so because the solution is so elegant that it can hardly be attributed to a scholiast. If the solution is not genuine, we have here an illustration of the truth that, however ingenious they are, Diophantus' solutions are not always the best imaginable (Loria, *Le scienze esatte nell' antica Grecia*, Libro v.). In this case the more elegant solution is the alternative one. Generalised, it is as follows. We have to find x, y, z so that

$$\left. \begin{array}{l} -x + y + z = \text{a square} \\ x - y + z = \text{a square} \\ x + y - z = \text{a square} \end{array} \right\},$$

and also

$$x + y + z = \text{a square}.$$

We have only to equate the first three expressions to squares a^2, b^2, c^2 such that $a^2 + b^2 + c^2 = $ a square, k^2 say, since the sum of the first three expressions is itself $x + y + z$.

The solution is then

$$x = \frac{1}{2}(b^2 + c^2), \quad y = \frac{1}{2}(c^2 + a^2), \quad z = \frac{1}{2}(a^2 + b^2).$$

The sum of the three $= \frac{5043}{2} = 2521\frac{1}{2}$, and

the three numbers are $120\frac{1}{2}$, $840\frac{1}{2}$, $1560\frac{1}{2}$.

8. Given one number, to find three others such that the sum of any pair of them added to the given number gives a square, and also the sum of the three added to the given number gives a square.

Given number 3.

Suppose first required number + second $= x^2 + 4x + 1$,

second + third $= x^2 + 6x + 6$,

sum of all three $= x^2 + 8x + 13$.

Therefore third $= 4x + 12$, second $= x^2 + 2x - 6$, first $= 2x + 7$.

Also   first + third $+3 =$ a square,

that is, $6x + 22 =$ square $= 100$, suppose.

Hence $x = 13$, and

the numbers are 33, 189, 64.

9. Given one number, to find three others such that the sum of any pair of them *minus* the given number gives a square, and also the sum of the three *minus* the given number gives a square.

Given number 3.

Suppose first of required numbers + second $= x^2 + 3$,

second + third $= x^2 + 2x + 4$,

sum of the three $= x^2 + 4x + 7$.

Therefore third $= 4x + 4$, second $= x^2 - 2x$, first $= 2x + 3$.

Lastly, first + third $-3 = 6x + 4 =$ a square $= 64$, say.

Therefore $x = 10$, and

(23, 80, 44) is a solution.

10. To find three numbers such that the product of any pair of them added to a given number gives a square.

Let the given number be 12. Take a square (say 25) and subtract 12. Take the difference (13) for the product of the first and second numbers, and let these numbers be $13x$, $\frac{1}{x}$ respectively.

Again subtract 12 from another square, say 16, and let the difference (4) be the product of the second and third numbers.

Therefore the third number $= 4x$.

The third condition gives $52x^2 + 12 =$ a square; now $52 = 4 \cdot 13$, and 13 is not a square; but, if it were a square, the equation could easily be solved[2].

---

[2] The equation $52x^2 + 12 = u^2$ can in reality be solved as it stands, by virtue of the fact that it has one obvious solution, namely $x = 1$. Another solution is found by substituting $y + 1$ for x, and so on. The value $x = 1$ itself gives (13, 1, 4) as a solution of the problem.

Thus we must find two numbers to replace 13 and 4 such that their product is a square, while either $+12$ is also a square.

Now the product is a square if both are squares; hence we must find two squares such that either $+12 =$ a square.

"This is easy[3] and, as we said, it makes the equation easy to solve."

The squares 4, $\frac{1}{4}$ satisfy the condition.

Retracing our steps, we now put $4x$, $\frac{1}{x}$ and $\frac{x}{4}$ for the numbers, and we have to solve the equation

$$x^2 + 12 = \text{square} = (x + 3)^2, \text{ say.}$$

Therefore $x = \frac{1}{2}$, and

$$\left(2, 2, \frac{1}{8}\right) \text{ is a solution[4].}$$

11. To find three numbers such that the product of any pair *minus* a given number gives a square.

Given number 10.

Put product of first and second $=$ a square $+10 = 4 + 10$, say, and let first $= 14x$, second $= \frac{1}{x}$.

Let product of second and third $=$ a square $+10 = 19$, say;

therefore third $= 19x$.

By the third condition, $266x^2 - 10$ must be a square; but 266 is not a square[5].

Therefore, as in the preceding problem, we must find two squares each of which exceeds a square by 10.

The squares $30\frac{1}{4}$, $12\frac{1}{4}$ satisfy these conditions[6].

Putting now $30\frac{1}{4}x$, $\frac{1}{x}$, $12\frac{1}{4}x$ for the numbers, we have, by the third condition, $370\frac{9}{16}x^2 - 10 =$ square [for $370\frac{9}{16}$ Diophantus writes $370\frac{1}{2}\frac{1}{16}$];

---

[3] The method is indicated in ii. 34. We have to find two pairs of squares differing by 12. (a) If we put $12 = 6 \cdot 2$, we have

$$\left\{\frac{1}{2}(6 - 2)\right\}^2 + 12 = \left\{\frac{1}{2}(6 + 2)\right\}^2,$$

and 16, 4 are squares differing by 12, or 4 is a square which when added to 12 gives a square. (b) If we put $12 = 4 \cdot 3$, we find $\{\frac{1}{2}(4 - 3)\}^2$ or $\frac{1}{4}$ to be a square which when added to 12 gives a square.

[4] Euler (*Algebra*, Part II. Art. 232) has an elegant solution of this problem in whole numbers. Let it be required to find x, y, z such that $xy + a$, $yz + a$, $zx + a$ are all squares. Suppose $xy + a = p^2$, and make $z = x + y + q$;

therefore $\quad xz + a = x^2 + xy + qx + a = x^2 + qx + p^2,$

and $\quad yz + a = xy + y^2 + qy + a = y^2 + qy + p^2;$

and the right hand expressions are both squares if $q = \pm 2p$, so that $z = x + y \pm 2p$.

We can therefore take any value for p such that $p^2 > a$, split $p^2 - a$ into factors, take those factors respectively for the values of x and y, and so find z.

E.g. suppose $a = 12$ and $p^2 = 25$, so that $xy = 13$; let $x = 1$, $y = 13$, and we have $z = 14 \pm 10 = 24$ or 4, and $(1, 13, 4)$, $(1, 13, 24)$ are solutions.

[5] As a matter of fact, the equation $266x^2 - 10 = u^2$ can be solved as it stands, since it has one obvious solution, namely $x = 1$. The value $x = 1$ gives $(14, 1, 19)$ as a solution of the problem.

[6] Tannery brackets the passage in the text in which these squares are found, on the ground that, as the solution was not given in the corresponding place of iii. 10, there was no necessity to give it here. 10 and 1 being factors of 10,

$$\left\{\frac{1}{2}(10 - 1)\right\}^2 + 10 = \left\{\frac{1}{2}(10 + 1)\right\}^2;$$

thus $30\frac{1}{4}$ is a square which exceeds a square by 10. Similarly $\{\frac{1}{2}(5 + 2)\}^2$ or $12\frac{1}{4}$ is such a square. The latter is found in the text by putting $m^2 - 10 =$ square $= (m - 2)^2$, whence $m = 3\frac{1}{2}$, and $m^2 = 12\frac{1}{4}$.

therefore $5929x^2 - 160 = $ square $= (77x - 2)^2$, say.

Therefore $x = \frac{41}{77}$, and

$$\text{the numbers are } \frac{1240\frac{1}{4}}{77}, \frac{77}{41}, \frac{502\frac{1}{4}}{77}.$$

12. To find three numbers such that the product of any two added to the third gives a square.

Take a square and subtract part of it for the third number; let $x^2 + 6x + 9$ be one of the sums, and 9 the third number.

Therefore product of first and second $= x^2 + 6x$; let first $=x$, so that second $= x + 6$.

By the two remaining conditions

$$\left. \begin{array}{c} 10x + 54 \\ 10x + 6 \end{array} \right\} \text{ are both squares.}$$

Therefore we have to find two squares differing by 48; "this is easy and can be done in an infinite number of ways."

The squares 16, 64 satisfy the condition. Equating these squares to the respective expressions, we obtain $x = 1$, and

the numbers are 1, 7, 9.

13. To find three numbers such that the product of any two *minus* the third gives a square.

First x, second $x + 4$; therefore product $= x^2 + 4x$, and we assume third $=4x$.

Therefore, by the other conditions,

$$\left. \begin{array}{c} 4x^2 + 15x \\ 4x^2 - x - 4 \end{array} \right\} \text{ are both squares.}$$

The difference $= 16x + 4 = 4(4x + 1)$, and we put

$$\left\{ \frac{1}{2}(4x + 5) \right\}^2 = 4x^2 + 15x.$$

Therefore $x = \frac{25}{20}$, and

$$\text{the numbers are } \frac{25}{20}, \frac{105}{20}, \frac{100}{20}.$$

14. To find three numbers such that the product of any two added to the square of the third gives a square[7].

---

[7] Wertheim gives a more general solution, as follows. If we take as the required numbers $X = \frac{1}{4}ax$, $Y = ax + b^2$, $Z = \frac{1}{4}b^2$, two conditions are already satisfied, namely $XY + Z^2 = $ a square, and $YZ + X^2 = $ a square.

It only remains to satisfy the condition $ZX + Y^2 = $ a square, or

$$a^2x^2 + \frac{33}{16}ab^2x + b^4 = \text{a square.}$$

Put

$$a^2x^2 + \frac{33}{16}ab^2x + b^4 = (ax + kb^2)^2,$$

and

$$x = \frac{16b^2(k^2 - 1)}{a(33 - 32k)},$$

where k remains undetermined.

H. D.

First x, second $4x + 4$, third 1. Two conditions are thus satisfied.

The third condition gives

$$x + (4x + 4)^2 = \text{a square} = (4x - 5)^2, \text{ say.}$$

Therefore $x = \frac{9}{73}$, and

the numbers (omitting the common denominator) are 9, 328, 73.

15. To find three numbers such that the product of any two added to the sum of those two gives a square[8].

[*Lemma.*] The product of the squares of any two consecutive numbers added to the sum of the said squares gives a square[9].

Let 4, 9 be two of the required numbers, x the third.

Therefore $\left.\begin{array}{r}10x + 9 \\ 5x + 4\end{array}\right\}$ are both squares.

The difference $= 5x + 5 = 5(x + 1)$.

Equating the square of half the sum of the factors to $10x + 9$, we have

$$\left\{\frac{1}{2}(x + 6)\right\}^2 = 10x + 9.$$

Therefore $x = 28$, and (4, 9, 28) is a solution.

Otherwise thus[10].

Assume first number to be x, second 3.

---

[8] The problem can of course be solved more elegantly, with our notation, thus. (The same remark applies to the next problem, iii. 16.)
If x, y, z are the required numbers, $xy + x + y$, etc. are to be squares. We may therefore write the conditions in the form

$$(y + 1)(z + 1) = \text{a square} + 1,$$
$$(z + 1)(x + 1) = \text{a square} + 1,$$
$$(x + 1)(y + 1) = \text{a square} + 1.$$

Assuming a^2, b^2, c^2 for the respective squares, and putting $\xi = x + 1, \eta = y + 1, \zeta = z + 1$, we have to solve

$$\eta\zeta = a^2 + 1,$$
$$\zeta\xi = b^2 + 1,$$
$$\xi\eta = c^2 + 1.$$

[This is practically the same problem as that in the Lemma to Dioph. v. 8.]
Multiplying the second and third equations and dividing by the first, we have

$$\xi = \sqrt{\{(b^2 + 1)(c^2 + 1)/(a^2 + 1)\}},$$

with similar expressions for η, ζ.

x, y, z are these expressions *minus* 1 respectively. a^2, b^2, c^2 must of course be so chosen that the resulting values of ξ, η, ζ may be rational. Cf. Euler, *Commentationes arithmeticae*, II.

[9] In fact, $a^2(a + 1)^2 + a^2 + (a + 1)^2 = \{a(a + 1) + 1\}^2$.

[10] This alternative solution would appear to be undoubtedly genuine. Diophantus has solved the equations

$$\left.\begin{array}{r}yz + y + z = u^2 \\ zx + z + x = v^2 \\ xy + x + y = w^2\end{array}\right\}.$$

Fermat shows how to solve the corresponding problem with *four* numbers instead of three. He uses for this purpose Diophantus' solution of v. 5, namely the problem of finding x^2, y^2, z^2, such that

$$\left.\begin{array}{lll}y^2z^2 + x^2 = r^2, & z^2x^2 + y^2 = s^2, & x^2y^2 + z^2 = t^2 \\ y^2z^2 + y^2 + z^2 = u^2, & z^2x^2 + z^2 + x^2 = v^2, & x^2y^2 + x^2 + y^2 = w^2\end{array}\right\}.$$

Diophantus finds $(\frac{25}{9}, \frac{64}{9}, \frac{196}{9})$ as a solution of the latter problem. Fermat takes these as the first three of the four numbers which are to satisfy the condition that the product of any two plus the sum of those two gives a square, and assumes x for the fourth. Three relations out of six are already satisfied, and the other three require

$$\left.\begin{array}{ll}\dfrac{25}{9}x + x + \dfrac{25}{9}, & \text{or} \quad \dfrac{34x}{9} + \dfrac{25}{9} \\[2mm] \dfrac{64}{9}x + x + \dfrac{64}{9}, & \text{or} \quad \dfrac{73x}{9} + \dfrac{64}{9} \\[2mm] \dfrac{196}{9}x + x + \dfrac{196}{9}, & \text{or} \quad \dfrac{205x}{9} + \dfrac{196}{9}\end{array}\right\}$$

to be made squares: a "triple-equation" to be solved by Fermat's method.

Therefore $4x + 3 =$ square $= 25$ say, whence $x = 5\frac{1}{2}$, and $5\frac{1}{2}$, 3 satisfy one condition.

Let the third be x, while $5\frac{1}{2}$, 3 are the first two.

Therefore $\left.\begin{array}{l} 4x + 3 \\ 6\frac{1}{2}x + 5\frac{1}{2} \end{array}\right\}$ must both be squares;

but, since the coefficients in one expression are respectively greater than those in the other, but neither of the ratios of corresponding coefficients is that of a square to a square, our suppositions will not serve the purpose; we cannot solve by our method.

Hence (to replace $5\frac{1}{2}$, 3) we must find two numbers such that their product + their sum =a square, and the ratio of the numbers increased by 1 respectively is the ratio of a square to a square.

Let these be y and $4y + 3$, which satisfy the latter condition; and, in order that the other may be satisfied, we must have

$$4y^2 + 8y + 3 = \text{square} = (2y - 3)^2, \text{ say.}$$

Therefore $$y = \tfrac{3}{10}.$$

Assume now $\frac{3}{10}$, $4\frac{1}{5}$, x for the three numbers.

Therefore $\left.\begin{array}{l} 5\frac{1}{5}x + 4\frac{1}{5} \\ \frac{13}{10}x + \frac{3}{10} \end{array}\right\}$ are both squares,

or, if we multiply by 25 and 100 respectively,

$$\left.\begin{array}{l} 130x + 105 \\ 130x + 30 \end{array}\right\} \text{ are both squares.}$$

The difference is $75 = 3 \cdot 25$, and the usual method of solution gives $x = \frac{7}{10}$.

$$\text{The numbers are } \frac{3}{10}, \frac{42}{10}, \frac{7}{10}.$$

16. To find three numbers such that the product of any two *minus* the sum of those two gives a square.

Fermat does not give the solution, but I had the curiosity to work it out in order to verify to what enormous numbers the method of the triple-equation leads, even in such comparatively simple cases.

We may of course neglect the denominator 9 and solve the equations

$$34x + 25 = u^2,$$
$$73x + 64 = v^2,$$
$$205x + 196 = w^2.$$

The method gives

$$x = -\frac{459185984964847787200}{631629004828419699201},$$

the denominator being equal to $(25132230399)^2$.

Verifying the correctness of the solution, we find that, in fact,

$$\frac{34}{25}x + 1 = \left(\frac{2505136897}{25132230399}\right)^2,$$

$$\frac{73}{64}x + 1 = \left(\frac{10351251901}{25132230399}\right)^2,$$

$$\frac{205}{196}x + 1 = \left(\frac{12275841601}{25132230399}\right)^2.$$

Strictly speaking, as the value found for x is negative, we ought to substitute $y - A$ for it (where $-A$ is the value found) in the three equations and start afresh. The portentous numbers which would thus arise must be left to the imagination.

Put x for the first, and any number for the second; we then fall into the same difficulty as in the last problem.

We have to find two numbers such that

(*a*) their product *minus* their sum = a square, and

(*b*) when each is diminished by 1, the remainders have the ratio of squares.

Now $4y + 1$, $y + 1$ satisfy the latter condition.

The former (*a*) requires that

$$4y^2 - 1 = \text{square} = (2y - 2)^2, \text{ say,}$$

which gives $y = \frac{5}{8}$.

Assume then $\frac{13}{8}$, $\frac{28}{8}$, x for the numbers.

Therefore $\left.\begin{array}{c} 2\frac{1}{2}x - 3\frac{1}{2} \\ \frac{5}{8}x - \frac{13}{8} \end{array}\right\}$ are both squares,

or, if we multiply by 4, 16 respectively,

$$\left.\begin{array}{c} 10x - 14 \\ 10x - 26 \end{array}\right\} \text{ are both squares.}$$

The difference is $12 = 2 \cdot 6$, and the usual method gives

$x = 3$.

The numbers are $\dfrac{13}{8}$, $3\frac{1}{2} = \dfrac{28}{8}$, $3 = \dfrac{24}{8}$.

17. To find two numbers such that their product added to both or to either gives a square.

Assume x, $4x - 1$ for the numbers, since

$$x(4x - 1) + x = 4x^2, \text{ a square.}$$

Therefore also $\left.\begin{array}{c} 4x^2 + 3x - 1 \\ 4x^2 + 4x - 1 \end{array}\right\}$ are both squares.

The difference is $x = 4x \cdot \frac{1}{4}$, and we find

$$x = \frac{65}{224}.$$

The numbers are $\dfrac{65}{224}$, $\dfrac{36}{224}$.

18. To find two numbers such that their product *minus* either, or *minus* the sum of both, gives a square[11].

[11] With this problem should be compared that in paragraph 42 of Part i. of the *Inventum Novum* of Jacobus de Billy (*Oeuvres de Fermat*, III.), where three conditions correspond to those of the above problem, and there is a fourth in addition. The problem is to find ξ, η ($\xi > \eta$) such that

$$\left.\begin{array}{c} \xi - \xi\eta \\ \eta - \xi\eta \\ \xi + \eta - \xi\eta \\ \xi - \eta - \xi\eta \end{array}\right\} \text{ are all squares.}$$

Suppose $\eta = x$, $\xi = 1 - x$; the first two conditions are thus satisfied. The other two give

$$x^2 - x + 1 = u^2,$$
$$x^2 - 3x + 1 = v^2.$$

Assume $x + 1$, $4x$ for the numbers, since

$$4x(x + 1) - 4x = \text{a square.}$$

Therefore also $\left.\begin{array}{c} 4x^2 + 3x - 1 \\ 4x^2 - x - 1 \end{array}\right\}$ are both squares.

The difference is $4x = 4x \cdot 1$, and we find

$$x = 1\tfrac{1}{4}.$$

The numbers are $2\tfrac{1}{4}$, 5.

19. To find four numbers such that the square of their sum *plus* or *minus* any one singly gives a square.

Since, in any right-angled triangle,

(sq. on hypotenuse) $\pm$ (twice product of perps.) = a square, we must seek four right-angled triangles [in rational numbers] having the same hypotenuse,

or we must find a square which is divisible into two squares in four different ways; and "we saw how to divide a square into two squares in an infinite number of ways." [ii. 8.]

Take right-angled triangles in the smallest numbers, (3, 4, 5) and (5, 12, 13); and multiply the sides of the first by the hypotenuse of the second and *vice versa*.

This gives the triangles (39, 52, 65) and (25, 60, 65); thus 65^2 is split up into two squares in *two* ways.

Again, 65 is "naturally" divided into two squares in two ways, namely into $7^2 + 4^2$ and $8^2 + 1^2$, "which is due to the fact that 65 is the product of 13 and 5, each of which numbers is the sum of two squares."

Separating the difference $2x$ into the factors $2x$, 1, we put, as usual,

$$\left(x + \frac{1}{2}\right)^2 = x^2 - x + 1,$$

whence $x = \dfrac{3}{8}$, and the numbers are $\dfrac{5}{8}, \dfrac{3}{8}$.

To find another value of x by means of the value thus found, we put $y + \dfrac{3}{8}$ in place of x in the double-equation, whence

$$y^2 - \frac{1}{4} + \frac{49}{64} = u^2,$$
$$y^2 - \frac{9}{4} + \frac{1}{64} = v^2.$$

Multiplying the lower expression by 49, we can solve in the usual way. Our expressions are now $y^2 - \tfrac{1}{4}y + \tfrac{49}{64}$ and $49y^2 - \tfrac{441}{4}y + \tfrac{49}{64}$, and the difference between them is $48y^2 - 110y$. The solution next mentioned by De Billy was clearly obtained by separating this difference into factors such that, when the square of half their difference is equated to $y^2 - \tfrac{1}{4}y + \tfrac{49}{64}$, the absolute terms cancel out. The factors are $\tfrac{440}{7}y$, $\tfrac{42}{55}y - \tfrac{7}{4}$, and we put

$$\left\{\left(\frac{220}{7} - \frac{21}{55}\right)y + \frac{7}{8}\right\}^2 = y^2 - \frac{1}{4}y + \frac{49}{64}.$$

This gives $y = -\tfrac{4045195}{71362992}$, whence $x = \tfrac{22715927}{71362992}$, and the numbers are

$$\frac{48647065}{71362992}, \frac{22715927}{71362992}.$$

A solution in smaller numbers is obtained by separating $48y^2 - 110y$ into factors such that the terms in x^2 in the resulting equation cancel out. The factors are $6y$, $8y - \tfrac{55}{3}$, and we put

$$\left(y - \frac{55}{6}\right)^2 = y^2 - \frac{1}{4}y + \frac{49}{64},$$

whence $y = \tfrac{47959}{10416}$ and $x = \tfrac{47959}{10416} + \tfrac{3}{8} = \tfrac{51865}{10416}$.

This would give a negative value for $1 - x$; but, owing to the symmetry of the original double-equation in x, since $x = \tfrac{51865}{10416}$ satisfies it, so does $x = \tfrac{10416}{51865}$; hence the numbers are $\tfrac{10416}{51865}$ and $\tfrac{41449}{51865}$: a solution also mentioned by De Billy.

Form now a right-angled triangle[12] from 7, 4. The sides are $(7^2 - 4^2, 2 \cdot 7 \cdot 4, 7^2 + 4^2)$ or (33, 56, 65).

Similarly, forming a right-angled triangle from 8, 1, we obtain $(2 \cdot 8 \cdot 1, 8^2 - 1^2, 8^2 + 1^2)$ or 16, 63, 65.

Thus 65^2 is split into two squares in *four* ways.

Assume now as the sum of the numbers $65x$ and

$$\text{as first number } 2 \cdot 39 \cdot 52x^2 = 4056x^2,$$
$$\text{,, second ,, } \quad 2 \cdot 25 \cdot 60x^2 = 3000x^2,$$
$$\text{,, third ,, } \quad 2 \cdot 33 \cdot 56x^2 = 3696x^2,$$
$$\text{,, fourth ,, } \quad 2 \cdot 16 \cdot 63x^2 = 2016x^2,$$

the coefficients of x^2 being four times the areas of the four right-angled triangles respectively.

The sum $12768x^2 = 65x$, and $x = \frac{65}{12768}$.

The numbers are

$$\frac{17136600}{163021824}, \frac{12675000}{163021824}, \frac{15615600}{163021824}, \frac{8517600}{163021824}.$$

20. To divide a given number into two parts and to find a square which, when either of the parts is subtracted from it, gives a square[13].

Given number 10, required square $x^2 + 2x + 1$.

Put for one of the parts $2x + 1$, and for the other $4x$.

The conditions are therefore satisfied if

$$6x + 1 = 10.$$

Therefore $x = 1\frac{1}{2}$;

$$\text{the parts are } (4, 6) \text{ and the square } 6\frac{1}{4}.$$

21. To divide a given number into two parts and to find a square which, when added to either of the parts, gives a square.

Given number 20, required square $x^2 + 2x + 1$.

If to the square there be added either $2x + 3$ or $4x + 8$, the result is a square.

Take $2x + 3$, $4x + 8$ as the parts of 20, and $6x + 11 = 20$, whence $x = 1\frac{1}{2}$.

Therefore the parts are (6, 14) and the square $6\frac{1}{4}$.

[12] If there are two numbers p, q, to "form a right-angled triangle" from them means to take the numbers $p^2 + q^2, p^2 - q^2, 2pq$. These are the sides of a right-angled triangle, since
$$(p^2 + q^2)^2 = (p^2 - q^2)^2 + (2pq)^2.$$
[13] This problem and the next are the same as ii. 15, 14 respectively. It may therefore be doubted whether the solutions here given are genuine, especially as interpolations from ancient commentaries occur most at the beginning and end of Books.

BOOK V

PROBLEMS 1–29

1. To find three numbers in geometrical progression such that each of them *minus* a given number gives a square.

Given number 12.

Find a *square* which exceeds 12 by a square. "This is easy [ii. 10]; $42\frac{1}{4}$ is such a number."

Let the first number be $42\frac{1}{4}$, the third x^2; therefore the middle number $= 6\frac{1}{2}x$.

Therefore $\left.\begin{array}{c} x^2 - 12 \\ 6\frac{1}{2}x - 12 \end{array}\right\}$ are both squares;

their difference $= x^2 - 6\frac{1}{2}x = x(x - 6\frac{1}{2})$; half the difference of the factors multiplied into itself $= \frac{169}{16}$; therefore, putting $6\frac{1}{2}x - 12 = \frac{169}{16}$, we have $x = \frac{361}{104}$,

and $\left(42\frac{1}{4}, \dfrac{2346\frac{1}{2}}{104}, \dfrac{130321}{10816}\right)$ is a solution.

2. To find three numbers in geometrical progression such that each of them when added to a given number gives a square.

Given number 20.

Take a square which when added to 20 gives a square, say 16.

Put for one of the extremes 16, and for the other x^2, so that the middle term $= 4x$.

Therefore $\left.\begin{array}{c} x^2 + 20 \\ 4x + 20 \end{array}\right\}$ are both squares.

Their difference is $x^2 - 4x = x(x - 4)$, and the usual method gives $4x + 20 = 4$, *which is absurd*, because the 4 ought to be some number greater than 20.

But the $4 = \frac{1}{4}(16)$, while the 16 is a square which when added to 20 makes a square; therefore, to replace 16, we must find some square greater than $4 \cdot 20$ and such that when increased by 20 it makes a square.

Now $81 > 80$; therefore, putting $(m + 9)^2$ for the required square, we have

$$(m + 9)^2 + 20 = \text{square} = (m - 11)^2, \text{ say};$$

therefore $m = \frac{1}{2}$, and the square $= (9\frac{1}{2})^2 = 90\frac{1}{4}$.

Assume now for the numbers $90\frac{1}{4}, 9\frac{1}{2}x, x^2$, and we have

$\left.\begin{array}{c} x^2 + 20 \\ 9\frac{1}{2}x + 20 \end{array}\right\}$ both squares.

The difference $= x(x - 9\frac{1}{2})$, and we put $9\frac{1}{2}x + 20 = \frac{361}{16}$.

Therefore $x = \frac{41}{152}$, and

$$\left(90\frac{1}{4}, \frac{389\frac{1}{2}}{152}, \frac{1681}{23104}\right) \text{ is a solution.}$$

3. Given one number, to find three others such that any one of them, or the product of any two of them, when added to the given number, gives a square.

Given number 5.

"*We have it in the Porisms* that if, of two numbers, each, as well as their product, when added to one and the same given number, severally make squares, the two numbers are obtained from the squares of consecutive numbers."

Take then the squares $(x + 3)^2, (x + 4)$[1], and, subtracting the given number 5 from each, put for the first number $x^2 + 6x + 4$, and for the second $x^2 + 8x + 11$, and let the third be twice their sum *minus* 1, or

$$4x^2 + 28x + 29.$$

Therefore $4x^2 + 28x + 34 = $ a square $= (2x - 6)^2$, say.

Hence $x = \frac{1}{26}$, and $\left(\frac{2861}{676}, \frac{7645}{676}, \frac{20336}{676}\right)$ is a solution[2].

4. Given one number, to find three others such that any one of them, or the product of any two, *minus* the given number gives a square.

Given number 6.

Take two consecutive squares $x^2, x^2 + 2x + 1$.

Adding 6 to each, we assume for the first number $x^2 + 6$, and for the second $x^2 + 2x + 7$.

For the third[3] we take twice the sum of the first and second *minus* 1, or $4x^2 + 4x + 25$.

---

[1] The Porism states that, if a be the given number, the numbers $x^2 - a, (x + 1)^2 - a$ satisfy the conditions.
In fact, their product $+a = \{x(x + 1)\}^2 - a(2x^2 + 2x + 1) + a^2 + a$
$= \{x(x + 1)\}^2 - 2ax(x + 1) + a^2 = \{x(x + 1) - a\}^2$.
Diophantus here adds, without explanation, that, if X, Y denote the above two numbers, we should assume for the third required number $Z = 2(X + Y) - 1$. We want *three* numbers such that *any two* satisfy the same conditions as X, Y. Diophantus takes for the third $Z = 2(X + Y) - 1$ because, as is easily seen, with this assumption two out of the three additional conditions are thereby satisfied.
For
$$Z = 2(X + Y) - 1 = 2(2x^2 + 2x + 1) - 4a - 1$$
$$= (2x + 1)^2 - 4a;$$
therefore
$$XZ + a = x^2(2x + 1)^2 - a\{(2x + 1)^2 + 4x^2\} + 4a^2 + a$$
$$= x^2(2x + 1)^2 - a \cdot 4x(2x + 1) + 4a^2$$
$$= \{x(2x + 1) - 2a\}^2,$$
while
$$YZ + a = (x + 1)^2(2x + 1)^2 - a\{(2x + 1)^2 + 4(x + 1)^2\} + 4a^2 + a$$
$$= (x + 1)^2(2x + 1)^2 - a(8x^2 + 12x + 4) + 4a^2$$
$$= \{(x + 1)(2x + 1) - 2a\}^2.$$
The only condition remaining is then
$$Z + a = \text{a square},$$
or
$$(2x + 1)^2 - 3a = \text{a square} = (2x - k)^2, \text{ say},$$
and x is found.
[2] Diophantus having solved the problem of finding three numbers ξ, η, ζ satisfying the six equations
$$\xi + a = r^2, \quad \eta\zeta + a = u^2,$$
$$\eta + a = s^2, \quad \zeta\xi + a = v^2,$$
$$\zeta + a = t^2, \quad \zeta\eta + a = w^2,$$
Fermat observes that we can deduce the solution of the problem
To find four numbers such that the product of any pair added to a given number produces a square.
Taking three numbers, as found by Diophantus, satisfying the above six conditions, we take $x + 1$ as the fourth number. We then have three conditions which remain to be satisfied. These give a "triple-equation" to be solved by Fermat's method.
[3] Diophantus makes this assumption for the same reason as in the last problem, v. 3. Note 1 above covers this case if we substitute $-a$ for a throughout.

266

Therefore third *minus* $6 = 4x^2 + 4x + 19 =$ square $= (2x - 6)^2$, say.
Therefore $x = \frac{17}{28}$,

$$\text{and} \left(\frac{4993}{784}, \frac{6729}{784}, \frac{22660}{784} \right) \text{ is a solution.}$$

[The same Porism is assumed as in the preceding problem but with a *minus* instead of a *plus*.]

5. To find three squares such that the product of any two added to the sum of those two, or to the remaining square, gives a square.

"We have it in the Porisms" that, if the squares on any two consecutive numbers be taken, and a third number be also taken which exceeds twice the sum of the squares by 2, we have three numbers such that the product of any two added to those two or to the remaining number gives a square.

Assume as the first square $x^2 + 2x + 1$, and as the second $x^2 + 4x + 4$, so that third number $= 4x^2 + 12x + 12$.

Therefore $x^2 + 3x + 3 =$ a square $= (x - 3)^2$, say, and $x = \frac{2}{3}$.

$$\text{Therefore} \left(\frac{25}{9}, \frac{64}{9}, \frac{196}{9} \right) \text{ is a solution.}$$

6. To find three numbers such that each *minus* 2 gives a square, and the product of any two *minus* the sum of those two, or *minus* the remaining number, gives a square.

Add 2 to each of three numbers found as in the Porism quoted in the preceding problem.

Let the numbers so obtained be $x^2 + 2$, $x^2 + 2x + 3$, $4x^2 + 4x + 6$.

All the conditions are now satisfied[4], except one, which gives

$$4x^2 + 4x + 6 - 2 = \text{a square.}$$

Divide by 4, and $x^2 + x + 1 =$ a square $= (x - 2)^2$, say.
Therefore $x = \frac{3}{5}$,

$$\text{and} \left(\frac{59}{25}, \frac{114}{25}, \frac{246}{25} \right) \text{ is a solution.}$$

Lemma I to the following problem.

To find two numbers such that their product added to the squares of both gives a square.

Suppose first number x, second any number (m), say 1.

Therefore $x \cdot 1 + x^2 + 1 = x^2 + x + 1 =$ a square $= (x - 2)^2$, say.

---

[4] The numbers are $x^2 + 2, (x + 1)^2 + 2, 2\{x^2 + (x + 1)^2 + 1\} + 2$; and if X, Y, Z denote these numbers respectively, it is easily verified that
$$XY - (X + Y) = (x^2 + x + 1)^2, \qquad XY - Z = (x^2 + x)^2,$$
$$XZ - (X + Z) = (2x^2 + x + 2)^2, \qquad XZ - Y = (2x^2 + x + 3)^2,$$
and
$$YZ - (Y + Z) = (2x^2 + 3x + 3)^2, \qquad YZ - X = (2x^2 + 3x + 4)^2.$$

267

Thus $x = \frac{3}{5}$, and

$$\left(\frac{3}{5},\ 1\right) \text{ is a solution, or } (3, 5).$$

Lemma II to the following problem.

To find three right-angled triangles (*i.e.* three right-angled triangles *in rational numbers*[5]) which have equal areas.

We must first find two numbers such that their product + the sum of their squares = a square, *e.g.* 3, 5, as in the preceding problem.

Now form right-angled triangles from the pairs of numbers[6]

$$(7, 3), (7, 5), (7, 3 + 5)$$

---

[5] All Diophantus' right-angled triangles must be understood to be right-angled triangles with sides expressible in rational numbers. In future I shall say "right-angled triangle" simply, for brevity.

[6] Diophantus here tacitly assumes that, if $ab + a^2 + b^2 = c^2$, and right-angled triangles be formed from (c, a), (c, b) and $(c, a + b)$ respectively, their areas are equal. The areas are of course $(c^2 - a^2)ca$, $(c^2 - b^2)cb$ and $\{(a + b)^2 - c^2\}(a + b)c$, and it is easy to see that each $= abc\,(a + b)$.

Nesselmann suggests that Diophantus discovered the property as follows. Let the triangles formed from (n, m), (q, m), (r, m) have their areas equal; therefore

$$n(m^2 - n^2) = q(m^2 - q^2) = r(r^2 - m^2).$$

It follows, first, since

$$m^2 n - n^3 = m^2 q - q^3,$$

that

$$m^2 = (n^3 - q^3)/(n - q) = n^2 + nq + q^2.$$

Again, given (q, m, n), to find r.

We have

$$q\,(m^2 - q^2) = r(r^2 - m^2),$$

and $m^2 - q^2 = n^2 + nq$, from above;

therefore

$$q\,(n^2 + nq) = r(r^2 - n^2 - nq - q^2),$$

or

$$q\,(n^2 + nr) + q^2(n + r) = r(r^2 - n^2).$$

Dividing by $r + n$, we have

$$qn + q^2 = r^2 - rn;$$

therefore

$$(q + r)n = r^2 - q^2,$$

and

$$r = q + n.$$

Fermat observes that, given any rational right-angled triangle, say z, b, d, where z is the hypotenuse, it is possible to find an infinite number of other rational right-angled triangles having the same area. Form a right-angled triangle from z^2, $2bd$; this gives the triangle $z^4 + 4b^2d^2$, $z^4 - 4b^2d^2$, $4z^2bd$. Divide each of these sides by $2z(b^2 - d^2)$, b being $> d$; and we have a triangle with the same area $(\frac{1}{2}bd)$ as the original triangle. Trying this method with Diophantus' first triangle $(40, 42, 58)$, we obtain as the new triangle

$$\frac{1412881}{1189},\quad \frac{1412880}{1189},\quad \frac{1681}{70}.$$

The method gives $(\frac{7}{10}, \frac{120}{7}, \frac{1201}{70})$ as a right-angled triangle with area equal to that of $(3, 4, 5)$.

Another method of finding other rational right-angled triangles having the same area as a given right-angled triangle is explained in the *Inventum Novum*, Part 1, paragraph 38 (*Œuvres de Fermat*, III.).

Let the given triangle be 3, 4, 5, so that it is required to find a new rational right-angled triangle with area 6.

Let 3, $x + 4$ be the perpendicular sides; therefore

the square of the hypotenuse $= x^2 + 8x + 25 =$ a square.

Again, the area is $\frac{3}{2}x + 6$; and, as this is to be 6, it must be six times a certain square; that is, $\frac{3}{2}x + 6$ divided by 6 must be a square, and this again multiplied by 36 must be a square; therefore

$$9x + 36 = \text{a square.}$$

Accordingly we have to solve the double-equation

$$\left.\begin{array}{l} x^2 + 8x + 25 = u^2 \\ 9x + 36 = v^2 \end{array}\right\}.$$

This gives

$$x = \frac{6725600}{2405601},$$

whence

$$x + 4 = \frac{2896804}{2405601}.$$

The triangle is thus found to be

$$3,\quad \frac{2896804}{2405601},\quad \frac{7776485}{2405601}.$$

The area is 6 times a certain square, namely $\frac{724201}{2405601}$, the root of which is $\frac{851}{1551}$.
Dividing each of the above sides by $\frac{851}{1551}$, we obtain a triangle with area 6, namely

$$\frac{4653}{851},\quad \frac{3404}{1551},\quad \frac{7776485}{1319901}.$$

Another solution of the double-equation, $x = -\frac{13959}{3600}$, giving $x + 4 = \frac{49}{400}$, leads to the same triangle $(\frac{7}{10}, \frac{120}{7}, \frac{1201}{70})$ as that obtained by Fermat's rule (see above).

The method of the *Inventum Novum* has a feature in common with the procedure in the ancient Greek problem reproduced and commented on by Heiberg and Zeuthen (*Bibliotheca Mathematica*, VIII₃, 1907/8), where it is required to find a rational right-angled triangle, having given the area, 5 feet, and where the 5 is multiplied by a square number containing 6 as a factor and such that the product "can form the area of a right-angled triangle." 36 is taken and the area becomes 180, which is the area of $(9, 40, 41)$. The sides of the latter triangle are then divided by 6, and we have the required triangle.

[*i.e.* the right-angled triangles $(7^2 + 3^2, 7^2 - 3^2, 2 \cdot 7 \cdot 3)$, etc.].

The triangles are (40, 42, 58), (24, 70, 74), (15, 112, 113), the area of each being 840.

7. To find three numbers such that the square of any one ± the sum of the three gives a square.

Since, in a right-angled triangle,

(hypotenuse)^2 ± twice product of perps. = a square, we make the three required numbers hypotenuses and the sum of the three four times the area.

Therefore we must find three right-angled triangles having the same area, *e.g.*, as in the preceding problem,

(40, 42, 58), (24, 70, 74), (15, 112, 113).

Reverting to the substantive problem, we put for the numbers $58x$, $74x$, $113x$; their sum $245x$ = four times the area of any one of the triangles = $3360x^2$.

Therefore $x = \frac{7}{96}$,

$$\text{and} \left(\frac{406}{96}, \frac{518}{96}, \frac{791}{96}\right) \text{ is a solution.}$$

Lemma to the following problem.

Given three squares, it is possible to find three numbers such that the products of the three pairs shall be respectively equal to those squares.

Squares 4, 9, 16.

First number x, so that the others are $\frac{4}{x}, \frac{9}{x}$; and $\frac{36}{x^2} = 16$.

Therefore $x = \frac{6}{4}$, and the numbers are $(1\frac{1}{2}, 2\frac{2}{3}, 6)$.

We observe that $x = \frac{6}{4}$, where 6 is the product of 2 and 3, and 4 is the side of 16.

Hence the following *rule.* Take the product of two sides (2, 3), divide by the side of the third square 4 [the result is the first number]; divide 4, 9 respectively by the result, and we have the second and third numbers.

8. To find three numbers such that the product of any two ± the sum of the three gives a square.

As in Lemma II to the 7th problem, we find three right-angled triangles with equal areas; the squares of their hypotenuses are 3364, 5476, 12769.

Now find, as in the last Lemma, three numbers such that the products of the three pairs are equal to these squares respectively, which we take because each ± 4. (area) or 3360 gives a square; the three numbers then are

$\frac{4292}{113}x$, $\frac{3277}{37}x$ $\left[\frac{380132}{4292}x \text{ Tannery}\right]$, $\frac{4181}{29}x$ $\left[\frac{618788}{4292}x \text{ Tannery}\right]$.

It remains that the sum of the three = $3360x^2$.

Therefore $\frac{32824806}{121249}x$ $\left[\frac{131299224}{484996}x \text{ Tannery}\right] = 3360x^2$.

Therefore $x = \frac{32824806}{407396640}$ $\left[\frac{131299224}{1629586560}\right.$ or $\frac{781543}{9699920}$ Tannery$\left.\right]$,

$$\left[\text{and the numbers are } \frac{781543}{255380}, \frac{781543}{109520}, \frac{781543}{67280}\right].$$

9. To divide unity into two parts such that, if the same given number be added to either part, the result will be a square.

Necessary condition. The given number must not be odd and the double of it $+1$ must not be divisible by any prime number which, when increased by 1, is divisible by 4 [*i.e.* any prime number of the form $4n - 1$].

Given number 6. Therefore 13 must be divided into two squares each of which >6. If then we divide 13 into two squares the difference of which <1, we solve the problem.

Take half of 13 or $6\frac{1}{2}$, and we have to add to $6\frac{1}{2}$ a small fraction which will make it a square,

or, multiplying by 4, we have to make $\frac{1}{x^2} + 26$ a square, *i.e.* $26x^2 + 1 = $ a square $= (5x + 1)^2$, say, whence $x = 10$.

That is, in order to make 26 a square, we must add $\frac{1}{100}$, or, to make $6\frac{1}{2}$ a square, we must add $\frac{1}{400}$, and

$$\frac{1}{400} + 6\frac{1}{2} = \left(\frac{51}{20}\right)^2.$$

Therefore we must divide 13 *into two squares such that their sides may be as nearly as possible equal to* $\frac{51}{20}$. [This is the παρισότητος ἀγωγή.]

Now $13 = 2^2 + 3^2$. Therefore we seek two numbers such that 3 *minus* the first $= \frac{51}{20}$, so that the first $= \frac{9}{20}$, and 2 *plus* the second $= \frac{51}{20}$, so that the second $= \frac{11}{20}$.

We write accordingly $(11x + 2)^2$, $(3 - 9x)^2$ for the required squares [substituting x for $\frac{1}{20}$].

The sum $= 202x^2 - 10x + 13 = 13$.

Therefore $x = \frac{5}{101}$, and the sides are $\frac{257}{101}, \frac{258}{101}$.

Subtracting 6 from the squares of each, we have, as the parts of unity,

$$\frac{4843}{10201}, \frac{5358}{10201}.$$

10. To divide unity into two parts such that, if we add different given numbers to each, the results will be squares.

Let the numbers[7] be 2, 6 and let them and the unit be represented in the figure, where $DA = 2$, $AB = 1$, $BE = 6$, and G is a point in AB so chosen that DG, GE are both squares.

[7] Loria, as well as Nesselmann, observes that Diophantus omits to state the necessary condition, namely that the sum of the two given numbers *plus* 1 must be the sum of two squares.

D   A G B   E

Now $DE = 9$. *Therefore we have to divide 9 into two squares such that one of them lies between 2 and 3.*

Let the latter square be x^2, so that the other is $9 - x^2$, where $3 > x^2 > 2$. Take two squares, one >2, the other <3 [the former being the smaller], say $\frac{289}{144}, \frac{361}{144}$. Therefore, if we can make x^2 lie between these, we shall solve the problem. We must have $x > \frac{17}{12}$ and $< \frac{19}{12}$.

Hence, in making $9 - x^2$ a square, we must find $x > \frac{17}{12}$ and $< \frac{19}{12}$. Put $9 - x^2 = (3 - mx)^2$, say, whence $x = \frac{6m}{(m^2 + 1)}$.

Therefore

$$\frac{17}{12} < \frac{6m}{m^2 + 1} < \frac{19}{12}.$$

The first inequality gives $72m > 17m^2 + 17$; and

$$36^2 - 17 \cdot 17 = 1007,$$

the square root of which[8] is not greater than 31; therefore $m \not> \frac{31 + 36}{17}$, *i.e.* $m \not> \frac{67}{17}$.

Similarly from the inequality $19m^2 + 19 > 72m$ we find

$$m \not< \frac{66}{19}.$$

Let $m = 3\frac{1}{2}$. Therefore $9 - x^2 = (3 - 3\frac{1}{2}x)^2$, and $x = \frac{84}{53}$. Therefore $x^2 = \frac{7056}{2809}$,

and the segments of 1 are $\left(\dfrac{1438}{2809}, \dfrac{1371}{2809} \right)$.

11. To divide unity into three parts such that, if we add the same number to each of the parts, the results are all squares.

Necessary condition. The given number must not be 2 or any multiple of 8 increased by 2.

Given number 3. Thus 10 is to be divided into three squares such that each >3. Take $\frac{1}{3}$ of 10, or $3\frac{1}{3}$, and find x so that $\frac{1}{9x^2} + 3\frac{1}{3}$ may be a square, or $30x^2 + 1 = $ a square $= (5x + 1)^2$, say.

Therefore $x = 2$, $x^2 = 4$, $\frac{1}{x^2} = \frac{1}{4}$, and

$$\frac{1}{36} + 3\frac{1}{3} = \frac{121}{36} = \text{a square}.$$

Therefore we have to divide 10 into three squares each of which is as near as possible to $\frac{121}{36}$. [παρισότητος ἀγωγή.]

---

[8] *I.e.* the *integral* part of the root is $\not> 31$. The limits taken in each case are *a fortiori* limits.

Now $10 = 3^2 + 1^2 =$ the sum of the three squares 9, $\frac{16}{25}$, $\frac{9}{25}$.

Comparing the sides 3, $\frac{4}{5}$, $\frac{3}{5}$ with $\frac{11}{6}$,

or (multiplying by 30) 90, 24, 18 with 55, we must make each side approach 55.

[Since then $\frac{55}{30} = 3 - \frac{35}{30} = \frac{4}{5} + \frac{31}{30} = \frac{3}{5} + \frac{37}{30}$], we put for the sides of the required numbers

$$3 - 35x, \quad 31x + \frac{4}{5}, \quad 37x + \frac{3}{5}.$$

The sum of the squares $= 3555x^2 - 116x + 10 = 10$.

Therefore $x = \frac{116}{3555}$,

and this solves the problem.

12. To divide unity into three parts such that, if three different given numbers be added to the parts respectively, the results are all squares.

Given numbers 2, 3, 4. Then I have to divide 10 into three squares such that the first >2, the second >3, and the third >4.

Let us add $\frac{1}{2}$ of unity to each, and we have to find three squares such that their sum is 10, while the first lies between 2, $2\frac{1}{2}$, the second between 3, $3\frac{1}{2}$, and the third between 4, $4\frac{1}{2}$.

It is necessary, first, to divide 10 (the sum of two squares) into two squares one of which lies between 2, $2\frac{1}{2}$; then, if we subtract 2 from the latter square, we have one of the required parts of unity.

Next divide the other square into two squares, one of which lies between 3, $3\frac{1}{2}$; subtracting 3 from the latter square, we have the second of the required parts of unity.

Similarly we can find the third part[9].

---

[9] Diophantus only thus briefly indicates the course of the solution. Wertheim solves the problem in detail after Diophantus' manner; and, as this is by no means too easy, I think it well to reproduce his solution.

I. It is first necessary to divide 10 into two squares one of which lies between 2 and 3. We use the παρισότητος ἀγωγή.

The first square must be in the neighbourhood of $2\frac{1}{2}$; and we seek a small fraction $\frac{1}{x^2}$ which when added to $2\frac{1}{2}$ gives a square: in other words, we must make $4(2\frac{1}{2} + \frac{1}{x^2})$ a square. This expression may be written $10 + (\frac{2}{x})^2$, and, to make this a square, we put

$$10y^2 + 1 = (3y + 1)^2, \text{ say,}$$

whence $y = 6$, $y^2 = 36$, $x^2 = 144$, so that $2\frac{1}{2} + \frac{1}{x^2} = \frac{361}{144} = (\frac{19}{12})^2$, which is an approximation to the first of the two squares the sum of which is 10.

The second of these squares approximates to $7\frac{1}{2}$, and we seek a small fraction $\frac{1}{x^2}$ such that $7\frac{1}{2} + \frac{1}{x^2}$ is a square, or $30 + (\frac{2}{x})^2 = 30 + (\frac{1}{y})^2$, say $=$ a square.

Put

$$30y^2 + 1 = (5y + 1)^2, \text{ say;}$$

therefore $y = 2$, $y^2 = 4$, $x^2 = 16$, so that $7\frac{1}{2} + \frac{1}{x^2} = \frac{121}{16} = (\frac{11}{4})^2 = (\frac{33}{12})^2$.

Now, since $10 = 1^2 + 3^2$, and $\frac{19}{12} = 1 + \frac{7}{12}$, while $\frac{33}{12} = 3 - \frac{3}{12}$, we put

$$(1 + 7x)^2 + (3 - 3x)^2 = 10, \qquad \text{[Cf. v. 9]}$$

so that

$$x = \frac{2}{29},$$

$$(1 + 7x)^2 = \left(1 + \frac{14}{29}\right)^2 = \left(\frac{43}{29}\right)^2 = \frac{1849}{841},$$

$$(3 - 3x)^2 = \left(3 - \frac{6}{29}\right)^2 = \left(\frac{81}{29}\right)^2 = \frac{6561}{841}.$$

Therefore the two squares into which 10 is divided are $\frac{1849}{841}$ and $\frac{6561}{841}$, and the first of these lies in fact between 2 and $2\frac{1}{2}$.

II. We have next to divide the square $\frac{6561}{841}$ into two squares, one of which, which we will call x^2, lies between 3 and 4. [The method of v. 10 is here applicable.]

13. To divide a given number into three parts such that the sum of any two of the parts gives a square.

Given number 10.

Since the sum of each pair of parts is a square less than 10, while the sum of the three pairs is twice the sum of the three parts or 20,

we have to divide 20 *into three squares each of which is* <10.

But 20 is the sum of two squares, 16 and 4;

and, if we put 4 for one of the required squares, we have to divide 16 into two squares, each of which is <10, or, in other words, into two squares, one of which lies between 6 and 10. This we learnt how to do[10] [v. 10].

We have, when this is done, three squares such that each is <10, while their sum is 20;

and by subtracting each of these squares from 10 we obtain the parts of 10 required.

Instead of 3, 4 take $\frac{49}{16}, \frac{64}{16}$ as the limits.

Therefore
$$\frac{49}{16} < x^2 < \frac{64}{16},$$

or
$$\frac{7}{4} < x < \frac{8}{4}.$$

And $\frac{6561}{841} - x^2$ must be a square $= \left(\frac{81}{29} - kx\right)^2$, say,

which gives
$$x = \frac{162k}{29(1 + k^2)};$$

k has now to be chosen such that

(1)
$$\frac{162k}{29(1 + k^2)} < \frac{7}{4},$$

from which it follows that
$$k < 2 \cdot 8 \ldots,$$

and (2)
$$\frac{162k}{29(1 + k^2)} < \frac{8}{4},$$

whence
$$k > 2 \cdot 3 \ldots.$$

We may therefore put
$$k = 2 \cdot 5.$$

Therefore
$$x = \frac{1620}{841}, \quad x^2 = \frac{2624400}{707281},$$

and
$$\frac{6561}{841} - x^2 = \frac{2893401}{707281}.$$

The three required squares into which 10 is divided are therefore
$$\frac{1849}{841}, \quad \frac{2624400}{707281}, \quad \frac{2893401}{707281}.$$

And if we subtract 2 from the first, 3 from the second and 4 from the third, we obtain as the required parts of unity
$$\frac{140447}{707281}, \quad \frac{502557}{707281}, \quad \frac{64277}{707281}.$$

[10] Wertheim gives a solution in full, thus.

Let the squares be x^2, $16 - x^2$, of which one, x^2, lies between 6 and 10.

Put instead of 6 and 10 the limits $\frac{25}{4}$ and 9, so that
$$\frac{5}{2} < x < 3.$$

To make $16 - x^2$ a square, we put
$$16 - x^2 = (4 - kx)^2,$$

whence
$$x = \frac{8k}{1 + k^2}.$$

Now (1) $\frac{8k}{1 + k^2} > \frac{5}{2}$, and (2) $\frac{8k}{1 + k^2} < 3$.

These conditions give, as limits for k, $2 \cdot 84 \ldots$ and $2 \cdot 21 \ldots$.

We may therefore *e.g.* put $k = 2\frac{1}{2}$.

Then
$$x = \frac{80}{29}, \quad x^2 = \frac{6400}{841}, \quad 16 - x^2 = \frac{7056}{841}.$$

The required three squares making up 20 are 4, $\frac{6400}{841}, \frac{7056}{841}$.

Subtracting these respectively from 10, we have the required parts of the given number 10, namely 6, $\frac{2010}{841}, \frac{1354}{841}$.

14. To divide a given number into four parts such that the sum of any three gives a square.

Given number 10.

Three times the sum of the parts = the sum of four squares. Therefore 30 has to be divided into four squares, each of which is <10.

(1) If we use the method of approximation ($\pi\alpha\rho\iota\sigma\acute{o}\tau\eta\varsigma$), we have to make each square approximate to $7\frac{1}{2}$; then, when the squares are found, we subtract each from 10, and so find the required parts.

(2) *Or*, observing that $30 = 16 + 9 + 4 + 1$, we take 4, 9 for two of the squares, and then divide 17 into two squares, each of which <10.

If then we divide 17 into two squares, one of which lies between $8\frac{1}{2}$ and 10, as we have learnt how to do[11] [cf. v. 10], the squares will satisfy the conditions.

We shall then have divided 30 into four squares, each of which is less than 10, two of them being 4, 9 and the other two the parts of 17 just found.

Subtracting each of the four squares from 10, we have the required parts of 10, two of which are 1 and 6.

15. To find three numbers such that the cube of their sum added to any one of them gives a cube.

Let the sum be x and the cubes $7x^3$, $26x^3$, $63x^3$.

Therefore $96x^3 = x$, or $96x^2 = 1$.

But 96 is not a square; we must therefore replace it by a square in order to solve the problem.

Now 96 is the sum of three numbers, each of which is 1 less than a cube;

therefore we have to find three numbers such that each of them is a cube less 1, and the sum of the three is a square.

[11] Wertheim gives a solution of this part of the problem.
As usual, we make $8\frac{1}{2} + \frac{1}{x^2}$, or $34 + \left(\frac{2}{x}\right)^2$, a square.
Putting $\frac{2}{x} = \frac{1}{y}$, we must make $34 + \left(\frac{1}{y}\right)^2$ a square.

Let
$$34y^2 + 1 = \text{a sqaure} = (6y - 1)^2,$$
and we obtain
$$y = 6, \ y^2 = 36, \ x^2 = 144.$$
Thus
$$8\frac{1}{2} + \frac{1}{144} = \frac{1225}{144} = \left(\frac{35}{12}\right)^2,$$
and $\frac{35}{12}$ is an approximation to the side of each of the required squares.
Next, since $17 = 1^2 + 4^2$, and $\frac{35}{12} = 1 + \frac{23}{12} = 4 - \frac{13}{12}$,
we put
$$17 = (1 + 23x)^2 + (4 - 13x)^2,$$
and we obtain
$$x = \frac{29}{349}.$$
The squares are then $(1 + 23x)^2 = \left(\frac{1016}{349}\right)^2 = \frac{1032256}{121801}$,
and
$$(4 - 13x)^2 = \left(\frac{1019}{349}\right)^2 = \frac{1038361}{121801}.$$
Subtracting each of these from 10, we have the third and fourth of the required parts of 10, namely
$$\frac{185754}{121801}, \ \frac{179649}{121801}.$$

Let the sides of the cubes[13] be $m + 1$, $2 - m$, 2, whence the numbers are $m^3 + 3m^2 + 3m$, $7 - 12m + 6m^2 - m^3$, 7; their sum $= 9m^2 - 9m + 14 =$ a square $= (3m - 4)^2$, say;

therefore $m = \frac{2}{15}$,

and the numbers are $\frac{1538}{3375}$, $\frac{18577}{3375}$, 7.

Reverting to the original problem, we put x for the sum of the numbers, and for the numbers respectively

$$\frac{1538}{3375}x^3, \quad \frac{18577}{3375}x^3, \quad 7x^3,$$

whence $\frac{43740}{3375}x^3 = x$,

that is (if we divide out by 15 and by x),

$$2916x^2 = 225, \text{ and } x = \frac{15}{54}.$$

The numbers are therefore found.

16. To find three numbers such that the cube of their sum *minus* any one of them gives a cube.

Let the sum be x, and the numbers $\frac{7}{8}x^3$, $\frac{26}{27}x^3$, $\frac{63}{64}x^3$.

Therefore $\frac{4877}{1728}x^3 = x$,

and, if $\frac{4877}{1728}$ were the ratio of a square to a square, the problem would be solved. But $\frac{4877}{1728} = 3 -$ (the sum of three cubes).

Therefore we must find three cubes, each of which <1, and such that $(3 -$ their sum$) =$ a square.

If, *a fortiori*, the sum of the three cubes is made <1, the square will be >2. Let[14] it be $2\frac{1}{4}$.

---

[13] If a^3, b^3, c^3 are the three cubes, so that $a^3 + b^3 + c^3 - 3$ has to be a square, Diophantus chooses c^3 arbitrarily (8) and then makes such assumptions for the sides of a^3, b^3, being linear expressions in m, that, in the expression to be turned into a square, the coefficient of m^3 vanishes, and that of m^2 is a square. If $a = m$, the condition is satisfied by putting $b = 3k^2 - m$, where k is any number.

[14] Bachet, finding no way of hitting upon $2\frac{1}{4}$ as the particular square to be taken in order that the difference between it and 3 may be separable into three cubes, and observing that he could not solve the problem if he took another arbitrary square between 2 and 3, e.g. $2\frac{7}{9}$, instead of $2\frac{1}{4}$, concluded that Diophantus must have hit upon $2\frac{1}{4}$, which does enable the problem to be solved, by accident.

Fermat would not admit this and considered that the method used by Diophantus for finding $2\frac{1}{4}$ as the square to be taken should not be difficult to discover. Fermat accordingly suggested a method as follows.

Let $x - 1$ be the side of the required square lying between 2 and 3. Then $3 - (x - 1)^2 = 2 + 2x - x^2$, and this has to be separated into three cubes. Fermat assumes for the sides of two of the required cubes two linear expressions in x such that, when the sum of their cubes is subtracted from $2 + 2x - x^2$, the result only contains terms in x^2 and x^3 or in x and units.

The first alternative is secured if the sides of the first and second cubes are $1 - \frac{1}{3}x$ and $1 + x$ respectively; for

$$2 + 2x - x^2 - \left(1 - \frac{1}{3}x\right)^3 - (1 + x)^3 = -4\frac{1}{3}x^2 - \frac{26}{27}x^3.$$

This latter expression has to be made a cube, for which purpose we put

$$-4\frac{1}{3}x^2 - \frac{26}{27}x^3 = -\frac{m^3x^3}{27}, \text{ say,}$$

which gives a value for x. We have only to see that this value makes $\frac{1}{3}x$ less than 1, and we can easily choose m so as to fulfil this condition.

[E.g. suppose $m = 5$, and we find $x = \frac{13}{11}$, so that

$$\frac{1}{3}x = \frac{13}{33}, \quad 1 - \frac{1}{3}x = \frac{20}{33}, \quad 1 + x = \frac{72}{33},$$

and the side of the third cube is $-\frac{65}{33}$.

The square $(x - 1)^2 = \left(\frac{2}{11}\right)^2$, and in fact $3 - \left\{\left(\frac{20}{33}\right)^3 + \left(\frac{72}{33}\right)^3 - \left(\frac{65}{33}\right)^3\right\} = \left(\frac{2}{11}\right)^2$.]

We then have three cubes which make up the excess of 3 over a certain square; but, while the first of these cubes is <1, the second is >1 and the third is negative. Hence we must, like Diophantus, proceed to transform the difference between the two latter cubes into the sum of two other cubes.

We have therefore to find three cubes the sum of which $= \frac{3}{4}$ or $\frac{162}{216}$;

that is, we have to divide 162 into three cubes.

But $162 = 125 + 64 - 27$;

and "we have it in the Porisms" that *the difference of two cubes can be transformed into the sum of two cubes.*

Having thus found the three cubes[15], we start again, and $x = 2\frac{1}{4}x^3$, so that $x = \frac{2}{3}$. The three numbers are thus determined.

17. To find three numbers such that each of them *minus* the cube of their sum gives a cube.

Let the sum be x and the numbers $2x^3$, $9x^3$, $28x^3$.

Therefore $39x^2 = 1$;

and we must replace 39 by a square which is the sum of three cubes $+3$;

therefore we must find three cubes such that their sum $+3$ is a square.

Let their sides[16] be m, $3 - m$, and any number, say 1.

Therefore $9m^2 + 31 - 27m =$ a square $= (3m - 7)^2$, say, so that $m = \frac{6}{5}$, and the sides of the cubes are $\frac{6}{5}$, $\frac{9}{5}$, 1.

Starting again, we put x for the sum, and for the numbers

$$\frac{341}{125}x^3, \quad \frac{854}{125}x^3, \quad \frac{250}{125}x^3,$$

whence $1445x^2 = 125$, $x^2 = \frac{25}{289}$, and $x = \frac{5}{17}$.

The required numbers are thus found.

18. To find three numbers such that their sum is a square and the cube of their sum added to any one of them gives a square.

Let the sum be x^2 and the numbers $3x^6$, $8x^6$, $15x^6$.

It follows that $26x^4 = 1$; and, if 26 were a fourth power, the problem would be solved.

To replace it by a fourth power, we have to find three numbers such that each increased by 1 gives a square, while the sum of the three gives a fourth power.

Let these numbers be $m^4 - 2m^2$, $m^2 + 2m$, $m^2 - 2m$ [the sum being m^4]; these are indeterminate numbers satisfying the conditions.

It will, however, be seen by trial that even Fermat's method is not quite general, for it will not, as a matter of fact, give the particular solution obtained by Diophantus in which the square is $2\frac{1}{4}$.

[15] Vieta's rule gives $4^3 - 3^3 = (\frac{303}{91})^3 + (\frac{40}{91})^3$. It follows that

$$\frac{3}{4} = \frac{162}{216} = \left(\frac{5}{6}\right)^3 + \left(\frac{101}{182}\right)^3 + \left(\frac{20}{273}\right)^3;$$

and, since $x^3 = \frac{8}{27}$, the required numbers are

$$\frac{91}{216} \cdot \frac{8}{27}, \quad \frac{4998267}{6028568} \cdot \frac{8}{27}, \quad \frac{20338417}{20346417} \cdot \frac{8}{27}.$$

[16] In this case, if one of the cubes is chosen arbitrarily and m^3 is another, we have only to put $(3k^2 - m)$ for the side of the third cube in order that, in the expression to be made a square, the term in m^3 may vanish, and the term in m^2 may be a square.

Putting any number, say 3, for m, we have as the required auxiliary numbers 63, 15, 3.

Starting again, we put x^2 for the sum and $3x^6$, $15x^6$, $63x^6$ for the required numbers,

and we have $81x^6 = x^2$, so that $x = \frac{1}{3}$.

The numbers are thus found $\left(\dfrac{3}{729}, \dfrac{15}{729}, \dfrac{63}{729} \right)$.

19. To find three numbers such that their sum is a square and the cube of their sum *minus* any one of them gives a square.

[There is obviously a lacuna in the text after this enunciation; for the next words are "And we have *again* to divide 2 *as before*,"
whereas there is nothing in our text to which they can refer, and the lines which follow are clearly no part of the solution of v. 19.

Bachet first noticed the probability that three problems intervened between v. 19 and v. 20, and he gave solutions of them. But he seems to have failed to observe that the eight lines or so in the text between the enunciation of v. 19 and the enunciation of v. 20 belonged to the solution of the last of the three missing problems. The first of the missing problems is connected with v. 18 and 19, making a natural trio with them, while the second and third similarly make with v. 20 a set of three. The enunciations were doubtless somewhat as follows.

19*a*. To find three numbers such that their sum is a square and any one of them *minus* the cube of their sum gives a square.

19*b*. To find three numbers such that their sum is a given number and the cube of their sum *plus* any one of them gives a square.

19*c*. To find three numbers such that their sum is a given number and the cube of their sum *minus* any one of them gives a square.

The words then in the text after the enunciation of v. 19 evidently belong to this last problem.]

The given sum is 2, the cube of which is 8.

We have to subtract each of the numbers from 8 and thereby make a square.

Therefore we have to divide 22 into three squares, each of which is greater than 6;

after which, by subtracting each of the squares from 8, we find the required numbers.

But we have already shown [cf. v. 11] how to divide 22 into three squares, each of which is greater than 6—and less than 8, Diophantus should have added.

[The above is explained by the fact that, by addition, three times the cube of the sum *minus* the sum itself is the sum of three squares, and three times the cube of the sum *minus* the sum = $3 \cdot 8 - 2 = 22$.][17]

20. To divide a given fraction into three parts such that any one of them *minus* the cube of their sum gives a square.

Given fraction $\frac{1}{4}$.

Therefore each part = $\frac{1}{64}$ + a square.

Therefore the sum of the three = $\frac{1}{4}$ = the sum of three squares $+\frac{3}{64}$.

Hence we have to divide $\frac{13}{64}$ into three squares, "which is easy[18]."

21. To find three squares such that their continued product added to any one of them gives a square.

Let the "solid content" $=x^2$.

We want now three squares, each of which increased by 1 gives a square.

They can be got from right-angled triangles[19] by dividing the square of one of the sides about the right angle by the square of the other.

Let the squares then be

$$\frac{9}{16}x^2, \quad \frac{25}{144}x^2, \quad \frac{64}{225}x^2.$$

The continued product = $\frac{14400}{518400}x^6 = x^2$, by hypothesis.

Therefore $\frac{120}{720}x^2 = 1$; and, if $\frac{120}{720}$ were a square, the problem would be solved.

As it is not, we must find three right-angled triangles such that, if b's are their bases, and p's are their perpendiculars, $p_1 p_2 p_3 b_1 b_2 b_3$ = a square;

and, if we assume one triangle arbitrarily (3, 4, 5), we have to make $12 p_1 p_2 b_1 b_2$ a square, or $3 p_1 b_1 / p_2 b_2$ a square.

"This is easy[20]," and the three triangles are (3, 4, 5), (9, 40, 41), (8, 15, 17) or similar to them.

---

[17] Wertheim adds a solution in Diophantus' manner. We have to find what small fraction of the form $\frac{1}{x^2}$ we have to add to $\frac{22}{3}$ or $\frac{66}{9}$, and therefore to 66, in order to make a square. In order that $66 + \frac{1}{x^2}$ may be a square, we put
$$66x^2 + 1 = \text{square} = (1 + 8x)^2, \text{ say,}$$
which gives
$$x = 8 \text{ and } x^2 = 64.$$
We have therefore to increase 66 by $\frac{1}{64}$, and therefore $7\frac{1}{3}$ by $\frac{1}{576}$, in order to make a square. And in fact $7\frac{1}{3} + \frac{1}{576} = (\frac{65}{24})^2$.
Next, since $22 = 3^2 + 3^2 + 2^2$, and $65 - 48 = 17$, while $72 - 65 = 7$, we put
$$22 = (3 - 7x)^2 + (3 - 7x)^2 + (2 + 17x)^2.$$
and
$$x = \frac{16}{387}.$$
Therefore the sides of the squares are $\frac{1049}{387}$, $\frac{1049}{387}$, $\frac{1046}{387}$,
the squares themselves $\frac{1100401}{149769}$, $\frac{1100401}{149769}$, $\frac{1094416}{149769}$,
and the required parts of 2 are $\frac{97751}{149769}$, $\frac{97751}{149769}$, $\frac{104036}{149769}$.
[18] As Wertheim observes, $\frac{13}{64} = \frac{9}{64} + \frac{1}{25} + \frac{9}{400}$, and the required fractions into which $\frac{1}{4}$ is divided are $\frac{250}{1600}$, $\frac{89}{1600}$, $\frac{61}{1600}$.
[19] If a, b be the perpendiculars, c the hypotenuse in a right-angled triangle,
$$\frac{a^2}{b^2} + 1 = \frac{c^2}{b^2} = \text{a square.}$$
Diophantus uses the triangles (3, 4, 5), (5, 12, 13), (8, 15, 17).
[20] Diophantus does not give the work here, but only the result. Bachet obtains it in this way. Suppose it required to find three rational right-angled triangles (b_1, p_1, b_1), (b_2, p_2, b_2) and (b_3, p_3, b_3) such that $p_1 p_2 p_3 / b_1 b_2 b_3$ is the ratio of a square to a square. One triangle (b_1, p_1, b_1) being chosen arbitrarily, form two others by putting
$$b_2 = b_1^2 + p_1^2, \quad p_2 = b_1^2 - p_1^2 = b_1^2, \quad b_2 = 2h_1 p_1,$$
$$b_3 = b_1^2 + b_1^2, \quad p_3 = h_1^2 - b_1^2 = p_1^2, \quad b_3 = 2h_1 b_1,$$

Starting again, we put for the squares

$$\frac{9}{16}x^2, \quad \frac{225}{64}x^2, \quad \frac{81}{1600}x^2.$$

Equating the product of these to x^2, we find x to be rational $[x = \frac{16}{9}$, and the squares are $\frac{16}{9}, \frac{100}{9}, \frac{4}{25}]$.

22. To find three squares such that their continued product *minus* any one of them gives a square.

Let the solid content be x^2, and let the numbers be obtained from right-angled triangles, being

$$\frac{16}{25}x^2, \quad \frac{25}{169}x^2, \quad \frac{64}{289}x^2.$$

Therefore the continued product $(\frac{4 \cdot 5 \cdot 8}{5 \cdot 13 \cdot 17})^2 x^6 = x^2$,

or $\qquad \left(\dfrac{4 \cdot 5 \cdot 8}{5 \cdot 13 \cdot 17}\right)^2 x^4 = \dfrac{25600}{1221025}x^4 = 1.$

If then $\frac{25600}{1221025}$ were a fourth power, *i.e.* if $\frac{4 \cdot 5 \cdot 8}{5 \cdot 13 \cdot 17}$ were a square, the problem would be solved.

We have therefore to find three right-angled triangles with hypotenuses h_1, h_2, h_3 respectively, and with p_1, p_2, p_3 as one of the perpendiculars in each respectively, such that

$$h_1 h_2 h_3 p_1 p_2 p_3 = \text{a square.}$$

and we have

$$\frac{p_1 p_2 p_3}{h_1 h_2 h_3} = \left(\frac{p_1}{2h_1}\right)^2 = \text{a square.}$$

If now $h_1 = 5$, $p_1 = 4$, $b_1 = 3$, the triangles (h_2, p_2, b_2) and (h_3, p_3, b_3) are $(41, 9, 40)$ and $(34, 16, 30)$ respectively. Dividing the sides of the latter throughout by 2 (which does not alter the ratio), we have Diophantus' second and third triangles $(9, 40, 41)$ and $(8, 15, 17)$.

Fermat, in his note on the problem, gives the following general rule for finding two right-angled triangles the areas of which are in the ratio $m : n$ ($m > n$).

Form (1) the greater triangle from $2m + n$, $m - n$, and the lesser from $m + 2n$, $m - n$,
or (2) the greater from $\quad 2m - n$, $m + n$, $\quad$ the lesser from $2n - m$, $m + n$,
or (3) the greater from $\quad\quad 6m$, $2m - n$, $\quad$ the lesser from $4m + n$, $4m - 2n$,
or (4) the greater from $m + 4n$, $2m - 4n$, $\quad$ the lesser from $\quad\quad 6n$, $m - 2n$.
The alternative (2) gives Diophantus' solution if we put $m = 3$, $n = 1$ and substitute $m - 2n$ for $2n - m$.

Fermat continues as follows: We can deduce a method of finding *three* right-angled triangles the areas of which are in the ratio of three given numbers, provided that the sum of two of these numbers is equal to four times the third. Suppose *e.g.* that m, n, q are three numbers such that $m + q = 4n$ ($m > q$). Now form the following triangles:

(1) from $m + 4n$, $2m - 4n$,
(2) from $\quad\quad 6n$, $m - 2n$,
(3) from $4n + q$, $4n - 2q$.

[If A_1, A_2, A_3 be the areas, we have, as a matter of fact,
$$A_1/m = A_2/n = A_3/q = -6m^3 + 36m^2 n + 144mn^2 - 384n^3.]$$

We can derive, says Fermat, a method of *finding three right-angled triangles the areas of which themselves form a right-angled triangle.* For we have only to find a triangle such that the sum of the base and hypotenuse is four times the perpendicular. This is easy, and the triangle will be similar to (17, 15, 8); the three triangles will then be formed

(1) from $17 + 4 \cdot 8$, $2 \cdot 17 - 4 \cdot 8$ $\quad$ or 49, 2,
(2) from $\quad\quad 6 \cdot 8$, $\quad 17 - 2 \cdot 8$ $\quad$ or 48, 1,
(3) from $4 \cdot 8 + 15$, $\quad 4 \cdot 8 - 2 \cdot 15$ $\quad$ or 47, 2.

[The areas of the three right-angled triangles are in fact 234906, 110544 and 207270, and these numbers form the sides of a right-angled triangle.]
Hence also we can derive a method of *finding three right-angled triangles the areas of which are in the ratio of three given squares such that the sum of two of them is equal to four times the third*, and we can also in the same way *find three right-angled triangles of the same area*; we can also *construct*, in an infinite number of ways, *two right-angled triangles the areas of which are in a given ratio*, by multiplying one of the terms of the ratio or the two terms by given squares, etc.

Assuming one of the triangles to be (3, 4, 5), so that *e.g.* $h_3p_3 = 5 \cdot 4 = 20$, we must have

$$5h_1p_1h_2p_2 = \text{ a square.}$$

This is satisfied if $h_1p_1 = 5h_2p_2$.

With a view to this we have first (cf. the last proposition) to find two right-angled triangles such that, if x_1, y_1 are the two *perpendiculars* in one and x_2, y_2 the two *perpendiculars* in the other, $x_1y_1 = 5x_2y_2$. From such a pair of triangles we can form two more right-angled triangles such that the product of the *hypotenuse* and *one perpendicular* in one is five times the product of the *hypotenuse* and *one perpendicular* in the other[21].

Since the triangles found satisfying the relation $x_1y_1 = 5x_2y_2$ are (5, 12, 13) and (3, 4, 5) respectively[22], we have in fact to find two new right-angled triangles from them, namely the triangles (h_1, p_1, b_1) and (h_2, p_2, b_2), such that

$$h_1p_1 = 30 \text{ and } h_2p_2 = 6,$$

the numbers 30 and 6 being the areas of the two triangles mentioned.

These triangles are $(6\frac{1}{2}, \frac{60}{13}, [\frac{119}{26}])$ and $(2\frac{1}{2}, \frac{12}{5}, [\frac{7}{10}])$ respectively.

Starting again, we take for the numbers

$$\frac{16}{25}x^2, \quad \frac{576}{625}x^2, \quad \frac{14400}{28561}x^2.$$

$[\frac{12}{5}$ divided by $2\frac{1}{2}$ gives $\frac{24}{25}$, and $\frac{60}{13}$ divided by $6\frac{1}{2}$ gives $\frac{120}{169}.]$

The product $= x^2$:

therefore, taking the square root, we have

$$\frac{4 \cdot 24 \cdot 120}{5 \cdot 25 \cdot 169}x^2 = 1,$$

so that $x = \frac{65}{48}$, and the required squares are found.

---

[21] Diophantus' procedure is only obscurely indicated in the Greek text. It was explained by Schulz in his edition (cf. Tannery in *Oeuvres de Fermat*, I.). Having given a rational right-angled triangle (z, x, y), Diophantus knows how to find a rational right-angled triangle (h, p, b) such that $hp = \frac{1}{2}xy$. We have in fact to put

$$h = \frac{1}{2}z, \; p = \frac{xy}{z}, \text{ whence } b^2 = h^2 - p^2 = \frac{1}{4}\left(\frac{z^4 - 4x^2y^2}{z^2}\right) = \left(\frac{x^2 - y^2}{2z}\right)^2.$$

Thus, having found two triangles (5, 12, 13) and (3, 4, 5) with areas in the ratio of 5 to 1 (see next paragraph of text with note thereon), Diophantus takes

$$h_1 = \frac{1}{2} \cdot 13 = 6\frac{1}{2}, \quad p_1 = \frac{5 \cdot 12}{13} = \frac{60}{13}, \text{ and similarly } h_2 = \frac{1}{2} \cdot 5 = 2\frac{1}{2}, \quad p_2 = \frac{3 \cdot 4}{5} = \frac{12}{5}.$$

Cossali (after Bachet) gives a formula for three right-angled triangles such that the solid content of the three hypotenuses has to the solid content of three perpendiculars (one in each triangle) the ratio of a square to a square; his triangles are

$$(1) \; i, b, p \; [i = \text{ipotenusa}], \quad (2) \frac{4p^2 + b^2}{b}, \frac{4p^2 - b^2}{b}, \frac{4bp}{b} = 4p,$$

$$(3) \frac{ib^2 + 4ip^2}{b}, \frac{b \cdot 4bp + p(4p^2 - b^2)}{b}, \frac{p \cdot 4bp - b(4p^2 - b^2)}{b} = b^2,$$

and, in fact,

$$\frac{i(b^2 + 4p^2)(ib^2 + 4ip^2)}{b^2} : p \cdot 4p \cdot b^2 = \frac{i^2(b^2 + 4p^2)^2}{4p^2b^2} : 4p^2b^2.$$

If $i = 5$, $b = 4$, $p = 3$, we can get from this triangle the triangles (13, 5, 12) and (65, 63, 16), and our equation is $\frac{3 \cdot 12 \cdot 16}{5 \cdot 13 \cdot 65}x^2 = 1$.

[22] These triangles can be obtained by putting $m = 5$, $n = 1$ in Fermat's *fourth* formula (note on last proposition). By that formula the triangles are formed from (9, 6) and (6, 3) respectively; and, dividing out by 3, we form the triangles from (3, 2) and (2, 1) respectively.

23. To find three squares such that each *minus* the product of the three gives a square.

Let the "solid content" be x^2, and let the squares be formed from right-angled triangles, as before.

If we take the same triangles as those found in the last problem and put for the three squares

$$\frac{25}{16}x^2, \quad \frac{625}{576}x^2, \quad \frac{28561}{14400}x^2,$$

each of these *minus* the continued product (x^2) gives a square.

It remains that their product $= x^2$;

this gives $x = \frac{48}{65}$, and the problem is solved.

24. To find three squares such that the product of any two increased by 1 gives a square.

Product of first and second $+1 =$ a square, and the third is a square; therefore "solid content" $+$ each $=$ a square.

The problem therefore reduces to v. 21 above[23].

25. To find three squares such that the product of any two *minus* 1 gives a square. This reduces, similarly, to v. 22 above.

26. To find three squares such that, if we subtract the product of any two of them from unity, the result is a square.

[23] De Billy in the *Inventum Novum*, Part II. paragraph 28 (*Oeuvres de Fermat*, III.), extends this problem, showing how to find *four* numbers, three of which (only) are squares, having the given property, *i.e.* to solve the equations

$$x_2^2 x_3^2 + 1 = r^2, \quad x_1^2 x_4 + 1 = u^2,$$
$$x_3^2 x_1^2 + 1 = s^2, \quad x_2^2 x_4 + 1 = v^2,$$
$$x_1^2 x_2^2 + 1 = t^2, \quad x_3^2 x_4 + 1 = w^2.$$

First seek three square numbers satisfying the conditions of Diophantus' problem v. 24, say $\left(\frac{9}{16}, \frac{25}{4}, \frac{256}{81}\right)$, the solution of v. 21 given in Bachet's edition. We have then to find a fourth number (x, say) such that

$$\left.\begin{array}{c} \dfrac{9}{16}x + 1 \\[2mm] \dfrac{25}{4}x + 1 \\[2mm] \dfrac{256}{81}x + 1 \end{array}\right\} \text{ are all squares.}$$

Substitute $\frac{16}{9}y^2 + \frac{32}{9}y$ for x, so as to make the first expression a square. We have then to solve the double-equation

$$\left.\begin{array}{c} \dfrac{100}{9}y^2 + \dfrac{200}{9}y + 1 = u^2 \\[2mm] \dfrac{4096}{729}y^2 + \dfrac{8192}{729}y + 1 = v^2 \end{array}\right\},$$

which can be solved by the ordinary method.

De Billy does not give the solution, but it may be easily supplied thus.

The difference
$$= \left(\frac{10}{3} + \frac{64}{27}\right)\left(\frac{10}{3} - \frac{64}{27}\right)(y^2 + 2y)$$
$$= \frac{154}{27}y\left(\frac{26}{27}y + \frac{2 \cdot 26}{27}\right).$$

Equating the square of half the sum of the factors to the larger expression, we have
$$\left(\frac{10}{3}y + \frac{26}{27}\right)^2 = \frac{100}{9}y^2 + \frac{200}{9}y + 1,$$

whence
$$y = -\frac{53}{11520}, \text{ and } y^2 + 2y = -\frac{1218311}{(11520)^2}.$$

Therefore $x = \frac{16}{9}(y^2 + 2y) = -\frac{1218311}{74649600}$, which satisfies the equations. In fact $\frac{9}{16}x + 1 = \left(\frac{11467}{11520}\right)^2$, $\frac{25}{4}x + 1 = \left(\frac{3275}{3456}\right)^2$, and $\frac{256}{81}x + 1 = \left(\frac{4733}{4860}\right)^2$.

But even here, as the value of x which we have found is negative, we ought, strictly speaking, to deduce a further value by substituting $y - \frac{1218311}{74649600}$ for x in the equations and solving again, which would of course lead to very large numbers.

This again reduces to an earlier problem, v. 23.

27. Given a number, to find three squares such that the sum of any two added to the given number makes a square.

Given number 15.

Let one of the required squares be 9;

I have then to find two other squares such that each $+24$ = a square, and their sum $+15$ = a square.

To find two squares, each of which $+24$ = a square, take two pairs of numbers which have 24 for their product[24].

Let one pair of factors be $\frac{4}{x}$, $6x$, and let the side of one square be half their difference or $\frac{2}{x} - 3x$.

Let the other pair of factors be $\frac{3}{x}$, $8x$, and let the side of the other square be half their difference or $\frac{1\frac{1}{2}}{x} - 4x$.

Therefore each of the squares $+24$ gives a square.

It remains that their sum $+15$ = a square;

therefore $\left(\frac{1\frac{1}{2}}{x} - 4x\right)^2 + \left(\frac{2}{x} - 3x\right)^2 + 15$ = a square,

or $\frac{6\frac{1}{4}}{x^2} + 25x^2 - 9$ = a square = $25x^2$, say.

Therefore $x = \frac{5}{6}$, and the problem is solved[25].

28. Given a number, to find three squares such that the sum of any two *minus* the given number makes a square.

Given number 13.

Let one of the squares be 25;

I have then to find two other squares such that each $+12$ = a square, and (sum of both) -13 = a square.

---

[24] The text adds the words "and [let us take] sides about the right angle in a right angled triangle." I think these words must be a careless interpolation: they are not wanted and give no sense; nor do they occur in the corresponding place in the next problem.

[25] Diophantus has found values of ξ, η, ζ satisfying the equations

$$\left.\begin{array}{c} \eta^2 + \zeta^2 + a = u^2 \\ \zeta^2 + \xi^2 + a = v^2 \\ \xi^2 + \eta^2 + a = w^2 \end{array}\right\}.$$

Fermat shows how to find *four* numbers (not squares) satisfying the corresponding conditions, namely that the sum of any two added to a shall give a square. Suppose $a = 15$.

Take three numbers satisfying the conditions of Diophantus' problem, say $9, \frac{1}{100}, \frac{529}{225}$.

Assume $x^2 - 15$ as the first of the four required numbers; and let the second be $6x + 9$ (because 9 is one of the square numbers taken and 6 is twice its side); for the same reason let the third number be $\frac{1}{5}x + \frac{1}{100}$ and the fourth $\frac{46}{15}x + \frac{529}{225}$.

Three of the conditions are now fulfilled since each of the last three numbers added to the first $(x^2 - 15)$ *plus* 15 gives a square. The three remaining conditions give the triple-equation

$$6\frac{1}{5}x + 9 + \frac{1}{100} + 15 = 6\frac{1}{5}x + \left(\frac{49}{10}\right)^2 = u^2,$$

$$\frac{136}{15}x + 9 + \frac{529}{225} + 15 = \frac{136}{15}x + \left(\frac{77}{15}\right)^2 = v^2,$$

$$\frac{49}{15}x + \frac{1}{100} + \frac{529}{225} + 15 = \frac{49}{15}x + \left(\frac{25}{6}\right)^2 = w^2.$$

Divide 12 into factors in two ways, and let the factors be $(3x, \frac{4}{x})$ and $(4x, \frac{3}{x})$.

Take as the sides of the squares half the differences of the factors, *i.e.* let the squares be

$$\left(1\tfrac{1}{2}x - \frac{2}{x}\right)^2, \left(2x - \frac{1\tfrac{1}{2}}{x}\right)^2.$$

Each of these $+12$ gives a square.

It remains that the sum of the squares $-13 = $ a square,

or $\dfrac{6\tfrac{1}{4}}{x^2} + 6\tfrac{1}{4}x^2 - 25 = $ a square $= \dfrac{6\tfrac{1}{4}}{x^2}$, say.

Therefore $x = 2$, and the problem is solved[26].

29. To find three squares such that the sum of their squares is a square.

Let the squares be x^2, 4, 9 respectively[27].

Therefore $x^4 + 97 = $ a square $= (x^2 - 10)^2$, say;

whence $x^2 = \tfrac{3}{20}$.

If the ratio of 3 to 20 were the ratio of a square to a square, the problem would be solved; but it is not.

Therefore *I have to find two squares* (p^2, q^2, *say*) *and a number* (m, *say*) *such that* $m^2 - p^4 - q^4$ *has to* $2m$ *the ratio of a square to a square.*

Let $p^2 = z^2$, $q^2 = 4$ and $m = z^2 + 4$.

Therefore $m^2 - p^4 - q^4 = (z^2 + 4)^2 - z^4 - 16 = 8z^2$.

Hence $8z^2/(2z^2 + 8)$, or $4z^2/(z^2 + 4)$, must be the ratio of a square to a square.

Put $z^2 + 4 = (z + 1)^2$, say;

therefore $z = 1\tfrac{1}{2}$, and the squares are $p^2 = 2\tfrac{1}{4}$, $q^2 = 4$, while $m = 6\tfrac{1}{4}$;

or, if we take 4 times each, $p^2 = 9$, $q^2 = 16$, $m = 25$.

Starting again, we put for the squares x^2, 9, 16;

then the sum of the squares $= x^4 + 337 = (x^2 - 25)^2$, and $x = \tfrac{12}{5}$.

The required squares are $\tfrac{144}{25}$, 9, 16.

30. [The enunciation of this problem is in the form of an epigram, the meaning of which is as follows.]

A man buys a certain number of measures (χόες) of wine, some at 8 drachmas, some at 5 drachmas each. He pays for them a *square* number of drachmas; and if

[26] Fermat observes that *four* numbers (not squares) with the property indicated can be found by the same procedure as that shown in the note to the preceding problem.

If a is the given number, put $x^2 + a$ for the first of the four required numbers.

If now k^2, l^2, m^2 represent three numbers satisfying the conditions of the present problem of Diophantus, put for the second of the required numbers $2kx + k^2$, for the third $2lx + l^2$, and for the fourth $2mx + m^2$. These satisfy three conditions, since each of the last three numbers added to the first ($x^2 + a$) less the number a gives a square. The remaining three conditions give a triple-equation.

[27] "Why," says Fermat, "does not Diophantus seek *two* fourth powers such that their sum is a square? This problem is in fact impossible, as by my method I am in a position to prove with all rigour." It is probable that Diophantus knew the fact without being able to prove it generally. That neither the sum nor the difference of two fourth powers can be a square was proved by Euler (*Commentationes arithmeticae*, 1., and *Algebra*, Part II. c. XIII.).

we add 60 to this number, the result is a square, the side of which is equal to the whole number of measures. Find how many he bought at each price.

Let x = the whole number of measures; therefore $x^2 - 60$ was the price paid, which is a square $= (x - m)^2$, say.

Now $\frac{1}{5}$ of the price of the five-drachma measures $+\frac{1}{8}$ of the price of the eight-drachma measures $=x$;

so that $x^2 - 60$, the total price, has to be divided into two parts such that $\frac{1}{5}$ of one $+\frac{1}{8}$ of the other $=x$.

We cannot have a real solution of this unless
$$x > \tfrac{1}{8}(x^2 - 60) \text{ and } < \tfrac{1}{5}(x^2 - 60).$$
Therefore $\qquad\qquad 5x < x^2 - 60 < 8x.$

(1) Since $\qquad\qquad x^2 > 5x + 60,$

$x^2 = 5x +$ a number greater than 60,

whence x is *not less than* 11.

(2) $\qquad\qquad\qquad x^2 < 8x + 60$

or $\qquad\qquad x^2 = 8x +$ some number less than 60,

whence x is *not greater than* 12.

Therefore $\qquad\qquad 11 < x < 12.$

Now (from above) $x = (m^2 + 60)/2m$;

therefore $22m < m^2 + 60 < 24m$.

Thus (1) $22m = m^2 +$ (some number less than 60),

and therefore m is *not less than* 19.

(2) $24m = m^2 +$ (some number greater than 60),

and therefore m is *less than* 21.

Hence we put $m = 20$, and
$$x^2 - 60 = (x - 20)^2,$$
so that $x = 11\frac{1}{2}$, $x^2 = 132\frac{1}{4}$, and $x^2 - 60 = 72\frac{1}{4}$.

Thus we have to divide $72\frac{1}{4}$ into two parts such that $\frac{1}{5}$ of one part *plus* $\frac{1}{8}$ of the other $= 11\frac{1}{2}$.

Let the first part be $5z$.

Therefore $\frac{1}{8}$ (second part) $= 11\frac{1}{2} - z$,

or second part $= 92 - 8z$;

therefore $5z + 92 - 8z = 72\frac{1}{4}$,

and $z = \frac{79}{12}$.

Therefore the number of five-drachma χόες $= \frac{79}{12}$.

,, ,, ,, ,, eight-drachma ,, $= \frac{59}{12}$.

René Descartes

(1596–1650)

HIS LIFE AND WORK

René Descartes was born into prominent families known for making significant contributions to French culture. His father, Joachim, came from a long line of medical doctors and was a councilor to the provincial Parlement of Brittany. His mother, Jeanne neé Brochard, came from a wealthy family with property in Poitou. They married in 1589 and already had a son named Pierre and a daughter named Jeanne when their third child, René, was born in 1596. They named him after his maternal grandfather who had died in 1596.

Descartes never knew his mother. She died in childbirth thirteen months after he was born. His father left the children in the care of a Madame Jeanne Sain who became Descartes' surrogate mother and remained so even when Descartes' father remarried in 1600.

If Jeanne Sain was Descartes' mother figure, then Father Etienne Charlet, the rector of the Jesuit College at La Fleche was his father figure. Descartes entered the College at La Fleche in 1606 and remained there for the next eight years. It had just opened two years earlier.

In contrast to the municipally run colleges in France, which were explicitly based on a humanist model, the Catholic colleges such as La Fleche continued to provide a thorough grounding in traditional Catholic thought and the classics of antiquity. Like most French colleges, La Fleche was a combination of a secondary school and a university. At La Fleche, the first year was devoted to preparatory classes, the next three years to grammar (in Latin and Greek) and the fifth year to rhetoric. All instruction was in Latin. Although the Jesuit fathers were very demanding of their students' academic work, they could have soft spots in their hearts. The fathers at La Fleche often allowed Descartes to stay in bed all morning because of his poor health.

Many students left Jesuit colleges like La Fleche after these first five years to enroll at a university and pursue a degree in one of the three major professions: law, medicine, and theology. In fact many of the secular universities refused to take students with more than five years at a Jesuit college fearing that they would be too indoctrinated with the radical teaching of the relatively new Jesuit order. Descartes was not one of these students. He was very pleased to learn that his father wished for him to stay at La Fleche for an additional three years.

At the beginning of these four years Descartes began his study of metaphysics, ethics, natural philosophy and dialectic. The curriculum in natural philosophy included the study of Euclid, Archimedes, and Diophantus as well as contemporary mathematics. While at La Fleche Descartes may have made a personal contact that proved to be just as valuable as the mathematical foundation he gained. One of Descartes' later correspondents, Marin Mersenne, was also a student at La Fleche. From the 1620s to the 1640s Mersenne acted as the hub of French scientific activity corresponding with anyone and everyone of any importance. His correspondence alone runs to more than 10,000 pages.

Descartes left the College at La Fleche in 1614 and enrolled at the nearby University of Poitiers to study law. He earned his law degree in two years. However, law seemed to hold little interest for him. There is no record of Descartes ever practicing law as his father probably wished. His real interest at Poitiers was medicine and he developed a keen interest in dissection.

Having lived in one region of France for the first twenty years of his life Descartes must have been eager to see other parts of Europe. Upon receiving his law degree he joined the army of Prince Maurice of Nassau as a gentleman officer. After two and a half years in the Netherlands, Descartes left the service of Prince Maurice and joined the army of Maximilian of Bavaria. Shortly after joining Maximilian Descartes traveled to Frankfurt where he witnessed preparations for the coronation of Ferdinand II as Holy Roman Emperor.

Descartes continued his military career until the age of thirty. He spent the first half of the 1620s traveling through Europe as the Thirty Years War raged. His correspondence records that he visited Bavaria, Bohemia, Germany, Italy, and Hungary, as well as Holland and France during these years. Although these were years of constant travel Descartes found the time to lay the groundwork for his work in metaphysics, epistemology, and natural philosophy. In 1618, while in the service of Maurice of Orange, Descartes met the Dutchman Isaac Beeckman, seven years his senior. Beeckman shared Descartes' interests in philosophy and mathematics.

Beeckman became another father figure to Descartes and Descartes became Beeckman's intellectual apprentice. Most importantly, Beeckman became a confidant to Descartes. Shortly before his 23rd birthday in 1619 Descartes outlined a scheme for a new mathematics in a letter to Beeckman:

> What I would like to present to the public is... a science with wholly new foundations, which will enable us to answer every question that can be put about any kind of quantity whatsoever, whether continuous or discontinuous, each according to its nature. In Arithmetic, certain quantities can be solved by means of rational numbers, others by using irrationals, and finally others can be imagined but not solved. In this way, I hope to demonstrate that in the case of continuous quantity, certain problems can be solved with straight lines and circles alone; that others can be solved only with curves other than circles, but which can be generated by a single motion and which can therefore be drawn using a new compass which I do not believe to be any less accurate than, and just as geometrical as, the ordinary compass which is used to draw circles; and finally other problems can be solved only with curves generated but motions not subordinated to one another, curves which are certainly only imaginary, such as the quadratrix, which is well known. I do not believe one can imagine anything which could not be solved along similar lines: indeed, I hope to show that particular kinds of questions can be resolved in one way and not another, so that there will remain almost nothing else to discover in Geometry. The task is infinite and could not be accomplished by one person. It is as incredible as it is ambitious. But I have seen a certain light in the dark chaos of this science, thanks to which the thickest clouds can be dispelled.

This served as a first draft for what would turn into part of Descartes' fourth *Rule for the Direction of the Mind.*

> It is evident to anyone with the most minimal education what pertains to mathematics and what does not in any context. When I

attended to the matter more closely, I came to see that the exclusive concern of mathematics is with questions of order or measure, and that it does not matter whether the measure in question involves numbers, shapes, stars, sounds or any other object whatsoever. This made me realize that there must be a general science that explains everything that can be raised concerning order and measure irrespective of the subject matter, and that this science should be termed mathesis universalis—*a venerable term with a well established meaning—for it covers everything that entitles these other sciences to be called branches of mathematics. How superior it is to these subordinate sciences both in usefulness and simplicity is clear from the fact that it covers all they deal with... Up to now, I have devoted all my energies to this* mathesis universalis *so that I might be able to tackle the more advanced sciences in due course.*

It would be nearly twenty years until Descartes actually had the time to develop his mathematics. After his military career, Descartes lived in Paris between 1625 and 1628. He found himself in a circle of thinkers who, like himself, rejected much of the Aristotelian orthodoxy still ascendant in the early 17th century. From time to time, he and his circle argued the merits of the heliocentric astronomy of Copernicus. However, they needed to couch their discussions in hypothetical terms. In 1624, the Theology Faculty at the Sorbonne in Paris had issued an edict threatening anyone with death for advocating views contrary to the approved teachings of the ancient authors.

Although he remained an ardent Catholic his entire life, Descartes found the clerically influenced intellectual atmosphere in Paris to be too threatening and in 1628 left for the more liberal environment in Protestant Holland. His decision to leave Paris may have also been hastened by a duel he fought over the honor of a lady. We know nothing of this lady except that Descartes compared her beauty to the beauty of truth! He must have left Paris in a great hurry. Descartes' correspondence records that he took nothing with him except for the Bible and the *Summa Theologica* of St. Thomas Aquinas.

Descartes settled in Amsterdam and soon made himself prominent in its intellectual circles. During the winter of 1634–5 Descartes became a fixture at the house of Elizabeth Stuart, Electress of the Palatine and Queen of Bohemia. She was the widow of the Frederick V of Bohemia whose armies had been routed at the Battle of the White Mountain in 1630 and was forced into an exile of poverty in the Netherlands. When Frederick died in 1632, his widow remained in the Netherlands on a pension provided by her brother, King Charles I of England. Descartes had a close intellectual relationship with the dowager Queen Elizabeth and later on with her daughter Princess Elizabeth of Bohemia.

While in Amsterdam, Descartes became friendly with Constatijn Huygens, secretary to Frederick Henry, Prince of Orange. Huygens came from a family of diplomats and he became an ardent supporter of Descartes. However, the Huygens family is best known for Constantijn's eldest son Christiaan (1629–1695) a contemporary of Newton who the English philosopher John Locke would describe as the "Huygenius." Descartes took an active interest in the education of the young Christiaan Huygens and Huygens developed Descartes' theory of vortices.

In the mid-1630s, Descartes had the only significant romantic relationship of his life with a woman named Hélène, a maid in the house where he was lodging. In July 1635 she gave birth to Descartes' daughter and they named the child Francine. The child was baptized as a Protestant and had only occasional contact with her father during her short life. She died of a fever that swept through Holland in 1640, just after her fifth birthday. Her death left Descartes greatly shaken. Only with her death could Descartes begin to understand what she really meant to him. True to form, he had no contact with Hélène following their child's death.

The Netherlands was also the birthplace of Descartes' greatest works: the *Discourse on Method*, the *Meditations on First Philosophy*, the *Principles of Philosophy*, and the *Geometry*, the work presented here.

We can easily see the power of Descartes' mathematical methods with his exposition of two very basic mathematical operations: determining products and square roots. Recall that to Euclid and his Greek successors all the way up through the sixteenth century, arithmetic propositions are stated in terms of geometrical figures such as line lengths, not because line lengths provide a good method for representing numbers, but because that is what numbers are! Thus, determining the product of two abstract values X and Y meant constructing a rectangle whose sides had length X and Y and realizing their product in the **area** of the rectangle! Similarly, the ancients would take the square root of an abstract value X by representing X as a two dimensional figure and finding a square of equal area. The side of that square realized the square root of X.

The crisis of the irrational forced the Greeks to resort to these geometric interpretations. Descartes would have none of these geometrical restrictions.

We should also note that those proportions that form a continuing sequence are to be understood in terms of a number of relations; others try to express these proportions in ordinary algebraic terms by means of several different dimensions and shapes. The first they call the root, the second the square, the third the cube, the fourth the biquadratic, and so on. These expressions have, I confess, long misled me.... All such names should be abandoned as they are liable to cause confusion in our thinking, For though a magnitude may be termed a cube or a biquadratic, it should never be represented to the imagination otherwise than as a line or a surface.... What, above all, requires to be noted is that the root, the square, the cube, etc, are merely magnitudes in continued proportion, which always implies the freely chose unit that we spoke of in the preceding Rule.

Thus, the cube of a quantity x, for example, is designated as x^3, not because it represents a geometric cube, a three dimensional figure, but because it represents a series of proportions with three relations.

$$1 : x = x : x^2 = x^2 : x^3$$

Instead of requiring geometric constructions for every one of the objects being considered, Descartes supposed that all of the given objects exist and shows how to construct the value sought as a line segment.

Here is how he determines the product of two quantities at the very beginning of his *Geometry*. He **supposes** that unity (i.e. the value 1) is given by the line segment **AB** and proceeds to multiply the values represented by the line segments **BD** (collinear with **AB**) and **BC**. He connects the points **A** and **C** and then constructs through **D** the line parallel to **AC** that intersects **BC** at **E**.

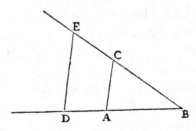

In so doing, Descartes has created two triangles **ABC** and **DBE** that are similar to each other. Consequently their sides are in proportion, so

BE : BC = BD : BA

but since **BA = 1** he has

BE : BC = BD

or

BE = BC*BD

the sought for solution.

Taking square roots is just as straightforward. He lets the line segment **GH** represent the quantity whose square root is to be determined and extends it from **G** to **F** so that **FG** equals unity. Having done that, Descartes bisects the line segment **FH** at the point **K** and constructs the semicircle centered on **K** having **KH** (= **KF**) as its radius. At the point **G** he erects the perpendicular to **FH** that intersects the circle at the point **I**.

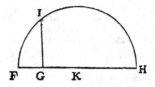

This time, Descartes has actually constructed three similar right triangles **IGF**, **HGI**, **HIF**. (The angles **IGF**, **HGI** and **HIF** are all right angles.) Since **IGF** and **HGI** are similar we have

$$\mathbf{GH : GI = GI : FG}$$

By supposition we have

$$\mathbf{FG = 1}$$

Therefore

$$\mathbf{GH = GI*GI}$$

demonstrating that **GI** is the square root of **GH**.

Descartes continued to live in the Netherlands until 1648 when Queen Christina of Sweden offered him a position at her court in Stockholm to establish an academy for scholars and give her lessons in ethics and theology. When Descartes accepted her offer, Christina sent a warship to take him to Sweden. Descartes had never been treated so royally and he probably expected to be treated just as royally at the Swedish court. Although his surroundings may have been royal, Descartes had not bothered to inquire about the hours he would be expected to keep.

Owing to his lifelong mediocre health, Descartes often remained in bed all morning. Much to his chagrin, he learned that this was not the style of the Swedish court. Queen Christina insisted that her lessons begin at 5 in the morning regardless of the weather. This required Descartes to leave his residence at 4:30 in the morning and brave the often-inclement northern climate. The winter of 1649–50 was exceptionally harsh and Descartes soon became ill with pneumonia. Preferring his own treatments to anything Christina's doctors might prescribe, Descartes tried to cure himself with a concoction of wine flavored tobacco designed to make him vomit up the phlegm. He soon got worse, falling into a delirium and dying two days later on February 11, 1650.

THE GEOMETRY OF RENÉ DESCARTES

BOOK I

PROBLEMS THE CONSTRUCTION OF WHICH REQUIRES ONLY STRAIGHT LINES AND CIRCLES

Any problem in geometry can easily be reduced to such terms that a knowledge of the lengths of certain straight lines is sufficient for its construction.[1] Just as arithmetic consists of only four or five operations, namely, addition, subtraction, multiplication, division and the extraction of roots, which may be considered a kind of division, so in geometry, to find required lines it is merely necessary to add or subtract other lines; or else, taking one line which I shall call unity in order to relate it as closely as possible to numbers,[2] and which can in general be chosen arbitrarily, and having given two other lines, to find a fourth line which shall be to one of the given lines as the other is to unity (which is the same as multiplication); or, again, to find a fourth line which is to one of the given lines as unity is to the other (which is equivalent to division); or, finally, to find one, two, or several mean proportionals between unity and some other line (which is the same as extracting the square root, cube root, etc., of the given line.[3] And I shall not hesitate to introduce these arithmetical terms into geometry, for the sake of greater clearness.

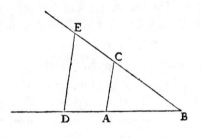

For example, let AB be taken as unity, and let it be required to multiply BD by BC. I have only to join the points A and C, and draw DE parallel to CA; then BE is the product of BD and BC.

If it be required to divide BE by BD, I join E and D, and draw AC parallel to DE; then BC is the result of the division.

---

1. Large collections of problems of this nature are contained in the following works: Vincenzo Riccati and Girolamo Saladino, *Institutiones Analyticae*, Bologna, 1765; Maria Gaetana Agnesi, *Istituzioni Analitiche*, Milan. 1748; Claude Rabuel, *Commentaires sur la Géométrie de M. Descartes*, Lyons, 1730 (hereafter referred to as Rabuel); and other books of the same period or earlier.

2. Van Schooten, in his Latin edition of 1683, has this note: "Per unitatem intellige lineam quandam determinatam, qua ad quamvis reliquarum linearum talem relationem habeat, qualem unitas ad certum aliquem numerum." *Geometria a Renato Des Cartes, una cum notis Florimondi de Beaune, opera atque studio Francisci à Schooten*, Amsterdam, 1683, (hereafter referred to as Van Schooten).

In general, the translation runs page for page with the facing original. On account of figures and footnotes, however, this plan is occasionally varied, but not in such a way as to cause the reader any serious inconvenience.

3. While in arithmetic the only exact roots obtainable are those of perfect powers, in geometry a length can be found which will represent exactly the square root of a given line, even though this line be not commensurable with unity. Of other roots, Descartes speaks later.

If the square root of GH is desired, I add, along the same straight line, FG equal to unity; then, bisecting FH at K, I describe the circle FIH about K as a center, and draw from G a perpendicular and extend it to I, and GI is the required root. I do not speak here of cube root, or other roots, since I shall speak more conveniently of them later.

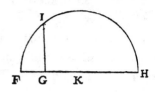

Often it is not necessary thus to draw the lines on paper, but it is sufficient to designate each by a single letter. Thus, to add the lines BD and GH, I call one a and the other b, and write $a + b$. Then $a - b$ will indicate that b is subtracted from a; ab that a is multiplied by b; $\frac{a}{b}$ that a is divided by b; aa or a^2 that a is multiplied by itself; a^3 that this result is multiplied by a, and so on, indefinitely.[4] Again, if I wish to extract the square root of $a^2 + b^2$, I write $\sqrt{a^2 + b^2}$; if I wish to extract the cube root of $a^3 - b^3 + ab^2$, I write $\sqrt[3]{a^3 - b^3 + ab^2}$, and similarly for other roots.[5] Here it must be observed that by a^2, b^3, and similar expressions, I ordinarily mean only simple lines, which, however, I name squares, cubes, etc., so that I may make use of the terms employed in algebra.[6]

It should also be noted that all parts of a single line should always be expressed by the same number of dimensions, provided unity is not determined by the conditions of the problem. Thus, a^3 contains as many dimensions as ab^2 or b^3, these being the component parts of the line which I have called $\sqrt[3]{a^3 - b^3 + ab^2}$. It is not, however, the same thing when unity is determined, because unity can always be understood, even where there are too many or too few dimensions; thus, if it be required to extract the cube root of $a^2b^2 - b$, we must consider the quantity a^2b^2 divided once by unity, and the quantity b multiplied twice by unity.[7]

Finally, so that we may be sure to remember the names of these lines, a separate list should always be made as often as names are assigned or changed. For example, we may write, AB = 1, that is AB is equal to 1;[8] GH = a, BD = b, and so on.

4. Descartes uses a^3, a^4, a^5, a^6, and so on, to represent the respective powers of a, but he uses both aa and a^2 without distinction. For example, he often has $aabb$, but he also uses $\dfrac{3a^2}{4b^2}$.

5. Descartes writes: $\sqrt{C \cdot a^3 - b^3 + abb}$.

6. At the time this was written, a^2 was commonly considered to mean the surface of a square whose side is a, and b^3 to mean the volume of a cube whose side is b; while b^4, b^5, . . . were unintelligible as geometric forms. Descartes here says that a^2 does not have this meaning, but means the line obtained by constructing a third proportional to 1 and a, and so on.

7. Descartes seems to say that each term must be of the third degree, and that therefore we must conceive of both a^2b^2 and b as reduced to the proper dimension.

8. Van Schooten adds "seu unitati." Descartes writes, AB ∝ 1. He seems to have been the first to use this symbol. Among the few writers who followed him, was Hudde (1633–1704). It is very commonly supposed that ∝ is a ligature representing the first two letters (or diphthong) of "æquare." See, for example, M. Aubry's note in W. W. R. Ball's *Récréations Mathématiques et Problèmes des Temps Anciens et Modernes*, French edition, Paris, 1909, Part III.

If, then, we wish to solve any problem, we first suppose the solution already effected,[9] and give names to all the lines that seem needful for its construction,—to those that are unknown as well as to those that are known.[10] Then, making no distinction between known and unknown lines, we must unravel the difficulty in any way that shows most naturally the relations between these lines, until we find it possible to express a single quantity in two ways.[11] This will constitute an equation, since the terms of one of these two expressions are together equal to the terms of the other.

We must find as many such equations as there are supposed to be unknown lines;[12] but if, after considering everything involved, so many cannot be found, it is evident that the question is not entirely determined. In such a case we may choose arbitrarily lines of known length for each unknown line to which there corresponds no equation.[13]

If there are several equations, we must use each in order, either considering it alone or comparing it with the others, so as to obtain a value for each of the unknown lines; and so we must combine them until there remains a single unknown line[14] which is equal to some known line, or whose square, cube, fourth power, fifth power, sixth power, etc., is equal to the sum or difference of two or more quantities,[15] one of which is known, while the others consist of mean proportionals between unity and this square, or cube, or fourth power, etc., multiplied by other known lines. I may express this as follows:

$$z = b,$$
$$\text{or } z^2 = -az + b^2,$$
$$\text{or } z^3 = az^2 + b^2z - c^3,$$
$$\text{or } z^4 = az^3 - c^3z + d^4, \text{ etc.}$$

That is, z, which I take for the unknown quantity, is equal to b; or, the square of z is equal to the square of b diminished by a multiplied by z; or, the cube of z is equal to a multiplied by the square of z, plus the square of b multiplied by z, diminished by the cube of c; and similarly for the others.

---

9. This plan, as is well known, goes back to Plato. It appears in the work of Pappus as follows: "In analysis we suppose that which is required to be already obtained, and consider its connections and antecedents, going back until we reach either something already known (given in the hypothesis), or else some fundamental principle (axiom or postulate) of mathematics." *Pappi Alexandrini Collectiones quae supersunt c libris manu scriptis edidit Latina interpellatione et commentariis instruxit Fredericus Hultsch*, Berlin, 1876–1878; Vol. II, (hereafter referred to as Pappus). See also Commandinus, *Pappi Alexandrini Mathematicae Collectiones*, Bologna, 1588, with later editions.

Pappus of Alexandria was a Greek mathematician who lived about 300 A.D. His most important work is a mathematical treatise in eight books, of which the first and part of the second are lost. This was made known to modern scholars by Commandinus. The work exerted a happy influence on the revival of geometry in the seventeenth century. Pappus was not himself a mathematician of the first rank, but he preserved for the world many extracts or analyses of lost works, and by his commentaries added to their interest.

10. Rabuel calls attention to the use of a, b, c, . . . for known, and x, y, z, . . . for unknown quantities.

11. That is, we must solve the resulting simultaneous equations.

12. Van Schooten gives two problems to illustrate this statement. Of these, the first is as follows: Given a line segment AB containing any point C, required to produce AB to D so that the rectangle AD.DB shall be equal to the square on CD. He lets AC $= a$, CB $= b$, and BD $= x$. Then AD $= a + b + x$, and CD $= b + x$, whence $ax + bx + x^2 = b^2 + 2bx + x^2$ and $x = \dfrac{b^2}{a-b}$.

13. Rabuel adds this note: "We may say that every indeterminate problem is an infinity of determinate problems, or that every problem is determined either by itself or by him who constructs it".

14. That is, a line represented by x, x^2, x^3, x^4, ...

15. In the older French, "le quarré, ou le cube, ou le quarré de quarré, ou le sursolide, ou le quarré de cube &c."

Thus, all the unknown quantities can be expressed in terms of a single quantity, whenever the problem can be constructed by means of circles and straight lines, or by conic sections, or even by some other curve of degree not greater than the third or fourth.[16]

But I shall not stop to explain this in more detail, because I should deprive you of the pleasure of mastering it yourself, as well as of the advantage of training your mind by working over it, which is in my opinion the principal benefit to be derived from this science. Because, I find nothing here so difficult that it cannot be worked out by any one at all familiar with ordinary geometry and with algebra, who will consider carefully all that is set forth in this treatise.[17]

I shall therefore content myself with the statement that if the student, in solving these equations, does not fail to make use of division wherever possible, he will surely reach the simplest terms to which the problem can be reduced.

And if it can be solved by ordinary geometry, that is, by the use of straight lines and circles traced on a plane surface,[18] when the last equation shall have been entirely solved there will remain at most only the square of an unknown quantity, equal to the product of its root by some known quantity, increased or diminished by some other quantity also known.[19] Then this root or unknown line can easily be found. For example, if I have $z^2 = az + b^2$,[20] I construct a right triangle NLM with one side LM, equal to b, the square root of the known quantity b^2, and the other side, LN, equal to $\frac{1}{2}a$, that is, to half the other known quantity which was multiplied by z, which I supposed to be the unknown line. Then prolonging MN, the hypotenuse[21]

---

16. Literally, "Only one or two degrees greater."

17. In the Introduction to the 1637 edition of *La Géométrie*, Descartes made the following remark: "In my previous writings I have tried to make my meaning clear to everybody; but I doubt if this treatise will be read by anyone not familiar with the books on geometry, and so I have thought it superfluous to repeat demonstrations contained in them." See *Oeuvres de Descartes*, edited by Charles Adam and Paul Tannery, Paris, 1897–1910, Vol. VI. In a letter written to Mersenne in 1637 Descartes says: "I do not enjoy speaking in praise of myself, but since few people can understand my geometry, and since you wish me to give you my opinion of it, I think it well to say that it is all I could hope for, and that in *La Dioptrique* and *Les Météores*, I have only tried to persuade people that my method is better than the ordinary one. I have proved this in my geometry, for in the beginning I have solved a question which, according to Pappus, could not be solved by any of the ancient geometers.

"Moreover, what I have given in the second book on the nature and properties of curved lines, and the method of examining them, is, it seems to me, as far beyond the treatment in the ordinary geometry, as the rhetoric of Cicero is beyond the a, b, c of children. . . .

"As to the suggestion that what I have written could easily have been gotten from Vieta, the very fact that my treatise is hard to understand is due to my attempt to put nothing in it that I believed to be known either by him or by any one else. . . . I begin the rules of my algebra with what Vieta wrote at the very end of his book, *De emendatione acquationum*. . . . Thus, I begin where he left off." *Oeuvres de Descartes, publiées par Victor Cousin*, Paris, 1824, Vol. VI (hereafter referred to as Cousin).

In another letter to Mersenne, written April 20, 1646, Descartes writes as follows: "I have omitted a number of things that might have made it (the geometry) clearer, but I did this intentionally, and would not have it otherwise. The only suggestions that have been made concerning changes in it are in regard to rendering it clearer to readers, but most of these are so malicious that I am completely disgusted with them." Cousin, Vol. IX.

In a letter to the Princess Elizabeth, Descartes says: "In the solution of a geometrical problem I take care, as far as possible, to use as lines of reference parallel lines or lines at right angles; and I use no theorems except those which assert that the sides of similar triangles are proportional, and that in a right triangle the square of the hypotenuse is equal to the sum of the squares of the sides. I do not hesitate to introduce several unknown quantities, so as to reduce the question to such terms that it shall depend only on these two theorems." Cousin, Vol. IX.

18. For a discussion of the possibility of constructions by the compasses and straight edge, see Jacob Steiner, *Die geometrischen Constructionen ausgeführt mittelst der geraden Linie und eines festen Kreises*, Berlin, 1833. For briefer treatments, consult Enriques, *Fragen der Elementar-Geometrie*, Leipzig, 1907; Klein, *Problems in Elementary Geometry*, trans. by Beman and Smith, Boston, 1897; Weber and Wellstein, *Encyklopädie der Elementaren Geometrie*, Leipzig, 1907. The work by Mascheroni, *La geometria del compasso*, Pavia, 1797, is interesting and well known.

19. That is, an expression of the form $z^2 = az \pm b$. "Esgal a ce qui se produit de l'Addition, ou soustraction de sa racine multipliée par quelque quantité connue, & de quelque autre quantité aussy connue."

20. Descartes proposes to show how a quadratic may be solved geometrically.

21. Descartes says "prolongeant MN la baze de ce triangle," because the hypotenuse was commonly taken as the base in earlier times.

of this triangle, to O, so that NO is equal to NL, the whole line OM is the required line z. This is expressed in the following way:[22]

$$z = \frac{1}{2}a + \sqrt{\frac{1}{4}a^2 + b^2}.$$

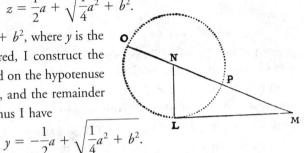

But if I have $y^2 = -ay + b^2$, where y is the quantity whose value is desired, I construct the same right triangle NLM, and on the hypotenuse MN lay off NP equal to NL, and the remainder PM is y, the desired root. Thus I have

$$y = -\frac{1}{2}a + \sqrt{\frac{1}{4}a^2 + b^2}.$$

In the same way, if I had

$$x^4 = -ax^2 + b^2,$$

PM would be x^2 and I should have

$$x = \sqrt{-\frac{1}{2}a + \sqrt{\frac{1}{4}a^2 + b^2}},$$

and so for other cases.

Finally, if I have $z^2 = az - b^2$, I make NL equal to $\frac{1}{2}a$ and LM equal to b as before; then, instead of joining the points M and N, I draw MQR parallel to LN, and with N as a center describe a circle through L cutting MQR in the points Q and R; then z, the line sought, is either MQ or MR, for in this case it can be expressed in two ways, namely:[23]

$$z = \frac{1}{2}a + \sqrt{\frac{1}{4}a^2 - b^2},$$

and

$$z = \frac{1}{2}a - \sqrt{\frac{1}{4}a^2 - b^2}.$$

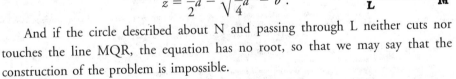

And if the circle described about N and passing through L neither cuts nor touches the line MQR, the equation has no root, so that we may say that the construction of the problem is impossible.

22. From the figure OM · PM = LM². If OM = z, PM = $z - a$, and since LM = b, we have $z(z-a) = b^2$ or $z^2 = az + b^2$. Again, MN = $\sqrt{\frac{1}{4}a^2 + b^2}$, whence OM = z = ON + MN = $\frac{1}{2}a + \sqrt{\frac{1}{4}a^2 + b^2}$. Descartes ignores the second root, which is negative.
23. Since MR · MQ = $\overline{LM}^2$, then if R = z, we have MQ = $a - z$, and so
$$z(a - z) = b^2 \text{ or } z^2 = az - b^2.$$
If, instead of this, MQ = z, then MR = $a - z$, and again, $z^2 = az - b^2$. Furthermore, letting O be the mid-point of QR,
$$MQ = OM - OQ = \frac{1}{2}a - \sqrt{\frac{1}{4}a^2 - b^2},$$
and
$$MR = MO + OR = \frac{1}{2}a + \sqrt{\frac{1}{4}a^2 - b^2}.$$

Descartes here gives both roots, since both are positive. If MR is tangent to the circle, that is, if $b = \frac{1}{2}a$, the roots will be equal; while if $b > \frac{1}{2}a$, the line MR will not meet the circle and both roots will be imaginary. Also, since RM · QM = $\overline{LM}^2$, $z_1 z_2 = b^2$, and RM + QM = $z_1 + z_2 = a$.

These same roots can be found by many other methods, I have given these very simple ones to show that it is possible to construct all the problems of ordinary geometry by doing no more than the little covered in the four figures that I have explained.[24] This is one thing which I believe the ancient mathematicians did not observe, for otherwise they would not have put so much labor into writing so many books in which the very sequence of the propositions shows that they did not have a sure method of finding all,[25] but rather gathered together those propositions on which they had happened by accident.

This is also evident from what Pappus has done in the beginning of his seventh book,[26] where, after devoting considerable space to an enumeration of the books on geometry written by his predecessors,[27] he finally refers to a question which he says that neither Euclid nor Apollonius nor any one else had been able to solve completely;[28] and these are his words:

"Quem autem dicit (Apollonius) in tertio libro locum ad tres, & quatuor lineas ab Euclide perfectum non esse, neque ipse perficere poterat, neque aliquis alius; sed neque paululum quid addere iis, quæ Euclides scripsit, per ea tantum conica, quæ usque ad Euclidis tempora præmonstrata sunt, &c."[29]

A little farther on, he states the question as follows:

"At locus ad tres, & quatuor lineas, in quo (Apollonius) magnifice se jactat, & ostentat, nulla habita gratia ei, qui prius scripserat, est hujusmodi.[30] *Si positione datis tribus rectis lineis ab uno & eodem puncto, ad tres lineas in datis angulis rectæ lineæ ducantur, & data sit proportio rectanguli contenti duabus ductis ad quadratum reliquæ: punctum contingit positione datum solidum locum, hoc est unam ex tribus conicis sectionibus. Et si ad quatuor rectas lineas positione datas in datis angulis lineæ ducantur; & rectanguli duabus ductis contenti ad contentum duabus reliquis proportio data sit; similiter punctum datum coni sectionem positione continget. Si quidem igitur ad duas tantum locus planus ostensus est. Quod si ad plures quam quatuor, punctum continget locos non adhuc cognitos, sed lineas tantum dictas; quales autem sint, vel quam habeant proprietatem, non*

24. It will be seen that Descartes considers only three types of the quadratic equation in z, namely, $z^2 + az - b^2 = 0$, $z^2 - az - b^2 = 0$, and $z^2 - az + b^2 = 0$. It thus appears that he has not been able to free himself from the old traditions to the extent of generalizing the meaning of the coefficients, — as negative and fractional as well as positive. He does not consider the type $z^2 + az + b^2 = 0$, because it has no positive roots.

25. "Qu'ils n'ont point eu la vraye methode pour les trouuer toutes."

26. See Note [9].

27. See Pappus, Vol. II. Pappus here gives a list of books that treat of analysis, in the following words: "Illorum librorum, quibus de loco, ἀναλνόμενος sive resoluto agitur, ordo hic est. Euclidis datorum liber unus, Apollonii de proportionis sectione libri duo, de spatii sectione duo, de sectione determinata duo, de tactionibus duo, Euclidis porismatum libri tres, Apollonii inclinationum libri duo, eiusdem locorum planorum duo, conicorum octo, Aristaci locorum solidorum libri duo." See also the Commandinus edition of Pappus, 1660 edition.

28. For the history of this problem. see Zeuthen: *Die Lehre von den Kegelschnitten im Alterthum*, Copenhagen, 1886. Also, Adam and Tannery, *Oeuvres de Descartes*, Vol. VI.

29. Pappus, Vol. II. Commandinus edition of 1660. Literally, "Moreover, he (Apollonius) says that the problem of the locus related to three or four lines was not entirely solved by Euclid, and that neither he himself, nor any one else has been able to solve it completely, nor were they able to add anything at all to those things which Euclid had written, by means of the conic sections only which had been demonstrated before Euclid." Descartes arrived at the solution of this problem four years before the publication of his geometry, after spending five or six weeks on it. See his letters, Cousin, Vol. VI, and Vol. VI.

30. Given as follows in the edition of Pappus by Hultsch, previously quoted: "Sed hic ad tres et quatuor lineas locus quo magnopere gloriatur simul addens ei qui conscripserit gratiam habendam esse, sic se habet."

constat: earum unam, neque primam, & quæ manifestissima videtur, composuerunt osten-dentes utilem esse. Propositiones autem ipsarum hæ sunt.

"Si ab aliquo puncto ad positione datas rectas lineas quinque ducantur rectæ lineæ in datis angulis, & data sit proportio solidi parallelepipedi rectanguli, quod tribus ductis lineis continetur ad solidum parallelepipedum rectangulum, quod continetur reliquis duabus, & data quapiam linea, punctum positione datam lineam continget. Si autem ad sex, & data sit proportio solidi tribus lincis contenti ad solidum, quod tribus reliquis continetur; rur-sus punctum continget positione datam lineam. Quod si ad plures quam sex, non adhuc habent dicere, an data sit proportio cujuspiam contenti quatuor lineis ad id quod reliquis continetur, quoniam non est aliquid contentum pluribus quam tribus dimensionibus." [31]

Here I beg you to observe in passing that the considerations that forced ancient writers to use arithmetical terms in geometry, thus making it impossible for them to proceed beyond a point where they could see clearly the relation between the two sub-jects, caused much obscurity and embarrassment, in their attempts at explanation.

Pappus proceeds as follows:

"Acquiescunt autem his, qui paulo ante talia interpretati sunt; neque unum aliquo pacto comprehensibile significantes quod his continetur. Licebit autem per conjunctas pro-portiones hæc, & dicere & demonstrare universe in dictis proportionibus, atque his in hunc modum. Si ab aliquo puncto ad positione datas rectas lineas ducantur rectæ lineæ in datis angulis, & data sit proportio conjuncta ex ea, quam habet una ductarum ad unam, & altera ad alteram, & alia ad aliam, & reliqua ad datam lineam, si sint septem; si vero octo, & reliqua ad reliquam: punctum continget positione datas lineas. Et similiter quotcumque sint impares vel pares multitudine, cum hæc, ut dixi, loco ad quatuor lin-eas respondeant, nullum igitur posuerunt ita ut linea nota sit, &c.[32]

---

31. This may be somewhat freely translated as follows: "The problem of the locus related to three or four lines, about which he (Apollonius) boasts so proudly, giving no credit to the writer who has preceded him, is of this nature: If three straight lines are given in position, and if straight lines be drawn from one and the same point, making given angles with the three given lines; and if there be given the ratio of the rectangle contained by two of the lines so drawn to the square of the other, the point lies on a solid locus given in position, namely, one of the three conic sections.

"Again, if lines be drawn making given angles with four straight lines given in position, and if the rectangle of two of the lines so drawn bears a given ratio to the rectangle of the other two; then, in like manner, the point lies on a conic section given in position. It has been shown that to only two lines there corresponds a plane locus. But if there be given more than four lines, the point generates loci not known up to the present time (that is, im-possible to determine by common methods), but merely called 'lines'. It is not clear what they are, or what their properties. One of them, not the first but the most manifest, has been examined, and this has proved to be helpful. (Paul Tannery, in the *Oeuvres de Descartes*, differs with Descartes in his translation of Pappus. He translates as follows: Et on n'a fait la synthèse d' aucune de ces lignes, ni montré qu'elle servit pour ces cieux, pas même pour celle qui semblerait la première et la plus indiquée.) These, however, are the propositions concerning them.

"If from any point straight lines be drawn making given angles with five straight lines given in position, and if the solid rectangular paral-lelepiped contained by three of the lines so drawn bears a given ratio to the solid rectangular parallelepiped contained by the other two and any given line whatever, the point lies on a 'line' given in position. Again, if there be six lines, and if the solid contained by three of the lines bears a given ratio to the solid contained by the other three lines, the point also lies on a 'line' given in position. But if there be more than six lines, we cannot say whether a ratio of something contained by four lines is given to that which is contained by the rest, since there is no figure of more than three dimensions."

32. This rather obscure passage may be translated as follows: "For in this are agreed those who formerly interpreted these things (that the dimen-sions of a figure cannot exceed three) in that they maintain that a figure that is contained by these lines is not comprehensible in any way. This is per-missible, however, both to say and to demonstrate generally by this kind of proportion, and in this manner: If from any point straight lines be drawn making given angles with straight lines given in position; and if there be given a ratio compounded of them, that is the ratio that one of the lines drawn has to one, the second has to a second, the third to a third, and so on to the given line if there be seven lines, or, if there be eight lines, of the last to a last, the point lies on the lines that are given in position. And similarly, whatever may be the odd or even number, since these, as I have said, correspond in position to the four lines; therefore they have not set forth any method so that a line may be known." The meaning of the passage appears from that which follows in the text.

The question, then, the solution of which was begun by Euclid and carried farther by Apollonius, but was completed by no one, is this:

Having three, four or more lines given in position, it is first required to find a point from which as many other lines may be drawn, each making a given angle with one of the given lines, so that the rectangle of two of the lines so drawn shall bear a given ratio to the square of the third (if there be only three); or to the rectangle of the other two (if there be four), or again, that the parallelepiped[33] constructed upon three shall bear a given ratio to that upon the other two and any given line (if there be five), or to the parallelepiped upon the other three (if there be six); or (if there be seven) that the product obtained by multiplying four of them together shall bear a given ratio to the product of the other three, or (if there be eight) that the product of four of them shall bear a given ratio to the product of the other four. Thus the question admits of extension to any number of lines.

Then, since there is always an infinite number of different points satisfying these requirements, it is also required to discover and trace the curve containing all such points.[34] Pappus says that when there are only three or four lines given, this line is one of the three conic sections, but he does not undertake to determine, describe, or explain the nature of the line required when the question involves a greater number of lines. He only adds that the ancients recognized one of them which they had shown to be useful, and which seemed the simplest, and yet was not the most important. This led me to try to find out whether, by my own method, I could go as far as they had gone.[35]

First, I discovered that if the question be proposed for only three, four, or five lines, the required points can be found by elementary geometry, that is, by the use of the ruler and compasses only, and the application of those principles that I have already explained, except in the case of five parallel lines. In this case, and in the cases where there are six, seven, eight, or nine given lines, the required points can always be found by means of the geometry of solid loci,[36] that is, by using some one of the three conic sections. Here, again, there is an exception in the case of nine parallel lines. For this and the cases of ten, eleven, twelve, or thirteen given lines, the required points may be found by means of a curve of degree next higher than that of the conic sections. Again, the case of thirteen parallel lines must be excluded, for which, as well as for the cases of fourteen, fifteen, sixteen, and seventeen lines, a curve of degree next higher than the preceding must be used; and so on indefinitely.

33. That is, continued product.
34. It is here that the essential feature of the work of Descartes may be said to begin.
35. Descartes gives here a brief summary of his solution, which he amplifies later.
36. This term was commonly applied by mathematicians of the seventeenth century to the three conic sections, while the straight line and circle were called plane loci, and other curves linear loci. See Fermat, *Isagoge ad Locos Planos et Solidos*, Toulouse, 1679.

Next, I have found that when only three or four lines are given, the required points lie not only all on one of the conic sections but sometimes on the circumference of a circle or even on a straight line.[37]

When there are five, six, seven, or eight lines, the required points lie on a curve of degree next higher than the conic sections, and it is impossible to imagine such a curve that may not satisfy the conditions of the problem; but the required points may possibly lie on a conic section, a circle, or a straight line. If there are nine, ten, eleven, or twelve lines, the required curve is only one degree higher than the preceding, but any such curve may meet the requirements, and so on to infinity.

Finally, the first and simplest curve after the conic sections is the one generated by the intersection of a parabola with a straight line in a way to be described presently.

I believe that I have in this way completely accomplished what Pappus tells us the ancients sought to do, and I will try to give the demonstration in a few words, for I am already wearied by so much writing.

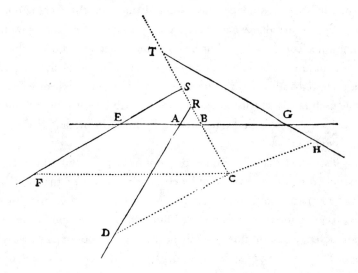

Let AB, AD, EF, GH, ... be any number of straight lines given in position,[38] and let it be required to find a point C, from which straight lines CB, CD, CF, CH, ... can be drawn, making given angles CBA, CDA, CFE, CHG, ... respectively, with the

---

37. Degenerate or limiting forms of the conic sections.

38. It should be noted that these lines are given in position but not in length. They thus become lines of reference or coördinate axes, and accordingly they play a very important part in the development of analytic geometry. In this connection we may quote as follows: "Among the predecessors of Descartes we reckon, besides Apollonius, especially Vieta, Oresme, Cavalieri, Roberval, and Fermat, the last the most distinguished in this field; but nowhere, even by, Fermat, had any attempt been made to refer several curves of different orders simultaneously to one system of coördinates, which at most possessed special significance for one of the curves. It is exactly this thing which Descartes systematically accomplished." Karl Fink, *A Brief History of Mathematics*, trans. by Beman and Smith, Chicago, 1903.

Heath calls attention to the fact that "the essential difference between the Greek and the modern method is that the Greeks did not direct their efforts to making the fixed lines of a figure as few as possible, but rather to expressing their equations between areas in as short and simple a form as possible." For further discussion see D. E. Smith, *History of Mathematics*, Boston, 1923–25, Vol. II (hereafter referred to as Smith).

given lines, and such that the product of certain of them is equal to the product of the rest, or at least such that these two products shall have a given ratio, for this condition does not make the problem any more difficult.

First, I suppose the thing done, and since so many lines are confusing, I may simplify matters by considering one of the given lines and one of those to be drawn (as, for example, AB and BC) as the principal lines, to which I shall try to refer all the others. Call the segment of the line AB between A and B, x, and call BC, y. Produce all the other given lines to meet these two (also produced if necessary) provided none is parallel to either of the principal lines. Thus, in the figure, the given lines cut AB in the points A, E, G, and cut BC in the points R, S, T.

Now, since all the angles of the triangle ARB are known,[39] the ratio between the sides AB and BR is known.[40] If we let AB : BR = $z : b$, since AB = x, we have RB = $\frac{bx}{z}$; and since B lies between C and R[41], we have CR = $y + \frac{bx}{z}$. (When R lies between C and B, CR is equal to $y - \frac{bx}{z}$, and when C lies between B and R, CR is equal to $-y + \frac{bx}{z}$) Again, the three angles of the triangle DRC are known,[42] and therefore the ratio between the sides CR and CD is determined. Calling this ratio $z : c$, since CR = $y + \frac{bx}{z}$, we have CD = $\frac{cy}{z} + \frac{b \cdot x}{z^2}$. Then, since the lines AB, AD, and EF are given in position, the distance from A to E is known. If we call this distance k, then EB = $k + x$; although EB = $k - x$ when B lies between E and A, and E = $-k + x$ when E lies between A and B. Now the angles of the triangle ESB

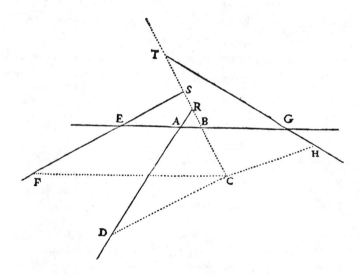

39. Since BC cuts AB and AD under given angles.
40. Since the ratio of the sines of the opposite angles is known.
41. In this particular figure, of course.
42. Since CB and CD cut AD under given angles.

being given, the ratio of BE to BS is known. We may call this ratio $z : d$. Then BS $= \frac{dk + dx}{z}$ and CS $= \frac{zy + dk + dx}{z}$.[43] When S lies between B and C we have CS $= \frac{zy - dk - dx}{z}$, and when C lies between B and S we have CS $= \frac{-zy + dk + dx}{z}$. The angles of the triangle FSC are known, and hence, also the ratio of CS to CF, or $z : e$. Therefore, CF $= \frac{ezy + dek + dex}{z^2}$. Likewise, AG or l is given, and BG $= l - x$. Also, in triangle BGT, the ratio of BG to BT, or $z : f$, is known. Therefore, BT $= \frac{fl - fx}{z}$ and CT $= \frac{zy + fl - fx}{z}$. In triangle TCH, the ratio of TC to CH, or $z : g$, is known,[44] whence CH $= \frac{gzy + fgl + fgx}{z^2}$.

And thus you see that, no matter how many lines are given in position, the length of any such line through C making given angles with these lines can always be expressed by three terms, one of which consists of the unknown quantity y multiplied or divided by some known quantity; another consisting of the unknown quantity x multiplied or divided by some other known quantity; and the third consisting of a known quantity.[45] An exception must be made in the case where the given lines are parallel either to AB (when the term containing x vanishes), or to CB (when the term containing y vanishes). This case is too simple to require further explanation.[46] The signs of the terms may be either $+$ or $-$ in every conceivable combination.[47]

You also see that in the product of any number of these lines the degree of any term containing x or y will not be greater than the number of lines (expressed by means of x and y) whose product is found. Thus, no term will be of degree higher than the second if two lines be multiplied together, nor of degree higher than the third, if there be three lines, and so on to infinity.

Furthermore, to determine the point C, but one condition is needed, namely, that the product of a certain number of lines shall be equal to, or (what is quite as

43. We have
$$CS = y + BS$$
$$= y + \frac{dk + dx}{z}$$
$$= \frac{zy + dk + dx}{z},$$
and similarly for the other cases considered below.

44. It should be noted that each ratio assumed has z as antecedent.

45. That is, an expression of the form $ax + by + c$, where a, b, c, are any real positive or negative quantities, integral or fractional (not zero, since this exception is considered later).

46. The following problem will serve as a very simple illustration: Given three parallel lines AB, CD, EF, so placed that AB is distant 4 units from CD, and CD is distant 3 units from EF; required to find a point P such that if PL, PM, PN

be drawn through P, making angles of 90°, 45°, 30°, respectively, with the parallels. Then $\overline{PM}^2 = PL \cdot PN$. Let PR $= y$, then PN $= 2y$, PM $= \sqrt{2}(y + 3)$, PL $= y + 7$. If $\overline{PM}^2 = PN \cdot PL$, we have $[\sqrt{2}(y + 3)]^2 = 2y(y + 7)$, whence $y = 9$. Therefore, the point P lies on the line XY parallel to EF and at a distance of 9 units from it.

47. Depending, of course, upon the relative positions of the given lines.

simple), shall bear a given ratio to the product of certain other lines. Since this condition can be expressed by a single equation in two unknown quantities,[48] we may give any value we please to either x or y and find the value of the other from this equation. It is obvious that when not more than five lines are given, the quantity x, which is not used to express the first of the lines can never be of degree higher than the second.[49]

Assigning a value to y, we have $x^2 = \pm ax \pm b^2$, and therefore x can be found with ruler and compasses, by a method already explained. If then we should take successively an infinite number of different values for the line y, we should obtain an infinite number of values for the line x, and therefore an infinity of different points, such as C, by means of which the required curve could be drawn.

This method can be used when the problem concerns six or more lines, if some of them are parallel to either AB or BC, in which case either x or y will be of only the second degree in the equation, so that the point C can be found with ruler and compasses.

On the other hand, if the given lines are all parallel even though a question should be proposed involving only five lines, the point C cannot be found in this way. For, since the quantity x does not occur at all in the equation, it is no longer allowable to give a known value to y. It is then necessary to find the value of y.[50] And since the term in y will now be of the third degree, its value can be found only by finding the root of a cubic equation, which cannot in general be done without the use of one of the conic sections.

And furthermore, if not more than nine lines are given, not all of them being parallel, the equation can always be so expressed as to be of degree not higher than the fourth. Such equations can always be solved by means of the conic sections in a way that I shall presently explain.

Again, if there are not more than thirteen lines, an equation of degree not higher than the sixth can be employed, which admits of solution by means of a curve just one degree higher than the conic sections by a method to be explained presently.[51]

This completes the first part of what I have to demonstrate here, but it is necessary, before passing to the second part, to make some general statements concerning the nature of curved lines.

48. That is, an indeterminate equation. "De plus, à cause que pour determiner le point C, il n'y a qu'une seule condition qui soit requise, à sçavoir que ce qui est produit par la multiplication d'un certain nombre de ces lignes soit égal, ou (ce qui n'est de rien plus mal-aisé) ait la proportion donnee, à ce qui est produit par la multiplication des autres; on peut prendre à discretion l'une des deux quantitez inconnuës x ou y, & chercher l'autre par cette Equation." Such variations in the texts of different editions are of no moment, but are occasionally introduced as matters of interest.

49. Since the product of three lines bears a given ratio to the product of two others and a given line, no term can be of higher degree than the third, and therefore, than the second in x.

50. That is, to solve the equation for y.

51. This line of reasoning may be extended indefinitely. Briefly, it means that for every two lines introduced the equation becomes one degree higher and the curve becomes correspondingly more complex.

GEOMETRY

BOOK II

ON THE NATURE OF CURVED LINES

The ancients were familiar with the fact that the problems of geometry may be divided into three classes, namely, plane, solid, and linear problems.[52] This is equivalent to saying that some problems require only circles and straight lines for their construction, while others require a conic section and still others require more complex curves.[53] I am surprised, however, that they did not go further, and distinguish between different degrees of these more complex curves, nor do I see why they called the latter mechanical, rather than geometrical.[54]

If we say that they are called mechanical because some sort of instrument[55] has to be used to describe them, then we must, to be consistent, reject circles and straight lines, since these cannot be described on paper without the use of compasses and a ruler, which may also be termed instruments. It is not because the other instruments, being more complicated than the ruler and compasses, are therefore less accurate, for if this were so they would have to be excluded from mechanics, in which accuracy of construction is even more important than in geometry. In the latter, exactness of reasoning alone[56] is sought, and this can surely be as thorough with reference to such lines as to simpler ones.[57] I cannot believe, either, that it was because they did not wish to make more than two postulates, namely, (1) a straight line can be drawn between any two points, and (2) about a given center a circle can be described passing through a given point. In their treatment of the conic sections they did not hesitate to introduce the assumption that any given cone can be cut by a given plane. Now to treat all the curves which I mean to introduce here, only one additional

---

52. Cf. Pappus, Vol. I, Proposition 5, Book III: "The ancients considered three classes of geometric problems, which they called plane, solid, and linear. Those which can be solved by means of straight lines and circumferences of circles are called plane problems, since the lines or curves by which they are solved have their origin in a plane. But problems whose solutions are obtained by the use of one or more of the conic sections are called solid problems, for the surfaces of solid figures (conical surfaces) have to be used. There remains a third class which is called linear because other 'lines' than those I have just described, having diverse and more involved origins, are required for their construction. Such lines are the spirals, the quadratrix, the conchoid, and the cissoid, all of which have many important properties."

53. Rabuel suggests dividing problems into classes, the first class to include all problems that can be constructed by means of straight lines, that is, curves whose equations are of the first degree; the second, those that require curves whose equations are of the second degree, namely, the circle and the conic sections, and so on.

54. Cf. *Encyclopédie ou Dictionnaire Raisonné des Sciences, des Arts et des Metiers, par une Société de gens de lettres, mis en ordre et publiées par M. Diderot, et quant à la Partie Mathematique par M. d'Alembert*, Lausanne and Berne, 1780. In substance as follows: *"Mechanical* is a mathematical term designating a construction not geometric, that is, that cannot be accomplished by geometric curves. Such are constructions depending upon the quadrature of the circle.

The term, mechanical curve, was used by Descartes to designate a curve that cannot be expressed by an algebraic equation." Leibniz and others call them transcendental.

55. "Machine."

56. An interesting question of modern education is here raised, namely, to what extent we should insist upon accuracy of construction even in elementary geometry.

57. Not only ancient writers but later ones, up to the time of Descartes, made the same distinction; for example, Vieta. Descartes's view has been universally accepted since his time.

assumption is necessary, namely, two or more lines can be moved, one upon the other, determining by their intersection other curves. This seems to me in no way more difficult.[58]

It is true that the conic sections were never freely received into ancient geometry,[59] and I do not care to undertake to change names confirmed by usage; nevertheless, it seems very clear to me that if we make the usual assumption that geometry is precise and exact, while mechanics is not;[60] and if we think of geometry as the science which furnishes a general knowledge of the measurement of all bodies, then we have no more right to exclude the more complex curves than the simpler ones, provided they can be conceived of as described by a continuous motion or by several successive motions, each motion being completely determined by those which precede; for in this way an exact knowledge of the magnitude of each is always obtainable.

Probably the real explanation of the refusal of ancient geometers to accept curves more complex than the conic sections lies in the fact that the first curves to which their attention was attracted happened to be the spiral,[61] the quadratrix,[62] and similar curves, which really do belong only to mechanics, and are not among those curves that I think should be included here, since they must be conceived of as described by two separate movements whose relation does not admit of exact determination. Yet they afterwards examined the conchoid,[63] the cissoid,[64] and a few others which should be accepted; but not knowing much about their properties they took no more account of these than of the others. Again, it may have been that, knowing as they did only a little about the conic sections,[65] and being still ignorant of many of the possibilities of the ruler and compasses, they dared not yet attack a matter of still greater difficulty. I hope that hereafter those who are clever enough to make use of the geometric methods herein suggested will find no great difficulty in applying them to plane or solid problems. I therefore think it proper to suggest to such a more extended line of investigation which will furnish abundant opportunities for practice.

Consider the lines AB, AD, AF, and so forth, which we may suppose to be described by means of the instrument YZ. This instrument consists of several rulers hinged together in such a way that YZ being placed along the line AN the angle XYZ can be increased or decreased in size, and when its sides are together the points

58. That is, in no way less obvious than the other postulates.

59. Because the ancients did not believe that the so-called constructions of the conic sections on a plane surface could be exact.

60. Since it is not possible to construct an ideal line, plane, and so on.

61. See Heath, *History of Greek Mathematics* (hereafter referred to as Heath). Cambridge, 2 vols., 1921. Also Cantor, *Vorlesungen über Geschichte der Mathematik*, Leipzig, Vol. I (2), and Vol. II (1) (hereafter referred to as Cantor).

62. See Heath, I, 225; Smith, Vol. II.

63. See Heath, I, 235, 238; Smith, Vol. II.

64. See Heath, I, 264; Smith, Vol. II.

65. They really knew much more than would be inferred from this statement. In this connection, see Taylor, *Ancient and Modern Geometry of Conics*, Cambridge, 1881.

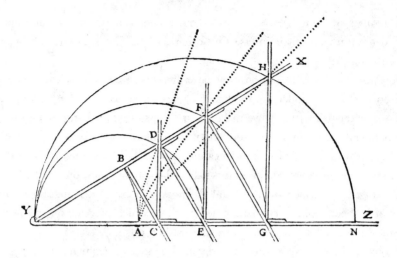

B, C, D, E, F, G, H, all coincide with A; but as the size of the angle is increased, the ruler BC, fastened at right angles to XY at the point B, pushes toward Z the ruler CD which slides along YZ always at right angles. In like manner, CD pushes DE which slides along YX always parallel to BC; DE pushes EF; EF pushes FG; FG pushes GH, and so on. Thus we may imagine an infinity of rulers, each pushing another, half of them making equal angles with YX and the rest with YZ.

Now as the angle XYZ is increased the point B describes the curve AB, which is a circle; while the intersections of the other rulers, namely, the points D, F, H describe other curves, AD, AF, AH, of which the latter are more complex than the first and this more complex than the circle. Nevertheless I see no reason why the description of the first[66] cannot be conceived as clearly and distinctly as that of the circle, or at least as that of the conic sections; or why that of the second, third,[67] or any other that can be thus described, cannot be as clearly conceived of as the first: and therefore I see no reason why they should not be used in the same way in the solution of geometric problems.[68]

66. That is, AD.

67. That is, AF and AH.

68. The equations of these curves may be obtained as follows: (1) Let YA = YB = a, YC = x, CD = y, YD = z; then $z : x = x : a$, whence $z = \dfrac{x^2}{a}$. Also $z^2 = x^2 + y^2$; therefore the equation of AD is $x^4 = a^2(x^2 + y^2)$. (2) Let YA = YB = a, YE = x, EF = y, YF = z. Then $z : x = x :$ YD, whence YD = $\dfrac{x^2}{z}$. Also

$$x : YD = YD : YC, \text{ whence } YC = \frac{x^4}{z^2} \div x = \frac{x^3}{z^2}.$$

But YD : YC = YC : a, and therefore

$$\frac{ax^2}{z} = \left(\frac{x^3}{z^2}\right)^2, \text{ or } z = \sqrt[3]{\frac{x^4}{a}}.$$

Also, $z^2 = x^2 + y^2$. Thus we get, as the equation of AF,

$$\sqrt[3]{\frac{x^8}{a^2}} = x^2 + y^2, \text{ or } x^8 = a^2(x^2 + y^2)^3.$$

(3) In the same way, it can be shown that the equation of AH is

$$x^{12} = a^2(x^2 + y^2)^5.$$

I could give here several other ways of tracing and conceiving a series of curved lines, each curve more complex than any preceding one,[69] but I think the best way to group together all such curves and then classify them in order, is by recognizing the fact that all points of those curves which we may call "geometric," that is, those which admit of precise and exact measurement, must bear a definite relation[70] to all points of a straight line, and that this relation must be expressed by means of a single equation.[71] If this equation contains no term of higher degree than the rectangle of two unknown quantities, or the square of one, the curve belongs to the first and simplest class,[72] which contains only the circle, the parabola, the hyperbola, and the ellipse; but when the equation contains one or more terms of the third or fourth degree[73] in one or both of the two unknown quantities[74] (for it requires two unknown quantities to express the relation between two points) the curve belongs to the second class; and if the equation contains a term of the fifth or sixth degree in either or both of the unknown quantities the curve belongs to the third class, and so on indefinitely.

Suppose the curve EC to be described by the intersection of the ruler GL and the rectilinear plane figure CNKL, whose side KN is produced indefinitely in the direction of C, and which, being moved in the same plane in such a way that its side[75] KL always coincides with some part of the line BA (produced in both directions), imparts to the ruler GL a rotary motion about G (the ruler being hinged to

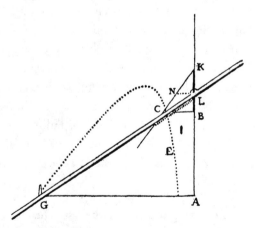

69. "Qui seroient de plus en plus composées par degrez à l'infini." The French quotations in the footnotes show a few variants in style in different editions.

70. That is, a relation exactly known, as, for example, that between two straight lines in distinction to that between a straight line and a curve, unless the length of the curve is known.

71. It will be recognized at once that this statement contains the fundamental concept of analytic geometry.

72. "Du premier & plus simple genre," an expression not now recognized. As now understood, the order or degree of a plane curve is the greatest number of points in which it can be cut by any arbitrary line, while the class is the greatest number of tangents that can be drawn to it from any arbitrary point in the plane.

73. Grouped together because an equation of the fourth degree can always be transformed into one of the third degree.

74. Thus Descartes includes such terms as x^2y, x^2y^2, . . . as well as x^3, y^4. . . .

75. "Diametre."

the figure CNKL at L).[76] If I wish to find out to what class this curve belongs, I choose a straight line, as AB, to which to refer all its points, and in AB I choose a point A at which to begin the investigation.[77] I say "choose this and that," because we are free to choose what we will, for, while it is necessary to use care in the choice in order to make the equation as short and simple as possible, yet no matter what line I should take instead of AB the curve would always prove to be of the same class, a fact easily demonstrated.[78]

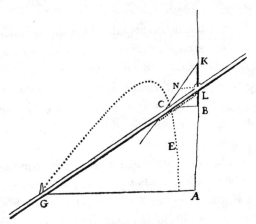

Then I take on the curve an arbitrary point, as C, at which we will suppose the instrument applied to describe the curve. Then I draw through C the line CB parallel to GA. Since CB and BA are unknown and indeterminate quantities, I shall call one of them y and the other x. To the relation between these quantities I must consider also the known quantities which determine the description of the curve, as GA, which I shall call a; KL, which I shall call b; and NL parallel to GA, which I shall call c. Then I say that as NL is to LK, or as c is to b, so CB, or y, is to BK, which is therefore equal to $\frac{b}{c}y$. Then BL is equal to $\frac{b}{c}y - b$, and AL is equal to $x + \frac{b}{c}y - b$. Moreover, as CB is to LB, that is, as y is to $\frac{b}{c}y - b$, so AG or a is to LA or $x + \frac{b}{c}y - b$. Multiplying the second by the third, we get $\frac{ab}{c}y - ab$ equal to

$$xy + \frac{b}{c}y^2 - by,$$

which is obtained by multiplying the first by the last. Therefore, the required equation is

$$y^2 = cy - \frac{cx}{b}y + ay - ac.$$

76. The instrument thus consists of three parts, (1) a ruler AK of indefinite length, fixed in a plane; (2) a ruler GL, also of indefinite length, fastened to a pivot, G, in the same plane, but not on AK; and (3) a rectilinear figure BKC, the side KC being indefinitely long, to which the ruler GL is hinged at L, and which is made to slide along the ruler GL.
77. That is, Descartes uses the point A as origin, and the line AB as axis of abscissas. He uses parallel ordinates, but does not draw the axis of ordinates.
78. That is, the nature of a curve is not affected by a transformation of coördinates.

From this equation we see that the curve EC belongs to the first class, it being, in fact, a hyperbola.[79]

If in the instrument used to describe the curve we substitute for the rectilinear figure CNK this hyperbola or some other curve of the first class lying in the plane CNKL, the intersection of this curve with the ruler GL will describe, instead of the hyperbola EC, another curve, which will be of the second class.

Thus, if CNK be a circle having its center at L, we shall describe the first conchoid of the ancients,[80] while if we use a parabola having KB as axis we shall describe the curve which, as I have already said, is the first and simplest of the curves required in the problem of Pappus, that is, the one which furnishes the solution when five lines are given in position.

If, instead of one of these curves of the first class, there be used a curve of the second class lying in the plane CNKL, a curve of the third class will be described; while if one of the third class be used, one of the fourth class will be obtained, and so on to infinity.[81] These statements are easily proved by actual calculation.

Thus, no matter how we conceive a curve to be described, provided it be one of those which I have called geometric, it is always possible to find in this manner an equation determining all its points. Now I shall place curves whose equations are of the fourth degree in the same class with those whose equations are of the third degree; and those whose equations are of the sixth degree[82] in the same class with

79. Cf. Briot and Bouquet, *Elements of Analytical Geometry of Two Dimensions*, trans. by J. H. Boyd, New York, 1896.
The two branches of the curve are determined by the position of the triangle CNKL with respect to the directrix AB.
Van Schooten, gives the following construction and proof: Produce AG to D, making DG = EA. Since E is a point of the curve obtained when GL coincides with GA, L with A, and C with N, then EA = NL. Draw DF parallel to KC. Now let GCE be a hyperbola through E whose asymptotes are DF and FA. To prove that this hyperbola is the curve given by the instrument described above, produce BC to cut DF in I, and draw DH parallel to AF meeting BC in H. Then KL:LN = DH:HI. But DH = AB = x, so we may write $b : c = x : $ HI, whence HI = $\dfrac{cx}{b}$, IB = $a + c - \dfrac{cx}{b}$,

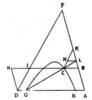

IC = $a + c - \dfrac{cx}{b} - y$. But in any hyperbola IC · BC = DE · EA, whence we have $(a + c - \dfrac{cx}{b} - y)y = ac$, or $y^2 = cy - \dfrac{cxy}{b} + ay - ac$. But this is the equation obtained above, which is therefore the equation of a hyperbola whose asymptotes are AF and FD.
Van Schooten, describes another similar instrument: Given a ruler AB pivoted at A, and another BD hinged to AB at B. Let AB rotate about A so that D moves along LK; then the curve generated by any point E of BE will be an ellipse whose semi-major axis is AB + BE and whose semi-minor axis is AB − BE.
80. See note 58.
81. Rabuel illustrates this, substituting for the curve CNKL the semicubical parabola, and showing that the resulting equation is of the fifth degree, and therefore, according to Descartes, of the third class. Rabuel also gives, a general method for finding the curve, no matter what figure is used for CNKL. Let GA = a, KL = b, AB = x, CB = y and KB = z; then LB = $z − b$, and AL = $x + z − b$. Now GA : AL = CB : BL, or
$a : x + z − b = y : z − b$, whence $z = \dfrac{xy − by + ab}{a − y}$.
This value of z is independent of the nature of the figure CNKL. But given any figure CNKL it is possible to obtain a second value for z from the nature of the curve. Equating these values of z we get the equation of the curve.
82. "Celles dont l'équation monte au quarré de cube."

those whose equations are of the fifth degree[83] and similarly for the rest. This classification is based upon the fact that there is a general rule for reducing to a cubic any equation of the fourth degree, and to an equation of the fifth degree[84] any equation of the sixth degree, so that the latter in each case need not be considered any more complex than the former.

It should be observed, however, with regard to the curves of any one class, that while many of them are equally complex so that they may be employed to determine the same points and construct the same problems, yet there are certain simpler ones whose usefulness is more limited. Thus, among the curves of the first class, besides the ellipse, the hyperbola, and the parabola, which are equally complex, there is also found the circle, which is evidently a simpler curve; while among those of the second class we find the common conchoid, which is described by means of the circle, and some others which, though less complicated[85] than many curves of the same class, cannot be placed in the first class.[86]

Having now made a general classification of curves, it is easy for me to demonstrate the solution which I have already given of the problem of Pappus. For, first, I have shown that when there are only three or four lines the equation which serves to determine the required points[87] is of the second degree. It follows that the curve containing these points must belong to the first class, since such an equation expresses the relation between all points of curves of Class I and all points of a fixed straight line. When there are not more than eight given lines the equation is at most a biquadratic, and therefore the resulting curve belongs to Class II or Class I. When there are not more than twelve given lines, the equation is of the sixth degree or lower, and therefore the required curve belongs to Class III or a lower class, and so on for other cases.

Now, since each of the given lines may have any conceivable position, and since any change in the position of a line produces a corresponding change in the values of the known quantities as well as in the signs $+$ and $-$ of the equation, it is clear that there is no curve of Class I that may not furnish a solution of this problem when it relates to four lines, and that there is no curve of Class II that may not furnish a solution when the problem relates to eight lines, none of Class III when it relates to twelve lines, etc. It follows that there is no geometric curve whose equation can be obtained that may not be used for some number of lines.[88]

83. "Celles dont elle ne monte qu'au sursolide."

84. "Au sursolide."

85. "Pas tant d'étenduë." Cf. Rabuel, "Pas tant d'étendue en leur puissance."

86. Various methods of tracing curves were used by writers of the seventeenth century. Among these there were not only the usual method of plotting a curve from its equation and that of using strings, pegs, etc., as in the popular construction of the ellipse, but also the method of using jointed rulers and that of using one curve from which to derive another, as for example the usual method of describing the cissoid.

87. That is, the equation of the required locus.

88. "En sorte qu'il n'y a pas une ligne courbe qui tombe sous le calcul & puisse être receuë en Geometrie, qui n'y soit utile pour quelque nombre de lignes."

It is now necessary to determine more particularly and to give the method of finding the curve required in each case, for only three or four given lines. This investigation will show that Class I contains only the circle and the three conic sections.

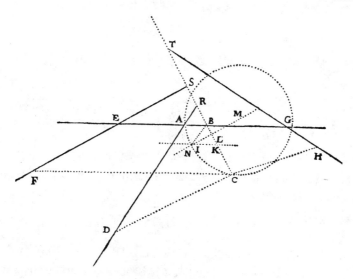

Consider again the four lines AB, AD, EF, and GH, given before, and let it be required to find the locus generated by a point C, such that, if four lines CB, CD, CF, and CH be drawn through it making given angles with the given lines, the product of CB and CF is equal to the product of CD and CH. This is equivalent to saying that if

$$CB = y,$$

$$CD = \frac{czy + bcx}{z^2},$$

$$CF = \frac{ezy + dek + dex}{z^2},$$

and

$$CH = \frac{gzy + fgl - fgx}{z^2}.$$

then the equation is

$$y^2 = \frac{(cfglz - dekz^2)y - (dez^2 + cfgz - bcgz)xy + bcfglx - bcfgx^2}{ez^3 - cgz^2}.$$

It is here assumed that ez is greater than cg; otherwise the signs $+$ and $-$ must all be changed.[89] If y is zero or less than nothing in this equation,[90] the point C being supposed to lie within the angle DAG, then C must be supposed to lie within one of the angles DAE, EAR, or RAG, and the signs must be changed to produce this

89. When ez is greater than cg, then $ez^3 - cgz^2$ is positive and its square root is therefore real.
90. Descartes uses "moindre que rien" for "negative."

result. If for each of these four positions y is equal to zero, then the problem admits of no solution in the case proposed.

Let us suppose the solution possible, and to shorten the work let us write $2m$ instead of $\frac{cfgz - dekz^2}{ez^3 - cgz^2}$, and $\frac{2n}{z}$ instead of $\frac{dez^2 + cfgz - bcgz}{ez^3 - cgz^2}$.

Then we have

$$y^2 = 2my - \frac{2n}{z}xy + \frac{bcfglx - bcfgx^2}{ez^3 - cgz^2},$$

of which the root[91] is

$$y = m - \frac{nx}{z} + \sqrt{m^2 - \frac{2mnx}{z} + \frac{n^2x^2}{z^2} + \frac{bcfglx - bcfgx^2}{ez^3 - cgz^2}}.$$

Again, for the sake of brevity, put $-\frac{2mn}{z} + \frac{bcfgl}{ez^3 - cgz^2}$ equal to o, and $\frac{n^2}{z^2} - \frac{bcfg}{ez^3 - cgz^2}$ equal to $\frac{p}{m}$; for these quantities being given, we can represent them in any way we please.[92] Then we have

$$y = m - \frac{n}{z}x + \sqrt{m^2 + ox + \frac{p}{m}x^2}.$$

This must give the length of the line BC, leaving AB or x undetermined. Since the problem relates to only three or four lines, it is obvious that we shall always have such terms, although some of them may vanish and the signs may all vary.[93]

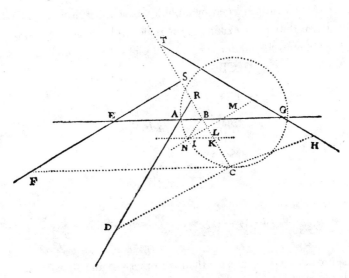

91. Descartes mentions here only one root; of course the other root would furnish a second locus.
92. In a letter to Mersenne (Cousin, Vol. VII), Descartes says: "In regard to the problem of Pappus, I have given only the construction and demonstration without putting in all the analysis; . . . in other words, I have given the construction as architects build structures, giving the specifications and leaving the actual manual labor to carpenters and masons."
93. Having obtained the value of BC algebraically, Descartes now proceeds to construct the length BC geometrically, term by term. He considers BC equal to BK + KL + LC, which is equal to BK − LK + LC which in turn is equal to

$$m - \frac{n}{z}x + \sqrt{m^2 + ox + \frac{p}{m}x^2}.$$

After this, I make KI equal and parallel to BA, and cutting off on BC a segment BK equal to m (since the expression for BC contains $+m$; if this were $-m$, I should have drawn IK on the other side of AB,[94] while if m were zero, I would not have drawn IK at all). Then I draw IL so that IK : KL = z : n; that is, so that if IK is equal to, x, KL is equal to $\frac{n}{z}x$. In the same way I know the ratio of KL to IL, which I may call $n : a$, so that if KL is equal to $\frac{n}{z}x$, IL is equal to $\frac{a}{z}x$. I take the point K between L and C, since the equation contains $-\frac{n}{z}x$; if this were $+\frac{n}{z}x$, I should take L between K and C;[95] while if $\frac{n}{z}x$ were equal to zero, I should not draw IL.

This being done, there remains the expression

$$LC = \sqrt{m^2 + ox + \frac{p}{m}x^2},$$

from which to construct LC. It is clear that if this were zero the point C would lie on the straight line IL;[96] that if it were a perfect square, that is if m^2 and $\frac{p}{m}x^2$ were both $+$[97] and o^2 was equal to $4pm$, or if m^2 and ox, or ox and $\frac{p}{m}x^2$, were zero, then the point C would lie on another straight line, whose position could be determined as easily as that of IL.[98]

If none of these exceptional cases occur,[99] the point C always lies on one of the three conic sections, or on a circle having its diameter in the line IL and having LC a line applied in order to this diameter,[100] or, on the other hand, having LC parallel to a diameter and IL applied in order.

In particular, if the term $\frac{p}{m}x^2$ is zero, the conic section is a parabola; if it is preceded by a plus sign, it is a hyperbola; and, finally, if it is preceded by a minus sign, it is an ellipse. An exception occurs when a^2m is equal to pz^2 and the angle ILC is a right angle,[101] in which case we get a circle instead of an ellipse.[102]

---

94. That is, take I on CB produced.

95. That is, on KB produced. C is not yet determined.

96. The equation of IL is $y = m - \frac{n}{z}x$.

97. There is considerable diversity in the treatment of this sentence in different editions. The Latin edition of 1683 has "Hoc est, ut, mm & $\frac{p}{m}xx$ signo + notalis." The French edition, Paris, 1705, has "C'est à dire que mm et $\frac{p}{m}xx$ étant marquez d'un même signe + ou −." Rabuel gives "C'est a dire que mm and $\frac{p}{m}xx$ étant marquez d'un même signe +." He adds the following note: "Il y a dans les Editions Françoises de Leyde, 1637, et de Paris, 1705, 'un meme signe + ou −', ce qui est une faute d'impression." The French edition, Paris, 1886, has "Etant marqués d'un meme signe + ou −."

98. Note the difficulty in generalization experienced even by Descartes.

99. "Mais lorsque cela n'est pas." In each case the equation giving the value of y is linear in x and y, and therefore represents a straight line. If the quantity under the radical sign and $\frac{n}{z}x$ are both zero, the line is parallel to AB. If the quantity under the radical sign and m are both zero, C lies in AL.

100. "An ordinate." The equivalent of "ordination application" was used in the 16th century translation of Apollonius. Hutton's Mathematical Dictionary, 1796, gives "applicate." "Ordinate applicate," was also used.

101. Rabuel adds "If $a^2m = pz^2$ or if $m = p$ the hyperbola is equilateral."

102. In this case the triangle ILK is a right triangle, whence $\overline{IK}^2 = \overline{LK}^2 + \overline{IC}^2$; but by hypothesis IL : IK : KL = a : z : n; then $a^2 + n^2 = z^2$. Now the equation of the curve is

$$y = m - \frac{n}{z}x + x\sqrt{m^2 + oz - \frac{p}{m}x^2},$$

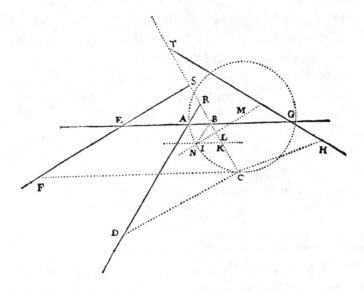

If the conic section is a parabola, its latus rectum is equal to $\frac{oz}{a}$ and its axis always lies along the line IL.[103] To find its vertex, N, make IN equal to $\frac{am^2}{oz}$, so that the point I lies between L and N if m^2 is positive and ox is positive; and L lies between I and N if m^2 is positive and ox negative; and N lies between I and L if m^2 is negative and ox positive. It is impossible that m^2 should be negative when the terms' are arranged as above. Finally, if m^2 is equal to zero, the points N and I must coincide. It is thus easy to determine this parabola, according to the first problem of the first book of Apollonius[104].

If, however, the required locus is a circle, an ellipse, or a hyperbola,[105] the point M, the center of the figure, must first be found. This will always lie on the line IL and may be found by taking IM equal to $\frac{aom}{2pz}$. If o is equal to zero M coincides with I. If the required locus is a circle or an ellipse, M and L must lie on the same side of I when the term ox is positive and on opposite sides when ox is negative. On the

and therefore the term in x^2 is

$$\left(\frac{n^2}{z^2} + \frac{p}{m}\right)x^2;$$

and if $a^2m = pz^2$, then $\frac{p}{m} = \frac{a^2}{z^2}$, and this term in x^2 becomes $\frac{a^2 + n^2}{z^2}x^2 = x^2$.

Therefore, the coefficients of x^2 and y^2 are unity and the locus is a circle.

103. This may be seen as follows: From the figure, and by the nature of the parabola $\overline{LC}^2 = LN \cdot p$ and $LN = IL + IN$. Let $IN = \phi$; then since $IL = \frac{a}{z}x$, we have $LN = \frac{a}{z}x + \phi$ and $LC = y - m + \frac{n}{z}x$; whence $(y - m + \frac{n}{z}x)^2 = (\frac{a}{z}x + \phi)p$. But $(y - m + \frac{n}{z}x)^2 = m^2 + ox$ from the equation of the parabola; therefore $\frac{a}{z}xp + \phi p = m^2 + ox$. Equating coefficients, we have $\frac{a}{z}p = o$; $p = \frac{oz}{a}$; $\phi p = m^2$; $\phi\frac{oz}{a} = m^2$; $\phi = \frac{am^2}{oz}$.

104. *Apollonii Pergacii Quae Graece exstant* edidit I. L. Heiberg, Leipzig, 1891. Vol. I, Liber I, Prop. LII. Hereafter referred to as Apollonius. This may be freely translated as follows: To describe in a plane a parabola, having given the parameter, the vertex, and the angle between an ordinate and the corresponding abscissa.

105. Central conics are thus grouped together by Descartes, the circle being treated as a special form of the ellipse, but being mentioned separately in all cases.

other hand, in the case of the hyperbola, M and L lie on the same side of I when *ox* is negative and on opposite sides when *ox* is positive.

The latus rectum of the figure must be

$$\sqrt{\frac{o^2 z^2}{a^2} + \frac{4mpz^2}{a^2}}$$

if m^2 is positive and the locus is a circle or an ellipse, or if m^2 is negative and the locus is a hyperbola. It must be

$$\sqrt{\frac{o^2 z^2}{a^2} - \frac{4mpz^2}{a^2}}$$

if the required locus is a circle or an ellipse and m^2 is negative, or if it is an hyperbola and o^2 is greater than $4mp$, m^2 being positive.

But if m^2 is equal to zero, the latus rectum is $\frac{oz}{a}$; and if *oz* is equal to zero[106], it is

$$\sqrt{\frac{4mpz^2}{a^2}}.$$

For the corresponding diameter a line must be found which bears the ratio $\frac{a^2 m}{pz^2}$ to the latus rectum; that is, if the latus rectum is

$$\sqrt{\frac{o^2 z^2}{a^2} + \frac{4mpz^2}{a^2}}$$

the diameter is

$$\sqrt{\frac{a^2 o^2 m^2}{p^2 z^2} + \frac{4a^2 m^3}{pz^2}}.$$

In every case, the diameter of the section lies along IM, and LC is one of its lines applied in order.[107] It is thus evident that, by making MN equal to half the diameter and taking N and L on the same side of M, the point N will be the vertex of this diameter.[108] It is then a simple matter to determine the curve, according to the second and third problems of the first book of Apollonius.[109]

When the locus is a hyperbola[110] and m^2 is positive, if o^2 is equal to zero or less than $4pm$ we must draw the line MOP from the center M parallel to LC, and draw CP parallel to LM, and take MO equal to

$$\sqrt{m^2 - \frac{o^2 m}{4p}};$$

while if *ox* is equal to zero, MO must be taken equal to *m*. Then considering O as the vertex of this hyperbola, the diameter being OP and the line applied in order

106. Some editions give, incorrectly, *ox* for *os*.
107. See note 100.
108. If the equation contains $-m^2$ and $+nx$, then n^2 must be greater than $4mp$, otherwise the problem is impossible.
109. Cf. Apollonius, Vol. I, Lib. I, Prop. LV: To describe a hyperbola, given the axis, the vertex, the parameter, and the angle between the axes. Also see Prop. LVI: To describe an ellipse, etc.
110. Cf. Letters of Descartes, Cousin, Vol. VIII.

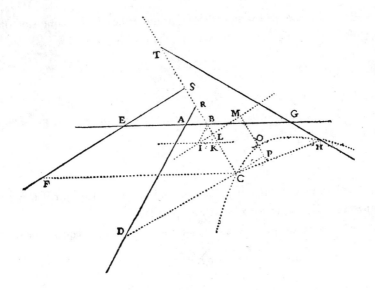

being CP, its latus rectum is

$$\sqrt{\frac{4a^4m^4}{p^2z^4} - \frac{a^4o^2m^3}{p^3z^4}}$$

and its diameter[111] is

$$\sqrt{4m^2 - \frac{o^2m}{p}}.$$

An exception must be made when ox is equal to zero, in which case the latus rectum is $\frac{2a^2m^2}{pz^2}$ and the diameter is $2m$. From these data the curve can be determined in accordance with the third problem of the first book of Apollonius.[112]

The demonstrations of the above statements are all very simple, for, forming the product[113] of the quantities given above as latus rectum, diameter, and segment of the diameter NL or OP, by the methods of Theorems 11, 12, and 13 of the first book of Apollonius, the result will contain exactly the terms which express the square of the line CP or CL, which is an ordinate of this diameter.

In this case take IM or $\frac{aom}{2pz}$ from NM or from its equal

$$\frac{am}{2pz}\sqrt{o^2 + 4mp}.$$

To the remainder IN add IL or $\frac{a}{z}x$, and we have

$$NL = \frac{a}{z}x - \frac{aom}{2pz} + \frac{am}{2pz}\sqrt{o^2 + 4mp}.$$

111. "Côté traversant."
112. See note 104.
113. "Composant un espace."

Multiplying this by

$$\frac{z}{a}\sqrt{o^2 + 4mp},$$

the latus rectum of the curve, we get

$$x\sqrt{o^2 + 4mp} - \frac{om}{2p}\sqrt{o^2 + 4mp} + \frac{mo^2}{2p} + 2m^2$$

for the rectangle, from which is to be subtracted a rectangle which is to the square of NL as the latus rectum is to the diameter. The square of NL is

$$\frac{a^2}{2^2}x^2 - \frac{a^2om}{pz^2}x + \frac{a^2m}{pz^2}x\sqrt{o^2 + 4mp} + \frac{a^2o^2m^2}{2p^2z^2} + \frac{a^2m^3}{pz^2} - \frac{a^2om^2}{2p^2z^2}\sqrt{o^2 + 4mp}.$$

Divide this by a^2m and multiply the quotient by pz^2, since these terms express the ratio between the diameter and the latus rectum. The result is

$$\frac{p}{m}x^2 - ox + x\sqrt{o^2 + 4mp} + \frac{o^2m}{2p} - \frac{om}{2p}\sqrt{o^2 + 4mp} + m^2.$$

This quantity being subtracted from the rectangle previously obtained, we get

$$\overline{CL}^2 = m^2 + ox - \frac{p}{m}x^2.$$

It follows that CL is an ordinate of an ellipse or circle applied to NL, the segment of the axis.

Suppose all the given quantities expressed numerically, as EA = 3, AG = 5, AB = BR, BS = $\frac{1}{2}$BE, GB = BT, CD = $\frac{3}{2}$CR, CF = 2CS, CH = $\frac{2}{3}$CT, the angle ABR = 60°; and let CB · CF = CD · CH. All these quanties must be known if the problem is to be entirely determined. Now let AB = x, and CB = y. By the method given above we shall obtain

$$y^2 = 2y - xy + 5x - x^2;$$

$$y = 1 - \frac{1}{2}x + \sqrt{1 + 4x - \frac{3}{4}x^2};$$

whence BK must be equal to 1, and KL must be equal to one-half KI; and since the angle IKL = angle ABR = 60° and angle KIL (which is one-half angle KIB or one-half angle IKL) is 30°, the angle ILK is a right angle. Since IK = AB = x, KL = $\frac{1}{2}x$, IL = $x\sqrt{\frac{3}{4}}$, and the quantity represented by z above is 1, we have $a = \sqrt{\frac{3}{4}}$, $m = 1$, $o = 4$, $p = \frac{3}{4}$, whence IM = $\sqrt{\frac{16}{3}}$, NM = $\sqrt{\frac{19}{3}}$; and since a^2m (which is $\frac{3}{4}$) is equal to pz^2, and the angle ILC is a right angle, it follows that the curve NC is a circle. A similar treatment of any of the other cases offers no difficulty.

Since all equations of degree not higher than the second are included in the discussion just given, not only is the problem of the ancients relating to three or four lines completely solved, but also the whole problem of what they called the composition of solid loci, and consequently that of plane loci, since they are included under

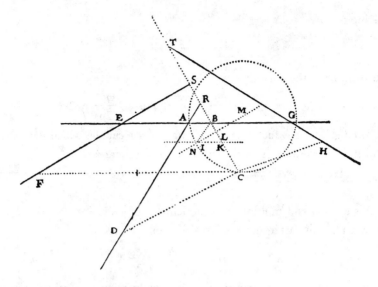

solid loci.[114] For the solution of any one of these problems of loci is nothing more than the finding of a point for whose complete determination one condition is wanting, the other conditions being such that (as in this example) all the points of a single line will satisfy them. If the line is straight or circular, it is said to be a plane locus; but if it is a parabola, a hyperbola, or an ellipse, it is called a solid locus. In every such case an equation can be obtained containing two unknown quantities and entirely analogous to those found above. If the curve upon which the required point lies is of higher degree than the conic sections, it may be called in the same way a supersolid locus,[115] and so on for other cases. If two conditions for the determination of the point are lacking, the locus of the point is a surface, which may be plane, spherical, or more complex. The ancients attempted nothing beyond the composition of solid loci, and it would appear that the sole aim of Apollonius in his treatise on the conic sections was the solution of problems of solid loci.

I have shown, further, that what I have termed the first class of curves contains no others besides the circle, the parabola, the hyperbola, and the ellipse. This is what I undertook to prove.

114. Since plane loci are degenerate cases of solid loci. The case in which neither x^2 nor y^2 but only xy occurs, and the case in which a constant term occurs, are omitted by Descartes. The various kinds of solid loci represented by the equation $y = \pm m \pm \dfrac{n}{z}x \pm \dfrac{n^2}{x} \pm \sqrt{\pm m^2 \pm ox \pm \dfrac{p}{m}x}$ may be summarized as follows: (1) If all the terms of the right member are zero except $\dfrac{n^2}{x}$, the equation represents an hyperbola referred to its asymptotes. (2) If $\dfrac{n^2}{x}$ is not present, there are several cases, as follows: (a) If the quantity under the radical sign is zero or a perfect square, the equation represents a straight line; (b) If this quantity is not a perfect square and if $\dfrac{p}{m}x^2 = 0$, the equation represents a parabola; (c) If it is not a perfect square and if $\dfrac{p}{m}x^2$ is negative, the equation represents a circle or an ellipse; (d) If $\dfrac{p}{m}x^2$ is positive, the equation represents a hyperbola.

115. "Un lieu sursolide."

If the problem of the ancients be proposed concerning five lines, all parallel, the required point will evidently always lie on a straight line. Suppose it be proposed concerning five lines with the following conditions:

(1) Four of these lines parallel and the fifth perpendicular to each of the others,

(2) The lines drawn from the required point to meet the given lines at right angles;

(3) The parallelepiped[116] composed of the three lines drawn to meet three of the parallel lines must be equal to that composed of three lines, namely, the one drawn to meet the fourth parallel, the one drawn to meet the perpendicular, and a certain given line.

This is, with the exception of the preceding one, the simplest possible case. The point required will lie on a curve generated by the motion of a parabola in the following way:

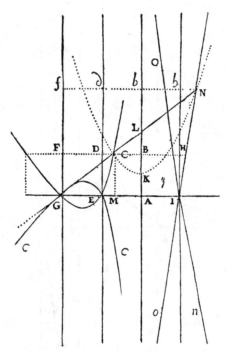

Let the required lines be AB, IH, ED, GF, and GA, and let it be required to find the point C, such that if CB, CF, CD, CH, and CM be drawn perpendicular respectively to the given lines, the parallelepiped of the three lines CF, CD, and CH shall be equal to that of the other two, CB and CM, and a third line AI. Let CB = y, CM = x, AI or AE or GE = a; whence if C lies between AB and DE, we have CF = $2a - y$, CD = $a - y$, and CH = $y + a$. Multiplying these three

116. That is, the product of the numerical measures of these lines.

together we get $y^3 - 2ay^2 - a^2y + 2a^3$ equal to the product of the other three, namely to *axy*.

I shall consider next the curve CEG, which I imagine to be described by the intersection of the parabola CKN (which is made to move so that its axis KL always lies along the straight line AB) with the ruler GL (which rotates about the point G in such a way that it constantly lies in the plane of the parabola and passes through the point L). I take KL equal to *a* and let the principal parameter, that is, the parameter corresponding to the axis of the given parabola, be also equal to *a*, and let GA = 2*a*, CB or MA = *y*, CM or AB = *x*. Since the triangles GMC and CBL are similar, GM (or 2*a* − *y*) is to MC (or *x*) as CB (or *y*) is to BL, which is therefore equal to $\frac{xy}{2a-y}$. Since KL is *a*, BK is $a - \frac{xy}{2a-y}$ or $\frac{2a^2 - ay - xy}{2a-y}$. Finally, since this same BK is a segment of the axis of the parabola, BK is to BC (its ordinate) as BC is to *a* (the latus rectum), whence we get $y^3 - 2ay^2 - a^2y + 2a^3 = axy$, and therefore C is the required point.

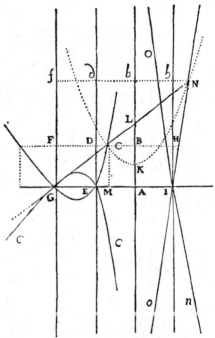

The point C can be taken on any part of the curve CEG or of its adjunct *c*EG*c*, which is described in the same way as the former, except that the vertex of the parabola is turned in the opposite direction; or it may lie on their counterparts[117] NI*o* and *n*IO, which are generated by the intersection of the line GL with the other branch of the parabola KN.

117. "En leurs contreposées."

Again, suppose that the given parallel lines AB, IH, ED, and GF are not equally distant from one another and are not perpendicular to GA, and that the lines through C are oblique to the given lines. In this case the point C will not always lie on a curve of just the same nature. This may even occur when no two of the given lines are parallel.

Next, suppose that we have four parallel lines, and a fifth line cutting them, such that the parallelepiped of three lines drawn through the point C (one to the cutting line and two to two of the parallel lines) is equal to the parallelepiped of two lines drawn through C to meet the other two parallels respectively and another given line. In this case the required point lies on a curve of different nature,[118] namely, a curve such that, all the ordinates to its axis being equal to the ordinates of a conic section, the segments of the axis between the vertex and the ordinates[119] bear the same ratio to a certain given line that this line bears to the segments of the axis of the conic section having equal ordinates.[120]

I cannot say that this curve is less simple than the preceding; indeed, I have always thought the former should be considered first, since its description and the determination of its equation are somewhat easier.

I shall not stop to consider in detail the curves corresponding to the other cases, for I have not undertaken to give a complete discussion of the subject; and having explained the method of determining an infinite number of points lying on any curve, I think I have furnished a way to describe them.

It is worthy of note that there is a great difference between this method[121] in which the curve is traced by finding several points upon it, and that used for the spiral and similar curves.[122] In the latter not any point of the required curve may be found at pleasure, but only such points as can be determined by a process simpler than that required for the composition of the curve. Therefore, strictly speaking, we do not find any one of its points, that is, not any one of those which are so peculiarly points of this curve that they cannot be found except by means of it. On the other hand, there is no point on these curves which supplies a solution for the proposed problem that cannot be determined by the method I have given.

But the fact that this method of tracing a curve by determining a number of its points taken at random applies only to curves that can be generated by a regular and continuous motion does not justify its exclusion from geometry. Nor should we reject the method[123] in which a string or loop of thread is used to determine the equality or

118. The general equation of this curve is $axy - xy^2 + 2a^2x = a^2y - ay^2$.
119. That is, the abscissas of points on the curve.
120. The thought, expressed in modern phraseology, is as follows: The curve is of such nature that the abscissa of any point on it is a third proportional to the abscissa of a point on a conic section whose ordinate is the same as that of the given point, and a given line.
121. That is, the method of analytic geometry.
122. That is, transcendental curves, called by Descartes "mechanical" curves.
123. Cf. the familiar "mechanical descriptions" of the conic sections.

difference of two or more straight lines drawn from each point of the required curve to certain other points,[124] or making fixed angles with certain other lines. We have used this method in "La Dioptrique"[125] in the discussion of the ellipse and the hyperbola.

On the other hand, geometry should not include lines that are like strings, in that they are sometimes straight and sometimes curved, since the ratios between straight and curved lines are not known, and I believe cannot be discovered by human minds,[126] and therefore no conclusion based upon such ratios can be accepted as rigorous and exact. Nevertheless, since strings can be used in these constructions only to determine lines whose lengths are known, they need not be wholly excluded.

When the relation between all points of a curve and all points of a straight line is known,[127] in the way I have already explained, it is easy to find the relation between the points of the curve and all other given points and lines; and from these relations to find its diameters, axes, center and other lines[128] or points which have especial significance for this curve, and thence to conceive various ways of describing the curve, and to choose the easiest.

By this method alone it is then possible to find out all that can be determined about the magnitude of their areas,[129] and there is no need for further explanation from me.

Finally, all other properties of curves depend only on the angles which these curves make with other lines. But the angle formed by two intersecting curves can be as easily measured as the angle between two straight lines, provided that a straight line can be drawn making right angles with one of these curves at its point of intersection with the other.[130] This is my reason for believing that I shall have given here a sufficient introduction to the study of curves when I have given a general method of drawing a straight line making right angles with a curve at an arbitrarily chosen point upon it. And I dare say that this is not only the most useful and most general problem in geometry that I know, but even that I have ever desired to know.

Let CE be the given curve, and let it be required to draw through C a straight line making right angles with CE. Suppose the problem solved, and let the required line be CP. Produce CP to meet the straight line GA, to whose points the points of CE are to be

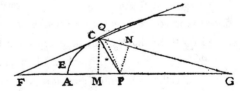

124. As for example, the foci, in the description of the ellipse.
125. This work was published at Leyden in 1637, together with Descartes's *Discours de la Methode*.
126. This is of course concerned with the problem of the rectification of curves. See Cantor, Vol. II (1). This statement, "ne pouvant être par les hommes" is a very noteworthy one, coming as it does from a philosopher like Descartes. On the philosophical question involved, consult such writers as Bertrand Russell.
127. Expressed by means of the equation of the curve.
128. For example, the equations of tangents, normals, etc.
129. For the history of the quadrature of curves, consult Cantor, Vol. II (1), Smith, *History*, Vol. II.
130. That is, the angle between two curves is defined as the angle between the normals to the curve at the point of intersection.

related.[131] Then, let MA = CB = y; and CM = BA = x. An equation must be found expressing the relation between x and y.[132] I let PC = s, PA = v, whence PM = $v - y$. Since PMC is a right triangle, we see that s^2, the square of the hypotenuse, is equal to $x^2 + v^2 - 2vy + y^2$, the sum of the squares of the two sides. That is to say, $x = \sqrt{s^2 - v^2 + 2vy - y^2}$ or $y = v + \sqrt{s^2 - x^2}$. By means of these last two equations, I can eliminate one of the two quantities x and y from the equation expressing the relation between the points of the curve CE and those of the straight line GA. If x is to be eliminated, this may easily be done by replacing x wherever it occurs by $\sqrt{s^2 - v^2 + 2vy - y^2}$, x^2 by the square of this expression, x^3 by its cube, etc., while if y is to be eliminated, y must be replaced by $v + \sqrt{s^2 - x^2}$, and y^2, y^3, . . . by the square of this expression, its cube, and so on. The result will be an equation in only one unknown quantity, x or y.

For example, if CE is an ellipse, MA the segment of its axis of which CM is an ordinate, r its latus rectum, and q its transverse axis,[133] then by Theorem 13, Book I, of Apollonius,[134] we have $x^2 = ry - \frac{r}{q}y^2$. Eliminating x^2 the resulting equation is

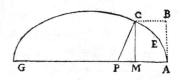

$$s^2 - v^2 + 2vy - y^2 = ry - \frac{r}{q}y^2, \text{ or } y^2 + \frac{qry - 2qvy + qv^2 - qs^2}{q - r} = 0.$$

In this case it is better to consider the whole as constituting a single expression than as consisting of two equal parts.[135]

If CE be the curve generated by the motion of a parabola already discussed, and if we represent GA by b, KL by c, and the parameter of the axis KL of the parabola by d, the equation expressing the relation between x and y is $y^3 - by^2 - cdy + bcd + dxy = 0$. Eliminating x, we have

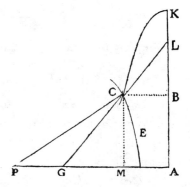

$$y^3 - by^2 - cdy + bcd + dy\sqrt{s^2 - v^2 + 2vy - y^2} = 0.$$

---

131. That is, the line GA is taken as one of the coordinate axes.

132. This will be the equation of the curve.

133. "Le traversant."

134. Apollonius: "Si conus per axem plano secatur autem alio quoque plano, quod cum utroque latere trianguli per axem posita concurrit, sed neque basi coni parallelum ducitur neque e contrario et si planum, in quo est basis coni, planumque secans concurrunt in recta perpendiculari aut ad basim trianguli per axem positi aut ad eam productam quælibet recta, quæ a sectione coni communi sectioni planorum parallela ducitur ad diametrum sectiones sumpta quadrata æqualis erit spatio adplicato rectæ cuidam, ad quam diametrus sectionis rationem habet, quam habet quadratum rectæ a vertice coni diametro sectionis parallelæ ductæ usque ad basim trianguli per rectangulum comprehensum rectis ab ea ad latera trianguli abscissis, latitudinem rectam ab ea e diametro ad verticem sectionis abscissam et figura deficiens simili similiterque posita rectangulo a diametro parametroque comprehenso; vocetur autem talis sectio ellipsis." Cf. *Apollonius of Perga*, edited by Sir T. L. Heath, Cambridge, 1896.

135. That is, to transpose all the terms to the left member.

Arranging the terms according to the powers of y by squaring,[136] this becomes

$$y^6 - 2by^5 + (b^2 - 2cd + d^2)y^4 + (4bcd - 2d^2v)y^3$$
$$+ (c^2d^2 - d^2s^2 + d^2v^2 - 2b^2cd)y^2 - 2bc^2d^2y + b^2c^2d^2 = 0,$$

and so for the other cases. If the points of the curve are not related to those of a straight line in the way explained, but are related in some other way,[137] such an equation can always be found.

Let CE be a curve which is so related to the points F, G, and A, that a straight line drawn from any point on it, as C, to F exceeds the line FA by a quantity which bears a given ratio to

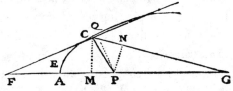

the excess of GA over the line drawn from the point C to G.[138] Let GA = b, AF = c, and taking an arbitrary point C on the curve let the quantity by which CF exceeds FA be to the quantity by which GA exceeds GC as d is to e. Then if we let z represent the undetermined quantity, FC = $c + z$ and GC = $b - \frac{e}{d}z$. Let MA = y, GM = $b - y$, and FM = $c + y$. Since CMG is a right triangle, taking the square of GM from the square of GC we have left the square of CM, or $\frac{e^2}{d^2}z^2 - \frac{2be}{d}z + 2by - y^2$. Again, taking the square of FM from the square of FC we have the square of CM expressed in another way, namely: $z^2 + 2cz - 2cy - y^2$. These two expressions being equal they will yield the value of y or MA, which is

$$\frac{d^2z^2 + 2cd^2z - e^2z^2 + 2bdez}{2bd^2 + 2cd^2}.$$

Substituting this value for y in the expression for the square of CM, we have

$$\overline{CM}^2 = \frac{bd^2z^2 + ce^2z^2 + 2bcd^2z - 2bcdez}{bd^2 + cd^2} - y^2.$$

If now we suppose the line PC to meet the curve at right angles at C, and let PC = s and PA = v as before, PM is equal to $v - y$; and since PCM is a right triangle, we have $s^2 - v^2 + 2vy - y^2$ for the square of CM. Substituting for y its value, and equating the values of the square of CM, we have

$$z^2 + \frac{2bcd^2z - 2bcdez - 2cd^2vz - 2bdevz - bd^2s^2 + bd^2v^2 - cd^2s^2 + cd^2v^2}{bd^2 + ce^2 + e^2v - d^2v} = 0$$

for the required equation.

Such an equation having been found[139] it is to be used, not to determine x, y, or z, which are known, since the point C is given, but to find v or s, which determine the required point P. With this in view, observe that if the point P fulfills the required

136. "En remettant en ordre ces termes par moyen de la multiplication."
137. "Mais en toute autre qu'on saurait imaginer."
138. That is the ratio of CF − FA to GA − CG is a constant.
139. Three such equations have been found by Descartes, namely those for the ellipse, the parabolic conchoid, and the curve just described.

conditions, the circle about P as center and passing through the point C will touch but not cut the curve CE; but if this point P be ever so little nearer to or farther from A than it should be, this circle must cut the curve not only at C but also in another point. Now if this circle cuts CE, the equation involving x and y as unknown quantities (supposing PA and PC known) must have two unequal roots. Suppose, for example, that the circle cuts the curve in the points C and E. Draw EQ parallel to CM. Then x and y may be used to represent EQ and QA respectively in just the same way as they were used to represent CM and MA; since PE is equal to PC (being radii of

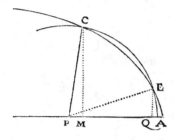

the same circle), if we seek EQ and QA (supposing PE and PA given) we shall get the same equation that we should obtain by seeking CM and MA (supposing PC and PA given). It follows that the value of x, or y, or any other such quantity, will be two-fold in this equation, that is, the equation will have two unequal roots. If the value of x be required, one of these roots will be CM and the other EQ; while if y be required, one root will be MA and the other QA. It is true that if E is not on the same side of the curve as C, only one of these will be a true root, the other being drawn in the opposite direction, or less than nothing.[140] The nearer together the points C and E are taken however, the less difference there is between the roots; and when the points coincide, the roots are exactly equal, that is to say, the circle through C will touch the curve CE at the point C without cutting it.

Furthermore, it is to be observed that when an equation has two equal roots, its left-hand member must be similar in form to the expression obtained by multiplying by itself the difference between the unknown quantity and a known quantity equal to it;[141] and then, if the resulting expression is not of as high a degree as the original equation, multiplying it by another expression which will make it of the same degree. This last step makes the two expressions correspond term by term.

For example, I say that the first equation found in the present discussion,[142] namely

$$y^2 + \frac{qry - 2qvy + qv^2 - qs^2}{q - r},$$

must be of the same form as the expression obtained by making $e = y$ and multiplying $y - e$ by itself, that is, as $y^2 - 2ey + e^2$. We may then compare the two

140. "Et l'autre sera renversée ou moindre que rien."
141. That is, the left-hand member will be the square of the binomial $x - a$ when $x = a$.
142. The original has "first equation," not "first member of the equation."

expressions term by term, thus: Since the first term, y^2, is the same in each, the second term,[143] $\frac{qry - 2qvy}{q - r}$, of the first is equal to $-2ey$, the second term of the second; whence, solving for v, or PA,

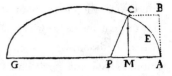

we have $v = e - \frac{r}{q}e + \frac{1}{2}r$; or, since we have assumed e equal to y, $v = y - \frac{r}{q}y + \frac{1}{2}r$. In the same way, we can find s from the third term, $e^2 = \frac{qv^2 - qs^2}{q - r}$; but since v completely determines P, which is all that is required, it is not necessary to go further.[144]

In the same way, the second equation found above,[145] namely,

$$y^6 - 2by^5 + (b^2 - 2cd + d^2)y^4 + (4bcd - 2d^2v)y^3$$
$$+ (c^2d^2 - 2b^2cd + d^2v^2 - d^2s^2)y^2 - 2bc^2d^2y + b^2c^2d^2,$$

must have the same form as the expression obtained by multiplying

$$y^2 - 2ey + e^2 \text{ by } y^4 + fy^3 + g^2y^2 + h^3y + k^4,$$

that is, as

$$y^6 + (f - 2e)y^5 + (g^2 - 2ef + e^2)y^4 + (h^3 - 2eg^2 + e^2f)y^3$$
$$+ (k^4 - 2eh^3 + e^2g^2)y^2 + (e^2h^3 - 2ek^4)y + e^2k^4.$$

From these two equations, six others may be obtained, which serve to determine the six quantities f, g, h, k, v, and s. It is easily seen that to whatever class the given curve may belong, this method will always furnish just as many equations as we necessarily have unknown quantities. In order to solve these equations, and ultimately to find v, which is the only value really wanted (the others being used only as means of finding v), we first determine f, the first unknown in the above expression, from the second term. Thus, $f = 2e - 2b$. Then in the last terms we can find k, the last unknown in the same expression, from which $k^4 = \frac{b^2c^2d^2}{e^2}$. From the third term we get the second quantity

$$g^2 = 3e^2 - 4be - 2cd + b^2 + d^2.$$

From the next to the last term we get h, the next to the last quantity, which is[146]

$$h^3 = \frac{2b^2c^2d^2}{e^3} - \frac{2bc^2d^2}{e^2}.$$

In the same way we should proceed in this order, until the last quantity is found.

Then from the corresponding term (here the fourth) we may find v, and we have

$$v = \frac{2e^3}{d^2} - \frac{3be^2}{d^2} + \frac{b^2e}{d^2} - \frac{2ce}{d} + e + \frac{2bc}{d} + \frac{bc^2}{e^2} - \frac{b^2c^2}{e^3};$$

143. That is, the second term in y.
144. That is, to construct PC we may lay off AP $= v$ and join P and C. If instead we use the value of e, taking C as center and a radius CP $= e$, we construct an arc cutting AG in P, and join P and C. To apply Descartes's method to the circle, for example, it is only necessary to observe that all parameters and diameters are equal, that is, $q = r$; and therefore the equation $v = y - \frac{r}{q}y + \frac{1}{2}r$ becomes $v = \frac{1}{2}q = \frac{1}{2}$ diameter. That is, the normal passes through the center and is a radius of the circle.
145. As before, Descartes uses "second equation" for "first member of the second equation."
146. Found from.

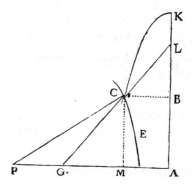

or putting y for its equal e, we get

$$v = \frac{2y^3}{d^2} - \frac{3by^2}{d^2} + \frac{b^2y}{d^2} - \frac{2cy}{d} + y + \frac{2bc}{d} + \frac{bc^2}{y^2} - \frac{b^2c^2}{y^3},$$

for the length of AP.

Again, the third[147] equation, namely,

$$z^2 + \frac{2bcd^2z - 2bcdez - 2cd^2vz - 2bdevz - bd^2s^2 + bd^2v^2 - cd^2s^2 + cd^2v^2}{bd^2 + ce^2 + e^2v - d^2v},$$

is of the same form as $z^2 - 2fz + f^2$ where $f = z$, so that $-2f$ or $-2z$ must be equal to

$$\frac{2bcd^2 - 2bcde - 2cd^2v - 2bdev}{bd^2 + ce^2 + e^2v - d^2v},$$

whence

$$v = \frac{bcd^2 - bcde + bd^2z + ce^2z}{cd^2 + bde - e^2z + d^2z}.$$

Therefore, if we take AP equal to the above value of v, all the terms of which are known, and join the point P thus determined to C, this line will cut the curve CE at right angles, which was required. I see no reason why this solution should not apply to every curve to which the methods of geometry are applicable.[148]

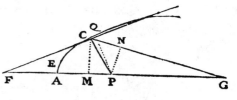

It should be observed regarding the expression taken arbitrarily to raise the original product to the required degree, as we just now took

$$y^4 + fy^3 + g^2y^2 + h^3y + k^4,$$

147. First member of the third equation.

148. Let us apply this method to the problem of constructing a normal to a parabola at a given point. As before, $s^2 = x^2 + v^2 - 2vy + y^2$. If we take as the equation of the parabola $x^2 = ry$, and substitute, we have

$$s^2 = ry + v^2 - 2vy + y^2 \text{ or } y^2 + (r - 2v)y + v^2 - s^2 = 0.$$

Comparing this with $y^2 - 2ey + e^2 = 0$, we have $r - 2v = -2e$; $v^2 - s^2 = e^2$; $v = \frac{r}{2} + e$. Since $e = y$, $v = \frac{r}{2} + y$. Let $AM = y$. and $v = AP$; then $AM - AP = MP = $ one-half the parameter.

that the signs + and − may be chosen at will, without producing different values of v or AP.[149] This is easily found to be the case, but if I should stop to demonstrate every theorem I use, it would require a much larger volume than I wish to write. I desire rather to tell you in passing that this method, of which you have here an example, of supposing two equations to be of the same form in order to compare them term by term and so to obtain several equations from one, will apply to an infinity of other problems and is not the least important feature of my general method.[150]

I shall not give the constructions for the required tangents and normals in connection with the method just explained, since it is always easy to find them, although it often requires some ingenuity to get short and simple methods of construction.

Given, for example, CD, the first conchoid of the ancients. Let A be its pole and BH the ruler, so that the segments of all straight lines, as CE and DB, converging toward A and included between the curve CD and the straight line BH are equal. Let it be required to find a line CG normal to the curve at the point C. In trying to find the point on BH through which CG must pass (according to the method just

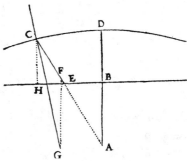

149. It will be observed that Descartes did not consider a coefficient, as a, in the general sense of a positive or a negative quantity, but that he always wrote the sign intended. In this sentence, however, he suggests some generalization.

150. The method may be used to draw a normal to a curve from a given point, to draw a tangent to a curve from a point without, and to discover points of inflexion, maxima, and minima. Compare Descartes's Letters, Cousin, Vol. VI. As an illustration, let it be required to find a point of inflexion on the first cubical parabola. Its equation is $y^3 = a^2 x$. Assume that D is a point of inflexion, and let CD = y, AC = x, PA = s, and AE = r.

Since triangle PAE is similar to triangle PCD we have $\dfrac{y}{x+s} = \dfrac{r}{s}$, whence $x = \dfrac{sy-rs}{r}$. Substituting in the equation of the curve, we have

$y^3 - \dfrac{a^2 s y}{r} + a^2 s = 0$. But if D is a point of inflexion this equation must have three equal roots, since at a point of inflexion there are three coincident

points of section. Compare the equation with

$$y^3 - 3ey^2 + 3e^2 y - e^3 = 0.$$

Then $3e^2 = 0$ and $e = 0$. But $e = y$, and therefore $y = 0$. Therefore the point of inflexion is (0, 0).

It will be of interest to compare the method of drawing tangents given by Fermat in *Methodus ad disquirendam maximam et minimam*, Toulouse, 1679, which is as follows: It is required to draw a tangent to the parabola BD from a point O without. From the nature of the parabola $\dfrac{CD}{DI} > \dfrac{BC^2}{OI^2}$, since O is without the curve. But by similar triangles $\dfrac{\overline{BC}^2}{\overline{OI}^2} = \dfrac{\overline{CE}^2}{\overline{IE}^2}$. Therefore $\dfrac{CD}{DI} > \dfrac{\overline{CE}^2}{\overline{IE}^2}$. Let CE = a, CI = e, and CD = d; then DI = $d - e$, and

$\dfrac{d}{d-e} > \dfrac{a^2}{(a-e)^2}$; whence

$$de^2 - 2ade > -a^2 e.$$

Dividing by e, we have $de - 2ad > -a^2$. Now if the line BO becomes tangent to the curve, the point B and O coincide, $de - 2ad = -a^2$, and e vanishes; then $2ad = a^2$ and $a = 2d$ in length. That is CE = 2CD.

explained), we would involve ourselves in a calculation as long as, or longer than any of those just given, and yet the resulting construction would be very simple. For we need only take CF on CA equal to CH, the perpendicular to BH; then through F draw FG parallel to BA and equal to EA, thus determining the point G, through which the required line CG must pass.

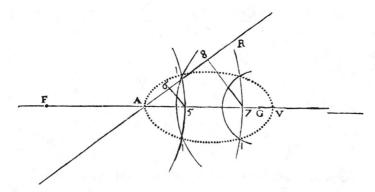

To show that a consideration of these curves is not without its use, and that they have diverse properties of no less importance than those of the conic sections I shall add a discussion of certain ovals which you will find very useful in the theory of catoptrics and dioptrics. They may be described in the following way: Drawing the two straight lines FA and AR intersecting at A under any angle, I choose arbitrarily a point F on one of them (more or less distant from A according as the oval is to be large or small). With F as center I describe a circle cutting FA at a point a little beyond A, as at the point 5. I then draw the straight line 56[151] cutting AR at 6, so that A6 is less than A5, and so that A6 is to A5 in any given ratio, as, for example, that which measures the refraction,[152] if the oval is to be used for dioptrics. This being done, I take an arbitrary point G in the line FA on the same side as the point 5, so that AF is to GA in any given ratio. Next, along the line A6 I lay off RA equal to GA, and with G as center and a radius equal to R6 I describe a circle. This circle will cut the first one in two points 1, 1,[153] through which the first of the required ovals must pass.

Next, with F as center I describe a circle which cuts FA as little nearer to or farther from A than the point 5, as, for example, at the point 7. I then draw 78 parallel to 56 and with G as center and a radius equal to R8 I describe another circle. This circle will cut the one through 7 in the points 1, 1[154] which are points of the same oval. We can thus find as many points as may be desired, by drawing lines parallel to 78 and describing circles with F and G as centers.

151. The confusion resulting from the use of Arabic figures to designate points is here apparent.
152. That is, the ratio corresponding to the index of refraction.
153. "Au point 1."
154. "Au point 1."

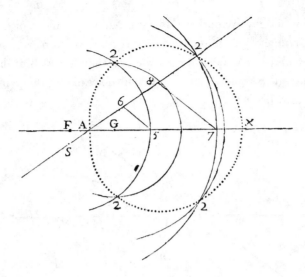

In the construction of the second oval the only difference is that instead of AR we must take AS on the other side of A, equal to AG, and that the radius of the circle about G cutting the circle about F and passing through 5 must be equal to the line S6; or if it is to cut the circle through 7 it must be equal to S8, and so on. In this way the circles intersect in the points 2, 2, which are points of this second oval A2X.

To construct the third and fourth ovals, instead of AG I take AH on the other side of A, that is, on the same side as F. It should be observed that this line AH must be greater than AF, which in any of these ovals may even be zero, in which case F and A coincide. Then, taking AR and AS each equal to AH, to describe the third oval, A3Y, I draw a circle about H as center with a radius equal to S6 and

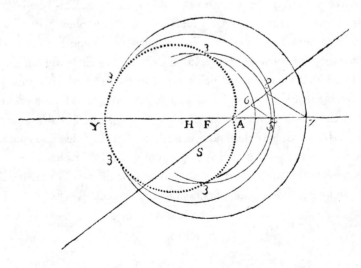

cutting in the point 3 the circle about F passing through 5, and another with a radius equal to S8 cutting the circle through 7 in the point also marked 3, and so on.

Finally, for the fourth oval, I draw circles about H as center with radii equal to R6, R8, and so on, and cutting the other circles in the points marked 4.[155]

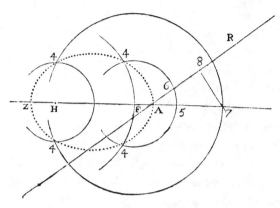

There are many other ways of describing these same ovals. For example, the first one, AV (provided we assume FA and AG equal) might be traced as follows: Divide the line FG at L so that FL:LG = A5:A6, that is, in the ratio corresponding to the index of refraction. Then bisecting AL at K, turn a ruler FE about the point F, pressing with the finger at C the cord EC, which, being attached at E to the end of the ruler, passes from C to K and then back to C and from C to G, where its other end is fastened. Thus the entire length of the cord is composed of GA + AL + FE − AF, and the point C will describe the first oval in a way similar to that in which the ellipse and hyperbola are described in *La Dioptrique*. But I cannot give any further attention to this subject.

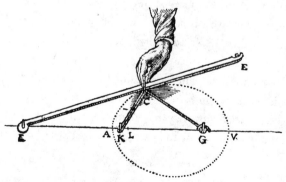

155. In all four ovals AF and AR or AF and AS intersect at A under any angle. F may coincide with A, and otherwise its distance from A determines the size of the oval. The ratio A5 : A6 is determined by the index of refraction of the material used. In the first two ovals, if A does not coincide with F it lies between F and G, and the ratio AF : AG is arbitrary. In the last two, if F does not coincide with A it lies between A and H, and the ratio AF : AH is arbitrary. In the first oval AR = AG and the points R, 6, 8 are on the same side of A. In the second oval AS = AG and S is on the opposite side of A from 6, 8. In the third oval AS = AH and S is on the opposite side of A from 6, 8. In the fourth oval AR = AH and R, 6, 8 are on the same side of A.

Athough these ovals seem to be of almost the same nature, they nevertheless belong to four different classes, each containing an infinity of sub-classes, each of which in turn contains as many different kinds as does the class of ellipses or of hyperbolas; the sub-classes depending upon the value of the ratio of A5 to A6. Then, as the ratio of AF to AG, or of AF to AH changes, the ovals of each sub-class change in kind, and the length of AG or AH determines the size of the oval.[156]

If A5 is equal to A6, the ovals of the first and third classes become straight lines; while among those of the second class we have all possible hyperbolas, and among those of the fourth all possible ellipses.[157]

In the case of each oval it is necessary further to consider two portions having different properties. In the first oval the portion toward A causes rays passing through the air from F to converge towards G upon meeting the convex surface 1A1 of a lens whose index of refraction, according to dioptrics, determines such ratios as that of A5 to A6, by means of which the oval is described.

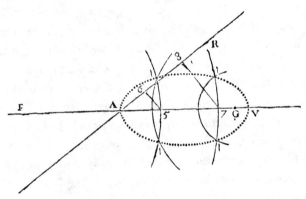

But the portion toward V causes all rays coming from G to converge toward F when they strike the concave surface of a mirror of the shape of 1V1 and of such material that it diminishes the velocity of these rays in the ratio of A5 to A6, for it is proved in dioptrics that in this case the angles of reflection will be unequal as well as the angles of refraction, and can be measured in the same way.

Now consider the second oval. Here, too, the portion 2A2 serves for reflections of which the angles may be assumed unequal. For if the surface of a mirror of the same material as in the case of the first oval be of this form, it will

156. Compare the changes in the ellipse and hyperbola as the ratio of the length of the transverse axis to the distance between the foci changes.
157. These theorems may be proved as follows: (1) Given the first oval, with A5 = A6; then RA = GA; FP = F5; GP = R6 = AR − R6 = GA − A5 = G5. Therefore FP + GP = F5 + G5. That is, the point P lies on the straight line FG. (2) Given the second oval, with A5 = A6; then F2 = F5 = FA + A5; G2 = S6 = SA + A6 = SA + A5; G2 − F2 = SA − FA = GA − FA = C. Therefore 2 lies on a hyperbola whose foci are F and G, and whose transverse axis is GA − FA. The proof for the third oval is analogous to (1) and that for the fourth to (2).
It may be noted that the first oval is the same curve as that described on page 327. For FP = F5, whence FP − AF = A5, and AR = AG; GP = R6; AG − GP = A6. If then A5 : A6 = d : e we have, as before,
$$FP − AF : AG − GP = d : e.$$

reflect all rays from G, making them seem to come from F. Observe, too, that if the line AG is considerably greater than AF, such a mirror will be convex in the center (toward A) and concave at each end; for such a curve would be heart-shaped rather than oval. The other part, X2, is useful for refracting lenses; rays which pass through the air toward F are refracted by a lens whose surface has this form.

The third oval is of use only for refraction, and causes rays traveling through the air toward F to move through the glass toward H, after they have passed through the surface whose form is A3Y3, which is convex throughout except toward A, where it is slightly concave, so that this curve is also heart-shaped. The difference between the two parts of this oval is that the one part is nearer F and farther from H, while the other is nearer H and farther from F.

Similarly, the last of these ovals is useful only in the case of reflection. Its effect is to make all rays coming from H and meeting the concave surface of a mirror of the same material as those previously discussed, and of the form A4Z4, converge towards F after reflection.

The points F, G and H may be called the "burning points"[158] of these ovals, to correspond to those of the ellipse and hyperbola, and they are so named in dioptrics.

I have not mentioned several other kinds of reflection and refraction that are effected[159] by these ovals; for being merely reverse or opposite effects they are easily deduced.

I must not, however, fail to prove the statements already made. For this purpose, take any point C on the first part of the first oval, and draw the straight line CP normal to the curve at

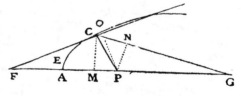

C. This can be done by the method given above, as follows:

Let AG = b, AF = c, FC = $c + z$. Suppose the ratio of d to e, which I always take here to measure the refractive power of the lens under consideration, to represent the ratio of A5 to A6 or similar lines used to describe the oval. Then

$$GC = b - \frac{e}{d}z,$$

whence

$$AP = \frac{bcd^2 - bcde + bd^2z + ce^2z}{bde + cd^2 + d^2z - e^2z}.$$

158. That is, the foci, from the Latin *focus*, "hearth." The word *focus* was first used in the geometric sense by Kepler, *Ad Vitellionem Paralipomena*, Frankfort, 1604. Chap. 4, Sect. 4.

159. "Reglées."

From P draw PQ perpendicular to FC, and PN perpendicular to GC.[160] Now if PO : PN = d : e, that is, if PQ : PN is equal to the same ratio as that between the lines which measure the refraction of the convex glass AC, then a ray passing from F to C must be refracted toward G upon entering the glass. This follows at once from dioptrics.

Now let us determine by calculation if it be true that PQ : PN = d : e. The right triangles PQF and CMF are similar, whence it follows that CF : CM = FP : PQ, and $\frac{FP \cdot CM}{CF}$ = PQ. Again, the right triangles PNG and CMG are similar, and therefore $\frac{GP \cdot CM}{CG}$ = PN. Now since the multiplication or division of two terms of a ratio by the same number does not alter the ratio, if $\frac{FP \cdot CM}{CF} : \frac{GP \cdot CM}{CG} = d : e$, then, dividing each term of the first ratio by CM and multiplying each by both CF and CG, we have FP $\cdot$ CG : GP $\cdot$ CF = d : e. Now by construction,

$$FP = c + \frac{bcd^2 - bcde + bd^2z + ce^2z}{cd^2 + bde - e^2z + d^2z},$$

or

$$FP = \frac{bcd^2 + c^2d^2 + bd^2z + cd^2z}{cd^2 + bde - e^2z + d^2z},$$

and

$$CG = b - \frac{e}{d}z.$$

Then

$$FP \cdot CG = \frac{b^2cd^2 + bc^2d^2 + b^2d^2z + bcd^2z - bcdez - c^2dez - bdez^2 - cdez^2}{cd^2 + bde - e^2z + d^2z}.$$

Then

$$GP = b - \frac{bcd^2 - bcde + bd^2z + ce^2z}{cd^2 + bde - e^2z + d^2z};$$

or

$$GP = \frac{b^2de + bcde - be^2z - ce^2z}{cd^2 + bde - e^2z + d^2z};$$

and CF = $c + z$. So that

$$GP \cdot CF = \frac{b^2cde + bc^2de + b^2dez + bcdez - bce^2z - c^2e^2z - be^2z^2 - ce^2z^2}{cd^2 + bde - e^2z + d^2z}.$$

The first of these products divided by d is equal to the second divided by e, whence it follows that PQ : PN = FP $\cdot$ CG : GP $\cdot$ CF = d : e, which was to be proved. This proof may be made to hold for the reflecting and refracting properties of any one of these ovals, by proper changes of the signs plus and minus; and as

---

160. Here PQ is the sine of the angle of incidence and PN is the sine of the angle of refraction. The ray FC is reflected along CG.

each can be investigated by the reader, there is no need for further discussion here.[161]

It now becomes necessary for me to supplement the statements made in my Dioptrique[162] to the effect that lenses of various forms serve equally well to cause rays coming from the same point and passing through them to converge to another point; and that among such lenses those which are convex on one side and concave on the other are more powerful burning-glasses than those which are convex on both sides; while, on the other hand, the latter make the better telescopes.[163] I shall describe and explain only those which I believe to have the greatest practical value, taking into consideration the difficulties of cutting. To complete the theory of the subject, I shall now have to describe again the form of lens which has one side of any desired degree of convexity or concavity, and which makes all the rays that are parallel or that come from a single point converge after passing through it; and also the form of lens having the same effect but being equally convex on both sides, or such that the convexity of one of its surfaces bears a given ratio to that of the other.

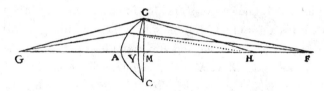

In the first place, let G, Y, C, and F be given points, such that rays coming from G or parallel to GA converge at F after passing through a concave lens. Let Y be the center of the inner surface of this lens and C its edge, and let the chord CMC be given, and also the altitude of the arc CYC. First we must determine which of these ovals can be used for a lens that will cause rays passing through it in the direction of H (a point as yet undetermined) to converge toward F after leaving it.

There is no change in the direction of rays by means of reflection or refraction which cannot be effected by at least one of these ovals; and it is easily seen that this particular result can be obtained by using either part of the third oval, marked 3A3

---

161. To obtain the equation of the first oval we may proceed as follows: Let AF = c; AG = b; FC = $c + z$; GC = $b - \frac{e}{d}z$. Let CM = x, AM = y. FM = $c + y$; GM = $b - y$. Draw PC normal to the curve at any point C. Let AP = v. Then $\overline{CF}^2 = \overline{CM}^2 + \overline{FM}^2$. Also, $c^2 + 2cz + z^2 = x^2 + c^2 + 2cy + y^2$, whence

$$z = -c + \sqrt{x^2 + c^2 + 2cy + y^2}.$$

Also, $\overline{CG}^2 = \overline{CM}^2 + \overline{GM}^2$, whence

$$b^2 - 2\frac{be}{d}z + \frac{e^2}{d^2}z^2 = x^2 + b^2 - 2by + y^2.$$

Substituting in this equation the value of z obtained above, squaring, and simplifying, we obtain:
$$[(d^2 - e^2)x^2 + (d^2 - e^2)y^2 - 2(e^2c + bd^2)y - 2ec(ec - bd)]^2 = 4e^2(bd + ec)^2(x^2 + c^2 + 2cy + y^2).$$

162. Descartes: *La Dioptrique*, published with *Discours de la Methode*, Leyden, 1637. See also Cousin, Vol. III.

163. "Lunetes." The laws of reflection were familiar to the geometers of the Platonic school, and burning-glasses, in the form of spherical glass shells filled with water, or balls of rock crystal are discussed by Pliny, Hist. Nat. xxxvi, 67 (25) and xxxvii, 10. Ptolemy, in his treatise on Optics, discussed reflection, refraction, and plane and concave mirrors.

or 3Y3, or the part of the second oval marked 2X2. Since the same method applied to each of these, we may in each case take Y, as the vertex, C as a point on the curve,[164] and F as one of the foci. It then remains to determine H, the other focus. This may be found by considering that the difference between FY and FC is to the difference between HY and HC as *d* is to *e*; that is, as the longer of the lines measuring the refractive power of the lens is to the shorter, as is evident from the manner of describing the ovals.

Since the lines FY and FC are given we know their difference; and then, since the ratio of the two differences is known, we know the difference between HY and HC.

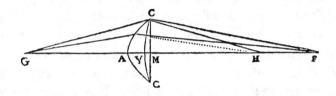

Again, since YM is known, we know the difference between MH and HC, and therefore CM. It remains to find MH, the side of the right triangle CMH. The other side of this triangle, CM, is known, and also the difference between the hypotenuse, CH and the required side, MH. We can therefore easily determine MH as follows:

Let $k = $ CH $-$ MH and $n = $ CM; then $\frac{n^2}{2k} - \frac{1}{2}k = $ MH, which determines the position of the point H.

If HY is greater than HF, the curve CY must be the first part of the third class of oval, which has already been designated by 3A3.

But suppose that HY is less than FY. This includes two cases: In the first, HY exceeds HF by such an amount that the ratio of their difference to the whole line FY is greater than the ratio of *e*, the smaller of the two lines that represent the refractive power, to *d*, the larger; that is, if HF $= c$, and HY $= c + h$, then *dh* is greater than $2ce + eh$. In this case CY must be the second part 3Y3 of the same oval of the third class.

In the second case *dh* is less than or equal to $2ce + eh$, and CY is the second part 2X2 of the oval of the second class.

Finally, if the points H and F coincide, FY $= $ FC and the curve YC is a circle.

It is also necessary to determine CAC, the other surface of the lens. If we suppose the rays falling on it to be parallel, this will be an ellipse having H as one of its foci, and the form is easily determined. If, however, we suppose the rays to come from the point G, the lens must have the form of the first part of an oval of the first class,

164. "Circonference."

the two foci of which are G and H and which passes through the point C. The point A is seen to be its vertex from the fact that the excess of GC over GA is to the excess of HA over HC as d is to e. For if k represents the difference between CH and HM, and x represents AM, then $x - k$ will represent the difference between AH and CH; and if g represents the difference between GC and GM, which are given, $g + x$ will represent the difference between GC and GA; and since $g + x : x - k = d : e$, we have $ge + ex = dx - dk$, or $AM = x = \frac{ge + dk}{d - e}$, which enables us to determine the required point A.

Again, suppose that only the points G, C, and F are given, together with the ratio of AM to YM; and let it be required to determine the form of the lens ACY which causes all the rays coming from the point G to converge to F.

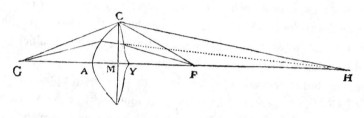

In this case, we can use two ovals, AC and CY, with foci G and H, and F and H respectively. To determine these let us suppose first that H, the focus common to both, is known. Then AM is determined by the three points G, C, and H in the way just now explained; that is if k represents the difference between CH and HM, and g the difference between GC and GM, and if AC be the first part of the oval of the first class, we have $AM = \frac{ge + dk}{d - e}$.

We may then find MY by means of the three points F, C, and H. If CY is the first part of an oval of the third class and we take y for MY and f for the difference between CF and FM, we have the difference between CF and FY equal to $f + y$; then let the difference between CH and HM equal k, and the difference between CH and HY equal $k + y$. Now $k + y : f + y = e : d$, since the oval is of the third class, whence $MY = \frac{fe - dk}{d - e}$. Therefore, $AM + MY = AY = \frac{ge + fe}{d - e}$, whence it follows that on whichever side the point H may lie, the ratio of the line AY to the excess of GC + CF over GF is always equal to the ratio of e, the smaller of the two lines representing the refractive power of the glass, to $d - e$, the difference of these two lines, which gives a very interesting theorem.[165]

The line AY being found, it must be divided in the proper ratio into AM and MY, and since M is known the points A and Y, and finally the point H, may be found by the preceding problem. We must first find whether the line AM thus found is greater than, equal to, or less than $\frac{ge}{d-e}$. If it is greater, AC must be the first part of one of the third class, as they have been considered here. If it is smaller, CY must be the first part of an oval of the first class and AC the first part of one of the third class. Finally, if AM is equal to $\frac{ge}{d-e}$, the curves AC and CY must both be hyperbolas.

These two problems can be extended to an infinity of other cases which I will not stop to deduce, since they have no practical value in dioptrics.

I might go farther and show how, if one surface of a lens is given and is neither entirely plane nor composed of conic sections or circles, the other surface can be so determined as to transmit all the rays from a given point to another point, also given. This is no more difficult than the problems I have just explained; indeed, it is much easier since the way is now open; I prefer, however, to leave this for others to work out, to the end that they may appreciate the more highly the discovery of those things here demonstrated, through having themselves to meet some difficulties.

In all this discussion I have considered only curves that can be described upon a plane surface, but my remarks can easily be made to apply to all those curves which can be conceived of as generated by the regular movement of the points of a body in three-dimensional space.[166] This can be done by dropping perpendiculars from each point of the curve under consideration upon two planes intersecting at right angles, for the ends of these perpendiculars will describe two other curves, one in each of the two planes, all points of which may be determined in the way already explained, and all of which may be related to those of a straight line common to the two planes; and by means of these the points of the three-dimensional curve will be entirely determined.

We can even draw a straight line at right angles to this curve at a given point, simply by drawing a straight line in each plane normal to the curve lying in that plane at the foot of the perpendicular drawn from the given point of the three-dimensional curve to that plane and then drawing two other planes, each passing through one of the straight lines and perpendicular to the plane containing it; the intersection of these two planes will be the required normal.

And so I think I have omitted nothing essential to an understanding of curved lines.

---

166. This is the hint which Descartes gives of the possibility of the extension of his theory to solid geometry. This extension was effected largely by Parent (1666–1716), Clairaut (1713–1765), and Van Schooten (d. 1661).

GEOMETRY

BOOK III

ON THE CONSTRUCTION OF SOLID AND SUPERSOLID PROBLEMS

While it is true that every curve which can be described by a continuous motion should be recognized in geometry, this does not mean that we should use at random the first one that we meet in the construction of a given problem. We should always choose with care the simplest curve that can be used in the solution of a problem, but it should be noted that the simplest means not merely the one most easily described, nor the one that leads to the easiest demonstration or construction of the problem, but rather the one of the simplest class that can be used to determine the required quantity.

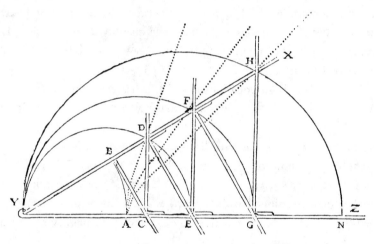

For example, there is, I believe, no easier method of finding any number of mean proportionals,[167] nor one whose demonstration is clearer, than the one which employs the curves described by the instrument XYZ, previously explained. Thus, if two mean proportionals between YA and YE be required, it is only necessary to describe a circle upon YE as diameter cutting the curve AD in D, and YD is then one of the required mean proportionals. The demonstration becomes obvious as soon as the instrument is applied to YD, since YA (or YB) is to YC as YC is to YD as YD is to YE.

Similarly, to find four mean proportionals between YA and YG, or six between YA and YN, it is only necessary to draw the circle YFG, which determines by its intersection with AF the line YF, one of the four mean proportionals; or the circle

---

167. For the history of this problem, see Heath, *History*, Vol. I.

YHN, which determines by its intersection with AH the line YH, one of the six mean proportionals, and so on.

But the curve AD is of the second class, while it is possible to find two mean proportionals by the use of the conic sections, which are curves of the first class.[168] Again, four or six mean proportionals can be found by curves of lower classes than AF and AH respectively. It would therefore be a geometric error to use these curves. On the other hand, it would be a blunder to try vainly to construct a problem by means of a class of lines simpler than its nature allows.[169]

Before giving the rules for the avoidance of both these errors, some general statements must be made concerning the nature of equations. An equation consists of several terms, some known and some unknown, some of which are together equal to the rest; or rather, all of which taken together are equal to nothing; for this is often the best form to consider.[170]

Every equation can have[171] as many distinct roots (values of the unknown quantity) as the number of dimensions of the unknown quantity in the equation.[172] Suppose, for example, $x = 2$ or $x - 2 = 0$, and again, $x = 3$, or $x - 3 = 0$. Multiplying together the two equations $x - 2 = 0$ and $x - 3 = 0$, we have $x^2 - 5x + 6 = 0$, or $x^2 = 5x - 6$. This is an equation in which x has the value 2 and at the same time[173] x has the value 3. If we next make $x - 4 = 0$ and multiply this by $x^2 - 5x + 6 = 0$, we have $x^3 - 9x^2 + 26x - 24 = 0$ another equation, in which x, having three dimensions, has also three values, namely, 2, 3, and 4.

It often happens, however, that some of the roots are false[174] or less than nothing. Thus, if we suppose x to represent the defect[175] of a quantity 5, we have $x + 5 = 0$ which, multiplied by $x^3 - 9x^2 + 26x - 24 = 0$, yields $x^4 - 4x^3 - 19x^2 + 106x - 120 = 0$, an equation having four roots, namely three true roots, 2, 3, and 4, and one false root, 5.[176]

It is evident from the above that the sum[177] of an equation having several roots is always divisible by a binomial consisting of the unknown quantity diminished by the value of one of the true roots, or plus the value of one of the false roots. In this way,[178] the degree of an equation can be lowered.

168. If we let x and y represent the two mean proportionals between a and b we have $a:x = x:y = y:b$, whence $z^2 = ay$; $y^2 = bx$, and $xy = ab$. Therefore x and y may be found by determining the intersections of two parabolas or of a parabola and a hyperbola.
169. Cf. Pappus, Book IV, Prop. 31, Vol. I. See also Guisnée, *Application de l'Algèbre a la Géométrie*, Paris, 1733, and L'Hospital, *Traité Analytique des Sections Coniques*, Paris, 1707.
170. The advantage of this arrangement had been recognized by several writers before Descartes.
171. It is worthy of note that Descartes writes "can have" ("peut-il y avoir"), not "must have," since he is considering only real positive roots.
172. That is, as the number denoting the degree of the equation.
173. "Tout ensemble,"—not quite the modern idea.
174. "Racines fausses," a term formerly used for "negative roots." Fibonacci, for example, does not admit negative quantities as roots of an equation. *Scritti de Leonardo Pisano*, published by Boncompagni, Rome, 1857. Cardan recognizes them, but calls them "æstimationes falsæ" or "fictæ," and attaches no special significance to them. See Cardan, *Ars Magna*, Nurnberg, 1545. Stifel called them "Numeri absurdi," as also in Rudolff's Coss, 1545.
175. "Le défaut." If $x = -5$, -5 is the "defect" of 5, that is, the remainder when 5 is subtracted from zero.
176. That is, three positive roots, 2, 3, and 4, and one negative root, -5.
177. "Somme," the left member when the right member is zero; that is, what we represent by $f(x)$ in the equation $f(x) = 0$.
178. That is, by performing the division.

On the other hand, if the sum of the terms of an equation[179] is not divisible by a binomial consisting of the unknown quantity plus or minus some other quantity, then this latter quantity is not a root of the equation. Thus the[180] above equation $x^4 - 4x^3 - 19x^2 + 106x - 120 = 0$ is divisible by $x - 2$, $x - 3$, $x - 4$ and $x + 5$,[181] but is not divisible by x plus or minus any other quantity. Therefore the equation can have only the four roots, 2, 3, 4, and 5.[182] We can determine also the number of true and false roots that any equation can have, as follows:[183] An equation can have as many true roots as it contains changes of sign, from $+$ to $-$ or from $-$ to $+$; and as many false roots as the number of times two $+$ signs or two $-$ signs are found in succession.

Thus, in the last equation, since $+x^4$ is followed by $-4x^3$, giving a change of sign from $+$ to $-$, and $-19x^2$ is followed by $+106x$ and $+106x$ by -120, giving two more changes, we know there are three true roots; and since $-4x^3$ is followed by $-19x^2$ there is one false root.

It is also easy to transform an equation so that all the roots that were false shall become true roots, and all those that were true shall become false. This is done by changing the signs of the second, fourth, sixth, and all even terms, leaving unchanged the signs of the first, third, fifth, and other odd terms. Thus, if instead of
$$+x^4 - 4x^3 - 19x^2 + 106x - 120 = 0$$
we write
$$+x^4 + 4x^3 - 19x^2 - 106x - 120 = 0$$
we get an equation having one true root, 5, and three false roots, 2, 3, and 4.[184]

If the roots of an equation are unknown and it be desired to increase or diminish each of these roots by some known number, we must substitute for the unknown quantity throughout the equation, another quantity greater or less by the given number. Thus, if it be desired to increase by 3 the value of each root of the equation
$$x^4 + 4x^3 - 19x^2 - 106x - 120 = 0$$
put y in the place of x, and let y exceed x by 3, so that $y - 3 = x$. Then for x^2 put the square of $y - 3$, or $y^2 - 6y + 9$; for x^3 put its cube, $y^3 - 9y^2 + 27y - 27$; and for x^4 put its fourth power,[185] or
$$y^4 - 12y^3 + 54y^2 - 108y + 81.$$

179. "Si la somme d'un équation."
180. First member of the equation. Descartes always speaks of dividing the equation.
181. Incorrectly given as $x - 5$ in some editions.
182. Where 5 would now be written -5. Descartes neither states nor explicitly assumes the fundamental theorem of algebra, namely, that every equation has at least one root.
183. This is the well known "Descartes's Rule of Signs." It was known however, before his time, for Harriot had given it in his *Artis analyticae praxis*, London, 1631. Cantor says Descartes may have learned it from Cardan's writings, but was the first to state it as a general rule. See Cantor, Vol. II(1).
184. In absolute value.
185. "Son quarré de quarré," that is, its fourth power.

Substituting these values in the above equation, and combining, we have

$$y^4 - 12y^3 + 54y^2 - 108y + 81$$
$$+ 4y^3 - 36y^2 + 108y - 108$$
$$- 19y^2 + 114y - 171$$
$$- 106y + 318$$
$$- 120$$
$$\overline{\qquad\qquad\qquad\qquad}$$
$$y^4 - 8y^3 - y^2 + 8y \qquad = 0,[186]$$

or

$$y^3 - 8y^2 - y + 8 = 0,$$

whose true root is now 8 instead of 5, since it has been increased by 3. If, on the other hand, it is desired to diminish by 3 the roots of the same equation, we must put $y + 3 = x$ and $y^2 + 6y + 9 = x^2$, and so on so that instead of $x^4 + 4x^3 - 19x^2 - 106x - 120 = 0$, we have

$$y^4 + 12y^3 + 54y^2 + 108y + 81$$
$$+ 4y^3 + 36y^2 + 108y + 108$$
$$- 19y^2 - 114y - 171$$
$$- 106y - 318$$
$$- 120$$
$$\overline{\qquad\qquad\qquad\qquad}$$
$$y^4 + 16y^3 + 71y^2 - 4y - 420 = 0.$$

It should be observed that increasing the true roots of an equation diminishes[187] the false roots by the same amount; and on the contrary diminishing the true roots increases the false roots; while diminishing either a true or a false root by a quantity equal to it makes the root zero; and diminishing it by a quantity greater than the root renders a true root false or a false root true.[188] Thus by increasing the true root 5 by 3, we diminish each of the false roots, so that the root previously 4 is now only 1, the root previously 3 is zero, and the root previously 2 is now a true root, equal to 1, since $-2 + 3 = +1$. This explains why the equation $y^3 - 8y^2 - y + 8 = 0$ has only three roots, two of them, 1 and 8, being true roots, and the third, also 1, being false; while the other equation $y^4 - 16y^3 + 71y^2 - 4y - 420 = 0$ has only one true root, 2, since $+5 - 3 = +2$, and three false roots, 5, 6, and 7.

Now this method of transforming the roots of an equation without determining their values yields two results which will prove useful: First, we can always remove the second term of an equation by diminishing its true roots by the known quantity of the second term divided by the number of dimensions of the first term, if

186. Descartes wrote this $y^4 - 8y^3 - y^2 + 8y$ * $\approx$ 0, indicating by a star the absence of a term in a completè polynomial.
187. In absolute value.
188. For example, the false root 5 diminished by 7 means $-(5 - 7) = +2$.

these two terms have opposite signs; or, if they have like signs, by increasing the roots by the same quantity.[189] Thus, to remove the second term of the equation $y^4 + 16y^3 + 71y^2 - 4y - 420 = 0$ I divide 16 by 4 (the exponent of y in y^4), the quotient being 4. I then make $z - 4 = y$ and write

$$
\begin{array}{r}
z^4 - 16z^3 + 96z^2 - 256z + 256 \\
+ 16z^3 - 192z^2 + 768z - 1024 \\
+ 71z^2 - 568z + 1136 \\
- 4z + 16 \\
- 420 \\
\hline
z^4 \qquad - 25z^2 - 60z - 36 = 0.
\end{array}
$$

The true root of this equation which was 2 is now 6, since it has been increased by 4, and the false roots, 5, 6, and 7, are only 1, 2, and 3, since each has been diminished by 4. Similarly, to remove the second terms of $x^4 - 2ax^3 + (2a^2 - c^2)x^2 - 2a^3x + a^4 = 0$; since $2a \div 4 = \frac{1}{2}a$ we must put $z + \frac{1}{2}a = x$ and write

$$
\begin{array}{r}
z^4 + 2az^3 + \dfrac{3}{2}a^2z^2 + \dfrac{1}{2}a^3z + \dfrac{1}{16}a^4 \\[4pt]
- 2az^3 - 3a^2z^2 - \dfrac{3}{2}a^3z - \dfrac{1}{4}a^4 \\[4pt]
+ 2a^2z^2 + 2a^3z + \dfrac{1}{2}a^4 \\[4pt]
- c^2z^2 - ac^2z - \dfrac{1}{4}a^2c^2 \\[4pt]
- 2a^3z - a^4 \\[4pt]
+ a^4 \\[4pt]
\hline
z^4 + \left(\dfrac{1}{2}a^2 - c^2\right)z^2 - (a^3 + ac^2)z + \dfrac{5}{16}a^4 - \dfrac{1}{4}a^2c^2 = 0.
\end{array}
$$

Having found the value of z, that of x is found by adding $\frac{1}{2}a$. Second, by increasing the roots by a quantity greater than any of the false roots[190] we make all the roots true. When this is done, there will be no two consecutive $+$ or $-$ terms; and further, the known quantity of the third term will be greater than the square of half that of the second term. This can be done even when the false roots are unknown, since approximate values can always be obtained for them and the roots can then be increased by a quantity as large as or larger than is required. Thus, given,

$$x^6 + nx^5 - 6n^2x^4 + 36n^3x^3 - 216n^4x^2 + 1296n^5x - 7776n^6 = 0,$$

---

189. That is, by diminishing the roots by a quantity equal to the coefficient of the second term divided by the exponent of the highest power of x, with the opposite sign.

190. In absolute value.

make $y - 6n = x$ and we have,

$$
\begin{aligned}
y^6 - 36n &\left\{ \begin{array}{l} y^5 \\ + n \end{array} \right.
\begin{array}{l} + 540n^2 \\ - 30n^2 \\ - 6n^2 \end{array} \left\{ \begin{array}{l} y^4 \end{array} \right.
\begin{array}{l} - 4320n^3 \\ + 360n^3 \\ + 144n^3 \\ + 36n^3 \end{array} \left\{ \begin{array}{l} y^3 \end{array} \right.
\begin{array}{l} + 19440n^4 \\ - 2160n^4 \\ - 1296n^4 \\ - 648n^4 \\ - 216n^4 \end{array} \left\{ \begin{array}{l} y^2 \end{array} \right. \\
&\begin{array}{l} - 46656n^5 \\ + 6480n^5 \\ + 5184n^5 \\ + 3888n^5 \\ + 2592n^5 \\ + 1296n^5 \end{array} \left\{ \begin{array}{l} y \end{array} \right.
\begin{array}{l} + 46656n^6 \\ - 7776n^6 \\ - 7776n^6 \\ - 7776n^6 \\ - 7776n^6 \\ - 7776n^6 \\ - 7776n^6 \end{array}
\end{aligned}
$$

$$
y^6 - 35ny^5 + 504n^2y^4 - 3780n^3y^3 + 15120n^4y^2 - 27216n^5y = 0
$$

Now it is evident that $504n^2$, the known quantity[191] of the third term, is larger than $\left(\frac{35}{2}n\right)^2$; that is, than the square of half that of the second term; and there is no case for which the true roots need be increased by a quantity larger in proportion to those given than for this one.

If it is undesirable to have the last term zero, as in this case, the roots must be increased just a little more, yet not too little, for the purpose. Similarly if it is desired to raise the degree of an equation, and also to have all its terms present, as if instead of $x^5 - b = 0$, we wish an equation of the sixth degree with no term zero, first, for $x^5 - b = 0$ write $x^6 - bx = 0$, and letting $y - a = x$ we have

$$ y^6 - 6ay^5 + 15a^2y^4 - 20a^3y^3 + 15a^4y^2 - (6a^5 + b)y + a^6 + ab = 0. $$

It is evident that, however small the quantity a, every term of this equation must be present.

We can also multiply or divide all the roots of an equation by a given quantity, without first determining their values. To do this, suppose the unknown quantity when multiplied or divided by the given number to be equal to a second unknown quantity. Then multiply or divide the known quantity of the second term by the given quantity, that in the third term by the square of the given quantity, that in the fourth term by its cube, and so on, to the end.

This device is useful in changing fractional terms of an equation, to whole numbers, and often in rationalizing the terms. Thus, given $x^3 - \sqrt{3}x^2 + \frac{26}{27}x - \frac{8}{27\sqrt{3}} = 0$, let there be required another equation in which all the terms are expressed in rational numbers. Let $y = \sqrt{3}$ and multiply the second term by $\sqrt{3}$, the third by 3, and the last by $3\sqrt{3}$. The resulting equation is $y^3 - 3y^2 + \frac{26}{9}y - \frac{8}{9} = 0$. Next let it be required to replace this equation by another in which the known quantities are expressed only by whole numbers. Let $z = 3y$. Multiplying 3 by 3, $\frac{26}{9}$ by 9, and $\frac{8}{9}$ by 27, we have

$$ z^3 - 9z^2 + 26z - 24 = 0. $$

191. I. e., the coefficient.

The roots of this equation are 2, 3, and 4; and hence the roots of the preceding equation are $\frac{2}{3}$, 1 and $\frac{4}{3}$, and those of the first equation are

$$\frac{2}{9}\sqrt{3}, \frac{1}{3}\sqrt{3}, \text{ and } \frac{4}{9}\sqrt{3}.$$

This method can also be used to make the known quantity of any term equal to a given quantity. Thus, given the equation

$$x^3 - b^2x + c^3 = 0,$$

let it be required to write an equation in which the coefficient of the third term.[192] namely b^2, shall be replaced by $3a^2$. Let

$$y = x\sqrt{\frac{3a^2}{b^2}}$$

and we have

$$y^3 - 3a^2y + \frac{3a^3c^3}{b^3}\sqrt{3} = 0.$$

Neither the true nor the false roots are always real; sometimes they are imaginary;[193] that is, while we can always conceive of as many roots for each equation as I have already assigned,[194] yet there is not always a definite quantity corresponding to each root so conceived of. Thus, while we may conceive of the equation $x^3 - 6x^2 + 13x - 10 = 0$ as having three roots, yet there is only one real root, 2, while the other two, however we may increase, diminish, or multiply them in accordance with the rules just laid down, remain always imaginary.

When the construction of a problem involves the solution of an equation in which the unknown quantity has three dimensions,[195] the following steps must be taken:

First, if the equation contains some fractional coefficients,[196] change them to whole numbers by the method explained above;[197] if it contains surds, change them as far as possible into rational numbers, either by multiplication or by one of several other methods easy enough to find. Second, by examining in order all the factors of the last term, determine whether the left member of the equation is divisible[198] by a binomial consisting of the unknown quantity plus or minus any one of these factors. If it is, the problem is plane, that is, it can be constructed by means of the ruler and compasses; for either the known quantity of the binomial is the required root[199]

192. Descartes wrote this equation $x * - bbx + c^3 \infty 0$, the star showing that a term is missing. Hence, he speaks of $-b^2x$ as the third term.
193. "Mais quelquefois seulement imaginaires." This is a rather interesting classification, signifying that we may have positive and negative roots that are imaginary. The use of the word "imaginary" in this sense begins here.
194. This seems to indicate that Descartes realized the fact that an equation of the nth degree has exactly n roots. Cf. Cantor, Vol. II (1).
195. That is, a cubic equation.
196. "Nombres rompues," the "numeri fracti" of the medieval Latin writers and "numeri rotti" of the Italians. The expression "broken numbers" was often used by early English writers.
197. That is, transform the equation into one having integral coefficients.
198. "Qui divise toute la somme."
199. That is, the root that satisfies the conditions of the problem.

or else, having divided the left member of the equation by the binomial, the quotient is of the second degree, and from this quotient the root can be found as explained in the first book.

Given, for example, $y^6 - 8y^4 - 124y^2 - 64 = 0$.[200] The last term, 64, is divisible by 1, 2, 4, 8, 16, 32, and 64; therefore we must find whether the left member is divisible by $y^2 - 1$, $y^2 + 1$, $y^2 - 2$, $y^2 + 2$, $y^2 - 4$, and so on. We shall find that it is divisible by $y^2 - 16$ as follows:

$$
\begin{array}{r}
+\, y^6 - 8y^4 - 124y^2 - 64 = 0 \\
-\, y^6 - 8y^4 - 4y^2 \\
\hline
0 - 16y^4 - 128y^2 \\
-\, 16 - 16 \\
\hline
+\, y^4 + 8y^2 + 4 = 0
\end{array}
\quad -\, 16
$$

Beginning with the last term, I divide -64 by -16 which gives $+4$; write this in the quotient; multiply $+4$ by $+y^2$ which gives $+4y^2$ and write in the dividend (for the opposite sign from that obtained by the multiplication must always be used). Adding $-124y^2$ and $-4y^2$ I have $-128y^2$. Dividing this by -16 I have $+8y^2$ in the quotient, and multiplying by y^2 I have $-8y^4$ to be added to the corresponding term, $-8y^4$, in the dividend. This gives $-16y^4$ which divided by -16 yields $+y^4$ in the quotient and $-y^6$ to be added to $+y^6$ which gives zero, and shows that the division is finished.

If, however, there is a remainder, or if any modified term is not exactly divisible by 16, then it is clear that the binomial is not a divisor.[201]

Similarly, given

$$
\left.
\begin{array}{l}
y^6 + a^2 \\
 - 2c^2
\end{array}
\right\}
\left.
\begin{array}{l}
y^4 - a^4 \\
 + c^4
\end{array}
\right\}
\left.
\begin{array}{l}
y^2 - a^6 \\
 - 2a^4c^2 \\
 - a^2c^4
\end{array}
\right\} = 0,
$$

the last term is divisible by a, a^2, $a^2 + c^2$, $a^3 + ac^2$, and so on, but only two of these need be considered, namely a^2 and $a^2 + c^2$. The others give a term in the quotient of lower or higher degree than the known quantity of the next to the last term, and thus render the division impossible.[202] Note that I am here considering y^6 as of the third degree, since there are no terms in y^5, y^3, or y. Trying the binomial

$$ y^2 - a^2 - c^2 = 0 $$

200. Descartes considers this equation as a function of y^2.

201. This is evidently a modified form of our modern "synthetic division," the basis of our "Remainder Theorem," and of Horner's Method of solving numerical equations, a method known to the Chinese in the thirteenth century. See Cantor, Vol. II(1). See also Smith and Mikami, *History of Japanese Mathematics*, Chicago, 1914; Smith, I.

202. This is not a general rule.

we find that the division can be performed as follows:

$$
\left.\begin{array}{r} + y^6 + a^2 \\ - y^6 - 2c^2 \end{array}\right\} y^4 \left.\begin{array}{r} - a^4 \\ + c^4 \end{array}\right\} y^2 \left.\begin{array}{r} - a^6 \\ - 2a^4 c^2 \end{array}\right\} = 0
$$

$$
\begin{array}{r} \left.\begin{array}{r} 0 - 2a^2 \\ + c^2 \end{array}\right\} y^4 \left.\begin{array}{r} - a^4 \\ - a^2 c^2 \end{array}\right\} y^2 \left.\begin{array}{r} - a^2 c^4 \\ - a^2 - c^2 \end{array}\right. \\ \hline - a^2 - c^2 \quad - a^2 - c^2 \end{array}
$$

$$
\left.\begin{array}{r} + y^4 \end{array}\right. \left.\begin{array}{r} + 2a^2 \\ - c^2 \end{array}\right\} y^2 \left.\begin{array}{r} + a^4 \\ + a^2 c^2 \end{array}\right\} = 0,
$$

This shows that $a^2 + c^2$ is the required root, which can easily be proved by multiplication.

But when no binomial divisor of the proposed equation can be found, it is certain that the problem depending upon it is solid,[203] and it is then as great a mistake to try to construct it by using only circles and straight lines as it is to use the conic sections to construct a problem requiring only circles; for any evidence of ignorance may be termed a mistake.

Again, given an equation in which the unknown quantity has four dimensions.[204] After removing any surds or fractions, see if a binomial having one term a factor of the last term of the expression will divide the left member. If such a binomial can be found, either the known quantity of the binomial is the required root, or,[205] after the division is performed, the resulting equation, which is of only three dimensions, must be treated in the same way. If no such binomial can be found, we must increase or diminish the roots so as to remove the second term, in the way already explained, and then reduce it to another of the third degree, in the following manner: Instead of

$$x^4 \pm px^2 \pm qx \pm r = 0$$

write

$$y^6 \pm 2py^4 + (p^2 \pm 4r)y^2 - q^2 = 0. [206]$$

For the ambiguous[207] sign put $+2p$ in the second expression if $+p$ occurs in the first; but if $-p$ occurs in the first, write $-2p$ in the second; and on the contrary, put $-4r$ if $+r$, and $+4r$ if $-r$ occurs; but whether the first expression contains $+q$ or $-q$ we always write $-q^2$ and $+p^2$ in the second, provided that x^4 and y^6 have the sign $+$; otherwise, we write $+q^2$ and $-p^2$. For example, given

$$x^4 - 4x^2 - 8x + 35 = 0$$

203. That is, that it involves a conic or some higher curve.
204. A biquadratic equation.
205. "Either, or," as in the original. It is like saying that the root of $x^2 - a^2 = 0$ is either $x = a$ or $x = -a$.
206. Descartes wrote substantially "Instead of

$$+x^{4^{*}} \cdot pxx \cdot qx \cdot r \propto 0$$

write

$$+y^6 \cdot 2py^4 + (pp \cdot 4r)yy - qq \propto 0."$$

The symbolism is characteristic of Descartes.
207. Descartes wrote "pour les signes $+$ ou $-$ que j'ai omis."

replace it by

$$y^6 - 8y^4 - 124y^2 - 64 = 0.$$

For since $p = -4$, we replace $2py^4$ by $-8y^4$; and since $r = 35$, we replace $(p^2 - 4r)y^2$ by $(16 - 140)y^2$ or $-124y^2$; and since $q = 8$, we replace $-q^2$ by -64.

Similarly, instead of

$$x^4 - 17x^2 - 20x - 6 = 0$$

we must write

$$y^6 - 34y^4 + 313y^2 - 400 = 0,$$

for 34 is twice 17, and 313 is the square of 17 increased by four times 6, and 400 is the square of 20.

In the same way, instead of

$$+z^4 + \left(\frac{1}{2}a^2 - c^2\right)z^2 - (a^3 + ac^2)z - \frac{5}{16}a^4 - \frac{1}{4}a^2c^2 = 0,$$

we must write

$$y^6 + (a^2 - 2c^2)y^4 + (c^4 - a^4)y^2 - a^6 - 2a^4c^2 - a^2c^4 = 0;$$

for

$$p = \frac{1}{2}a^2 - c^2, \, p^2 = \frac{1}{4}a^4 - a^2c^2 + c^4, \, 4r = -\frac{5}{4}a^4 + a^2c^2.$$

And, finally,

$$-q^2 = -a^6 - 2a^4c^2 - a^2c^4.$$

When the equation has been reduced to three dimensions, the value of y^2 is found by the method already explained. If this cannot be done it is useless to pursue the question further, for it follows inevitably that the problem is solid. If, however, the value of y^2 can be found, we can by means of it separate the preceding equation into two others, each of the second degree, whose roots will be the same as those of the original equation. Instead of $+x^4 \pm px^2 \pm qx \pm r = 0$, write the two equations

$$+x^2 - yx + \frac{1}{2}y^2 \pm \frac{1}{2}p \pm \frac{q}{2y} = 0$$

and

$$+x^2 + yx + \frac{1}{2}y^2 \pm \frac{1}{2}p \pm \frac{q}{2y} = 0.$$

For the ambiguous signs write $+\frac{1}{2}p$ in each new equation, when p has a positive sign, and $-\frac{1}{2}p$ when p has a negative sign, but write $+\frac{q}{2y}$ when we have $-yx$, and $-\frac{q}{2y}$ when we have $+yx$, provided q has a positive sign, and the opposite when q has a negative sign. It is then easy to determine all the roots of the proposed equation, and consequently to construct the problem of which it contains the solution, by the exclusive use of circles and straight lines. For example, writing $y^6 - 34y^4 + 313y^2 - 400 = 0$

348

instead of $x^4 - 17x^2 - 20x - 6 = 0$ we find that $y^2 = 16$; then, instead of the original equation

$$+x^4 - 17x^2 - 20x - 6 = 0$$

write the two equations $+x^2 - 4x - 3 = 0$ and $+x^2 + 4x + 2 = 0$. For, $y = 4$, $\frac{1}{2}y^2 = 8$, $p = 17$, $q = 20$, and therefore

$$+\frac{1}{2}y^2 - \frac{1}{2}p - \frac{q}{2y} = -3$$

and

$$+\frac{1}{2}y^2 - \frac{1}{2}p + \frac{q}{2y} = +2.$$

Obtaining the roots of these two equations, we get the same results as if we had obtained the roots of the equation containing x^4, namely, one true root, $\sqrt{7} + 2$, and three false ones, $\sqrt{7} - 2$, $2 + \sqrt{2}$, and $2 - \sqrt{2}$. Again, given $x^4 - 4x^2 - 8x + 35 = 0$, we have $y^6 - 8y^4 - 124y^2 - 64 = 0$, and since the root of the latter equation is 16, we must write $x^2 - 4x + 5 = 0$ and $x^2 + 4x + 7 = 0$. For in this case,

$$+\frac{1}{2}y^2 - \frac{1}{2}p - \frac{q}{2y} = 5$$

and

$$+\frac{1}{2}y^2 - \frac{1}{2}p + \frac{q}{2y} = 7.$$

Now these two equations have no roots either true or false,[208] whence we know that the four roots of the original equation are imaginary; and that the problem whose solution depends upon this equation is plane, but that its construction is impossible, because the given quantities cannot be united.[209]

Similarly, given

$$z^4 + \left(\frac{1}{2}a^2 - c^2\right)z^2 - (a^3 + ac^2)z + \frac{5}{16}a^4 - \frac{1}{4}a^2c^2 = 0,$$

since we have found $y^2 = a^2 + c^2$, we must write

$$z^2 - \sqrt{a^2 + c^2}\,z + \frac{3}{4}a^2 - \frac{1}{2}a\sqrt{a^2 + c^2} = 0,$$

and

$$z^2 + \sqrt{a^2 + c^2}\,z + \frac{3}{4}a^2 + \frac{1}{2}a\sqrt{a^2 + c^2} = 0.$$

For $y = \sqrt{a^2 + c^2}$ and $+\frac{1}{2}y^2 + \frac{1}{2}p = \frac{3}{4}a^2$, and $\frac{q}{2y} = \frac{1}{2}a\sqrt{a^2 + c^2}$, then we have

$$z = \frac{1}{2}\sqrt{a^2 + c^2} + \sqrt{-\frac{1}{2}a^2 + \frac{1}{4}c^2 + \frac{1}{2}a\sqrt{a^2 + c^2}}$$

208. That is, all its roots are imaginary.
209. That is, the given quantities cannot be taken together in the same problem.

or

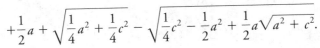

$$z = \frac{1}{2}\sqrt{a^2 + c^2} - \sqrt{-\frac{1}{2}a^2 + \frac{1}{4}c^2 + \frac{1}{2}a\sqrt{a^2 + c^2}}.$$

Now we already have $z + \frac{1}{2}a = x$, and therefore x, the quantity in the search for which we have performed all these operations, is

$$+\frac{1}{2}a + \sqrt{\frac{1}{4}a^2 + \frac{1}{4}c^2} - \sqrt{\frac{1}{4}c^2 - \frac{1}{2}a^2 + \frac{1}{2}a\sqrt{a^2 + c^2}}.$$

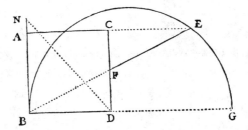

To emphasize the value of this rule, I shall apply it to a problem. Given the square AD and the line BN, to prolong the side AC to E, so that EF, laid off from E on EB, shall be equal to NB.

Pappus showed that if BD is produced to G, so that DG = DN, and a circle is described on BG as diameter, the required point E will be the intersection of the straight line AC (produced) with the circumference of this circle.[210]

Those not familiar with this construction would not be likely to discover it, and if they applied the method suggested here they would never think of taking DG for the unknown quantity rather than CF or FD, since either of these would much more easily lead to an equation. They would thus get an equation which could not easily be solved without the rule which I have just explained.

For, putting a for BD or CD, c for EF and x for DF, we have CF = $a - x$, and, since CF is to FE as FD is to BF, we have

$$a - x : c = x : BF,$$

whence BF $= \frac{cx}{a - x}$. Now, in the right triangle BDF whose sides are x and a, $x^2 + a^2$, the sum of their squares, is equal to the square of the hypotenuse, which is $\frac{c^2x^2}{x^2 - 2ax + a^2}$ Multiplying both sides by

$$x^2 - 2ax + a^2$$

we get the equation,

$$x^4 - 2ax^3 + 2a^2x^2 - 2a^3x + a^4 = c^2x^2,$$

210. Pappus Lib. VII, Prop. 72, Vol. II. The following is in substance the proof given by Pappus. He first gives an elaborate proof of the following lemma: Given a square ABCD, and E a point in AC produced, EG perpendicular to BE at E, meeting BD produced in G, and F the point of intersection of BE and CD. Then $\overline{CD}^2 + FE^2 = \overline{DG}^2$. Then he proceeds as follows: By the construction given in the problem, $\overline{DN}^2 = BD^2 + \overline{BN}^2$. By the lemma, $\overline{DG}^2 = \overline{CD}^2 + \overline{FE}^2$. By construction, BD = CD and DG = DN. Therefore, FE = BN.

or

$$x^4 - 2ax^3 + (2a^2 - c^2)x^2 - 2a^3x + a^4 = 0,$$

and by the preceding rule we know that its root, which is the length of the line DF, is

$$\frac{1}{2}a + \sqrt{\frac{1}{4}a^2 + \frac{1}{4}c^2} - \sqrt{\frac{1}{4}c^2 - \frac{1}{2}a^2 + \frac{1}{2}a\sqrt{a^2 + c^2}}.$$

If, on the other hand, we consider BF, CE, or BE as the unknown quantity, we obtain an equation of the fourth degree, but much easier to solve, and quite simply obtained.[211]

Again, if DG were used, the equation would be much more difficult to obtain, but its solution would be very simple. I state this simply to warn you that, when the proposed problem is not solid, if one method of attack yields a very complicated equation a much simpler one can usually be found by some other method.

I might add several different rules for the solution of cubic and biquadratic equations but they would be superfluous, since the construction of any plane problem can be found by means of those already given.

I could also add rules for equations of the fifth, sixth, and higher degrees, but I prefer to consider them all together and to state the following general rule:

First, try to put the given equation into the form of an equation of the same degree obtained by multiplying together two others, each of a lower degree. If, after all possible ways of doing this have been tried, none has been sucessful, then it is certain that the given equation cannot be reduced to a simpler one; and, consequently, if it is of the third or fourth degree, the problem depending upon it is solid; if of the fifth or sixth, the problem is one degree more complex, and so on. I have also omitted here the demonstration of most of my statements, because they seem to me so easy that if you take the trouble to examine them systematically the demonstrations will present themselves to you and it will be of much more value to you to learn them in that way than by reading them.

Now, when it is clear that the proposed problem is solid, whether the equation upon which its solution depends is of the fourth degree or only of the third. its roots can always be found by any one of the three conic sections, or even by some part of one of them, however small, together with only circles and straight lines. I shall content myself with giving here a general rule for finding them all by means of a parabola, since that is in some respects the simplest of these curves.

First, remove the second term of the proposed equation, if this is not already zero, thus reducing it to the form $z^3 = \pm apz \pm a^2q$, if the given equation is of the

211. Taking BF as the unknown quantity, the resulting equation is
$$x^4 + 2cx^3 + (c^2 - 2a^2)x^2 - 2a^2cx - a^2c^2 = 0.$$

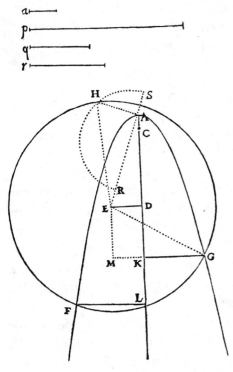

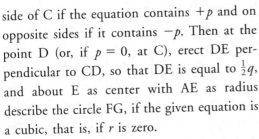

third degree, or $z^4 = \pm apz^2 \pm a^2qz \pm a^3r$, if it is of the fourth degree. By choosing a as the unit, the former may be written $z^3 = \pm pz \pm q$ and the latter $z^4 = \pm pz^2 \pm qz \pm r$. Suppose that the parabola FAG is already described; let ACDKL be its axis, a, or 1 which equals 2AC, its latus rectum (C being within the parabola), and A its vertex. Lay off CD equal to $\frac{1}{2}p$ so that the points D and A lie on the same

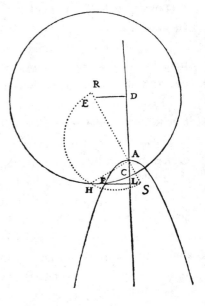

side of C if the equation contains $+p$ and on opposite sides if it contains $-p$. Then at the point D (or, if $p = 0$, at C), erect DE perpendicular to CD, so that DE is equal to $\frac{1}{2}q$, and about E as center with AE as radius describe the circle FG, if the given equation is a cubic, that is, if r is zero.

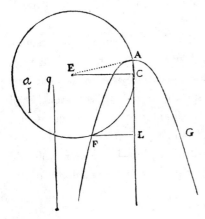

If the equation contains $+r$, on one side of AE produced, lay off AR equal to r, and on the other side lay off AS equal to the latus rectum of the parabola, that is, to 1, and describe a circle on RS as diameter. Then if AH is drawn perpendicular to AE it will meet the circle RHS in the point H, through which the other circle FHG must pass.

If the equation contains $-r$, construct a circle upon AE as diameter and in it inscribe

AI, a line equal to AH;[212] then the first circle must pass through the point I.

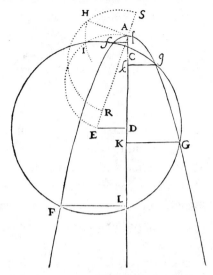

Now the circle FG can cut or touch the parabola in 1, 2, 3, or 4 points; and if perpendiculars are drawn from these points upon the axis they will represent all the roots of the equation, both true and false. If the quantity q is positive the true roots will be those perpendiculars, such as FL, on the same side of the parabola, as E,[213] the center of the circle; while the others, as GK, will be the false roots. On the other hand, if q is negative, the true roots will be those on the opposite side, and the false or negative roots[214] will be those on the same side as E, the center of the circle. If the circle neither cuts nor touches the parabola at any point, it is an indication that the equation has neither a true nor a false root, but that all the roots are imaginary.[215]

This rule is evidently as general and complete as could possibly be desired. Its demonstration is also very easy. If the line GK thus constructed be represented by z, then AK is z^2, since by the nature of the parabola, GK is the mean proportional between AK and the latus rectum, which is 1. Then if AC or $\frac{1}{2}$, and CD or $\frac{1}{2}p$, be subtracted from AK, the remainder is DK or EM, which is equal to $z^2 - \frac{1}{2}p - \frac{1}{2}$ of which the square is

$$z^4 - pz^2 - z^2 + \frac{1}{4}p^2 + \frac{1}{2}p + \frac{1}{4}.$$

And since DE = KM = $\frac{1}{2}q$, the whole line GM = $z + \frac{1}{2}q$, and the square of GM equals $z^2 + qz + \frac{1}{4}q^2$. Adding these two squares we have

$$z^4 - pz^2 + qz + \frac{1}{4}q^2 + \frac{1}{4}p^2 + \frac{1}{2}p + \frac{1}{4}$$

for the square of GE, since GE is the hypotenuse of the right triangle EMG.

But GE is the radius of the circle FG and can therefore be expressed in another way. For since ED = $\frac{1}{2}q$, and AD = $\frac{1}{2}p + \frac{1}{2}$, and ADE is a right angle, we have

$$EA = \sqrt{\frac{1}{4}q^2 + \frac{1}{4}p^2 + \frac{1}{2}p + \frac{1}{4}}.$$

212. That is, draw a chord equal to AH.
213. That is, on the same side of the axis of the parabola.
214. "Les fausses ou moindres que rien." This is the first time Descartes has directly used this synonym.
215. It may be noted that Descartes considers the cubic as a quartic having zero as one of its roots. Therefore, the circle always cuts the parabola at the vertex. It must then cut it in another point, since the cubic must have one real root. It may or may not cut it in two other points. It may cut it in two coincident points at the vertex, in which case the equation reduces to a quadratic.

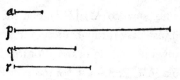

Then, since HA is the mean proportional between AS or 1 and AR or r, HA = $\sqrt{r}$; and since EAH is a right angle, the square of HE or of EG is

$$\frac{1}{4}q^2 + \frac{1}{4}p^2 + \frac{1}{2}p + \frac{1}{4} + r,$$

and we can form an equation from this expression and the one already obtained. This equation will be of the form $z^4 = pz^2 - qz + r$, and therefore the line GK, or z, is the root of this equation, which was to be proved. If you will apply this method in all the other cases, with the proper changes of sign, you will be convinced of its usefulness, without my writing anything further about it.

Let us apply it to the problem of finding two mean proportionals between the lines a and q. It is evident that if we represent one of the mean

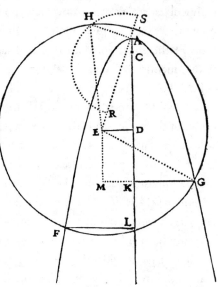

proportionals by z, then $a:z = z:\frac{z^2}{a} = \frac{z^2}{a}:\frac{z^3}{a^2}$. Thus we have an equation between q and $\frac{z^3}{a^2}$, namely, $z^3 = a^2q$.

Describe the parabola FAG with its axis along AC, and with AC equal to $\frac{1}{2}a$, that is, to half the latus rectum. Then erect CE equal to $\frac{1}{2}q$ and perpendicular to AC at C, and describe the circle AF about E as center, passing through A. Then FL and LA are the required mean proportionals.[216]

Again, let it be required to divide the angle NOP, or rather, the circular arc NQTP, into three equal parts. Let NO = 1 be the radius of

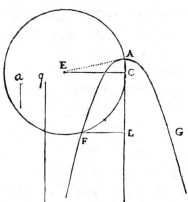

216. This may be shown as follows: Draw FM $\perp$ to EC; let FL = z. From the nature of the parabola, $\overline{FL}^2 = a \cdot AL$; $AL = \frac{z^2}{a}$;

$\overline{EC}^2 + \overline{CA}^2 = \overline{EA}^2$; $\overline{EM}^2 + \overline{FM}^2 = \overline{EF}^2$; $\overline{EA}^2 = \frac{q^2}{4} + \frac{a^2}{4}$; $\overline{EM}^2 = (EC - FL)^2 = \left(\frac{1}{2}q - z\right)^2$; $\overline{FM}^2 = \overline{CL}^2 = (AL - AC)^2 = \left(\frac{z^2}{a} - \frac{a}{2}\right)^2$;

$\overline{EF}^2 = \frac{q^2}{4} - qz + z^2 + \frac{z^4}{a^2} - z^2 + \frac{a^2}{4}$. But EF = EA.

$$\therefore \frac{q^2}{4} + \frac{a^2}{4} = \frac{q^2}{4} - qz + z^2 + \frac{z^4}{a^2} - z^2 + \frac{a^2}{4},$$

whence $z^3 = a^2q$.

the circle, NP = q be the chord subtending the given arc, and NQ = z be the chord subtending one-third of that arc; then the equation is $z^3 = 3z - q$. For, drawing NQ, OQ and OT, and drawing QS parallel to TO, it is obvious that NO is to NQ as NQ is to QR as QR is to RS. Since NO = 1 and NQ = z, then QR = z^2 and RS = z^3; and since NP or q lacks only RS or z^3 of being three times NQ or z, we have $q = 3z - z^3$ or $z^3 = 3z - q$.[217]

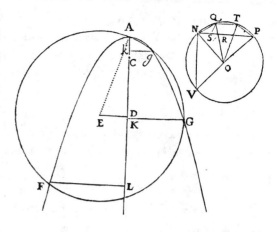

Describe the parabola FAG so that CA, one-half its latus rectum, shall be equal to $\frac{1}{2}$; take CD = $\frac{3}{2}$ and the perpendicular DE = $\frac{1}{2}q$; then describe the circle FAgG about E as center, passing through A. This circle cuts the parabola in three points, F, g, and G, besides the vertex, A. This shows that the given equation has three roots, namely, the two true roots, GK and gk, and one false root, FL.[218] The smaller of the two roots, gk, must be taken as the length of the required line NQ, for the other root, GK, is equal to NV, the chord subtended by one-third the arc VNP, which, together with the arc NQP constitutes the circle; and the false root, FL, is equal to the sum of QN and NV, as may easily be shown.[219]

217. ∠NOQ is measured by arc NQ;
  ∠QNS is measured by $\frac{1}{2}$ arc QP or arc NQ;
  ∠SQR = ∠QOT is measured by arc QT or NQ;
  ∴ ∠OQN = ∠NQR = ∠QSR.
  ∴ NO:NQ = NQ:QR = QR:RS.
  QR = z^2; RS = z^3. Let OT cut NP at M.
  NP = 2NR + MR = 2NQ + MR
    = 2NQ + MS − RS
    = 2NQ + QT − RS
    = 3NQ − RS.
  Or $q = 3z - z^3$.
218. G and g being on the opposite side of the axis from E, and F being on the same side.

219. Let AB = b; EB = MR = mk = NL = c; AK = t; Ak = s, AL = r; KG = y; kg = z, FL = v. Then GM = $y + c$, $gm = z + c$, FN = $v - c$, GK$^2 = a \cdot$ AK, $at = y^2$, $t = \frac{y^2}{a}$, $\overline{gk}^2 = a \cdot A k$, $as = z^2$, $s = \frac{z^2}{a}$, $\overline{FL}^2 = a \cdot$ AL, $ar = v^2$, $r = \frac{v^2}{a}$,

$$ME = AB - AK = b - \frac{y^2}{a}$$

$$mE = b - \frac{z^2}{a}$$

$$EN = \frac{v^2}{a} - b$$

$$\overline{EA}^2 = \overline{AB}^2 + \overline{BE}^2$$

$$\overline{EG}^2 = \overline{EM}^2 + \overline{MG}^2$$

It is unnecessary for me to give other examples here, for all problems that are only solid can be reduced to such forms as not to require this rule for their construction except when they involve the finding of two mean proportionals or the trisection of an angle. This will be obvious if it is noted that the most difficult of these problems can be expressed by equations of the third or fourth degree; that all equations of the fourth degree can be reduced to quadratic equations by means of other equations not exceeding the third degree; and finally, that the second terms of these equations can be removed; so that every such equation can be reduced to one of the following forms:

$$z^3 = -pz + q \qquad\qquad z^3 = +pz + q \qquad\qquad z^3 = +pz - q$$

Now, if we have $z^3 = -pz + q$, the rule, attributed by Cardan[220] to one Scipio Ferreus, gives us the root

$$\sqrt[3]{\frac{1}{2}q + \sqrt{\frac{1}{4}q^2 + \frac{1}{27}p^3}} - \sqrt[3]{-\frac{1}{2}q + \sqrt{\frac{1}{4}q^2 + \frac{1}{27}p^3}}.[221]$$

Similarly, when we have $z^3 = +pz + q$ where the square of half the last term is greater than the cube of one-third the coefficient of the next to the last term, the corresponding rule gives us the root

$$\sqrt[3]{\frac{1}{2}q + \sqrt{\frac{1}{4}q^2 - \frac{1}{27}p^3}} + \sqrt[3]{\frac{1}{2}q - \sqrt{\frac{1}{4}q^2 - \frac{1}{27}p^3}}.$$

It is now clear that all problems of which the equations can be reduced to either of these two forms can be constructed without the use of the conic sections except to extract the cube roots of certain known quantities, which process is equivalent to finding two mean proportionals between such a quantity and unity. Again, if we have $z^3 = +pz + q$, where the square of half the last term is not greater than the cube of one-third the coefficient of the next to the last term, describe the circle NQPV with radius NO equal to $\sqrt{\frac{1}{3}p}$, that is to the mean proportional between unity and one-third the known quantity p. Then take NP $= \frac{3q}{p}$, that is, such that NP is to q,

$$\overline{EG}^2 = b^2 - \frac{2by^2}{a} + \frac{y^4}{a^2} + y^2 + 2cy + c^2$$

$$2ab = \frac{y^3 + 2a^2c + a^2y}{y} \qquad\qquad\qquad\qquad 2ab = \frac{z^3 + 2a^2c + a^2z}{z}$$

$$\frac{y^3 + 2a^2c + a^2y}{y} = \frac{z^3 + 2a^2c + a^2z}{z}$$

$$2a^2c = z^2y + zy^2$$

Similarly,

$$2a^2c = v^2y - vy^2$$

$$z^2y + zy^2 = v^2y - vy^2 \qquad\qquad v^2 - z^2 = vy + zy$$

$$v = y + z \qquad\qquad\qquad\qquad\qquad \text{FL} = \text{KG} + kg$$

$v - z = y$

220. Cardan; Liber X, Cap. XI, fol. 29: "Scipio Ferreus Bononiensis iam annis ab hinc triginta fermè capitulum hoc inuenit, tradidit uero Anthonio Mariæ Florido Veneto, qui cù in certamen cù Nicolao Tartalea Brixellense aliquando uenisset, occasionem dedit, ut Nocolaus inuenerit & ipse, qui cum nobis rogantibus tradidisset, suppressa demonstratione, freti hoc auxilio, demonstrationem quæliuimus, eamque in modos, quod diffcillimum fuit, redactam sic subjecimus."

See also Cantor, Vol. II (1); Smith, Vol. II.

221. Descartes wrote this:

$$\sqrt{C \cdot + \frac{1}{2}q + \sqrt{\frac{1}{4}qq + \frac{1}{27}p^3}} + \sqrt{C \cdot \frac{1}{2}q + \sqrt{\frac{1}{4}qq + \frac{1}{27}p^3}}.$$

the other known quantity, as 1 is to $\frac{1}{3}p$, and inscribe NP in the circle. Divide each of the two arcs NQP and NVP into three equal parts, and the required root is the sum of NQ, the chord subtending one-third the first arc, and NV, the chord subtending one-third of the second arc.[222]

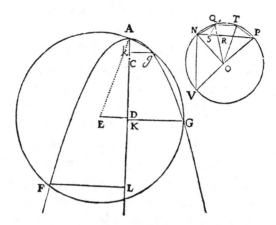

Finally, suppose that we have $z^3 = pz - q$. Construct the circle NQPV whose radius NO is equal to $\sqrt{\frac{1}{3}p}$, and let NP, equal to $\frac{3q}{b}$, be inscribed in this circle; then NQ, the chord of one-third the arc NQP, will be the first of the required roots, and NV, the chord of one-third the other arc, will be the second.

An exception must be made in the case in which the square of half the last term is greater than the cube of one-third the coefficient of the next to the last term;[223] for then the line NP cannot be inscribed in the circle, since it is longer than the diameter. In this case, the two roots that were true are merely imaginary, and the only real root is the one previously false, which according to Cardan's rule is

$$\sqrt[3]{\frac{1}{2}q + \sqrt{\frac{1}{4}q^2 - \frac{1}{27}p^3}} + \sqrt[3]{\frac{1}{2}q - \sqrt{\frac{1}{4}q^2 - \frac{1}{27}p^3}}.$$

Furthermore it should be remarked that this method of expressing the roots by means of the relations which they bear to the sides of certain cubes whose contents only are known[224] is in no respect clearer or simpler than the method of expressing them by means of the relations which they bear to the chords of certain arcs (or portions of circles), when arcs three times as long are known. And the roots of the cubic equations which cannot be solved by Cardan's method can be expressed as clearly as any others, or more clearly than the others, by the method given here.

---

222. It may be noted that the equation $z^3 = 3z - q$ may be obtained from the equation $z^3 = 3z + q$ by transforming the latter into an equation whose roots have the opposite signs. Then the true roots of $z^3 = 3z - q$ are the false roots of $z^3 = 3z + q$ and vice-versa. Therefore FL = NQ + NP is now the true root.

223. The so-called irreducible case.

224. Descartes here makes use of the geometrical conception of finding the cube root of a given quantity.

For example, grant that we may consider a root of the equation $z^3 = -qz + p$ known, because we know that it is the sum of two lines of which one is the side of a cube whose volume is $\frac{1}{2}q$ increased by the side of a square whose area is $\frac{1}{4}q^2 - \frac{1}{27}p^3$, and the other is the side of another cube whose volume is the difference between $\frac{1}{2}q$ and the side of a square whose area is $\frac{1}{4}q^2 - \frac{1}{27}p^3$. This is as much knowledge of the roots as is furnished by Cardan's method. There is no doubt that the value of the root of the equation $z^3 = +qz - p$ is quite as well known and as clearly conceived when it is considered as the length of a chord inscribed in a circle of radius $\sqrt{\frac{1}{3}p}$ and subtending an arc that is one-third the arc subtended by a chord of length $\frac{3q}{p}$.

Indeed, these terms are much less complicated than the others, and they might be made even more concise by the use of some particular symbol to express such chords,[225] just as the symbol $\sqrt[3]{}$ [226] is used to represent the side of a cube.

By methods similar to those already explained, we can express the roots of any biquadratic equation, and there seems to me nothing further to be desired in the matter; for by their very nature these roots cannot be expressed in simpler terms, nor can they be determined by any constuction that is at the same time easier and more general.

It is true that I have not yet stated my grounds for daring to declare a thing possible or impossible, but if it is remembered that in the method I use all problems which present themselves to geometers reduce to a single type, namely, to the question of finding the values of the roots of an equation, it will be clear that a list can be made of all the ways of finding the roots, and that it will then be easy to prove our method the simplest and most general. Solid problems in particular cannot, as I have already said, be constructed without the use of a curve more complex than the circle. This follows at once from the fact that they all reduce to two constructions, namely, to one in which two mean proportionals are to be found between two given lines, and one in which two points are to be found which divide a given arc into three equal parts. Inasmuch as the curvature of a circle depends only upon a simple relation between the center and all points on the circumference, the circle can only be used to determine a single point between two extremes, as, for example, to find one mean proportional between two given lines or to bisect a given arc; while, on the other hand, since the curvature of the conic sections always depends upon two different things,[227] it can be used to determine two different points.

For a similar reason, it is impossible that any problem of degree more complex than the solid, involving the finding of four mean proportionals or the division of an angle into five equal parts, can be constructed by the use of one of the conic sections.

225. This is another indication of the tendency of Descartes's age toward symbolism. This suggestion was never adopted.
226. In Descartes's notation, $\sqrt{C}$.
227. As, for example, the distance of any point from the two foci. Descartes does not say "all points on the circumference," but "toutes ses parties."

I therefore believe that I shall have accomplished all that is possible when I have given a general rule for constructing problems by means of the curve described by the intersection of a parabola and a straight line, as previously explained; for I am convinced that there is nothing of a simpler nature that will serve this purpose. You have seen, too, that this curve directly follows the conic sections in that question to which the ancients devoted so much attention, and whose solution presents in order all the curves that should be received into geometry.

When quantities required for the construction of these problems are to be found, you already know how an equation can always be formed that is of no higher degree than the fifth or sixth. You also know how by increasing the roots of this equation we can make them all true, and at the same time have the coefficient of the third term greater than the square of half that of the second term. Also, if it is not higher than the fifth degree it can always be changed into an equation of the sixth degree in which every term is present.

Now to overcome all these difficulties by means of a single rule, I shall consider all these directions applied and the equation thereby reduced to the form:

$$y^6 - py^5 + qy^4 - ry^3 + sy^2 - ty + u = 0$$

in which q is greater than the square of $\frac{1}{2}p$.

Produce BK indefinitely in both directions, and at B draw AB perpendicular to BK and equal to $\frac{1}{2}p$. In a separate plane[228] describe the parabola CDF whose principal parameter is

$$\sqrt{\frac{t}{\sqrt{u}} + q - \frac{1}{4}p^2}$$

which we shall represent by n.

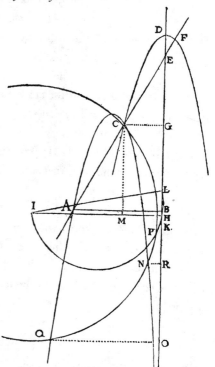

Now place the plane containing the parabola on that containing the lines AB and BK, in such a way that the axis DE of the parabola falls along the line BK. Take a point E such that DE $= \frac{2\sqrt{u}}{pn}$ and place a ruler so as to connect this point E and the point A of the lower plane. Hold the ruler so that it always connects these points, and slide the parabola up or down, keeping its axis always along BK. Then the

228. This does not mean in a fixed plane intersecting the first, but, for example, on another piece of paper.

point C, the intersection of the parabola and the ruler, will describe the curve ACN, which is to be used in the construction of the proposed problem.

Having thus described the curve, take a point L in the line BK on the concave side of the parabola, and such that BL = DE = $\frac{2\sqrt{u}}{pn}$; then lay off on BK, toward B, LH equal to $\frac{t}{2n\sqrt{u}}$, and from H draw HI perpendicular to LH and on the same side as the curve ACN. Take HI equal to

$$\frac{r}{2n^2} + \frac{\sqrt{u}}{n^2} + \frac{pt}{4n^2\sqrt{u}}$$

which we may, for the sake of brevity, set equal to $\frac{m}{n^2}$. Join L and I, and describe the circle LPI on LI as diameter; then inscribe in this circle the line LP equal to $\sqrt{\frac{s + p\sqrt{u}}{n^2}}$. Finally, describe the circle PCN about I as center and passing through P. This circle will cut or touch the curve ACN in as many points as the equation has roots; and hence the perpendiculars CG, NR, QO, and so on, dropped from these points upon BK, will be the required roots. This rule never fails nor does it admit of any exceptions.

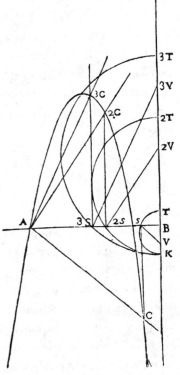

For if the quantity s were so large in proportion to the others, p, q, r, t, u, that the line LP was greater than the diameter of the circle LI,[229] so that LP could not be inscribed in it, every root of the proposed equation would be imaginary; and the same would be true if the circle IP[230] were so small that it did not cut the curve ACN at any point. The circle IP will in general cut the curve ACN in six different points, so that the equation can have six distinct roots.[231] But if it cuts it in fewer points, this indicates that some of the roots are equal or else imaginary.

If, however, this method of tracing the curve ACN by the translation of a parabola seems to you awkward, there are many other ways of describing it. We might take AB and BL as before, and BK equal to the latus rectum of the parabola, and describe the semicircle KST with its center in BK and cutting AB in some point S. Then from the point T where it

229. That is, the circle IPL, of which the diameter is t.
230. That is, the circle PCN.
231. The points determining these roots must be points of intersection of the circle with the main branch of the curve obtained, that is, of the branch ACN.

ends, take TV toward K equal to BL and join
S and V. Draw AC through A parallel to SV,
and draw SC through S parallel to BK; then
C, the intersection of AC and SC will be one
point of the required curve. In this way we can
find as many points of the curve as may be
desired.

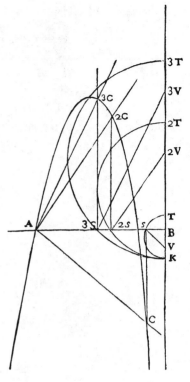

The demonstration of all this is very simple.
Place the ruler AE and the parabola FD so that
both pass through the point C. This can always
be done, since C lies on the curve ACN which
is described by the intersection of the parabola
and the ruler. If we let CG = y, GD will equal
$\frac{y^2}{n}$, since the latus rectum n is to CG as CG is
to GD. Then DE = $\frac{2\sqrt{u}}{pn}$, and subtracting DE
from GD we have GE = $\frac{y^2}{n} - \frac{2\sqrt{u}}{pn}$. Since AB is
to BE as CG is to GE, and AB is equal to $\frac{1}{2}p$,
therefore BE = $\frac{py}{2n} - \frac{\sqrt{u}}{ny}$. Now let C be a point
on the curve generated by the intersection of the
line SC, which is parallel to BK, and AC, which
is parallel to SV. Let SB = CG = y, and BK = n, the latus rectum of the parabola.
Then BT = $\frac{y^2}{n}$, for KB is to BS as BS is to BT, and since TV = BL = $\frac{2\sqrt{u}}{pn}$ we have
BV = $\frac{y^2}{n} - \frac{2\sqrt{u}}{pn}$. Also SB is to BV as AB is to BE, whence BE = $\frac{py}{2n} - \frac{\sqrt{u}}{ny}$ as
before. It is evident, therefore, that one and the same curve is described by these two
methods.

Furthermore, BL = DE, and therefore DL = BE; also LH = $\frac{t}{2n\sqrt{u}}$

and
$$DL = \frac{py}{2n} - \frac{\sqrt{u}}{ny}$$

therefore
$$DH = LH + DL = \frac{py}{2n} - \frac{\sqrt{u}}{ny} + \frac{t}{2n\sqrt{u}}.$$

Also, since GD = $\frac{y^2}{n}$,

$$GH = DH - GD = \frac{py}{2n} - \frac{\sqrt{u}}{ny} + \frac{t}{2n\sqrt{u}} - \frac{y^2}{n}$$

which may be written

$$GH = \frac{-y^3 + \frac{1}{2}py^2 + \frac{ty}{2\sqrt{u}} - \sqrt{u}}{ny}$$

and the square of GH is equal to

$$\dfrac{y^6 - py^5 + \left(\dfrac{1}{4}p^2 - \dfrac{t}{\sqrt{u}}\right)y^4 + \left(2\sqrt{u} + \dfrac{pt}{2\sqrt{u}}\right)y^3 + \left(\dfrac{t^2}{4u} - p\sqrt{u}\right)y^2 - ty + u}{n^2 y^2}.$$

Whatever point of the curve is taken as C, whether toward N or toward Q, it will always be possible to express the square of the segment of BH between the point H and the foot of the perpendicular from C to BH in these same terms connected by these same signs.

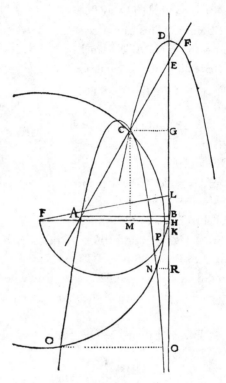

Again, IH $= \frac{m}{n^2}$, LH $= \frac{t}{2n\sqrt{u}}$, whence

$$\text{IL} = \sqrt{\dfrac{m^2}{n^4} + \dfrac{t^2}{4n^2 u}},$$

since the angle IHL is a right angle; and since

$$\text{LP} = \sqrt{\dfrac{s}{n^2} + \dfrac{p\sqrt{u}}{n^2}}$$

and the angle IPL is a right angle,

$$\text{IC} = \text{IP} = \sqrt{\dfrac{m^2}{n^4} + \dfrac{t^2}{4n^2 u} - \dfrac{s}{n^2} - \dfrac{p\sqrt{u}}{n^2}}.$$

Now draw CM perpendicular to IH, and

$$IM = HI - HM = HI - CG = \frac{m}{n^2} - y;$$

whence the square of IM is $\frac{m^2}{n^4} - \frac{2my}{n^2} + y^2$.

Taking this from the square of IC there remains the square of CM, or

$$\frac{t^2}{4n^2u} - \frac{s}{n^2} - \frac{p\sqrt{u}}{n^2} + \frac{2my}{n^2} - y^2,$$

and this is equal to the square of GH, previously found. This may be written

$$\frac{-n^2y^4 + 2my^3 - p\sqrt{u}\,y^2 - sy^2 + \dfrac{t^2}{4u}y^2}{n^2y^2}.$$

Now, putting

$$\frac{t}{\sqrt{u}}y^4 + qy^4 - \frac{1}{4}p^2y^4$$

for n^2y^4, and

$$ry^3 + 2\sqrt{u}\,y^3 + \frac{pt}{2\sqrt{u}}y^3$$

for $2my^3$, and multiplying both members by n^2y^2, we have

$$y^6 - py^5 + \left(\frac{1}{4}p^2 - \frac{t}{\sqrt{u}}\right)y^4 + \left(2\sqrt{u} + \frac{pt}{2\sqrt{u}}\right)y^3 + \left(\frac{t^2}{4u} - p\sqrt{u}\right)y^2 - ty + u$$

equals

$$\left(\frac{1}{4}p^2 - q - \frac{t}{\sqrt{u}}\right)y^4 + \left(r + 2\sqrt{u} + \frac{pt}{2\sqrt{u}}\right)y^3 + \left(\frac{t^2}{4u} - s - p\sqrt{u}\right)y^2,$$

or

$$y^6 - py^5 + qy^4 - ry^3 + sy^2 - ty + u = 0,$$

whence it appears that the lines CG, NR, QO, etc., are the roots of this equation.

If then it be desired to find four mean proportionals between the lines a and b, if we let x be the first, the equation is $x^5 - a^4b = 0$ or $x^6 - a^4bx = 0$. Let $y - a = x$, and we get

$$y^6 - 6ay^5 + 15a^2y^4 - 20a^3y^3 + 15a^4y^2 - (6a^5 + a^4b)y + a^6 + a^5b = 0.$$

Therefore, we must take AB $= 3a$, and BK, the latus rectum of the parabola must be

$$\sqrt{\frac{6a^3 + a^2b}{\sqrt{a^2 + ab}} + 6a^2}$$

which I shall call n, and DE or BL will be

$$\frac{2a}{3n}\sqrt{a^2 + ab}.$$

Then having described the curve ACN, we must have

$$LH = \frac{6a^3 + a^2b}{2n\sqrt{a^2 + ab}}$$

and

$$HI = \frac{10a^3}{n^2} + \frac{a^2}{n^2}\sqrt{a^2 + ab} + \frac{18a^4 + 3a^3b}{2n^2\sqrt{a^2 + ab}},$$

and

$$LP = \frac{a}{n}\sqrt{15a^2 + 6a\sqrt{a^2 + ab}}.$$

For the circle about I as center will pass through the point P thus found, and cut the curve in the two points C and N. If we draw the perpendiculars NR and CG, and subtract NR, the smaller, from CG, the greater, the remainder will be x, the first of the four required mean proportionals.[232]

This method applies as well to the division of an angle into five equal parts, the inscription of a regular polygon of eleven or thirteen sides in a circle, and an infinity of other problems. It should be remarked, however, that in many of these problems it may happen that the circle cuts the parabola of the second class so obliquely[233] that it is hard to determine the exact point of intersection. In such cases this construction is not of practical value.[234] The difficulty could easily be overcome by forming other rules analogous to these, which might be done in a thousand different ways.

But it is not my purpose to write a large book. I am trying rather to include much in a few words, as will perhaps be inferred from what I have done, if it is considered that, while reducing to a single construction all the problems of one class, I have at the same time given a method of transforming them into an infinity of others, and thus of solving each in an infinite number of ways; that, furthermore, having constructed all plane problems by the cutting of a circle by a straight line, and all solid problems by the cutting of a circle by a parabola; and, finally, all that are but one degree more complex by cutting a circle by a curve but one degree higher than the parabola, it is only necessary to follow the same general method to construct all problems, more and more complex, ad infinitum; for in the case of a mathematical progression, whenever the first two or three terms are given, it is easy to find the rest.

I hope that posterity will judge me kindly, not only as to the things which I have explained, but also as to those which I have intentionally omitted so as to leave to others the pleasure of discovery.

232. The two roots of the above equation in y are NR and CG. But we know that a is one of the roots of this equation, and therefore NR, the shorter length, must be a, and CG must be y. Then $x = y - a = $ CG $-$ NR, the first of the required mean proportionals.

233. That is, makes so small an angle with it.

234. This is especially noticeable when there are six real positive roots.

Isaac Newton

(1642–1727)

HIS LIFE AND WORK

Galileo died on January 8, 1642, exactly three hundred years before the day I was born. Isaac Newton was born on Christmas Day of that year in the English industrial town of Woolsthorpe, Lincolnshire. He would later become Lucasian Professor of Mathematics at Cambridge University, the chair I now hold.

Newton's mother did not expect him to live long, as he was born very prematurely; he would later describe himself as having been so small at birth he could fit into a quart pot. Newton's yeoman father, also named Isaac, had died three months earlier, and when Newton reached two years of age, his mother, Hannah Ayscough, remarried, wedding Barnabas Smith, a rich clergyman from North Witham. Apparently there was no place in the new Smith family for the young Newton, and he was placed in the care of his grandmother, Margery Ayscough. The specter of this abandonment, coupled with the tragedy of never having known his father, haunted Newton for the rest of his life. He despised his stepfather; in journal entries for 1662, Newton recalled "threatening my father and mother Smith to burne them and the house over them."

Much like his adulthood, Newton's childhood was filled with episodes of harsh, vindictive attacks, not only against perceived enemies but also against friends and family. He also displayed the kind of curiosity early on that would define his life's

365

achievements, taking an interest in mechanical models and architectural drawing. Newton spent countless hours building clocks, flaming kites, sundials, and miniature mills (powered by mice), as well as drawing elaborate sketches of animals and ships. At the age of five, he attended schools at Skillington and Stoke but was considered one of the poorest students, with his teachers' reports stating he was "inattentive" and "idle." Despite his curiosity and demonstrable passion for learning, he was unable to apply himself to schoolwork.

By the time Newton reached the age of ten, Barnabas Smith had passed away and Hannah had come into a considerable sum from Smith's estate. Isaac and his grandmother began living with Hannah, a half-brother, and two half-sisters. Because his work at school was uninspiring, including his studies of mathematics, Hannah decided that Isaac would be better off managing the farm and estate, and she pulled him out of the Free Grammar School in Grantham. Unfortunately for her, Newton had even less skill or interest in managing the family estate than he had in schoolwork. Hannah's brother William, a clergyman, decided that it would be best for the family if the absent-minded Isaac returned to school to finish his education.

This time, Newton lived with the headmaster of the Free Grammar School, John Stokes, and he seemed to turn a corner in his education. One story has it that a blow to the head, administered by a schoolyard bully, somehow enlightened him, enabling the young Newton to reverse the negative course of his educational promise. Now demonstrating intellectual aptitude and curiosity, Newton began preparing for further study at a university. He decided to attend Trinity College, his uncle William's alma mater, at Cambridge University.

At Trinity, Newton became a subsizar, receiving an allowance toward the cost of his education in exchange for performing various chores such as waiting tables and cleaning rooms for the faculty. But by 1664, he was elected scholar, which status guaranteed him financial support and freed him from menial duties. When the university closed because of the bubonic plague in 1665, Newton retreated to Lincolnshire. In the eighteen months he spent at home during the plague, he devoted himself to mechanics and mathematics and began to concentrate on optics and gravitation. This "annus mirabilis" (miraculous year), as Newton called it, was one of the most productive and fruitful periods of his life. It is also around this time that an apple, according to legend, fell onto Newton's head, awakening him from a nap under a tree and spurring him on to define the laws of gravity. However far-fetched the tale, Newton himself wrote that a falling apple had "occasioned" his foray into gravitational contemplation, and he is believed to have performed his pendulum experiments then. "I was in the prime of my age for invention," Newton later recalled, "and minded Mathematicks and Philosophy more than at any time since."

When he returned to Cambridge, Newton studied the philosophy of Aristotle and Descartes, as well as the science of Thomas Hobbes and Robert Boyle. He was taken by the mechanics of Copernicus and Galileo's astronomy, in addition to Kepler's optics. We have little direct information on Newton's mathematical education prior to his matriculation at Cambridge. Newton's first tutor at Cambridge was Benjamin Pulleyn, who later became Regius Professor of Greek. Newton soon came under the tutelage of Isaac Barrow, an outstanding mathematician and one of the founders of the Royal Society, who tutored Newton as young Isaac sped through Euclid's *Elements*. Following this, Newton soon mastered works on algebra by William Oughtred (1574–1660) and François Viète (1540–1603) and, most importantly, Descartes' *Geometrie*.

Around this time, Newton began his prism experiments in light refraction and dispersion, possibly in his room at Trinity or at home in Woolsthorpe. A development at the university that clearly had a profound influence on Newton's future—was the arrival of Isaac Barrow, who had been named the Lucasian Professor of Mathematics. Barrow recognized Newton's extraordinary mathematical talents, and when he resigned his professorship in 1669 to pursue theology, he recommended the twenty-seven-year-old Newton as his replacement.

Newton's first studies as Lucasian Professor were centered in the field of optics. He set out to prove that white light was composed of a mixture of various types of light, each producing a different color of the spectrum when refracted by a prism. His series of elaborate and precise experiments to prove that light was composed of minute particles drew the ire of scientists such as Hooke, who contended that light traveled in waves. Hooke challenged Newton to offer further proof of his eccentric optical theories. Newton's way of responding was one he did not outgrow as he matured. He withdrew, set out to humiliate Hooke at every opportunity, and refused to publish his book, *Opticks,* until after Hooke's death in 1703.

Early in his tenure as Lucasian Professor, Newton was well along in his study of pure mathematics, but he shared his work with few of his colleagues. Already by 1666, he had discovered general methods of solving problems of curvature—what he termed "theories of fluxions and inverse fluxions." The discovery set off a dramatic feud with supporters of the German mathematician and philosopher Gottfried Wilhelm Leibniz, who more than a decade later published his findings on differential and integral calculus. Both men arrived at roughly the same mathematical principles, but Leibniz published his work before Newton. Newton's supporters claimed that Leibniz had seen the Lucasian Professor's papers years before, and a heated argument between the two camps, known as the Calculus Priority Dispute, did not end until Leibniz died in 1716. Newton's vicious attacks, which often spilled over to touch on views about God and the universe, as well as his accusations of plagiarism, left Leibniz impoverished and disgraced.

Most historians of science believe that the two men in fact arrived at their ideas independently and that the dispute was pointless. Newton's vitriolic aggression toward Leibniz took a physical and emotional toll on Newton as well. He soon found himself involved in another battle, this time over his theory of color and with the English Jesuits, and in 1678 he suffered a severe mental breakdown. The next year, his mother passed away, and Newton began to distance himself from others. In secret he delved into alchemy, a field widely regarded already in Newton's time as fruitless. This episode in the scientist's life has been a source of embarrassment to many Newton scholars. Only long after Newton died did it become apparent that his interest in chemical experiments was related to his later research in celestial mechanics and gravitation.

Newton had already begun forming theories about motion by 1666, but he was as yet unable to adequately explain the mechanics of circular motion. Some fifty years earlier, the German mathematician and astronomer Johannes Kepler had proposed three laws of planetary motion, which accurately described how the planets moved in relation to the sun, but he could not explain why the planets moved as they did. The closest Kepler came to understanding the forces involved was to say that the sun and the planets were "magnetically" related.

Newton set out to discover the cause of the planets' elliptical orbits. By applying his own law of centripetal force to Kepler's third law of planetary motion, (the law of harmonies) he deduced the inverse-square law, which states that the force of gravity between any two objects is inversely proportional to the square of the distance between the object's centers. Newton was thereby coming to recognize that gravitation is universal—that one and the same force causes an apple to fall to the ground and the moon to race around the earth. He then set out to test the inverse-square relation against known data. He accepted Galileo's estimate that the moon's distance from the earth is sixty earth radii, but the inaccuracy of his own estimate of the earth's diameter made it impossible to complete the test to his satisfaction. Ironically, it was an exchange of letters in 1679 with his old adversary Hooke that renewed his interest in the problem. This time, he turned his attention to Kepler's second law, the law of equal areas, which Newton was able to prove held true because of centripetal force. Hooke, too, was attempting to explain the planetary orbits, and some of his letters on that account were of particular interest to Newton.

At an infamous gathering in 1684, three members of the Royal Society—Robert Hooke, Edmond Halley, and Christopher Wren, the noted architect of St. Paul's

Cathedral—engaged in a heated discussion about the inverse-square relation governing the motions of the planets. In the early 1670s, the talk in the coffee-houses of London and other intellectual centers was that gravity emanated from the sun in all directions and fell off at a rate inverse to the square of the distance, thus becoming more and more diluted over the surface of the sphere as that surface expands. The 1684 meeting was, in effect, the birth of *Principia*. Hooke declared that he had derived from Kepler's law of ellipses the proof that gravity was an ema-nating force, but would withhold it from Halley and Wren until he was ready to make it public. Furious, Halley went to Cambridge, told Newton Hooke's claim, and proposed the following problem. "What would be the form of a planet's orbit about the sun if it were drawn towards the sun by a force that varied inversely as the square of the distance?" Newton's response was staggering. "It would be an ellipse," he answered immediately, and then told Halley that he had solved the prob-lem four years earlier but had misplaced the proof in his office.

At Halley's request, Newton spent three months reconstituting and improving the proof. Then, in a burst of energy sustained for eighteen months, during which he was so caught up in his work that he often forgot to eat, Newton further devel-oped these ideas until their presentation filled three volumes. He chose to title the work *Philosophiae naturalis principia mathematica,* in deliberate contrast with Descartes' *Principia philosophiae*. The three books of Newton's *Principia* provided the link between Kepler's laws and the physical world. Halley reacted with "joy and amazement" to Newton's discoveries. To Halley, it seemed the Lucasian Professor had succeeded where all others had failed and he personally financed publication of the massive work as a masterpiece and a gift to humanity.

Where Galileo had shown that objects were "pulled" toward the center of the earth, Newton was able to prove that this same force, gravity, affected the orbits of the planets. He was also familiar with Galileo's work on the motion of projectiles, and he asserted that the moon's orbit around the earth adhered to the same princi-ples. Newton demonstrated that gravity explained and predicted the moon's motions as well as the rising and falling of the tides on earth. Book 1 of *Principia* encom-passes Newton's three laws of motion:

1. Every body perseveres in its state of resting, or uniformly moving in a right line, unless it is compelled to change that state by forces impressed upon it.

2. The change of motion is proportional to the motive force impressed; and is made in the direction of the right line in which that force is impressed.

3. To every action there is always opposed an equal reaction; or, the mutual actions of two bodies upon each other are always equal, and directed to contrary directions.

Book 2 began for Newton as something of an afterthought to Book 1; it was not included in the original outline of the work. It is essentially a treatise on fluid mechanics, and it allowed Newton room to display his mathematical ingenuity. Toward the end of the book, Newton concludes that the vortices invoked by Descartes to explain the motions of planets do not hold up to scrutiny, for the motions could be performed in free space without vortices. How that is so, Newton wrote, "may be understood by the first Book; and I shall now more fully treat of it in the following Book."

In Book 3, subtitled *System of the World,* by applying the laws of motion from Book 1 to the physical world Newton concluded, "there is a power of gravity tending to all bodies, proportional to the several quantities of matter which they contain." He thus demonstrated that his law of universal gravitation could explain the motions of the six known planets, as well as moons, comets, equinoxes, and tides. The law states that all matter is mutually attracted with a force directly proportional to the product of their masses and inversely proportional to the square of the distance between them. Newton, by a single set of laws, had united the earth with all that could be seen in the skies. In the first two "Rules of Reasoning" from Book 3, Newton wrote:

> *We are to admit no more causes of natural things than such as are both true and sufficient to explain their appearances. Therefore, to the same natural effects we must, as far as possible, assign the same causes.*

It is the second rule that actually unifies heaven and earth. An Aristotelian would have asserted that heavenly motions and terrestrial motions are manifestly not the same natural effects and that Newton's second rule could not, therefore, be applied. Newton saw things otherwise. *Principia* was moderately praised on its publication in 1687, but only about five hundred copies of the first edition were printed. However, Newton's nemesis, Robert Hooke, had threatened to spoil any coronation Newton might have enjoyed. After Book 2 appeared, Hooke publicly claimed that the letters he had written in 1679 had provided scientific ideas that were vital to Newton's discoveries. His claims, though not without merit, were abhorrent to Newton, who vowed to delay or even abandon publication of Book 3. Newton ultimately relented and published the final book of *Principia,* but not before painstakingly removing from it every mention of Hooke's name.

Newton's work on integral and differential calculus can be found in his notebooks from the mid-1660s. However, Newton never published a purely mathematical text on his own. Only in the second half of the twentieth century have an extensive portion of his mathematical papers been published. Newton gave his contemporaries a glimpse of his discoveries in the calculus in Book I, Section 1, of the *Principia.* He titled this section "The method of first and last ratios of quantities, by the help of which we demonstrate the propositions that follow." In this section, included in the present volume, Newton presents eleven "lemmas," a Greek term for "subsidiary proposition" on first and last ratios that will allow him to make interchangeable use of figures constructed with curves and corresponding figures constructed from straight lines.

In the first lemma Newton proves that

> *Quantities, and the ratios of quantities, which in any finite time converge*
> *continually to equality, and before the end of that time approach nearer*
> *to each other than by any given difference, become ultimately equal.*

He proves this in a very straightforward manner that anticipates Weierstrass's epsilon-delta method two centuries later. If the quantities and the ratios do not ultimately become equal, then they will have some finite ultimate difference D and they cannot ultimately approach each closer than that difference D! Notice that Newton frames this statement in terms of change over time. Given the setting in the *Principia,* a work on physical science, this is hardly surprising.

Two millennia before Newton, Archimedes had proven theorems about the area of particular geometric objects such as the circle by inscribing and circumscribing polygons about the object. In the second lemma, Newton takes a cue from Archimedes and extends this method to an arbitrary curve by inscribing and circumscribing rectangles about portions of the curve and demonstrating that the inscribed and circumscribed figures composed of these inscribed and circumscribed rectangles have areas that have an ultimate ratio of equality.

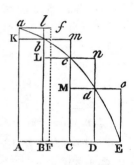

Newton has the reader consider an arbitrary curve in relation to a straight line represented here by AE that he divides into equal parts AB, BC, CD, and so forth, that will be diminished *ad infinitum.* He then constructs the rectangles such as AKbB inscribed within a segment of the curve and AalB circumscribed about the same segment of the curve. He notes that the difference in the areas of these rectangles is the area of the rectangle aKbl and that the sum of the areas of these "difference" rectangles is simply the area of the first circumscribed rectangle AalB! And then Newton clinches the proof noting that the length of the base AB of the rectangle AalB is being diminished *ad infinitum* so that the area of the rectangle AalB "becomes less than any given space." Hence, the areas of the inscribed and circumscribed figures ultimately "become ultimately equal one to the other" and the area described by the curvilinear figure as well!

Newton put his lemmas to immediate use in the first proposition he demonstrates in the *Principia:* Kepler's Area Law!

In a Scholium at the end of Section 1, Newton addresses concerns that would soon be voiced by critics that quantities that themselves vanish to zero cannot have an ultimate proportion as they vanish to zero. He responds by comparing ultimate ratios to a body's velocity at a particular point in space. In an argument having its roots in the ancient Greeks, Newton points out that the body is certainly not at rest for the moment that it is at the particular place. In contrast, it has a definite ultimate velocity, for the moment that it is at the particular place, just as quantities that vanish to zero can have an ultimate ratio to each other as the vanish to zero.

> *For those ultimate ratios with which quantities vanish are not truly the ratios of ultimate quantities, but limits towards which they approach nearer than by any given difference, but never go beyond, nor in effect attain to, till the quantities are diminished* in infinitum.

Newton began the eighteenth century in a government post as warden of the Royal Mint, where he utilized his work in alchemy to determine methods for reestablishing the integrity of the English currency. As president of the Royal Society, he continued to battle perceived enemies with inexorable determination, in particular carrying on his longstanding feud with Leibniz over their competing claims to have invented calculus. He was knighted by Queen Anne in 1705, and lived to see publication of the second and third editions of the *Principia.*

Newton occasionally claimed that he had derived many of the major propositions of the *Principia* using his discoveries in the calculus. He made one such claim in an unpublished preface to the *Principia* that he drafted around 1715 and finally went into print with such a claim in an anonymously published 1722 review of a book assessing his and Leibniz's claims to have invented calculus. He wrote:

> *By the help of the new analysis Mr. Newton found out most of the propositions of his* Principia Philosophiae: *but because the ancients for making things certain admitted nothing into geometry before it was demonstrated synthetically, he demonstrated the propositions synthetically, that the system of the heavens might be founded upon good geometry. And this makes it difficult now for unskillful men to see the analysis by which the propositions were found out.*

Recent scholarly analysis of Newton's notebooks have revealed that there is not a shred of evidence for his extravagant claims made to secure priority for discovering the calculus for himself and not for his arch-rival Leibniz.

Isaac Newton died in March 1727, after bouts of pulmonary inflammation and gout. As was his wish, Newton had no rival in the field of science. The man who apparently formed no romantic attachments with women (some historians have speculated on possible relationships with men, such as the Swiss natural philosopher Nicolas Fatio de Duillier) cannot, however, be accused of a lack of passion for his work. The poet Alexander Pope, a contemporary of Newton's, most elegantly described the great thinker's gift to humanity:

Nature and Nature's laws lay hid in night:
God said, "Let Newton be! and all was light."

For all the petty arguments and undeniable arrogance that marked his life, toward its end, Isaac Newton was remarkably poignant in assessing his accomplishments: "I do not know how I may appear to the world, but to myself I seem to have been only like a boy, playing on the sea-shore, and diverting myself, in now and then finding a smoother pebble or prettier shell than ordinary, whilst the great ocean of truth lay all undiscovered before me."

SELECTIONS FROM *PRINCIPIA*
BOOK I
OF THE MOTION OF BODIES
SECTION I

Of the method of first and last ratios of quantities, by the help whereof we demonstrate the propositions that follow.

LEMMA I

QUANTITIES, AND THE RATIOS OF QUANTITIES, WHICH IN ANY FINITE TIME CONVERGE CONTINUALLY TO EQUALITY, AND BEFORE THE END OF THAT TIME APPROACH NEARER THE ONE TO THE OTHER THAN BY ANY GIVEN DIFFERENCE, BECOME ULTIMATELY EQUAL.

If you deny it, suppose them to be ultimately unequal, and let D be their ultimate difference. Therefore they cannot approach nearer to equality than by that given difference D; which is against the supposition.

LEMMA II

IF IN ANY FIGURE AacE, TERMINATED BY THE RIGHT LINES Aa, AE, AND THE CURVE acE, THERE BE INSCRIBED ANY NUMBER OF PARALLELOGRAMS Ab, Bc, Cd, &c., COMPREHENDED UNDER EQUAL BASES AB, BC, CD, &c., AND THE SIDES, Bb, Cc, Dd, &c., PARALLEL TO ONE SIDE Aa OF THE FIGURE; AND THE PARALLELOGRAMS aKbl, bLcm, cMdn, &c., ARE COMPLETED. THEN IF THE BREADTH OF THOSE PARALLELOGRAMS BE SUPPOSED TO BE DIMINISHED, AND THEIR NUMBER TO BE AUGMENTED IN INFINITUM; I SAY, THAT THE ULTIMATE RATIOS WHICH THE INSCRIBED FIGURE AKbLcMdD, THE CIRCUMSCRIBED FIGURE A$albmendo$E, AND CURVILINEAR FIGURE A$abcd$E, WILL HAVE TO ONE ANOTHER, ARE RATIOS OF EQUALITY.

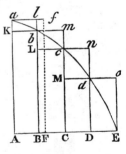

For the difference of the inscribed and circumscribed figures is the sum of the parallelograms Kl, Lm, Mn, Do, that is (from the equality of all their bases), the rectangle under one of their bases Kb and the sum of their altitudes Aa, that is, the rectangle ABla. But this rectangle, because its breadth AB is supposed diminished in infinitum, becomes less than any given space. And therefore (by Lem. I) the figures inscribed and circumscribed become ultimately equal one to the other; and much more will the intermediate curvilinear figure be ultimately equal to either.

Q.E.D.

LEMMA III

The same ultimate ratios are also ratios of equality, when the breadths, AB, BC, DC, &c., of the parallelograms are unequal, and are all diminished in infinitum.

For suppose AF equal to the greatest breadth, and complete the parallelogram FA*af*. This parallelogram will be greater than the difference of the inscribed and circumscribed figures; but, because its breadth AF is diminished in infinitum, it will become less than any given rectangle.

Q.E.D.

COR. 1. Hence the ultimate sum of those evanescent parallelograms will in all parts coincide with the curvilinear figure.

COR. 2. Much more will the rectilinear figure comprehended under the chords of the evanescent arcs *ab*, *bc*, *cd*, &c., ultimately coincide with the curvilinear figure.

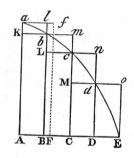

COR. 3. And also the circumscribed rectilinear figure comprehended under the tangents of the same arcs.

COR. 4. And therefore these ultimate figures (as to their perimeters E) are not rectilinear, but curvilinear limits of rectilinear figures.

LEMMA IV

IN TWO FIGURES A*ac*E, P*pr*T, YOU INSCRIBE (AS BEFORE) TWO RANKS OF PARALLELOGRAMS, AN EQUAL NUMBER IN EACH RANK, AND, WHEN THEIR BREADTHS ARE DIMINISHED IN INFINITUM, THE ULTIMATE RATIOS OF THE PARALLELOGRAMS IN ONE FIGURE TO THOSE IN THE OTHER, EACH TO EACH RESPECTIVELY, ARE THE SAME; I SAY, THAT THOSE TWO FIGURES A*ac*E, P*pr*T, ARE TO ONE ANOTHER IN THAT SAME RATIO.

For as the parallelograms in the one are severally to P the parallelograms in the other, so (by composition) is the sum of all in the one to the sum of all in the other; and so is the one figure to the other; because (by Lem. III) the former figure to the former sum, and the latter figure to the latter sum, are both in the ratio of equality.

Q.E.D.

COR. Hence if two quantities of any kind are any how divided into an equal number of parts, and those parts, when their number is augmented, and their magnitude diminished in infinitum, have a given ratio one to the other, the first to the first, the second to the second, and so on in order, the whole quantities

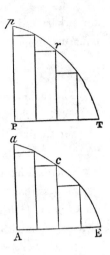

will be one to the other in that same given ratio. For if, in the figures of this Lemma, the parallelograms are taken one to the other in the ratio of the parts, the sum of the parts will always be as the sum of the parallelograms; and therefore supposing the number of the parallelograms and parts to be augmented, and their magnitudes diminished in infinitum, those sums will be in the ultimate ratio of the parallelogram in the one figure to the correspondent parallelogram in the other; that is (by the supposition), in the ultimate ratio of any part of the one quantity to the correspondent part of the other.

LEMMA V

IN SIMILAR FIGURES, ALL SORTS OF HOMOLOGOUS SIDES, WHETHER CURVILINEAR OR RECTILINEAR, ARE PROPORTIONAL; AND THE AREAS ARE IN THE DUPLICATE RATIO OF THE HOMOLOGOUS SIDES.

LEMMA VI

IF ANY ARC ACB, GIVEN IN POSITION IS SUBTENDED BY ITS CHORD AB, AND IN ANY POINT A, IN THE MIDDLE OF THE CONTINUED CURVATURE, IS TOUCHED BY A RIGHT LINE AD, PRODUCED BOTH WAYS; THEN IF THE POINTS A AND B APPROACH ONE ANOTHER AND MEET, I SAY, THE ANGLE BAD, CONTAINED BETWEEN THE CHORD AND THE TANGENT, WILL BE DIMINISHED IN INFINITUM, AND ULTIMATELY WILL VANISH.

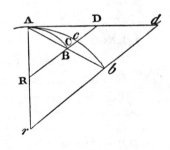

For if that angle does not vanish, the arc ACB will contain with the tangent AD an angle equal to a rectilinear angle; and therefore the curvature at the point A will not be continued, which is against the supposition.

LEMMA VII

THE SAME THINGS BEING SUPPOSED, I SAY THAT THE ULTIMATE RATIO OF THE ARC, CHORD, AND TANGENT, ANY ONE TO ANY OTHER, IS THE RATIO OF EQUALITY.

For while the point B approaches towards the point A, consider always AB and AD as produced to the remote points b and d, and parallel to the secant BD draw bd: and let the arc Acb be always similar to the arc ACB. Then, supposing the points

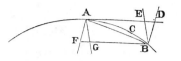

A and B to coincide, the angle dAb will vanish, by the preceding Lemma; and therefore the right lines Ab, Ad (which are always finite), and the intermediate arc Acb, will coincide, and become equal among themselves. Wherefore, the rightlines AB, AD, and the intermediate arc ACB (which are always proportional to the former), will vanish, and ultimately acquire the ratio of equality.

Q.E.D.

COR. 1. Whence if through B we draw BF parallel to the tangent, always cutting any right line AF passing through A in F, this line BF will be ultimately in the ratio of equality with the evanescent arc ACB; because, completing the parallelogram AFBD, it is always in a ratio of equality with AD.

COR. 2. And if through B and A more right lines are drawn, as BE, BD, AF, AG, cutting the tangent AD and its parallel BF; the ultimate ratio of all the abscissas AD, AE, BF, BG, and of the chord and arc AB, any one to any other, will be the ratio of equality.

COR. 3. And therefore in all our reasoning about ultimate ratios, we may freely use any one of those lines for any other.

LEMMA VIII

IF THE RIGHT LINES AR, BR, WITH THE ARC ACB, THE CHORD AB, AND THE TANGENT AD, CONSTITUTE THREE TRIANGLES RAB, RACB, RAD, AND THE POINTS A AND B APPROACH AND MEET: I SAY, THAT THE ULTIMATE FORM OF THESE EVANESCENT TRIANGLES IS THAT OF SIMILITUDE, AND THEIR ULTIMATE RATIO THAT OF EQUALITY.

For while the point B approaches towards the point A, consider always AB, AD, AR, as produced to the remote points b, d, and r, and rbd as drawn parallel to RD, and let the arc Acb be always similar to the arc ACB. Then supposing the points A and B to coincide. The angle bAd will vanish; and therefore the three triangles rAb, $rAci$, rAd (which are always finite), will coincide, and on that account become both similar and equal. And therefore the triangles RAB, RACB, RAD which are always similar and proportional to these, will ultimately become both similar and equal among themselves.

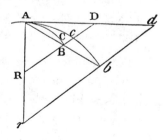

Q.E.D.

COR. And hence in all reasonings about ultimate ratios, we may indifferently use any one of those triangles for any other.

LEMMA IX

IF A RIGHT LINE AE, AND A CURVED LINE ABC, BOTH GIVEN BY POSITION, CUT EACH OTHER IN A GIVEN ANGLE, A; AND TO THAT RIGHT LINE, IN ANOTHER GIVEN ANGLE, BD, CE ARE ORDINATELY APPLIED, MEETING, THE CURVE IN B, C; AND THE POINTS B AND C TOGETHER APPROACH TOWARDS AND MEET IN THE POINT A: I SAY, THAT THE AREAS OF THE TRIANGLES ABD, ACE, WILL ULTIMATELY BE ONE TO THE OTHER IN THE DUPLICATE RATIO OF THE SIDES.

For while the points B, C, approach towards the point A, suppose always AD to be produced to the remote points *d* and *e*, so as A*d*, A*e* may be proportional to AD, AE; and the ordinates *db*, *ec*, to be drawn parallel to the ordinates DB and EC, and meeting AB and AC produced in *b* and *c*. Let the curve A*bc* be similar to the curve ABC, and draw the right line A*g* so as to touch both curves in A, and cut the ordinates DB, EC, *db*, *ec*, in F, G, *f*, *g*. Then, supposing the length A*e* to remain the same, let the points B and C meet in the point A; and the angle *c*A*g* vanishing, the curvilinear areas A*bd*, A*ce* will coincide with the rectilinear areas A*fd*, A; 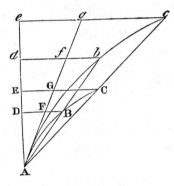 and therefore (by Lem. V) will be one to the other in the duplicate ratio of the sides A*d*, A*e*. But the areas ABD, ACE are always proportional to these areas; and so the sides AD, AE are to these sides. And therefore the areas ABD, ACE are ultimately one to the other in the duplicate ratio of the sides AD, AE.   Q.E.D.

LEMMA X

THE SPACES WHICH A BODY DESCRIBES BY ANY FINITE FORCE URGING IT, WHETHER THAT FORCE IS DETERMINED AND IMMUTABLE, OR IS CONTINUALLY AUGMENTED OR CONTINUALLY DIMINISHED, ARE IN THE VERY BEGINNING OF THE MOTION ONE TO THE OTHER IN THE DUPLICATE RATIO OF THE TIMES.

Let the times be represented by the lines AD, AE, and the velocities generated in those times by the ordinates DB, EC. The spaces described with these velocities will be as the areas ABD, ACE, described by those ordinates, that is, at the very beginning of the motion (by Lem. IX), in the duplicate ratio of the times AD, AE.   Q.E.D.

COR. 1. And hence one may easily infer, that the errors of bodies describing similar parts of similar figures in proportional times, are nearly as the squares of the times in which they are generated; if so be these errors are generated by any equal forces similarly applied to the bodies, and measured by the distances of the bodies from those places of the similar figures, at which, without the action of those form, the bodies would have arrived in those proportional times.

COR. 2. But the errors that are generated by proportional forces, similarly applied to the bodies at similar parts of the similar figures, are as the forces and the squares of the times conjunctly.

COR. 3. The same thing is to be understood of any spaces whatsoever described by bodies urged with different forces; all which, in the very beginning of the motion, are as the forces and the squares of the times conjunctly.

COR. 4. And therefore the forces are as the spaces described in the very beginning of the motion directly, and the squares of the times inversely.

COR. 5. And the squares of the times are as the spaces described directly, and the forces inversely.

SCHOLIUM

If in comparing indetermined quantities of different sorts one with another, any one is said to be as any other directly or inversely, the meaning is, that the former is augmented or diminished in the same ratio with the latter, or with its reciprocal. And if any one is said to be as any other two or more directly or inversely, the meaning is, that the first is augmented or diminished in the ratio compounded of the ratios in which the others, or the reciprocals of the others, are augmented or diminished. As if A is said to be as B directly, and C directly, and D inversely, the meaning is, that A is augmented or diminished in the same ratio with $B \times C \times \frac{1}{D}$, that is to say, that A and $\frac{BC}{D}$ are one to the other in a given ratio.

LEMMA XI

THE EVANESCENT SUBTENSE OF THE ANGLE OF CONTACT, IN ALL
CURVES WHICH AT THE POINT OF CONTACT HAVE A FINITE
CURVATURE, IS ULTIMATELY IN THE DUPLICATE RATIO OF
THE SUBTENSE OF THE CONTERMINATE ARC.

CASE 1. Let AB be that arc, AD its tangent, BD the subtense of the angle of contact perpendicular on the tangent, AB the subtense of the arc. Draw BG perpendicular to the subtense AB, and AG to the tangent AD, meeting in G; then let the points D, B, and G, approach to the points d, b, and g, and suppose J to be the ultimate intersection of the lines BG, AG, when the points D, B, have come to A. It is evident that the distance GJ may be less than any assignable. But (from the nature of the circles passing through q the points A, B, G, A, b, g,) $AB^2 = AG \times BD$, and $A^2 = Ag \times bd$; and therefore the ratio of AB^2 to Ab^2 is compounded of the ratios of AG to Ag, and of Bd to bd. But because GJ may be assumed of less length than any assignable, the ratio of AG to Ag may be such as to differ from the ratio of equality by less than any assignable difference; and therefore the ratio of AB^2 to Ab^2 may be such as to differ from the ratio of BD to bd by less than any assignable difference. Therefore, by Lem. I, the ultimate ratio of AB^2 to Ab^2 is the same with the ultimate ratio of BD to bd. Q.E.D.

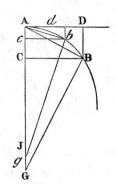

CASE 2. Now let BD be inclined to AD in any given angle, and the ultimate ratio of BD to bd will always be the same as before and therefore the same with the ratio of AB^2 to Ab^2. Q.E.D.

CASE 3. And if we suppose the angle D not to be given, but that the right line BD converges to a given point, or is determined by any other condition whatever; nevertheless the angles D, d, being determined by the same law, will always draw nearer to equality, and approach nearer to each other than by any assigned difference, and therefore, by Lem. I, will at last be equal; and therefore the lines BD, bd are in the same ratio to each other as before. Q.E.D.

COR. 1. Therefore since the tangents AD, Ad, the arcs AB, Ab, and their sines, BC, bc, become ultimately equal to the chords AB, Ab, their squares will ultimately become as the subtenses BD, bd.

COR. 2. Their squares are also ultimately as the versed sines of the arcs, bisecting the chords, and converging to a given point. For those versed sines are as the subtenses BD, bd.

COR. 3. And therefore the versed sine is in the duplicate ratio of the time in which a body will describe the arc with a given velocity.

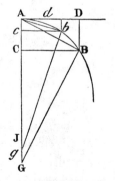

COR. 4. The rectilinear triangles ADB, Adb are ultimately in the triplicate ratio of the sides AD, Ad, and in a sesquiplicate ratio of the sides DB, db; as being in the ratio compounded of the sides AD to DB, and of Ad to db. So also the triangles ABC, Abc are ultimately in the triplicate ratio of the sides BC, bc. What I call the sesquiplicate ratio is the subduplicate of the triplicate, as being compounded of the simple and subduplicate ratio.

COR. 5. And because DB, db are ultimately parallel and in the duplicate ratio of the lines AD, Ad, the ultimate curvilinear areas ADB, Adb will be (by the nature of the parabola) two thirds of the rectilinear triangles ADB, Adb; and the segments AB, Ab will be one third of the same triangles. And thence those areas and those segments will be in the triplicate ratio as well of the tangents AD, A*d*, as of the chords and arcs AB, AB.

SCHOLIUM

But we have all along supposed the angle of contact to be neither infinitely greater nor infinitely less than the angles of contact made by circles and their tangents; that is, that the curvature at the point A is neither infinitely small nor infinitely great, or that the interval AJ is of a finite magnitude. For DB may be taken as AD^3: in which case no circle can be drawn through the point A, between the tangent AD and the curve AB, and therefore the angle of contact will be infinitely less than those of circle. And by a like reasoning, if DB be made successfully as AD^4, AD^5, AD^6, AD^7 &c., we shall have a series of angles of contact, proceeding in infinitum, wherein every succeeding term is infinitely less than the preceding. And if DB be made successively as AD^2, $AD^{3/2}$ $AD^{4/3}$ $AD^{5/4}$ $AD^{6/5}$ AD^7, &c., we shall have another infinite series of angles of contact, the first of which is of the same sort with those of circles, the second infinitely greater, and every succeeding one infinitely greater than the preceding. But between any two of these angles another series of intermediate angles of contact may be interposed, proceeding both ways in infinitum, wherein every succeeding angle shall be infinitely greater or infinitely less than the preceding. As if between the terms AD^2 and AD^3 there were interposed the series $AD^{13/6}$, $AD^{11/5}$ $AD^{9/4}$ $AD^{7/3}$ $AD^{5/2}$ $AD^{8/3}$ $AD^{11/4}$ $AD^{14/5}$ $AD^{17/6}$, &c. And again, between any two angles of this series, a new series of intermediate angles may be interposed, differing from one another by infinite intervals. Nor is nature confined to any bounds.

Those things which have been demonstrated of curve lines, and the superfices which they comprehend, may be easily applied to the curve superfices and contents of solids. These Lemmas are premised to avoid the tediousness of deducing perplexed demonstrations ad absurdum, according to the method of the ancient geometers. For demonstrations are more contracted by the method of indivisibles: but because the hypothesis of indivisibles seems somewhat harsh, and therefore that method is reckoned less geometrical, I chose rather to reduce the demonstrations of the following propositions to the first and last sums and ratios of nascent and evanescent quantities, that is, to the limits of those sums and ratios; and so to premise, as short as I could, the demonstrations of those limits. For hereby the same thing is performed as by the method of indivisibles; and now those principles being demonstrated, we may use them with more safety. Therefore if hereafter I should happen to consider quantities as made up of particles, or should use little curve lines for right ones, I would not be understood to mean indivisibles, but evanescent divisible quantities;

not the sums and ratios of determinate parts, but always the limits of sums and ratios; and that the force of such demonstrations always depends on the method laid down in the foregoing Lemmas.

Perhaps it may be objected, that there is no ultimate proportion, of evanescent quantities; because the proportion, before the quantities have vanished, is not the ultimate, and when they are vanished, is none. But by the same argument, it may be alledged, that a body arriving at a certain place, and there stopping, has no ultimate velocity: because the velocity, before the body comes to the place, is not its ultimate velocity; when it has arrived, is none. But the answer is easy; for by the ultimate velocity is meant that with which the body is moved, neither before it arrives at its last place and the motion ceases, nor after, but at the very instant it arrives; that is, that velocity with which the body arrives at its last place, and with which the motion ceases. And in like manner, by the ultimate ratio of evanescent quantities is to be understood the ratio of the quantities not before they vanish, nor afterwards, but with which they vanish. In like manner the first ratio of nascent quantities is that with which they begin to be. And the first or last sum is that with which they begin and cease to be (or to be augmented or diminished). There is a limit which the velocity at the end of the motion may attain, but not exceed. This is the ultimate velocity. And there is the like limit in all quantities and proportions that begin and cease to be. And since such limits are certain and definite, to determine the same is a problem strictly geometrical. But whatever is geometrical we may be allowed to use in determining and demonstrating any other thing that is likewise geometrical.

It may also be objected, that if the ultimate ratios of evanescent quantities are given, their ultimate magnitudes will be also given: and so all quantities will consist of indivisibles, which is contrary to what Euclid has demonstrated concerning incommensurables, in the 10th Book of his Elements. But this objection is founded on a false supposition. For those ultimate ratios with which quantities vanish are not truly the ratios of ultimate quantities, but limits towards which the ratios of quantities decreasing without limit do always converge; and to which they approach nearer than by any given difference, but never go beyond, nor in effect attain to, till the quantities are diminished in infinitum. This thing will appear more evident in quantities infinitely great. If two quantities, whose difference is given, be augmented in infinitum, the ultimate ratio of these quantities will be given, to wit, the ratio of equality; but it does not from thence follow, that the ultimate or greatest quantities themselves, whose ratio that is, will be given. Therefore if in what follows, for the sake of being more easily understood, I should happen to mention quantities as least, or evanescent, or ultimate, you are not to suppose that quantities of any determinate magnitude are meant, but such as are conceived to be always diminished without end.

Pierre Simon de Laplace

(1749–1827)

HIS LIFE AND WORK

ritics of contemporary Darwinian biology often point to highly complex biological systems such as blood clotting, that they describe as "irreducibly complex," so complex that their evolution cannot be explained by Darwinian evolution. They cite this as evidence of intelligent design and, therefore, an intelligent designer. They would do well to cast their glance back two centuries to the accomplishments of Pierre Simon de Laplace, perhaps the greatest mathematical physicist of all times.

Isaac Newton's *Principia* unified and simplified all of physical science as it was then known to man. But it still left many questions unanswered, perhaps even unanswerable by physical science. Shortly after its publication in 1687, the Reverend Richard Bentley, one of Newton's earliest supporters and popularizers, presented the master with a question not considered in the *Principia*. Could the solar system, with the planets all revolving around the sun in the same direction in almost the same plane, be formed out of an initial uniform distribution of matter by the action of only natural causes or was it evidence of design?

Newton believed that his system could in no way explain these obvious regularities in the heavens. He replied that most definitely they could not result from the action of only natural causes. The cause had "to be not blind and fortuitous, but very skilled in Mechanics and Geometry."

And so matters stood for nearly the entire eighteenth century until Pierre Simon de Laplace blazed his way across the firmament of French science.

Pierre Simon de Laplace was born in a small town in Normandy in 1749. It is entirely possible that some of his distant uncles were among the Normans that invaded England in 1066 with William the Conqueror, but we can never know for sure. We do know that Laplace's father was a successful cider trader and that his mother came from a prosperous farming family—a comfortable background to be sure, but one that might have suited a son to a career in the clergy or the military.

That indeed is the educational path on which Laplace embarked, studying first at local Benedictine schools. At age 16, he matriculated at the nearby University of Caen intending to pursue a theology degree. However, Laplace soon discovered the joys of mathematics. After two years, he left Caen for Paris, the center of French science. In Paris, Laplace quickly impressed the mathematician Jean Le Rond d'Alembert who obtained a teaching position for his young protégé at the École Militaire. In that position, Laplace taught geometry and trigonometry to cadets from good families, but with little interest in the subject. Nevertheless, the position allowed Laplace to remain in Paris, the center of French, indeed global, mathematics.

In 1770, Laplace began a rapid outpouring of papers on a wide variety of topics in pure and applied mathematics, drawing wide attention to himself. The most important papers focused on outstanding problems in planetary theory. The orbits of the two largest planets Jupiter and Saturn sometimes lagged behind and sometimes ran ahead of their predicted position. Laplace sought to explain how the planets influenced each other in their orbits. This is a more difficult problem than the three-body problem which even today can only be solved by successive approximations. Laplace demonstrated that perturbations were not cumulative, as Newton feared, but periodic. God did not need to intervene to keep the Solar System from collapsing.

Given this heady success, Laplace first sought election to the prestigious Acadèmie des Sciences at the tender age of 22. Undismayed by his initial failure in 1771, he tried again the next year with the same result; losing the election to someone he considered to be a mediocre candidate who surpassed Laplace in only one dimension, age! Laplace never hesitated to let anyone know the high opinion he held of himself. Even in his early twenties he considered himself the best mathematician in France. His mentors may not have held quite such a high opinion of the young Laplace, but they certainly recognized his talent. In 1773, d'Alembert and Lagrange conspired to have Laplace elected first to the Berlin Academy of Science and then as a special adjunct to the Paris Academy.

With a place in the French scientific establishment secured, Laplace continued publishing profusely and teaching as necessary. In 1784, the army appointed him an examiner for the Royal Artillery Corps. This appointment secured Laplace an additional income and also gave him visibility to high government officials. But perhaps the most important relationship Laplace made from this appointment was with one of his first students, a sixteen-year-old from Corsica named Napolean Bonaparte!

In 1788, at the age of thirty-nine, Laplace married Marie-Charlotte de Courty de Romanges, the daughter of a well-to-do squire from Besançon in the French Alps. She was not much more than half his age. Within a year their marriage was blessed with the arrival of a son, Charles-Émile, who went on to have a distinguished military career but died childless in 1874. Madame Laplace gave birth to a daughter, Sophie-Suzanne, in 1792. Sophie married the Marquis de Partes, a minor Napoleonic noble. She died giving birth to Laplace's only grandchild, a girl also named Sophie. Young Sophie later married the Comte de Colbert Chabannais, the scion of a major aristocratic family. They had many children, and Laplace's male descendants now bear the family name Colbert-Laplace.

Engrossed as he was in science, Laplace had little time for the tumult of politics in the late 1780s and early 1790s. He took no part in the affairs of the French Revolution during its most radical phase in the early 1790s, except to participate in the committee that devised the metric system, part of a systematic attempt to overthrow the shackles of the Old Regime. One camp argued that the fundamental unit of length should be defined in terms of the earth's equatorial circumference. Laplace argued, instead, that given the role of the right angle in geometry, the fundamental unit of length should be based on the length of the quadrant from the North Pole to the equator. Thanks to Laplace's successful argument, the meter was defined as $\frac{1}{10,000}$ of the distance from the pole to the equator.

By the end of 1793 the political atmosphere in Paris became too intemperate for Laplace. Along with many other leading scientists, including Lavoisier and Coulomb, Laplace was purged from the committee devising the metric system. The radical republicans loudly announced that such responsibilities could only be entrusted to men known for "their Republican virtues and hatred of kings."

Literally fearing for his neck, Laplace and his family fled from Paris to the country town of Melun, thirty miles away. In retrospect, Laplace had correctly judged the probabilities of the possible outcomes. His friend and colleague, the chemist Antoine Lavoisier, chose to remain in Paris, where he lost his head to the guillotine in the spring of 1794.

A semblance of order returned to Paris in July 1794 following the overthrow of the Jacobin dictatorship and its lead henchman Robespierre. Shortly thereafter the new government recalled Laplace to Paris to help French science recover from the radical excesses of the Revolution. Laplace zestfully attended to his new role, naturally gravitating to a leadership role in the refounding of the great schools of learning, the Écoles Normale, Polytechnique, and Centrale des Travaux Publics. The new government also reincarnated the Acadèmie des Sciences as the Institut de France which elected Laplace its vice president in December 1795 and then, in April 1796, its president.

The years following his fiftieth birthday saw the publication of Laplace's greatest works. In 1799, he published the first of what would eventually be five volumes of his epoch work, *Mécanique céleste* (*Celestial Mechanics*). The last volume was published in 1825.

During the course of the eighteenth century, astronomers discovered the planet Uranus as well as several additional moons of Saturn. All of the known planets and moons revolved in the same direction and almost in the sun's equatorial plane. The sun, planets, and the moons all rotated in the same direction as the earth. If the commonality in all twenty-nine motions was merely the result of chance, then the probability of that chance would be exceedingly small, $(\frac{1}{2})^{29}$, about 1 in 100 billion!

Laplace recognized that the regularity in the heavens had to have a physical cause and he found it. He achieved what Newton had thought impossible. In the *Mécanique céleste*, Laplace demonstrated how the solar system could have arisen from a whirling solar atmosphere with planets whose moons formed as rotating balls of gas that cooled and condensed, with the revolutions necessarily in the same direction in almost the same plane about the sun.

Laplace's work on probability began as early as 1780. In 1812, he published the first edition of his masterpiece in that area, the *Théorie analytique des probabilités* (*Analytical Theory of Probabilities*). Prior to Laplace, all mathematicians working in the area of probability had presumed that one could have knowledge of absolute probabilities of chance events. Laplace disagreed. Laplace held that because the world is determined, there can be no probabilities in things. Probability results from our lack of knowledge. Laplace wrote that probability

> . . . consists in reducing all events of the same kind to a certain number of equally possible cases, that is to say, those such that we are equally undecided about their existence.

In so doing, Laplace expanded the scope of probability theory to include the probability of past causes of known events, as well as future outcomes. In the *Analytical*

Theory, Laplace developed formulae for assigning degrees of rational belief to propositions about past events.

The concluding problem from the first chapter is a good example of the sort of problem that attracted Laplace. In this problem, Laplace begins with a ring of urns. One urn contains only black balls. Another urn contains only white balls. All of the other urns contain widely varied mixtures of black and white balls. Laplace considers the following process: withdraw a ball from one urn, place it in the next urn, shake that urn well, withdraw a ball from that urn and place it in the next urn. Laplace demonstrated that eventually there would be the same ratio of black to white balls on all of the urns. In so doing, Laplace actually gave an example of the tendency of natural forces to impose order on even the most chaotic of systems, akin to his demonstration of the solar system's order.

The critical success of the *Analytical Theory of Probabilities* inspired Laplace to compose a popular version, the *Essai philosophique sur les probabilités* (*A Philosophical Essay on Probabilities*), first published in 1814 and reproduced here. Many commentators consider this essay the most complete perfect argument ever made for a deterministic interpretation of the universe.

The *Essay* begins with the statement of seven principles.

1. *Probability is the ratio of the number of favorable cases to the number of all possible cases.*

2. *If the cases are not all equally possible we must first determine their respective possibilities.*

3. *If events are independent of one another the probability of their combined existence is the product of their respective probabilities.*

4. *When two events depend on each other, the probability of the compound event is the product of the first event and the probability that, this event having occurred, the second event will occur.*

5. *If we calculate* a priori *the probability of the occurred event and the probability of an event composed of that one and a second one that is expected, the second probability divided by the first will be the probability of the event expected, drawn from the observed event.*

6. *Each of the causes to which an observed event may be attributed is indicated with just as much likelihood as there is probability that the event will take place, supposing the event to be constant.*

7. *The probability of a future event is the sum of the products of the probability of each cause, drawn from the event observed, by the probability that, this cause existing, the future event will occur.*

Not surprisingly, the *Essay* includes many problems involving coin tossing. Laplace wrote:

> *One regards two events as equally probable when one can see no reason that would make one more probable than the other, because, even though there is an unequal possibility between them, we know not which way, and this uncertainty makes us look on each as if it were as probable as the other.*

Thus, even if we are told that a coin is biased, we have no way of knowing the bias, that is if we are ignorant of the bias, we begin with the assumption that it is a fair cause. When considering a run of twenty successive heads during a much longer run of coin tosses, Laplace asks whether the twenty successive heads are to be considered a chance event or the result of a cause. During the analysis Laplace writes:

> *We look upon a thing as the effect of chance when we see nothing regular in it, nothing that manifests design, and when furthermore, we are ignorant of the causes that brought it about. Thus, chance has no reality in itself. It is nothing but a term for expressing our ignorance of the way in which the various aspects of a phenomenon are interconnected and related to the rest of nature.*

Is the run of twenty heads a matter of chance or a manifestation of design? To the common person, it appears to be a regularity, but is it? Consider how this problem reappears when playing a lottery in which six numbers get picked out of fifty. Many people would avoid playing the set {01,02,03,04,05,06} supposing that that set of six numbers shows much more regularity than the recent winning set of numbers {06,13,15,15,32,36}. But, Laplace would argue, a thorough analysis of the lottery process would reveal the independence of each number's selection. Thus, the set {01,02,03,04,05,06} shows neither more nor less regularity than the set {06,13,15, 15,32,36}. Given the fact that a lottery payout depends on the number of winners, the very fact that many people would avoid playing {01,02,03,04,05,06} is, in fact, a very good reason to play it. There would be less likelihood of anyone else having the same winning set of numbers!

In his later years, Laplace directed his focus to a wide variety of topics in mathematical physics: capillary action, double refraction, the velocity of sound, the shape and heat of the earth, and elastic fluids.

Laplace had no firm political views, other than making sure to keep his head firmly connected to the rest of his body so that he could pursue his scientific work. Napoleon had made him a member of the Senate. But in 1814, Laplace voted in favor of restoring the monarchy and exiling his former pupil and patron Napoleon to exile in Elba. When Napoleon briefly returned to power in 1815, Laplace once again exited himself from Paris. After Napoleon's defeat at Waterloo and exile to St Helena, Laplace became a staunch supporter of the Bourbon monarchy. In return, Louis XVIII elevated Laplace to a peerage, making him a Marquis in 1817.

Shortly before his seventy-eighth birthday in 1827, Laplace died following two years of enfeeblement. In his funeral elegy, Fourier remarked that Laplace retained his extraordinary powers of memory and critical thinking until almost the end of his life. Fourier also remarked:

> *Laplace was born to perfect everything, to deepen everything, to push back all the boundaries, to solve what was thought to be insoluble.*

In his funeral elegy, the mathematician Simeon Denis Poisson remarked:

> *For Laplace, mathematical analysis was an instrument that he bent to his purposes for the most varied applications, but always subordinating the method itself to the content of each question.*

Surely the *Mécanique céleste* was Laplace's supreme achievement in the application of mathematical analysis. Laplace dedicated the third volume of his *Mécanique céleste* to Napoleon. After browsing through the vast volume, Napoleon commented that he saw no mention of God. On hearing this, Laplace responded "Sire, there is no need of that hypothesis!" Darwin's modern critics might do well to remember that.

A PHILOSOPHICAL ESSAY ON PROBABILITIES

CHAPTER I

INTRODUCTION

THIS philosophical essay is the development of a lecture on probabilities which I delivered in 1795 to the normal schools whither I had been called, by a decree of the national convention, as professor of mathematics with Lagrange. I have recently published upon the same subject a work entitled *The Analytical Theory of Probabilities*. I present here without the aid of analysis the principles and general results of this theory, applying them to the most important questions of life, which are indeed for the most part only problems of probability. Strictly speaking it may even be said that nearly all our knowledge is problematical; and in the small number of things which we are able to know with certainty, even in the mathematical sciences themselves, the principal means for ascertaining truth—induction, and analogy—are based on probabilities; so that the entire system of human knowledge is connected with the theory set forth in this essay. Doubtless it will be seen here with interest that in considering, even in the eternal principles of reason, justice, and humanity, only the favorable chances which are constantly attached to them, there is a great advantage in following these principles and serious inconvenience in departing from them: their chances, like those favorable to lotteries, always end by prevailing in the midst of the vacillations of hazard. I hope that the reflections given in this essay may merit the attention of philosophers and direct it to a subject so worthy of engaging their minds.

CHAPTER II

CONCERNING PROBABILITY

ALL events, even those which on account of their insignificance do not seem to follow the great laws of nature, are a result of it just as necessarily as the revolutions of the sun. In ignorance of the ties which unite such events to the entire system of the universe, they have been made to depend upon final causes or upon hazard, according as they occur and are repeated with regularity, or appear without regard to order; but these imaginary causes have gradually receded with the widening bounds of knowledge and disappear entirely before sound philosophy, which sees in them only the expression of our ignorance of the true causes.

Present events are connected with preceding ones by a tie based upon the evident principle that a thing cannot occur without a cause which produces it. This axiom, known by the name of *the principle of sufficient reason*, extends even to actions which are considered indifferent; the freest will is unable without a determinative motive to give them birth; if we assume two positions with exactly similar circumstances and find that the will is active in the one and inactive in the other, we say that its choice is an effect without a cause. It is then, says Leibnitz, the blind chance of the Epicureans. The contrary opinion is an illusion of the mind, which, losing sight of the evasive reasons of the choice of the will in indifferent things, believes that choice is determined of itself and without motives.

We ought then to regard the present state of the universe as the effect of its anterior state and as the cause of the one which is to follow. Given for one instant an intelligence which could comprehend all the forces by which nature is animated and the respective situation of the beings who compose it—an intelligence sufficiently vast to submit these data to analysis—it would embrace in the same formula the movements of the greatest bodies of the universe and those of the lightest atom; for it, nothing would be uncertain and the future, as the past, would be present to its eyes. The human mind offers, in the perfection which it has been able to give to astronomy, a feeble idea of this intelligence. Its discoveries in mechanics and geometry, added to that of universal gravity, have enabled it to comprehend in the same analytical expressions the past and future states of the system of the world. Applying the same method to some other objects of its knowledge, it has succeeded in referring to general laws observed phenomena and in foreseeing those which given circumstances ought to produce. All these efforts in the search for truth tend to lead it back continually to the vast intelligence which we have just mentioned, but from which it will always remain infinitely removed. This tendency, peculiar to the human race, is that which renders it superior to animals; and their progress in this respect distinguishes nations and ages and constitutes their true glory.

Let us recall that formerly, and at no remote epoch, an unusual rain or an extreme drought, a comet having in train a very long tail, the eclipses, the aurora borealis, and in general all the unusual phenomena were regarded as so many signs of celestial wrath. Heaven was invoked in order to avert their baneful influence. No one prayed to have the planets and the sun arrested in their courses: observation had soon made apparent the futility of such prayers. But as these phenomena, occurring and disappearing at long intervals, seemed to oppose the order of nature, it was supposed that Heaven, irritated by the crimes of the earth, had created them to announce its vengeance. Thus the long tail of the comet of 1456 spread terror through Europe, already thrown into consternation by the rapid successes of the Turks, who had just overthrown the Lower Empire. This star after four revolutions has excited among us a very different interest. The knowledge of the laws of the system of the world acquired in the interval had dissipated the fears begotten by the ignorance of the true relationship of man to the universe; and Halley, having recognized the identity of this comet with those of the years 1531, 1607, and 1682, announced its next return for the end of the year 1758 or the beginning of the year 1759. The learned world awaited with impatience this return which was to confirm one of the greatest discoveries that have been made in the sciences, and fulfil the prediction of Seneca when he said, in speaking of the revolutions of those stars which fall from an enormous height: "The day will come when, by study pursued through several ages, the things now concealed will appear with evidence; and posterity will be astonished that truths so clear had escaped us." Clairaut then undertook to submit to analysis the perturbations which the comet had experienced by the action of the two great planets, Jupiter and Saturn; after immense calculations he fixed its next passage at the perihelion toward the beginning of April, 1759, which was actually verified by observation. The regularity which astronomy shows us in the movements of the comets doubtless exists also in all phenomena.

The curve described by a simple molecule of air or vapor is regulated in a manner just as certain as the planetary orbits; the only difference between them is that which comes from our ignorance.

Probability is relative, in part to this ignorance, in part to our knowledge. We know that of three or a greater number of events a single one ought to occur; but nothing induces us to believe that one of them will occur rather than the others. In this state of indecision it is impossible for us to announce their occurrence with certainty. It is, however, probable that one of these events, chosen at will, will not occur because we see several cases equally possible which exclude its occurrence, while only a single one favors it.

The theory of chance consists in reducing all the events of the same kind to a certain number of cases equally possible, that is to say, to such as we may be equally undecided about in regard to their existence, and in determining the number of cases favorable to the event whose probability is sought. The ratio of this number to that of all the cases possible is the measure of this probability, which is thus simply a fraction whose numerator is the number of favorable cases and whose denominator is the number of all the cases possible.

The preceding notion of probability supposes that, in increasing in the same ratio the number of favorable cases and that of all the cases possible, the probability remains the same. In order to convince ourselves let us take two urns, A and B, the first containing four white and two black balls, and the second containing only two white balls and one black one. We may imagine the two black balls of the first urn attached by a thread which breaks at the moment when one of them is seized in order to be drawn out, and the four white balls thus forming two similar systems. All the chances which will favor the seizure of one of the balls of the black system will lead to a black ball. If we conceive now that the threads which unite the balls do not break at all, it is clear that the number of possible chances will not change any more than that of the chances favorable to the extraction of the black balls; but two balls will be drawn from the urn at the same time; the probability of drawing a black ball from the urn A will then be the same as at first. But then we have obviously the case of urn B with the single difference that the three balls of this last urn would be replaced by three systems of two balls invariably connected.

When all the cases are favorable to an event the probability changes to certainty and its expression becomes equal to unity. Upon this condition, certainty and probability are comparable, although there may be an essential difference between the two states of the mind when a truth is rigorously demonstrated to it, or when it still perceives a small source of error.

In things which are only probable the difference of the data, which each man has in regard to them, is one of the principal causes of the diversity of opinions which prevail in regard to the same objects. Let us suppose, for example, that we have three urns, A, B, C, one of which contains only black balls while the two others contain only white balls; a ball is to be drawn from the urn C and the probability is demanded that this ball will be black. If we do not know which of the three urns contains black balls only, so that there is no reason to believe that it is C rather than B or A, these three hypotheses will appear equally possible, and since a black ball can be drawn only in the first hypothesis, the probability of drawing it is equal to one third. If it is known that the urn A contains white balls only, the indecision then extends only to the urns B and C, and the probability that the ball drawn from the urn C will be black is one half. Finally this probability changes to certainty if we are assured that the urns A and B contain white balls only.

It is thus that an incident related to a numerous assembly finds various degrees of credence, according to the extent of knowledge of the auditors. If the man who reports it is fully convinced of it and if, by his position and character, he inspires great confidence, his statement, however extraordinary it may be, will have for the auditors who lack information the same degree of probability as an ordinary statement made by the same man, and they will have entire faith in it. But if some one of them knows that the same incident is rejected by other equally trustworthy men, he will be in doubt and the incident will be discredited by the enlightened auditors, who will reject it whether it be in regard to facts well averred or the immutable laws of nature.

It is to the influence of the opinion of those whom the multitude judges best informed and to whom it has been accustomed to give its confidence in regard to the most important matters of life that the propagation of those errors is due which in times of ignorance have covered the face of the earth. Magic and astrology offer us two great examples. These errors inculcated in infancy, adopted without examination, and having for a basis only universal credence, have maintained themselves during a very long time; but at last the progress of science has destroyed them in the minds of enlightened men, whose opinion consequently has caused them to disappear even among the common people, through the power of imitation and habit which had so generally spread them abroad. This power, the richest resource of the moral world, establishes and conserves in a whole nation ideas entirely contrary to those which it upholds elsewhere with the same authority. What indulgence ought we not then to have for opinions different from ours, when this difference often depends only upon the various points of view where circumstances have placed us! Let us enlighten those whom we judge insufficiently instructed; but first let us examine critically our own opinions and weigh with impartiality their respective probabilities.

The difference of opinions depends, however, upon the manner in which the influence of known data is determined. The theory of probabilities holds to considerations so delicate that it is not surprising that with the same data two persons arrive at different results, especially in very complicated questions. Let us examine now the general principles of this theory.

CHAPTER III

THE GENERAL PRINCIPLES OF THE CALCULUS
OF PROBABILITIES

First Principle.—The first of these principles is the definition itself of probability, which, as has been seen, is the ratio of the number of favorable cases to that of all the cases possible.

Second Principle.—But that supposes the various cases equally possible. If they are not so, we will determine first their respective possibilities, whose exact appreciation is one of the most delicate points of the theory of chance. Then the probability will be the sum of the possibilities of each favorable case. Let us illustrate this principle by an example.

GOD CREATED THE INTEGERS

Let us suppose that we throw into the air a large and very thin coin whose two large opposite faces, which we will call heads and tails, are perfectly similar. Let us find the probability of throwing heads at least one time in two throws. It is clear that four equally possible cases may arise, namely, heads at the first and at the second throw; heads at the first throw and tails at the second; tails at the first throw and heads at the second; finally, tails at both throws. The first three cases are favorable to the event whose probability is sought; consequently this probability is equal to $\frac{3}{4}$; so that it is a bet of three to one that heads will be thrown at least once in two throws.

We can count at this game only three different cases, namely, heads at the first throw, which dispenses with throwing a second time; tails at the first throw and heads at the second; finally, tails at the first and at the second throw. This would reduce the probability to $\frac{2}{3}$ if we should consider with d'Alembert these three cases as equally possible. But it is apparent that the probability of throwing heads at the first throw is $\frac{1}{2}$, while that of the two other cases is $\frac{1}{4}$, the first case being a simple event which corresponds to two events combined: heads at the first and at the second throw, and heads at the first throw, tails at the second. If we then, conforming to the second principle, add the possibility $\frac{1}{2}$ of heads at the first throw to the possibility $\frac{1}{4}$ of tails at the first throw and heads at the second, we shall have $\frac{3}{4}$ for the probability sought, which agrees with what is found in the supposition when we play the two throws. This supposition does not change at all the chance of that one who bets on this event; it simply serves to reduce the various cases to the cases equally possible.

Third Principle.—One of the most important points of the theory of probabilities and that which lends the most to illusions is the manner in which these probabilities increase or diminish by their mutual combination. If the events are independent of one another, the probability of their combined existence is the product of their respective probabilities. Thus the probability of throwing one ace with a single die is $\frac{1}{6}$; that of throwing two aces in throwing two dice at the same time is $\frac{1}{36}$. Each face of the one being able to combine with the six faces of the other, there are in fact thirty-six equally possible cases, among which one single case gives two aces. Generally the probability that a simple event in the same circumstances will occur consecutively a given number of times is equal to the probability of this simple event raised to the power indicated by this number. Having thus the successive powers of a fraction less than unity diminishing without ceasing, an event which depends upon a series of very great probabilities may become extremely improbable. Suppose then an incident be transmitted to us by twenty witnesses in such manner that the first has transmitted it to the second, the second to the third, and so on. Suppose again that the probability of each testimony be equal to the fraction $\frac{9}{10}$; that of the incident resulting from the testimonies will be less than $\frac{1}{8}$. We cannot better compare this diminution of the probability than with the extinction of the light of objects by the interposition of several pieces of glass. A relatively small number of pieces suffices to take away the view of an object that a single piece allows us to perceive in a distinct manner. The historians do not appear to have paid sufficient attention to this degradation of the probability of events when seen across a great number of successive generations; many historical events reputed as certain would be at least doubtful if they were submitted to this test.

In the purely mathematical sciences the most distant consequences participate in the certainty of the principle from which they are derived. In the applications of analysis to physics the results have all the certainty of facts or experiences. But in the moral sciences, where each inference is deduced from that which precedes it only in a probable manner, however probable these deductions may be, the chance of error increases with their number and ultimately surpasses the chance of truth in the consequences very remote from the principle.

Fourth Principle.—When two events depend upon each other, the probability of the compound event is the product of the probability of the first event and the probability that, this event having occurred, the second will occur. Thus in the preceding case of the three urns A, B, C, of which two contain only white balls and one contains only black balls, the probability of drawing a white ball from the urn C is $\frac{2}{8}$, since of the three urns only two contain balls of that color. But when a white ball has been drawn from the urn C, the indecision relative to that one of the urns which contain only black balls extends only to the urns A and B; the probability of drawing a white ball from the urn B is $\frac{1}{2}$; the product of $\frac{2}{8}$ by $\frac{1}{2}$, or $\frac{1}{8}$, is then the probability of drawing two white balls at one time from the urns B and C.

We see by this example the influence of past events upon the probability of future events. For the probability of drawing a white ball from the urn B, which primarily is $\frac{2}{8}$, becomes $\frac{1}{2}$ when a white ball has been drawn from the urn C; it would change to certainty if a black ball had been drawn from the same urn. We will determine this influence by means of the following principle, which is a corollary of the preceding one.

Fifth Principle.—If we calculate à priori the probability of the occurred event and the probability of an event composed of that one and a second one which is expected, the second probability divided by the first will be the probability of the event expected, drawn from the observed event.

Here is presented the question raised by some philosophers touching the influence of the past upon the probability of the future. Let us suppose at the play of heads and tails that heads has occurred oftener than tails. By this alone we shall be led to believe that in the constitution of the coin there is a secret cause which favors it. Thus in the conduct of life constant happiness is a proof of competency which should induce us to employ preferably happy persons. But if by the unreliability of circumstances we are constantly brought back to a state of absolute indecision, if, for example, we change the coin at each throw at the play of heads and tails, the past can shed no light upon the future and it would be absurd to take account of it.

Sixth Principle.—Each of the causes to which an observed event may be attributed is indicated with just as much likelihood as there is probability that the event will take place, supposing the event to be constant. The probability of the existence of any one of these causes is then a fraction whose numerator is the probability of the event resulting from this cause and whose denominator is the sum of the similar probabilities relative to all the causes; if these various causes, considered *à priori*, are unequally probable, it is necessary, in place of the probability of the event resulting from each cause, to employ the product of this probability by the possibility of the cause itself. This is the fundamental principle of this branch of the analysis of chances which consists in passing from events to causes.

This principle gives the reason why we attribute regular events to a particular cause. Some philosophers have thought that these events are less possible than others and that at the play of heads and tails, for example, the combination in which heads occurs twenty successive times is less easy in its nature than those where heads and tails are mixed in an irregular manner. But this opinion supposes that past events have an influence on the possibility of future events, which is not at all admissible. The regular combinations occur more rarely only because they are less numerous. If we seek a cause wherever we perceive symmetry, it is not that we regard a symmetrical event as less possible than the others, but, since this event ought to be the effect of a regular cause or that of chance, the first of these suppositions is more probable than the second. On a table we see letters arranged in this order, *Constantinople*, and we judge that this arrangement is not the result of chance, not because it is less possible than the others, for if this word were not employed in any language we should not suspect it came from any particular cause, but this word being in use among us, it is incomparably more probable that some person has thus arranged the aforesaid letters than that this arrangement is due to chance.

This is the place to define the word *extraordinary*. We arrange in our thought all possible events in various classes; and we regard as *extraordinary* those classes which include a very small number. Thus at the play of heads and tails the occurrence of heads a hundred successive times appears to us extraordinary because of the almost infinite number of combinations which may occur in a hundred throws; and if we divide the combinations into regular series containing an order easy to comprehend, and into irregular series, the latter are incomparably more numerous. The drawing of a white ball from an urn which among a million balls contains only one of this color, the others being black, would appear to us likewise extraordinary, because we form only two classes of events relative to the two colors. But the drawing of the number 475813, for example, from an urn that contains a million numbers seems to us an ordinary event; because, comparing individually the numbers with one another without dividing them into classes, we have no reason to believe that one of them will appear sooner than the others.

From what precedes, we ought generally to conclude that the more extraordinary the event, the greater the need of its being supported by strong proofs. For those who attest it, being able to deceive or to have been deceived, these two causes are as much more probable as the reality of the event is less. We shall see this particularly when we come to speak of the probability of testimony.

Seventh Principle.—The probability of a future event is the sum of the products of the probability of each cause, drawn from the event observed, by the probability that, this cause existing, the future event will occur. The following example will illustrate this principle.

Let us imagine an urn which contains only two balls, each of which may be either white or black. One of these balls is drawn and is put back into the urn before proceeding to a new draw. Suppose that in the first two draws white balls have been drawn; the probability of again drawing a white ball at the third draw is required.

Only two hypotheses can be made here: either one of the balls is white and the other black, or both are white. In the first hypothesis the probability of the event observed is $\frac{1}{4}$; it is unity or certainty in the second. Thus in regarding these hypotheses as so many causes, we shall have for the sixth principle $\frac{1}{5}$ and $\frac{4}{5}$ for their respective probabilities. But if the first hypothesis occurs, the probability of drawing a white ball at the third draw is $\frac{1}{2}$; it is equal to certainty in the second hypothesis; multiplying then the last probabilities by those of the corresponding hypotheses, the sum of the products, or $\frac{9}{10}$, will be the probability of drawing a white ball at the third draw.

When the probability of a single event is unknown we may suppose it equal to any value from zero to unity. The probability of each of these hypotheses, drawn from the event observed, is, by the sixth principle, a fraction whose numerator is the probability of the event in this hypothesis and whose denominator is the sum of the similar probabilities relative to all the hypotheses. Thus the probability that the possibility of the event is comprised within given limits is the sum of the fractions comprised within these limits. Now if we multiply each fraction by the probability of the future event, determined in the corresponding hypothesis, the sum of the products relative to all the hypotheses will be, by the seventh principle, the probability of the future event drawn from the event observed. Thus we find that an event having occurred successively any number of times, the probability that it will happen again the next time is equal to this number increased by unity divided by the same number, increased by two units. Placing the most ancient epoch of history at five thousand years ago, or at 182623 days, and the sun having risen constantly in the interval at each revolution of twenty-four hours, it is a bet of 1826214 to one that it will rise again to-morrow. But this number is incomparably greater for him who, recognizing in the totality of phenomena the principal regulator of days and seasons, sees that nothing at the present moment can arrest the course of it.

Buffon in his *Political Arithmetic* calculates differently the preceding probability. He supposes that it differs from unity only by a fraction whose numerator is unity and whose denominator is the number 2 raised to a power equal to the number of days which have elapsed since the epoch. But the true manner of relating past events with the probability of causes and of future events was unknown to this illustrious writer.

CHAPTER IV

CONCERNING HOPE

THE probability of events serves to determine the hope or the fear of persons interested in their existence. The word *hope* has various acceptations; it expresses generally the advantage of that one who expects a certain benefit in suppositions which are only probable. This advantage in the theory of chance is a product of the sum hoped for by the probability of obtaining it; it is the partial sum which ought to result when we do not wish to run the risks of the event in supposing that the division is made proportional to the probabilities. This division is the only equitable one when all strange circumstances are eliminated; because an equal degree of probability gives an equal right to the sum hoped for. We will call this advantage *mathematical hope.*

Eighth Principle.—When the advantage depends on several events it is obtained by taking the sum of the products of the probability of each event by the benefit attached to its occurrence.

Let us apply this principle to some examples. Let us suppose that at the play of heads and tails Paul receives two francs if he throws heads at the first throw and five francs if he throws it only at the second. Multiplying two francs by the probability $\frac{1}{2}$ of the first case, and five francs by the probability $\frac{1}{4}$ of the second case, the sum of the products, or two and a quarter francs, will be Paul's advantage. It is the sum which he ought to give in advance to that one who has given him this advantage; for, in order to maintain the equality of the play, the throw ought to be equal to the advantage which it procures.

If Paul receives two francs by throwing heads at the first and five francs by throwing it at the second throw, whether he has thrown it or not at the first, the probability of throwing heads at the second throw being $\frac{1}{2}$, multiplying two francs and five francs by $\frac{1}{2}$ the sum of these products will give three and one half francs for Paul's advantage and consequently for his stake at the game.

Ninth Principle.—In a series of probable events of which the ones produce a benefit and the others a loss, we shall have the advantage which results from it by making a sum of the products of the probability of each favorable event by the benefit which it procures, and subtracting from this sum that of the products of the probability of each unfavorable event by the loss which is attached to it. If the second sum is greater than the first, the benefit becomes a loss and hope is changed to fear.

Consequently we ought always in the conduct of life to make the product of the benefit hoped for, by its probability, at least equal to the similar product relative to the loss. But it is necessary, in order to attain this, to appreciate exactly the advantages, the losses, and their respective probabilities. For this a great accuracy of mind, a delicate judgment, and a great experience in affairs is necessary; it is necessary to know how to guard one's self against prejudices, illusions of fear or hope, and erroneous ideas, ideas of fortune and happiness, with which the majority of people feed their self-love.

The application of the preceding principles to the following question has greatly exercised the geometricians. Paul plays at heads and tails with the condition of receiving two francs if he throws heads at the first throw, four francs if he throws it only at the second throw, eight francs if he throws it only at the third, and so on. His stake at the play ought to be, according to the eighth principle, equal to the number of throws, so that if the game continues to infinity the stake ought to be infinite. However, no reasonable man would wish to risk at this game even a small sum, for example five francs. Whence comes this difference between the result of calculation and the indication of common sense? We soon recognize that it amounts to this: that the moral advantage which a benefit procures for us is not proportional to this benefit and that it depends upon a thousand circumstances, often very difficult to define, but of which the most general and most important is that of fortune.

Indeed it is apparent that one franc has much greater value for him who possesses only a hundred than for a millionaire. We ought then to distinguish in the hoped-for benefit its absolute from its relative value. But the latter is regulated by the motives which make it desirable, whereas the first is independent of them. The general principle for appreciating this relative value cannot be given, but here is one proposed by Daniel Bernoulli which will serve in many cases.

Tenth Principle.—The relative value of an infinitely small sum is equal to its absolute value divided by the total benefit of the person interested. This supposes that every one has a certain benefit whose value can never be estimated as zero. Indeed even that one who possesses nothing always gives to the product of his labor and to his hopes a value at least equal to that which is absolutely necessary to sustain him.

If we apply analysis to the principle just propounded, we obtain the following rule: Let us designate by unity the part of the fortune of an individual, independent of his expectations. If we determine the different values that this fortune may have by virtue of these expectations and their probabilities, the product of these values raised respectively to the powers indicated by their probabilities will be the physical fortune which would procure for the individual the same moral advantage which he receives from the part of his fortune taken as unity and from his expectations; by subtracting unity from the product, the difference will be the increase of the physical fortune due to expectations: we will call this increase *moral hope*. It is easy to see that it coincides with mathematical hope when the fortune taken as unity becomes infinite in reference to the variations which it receives from the expectations. But when these variations are an appreciable part of this unity the two hopes may differ very materially among themselves.

This rule conduces to results conformable to the indications of common sense which can by this means be appreciated with some exactitude. Thus in the preceding question it is found that if the fortune of Paul is two hundred francs, he ought not reasonably to stake more than nine francs. The same rule leads us again to distribute the danger over several parts of a benefit expected rather than to expose the entire benefit to this danger. It results similarly that at the fairest game the loss is always greater than the gain. Let us suppose, for example, that a player having a fortune of one hundred francs risks fifty at the play of heads and tails; his fortune after his stake at the play will be reduced to eighty-seven francs, that is to say, this last sum would procure for the player the same moral advantage as the state of his fortune after the stake. The play is then disadvantageous even in the case where the stake is equal to the product of the sum hoped for, by its probability. We can judge by this of the immorality of games in which the sum hoped for is below this product. They subsist only by false reasonings and by the cupidity which they excite and which, leading the people to sacrifice their necessaries to chimerical hopes whose improbability they are not in condition to appreciate, are the source of an infinity of evils.

The disadvantage of games of chance, the advantage of not exposing to the same danger the whole benefit that is expected, and all the similar results indicated by common sense, subsist, whatever may be the function of the physical fortune which for each individual expresses his moral fortune. It is enough that the proportion of the increase of this function to the increase of the physical fortune diminishes in the measure that the latter increases.

CHAPTER V

CONCERNING THE ANALYTICAL METHODS
OF THE CALCULUS OF PROBABILITIES

THE application of the principle which we have just expounded to the various questions of probability requires methods whose investigation has given birth to several methods of analysis and especially to the theory of combinations and to the calculus of finite differences.

If we form the product of the binomials, unity plus the first letter, unity plus the second letter, unity plus the third letter, and so on up to n letters, and subtract unity from this developed product, the result will be the sum of the combination of all these letters taken one by one, two by two, three by three, etc., each combination having unity for a coefficient. In order to have the number of combinations of these n letters taken s by s times, we shall observe that if we suppose these letters equal among themselves, the preceding product will become the nth power of the binomial one plus the first letter; thus the number of combinations of n letters taken s by s times will be the coefficient of the sth power of the first letter in the development in this binomial; and this number is obtained by means of the known binomial formula.

Attention must be paid to the respective situations of the letters in each combination, observing that if a second letter is joined to the first it may be placed in the first or second position which gives two combinations. If we join to these combinations a third letter, we can give it in each combination the first, the second, and the third rank which forms three combinations relative to each of the two others, in all six combinations. From this it is easy to conclude that the number of arrangements of which s letters are susceptible is the product of the numbers from unity to s. In order to pay regard to the respective positions of the letters it is necessary then to multiply by this product the number of combinations of n letters s by s times, which is tantamount to taking away the denominator of the coefficient of the binomial which expresses this number.

Let us imagine a lottery composed of n numbers, of which r are drawn at each draw. The probability is demanded of the drawing of s given numbers in one draw. To arrive at this let us form a fraction whose denominator will be the number of all the cases possible or of the combinations of n letters taken r by r times, and whose numerator will be the number of all the combinations which contain the given s numbers. This last number is evidently that of the combinations of the other numbers taken n less s by n less s times. This fraction will be the required probability, and we shall easily find that it can be reduced to a fraction whose numerator is the number of combinations of r numbers taken s by s times, and whose denominator is the number of combinations of n numbers taken similarly s by s times. Thus in the lottery of France, formed as is known of 90 numbers of which five are drawn at each draw, the probability of drawing a given combination is $\frac{5}{90}$, or $\frac{1}{18}$; the lottery ought then for the equality of the play to give eighteen times the stake. The total number of combinations two by two of the 90 numbers is 4005, and that of the combinations two by two of 5 numbers is 10. The probability of the drawing of a given pair is then $\frac{1}{4005}$, and the lottery ought to give four hundred and a half times the stake; it ought to give 11748 times for a given tray, 511038 times for a quaternary, and 43949268 times for a quint. The lottery is far from giving the player these advantages.

Suppose in an urn a white balls, b black balls, and after having drawn a ball it is put back into the urn; the probability is asked that in n number of draws m white balls and $n - m$ black balls will be drawn. It is clear that the number of cases that may occur at each drawing is $a + b$. Each case of the second drawing being able to combine with all the cases of the first, the number of possible cases in two drawings is the square of the binomial $a + b$. In the development of this square, the square of a expresses the number of cases in which a white ball is twice drawn, the double product of a by b expresses the number of cases in which a white ball and a black ball are drawn. Finally, the square of b expresses the number of cases in which two black balls are drawn. Continuing thus, we see generally that the nth power of the binomial $a + b$ expresses the number of all the cases possible in n draws; and that in the development of this power the term multiplied by the mth power of a expresses the number of cases in which m white balls and $n - m$ black balls may be drawn. Dividing then this term by the entire power of the binomial, we shall have the probability of drawing m white balls and $n - m$ black balls. The ratio of the numbers a and $a + b$ being the probability of drawing one white ball at one draw; and the ratio of the numbers b and $a + b$ being the probability of drawing one black ball; if we call these probabilities p and q, the probability of drawing m white balls in n draws will be the term multiplied by the mth power of p in the development of the nth power of the binomial $p + q$; we may see that the sum $p + q$ is unity. This remarkable property of the binomial is very useful in the theory of probabilities. But the most general and direct method of resolving questions of probability consists in making them depend upon equations of differences. Comparing the successive conditions of the function which expresses the probability when we increase the variables by their respective differences, the proposed question often furnishes a very simple proportion between the conditions. This proportion is what is called *equation of ordinary or partial differentials; ordinary* when there is only one variable, *partial* when there are several. Let us consider some examples of this.

Three players of supposed equal ability play together on the following conditions: that one of the first two players who beats his adversary plays the third, and if he beats him the game is finished. If he is beaten, the victor plays against the second until one of the players has defeated consecutively the two others, which ends the game. The probability is demanded that the game will be finished in a certain number n of plays. Let us find the probability that it will end precisely at the nth play. For that the player who wins ought to enter the game at the play $n - 1$ and win it thus at the following play. But if in place of winning the play $n - 1$ he should be beaten by his adversary who had just beaten the other player, the game would end at this play. Thus the probability that one of the players will enter the game at the play $n - 1$ and will win it is equal to the probability that the game will end precisely with this play; and as this player ought to win the following play in order that the game may be finished at the nth play, the probability of this last case will be only one half of the preceding one. This probability is evidently a function of the number n; this function is then equal to the half of the same function when n is diminished by unity. This equality forms one of those equations called *ordinary finite differential equations.*

We may easily determine by its use the probability that the game will end precisely at a certain play. It is evident that the play cannot end sooner than at the second play; and for this it is necessary that that one of the first two players who has beaten his adversary should beat at the second play the third player; the probability that the game will end at this play is $\frac{1}{2}$. Hence by virtue of the preceding equation we conclude that the successive probabilities of the end of the game are $\frac{1}{4}$ for the third play, $\frac{1}{8}$ for the fourth play, and so on; and in general $\frac{1}{2}$ raised to the power $n - 1$ for the nth play. The sum of all these powers of $\frac{1}{2}$ is unity less the last of these powers; it is the probability that the game will end at the latest in n plays.

Let us consider again the first problem more difficult which may be solved by probabilities and which Pascal proposed to Fermat to solve. Two players, A and B, of equal skill play together on the conditions that the one who first shall beat the other a given number of times shall win the game and shall take the sum of the stakes at the game; after some throws the players agree to quit without having finished the game: we ask in what manner the sum ought to be divided between them. It is evident that the parts ought to be proportional to the respective probabilities of winning the game. The question is reduced then to the determination of these probabilities. They depend evidently upon the number of points which each player lacks of having attained the given number. Hence the probability of A is a function of the two numbers which we will call *indices*. If the two players should agree to play one throw more (an agreement which does not change their condition, provided that after this new throw the division is always made proportionally to the new probabilities of winning the game), then either A would win this throw and in that case the number of points which he lacks would be diminished by unity, or the player B would win it and in that case the number of points lacking to this last player would be less by unity. But the probability of each of these cases is $\frac{1}{2}$; the function sought is then equal to one half of this function in which we diminish by unity the first index plus the half of the same function in which the second variable is diminished by unity. This equality is one of those equations called *equations of partial differentials*.

We are able to determine by its use the probabilities of A by dividing the smallest numbers, and by observing that the probability or the function which expresses it is equal to unity when the player A does not lack a single point, or when the first index is zero, and that this function becomes zero with the second index. Supposing thus that the player A lacks only one point, we find that his probability is $\frac{1}{2}$, $\frac{3}{4}$, $\frac{7}{8}$, etc., according as B lacks one point, two, three, etc. Generally it is then unity less the power of $\frac{1}{2}$, equal to the number of points which B lacks. We will suppose then that the player A lacks two points and his probability will be found equal to $\frac{1}{4}$, $\frac{1}{2}$, $\frac{11}{16}$, etc., according as B lacks one point, two points, three points, etc. We will suppose again that the player A lacks three points, and so on.

This manner of obtaining the successive values of a quantity by means of its equation of differences is long and laborious. The geometricians have sought methods to obtain the general function of indices that satisfies this equation, so that for any particular case we need only to substitute in this function the corresponding values of the indices. Let us consider this subject in a general way. For this purpose let us conceive a series of terms arranged along a horizontal line so that each of them is derived from the preceding one according to a given law. Let us suppose this law expressed by an equation among several consecutive terms and their index, or the number which indicates the rank that they occupy in the series. This equation I call the *equation of finite differences by a single index*. The order or the degree of this equation is the difference of rank of its two extreme terms. We are able by its use to determine successively the terms of the series and to continue it indefinitely; but for that it is necessary to know a number of terms of the series equal to the degree of the equation. These terms are the arbitrary constants of the expression of the general term of the series or of the integral of the equation of differences.

Let us imagine now below the terms of the preceding series a second series of terms arranged horizontally; let us imagine again below the terms of the second series a third horizontal series, and so on to infinity; and let us suppose the terms of all these series connected by a general equation among several consecutive terms, taken as much in the horizontal as in the vertical sense, and the numbers which indicate their rank in the two senses. This equation is called the *equation of partial finite differences by two indices.*

Let us imagine in the same way below the plan of the preceding series a second plan of similar series, whose terms should be placed respectively below those of the first plan; let us imagine again below this second plan a third plan of similar series, and so on to infinity; let us suppose all the terms of these series connected by an equation among several consecutive terms taken in the sense of length, width, and depth, and the three numbers which indicate their rank in these three senses. This equation I call the *equation of partial finite differences by three indices.*

Finally, considering the matter in an abstract way and independently of the dimensions of space, let us imagine generally a system of magnitudes, which should be functions of a certain number of indices, and let us suppose among these magnitudes, their relative differences to these indices and the indices themselves, as many equations as there are magnitudes; these equations will be partial finite differences by a certain number of indices.

. We are able by their use to determine successively these magnitudes. But in the same manner as the equation by a single index requires for it that we known a certain number of terms of the series, so the equation by two indices requires that we know one or several lines of series whose general terms should be expressed each by an arbitrary function of one of the indices. Similarly the equation by three indices requires that we know one or several plans of series, the general terms of which should be expressed each by an arbitrary function of two indices, and so on. In all these cases we shall be able by successive eliminations to determine a certain term of the series. But all the equations among which we eliminate being comprised in the same system of equations, all the expressions of the successive terms which we obtain by these eliminations ought to be comprised in one general expression, a function of the indices which determine the rank of the term. This expression is the integral of the proposed equation of differences, and the search for it is the object of integral calculus.

Taylor is the first who in his work entitled *Metodus incrementorum* has considered linear equations of finite differences. He gives the manner of integrating those of the first order with a coefficient and a last term, functions of the index. In truth the relations of the terms of the arithmetical and geometrical progressions which have always been taken into consideration are the simplest cases of linear equations of differences; but they had not been considered from this point of view. It was one of those which, attaching themselves to general theories, lead to these theories and are consequently veritable discoveries.

About the same time Moivre was considering under the name of recurring series the equations of finite differences of a certain order having a constant coefficient. He succeeded in integrating them in a very ingenious manner. As it is always interesting to follow the progress of inventors, I shall expound the method of Moivre by applying it to a recurring series whose relation among three consecutive terms is given. First he considers the relation among the consecutive terms of a geometrical progression or the equation of two terms which expresses it. Referring it to terms less than unity, he multiplies it in this state by a constant factor and subtracts the product from the first equation. Thus he obtains an equation among three consecutive terms of the geometrical progression. Moivre considers next a second progression whose ratio of terms is the same factor which he has just used. He diminishes similarly by unity the index of the terms of the equation of this new progression. In this condition he multiplies it by the ratio of the terms of the first progression, and he subtracts the product from the equation of the second progression, which gives him among three consecutive terms of this progression a relation entirely similar to that which he has found for the first progression. Then he observes that if one adds term by term the two progressions, the same ratio exists among any three of these consecutive terms. He compares the coefficients of this ratio to those of the relation of the terms of the proposed recurrent series, and he finds for determining the ratios of the two geometrical progressions an equation of the second degree, whose roots are these ratios. Thus Moivre decomposes the recurrent series into two geometrical progressions, each multiplied by an arbitrary constant which he determines by means of the first two terms of the recurrent series. This ingenious process is in fact the one that d'Alembert has since employed for the integration of linear equations of infinitely small differences with constant coefficients, and Lagrange has transformed into similar equations of finite differences.

Finally, I have considered the linear equations of partial finite differences, first under the name of *recurrorecurrent* series and afterwards under their own name. The most general and simplest manner of integrating all these equations appears to me that which I have based upon the consideration of discriminant functions, the idea of which is here given.

If we conceive a function V of a variable t developed according to the powers of this variable, the coefficient of any one of these powers will be a function of the exponent or index of this power, which index I shall call x. V is what I call the discriminant function of this coefficient or of the function of the index.

Now if we multiply the series of the development of V by a function of the same variable, such, for example, as unity plus two times this variable, the product will be a new discriminant function in which the coefficient of the power x of the variable t will be equal to the coefficient of the same power in V plus twice the coefficient of the power less unity. Thus the function of the index x in the product will be equal to the function of the index x in V plus twice the same function in which the index is diminished by unity. This function of the index x is thus a derivative of the function of the same index in the development of V, a function which I shall call the *primitive function* of the index. Let us designate the derivative function by the letter δ placed before the primitive function. The derivation indicated by this letter will depend upon the multiplier of V, which we will call T and which we will suppose developed like V by the ratio to the powers of the variable t. If we multiply anew by T the product of V by T, which is equivalent to multiplying V by T^2, we shall form a third discriminant function, in which the coefficient of the xth power of t will be a derivative similar to the corresponding coefficient of the preceding product; it may be expressed by the same character δ placed before the preceding derivative, and then this character will be written twice before the primitive function of x. But in place of writing it thus twice we give it 2 for an exponent.

Continuing thus, we see generally that if we multiply V by the nth power of T, we shall have the coefficient of the xth power of t in the product of V by the nth power of T by placing before the primitive function the character δ with n for an exponent.

Let us suppose, for example, that T be unity divided by t; then in the product of V by T the coefficient of the xth power of t will be the coefficient of the power greater by unity in V; this coefficient in the product of V by the nth power of T will then be the primitive function in which x is augmented by n units.

Let us consider now a new function Z of t, developed like V and T according to the powers of t; let us designate by the character Δ placed before the primitive function the coefficient of the xth power of t in the product of V by Z; this coefficient in the product of V by the nth power of Z will be expressed by the character Δ affected by the exponent n and placed before the primitive function of x.

If, for example, Z is equal to unity divided by t less one, the coefficient of the xth power of t in the product of V by Z will be the coefficient of the $x + 1$ power of t in V less the coefficient of the xth power. It will be then the finite difference of the primitive function of the index x. Then the character Δ indicates a finite difference of the primitive function in the case where the index varies by unity; and the nth power of this character placed before the primitive function will indicate the finite nth difference of this function. If we suppose that T be unity divided by t, we shall have T equal to the binomial $Z + 1$. The product of V by the nth power of T will then be equal to the product of V by the nth power of the binomial $Z + 1$. Developing this power in the ratio of the powers of Z, the product of V by the various terms of this development will be the discriminant functions of these same terms in which we substitute in place of the powers of Z the corresponding finite differences of the primitive function of the index.

Now the product of V by the nth power of T is the primitive function in which the index x is augmented by n units; repassing from the discriminant functions to their coefficients, we shall have this primitive function thus augmented equal to the development of the nth power of the binomial $Z + 1$, provided that in this development we substitute in place of the powers of Z the corresponding differences of the primitive function and that we multiply the independent term of these powers by the primitive function. We shall thus obtain the primitive function whose index is augmented by any number n by means of its differences.

Supposing that T and Z always have the preceding values, we shall have Z equal to the binomial $T - 1$; the product of V by the nth power of Z will then be equal to the product of V by the development of the nth power of the binomial $T - 1$. Repassing from the discriminant functions to their coefficients as has just been done, we shall have the nth difference of the primitive function expressed by the development of the nth power of the binomial $T - 1$, in which we substitute for the powers of T this same function whose index is augmented by the exponent of the power, and for the independent term of t, which is unity, the primitive function, which gives this difference by means of the consecutive terms of this function.

Placing δ before the primitive function expressing the derivative of this function, which multiplies the x power of t in the product of V by T, and Δ expressing the same derivative in the product of V by Z, we are led by that which precedes to this general result: whatever may be the function of the variable t represented by T and Z, we may, in the development of all the identical equations susceptible of being formed among these functions, substitute the characters δ and Δ in place of T and Z, provided that we write the primitive function of the index in series with the powers and with the products of the powers of the characters, and that we multiply by this function the independent terms of these characters.

We are able by means of this general result to transform any certain power of a difference of the primitive function of the index x, in which x varies by unity, into a series of differences of the same function in which x varies by a certain number of units and reciprocally. Let us suppose that T be the i power of unity divided by $t - 1$, and that Z be always unity divided by $t - 1$; then the coefficient of the

x power of t in the product of V by T will be the coefficient of the $x + i$ power of t in V less the coefficient of the x power of t; it will then be the finite difference of the primitive function of the index x in which we vary this index by the number i. It is easy to see that T is equal to the difference between the i power of the binomial $Z + 1$ and unity. The nth power of T is equal to the nth power of this difference. If in this equality we substitute in place of T and Z the characters δ and Δ, and after the development we place at the end of each term the primitive function of the index x, we shall have the nth difference of this function in which x varies by i units expressed by a series of differences of the same function in which x varies by unity. This series is only a transformation of the difference which it expresses and which is identical with it; but it is in similar transformations that the power of analysis resides.

The generality of analysis permits us to suppose in this expression that n is negative. Then the negative powers of δ and Δ indicate the integrals. Indeed the nth difference of the primitive function having for a discriminant function the product of V by the nth power of the binomial one divided by t less unity, the primitive function which is the nth integral of this difference has for a discriminant function that of the same difference multiplied by the nth power taken less than the binomial one divided by t minus one, a power to which the same power of the character Δ corresponds; this power indicates then an integral of the same order, the index x varying by unity; and the negative powers of δ indicate equally the integrals x varying by i units. We see, thus, in the clearest and simplest manner the rationality of the analysis observed among the positive powers and differences, and among the negative powers and the integrals.

If the function indicated by δ placed before the primitive function is zero, we shall have an equation of finite differences, and V will be the discriminant function of its integral. In order to obtain this discriminant function we shall observe that in the product of V by T all the powers of t ought to disappear except the powers inferior to the order of the equation of differences; V is then equal to a fraction whose denominator is T and whose numerator is a polynomial in which the highest power of t is less by unity than the order of the equation of differences. The arbitrary coefficients of the various powers of t in this polynomial, including the power zero, will be determined by as many values of the primitive function of the index when we make successively x equal to zero, to one, to two, etc. When the equation of differences is given we determine T by putting all its terms in the first member and zero in the second; by substituting in the first member unity in place of the function which has the largest index; the first power of t in place of the primitive function in which this index is diminished by unity; the second power of t for the primitive function where this index is diminished by two units, and so on. The coefficient of

410

the xth power of t in the development of the preceding expression of V will be the primitive function of x or the integral of the equation of finite differences. Analysis furnishes for this development various means, among which we may choose that one which is most suitable for the question proposed; this is an advantage of this method of integration.

Let us conceive now that V be a function of the two variables t and t' developed according to the powers and products of these variables; the coefficient of any product of the powers x and x' of t and t' will be a function of the exponents or indices x and x' of these powers; this function I shall call the *primitive function* of which V is the discriminant function.

Let us multiply V by a function T of the two variables t and t' developed like V in ratio of the powers and the products of these variables; the product will be the discriminant function of a derivative of the primitive function; if T, for example, is equal to the variable t plus the variable t' minus two, this derivative will be the primitive function of which we diminish by unity the index x plus this same primitive function of which we diminish by unity the index x' less two times the primitive function. Designating whatever T may be by the character δ placed before the primitive function, this derivative, the product of V by the nth power of T, will be the discriminant function of the derivative of the primitive function before which one places the nth power of the character δ. Hence result the theorems analogous to those which are relative to functions of a single variable.

Suppose the function indicated by the character δ be zero; one will have an equation of partial differences. If, for example, we make as before T equal to the variable t plus the variable $t' - 2$, we have zero equal to the primitive function of which we diminish by unity the index x plus the same function of which we diminish by unity the index x' minus two times the primitive function. The discriminant function V of the primitive function or of the integral of this equation ought then to be such that its product by T does not include at all the products of t by t'; but V may include separately the powers of t and those of t', that is to say, an arbitrary function of t and an arbitrary function of t'; V is then a fraction whose numerator is the sum of these two arbitrary functions and whose denominator is T. The coefficient of the product of the xth power of t by the x' power of t' in the development of this fraction will then be the integral of the preceding equation of partial differences. This method of integrating this kind of equations seems to me the simplest and the easiest by the employment of the various analytical processes for the development of rational fractions.

More ample details in this matter would be scarcely understood without the aid of calculus.

Considering equations of infinitely small partial differences as equations of finite partial differences in which nothing is neglected, we are able to throw light upon the obscure points of their calculus, which have been the subject of great discussions among geometricians. It is thus that I have demonstrated the possibility of introducing discontinued functions in their integrals, provided that the discontinuity takes place only for the differentials of the order of these equations or of a superior order. The transcendent results of calculus are, like all the abstractions of the understanding, general signs whose true meaning may be ascertained only by repassing by metaphysical analysis to the elementary ideas which have led to them; this often presents great difficulties, for the human mind tries still less to transport itself into the future than to retire within itself. The comparison of infinitely small differences with finite differences is able similarly to shed great light upon the metaphysics of infinitesimal calculus.

It is easily proven that the finite nth difference of a function in which the increase of the variable is E being divided by the nth power of E, the quotient reduced in series by ratio to the powers of the increase E is formed by a first term independent of E. In the measure that E diminishes, the series approaches more and more this first term from which it can differ only by quantities less than any assignable magnitude. This term is then the limit of the series and expresses in differential calculus the infinitely small nth difference of the function divided by the nth power of the infinitely small increase.

Considering from this point of view the infinitely small differences, we see that the various operations of differential calculus amount to comparing separately in the development of identical expressions the finite terms or those independent of the increments of the variables which are regarded as infinitely small; this is rigorously exact, these increments being indeterminant. Thus differential calculus has all the exactitude of other algebraic operations.

The same exactitude is found in the applications of differential calculus to geometry and mechanics. If we imagine a curve cut by a secant at two adjacent points, naming E the interval of the ordinates of these two points, E will be the increment of the abscissa from the first to the second ordinate. It is easy to see that the corresponding increment of the ordinate will be the product of E by the first ordinate divided by its subsecant; augmenting then in this equation of the curve the first ordinate by this increment, we shall have the equation relative to the second ordinate. The difference of these two equations will be a third equation which, developed by the ratio of the powers of E and divided by E, will have its first term independent of E, which will be the limit of this development. This term, equal to zero, will give then the limit of the subsecants, a limit which is evidently the subtangent.

This singularly happy method of obtaining the subtangent is due to Fermat, who has extended it to transcendent curves. This great geometrician expresses by the character E the increment of the abscissa; and considering only the first power of this increment, he determines exactly as we do by differential calculus the subtangents of the curves, their points of inflection, the *maxima* and *minima* of their ordinates, and in general those of rational functions. We see likewise by his beautiful solution of the problem of the refraction of light inserted in the *Collection of the Letters of Descartes* that he knows how to extend his methods to irrational functions in freeing them from irrationalities by the elevation of the roots to powers. Fermat should be regarded, then, as the true discoverer of Differential Calculus. Newton has since rendered this calculus more analytical in his *Method of Fluxions*, and simplified and generalized the processes by his beautiful theorem of the binomial. Finally, about the same time Leibnitz has enriched differential calculus by a notation which, by indicating the passage from the finite to the infinitely small, adds to the advantage of expressing the general results of calculus that of giving the first approximate values of the differences and of the sums of the quantities; this notation is adapted of itself to the calculus of partial differentials.

We are often led to expressions which contain so many terms and factors that the numerical substitutions are impracticable. This takes place in questions of probability when we consider a great number of events. Meanwhile it is necessary to have the numerical value of the formulæ in order to know with what probability the results are indicated, which the events develop by multiplication. It is necessary especially to have the law according to which this probability continually approaches certainty, which it will finally attain if the number of events were infinite. In order to obtain this law I considered that the definite integrals of differentials multiplied by the factors raised to great powers would give by integration the formulæ composed of a great number of terms and factors. This remark brought me to the idea of transforming into similar integrals the complicated expressions of analysis and the integrals of the equation of differences. I fulfilled this condition by a method which gives at the same time the function comprised under the integral sign and the limits of the integration. It offers this remarkable thing, that the function is the same discriminant function of the expressions and the proposed equations; this attaches this method to the theory of discriminant functions of which it is thus the complement. Further, it would only be a question of reducing the definite integral to a converging series. This I have obtained by a process which makes the series converge with as much more rapidity as the formula which it represents is more complicated, so that it is more exact as it becomes more necessary. Frequently the series has for a factor the square root of the ratio of the circumference to the diameter; sometimes it depends upon other transcendents whose number is infinite.

An important remark which pertains to great generality of analysis, and which permits us to extend this method to formulæ and to equations of difference which the theory of probability presents most frequently, is that the series to which one comes by supposing the limits of the definite integrals to be real and positive take place equally in the case where the equation which determines these limits has only negative or imaginary roots. These passages from the positive to the negative and from the real to the imaginary, of which I first have made use, have led me further to the values of many singular definite integrals, which I have accordingly demonstrated directly. We may then consider these passages as a means of discovery parallel to induction and analogy long employed by geometricians, at first with an extreme reserve, afterwards with entire confidence, since a great number of examples has justified its use. In the mean time it is always necessary to confirm by direct demonstrations the results obtained by these divers means.

I have named the ensemble of the preceding methods the *Calculus of Discriminant Functions*; this calculus serves as a basis for the work which I have published under the title of the *Analytical Theory of Probabilities*. It is connected with the simple idea of indicating the repeated multiplications of a quantity by itself or its entire and positive powers by writing toward the top of the letter which expresses it the numbers which mark the degrees of these powers.

This notation, employed by Descartes in his *Geometry* and generally adopted since the publication of this important work, is a little thing, especially when compared with the theory of curves and variable functions by which this great geometrician has established the foundations of modern calculus. But the language of analysis, most perfect of all, being in itself a powerful instrument of discoveries, its notations, especially when they are necessary and happily conceived, are so many germs of new calculi. This is rendered appreciable by this example.

Wallis, who in his work entitled *Arithmetica Infinitorum*, one of those which have most contributed to the progress of analysis, has interested himself especially in following the thread of induction and analogy, considered that if one divides the exponent of a letter by two, three, etc., the quotient will be accordingly the Cartesian notation, and when division is possible the exponent of the square, cube, etc., root of the quantity which represents the letter raised to the dividend exponent. Extending by analogy this result to the case where division is impossible, he considered a quantity raised to a fractional exponent as the root of the degree indicated by the denominator of this fraction— namely, of the quantity raised to a power indicated by the numerator. He observed then that, according to the Cartesian notation, the multiplication of two powers of the same letter amounts to adding their exponents, and that their division amounts to subtracting the exponents of the power of the divisor from that of the power of the dividend, when the second of these exponents is greater than the first. Wallis extended this result to the case where the first exponent is equal to or greater than the second, which makes the difference zero or negative. He supposed then that a negative exponent indicates unity divided by the quantity raised to the same exponent taken positively. These remarks led him to integrate generally the monomial differentials, whence he inferred the definite integrals of a particular kind of binomial differentials whose exponent is a positive integral number. The observation then of the law of the numbers which express these integrals, a series of interpolations and happy inductions where one perceives the germ of the calculus of definite integrals which has so much exercised geometricians and which is one of the fundaments of my new *Theory of Probabilities*, gave him the ratio of the area of the circle to the square of its diameter expressed by an infinite product, which, when one stops it, confines this ratio to limits more and more converging; this is one of the most singular results in analysis. But it is remarkable that Wallis, who had so well considered the fractional exponents of radical powers, should have continued to note these powers as had been done before him. Newton in his *Letters to Oldembourg*, if I am not mistaken, was the first to employ the notation of these powers by fractional exponents. Comparing by the way of induction, of which Wallis had made such a beautiful use, the exponents of the powers of the binomial with the coefficients of the terms of its development in the case where this exponent is integral and positive, he determined the law of these coefficients and extended it by analogy to fractional and negative powers. These various results, based upon the notation of Descartes, show his influence on the progress of analysis. It has still the advantage of giving the simplest and fairest idea of logarithms, which are indeed only the exponents of a magnitude whose successive powers, increasing by infinitely small degrees, can represent all numbers.

But the most important extension that this notation has received is that of variable exponents, which constitutes exponential calculus, one of the most fruitful branches of modern analysis. Leibnitz was the first to indicate the transcendents by variable exponents, and thereby he has completed the system of elements of which a finite function can be composed; for every finite explicit function of a variable may be reduced in the last analysis to simple magnitudes, combined by the method of addition, subtraction, multiplication, and division and raised to constant or variable powers. The roots of the equations formed from these elements are the implicit functions of the variable. It is thus that a variable has for a logarithm the exponent of the power which is equal to it in the series of the powers of the number whose hyperbolic logarithm is unity, and the logarithm of a variable of it is an implicit function.

Leibnitz thought to give to his differential character the same exponents as to magnitudes; but then in place of indicating the repeated multiplications of the same magnitude these exponents indicate the repeated differentiations of the same function. This new extension of the Cartesian notation led Leibnitz to the analogy of positive powers with the differentials, and the negative powers with the integrals. Lagrange has followed this singular analogy in all its developments; and by series of inductions which may be regarded as one of the most beautiful applications which have ever been made of the method of induction he has arrived at general formulæ which are as curious as useful on the transformations of differences and of integrals the ones into the others when the variables have divers finite increments and when these increments are infinitely small. But he has not given the demonstrations of it which appear to him difficult. The theory of discriminant functions extends the Cartesian notations to some of its characters; it shows with proof the analogy of the powers and operations indicated by these characters; so that it may still be regarded as the exponential calculus of characters. All that concerns the series and the integration of equations of differences springs from it with an extreme facility.

PART II
APPLICATIONS OF THE CALCULUS OF PROBABILITIES
CHAPTER VI
GAMES OF CHANCE

THE combinations which games present were the object of the first investigations of probabilities. In an infinite variety of these combinations many of them lend themselves readily to calculus; others require more difficult calculi; and the difficulties increasing in the measure that the combinations become more complicated, the desire to surmount them and curiosity have excited geometricians to perfect more and more this kind of analysis. It has been seen already that the benefits of a lottery are easily determined by the theory of combinations. But it is more difficult to know in how many draws one can bet one against one, for example that all the numbers will be drawn, n being the number of numbers, r that of the numbers drawn at each draw, and i the unknown number of draws. The expression of the probability of drawing all the numbers depends upon the nth finite difference of the i power of a product of r consecutive numbers. When the number n is considerable the search for the value of i which renders this probability equal to $\frac{1}{2}$ becomes impossible at least unless this difference is converted into a very converging series. This is easily done by the method here below indicated by the approximations of functions of very large numbers. It is found thus since the lottery is composed of ten thousand numbers, one of which is drawn at each draw, that there is a disadvantage in betting one against one that all the numbers will be drawn in 95767 draws and an advantage in making the same bet for 95768 draws. In the lottery of France this bet is disadvantageous for 85 draws and advantageous for 86 draws.

Let us consider again two players, A and B, playing together at heads and tails in such a manner that at each throw if heads turns up A gives one counter to B, who gives him one if tails turns up; the number of counters of B is limited, while that of A is unlimited, and the game is to end only when B shall have no more counters. We ask in how many throws one should bet one to one that the game will end. The expression of the probability that the game will end in an i number of throws is given by a series which comprises a great number of terms and factors if the number of counters of B is considerable; the search for the value of the unknown i which renders this series $\frac{1}{2}$ would then be impossible if we did not reduce the same to a very convergent series. In applying to it the method of which we have just spoken, we find a very simple expression for the unknown from which it results that if, for example, B has a hundred counters, it is a bet of a little less than one against one that the game will end in 23780 throws, and a bet of a little more than one against one that it will end in 23781 throws.

These two examples added to those we have already given are sufficient to shows how the problems of games have contributed to the perfection of analysis.

CHAPTER VII

CONCERNING THE UNKNOWN INEQUALITIES WHICH MAY EXIST AMONG CHANCES WHICH ARE SUPPOSED EQUAL

INEQUALITIES of this kind have upon the results of the calculation of probabilities a sensible influence which deserves particular attention. Let us take the game of heads and tails, and let us suppose that it is equally easy to throw the one or the other side of the coin. Then the probability of throwing heads at the first throw is $\frac{1}{2}$ and that of throwing it twice in succession is $\frac{1}{4}$. But if there exist in the coin an inequality which causes one of the faces to appear rather than the other without knowing which side is favored by this inequality, the probability of throwing heads at the first throw will always be $\frac{1}{2}$; because of our ignorance of which face is favored by the inequality the probability of the simple event is increased if this inequality is favorable to it, just so much is it diminished if the inequality is contrary to it. But in this same ignorance the probability of throwing heads twice in succession is increased. Indeed this probability is that of throwing heads at the first throw multiplied by the probability that having thrown it at the first throw it will be thrown at the second; but its happening at the first throw is a reason for belief that the inequality of the coin favors it; the unknown inequality increases, then, the probability of throwing heads at the second throw; it consequently increases the product of these two probabilities. In order to submit this matter to calculus let us suppose that this inequality increases by a twentieth the probability of the simple event which it favors. If this event is heads, its probability will be $\frac{1}{2}$ plus $\frac{1}{20}$, or $\frac{11}{20}$, and the probability of throwing it twice in succession will be the square of $\frac{11}{20}$, or $\frac{121}{400}$. If the favored event is tails, the probability of heads, will be $\frac{1}{2}$ minus $\frac{1}{20}$, or $\frac{9}{20}$, and the probability of throwing it twice in succession will be $\frac{81}{400}$. Since we have at first no reason for believing that the inequality favors one of these events rather than the other, it is clear that in order to have the probability of the compound event heads heads it is necessary to add the two preceding probabilities and take the half of their sum, which gives $\frac{101}{400}$ for this probability, which exceeds $\frac{1}{4}$ by $\frac{1}{400}$ or by the square of the favor $\frac{1}{20}$ that the inequality adds to the possibilities of the event which it favors. The probability of throwing tails tails is similarly $\frac{101}{400}$, but the probability of throwing heads tails or tails heads is each $\frac{99}{400}$; for the sum of these four probabilities ought to equal certainty or unity. We find thus generally that the constant and unknown causes which favor simple events which are judged equally possible always increase the probability of the repetition of the same simple event.

In an even number of throws heads and tails ought both to happen either an even number of times or odd number of times. The probability of each of these cases is $\frac{1}{2}$ if the possibilities of the two faces are equal; but if there is between them an unknown inequality, this inequality is always favorable to the first case.

Two players whose skill is supposed to be equal play on the conditions that at each throw that one who loses gives a counter to his adversary, and that the game continues until one of the players has no more counters. The calculation of the probabilities shows us that for the equality of the play the throws of the players ought to be an inverse ratio to their counters. But if there is between the players a small unknown inequality, it favors that one of the players who has the smallest number of counters. His probability of winning the game increases if the players agree to double or triple their counters; and it will be $\frac{1}{2}$ or the same as the probability of the other player in the case where the number of their counters should become infinite, preserving always the same ratio.

One may correct the influence of these unknown inequalities by submitting them themselves to the chances of hazard. Thus at the play of heads and tails, if one has a second coin which is thrown each time with the first and one agrees to name constantly heads the face turned up by the second coin, the probability of throwing heads twice in succession with the first coin will approach much nearer $\frac{1}{4}$ than in the case of a single coin. In this last case the difference is the square of the small increment of possibility that the unknown inequality gives to the face of the first coin which it favors; in the other case this difference is the quadruple product of this square by the corresponding square relative to the second coin.

Let there be thrown into an urn a hundred numbers from 1 to 100 in the order of numeration, and after having shaken the urn in order to mix the numbers one is drawn; it is clear that if the mixing has been well done the probabilities of the drawing of the numbers will be the same. But if we fear that there is among them small differences dependent upon the order according to which the numbers have been thrown into the urn, we shall diminish considerably these differences by throwing into a second urn the numbers according to the order of their drawing from the first urn, and by shaking then this second urn in order to mix the numbers. A third urn, a fourth urn, etc., would diminish more and more these differences already inappreciable in the second urn.

CHAPTER VIII

CONCERNING THE LAWS OF PROBABILITY WHICH RESULT FROM THE INDEFINITE MULTIPLICATION OF EVENTS

AMID the variable and unknown causes which we comprehend under the name of *chance*, and which render uncertain and irregular the march of events, we see appearing, in the measure that they multiply, a striking regularity which seems to hold to a design and which has been considered as a proof of Providence. But in reflecting upon this we soon recognize that this regularity is only the development of the respective possibilities of simple events which ought to present themselves more often when they are more probable. Let us imagine, for example, an urn which contains white balls and black balls; and let us suppose that each time a ball is drawn it is put back into the urn before proceeding to a new draw. The ratio of the number of the white balls drawn to the number of black balls drawn will be most often very irregular in the first drawings; but the variable causes of this irregularity produce effects alternately favorable and unfavorable to the regular march of events which destroy each other mutually in the totality of a great number of draws, allowing us to perceive more and more the ratio of white balls to the black balls contained in the urn, or the respective possibilities of drawing a white ball or black ball at each draw. From this results the following theorem.

The probability that the ratio of the number of white balls drawn to the total number of balls drawn does not deviate beyond a given interval from the ratio of the number of white balls to the total number of balls contained in the urn, approaches indefinitely to certainty by the indefinite multiplication of events, however small this interval.

This theorem indicated by common sense was difficult to demonstrate by analysis. Accordingly the illustrious geometrician Jacques Bernouilli, who first has occupied himself with it, attaches great importance to the demonstrations he has given. The calculus of discriminant functions applied to this matter not only demonstrates with facility this theorem, but still more it gives the probability that the ratio of the events observed deviates only in certain limits from the true ratio of their respective possibilities.

One may draw from the preceding theorem this consequence which ought to be regarded as a general law, namely, that the ratios of the acts of nature are very nearly constant when these acts are considered in great number. Thus in spite of the variety of years the sum of the productions during a considerable number of years is sensibly the same; so that man by useful foresight is able to provide against the irregularity of the seasons by spreading out equally over all the seasons the goods which nature distributes in an unequal manner. I do not except from the above law results due to moral causes. The ratio of annual births to the population, and that of marriages to births, show only small variations; at Paris the number of annual births is almost the same, and I have heard it said at the post-office in ordinary seasons the number of letters thrown aside on account of defective addresses changes little each year; this has likewise been observed at London.

It follows again from this theorem that in a series of events indefinitely prolonged the action of regular and constant causes ought to prevail in the long run over that of irregular causes. It is this which renders the gains of the lotteries just as certain as the products of agriculture; the chances which they reserve assure them a benefit in the totality of a great number of throws. Thus favorable and numerous chances being constantly attached to the observation of the eternal principles of reason, of justice, and of humanity which establish and maintain societies, there is a great advantage in conforming to these principles and of grave inconvenience in departing from them. If one consult histories and his own experience, one will see all the facts come to the aid of this result of calculus. Consider the happy effects of institutions founded upon reason and the natural rights of man among the peoples who have known how to establish and preserve them. Consider again the advantages which good faith has procured for the governments who have made it the basis of their conduct and how they have been indemnified for the sacrifices which a scrupulous exactitude in keeping their engagements has cost them. What immense credit at home! What preponderance abroad! On the contrary, look into what an abyss of misfortunes nations have often been precipitated by the ambition and the perfidy of their chiefs. Every time that a great power intoxicated by the love of conquest aspires to universal domination the sentiment of independence produces among the menaced nations a coalition of which it becomes almost always the victim. Similarly in the midst of the variable causes which extend or restrain the divers states, the natural limits acting as constant causes ought to end by prevailing. It is important then to the stability as well as to the happiness of empires not to extend them beyond those limits into which they are led again without cessation by the action of the causes; just as the waters of the seas raised by violent tempests fall again into their basins by the force of gravity. It is again a result of the calculus of probabilities confirmed by numerous and melancholy experiences. History treated from the point of view of the influence of constant causes would unite to the interest of curiosity that of offering to man most useful lessons. Sometimes we attribute the inevitable results of these causes to the accidental circumstances which have produced their action. It is, for example, against the nature of things that one people should ever be governed by another when a vast sea or a great distance separates them. It may be affirmed that in the long run this constant cause, joining itself without ceasing to the variable causes which act in the same way and which the course of time develops, will end by finding them sufficiently strong to give to a subjugated people its natural independence or to unite it to a powerful state which may be contiguous.

In a great number of cases, and these are the most important of the analysis of hazards, the possibilities of simple events are unknown and we are forced to search in past events for the indices which can guide us in our conjectures about the causes upon which they depend. In applying the analysis of discriminant functions to the principle elucidated above on the probability of the causes drawn from the events observed, we are led to the following theorem.

When a simple event or one composed of several simple events, as, for instance, in a game, has been repeated a great number of times the possibilities of the simple events which render most probable that which has been observed are those that observation indicates with the greatest probability; in the measure that the observed event is repeated this probability increases and would end by amounting to certainty if the numbers of repetitions should become infinite.

There are two kinds of approximations: the one is relative to the limits taken on all sides of the possibilities which give to the past the greatest probability; the other approximation is related to the probability that these possibilities fall within these limits. The repetition of the compound event increases more and more this probability, the limits remaining the same; it reduces more and more the interval of these limits, the probability remaining the same; in infinity this interval becomes zero and the probability changes to certainty.

If we apply this theorem to the ratio of the births of boys to that of girls observed in the different countries of Europe, we find that this ratio, which is everywhere about equal to that of 22 to 21, indicates with an extreme probability a greater facility in the birth of boys. Considering further that it is the same at Naples and at St. Petersburg, we shall see that in this regard the influence of climate is without effect. We might then suspect, contrary to the common belief, that this predominance of masculine births exists even in the Orient. I have consequently invited the French scholars sent to Egypt to occupy themselves with this interesting question; but the difficulty in obtaining exact information about the births has not permitted them to solve it. Happily, M. de Humboldt has not neglected this matter among the innumerable new things which he has observed and collected in America with so much sagacity, constancy, and courage. He has found in the tropics the same ratio of the births as we observe in Paris; this ought to make us regard the greater number of masculine births as a general law of the human race. The laws which the different kinds of animals follow in this regard seem to me worthy of the attention of naturalists.

The fact that the ratio of births of boys to that of girls differs very little from unity even in the great number of the births observed in a place would offer in this regard a result contrary to the general law, without which we should be right in concluding that this law did not exist. In order to arrive at this result it is necessary to employ great numbers and to be sure that it is indicated by great probability. Buffon cites, for example, in his *Political Arithmetic* several communities of Bourgogne where the births of girls have surpassed those of boys. Among these communities that of Carcelle-le-Grignon presents in 2009 births during five years 1026 girls and 983 boys. Although these numbers are considerable, they indicate, however, only a greater possibility in the births of girls with a probability of $\frac{9}{10}$, and this probability, smaller than that of not throwing heads four times in succession in the game of heads and tails, is not sufficient to investigate the cause for this anomaly, which, according to all probability, would disappear if one should follow during a century the births in this community.

The registers of births, which are kept with care in order to assure the condition of the citizens, may serve in determining the population of a great empire without recurring to the enumeration of its inhabitants—a laborious operation and one difficult to make with exactitude. But for this it is necessary to know the ratio of the population to the annual births. The most precise means of obtaining it consists, first, in choosing in the empire districts distributed in an almost equal manner over its whole surface, so as to render the general result independent of local circumstances;

second, in enumerating with care for a given epoch the inhabitants of several communities in each of these districts; third, by determining from the statement of the births during several years which precede and follow this epoch the mean number corresponding to the annual births. This number, divided by that of the inhabitants, will give the ratio of the annual births to the population in a manner more and more accurate as the enumeration becomes more considerable. The government, convinced of the utility of a similar enumeration, has decided at my request to order its execution. In thirty districts spread out equally over the whole of France, communities have been chosen which would be able to furnish the most exact information. Their enumerations have given 2037615 individuals as the total number of their inhabitants on the 23d of September, 1802. The statement of the births in these communities during the years 1800, 1801, and 1802 have given:

Births.	Marriages.	Deaths.
110312 boys	46037	103659 men
105287 girls		99443 women

The ratio of the population to annual births is then $28\frac{352845}{1000000}$; it is greater than had been estimated up to this time. Multiplying the number of annual births in France by this ratio, we shall have the population of this kingdom. But what is the probability that the population thus determined will not deviate from the true population beyond a given limit? Resolving this problem and applying to its solution the preceding data, I have found that, the number of annual births in France being supposed to be 1000000, which brings the population to 28352845 inhabitants, it is a bet of almost 300000 against 1 that the error of this result is not half a million.

The ratio of the births of boys to that of girls which the preceding statement offers is that of 22 to 21; and the marriages are to the births as 3 is to 4.

At Paris the baptisms of children of both sexes vary a little from the ratio of 22 to 21. Since 1745, the epoch in which one has commenced to distinguish the sexes upon the birth-registers, up to the end of 1784, there have been baptized in this capital 393386 boys and 377555 girls. The ratio of the two numbers is almost that of 25 to 24; it appears then at Paris that a particular cause approximates an equality of baptisms of the two sexes. If we apply to this matter the calculus of probabilities, we find that it is a bet of 238 to 1 in favor of the existence of this cause, which is sufficient to authorize the investigation. Upon reflection it has appeared to me that the difference observed holds to this, that the parents in the country and the provinces, finding some advantage in keeping the boys at home, have sent to the Hospital for Foundlings in Paris fewer of them relative to the number of girls according to the ratio of births of the two sexes. This is proved by the statement of the registers of this hospital. From the beginning of 1745 to the end of 1809 there were entered 163499 boys and 159405 girls. The first of these numbers exceeds only by $\frac{1}{38}$ the second, which it ought to have surpassed at least by $\frac{1}{24}$. This confirms the existence of the assigned cause, namely, that the ratio of births of boys to those of girls is at Paris that of 22 to 21, no attention having been paid to foundlings.

The preceding results suppose that we may compare the births to the drawings of balls from an urn which contains an infinite number of white balls and black balls so mixed that at each draw the chances of drawing ought to be the same for each ball; but it is possible that the variations of the same seasons in different years may have some influence upon the annual ratio of the births of boys to those of girls. The Bureau of Longitudes of France publishes each year in its annual the tables of the annual movement of the population of the kingdom. The tables already published commence in 1817; in that year and in the five following years there were born 2962361 boys and 2781997 girls, which gives about $\frac{16}{15}$ for the ratio of the births of boys to that of girls. The ratios of each year vary little from this mean result; the smallest ratio is that of 1822, where it was only $\frac{17}{16}$; the greatest is of the year 1817, when it was $\frac{15}{14}$. These ratios vary appreciably from the ratio of $\frac{22}{21}$ found above. Applying to this deviation the analysis of probabilities in the hypothesis of the comparison of births to the drawings of balls from an urn, we find that it would be scarcely probable. It appears, then, to indicate that this hypothesis, although closely approximated, is not rigorously exact. In the number of births which we have just stated there are of natural children 200494 boys and 190698 girls. The ratio of masculine and feminine births was then in this regard $\frac{20}{19}$, smaller than the mean ratio of $\frac{16}{15}$. This result is in the same sense as that of the births of foundlings; and it seems to prove that in the class of natural children the births of the two sexes approach more nearly equality than in the class of legitimate children. The difference of the climates from the north to the south of France does not appear to influence appreciably the ratio of the births of boys and girls. The thirty most southern districts have given $\frac{16}{15}$ for this ratio, the same as that of entire France.

The constancy of the superiority of the births of boys over girls at Paris and at London since they have been observed has appeared to some scholars to be a proof of Providence, without which they have thought that the irregular causes which disturb without ceasing the course of events ought several times to have rendered the annual births of girls superior to those of boys.

But this proof is a new example of the abuse which has been so often made of final causes which always disappear on a searching examination of the questions when we have the necessary data to solve them. The constancy in question is a result of regular causes which give the superiority to the births of boys and which extend it to the anomalies due to hazard when the number of annual births is considerable. The investigation of the probability that this constancy will maintain itself for a long time belongs to that branch of the analysis of hazards which passes from past events to the probability of future events; and taking as a basis the births observed from 1745 to 1784, it is a bet of almost 4 against 1 that at Paris the annual births of boys will constantly surpass for a century the births of girls; there is then no reason to be astonished that this has taken place for a half-century.

Let us take another example of the development of constant ratios which events present in the measure that they are multiplied. Let us imagine a series of urns arranged circularly, and each containing a very great number of white balls and black balls; the ratio of white balls to the black in the urns being originally very different and such, for example, that one of these urns contains only white balls, while another contains only black balls. If one draws a ball from the first urn in order to put it into the second, and, after having shaken the second urn in order to mix well the new ball with the others, one draws a ball to put it into the third urn, and so on to the last urn, from which is drawn a ball to put into the first, and if this series is recommenced continually, the analysis of probability shows us that the ratios of the white balls to the black in these urns will end by being the same and equal to the ratio of the sum of all the white balls to the sum of all the black balls contained in the urns. Thus by this regular mode of change the primitive irregularity of these ratios disappears eventually in order to make room for the most simple order. Now if among these urns one intercalate new ones in which the ratio of the sum of the white balls to the sum of the black balls which they contain differs from the preceding, continuing indefinitely in the totality of the urns the drawings which we have just indicated, the simple order established in the old urns will be at first disturbed, and the ratios of the white balls to the black balls will become irregular; but little by little this irregularity will disappear in order to make room for a new order, which will finally be that of the equality of the ratios of the white balls to the black balls contained in the urns. We may apply these results to all the combinations of nature in which the constant forces by which their elements are animated establish regular modes of action, suited to bring about in the very heart of chaos systems governed by admirable laws.

The phenomena which seem the most dependent upon hazard present, then, when multiplied a tendency to approach without ceasing fixed ratios, in such a manner that if we conceive on all sides of each of these ratios an interval as small as desired, the probability that the mean result of the observations falls within this interval will end by differing from certainty only by a quantity greater than an assignable magnitude. Thus by the calculations of probabilities applied to a great number of observations we may recognize the existence of these ratios. But before seeking the causes it is necessary, in order not to be led into vain speculations, to assure ourselves that they are indicated by a probability which does not permit us to regard them as anomalies due to hazard. The theory of discriminant functions gives a very simple expression for this probability, which is obtained by integrating the product of the differential of the quantity of which the result deduced from a great number of observations varies from the truth by a constant less than unity, dependent upon the nature of the problem, and raised to a power whose exponent is the ratio of the square of this variation to the number of observations. The integral taken between the limits given and divided by the same integral, applied to a positive and negative infinity, will express the probability that the variation from the truth is comprised between these limits. Such is the general law of the probability of results indicated by a great number of observations.

CHAPTER IX

The Application of the Calculus of Probabilities to Natural Philosophy

The phenomena of nature are most often enveloped by so many strange circumstances, and so great a number of disturbing causes mix their influence, that it is very difficult to recognize them. We may arrive at them only by multiplying the observations or the experiences, so that the strange effects finally destroy reciprocally each other, the mean results putting in evidence those phenomena and their divers elements. The more numerous the number of observations and the less they vary among themselves the more their results approach the truth. We fulfil this last condition by the choice of the methods of observations, by the precision of the instruments, and by the care which we take to observe closely; then we determine by the theory of probabilities the most advantageous mean results or those which give the least value of the error. But that is not sufficient; it is further necessary to appreciate the probability that the errors of these results are comprised in the given limits; and without this we have only an imperfect knowledge of the degree of exactitude obtained. Formulæ suitable to these matters are then true improvements of the method of sciences, and it is indeed important to add them to this method. The analysis which they require is the most delicate and the most difficult of the theory of probabilities; it is one of the principal objects of the work which I have published upon this theory, and in which I have arrived at formulæ of this kind which have the remarkable advantage of being independent of the law of the probability of errors and of including only the quantities given by the observations themselves and their expressions.

Each observation has for an analytic expression a function of the elements which we wish to determine; and if these elements are nearly known, this function becomes a linear function of their corrections. In equating it to the observation itself there is formed *an equation of condition.* If we have a great number of similar equations, we combine them in such a manner as to obtain as many final equations as there are elements whose corrections we determine then by resolving these equations. But what is the most advantageous manner of combining equations of condition in order to obtain final equations? What is the law of the probabilities of errors of which the elements are still susceptible that we draw from them? This is made clear to us by the theory of probabilities. The formation of a final equation by means of the equation of condition amounts to multiplying each one of these by an indeterminate factor and by uniting the products; it is necessary to choose the system of factors which gives the smallest opportunity for error. But it is apparent that if we multiply the possible errors of an element by their respective probabilities, the most advantageous system will be that in which the sum of these products all, taken, positively is a *minimum*; for a positive or a negative error ought to be considered as a loss. Forming, then, this sum of products, the condition of the *minimum* will determine the system of factors which it is expedient to adopt, or the most advantageous system. We find thus that this system is that of the coefficients of the elements in each equation of condition; so that we form a first final equation by multiplying respectively each equation of condition by its coefficient of the first element and by uniting all these equations thus multiplied. We form a second final equation by employing in the same manner the coefficients of the second element, and so on. In this manner the elements and the laws of the phenomena obtained in the collection of a great number of observations are developed with the most evidence.

The probability of the errors which each element still leaves to be feared is proportional to the number whose hyperbolic logarithm is unity raised to a power equal to the square of the error taken as a minus quantity and multiplied by a constant coefficient which may be considered as the modulus of the probability of the errors; because, the error remaining the same, its probability decreases with rapidity when the former increases; so that the element obtained weighs, if I may thus speak toward the truth, as much more as this modulus is greater. I would call for this reason this modulus the *weight* of the element or of the result. This weight is the greatest possible in the system of factors—the most advantageous; it is this which gives to this system superiority over others. By a remarkable analogy of this weight with those of bodies compared at their common centre of gravity it results that if the same element is given by divers systems, composed each of a great number of observations, the most advantageous, the mean result of their totality is the sum of the products of each partial result by its weight. Moreover, the total weight of the results of the divers systems is the sum of their partial weights; so that the probability of the errors of the mean result of their totality is proportional to the number which has unity for an hyperbolic logarithm raised to a power equal to the square of the error taken as minus and multiplied by the sum of the weights. Each weight depends in truth upon the law of the probability of error of each system, and almost always this law is unknown; but happily I have been able to eliminate the factor which contains it by means of the sum of the squares of the variations of the observations in this system from their mean result. It would then be desirable in order to complete our knowledge of the results obtained by the totality of a great number of observations that we write by the side of each result the weight which corresponds to it; analysis furnishes for this object both general and simple methods. When we have thus obtained the exponential which represents the law of the probability of errors, we shall have the probability that the error of the result is included within given limits by taking within the limits the integral of the product of this exponential by the differential of the error and multiplying it by the square root of the weight of the result divided by the circumference whose diameter is unity. Hence it follows that for the same probability the errors of the results are reciprocal to the square roots of their weights, which serves to compare their respective precision.

In order to apply this method with success it is necessary to vary the circumstances of the observations or the experiences in such a manner as to avoid the constant causes of error. It is necessary that the observations should be numerous, and that they should be so much the more so as there are more elements to determine; for the weight of the mean result increases as the number of observations divided by the number of the elements. It is still necessary that the elements follow in these observations a different course; for if the course of the two elements were exactly the same, which would render their coefficients proportional in equation of conditions, these elements would form only a single unknown quantity and it would be impossible to distinguish them by these observations. Finally it is necessary that the observations should be precise; this condition, the first of all, increases greatly the weight of the result the expression of which has for a divisor the sum of the squares of the deviations of the observations from this result. With these precautions we shall be able to make use of the preceding method and measure the degree of confidence which the results deduced from a great number of observations merit.

The rule which we have just given to conclude equations of condition, final equations, amount to rendering a minimum the sum of the squares of the errors of observations; for each equation of condition becomes exact by substituting in it the observation plus its error; and if we draw from it the expression of this error, it is easy to see that the condition of the *minimum* of the sum of the squares of these expressions gives the rule in question. This rule is the more precise as the observations are more numerous; but even in the case where their number is small it appears natural to employ the same rule which in all cases offers a simple means of obtaining without groping the corrections which we seek to determine. It serves further to compare the precision of the divers astronomical tables of the same star. These tables may always be supposed as reduced to the same form, and then they differ only by the epochs, the mean movements and the coefficients of the arguments; for if one of them contains a coefficient which is not found in the others, it is clear that this amounts to supposing zero in them as the coefficient of this argument. If now we rectify these tables by the totality of the good observations, they would satisfy the condition that the sum of the squares of the errors should be a minimum; the tables which, compared to a considerable number of observations, approach nearest this condition merit then the preference.

It is principally in astronomy that the method explained above may be employed with advantage. The astronomical tables owe the truly astonishing exactitude which they have attained to the precision of observations and of theories, and to the use of equations of conditions which cause to concur a great number of excellent observations in the correction of the same element. But it remains to determine the probability of the errors that this correction leaves still to be feared; and the method which I have just explained enables us to recognize the probability of these errors. In order to give some interesting applications of it I have profited by the immense work which M. Bouvard has just finished on the movements of Jupiter and Saturn, of which he has formed very precise tables. He has discussed with the greatest care the oppositions and quadratures of these two planets observed by Bradley and by the astronomers who have followed him down to the last years; he has concluded the corrections of the elements of their movement and their masses compared to that of the sun taken as unity. His calculations give him the mass of Saturn equal to the 3512th part of that of the sun. Applying to them my formulæ of probability, I find that it is a bet of 11,000 against one that the error of this result is not $\frac{1}{100}$ of its value, or that which amounts to almost the same—that after a century of new observations added to the preceding ones, and examined in the same manner, the new result will not differ by $\frac{1}{100}$ from that of M. Bouvard. This wise astronomer finds again the mass of Jupiter equal to the 1071th part of the sun; and my method of probability gives a bet of 1,000,000 to one that this result is not $\frac{1}{100}$ in error.

This method may be employed again with success in geodetic operations. We determine the length of the great arc on the surface of the earth by triangulation, which depends upon a base measured with exactitude. But whatever precision may be brought to the measure of the angles, the inevitable errors can, by accumulating, cause the value of the arc concluded from a great number of triangles to deviate appreciably from the truth. We recognize this value, then, only imperfectly unless the probability that its error is comprised within given limits can be assigned. The error of a geodetic result is a function of the errors of the angles of each triangle. I have given in the work cited general formulæ in order to obtain the probability of the values of one or of several linear functions of a great number of partial errors of which we know the law of probability; we may then by means of these formulæ determine the probability that the error of a geodetic result is contained within the assigned limits, whatever may be the law of the probability of partial errors. It is moreover more necessary to render ourselves independent of the law, since the most simple laws themselves are always infinitely less probable, seeing the infinite number of those which may exist in nature. But the unknown law of partial errors introduces into the formulæ an indeterminant which does not permit of reducing them to numbers unless we are able to eliminate it. We have seen that in astronomical questions, where each observation furnishes an equation of condition for obtaining the elements, we eliminate this determinant by means of the sum of the squares of the remainders when the most probable values of the elements have been substituted in each equation. Geodetic questions not offering similar equations, it is necessary to seek another means of elimination. The quantity by which the sum of the angles of each observed triangle surpasses two right angles plus the spherical excess furnishes this means. Thus we replace by the sum of the squares of these quantities the sum of the squares of the remainders of the equations of condition; and we may assign in numbers the probability that the error of the final result of a series of geodetic operations will not exceed a given quantity. But what is the most advantageous manner of dividing among the three angles of each triangle the observed sum of their errors? The analysis of probabilities renders it apparent that each angle ought to be diminished by a third of this sum, provided that the weight of a geodetic result be the greatest possible, which renders the same error less probable. There is then a great advantage in observing the three angles of each triangle and of correcting them as we have just said. Simple common sense indicates this advantage; but the calculation of probabilities alone is able to appreciate it and to render apparent that by this correction it becomes the greatest possible.

In order to assure oneself of the exactitude of the value of a great arc which rests upon a base measured at one of its extremities one measures a second base toward the other extremity; and one concludes from one of these bases the length of the other. If this length varies very little from the observation, there is all reason to believe that the chain of triangles which unites these bases is very nearly exact and likewise the value of the large arc which results from it. One corrects, then, this value by modifying the angles of the triangles in such a manner that the base is calculated according to the bases measured. But this may be done in an infinity of ways, among which is preferred that of which the geodetic result has the greatest weight, inasmuch as the same error becomes less probable. The analysis of probabilities gives formulæ for obtaining directly the most advantageous correction which results from the measurements of the several bases and the laws of probability which the multiplicity of the bases makes—laws which become very rapidly decreasing by this multiplicity.

Generally the errors of the results deduced from a great number of observations are the linear functions of the partial errors of each observation. The coefficients of these functions depend upon the nature of the problem and upon the process followed in order to obtain the results. The most advantageous process is evidently that in which the same error in the results is less probable than according to any other process. The application of the calculus of probabilities to natural philosophy consists, then, in determining analytically the probability of the values of these functions and in choosing their indeterminant coefficients in such a manner that the law of this probability should be most rapidly descending. Eliminating, then, from the formulæ by the data of the question the factor which is introduced by the almost always unknown law of the probability of partial errors, we may be able to evaluate numerically the probability that the errors of the results do not exceed a given quantity. We shall thus have all that may be desired touching the results deduced from a great number of observations.

Very approximate results may be obtained by other considerations. Suppose, for example, that one has a thousand and one observations of the same quantity; the arithmetical mean of all these observations is the result given by the most advantageous method. But one would be able to choose the result according to the condition that the sum of the variations from each partial value all taken positively should be a *minimum*. It appears indeed natural to regard as very approximate the result which satisfies this condition. It is easy to see that if one disposes the values given by the observations according to the order of magnitude, the value which will occupy the mean will fulfil the preceding condition, and calculus renders it apparent that in the case of an infinite number of observations it would coincide with the truth; but the result given by the most advantageous method is still preferable.

We see by that which precedes that the theory of probabilities leaves nothing arbitrary in the manner of distributing the errors of the observations; it gives for this distribution the most advantageous formulæ which diminishes as much as possible the errors to be feared in the results.

The consideration of probabilities can serve to distinguish the small irregularities of the celestial movements enveloped in the errors of observations, and to repass to the cause of the anomalies observed in these movements.

In comparing all the observations it was Ticho-Brahé who recognized the necessity of applying to the moon an equation of time different from that which had been applied to the sun and to the planets. It was similarly the totality of a great number of observations which made Mayer recognize that the coefficient of the inequality of the precession ought to be diminished a little for the moon. But since this diminution, although confirmed and even augmented by Mason, did not appear to result from universal gravitation, the majority of astronomers neglect it in their calculations. Having submitted to the calculation of probabilities a considerable number of lunar observations chosen for this purpose and which M. Bouvard consented to examine at my request, it appeared to me to be indicated with so strong a probability that I believed the cause of it ought to be investigated. I soon saw that it would be only the ellipticity of the terrestrial spheroid, neglected up to that time in the theory of the lunar movement as being able to produce only imperceptible terms. I concluded that these terms became perceptible by the successive integrations of differential equations. I determined then those terms by a particular analysis, and I discovered first the inequality of the lunar movement in latitude which is proportional to the sine of the longitude of the moon, which no astronomer before had suspected. I recognized then by means of this inequality that another exists in the lunar movement in longitude which produces the diminution observed by Mayer in the equation of the precession applicable to the moon. The quantity of this diminution and the coefficient of the preceding inequality in latitude are very appropriate to fix the oblateness of the earth. Having communicated my researches to M. Burg, who was occupied at that time in perfecting the tables of the moon by the comparison of all the good observations, I requested him to determine with a particular care these two quantities. By a very remarkable agreement the values which he has found give to the earth the same oblateness, $\frac{1}{305}$, which differs little from the mean derived from the measurements of the degrees of the meridian and the pendulum; but those regarded from the point of view of the influence of the errors of the observations and of the perturbing causes in these measurements, did not appear to me exactly determined by these lunar inequalities.

It was again by the consideration of probabilities that I recognized the cause of the secular equation of the moon. The modern observations of this star compared to the ancient eclipses had indicated to astronomers an acceleration in the lunar movement; but the geometricians, and particularly Lagrange, having vainly sought in the perturbations which this movement experienced the terms upon which this acceleration depends, reject it. An attentive examination of the ancient and modern observations and of the intermediary eclipses observed by the Arabians convinced me that it was indicated with a great probability. I took up again then from this point of view the lunar theory, and I recognized that the secular equation of the moon is due to the action of the sun upon this satellite, combined with the secular variation of the eccentricity of the terrestrial orb; this brought me to the discovery of the secular equations of the movements of the nodes and of the perigees of the lunar orbit, which equations had not been even suspected by astronomers. The very remarkable agreement of this theory with all the ancient and modern observations has brought it to a very high degree of evidence.

The calculus of probabilities has led me similarly to the cause of the great irregularities of Jupiter and Saturn. Comparing modern observations with ancient, Halley found an acceleration in the movement of Jupiter and a retardation in that of Saturn. In order to conciliate the observations he reduced the movements to two secular equations of contrary signs and increasing as the squares of the times passed since 1700. Euler and Lagrange submitted to analysis the alterations which the mutual attraction of these two planets ought to produce in these movements. They found in doing this the secular equations; but their results were so different that one of the two at least ought to be erroneous. I determined then to take up again this important problem of *celestial mechanics*, and I recognized the invariability of the mean planetary movements, which nullified the secular equations introduced by Halley in the tables of Jupiter and Saturn. Thus there remain, in order to explain the great irregularity of these planets, only the attractions of the comets to which many astronomers had effective recourse, or the existence of an irregularity over a long period produced in the movements of the two planets by their reciprocal action and affected by contrary signs for each of them. A theorem which I found in regard to the inequalities of this kind rendered this inequality very probable. According to this theorem, if the movement of Jupiter is accelerated, that of Saturn is retarded, which has already conformed to what Halley had noticed; moreover, the acceleration of Jupiter resulting from the same theorem is to the retardation of Saturn very nearly in the ratio of the secular equations proposed by Halley. Considering the mean movements of Jupiter and Saturn I was enabled easily to recognize that two times that of Jupiter differed only by a very small quantity from five times that of Saturn. The period of an irregularity which would have for an argument this difference would be about nine centuries. Indeed its coefficient would be of the order of the cubes of the eccentricities of the orbits; but I knew that by virtue of successive integrations it acquired for divisor the square of the very small multiplier of the time in the argument of this inequality which is able to give it a great value; the existence of this inequality appeared to me then very probable. The following observation increased then its probability. Supposing its argument zero toward the epoch of the observations of Ticho-Brahé, I saw that Halley ought to have found by the comparison of modern with ancient observations the alterations which he had indicated; while the comparison of the modern observations among themselves ought to offer contrary alterations similar to those which Lambert had concluded from this comparison. I did not then hesitate at all to undertake this long and tedious calculation necessary to assure myself of this inequality. It was entirely confirmed by the result of this calculation, which moreover made me recognize a great number of other inequalities of which the totality has inclined the tables of Jupiter and Saturn to the precision of the same observations.

It was again by means of the calculus of probabilities that I recognized the remarkable law of the mean movements of the three first satellites of Jupiter, according to which the mean longitude of the first minus three times that of the second plus two times that of the third is rigorously equal to the half-circumference. The approximation with which the mean movements of these stars satisfy this law since their discovery indicates its existence with an extreme probability. I sought then the cause of it in their mutual action. The searching examination of this action convinced me that it was sufficient if in the beginning the ratios of their mean movements had approached this law within certain limits, because their mutual action had established and maintained it rigorously. Thus these three bodies will balance one another eternally in space according to the preceding law unless strange causes, such as comets, should change suddenly their movements about Jupiter.

Accordingly it is seen how necessary it is to be attentive to the indications of nature when they are the result of a great number of observations, although in other respects they may be inexplicable by known means. The extreme difficulty of problems relative to the system of the world has forced geometricians to recur to the approximation which always leaves room for the fear that the quantities neglected may have an appreciable influence. When they have been warned of this influence by the observations, they have recurred to their analysis; in rectifying it they have always found the cause of the anomalies observed; they have determined the laws and often they have anticipated the observations in discovering the inequalities which it had not yet indicated. Thus one may say that nature itself has concurred in the analytical perfection of the theories based upon the principle of universal gravity; and this is to my mind one of the strongest proofs of the truth of this admirable principle.

In the cases which I have just considered the analytical solution of the question has changed the probability of the causes into certainty. But most often this solution is impossible and it remains only to augment more and more this probability. In the midst of numerous and incalculable modifications which the action of the causes receives then from strange circumstances these causes conserve always with the effects observed the proper ratios to make them recognizable and to verify their existence. Determining these ratios and comparing them with a great number of observations if one finds that they constantly satisfy it, the probability of the causes may increase to the point of equalling that of facts in regard to which there is no doubt. The investigation of these ratios of causes to their effects is not less useful in natural philosophy than the direct solution of problems whether it be to verify the reality of these causes or to determine the laws from their effects; since it may be employed in a great number of questions whose direct solution is not possible, it replaces it in the most advantageous manner. I shall discuss here the application which I have made of it to one of the most interesting phenomena of nature, the flow and the ebb of the sea.

Pliny has given of this phenomenon a description remarkable for its exactitude, and in it one sees that the ancients had observed that the tides of each month are greatest toward the syzygies and smallest toward the quadratures; that they are higher in the perigees than in the apogees of the moon, and higher in the equinoxes than in the solstices. They concluded from this that this phenomenon is due to the action of the sun and moon upon the sea. In the preface of his work *De Stella Martis* Kepler admits a tendency of the waters of the sea toward the moon; but, ignorant of the law of this tendency, he was able to give on this subject only a probable idea. Newton converted into certainty the probability of this idea by attaching it to his great principle of universal gravity. He gave the exact expression of the attractive forces which produced the flood and the ebb of the sea; and in order to determine the effects he supposed that the sea takes at each instant the position of equilibrium which is agreeable to these forces. He explained in this manner the principal phenomena of the tides; but it followed from this theory that in our ports the two tides of the same day would be very unequal if the sun and the moon should have a great declination. At Brest, for example, the evening tide would be in the syzygies of the solstices about eight times greater than the morning tide, which is certainly contrary to the observations which prove that these two tides are very nearly equal. This result from the Newtonian theory might hold to the supposition that the sea is agreeable at each instant to a position of equilibrium, a supposition which is not at all admissible. But the investigation of the true figure of the sea presents great difficulties. Aided by the discoveries which the geometricians had just made in the theory of the movement of fluids and in the calculus of partial differences, I undertook this investigation, and

I gave the differential equations of the movement of the sea by supposing that it covers the entire earth. In drawing thus near to nature I had the satisfaction of seeing that my results approached the observations, especially in regard to the little difference which exists in our ports between the two tides of the solstitial syzygies of the same day. I found that they would be equal if the sea had everywhere the same depth; I found further that in giving to this depth convenient values one was able to augment the height of the tides in a port conformably to the observations. But these investigations, in spite of their generality, did not satisfy at all the great differences which even adjacent ports present in this regard and which prove the influence of local circumstances. The impossibility of knowing these circumstances and the irregularity of the basin of the seas and that of integrating the equations of partial differences which are relative has compelled me to make up the deficiency by the method I have indicated above. I then endeavored to determine the greatest ratios possible among the forces which affect all the molecules of the sea; and their effects observable in our ports. For this I made use of the following principle, which may be applied to many other phenomena.

"The state of the system of a body in which the primitive conditions of the movement have disappeared by the resistances which this movement meets is periodic as the forces which animate it."

Combining this principle with that of the coexistence of very small oscillations, I have found an expression of the height of the tides whose arbitraries contain the effect of local circumstances of each port and are reduced to the smallest number possible; it is only necessary to compare it to a great number of observations.

Upon the invitation of the Academy of Sciences, observations were made at the beginning of the last century at Brest upon the tides, which were continued during six consecutive years. The situation of this port is very favorable to this sort of observations; it communicates with the sea by a canal which empties into a vast roadstead at the far end of which the port has been constructed. The irregularities of the sea extend thus only to a small degree into the port, just as the oscillations which the irregular movement of a vessel produces in a barometer are diminished by a throttling made in the tube of this instrument. Moreover, the tides being considerable at Brest, the accidental variations caused by the winds are only feeble; likewise we notice in the observations of these tides, however little we multiply them, a great regularity which induced me to propose to the government to order in this port a new series of observations of the tides, continued during a period of the movement of the nodes of the lunar orbit. This has been done. The observations began June 1, 1806; and since this time they have been made every day without interruption. I am indebted to the indefatigable zeal of M. Bouvard, for all that interests astronomy, the immense calculations which the comparison of my analysis with the observations has demanded. There have been used about six thousand observations, made during the year 1807 and the fifteen years following. It results from this comparison that my formulæ represent with a remarkable precision all the varieties of the tides relative to the digression of the moon, from the sun, to the declination of these stars, to their distances from the earth, and to the laws of variation at the *maximum* and *minimum* of each of these elements. There results from this accord a probability that the flow and the ebb of the sea is due to the attraction of the sun and moon, so approaching certainty that it ought to leave room for no reasonable doubt. It changes into certainty when we consider that this attraction is derived from the law of universal gravity demonstrated by all the celestial phenomena.

The action of the moon upon the sea is more than double that of the sun. Newton and his successors in the development of this action have paid attention only to the terms divided by the cube of the distance from the moon to the earth, judging that the effects due to the following terms ought to be inappreciable. But the calculation of probabilities makes it clear to us that the smallest effects of regular causes may manifest themselves in the results of a great number of observations arranged in the order most suitable to indicate them. This calculation again determines their probability and up to what point it is necessary to multiply the observations to make it very great. Applying it to the numerous observations discussed by M. Bouvard I recognized that at Brest the action of the moon upon the sea is greater in the full moons than in the new moons, and greater when the moon is austral than when it is boreal—phenomena which can result only from the terms of the lunar action divided by the fourth power of the distance from the moon to the earth.

To arrive at the ocean the action of the sun and the moon traverses the atmosphere, which ought consequently to feel its influence and to be subjected to movements similar to those of the sea.

These movements produce in the barometer periodic oscillations. Analysis has made it clear to me that they are inappreciable in our climates. But as local circumstances increase considerably the tides in our ports, I have inquired again if similar circumstances have made appreciable these oscillations of the barometer. For this I have made use of the meteorological observations which have been made every day for many years at the royal observatory. The heights of the barometer and of the thermometer are observed there at nine o'clock in the morning, at noon, at three o'clock in the afternoon, and at eleven o'clock in the evening. M. Bouvard has indeed wished to take up the consideration of observations of the eight years elapsed from October 1, 1815, to October 1, 1823, on the registers. In disposing the observations in the manner most suitable to indicate the lunar atmospheric flood at Paris, I find only one eighteenth of a millimeter for the extent of the corresponding oscillation of the barometer. It is this especially which has made us feel the necessity of a method for determining the probability of a result, and without this method one is forced to present as the laws of nature the results of irregular causes which has often happened in meteorology. This method applied to the preceding result shows the uncertainty of it in spite of the great number of observations employed, which it would be necessary to increase tenfold in order to obtain a result sufficiently probable.

The principle which serves as a basis for my theory of the tides may be extended to all the effects of hazard to which variable causes are joined according to regular laws. The action of these causes produces in the mean results of a great number of effects varieties which follow the same laws and which one may recognize by the analysis of probabilities. In the measure which these effects are multiplied those varieties are manifested with an ever-increasing probability, which would approach certainty if the number of the effects of the results should become infinite. This theorem is analogous to that which I have already developed upon the action of constant causes. Every time, then, that a cause whose progress is regular can have influence upon a kind of events, we may seek to discover its influence by multiplying the observations and arranging them in the most suitable order to indicate it. When this influence appears to manifest itself the analysis of probabilities determines the probability of its existence and that of its intensity; thus the variation of the temperature from day to night modifying the pressure of the atmosphere and consequently the height of the barometer, it is natural to think that the multiplied observations of these heights ought to show the influence of the solar heat. Indeed there has long been recognized at the equator, where this influence appears to be greatest, a small diurnal variation in the height of the barometer of which the *maximum* occurs about nine o'clock in the morning and the *minimum* about three o'clock in the afternoon. A second *maximum* occurs about eleven o'clock in the evening and a second *minimum* about four o'clock in the morning. The oscillations of the night are less than those of the day, the extent of which is about two millimeters. The inconstancy of our climate has not taken this variation from our observers, although it may be less appreciable than in the tropics. M. Ramond has recognized and determined it at Clermont, the chief place of the district of Puy-de-Dôme, by a series of precise observations made during several years; he has even found that it is smaller in the months of winter than in other months. The numerous observations which I have discussed in order to estimate the influence of attractions of the sun and the moon upon the barometric heights at Paris have served me in determining their diurnal variation. Comparing the heights at nine o'clock in the morning with those of the same days at three o'clock in the afternoon, this variation is manifested with so much evidence that its mean value each month has been constantly positive for each of the seventy-two months from January 1, 1817, to January 1, 1823; its mean value in these seventy-two months has been almost .8 of a millimeter, a little less than at Clermont and much less than at the equator. I have recognized that the mean result of the diurnal variations of the barometer from 9 o'clock A.M. to 3 P.M. has been only .5428 millimeter in the three months of November, December, January, and that it has risen to 1.0563 millimeters in the three following months, which coincides with the observations of M. Ramond. The other months offer nothing similar.

In order to apply to these phenomena the calculation of these probabilities, I commenced by determining the law of the probability of the anomalies of the diurnal variation due to hazard. Applying it then to the observations of this phenomenon, I found that it was a bet of more than 300,000 against one that a regular cause produced it. I do not seek to determine this cause; I content myself with stating its existence. The period of the diurnal variation regulated by the solar day indicates evidently that this variation is due to the action of the sun. The extreme smallness of the attractive action of the sun upon the atmosphere is proved by the smallness of the effects due to the united attractions of the sun and the moon. It is then by the action of its heat that the sun produces the diurnal variation of the barometer; but it is impossible to subject to calculus the effects of its action on the height of the barometer and upon the winds. The diurnal variation of the magnetic needle is certainly a result of the action of the sun. But does this star act here as in the diurnal variation of the barometer by its heat or by its influence upon electricity and upon magnetism, or finally by the union of these influences? A long series of observations made in different countries will enable us to apprehend this.

One of the most remarkable phenomena of the system of the world is that of all the movements of rotation and of revolution of the planets and the satellites in the sense of the rotation of the sun and about in the same plane of its equator. A phenomenon so remarkable is not the effect of hazard: it indicates a general cause which has determined all its movements. In order to obtain the probability with which this cause is indicated we shall observe that the planetary system, such as we know it to-day, is composed of eleven planets and of eighteen satellites at least, if we attribute with Herschel six satellites to the planet Uranus. The movements of the rotation of the sun, of six planets, of the moon, of the satellites of Jupiter, of the ring of Saturn, and of one of its satellites have been recognized. These movements form with those of revolution a totality of forty-three movements directed in the same sense; but one finds by the analysis of probabilities that it is a bet of more than 4000000000000 against one that this disposition is not the result of hazard; this forms a probability indeed superior to that of historical events in regard to which no doubt exists. We ought then to believe at least with equal confidence that a primitive cause has directed the planetary movements, especially if we consider that the inclination of the greatest number of these movements at the solar equator is very small.

Another equally remarkable phenomenon of the solar system is the small degree of the eccentricity of the orbs of the planets and the satellites, while those of the comets are very elongated, the orbs of the system not offering any intermediate shades between a great and a small eccentricity. We are again forced to recognize here the effect of a regular cause; chance has certainly not given an almost circular form to the orbits of all the planets and their satellites; it is then that the cause which has determined the movements of these bodies has rendered them almost circular. It is necessary, again, that the great eccentricities of the orbits of the comets should result from the existence of this cause without its having influenced the direction of their movements; for it is found that there are almost as many retrograde comets as direct comets, and that the mean inclination of all their orbits to the ecliptic approaches very nearly half a right angle, as it ought to be if the bodies had been thrown at hazard.

Whatever may be the nature of the cause in question, since it has produced or directed the movement of the planets, it is necessary that it should have embraced all the bodies and considered all the distances which separate them, it can have been only a fluid of an immense extension. Therefore in order to have given them in the same sense an almost circular movement about the sun it is necessary that this fluid should have surrounded this star as an atmosphere. The consideration of the planetary movements leads us then to think that by virtue of an excessive heat the atmosphere of the sun was originally extended beyond the orbits of all the planets, and that it has contracted gradually to its present limits.

In the primitive state where we imagine the sun it resembled the nebulæ that the telescope shows us composed of a nucleus more or less brilliant surrounded by a nebula which, condensing at the surface, ought to transform it some day into a star. If one conceives by analogy all the stars formed in this manner, one can imagine their anterior state of nebulosity itself preceded by other stars in which the nebulous matter was more and more diffuse, the nucleus being less and less luminous and dense. Going back, then, as far as possible, one would arrive at a nebulosity so diffuse that one would be able scarcely to suspect its existence.

Such is indeed the first state of the nebulæ which Herschel observed with particular care by means of his powerful telescopes, and in which he has followed the progress of condensation, not in a single one, these stages not becoming appreciable to us except after centuries, but in their totality, just about as one can in a vast forest follow the increase of the trees by the individuals of the divers ages which the forest contains. He has observed from the beginning nebulous matter spread out in divers masses in the different parts of the heavens, of which it occupies a great extent. He has seen in some of these masses this matter slightly condensed about one or several faintly luminous nebulæ. In the other nebulæ these nuclei shine, moreover, in proportion to the nebulosity which surrounds them. The atmospheres of each nucleus becoming separated by an ulterior condensation, there result the multifold nebulæ formed of brilliant nuclei very adjacent and surrounded each by an atmosphere; sometimes the nebulous matter, by condensing in a uniform manner, has produced the nebulæ which are called *planetary*. Finally a greater degree of condensation transforms all these nebulæ into stars. The nebulæ classed according to this philosophic view indicate with an extreme probability their future transformation into stars and the anterior state of nebulosity of existing stars. The following considerations come to the aid of proofs drawn from these analogies.

For a long time the particular disposition of certain stars visible to the naked eye has struck the attention of philosophical observers. Mitchel has already remarked how improbable it is that the stars of the Pleiades, for example, should have been confined in the narrow space which contain them by the chances of hazard alone, and he has concluded from this that this group of stars and the similar groups that the heaven presents us are the results of a primitive cause or of a general law of nature. These groups are a necessary result of the condensation of the nebulæ at several nuclei; it is apparent that the nebulous matter being attracted continuously by the divers nuclei, they ought to form in time a group of stars equal to that of the Pleiades. The condensation of the nebulæ at two nuclei forms similarly very adjacent stars, revolving the one about the other, equal to those whose respective movements Herschel has already considered. Such are, further, the 61st of the Swan and its following one in which Bessel has just recognized particular movements so considerable and so little different that the proximity of these stars to one another and their movement about the common centre of gravity ought to leave no doubt. Thus one descends by degrees from the condensation of nebulous matter to the consideration of the sun surrounded formerly by a vast atmosphere, a consideration to which one repasses, as has been seen, by the examination of the phenomena of the solar system. A case so remarkable gives to the existence of this anterior state of the sun a probability strongly approaching certainty.

But how has the solar atmosphere determined the movements of rotation and revolution of the planets and the satellites? If these bodies had penetrated deeply the atmosphere its resistance would have caused them to fall upon the sun; one is then led to believe with much probability that the planets have been formed at the successive limits of the solar atmosphere which, contracting by the cold, ought to have abandoned in the plane of its equator zones of vapors which the mutual attraction of their molecules has changed into divers spheroids. The satellites have been similarly formed by the atmospheres of their respective planets.

I have developed at length in my *Exposition of the System of the World* this hypothesis, which appears to me to satisfy all the phenomena which this system presents us. I shall content myself here with considering that the angular velocity of rotation of the sun and the planets being accelerated by the successive condensation of their atmospheres at their surfaces, it ought to surpass the angular velocity of revolution of the nearest bodies which revolve about them. Observation has indeed confirmed this with regard to the planets and satellites, and even in ratio to the ring of Saturn, the duration of whose revolution is .438 days, while the duration of the rotation of Saturn is .427 days.

In this hypothesis the comets are strangers to the planetary system. In attaching their formation to that of the nebulæ they may be regarded as small nebulæ at the nuclei, wandering from systems to solar systems, and formed by the condensation of the nebulous matter spread out in such great profusion in the universe. The comets would be thus, in relation to our system, as the aerolites are relatively to the Earth, to which they would appear strangers. When these stars become visible to us they offer so perfect resemblance to the nebulæ that they are often confounded with them; and it is only by their movement, or by the knowledge of all the nebulæ confined to that part of the heavens where they appear, that we succeed in distinguishing them. This supposition explains in a happy manner the great extension which the heads and tails of comets take in the measure that they approach the sun, and the extreme rarity of these tails which, in spite of their immense depth, do not weaken at all appreciably the light of the stars which we look across.

When the little nebulæ come into that part of space where the attraction of the sun is predominant, and which we shall call the *sphere of activity* of this star, it forces them to describe elliptic or hyperbolic orbits. But their speed being equally possible in all directions they ought to move indifferently in all the senses and under all inclinations of the elliptic, which is conformable to that which has been observed.

The great eccentricity of the cometary orbits results again from the preceding hypothesis. Indeed if these orbits are elliptical they are very elongated, since their great axes are at least equal to the radius of the sphere of activity of the sun. But these orbits may be hyperbolic; and if the axes of these hyperbolæ are not very large in proportion to the mean distance from the sun to the earth, the movement of the comets which describe them will appear sensibly hyperbolic. However, of the hundred comets of which we already have the elements, not one has appeared certainly to move in an hyperbola; it is necessary, then, that the chances which give an appreciable hyperbola should be extremely rare in proportion to the contrary chances.

The comets are so small that, in order to become visible, their perihelion distance ought to be inconsiderable. Up to the present this distance has surpassed only twice the diameter of the terrestrial orbit, and most often it has been below the radius of this orbit. It is conceived that, in order to approach so near the sun, their speed at the moment of their entrance into its sphere of activity ought to have a magnitude and a direction confined within narrow limits. In determining by the analysis of probabilities the ratio of the chances which, in these limits, give an appreciable hyperbola, to the chances which give an orbit which may be confounded with a parabola, I have found that it is a bet of at least 6000 against one that a nebula which penetrates into the activity of the sun in such a manner as to be observed will describe either a very elongated ellipse or an hyperbola. By the magnitude of its axis, the latter will be appreciably confounded with a parabola in the part which is observed; it is then not surprising that, up to this time, hyperbolic movements have not been recognized.

The attraction of the planets, and, perhaps further, the resistance of the ethereal centres, ought to have changed many cometary orbits in the ellipses whose great axis is less than the radius of the sphere of activity of the sun, which augments the chances of the elliptical orbits. We may believe that this change has taken place with the comet of 1759, and with the comet whose duration is only twelve hundred days, and which will reappear without ceasing in this short interval, unless the evaporation which it meets at each of its returns to the perihelion ends by rendering it invisible.

We are able further, by the analysis of probabilities, to verify the existence or the influence of certain causes whose action is believed to exist upon organized beings. Of all the instruments that we are able to employ in order to recognize the imperceptible agents of nature the most sensitive are the nerves, especially when particular causes increase their sensibility. It is by their aid that the feeble electricity which the contact of two heterogeneous metals develops has been discovered; this has opened a vast field to the researches of physicists and chemists. The singular phenomena which results from extreme sensibility of the nerves in some individuals have given birth to divers opinions about the existence of a new agent which has been named *animal magnetism*, about the action on ordinary magnetism, and about the influence of the sun and moon in some nervous affections, and finally, about the impressions which the proximity of metals or of running water makes felt. It is natural to think that the action of these causes is very feeble, and that it may be easily disturbed by accidental circumstances; thus because in some cases it is not manifested at all its existence ought not to be denied. We are so far from recognizing all the agents of nature and their divers modes of action that it would be unphilosophical to deny the phenomena solely because they are inexplicable in the present state of our knowledge. But we ought to examine them with an attention as much the more scrupulous as it appears the more difficult to admit them; and it is here that the calculation of probabilities becomes indispensable in determining to just what point it is necessary to multiply the observations or the experiences in order to obtain in favor of the agents which they indicate, a probability superior to the reasons which can be obtained elsewhere for not admitting them.

The calculation of probabilities can make appreciable the advantages and the inconveniences of the methods employed in the speculative sciences. Thus in order to recognize the best of the treatments in use in the healing of a malady, it is sufficient to test each of them on an equal number of patients, making all the conditions exactly similar; the superiority of the most advantageous treatment will manifest itself more and more in the measure that the number is increased; and the calculation will make apparent the corresponding probability of its advantage and the ratio according to which it is superior to the others.

CHAPTER X

APPLICATION OF THE CALCULUS OF PROBABILITIES
TO THE MORAL SCIENCES

WE have just seen the advantages of the analysis of probabilities in the investigation of the laws of natural phenomena whose causes are unknown or so complicated that their results cannot be submitted to calculus. This is the case of nearly all subjects of the moral sciences. So many unforeseen causes, either hidden or inappreciable, influence human institutions that it is impossible to judge *à priori* the results. The series of events which time brings about develops these results and indicates the means of remedying those that are harmful. Wise laws have often been made in this regard; but because we had neglected to conserve the motives many have been abrogated as useless, and the fact that vexatious experiences have made the need felt anew ought to have reëstablished them.

It is very important to keep in each branch of the public administration an exact register of the results which the various means used have produced, and which are so many experiences made on a large scale by governments. Let us apply to the political and moral sciences the method founded upon observation and upon calculus, the method which has served us so well in the natural sciences. Let us not offer in the least a useless and often dangerous resistance to the inevitable effects of the progress of knowledge; but let us change only with an extreme circumspection our institutions and the usages to which we have already so long conformed. We should know well by the experience of the past the difficulties which they present; but we are ignorant of the extent of the evils which their change can produce. In this ignorance the theory of probability directs us to avoid all change; especially is it necessary to avoid the sudden changes which in the moral world as well as in the physical world never operate without a great loss of vital force.

Already the calculus of probabilities has been applied with success to several subjects of the moral sciences. I shall present here the principal results.

CHAPTER XI

CONCERNING THE PROBABILITIES OF TESTIMONIES

THE majority of our opinions being founded on the probability of proofs it is indeed important to submit it to calculus. Things it is true often become impossible by the difficulty of appreciating the veracity of witnesses and by the great number of circumstances which accompany the deeds they attest; but one is able in several cases to resolve the problems which have much analogy with the questions which are proposed and whose solutions may be regarded as suitable approximations to guide and to defend us against the errors and the dangers of false reasoning to which we are exposed. An approximation of this kind, when it is well made, is always preferable to the most specious reasonings. Let us try then to give some general rules for obtaining it.

A single number has been drawn from an urn which contains a thousand of them. A witness to this drawing announces that number 79 is drawn; one asks the probability of drawing this number. Let us suppose that experience has made known that this witness deceives one time in ten, so that the probability of his testimony is $\frac{1}{10}$. Here the event observed is the witness attesting that number 79 is drawn. This event may result from the two following hypotheses, namely: that the witness utters the truth or that he deceives. Following the principle that has been expounded on the probability of causes drawn from events observed it is necessary first to determine *à priori* the probability of the event in each hypothesis. In the first, the probability that the witness will announce number 79 is the probability itself of the drawing of this number, that is to say, $\frac{1}{1000}$. It is necessary to multiply it by the probability $\frac{6}{10}$ of the veracity of the witness; one will have then $\frac{9}{10000}$ for the probability of the event observed in this hypothesis. If the witness deceives, number 79 is not drawn, and the probability of this case is $\frac{999}{1000}$. But to announce the drawing of this number the witness has to choose it among the 999 numbers not drawn; and as he is supposed to have no motive of preference for the ones rather than the others, the probability that he will choose number 79 is $\frac{1}{999}$; multiplying, then, this probability by the preceding one, we shall have $\frac{1}{1000}$ for the probability that the witness will announce number 79 in the second hypothesis. It is necessary again to multiply this probability by $\frac{1}{10}$ of the hypothesis itself, which gives $\frac{1}{10000}$ for the probability of the event relative to this hypothesis. Now if we form a fraction whose numerator is the probability relative to the first hypothesis, and whose denominator is the sum of the probabilities relative to the two hypotheses, we shall have, by the sixth principle, the probability of the first hypothesis, and this probability will be $\frac{9}{10}$; that is to say, the veracity itself of the witness. This is likewise the probability of the drawing of number 79. The probability of the falsehood of the witness and of the failure of drawing this number is $\frac{1}{10}$.

If the witness, wishing to deceive, has some interest in choosing number 79 among the numbers not drawn,—if he judges, for example, that having placed upon this number a considerable stake, the announcement of its drawing will increase his credit, the probability that he will choose this number will no longer be as at first, $\frac{1}{999}$, it will then be $\frac{1}{2}$, $\frac{1}{8}$, etc., according to the interest that he will have in announcing its drawing. Supposing it to be $\frac{1}{9}$, it will be necessary to multiply by this fraction the probability $\frac{999}{1000}$ in order to get in the hypothesis of the falsehood the probability of the event observed, which it is necessary still to multiply by $\frac{1}{10}$, which gives $\frac{111}{10000}$ for the probability of the event in the second hypothesis. Then the probability of the first hypothesis, or of the drawing of number 79, is reduced by the preceding rule to $\frac{9}{120}$. It is then very much decreased by the consideration of the interest which the witness may have in announcing the drawing of number 79. In truth this same interest increases the probability $\frac{9}{10}$ that the witness will speak the truth if number 79 is drawn. But this probability cannot exceed unity or $\frac{10}{10}$; thus the probability of the drawing of number 79 will not surpass $\frac{10}{121}$. Common sense tells us that this interest ought to inspire distrust, but calculus appreciates the influence of it.

The probability à priori of the number announced by the witness is unity divided by the number of the numbers in the urn; it is changed by virtue of the proof into the veracity itself of the witness; it may then be decreased by the proof. If, for example, the urn contains only two numbers, which gives $\frac{1}{2}$ for the probability à priori of the drawing of number 1, and if the veracity of a witness who announces it is $\frac{4}{10}$, this drawing becomes less probable. Indeed it is apparent, since the witness has then more inclination towards a falsehood than towards the truth, that his testimony ought to decrease the probability of the fact attested every time that this probability equals or surpasses $\frac{1}{2}$. But if there are three numbers in the urn the probability à priori of the drawing of number 1 is increased by the affirmation of a witness whose veracity surpasses $\frac{1}{3}$.

Suppose now that the urn contains 999 black balls and one white ball, and that one ball having been drawn a witness of the drawing announces that this ball is white.

The probability of the event observed, determined *à priori* in the first hypothesis, will be here, as in the preceding question, equal to $\frac{9}{10000}$. But in the hypothesis where the witness deceives, the white ball is not drawn and the probability of this case is $\frac{999}{1000}$. It is necessary to multiply it by the probability $\frac{1}{10}$ of the falsehood, which gives $\frac{999}{10000}$ for the probability of the event observed relative to the second hypothesis. This probability was only $\frac{1}{10000}$ in the preceding question; this great difference results from this— that a black ball having been drawn the witness who wishes to deceive has no choice at all to make among the 999 balls not drawn in order to announce the drawing of a white ball. Now if one forms two fractions whose numerators are the probabilities relative to each hypothesis, and whose common denominator is the sum of these probabilities, one will have $\frac{9}{1008}$ for the probability of the first hypothesis and of the drawing of a white ball, and $\frac{999}{1008}$ for the probability of the second hypothesis and of the drawing of a black ball. This last probability strongly approaches certainty; it would approach it much nearer and would become $\frac{999999}{1000008}$ if the urn contained a million balls of which one was white, the drawing of a white ball becoming then much more extraordinary. We see thus how the probability of the falsehood increases in the measure that the deed becomes more extraordinary.

We have supposed up to this time that the witness was not mistaken at all; but if one admits, however, the chance of his error the extraordinary incident becomes more improbable. Then in place of the two hypotheses one will have the four following ones, namely: that of the witness not deceiving and not being mistaken at all; that of the witness not deceiving at all and being mistaken; the hypothesis of the witness deceiving and not being mistaken at all; finally, that of the witness deceiving and being mistaken. Determining *à priori* in each of these hypotheses the probability of the event observed, we find by the sixth principle the probability that the fact attested is false equal to a fraction whose numerator is the number of black balls in the urn multiplied by the sum of the probabilities that the witness does not deceive at all and is mistaken, or that he deceives and is not mistaken, and whose denominator is this numerator augmented by the sum of the probabilities that the witness does not deceive at all and is not mistaken at all, or that he deceives and is mistaken at the same time. We see by this that if the number of black balls in the urn is very great, which renders the drawing of the white ball extraordinary, the probability that the fact attested is not true approaches most nearly to certainty.

Applying this conclusion to all extraordinary deeds it results from it that the probability of the error or of the falsehood of the witness becomes as much greater as the fact attested is more extraordinary. Some authors have advanced the contrary on this basis that the view of an extraordinary fact being perfectly similar to that of an ordinary fact the same motives ought to lead us to give the witness the same credence when he affirms the one or the other of these facts. Simple common sense rejects such a strange assertion; but the calculus of probabilities, while confirming the findings of common sense, appreciates the greatest improbability of testimonies in regard to extraordinary facts.

These authors insist and suppose two witnesses equally worthy of belief, of whom the first attests that he saw an individual dead fifteen days ago whom the second witness affirms to have seen yesterday full of life. The one or the other of these facts offers no improbability. The reservation of the individual is a result of their combination; but the testimonies do not bring us at all directly to this result, although the credence which is due these testimonies ought not to be decreased by the fact that the result of their combination is extraordinary.

But if the conclusion which results from the combination of the testimonies was impossible one of them would be necessarily false; but an impossible conclusion is the limit of extraordinary conclusions, as error is the limit of improbable conclusions; the value of the testimonies which becomes zero in the case of an impossible conclusion ought then to be very much decreased in that of an extraordinary conclusion. This is indeed confirmed by the calculus of probabilities.

In order to make it plain let us consider two urns, A and B, of which the first contains a million white balls and the second a million black balls. One draws from one of these urns a ball, which he puts back into the other urn, from which one then draws a ball. Two witnesses, the one of the first drawing, the other of the second, attest that the ball which they have seen drawn is white without indicating the urn from which it has been drawn. Each testimony taken alone is not improbable; and it is easy to see that the probability of the fact attested is the veracity itself of the witness. But it follows from the combination of the testimonies that a white ball has been extracted from the urn A at the first draw, and that then placed in the urn B it has reappeared at the second draw, which is very extraordinary; for this second urn, containing then one white ball among a million black balls, the probability of drawing the white ball is $\frac{1}{1000001}$. In order to determine the diminution which results in the probability of the thing announced by the two witnesses we shall notice that the event observed is here the affirmation by each of them that the ball which he has seen extracted is white. Let us represent by $\frac{9}{10}$ the probability that he announces the truth, which can occur in the present case when the witness does not deceive and is not mistaken at all, and when he deceives and is mistaken at the same time. One may form the four following hypotheses:

1st. The first and second witness speak the truth. Then a white ball has at first been drawn from the urn A, and the probability of this event is $\frac{1}{2}$, since the ball drawn at the first draw may have been drawn either from the one or the other urn. Consequently the ball drawn, placed in the urn B, has reappeared at the second draw;

the probability of this event is $\frac{1}{1000001}$, the probability of the fact announced is then $\frac{1}{2000002}$. Multiplying it by the product of the probabilities $\frac{9}{10}$ and $\frac{9}{10}$ that the witnesses speak the truth one will have $\frac{81}{200000200}$ for the probability of the event observed in this first hypothesis.

2d. The first witness speaks the truth and the second does not, whether he deceives and is not mistaken or he does not deceive and is mistaken. Then a white ball has been drawn from the urn A at the first draw, and the probability of this event is $\frac{1}{2}$. Then this ball having been placed in the urn B a black ball has been drawn from it: the probability of such drawing is $\frac{1000000}{1000001}$; one has then $\frac{1000000}{2000002}$ for the probability of the compound event. Multiplying it by the product of the two probabilities $\frac{9}{10}$ and $\frac{1}{10}$ that the first witness speaks the truth and that the second does not, one will have $\frac{9000000}{200000200}$ for the probability for the event observed in the second hypothesis.

3d. The first witness does not speak the truth and the second announces it. Then a black ball has been drawn from the urn B at the first drawing, and after having been placed in the urn A a white ball has been drawn from this urn. The probability of the first of these events is $\frac{1}{2}$ and that of the second is $\frac{1000000}{1000001}$; the probability of the compound event is then $\frac{1000000}{2000002}$. Multiplying it by the product of the probabilities $\frac{1}{10}$ and $\frac{9}{10}$ that the first witness does not speak the truth and that the second announces it, one will have $\frac{9000000}{200000200}$ for the probability of the event observed relative to this hypothesis.

4th. Finally, neither of the witnesses speaks the truth. Then a black ball has been drawn from the urn B at the first draw; then having been placed in the urn A it has reappeared at the second drawing: the probability of this compound event is $\frac{1}{2000002}$. Multiplying it by the product of the probabilities $\frac{1}{10}$ and $\frac{1}{10}$ that each witness does not speak the truth one will have $\frac{1}{20000200}$ for the probability of the event observed in this hypothesis.

Now in order to obtain the probability of the thing announced by the two witnesses, namely, that a white ball has been drawn at each draw, it is necessary to divide the probability corresponding to the first hypothesis by the sum of the probabilities relative to the four hypotheses; and then one has for this probability $\frac{81}{18000082}$, an extremely small fraction.

If the two witnesses affirm the first, that a white ball has been drawn from one of the two urns A and B; the second that a white ball has been likewise drawn from one of the two urns A′ and B′, quite similar to the first ones, the probability of the thing announced by the two witnesses will be the product of the probabilities of their testimonies, or $\frac{81}{100}$; it will then be at least a hundred and eighty thousand times greater than the preceding one. One sees by this how much, in the first case, the reappearance at the second draw of the white ball drawn at the first draw, the extraordinary conclusion of the two testimonies decreases the value of it.

We would give no credence to the testimony of a man who should attest to us that in throwing a hundred dice into the air they had all fallen on the same face. If we had ourselves been spectators of this event we should believe our own eyes only after having carefully examined all the circumstances, and after having brought in the testimonies of other eyes in order to be quite sure that there had been neither hallucination nor deception. But after this examination we should not hesitate to admit it in spite of its extreme improbability; and no one would be tempted, in order to explain it, to recur to a denial of the laws of vision. We ought to conclude from it that the probability of the constancy of the laws of nature is for us greater than this, that the event in question has not taken place at all—a probability greater than that of the majority of historical facts which we regard as incontestable. One may judge by this the immense weight of testimonies necessary to admit a suspension of natural laws, and how improper it would be to apply to this case the ordinary rules of criticism. All those who without offering this immensity of testimonies support this when making recitals of events contrary to those laws, decrease rather than augment the belief which they wish to inspire; for then those recitals render very probable the error or the falsehood of their authors. But that which diminishes the belief of educated men increases often that of the uneducated, always greedy for the wonderful.

There are things so extraordinary that nothing can balance their improbability. But this, by the effect of a dominant opinion, can be weakened to the point of appearing inferior to the probability of the testimonies; and when this opinion changes an absurd statement admitted unanimously in the century which has given it birth offers to the following centuries only a new proof of the extreme influence of the general opinion upon the more enlightened minds. Two great men of the century of Louis XIV.—Racine and Pascal—are striking examples of this. It is painful to see with what complaisance Racine, this admirable painter of the human heart and the most perfect poet that has ever lived, reports as miraculous the recovery of Mlle. Perrier, a niece of Pascal and a day pupil at the monastery of Port-Royal; it is painful to read the reasons by which Pascal seeks to prove that this miracle should be necessary to religion in order to justify the doctrine of the monks of this abbey, at that time persecuted by the Jesuits. The young Perrier had been afflicted for three years and a half by a lachrymal fistula; she touched her afflicted eye with a relic which was pretended to be one of the thorns of the crown of the Saviour and she had faith in instant recovery. Some days afterward the physicians and the surgeons attest the recovery, and they declare that nature and the remedies have had no part in it. This event, which took place in 1656, made a great sensation, and "all Paris rushed," says Racine, "to Port-Royal. The crowd increased from day to day, and God himself seemed to take pleasure in authorizing the devotion of the people by the number of miracles which were performed in this church." At this time miracles and sorcery did not yet appear improbable, and one did not hesitate at all to attribute to them the singularities of nature which could not be explained otherwise.

This manner of viewing extraordinary results is found in the most remarkable works of the century of Louis XIV.; even in the Essay on the Human Understanding by the philosopher Locke, who says, in speaking of the degree of assent: "Though the common experience and the ordinary course of things have justly a mighty influence on the minds of men, to make them give or refuse credit to anything proposed to their belief; yet there is one case, wherein the strangeness of the fact lessens not the assent to a fair testimony of it. For where such supernatural events are suitable to ends aimed at by him who has the power to change the course of nature, there, under such circumstances, they may be the fitter to procure belief, by how much the more they are beyond or contrary to ordinary observation." The true principles of the probability of testimonies having been thus misunderstood by philosophers to whom reason is principally indebted for its progress, I have thought it necessary to present at length the results of calculus upon this important subject.

There comes up naturally at this point the discussion of a famous argument of Pascal, that Craig, an English mathematician, has produced under a geometric form. Witnesses declare that they have it from Divinity that in conforming to a certain thing one will enjoy not one or two but an infinity of happy lives. However feeble the probability of the proofs may be, provided that it be not infinitely small, it is clear that the advantage of those who conform to the prescribed thing is infinite since it is the product of this probability and an infinite good; one ought not to hesitate then to procure for oneself this advantage.

This argument is based upon the infinite number of happy lives promised in the name of the Divinity by the witnesses; it is necessary then to prescribe them, precisely because they exaggerate their promises beyond all limits, a consequence which is repugnant to good sense. Also calculus teaches us that this exaggeration itself enfeebles the probability of their testimony to the point of rendering it infinitely small or zero. Indeed this case is similar to that of a witness who should announce the drawing of the highest number from an urn filled with a great number of numbers, one of which has been drawn and who would have a great interest in announcing the drawing of this number. One has already seen how much this interest enfeebles his testimony. In evaluating only at $\frac{1}{2}$ the probability that if the witness deceives he will choose the largest number, calculus gives the probability of his announcement as smaller than a fraction whose numerator is unity and whose denominator is unity plus the half of the product of the number of the numbers by the probability of falsehood considered *à priori* or independently of the announcement. In order to compare this case to that of the argument of Pascal it is sufficient to represent by the numbers in the urn all the possible numbers of happy lives which the number of these numbers renders infinite; and to observe that if the witnesses deceive they have the greatest interest, in order to accredit their falsehood, in promising an eternity of happiness. The expression of the probability of their testimony becomes then infinitely small. Multiplying it by the infinite number of happy lives promised, infinity would disappear from the product which expresses the advantage resultant from this promise which destroys the argument of Pascal.

Let us consider now the probability of the totality of several testimonies upon an established fact. In order to fix our ideas let us suppose that the fact be the drawing of a number from an urn which contains a hundred of them, and of which one single number has been drawn. Two witnesses of this drawing announce that number 2 has been drawn, and one asks for the resultant probability of the totality of these testimonies. One may form these two hypotheses: the witnesses speak the truth; the witnesses deceive. In the first hypothesis the number 2 is drawn and the probability of this event is $\frac{1}{100}$. It is necessary to multiply it by the product of the veracities of the witnesses, veracities which we will suppose to be $\frac{9}{10}$ and $\frac{7}{10}$: one will have then $\frac{63}{10000}$ for the probability of the event observed in this hypothesis. In the second, the number 2 is not drawn and the probability of this event is $\frac{99}{100}$. But the agreement of the witnesses requires then that in seeking to deceive they both choose the number 2 from the 99 numbers not drawn: the probability of this choice if the witnesses do not have a secret agreement is the product of the fraction $\frac{1}{99}$ by itself; it becomes necessary then to multiply these two probabilities together, and by the product of the probabilities $\frac{1}{10}$ and $\frac{3}{10}$ that the witnesses deceive; one will have thus $\frac{1}{330000}$ for the probability of the event observed in the second hypothesis. Now one will have the probability of the fact attested or of the drawing of number 2 in dividing the probability relative to the first hypothesis by the sum of the probabilities relative to the two hypotheses; this probability will be then $\frac{2079}{2080}$, and the probability of the failure to draw this number and of the falsehood of the witnesses will be $\frac{1}{2080}$.

If the urn should contain only the numbers 1 and 2 one would find in the same manner $\frac{21}{22}$ for the probability of the drawing of number 2, and consequently $\frac{1}{22}$ for the probability of the falsehood of the witnesses, a probability at least ninety-four times larger than the preceding one. One sees by this how much the probability of the falsehood of the witnesses diminishes when the fact which they attest is less probable in itself. Indeed one conceives that then the accord of the witnesses, when they deceive, becomes more difficult, at least when they do not have a secret agreement, which we do not suppose here at all.

In the preceding case where the urn contained only two numbers the *à priori* probability of the fact attested is $\frac{1}{2}$, the resultant probability of the testimonies is the product of the veracities of the witnesses divided by this product added to that of the respective probabilities of their falsehood.

It now remains for us to consider the influence of time upon the probability of facts transmitted by a traditional chain of witnesses. It is clear that this probability ought to diminish in proportion as the chain is prolonged. If the fact has no probability itself, such as the drawing of a number from an urn which contains an infinity of them, that which it acquires by the testimonies decreases according to the continued product of the veracity of the witnesses. If the fact has a probability in itself; if, for example, this fact is the drawing of the number 2 from an urn which contains an infinity of them, and of which it is certain that one has drawn a single number; that which the traditional chain adds to this probability decreases, following a continued product of which the first factor is the ratio of the number of numbers in the urn less one to the same number, and of which each other factor is the veracity of each witness diminished by the ratio of the probability of his falsehood to the number of the numbers in the urn less one; so that the limit of the probability of the fact is that of this fact considered à priori, or independently of the testimonies, a probability equal to unity divided by the number of the numbers in the urn.

The action of time enfeebles then, without ceasing, the probability of historical facts just as it changes the most durable monuments. One can indeed diminish it by multiplying and conserving the testimonies and the monuments which support them. Printing offers for this purpose a great means, unfortunately unknown to the ancients. In spite of the infinite advantages which it procures the physical and moral revolutions by which the surface of this globe will always be agitated will end, in conjunction with the inevitable effect of time, by rendering doubtful after thousands of years the historical facts regarded to-day as the most certain.

Craig has tried to submit to calculus the gradual enfeebling of the proofs of the Christian religion; supposing that the world ought to end at the epoch when it will cease to be probable, he finds that this ought to take place 1454 years after the time when he writes. But his analysis is as faulty as his hypothesis upon the duration of the moon is bizarre.

CHAPTER XII

CONCERNING THE SELECTIONS AND
THE DECISIONS OF ASSEMBLIES

THE probability of the decisions of an assembly depends upon the plurality of votes, the intelligence and the impartiality of the members who compose it. So many passions and particular interests so often add their influence that it is impossible to submit this probability to calculus. There are, however, some general results dictated by simple common sense and confirmed by calculus. If, for example, the assembly is poorly informed about the subject submitted to its decision, if this subject requires delicate considerations, or if the truth on this point is contrary to established prejudices, so that it would be a bet of more than one against one that each voter will err; then the decision of the majority will be probably wrong, and the fear of it will be the better based as the assembly is more numerous. It is important then, in public affairs, that assemblies should have to pass upon subjects within reach of the greatest number; it is important for them that information be generally diffused and that good works founded upon reason and experience should enlighten those who are called to decide the lot of their fellows or to govern them, and should forewarn them against false ideas and the prejudices of ignorance. Scholars have had frequent occasion to remark that first conceptions often deceive and that the truth is not always probable.

It is difficult to understand and to define the desire of an assembly in the midst of a variety of opinions of its members. Let us attempt to give some rules in regard to this matter by considering the two most ordinary cases: the election among several candidates, and that among several propositions relative to the same subject.

When an assembly has to choose among several candidates who present themselves for one or for several places of the same kind, that which appears simplest is to have each voter write upon a ticket the names of all the candidates according to the order of merit that he attributes to them. Supposing that he classifies them in good faith, the inspection of these tickets will give the results of the elections in such a manner that the candidates may be compared among themselves; so that new elections can give nothing more in this regard. It is a question now to conclude the order of preference which the tickets establish among the candidates. Let us imagine that one gives to each voter an urn which contains an infinity of balls by means of which he is able to shade all the degrees of merit of the candidates; let us conceive again that he draws from his urn a number of balls proportional to the merit of each candidate, and let us suppose this number written upon a ticket at the side of the name of the candidate. It is clear that by making a sum of all the numbers relative to each candidate upon each ticket, that one of all the candidates who shall have the largest sum will be the candidate whom the assembly prefers; and that in general the order of preference of the candidates will be that of the sums relative to each of them. But the tickets do not mark at all the number of balls which each voter gives to the candidates; they indicate solely that the first has more of them than the second, the second more than the third, and so on. In supposing then at first upon a given ticket a certain number of balls all the combinations of the inferior numbers which fulfil the preceding conditions are equally admissible; and one will have the number of balls relative to each candidate by making a sum of all the numbers which each combination gives him and dividing it by the entire number of combinations. A very simple analysis shows that the numbers which must be written upon each ticket at the side of the last name, of the one before the last, etc., are proportional to the terms of the arithmetical progression 1, 2, 3, etc. Writing then thus upon each ticket the terms of this progression, and adding the terms relative to each candidate upon these tickets, the divers sums will indicate by their magnitude the order of their preference which ought to be established among the candidates. Such is the mode of election which The Theory of Probabilities indicates. Without doubt it would be better if each voter should write upon his ticket the names of the candidates in the order of merit which he attributes to them. But particular interests and many strange considerations of merit would affect this order and place sometimes in the last rank the candidate most formidable to that one whom one prefers, which gives too great an advantage to the candidates of mediocre merit. Likewise experience has caused the abandonment of this mode of election in the societies which had adopted it.

The election by the absolute majority of the suffrages unites to the certainty of not admitting any one of the candidates whom this majority rejects, the advantage of expressing most often the desire of the assembly. It always coincides with the preceding mode when there are only two candidates. Indeed it exposes an assembly to the inconvenience of rendering elections interminable. But experience has shown that this inconvenience is nil, and that the general desire to put an end to elections soon unites the majority of the suffrages upon one of the candidates.

The choice among several propositions relative to the same object ought to be subjected, seemingly, to the same rules as the election among several candidates. But there exists between the two cases this difference, namely, that the merit of a candidate does not exclude that of his competitors; but if it is necessary to choose among propositions which are contrary, the truth of the one excludes the truth of the others. Let us see how one ought then to view this question.

Let us give to each voter an urn which contains an infinite number of balls, and let us suppose that he distributes them upon the divers propositions according to the respective probabilities which he attributes to them. It is clear that the total number of balls expressing certainty, and the voter being by the hypothesis assured that one of the propositions ought to be true, he will distribute this number at length upon the propositions. The problem is reduced then to this, namely, to determine the combinations in which the balls will be distributed in such a manner that there may be more of them upon the first proposition of the ticket than upon the second, more upon the second than upon the third, etc.; to make the sums of all the numbers of balls relative to each proposition in the divers combinations, and to divide this sum by the number of combinations; the quotients will be the numbers of balls that one ought to attribute to the propositions upon a certain ticket. One finds by analysis that in going from the last proposition these quotients are among themselves as the following quantities: first, unity divided by the number of propositions; second, the preceding quantity, augmented by unity, divided by the number of propositions less one; third, this second quantity, augmented by unity, divided by the number of propositions less two, and so on for the others. One will write then upon each ticket these quantities at the side of the corresponding propositions, and adding the relative quantities to each proposition upon the divers tickets the sums will indicate by their magnitude the order of preference which the assembly gives to these propositions.

Let us speak a word about the manner of renewing assemblies which should change in totality in a definite number of years. Ought the renewal to be made at one time, or is it advantageous to divide it among these years? According to the last method the assembly would be formed under the influence of the divers opinions dominant during the time of its renewal; the opinion which obtained then would be probably the mean of all these opinions. The assembly would receive thus at the time the same advantage that is given to it by the extension of the elections of its members to all parts of the territory which it represents. Now if one considers what experience has only too clearly taught, namely, that elections are always directed in the greatest degree by dominant opinions, one will feel how useful it is to temper these opinions, the ones by the others, by means of a partial renewal.

CHAPTER XIII

CONCERNING THE PROBABILITY OF THE
JUDGMENTS OF TRIBUNALS

ANALYSIS confirms what simple common sense teaches us, namely, the correctness of judgments is as much more probable as the judges are more numerous and more enlightened. It is important then that tribunals of appeal should fulfil these two conditions. The tribunals of the first instance standing in closer relation to those amenable offer to the higher tribunal the advantage of a first judgment already probable, and with which the latter often agree, be it in compromising or in desisting from their claims. But if the uncertainty of the matter in litigation and its importance determine a litigant to have recourse to the tribunal of appeals, he ought to find in a greater probability of obtaining an equitable judgment greater security for his fortune and the compensation for the trouble and expense which a new procedure entails. It is this which had no place in the institution of the reciprocal appeal of the tribunals of the district, an institution thereby very prejudicial to the interest of the citizens. It would be perhaps proper and conformable to the calculus of probabilities to demand a majority of at least two votes in a tribunal of appeal in order to invalidate the sentence of the lower tribunal. One would obtain this result if the tribunal of appeal being composed of an even number of judges the sentence should stand in the case of the equality of votes.

I shall consider particularly the judgments in criminal matters.

In order to condemn an accused it is necessary without doubt that the judges should have the strongest proofs of his offence. But a moral proof is never more than a probability; and experience has only too clearly shown the errors of which criminal judgments, even those which appear to be the most just, are still susceptible. The impossibility of amending these errors is the strongest argument of the philosophers who have wished to proscribe the penalty of death. We should then be obliged to abstain from judging if it were necessary for us to await mathematical evidence. But the judgment is required by the danger which would result from the impunity of the crime. This judgment reduces itself, if I am not mistaken, to the solution of the following question: Has the proof of the offence of the accused the high degree of probability necessary so that the citizens would have less reason to doubt the errors of the tribunals, if he is innocent and condemned, than they would have to fear his new crimes and those of the unfortunate ones who would be emboldened by the example of his impunity if he were guilty and acquitted? The solution of this question depends upon several elements very difficult to ascertain. Such is the eminence of danger which would threaten society if the criminal accused should remain unpunished. Sometimes this danger is so great that the magistrate sees himself constrained to waive forms wisely established for the protection of innocence. But that which renders almost always this question insoluble is the impossibility of appreciating exactly the probability of the offence and of fixing that which is necessary for the condemnation of the accused. Each judge in this respect is forced to rely upon his own judgment. He forms his opinion by comparing the divers testimonies and the circumstances by which the offence is accompanied, to the results of his reflections and his experiences, and in this respect a long habitude of interrogating and judging accused persons gives great advantage in ascertaining the truth in the midst of indices often contradictory.

The preceding question depends again upon the care taken in the investigation of the offence; for one demands naturally much stronger proofs for imposing the death penalty than for inflicting a detention of some months. It is a reason for proportioning the care to the offence, great care taken with an unimportant case inevitably clearing many guilty ones. A law which gives to the judges power of moderating the care in the case of attenuating circumstances is then conformable at the same time to principles of humanity towards the culprit, and to the interest of society. The product of the probability of the offence by its gravity being the measure of the danger to which the acquittal of the accused can expose society, one would think that the care taken ought to depend upon this probability. This is done indirectly in the tribunals where one retains for some time the accused against whom there are very strong proofs, but insufficient to condemn him; in the hope of acquiring new light one does not place him immediately in the midst of his fellow citizens, who would not see him again without great alarm. But the arbitrariness of this measure and the abuse which one can make of it have caused its rejection in the countries where one attaches the greatest price to individual liberty.

Now what is the probability that the decision of a tribunal which can condemn only by a given majority will be just, that is to say, conform to the true solution of the question proposed above? This important problem well solved will give the means of comparing among themselves the different tribunals. The majority of a single vote in a numerous tribunal indicates that the affair in question is very doubtful; the condemnation of the accused would be then contrary to the principles of humanity, protectors of innocence. The unanimity of the judges would give very strong probability of a just decision; but in abstaining from it too many guilty ones would be acquitted. It is necessary, then, either to limit the number of judges, if one wishes that they should be unanimous, or increase the majority necessary for a condemnation, when the tribunal becomes more numerous. I shall attempt to apply calculus to this subject, being persuaded that it is always the best guide when one bases it upon the data which common sense suggests to us.

The probability that the opinion of each judge is just enters as the principal element into this calculation. If in a tribunal of a thousand and one judges, five hundred and one are of one opinion, and five hundred are of the contrary opinion, it is apparent that the probability of the opinion of each judge surpasses very little $\frac{1}{2}$; for supposing it obviously very large a single vote of difference would be an improbable event. But if the judges are unanimous, this indicates in the proofs that degree of strength which entails conviction; the probability of the opinion of each judge is then very near unity or certainty, provided that the passions or the ordinary prejudices do not affect at the same time all the judges. Outside of these cases the ratio of the votes for or against the accused ought alone to determine this probability. I suppose thus that it can vary from $\frac{1}{2}$ to unity, but that it cannot be below $\frac{1}{2}$. If that were not the case the decision of the tribunal would be as insignificant as chance; it has value only in so far as the opinion of the judge has a greater tendency to truth than to error. It is thus by the ratio of the numbers of votes favorable, and contrary to the accused, that I determine the probability of this opinion.

These data suffice to ascertain the general expression of the probability that the decision of a tribunal judging by a known majority is just. In the tribunals where of eight judges five votes would be necessary for the condemnation of an accused, the probability of the error to be feared in the justice of the decision would surpass $\frac{1}{4}$. If the tribunal should be reduced to six members who are able to condemn only by a plurality of four votes, the probability of the error to be feared would be below $\frac{1}{4}$. There would be then for the accused an advantage in this reduction of the tribunal. In both cases the majority required is the same and is equal to two. Thus the majority remaining constant, the probability of error increases with the number of judges; this is general whatever may be the majority required, provided that it remains the same. Taking, then, for the rule the arithmetical ratio, the accused finds himself in a position less and less advantageous in the measure that the tribunal becomes more numerous. One might believe that in a tribunal where one might demand a majority of twelve votes, whatever the number of the judges was, the votes of the minority, neutralizing an equal number of votes of the majority, the twelve remaining votes would represent the unanimity of a jury of twelve members, required in England for the condemnation of an accused; but one would be greatly mistaken. Common sense shows that there is a difference between the decision of a tribunal of two hundred and twelve judges, of which one hundred and twelve condemn the accused, while one hundred acquit him, and that of a tribunal of twelve judges unanimous for condemnation. In the first case the hundred votes favorable to the accused warrant in thinking that the proofs are far from attaining the degree of strength which entails conviction; in the second case, the unanimity of the judges leads to the belief that they have attained this degree. But simple common sense does not suffice at all to appreciate the extreme difference of the probability of error in the two cases. It is necessary then to recur to calculus, and one finds nearly one fifth for the probability of error in the first case, and only $\frac{1}{8192}$ for this probability in the second case, a probability which is not one thousandth of the first. It is a confirmation of the principle that the arithmetical ratio is unfavorable to the accused when the number of judges increases. On the contrary, if one takes for a rule the geometrical ratio, the probability of the error of the decision diminishes when the number of judges increases. For example, in the tribunals which can condemn only by a plurality of two thirds of the votes, the probability of the error to be feared is nearly one fourth if the number of the judges is six; it is below $\frac{1}{7}$ if this number is increased to twelve. Thus one ought to be governed neither by the arithmetical ratio nor by the geometrical ratio if one wishes that the probability of error should never be above nor below a given fraction.

But what fraction ought to be determined upon? It is here that the arbitrariness begins and the tribunals offer in this regard the greatest variety. In the special tribunals where five of the eight votes suffice for the condemnation of the accused, the probability of the error to be feared in regard to justice of the judgment is $\frac{65}{256}$, or more than $\frac{1}{4}$. The magnitude of this fraction is dreadful; but that which ought to reassure us a little is the consideration that most frequently the judge who acquits an accused does not regard him as innocent; he pronounces solely that it is not attained by proofs sufficient for condemnation. One is especially reassured by the pity which nature has placed in the heart of man and which disposes the mind to see only with reluctance a culprit in the accused submitted to his judgment. This sentiment, more active in those who have not the habitude of criminal judgments, compensates for the inconveniences attached to the inexperience of the jurors. In a jury of twelve members, if the plurality demanded for the condemnation is eight of twelve votes, the probability of the error to be feared $\frac{1093}{8192}$, or a little more than one eighth, it is almost $\frac{1}{22}$ if this plurality consists of nine votes. In the case of unanimity the probability of the error to be feared is $\frac{1}{8192}$, that is to say, more than a thousand times less than in our juries. This supposes that the unanimity results only from proofs favorable or contrary to the accused; but motives that are entirely strange, ought oftentimes to concur in producing it, when it is imposed upon the jury as a necessary condition of its judgment. Then its decisions depending upon the temperament, the character, the habits of the jurors, and the circumstances in which they are placed, they are sometimes contrary to the decisions which the majority of the jury would have made if they had listened only to the proofs; this seems to me to be a great fault of this manner of judging.

The probability of the decision is too feeble in our juries, and I think that in order to give a sufficient guarantee to innocence, one ought to demand at least a plurality of nine votes in twelve.

CHAPTER XIV

CONCERNING TABLES OF MORTALITY, AND OF MEAN DURATIONS OF LIFE, OF MARRIAGES, AND OF ASSOCIATIONS

THE manner of preparing tables of mortality is very simple. One takes in the civil registers a great number of individuals whose birth and death are indicated. One determines how many of these individuals have died in the first year of their age, how many in the second year, and so on. It is concluded from these the number of individuals living at the commencement of each year, and this number is written in the table at the side of that which indicates the year. Thus one writes at the side of zero the number of births; at the side of the year 1 the number of infants who have attained one year; at the side of the year 2 the number of infants who have attained two years, and so on for the rest. But since in the first two years of life the mortality is very great, it is necessary for the sake of greater exactitude to indicate in this first age the number of survivors at the end of each half year.

If we divide the sum of the years of the life of all the individuals inscribed in a table of mortality by the number of these individuals we shall have the mean duration of life which corresponds to this table. For this, we will multiply by a half year the number of deaths in the first year, a number equal to the difference of the numbers of individuals inscribed at the side of the years 0 and 1. Their mortality being distributed over the entire year the mean duration of their life is only a half year. We will multiply by a year and a half the number of deaths in the second year; by two years and a half the number of deaths in the third year; and so on. The sum of these products divided by the number of births will be the mean duration of life. It is easy to conclude from this that we will obtain this duration, by making the sum of the numbers inscribed in the table at the side of each year, dividing it by the number of births and subtracting one half from the quotient, the year being taken as unity. The mean duration of life that remains, starting from any age, is determined in the same manner, working upon the number of individuals who have arrived at this age, as has just been done with the number of births. But it is not at the moment of birth that the mean duration of life is the greatest; it is when one has escaped the dangers of infancy and it is then about forty-three years. The probability of arriving at a certain age, starting from a given age is equal to the ratio of the two numbers of individuals indicated in the table at these two ages.

The precision of these results demands that for the formation of tables we should employ a very great number of births. Analysis gives then very simple formulæ for appreciating the probability that the numbers indicated in these tables will vary from the truth only within narrow limits. We see by these formulæ that the interval of the limits diminishes and that the probability increases in proportion as we take into consideration more births; so that the tables would represent exactly the true law of mortality if the number of births employed were infinite.

A table of mortality is then a table of the probability of human life. The ratio of the individuals inscribed at the side of each year to the number of births is the probability that a new birth will attain this year. As we estimate the value of hope by making a sum of the products of each benefit hoped for, by the probability of obtaining it, so we can equally evaluate the mean duration of life by adding the products of each year by half the sum of the probabilities of attaining the commencement and the end of it, which leads to the result found above. But this manner of viewing the mean duration of life has the advantage of showing that in a stationary population, that is to say, such that the number of births equals that of deaths, the mean duration of life is the ratio itself of the population to the annual births; for the population being supposed stationary, the number of individuals of an age comprised between two consecutive years of the table is equal to the number of annual births, multiplied by half the sum of the probabilities of attaining these years; the sum of all these products will be then the entire population. Now it is easy to see that this sum, divided by the number of annual births, coincides with the mean duration of life as we have just defined it.

It is easy by means of a table of mortality to form the corresponding table of the population supposed to be stationary. For this we take the arithmetical means of the numbers of the table of mortality corresponding to the ages zero and one year, one and two years, two and three years, etc. The sum of all these means is the entire population; it is written at the side of the age zero. There is subtracted from this sum the first mean and the remainder is the number of individuals of one year and upwards; it is written at the side of the year 1. There is subtracted from this first remainder the second mean; this second remainder is the number of individuals of two years and upwards; it is written at the side of the year 2, and so on.

So many variable causes influence mortality that the tables which represent it ought to be changed according to place and time. The divers states of life offer in this regard appreciable differences relative to the fatigues and the dangers inseparable from each state and of which it is indispensable to keep account in the calculations founded upon the duration of life. But these differences have not been sufficiently observed. Some day they will be and then will be known what sacrifice of life each profession demands and one will profit by this knowledge to diminish the dangers.

The greater or less salubrity of the soil, its elevation, its temperature, the customs of the inhabitants, and the operations of governments have a considerable influence upon mortality. But it is always necessary to precede the investigation of the cause of the differences observed by that of the probability with which this cause is indicated. Thus the ratio of the population to annual births, which one has seen raised in France to twenty-eight and one third, is not equal to twenty-five in the ancient duchy of Milan. These ratios, both established upon a great number of births, do not permit of calling into question the existence among the Milanese of a special cause of mortality, which it is of moment for the government of our country to investigate and remove.

The ratio of the population to the births would increase again if we could diminish and remove certain dangerous and widely spread maladies. This has happily been done for the smallpox, at first by the inoculation of this disease, then in a manner much more advantageous, by the inoculation of vaccine, the inestimable discovery of Jenner, who has thereby become one of the greatest benefactors of humanity.

The smallpox has this in particular, namely, that the same individual is not twice affected by it, or at least such cases are so rare that they may be abstracted from the calculation. This malady, from which few escaped before the discovery of vaccine, is often fatal and causes the death of one seventh of those whom it attacks. Sometimes it is mild, and experience has taught that it can be given this latter character by inoculating it upon healthy persons, prepared for it by a proper diet and in a favorable season. Then the ratio of the individuals who die to the inoculated ones is not one three hundredth. This great advantage of inoculation, joined to those of not altering the appearance and of preserving from the grievous consequences which the natural smallpox often brings, caused it to be adopted by a great number of persons. The practice was strongly recommended, but it was strongly combated, as is nearly always the case in things subject to inconvenience. In the midst of this dispute Daniel Bernoulli proposed to submit to the calculus of probabilities the influence of inoculation upon the mean duration of life. Since precise data of the mortality produced by the smallpox at the various ages of life were lacking, he supposed that the danger of having this malady and that of dying of it are the same at every age. By means of these suppositions he succeeded by a delicate analysis in converting an ordinary table of mortality into that which would be used if smallpox did not exist, or if it caused the death of only a very small number of those affected, and he concludes from it that inoculation would augment by three years at least the mean duration of life, which appeared to him beyond doubt the advantage of this operation. D'Alembert attacked the analysis of Bernoulli: at first in regard to the uncertainty of his two hypotheses, then in regard to its insufficiency in this, that no comparison was made of the immediate danger, although very small, of dying of inoculation, to the very great but very remote danger of succumbing to natural smallpox. This consideration, which disappears when one considers a great number of individuals, is for this reason immaterial for governments and the advantages of inoculation for them still remain; but it is of great weight for the father of a family who must fear, in having his children inoculated, to see that one perish whom he holds most dear and to be the cause of it. Many parents were restrained by this fear, which the discovery of vaccine has happily dissipated. By one of those mysteries which nature offers to us so frequently, vaccine is a preventive of smallpox just as certain as variolar virus, and there is no danger at all; it does not expose to any malady and demands only very little care. Therefore the practice of it has spread quickly; and to render it universal it remains only to overcome the natural inertia of the people, against which it is necessary to strive continually, even when it is a question of their dearest interests.

The simplest means of calculating the advantage which the extinction of a malady would produce consists in determining by observation the number of individuals of a given age who die of it each year and subtracting this number from the number of deaths at the same age. The ratio of the difference to the total number of individuals of the given age would be the probability of dying in the year at this age if the malady did not exist. Making, then, a sum of these probabilities from birth up to any given age, and subtracting this sum from unity, the remainder will be the probability of living to that age corresponding to the extinction of the malady. The series of these probabilities will be the table of mortality relative to this hypothesis, and we may conclude from it, by what precedes, the mean duration of life. It is thus that Duvilard has found that the increase of the mean duration of life, due to inoculation with vaccine, is three years at the least. An increase so considerable would produce a very great increase in the population if the latter, for other reasons, were not restrained by the relative diminution of subsistences.

It is principally by the lack of subsistences that the progressive march of the population is arrested. In all kinds of animals and vegetables, nature tends without ceasing to augment the number of individuals until they are on a level of the means of subsistence. In the human race moral causes have a great influence upon the population. If easy clearings of the forest can furnish an abundant nourishment for new generations, the certainty of being able to support a numerous family encourages marriages and renders them more productive. Upon the same soil the population and the births ought to increase at the same time simultaneously in geometric progression. But when clearings become more difficult and more rare then the increase of population diminishes; it approaches continually the variable state of subsistences, making oscillations about it just as a pendulum whose periodicity is retarded by changing the point of suspension, oscillates about this point by virtue of its own weight. It is difficult to evaluate the *maximum* increase of the population; it appears after observations that in favorable circumstances the population of the human race would be doubled every fifteen years. We estimate that in North America the period of this doubling is twenty-two years. In this state of things, the population, births, marriages, mortality, all increase according to the same geometric progression of which we have the constant ratio of consecutive terms by the observation of annual births at two epochs.

By means of a table of mortality representing the probabilities of human life, we may determine the duration of marriages. Supposing in order to simplify the matter that the mortality is the same for the two sexes, we shall obtain the probability that the marriage will subsist one year, or two, or three, etc., by forming a series of fractions whose common denominator is the product of the two numbers of the table corresponding to the ages of the consorts, and whose numerators are the successive products of the numbers corresponding to these ages augmented by one, by two, by three, etc., years. The sum of these fractions augmented by one half will be the mean duration of marriage, the year being taken as unity. It is easy to extend the same rule to the mean duration of an association formed of three or of a greater number of individuals.

CHAPTER XV

CONCERNING THE BENEFITS OF INSTITUTIONS WHICH DEPEND UPON THE PROBABILITY OF EVENTS

LET us recall here what has been said in speaking of hope. It has been seen that in order to obtain the advantage which results from several simple events, of which the ones produce a benefit and the others a loss, it is necessary to add the products of the probability of each favorable event by the benefit which it procures, and subtract from their sum that of the products of the probability of each unfavorable event by the loss which is attached to it. But whatever may be the advantage expressed by the difference of these sums, a single event composed of these simple events does not guarantee against the fear of experiencing a loss. One imagines that this fear ought to decrease when one multiplies the compound event. The analysis of probabilities leads to this general theorem.

By the repetition of an advantageous event, simple or compound, the real benefit becomes more and more probable and increases without ceasing; it becomes certain in the hypothesis of an infinite number of repetitions; and dividing it by this number the quotient or the mean benefit of each event is the mathematical hope itself or the advantage relative to the event. It is the same with a loss which becomes certain in the long run, however small the disadvantage of the event may be.

This theorem upon benefits and losses is analogous to those which we have already given upon the ratios which are indicated by the indefinite repetition of events simple or compound; and, like them, it proves that regularity ends by establishing itself even in the things which are most subordinated to that which we name *hazard*.

When the events are in great number, analysis gives another very simple expression of the probability that the benefit will be comprised within determined limits. This is the expression which enters again into the general law of probability given above in speaking of the probabilities which result from the indefinite multiplication of events.

The stability of institutions which are based upon probabilities depends upon the truth of the preceding theorem. But in order that it may be applied to them it is necessary that those institutions should multiply these advantageous events for the sake of numerous things.

There have been based upon the probabilities of human life divers institutions, such as life annuities and tontines. The most general and the most simple method of calculating the benefits and the expenses of these institutions consists in reducing these to actual amounts. The annual interest of unity is that which is called *the rate of interest*. At the end of each year an amount acquires for a factor unity plus the rate of interest; it increases then according to a geometrical progression of which this factor is the ratio. Thus in the course of time it becomes immense. If, for example, the rate of interest is $\frac{1}{20}$ or five per cent, the capital doubles very nearly in fourteen years, quadruples in twenty-nine years, and in less than three centuries it becomes two million times larger.

An increase so prodigious has given birth to the idea of making use of it in order to pay off the public debt. One forms for this purpose a sinking fund to which is devoted an annual fund employed for the redemption of public bills and without ceasing increased by the interest of the bills redeemed. It is clear that in the long run this fund will absorb a great part of the national debt. If, when the needs of the State make a loan necessary, a part of this loan is devoted to the increasing of the annual sinking fund, the variation of public bills will be less; the confidence of the lenders and the probability of retiring without loss of capital loaned when one desires will be augmented and will render the conditions of the loan less onerous. Favorable experiences have fully confirmed these advantages. But the fidelity in engagements and the stability, so necessary to the success of such institutions, can be guaranteed only by a government in which the legislative power is divided among several independent powers. The confidence which the necessary coöperation of these powers inspires, doubles the strength of the State, and the sovereign himself gains then in legal power more than he loses in arbitrary power.

It results from that which precedes that the actual capital equivalent to a sum which is to be paid only after a certain number of years is equal to this sum multiplied by the probability that it will be paid at that time and divided by unity augmented by the rate of interest and raised to a power expressed by the number of these years.

It is easy to apply this principle to life annuities upon one or several persons, and to savings banks, and to assurance societies of any nature. Suppose that one proposes to form a table of life annuities according to a given table of mortality. A life annuity payable at the end of five years, for example, and reduced to an actual amount is, by this principle, equal to the product of the two following quantities, namely, the annuity divided by the fifth power of unity augmented by the rate of interest and the probability of paying it. This probability is the inverse ratio of the number of individuals inscribed in the table opposite to the age of that one who settles the annuity to the number inscribed opposite to this age augmented by five years. Forming, then, a series of fractions whose denominators are the products of the number of persons indicated in the table of mortality as living at the age of that one who settles the annuity, by the successive powers of unity augmented by the rate of interest, and whose numerators are the products of the annuity by the number of persons living at the same age augmented successively by one year, by two years, etc., the sum of these fractions will be the amount required for the life annuity at that age.

Let us suppose that a person wishes by means of a life annuity to assure to his heirs an amount payable at the end of the year of his death. In order to determine the value of this annuity, one may imagine that the person borrows in life at a bank this capital and that he places it at perpetual interest in the same bank. It is clear that this same capital will be due by the bank to his heirs at the end of the year of his death; but he will have paid each year only the excess of the life interest over the perpetual interest. The table of life annuities will then show that which the person ought to pay annually to the bank in order to assure this capital after his death.

Maritime assurance, that against fire and storms, and generally all the institutions of this kind, are computed on the same principles. A merchant having vessels at sea wishes to assure their value and that of their cargoes against the dangers that they may run; in order to do this, he gives a sum to a company which becomes responsible to him for the estimated value of his cargoes and his vessels. The ratio of this value to the sum which ought to be given for the price of the assurance depends upon the dangers to which the vessels are exposed and can be appreciated only by numerous observations upon the fate of vessels which have sailed from port for the same destination.

If the persons assured should give to the assurance company only the sum indicated by the calculus of probabilities, this company would not be able to provide for the expenses of its institution; it is necessary then that they should pay a sum much greater than the cost of such insurance. What then is their advantage? It is here that the consideration of the moral disadvantage attached to an uncertainty becomes necessary. One conceives that the fairest game becomes, as has already been seen, disadvantageous, because the player exchanges a certain stake for an uncertain benefit; assurance by which one exchanges the uncertain for the certain ought to be advantageous. It is indeed this which results from the rule which we have given above for determining moral hope and by which one sees moreover how far the sacrifice may extend which ought to be made to the assurance company by reserving always a moral advantage. This company can then in procuring this advantage itself make a great benefit, if the number of the assured persons is very large, a condition necessary to its continued existence. Then its benefits become certain and the mathematical and moral hopes coincide; for analysis leads to this general theorem, namely, that if the expectations are very numerous the two hopes approach each other without ceasing and end by coinciding in the case of an infinite number.

We have said in speaking of mathematical and moral hopes that there is a moral advantage in distributing the risks of a benefit which one expects over several of its parts. Thus in order to send a sum of money to a distant part it is much better to send it on several vessels than to expose it on one. This one does by means of mutual assurances. If two persons, each having the same sum upon two different vessels which have sailed from the same port to the same destination, agree to divide equally all the money which may arrive, it is clear that by this agreement each of them divides equally between the two vessels the sum which he expects. Indeed this kind of assurance always leaves uncertainty as to the loss which one may fear. But this uncertainty diminishes in proportion as the number of policy-holders increases; the moral advantage increases more and more and ends by coinciding with the mathematical advantage, its natural limit. This renders the association of mutual assurances when it is very numerous more advantageous to the assured ones than the companies of assurance which, in proportion to the benefit that they give, give a moral advantage always inferior to the mathematical advantage. But the surveillance of their administration can balance the advantage of the mutual assurances. All these results are, as has already been seen, independent of the law which expresses the moral advantage.

One may look upon a free people as a great association whose members secure mutually their properties by supporting proportionally the charges of this guaranty. The confederation of several peoples would give to them advantages analogous to those which each individual enjoys in the society. A congress of their representatives would discuss objects of a utility common to all and without doubt the system of weights, measures, and moneys proposed by the French scientists would be adopted in this congress as one of the things most useful to commerical relations.

Among the institutions founded upon the probabilities of human life the better ones are those in which, by means of a light sacrifice of his revenue, one assures his existence and that of his family for a time when one ought to fear to be unable to satisfy their needs. As far as games are immoral, so far these institutions are advantageous to customs by favoring the strongest bents of our nature. The government ought then to encourage them and respect them in the vicissitudes of public fortune;

since the hopes which they present look toward a distant future, they are able to prosper only when sheltered from all inquietude during their existence. It is an advantage that the institution of a representative government assures them.

Let us say a word about loans. It is clear that in order to borrow perpetually it is necessary to pay each year the product of the capital by the rate of interest. But one may wish to discharge this principal in equal payments made during a definite number of years, payments which are called *annuities* and whose value is obtained in this manner. Each annuity in order to be reduced at the actual moment ought to be divided by a power of unity augmented by the rate of interest equal to the number of years after which this annuity ought to be paid. Forming then a geometric progression whose first term is the annuity divided by unity augmented by the rate of interest, and whose last term is this annuity divided by the same quantity raised to a power equal to the number of years during which the payment should have been made, the sum of this progression will be equivalent to the capital borrowed, which will determine the value of the annuity. A sinking fund is at bottom only a means of converting into annuities a perpetual rent with the sole difference that in the case of a loan by annuities the interest is supposed constant, while the interest of funds acquired by the sinking fund is variable. If it were the same in both cases, the annuity corresponding to the funds acquired would be formed by these funds and from this annuity the State contributes annually to the sinking fund.

If one wishes to make a life loan it will be observed that the tables of life annuities give the capital required to constitute a life annuity at any age, a simple proportion will give the rent which one ought to pay to the individual from whom the capital is borrowed. From these principles all the possible kinds of loans may be calculated.

The principles which we have just expounded concerning the benefits and the losses of institutions may serve to determine the mean result of any number of observations already made, when one wishes to regard the deviations of the results corresponding to divers observations. Let us designate by x the correction of the least result and by x augmented successively by q, q', q'', etc., the corrections of the following results. Let us name e, e', e'', etc., the errors of the observations whose law of probability we will suppose known. Each observation being a function of the result, it is easy to see that by supposing the correction x of this result to be very small, the error e of the first observation will be equal to the product of x by a determined coefficient. Likewise the error e' of the second observation will be the product of the sum q plus x, by a determined coefficient, and so on. The probability of the error e being given by a known function, it will be expressed by the same function of the first of the preceding products. The probability of e' will be expressed by the same function of the second of these products, and so on of the others. The probability of the simultaneous existence of the errors e, e', e'', etc., will be then proportional to the product of these divers functions, a product which will be a function of x. This being granted, if one conceives a curve whose abscissa is x, and whose corresponding ordinate is this product, this curve will represent the probability of the divers values of x, whose limits will be determined by the limits of the errors e, e', e'', etc. Now let us designate by X the abscissa which it is necessary to choose; X diminished by x will be the error which would be committed if the abscissa x were the true correction. This error, multiplied by the probability of x or by the corresponding ordinate of the curve, will be the product of the loss by its probability, regarding, as one should, this error as a loss attached to the choice X. Multiplying this product by the differential of x the integral taken from the first extremity of the curve to X will be the disadvantage of X resulting from the values of x inferior to X. For the values of x superior to X, x less X would be the error of X if x were the true correction; the integral of the product of x by the corresponding ordinate of the curve and by the differential of x will be then the disadvantage of X resulting from the values x superior to x, this integral being taken from x equal to X up to the last extremity of the curve. Adding this disadvantage to the preceding one, the sum will be the disadvantage attached to the choice of X. This choice ought to be determined by the condition that this disadvantage be a *minimum*; and a very simple calculation shows that for this, X ought to be the abscissa whose ordinate divides the curve into two equal parts, so that it is thus probable that the true value of x falls on neither the one side nor the other of X.

Celebrated geometricians have chosen for X the most probable value of x and consequently that which corresponds to the largest ordinate of the curve; but the preceding value appears to me evidently that which the theory of probability indicates.

CHAPTER XVI

CONCERNING ILLUSIONS IN THE ESTIMATION OF PROBABILITIES

THE mind has its illusions as the sense of sight; and in the same manner that the sense of feeling corrects the latter, reflection and calculation correct the former. Probability based upon a daily experience, or exaggerated by fear and by hope, strikes us more than a superior probability but it is only a simple result of calculus. Thus we do not fear in return for small advantages to expose our life to dangers much less improbable than the drawing of a quint in the lottery of France; and yet no one would wish to procure for himself the same advantages with the certainty of losing his life if this quint should be drawn.

Our passions, our prejudices, and dominating opinions, by exaggerating the probabilities which are favorable to them and by attenuating the contrary probabilities, are the abundant sources of dangerous illusions.

Present evils and the cause which produced them effect us much more than the remembrance of evils produced by the contrary cause; they prevent us from appreciating with justice the inconveniences of the ones and the others, and the probability of the proper means to guard ourselves against them. It is this which leads alternately to despotism and to anarchy the people who are driven from the state of repose to which they never return except after long and cruel agitations.

This vivid impression which we receive from the presence of events, and which allows us scarcely to remark the contrary events observed by others, is a principal cause of error against which one cannot sufficiently guard himself.

It is principally at games of chance that a multitude of illusions support hope and sustain it against unfavorable chances. The majority of those who play at lotteries do not know how many chances are to their advantage, how many are contrary to them. They see only the possibility by a small stake of gaining a considerable sum, and the projects which their imagination brings forth, exaggerate to their eyes the probability of obtaining it; the poor man especially, excited by the desire of a better fate, risks at play his necessities by clinging to the most unfavorable combinations which promise him a great benefit. All would be without doubt surprised by the immense number of stakes lost if they could know of them; but one takes care on the contrary to give to the winnings a great publicity, which becomes a new cause of excitement for this funereal play.

When a number in the lottery of France has not been drawn for a long time the crowd is eager to cover it with stakes. They judge since the number has not been drawn for a long time that it ought at the next drawing to be drawn in preference to others. So common an error appears to me to rest upon an illusion by which one is carried back involuntarily to the origin of events. It is, for example, very improbable that at the play of heads and tails one will throw heads ten times in succession. This improbability which strikes us indeed when it has happened nine times, leads us to believe that at the tenth throw tails will be thrown. But the past indicating in the coin a greater propensity for heads than for tails renders the first of the events more probable than the second; it increases as one has seen the probability of throwing heads at the following throw. A similar illusion persuades many people that one can certainly win in a lottery by placing each time upon the same number, until it is drawn, a stake whose product surpasses the sum of all the stakes. But even when similar speculations would not often be stopped by the impossibility of sustaining them they would not diminish the mathematical disadvantage of speculators and they would increase their moral disadvantage, since at each drawing they would risk a very large part of their fortune.

I have seen men, ardently desirous of having a son, who could learn only with anxiety of the births of boys in the month when they expected to become fathers. Imagining that the ratio of these births to those of girls ought to be the same at the end of each month, they judged that the boys already born would render more probable the births next of girls. Thus the extraction of a white ball from an urn which contains a limited number of white balls and of black balls increases the probability of extracting a black ball at the following drawing. But this ceases to take place when the number of balls in the urn is unlimited, as one must suppose in order to compare this case with that of births. If, in the course of a month, there were born many more boys than girls, one might suspect that toward the time of their conception a general cause had favored masculine conception, which would render more probable the birth next of a boy. The irregular events of nature are not exactly comparable to the drawing of the numbers of a lottery in which all the numbers are mixed at each drawing in such a manner as to render the chances of their drawing perfectly equal. The frequency of one of these events seems to indicate a cause slightly favoring it, which increases the probability of its next return, and its repetition prolonged for a long time, such as a long series of rainy days, may develop unknown causes for its change; so that at each expected event we are not, as at each drawing of a lottery, led back to the same state of indecision in regard to what ought to happen. But in proportion as the observation of these events is multiplied, the comparison of their results with those of lotteries becomes more exact.

By an illusion contrary to the preceding ones one seeks in the past drawings of the lottery of France the numbers most often drawn, in order to form combinations upon which one thinks to place the stake to advantage. But when the manner in which the mixing of the numbers in this lottery is considered, the past ought to have no influence upon the future. The very frequent drawings of a number are only the anomalies of chance; I have submitted several of them to calculation and have constantly found that they are included within the limits which the supposition of an equal possibility of the drawing of all the numbers allows us to admit without improbability.

In a long series of events of the same kind the single chances of hazard ought sometimes to offer the singular veins of good luck or bad luck which the majority of players do not fail to attribute to a kind of fatality. It happens often in games which depend at the same time upon hazard and upon the competency of the players, that that one who loses, troubled by his loss, seeks to repair it by hazardous throws which he would shun in another situation; thus he aggravates his own ill luck and prolongs its duration. It is then that prudence becomes necessary and that it is of importance to convince oneself that the moral disadvantage attached to unfavorable chances is increased by the ill luck itself.

The opinion that man has long been placed in the centre of the universe, considering himself the special object of the cares of nature, leads each individual to make himself the centre of a more or less extended sphere and to believe that hazard has preference for him. Sustained by this belief, players often risk considerable sums at games when they know that the chances are unfavorable. In the conduct of life a similar opinion may sometimes have advantages; but most often it leads to disastrous enterprises. Here as everywhere illusions are dangerous and truth alone is generally useful.

One of the great advantages of the calculus of probabilities is to teach us to distrust first opinions. As we recognize that they often deceive when they may be submitted to calculus, we ought to conclude that in other matters confidence should be given only after extreme circumspection. Let us prove this by example.

An urn contains four balls, black and white, but which are not all of the same color. One of these balls has been drawn whose color is white and which has been put back in the urn in order to proceed again to similar drawings. One demands the probability of extracting only black balls in the four following drawings.

If the white and black were in equal number this probability would be the fourth power of the probability $\frac{1}{2}$ of extracting a black ball at each drawing; it would be then $\frac{1}{16}$. But the extraction of a white ball at the first drawing indicates a superiority in the number of white balls in the urn; for if one supposes in the urn three white balls and one black the probability of extracting a white ball is $\frac{3}{4}$; it is $\frac{2}{4}$ if one supposes two white balls and two black; finally it is reduced to $\frac{1}{4}$ if one supposes three black balls and one white. Following the principle of the probability of causes drawn from events the probabilities of these three suppositions are among themselves as the quantities $\frac{3}{4}, \frac{2}{4}, \frac{1}{4}$; they are consequently equal to $\frac{3}{6}, \frac{2}{6}, \frac{1}{6}$. It is thus a bet of 5 against 1 that the number of black balls is inferior, or at the most equal, to that of the white. It seems then that after the extraction of a white ball at the first drawing, the probability of extracting successively four black balls ought to be less than in the case of the equality of the colors or smaller than one sixteenth. However, it is not, and it is found by a very simple calculation that this probability is greater than one fourteenth. Indeed it would be the fourth power of $\frac{1}{4}$, of $\frac{2}{4}$, and of $\frac{3}{4}$ in the first, the second, and the third of the preceding suppositions concerning the colors of the balls in the urn. Multiplying respectively each power by the probability of the corresponding supposition, or by $\frac{3}{6}, \frac{2}{6}$, and $\frac{1}{6}$, the sum of the products will be the probability of extracting successively four black balls. One has thus for this probability $\frac{29}{384}$, a fraction greater than $\frac{1}{14}$. This paradox is explained by considering that the indication

473

of the superiority of white balls over the black ones at the first drawing does not exclude at all the superiority of the black balls over the white ones, a superiority which excludes the supposition of the equality of the colors. But this superiority, though but slightly probable, ought to render the probability of drawing successively a given number of black balls greater than in this supposition if the number is considerable; and one has just seen that this commences when the given number is equal to four. Let us consider again an urn which contains several white and black balls. Let us suppose at first that there is only one white ball and one black. It is then an even bet that a white ball will be extracted in one drawing. But it seems for the equality of the bet that one who bets on extracting the white ball ought to have two drawings if the urn contains two black and one white, three drawings if it contains three black and one white, and so on; it is supposed that after each drawing the extracted ball is placed again in the urn.

We are convinced easily that this first idea is erroneous. Indeed in the case of two black and one white ball, the probability of extracting two black in two drawings is the second power of $\frac{2}{3}$ or $\frac{4}{9}$; but this probability added to that of drawing a white ball in two drawings is certainty or unity, since it is certain that two black balls or at least one white ball ought to be drawn; the probability in this last case is then $\frac{5}{9}$, a fraction greater than $\frac{1}{2}$. There would still be a greater advantage in the bet of drawing one white ball in five draws when the urn contains five black and one white ball; this bet is even advantageous in four drawings; it returns then to that of throwing six in four throws with a single die.

The Chevalier de Meré, who caused the invention of the calculus of probabilities by encouraging his friend Pascal, the great geometrician, to occupy himself with it, said to him "that he had found error in the numbers by this ratio. If we undertake to make six with one die there is an advantage in undertaking it in four throws, as 671 to 625. If we undertake to make two sixes with two dice, there is a disadvantage in undertaking in 24 throws. At least 24 is to 36, the number of the faces of the two dice, as 4 is to 6, the number of faces of one die." "This was," wrote Pascal to Fermat, "his great scandal which caused him to say boldly that the propositions were not constant and that arithmetic was demented. . . . He has a very good mind, but he is not a geometrician, which is, as you know, a great fault." The Chevalier de Meré, deceived by a false analogy, thought that in the case of the equality of bets the number of throws ought to increase in proportion to the number of all the chances possible, which is not exact, but which approaches exactness as this number becomes larger.

One has endeavored to explain the superiority of the births of boys over those of girls by the general desire of fathers to have a son who would perpetuate the name. Thus by imagining an urn filled with an infinity of white and black balls in equal number, and supposing a great number of persons each of whom draws a ball from this urn and continues with the intention of stopping when he shall have extracted a white ball, one has believed that this intention ought to render the number of white balls extracted superior to that of the black ones. Indeed this intention gives necessarily after all the drawings a number of white balls equal to that of persons, and it is possible that these drawings would never lead a black ball. But it is easy to see that this first notion is only an illusion; for if one conceives that in the first drawing all the persons draw at once a ball from the urn, it is evident that their intention can have no influence upon the color of the balls which ought to appear at this drawing. Its unique effect will be to exclude from the second drawing the persons who shall have drawn a white one at the first. It is likewise apparent that the intention of the persons who shall take part in the new drawing will have no influence upon the color of the balls which shall be drawn, and that it will be the same at the following drawings. This intention will have no influence then upon the color of the balls extracted in the totality of drawings; it will, however, cause more or fewer to participate at each drawing. The ratio of the white balls extracted to the black ones will differ thus very little from unity. It follows that the number of persons being supposed very large, if observation gives between the colors extracted a ratio which differs sensibly from unity, it is very probable that the same difference is found between unity and the ratio of the white balls to the black contained in the urn.

I count again among illusions the application which Liebnitz and Daniel Bernoulli have made of the calculus of probabilities to the summation of series. If one reduces the fraction whose numerator is unity and whose denominator is unity plus a variable, in a series prescribed by the ratio to the powers of this variable, it is easy to see that in supposing the variable equal to unity the fraction becomes $\frac{1}{2}$, and the series becomes plus one, minus one, plus one, minus one, etc. In adding the first two terms, the second two, and so on, the series is transformed into another of which each term is zero. Grandi, an Italian Jesuit, concluded from this the possibility of the creation; because the series being always $\frac{1}{2}$, he saw this fraction spring from an infinity of zeros or from nothing. It was thus that Liebnitz believed he saw the image of creation in his binary arithmetic where he employed only the two characters; unity and zero. He imagined, since God can be represented by unity and nothing by zero, that the Supreme Being had drawn from nothing all beings; as unity with zero expresses all the numbers in this system of arithmetic. This idea was so pleasing to Liebnitz that he communicated it to the Jesuit Grimaldi, president of the tribunal of mathematics in China, in the hope that this emblem of creation would convert to Christianity the emperor there who particularly loved the sciences. I report this incident only to show to what extent the prejudices of infancy can mislead the greatest men.

Liebnitz, always led by a singular and very loose metaphysics, considered that the series plus one, minus one, plus one, etc., becomes unity or zero according as one stops at a number of terms odd or even; and as in infinity there is no reason to prefer the even number to the odd, one ought following the rules of probability, to take the half of the results relative to these two kinds of numbers, and which are zero and unity, which gives $\frac{1}{2}$ for the value of the series. Daniel Bernoulli has since extended this reasoning to the summation of series formed from periodic terms. But all these series have no values properly speaking; they get them only in the case where their terms are multiplied by the successive powers of a variable less than unity. Then these series are always convergent, however small one supposes the difference of the variable from unity; and it is easy to demonstrate that the values assigned by Bernoulli, by virtue of the rule of probabilities, are the same values of the generative fraction of the series, when one supposes in these fractions the variable equal to unity. These values are again the limits which the series approach more and more, in proportion as the variable approaches unity. But when the variable is exactly equal to unity the series cease to be convergent; they have values only as far as one arrests them. The remarkable ratio of this application of the calculus of probabilities with the limits of the values of periodic series supposes that the terms of these series are multiplied by all the consecutive powers of the variable. But this series may result from the development of an infinity of different fractions in which this did not occur. Thus the series plus one, minus one, plus one, etc., may spring from the development of a fraction whose numerator is unity plus the variable, and whose denominator is this numerator augmented by the square of the variable. Supposing the variable equal to unity, this development changes, in the series proposed, and the generative fraction becomes equal to $\frac{2}{8}$; the rules of probabilities would give then a false result, which proves how dangerous it would be to employ similar reasoning, especially in the mathematical sciences, which ought to be especially distinguished by the rigor of their operations.

We are led naturally to believe that the order according to which we see things renewed upon the earth has existed from all times and will continue always. Indeed if the present state of the universe were exactly similar to the anterior state which has produced it, it would give birth in its turn to a similar state; the succession of these states would then be eternal. I have found by the application of analysis to the law of universal gravity that the movement of rotation and of revolution of the planets and satellites, and the position of the orbits and of their equators are subjected only to periodic inequalities. In comparing with ancient eclipses the theory of the secular equation of the moon I have found that since Hipparchus the duration of the day has not varied by the hundredth of a second, and that the mean temperature of the earth has not diminished the one-hundredth of a degree. Thus the stability of actual order appears established at the same time by theory and by observations. But this order is effected by divers causes which an attentive examination reveals, and which it is impossible to submit to calculus.

The actions of the ocean, of the atmosphere, and of meteors, of earthquakes, and the eruptions of volcanoes, agitate continually the surface of the earth and ought to effect in the long run great changes. The temperature of climates, the volume of the atmosphere, and the proportion of the gases which constitute it, may vary in an inappreciable manner. The instruments and the means suitable to determine these variations being new, observation has been unable up to this time to teach us anything in this regard. But it is hardly probable that the causes which absorb and renew the gases constituting the air maintain exactly their respective proportions. A long series of centuries will show the alterations which are experienced by all these elements so essential to the conservation of organized beings. Although historical monuments do not go back to a very great antiquity they offer us nevertheless sufficiently great changes which have come about by the slow and continued action of natural agents. Searching in the bowels of the earth one discovers numerous débris of former nature, entirely different from the present. Moreover, if the entire earth was in the beginning fluid, as everything appears to indicate, one imagines that in passing from that state to the one which it has now, its surface ought to have experienced prodigious changes. The heavens itself in spite of the order of its movements, is not unchangeable. The resistance of light and of other ethereal fluids, and the attraction of the stars ought, after a great number of centuries, to alter considerably the planetary movements. The variations already observed in the stars and in the form of the nebulæ give us a presentiment of those which time will develop in the system of these great bodies. One may represent the successive states of the universe by a curve, of which time would be the abscissa and of which the ordinates are the divers states. Scarcely knowing an element of this curve we are far from being able to go back to its origin; and if in order to satisfy the imagination, always restless from our ignorance of the cause of the phenomena which interest it, one ventures some conjectures it is wise to present them only with extreme reserve.

There exists in the estimation of probabilities a kind of illusions, which depending especially upon the laws of the intellectual organization demands, in order to secure oneself against them, a profound examination of these laws. The desire to penetrate into the future and the ratios of some remarkable events, to the predictions of astrologers, of diviners and soothsayers, to presentiments and dreams, to the numbers and the days reputed lucky or unlucky, have given birth to a multitude of prejudices still very widespread. One does not reflect upon the great number of non-coincidences which have made no impression or which are unknown. However, it is necessary to be acquainted with them in order to appreciate the probability of the causes to which the coincidences are attributed. This knowledge would confirm without doubt that which reason tells us in regard to these prejudices. Thus the philosopher of antiquity to whom is shown in a temple, in order to exalt the power of the god who is adored there, the *ex veto* of all those who after having invoked it were saved from shipwreck, presents an incident consonant with the calculus of probabilities, observing that he does not see inscribed the names of those who, in spite of this invocation, have perished. Cicero has refuted all these prejudices with much reason and eloquence in his *Treatise on Divination*, which he ends by a passage which I shall cite; for one loves to find again among the ancients the thunderbolts of reason, which, after having dissipated all the prejudices by its light, shall become the sole foundation of human institutions.

"It is necessary," says the Roman orator, "to reject divination by dreams and all similar prejudices. Widespread superstition has subjugated the majority of minds and has taken possession of the feebleness of men. It is this we have expounded in our books upon the nature of the gods and especially in this work, persuaded that we shall render a service to others and to ourselves if we succeed in destroying superstition. However (and I desire especially in this regard my thought be well comprehended), in destroying superstition I am far from wishing to disturb religion. Wisdom enjoins us to maintain the institutions and the ceremonies of our ancestors, touching the cult of the gods. Moreover, the beauty of the universe and the order of celestial things force us to recognize some superior nature which ought to be remarked and admired by the human race. But as far as it is proper to propagate religion, which is joined to the knowledge of nature, so far it is necessary to work toward the extirpation of superstition, for it torments one, importunes one, and pursues one continually and in all places. If one consult a diviner or a soothsayer, if one immolates a victim, if one regards the flight of a bird, if one encounters a Chaldean or an aruspex, if it lightens, if it thunders, if the thunderbolt strikes, finally, if there is born or is manifested a kind of prodigy, things one of which ought often to happen, then superstition dominates and leaves no repose. Sleep itself, this refuge of mortals in their troubles and their labors, becomes by it a new source of inquietude and fear."

All these prejudices and the terrors which they inspire are connected with physiological causes which continue sometimes to operate strongly after reason has disabused us of them. But the repetition of acts contrary to these prejudices can always destroy them.

CHAPTER XVII

CONCERNING THE VARIOUS MEANS OF APPROACHING CERTAINTY

INDUCTION, analogy, hypotheses founded upon facts and rectified continually by new observations, a happy tact given by nature and strengthened by numerous comparisons of its indications with experience, such are the principal means for arriving at truth.

If one considers a series of objects of the same nature one perceives among them and in their changes ratios which manifest themselves more and more in proportion as the series is prolonged, and which, extending and generalizing continually, lead finally to the principle from which they were derived. But these ratios are enveloped by so many strange circumstances that it requires great sagacity to disentangle them and to recur to this principle: it is in this that the true genius of sciences consists. Analysis and natural philosophy owe their most important discoveries to this fruitful means, which is called *induction*. Newton was indebted to it for his theorem of the binomial and the principle of universal gravity. It is difficult to appreciate the probability of the results of induction, which is based upon this that the simplest ratios are the most common; this is verified in the formulæ of analysis and is found again in natural phenomena, in crystallization, and in chemical combinations. This simplicity of ratios will not appear astonishing if we consider that all the effects of nature are only mathematical results of a small number of immutable laws.

Yet induction, in leading to the discovery of the general principles of the sciences, does not suffice to establish them absolutely. It is always necessary to confirm them by demonstrations or by decisive experiences; for the history of the sciences shows us that induction has sometimes led to inexact results. I shall cite, for example, a theorem of Fermat in regard to prime numbers. This great geometrician, who had meditated profoundly upon this theorem, sought a formula which, containing only prime numbers, gave directly a prime number greater than any other number assignable. Induction led him to think that two, raised to a power which was itself a power of two, formed with unity a prime number. Thus, two raised to the square plus one, forms the prime number five; two raised to the second power of two, or sixteen, forms with one the prime number seventeen. He found that this was still true for the eighth and the sixteenth power of two augmented by unity; and this induction, based upon several arithmetical considerations, caused him to regard this result as general. However, he avowed that he had not demonstrated it. Indeed, Euler recognized that this does not hold for the thirty-second power of two, which, augmented by unity, gives 4,294,967,297, a number divisible by 641.

We judge by induction that if various events, movements, for example, appear constantly and have been long connected by a simple ratio, they will continue to be subjected to it; and we conclude from this, by the theory of probabilities, that this ratio is due, not to hazard, but to a regular cause. Thus the equality of the movements of the rotation and the revolution of the moon; that of the movements of the nodes of the orbit and of the lunar equator, and the coincidence of these nodes; the singular ratio of the movements of the first three satellites of Jupiter, according to which the mean longitude of the first satellite, less three times that of the second, plus two times that of the third, is equal to two right angles; the equality of the interval of the tides to that of the passage of the moon to the meridian; the return of the greatest tides with the syzygies, and of the smallest with the quadratures; all these things, which have been maintained since they were first observed, indicate with an extreme probability, the existence of constant causes which geometricians have happily succeeded in attaching to the law of universal gravity, and the knowledge of which renders certain the perpetuity of these ratios.

The chancellor Bacon, the eloquent promoter of the true philosophical method, has made a very strange misuse of induction in order to prove the immobility of the earth. He reasons thus in the *Novum Organum*, his finest work: "The movement of the stars from the orient to the occident increases in swiftness, in proportion to their distance from the earth. This movement is swiftest with the stars; it slackens a little with Saturn, a little more with Jupiter, and so on to the moon and the highest comets. It is still perceptible in the atmosphere, especially between the tropics, on account of the great circles which the molecules of the air describe there; finally, it is almost inappreciable with the ocean; it is then nil for the earth." But this induction proves only that Saturn, and the stars which are inferior to it, have their own movements, contrary to the real or apparent movement which sweeps the whole celestial sphere from the orient to the occident, and that these movements appear slower with the more remote stars, which is conformable to the laws of optics. Bacon ought to have been struck by the inconceivable swiftness which the stars require in order to accomplish their diurnal revolution, if the earth is immovable, and by the extreme simplicity with which its rotation explains how bodies so distant, the ones from the others, as the stars, the sun, the planets, and the moon, all seem subjected to this revolution. As to the ocean and to the atmosphere, he ought not to compare their movement with that of the stars which are detached from the earth; but since the air and the sea make part of the terrestrial globe, they ought to participate in its movement or in its repose. It is singular that Bacon, carried to great prospects by his genius, was not won over by the majestic idea which the Copernican system of the universe offers. He was able, however, to find in favor of that system, strong analogies in the discoveries of Galileo, which were continued by him. He has given for the search after truth the precept, but not the example. But by insisting, with all the force of reason and of eloquence, upon the necessity of abandoning the insignificant subtilties of the school, in order to apply oneself to observations and to experiences, and by indicating the true method of ascending to the general causes of phenomena, this great philosopher contributed to the immense strides which the human mind made in the grand century in which he terminated his career.

Analogy is based upon the probability, that similar things have causes of the same kind and produce the same effects. This probability increase as the similitude becomes more perfect. Thus we judge without doubt that beings provided with the same organs, doing the same things, experience the same sensations, and are moved by the same desires. The probability that the animals which resemble us have sensations analogous to ours, although a little inferior to that which is relative to individuals of our species, is still exceedingly great; and it has required all the influence of religious prejudices to make us think with some philosophers that animals are mere automatons. The probability of the existence of feeling decreases in the same proportion as the similitude of the organs with ours diminishes, but it is always very great, even with insects. In seeing those of the same species execute very complicated things exactly in the same manner from generation to generation, and without having learned them, one is led to believe that they act by a kind of affinity analogous to that which brings together the molecules of crystals, but which, together with the sensation attached to all animal organization, produces, with the regularity of chemical combinations, combinations that are much more singular; one might, perhaps, name this mingling of elective affinities and sensations *animal affinity*. Although there exists a great analogy between the organization of plants and that of animals, it does not seem to me sufficient to extend to vegetables the sense of feeling; but nothing authorizes us in denying it to them.

Since the sun brings forth, bythe beneficent action of its light and of its heat, the animals and plants which cover the earth, we judge by analogy that it produces similar effects upon the other planets; for it is not natural to think that the cause whose activity we see developed in so many ways should be sterile upon so great a planet as Jupiter, which, like the terrestrial globe, has its days, its nights, and its years, and upon which observations indicate changes which suppose very active forces. Yet this would be giving too great an extension to analogy to conclude from it the similitude of the inhabitants of the planets and of the earth. Man, made for the temperature which he enjoys, and for the element which he breathes, would not be able, according to all appearance, to live upon the other planets. But ought there not to be an infinity of organization relative to the various constitutions of the globes of this universe? If the single difference of the elements and of the climates make so much variety in terrestrial productions, how much greater the difference ought to be among those of the various planets and of their satellites! The most active imagination can form no idea of it; but their existence is very probable.

We are led by a strong analogy to regard the stars as so many suns endowed, like ours, with an attractive power proportional to the mass and reciprocal to the square of the distances; for this power being demonstrated for all the bodies of the solar system, and for their smallest molecules, it appears to appertain to all matter. Already the movements of the small stars, which have been called *double*, on account of their being binary, appear to indicate it; a century at most of precise observations, by verifying their movements of revolution, the ones about the others, will place beyond doubt their reciprocal attractions.

The analogy which leads us to make each star the centre of a planetary system is far less strong than the preceding one; but it acquires probability by the hypothesis which has been proposed in regard to the formation of the stars and of the sun; for in this hypothesis each star, having been like the sun, primitively environed by a vast atmosphere, it is natural to attribute to this atmosphere the same effects as to the solar atmosphere, and to suppose that it has produced, in condensing, planets and satellites.

A great number of discoveries in the sciences is due to analogy. I shall cite as one of the most remarkable, the discovery of atmospheric electricity, to which one has been led by the analogy of electric phenomena with the effects of thunder.

The surest method which can guide us in the search for truth, consists in rising by induction from phenomena to laws and from laws to forces. Laws are the ratios which connect particular phenomena together: when they have shown the general principle of the forces from which they are derived, one verifies it either by direct experiences, when this is possible, or by examination if it agrees with known phenomena; and if by a rigorous analysis we see them proceed from this principle, even in their small details, and if, moreover, they are quite varied and very numerous, then science acquires the highest degree of certainty and of perfection that it is able to attain. Such, astronomy has become by the discovery of universal gravity. The history of the sciences shows that the slow and laborious path of induction has not always been that of inventors. The imagination, impatient to arrive at the causes, takes pleasure in creating hypotheses, and often it changes the facts in order to adapt them to its work; then the hypotheses are dangerous. But when one regards them only as the means of connecting the phenomena in order to discover the laws; when, by refusing to attribute them to a reality, one rectifies them continually by new observations, they are able to lead to the veritable causes, or at least put us in a position to conclude from the phenomena observed those which given circumstances ought to produce.

If we should try all the hypotheses which can be formed in regard to the cause of phenomena we should arrive, by a process of exclusion, at the true one. This means has been employed with success; sometimes we have arrived at several hypotheses which explain equally well all the facts known, and among which scholars are divided, until decisive observations have made known the true one. Then it is interesting, for the history of the human mind, to return to these hypotheses, to see how they succeed in explaining a great number of facts, and to investigate the changes which they ought to undergo in order to agree with the history of nature. It is thus that the system of Ptolemy, which is only the realization of celestial appearances, is transformed into the hypothesis of the movement of the planets about the sun, by rendering equal and parallel to the solar orbit the circles and the epicycles which he causes to be described annually, and the magnitude of which he leaves undetermined. It suffices, then, in order to change this hypothesis into the true system of the world, to transport the apparent movement of the sun in a sense contrary to the earth.

It is almost always impossible to submit to calculus the probability of the results obtained by these various means; this is true likewise for historical facts. But the totality of the phenomena explained, or of the testimonies, is sometimes such that without being able to appreciate the probability we cannot reasonably permit ourselves any doubt in regard to them. In the other cases it is prudent to admit them only with great reserve.

CHAPTER XVIII
HISTORICAL NOTICE CONCERNING THE CALCULUS
OF PROBABILITIES

LONG ago were determined, in the simplest games, the ratios of the chances which are favorable or unfavorable to the players; the stakes and the bets were regulated according to these ratios. But no one before Pascal and Fermat had given the principles and the methods for submitting this subject to calculus, and no one had solved the rather complicated questions of this kind. It is, then, to these two great geometricians that we must refer the first elements of the science of probabilities, the discovery of which can be ranked among the remarkable things which have rendered illustrious the seventeenth century—the century which has done the greatest honor to the human mind. The principal problem which they solved by different methods, consists, as we have seen, in distributing equitably the stake among the players, who are supposed to be equally skilful and who agree to stop the game before it is finished, the condition of play being that, in order to win the game, one must gain a given number of points different for each of the players. It is clear that the distribution should be made proportionally to the respective probabilities of the players of winning this game, the probabilities depending upon the numbers of points which are still lacking. The method of Pascal is very ingenious, and is at bottom only the equation of partial differences of this problem applied in determining the successive probabilities of the players, by going from the smallest numbers to the following ones. This method is limited to the case of two players; that of Fermat, based upon combinations, applies to any number of players. Pascal believed at first that it was, like his own, restricted to two players; this brought about between them a discussion, at the conclusion of which Pascal recognized the generality of the method of Fermat.

Huygens united the divers problems which had already been solved and added new ones in a little treatise, the first that has appeared on this subject and which has the title *De Ratiociniis in ludo aleæ*. Several geometricians have occupied themselves with the subject since: Hudde, the great pensionary, Witt in Holland, and Halley in England, applied calculus to the probabilities of human life, and Halley published in this field the first table of mortality. About the same time Jacques Bernoulli proposed to geometricians various problems of probability, of which he afterwards gave solutions. Finally he composed his beautiful work entitled *Ars conjectandi*, which appeared seven years after his death, which occurred in 1706. The science of probabilities is more profoundly investigated in this work than in that of Huygens. The author gives a general theory of combinations and series, and applies it to several difficult questions concerning hazards. This work is still remarkable on account of the justice and the cleverness of view, the employment of the formula of the binomial in this kind of questions, and by the demonstration of this theorem, namely, that in multiplying indefinitely the observations and the experiences, the ratio of the events of different natures approaches that of their respective probabilities in the limits whose interval becomes more and more narrow in proportion as they are multiplied, and become less than any assignable quantity. This theorem is very useful for obtaining by observations the laws and the causes of phenomena. Bernoulli attaches, with reason, a great importance to his demonstration, upon which he has said to have meditated for twenty years.

In the interval, from the death of Jacques Bernoulli to the publication of his work, Montmort and Moivre produced two treatises upon the calculus of probabilities. That of Montmort has the title *Essai sur les Jeux de hasard*; it contains numerous applications of this calculus to various games. The author has added in the second edition some letters in which Nicolas Bernoulli gives the ingenious solutions of several difficult problems. The treatise of Moivre, later than that of Montmort, appeared at first in the *Transactions philosophiques* of the year 1711. Then the author published it separately, and he has improved it successively in three editions. This work is principally based upon the formula of the binomial and the problems which it contains have, like their solutions, a grand generality. But its distinguishing feature is the theory of recurrent series and their use in this subject. This theory is the integration of linear equations of finite differences with constant coefficients, which Moivre made in a very happy manner.

In his work, Moivre has taken up again the theory of Jacques Bernoulli in regard to the probability of results determined by a great number of observations. He does not content himself with showing, as Bernoulli does, that the ratio of the events which ought to occur approaches without ceasing that of their respective probabilities; but he gives besides an elegant and simple expression of the probability that the difference of these two ratios is contained within the given limits. For this purpose he determines the ratio of the greatest term of the development of a very high power of the binomial to the sum of all its terms, and the hyperbolic logarithm of the excess of this term above the terms adjacent to it.

The greatest term being then the product of a considerable number of factors, his numerical calculus becomes impracticable. In order to obtain it by a convergent approximation, Moivre makes use of a theorem of Stirling in regard to the mean term of the binomial raised to a high power, a remarkable theorem, especially in this, that it introduces the square root of the ratio of the circumference to the radius in an expression which seemingly ought to be irrelevant to this transcendent. Moreover, Moivre was greatly struck by this result, which Stirling had deduced from the expression of the circumference in infinite products; Wallis had arrived at this expression by a singular analysis which contains the germ of the very curious and useful theory of definite intergrals.

Many scholars, among whom one ought to name Deparcieux, Kersseboom, Wargentin, Dupré de Saint-Maure, Simpson, Sussmilch, Messène, Moheau, Price, Bailey, and Duvillard, have collected a great amount of precise data in regard to population, births, marriages, and mortality. They have given formulæ and tables relative to life annuities, tontines, assurances, etc. But in this short notice I can only indicate these useful works in order to adhere to original ideas. Of this number special mention is due to the mathematical and moral hopes and to the ingenious principle which Daniel Bernoulli has given for submitting the latter to analysis. Such is again the happy application which he has made of the calculus of probabilities to inoculation. One ought especially to include, in the number of these original ideas, direct consideration of the possibility of events drawn from events observed. Jacques Bernoulli and Moivre supposed these possibilities known, and they sought the probability that the result of future experiences will more and more nearly represent them. Bayes, in the *Transactions philosophiques* of the year 1763, sought directly the probability that the possibilities indicated by past experiences are comprised within given limits; and he has arrived at this in a refined and very ingenious manner, although a little perplexing. This subject is connected with the theory of the probability of causes and future events, concluded from events observed. Some years later I expounded the principles of this theory with a remark as to the influence of the inequalities which may exist among the chances which are supposed to be equal. Although it is not known which of the simple events these inequalities favor, nevertheless this ignorance itself often increases the probability of compound events.

In generalizing analysis and the problems concerning probabilities, I was led to the calculus of partial finite differences, which Lagrange has since treated by a very simple method, elegant applications of which he has used in this kind of problems. The theory of generative functions which I published about the same time includes these subjects among those it embraces, and is adapted of itself and with the greatest generality to the most difficult questions of probability. It determines again, by very convergent approximations, the values of the functions composed of a great number of terms and factors; and in showing that the square root of the ratio of the circumference to the radius enters most frequently into these values, it shows that an infinity of other transcendents may be introduced.

Testimonies, votes, and the decisions of electoral and deliberative assemblies, and the judgments of tribunals, have been submitted likewise to the calculus of probabilities. So many passions, divers interests, and circumstances complicate the questions relative to the subjects, that they are almost always insoluble. But the solution of very simple problems which have a great analogy with them, may often shed upon difficult and important questions great light, which the surety of calculus renders always preferable to the most specious reasonings.

One of the most interesting applications of the calculus of probabilities concerns the mean values which must be chosen among the results of observations. Many geometricians have studied the subject, and Lagrange has published in the *Mémoires de Turin* a beautiful method for determining these mean values when the law of the errors of the observations is known. I have given for the same purpose a method based upon a singular contrivance which may be employed with advantage in other questions of analysis; and this, by permitting indefinite extension in the whole course of a long calculation of the functions which ought to be limited by the nature of the problem, indicates the modifications which each term of the final result ought to receive by virtue of these limitations. It has already been seen that each observation furnishes an equation of condition of the first degree, which may always be disposed of in such a manner that all its terms be in the first member, the second being zero. The use of these equations is one of the principal causes of the great precision of our astronomical tables, because an immense number of excellent observations has thus been made to concur in determining their elements. When there is only one element to be determined Côtes prescribed that the equations of condition should be prepared in such a manner that the coefficient of the unknown element be positive in each of them; and that all these equations should be added in order to form a final equation, whence is derived the value of this element. The rule of Côtes was followed by all calculators, but since he failed to determine several elements, there was no fixed rule for combining the equations of condition in such a manner as to obtain the necessary final equations; but one chose for each element the observations most suitable to determine it. It was in order to obviate these gropings that Legendre and Gauss concluded to add the squares of the first members of the equations of condition, and to render the sum a minimum, by varying each unknown element; by this means is obtained directly as many final equations as there are elements. But do the values determined by these equations merit the preference over all those which may be obtained by other means? This question, the calculus of probabilities alone was able to answer. I applied it, then, to this subject, and obtained by a delicate analysis a rule which includes the preceding method, and which adds to the advantage of giving, by a regular process, the desired elements that of obtaining them with the greatest show of evidence from the totality of observations, and of determining the values which leave only the smallest possible errors to be feared.

However, we have only an imperfect knowledge of the results obtained, as long as the law of the errors of which they are susceptible is unknown; we must be able to assign the probability that these errors are contained within given limits, which amounts to determining that which I have called the *weight* of a result. Analysis leads to general and simple formulæ for this purpose. I have applied this analysis to the results of geodetic observations. The general problem consists in determining the probabilities that the values of one or of several linear functions of the errors of a very great number of observations are contained within any limits.

The law of the possibility of the errors of observations introduces into the expressions of these probabilities a constant, whose value seems to require the knowledge of this law, which is almost always unknown. Happily this constant can be determined from the observations.

In the investigation of astronomical elements it is given by the sum of the squares of the differences between each observation and the calculated one. The errors equally probable being proportional to the square root of this sum, one can, by the comparison of these squares, appreciate the relative exactitude of the different tables of the same star. In geodetic operations these squares are replaced by the squares of the errors of the sums observed of the three angles of each triangle. The comparison of the squares of these errors will enable us to judge of the relative precision of the instruments with which the angles have been measured. By this comparison is seen the advantage of the repeating circle over the instruments which it has replaced in geodesy.

There often exists in the observations many sources of errors: thus the positions of the stars being determined by means of the meridian telescope and of the circle, both susceptible of errors whose law of probability ought not to be supposed the same, the elements that are deduced from these positions are affected by these errors. The equations of condition, which are made to obtain these elements, contain the errors of each instrument and they have various coefficients. The most advantageous system of factors by which these equations ought to be multiplied respectively, in order to obtain, by the union of the products, as many final equations as there are elements to be determined, is no longer that of the coefficients of the elements in each equation of condition. The analysis which I have used leads easily, whatever the number of the sources of error may be, to the system of factors which gives the most advantageous results, or those in which the same error is less probable than in any other system. The same analysis determines the laws of probability of the errors of these results. These formulæ contain as many unknown constants as there are sources of error, and they depend upon the laws of probability of these errors. It has been seen that, in the case of a single source, this constant can be determined by forming the sum of the squares of the residuals of each equation of condition, when the values found for these elements have been substituted. A similar process generally gives values of these constants, whatever their number may be, which completes the application of the calculus of probabilities to the results of observations.

I ought to make here an important remark. The small uncertainty that the observations, when they are not numerous, leave in regard to the values of the constants of which I have just spoken, renders a little uncertain the probabilities determined by analysis. But it almost always suffices to know if the probability, that the errors of the results obtained are comprised within narrow limits, approaches closely to unity; and when it is not, it suffices to know up to what point the observations should be multiplied, in order to obtain a probability such that no reasonable doubt remains in regard to the correctness of the results. The analytic formulæ of probabilities satisfy perfectly this requirement; and in this connection they may be viewed as the necessary complement of the sciences, based upon a totality of observations susceptible of error. They are likewise indispensable in solving a great number of problems in the natural and moral sciences. The regular causes of phenomena are most frequently either unknown, or too complicated to be submitted to calculus;

again, their action is often disturbed by accidental and irregular causes; but its impression always remains in the events produced by all these causes, and it leads to modifications which only a long series of observations can determine. The analysis of probabilities develops these modifications; it assigns the probability of their causes and it indicates the means of continually increasing this probability. Thus in the midst of the irregular causes which disturb the atmosphere, the periodic changes of solar heat, from day to night, and from winter to summer, produce in the pressure of this great fluid mass and in the corresponding height of the barometer, the diurnal and annual oscillations; and numerous barometric observations have revealed the former with a probability at least equal to that of the facts which we regard as certain. Thus it is again that the series of historical events shows us the constant action of the great principles of ethics in the midst of the passions and the various interests which disturb societies in every way. It is remarkable that a science, which commenced with the consideration of games of chance, should be elevated to the rank of the most important subjects of human knowledge.

I have collected all these methods in my *Théorie analytique des Probabilités*, in which I have proposed to expound in the most general manner the principles and the analysis of the calculus of probabilities, likewise the solutions of the most interesting and most difficult problems which calculus presents.

It is seen in this essay that the theory of probabilities is at bottom only common sense reduced to calculus; it makes us appreciate with exactitude that which exact minds feel by a sort of instinct without being able ofttimes to give a reason for it. It leaves no arbitrariness in the choice of opinions and sides to be taken; and by its use can always be determined the most advantageous choice. Thereby it supplements most happily the ignorance and the weakness of the human mind. If we consider the analytical methods to which this theory has given birth; the truth of the principles which serve as a basis; the fine and delicate logic which their employment in the solution of problems requires; the establishments of public utility which rest upon it; the extension which it has received and which it can still receive by its application to the most important questions of natural philosophy and the moral science; if we consider again that, even in the things which cannot be submitted to calculus, it gives the surest hints which can guide us in our judgments, and that it teaches us to avoid the illusions which ofttimes confuse us, then we shall see that there is no science more worthy of our meditations, and that no more useful one could be incorporated in the system of public instruction.

Jean Baptiste Joseph Fourier

(1768–1830)

HIS LIFE AND WORK

Introduction of a new tool into a field of intellectual endeavor often sparks a spate of remarkable results in that field. Jean Francois Champollion's decipherment of the Rosetta stone in 1822 provided such a spark to the field of Egyptology. Researchers were suddenly able to translate the hieroglyphics found in Egyptian antiquities. In that same year, Jean Baptiste Joseph Fourier published his groundbreaking work *The Analytical Theory of Heat (Théorie analytique de la chaleur)*, which paved the way for significant breakthroughs in pure mathematics and mathematical physics during the remainder of the nineteenth century.

It is no coincidence that these new tools appeared at the same time. Champollion was a protégé of Fourier, who was the leader of the French scientific expedition that discovered the Rosetta stone in 1799. During his lifetime, Fourier was better known as a serious Egyptologist and able government official than as a mathematical physicist. Today, we remember him by the mathematical tools that bear his name.

The vernal equinox of 1768 heralded the arrival of another spring season in France. It also marked the birth of Jean Baptiste Joseph Fourier in the town of Auxerre, about one hundred miles southeast of Paris. Fourier was the ninth of twelve surviving children born to the tailor Joseph Fourier and his second wife, the former Edmie Germaine Lebegue. Monsieur Fourier senior had also sired three children by his first wife.

Little is known of Fourier's early childhood beyond the fact that his father and mother died in quick succession before his ninth birthday. A Madame Moitton rescued the orphaned Fourier from total misfortune by paying his tuition at the École Royale Militaire, the local military school run by the Benedictine order. It was here that Fourier first discovered the joys of mathematics, devoting all of his spare time to his newfound joy, as he prepared for a career as a military officer. After graduating from the local military school in 1787, Fourier hoped to continue his education at the Benedictine College de Montaigu in Paris, but for some reason, his application was rejected.

Finding his way to a military career blocked, Fourier pursued another calling common to bright young Frenchmen without other prospects: the priesthood. He became a novice at the Benedictine abbey of St Benoit-sur-Loire in the town of Fleury, near the city of Orleans. In short time, the head abbot recognized Fourier's intellectual brilliance and put him in charge of the abbey's teaching program, a heavy load for someone also studying for the priesthood. Soon, however, Fourier began to doubt whether he truly had a vocation to the priesthood. He yearned to work more intensely with mathematics and physics, as is evidenced by this note to a friend: "Yesterday was my 21st birthday, at that age Newton and Pascal had already acquired many claims to immortality."

Fourier's first significant mathematical research was conducted at the abbey. In 1789, he submitted a paper on a problem in the theory of equations to the Academy of Sciences in Paris. The noted mathematicians Adrien-Marie Legendre and Gaspard Monge recommended the paper for publication in the fall. However, the paper's publication was thwarted when, on July 14, 1789, a mob stormed the Bastille and the French Revolution began.

The cancelled publication was the least of the impacts the French Revolution would have on Fourier. Shortly after the storming of the Bastille, Fourier left the abbey. He returned to his hometown where he began teaching philosophy, history, rhetoric, and mathematics at the École Royale Militaire.

The revolutionary milieu worked its effects on the young Fourier who, in his later years, wrote:

As the natural ideas of equality developed it was possible to conceive the sublime hope of establishing among us a free government, exempt from kings and priests, and to free from this double yoke the long-usurped soil of Europe. I readily became enamored of this cause, in my opinion the greatest and most beautiful which any nation has ever undertaken.

Fourier matched his actions with his passions and joined Auxerre's municipal Committee on Surveillance, which was charged with enforcing government decrees. More ominously, Fourier also joined the local Popular Society (*Société Populaire*), which had aligned itself with the revolutionary Jacobin party. The leading Jacobins, most notably Robespierre, believed that an energetic, centralized state was essential to consolidate the advances achieved by the Revolution. By the middle of 1793, the Jacobins purged members of the more tolerant Girondist party from the National Convention.

Fourier became known as a prominent speaker at the meetings of Auxerre's Popular Society. At one meeting, he spoke so persuasively on the need for men to serve in the French army that the area's quota was filled exclusively by volunteers. He also strove to protect Auxerre's citizens from the Reign of Terror that Robespierre and his Committee of Public Safety inflicted on France. In 1794, Fourier's stinging criticism of corruption among his fellow local officials led the Committee to issue a decree for his arrest and execution by the guillotine.

Fourier went to Paris to plead his case, but it was to no avail. When he returned home, the local Surveillance Committee issued the order for his arrest and execution but the public outcry in his support was so great that order was repealed. Eight days later, however, Fourier was arrested and imprisoned after the order was reinstated under pressure from the Committee of Public Safety in Paris.

Once again, Auxerre's citizenry rallied to Fourier's support. They sent a delegation to Paris to meet with Claude-Louis St Just, Robespierre's colleague on the Committee of Public Safety. After much effort, Fourier's supporters convinced St Just to order his release. Soon, however, none of this would matter. The Reign of Terror suddenly ended when Robespierre, St Just, and other leading Jacobins were arrested and sent to the guillotine. A general amnesty followed and Fourier was released.

The Jacobin-controlled National Convention left a legacy in the area of higher education that had a profoundly positive impact on Fourier's career. In 1794, it decreed the establishment of the École Normale in Paris to teach university-level courses appropriate to the new revolutionary environment. The Convention intended the École Normale to be different from the specialized institutions of higher learning, such as the School of Mines, which the monarchy had established. In sharp contrast, the École Normale offered a wide range of courses taught by the best scholars in France. Its mathematics faculty included the renowned mathematicians Joseph Louis LaGrange, Pierre Simon Laplace, and Monge. Fourier's home district quickly chose him to be a student at the École Normale.

The École Normale had an even shorter life than did many of the French Revolution's creations. It opened its doors in January 1795 and closed them that May. Fortunately for Fourier, this was long enough for him to make a very significant impression on Monge.

In 1794, the Convention established the École Polytechnique to provide training in engineering and applied sciences for students who would be directed toward the military. When the École Normale closed its doors, many science and engineering students transferred to the École Polytechnique. Although Fourier wished to do so as well, the École Polytechnique had an upper age limit of twenty for entering students, and he was twenty-five. Thanks to Monge's assistance, Fourier overcame this hurdle.

At first, Monge tried to secure a teaching position for Fourier, but the school had no budget for additional teaching staff. Monge then secured Fourier an appointment as *administrateur de police*, which made him responsible for maintaining discipline in the lecture hall. Fourier's diary reveals that he assisted Monge in a course on the use of science and mathematics in military contexts.

Fourier's earlier Jacobin connections haunted him one final time while he was at the École Polytechnique. The government that replaced Robespierre was, in many respects, no better than its predecessor. It suspected conspiracies around every corner. It arrested Fourier because of his Jacobin connections and connection with Robespierre even though Robespierre himself had ordered Fourier's arrest. By August, the government agreed to release Fourier after hearing pleas from LaGrange, Laplace, Monge, and Fourier's students.

In 1797, Fourier succeeded Laplace as professor of analysis and mechanics even though he had completed hardly any original research in science or mathematics. This would soon change. He published his first mathematical result in 1798, a new proof of Descartes' rule of signs. In the polynomial:

$$f(x) = x^m + a_1 x^{m-1} + \ldots + a_m$$

two successive coefficients are called combinations. If they have the same sign, the combination is called a permanence. If they have different signs, the combination is called a variation. Descartes' rule says that the number of positive roots does not exceed the number of variations and the number of negative roots does not exceed the number of permanences. Fourier gave a vastly simplified proof.

Fourier was flush with the success of his first publication when his research and teaching career soon received an unanticipated interruption. In early 1798, the French government named Napoleon Bonaparte to head a military expedition into Egypt. Napoleon asked Monge and the chemist Claude-Louis Berthollet to choose the scientific members of the expedition. In May, Monge chose his pupil and colleague Fourier. Three months later, Fourier was named perpetual secretary of the Institut d'Egypte.

Napoleon took the expedition to Egypt in late July 1798 and then left in August 1799 to return to Paris to seize power as First Consul. A few months before he left, Napoleon named Fourier leader of one of two scientific expeditions that spent the autumn of 1799 investigating the monuments and inscriptions in Upper Egypt. Fourier's expedition team discovered the Rosetta stone in July 1799.

Fourier remained in Egypt with the expedition for two more years until the Anglo-French peace treaty of 1801 forced the French to withdraw (and relinquish the Rosetta stone to the English). With the Egyptian expedition drawing to an end, Fourier hoped to return to his teaching post at the École Polytechnique. Napoleon had other plans for him.

In the few months they had spent together on the expedition, Napoleon had become impressed with Fourier's diplomatic and military talents. By 1801, Napoleon had assumed absolute power in France and needed no one else's approval to appoint Fourier to the post of prefect of the Isere department, which was based in the city of Grenoble, near the Italian border. The post carried a heavy load. The prefecture had to carry out the orders of the central government while at the same time being responsive to the needs of its own citizens. Fourier carried out his appointment with intensity and integrity. While many prefects used their position to place cronies from Paris or their home district into positions of local power, Fourier did just the opposite. He made sure to appoint deserving local citizens to important posts, among them Jean Francois Champollion, who would later decipher the Rosetta stone.

When Napoleon made a grand visit to Grenoble in February 1804, just three months before crowning himself Emperor, he made Fourier a *chevalier* of the newly formed Legion of Honor. Five years later, Napoleon elevated Fourier to the title of Baron.

While diligently acting as prefect, Fourier yearned to be free of governmental responsibilities so that he could pursue his scientific interests. In 1807, Monge and Berthollet tried to persuade Napoleon to release Fourier from his duties but Napoleon would hear nothing of the kind. Fourier toiled on with his governmental duties even as he began to suffer bad attacks of rheumatism in Grenoble's chilling winds.

He briefly considered going into exile as a way to rid himself of his onerous governmental duties and find a healthier climate.

Somehow, Fourier found enough time in 1807 to work out his theory of heat diffusion. At the very end of the year, he presented an essay entitled *On the Propagation of Heat in Solid Bodies* to the Institut de France. This essay was a precursor to *The Analytical Theory of Heat* that would appear fifteen years later. The Institut appointed the mathematicians Sylvestre Francois Lacroix, Monge, Laplace, and LaGrange to a committee to determine whether the work was worthy of publication. The first three members approved publication. LaGrange withheld his approval because he objected to Fourier's expansion of functions by the trigonometrical series now known as the *Fourier series*, the very piece of mathematics reproduced here, for which Fourier is best remembered.

Finding few rewards to his scientific researches, Fourier turned his attention to the multivolume *Description of Egypt*, for which he was the general editor. It ultimately ran to twenty volumes and took twenty years (1808–1827) to complete. On New Year's Day 1810, the Institut de France announced that it would offer a prize for the best work on the mathematical theory of heat and its confirmation. In addition, the Institut pledged to publish the work in its house journal or independently, if the reviewers found the winning entry deserving.

On September 23, 1811, a week before the close of entries, the Institut received an entry from a Monsieur Antoine Cardon-Michaels of Borgues, who submitted a 21-page essay describing heat as "a symbol of man's return to the fire." Fourier's 215-page manuscript arrived five days later. It was a revision of his 1807 memoir.

LaGrange, LaPlace, and Monge served on the review committee along with Etienne Louis Malus and Rene Just Hauy. Once again, LaGrange objected to Fourier's work, but he had to agree that Fourier's work was the best entry submitted. LaGrange was able to block independent publication of Fourier's essay and force its publication to a regular issue of the Institut's house journal. As it turned out, no issues of the house journal, the *Memoires presentes par divers savants*, were published between 1811 and 1827.

Even though LaGrange, and his opposition to Fourier's mathematics, died in April 1813, Fourier realized that the only way to guarantee publication of his work would be to revise it into book form and publish it on his own. During his period of revising, Napoleon abdicated in April 1814. Although Fourier's allegiance to Napoleon put him at risk of prosecution by the newly restored monarchy, the new government recognized that he had always acted with integrity while prefect of Isere. Because of this, he was asked him to continue in that office.

Napoleon's return from exile in Elba a year later severely tested Fourier's allegiance. He prepared the city's defenses against Napoleon. Fourier fled Grenoble just before Napoleon forced open its gates on March 7, 1815. The next day, Napoleon issued a decree removing Fourier from the prefecture — about eight years too late for Fourier's liking! However, the two old comrades met the very next day outside the gates of Grenoble and Napoleon immediately named Fourier prefect of the nearby Rhone department, based in its capital the city of Lyon. Fourier had little choice but to accept the appointment, otherwise Napoleon might have imprisoned him. Fourier soon resigned when he received an order from Paris that would have forced him to remove all officials who had expressed Royalist sympathies.

Fourier doubted Napoleon's prospects of remaining in power. Sure enough, on June 18, the combined English and Prussian armies defeated Napoleon at the Battle of Waterloo. Four days later, Napoleon abdicated and was exiled to St Helena in the South Atlantic. On July 8, King Louis XVIII reentered Paris and the Bourbon monarchy was reestablished.

Fourier's flip-flop between the monarchy and Napoleon put him in a bad light with the new government. He was finally free of his governmental duties but had very few prospects for an income. He applied to the new government for a pension. At first, the government was inclined to grant the pension but it ended up refusing to do so.

Once again, a former student and colleague came to Fourier's aid. The Comte de Chabrol de Volvic, a former student and a former colleague on the Egyptian expedition, had become the new government's prefect of the Seine department, which included the capital city of Paris. He offered Fourier the directorship of the Bureau of Statistics for the department, a position that could provide him with a reasonable income. Fourier remained in this position for the rest of his life.

Initially, the new king refused to approve Fourier's election to the scientific section of the Insitut of France. (The King had Monge expelled because of his ties to Napoleon.) In 1817, however, the King relented and Fourier was elected to the Acadmie des Sciences. In 1822, Fourier became one of its perpetual secretaries.

The year 1822 also marked the publication of the crowning achievement of Fourier's career: *The Analytical Theory of Heat*. Fourier's main concern in this work is the diffusion of heat in solid bodies. Our main interest is Chapter III presented in this volume. In this chapter, Fourier demonstrates that an arbitrary function $f(x)$ can be expressed in the interval $[-\pi, +\pi]$ as the infinite trigonometric series:

$$\frac{1}{2}a_0 + \sum_{n=1}^{\infty} a_n \cos(nx) + \sum_{n=1}^{\infty} b_n \sin(nx).$$

The great Swiss mathematician Daniel Bernoulli first suggested such an infinite series representation in the late 1740s as he was developing a mathematical analysis of vibrating musical strings. However, Bernoulli was unable to determine the values of the coefficients of the series. Determining their values was Fourier's great accomplishment. As early as his 1807 essay, Fourier had demonstrated the following:

$$a_0 = \frac{1}{\pi} \int_{-\pi}^{\pi} f(x)\, dx,$$

$$a_n = \frac{1}{\pi} \int_{-\pi}^{\pi} f(x) \cos(nx)\, dx,$$

$$b_n = \frac{1}{\pi} \int_{-\pi}^{\pi} f(x) \sin(nx)\, dx.$$

Briefly, Fourier determines these values as follows. He begins by supposing

$$f(x) = \frac{1}{2}a_0 + \sum_{n=1}^{\infty} a_n \cos(nx) + \sum_{n=1}^{\infty} b_n \sin(nx),$$

and proceeds as follows:

- To determine the value of the coefficient a_0, Fourier integrates both sides of the equation over the interval $[-\pi, +\pi]$.
- To determine the values of the coefficients a_n, Fourier multiplies both sides of the equation by $\cos(mx)$ and then integrates over the interval $[-\pi, +\pi]$.
- To determine the values of the coefficients b_n Fourier multiplies both sides of the equation by $\sin(mx)$ and then integrates over the interval $[-\pi, +\pi]$.

Fourier made two key assumptions. First, he assumed the integrability of the function $f(x)$. Second, he assumed that the integral of the infinite sums; for example,

$$\int_{-\pi}^{\pi} \left[\sum_{n=1}^{\infty} a_n \cos(nx) + \sum_{n=1}^{\infty} b_n \sin(nx) + \frac{1}{2}a_0 \right] dx$$

was identical to the infinite sum of the integrals; for example,

$$\sum_{n=1}^{\infty} \int_{-\pi}^{\pi} [a_n \cos(nx) + b_n \sin(nx)]dx + \frac{1}{2}a_0 \int_{-\pi}^{\pi} dx.$$

(This is an associative property of definite integrals over infinite sums.)
During the rest of the nineteenth-century, mathematicians spent a good deal of effort determining the range of functions for which these assumptions are valid.

Fourier was in his mid-fifties when he finally published *The Analytical Theory of Heat*. He was beginning to suffer from ill health, especially chronic rheumatism. He might also have been suffering from malaria, which he might have caught during his time in Egypt. He never married, although his acquaintances said he was extremely fond of the company of intelligent women, among them Sophie Germain, one of the first prominent female mathematicians.

On May 4, 1830, Fourier was stricken with a severe attack of rheumatism. He died on May 16, 1830, of nervous angina and heart problems, according to his doctors.

Fourier's accomplishment in physics revolved around his solution of the problem of heat diffusion in sold bodies. This problem stood outside the scope of rational and celestial mechanics that had dominated physics since the days Isaac Newton. From this perspective, Fourier's work formed a major achievement in the field of mathematical physics. In the course of this work, he developed a new mathematical technique that would not likely have arisen otherwise. It paved the way for many advances in both mathematical physics and pure mathematics during the remainder of the nineteenth century and remains a primary analytical tool in both areas.

The great British mathematical physicist Lord Kelvin would later write of him:

> *Fourier's Theorem is not only one of the most beautiful results of modern analysis, but it is said to furnish an indispensable instrument in the treatment of nearly every recondite question in modern physics. . . . Fourier is a mathematical poem.*

SELECTION FROM *THE ANALYTICAL THEORY OF HEAT*

CHAPTER III

PROPAGATION OF HEAT IN AN INFINITE RECTANGULAR SOLID

SECTION I

Statement of the Problem

163.   PROBLEMS relative to the uniform propagation, or to the varied movement of heat in the interior of solids, are reduced, by the foregoing methods, to problems of pure analysis, and the progress of this part of physics will depend in consequence upon the advance which may be made in the art of analysis. The differential equations which we have proved contain the chief results of the theory; they express, in the most general and most concise manner, the necessary relations of numerical analysis to a very extensive class of phenomena; and they connect for ever with mathematical science one of the most important branches of natural philosophy.

It remains now to discover the proper treatment of these equations in order to derive their complete solutions and an easy application of them. The following problem offers the first example of analysis which leads to such solutions; it appeared to us better adapted than any other to indicate the elements of the method which we have followed.

164.   Suppose a homogeneous solid mass to be contained between two planes B and C vertical, parallel, and infinite, and to be divided into two parts by a plane A perpendicular to the other two (fig. 7); we proceed to consider the temperatures of the mass BAC bounded by the three infinite planes A, B, C. The other part $B'AC'$ of the infinite solid is supposed to be a constant source of heat, that is to say, all its points are maintained at the temperature 1, which cannot alter. The two lateral solids bounded, one by the plane C and the plane A produced, the other by the plane B and the plane A produced, have at all points the constant temperature 0, some external cause maintaining them always at that temperature; lastly, the molecules of the solid bounded by A, B and C have the initial temperature 0. Heat will pass continually from the source A into the solid BAC, and will be propagated there in the longitudinal direction, which is infinite, and at the same time will turn towards the cool masses B and C, which will

500

Fig. 7.

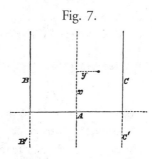

absorb great part of it. The temperatures of the solid *BAC* will be raised gradually: but will not be able to surpass nor even to attain a maximum of temperature, which is different for different points of the mass. It is required to determine the final and constant state to which the variable state continually approaches.

If this final state were known, and were then formed, it would subsist of itself, and this is the property which distinguishes it from all other states. Thus the actual problem consists in determining the permanent temperatures of an infinite rectangular solid, bounded by two masses of ice *B* and *C*, and a mass of boiling water *A*; the consideration of such simple and primary problems is one of the surest modes of discovering the laws of natural phenomena, and we see, by the history of the sciences, that every theory has been formed in this manner.

165.　To express more briefly the same problem, suppose a rectangular plate *BAC*, of infinite length, to be heated at its base *A*, and to preserve at all points of the base a constant temperature 1, whilst each of the two infinite sides *B* and *C*, perpendicular to the base *A*, is submitted also at every point to a constant temperature 0; it is required to determine what must be the stationary temperature at any point of the plate.

It is supposed that there is no loss of heat at the surface of the plate, or, which is the same thing, we consider a solid formed by superposing an infinite number of plates similar to the preceding: the straight line *Ax* which divides the plate into two equal parts is taken as the axis of *x*, and the co-ordinates of any point *m* are *x* and *y*; lastly, the width *A* of the plate is represented by $2l$, or, to abridge the calculation, by π, the value of the ratio of the diameter to the circumference of a circle.

Imagine a point *m* of the solid plate *BAC*, whose co-ordinates are *x* and *y*, to have the actual temperature *v*, and that the quantities *v*, which correspond to different points, are such that no change can happen in the temperatures, provided that the temperature of every point of the base *A* is always 1, and that the sides *B* and *C* retain at all their points the temperature 0.

If at each point *m* a vertical co-ordinate be raised, equal to the temperature *v*, a curved surface would be formed which would extend above the plate and be prolonged to infinity. We shall endeavour to find the nature of this surface, which passes through a line drawn above the axis of y at a distance equal to unity, and which cuts the horizontal plane of xy along two infinite straight lines parallel to x.

166.   To apply the general equation

$$\frac{dv}{dt} = \frac{K}{CD}\left(\frac{d^2v}{dx^2} + \frac{d^2v}{dy^2} + \frac{d^2v}{dz^2}\right),$$

we must consider that, in the case in question, abstraction is made of the co-ordinate z, so that the term $\frac{d^2v}{dz^2}$ must be omitted; with respect to the first member $\frac{dv}{dt}$, it vanishes, since we wish to determine the stationary temperatures; thus the equation which belongs to the actual problem, and determines the properties of the required curved surface, is the following:

$$\frac{d^2v}{dx^2} + \frac{d^2v}{dy^2} = 0 \dots\dots\dots\dots\dots\dots\dots\dots(a).$$

The function of x and y, $\phi(x, y)$, which represents the permanent state of the solid BAC, must, 1st, satisfy the equation (a); 2nd, become nothing when we substitute $-\frac{1}{2}\pi$ or $+\frac{1}{2}\pi$ for y, whatever the value of x may be; 3rd, must be equal to unity when we suppose $x = 0$ and y to have any value included between $-\frac{1}{2}\pi$ and $+\frac{1}{2}\pi$.

Further, this function $\phi(x, y)$ ought to become extremely small when we give to x a very large value, since all the heat proceeds from the source A.

167.   In order to consider the problem in its elements, we shall in the first place seek for the simplest functions of x and y, which satisfy equation (a); we shall then generalise the value of v in order to satisfy all the stated conditions. By this method the solution will receive all possible extension, and we shall prove that the problem proposed admits of no other solution.

Functions of two variables often reduce to less complex expressions, when we attribute to one of the variables or to both of them infinite values; this is what may be remarked in algebraic functions which, in this particular case, take the form of the product of a function of x by a function of y.

We shall examine first if the value of v can be represented by such a product; for the function v must represent the state of the plate throughout its whole extent, and consequently that of the points whose co-ordinate x is infinite. We shall then write $v = F(x)f(y)$; substituting in equation (a) and denoting $\frac{d^2F(x)}{dx^2}$ by $F''(x)$ and $\frac{d^2f(y)}{dy^2}$ by $f''(y)$, we shall have

$$\frac{F''(x)}{F(x)} + \frac{f''(y)}{f(y)} = 0;$$

we then suppose $\frac{F''(x)}{F(x)} = m$ and $\frac{f''(y)}{f(y)} = -m$, m being any constant quantity, and as it is proposed only to find a particular value of v, we deduce from the preceding equations $F(x) = e^{-mx}$, $f(y) = \cos my$.

168.   We could not suppose m to be a negative number, and we must necessarily exclude all particular values of v, into which terms such as e^{mx} might enter, m
being a positive number, since the temperature v cannot become infinite when x is
infinitely great. In fact, no heat being supplied except from the constant source A,
only an extremely small portion can arrive at those parts of space which are very far
removed from the source. The remainder is diverted more and more towards the infinite edges B and C, and is lost in the cold masses which bound them.

The exponent m which enters into the function $e^{-mx} \cos my$ is unknown, and
we may choose for this exponent any positive number: but, in order that v may
become nul on making $y = -\frac{1}{2}\pi$ or $y = +\frac{1}{2}\pi$, whatever x may be, m must be taken
to be one of the terms of the series, 1, 3, 5, 7, &c.; by this means the second condition will be fulfilled.

169.   A more general value of v is easily formed by adding together several terms
similar to the preceding, and we have

$$v = ae^{-x} \cos y + be^{-3x} \cos 3y + ce^{-5x} \cos 5y + de^{-7x} \cos 7y + \&c. \quad ....(b).$$

It is evident that the function v denoted by $\phi(x, y)$ satisfies the equation
$\frac{d^2v}{dx^2} + \frac{d^2v}{dy^2} = 0$, and the condition $\phi(x, \pm\frac{1}{2}\pi) = 0$. A third condition remains to be
fulfilled, which is expressed thus, $\phi(0, y) = 1$, and it is essential to remark that this
result must exist when we give to y any value whatever included between $-\frac{1}{2}\pi$ and
$+\frac{1}{2}\pi$. Nothing can be inferred as to the values which the function $\phi(0, y)$ would
take, if we substituted in place of y a quantity not included between the limits $-\frac{1}{2}\pi$
and $+\frac{1}{2}\pi$. Equation (b) must therefore be subject to the following condition:

$$1 = a \cos y + b \cos 3y + c \cos 5y + d \cos 7y + \&c.$$

The coefficients, a, b, c, d, &c., whose number is infinite, are determined by
means of this equation.

The second member is a function of y, which is equal to 1 so long as the vari-
able y is included between the limits $-\frac{1}{2}\pi$ and $+\frac{1}{2}\pi$. It may be doubted whether
such a function exists, but this difficulty will be fully cleared up by the sequel.

170.   Before giving the calculation of the coefficients, we may notice the effect
represented by each one of the terms of the series in equation (b).

Suppose the fixed temperature of the base A, instead of being equal to unity at
every point, to diminish as the point of the line A becomes more remote from the
middle point, being proportional to the cosine of that distance; in this case it will
easily be seen what is the nature of the curved surface, whose vertical ordinate
expresses the temperature v or $\phi(x, y)$. If this surface be cut at the origin by a plane
perpendicular to the axis of x, the curve which bounds the section will have for its
equation $v = a \cos y$; the values of the coefficients will be the following:

$$a = \alpha, b = 0, c = 0, d = 0,$$

and so on, and the equation of the curved surface will be

$$v = ae^{-x}\cos y.$$

If this surface be cut at right angles to the axis of y, the section will be a logarithmic spiral whose convexity is turned towards the axis; if it be cut at right angles to the axis of x, the section will be a trigonometric curve whose concavity is turned towards the axis.

It follows from this that the function $\frac{d^2v}{dx^2}$ is always positive, and $\frac{d^2v}{dy^2}$ is always negative. Now the quantity of heat which a molecule acquires in consequence of its position between two others in the direction of x is proportional to the value of $\frac{d^2v}{dx^2}$ (Art. 123): it follows then that the intermediate molecule receives from that which precedes it, in the direction of x, more heat than it communicates to that which follows it. But, if the same molecule be considered as situated between two others in the direction of y, the function $\frac{d^2v}{dy^2}$ being negative, it appears that the intermediate molecule communicates to that which follows it more heat than it receives from that which precedes it. Thus it follows that the excess of the heat which it acquires in the direction of x, is exactly compensated by that which it loses in the direction of y, as the equation $\frac{d^2v}{dx^2} + \frac{d^2v}{dy^2} = 0$ denotes. Thus then the route followed by the heat which escapes from the source A becomes known. It is propagated in the direction of x, and at the same time it is decomposed into two parts, one of which is directed towards one of the edges, whilst the other part continues to separate from the origin, to be decomposed like the preceding, and so on to infinity. The surface which we are considering is generated by the trigonometric curve which corresponds to the base A, moved with its plane at right angles to the axis of x along that axis, each one of its ordinates decreasing indefinitely in proportion to successive powers of the same fraction.

Analogous inferences might be drawn, if the fixed temperatures of the base A were expressed by the term

$$b \cos 3y \quad \text{or} \quad c \cos 5y, \&\text{c.};$$

and in this manner an exact idea might be formed of the movement of heat in the most general case; for it will be seen by the sequel that the movement is always compounded of a multitude of elementary movements, each of which is accomplished as if it alone existed.

SECTION II

First Example of the Use of Trigonometric Series in the Theory of Heat

171. Take now the equation

$$1 = a \cos y + b \cos 3y + c \cos 5y + d \cos 7y + \&\text{c.},$$

in which the coefficients a, b, c, d, &c. are to be determined. In order that this equation may exist, the constants must necessarily satisfy the equations which are obtained by successive differentiations; whence the following results,

$$1 = a \cos y + \quad b \cos 3y + \quad c \cos 5y + \quad d \cos 7y + \text{&c.},$$
$$0 = a \sin y + \quad 3b \sin 3y + \quad 5c \sin 5y + \quad 7d \sin 7y + \text{&c.},$$
$$0 = a \cos y + \quad 3^2 b \cos 3y + \quad 5^2 c \cos 5y + \quad 7^2 d \cos 7y + \text{&c.},$$
$$0 = a \sin y + \quad 3^3 b \sin 3y + \quad 5^3 c \sin 5y + \quad 7^3 d \sin 7y + \text{&c.},$$

and so on to infinity.

These equations necessarily hold when $y = 0$, thus we have

$$1 = a + \quad b + \quad c + \quad d + \quad e + \quad f + \quad g + \ldots \text{&c.},$$
$$0 = a + 3^2 b + 5^2 c + 7^2 d + 9^2 e + 11^2 f + \ldots \text{&c.},$$
$$0 = a + 3^4 b + 5^4 c + 7^4 d + 9^4 e + \ldots \text{&c.},$$
$$0 = a + 3^6 b + 5^6 c + 7^6 d + \ldots \text{&c.},$$
$$0 = a + 3^8 b + 5^8 c + \ldots \text{&c.},$$
$$\text{&c.}$$

The number of these equations is infinite like that of the unknowns a, b, c, d, e, ... &c. The problem consists in eliminating all the unknowns, except one only.

172. In order to form a distinct idea of the result of these eliminations, the number of the unknowns a, b, c, d, ... &c., will be supposed at first definite and equal to m. We shall employ the first m equations only, suppressing all the terms containing the unknowns which follow the m first. If in succession m be made equal to 2, 3, 4, 5, and so on, the values of the unknowns will be found on each one of these hypotheses. The quantity a, for example, will receive one value for the case of two unknowns, others for the cases of three, four, or successively a greater number of unknowns. It will be the same with the unknown b, which will receive as many different values as there have been cases of elimination; each one of the other unknowns is in like manner susceptible of an infinity of different values. Now the value of one of the unknowns, for the case in which their number is infinite, is the limit towards which the values which it receives by means of the successive eliminations tend. It is required then to examine whether, according as the number of unknowns increases, the value of each one of a, b, c, d ... &c. does not converge to a finite limit which it continually approaches.

Suppose the six following equations to be employed:

$$1 = a + \quad b + \quad c + \quad d + \quad e + \quad f + \text{&c.},$$
$$0 = a + 3^2 b + \quad 5^2 c + \quad 7^2 d + \quad 9^2 e + \quad 11^2 f + \text{&c.},$$
$$0 = a + 3^4 b + \quad 5^4 c + \quad 7^4 d + \quad 9^4 e + \quad 11^4 f + \text{&c.},$$
$$0 = a + 3^6 b + \quad 5^6 c + \quad 7^6 d + \quad 9^6 e + \quad 11^6 f + \text{&c.},$$
$$0 = a + 3^8 b + \quad 5^8 c + \quad 7^8 d + \quad 9^8 e + \quad 11^8 f + \text{&c.},$$
$$0 = a + 3^{10} b + 5^{10} c + 7^{10} d + 9^{10} e + 11^{10} f + \text{&c.}$$

The five equations which do not contain f are:

$$11^2 = a(11^2 - 1^2) + b(11^2 - 3^2) + c(11^2 - 5^2) + d(11^2 - 7^2) + e(11^2 - 9^2),$$

$$0 = a(11^2 - 1^2) + 3^2 b(11^2 - 3^2) + 5^2 c(11^2 - 5^2) + 7^2 d(11^2 - 7^2) + 9^2 e(11^2 - 9^2),$$

$$0 = a(11^2 - 1^2) + 3^4 b(11^2 - 3^2) + 5^4 c(11^2 - 5^2) + 7^4 d(11^2 - 7^2) + 9^4 e(11^2 - 9^2),$$

$$0 = a(11^2 - 1^2) + 3^6 b(11^2 - 3^2) + 5^6 c(11^2 - 5^2) + 7^6 d(11^2 - 7^2) + 9^6 e(11^2 - 9^2),$$

$$0 = a(11^2 - 1^2) + 3^8 b(11^2 - 3^2) + 5^8 c(11^2 - 5^2) + 7^8 d(11^2 - 7^2) + 9^8 e(11^2 - 9^2).$$

Continuing the elimination we shall obtain the final equation in a, which is:

$$a(11^2 - 1^2)(9^2 - 1^2)(7^2 - 1^2)(5^2 - 1^2)(3^2 - 1^2) = 11^2 \cdot 9^2 \cdot 7^2 \cdot 5^2 \cdot 3^2 \cdot 1^2.$$

173. If we had employed a number of equations greater by unity, we should have found, to determine a, an equation analogous to the preceding, having in the first member one factor more, namely, $13^2 - 1^2$, and in the second member 13^2 for the new factor. The law to which these different values of a are subject is evident, and it follows that the value of a which corresponds to an infinite number of equations is expressed thus:

$$a = \frac{3^2}{3^2 - 1^2} \cdot \frac{5^2}{5^2 - 1^2} \cdot \frac{7^2}{7^2 - 1^2} \cdot \frac{9^2}{9^2 - 1^2} \cdot \frac{11^2}{11^2 - 1^2} \cdot \&c.,$$

or

$$a = \frac{3 \cdot 3}{2 \cdot 4} \cdot \frac{5 \cdot 5}{4 \cdot 6} \cdot \frac{7 \cdot 7}{6 \cdot 8} \cdot \frac{9 \cdot 9}{8 \cdot 10} \cdot \frac{11 \cdot 11}{10 \cdot 12} \cdot \&c.$$

Now the last expression is known and, in accordance with Wallis' Theorem, we conclude that $a = \frac{4}{\pi}$. It is required then only to ascertain the values of the other unknowns.

174. The five equations which remain after the elimination of f may be compared with the five simpler equations which would have been employed if there had been only five unknowns. The last equations differ from the equations of Art. 172, in that in them e, d, c, b, a are found to be multiplied respectively by the factors

$$\frac{11^2 - 9^2}{11^2}, \quad \frac{11^2 - 7^2}{11^2}, \quad \frac{11^2 - 5^2}{11^2}, \quad \frac{11^2 - 3^2}{11^2}, \quad \frac{11^2 - 1^2}{11^2}.$$

It follows from this that if we had solved the five linear equations which must have been employed in the case of five unknowns, and had calculated the value of each unknown, it would have been easy to derive from them the value of the unknowns of the same name corresponding to the case in which six equations should have been employed. It would suffice to multiply the values of e, d, c, b, a, found in the first case, by the known factors. It will be easy in general to pass from the value of one of these quantities, taken on the supposition of a certain number of equations and unknowns, to the value of the same quantity, taken in the case in which there should have been one unknown and one equation more. For example, if the value of e, found on the hypothesis of five equations and five unknowns, is represented by E, that of the same quantity, taken in the case of one unknown more, will be $E\frac{11^2}{11^2 - 9^2}$. The same value, taken in the case of seven unknowns, will be, for the same reason,

$$E\frac{11^2}{11^2 - 9^2} \cdot \frac{13^2}{13^2 - 9^2},$$

and in the case of eight unknowns it will be

$$E\frac{11^2}{11^2 - 9^2} \cdot \frac{13^2}{13^2 - 9^2} \cdot \frac{15^2}{15^2 - 9^2},$$

and so on. In the same manner it will suffice to know the value of b, corresponding to the case of two unknowns, to derive from it that of the same letter which corresponds to the cases of three, four, five unknowns, &c. We shall only have to multiply this first value of b by

$$\frac{5^2}{5^2 - 3^2} \cdot \frac{7^2}{7^2 - 3^2} \cdot \frac{9^2}{9^2 - 3^2} \cdots \&c.$$

Similarly if we knew the value of c for the case of three unknowns, we should multiply this value by the successive factors

$$\frac{7^2}{7^2 - 5^2} \cdot \frac{9^2}{9^2 - 5^2} \cdot \frac{11^2}{11^2 - 5^2} \cdots \&c.$$

We should calculate the value of d for the case of four unknowns only, and multiply this value by

$$\frac{9^2}{9^2 - 7^2} \cdot \frac{11^2}{11^2 - 7^2} \cdot \frac{13^2}{13^2 - 7^2} \cdots \&c.$$

The calculation of the value of a is subject to the same rule, for if its value be taken for the case of one unknown, and multiplied successively by

$$\frac{3^2}{3^2 - 1^2}, \quad \frac{5^2}{5^2 - 1^2}, \quad \frac{7^2}{7^2 - 1^2}, \quad \frac{9^2}{9^2 - 1^2},$$

the final value of this quantity will be found.

175. The problem is therefore reduced to determining the value of a in the case of one unknown, the value of b in the case of two unknowns, that of c in the case of three unknowns, and so on for the other unknowns.

It is easy to conclude, by inspection only of the equations and without any calculation, that the results of these successive eliminations must be

$$a = 1,$$

$$b = \frac{1^2}{1^2 - 3^2},$$

$$c = \frac{1^2}{1^2 - 5^2} \cdot \frac{3^2}{3^2 - 5^2},$$

$$d = \frac{1^2}{1^2 - 7^2} \cdot \frac{3^2}{3^2 - 7^2} \cdot \frac{5^2}{5^2 - 7^2},$$

$$e = \frac{1^2}{1^2 - 9^2} \cdot \frac{3^2}{3^2 - 9^2} \cdot \frac{5^2}{5^2 - 9^2} \cdot \frac{7^2}{7^2 - 9^2}.$$

176. It remains only to multiply the preceding quantities by the series of products which ought to complete them, and which we have given (Art. 174). We shall have consequently, for the final values of the unknowns *a, b, c, d, e, f*, &c., the following expressions:

$$a = 1 \cdot \frac{3^2}{3^2 - 1^2} \cdot \frac{5^2}{5^2 - 1^2} \cdot \frac{7^2}{7^2 - 1^2} \cdot \frac{9^2}{9^2 - 1^2} \cdot \frac{11^2}{11^2 - 1^2} \text{ &c.,}$$

$$b = \frac{1^2}{1^2 - 3^2} \cdot \frac{5^2}{5^2 - 3^2} \cdot \frac{7^2}{7^2 - 3^2} \cdot \frac{9^2}{9^2 - 3^2} \cdot \frac{11^2}{11^2 - 3^2} \text{ &c.,}$$

$$c = \frac{1^2}{1^2 - 5^2} \cdot \frac{3^2}{3^2 - 5^2} \cdot \frac{7^2}{7^2 - 5^2} \cdot \frac{9^2}{9^2 - 5^2} \cdot \frac{11^2}{11^2 - 5^2} \text{ &c.,}$$

$$d = \frac{1^2}{1^2 - 7^2} \cdot \frac{3^2}{3^2 - 7^2} \cdot \frac{5^2}{5^2 - 7^2} \cdot \frac{9^2}{9^2 - 7^2} \cdot \frac{11^2}{11^2 - 7^2} \text{ &c.,}$$

$$e = \frac{1^2}{1^2 - 9^2} \cdot \frac{3^2}{3^2 - 9^2} \cdot \frac{5^2}{5^2 - 9^2} \cdot \frac{7^2}{7^2 - 9^2} \cdot \frac{11^2}{11^2 - 9^2} \cdot \frac{13^2}{13^2 - 9^2} \text{ &c.,}$$

$$f = \frac{1^2}{1^2 - 11^2} \cdot \frac{3^2}{3^2 - 11^2} \cdot \frac{5^2}{5^2 - 11^2} \cdot \frac{7^2}{7^2 - 11^2} \cdot \frac{9^2}{9^2 - 11^2} \cdot \frac{13^2}{13^2 - 11^2} \text{ &c.,}$$

or,

$$a = +1 \cdot \frac{3 \cdot 3}{2 \cdot 4} \cdot \frac{5 \cdot 5}{4 \cdot 6} \cdot \frac{7 \cdot 7}{6 \cdot 8} \text{ &c.,}$$

$$b = -\frac{1 \cdot 1}{2 \cdot 4} \cdot \frac{5 \cdot 5}{2 \cdot 8} \cdot \frac{7 \cdot 7}{4 \cdot 10} \cdot \frac{9 \cdot 9}{6 \cdot 12} \text{ &c.,}$$

$$c = +\frac{1 \cdot 1}{4 \cdot 6} \cdot \frac{3 \cdot 3}{2 \cdot 8} \cdot \frac{7 \cdot 7}{2 \cdot 12} \cdot \frac{9 \cdot 9}{4 \cdot 14} \cdot \frac{11 \cdot 11}{6 \cdot 16} \text{ &c.,}$$

$$d = -\frac{1 \cdot 1}{6 \cdot 8} \cdot \frac{3 \cdot 3}{4 \cdot 10} \cdot \frac{5 \cdot 5}{2 \cdot 12} \cdot \frac{9 \cdot 9}{2 \cdot 16} \cdot \frac{11 \cdot 11}{4 \cdot 18} \text{ &c.,}$$

$$e = +\frac{1 \cdot 1}{8 \cdot 10} \cdot \frac{3 \cdot 3}{6 \cdot 12} \cdot \frac{5 \cdot 5}{4 \cdot 14} \cdot \frac{7 \cdot 7}{2 \cdot 16} \cdot \frac{11 \cdot 11}{2 \cdot 20} \cdot \frac{13 \cdot 13}{4 \cdot 22} \text{ &c.,}$$

$$f = -\frac{1 \cdot 1}{10 \cdot 12} \cdot \frac{3 \cdot 3}{8 \cdot 14} \cdot \frac{5 \cdot 5}{6 \cdot 16} \cdot \frac{7 \cdot 7}{4 \cdot 18} \cdot \frac{9 \cdot 9}{2 \cdot 20} \cdot \frac{13 \cdot 13}{2 \cdot 24} \cdot \frac{15 \cdot 15}{4 \cdot 26} \text{ &c.}$$

The quantity $\frac{1}{2}\pi$ or a quarter of the circumference is equivalent, according to Wallis' Theorem, to

$$\frac{2 \cdot 2}{1 \cdot 3} \cdot \frac{4 \cdot 4}{3 \cdot 5} \cdot \frac{6 \cdot 6}{5 \cdot 7} \cdot \frac{8 \cdot 8}{7 \cdot 9} \cdot \frac{10 \cdot 10}{9 \cdot 11} \cdot \frac{12 \cdot 12}{11 \cdot 13} \cdot \frac{14 \cdot 14}{13 \cdot 15} \&c.$$

If now in the values of a, b, c, d, &c., we notice what are the factors which must be joined on to numerators and denominators to complete the double series of odd and even numbers, we find that the factors to be supplied are:

$$\left. \begin{array}{l} \text{for } b \; \dfrac{3 \cdot 3}{6}, \\[2mm] \text{for } c \; \dfrac{5 \cdot 5}{10}, \\[2mm] \text{for } d \; \dfrac{7 \cdot 7}{14}, \\[2mm] \text{for } e \; \dfrac{9 \cdot 9}{18}, \\[2mm] \text{for } f \; \dfrac{11 \cdot 11}{22}, \end{array} \right\} \text{ whence we conclude } \left\{ \begin{array}{l} a = 2 \cdot \dfrac{2}{\pi}, \\[2mm] b = -2 \cdot \dfrac{2}{3\pi}, \\[2mm] c = 2 \cdot \dfrac{2}{5\pi}, \\[2mm] d = -2 \cdot \dfrac{2}{7\pi}, \\[2mm] e = 2 \cdot \dfrac{2}{9\pi}, \\[2mm] f = -2 \cdot \dfrac{2}{11\pi}.[1] \end{array} \right.$$

177. Thus the eliminations have been completely effected, and the coefficients a, b, c, d, &c., determined in the equation

$$1 = a \cos y + b \cos 3y + c \cos 5y + d \cos 7y + e \cos 9y + \&c.$$

The substitution of these coefficients gives the following equation:

$$\frac{\pi}{4} = \cos y - \frac{1}{3} \cos 3y + \frac{1}{5} \cos 5y - \frac{1}{7} \cos 7y + \frac{1}{9} \cos 9y - \&c.[2]$$

The second member is a function of y, which does not change in value when we give to the variable y a value included between $-\frac{1}{2}\pi$ and $+\frac{1}{2}\pi$. It would be easy to prove that this series is always convergent, that is to say that writing instead of y any number whatever, and following the calculation of the coefficients, we approach more and more to a fixed value, so that the difference of this value from the sum of the calculated terms becomes less than any assignable magnitude. Without stopping for a proof, which the reader may supply, we remark that the fixed value which is continually approached is $\frac{1}{4}\pi$, if the value attributed to y is included between 0 and $\frac{1}{2}\pi$, but that it is $-\frac{1}{4}\pi$, if y is included between $\frac{1}{2}\pi$ and $\frac{3}{2}\pi$; for, in this second interval, each term of the series changes in sign. In general the limit of the series is alternately

---

[1] It is a little better to deduce the value of b in a, of c in b, &c. [R. L. E.]

[2] The coefficients a, b, c, &c., might be determined, according to the methods of Section VI., by multiplying both sides of the first equation by $\cos y$, $\cos 3y$, $\cos 5y$, &c., respectively, and integrating from $-\frac{1}{2}\pi$ to $+\frac{1}{2}\pi$, as was done by D. F. Gregory, *Cambridge Mathematical Journal*, Vol. I. [A. F.]

positive and negative; in other respects, the convergence is not sufficiently rapid to produce an easy approximation, but it suffices for the truth of the equation.

178. The equation

$$y = \cos x - \frac{1}{3}\cos 3x + \frac{1}{5}\cos 5x - \frac{1}{7}\cos 7x + \&c.$$

belongs to a line which, having x for abscissa and y for ordinate, is composed of separated straight lines, each of which is parallel to the axis, and equal to the circumference. These parallels are situated alternately above and below the axis, at the distance $\frac{1}{4}\pi$, and joined by perpendiculars which themselves make part of the line. To form an exact idea of the nature of this line, it must be supposed that the number of terms of the function

$$\cos x - \frac{1}{3}\cos 3x + \frac{1}{5}\cos 5x - \&c.$$

has first a definite value. In the latter case the equation

$$y = \cos x - \frac{1}{3}\cos 3x + \frac{1}{5}\cos 5x - \&c.$$

belongs to a curved line which passes alternately above and below the axis, cutting it every time that the abscissa x becomes equal to one of the quantities

$$0, \pm\frac{1}{2}\pi, \pm\frac{3}{2}\pi, \pm\frac{5}{2}\pi, \&c.$$

According as the number of terms of the equation increases, the curve in question tends more and more to coincidence with the preceding line, composed of parallel straight lines and of perpendicular lines; so that this line is the limit of the different curves which would be obtained by increasing successively the number of terms.

SECTION III

Remarks on These Series

179. We may look at the same equations from another point of view, and prove directly the equation

$$\frac{\pi}{4} = \cos x - \frac{1}{3}\cos 3x + \frac{1}{5}\cos 5x - \frac{1}{7}\cos 7x + \frac{1}{9}\cos 9x - \&c.$$

The case where x is nothing is verified by Leibnitz' series,

$$\frac{\pi}{4} = 1 - \frac{1}{3} + \frac{1}{5} - \frac{1}{7} + \frac{1}{9} - \&c.$$

We shall next assume that the number of terms of the series

$$\cos x - \frac{1}{3}\cos 3x + \frac{1}{5}\cos 5x - \frac{1}{7}\cos 7x + \&c.$$

instead of being infinite is finite and equal to m. We shall consider the value of the finite series to be a function of x and m. We shall express this function by a series arranged according to negative powers of m; and it will be found that the value of the function approaches more nearly to being constant and independent of x, as the number m becomes greater.

Let y be the function required, which is given by the equation $y = \cos x - \frac{1}{3} \cos 3x + \frac{1}{5} \cos 5x - \frac{1}{7} \cos 7x + \ldots - \frac{1}{2m-1} \cos (2m-1)x$, m, the number of terms, being supposed even. This equation differentiated with respect to x gives

$$-\frac{dy}{dx} = \sin x - \sin 3x + \sin 5x - \sin 7x + \ldots$$
$$+ \sin (2m-3)x - \sin (2m-1)x;$$

multiplying by $2 \sin 2x$, we have

$$-2\frac{dy}{dx} \sin 2x = 2 \sin x \sin 2x - 2 \sin 3x \sin 2x + 2 \sin 5x \sin 2x \ldots$$
$$+ 2 \sin (2m-3)x \sin 2x - 2 \sin (2m-1)x \sin 2x.$$

Each term of the second member being replaced by the difference of two cosines, we conclude that

$$-2\frac{dy}{dx} \sin 2x = \cos (-x) - \cos 3x$$
$$- \cos x + \cos 5x$$
$$+ \cos 3x - \cos 7x$$
$$- \cos 5x + \cos 9x$$
$$\cdots\cdots\cdots\cdots\cdots\cdots\cdots\cdots$$
$$+ \cos (2m-5)x - \cos (2m-1)x$$
$$- \cos (2m-3)x + \cos (2m+1)x.$$

The second member reduces to

$$\cos (2m+1)x - \cos (2m-1)x, \quad \text{or} \quad -2 \sin 2mx \sin x;$$

hence

$$y = \frac{1}{2}\int\left(dx\,\frac{\sin 2mx}{\cos x}\right).$$

180. We shall integrate the second member by parts, distinguishing in the integral between the factor $\sin 2mx\,dx$ which must be integrated successively, and the factor $\frac{1}{\cos x}$ or $\sec x$ which must be differentiated successively; denoting the results of these differentiations by $\sec' x$, $\sec'' x$, $\sec''' x$, $\ldots$ &c., we shall have

$$2y = \text{const.} - \frac{1}{2m} \cos 2mx \sec x + \frac{1}{2^2 m^2} \sin 2mx \sec' x$$
$$+ \frac{1}{2^3 m^3} \cos 2mx \sec'' x + \&c.;$$

thus the value of y or

$$\cos x - \frac{1}{3}\cos 3x + \frac{1}{5}\cos 5x - \frac{1}{7}\cos 7x + \ldots - \frac{1}{2m-1}\cos(2m-1)x,$$

which is a function of x and m, becomes expressed by an infinite series; and it is evident that the more the number m increases, the more the value of y tends to become constant. For this reason, when the number m is infinite, the function y has a definite value which is always the same, whatever be the positive value of x, less than $\frac{1}{2}\pi$. Now, if the arc x be supposed nothing, we have

$$y = 1 - \frac{1}{3} + \frac{1}{5} - \frac{1}{7} + \frac{1}{9} - \&c.,$$

which is equal to $\frac{1}{4}\pi$. Hence generally we shall have

$$\frac{1}{4}\pi = \cos x - \frac{1}{3}\cos 3x + \frac{1}{5}\cos 5x - \frac{1}{7}\cos 7x + \frac{1}{9}\cos 9x - \&c. \ldots\ldots\ldots (b).$$

181.   If in this equation we assume $x = \frac{1}{2}\frac{\pi}{2}$, we find

$$\frac{\pi}{2\sqrt{2}} = 1 + \frac{1}{3} - \frac{1}{5} - \frac{1}{7} + \frac{1}{9} + \frac{1}{11} - \frac{1}{13} - \frac{1}{15} + \ldots \&c.;$$

by giving to the arc x other particular values, we should find other series, which it is useless to set down, several of which have been already published in the works of Euler. If we multiply equation (b) by dx, and integrate it, we have

$$\frac{\pi x}{4} = \sin x - \frac{1}{3^2}\sin 3x + \frac{1}{5^2}\sin 5x - \frac{1}{7^2}\sin 7x + \&c.$$

Making in the last equation $x = \frac{1}{2}\pi$, we find

$$\frac{\pi^2}{8} = 1 + \frac{1}{3^2} + \frac{1}{5^2} + \frac{1}{7^2} + \frac{1}{9^2} + \&c.,$$

a series already known. Particular cases might be enumerated to infinity; but it agrees better with the object of this work to determine, by following the same process, the values of the different series formed of the sines or cosines of multiple arcs.

182.   Let

$$y = \sin x - \frac{1}{2}\sin 2x + \frac{1}{3}\sin 3x - \frac{1}{4}\sin 4x \ldots$$

$$+ \frac{1}{m-1}\sin(m-1)x - \frac{1}{m}\sin mx,$$

m being any even number. We derive from this equation

$$\frac{dy}{dx} = \cos x - \cos 2x + \cos 3x - \cos 4x \ldots + \cos(m-1)x - \cos mx;$$

multiplying by $2\sin x$, and replacing each term of the second member by the difference of two sines, we shall have

$$2 \sin x \frac{dy}{dx} = \sin(x + x) - \sin(x - x)$$

$$\begin{aligned} &- \sin(2x + x) + \sin(2x - x) \\ &+ \sin(3x + x) - \sin(3x - x) \end{aligned}$$

$$\cdots\cdots\cdots\cdots\cdots\cdots$$

$$\begin{aligned} &+ \sin\{(m-1)x - x\} - \sin\{(m+1)x - x\} \\ &- \sin(mx + x) + \sin(mx - x); \end{aligned}$$

and, on reduction,

$$2 \sin x \frac{dy}{dx} = \sin x + \sin mx - \sin(mx + x):$$

the quantity $\qquad \sin mx - \sin(mx + x),$

or $\qquad \sin\left(mx + \tfrac{1}{2}x - \tfrac{1}{2}x\right) - \sin\left(mx + \tfrac{1}{2}x + \tfrac{1}{2}x\right),$

is equal to $\qquad -2 \sin \tfrac{1}{2}x \cos\left(mx + \tfrac{1}{2}x\right);$

we have therefore

$$\frac{dy}{dx} = \frac{1}{2} - \frac{\sin \tfrac{1}{2}x}{\sin x} \cos\left(mx + \tfrac{1}{2}x\right),$$

or

$$\frac{dy}{dx} = \frac{1}{2} - \frac{\cos\left(mx + \tfrac{1}{2}x\right)}{2 \cos \tfrac{1}{2}x};$$

whence we conclude

$$y = \frac{1}{2}x - \int dx \, \frac{\cos\left(mx + \tfrac{1}{2}x\right)}{2 \cos \tfrac{1}{2}x}.$$

If we integrate this by parts, distinguishing between the factor $\dfrac{1}{\cos \tfrac{1}{2}x}$ or $\sec \tfrac{1}{2}x$, which must be successively differentiated, and the factor $\cos\left(mx + \tfrac{1}{2}x\right)$, which is to be integrated several times in succession, we shall form a series in which the powers of $m + \tfrac{1}{2}$ enter into the denominators. As to the constant it is nothing, since the value of y begins with that of x.

It follows from this that the value of the finite series

$$\sin x - \frac{1}{2}\sin 2x + \frac{1}{3}\sin 3x - \frac{1}{5}\sin 5x + \frac{1}{7}\sin 7x - \ldots - \frac{1}{m}\sin mx$$

differs very little from that of $\tfrac{1}{2}x$, when the number of terms is very great; and if this number is infinite, we have the known equation

$$\frac{1}{2}x = \sin x - \frac{1}{2}\sin 2x + \frac{1}{3}\sin 3x - \frac{1}{4}\sin 4x + \frac{1}{5}\sin 5x - \&c.$$

From the last series, that which has been given above for the value of $\tfrac{1}{4}\pi$ might also be derived.

183.   Let now

$$y = \frac{1}{2} \cos 2x - \frac{1}{4} \cos 4x + \frac{1}{6} \cos 6x - \ldots$$

$$+ \frac{1}{2m - 2} \cos(2m - 2)x - \frac{1}{2m} \cos 2mx.$$

Differentiating, multiplying by $2 \sin 2x$, substituting the differences of cosines, and reducing, we shall have

$$2\frac{dy}{dx} = -\tan x + \frac{\sin(2m + 1)x}{\cos x},$$

or

$$2y = c - \int dx \tan x + \int dx \frac{\sin(2m + 1)x}{\cos x};$$

integrating by parts the last term of the second member, and supposing m infinite, we have $y = c + \frac{1}{2} \log \cos x$. If in the equation

$$y = \frac{1}{2} \cos 2x - \frac{1}{4} \cos 4x + \frac{1}{6} \cos 6x - \frac{1}{8} \cos 8x + \ldots \&c.$$

we suppose x nothing, we find

$$y = \frac{1}{2} - \frac{1}{4} + \frac{1}{6} - \frac{1}{8} + \ldots \&c. = \frac{1}{2} \log 2;$$

therefore

$$y = \frac{1}{2} \log 2 + \frac{1}{2} \log \cos x.$$

Thus we meet with the series given by Euler,

$$\log(2 \cos \tfrac{1}{2}x) = \cos x - \frac{1}{2} \cos 2x + \frac{1}{3} \cos 3x - \frac{1}{4} \cos 4x + \&c.$$

184.   Applying the same process to the equation

$$y = \sin x + \frac{1}{3} \sin 3x + \frac{1}{5} \sin 5x + \frac{1}{7} \sin 7x + \&c.,$$

we find the following series, which has not been noticed,

$$\frac{1}{4} \pi = \sin x + \frac{1}{3} \sin 3x + \frac{1}{5} \sin 5x + \frac{1}{7} \sin 7x + \frac{1}{9} \sin 9x + \&c.[3]$$

It must be observed with respect to all these series, that the equations which are formed by them do not hold except when the variable x is included between certain limits. Thus the function

$$\cos x - \frac{1}{3} \cos 3x + \frac{1}{5} \cos 5x - \frac{1}{7} \cos 7x + \&c.$$

---

[3] This may be derived by integration from 0 to π as in Art. 222. [R. L. E.]

is not equal to $\frac{1}{4}\pi$, except when the variable x is contained between the limits which we have assigned. It is the same with the series

$$\sin x - \frac{1}{2}\sin 2x + \frac{1}{3}\sin 3x - \frac{1}{4}\sin 4x + \frac{1}{5}\sin 5x - \&c.$$

This infinite series, which is always convergent, has the value $\frac{1}{2}x$ so long as the arc x is greater than 0 and less than π. But it is not equal to $\frac{1}{2}x$, if the arc exceeds π; it has on the contrary values very different from $\frac{1}{2}x$; for it is evident that in the interval from $x = \pi$ to $x = 2\pi$, the function takes with the contrary sign all the values which it had in the preceding interval from $x = 0$ to $x = \pi$. This series has been known for a long time, but the analysis which served to discover it did not indicate why the result ceases to hold when the variable exceeds π.

The method which we are about to employ must therefore be examined attentively, and the origin of the limitation to which each of the trigonometrical series is subject must be sought.

185. To arrive at it, it is sufficient to consider that the values expressed by infinite series are not known with exact certainty except in the case where the limits of the sum of the terms which complete them can be assigned; it must therefore be supposed that we employ only the first terms of these series, and the limits between which the remainder is included must be found.

We will apply this remark to the equation

$$y = \cos x - \frac{1}{3}\cos 3x + \frac{1}{5}\cos 5x - \frac{1}{7}\cos 7x \ldots$$

$$+ \frac{\cos (2m - 3)x}{2m - 3} - \frac{\cos (2m - 1)x}{2m - 1}.$$

The number of terms is even and is represented by m; from it is derived the equation $\frac{2dy}{dx} = \frac{\sin 2mx}{\cos x}$, whence we may infer the value of y, by integration by parts. Now the integral $\int uvdx$ may be resolved into a series composed of as many terms as may be desired, u and v being functions of x. We may write, for example,

$$\int uvdx = c + u\int vdx - \frac{du}{dx}\int dx\int vdx + \frac{d^2u}{dx^2}\int dx\int dx\int vdx$$

$$- \int \left\{d\left(\frac{d^2u}{dx^2}\right)\right\}\int dx\int dx\int vdx\right\},$$

an equation which is verified by differentiation.

Denoting $\sin 2mx$ by v and $\sec x$ by u, it will be found that

$$2y = c - \frac{1}{2m}\sec x \cos 2mx + \frac{1}{2^2 m^2}\sec' x \sin 2mx + \frac{1}{2^3 m^3}\sec'' x \cos 2mx$$

$$- \int \left(d\frac{\sec'' x}{2^3 m^3} \cdot \cos 2mx\right).$$

186. It is required now to ascertain the limits between which the integral $\frac{1}{2^3m^3}\int\{d(\sec''x)\cos 2mx\}$ which completes the series is included. To form this integral an infinity of values must be given to the arc x, from 0, the limit at which the integral begins, up to x, which is the final value of the arc; for each one of these values of x the value of the differential $d(\sec''x)$ must be determined, and that of the factor $\cos 2mx$, and all the partial products must be added: now the variable factor $\cos 2mx$ is necessarily a positive or negative fraction; consequently the integral is composed of the sum of the variable values of the differential $d(\sec''x)$, multiplied respectively by these fractions. The total value of the integral is then less than the sum of the differentials $d(\sec''x)$, taken from $x = 0$ up to x, and it is greater than this sum taken negatively: for in the first case we replace the variable factor $\cos 2mx$ by the constant quantity 1, and in the second case we replace this factor by -1: now the sum of the differentials $d(\sec''x)$, or which is the same thing, the integral $\int d(\sec''x)$, taken from $x = 0$, is $\sec''x - \sec''0$; $\sec''x$ is a certain function of x, and $\sec''0$ is the value of this function, taken on the supposition that the arc x is nothing.

The integral required is therefore included between

$$+(\sec''x - \sec''0) \quad \text{and} \quad -(\sec''x - \sec''0);$$

that is to say, representing by k an unknown fraction positive or negative, we have always

$$\int\{d(\sec''x)\cos 2mx\} = k(\sec''x - \sec''0).$$

Thus we obtain the equation

$$2y = c - \frac{1}{2m}\sec x \cos 2mx + \frac{1}{2^2m^2}\sec'x \sin 2mx + \frac{1}{2^3m^3}\sec''x \cos 2mx$$

$$- \frac{k}{2^3m^3}(\sec''x - \sec''0),$$

in which the quantity $\frac{k}{2^3m^3}(\sec''x - \sec''0)$ expresses exactly the sum of all the last terms of the infinite series.

187. If we had investigated two terms only we should have had the equation

$$2y = c - \frac{1}{2m}\sec x \cos 2mx + \frac{1}{2^2m^2}\sec'x \sin 2mx + \frac{k}{2^2m^2}(\sec'x - \sec'0).$$

From this it follows that we can develope the value of y in as many terms as we wish, and express exactly the remainder of the series; we thus find the set of equations

$$2y = c - \frac{1}{2m}\sec x \cos 2mx + \frac{k}{2^2m^2}(\sec x - \sec 0),$$

$$2y = c - \frac{1}{2m}\sec x \cos 2mx + \frac{1}{2^2m^2}\sec'x \sin 2mx - \frac{k}{2^2m^2}(\sec'x - \sec'0),$$

$$2y = c - \frac{1}{2m}\sec x \cos 2mx + \frac{1}{2^2 m^2}\sec' x \sin 2mx + \frac{1}{2^3 m^3}\sec'' x \cos 2mx$$

$$- \frac{k}{2^3 m^3}(\sec'' x - \sec'' 0).$$

The number k which enters into these equations is not the same for all, and it represents in each one a certain quantity which is always included between 1 and -1; m is equal to the number of terms of the series

$$\cos x - \frac{1}{3}\cos 3x + \frac{1}{5}\cos 5x - \ldots - \frac{1}{2m - 1}\cos(2m - 1)x,$$

whose sum is denoted by y.

188.   These equations could be employed if the number m were given, and however great that number might be, we could determine as exactly as we pleased the variable part of the value of y. If the number m be infinite, as is supposed, we consider the first equation only; and it is evident that the two terms which follow the constant become smaller and smaller; so that the exact value of $2y$ is in this case the constant c; this constant is determined by assuming $x = 0$ in the value of y, whence we conclude

$$\frac{\pi}{4} = \cos x - \frac{1}{3}\cos 3x + \frac{1}{5}\cos 5x - \frac{1}{7}\cos 7x + \frac{1}{9}\cos 9x - \&c.$$

It is easy to see now that the result necessarily holds if the arc x is less than $\frac{1}{2}\pi$. In fact, attributing to this arc a definite value X as near to $\frac{1}{2}\pi$ as we please, we can always give to m a value so great, that the term $\frac{k}{2m}(\sec x - \sec 0)$, which completes the series, becomes less than any quantity whatever; but the exactness of this conclusion is based on the fact that the term $\sec x$ acquires no value which exceeds all possible limits, whence it follows that the same reasoning cannot apply to the case in which the arc x is not less than $\frac{1}{2}\pi$.

The same analysis could be applied to the series which express the values of $\frac{1}{2}x$, log cos x, and by this means we can assign the limits between which the variable must be included, in order that the result of analysis may be free from all uncertainty; moreover, the same problems may be treated otherwise by a method founded on other principles.

189.   The expression of the law of fixed temperatures in a solid plate supposed the knowledge of the equation

$$\frac{\pi}{4} = \cos x - \frac{1}{3}\cos 3x + \frac{1}{5}\cos 5x - \frac{1}{7}\cos 7x + \frac{1}{9}\cos 9x - \&c.$$

A simpler method of obtaining this equation is as follows:

If the sum of two arcs is equal to $\frac{1}{2}\pi$, a quarter of the circumference, the product of their tangent is 1; we have therefore in general

$$\frac{1}{2}\pi = \text{arc tan } u + \text{arc tan } \frac{1}{u} \quad \dots\dots\dots\dots\dots\dots(c);$$

the symbol arc tan u denotes the length of the arc whose tangent is u, and the series which gives the value of that arc is well known; whence we have the following result:

$$\frac{1}{2}\pi = u + \frac{1}{u} - \frac{1}{3}\left(u^3 + \frac{1}{u^3}\right) + \frac{1}{5}\left(u^5 + \frac{1}{u^5}\right) - \frac{1}{7}\left(u^7 + \frac{1}{u^7}\right)$$

$$+ \frac{1}{9}\left(u^9 + \frac{1}{u^9}\right) - \&c. \quad \dots\dots\dots\dots\dots(d).$$

If now we write $e^{x\sqrt{-1}}$ instead of u in equation (c), and in equation (d), we shall have

$$\frac{1}{2}\pi = \text{arc tan } e^{x\sqrt{-1}} + \text{arc tan } e^{-x\sqrt{-1}},$$

and $\frac{1}{4}\pi = \cos x - \frac{1}{3}\cos 3x + \frac{1}{5}\cos 5x - \frac{1}{7}\cos 7x + \frac{1}{9}\cos 9x - \&c.$

The series of equation (d) is always divergent, and that of equation (b) (Art. 180) is always convergent; its value is $\frac{1}{4}\pi$ or $-\frac{1}{4}\pi$.

SECTION IV

General Solution

190. We can now form the complete solution of the problem which we have proposed; for the coefficients of equation (b) (Art. 169) being determined, nothing remains but to substitute them, and we have

$$\frac{\pi v}{4} = e^{-x}\cos y - \frac{1}{3}e^{-3x}\cos 3y + \frac{1}{5}e^{-5x}\cos 5y - \frac{1}{7}e^{-7x}\cos 7y + \&c. \quad \dots\dots(\alpha).$$

This value of v satisfies the equation $\frac{d^2v}{dx^2} + \frac{d^2v}{dy^2} = 0$; it becomes nothing when we give to y a value equal to $\frac{1}{2}\pi$ or $-\frac{1}{2}\pi$; lastly, it is equal to unity when x is nothing and y is included between $-\frac{1}{2}\pi$ and $+\frac{1}{2}\pi$. Thus all the physical conditions of the problem are exactly fulfilled, and it is certain that, if we give to each point of the plate the temperature which equation (α) determines, and if the base A be maintained at the same time at the temperature 1, and the infinite edges B and C at the temperature 0, it would be impossible for any change to occur in the system of temperatures.

191. The second member of equation (α) having the form of an exceedingly convergent series, it is always easy to determine numerically the temperature of a point whose co-ordinates x and y are known. The solution gives rise to various results which it is necessary to remark, since they belong also to the general theory.

If the point m, whose fixed temperature is considered, is very distant from the origin A, the value of the second member of the equation (α) will be very nearly equal to $e^{-x} \cos y$; it reduces to this term if x is infinite.

The equation $v = \frac{4}{\pi} e^{-x} \cos y$ represents also a state of the solid which would be preserved without any change, if it were once formed; the same would be the case with the state represented by the equation $v = \frac{4}{3\pi} e^{-3x} \cos 3y$, and in general each term of the series corresponds to a particular state which enjoys the same property. All these partial systems exist at once in that which equation (α) represents; they are superposed, and the movement of heat takes place with respect to each of them as if it alone existed. In the state which corresponds to any one of these terms, the fixed temperatures of the points of the base A differ from one point to another, and this is the only condition of the problem which is not fulfilled; but the general state which results from the sum of all the terms satisfies this special condition.

According as the point whose temperature is considered is more distant from the origin, the movement of heat is less complex: for if the distance x is sufficiently great, each term of the series is very small with respect to that which precedes it, so that the state of the heated plate is sensibly represented by the first three terms, or by the first two, or by the first only, for those parts of the plate which are more and more distant from the origin.

The curved surface whose vertical ordinate measures the fixed temperature v, is formed by adding the ordinates of a multitude of particular surfaces whose equations are

$$\frac{\pi v_1}{4} = e^{-x} \cos y, \quad \frac{\pi v_2}{4} = -\frac{1}{3} e^{-3x} \cos 3y, \quad \frac{\pi v_3}{4} = \frac{1}{5} e^{-5x} \cos 5y, \text{ \&c.}$$

The first of these coincides with the general surface when x is infinite, and they have a common asymptotic sheet.

If the difference $v - v_1$ of their ordinates is considered to be the ordinate of a curved surface, this surface will coincide, when x is infinite, with that whose equation is $\frac{1}{4}\pi v_2 = -\frac{1}{3} e^{-3x} \cos 3y$. All the other terms of the series produce similar results.

The same results would again be found if the section at the origin, instead of being bounded as in the actual hypothesis by a straight line parallel to the axis of y, had any figure whatever formed of two symmetrical parts. It is evident therefore that the particular values

$$ae^{-x} \cos y, \quad be^{-3x} \cos 3y, \quad ce^{-5x} \cos 5y, \text{ \&c.,}$$

have their origin in the physical problem itself, and have a necessary relation to the phenomena of heat. Each of them expresses a simple mode according to which heat is established and propagated in a rectangular plate, whose infinite sides retain a constant temperature. The general system of temperatures is compounded always of a multitude of simple systems, and the expression for their sum has nothing arbitrary but the coefficients a, b, c, d, &c.

192.   Equation (α) may be employed to determine all the circumstances of the permanent movement of heat in a rectangular plate heated at its origin. If it be asked, for example, what is the expenditure of the source of heat, that is to say, what is the quantity which, during a given time, passes across the base A and replaces that which flows into the cold masses B and C; we must consider that the flow perpendicular to the axis of y is expressed by $-K\frac{dv}{dx}$. The quantity which during the instant dt flows across a part dy of the axis is therefore

$$-K\frac{dv}{dx}\,dy\,dt;$$

and, as the temperatures are permanent, the amount of the flow, during unit of time, is $-K\frac{dv}{dx}\,dy$. This expression must be integrated between the limits $y = -\frac{1}{2}\pi$ and $y = +\frac{1}{2}\pi$, in order to ascertain the whole quantity which passes the base, or which is the same thing, must be integrated from $y = 0$ to $y = \frac{1}{2}\pi$, and the result doubled. The quantity $\frac{dv}{dx}$ is a function of x and y, in which x must be made equal to 0, in order that the calculation may refer to the base A, which coincides with the axis of y. The expression for the expenditure of the source of heat is therefore $2\int(-K\frac{dv}{dx}\,dy)$. The integral must be taken from $y = 0$ to $y = \frac{1}{2}\pi$; if, in the function $\frac{dv}{dx}$, x is not supposed equal to 0, but $x = x$, the integral will be a function of x which will denote the quantity of heat which flows in unit of time across a transverse edge at a distance x from the origin.

193.   If we wish to ascertain the quantity of heat which, during unit of time, passes across a line drawn on the plate parallel to the edges B and C, we employ the expression $-K\frac{dv}{dy}$, and, multiplying it by the element dx of the line drawn, integrate with respect to x between the given boundaries of the line; thus the integral $\int(-K\frac{dv}{dy}\,dx)$ shews how much heat flows across the whole length of the line; and if before or after the integration we make $y = \frac{1}{2}\pi$, we determine the quantity of heat which, during unit of time, escapes from the plate across the infinite edge C. We may next compare the latter quantity with the expenditure of the source of heat; for the source must necessarily supply continually the heat which flows into the masses B and C. If this compensation did not exist at each instant, the system of temperatures would be variable.

194.   Equation (α) gives

$$-K\frac{dv}{dx} = \frac{4K}{\pi}\left(e^{-x}\cos y - e^{-3x}\cos 3y + e^{-5x}\cos 5y - e^{-7x}\cos 7y + \text{&c.}\right);$$

multiplying by dy, and integrating from $y = 0$, we have

$$\frac{4K}{\pi}\left(e^{-x}\sin y - \frac{1}{3}e^{-3x}\sin 3y + \frac{1}{5}e^{-5x}\sin 5y - \frac{1}{7}e^{-7x}\sin 7y + \&c.\right).$$

If y be made $= \frac{1}{2}\pi$, and the integral doubled, we obtain

$$\frac{8K}{\pi}\left(e^{-x} + \frac{1}{3}e^{-3x} + \frac{1}{5}e^{-5x} + \frac{1}{7}e^{-7x} + \&c.\right)$$

as the expression for the quantity of heat which, during unit of time, crosses a line parallel to the base, and at a distance x from that base.

From equation (α) we derive also

$$-K\frac{dv}{dy} = \frac{4K}{\pi}\left(e^{-x}\sin y - e^{-3x}\sin 3y + e^{-5x}\sin 5y - e^{-7x}\sin 7y + \&c.\right):$$

hence the integral $\int -K\left(\frac{dv}{dy}\right)dx$, taken from $x = 0$, is

$$\frac{4K}{\pi}\{(1 - e^{-x})\sin y - (1 - e^{-3x})\sin 3y + (1 - e^{-5x})\sin 5y$$

$$-(1 - e^{-7}x)\sin 7y + \&c.\}.$$

If this quantity be subtracted from the value which it assumes when x is made infinite, we find

$$\frac{4K}{\pi}\left(e^{-x}\sin y - \frac{1}{3}e^{-3x}\sin 3y + \frac{1}{5}e^{-5x}\sin 5y - \&c.\right);$$

and, on making $y = \frac{1}{2}\pi$, we have an expression for the whole quantity of heat which crosses the infinite edge C, from the point whose distance from the origin is x up to the end of the plate; namely,

$$\frac{4K}{\pi}\left(e^{-x} + \frac{1}{3}e^{-3x} + \frac{1}{5}e^{-5x} + \frac{1}{7}e^{-7x} + \&c.\right),$$

which is evidently equal to half the quantity which in the same time passes beyond the transverse line drawn on the plate at a distance x from the origin. We have already remarked that this result is a necessary consequence of the conditions of the problem; if it did not hold, the part of the plate which is situated beyond the transverse line and is prolonged to infinity would not receive through its base a quantity of heat equal to that which it loses through its two edges; it could not therefore preserve its state, which is contrary to hypothesis.

195. As to the expenditure of the source of heat, it is found by supposing $x = 0$ in the preceding expression; hence it assumes an infinite value, the reason for which is evident if it be remarked that, according to hypothesis, every point of the line A has and retains the temperature 1: parallel lines which are very near to this base have also a temperature very little different from unity: hence, the extremities of all these lines contiguous to the cold masses B and C communicate to them a quantity of heat incomparably greater than if the decrease of temperature were

continuous and imperceptible. In the first part of the plate, at the ends near to B or C, a *cataract* of heat, or an infinite flow, exists. This result ceases to hold when the distance x becomes appreciable.

196.   The length of the base has been denoted by π. If we assign to it any value $2l$, we must write $\frac{1}{2}\pi\frac{y}{l}$ instead of y, and multiplying also the values of x by $\frac{\pi}{2l}$, we must write $\frac{1}{2}\pi\frac{x}{l}$ instead of x. Denoting by A the constant temperature of the base, we must replace v by $\frac{v}{A}$. These substitutions being made in the equation (α), we have

$$v = \frac{4A}{\pi}\left(e^{-\frac{\pi x}{2l}}\cos\frac{\pi y}{2l} - \frac{1}{3}e^{-\frac{3\pi x}{2l}}\cos 3\frac{\pi y}{2l} + \frac{1}{5}e^{-\frac{5\pi x}{2l}}\cos 5\frac{\pi y}{2l} \right.$$
$$\left. - \frac{1}{7}e^{-\frac{7\pi x}{2l}}\cos 7\frac{\pi y}{2l} + \&\text{c.} \right) \dots\dots(\beta).$$

This equation represents exactly the system of permanent temperature in an infinite rectangular prism, included between two masses of ice B and C, and a constant source of heat.

197.   It is easy to see either by means of this equation, or from Art. 171, that heat is propagated in this solid, by separating more and more from the origin, at the same time that it is directed towards the infinite faces B and C. Each section parallel to that of the base is traversed by a wave of heat which is renewed at each instant with the same intensity: the intensity diminishes as the section becomes more distant from the origin. Similar movements are effected with respect to any plane parallel to the infinite faces; each of these planes is traversed by a constant wave which conveys its heat to the lateral masses.

The developments contained in the preceding articles would be unnecessary, if we had not to explain an entirely new theory, whose principles it is requisite to fix. With that view we add the following remarks.

198.   Each of the terms of equation (α) corresponds to only one particular system of temperatures, which might exist in a rectangular plate heated at its end, and whose infinite edges are maintained at a constant temperature. Thus the equation $v = e^{-x}\cos y$ represents the permanent temperatures, when the points of the base A are subject to a fixed temperature, denoted by $\cos y$. We may now imagine the heated plate to be part of a plane which is prolonged to infinity in all directions, and denoting the co-ordinates of any point of this plane by x and y, and the temperature of the same point by v, we may apply to the entire plane the equation $v = e^{-x}\cos y$; by this means, the edges B and C receive the constant temperature 0; but it is not the same with contiguous parts BB and CC; they receive and keep lower temperatures. The base A has at every point the permanent temperature denoted by $\cos y$, and the contiguous parts AA have higher temperatures. If we construct the curved surface whose vertical ordinate is equal to the permanent temperature at each point

of the plane, and if it be cut by a vertical plane passing through the line A or parallel to that line, the form of the section will be that of a trigonometrical line whose ordinate represents the infinite and periodic series of cosines. If the same curved surface be cut by a vertical plane parallel to the axis of x, the form of the section will through its whole length be that of a logarithmic curve.

199. By this it may be seen how the analysis satisfies the two conditions of the hypothesis, which subjected the base to a temperature equal to cos y, and the two sides B and C to the temperature 0. When we express these two conditions we solve in fact the following problem: If the heated plate formed part of an infinite plane, what must be the temperatures at all the points of the plane, in order that the system may be self-permanent, and that the fixed temperatures of the infinite rectangle may be those which are given by the hypothesis?

We have supposed in the foregoing part that some external causes maintained the faces of the rectangular solid, one at the temperature 1, and the two others at the temperature 0. This effect may be represented in different manners; but the hypothesis proper to the investigation consists in regarding the prism as part of a solid all of whose dimensions are infinite, and in determining the temperatures of the mass which surrounds it, so that the conditions relative to the surface may be always observed.

200. To ascertain the system of permanent temperatures in a rectangular plate whose extremity A is maintained at the temperature 1, and the two infinite edges at the temperature 0, we might consider the changes which the temperatures undergo, from the initial state which is given, to the fixed state which is the object of the problem. Thus the variable state of the solid would be determined for all values of the time, and it might then be supposed that the value was infinite.

The method which we have followed is different, and conducts more directly to the expression of the final state, since it is founded on a distinctive property of that state. We now proceed to shew that the problem admits of no other solution than that which we have stated. The proof follows from the following propositions.

201. If we give to all the points of an infinite rectangular plate temperatures expressed by equation (α), and if at the two edges B and C we maintain the fixed temperature 0, whilst the end A is exposed to a source of heat which keeps all points of the line A at the fixed temperature 1; no change can happen in the state of the solid. In fact, the equation $\frac{d^2v}{dx^2} + \frac{d^2v}{dy^2} = 0$ being satisfied, it is evident (Art. 170) that the quantity of heat which determines the temperature of each molecule can be neither increased nor diminished.

The different points of the same solid having received the temperatures expressed by equation (α) or $v = \phi(x, y)$, suppose that instead of maintaining the edge A at

523

the temperature 1, the fixed temperature 0 be given to it as to the two lines B and C; the heat contained in the plate BAC will flow across the three edges A, B, C, and by hypothesis it will not be replaced, so that the temperatures will diminish continually, and their final and common value will be zero. This result is evident since the points infinitely distant from the origin A have a temperature infinitely small from the manner in which equation (α) was formed.

The same effect would take place in the opposite direction, if the system of temperatures were $v = -\phi(x, y)$, instead of being $v = \phi(x, y)$; that is to say, all the initial negative temperatures would vary continually, and would tend more and more towards their final value 0, whilst the three edges A, B, C preserved the temperature 0.

202.   Let $v = f(x, y)$ be a given equation which expresses the initial temperature of points in the plate BAC, whose base A is maintained at the temperature 1, whilst the edges B and C preserve the temperature 0.

Let $v = F(x, y)$ be another given equation which expresses the initial temperature of each point of a solid plate BAC exactly the same as the preceding, but whose three edges B, A, C are maintained at the temperature 0.

Suppose that in the first solid the variable state which succeeds to the final state is determined by the equation $v = \phi(x, y, t)$, t denoting the time elapsed, and that the equation $v = \Phi(x, y, t)$ determines the variable state of the second solid, for which the initial temperatures are $F(x, y)$.

Lastly, suppose a third solid like each of the two preceding: let $v = f(x, y) + F(x, y)$ be the equation which represents its initial state, and let 1 be the constant temperature of the base A, 0 and 0 those of the two edges B and C.

We proceed to shew that the variable state of the third solid is determined by the equation $v = \phi(x, y, t) + \Phi(x, y, t)$.

In fact, the temperature of a point m of the third solid varies, because that molecule, whose volume is denoted by M, acquires or loses a certain quantity of heat Δ. The increase of temperature during the instant dt is

$$\frac{\Delta}{cM}\, dt,$$

the coefficient c denoting the specific capacity with respect to volume. The variation of the temperature of the same point in the first solid is $\frac{d}{cM}\, dt$, and $\frac{D}{cM}\, dt$ in the second, the letters d and D representing the quantity of heat positive or negative which the molecule acquires by virtue of the action of all the neighbouring molecules. Now it is easy to perceive that Δ is equal to $d + D$. For proof it is sufficient to consider the quantity of heat which the point m receives from another point m' belonging to the interior of the plate, or to the edges which bound it.

The point m_1, whose initial temperature is denoted by f_1, transmits, during the instant dt, to the molecule m, a quantity of heat expressed by $q_1(f_1 - f)dt$, the factor q_1 representing a certain function of the distance between the two molecules. Thus the whole quantity of heat acquired by m is $\Sigma q_1(f_1 - f)dt$, the sign Σ expressing the sum of all the terms which would be found by considering the other points m_2, m_3, m_4 &c. which act on m; that is to say, writing q_2, f_2 or q_3, f_3, or q_4, f_4 and so on, instead of q_1, f_1. In the same manner $\Sigma q_1(F_1 - F)dt$ will be found to be the expression of the whole quantity of heat acquired by the same point m of the second solid; and the factor q_1 is the same as in the term $\Sigma q_1(f_1 - f)dt$, since the two solids are formed of the same matter, and the position of the points is the same; we have then

$$d = \Sigma q_1(f_1 - f)dt \quad \text{and} \quad D = \Sigma q_1(F_1 - F)dt.$$

For the same reason it will be found that

$$\Delta = \Sigma q_1\{f_1 + F_1 - (f + F)\}dt;$$

hence

$$\Delta = d + D \quad \text{and} \quad \frac{\Delta}{cM} = \frac{d}{cM} + \frac{D}{cM}.$$

It follows from this that the molecule m of the third solid acquires, during the instant dt, an increase of temperature equal to the sum of the two increments which the same point would have gained in the two first solids. Hence at the end of the first instant, the original hypothesis will again hold, since any molecule whatever of the third solid has a temperature equal to the sum of those which it has in the two others. Thus the same relation exists at the beginning of each instant, that is to say, the variable state of the third solid can always be represented by the equation

$$v = \phi(x, y, t) + \Phi(x, y, t).$$

203.   The preceding proposition is applicable to all problems relative to the uniform or varied movement of heat. It shews that the movement can always be decomposed into several others, each of which is effected separately as if it alone existed. This superposition of simple effects is one of the fundamental elements in the theory of heat. It is expressed in the investigation, by the very nature of the general equations, and derives its origin from the principle of the communication of heat.

Let now $v = \phi(x, y)$ be the equation (α) which expresses the permanent state of the solid plate BAC, heated at its end A, and whose edges B and C preserve the temperature 0; the initial state of the plate is such, according to hypothesis, that all its points have a nul temperature, except those of the base A, whose temperature is 1. The initial state can then be considered as formed of two others, namely: a first, in which the initial temperatures are $-\phi(x, y)$, the three edges being maintained at the temperature 0, and a second state, in which the initial temperatures are $+\phi(x, y)$, the two edges B and C preserving the temperature 0, and the base A the temperature 1; the superposition of these two states produces the initial state which results from the hypothesis. It remains then only to examine the movement of heat in each one of the two partial states. Now, in the second, the system of temperatures can undergo no change; and in the first, it has been remarked in Article 201 that the temperatures vary continually, and end with being nul. Hence the final state, properly so called, is that which is represented by $v = \phi(x, y)$ or equation (α).

If this state were formed at first it would be self-existent, and it is this property which has served to determine it for us. If the solid plate be supposed to be in another initial state, the difference between the latter state and the fixed state forms a partial state, which imperceptibly disappears. After a considerable time, the difference has nearly vanished, and the system of fixed temperatures has undergone no change. Thus the variable temperatures converge more and more to a final state, independent of the primitive heating.

204. We perceive by this that the final state is unique; for, if a second state were conceived, the difference between the second and the first would form a partial state, which ought to be self-existent, although the edges A, B, C were maintained at the temperature 0. Now the last effect cannot occur; similarly if we supposed another source of heat independent of that which flows from the origin A; besides, this hypothesis is not that of the problem we have treated, in which the initial temperatures are nul. It is evident that parts very distant from the origin can only acquire an exceedingly small temperature.

Since the final state which must be determined is unique, it follows that the problem proposed admits no other solution than that which results from equation (α). Another form may be given to this result, but the solution can be neither extended nor restricted without rendering it inexact.

The method which we have explained in this chapter consists in forming first very simple particular values, which agree with the problem, and in rendering the solution more general, to the intent that v or $\phi(x, y)$ may satisfy three conditions, namely:

$$\frac{d^2v}{dx^2} + \frac{d^2v}{dy^2} = 0, \quad \phi(x, 0) = 1, \quad \phi(x, \pm\tfrac{1}{2}\pi) = 0.$$

It is clear that the contrary order might be followed, and the solution obtained would necessarily be the same as the foregoing. We shall not stop over the details, which are easily supplied, when once the solution is known. We shall only give in the following section a remarkable expression for the function $\phi(x, y)$ whose value was developed in a convergent series in equation (α).

SECTION V

Finite Expression of the Result of the Solution

205. The preceding solution might be deduced from the integral of the equation $\frac{d^2v}{dx^2} + \frac{d^2v}{dy^2} = 0$,[4] which contains imaginary quantities, under the sign of the arbitrary functions. We shall confine ourselves here to the remark that the integral

$$v = \phi(x + y\sqrt{-1}) + \psi(x - y\sqrt{-1}),$$

has a manifest relation to the value of v given by the equation

$$\frac{\pi v}{4} = e^{-x}\cos y - \frac{1}{3}e^{-3x}\cos 3y + \frac{1}{5}e^{-5x}\cos 5y - \&c.$$

In fact, replacing the cosines by their imaginary expressions, we have

$$\frac{\pi v}{2} = e^{-(x-y\sqrt{-1})} - \frac{1}{3}e^{-3(x-y\sqrt{-1})} + \frac{1}{5}e^{-5(x-y\sqrt{-1})} - \&c.$$

$$+ e^{-(x+y\sqrt{-1})} - \frac{1}{3}e^{-3(x+y\sqrt{-1})} + \frac{1}{5}e^{-5(x+y\sqrt{-1})} - \&c.$$

The first series is a function of $x - y\sqrt{-1}$, and the second series is the same function of $x + y\sqrt{-1}$.

Comparing these series with the known development of arc tan z in functions of z its tangent, it is immediately seen that the first is arc tan $e^{-(x-y\sqrt{-1})}$, and the second is arc tan $e^{-(x+y\sqrt{-1})}$; thus equation (α) takes the finite form

$$\frac{\pi v}{2} = \text{arc tan } e^{-(x+y\sqrt{-1})} + \text{arc tan } e^{-(x-y\sqrt{-1})} \quad \dots\dots\dots\dots\dots(B).$$

In this mode it conforms to the general integral

$$v = \phi(x + y\sqrt{-1}) + \psi(x - y\sqrt{-1}) \quad \dots\dots\dots\dots\dots(A),$$

the function $\phi(z)$ is arc tan e^{-z}, and similarly the function $\psi(z)$.

If in equation (B) we denote the first term of the second member by p and the second by q, we have

$$\frac{1}{2}\pi v = p + q, \ \tan p = e^{-(x+y\sqrt{-1})}, \ \tan q = e^{-(x-y\sqrt{-1})};$$

whence $\quad \tan(p + q) \quad$ or $\quad \dfrac{\tan p + \tan q}{1 - \tan p \tan q} = \dfrac{2e^{-x}\cos y}{1 - e^{-2x}} = \dfrac{2\cos y}{e^x - e^{-x}};$

whence we deduce the equation $\quad \dfrac{1}{2}\pi v = \text{arc tan}\left(\dfrac{2\cos y}{e^x - e^{-x}}\right) \quad \dots\dots\dots\dots\dots(C).$

This is the simplest form under which the solution of the problem can be presented.

[4] D. F. Gregory derived the solution from the form

$$v = \cos\left(y\frac{d}{dx}\right)\phi(x) + \sin\left(y\frac{d}{dx}\right)\psi(x).$$

Camb. Math. Journal, Vol. I. [A. F.]

206. This value of v or $\phi(x, y)$ satisfies the conditions relative to the ends of the solid, namely, $\phi(x, \pm \frac{1}{2}\pi) = 0$, and $\phi(0, y) = 1$; it satisfies also the general equation $\frac{d^2v}{dx^2} + \frac{d^2v}{dy^2} = 0$, since equation (C) is a transformation of equation (B). Hence it represents exactly the system of permanent temperatures; and since that state is unique, it is impossible that there should be any other solution, either more general or more restricted.

The equation (C) furnishes, by means of tables, the value of one of the three unknowns v, x, y, when two of them are given; it very clearly indicates the nature of the surface whose vertical ordinate is the permanent temperature of a given point of the solid plate. Finally, we deduce from the same equation the values of the differential coefficients $\frac{dv}{dx}$ and $\frac{dv}{dy}$ which measure the velocity with which heat flows in the two orthogonal directions; and we consequently know the value of the flow in any other direction.

These coefficients are expressed thus,

$$\frac{dv}{dx} = -2\cos y \left(\frac{e^x + e^{-x}}{e^{2x} + 2\cos 2y + e^{-2x}} \right),$$

$$\frac{dv}{dy} = -2\sin y \left(\frac{e^x - e^{-x}}{e^{2x} + 2\cos 2y + e^{-2x}} \right).$$

It may be remarked that in Article 194 the value of $\frac{dv}{dx}$, and that of $\frac{dv}{dy}$ are given by infinite series, whose sums may be easily found, by replacing the trigonometrical quantities by imaginary exponentials. We thus obtain the values of $\frac{dv}{dx}$ and $\frac{dv}{dy}$ which we have just stated.

The problem which we have now dealt with is the first which we have solved in the theory of heat, or rather in that part of the theory which requires the employment of analysis. It furnishes very easy numerical applications, whether we make use of the trigonometrical tables or convergent series, and it represents exactly all the circumstances of the movement of heat. We pass on now to more general considerations.

SECTION VI

Development of an Arbitrary Function in Trigonometric Series

207. The problem of the propagation of heat in a rectangular solid has led to the equation $\frac{d^2v}{dx^2} + \frac{d^2v}{dy^2} = 0$; and if it be supposed that all the points of one of the faces of the solid have a common temperature, the coefficients a, b, c, d, etc. of the series

$$a\cos x + b\cos 3x + c\cos 5x + d\cos 7x + \ldots \&c.,$$

must be determined so that the value of this function may be equal to a constant whenever the arc x is included between $-\frac{1}{2}\pi$ and $+\frac{1}{2}\pi$. The value of these coefficients has just been assigned; but herein we have dealt with a single case only of a more general problem, which consists in developing any function whatever in an infinite series of sines or cosines of multiple arcs. This problem is connected with the theory of partial differential equations, and has been attacked since the origin of that analysis. It was necessary to solve it, in order to integrate suitably the equations of the propagation of heat; we proceed to explain the solution.

We shall examine, in the first place, the case in which it is required to reduce into a series of sines of multiple arcs, a function whose development contains only odd powers of the variable. Denoting such a function by $\phi(x)$, we arrange the equation

$$\phi(x) = a \sin x + b \sin 2x + c \sin 3x + d \sin 4x + \ldots \&c.,$$

in which it is required to determine the value of the coefficients a, b, c, d, &c. First we write the equation

$$\phi(x) = x\phi'(0) + \frac{x^2}{\underline{2}}\phi''(0) + \frac{x^3}{\underline{3}}\phi'''(0) + \frac{x^4}{\underline{4}}\phi^{iv}(0) + \frac{x^5}{\underline{5}}\phi^{v}(0) + \ldots \&c.,$$

in which $\phi'(0)$, $\phi''(0)$, $\phi'''(0)$, $\phi^{iv}(0)$, &c. denote the values taken by the coefficients

$$\frac{d\phi(x)}{dx}, \quad \frac{d^2\phi(x)}{dx^2}, \quad \frac{d^3\phi(x)}{dx^3}, \quad \frac{d^4\phi(x)}{dx^4}, \&c.,$$

when we suppose $x = 0$ in them. Thus, representing the development according to powers of x by the equation

$$\phi(x) = Ax - B\frac{x^3}{\underline{3}} + C\frac{x^5}{\underline{5}} - D\frac{x^7}{\underline{7}} + E\frac{x^9}{\underline{9}} - \&c.,$$

we have

$$\phi(0) = 0, \quad \text{and} \quad \phi'(0) = A,$$
$$\phi''(0) = 0, \quad \phi'''(0) = -B,$$
$$\phi^{iv}(0) = 0, \quad \phi^{v}(0) = C,$$
$$\phi^{vi}(0) = 0, \quad \phi^{vii}(0) = -D$$
$$\&c. \qquad \&c.$$

If now we compare the preceding equation with the equation

$$\phi(x) = a \sin x + b \sin 2x + c \sin 3x + d \sin 4x + e \sin 5x + \&c.,$$

developing the second member with respect to powers of x, we have the equations

$$A = a + \ 2b + \ 3c + \ 4d + \ 5e + \&c.,$$
$$B = a + 2^3b + 3^3c + 4^3d + 5^3e + \&c.,$$
$$C = a + 2^5b + 3^5c + 4^5d + 5^5e + \&c.,$$
$$D = a + 2^7b + 3^7c + 4^7d + 5^7e + \&c.,$$
$$E = a + 2^9b + 3^9c + 4^9d + 5^9e + \&c. \ \ldots\ldots\ldots\ldots\ldots(a).$$

These equations serve to find the coefficients a, b, c, d, e, &c., whose number is infinite. To determine them, we first regard the number of unknowns as finite and equal to m; thus we suppress all the equations which follow the first m equations, and we omit from each equation all the terms of the second member which follow the first m terms which we retain. The whole number m being given, the coefficients a, b, c, d, e, &c. have fixed values which may be found by elimination. Different values would be obtained for the same quantities, if the number of the equations and that of the unknowns were greater by one unit. Thus the value of the coefficients varies as we increase the number of the coefficients and of the equations which ought to determine them. It is required to find what the limits are towards which the values of the unknowns converge continually as the number of equations increases. These limits are the true values of the unknowns which satisfy the preceding equations when their number is infinite.

208. We consider then in succession the cases in which we should have to determine one unknown by one equation, two unknowns by two equations, three unknowns by three equations, and so on to infinity.

Suppose that we denote as follows different systems of equations analogous to those from which the values of the coefficients must be derived:

$$a_1 = A_1, \qquad a_2 + 2b_2 = A_2, \qquad a_3 + 2b_3 + 3c_3 = A_3,$$
$$a_2 + 2^3 b_2 = B_2, \qquad a_3 + 2^3 b_3 + 3^3 c_3 = B_3,$$
$$a_3 + 2^5 b_3 + 3^5 c_3 = C_3,$$

$$a_4 + 2b_4 + 3c_4 + 4d_4 = A_4,$$
$$a_4 + 2^3 b_4 + 3^3 c_4 + 4^3 d_4 = B_4,$$
$$a_4 + 2^5 b_4 + 3^5 c_4 + 4^5 d_4 = C_4,$$
$$a_4 + 2^7 b_4 + 3^7 c_4 + 4^7 d_4 = D_4,$$

$$a_5 + 2b_5 + 3c_5 + 4d_5 + 5e_5 = A_5,$$
$$a_5 + 2^3 b_5 + 3^3 c_5 + 4^3 d_5 + 5^3 e_5 = B_5,$$
$$a_5 + 2^5 b_5 + 3^5 c_5 + 4^5 d_5 + 5^5 e_5 = C_5,$$
$$a_5 + 2^7 b_5 + 3^7 c_5 + 4^7 d_5 + 5^7 e_5 = D_5,$$
$$a_5 + 2^9 b_5 + 3^9 c_5 + 4^9 d_5 + 5^9 e_5 = E_5,$$

$$\text{\&c.} \qquad\qquad \text{\&c.} \dots\dots\dots\dots\dots\dots\dots\dots(b).$$

If now we eliminate the last unknown e_5 by means of the five equations which contain A_5, B_5, C_5, D_5, E_5, &c., we find

$$a_5(5^2 - 1^2) + 2b_5(5^2 - 2^2) + 3c_5(5^2 - 3^2) + 4d_5(5^2 - 4^2) = 5^2 A_5 - B_5,$$
$$a_5(5^2 - 1^2) + 2^3 b_5(5^2 - 2^2) + 3^3 c_5(5^2 - 3^2) + 4^3 d_5(5^2 - 4^2) = 5^2 B_5 - C_5,$$
$$a_5(5^2 - 1^2) + 2^5 b_5(5^2 - 2^2) + 3^5 c_5(5^2 - 3^2) + 4^5 d_5(5^2 - 4^2) = 5^2 C_5 - D_5,$$
$$a_5(5^2 - 1^2) + 2^7 b_5(5^2 - 2^2) + 3^7 c_5(5^2 - 3^2) + 4^7 d_5(5^2 - 4^2) = 5^2 D_5 - E_5.$$

We could have deduced these four equations from the four which form the preceding system, by substituting in the latter instead of

$$a_4, \quad (5^2 - 1^2)a_5,$$
$$b_4, \quad (5^2 - 2^2)b_5,$$
$$c_4, \quad (5^2 - 3^2)c_5,$$
$$d_4, \quad (5^2 - 4^2)d_5;$$

and instead of
$$A_4, \quad 5^2 A_5 - B_5,$$
$$B_4, \quad 5^2 B_5 - C_5,$$
$$C_4, \quad 5^2 C_5 - D_5,$$
$$D_4, \quad 5^2 D_5 - E_5.$$

By similar substitutions we could always pass from the case which corresponds to a number m of unknowns to that which corresponds to the number $m + 1$. Writing in order all the relations between the quantities which correspond to one of the cases and those which correspond to the following case, we shall have

$$a_1 = a_2(2^2 - 1),$$
$$a_2 = a_3(3^2 - 1), \quad b_2 = b_3(3^2 - 2^2),$$
$$a_3 = a_4(4^2 - 1), \quad b_3 = b_4(4^2 - 2^2), \quad c_3 = c_4(4^2 - 3^2),$$
$$a_4 = a_5(5^2 - 1), \quad b_4 = b_5(5^2 - 2^2), \quad c_4 = c_5(5^2 - 3^2), \quad d_4 = d_5(5^2 - 4^2),$$
$$a_5 = a_6(6^2 - 1), \quad b_5 = b_6(6^2 - 2^2), \quad c_5 = c_6(6^2 - 3^2), \quad d_5 = d_6(6^2 - 4^2),$$
$$e_5 = e_6(6^2 - 5^2),$$

&c.   &c. ....(c).

We have also
$$A_1 = 2A_2 - B_2,$$
$$A_2 = 3A_3 - B_3, \quad B_2 = 3B_3 - C_3,$$
$$A_3 = 4A_4 - B_4, \quad B_3 = 4B_4 - C_4, \quad C_3 = 4C_4 - D_4,$$
$$A_4 = 5A_5 - B_5, \quad B_4 = 5B_5 - C_5, \quad C_4 = 5C_5 - D_5, \quad D_4 = 5D_5 - E_5,$$

&c.   &c. ....(d).[5]

From equations (c) we conclude that on representing the unknowns, whose number is infinite, by a, b, c, d, e, &c., we must have

$$a = \frac{a_1}{(2^2 - 1)(3^2 - 1)(4^2 - 1)(5^2 - 1)\ldots},$$

$$b = \frac{b_2}{(3^2 - 2^2)(4^2 - 2^2)(5^2 - 2^2)(6^2 - 2^2)\ldots},$$

$$c = \frac{c_3}{(4^2 - 3^2)(5^2 - 3^2)(6^2 - 3^2)(7^2 - 3^2)\ldots},$$

$$d = \frac{d_4}{(5^2 - 4^2)(6^2 - 4^2)(7^2 - 4^2)(8^2 - 4^2)\ldots},$$

&c.   &c. ....(e).

209.   It remains then to determine the values of a_1, b_2, c_3, d_4, e_5, &c.; the first is given by one equation, in which A_1 enters; the second is given by two equations into which $A_2 B_2$ enter; the third is given by three equations, into which $A_3 B_3 C_3$ enter;

[5] the numerals should be squared in line 2, 3, 4, 5.

and so on. It follows from this that if we knew the values of

$$A_1, \quad A_2B_2, \quad A_3B_3C_3, \quad A_4B_4C_4D_4 \ldots, \&c.,$$

we could easily find a_1 by solving one equation, a_2b_2 by solving two equations, $a_3b_3c_3$ by solving three equations, and so on: after which we could determine a, b, c, d, e, &c. It is required then to calculate the values of

$$A_1, \quad A_2B_2, \quad A_3B_3C_3, \quad A_4B_4C_4D_4, \quad A_5B_5C_5D_5E_5 \ldots, \&c.,$$

by means of equations (d). 1st, we find the value of A_1 in terms of A_2 and B_2; 2nd, by two substitutions we find this value of A_1 in terms of $A_3B_3C_3$; 3rd, by three substitutions we find the same value of A_1 in terms of $A_4B_4C_4D_4$, and so on. The successive values of A_1 are

$$A_1 = A_2 2^2 - B_2,$$
$$A_1 = A_3 2^2 \cdot 3^2 - B_3(2^2 + 3^2) + C_3,$$
$$A_1 = A_4 2^2 \cdot 3^2 \cdot 4^2 - B_4(2^2 \cdot 3^2 + 2^2 \cdot 4^2 + 3^2 \cdot 4^2) + C_4(2^2 + 3^2 + 4^2) - D_4,$$
$$A_1 = A_5 2^2 \cdot 3^2 \cdot 4^2 \cdot 5^2 - B_5(2^2 \cdot 3^2 \cdot 4^2 + 2^2 \cdot 3^2 \cdot 5^2 + 2^2 \cdot 4^2 \cdot 5^2 + 3^2 \cdot 4^2 \cdot 5^2)$$
$$+ C_5(2^2 \cdot 3^2 + 2^2 \cdot 4^2 + 2^2 \cdot 5^2 + 3^2 \cdot 4^2 + 3^2 \cdot 5^2 + 4^2 \cdot 5^2)$$
$$- D_5(2^2 + 3^2 + 4^2 + 5^2) + E_5, \&c.,$$

the law of which is readily noticed. The last of these values, which is that which we wish to determine, contains the quantities A, B, C, D, E, &c., with an infinite index, and these quantities are known; they are the same as those which enter into equations (a).

Dividing the ultimate value of A_1 by the infinite product

$$2^2 \cdot 3^2 \cdot 4^2 \cdot 5^2 \cdot 6^2 \ldots \&c.,$$

we have

$$A - B\left(\frac{1}{2^2} + \frac{1}{3^2} + \frac{1}{4^2} + \frac{1}{5^2} + \&c.\right) + C\left(\frac{1}{2^2 \cdot 3^2} + \frac{1}{2^2 \cdot 4^2} + \frac{1}{3^2 \cdot 4^2} + \&c.\right)$$
$$- D\left(\frac{1}{2^2 \cdot 3^2 \cdot 4^2} + \frac{1}{2^2 \cdot 3^2 \cdot 5^2} + \frac{1}{3^2 \cdot 4^2 \cdot 5^2} + \&c.\right)$$
$$+ E\left(\frac{1}{2^2 \cdot 3^2 \cdot 4^2 \cdot 5^2} + \frac{1}{2^2 \cdot 3^2 \cdot 4^2 \cdot 6^2} + \&c.\right) + \&c.$$

The numerical coefficients are the sums of the products which could be formed by different combinations of the fractions

$$\frac{1}{1^2}, \quad \frac{1}{2^2}, \quad \frac{1}{3^2}, \quad \frac{1}{5^2}, \quad \frac{1}{6^2} \ldots \&c.,$$

after having removed the first fraction $\frac{1}{1^2}$. If we represent the respective sums of products by $P_1, Q_1, R_1, S_1, T_1, \ldots$ &c., and if we employ the first of equations (e) and the first of equations (b), we have, to express the value of the first coefficient a, the equation

$$\frac{a(2^2 - 1)(3^2 - 1)(4^2 - 1)(5^2 - 1) \cdots}{2^3 \cdot 3^2 \cdot 4^2 \cdot 5^2 \ldots} = A - BP_1 + CQ_1 - DR_1 + ES_1 - \&c.,$$

now the quantities $P_1, Q_1, R_1, S_1, T_1 \ldots$ &c. may be easily determined, as we shall see lower down; hence the first coefficient a becomes entirely known.

210. We must pass on now to the investigation of the following coefficients b, c, d, e, &c., which from equations (e) depend on the quantities b_2, c_3, d_4, e_5, &c. For this purpose we take up equations (b), the first has already been employed to find the value of a_1, the two following give the value of b_2, the three following the value of c_3, the four following the value of d_4, and so on.

On completing the calculation, we find by simple inspection of the equations the following results for the values of b_2, c_3, d_4, &c.

$$2b_2(1^2 - 2^2) = A_2 1^2 - B_2,$$
$$3c_3(1^2 - 3^2)(2^2 - 3^2) = A_3 1^2 \cdot 2^2 - B_3(1^2 + 2^2) + C_3,$$
$$4d_4(1^2 - 4^2)(2^2 - 4^2)(3^2 - 4^2)$$
$$= A_4 1^2 \cdot 2^2 \cdot 3^2 - B_4(1^2 \cdot 2^2 + 1^2 \cdot 3^2 + 2^2 \cdot 3^2) + C_4(1^2 + 2^2 + 3^2) - D_4,$$
&c.

It is easy to perceive the law which these equations follow; it remains only to determine the quantities $A_2 B_2, A_3 B_3 C_3, A_4 B_4 C_4$, &c.

Now the quantities $A_2 B_2$ can be expressed in terms of $A_3 B_3 C_3$, the latter in terms of $A_4 B_4 C_4 D_4$. For this purpose it suffices to effect the substitutions indicated by equations (d); the successive changes reduce the second members of the preceding equations so as to contain only the $ABCD$, &c. with an infinite suffix, that is to say, the known quantities $ABCD$, &c. which enter into equations (a); the coefficients become the different products which can be made by combining the squares of the numbers $1, 2, 3, 4, 5$ to infinity. It need only be remarked that the first of these squares 1^2 will not enter into the coefficients of the value of a_1; that the second 2^2 will not enter into the coefficients of the value of b_2; that the third square 3^2 will be omitted only from those which serve to form the coefficients of the value of c_3; and so of the rest to infinity. We have then for the values of $b_2 c_3 d_4 e_5$, &c., and consequently for those of $bcde$, &c., results entirely analogous to that which we have found above for the value of the first coefficient a_1.

211. If now we represent by P_2, Q_2, R_2, S_2, &c., the quantities

$$\frac{1}{1^2} + \frac{1}{3^2} + \frac{1}{4^2} + \frac{1}{5^2} + \ldots,$$

$$\frac{1}{1^2 \cdot 3^2} + \frac{1}{1^2 \cdot 4^2} + \frac{1}{1^2 \cdot 5^2} + \frac{1}{3^2 \cdot 4^2} + \frac{1}{3^2 \cdot 5^2} + \ldots,$$

$$\frac{1}{1^2 \cdot 3^2 \cdot 4^2} + \frac{1}{1^2 \cdot 3^2 \cdot 5^2} + \frac{1}{3^2 \cdot 4^2 \cdot 5^2} + \ldots$$

$$\frac{1}{1^2 \cdot 3^2 \cdot 4^2 \cdot 5^2} + \frac{1}{1^2 \cdot 4^2 \cdot 5^2 \cdot 6^2} + \ldots, \qquad \text{&c.,}$$

which are formed by combinations of the fractions $\frac{1}{1^2}, \frac{1}{2^2}, \frac{1}{3^2}, \frac{1}{4^2}, \frac{1}{5^2} \ldots$ &c. to infinity, omitting $\frac{1}{2^2}$ the second of these fractions we have, to determine the value of b_2, the equation

$$2b_2 \frac{1^2 - 2^2}{1^2 \cdot 3^2 \cdot 4^2 \cdot 5^2 \ldots} = A - BP_2 + CQ_2 - DR_2 + ES_2 - \&\text{c.}$$

Representing in general by $P_n \, Q_n \, R_n \, S_n \ldots$ the sums of the products which can be made by combining all the fractions $\frac{1}{1^2}, \frac{1}{2^2}, \frac{1}{3^2}, \frac{1}{4^2}, \frac{1}{5^2} \ldots$ to infinity, after omitting the fraction $\frac{1}{n^2}$ only; we have in general to determine the quantities $a_1, b_2, c_3, d_4, e_5 \ldots$, &c., the following equations:

$$A - BP_1 + CQ_1 - DR_1 + ES_1 - \&\text{c.} = a_1 \frac{1}{2^2 \cdot 3^2 \cdot 4^2 \cdot 5^2 \ldots},$$

$$A - BP_2 + CQ_2 - DR_2 + ES_2 - \&\text{c.} = 2b_2 \frac{(1^2 - 2^2)}{1^2 \cdot 3^2 \cdot 4^2 \cdot 5^2 \ldots},$$

$$A - BP_3 + CQ_3 - DR_3 + ES_3 - \&\text{c.} = 3c_3 \frac{(1^2 - 3^2)(2^2 - 3^2)}{1^2 \cdot 2^2 \cdot 4^2 \cdot 5^2 \cdot 6^2 \ldots},$$

$$A - BP_4 + CQ_4 - DR_4 + ES_4 - \&\text{c.} = 4d_4 \frac{(1^2 - 4^2)(2^2 - 4^2)(3^2 - 4^2)}{1^2 \cdot 2^2 \cdot 3^2 \cdot 5^2 \cdot 6^2 \ldots},$$

&c.

212.   If we consider now equations (e) which give the values of the coefficients a, b, c, d, &c., we have the following results:

$$a \frac{(2^2 - 1^2)(3^2 - 1^2)(4^2 - 1^2)(5^2 - 1^2) \ldots}{2^2 \cdot 3^2 \cdot 4^2 \cdot 5^2 \ldots}$$
$$= A - BP_1 + CQ_1 - DR_1 + ES_1 - \&\text{c.},$$

$$2b \frac{(1^2 - 2^2)(3^2 - 2^2)(4^2 - 2^2)(5^2 - 2^2) \ldots}{1^2 \cdot 3^2 \cdot 4^2 \cdot 5^2 \ldots}$$
$$= A - BP_2 + CQ_2 - DR_2 + ES_2 - \&\text{c.},$$

$$3c \frac{(1^2 - 3^2)(2^2 - 3^2)(4^2 - 3^2)(5^2 - 3^2) \ldots}{1^2 \cdot 2^2 \cdot 4^2 \cdot 5^2 \ldots}$$
$$= A - BP_3 - CQ_3 - DR_3 + ES_3 - \&\text{c.},$$

$$4d \frac{(1^2 - 4^2)(2^2 - 4^2)(3^2 - 4^2)(5^2 - 4^2) \ldots}{1^2 \cdot 2^2 \cdot 3^2 \cdot 5^2 \ldots}$$
$$= A - BP_4 + CQ_4 - DR_4 + ES_4 - \&\text{c.},$$

&c.

Remarking the factors which are wanting to the numerators and denominators to complete the double series of natural numbers, we see that the fraction is reduced, in the first equation to $\frac{1}{1} \cdot \frac{1}{2}$; in the second to $-\frac{2}{2} \cdot \frac{2}{4}$; in the third to $\frac{3}{3} \cdot \frac{3}{6}$;

in the fourth to $-\frac{4}{4} \cdot \frac{4}{8}$; so that the products which multiply a, $2b$, $3c$, $4d$, &c., are alternately $\frac{1}{2}$ and $-\frac{1}{2}$. It is only required then to find the values of $P_1 Q_1 R_1 S_1$, $P_2 Q_2 R_2 S_2$, $P_3 Q_3 R_3 S_3$, &c.

To obtain them we may remark that we can make these values depend upon the values of the quantities $PQRST$, &c., which represent the different products which may be formed with the fractions $\frac{1}{1^2}$, $\frac{1}{2^2}$, $\frac{1}{3^2}$, $\frac{1}{4^2}$, $\frac{1}{5^2}$, $\frac{1}{6^2}$, &c., without omitting any.

With respect to the latter products, their values are given by the series for the developments of the sine. We represent then the series

$$\frac{1}{1^2} + \frac{1}{2^2} + \frac{1}{3^2} + \frac{1}{4^2} + \frac{1}{5^2} + \&c.$$

$$\frac{1}{1^2 \cdot 2^2} + \frac{1}{1^2 \cdot 3^2} + \frac{1}{1^2 \cdot 4^2} + \frac{1}{2^2 \cdot 3^2} + \frac{1}{2^2 \cdot 4^2} + \frac{1}{3^2 \cdot 4^2} + \&c.$$

$$\frac{1}{1^2 \cdot 2^2 \cdot 3^2} + \frac{1}{1^2 \cdot 2^2 \cdot 4^2} + \frac{1}{1^2 \cdot 3^2 \cdot 4^2} + \frac{1}{2^2 \cdot 3^2 \cdot 4^2} + \&c.$$

$$\frac{1}{1^2 \cdot 2^2 \cdot 3^2 \cdot 4^2} + \frac{1}{2^2 \cdot 3^2 \cdot 4^2 \cdot 5^2} + \frac{1}{1^2 \cdot 2^2 \cdot 3^2 \cdot 5^2} + \&c.$$

by P, Q, R, S, &c.

The series sin $\quad x = x - \dfrac{x^3}{\lfloor 3} + \dfrac{x^5}{\lfloor 5} - \dfrac{x^7}{\lfloor 7} + \&c.$

furnishes the values of the quantities P, Q, R, S, &c. In fact, the value of the sine being expressed by the equation

$$\sin x = x \left(1 - \frac{x^2}{1^2 \pi^2}\right)\left(1 - \frac{x^2}{2^2 \pi^2}\right)\left(1 - \frac{x^2}{3^2 \pi^2}\right)\left(1 - \frac{x^2}{4^2 \pi^2}\right)\left(1 - \frac{x^2}{5^2 \pi^2}\right) \&c.$$

we have

$$1 - \frac{x^2}{\lfloor 3} + \frac{x^4}{\lfloor 5} - \frac{x^6}{\lfloor 7} + \&c.$$

$$= \left(1 - \frac{x^2}{1^2 \pi^2}\right)\left(1 - \frac{x^2}{2^2 \pi^2}\right)\left(1 - \frac{x^2}{3^2 \pi^2}\right)\left(1 - \frac{x^2}{4^2 \pi^2}\right) \&c.$$

Whence we conclude at once that

$$P = \frac{\pi^2}{\lfloor 3}, \quad Q = \frac{\pi^4}{\lfloor 5}, \quad R = \frac{\pi^6}{\lfloor 7}, \quad S = \frac{\pi^8}{\lfloor 9}, \&c.$$

213. Suppose now that P_n, Q_n, R_n, S_n, &c., represent the sums of the different products which can be made with the fractions $\frac{1}{1^2}$, $\frac{1}{2^2}$, $\frac{1}{3^2}$, $\frac{1}{4^2}$, $\frac{1}{5^2}$, &c., from which the fraction $\frac{1}{n^2}$ has been removed, n being any integer whatever; it is required to determine P_n, Q_n, R_n, S_n, &c., by means of P, Q, R, S, &c. If we denote by

$$1 - qP_n + q^2 Q_n - q^3 R_n + q^4 S_n - \&c.,$$

the products of the factors

$$\left(1 - \frac{q}{1^2}\right)\left(1 - \frac{q}{2^2}\right)\left(1 - \frac{q}{3^2}\right)\left(1 - \frac{q}{4^2}\right) \&c.,$$

among which the factor $\left(1 - \frac{q}{n^2}\right)$ only has been omitted; it follows that on multiplying by $\left(1 - \frac{q}{n^2}\right)$ the quantity

$$1 - qP_n + q^2Q_n - q^3R_n + q^4S_n - \&c.,$$

we obtain $\qquad 1 - qP + q^2Q - q^3R + q^4S - \&c.$

This comparison gives the following relations:

$$P_n + \frac{1}{n^2} = P,$$

$$Q_n + \frac{1}{n^2}P_n = Q,$$

$$R_n + \frac{1}{n^2}Q_n = R,$$

$$S_n + \frac{1}{n^2}R_n = S,$$

$$\&c.;$$

or

$$P_n = P - \frac{1}{n^2},$$

$$Q_n = Q - \frac{1}{n^2}P + \frac{1}{n^4},$$

$$R_n = R - \frac{1}{n^2}Q + \frac{1}{n^4}P - \frac{1}{n^6},$$

$$S_n = S - \frac{1}{n^2}R + \frac{1}{n^4}Q - \frac{1}{n^6}P + \frac{1}{n^8},$$

$$\&c.$$

Employing the known values of P, Q, R, S, and making n equal to 1, 2, 3, 4, 5, &c. successively, we shall have the values of $P_1Q_1R_1S_1$, &c.; those of $P_2Q_2R_2S_2$, &c.; those of $P_3Q_3R_3S_3$, &c.

214. From the foregoing theory it follows that the values of a, b, c, d, e, &c., derived from the equations

$$a + 2b + 3c + 4d + 5e + \&c. = A,$$
$$a + 2^3b + 3^3c + 4^3d + 5^3e + \&c. = B,$$
$$a + 2^5b + 3^5c + 4^5d + 5^5e + \&c. = C,$$
$$a + 2^7b + 3^7c + 4^7d + 5^7e + \&c. = D,$$
$$a + 2^9b + 3^9c + 4^9d + 5^9e + \&c. = E,$$
$$\&c.,$$

are thus expressed,

$$\frac{1}{2}a = A - B\left(\frac{\pi^2}{\lfloor 3} - \frac{1}{1^2}\right) + C\left(\frac{\pi^4}{\lfloor 5} - \frac{1}{1^2}\frac{\pi^2}{\lfloor 3} + \frac{1}{1^4}\right)$$
$$- D\left(\frac{\pi^6}{\lfloor 7} - \frac{1}{1^2}\frac{\pi^4}{\lfloor 5} + \frac{1}{1^4}\frac{\pi^2}{\lfloor 3} - \frac{1}{1^6}\right)$$
$$+ E\left(\frac{\pi^8}{\lfloor 9} - \frac{1}{1^2}\frac{\pi^6}{\lfloor 7} + \frac{1}{1^4}\frac{\pi^4}{\lfloor 5} - \frac{1}{1^6}\frac{\pi^2}{\lfloor 3} + \frac{1}{1^8}\right) - \&c.;$$

$$-\frac{1}{2}2b = A - B\left(\frac{\pi^2}{\lfloor 3} - \frac{1}{2^2}\right) + C\left(\frac{\pi^4}{\lfloor 5} - \frac{1}{2^2}\frac{\pi^4}{\lfloor 3} + \frac{1}{2^4}\right)$$
$$- D\left(\frac{\pi^6}{\lfloor 7} - \frac{1}{2^2}\frac{\pi^4}{\lfloor 5} + \frac{1}{2^4}\frac{\pi^2}{\lfloor 3} - \frac{1}{2^6}\right)$$
$$+ E\left(\frac{\pi^8}{\lfloor 9} - \frac{1}{2^2}\frac{\pi^6}{\lfloor 7} + \frac{1}{2^4}\frac{\pi^4}{\lfloor 5} - \frac{1}{2^6}\frac{\pi^2}{\lfloor 3} + \frac{1}{2^8}\right) - \&c.;$$

$$\frac{1}{2}3c = A - B\left(\frac{\pi^2}{\lfloor 3} - \frac{1}{3^2}\right) + C\left(\frac{\pi^4}{\lfloor 5} - \frac{1}{3^2}\frac{\pi^2}{\lfloor 3} + \frac{1}{3^4}\right)$$
$$- D\left(\frac{\pi^6}{\lfloor 7} - \frac{1}{3^2}\frac{\pi^4}{\lfloor 5} + \frac{1}{3^4}\frac{\pi^2}{\lfloor 3} - \frac{1}{3^6}\right)$$
$$+ E\left(\frac{\pi^8}{\lfloor 9} - \frac{1}{3^2}\frac{\pi^6}{\lfloor 7} + \frac{1}{3^4}\frac{\pi^4}{\lfloor 5} - \frac{1}{3^6}\frac{\pi^2}{\lfloor 3} + \frac{1}{3^8}\right) - \&c.;$$

$$-\frac{1}{2}4d = A - B\left(\frac{\pi^2}{\lfloor 3} - \frac{1}{4^2}\right) + C\left(\frac{\pi^4}{\lfloor 5} - \frac{1}{4^2}\frac{\pi^4}{\lfloor 3} + \frac{1}{4^4}\right)$$
$$- D\left(\frac{\pi^6}{\lfloor 7} - \frac{1}{4^2}\frac{\pi^4}{\lfloor 5} + \frac{1}{4^4}\frac{\pi^2}{\lfloor 3} - \frac{1}{4^6}\right)$$
$$+ E\left(\frac{\pi^8}{\lfloor 9} - \frac{1}{4^2}\frac{\pi^6}{\lfloor 7} + \frac{1}{4^4}\frac{\pi^4}{\lfloor 5} - \frac{1}{4^6}\frac{\pi^2}{\lfloor 3} + \frac{1}{4^8}\right) - \&c.;$$

&c.

215. Knowing the values of a, b, c, d, e, &c., we can substitute them in the proposed equation

$$\phi(x) = a \sin x + b \sin 2x + c \sin 3x + d \sin 4x + e \sin 5x + \&c.,$$

and writing also instead of the quantities A, B, C, D, E, &c., their values $\phi'(0)$, $\phi'''(0)$, $\phi^{v}(0)$, $\phi^{vii}(0)$, $\phi^{ix}(0)$, &c., we have the general equation

$$\frac{1}{2}\phi(x) = \sin x \left\{ \phi'(0) + \phi'''(0)\left(\frac{\pi^2}{\lfloor 3} - \frac{1}{1^2}\right) + \phi^{v}(0)\left(\frac{\pi^4}{\lfloor 5} - \frac{1}{1^2}\frac{\pi^2}{\lfloor 3} + \frac{1}{1^4}\right) \right.$$
$$\left. + \phi^{vii}(0)\left(\frac{\pi^6}{\lfloor 7} - \frac{1}{1^2}\frac{\pi^4}{\lfloor 5} + \frac{1}{1^4}\frac{\pi^2}{\lfloor 3} - \frac{1}{1^6}\right) + \&c. \right\};$$

$$-\frac{1}{2}\sin 2x \left\{ \phi'(0) + \phi'''(0)\left(\frac{\pi^2}{\lfloor 3} - \frac{1}{2^2}\right) + \phi^v(0)\left(\frac{\pi^4}{\lfloor 5} - \frac{1}{2^2}\frac{\pi^2}{\lfloor 3} + \frac{1}{2^4}\right) \right.$$

$$\left. + \phi^{vii}(0)\left(\frac{\pi^6}{\lfloor 7} - \frac{1}{2^2}\frac{\pi^4}{\lfloor 5} + \frac{1}{2^4}\frac{\pi^2}{\lfloor 3} - \frac{1}{2^6}\right) + \&c. \right\};$$

$$+\frac{1}{3}\sin 3x \left\{ \phi'(0) + \phi'''(0)\left(\frac{\pi^2}{\lfloor 3} - \frac{1}{3^2}\right) + \phi^v(0)\left(\frac{\pi^4}{\lfloor 5} - \frac{1}{3^2}\frac{\pi^2}{\lfloor 3} + \frac{1}{3^4}\right) \right.$$

$$\left. + \phi^{vii}(0)\left(\frac{\pi^6}{\lfloor 7} - \frac{1}{3^2}\frac{\pi^4}{\lfloor 5} + \frac{1}{3^4}\frac{\pi^2}{\lfloor 3} - \frac{1}{3^6}\right) + \&c. \right\};$$

$$-\frac{1}{4}\sin 4x \left\{ \phi'(0) + \phi'''(0)\left(\frac{\pi^2}{\lfloor 3} - \frac{1}{4^2}\right) + \phi^v(0)\left(\frac{\pi^4}{\lfloor 5} - \frac{1}{4^2}\frac{\pi^2}{\lfloor 3} + \frac{1}{4^4}\right) \right.$$

$$\left. + \phi^{vii}(0)\left(\frac{\pi^6}{\lfloor 7} - \frac{1}{4^2}\frac{\pi^4}{\lfloor 5} + \frac{1}{4^4}\frac{\pi^2}{\lfloor 3} - \frac{1}{4^6}\right) + \&c. \right\};$$

$$+ \&c. \tag{A}$$

We may make use of the preceding series to reduce into a series of sines of multiple arcs any proposed function whose development contains only odd powers of the variable.

216.   The first case which presents itself is that in which $\phi(x) = x$; we find then $\phi'(0) = 1, \phi'''(0) = 0, \phi^v(0) = 0$, &c., and so for the rest. We have therefore the series

$$\frac{1}{2}x = \sin x - \frac{1}{2}\sin 2x + \frac{1}{3}\sin 3x - \frac{1}{4}\sin 4x + \&c.,$$

which has been given by Euler.

If we suppose the proposed function to be x^3, we shall have

$$\phi'(0) = 0, \phi'''(0) = \lfloor 3, \phi^v(0) = 0, \phi^{vii}(0) = 0, \&c.,$$

which gives the equation

$$\frac{1}{2}x^3 = \left(\pi^2 - \frac{\lfloor 3}{1^2}\right)\sin x - \left(\pi^2 - \frac{\lfloor 3}{2^2}\right)\frac{1}{2}\sin 2x + \left(\pi^2 - \frac{\lfloor 3}{3^2}\right)\frac{1}{3}\sin 3x + \&c.$$

We should arrive at the same result, starting from the preceding equation,

$$\frac{1}{2}x = \sin x - \frac{1}{2}\sin 2x + \frac{1}{3}\sin 3x - \frac{1}{4}\sin 4x + \&c.$$

In fact, multiplying each member by dx, and integrating, we have

$$C - \frac{x^2}{4} = \cos x - \frac{1}{2^2}\cos 2x + \frac{1}{3^2}\cos 3x - \frac{1}{4^2}\cos 4x + \&c.;$$

the value of the constant C is

$$1 - \frac{1}{2^2} + \frac{1}{3^2} - \frac{1}{4^2} + \frac{1}{5^2} - \&c.;$$

a series whose sum is known to be $\frac{1}{2}\frac{\pi^2}{\underline{|3}}$. Multiplying by dx the two members of the equation

$$\frac{1}{2}\frac{\pi^2}{\underline{|3}} - \frac{x^2}{4} = \cos x - \frac{1}{2^2}\cos 2x + \frac{1}{3^2}\cos 3x - \&c.,$$

and integrating we have

$$\frac{1}{2}\frac{\pi^2 x}{\underline{|3}} - \frac{1}{2}\frac{x^3}{\underline{|3}} = \sin x - \frac{1}{2^2}\sin 2x + \frac{1}{3^2}\sin 3x - \&c.$$

If now we write instead of x its value derived from the equation

$$\frac{1}{2}x = \sin x - \frac{1}{2}\sin 2x + \frac{1}{3}\sin 3x - \frac{1}{4}\sin 4x + \&c.,$$

we shall obtain the same equation as above, namely,

$$\frac{1}{2}\frac{x^3}{\underline{|3}} = \sin x\left(\frac{\pi^2}{\underline{|3}} - \frac{1}{1^2}\right) - \frac{1}{2}\sin 2x\left(\frac{\pi^2}{\underline{|3}} - \frac{1}{2^2}\right) + \frac{1}{3}\sin 3x\left(\frac{\pi^2}{\underline{|3}} - \frac{1}{3^2}\right) - \&c.$$

We could arrive in the same manner at the development in series of multiple arcs of the powers x^5, x^7, x^9, &c., and in general every function whose development contains only odd powers of the variable.

217. Equation (A), (Art. 215), can be put under a simpler form, which we may now indicate. We remark first, that part of the coefficient of $\sin x$ is the series

$$\phi'(0) + \frac{\pi^2}{\underline{|3}}\phi'''(0) + \frac{\pi^4}{\underline{|5}}\phi^v(0) + \frac{\pi^6}{\underline{|7}}\phi^{vii}(0) + \&c.,$$

which represents the quantity $\frac{1}{\pi}\phi(\pi)$. In fact, we have, in general,

$$\phi(x) = \phi(0) + x\phi'(0) + \frac{x^2}{\underline{|2}}\phi''(0) + \frac{x^3}{\underline{|3}}\phi'''(0) + \frac{x^4}{\underline{|4}}\phi^{iv}(0) + \&c.$$

Now, the function $\phi(x)$ containing by hypothesis only odd powers, we must have $\phi(0) = 0$, $\phi''(0) = 0$, $\phi^{iv}(0) = 0$, and so on. Hence

$$\phi(x) = x\phi'(0) + \frac{x^3}{\underline{|3}}\phi'''(0) + \frac{x^5}{\underline{|5}}\phi^v(0) + \&c.;$$

a second part of the coefficient of $\sin x$ is found by multiplying by $-\frac{1}{1^2}$ the series

$$\phi'''(0) + \frac{\pi^3}{\underline{|3}}\phi^v(0) + \frac{\pi^4}{\underline{|5}}\phi^{vii}(0) + \frac{\pi^6}{\underline{|7}}\phi^{ix}(0) + \&c.,$$

whose value is $\frac{1}{\pi}\phi''(\pi)$. We can determine in this manner the different parts of the coefficient of $\sin x$, and the components of the coefficients of $\sin 2x$, $\sin 3x$, $\sin 4x$, &c. We may employ for this purpose the equations:

$$\phi'(0) + \frac{\pi^2}{\underline{|3}}\phi'''(0) + \frac{\pi^4}{\underline{|5}}\phi^v(0) + \&c. = \frac{1}{\pi}\phi(\pi);$$

$$\phi'''(0) + \frac{\pi^2}{\underline{|3}}\phi^{v}(0) + \frac{\pi^4}{\underline{|5}}\phi^{vii}(0) + \&c. = \frac{1}{\pi}\phi''(\pi);$$

$$\phi^{v}(0) + \frac{\pi^2}{\underline{|3}}\phi^{vii}(0) + \frac{\pi^4}{\underline{|5}}\phi^{ix}(0) + \&c. = \frac{1}{\pi}\phi^{iv}(\pi);$$

By means of these reductions equation (A) takes the following form:

$$\frac{1}{2}\pi\phi(x) = \sin x \left\{ \phi(\pi) - \frac{1}{1^2}\phi''(\pi) + \frac{1}{1^4}\phi^{iv}(\pi) - \frac{1}{1^6}\phi^{vi}(\pi) + \&c. \right\}$$

$$-\frac{1}{2}\sin 2x \left\{ \phi(\pi) - \frac{1}{2^2}\phi''(\pi) + \frac{1}{2^4}\phi^{iv}(\pi) - \frac{1}{2^6}\phi^{vi}(\pi) + \&c. \right\}$$

$$+\frac{1}{3}\sin 3x \left\{ \phi(\pi) - \frac{1}{3^2}\phi''(\pi) + \frac{1}{3^4}\phi^{iv}(\pi) - \frac{1}{3^6}\phi^{vi}(\pi) + \&c. \right\}$$

$$-\frac{1}{4}\sin 4x \left\{ \phi(\pi) - \frac{1}{4^2}\phi''(\pi) + \frac{1}{4^4}\phi^{iv}(\pi) - \frac{1}{4^6}\phi^{vi}(\pi) + \&c. \right\}$$

$$+ \&c. \qquad\qquad\qquad\qquad\qquad\qquad\qquad\qquad\qquad\qquad (B);$$

or this,

$$\frac{1}{2}\pi\phi(x) = \phi(\pi) \left\{ \sin x - \frac{1}{2}\sin 2x + \frac{1}{3}\sin 3x - \&c. \right\}$$

$$-\phi''(\pi) \left\{ \sin x - \frac{1}{2^3}\sin 2x + \frac{1}{3^3}\sin 3x - \&c. \right\}$$

$$+\phi^{iv}(\pi) \left\{ \sin x - \frac{1}{2^5}\sin 2x + \frac{1}{3^5}\sin 3x - \&c. \right\}$$

$$-\phi^{vi}(\pi) \left\{ \sin x - \frac{1}{2^7}\sin 2x + \frac{1}{3^7}\sin 3x - \&c. \right\}$$

$$+ \&c. \qquad\qquad\qquad\qquad\qquad\qquad\qquad\qquad\qquad\qquad (C).$$

218. We can apply one or other of these formulæ as often as we have to develope a proposed function in a series of sines of multiple arcs. If, for example, the proposed function is $e^x - e^{-x}$, whose development contains only odd powers of x, we shall have

$$\frac{1}{2}\pi\frac{e^x - e^{-x}}{e^\pi - e^{-\pi}} = \left(\sin x - \frac{1}{2}\sin 2x + \frac{1}{3}\sin 3x - \&c. \right)$$

$$- \left(\sin x - \frac{1}{2^3}\sin 2x + \frac{1}{3^3}\sin 3x - \&c. \right)$$

$$+ \left(\sin x - \frac{1}{2^5}\sin 2x + \frac{1}{3^5}\sin 3x - \&c. \right)$$

$$- \left(\sin x - \frac{1}{2^7}\sin 2x + \frac{1}{3^7}\sin 3x - \&c. \right)$$

$$+ \&c.$$

Collecting the coefficients of $\sin x$, $\sin 2x$, $\sin 3x$, $\sin 4x$, &c., and writing, instead of $\frac{1}{n} - \frac{1}{n^3} + \frac{1}{n^5} - \frac{1}{n^7} + $ etc., its value $\frac{n}{n^2 + 1}$, we have

$$\frac{1}{2}\,\pi\,\frac{(e^x - e^{-x})}{e^\pi - e^{-\pi}} = \frac{\sin x}{1 + \frac{1}{1}} - \frac{\sin 2x}{2 + \frac{1}{2}} + \frac{\sin 3x}{3 + \frac{1}{3}} - \&c.$$

We might multiply these applications and derive from them several remarkable series. We have chosen the preceding example because it appears in several problems relative to the propagation of heat.

219. Up to this point we have supposed that the function whose development is required in a series of sines of multiple arcs can be developed in a series arranged according to powers of the variable x, and that only odd powers enter into that series. We can extend the same results to any functions, even to those which are discontinuous and entirely arbitrary. To establish clearly the truth of this proposition, we must follow the analysis which furnishes the foregoing equation (B), and examine what is the nature of the coefficients which multiply $\sin x$, $\sin 2x$, $\sin 3x$, &c. Denoting by s the quantity which multiplies $\frac{1}{n}\sin nx$ in this equation when n is odd, and $-\frac{1}{n}\sin nx$ when n is even, we have

$$s = \phi\,(\pi) - \frac{1}{n^2}\,\phi''\,(\pi) + \frac{1}{n^4}\,\phi^{iv}(\pi) - \frac{1}{n^6}\,\phi^{vi}(\pi) + \&c.$$

Considering s as a function of π, differentiating twice, and comparing the results, we find $s + \frac{1}{n^2}\frac{d^2s}{d\pi^2} = \phi(\pi)$; an equation which the foregoing value of s must satisfy.

Now the integral of the equation $s + \frac{1}{n^2}\frac{d^2s}{dx^2} = \phi(x)$, in which s is considered to be a function of x, is

$$s = a \cos nx + n \sin nx \int \cos nx\, \phi\,(x)\, dx - n \cos nx \int \sin nx\, \phi\,(x)\, dx.$$

If n is an integer, and the value of x is equal to π, we have $s = \pm n \int \phi(x) \sin nx\, dx$. The sign $+$ must be chosen when n is odd, and the sign $-$ when that number is even. We must make x equal to the semi-circumference π, after the integration indicated; the result may be verified by developing the term $\int \phi(x) \sin nx\, dx$, by means of integration by parts, remarking that the function $\phi(x)$ contains only odd powers of the variable x, and taking the integral from $x = 0$ to $x = \pi$.

We conclude at once that the term is equal to

$$\pm \left\{ \phi(\pi) - \frac{1}{n^2}\,\phi''\,(\pi) + \frac{1}{n^4}\,\phi^{iv}(\pi) - \frac{1}{n^6}\,\phi^{vi}(\pi) + \frac{1}{n^8}\,\phi^{viii}(\pi) - \&c. \right\}.$$

If we substitute this value of $\frac{s}{n}$ in equation (B), taking the sign $+$ when the term of this equation is of odd order, and the sign $-$ when n is even, we shall have in general $\int \phi(x) \sin nx\, dx$ for the coefficient of $\sin nx$; in this manner we arrive at a

541

very remarkable result expressed by the following equation:

$$\frac{1}{2}\pi\,\phi(x) = \sin x \int \sin x\phi(x)\,dx + \sin 2x \int \sin 2x\phi(x)\,dx + \&c.$$

$$+ \sin ix \int \sin ix\,\phi(x)\,dx + \&c. \dots\dots\dots (D),$$

the second member will always give the development required for the function $\phi(x)$, if we integrate from $x = 0$ to $x = \pi$.[6]

220. We see by this that the coefficients a, b, c, d, e, f, &c., which enter into the equation

$$\frac{1}{2}\pi\,\phi(x) = a\sin x + b\sin 2x + c\sin 3x + d\sin 4x + \&c.,$$

and which we found formerly by way of successive eliminations, are the values of definite integrals expressed by the general term $\int \sin ix\phi(x)\,dx$, i being the number of the term whose coefficient is required. This remark is important, because it shews how even entirely arbitrary functions may be developed in series of sines of multiple arcs. In fact, if the function $\phi(x)$ be represented by the variable ordinate of any curve whatever whose abscissa extends from $x = 0$ to $x = \pi$, and if on the same part of the axis the known trigono-metric curve, whose ordinate is $y = \sin x$, be constructed, it is easy to represent the value of any integral term. We must suppose that for each abscissa x, to which corresponds one value of $\phi(x)$, and one value of $\sin x$, we multiply the latter value by the first, and at the same point of the axis raise an ordinate equal to the product $\phi(x)\sin x$. By this contin-uous operation a third curve is formed, whose ordinates are those of the trigonometric curve, reduced in proportion to the ordinates of the arbitary curve which represents $\phi(x)$. This done, the area of the reduced curve taken from $x = 0$ to $x = \pi$ gives the exact value of the coefficient of $\sin x$; and whatever the given curve may be which corresponds to $\phi(x)$, whether we can assign to it an analytical equation, or whether it depends on no regular law, it is evident that it always serves to reduce in any manner whatever the trigonometric curve; so that the area of the reduced curve has, in all possible cases, a definite value, which is the value of the coefficient of $\sin x$ in the development of the function. The same is the case with the following coefficient b, or $\int \phi(x)\sin 2x\,dx$.

In general, to construct the values of the coefficients a, b, c, d, &c., we must imagine that the curves, whose equations are

$$y = \sin x, \quad y = \sin 2x, \quad y = \sin 3x, \quad y = \sin 4x, \&c.,$$

---

[6] Lagrange had already shown (*Miscellanea Taurinensia*, Tom. III., 1766,) that the function y given by the equation

$$y = 2\left(\sum_{r=1}^{r=n} Y_r \sin X_r \pi \Delta X\right)\sin x\pi + 2\left(\sum_{r=1}^{r=n} Y_r \sin 2X_r \pi \Delta X\right)\sin 2x\pi + 2\left(\sum_{r=1}^{r=n} Y_r \sin 3X_r \pi \Delta X\right)\sin 3x\pi + \dots + 2\left(\sum_{r=1}^{r=n} Y_r \sin nX_r \pi \Delta X\right)\sin nx\pi$$

receives the values $Y_1, Y_2, Y_3 \dots Y_n$ corresponding to the values $X_1, X_2, X_3 \dots X_n$ of x, where $X_r = \frac{r}{n+1}$, and $\Delta X = \frac{1}{n+1}$.

Lagrange however abstained from the transition from this summation-formula to the integration-formula given by Fourier.

Cf. Riemann's *Gesammelte Mathematische Werke*, Leipzig, 1876, of his historical criticism, *Ueber die Darstellbarkeit einer Function durch eine Trigonometrische Reihe.* [A. F.]

have been traced for the same interval on the axis of x, from $x = 0$ to $x = \pi$; and then that we have changed these curves by multiplying all their ordinates by the corresponding ordinates of a curve whose equation is $y = \phi(x)$. The equations of the reduced curves are

$$y = \sin x \, \phi(x), \quad y = \sin 2x \, \phi(x), \quad y = \sin 3x \, \phi(x), \text{ \&c.}$$

The areas of the latter curves, taken from $x = 0$ to $x = \pi$, are the values of the coefficients a, b, c, d, \&c., in the equation

$$\frac{1}{2}\pi \phi(x) = a \sin x + b \sin 2x + c \sin 3x + d \sin 4x + \text{\&c.}$$

221. We can verify the foregoing equation (D), (Art. 220), by determining directly the quantities $a_1, a_2, a_3, \ldots a_j$, \&c., in the equation

$$\phi(x) = a_1 \sin x + a_2 \sin 2x + a_3 \sin 3x + \ldots a_j \sin jx + \text{\&c.;}$$

for this purpose, we multiply each member of the latter equation by $\sin ix \, dx$, i being an integer, and take the integral from $x = 0$ to $x = \pi$, whence we have

$$\int \phi(x) \sin ix \, dx = a_1 \int \sin x \sin ix \, dx + a_2 \int \sin 2x \sin ix \, dx + \text{\&c.}$$

$$+ a_j \int \sin jx \sin ix \, dx + \ldots \text{\&c.}$$

Now it can easily be proved, 1st, that all the integrals, which enter into the second member, have a nul value, except only the term $a_i \int \sin ix \sin ix \, dx$; 2nd, that the value of $\int \sin ix \sin ix \, dx$ is $\frac{1}{2}\pi$; whence we derive the value of a_i, namely

$$\frac{2}{\pi} \int \phi(x) \sin ix \, dx.$$

The whole problem is reduced to considering the value of the integrals which enter into the second member, and to demonstrating the two preceding propositions. The integral

$$2 \int \sin jx \sin ix \, dx,$$

taken from $x = 0$ to $x = \pi$, in which i and j are integers, is

$$\frac{1}{i-j} \sin(i-j)x - \frac{1}{i+j} \sin(i+j)x + C.$$

Since the integral must begin when $x = 0$ the constant C is nothing, and the numbers i and j being integers, the value of the integral will become nothing when $x = p$; it follows that each of the terms, such as

$$a_1 \int \sin x \sin ix \, dx, \quad a_2 \int \sin 2x \sin ix \, dx, \quad a_3 \int \sin 3x \sin ix \, dx, \quad \text{\&c.,}$$

vanishes, and that this will occur as often as the numbers i and j are different. The same is not the case when the numbers i and j are equal, for the term $\frac{1}{i-j}\sin(i-j)x$ to which the integral reduces, becomes $\frac{0}{0}$, and its value is π. Consequently we have

$$2\int \sin ix \sin ix \, dx = \pi;$$

thus we obtain, in a very brief manner, the values of $a_1, a_2, a_3, \ldots a_i$, &c., namely,

$$a_1 = \frac{2}{\pi}\int \phi(x)\sin x \, dx, \qquad a_2 = \frac{2}{\pi}\int \phi(x)\sin 2x \, dx,$$

$$a_3 = \frac{2}{\pi}\int \phi(x)\sin 3x \, dx, \qquad a_i = \frac{2}{\pi}\int \phi(x)\sin ix \, dx.$$

Substituting these we have

$$\frac{1}{2}\pi\phi(x) = \sin x\int \phi(x)\sin x \, dx + \sin 2x\int \phi(x)\sin 2x \, dx + \&c.$$

$$+ \sin ix\int \phi(x)\sin ix \, dx + \&c.$$

222. The simplest case is that in which the given function has a constant value for all values of the variable x included between 0 and π; in this case the integral $\int \sin ix \, dx$ is equal to $\frac{2}{i}$, if the number i is odd, and equal to 0 if the number i is even.

Hence we deduce the equation

$$\frac{1}{4}\pi = \sin x + \frac{1}{3}\sin 3x + \frac{1}{5}\sin 5x + \frac{1}{7}\sin 7x + \&c., \tag{A}$$

which has been found before.

It must be remarked that when a function $\phi(x)$ has been developed in a series of sines of multiple arcs, the value of the series

$$a\sin x + b\sin 2x + c\sin 3x + d\sin 4x + \&c.$$

is the same as that of the function $\phi(x)$ so long as the variable x is included between 0 and π; but this equality ceases in general to hold good when the value of x exceeds the number π.

Suppose the function whose development is required to be x, we shall have, by the preceding theorem,

$$\frac{1}{2}\pi x = \sin x\int x\sin x \, dx + \sin 2x\int x\sin 2x \, dx$$

$$+ \sin 3x\int x\sin 3x \, dx + \&c.$$

The integral $\int_0^\pi x \sin i x\, dx$ is equal to $\pm\frac{\pi}{i}$; the indices 0 and π, which are connected with the sign $\int$, shew the limits of the integral; the sign $+$ must be chosen when i is odd, and the sign $-$ when i is even. We have then the following equation,

$$\frac{1}{2}x = \sin x - \frac{1}{2}\sin 2x + \frac{1}{3}\sin 3x - \frac{1}{4}\sin 4x + \frac{1}{5}\sin 5x - \&c.$$

223.   We can develope also in a series of sines of multiple arcs functions different from those in which only odd powers of the variable enter. To instance by an example which leaves no doubt as to the possibility of this development, we select the function $\cos x$, which contains only even powers of x, and which may be developed under the following form:

$$a \sin x + b \sin 2x + c \sin 3x + d \sin 4x + e \sin 5x + \&c.,$$

although in this series only odd powers of the variable enter.

We have, in fact, by the preceding theorem,

$$\frac{1}{2}\pi \cos x = \sin x \int \cos x \sin x\, dx + \sin 2x \int \cos x \sin 2x\, dx$$

$$+ \sin 3x \int \cos x \sin 3x\, dx + \&c.$$

The integral $\int \cos x \sin i x\, dx$ is equal to zero when i is an odd number, and to $\frac{2i}{i^2 - 1}$ when i is an even number. Supposing successively $i = 2, 4, 6, 8$, etc., we have the always convergent series

$$\frac{1}{4}\pi \cos x = \frac{2}{1 \cdot 3}\sin 2x + \frac{4}{3 \cdot 5}\sin 4x + \frac{6}{5 \cdot 7}\sin 6x + \&c.;$$

or,

$$\cos x = \frac{2}{\pi}\left\{\left(\frac{1}{1} + \frac{1}{3}\right)\sin 2x + \left(\frac{1}{3} + \frac{1}{5}\right)\sin 4x + \left(\frac{1}{5} + \frac{1}{7}\right)\sin 6x + \&c.\right\}.$$

This result is remarkable in this respect, that it exhibits the development of the cosine in a series of functions, each one of which contains only odd powers. If in the preceding equation x be made equal to $\frac{1}{4}\pi$, we find

$$\frac{1}{4}\frac{\pi}{\sqrt{2}} = \frac{1}{2}\left(\frac{1}{1} + \frac{1}{3} - \frac{1}{5} - \frac{1}{7} + \frac{1}{9} + \frac{1}{11} - \&c.\right).$$

This series is known (*Introd. ad analysin. infinit.* cap. x.).

224.   A similar analysis may be employed for the development of any function whatever in a series of cosines of multiple arcs.

Let $\phi(x)$ be the function whose development is required, we may write

$$\phi(x) = a_0 \cos 0x + a_1 \cos x + a_2 \cos 2x + a_3 \cos 3x + \&c.$$

$$+ a_i \cos i x + \&c. \ldots\ldots (m).$$

If the two members of this equation be multiplied by cos *jx*, and each of the terms of the second member integrated from $x = 0$ to $x = \pi$; it is easily seen that the value of the integral will be nothing, save only for the term which already contains cos *jx*. This remark gives immediately the coefficient a_j; it is sufficient in general to consider the value of the integral

$$\int \cos jx \cos ix \, dx,$$

taken from $x = 0$ to $x = \pi$, supposing j and i to be integers. We have

$$\int \cos jx \cos ix \, dx = \frac{1}{2(j+i)} \sin(j+i)x + \frac{1}{2(j-i)} \sin(j-i)x + c.$$

This integral, taken from $x = 0$ to $x = \pi$, evidently vanishes whenever j and i are two different numbers. The same is not the case when the two numbers are equal. The last term

$$\frac{1}{2(j-i)} \sin(j-i)x$$

becomes $\frac{0}{0}$, and its value is $\frac{1}{2}\pi$, when the arc x is equal to π. If then we multiply the two terms of the preceding equation (*m*) by cos *ix*, and integrate it from 0 to π, we have

$$\int \phi(x) \cos ix \, dx = \frac{1}{2}\pi a_i,$$

an equation which exhibits the value of the coefficient a_i.

To find the first coefficient a_0, it may be remarked that in the integral

$$\frac{1}{2(j+i)} \sin(j+i)x + \frac{1}{2(j-i)} \sin(j-i)x,$$

if $j = 0$ and $i = 0$ each of the terms becomes $\frac{0}{0}$, and the value of each term is $\frac{1}{2}\pi$; thus the integral $\int \cos jx \cos ix\, dx$ taken from $x = 0$ to $x = \pi$ is nothing when the two integers j and i are different: it is $\frac{1}{2}\pi$ when the two numbers j and i are equal but different from zero; it is equal to π when j and i are each equal to zero; thus we obtain the following equation,

$$\frac{1}{2}\pi\phi(x) = \frac{1}{2}\int_0^\pi \phi(x)\,dx + \cos x \int_0^\pi \phi(x)\cos x\,dx + \cos 2x \int_0^\pi \phi(x)\cos 2x\,dx$$

$$+ \cos 3x \int_0^\pi \phi(x)\cos 3x\,dx + \&c. \quad (n)[7].$$

This and the preceding theorem suit all possible functions, whether their character can be expressed by known methods of analysis, or whether they correspond to curves traced arbitrarily.

---

[7] The process analogous to (*A*) in Art. 222 fails here; yet we see, Art. 177, that an analogous result exists. [R. L. E.]

225.   If the proposed function whose development is required in cosines of multiple arcs is the variable x itself; we may write down the equation

$$\frac{1}{2}\pi x = a_0 + a_1 \cos x + a_2 \cos 2x + a_3 \cos 3x + \ldots + a_i \cos ix + \&c.,$$

and we have, to determine any coefficient whatever a_i, the equation $a_i = \int_0^\pi x \cos ix \, dx$. This integral has a nul value when i is an even number, and is equal to $-\frac{2}{i^2}$ when i is odd. We have at the same time $a_0 = \frac{1}{4}\pi^2$. We thus form the following series,

$$x = \frac{1}{2}\pi - 4\frac{\cos x}{\pi} - 4\frac{\cos 3x}{3^2\pi} - 4\frac{\cos 5x}{5^2\pi} - 4\frac{\cos 7x}{7^2\pi} - \&c.$$

We may here remark that we have arrived at three different developments for x, namely,

$$\frac{1}{2}x = \sin x - \frac{1}{2}\sin 2x + \frac{1}{3}\sin 3x - \frac{1}{4}\sin 4x + \frac{1}{5}\sin 5x - \&c.,$$

$$\frac{1}{2}x = \frac{2}{\pi}\sin x - \frac{2}{3^2\pi}\sin 3x + \frac{2}{5^2\pi}\sin 5x - \&c. \text{ (Art. 181)},$$

$$\frac{1}{2}x = \frac{1}{4}\pi - \frac{2}{\pi}\cos x - \frac{2}{3^2\pi}\cos 3x - \frac{2}{5^2\pi}\cos 5x - \&c.$$

It must be remarked that these three values of $\frac{1}{2}x$ ought not to be considered as equal; with reference to all possible values of x, the three preceding developments have a common value only when the variable x is included between 0 and $\frac{1}{2}\pi$. The construction of the values of these three series, and the comparison of the lines whose ordinates are expressed by them, render sensible the alternate coincidence and divergence of values of these functions.

To give a second example of the development of a function in a series of cosines of multiple arcs, we choose the function $\sin x$, which contains only odd powers of the variable, and we may suppose it to be developed in the form

$$a + b \cos x + c \cos 2x + d \cos 3x + \&c.$$

Applying the general equation to this particular case, we find, as the equation required,

$$\frac{1}{4}\pi \sin x = \frac{1}{2} - \frac{\cos 2x}{1 \cdot 3} - \frac{\cos 4x}{3 \cdot 5} - \frac{\cos 6x}{5 \cdot 7} - \&c.$$

Thus we arrive at the development of a function which contains only odd powers in a series of cosines in which only even powers of the variable enter. If we give to x the particular value $\frac{1}{2}\pi$, we find

$$\frac{1}{4}\pi = \frac{1}{2} + \frac{1}{1 \cdot 3} - \frac{1}{3 \cdot 5} + \frac{1}{5 \cdot 7} - \frac{1}{7 \cdot 9} + \&c.$$

Now, from the known equation,

$$\frac{1}{4}\pi = 1 - \frac{1}{3} + \frac{1}{5} - \frac{1}{7} + \frac{1}{9} - \frac{1}{11} + \&c.,$$

we derive

$$\frac{1}{8}\pi = \frac{1}{1 \cdot 3} + \frac{1}{5 \cdot 7} + \frac{1}{9 \cdot 11} + \frac{1}{13 \cdot 15} + \&c.,$$

and also

$$\frac{1}{8}\pi = \frac{1}{2} - \frac{1}{3 \cdot 5} - \frac{1}{7 \cdot 9} - \frac{1}{11 \cdot 13} - \&c.$$

Adding these two results we have, as above,

$$\frac{1}{4}\pi = \frac{1}{2} + \frac{1}{1 \cdot 3} - \frac{1}{3 \cdot 5} + \frac{1}{5 \cdot 7} - \frac{1}{7 \cdot 9} + \frac{1}{9 \cdot 11} - \&c.$$

226. The foregoing analysis giving the means of developing any function whatever in a series of sines or cosines of multiple arcs, we can easily apply it to the case in which the function to be developed has definite values when the variable is included between certain limits and has real values, or when the variable is included between other limits. We stop to examine this particular case, since it is presented in physical questions which depend on partial differential equations, and was proposed formerly as an example of functions which cannot be developed in sines or cosines of multiple arcs. Suppose then that we have reduced to a series of this form a function whose value is constant, when x is included between 0 and α, and all of whose values are nul when x is included between α and π. We shall employ the general equation (D), in which the integrals must be taken from $x = 0$ to $x = \pi$. The values of $\phi(x)$ which enter under the integral sign being nothing from $x = \alpha$ to $x = \pi$, it is sufficient to integrate from $x = 0$ to $x = \alpha$. This done, we find, for the series required, denoting by h the constant value of the function,

$$\frac{1}{2}\pi\phi(x) = h\left\{(1 - \cos\alpha)\sin x + \frac{1 - \cos 2\alpha}{2}\sin 2x \right.$$
$$\left. + \frac{1 - \cos 3\alpha}{3}\sin 3x + \&c.\right\}.$$

If we make $h = \frac{1}{2}\pi$, and represent the versed sine of the arc x by versin x, we have

$$\phi(x) = \text{versin } \alpha \sin x + \frac{1}{2}\text{versin } 2\alpha \sin 2x + \frac{1}{3}\text{versin } 3\alpha \sin 3x + \&c.[8]$$

This series, always convergent, is such that if we give to x any value whatever included between 0 and α, the sum of its terms will be $\frac{1}{2}\pi$; but if we give to x

[8]In what cases a function, arbitrary between certain limits, can be developed in a series of cosines, and in what cases in a series of sines, has been shown by Sir W. Thomson, *Camb. Math. Journal*, Vol. II., in an article signed P. Q. R., *On Fourier's Expansions of Functions in Trigonometrical Series*. [A. F.]

any value whatever greater than α and less than π, the sum of the terms will be nothing.

In the following example, which is not less remarkable, the values of $\phi(x)$ are equal to $\sin \frac{\pi x}{\alpha}$ for all values of x included between 0 and α, and nul for values of x between α and π. To find what series satisfies this condition, we shall employ equation (D).

The integrals must be taken from $x = 0$ to $x = \pi$; but it is sufficient, in the case in question, to take these integrals from $x = 0$ to $x = \alpha$, since the values of $\phi(x)$ are supposed nul in the rest of the interval. Hence we find

$$\phi(x) = 2\alpha \left\{ \frac{\sin \alpha \sin x}{\pi^2 - \alpha^2} + \frac{\sin 2\alpha \sin 2x}{\pi^2 - 2^2\alpha^2} + \frac{\sin 3\alpha \sin 3x}{\pi^2 - 3^2\alpha^2} + \&c. \right\}.$$

If α be supposed equal to π, all the terms of the series vanish, except the first, which becomes $\frac{0}{0}$, and whose value is $\sin x$; we have then $\phi(x) = \sin x$.

227.   The same analysis could be extended to the case in which the ordinate represented by $\phi(x)$ was that of a line composed of different parts, some of which might be arcs of curves and others straight lines. For example, let the value of the function, whose development is equired in a series of cosines of multiple arcs, be $\left(\frac{\pi}{2}\right)^2 - x^2$, from $x = 0$ to $x = \frac{1}{2}\pi$, and be nothing from $x = \frac{1}{2}\pi$ to $x = \pi$. We shall employ the general equation (n), and effecting the integrations within the given limits, we find that the general term[9] $\int \left[\left(\frac{\pi}{2}\right)^2 - x^2\right] \cos ix\, dx$ is equal to $\frac{2}{i^3}$ when i is even, to $+\frac{\pi}{i^2}$ when i is the double of an odd number, and to $-\frac{\pi}{i^2}$ when i is four times an odd number. On the other hand, we find $\frac{1}{3}\frac{\pi^3}{2^3}$ for the value of the first term $\frac{1}{2}\int \phi(x)dx$. We have then the following development:

$$\frac{1}{2}\phi(x) = \frac{1}{2 \cdot 3}\left(\frac{\pi}{2}\right)^2 + \frac{2}{\pi}\left\{ \frac{\cos x}{1^3} - \frac{\cos 3x}{3^3} + \frac{\cos 5x}{5^3} - \frac{\cos 7x}{7^3} + \&c. \right\}$$
$$+ \frac{\cos 2x}{2^2} - \frac{\cos 4x}{4^2} + \frac{\cos 6x}{6^2} - \&c.$$

The second member is represented by a line composed of parabolic arcs and straight lines.

228.   In the same manner we can find the development of a function of x which expresses the ordinate of the contour of a trapezium. Suppose $\phi(x)$ to be equal to x from $x = 0$ to $x = \alpha$, that the function is equal to α from $x = \alpha$ to $x = \pi - \alpha$, and lastly equal to $\pi - x$, from $x = \pi - \alpha$ to $x = \pi$. To reduce it to a series of sines of multiple arcs, we employ the general equation (D). The general term $\int \phi(x) \sin ix\, dx$ is composed of three different parts, and we have, after the reductions, $\frac{2}{i^2}\sin i\alpha$ for the coefficient of $\sin ix$, when i is an odd number;

---

[9] $\int \left[\left(\frac{\pi}{2}\right)^2 - x^2\right] \cos ix\, dx = \left(\frac{\pi}{2}\right)^2 \frac{\sin ix}{i} - \frac{x^2}{i}\sin ix - \frac{2}{i^2}x \cos ix + 2\frac{\sin ix}{i^3}$ [R.L.E.]

but the coefficient vanishes when i is an even number. Thus we arrive at the equation

$$\frac{1}{2}\pi\,\phi(x) = 2\left\{ \sin\alpha \sin x + \frac{1}{3^2}\sin 3\alpha \sin 3x + \frac{1}{5^2}\sin 5\alpha \sin 5x \right.$$
$$\left. + \frac{1}{7^2}\sin 7\alpha \sin 7x + \&c. \right\} \quad (\lambda).[10]$$

If we supposed $\alpha = \frac{1}{2}\pi$, the trapezium would coincide with an isosceles triangle, and we should have, as above, for the equation of the contour of this triangle,

$$\frac{1}{2}\pi\phi(x) = 2\left(\sin x - \frac{1}{3^2}\sin 3x + \frac{1}{5^2}\sin 5x - \frac{1}{7^2}\sin 7x + \&c. \right),[11]$$

a series which is always convergent whatever be the value of x. In general, the trigonometric series at which we have arrived, in developing different functions are always convergent, but it has not appeared to us necessary to demonstrate this here; for the terms which compose these series are only the coefficients of terms of series which give the values of the temperature; and these coefficients are affected by certain exponential quantities which decrease very rapidly, so that the final series are very convergent. With regard to those in which only the sines and cosines of multiple arcs enter, it is equally easy to prove that they are convergent, although they represent the ordinates of discontinuous lines. This does not result solely from the fact that the values of the terms diminish continually; for this condition is not sufficient to establish the convergence of a series. It is necessary that the values at which we arrive on increasing continually the number of terms, should approach more and more a fixed limit, and should differ from it only by a quantity which becomes less than any given magnitude: this limit is the value of the series. Now we may prove rigorously that the series in question satisfy the last condition.

229. Take the preceding equation (λ) in which we can give to x any value whatever; we shall consider this quantity as a new ordinate, which gives rise to the following construction.

Fig. 8.

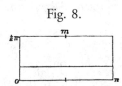

Having traced on the plane of x and y (see fig. 8) a rectangle whose base $O\pi$ is equal to the semi-circumference, and whose height is $\frac{1}{2}\pi$; on the middle point m

---

[10] The accuracy of this and other series given by Fourier is maintained by Sir W. Thomson in the article quoted in note 9.

[11] Expressed in cosines between the limits 0 and π,

$$\frac{1}{2}\pi\phi(x) = \frac{\pi^2}{8} - \left(\cos 2x + \frac{1}{3^2}\cos 6x + \frac{1}{5^2}\cos 10x + \&c. \right).$$

of the side parallel to the base, let us raise perpendicularly to the plane of the rectangle a line equal to $\frac{1}{2}\pi$, and from the upper end of this line draw straight lines to the four corners of the rectangle. Thus will be formed a quadrangular pyramid. If we now measure from the point O on the shorter side of the rectangle, any line equal to α, and through the end of this line draw a plane parallel to the base $O\pi$, and perpendicular to the plane of the rectangle, the section common to this plane and to the solid will be the trapezium whose height is equal to α. The variable ordinate of the contour of this trapezium is equal, as we have just seen, to

$$\frac{4}{\pi}\left(\sin\alpha \sin x + \frac{1}{3^2}\sin 3\alpha \sin 3x + \frac{1}{5^2}\sin 5\alpha \sin 5x + \&c.\right).$$

It follows from this that calling x, y, z the co-ordinates of any point whatever of the upper surface of the quadrangular pyramid which we have formed, we have for the equation of the surface of the polyhedron, between the limits

$$x = 0, \ x = \pi, \ y = 0, \ y = \tfrac{1}{2}\pi,$$

$$\frac{1}{2}\pi z = \frac{\sin x \sin y}{1^2} + \frac{\sin 3x \sin 3y}{3^2} + \frac{\sin 5x \sin 5y}{5^2} + \&c.$$

This convergent series gives always the value of the ordinate z or the distance of any point whatever of the surface from the plane of x and y.

The series formed of sines or cosines of multiple arcs are therefore adapted to represent, between definite limits, all possible functions, and the ordinates of lines or surfaces whose form is discontinuous. Not only has the possibility of these developments been demonstrated, but it is easy to calculate the terms of the series; the value of any coefficient whatever in the equation

$$\phi(x) = a_1 \sin x + a_2 \sin 2x + a_3 \sin 3x + \ldots + a_i \sin ix + \text{etc.},$$

is that of a definite integral, namely,

$$\frac{2}{\pi}\int \phi(x) \ \sin ix \ dx.$$

Whatever be the function $\phi(x)$, or the form of the curve which it represents, the integral has a definite value which may be introduced into the formula. The values of these definite integrals are analogous to that of the whole area $\int \phi(x)\,dx$ included between the curve and the axis in a given interval, or to the values of mechanical quantities, such as the ordinates of the centre of gravity of this area or of any solid whatever. It is evident that all these quantities have assignable values, whether the figure of the bodies be regular, or whether we give to them an entirely arbitrary form.

230. If we apply these principles to the problem of the motion of vibrating strings, we can solve difficulties which first appeared in the researches of Daniel Bernoulli. The solution given by this geometrician assumes that any function whatever may always be developed in a series of sines or cosines of multiple arcs.

Now the most complete of all the proofs of this proposition is that which consists in actually resolving a given function into such a series with determined coefficients.

In researches to which partial differential equations are applied, it is often easy to find solutions whose sum composes a more general integral; but the employment of these integrals requires us to determine their extent, and to be able to distinguish clearly the cases in which they represent the general integral from those in which they include only a part. It is necessary above all to assign the values of the constants, and the difficulty of the application consists in the discovery of the coefficients. It is remarkable that we can express by convergent series, and, as we shall see in the sequel, by definite integrals, the ordinates of lines and surfaces which are not subject to a continuous law[12]. We see by this that we must admit into analysis functions which have equal values, whenever the variable receives any values whatever included between two given limits, even though on substituting in these two functions, instead of the variable, a number included in another interval, the results of the two substitutions are not the same. The functions which enjoy this property are represented by different lines, which coincide in a definite portion only of their course, and offer a singular species of finite osculation. These considerations arise in the calculus of partial differential equations; they throw a new light on this calculus, and serve to facilitate its employment in physical theories.

231. The two general equations which express the development of any function whatever, in cosines or sines of multiple arcs, give rise to several remarks which explain the true meaning of these theorems, and direct the application of them.

If in the series

$$a + b \cos x + c \cos 2x + d \cos 3x + e \cos 4x + \&c.,$$

we make the value of x negative, the series remains the same; it also preserves its value if we augment the variable by any multiple whatever of the circumference 2π. Thus in the equation

$$\frac{1}{2}\pi\phi(x) = \frac{1}{2}\int \phi(x)\,dx + \cos x \int \phi(x)\,\cos x\,dx + \cos 2x \int \phi(x)\,\cos 2x\,dx$$

$$+ \cos 3x \int \phi(x)\,\cos 3x\,dx + \&c. \quad\ldots\ldots\ldots\ldots\ldots(\nu),$$

the function ϕ is periodic, and is represented by a curve composed of a multitude of equal arcs, each of which corresponds to an interval equal to 2π on the axis of the abscissæ. Further, each of these arcs is composed of two symmetrical branches, which correspond to the halves of the interval equal to 2π.

---

[12]Demonstrations have been supplied by Poisson, Deflers, Dirichlet, Dirksen, Bessel, Hamilton, Boole, De Morgan, Stokes. See note, p. 559.
[A. F.]

Suppose then that we trace a line of any form whatever $\phi\phi\alpha$ (see fig. 9.), which corresponds to an interval equal to π.

Fig. 9.

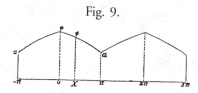

If a series be required of the form.

$$a + b\cos x + c\cos 2x + d\cos 3x + \&c.,$$

such that, substituting for x any value whatever X included between 0 and π, we find for the value of the series that of the ordinate $X\phi$, it is easy to solve the problem: for the coefficients given by the equation (ν) are

$$\frac{1}{\pi}\int \phi(x)\,dx, \quad \frac{2}{\pi}\int \phi(x)\cos 2x\,dx, \quad \frac{2}{\pi}\int \phi(x)\cos 3x\,dx, \quad \&c.$$

These integrals, which are taken from $x = 0$ to $x = \pi$, having always measurable values like that of the area $0\phi\alpha\pi$, and the series formed by these coefficients being always convergent, there is no form of the line $\phi\phi\alpha$, for which the ordinate $X\phi$ is not exactly represented by the development

$$a + b\cos x + c\cos 2x + d\cos 3x + e\cos 4x + \&c.$$

The arc $\phi\phi\alpha$ is entirely arbitrary; but the same is not the case with other parts of the line, they are, on the contrary, determinate; thus the arc $\phi\alpha$ which corresponds to the interval from 0 to $-\pi$ is the same as the arc ϕa; and the whole arc $\alpha\phi a$ is repeated on consecutive parts of the axis, whose length is 2π.

We may vary the limits of the integrals in equation (ν). If they are taken from $x = -\pi$ to $x = \pi$ the result will be doubled: it would also be doubled if the limits of the integrals were 0 and 2π, instead of being 0 and π. We denote in general by the sign $\int_a^b$ an integral which begins when the variable is equal to a, and is completed when the variable is equal to b; and we write equation (n) under the following form:

$$\frac{1}{2}\pi\phi(x) = \frac{1}{2}\int_0^\pi \phi(x)\,dx + \cos x\int_0^\pi \phi(x)\cos x\,dx + \cos 2x\int_0^\pi \phi(x)\cos 2x\,dx$$

$$+ \cos 3x\int_0^\pi \phi(x)\cos 3x\,dx + \text{etc.} \quad\ldots\ldots\ldots\ldots(\nu).$$

Instead of taking the integrals from $x = 0$ to $x = \pi$, we might take them from $x = 0$ to $x = 2\pi$, or from $x = -\pi$ to $x = \pi$; but in each of these two cases, $\pi\phi(x)$ must be written instead of $\frac{1}{2}\pi\phi(x)$ in the first member of the equation.

232. In the equation which gives the development of any function whatever in sines of multiple arcs, the series changes sign and retains the same absolute value

when the variable x becomes negative; it retains its value and its sign when the variable is increased or diminished by any multiple whatever of the circumference 2π.

Fig. 10.

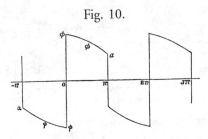

The arc $\phi\phi a$ (see fig. 10), which corresponds to the interval from 0 to π is arbitrary; all the other parts of the line are determinate. The arc $\phi\phi\alpha$, which corresponds to the interval from 0 to $-\pi$, has the same form as the given arc $\phi\phi a$; but it is in the opposite position. The whole arc $\alpha\phi\phi\phi\phi a$ is repeated in the interval from π to 3π, and in all similar intervals. We write this equation as follows:

$$\frac{1}{2}\pi\phi\,(x) = \sin x \int_0^\pi \phi(x)\sin x dx + \sin 2x \int_0^\pi \phi(x)\sin 2x dx$$
$$+ \sin 3x \int_0^\pi \phi(x)\sin 3x dx + \&c. \ \dots\dots(\mu).$$

We might change the limits of the integrals and write

$$\int_0^{2\pi} \text{ or } \int_{-\pi}^{+\pi} \text{ instead of } \int_0^\pi \ ;$$

but in each of these two cases it would be necessary to substitute in the first member $\pi\phi\,(x)$ for $\frac{1}{2}\pi\phi\,(x)$.

233.   The function $\phi\,(x)$ developed in cosines of multiple arcs, is represented by a line formed of two equal arcs placed symmetrically on each side of the axis of y,

Fig. 11.

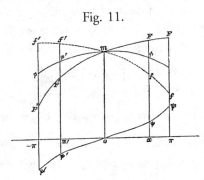

in the interval from $-\pi$ to $+\pi$ (see fig. 11); this condition is expressed thus,

$$\phi\,(x) = \phi\,(-x).$$

The line which represents the function $\psi(x)$ is, on the contrary, formed in the same interval of two opposed arcs, which is what is expressed by the equation

$$\psi(x) = -\psi(-x).$$

Any function whatever $F(x)$, represented by a line traced arbitrarily in the interval from $-\pi$ to $+\pi$, may always be divided into two functions such as $\phi(x)$ and $\psi(x)$. In fact, if the line $F'F'mFF$ represents the function $F(x)$, and we raise at the point o the ordinate om, we can draw through the point m to the right of the axis om the arc mff similar to the arc $mF'F$ of the given curve, and to the left of the same axis we may trace the arc $mf'f'$ similar to the arc mFF; we must then draw through the point m a line $\phi'\phi'm\phi\phi$ which shall divide into two equal parts the difference between each ordinate xF or $x'f'$ and the corresponding ordinate xf or $x'F'$. We must draw also the line $\psi'\psi'0\psi\psi$, whose ordinate measures the half-difference between the ordinate of $F'F'mFF$ and that *of $f'f'mff$.* This done the ordinate of the lines $F'F'mFF$, and $f'f'mff$ being denoted by $F(x)$ and $f(x)$ respectively, we evidently have $f(x) = F(-x)$; denoting also the ordinate of $\phi'\phi'm\phi\phi$ by $\phi(x)$, and that of $\psi'\psi'0\psi\psi$ by $\psi(x)$, we have

$$F(x) = \phi(x) + \psi(x) \quad \text{and} \quad f(x) = \phi(x) - \psi(x) = F(-x),$$

hence

$$\phi(x) = \frac{1}{2}F(x) + \frac{1}{2}F(-x) \quad \text{and} \quad \psi(x) = \frac{1}{2}F(x) - \frac{1}{2}F(-x),$$

whence we conclude that

$$\phi(x) = \phi(-x) \quad \text{and} \quad \psi(x) = -\psi(-x),$$

which the construction makes otherwise evident.

Thus the two functions $\phi(x)$ and $\psi(x)$, whose sum is equal to $F(x)$ may be developed, one in cosines of multiple arcs, and the other in sines.

If to the first function we apply equation (v), and to the second the equation (μ), taking the integrals in each case from $x = -\pi$ to $x = \pi$, and adding the two results, we have

$$\pi\big[\phi(x) + \psi(x)\big] = \pi F(x)$$
$$= \frac{1}{2}\int \phi(x)\,dx + \cos x \int \phi(x)\cos x\,dx + \cos 2x \int \phi(x)\cos 2x\,dx + \&c.$$
$$+ \sin x \int \psi(x)\sin x\,dx + \sin 2x \int \psi(x)\sin 2x\,dx + \&c.$$

The integrals must be taken from $x = -\pi$ to $x = \pi$. It may now be remarked, that in the integral $\int_{-\pi}^{+\pi}\phi(x)\cos x\,dx$ we could, without changing its value, write $\phi(x) + \psi(x)$ instead of $\phi(x)$: for the function $\cos x$ being composed, to right and left of the axis of x, of two similar parts, and the function $\psi(x)$ being, on the contrary, formed of two opposite parts, the integral $\int_{-\pi}^{+\pi}\psi(x)\cos x\,dx$ vanishes. The same would be the case if we wrote $\cos 2x$ or $\cos 3x$, and in general $\cos ix$ instead of $\cos x$, i being any integer from 0 to infinity. Thus the integral $\int_{-\pi}^{+\pi}\phi(x)\cos ix\, dx$ is the same as the integral

$$\int_{-\pi}^{+\pi}\left[\psi(x) + \psi(x)\right]\cos ix\, dx, \quad \text{or} \quad \int_{-\pi}^{+\pi} F(x)\cos ix\, dx.$$

It is evident also that the integral $\int_{-\pi}^{+\pi}\psi(x)\sin ix\, dx$ is equal to the integral $\int_{-\pi}^{+\pi} F(x)\sin ix\, dx$, since the integral $\int_{-\pi}^{+\pi}\phi(x)\sin ix\, dx$ vanishes. Thus we obtain the following equation (p), which serves to develope any function whatever in a series formed of sines and cosines of multiple arcs:

$$\pi F(x) = \frac{1}{2}\int F(x)\, dx \tag{p}$$

$$+ \cos x \int F(x)\cos x\, dx + \cos 2x \int F(x)\cos 2x\, dx + \&c.$$

$$+ \sin x \int F(x)\sin x\, dx + \sin 2x \int F(x)\sin 2x\, dx + \&c.$$

234.   The function $F(x)$, which enters into this equation, is represented by a line $F'F'FF$, of any form whatever. The arc $F'F'FF$, which corresponds to the interval from $-\pi$ to $+\pi$, is arbitrary; all the other parts of the line are determinate, and the arc $F'F'FF$ is repeated in each consecutive interval whose length is 2π. We shall make frequent applications of this theorem, and of the preceding equations (μ) and (ν).

If it be supposed that the function $F(x)$ in equation (p) is represented, in the interval from $-\pi$ to $+\pi$, by a line composed of two equal arcs symmetrically placed, all the terms which contain sines vanish, and we find equation (ν). If, on the contrary, the line which represents the given function $F(x)$ is formed of two equal arcs opposed in position, all the terms which do not contain sines disappear, and we find equation (μ). Submitting the function $F(x)$ to other conditions, we find other results.

If in the general equation (p) we write, instead of the variable x, the quantity $\frac{\pi x}{r}$, x denoting another variable, and $2r$ the length of the interval which includes the arc which represents $F(x)$; the function becomes $F(\frac{\pi x}{r})$, which we may denote by $f(x)$. The limits $x = -\pi$ and $x = \pi$ become $\frac{\pi x}{r} = -\pi$, $\frac{\pi x}{r} = \pi$; we have therefore, after the substitution,

$$rf(x) = \frac{1}{2}\int_{-r}^{+r} f(x)\, dx \tag{P.}$$

$$+ \cos\pi\frac{x}{r}\int f(x)\cos\frac{\pi x}{r}\, dx + \cos\frac{2\pi x}{r}\int f(x)\cos\frac{2\pi x}{r}\, dx + \text{etc.}$$

$$+ \sin\pi\frac{x}{r}\int f(x)\sin\frac{\pi x}{r}dx + \sin\frac{2\pi x}{r}\int f(x)\sin\frac{2\pi x}{r}dx + \text{etc.}$$

All the integrals must be taken like the first from $x = -r$ to $x = +r$. If the same substitution be made in the equations (ν) and (μ), we have

$$\frac{1}{2}rf(x) = \frac{1}{2}\int_0^r f(x)\,dx + \cos\frac{\pi x}{r}\int f(x)\,\cos\frac{\pi x}{r}\,dx$$

$$+ \cos\frac{2\pi x}{r}\int f(x)\,\cos\frac{2\pi x}{r}dx + \&c.\dots\dots\dots\dots\dots(N),$$

and

$$\frac{1}{2}rf(x) = \sin\frac{\pi x}{r}\int_0^r f(x)\,\sin\frac{\pi x}{r}\,dx$$

$$+ \sin\frac{2\pi x}{r}\int f(x)\,\sin\frac{2\pi x}{r}dx + \&c.\dots\dots\dots\dots\dots(M).$$

In the first equation (P) the integrals might be taken from from $x = 0$ to $x = 2r$, and representing by X the whole interval $2r$, we should have[13]

$$\frac{1}{2}Xf(x) = \frac{1}{2}\int f(x)\,dx \qquad\qquad\qquad\qquad\qquad\text{(II)}$$

$$+ \cos\frac{2\pi x}{X}\int f(x)\,\cos\frac{2\pi x}{X}dx + \cos\frac{4\pi x}{X}\int f(x)\,\cos\frac{4\pi x}{X}dx + \&c.$$

$$+ \sin\frac{2\pi x}{X}\int f(x)\,\sin\frac{2\pi x}{X}dx + \sin\frac{4\pi x}{X}\int f(x)\,\sin\frac{4\pi x}{X}dx + \&c.$$

235. It follows from that which has been proved in this section, concerning the development of functions in trigonometrical series, that if a function $f(x)$ be proposed, whose value in a definite interval from $x = 0$ to $x = X$ is represented by the ordinate of a curved line arbitrarily drawn; we can always develope this function in a series which contains only sines or only cosines, or the sines and cosines of multiple arcs, or the cosines only of odd multiples. To ascertain the terms of these series we must employ equations (M), (N), (P).

The fundamental problems of the theory of heat cannot be completely solved, without reducing to this form the functions which represent the initial state of the temperatures.

These trigonometric series, arranged according to cosines or sines of multiples of arcs, belong to elementary analysis, like the series whose terms contain the

---

[13] It has been shown by Mr J. O'Kinealy that if the values of the arbitrary function $f(x)$ be imagined to recur for every range of x over successive intervals λ, we have the symbolical equation

$$(e^{\lambda\frac{d}{dx}} - 1)\,f(x) = 0;$$

and the roots of the auxiliary equation being

$$\pm n\frac{2\pi\sqrt{-1}}{\lambda}, n = 0, 1, 2, 3\dots\infty,$$

it follows that

$$f(x) = A_0 + A_1\cos\frac{2\pi x}{\lambda} + A_2\cos 2\frac{2\pi x}{\lambda} + A_3\cos 3\frac{2\pi x}{\lambda} + \&c.$$

$$+ B_1\sin\frac{2\pi x}{\lambda} + B_2\sin 2\frac{2\pi x}{\lambda} + B_3\sin 3\frac{2\pi x}{\lambda} + \&c.$$

The coefficients being determined in Fourier's manner by multiplying both sides by $\frac{\cos}{\sin}\,n\frac{2\pi x}{\lambda}$ and integrating from 0 to λ. (*Philosophical Magazine*, August 1874). [A. F.]

successive powers of the variable. The coefficients of the trigonometric series are definite areas, and those of the series of powers are functions given by differentiation, in which, moreover, we assign to the variable a definite value. We could have added several remarks concerning the use and properties of trigonometrical series; but we shall limit ourselves to enunciating briefly those which have the most direct relation to the theory with which we are concerned.

1st. The series arranged according to sines or cosines of multiple arcs are always convergent; that is to say, on giving to the variable any value whatever that is not imaginary, the sum of the terms converges more and more to a single fixed limit, which is the value of the developed function.

2nd. If we have the expression of a function $f(x)$ which corresponds to a given series

$$a + b \cos x + c \cos 2x + d \cos 3x + e \cos 4x + \&\text{c.},$$

and that of another function $\phi(x)$, whose given development is

$$\alpha + \beta \cos x + \gamma \cos 2x + \delta \cos 3x + \epsilon \cos 4x + \&\text{c.},$$

it is easy to find in real terms the sum of the compound series

$$a\alpha + b\beta + c\gamma + d\delta + e\epsilon + \&\text{c.},[14]$$

and more generally that of the series

$$a\alpha + b\beta \cos x + c\gamma \cos 2x + d\delta \cos 3x + e\epsilon \cos 4x + \&\text{c.},$$

which is formed by comparing term by term the two given series. This remark applies to any number of series.

3rd. The series (P) (Art. 234) which gives the development of a function $F(x)$ in a series of sines and cosines of multiple arcs, may be arranged under the form

$$\pi F(x) = \frac{1}{2}\int F(\alpha)\,d\alpha$$

$$+ \cos x \int F(\alpha) \cos \alpha\,d\alpha + \cos 2x \int F(\alpha) \cos 2\alpha\,d\alpha + \&\text{c.}$$

$$+ \sin x \int F(\alpha) \sin \alpha\,d\alpha + \sin 2x \int F(\alpha) \sin 2\alpha\,d\alpha + \&\text{c.}$$

α being a new variable which disappears after the integrations.

We have then

$$\pi F(x) = \int_{-\pi}^{+\pi} F(\alpha)\,d\alpha \left\{ \frac{1}{2} \right.$$

$$+ \cos x \cos \alpha + \cos 2x \cos 2\alpha + \cos 3x \cos 3\alpha + \&\text{c.}$$

$$\left. + \sin x \sin \alpha + \sin 2x \sin 2\alpha + \sin 3x \sin 3\alpha + \&\text{c.} \right\},$$

[14] We shall have
$$\int_0^\pi \psi(x)\phi(x)\,dx = a\alpha\pi + \tfrac{1}{2}\pi\{b\beta + c\gamma + \ldots\}. \qquad [\text{R. L. E.}]$$

or

$$F\left(x\right) = \frac{1}{\pi} \int_{-\pi}^{+\pi} F\left(\alpha\right) d\alpha \left\{\frac{1}{2} + \cos\left(x - \alpha\right) + \cos 2\left(x - \alpha\right) + \&c.\right\}.$$

Hence, denoting the sum of the preceding series by

$$\sum \cos i\left(x - \alpha\right)$$

taken from $i = 1$ to $i = \infty$, we have

$$F\left(x\right) = \frac{1}{\pi} \int F\left(\alpha\right) d\alpha \left\{\frac{1}{2} + \sum \cos i\left(x - \alpha\right)\right\}.$$

The expression $\frac{1}{2} + \sum \cos i\left(x - \alpha\right)$ represents a function of x and α, such that if it be multiplied by any function whatever $F(\alpha)$, and integrated with respect to α between the limits $\alpha = -\pi$ and $\alpha = \pi$, the proposed function $F(\alpha)$ becomes changed into a like function of x multiplied by the semi-circumference π. It will be seen in the sequel what is the nature of the quantities, such as $\frac{1}{2} + \sum \cos i\left(x - \alpha\right)$, which enjoy the property we have just enunciated.

4th. If in the equations (M), (N), and (P) (Art. 234), which on being divided by r give the development of a function $f\left(x\right)$, we suppose the interval r to become infinitely large, each term of the series is an infinitely small element of an integral; the sum of the series is then represented by a definite integral. When the bodies have determinate dimensions, the arbitrary functions which represent the initial temperatures, and which enter into the integrals of the partial differential equations, ought to be developed in series analogous to those of the equations (M), (N), (P); but these

---

Note on Section VI. On the subject of the development of a function whose values are arbitrarily assigned between certain limits, in series of sines and cosines of multiple arcs, and on questions connected with the values of such series at the limits, on the convergency of the series, and on the discontinuity of their values, the principal authorities are

Poisson, *Théorie mathématique de la Chaleur,* Paris, 1835, Chap. VII. Arts. 92–102, *Sur la manière d'exprimer les fonctions arbitraires par des séries de quantités périodiques.* Or, more briefly, in his *Traité de Mécanique,* Arts. 325–328. Poisson's original memoirs on the subject were published in the *Journal de l'École Polytechnique,* Cahier 18, year 1820, and Cahier 19, year 1823.

De Morgan, *Differential and Integral Calculus.* London, 1842. The proofs of the developments appear to be original. In the verification of the developments the author follows Poisson's methods.

Stokes, *Cambridge Philosophical Transactions,* 1847, Vol. VIII. *On the Critical values of the sums of Periodic Series.* Section I. *Mode of ascertaining the nature of the discontinuity of a function which is expanded in a series of sines or cosines, and of obtaining the developments of the derived functions.* Graphically illustrated.

Thomson and Tait, *Natural Philosophy,* Oxford, 1867, Vol. I. Arts. 75–77.

Donkin, *Acoustics,* Oxford, 1870, Arts. 72–79, and Appendix to Chap. IV.

Matthieu, *Cours de Physique Mathématique,* Paris, 1873.

Entirely different methods of discussion, not involving the introduction of arbitrary multipliers to the successive terms of the series were originated by

Dirichlet, *Crelle's Journal,* Berlin, 1829, Band IV. *Sur la convergence des séries trigonométriques qui servent à représenter une fonction arbitraire entre les limites données.* The methods of this memoir thoroughly deserve attentive study, but are not yet to be found in English text-books. Another memoir, of greater length, by the same author appeared in Dove's *Repertorium der Physik,* Berlin, 1837, Band I. *Ueber die Darstellung ganz willkührlicher Functionen durch Sinus- und Cosinusreihen.* Von G. Lejeune Dirichlet.

Other methods are given by

Dirksen, *Crelle's Journal,* 1829, Band IV. *Ueber die Convergenzeiner nach den Sinussen und Cosinussen der Vielfachen eines Winkels fortschreitenden Reihe.*

Bessel, *Astronomische Nachrichten,* Altona, 1839. *Ueber den Ausdruck einer Function ϕ (x) durch Cosinusse und Sinusse der Vielfachen von x.*

The writings of the last three authors are criticised by Riemann, *Gesammelte Mathematische Werke,* Leipzig, 1876. *Ueber die Darstellbarkeit einer Function durch eine Trigonometrische Reihe.*

On *Fluctuating Functions* and their properties, a memoir was published by Sir W. R. Hamilton, *Transactions of the Royal Irish Academy,* 1843, Vol. XIX. The introductory and concluding remarks may at this stage be studied.

The writings of Deflers, Boole, and others, on the subject of the expansion of an arbitrary function by means of a double integral (*Fourier's Theorem*) will be alluded to in the notes on Chap. IX. Arts. 361, 362. [A. F.]

functions take the form of definite integrals, when the dimensions of the bodies are not determinate, as will be explained in the course of this work, in treating of the free diffusion of heat (Chapter IX.).

SECTION VII

Application to the Actual Problem

236.   We can now solve in a general manner the problem of the propagation of heat in a rectangular plate *BAC*, whose end *A* is constantly heated, whilst its two infinite edges *B* and *C* are maintained at the temperature 0.

Suppose the initial temperature at all points of the slab *BAC* to be nothing, but that the temperature at each point *m* of the edge *A* is preserved by some external cause, and that its fixed value is a function $f(x)$ of the distance of the point *m* from the end *O* of the edge *A* whose whole length is *r*; let *v* be the constant temperature of the point *m* whose co-ordinates are *x* and *y*, it is required to determine *v* as a function of *x* and *y*.

The value $v = ae^{-my} \sin mx$ satisfies the equation

$$\frac{d^2v}{dx^2} + \frac{d^2v}{dy^2} = 0;$$

a and *m* being any quantities whatever. If we take $m = i\frac{\pi}{r}$, *i* being an integer, the value $ae^{-i\pi\frac{y}{r}} \sin \frac{i\pi x}{r}$ vanishes, when $x = r$, whatever the value of *y* may be. We shall therefore assume, as a more general value of *v*,

$$v = a_1 e^{-\pi\frac{y}{r}} \sin \frac{\pi x}{r} + a_2 e^{-2\pi\frac{y}{r}} \sin \frac{2\pi x}{r} + a_3 e^{-3\pi\frac{y}{r}} \sin \frac{3\pi x}{r} + \&c.$$

If *y* be supposed nothing, the value of *v* will by hypothesis be equal to the known function $f(x)$. We then have

$$f(x) = a_1 \sin \frac{\pi x}{r} + a_2 \sin \frac{2\pi x}{r} + a_3 \sin \frac{3\pi x}{r} + \&c.$$

The coefficients a_1, a_2, a_3, &c. can be determined by means of equation (M), and on substituting them in the value of *v* we have

$$\frac{1}{2}rv = e^{-\pi\frac{y}{r}} \sin \frac{\pi x}{r} \int f(x) \sin \frac{\pi x}{r} dx + e^{-2\pi\frac{y}{r}} \sin \frac{2\pi x}{r} \int f(x) \sin \frac{2\pi x}{r} dx$$

$$+ e^{-3\pi\frac{y}{r}} \sin \frac{3\pi x}{r} \int f(x) \sin \frac{3\pi x}{r} dx + \&c.$$

237.   Assuming $r = \pi$ in the preceding equation, we have the solution under a more simple form, namely

$$\frac{1}{2}\pi v = e^{-y} \sin x \int f(x) \sin x dx + e^{-2y} \sin 2x \int f(x) \sin 2x dx$$

$$+ e^{-3y} \sin 3x \int f(x) \sin 3x dx + \&c. \dots\dots\dots(a),$$

or

$$\frac{1}{2}\pi v = \int_0^\pi f(\alpha)\,d\alpha\,(e^{-y}\sin x \sin \alpha + e^{-2y}\sin 2x \sin 2\alpha + e^{-3y}\sin 3x \sin 3\alpha + \&c.)$$

α is a new variable, which disappears after integration.

If the sum of the series be determined, and if it be substituted in the last equation, we have the value of v in a finite form. The double of the series is equal to

$$e^{-y}[\cos(x-\alpha) - \cos(x+\alpha)] + e^{-2y}[\cos 2(x-\alpha) - \cos 2(x+\alpha)]$$
$$+ e^{-3y}[\cos 3(x-\alpha) - \cos 3(x+\alpha)] + \&c.;$$

denoting by $F(y,p)$ the sum of the infinite series

$$e^{-y}\cos p + e^{-2y}\cos 2p + e^{-3y}\cos 3p + \&c.,$$

we find

$$\pi v = \int_0^\pi f(\alpha)\,d\alpha\,\{F(y, x-\alpha) - F(y, x+\alpha)\}.$$

We have also

$$2F(y,p) = \left\{\begin{array}{l} e^{-(y+p\sqrt{-1})} + e^{-2(y+p\sqrt{-1})} + e^{-3(y+p\sqrt{-1})} + \&c. \\ + e^{-(y-p\sqrt{-1})} + e^{-2(y-p\sqrt{-1})} + e^{-3(y-p\sqrt{-1})} + \&c. \end{array}\right.$$

$$= \frac{e^{-(y+p\sqrt{-1})}}{1-e^{-(y+p\sqrt{-1})}} + \frac{e^{-(y-p\sqrt{-1})}}{1-e^{-(y-p\sqrt{-1})}},$$

or

$$F(y,p) = \frac{\cos p - e^{-y}}{e^y - 2\cos p + e^{-y}},$$

whence

$$\pi v = \int_0^\pi f(\alpha)\,d\alpha\left\{\frac{\cos(x-\alpha) - e^{-y}}{e^y - 2\cos(x-\alpha) + e^{-y}} - \frac{\cos(x+\alpha) - e^{-y}}{e^y - 2\cos(x+\alpha) + e^{-y}}\right\},$$

or

$$\pi v = \int_0^\pi f(\alpha)\,d\alpha\left\{\frac{2(e^y - e^{-y})\sin x \sin \alpha}{[e^y - 2\cos(x-\alpha) + e^{-y}][e^y - 2\cos(x+\alpha) + e^{-y}]}\right\},$$

or, decomposing the coefficient into two fractions,

$$\pi v = \frac{e^y - e^{-y}}{2}\int_0^\pi f(\alpha)\,d\alpha\left\{\frac{1}{e^y - 2\cos(x-\alpha) + e^{-y}}\right.$$

$$\left. - \frac{1}{e^y - 2\cos(x+\alpha) + e^{-y}}\right\}.$$

This equation contains, in real terms under a finite form, the integral of the equation $\frac{d^2v}{dx^2} + \frac{d^2v}{dy^2} = 0$, applied to the problem of the uniform movement of heat in a rectangular solid, exposed at its extremity to the constant action of a single source of heat.

It is easy to ascertain the relations of this integral to the general integral, which has two arbitrary functions; these functions are by the very nature of the problem determinate, and nothing arbitrary remains but the function $f(\alpha)$, considered between the limits $\alpha = 0$ and $\alpha = \pi$. Equation (a) represents, under a simple form, suitable for numerical applications, the same value of v reduced to a convergent series.

If we wished to determine the quantity of heat which the solid contains when it has arrived at its permanent state, we should take the integral $\int dx \int dy\, v$ from $x = 0$ to $x = \pi$, and from $y = 0$ to $y = \infty$; the result would be proportional to the quantity required. In general there is no property of the uniform movement of heat in a rectangular plate, which is not exactly represented by this solution.

We shall next regard problems of this kind from another point of view, and determine the varied movement of heat in different bodies.

Carl Friedrich Gauss

(1777–1855)

HIS LIFE AND WORK

In 1940, the eminent British mathematician G. H. Hardy wrote:

> *317 is a prime, not because we think so, or because our minds are shaped in one way rather than another, but because it is so, because mathematical reality is built that way.*

Could this attitude toward mathematics explain why Carl Friedrich Gauss, unquestionably the greatest mathematician of all time, withheld from publication his work in non-Euclidean geometry while seeking an empirical verification he would never find? Perhaps, but we will never know with certainty. What we do know with certainty is the enormity of Gauss's mathematical gifts and accomplishments.

Gauss's talents were obvious as soon as he stepped into a classroom at the age of seven. When the class began to be unruly, the teacher, J. G. Büttner, assigned them the task of adding up all of the integers from 1 to 100. As his classmates struggled to fit their calculations on their individual slates, Gauss wrote down the answer immediately: 5,050. As soon as the problem was stated, Gauss recognized that the set of integers from 1 to 100 was identical to 50 pairs of integers each adding up to 101: ({1,100}, {2,99}, . . . , {50,51}).

Herr Büttner approached Gauss's parents to persuade them to let their son stay after school for special math instruction. Gauss's parents were at first skeptical. They had recognized their son's calculating ability when, at the age of three, he corrected a mistake his father made in paying out wages to men who worked him. However, the elder Gausses had very limited horizons. Gauss's father, Gebhard, born in 1744, was a gardener, laborer, and foreman who came from a long line of poor, unlettered workmen who had had little success escaping their peasant roots to the lower middle class. His mother, born Dorothea Benze in 1743, was a maid before becoming Gebhard's second wife in 1776. Their only son Carl Friedrich was born a year later in the city of Brunswick, the capital of a Duchy of the same name.

In one respect, Gauss was very fortunate that his parents had such limited horizons and did not abuse his talents. Otherwise, they might have taken him on tour and exhibited him as a calculating prodigy, in much the same way that Mozart's father had taken young Wolfgang on tour as a musical prodigy. By his early teens, Gauss had worked out two methods to compute square roots to as many as fifty decimal places. He is reputed to have read tables of logarithms and found slight errors many places to the right of the decimal point. This computational facility would serve Gauss well later in his life.

Fortunately, he was to have as close to a regular education as possible for someone of his gifts. At the age of eleven, Gauss entered the local Gymnasium and received a thorough education in the classics. His real math education came in one-on-one instruction after school and on his own as he devoured the contents of works like Newton's *Principia* and Bernoulli's *Ars conjectandi* in his spare time. Gauss achieved such a distinguished record at the Gymnasium that at the age of fifteen, Carl Wilhelm Ferdinand, Duke of Brunswick, gave Gauss a stipend to continue his education at the Brunswick Collegium Carolinum.

Gauss entered the Collegium already possessed of a scientific and classical education worthy of a graduate. Three years later, he left the Collegium in 1795 having done enough mathematics to fill a career. During these years, Gauss gave the first proposals for approximations for $\pi(n)$, the function counting the number of prime numbers less than n. Gauss first proposed the following:

$$\pi(n) \sim \frac{n}{\ln n}.$$

(**ln** n is the natural logarithm of n) and then refined this to $\pi(n) = \mathrm{Li}(n)$, where

$$\mathrm{Li}\,(n) = \int_2^n \frac{dx}{\ln x}$$

Ever the calculator, Gauss computed the number of primes and tested the formula up to 3,000,000.

Gauss sped through the curriculum at the Collegium and enrolled at the University of Göttingen, sixty miles from Brunswick, rather than the Duchy's official university in nearby Helmstedt, most likely due to Göttingen's superior mathematics library. Surprisingly, Göttingen's records show that Gauss borrowed far more books in the humanities than in mathematics. Gauss's notebooks tell us that he had a much higher regard for his classics professor than for his mathematics professor. However, mathematical accomplishment quickly won out.

The Greeks had known that regular polygons with 3 or 5 or 15 sides could be constructed with a straight edge and compass. So could any regular polygon having a number of sides that was a power of 2 times 3, 5, or 15, and that is where the boundary of the field stood for two millennia until 1796 when Gauss discovered that a seventeen-sided regular polygon could also be constructed with the classical geometrical tools: the straight edge and compass. He quickly generalized his result to any regular polygon with a number of sides that is a product of a power of 2 and any number of Fermat primes. A Fermat prime is a prime number of the form $2^N + 1$, where N is itself a power of 2. Fermat (1601–1665) thought that all numbers of the form $2^N + 1$, where N is a power of 2, are prime numbers, and it is easy to see that 3, 5, 17, and 257 have this property. With only a little bit of brute force you can demonstrate, as Fermat must have done, that 65,537 ($= 2^{16} + 1$) is a prime. The next candidate Fermat prime is 4,294,967,297 ($= 2^{32} + 1$). We cannot fault Fermat for not finding 641 as its smallest prime divisor. It took a century for the great mathematician Leonhard Euler to find that. In his notebooks, Gauss speculated that there are no other Fermat primes. To date, none have been found to exist.

Gauss was so overjoyed at this result that it convinced him to pursue a career in mathematics. After two years at Göttingen, he realized that no one on the faculty could really be of any assistance to him, so he went home to Brunswick to write his doctoral dissertation. For his topic he chose the fundamental theorem of algebra, that every polynomial equation of degree n with complex coefficients has exactly n roots in the complex numbers. His dissertation was the first of what would be four proofs throughout his career.

Freed of the need to write a set piece, Gauss turned his attention to number theory. Number theory goes back to the Greeks with Euclid's proofs of the infinity of prime numbers and the form of even perfect numbers being two of the earliest results in the field. From time to time, new results were added or new conjectures made. In the seventeenth century, the French mathematician Pierre de Fermat, a contemporary of Descartes, made his famous conjecture that the equation $x^n + y^n = z^n$ has no nontrivial solutions in integers for $n > 2$. The Arabs had proved this for

$n = 3$ and $n = 4$, however, the complete proof of Fermat's Last Theorem did not come until 1992. In the fifty years before Gauss, Lagrange had provided the first proof that every integer can be expressed as the sum of no more than four squares and Goldbach had conjectured that every even number other than 2 can be expressed as the sum of two primes. In 1770, the English mathematician published the statement of a theorem that had first been proposed by his former student John Wilson.

The integer p is a prime if and only if p evenly divides $(p - 1)! + 1$.

Recall that $n! = 1 \cdot 2 \cdot 3 \cdot 4 \cdot \ldots \cdot (n - 1) \cdot n$. Given the speed with which $n!$ (n factorial) grows, Waring despaired of ever having a notation that would enable a proof of the conjecture whose truth could barely be established in particular cases.

When Gauss began work on his epochal *Disquisitiones arithmeticae* (*Arithmetic disquisitions*) excerpted here, number theory was merely a collection of isolated results. Upon hearing of Waring's despair, Gauss is supposed to have remarked that mathematics is concerned with notions rather than with notations. In the *Disquisitiones*, he introduced the notion of congruence and in so doing unified number theory. Two integers x and y are said to be *congruent* modulo the integer z if and only if $(x - y)$ is evenly divisible by z. Following Gauss, we express this congruence relationship as

$$x \equiv y \ (mod \ z)$$

This powerful analytical method was the foundation of the *Disquisitiones arithmeticae*, a compendium of Gauss's result in number theory, which he published at the age of twenty-four. It is divided into seven sections:

1. congruences in general
2. congruences of the first degree
3. residues of powers
4. congruences of the second degree
5. quadratic forms
6. applications
7. division of the circle

With the notion of congruence and Gauss's compact notation for it, he could easily state Wilson's theorem as

The integer p is a prime if and only if $(p - 1)! \equiv -1 (mod \ p)$

and he could also state and prove the complement to Wilson's theorem:

If n is a composite number other than 4, then $(n - 1)! \equiv 0 \ (mod \ n)$.

These results only hinted at the power of congruence as a mathematical tool. In 1795, while still at the Collegium, Gauss had discovered the law of quadratic reciprocity, which had only been stated and incompletely proven a decade earlier by the thirty-three-year-old French mathematician Adrien-Marie Legendre. In the language of congruences, the law of quadratic reciprocity states:

If p and q are primes not both congruent to 3 modulo 4 then either
$$\text{both } x^2 \equiv p \,(\text{mod } q) \text{ and } x^2 \equiv q \,(\text{mod } p) \text{ are solvable}$$
or
$$\text{neither } x^2 \equiv p \,(\text{mod } q) \text{ nor } x^2 \equiv q \,(\text{mod } p) \text{ is solvable.}$$

If p and q are primes both congruent to 3 modulo 4 then exactly one of the equations
$$x^2 \equiv p \,(\text{mod } q) \text{ is } x^2 \equiv q \,(\text{mod } p).$$

Notice that this theorem comes nowhere close to resolving the question of whether p is in fact a quadratic residue modulo q *in isolation*! It merely allows one difficult numeric calculation, such as whether 257 is a prime modulo 65,537, to be replaced by the easier numeric calculation of whether 65,537 is a prime modulo 257. (The first step is to reduce 65,537 to its residue modulo 257, which is 2, and then determine whether 2 is a quadratic residue modulo 257.)

Gauss considered the law of quadratic reciprocity to be the *theorem aureum*— the golden theorem — or the *gemma arithmeticae* — the gem of arithmetic. He considered arithmetic itself, as he called number theory, to be the *queen of mathematics*, which he called the *queen of sciences*.

Early in 1801, the Duke of Brunswick raised Gauss's stipend. However, Gauss felt as though he had done little to merit the raise. With the *Disquisitiones arithmeticae* in press, Gauss sought a new challenge and turned his attention to planetary theory. In January 1801, the Italian astronomer Joseph Piazzi had briefly observed what he thought to be a new planet before losing track of it. Gauss spent much of 1801 improving the theory to planetary perturbation to utilize true ellipses rather than circular approximations. Gauss predicted the mysterious body was an asteroid rather than a new planet. At the end of the year, astronomers found the asteroid Ceres exactly where Gauss's improved method said it should be. The discovery of Ceres earned Gauss genuine international fame. In January 1802, the Academy of Sciences in the Russian capital of St Petersburg elected him a corresponding member. Gauss felt as though he had indeed merited the increase in the Duke's stipend.

The Duke raised Gauss's stipend again in 1803. The increased income may have prompted Gauss to think about his personal situation. In 1805, Gauss surprised everyone around him by announcing his engagement to Johanna Osthoff after a year-long courtship. He wrote his friend Bolyai, "For three days now this angel, almost too heavenly for our earth, has been my fiancée . . . Life lies before me like an eternal spring with radiant colors." They married on October 9, 1805. For a brief time, Gauss did seem to be in the spring of his life. His patron rewarded him with a hefty increase following his marriage, perhaps motivated by an offer of a position in St. Petersburg. The Gauss's first child Joseph, named for the discoverer of Ceres, was born on August 21, 1806. A daughter christened Wilhelmina followed a year and a half later on February 29, 1808.

Unfortunately for Gauss, his blissful spring would not last long. In November 1806, Duke Ferdinand died from a wound he suffered losing the Battle of Auerstädt to Napoleon the previous month. Following Napoleon's victories in Germany, Göttingen found itself in the Kingdom of Westphalia a French vassal state. As a professor, Gauss had to pay 2,000 francs tax, a small fortune in those days. The astronomer Olbers sent Gauss the necessary funds to pay the tax; however, Gauss refused his benevolence. Then Gauss received a letter from the French mathematician Laplace who said he considered it an honor to pay the tax and lift the obligation from Gauss. Once again Gauss declined the offer, but not out of any animosity for the Frenchman. Gauss had a great respect for Laplace and Laplace for Gauss. When Laplace was once asked who was the greatest mathematician in Germany, he immediately replied with the name of Pfaff, who had nominally supervised Gauss's doctoral dissertation. When asked why he hadn't named Gauss, Laplace immediately replied, "Gauss is the greatest mathematician in the world!" In the end, an anonymous donor sent Gauss the money to pay the tax. Unable to repay the donor, Gauss made regular donations to charity with interest as if paying of a loan for the amount donated.

Gauss would not have a long marriage to Johanna. She died a month after giving birth to their third child, Ludwig, in 1809. Sadly, Ludwig died five months later. Within a year of Johanna's death, Gauss married Minna Waldeck, Johanna's best friend. We can only guess that Gauss may have been motivated by a need to find a stepmother for his three children. Gauss had three more children by Minna before she became ill, first with tuberculosis and then with what was diagnosed as hysterical neurosis. Gauss and Minna were perpetually unhappy until her death in 1831.

At the beginning of the nineteenth century, Paris remained the center of the mathematical world. Göttingen was at best a remote outpost. Gauss had occasional correspondence with the mathematical giants in France, but he never troubled himself to visit Paris. There were no mathematicians in Germany who came close to him in stature during the prime of his career and few worthy of correspondence. Given the mediocre state of affairs in the German mathematics community, it is not surprising that Gauss chose to serve as Professor of Astronomy and Director of the Observatory at Göttingen. Had he been Professor of Mathematics, he would have been required to spend his time teaching Mathematics to indifferent undergraduates.

Not having mathematics classes to teach may have given Gauss the free time to pursue the consequences of denying Euclid's parallel postulate, research he had begun in his student days and then put aside. He seems to have done this work almost against his will, hardly bearing the thought that the parallel postulate might not be true. He was never able to allow himself to publish this work. We know it only through his notebooks.

Gauss's contemporaries considered him to be a mathematical scientist with strong interests in applied and even empirical mathematics. His interest in geodesy, the mathematics of surveying and representing the land, is a good example of his strong empirical leanings. He initially worked on surveying problems in his early twenties, then set that interest aside for nearly two decades. In 1817, at the age of forty he returned to this subject, taking responsibility for a survey of the state of Hanover. For several years Gauss spent his summers surveying the land and spending much of the rest of the year analyzing the data. Dissatisfied with the standard geodetic measurement techniques based on sighting lamps or flares, Gauss invented a new method using a device called a *heliotrope*. It employed mirrors to deflect light rays to small aperture telescopes.

Gauss employed his geodetic work to find empirical support for non-Euclidean geometry. As part of one of his surveys he measured the angles of the triangle formed by the mountain tops of the northern German peaks Hohenhagen, Brocken, and Inselsberg. The measurement of this triangle, with sides between 45 and 70 miles long, proved inconclusive. Gauss calculated the sum of the angles of the triangle to be $180° \ 0' \ 15''$, 1 part in 43,200, close enough to $180°$, to be due to measuring error. Thanks to Einstein's theory of general relativity, we now know that the sum of the angles of such a triangle exceeds $10^{-17}''$, 1 part in 10^{21}!

Without any empirical data, Gauss chose to not to publish any of his work on non-Euclidean geometry. However, when Gauss received word in 1831 that the Hungarian mathematician Johan Bolyai had published work demonstrating the consistency of a non-Euclidean geometry, he responded to Johan's father Wolfgang, an old friend:

> *To praise it would amount to praising myself. For the entire content of the work . . . coincides almost exactly with my own meditations which have occupied my mind for the past thirty or thirty-five years.*

Gauss thought that he was merely reporting a fact. Both Bolyais took it as a great offense and as an attempt to steal priority.

In 1824, Gauss received a substantial salary increase, his first raise since 1807. A year later, he received a large bonus for the surveying work. This newfound financial bounty came at a very fortunate point in his life, just as Gauss was beginning to suffer from asthma and a heart ailment. By 1825, the physical burden of the summer surveys became too much for him to bear and he satisfied himself with supervising the surveys and doing all of the calculations. It has been estimated that he handled more than a million pieces of numeric data by himself. Given his computational talents, it is hardly surprising that even late in his life Gauss would find new fields for his talents. In the 1840s, he took on the task of putting the university's pension fund on a sound actuarial basis. He must have been especially good at investments. When he died, his estate was equal to two hundred times his annual income.

Over the years, Gauss began to attract a handful of students to his occasional mathematical lectures. Anyone who attended his lectures on number theory in 1809, the theory of curved surfaces in 1827, or the method of least squares in 1851 should have considered himself exceptionally fortunate. Bernhard Riemann and Richard Dedekind, each included in this book, were among the fortunate. Their generation established Göttingen as the center of the mathematical universe.

Gauss resisted his physical ailments until he was well into his seventies. He finally died on February 23, 1855, two months short of his seventy-eighth birthday.

In his will, Gauss stipulated that a seventeen-sided regular polygon be carved into his gravestone. However, that was not to be. The mason charged with the task thought that viewers would confuse a seventeen-sided regular polygon with a circle, so he carved a seventeen-pointed star. Although the mason may not have followed Gauss's instructions, he did mark Gauss as a star, the greatest star in the mathematical firmament.

Selections from *Disquisitiones Arithmeticae* (Arithmetic Disquisitions)

SECTION III

Residues of Powers

▶ 45. THEOREM. *In any geometric progression* 1, *a*, *aa*, a^3, *etc., outside of the first term* 1, *there is still another term* a^t *which is congruent to unity relative to the modulus p when p is prime relative to a; and the exponent t is* $<p$.

Demonstration. Since the modulus p is prime relative to a and hence to any power of a, no term of the progression will be $\equiv 0$ (mod. p) but each of them will be congruent to one of the numbers 1, 2, 3, ... $p - 1$. Since the number of these is $p - 1$ it is clear that if we consider more than $p - 1$ terms of the progression, not all can have different least residues. So among the terms 1, a, aa, a^3, ... a^{p-1} there must be at least one congruent pair. Let therefore $a^m \equiv a^n$ with $m > n$. Dividing by a^n we get $a^{m-n} \equiv 1$ (art. 22) with $m - n < p$ and >0. Q.E.D.

Example. In the progession 2, 4, 8, etc. the first term which is congruent to unity relative to the modulus 13 is $2^{12} = 4096$. But relative to the modulus 23 in the same progression we have $2^{11} = 2048 \equiv 1$. Similarly 15625, the sixth power of the number 5, is congruent to unity relative to the modulus 7 but relative to 11 it is 3125, the fifth power. In some cases therefore the power is less than $p - 1$, but in others it is necessary to go to the $p - $ 1st power itself.

▶ 46. When the progression is continued beyond the term which is congruent to unity, the same residues that we had from the beginning will appear again. Thus if $a^t \equiv 1$, we will have $a^{t+1} \equiv a$, $a^{t+2} \equiv aa$, etc. until we come to the term a^{2t}. Its least residue will again $\equiv 1$, and the *period* of the residues will begin again. Thus we have a period of t residues which as long as it is finite will always be repeated from the beginning; and no residues other than the ones appearing in this period can occur in the whole progression. In general we will have $a^{mt} \equiv 1$ and $a^{mt+n} \equiv a^n$. Following our notation this can be represented thus:

$$\text{if} \quad r \equiv \rho \quad (\text{mod. } t) \qquad \text{then} \qquad a^r \equiv a^\rho \quad (\text{mod. } p)$$

▶ 47. This theorem facilitates our finding the residues of powers, regardless of the size of the exponents, at the same time that we find the power that is congruent to unity. If, e.g., we want to find the residue resulting from division of 3^{1000} by 13 we have $t \equiv 3$ because $3^3 \equiv 1$ (mod. 13); so since $1000 \equiv 1$ (mod. 3), $3^{1000} \equiv 3$ (mod. 13).

▶ 48. When a^t is the *lowest* power congruent to unity (except $a^0 = 1$, which case we are not considering here) all the t terms that comprise the period of the residues will be different from one another, as is clear from the demonstration in article 45. From article 46 we see that the proposition can be converted; that is, if $a^m \equiv a^n$ (mod. p), we have $m \equiv n$ (mod. t). For if m, n are noncongruent relative to the modulus t, their least residues μ, ν will be different. But $a^\mu \equiv a^m$, $a^\nu \equiv a^n$; so $a^\mu \equiv a^\nu$, i.e. not all the powers less than a^t will be noncongruent, contrary to our hypothesis.

If therefore $a^k \equiv 1$ (mod. p) then $k \equiv 0$ (mod. t); i.e. k is divisible by t.

Thus far we have spoken about moduli which are prime relative to some a. Now we will consider separately moduli which are absolutely prime numbers; we will then develop a more general investigation based on these.

▶ 49. THEOREM. *If p is a prime number that does not divide a, and if a^t is the lowest power of a that is congruent to unity relative to the modulus p, the exponent t will either $= p - 1$ or be a factor of this number.*

See article 45 for examples.

Demonstration. We have already seen that t either $= p - 1$ or $< p - 1$. It remains to show that in the latter case t will always be a factor of $p - 1$.

I. Collect the least positive residues of all the terms 1, a, aa, ... a^{t-1} and call them α, α', α'', etc. Thus $\alpha = 1$, $\alpha' \equiv a$, $\alpha'' \equiv aa$, etc. Obviously all these are different for if two terms a^m, a^n had the same residue, we would have (supposing $m > n$) $a^{m-n} \equiv 1$ with $m - n < t$, Q.E.A., since by hypothesis no power lower than a^t can be congruent to unity. Further, all the α, α', α'', etc. are contained in the series of numbers $1, 2, 3, \ldots p - 1$ without, however, exhausting the series, since $t < p - 1$. Let (A) designate the complex of all the α, α', α'', etc. (A) therefore will have t terms.

II. Pick any number β from $1, 2, 3, \ldots p - 1$ which is not contained in (A). Multiply β by all the α, α', α'', etc. and produce the least residues of these β, β', β'', etc. There will also be t of them. But all of these will be different from one another as well as from all the α, α', α'', etc. If the *former* assertion were false, we would have $\beta a^m \equiv \beta a^n$, and dividing by β, $a^m \equiv a^n$, contrary to what we have just shown. If the *latter* were false, we would have $\beta a^m \equiv a^n$; therefore when $m < n$, $\beta \equiv a^{n-m}$ (i.e. β is congruent to one of the α, α', α'', etc., contrary to the hypothesis). If finally $m > n$, multiplying by a^{t-m} we have $\beta a^t \equiv a^{t+n-m}$ or, since $a^t \equiv 1$, $\beta \equiv a^{t+n-m}$, which is the same absurdity. Let (B) designate the complex of numbers β, β', β'', etc. The number of these is t and so we now have $2t$ numbers from $1, 2, 3, \ldots p - 1$. And if (A) and (B) include all these numbers, $(p - 1)/2 = t$, and the theorem is proved.

III. But if there are some numbers left, let one of them be γ. Multiply all the α, α', α'', etc. by γ and let the least residues of the products be γ, γ', γ'', etc. Designate by (C) the complex of all of these. (C) therefore will have t of the numbers $1, 2, 3, \ldots p - 1$, all of them different from each other and from the numbers contained in (A) and (B). The first two assertions can be demonstrated in the same way as in II. For the third, if we had $\gamma a^m \equiv \beta a^n$, then $\gamma \equiv \beta a^{n-m}$ or $\gamma \equiv \beta a^{t+n-m}$, according as m is $<n$ or $>n$. In either case γ would be congruent to one of the numbers in (B), contrary to the hypothesis. So we have $3t$ numbers of the $1, 2, 3, \ldots p - 1$ and if there is none remaining, $t = (p - 1)/3$ and the theorem is proved.

IV. If, however, there are some remaining, we proceed to a fourth complex of numbers (D) etc. It is clear that since the number of the $1, 2, 3, \ldots p - 1$ is finite, they will be finally exhausted. So $p - 1$ is a multiple of t and t is a factor of the number $p - 1$.   Q.E.D.

▶ 50. Since $(p - 1)/t$ is therefore an integer, it follows that by raising each side of the congruence $a^t \equiv 1$ to the power $(p - 1)/t$ we have $a^{p-1} \equiv 1$, or $a^{p-1} - 1$ *will always be divisible by p when p is a prime and does not divide a.*

This theorem is worthy of attention both because of its elegance and its great usefulness. It is usually called Fermat's theorem after its discoverer. (See Fermat, *Opera Mathem.*,[1]) Euler first published a proof in his dissertation entitled "Theorematum quorundam ad numeros primos spectantium demonstratio," (*Comm. acad. Petrop.*, 8 [1736], 1741, 141).[2] The proof is based on the expansion of $(a + 1)^p$. From the form of the coefficients it is easily seen that $(a + 1)^p - a^p - 1$ will always be divisible by p and consequently so will $(a + 1)^p - (a + 1)$ whenever $a^p - a$ is divisible by p. Now since $1^p - 1$ is always divisible by p so also will $2^p - 2$, $3^p - 3$, and in general $a^p - a$. And since p does not divide a, $a^{p-1} - 1$ will be divisible by p. The illustrious Lambert gave a similar demonstration in *Nova acta erudit.*, 1769.[3] But since the expansion of a binomial power seemed quite alien to the theory of numbers, Euler gave another demonstration that appears in *Novi comm. acad. Petrop.*[4] (8 [1760–61], 1763), which is more in harmony with what we have done in the preceding article. We will offer still others later. Here we will add another deduction based on principles similar to those of Euler. The following

[1] The reference is to a letter of Fermat to Frénicle of 18 October 1640 (Bernard Frénicle de Bessy [1605–75]).

[2] In a previous commentary (*Comm. acad. Petrop.*, 6 [1723–33], 1738 ["Observationes de theoremate quodam Fermatiano aliisque ad numeros primos spectantibus"]), this great man had not yet reached this result. In the famous controversy between Maupertuis and König on the principle of the least action, a controversy that led to strange digressions, König claimed to have at hand a manuscript of Leibniz containing a demonstration of this theorem very like Euler's (König, *Appel au public*). We do not wish to deny this testimony, but certainly Leibniz never published his discovery. Cf. *Hist. de l'Ac. de Prusse*, 6 [1750], 1752. [The reference is "Lettre de M. Euler à M. Merian (traduit du Latin)" which was sent from Berlin Sept. 3, 1752.]

[3] "Adnotata quaedam de numeris eorumque anatomia."

[4] "Solutio problematis de investigatione trium numerorum, quorum tan summa quam productum necnon summa productorum ex binis sint numeri quadrati."

proposition of which our theorem is only a particular case will also prove useful for other investigations.

▶ 51. *If p is a prime number the pth power of the polynomial a + b + c + etc. relative to the modulus p is*

$$\equiv a^p + b^p + c^p + \text{etc.}$$

Demonstration. It is clear that the *pth* power of the polynomial $a + b + c +$ etc. is composed of terms of the form $xa^\alpha b^\beta c^\gamma$ etc. where $\alpha + \beta + \gamma +$ etc. $= p$ and x is the number of permutations of p things among which a, b, c, etc. appear respectively α, β, γ, etc. times. But in article 41 above we showed that this number is always divisible by p unless all the objects are the same; i.e. unless some one of the numbers α, β, γ, etc. $= p$ and all the rest $= 0$. It follows that all the terms of $(a + b + c + \text{etc.})^p$ except a^p, b^p, c^p, etc. are divisible by p; and when we treat a congruence relative to the modulus p we can safely omit them and thereby get

$$(a + b + c + \text{etc.})^p \equiv a^p + b^p + c^p + \text{etc.} \qquad \text{Q.E.D.}$$

If all the quantities a, b, c, etc. $= 1$, their sum will $= k$ and we will have $k^p \equiv k$, as in the preceding article.

▶ 52. Suppose we are given numbers which are to be made congruent to unity by raising to a power. We know that for the exponent involved to be of lowest degree it must be a divisor of $p - 1$. The question arises whether all the divisors of $p - 1$ enjoy this property. And if we take all numbers not divisible by p and classify them according to the exponent (in the lowest degree) which makes them congruent to unity, how many are there for each exponent? We observe first that it is sufficient to consider all positive numbers from 1 to $p - 1$; for, manifestly, numbers congruent to each other have to be raised to the same power to become congruent to unity, and therefore each number should be related to the same exponent as its least positive residue. Thus we must find out how in this respect the numbers $1, 2, 3, \ldots p - 1$ should be distributed among the individual factors of the number $p - 1$. In brief, if d is one of the divisors of the number $p - 1$ (among these 1 and $p - 1$ itself must be included), we will designate by ψd the number of positive numbers less than p whose dth power is the lowest one congruent to unity.

▶ 53. To make this easier to understand, we give an example. For $p = 19$ the numbers $1, 2, 3, \ldots 18$ will be distributed among the divisors of 18 in the following way:

1	1
2	18
3	7, 11
6	8, 12
9	4, 5, 6, 9, 16, 17
18	2, 3, 10, 13, 14, 15

Thus in this case $\psi 1 = 1$, $\psi 2 = 1$, $\psi 3 = 2$, $\psi 6 = 2$, $\psi 9 = 6$, $\psi 18 = 6$. A little attention shows that with each exponent there are associated as many numbers as there are numbers relatively prime to the exponent not greater than it. In other words in this case (if we keep the symbol of article 39) $\psi d = \phi d$. Now we will show that this observation is true in general.

I. Suppose we have a number a *belonging* to the exponent d (i.e. its dth power is congruent to unity and all lower powers are noncongruent). All its powers $a^2, a^3, a^4, \ldots a^d$ or their least residues will have the same property (their dth power will be congruent to unity). Since this can be expressed by saying that the least residues of the numbers $a, a^2, a^3, \ldots a^d$ (which are all different) are roots of the congruence $x^d \equiv 1$, and since this can have no more than d roots, it is clear that except for the least residues of the numbers $a, a^2, a^3, \ldots a^d$ there can be no other numbers between 1 and $p - 1$ inclusively whose power to the exponent d is congruent to unity. So all numbers belonging to the exponent d are found among the least residues of the numbers $a, a^2, a^3, \ldots a^d$. What they are and how many there are, we find as follows. If k is a number relatively prime to d, all the powers of a^k whose exponents are $<d$ will not be congruent to unity; for let $1/k$ (mod. d) $\equiv m$ (see art. 31) and we will have $a^{km} \equiv a$. And if the eth power of a^k were congruent to unity and $e < d$, we would also have $a^{kme} \equiv 1$ and hence $a^e \equiv 1$, contrary to the hypothesis. Manifestly the least residue of a^k belongs to the exponent d. If, however, k has a divisor δ in common with d, the least residue of a^k will not belong to the exponent d, because in that case the (d/δ)th power is already congruent to unity (for kd/δ is divisible by d; that is $\equiv 0$ (mod. d) and $a^{kd/\delta} \equiv 1$). We conclude that there are as many numbers belonging to the exponent d as there are numbers relatively prime to d among $1, 2, 3, \ldots d$. But it must be remembered that this conclusion depends on the supposition that we already have one number a belonging to the exponent d. So the doubt remains as to whether it could happen that no number at all pertains to some exponents. The conclusion is therefore limited to the statement that either $\psi d = 0$ or $\psi d = \phi d$.

▶ 54. II. Let d, d', d'', etc. be all the divisors of the number $p - 1$. Since all the numbers $1, 2, 3, \ldots p - 1$ are distributed among these,

$$\psi d + \psi d' + \psi d'' + \text{etc.} = p - 1$$

But in article 40 we showed that

$$\phi d + \phi d' + \phi d'' + \text{etc.} = p - 1$$

and from the preceding article it follows that ψd is equal to or less than ϕd but not greater. Similarly for $\psi d'$ and $\phi d'$ etc. So if one or more from among the terms ψd, $\psi d'$, $\psi d''$, etc. were each smaller than the corresponding term from among the ϕd, $\phi d'$, $\phi d''$, etc., the first sum could not be equal to the second. Thus we conclude finally that ψd *is always equal to* ϕd and so does not depend on the size of $p - 1$.

▶ 55. There is a particular case of the preceding proposition which merits special attention. *There always exist numbers with the property that no power less than the $p - 1$st is congruent to unity*, and there are as many of them between 1 and $p - 1$ as there are numbers less than $p - 1$ and relatively prime to $p - 1$. Since the demonstration of this theorem is less obvious than it first seems and because of the importance of the theorem itself we will use a method somewhat different from the preceding. Such diversity of methods is helpful in shedding light on more obscure points. Let $p - 1$ be resolved into its prime factors so that $p - 1 = a^\alpha b^\beta c^\gamma$ etc.; a, b, c, etc. are unequal prime numbers. We will complete the demonstration by means of the following.

I. We can always find a number A (or several of them) which belongs to the exponent a^α, and likewise the numbers B, C, etc. belonging respectively to the exponents b^β, c^γ, etc.

II. The product of all the numbers A, B, C, etc. (or the least residue of this product) belongs to the exponent $p - 1$. We show this as follows.

1) Let g be some number from $1, 2, 3, \dots p - 1$ which does *not* satisfy the congruence $x^{(p-1)/a} \equiv 1$ (mod. p). Since the degree of the congruence is $<p - 1$ not all these numbers can satisfy it. Now if the $(p - 1)/a^\alpha$ power of $g \equiv h$, then h or its least residue belongs to the exponent a^α.

For it is clear that the a^α power of h is congruent to the $p - 1$ power of g, i.e. to unity. But the $a^{\alpha - 1}$ power of h is congruent to the $(p - 1)/a$ power of g; i.e. it is not congruent to unity. Much less can the $a^{\alpha - 2}$, $a^{\alpha - 3}$, etc. powers of h be congruent to unity. But the exponent of the smallest power of h which is congruent to unity (that is, the exponent to which h belongs) must divide a^α (art. 48). And since a^α is divisible only by itself and lower powers of a, a^α will necessarily be the exponent to which h belongs. Q.E.D. By a similar method we show that there exist numbers belonging to the exponents b^β, c^γ, etc.

2) If we suppose that the product of all the A, B, C, etc. does not belong to the exponent $p - 1$ but to a smaller number t, then t divides $p - 1$ (art. 48); that is $(p - 1)/t$ will be an integer greater than unity. It is easily seen that this quotient is one of the prime numbers a, b, c, etc. or at least is divisible by one of them (art. 17), e.g. by a. From this point on the reasoning is the same as above. Thus t divides $(p - 1)/a$ and the product ABC etc. raised to the $(p - 1)/a$ power will also be congruent to unity (art. 46). But it is evident that all the numbers B, C, etc. (A excepted) will become congruent to unity if raised to the $(p - 1)/a$ power because the exponents b^β, c^γ, etc. to which they each belong divide $(p - 1)/a$. Thus we have

$$A^{(p-1)/a} B^{(p-1)/a} C^{(p-1)/a} \text{ etc.} \equiv A^{(p-1)/a} \equiv 1$$

It follows that the exponent to which A belongs ought to divide $(p-1)/a$ (art. 48); i.e. $(p-1)/a^{\alpha+1}$ is an integer. But $(p-1)/a^{\alpha+1} = (b^\beta c^\gamma$ etc.$)/a$ cannot be an integer (art. 15). We conclude therefore that our supposition is inconsistent, i.e. that the product ABC etc. really belongs to the exponent $p-1$.   Q.E.D.

The second demonstration seems a little longer than the first, but the first is less direct than the second.

▶ 56. This theorem furnishes an outstanding example of the need for circumspection in number theory so that we do not accept fallacies as certainties. Lambert in the dissertation that we praised above, *Nova acta erudit.* 1769, makes mention of this proposition but does not mention the need for demonstrating it. No one has attempted the demonstration except Euler in *Novi comm. Acad. Petrop.*, (*18* [1773], 1774), "*Demonstrationes circa residua ex divisione potestatum per numeros primos resultantia.*" See especially his article 37 where he speaks at great length of the need for demonstration. But the demonstration which this shrewdest of men presents has two defects. One is that his article 31 ff. tacitly supposes that the congruence $x^n \equiv 1$ (translating his argument into our notation) really has n different roots, although nothing more had been shown previously than that it could not have *more* than n roots; the other is that the formula of article 34 was derived only by induction.

▶ 57. Along with Euler we will call numbers belonging to the exponent $p-1$ *primitive roots*. Therefore if a is a primitive root the least residues of the powers $a, a^2, a^3, \ldots a^{p-1}$ will all be different. It is then easy to deduce that among them we will find all the numbers $1, 2, 3, \ldots p-1$, since each has the same number of elements. This means that any number not divisible by p is congruent to some power of a. This remarkable property is of great usefulness, and it can considerably reduce the arithmetic operations relative to congruences in much the same way that the introduction of logarithms reduces the operations in ordinary arithmetic. We will arbitrarily choose some primitive root a as a *base* to which we will refer all numbers not divisible by p. And if $a^e \equiv b$ (mod. p), we will call e the *index* of b. For example, if relative to the modulus 19 we take the primitive root 2 as base, we have

numbers 1. 2. 3. 4. 5. 6. 7. 8. 9. 10. 11. 12. 13. 14. 15. 16. 17. 18

indices 0. 1. 13. 2. 16. 14. 6. 3. 8. 17. 12. 15. 5. 7. 11. 4. 10. 9

For the rest, it is clear that with the base unchanged each number has many indices but that they are all congruent relative to the modulus $p-1$; so when there is question of indices, those that are congruent relative to the modulus $p-1$ will be regarded as equivalent in the same way that numbers are regarded as equivalent when they are congruent relative to the modulus p.

▶ 58. Theorems pertaining to indices are completely analogous to those that refer to logarithms.

The index of the product of any number of factors is congruent to the sum of the indices of the individual factors relative to the modulus p − 1.

The index of the power of a number is congruent relative to the modulus p − 1 to the product of the index of the number by the exponent of the power.

Because of their simplicity we omit demonstrations of these theorems.

It is clear from the above that if we wish to construct a table that gives the indices for all numbers relative to different moduli, we can omit from it all numbers larger than the modulus and also all composite numbers. An example of this kind of table appears at the end of this work (Table 1). In the first vertical column are arranged prime numbers and the powers of prime numbers from 3 to 97. These are to be regarded as moduli. Next to each of these in the following column are the numbers chosen as base; then follow the indices of successive prime numbers arranged in blocks of five. At the top of the columns the prime numbers are again arranged in the same order so that it is easy to find the index corresponding to a given prime number relative to a given modulus.

For example, if $p = 67$, the index for the number 60 with 12 as base $\equiv$ 2 Ind. 2 + Ind. 3 + Ind. 5 (mod. 66) $\equiv$ 58 + 9 + 39 $\equiv$ 40

▶ 59. *The index of the value of an expression like a/b (mod. p) (see art. 31) is congruent relative to the modulus p − 1 to the difference between the index of the numerator a and the index of the denominator b, provided a, b are not divisible by p.*

Let c be such a value. We have $bc \equiv a$ (mod. p) and so

$$\text{Ind. } b + \text{Ind. } c \equiv \text{Ind. } a \,(\text{mod. } p - 1)$$

and

$$\text{Ind. } c \equiv \text{Ind. } a - \text{Ind. } b$$

Then if we have two tables, one of which gives the index for any number relative to any prime modulus and the other gives the number belonging to a given index, all congruences of the first degree can be easily solved because they can be reduced to congruences whose modulus is a prime number (art. 30). For example, a given congruence

$$29x + 7 \equiv 0 \,(\text{mod. } 47) \qquad \text{becomes} \qquad x \equiv \frac{-7}{29} \,(\text{mod. } 47)$$

and so

$$\text{Ind. } x \equiv \text{Ind. } -7 - \text{Ind. } 29 \equiv \text{Ind. } 40 - \text{Ind. } 29 \equiv 15 - 43$$
$$\equiv 18 \,(\text{mod. } 46)$$

Now 3 is the number whose index is 18. So $x \equiv 3$ (mod. 47). We have not added the second table, but in Section VI we shall see how to replace it with another.

▶ 60. In article 31 we designated by a special sign roots of congruences of the first degree, so in what follows we will represent simple congruences of higher degrees by a special sign. Just as $\sqrt[n]{A}$ indicates the root of the equation $x^n = A$, so with the modulus added $\sqrt[n]{A}$ (mod. p) will denote any root of the congruence $x^n \equiv A$ (mod. p). We will say that the expression $\sqrt[n]{A}$ (mod. p) has as many values as it has values that are noncongruent relative to p, since all those that are congruent relative to p are considered equivalent (art. 26). It is clear that if A, B are congruent relative to p, the expressions $\sqrt[n]{A}$, $\sqrt[n]{B}$ (mod. p) will be equivalent.

Now if we are given $\sqrt[n]{A} \equiv x$ (mod. p), we will have n Ind. $x \equiv$ Ind. A (mod. $p - 1$). According to the rules of the preceding section, from this congruence we can deduce the values of Ind. x and from these the corresponding values of x. It is easily seen that x will have as many values as there are roots of the congruence n Ind. $x \equiv$ Ind. A (mod. $p - 1$). And manifestly $\sqrt[n]{A}$ will have only one value when n is prime relative to $p - 1$. But when n, $p - 1$ have a greatest common divisor δ, Ind. x will have δ noncongruent values relative to $p - 1$, and $\sqrt[n]{A}$ the same number of noncongruent values relative to p, provided Ind. A is divisible by δ. If this condition is lacking, $\sqrt[n]{A}$ will have no real value.

Example. We are looking for the values of the expression $\sqrt[15]{11}$ (mod. 19). We must therefore solve the congruence 15 Ind. $x \equiv$ Ind. $11 = 6$ (mod. 18), and we find three values of Ind. $x \equiv 4, 10, 16$ (mod. 18). The corresponding values of x are 6, 9, 4.

▶ 61. No matter how expeditious this method is when we have the necessary tables, we should not forget that it is indirect. It will be useful therefore to determine how powerful the direct methods are. We will consider here what we can deduce from previous sections; other considerations that demand more profound investigation will be reserved for Section VIII.[5] We begin with the simplest case where $A = 1$. That is, we will look for the roots of the congruence $x^n \equiv 1$ (mod. p). Here after taking some primitive root as base we must have n Ind. $x \equiv 0$ (mod. $p - 1$). Now if n is prime relative to $p - 1$ this congruence will have only one root; that is, Ind. $x \equiv 0$ (mod. $p - 1$). In this case then $\sqrt[n]{1}$ (mod. p) has a unique value, namely $\equiv 1$. But when the numbers n, $p - 1$ have a (greatest) common divisor δ, the complete solution of the congruence n Ind. $x \equiv 0$ (mod. $p - 1$) will be Ind. $x \equiv 0$ (mod. $(p - 1)/\delta$) [see art. 29]; i.e. Ind. x relative to the modulus $p - 1$ should be congruent to one of the numbers

$$0, \quad \frac{p - 1}{\delta}, \quad \frac{2(p - 1)}{\delta}, \quad \frac{3(p - 1)}{\delta}, \quad \ldots, \quad \frac{(\delta - 1)(p - 1)}{\delta}$$

[5]Section VIII was not published.

that is, it will have δ values which are noncongruent relative to the modulus $p - 1$; and so in this case x also will have δ different values (noncongruent relative to the modulus p). Thus we see that the expression $\sqrt[\delta]{1}$ also has δ different values whose indices are absolutely the same as the previous ones. For this reason the expression $\sqrt[\delta]{1}$ (mod. p) is completely equivalent to $\sqrt[n]{1}$ (mod. p); i.e. the congruence $x^\delta \equiv 1$ (mod. p) has the same roots as $x^n \equiv 1$ (mod. p). The former, however, will be of lower degree if δ and n are not equal.

Example. $\sqrt[15]{1}$ (mod. 19) has three values because 3 is the greatest common divisor of the numbers 15, 18 and they will also be the values of the expression $\sqrt[3]{1}$ (mod. 19). They are 1, 7, 11.

▶ 62. By this reduction we know that other congruences of the form $x^n \equiv 1$ need not be solved unless n is a divisor of the number $p - 1$. We will see later that congruences of this form can always be reduced; what we showed so far is not sufficient to indicate this. There is one case, however, that we can dispose of here, i.e. when $n = 2$. Clearly the values of the expression $\sqrt[2]{1}$ will be $+1$ and -1, since there cannot be more than two, and $+1$ and -1 are always noncongruent unless the modulus $=2$, in which case $\sqrt[2]{1}$ can have only one value. It follows that $+1$ and -1 will also be values of the expression $\sqrt[2m]{1}$ when m is prime relative to $(p - 1)/2$. This always happens when the modulus is such that $(p - 1)/2$ is an absolutely prime number (except when $p - 1 = 2m$ in which case all the numbers $1, 2, 3, \ldots p - 1$ are roots)—e.g. when $p = 3, 5, 7, 11, 23, 47, 59, 83, 107$, etc. As a corollary we note that the index of -1 is always $\equiv (p - 1)/2$ (mod. $p - 1$), no matter what primitive root is taken as base. This is so because 2 Ind. $(-1) \equiv 0$ (mod. $p - 1$), and so Ind. (-1) will be either $\equiv 0$ or $\equiv (p - 1)/2$ (mod. $p - 1$). However 0 is always an index of $+1$, and $+1$ and -1 must always have different indices (except for the case $p = 2$ which we need not consider here).

▶ 63. We showed in article 60 that the expression $\sqrt[n]{A}$ (mod. p) has δ different values, or none at all if δ is the greatest common divisor of the numbers $n, p - 1$. Now in the same way that $\sqrt[n]{A}$ and $\sqrt[n]{1}$ are equivalent when $A \equiv 1$, we will give a more general proof showing that $\sqrt[n]{A}$ can always be reduced to another expression, $\sqrt[\delta]{B}$, to which it will be equivalent. If we denote some value of this by x we will have $x^n \equiv A$; further let t be a value of the expression δ/n (mod. $p - 1$). These values are real as is clear from article 31. Now $x^{tn} \equiv A^t$, but $x^{tn} \equiv x^\delta$ because $tn \equiv \delta$ (mod. $p - 1$). Therefore $x^\delta \equiv A^t$, and so any value of $\sqrt[n]{A}$ will also be a value of $\sqrt[\delta]{A^t}$. As often therefore as $\sqrt[n]{A}$ has real values, it will be completely equivalent to the expression $\sqrt[\delta]{A^t}$. This is true because the former cannot have different values than the latter nor can there be fewer of them except when $\sqrt[n]{A}$ has no real values. In this case, however, it can happen that $\sqrt[\delta]{A^t}$ has real values.

Example. If we are looking for the values of the expression $\sqrt[21]{2}$ (mod. 31), the greatest common divisor of the numbers 21 and 30 is 3, and 3 is a value of the expression 3/21 (mod. 30); so if $\sqrt[21]{2}$ has real values it will be equivalent to the expression $\sqrt[3]{2^3}$ or $\sqrt[3]{8}$; and we find in fact that the values of the latter which are 2, 10, 19 satisfy the former also.

▶ 64. In order to avoid trying this operation in vain, we should investigate the rule for determining whether $\sqrt[n]{A}$ has real values. If we have a table of indices, it is easy, for from article 60 it is clear that we have real values if the index of A with any primitive root taken as base is divisible by δ. In the contrary case there will be no real values. But we can still determine this without such a table. Let the index of A be $=k$. If this is divisible by δ, then $k(p-1)/\delta$ will be divisible by $p-1$ and vice versa. But the index of the number $A^{(p-1)/\delta}$ will be $k(p-1)/\delta$. So if $\sqrt[n]{A}$ (mod. p) has real values, $A^{(p-1)/\delta}$ will be congruent to unity; if not, it will be noncongruent. So in the example of the preceding article we have $2^{10} = 1024 \equiv 1$ (mod. 31) and we conclude that $\sqrt[21]{2}$ (mod. 31) has real values. In the same way we see that $\sqrt[2]{-1}$ (mod. p) will always have a pair of real values when p is of the form $4m + 1$ but none when p is of the form $4m + 3$, because $(-1)^{2m} = 1$ and $(-1)^{2m+1} = -1$. This elegant theorem which is commonly enunciated thus: *If p is a prime number of the form $4m + 1$, a square a^2 can be found such that $a^2 + 1$ is divisible by p; but if p is of the form $4m - 1$ such a square cannot be found,* . . . was demonstrated in this manner by Euler in *Novi comm. acad. Petrop.* (*18* [1773], 1774). He had given another demonstration much before that in 1760 (*Novi comm. acad. Petrop.*, 5 [1754–55]). In a previous dissertation[6] (*4* [1752–53], 1758, 25) he had not yet arrived at the result. Later also Lagrange gave a demonstration of the theorem in *Nouv. mém. Acad. Berlin*, 1775.[7] We will give another demonstration in the following section which is properly devoted to this subject.

▶ 65. After discussing how to reduce expressions $\sqrt[n]{A}$ (mod. p) to others in which n is a divisor of $p - 1$, and finding a criterion by which we can determine whether or not they admit of real values, we will consider expressions $\sqrt[n]{A}$ (mod. p) in which n is already a divisor of $p - 1$. First we will show what relation the different values of the expression have among themselves; we will then treat of certain devices by means of which one value can very often be found.

First. When $A \equiv 1$ and r is one of the values for the expressions $\sqrt[n]{1}$ (mod. p) or $r^n \equiv 1$ (mod. p), all the powers of this r will also be values of the expression; there will be as many different ones as there are unities in the exponent to which

[6]"De numeris qui sunt aggregata duorum quadratorum."
[7]"Suite des Recherches d'arithmétique imprimées dans le volume de l'année 1773."

r belongs (art. 48). Therefore if r is a value belonging to the exponent n, all the powers $r, r^2, r^3, \ldots r^n$ (where *unity* can replace the last) involve values of the expression $\sqrt[n]{1}$ (mod. p). We will explain in Section VIII what aids there are for finding such values that belong to the exponent n.

Second. When A is not congruent to unity and one value z of the expression $\sqrt[n]{A}$ (mod. p) is known, the others can be found as follows. Let the values of the expression $\sqrt[n]{1}$ be

$$1, r, r^2, \ldots, r^{n-1}$$

(as we have just shown). All the values of the expression $\sqrt[n]{A}$ will be

$$z, zr, zr^2, \ldots, zr^{n-1}$$

It is evident that all these satisfy the congruence $x^n \equiv A$ because if one of them is $\equiv zr^k$, its nth power $z^n r^{nk}$ will be congruent to A. This is obvious because $r^n \equiv 1$ and $z^n \equiv A$. And from article 23 it is easy to see that all these values are different. The expression $\sqrt[n]{A}$ can have none except these n values. So, e.g., if one value of the expression $\sqrt[2]{A}$ is z, another will be $-z$. From the preceding we must conclude that it is not possible to find all values of the expression $\sqrt[n]{A}$ without at the same time establishing all the values of the expression $\sqrt[n]{1}$.

▶ 66. The second thing that we proposed to do was to find out when one value of the expression $\sqrt[n]{A}$ (mod. p) can be determined directly (presupposing, of course, that n is a divisor of $p-1$). This happens when some value is congruent to a power of A. This is a rather frequent occurrence and warrants a moment's consideration. *If this value exists* let it be z, that is $z \equiv A^k$ and $z \equiv z^n$ (mod. p). As a result $A \equiv A^{kn}$; and so if we can find a number k such that $A \equiv A^{kn}$, A^k will be the value we seek. But this condition is the same as saying that $1 \equiv kn$ (mod. t) where t is the exponent to which A belongs (art. 46, 48). But for such a congruence to be possible it is necessary that n and t be relatively prime. In which case we will have $k \equiv 1/n$ (mod. t); if however t and n have a common divisor, there can be no value of z that is congruent to a power of A.

▶ 67. Since to arrive at this solution it is necessary to know t, let us see how we proceed if it is not known. First, it is obvious that t must divide $(p-1)/n$ if $\sqrt[n]{A}$ (mod. p) is to have real values as we always suppose here. Let y be some one of these values; we will then have $y^{p-1} \equiv 1$ and $y^n \equiv A$ (mod. p); and by raising both sides of the latter congruence to the $(p-1)/n$ power we get $A^{(p-1)/n} \equiv 1$; and so $(p-1)/n$ is divisible by t (art. 48). Now if $(p-1)/n$ is prime relative to n, the congruence of the preceding article $kn \equiv 1$ can also be solved relative to the modulus $(p-1)/n$, and obviously any value of k satisfying the congruence relative to this modulus will satisfy it also relative to the modulus t which divides $(p-1)/n$ (art. 5). Thus we have found what we wanted. If $(p-1)/n$ is not prime relative to

n, eliminate from $(p-1)/n$ all the prime factors which at the same time divide n. Thus we get the number $(p-1)/nq$ which is prime relative to n. Here we use q to indicate the product of all the prime factors that have been eliminated. Now if the condition of the previous article holds, that is if t is prime relative to n, then t will also be prime relative to q and so will divide $(p-1)/nq$. So if we solve the congruence $kn \equiv 1$ (mod. $(p-1)/nq$), (which can be done because n is prime relative to $(p-1)/nq$), the value of k will also satisfy the congruence relative to the modulus t, which is what we were looking for. This whole device consists in discovering a number to take the place of the t which we do not know. But we must remember that when $(p-1)/n$ is not prime relative to n we have supposed the condition of the previous article. If this is not true, all the conclusions will be erroneous; and if by following the given rules without care we find a value for z whose nth power is not congruent to A, we know that the condition is lacking and the method cannot be used.

▶ 68. But even in this case it is often advantageous to have done the work involved, and it is worth investigating the relationship between this false value and the true ones. Suppose therefore that k, z have been duly determined but that z^n is not $\equiv A$ (mod. p). Then if we can only determine the values of the expression $\sqrt[n]{(A/z^n)}$ (mod. p), we will obtain the values of $\sqrt[n]{A}$ by multiplying each of them by z. For if v is a value of $\sqrt[n]{(A/z^n)}$, we will have $(vz)^n \equiv A$. But the expression $\sqrt[n]{(A/z^n)}$ is simpler than $\sqrt[n]{A}$ because very often A/z^n (mod. p) belongs to a lower exponent than A. More precisely, if the greatest common divisor of the numbers t, q is d, A/z^n (mod. p) will belong to the exponent d as we have seen. Substituting for z, we get $A/z^n \equiv 1/A^{kn-1}$ (mod. p). But $kn-1$ is divisible by $(p-1)/nq$, and $(p-1)/n$ by t (preceding article); that is $(p-1)/nd$ is divisible by t/d. But t/d and q/d are relatively prime (hypothesis), so also $(p-1)/nd$ is divisible by tq/dd or $(p-1)/nq$ by t/d. And thus $kn-1$ will be divisible by t/d and $(kn-1)d$ by t. The final result gives us $A^{(kn-1)d} \equiv 1$ (mod. p), and we deduce that A/z^n raised to the dth power is congruent to unity. It is easy to show that A/z^n cannot belong to any exponent less than d, but since this is not required for our purpose we will not dwell on it. We can be certain therefore that A/z^n (mod. p) always belongs to a smaller exponent than A except when t divides q and so $d = t$.

But what is the advantage in having A/z^n belong to a smaller exponent than A? There are more numbers that can be A than can be A/z^n and when we want to solve many expressions of the form $\sqrt[n]{A}$ relative to the same modulus, we have the advantage of being able to derive several results from the one computation. Thus, e.g., we will always be able to determine at least one value of the expression $\sqrt[n]{A}$ (mod. 29) if we know the values of the expression $\sqrt[2]{-1}$ (mod. 29) [these are ± 12]. From the preceding article it is easy to see how we can determine one value of this expression directly when t is odd, and that d will be $=2$ when t is even; except for -1 no number belongs to the exponent 2.

Examples. We want to solve $\sqrt[3]{31}$ (mod. 37). Here $p - 1 = 36$, $n = 3$, $(p - 1)/3 = 12$, and so $q = 3$. Now we want $3k \equiv 1$ (mod. 4), and we get this by letting $k = 3$. Thus $z \equiv 31^3$ (mod. 37) $\equiv 6$, and it is true that $6^3 \equiv 31$ (mod. 37). If the values of the expression $\sqrt[3]{1}$ (mod. 37) are known, the remaining values of the expression $\sqrt[3]{6}$ can be determined. The values of $\sqrt[3]{1}$ (mod. 37) are 1,10, 26. Multiplying these by 6 we get for the other two $\sqrt[3]{31} \equiv 23$, 8.

If however we are looking for the value of the expression $\sqrt[2]{3}$ (mod. 37), $n = 2$, $(p - 1)/n = 18$, and so $q = 2$. Since we want $2k \equiv 1$ (mod. 9), $k \equiv 5$ (mod. 9). And $z \equiv 3^5 \equiv 21$ (mod. 37); but 21^2 is not $\equiv 3$ whereas it is $\equiv 34$; on the other hand $3/34$ (mod. 37) $\equiv -1$ and $\sqrt[2]{-1}$ (mod. 37) $\equiv \pm 6$; thus we find the true values $\pm 6 \cdot 21 \equiv \pm 15$.

This is practically all we can say about the solution of such expressions. It is clear that direct methods are often rather long, but this is true of almost all direct methods in the theory of numbers; nevertheless their usefulness should be demonstrated. It is beyond the purpose of our investigation, however, to explain one by one the particular artifices that become familiar to anyone working in this field.

▶ 69. We return now to consideration of the roots we call primitive. We have shown that when we take any primitive root as a base all numbers whose indices are prime relative to $p - 1$ will also be primitive roots and that these are the only ones; we therefore know at the same time the number of the primitive roots (see art. 53). In general it is left to our own discretion to decide which root to use as a base. Here, as in logarithmic calculus, we can have many different systems.[8] Let us see how they are connected. Let a, b be two primitive roots and m another number. When we take a as our base, the index of the number $b \equiv \beta$ and the index of the number $m \equiv \mu$ (mod. $p - 1$). But when we take b as our base, the index of the number a is $\equiv \alpha$ and of the number m is $\equiv \nu$ (mod. $p - 1$). Now $\alpha^\beta \equiv b$ and $a^{\alpha\beta} \equiv b^\alpha \equiv a$ (mod. p) [by hypothesis] so $\alpha\beta \equiv 1$ (mod. $p - 1$). By a similar process we find that $\nu \equiv \alpha\mu$, and $\mu \equiv \beta\nu$ (mod. $p - 1$). Therefore if we have a table of indices constructed for the base a it is easy to convert it into another in which the base is b. For if for the base a the index of b is $\equiv \beta$, for the base b the index of a will be $\equiv 1/\beta$ (mod. $p - 1$) and by multiplying all the indices of the table by this number, we will find all the indices for the base b.

---

[8]But it differs in that for logarithms the number of systems is infinite; here there are only as many as there are primitive roots. For it is manifest that congruent bases generate the same system.

▶ 70. Although a given number can have varying indices according as different primitive roots are taken as bases, they all agree in this—that the greatest common divisor for each of them and $p - 1$ will be the same. For if for the base a, the index of a given number is m, and for the base b, the index is n and if the greatest common divisors of these numbers and $p - 1$ are μ, ν and unequal, then one will be greater; e.g. $\mu > \nu$ and so μ will not divide n. Now assuming b for the base, let α be the index of a and we will have (preceding article) $n \equiv \alpha m$ (mod. $p - 1$) and μ will divide n. Q.E.A.

We can also see that this greatest divisor common to the indices of a given number and to $p - 1$ does not depend on the base from the fact that it is equal to $(p - 1)/t$. Here t is the exponent to which the number belongs whose indices we are considering. For if the index for any base is k, t will be the smallest number (except zero) which multiplied by k gives a multiple of $p - 1$ (see art. 48, 58), that is, the least value of the expression $0/k$ (mod. $p - 1$) except zero. That this is equal to the greatest common divisor of the numbers k and $p - 1$ is derived with no difficulty from article 29.

▶ 71. It is always permissible to choose a base so that the number pertaining to the exponent t matches some predetermined index whose greatest divisor in common with $p - 1$ is $= (p - 1)/t$. Let us designate by d this divisor, let the index proposed be $\equiv dm$, and let the index of the proposed number be $\equiv dn$. Here a primitive root a is selected for the base; m, n will be prime relative to $(p - 1)/d$ or to t. Then if ε is a value of the expression dn/dm (mod. $p - 1$) and at the same time is prime relative to $p - 1$, a^ε will be a primitive root. With this base the proposed number will produce the index dm just as we wanted (for $a^{\varepsilon dm} \equiv a^{dn} \equiv$ the proposed number). To prove that the expression dn/dm (mod. $p - 1$) has values relatively prime to $p - 1$ we proceed as follows. This expression is equivalent to n/m (mod. $(p - 1)/d$), or to n/m (mod. t) [see art. 31, 2] and all its values will be prime relative to t; for if any value e should have a common divisor with t, this divisor will also divide me and so also n, since me is congruent to n relative to t. But this is contrary to the hypothesis which demands that n be prime relative to t. When therefore all the prime divisors of $p - 1$ divide t, *all* the values of the expression n/m (mod. t) will be prime relative to $p - 1$ and their number will $= d$. When however $p - 1$ has other divisors f, g, h, etc., which are prime and do not divide t, one value of the expression n/m (mod. t) $\equiv e$. But then since all the numbers t, f, g, h, etc. are prime relative to each other, a number ε can be found which is congruent to e relative to t, and relative to f, g, h, etc. is congruent to other numbers which are prime relative to these (art. 32). Such a number will be divisible by no prime factor of $p - 1$ and so will be prime relative to $p - 1$ as was desired. From the theory of combinations we easily deduce that the number of such values will be equal to

$$\frac{(p-1)\cdot(f-1)\cdot(g-1)\cdot(h-1)\cdot \text{etc.}}{t \quad \cdot \quad f \quad \cdot \quad g \quad \cdot \quad h \quad \cdot \text{etc.}}$$

but in order not to extend this digression too far, we omit the demonstration. It is not necessary to our purpose in any case.

▶ 72. Although in general the choice of a primitive root for the base is entirely arbitrary, at times some bases will prove to have special advantages over others. In Table 1 we have always taken 10 as the base when it was a primitive root; in other cases we have always chosen the base so that the index of the number 10 would be the smallest possible; i.e. we let it $= p - 1/t$ with t the exponent to which 10 belongs. We will indicate the advantage of this in Section VI where the same table is used for other purposes. But here too there still remains some freedom of choice, as we have seen in the preceding article. And so we have always chosen as base the *smallest* primitive root that satisfies the conditions. Thus for $p = 73$ where $t = 8$ and $d = 9$, a^ε has $72/8 \cdot 2/3$; i.e. 6 values, which are 5, 14, 20, 28, 39, 40. We chose the smallest, 5, as base.

▶ 73. Finding primitive roots is reduced for the most part to trial and error. If we add what we have said in article 55 to what we have indicated above about the solution of the congruence $x^n \equiv 1$, we will have all that can be done by direct methods. Euler (*Opuscula Analytica*, 1[9]) admits that it is extremely difficult to assign these numbers and that their nature is one of the deepest mysteries of numbers. But they can be determined easily enough by the following method. Skillful mathematicians know how to reduce tedious calculations by a variety of devices, and here experience is a better teacher than precept.

1) Arbitrarily select a number a which is prime relative to p (we always use this letter to designate the modulus; usually the calculation is simpler if we choose the smallest possible—e.g. the number 2). Next determine its period (art. 46), i.e. the least residues of its powers, until we come to the power a^t whose least residue is 1.[10] If $t = p - 1$, a is a primitive root.

2) If however $t < p - 1$, choose another number b that is not contained in the period of a and in the same way investigate its period. If we designate by u the exponent to which b belongs, we see that u cannot be equal to t or to a factor of t; for in either case we would have $b^t \equiv 1$ which cannot be, since the period of a includes all numbers whose t power is congruent to unity (art. 53). Now if $u = p - 1$, b

---

[9] "Disquisitio accuratior circa residua ex divisione quadratorum altiorumque potestatum per numeros primos relicta."
[10] It is not necessary to know these powers, since the least residue can be easily obtained from the least residue of the preceding power.

would be a primitive root; however if u is not $= p - 1$ but is a multiple of t, we have gained this much—we know we can find a number belonging to a higher exponent and so we will be closer to our goal, which is to find a number belonging to the *maximum* exponent. Now if u does not $= p - 1$ and is not a multiple of t, we can nevertheless find a number belonging to an exponent greater than t, u, namely to the exponent equal to the least common multiple of t, u. Let this number $= y$ and resolve y into two factors m, n which are relatively prime and such that one of them divides t and the other divides u.[11] As a result, the t/m power of a will $\equiv A$, the u/n power of b will $\equiv B$ (mod. p), and the product AB will be the number belonging to the exponent y. This is clear, since A belongs to the exponent m and B belongs to the exponent n and, since m, n are relatively prime, the product AB will belong to mn. We can prove this in practically the same way as in article 55, Section II.

3) Now if $y = p - 1$, AB will be a primitive root; if not, we proceed as before, using another number not occurring in the period of AB. This will either be a primitive root, or it will belong to an exponent greater than y, or with its help (as before) we can find a number belonging to an exponent greater than y. Since the numbers we get by repeating this operation belong to constantly increasing exponents, we must finally discover a number that belongs to the *maximum* exponent. This will be a primitive root.
Q.E.F.

▶ 74. This becomes clearer by an example. Let $p = 73$ and let us look for a primitive root. We will try first the number 2 whose period is

1. 2. 4. 8. 16. 32. 64. 55. 37. 1 etc.
0. 1. 2. 3. 4. 5. 6. 7. 8. 9 etc.

Since the power of the exponent 9 is therefore congruent to unity, 2 is not a primitive root. We will try another number that does not occur in the period—e.g. 3. Its period will be

1. 3. 9. 27. 8. 24. 72. 70. 64. 46. 65. 49. 1 etc.
0. 1. 2. 3. 4. 5. 6. 7. 8. 9. 10. 11. 12 etc.

So 3 is not a primitive root. However the least common multiple of the exponents to which 2, 3 belong (i.e. the numbers 9, 12) is 36 which is resolved into the factors 9 and 4 as we saw in the previous article. Raising 2 to the power 9/9, 3 to the power 3, the product of these is 54 which belongs to the exponent 36. If finally we compute the period of 54 and try a number that is not contained in it (e.g. 5), we find that this is a primitive root.

---

[11]From article 18 we see how this can be done without difficulty. Resolve y into factors which are either different prime numbers or powers of different prime numbers. Each of these will divide t or u (or both of them). Assign each of them either to t or to u according to which one it divides. If one of them divides both, it can be arbitrarily assigned. Let the product of those assigned to $t = m$, the product of the others $= n$. Obviously m divides t, n divides u, and $mn = y$.

▶ 75. Before abandoning this line of argument we will present certain propositions which because of their simplicity are worth attention.

The product of all the terms of the period of any number is ≡1 *when their number or the exponent to which the number belongs is odd and* ≡−1 *when the exponent is even.*

Example. For the modulus 13, the period of the number 5 consists of the terms 1, 5, 12, 8, and its product 480 ≡ −1 (mod. 13).

Relative to the same modulus the period of the number 3 consists of the terms 1, 3, 9, and the product 27 ≡ 1 (mod. 13).

Demonstration. Let the exponent to which a number belongs be t and the index of the number $(p − 1)/t$. This can always be done if we choose the right base (art. 71). Then the index of the product of all the terms of the period will be

$$\equiv (1 + 2 + 3 + \text{etc.} + t − 1)\frac{p − 1}{t} = \frac{(t − 1)(p − 1)}{2}$$

i.e. ≡0 (mod. $p − 1$) when t is odd, and ≡$(p − 1)/2$ when t is even; in the former case the product will be ≡1 (mod. p) and in the latter ≡−1 (mod. p) (art. 62).
Q.E.D.

▶ 76. If the number in the preceding theorem is a primitive root, its period will include all the numbers 1, 2, 3, . . . , $p − 1$. Its product will always ≡−1 (for $p − 1$ will always be even except when $p = 2$; in this case −1 and +1 are equivalent). This elegant theorem is usually worded thus: *The product of all numbers less than a given prime number is divisible by this prime number if unity is added to the product.* It was first published by Waring and attributed to Wilson: Waring, *Meditationes Algebraicae* (3d ed., Cambridge, 1782).[12] But neither of them was able to prove the theorem, and Waring confessed that the demonstration was made more difficult because no *notation* can be devised to express a prime number. But in our opinion truths of this kind should be drawn from the ideas involved rather than from notations. Afterward Lagrange gave a proof (*Nouv. mém. Acad. Berlin,* 1771[13]). He does this by considering the coefficients that arise in expanding the product

$$(x + 1)(x + 2)(x + 3)\ldots(x + p − 1)$$

By letting this product

$$\equiv x^{p−1} + Ax^{p−2} + Bx^{p−3} + \text{etc.} + Mx + N$$

the coefficients A, B, etc., M will be divisible by p, and N will be $=1 \cdot 2 \cdot 3 \cdot \ldots \cdot p − 1$. If $x = 1$ the product will be divisible by p; but then it will be ≡$1 + N$ (mod. p) and so $1 + N$ will necessarily be divisible by p.

---

[12]In the first edition of 1770, to which Lagrange refers.
[13]"Démonstration d'un théoreme noveau concernant les nombres premiers."

Finally, Euler, in *Opuscula Analytica, 1*[14], gave a proof that agrees with what we have shown. Since such distinguished mathematicians did not consider this theorem unworthy of their attention, we are emboldened to add still another proof.

▶ 77. When relative to the modulus p the product of two numbers a, b is congruent to unity we will call the numbers *associates* as Euler did. Then according to the preceding section any positive number less than p will have a positive associate less than p and it will be unique. It is easy to prove that of the numbers $1, 2, 3, \ldots, p - 1$, 1 and $p - 1$ are the unique numbers which are associates of one another, for associate numbers will be roots of the congruence $x^2 \equiv 1$. And since this congruence is of the second degree it cannot have more than two roots; i.e. only 1 and $p - 1$. Suppressing these, the remaining numbers $2, 3, \ldots, p - 2$ will be associated in pairs. Their product then will $\equiv 1$ and the product of all of them $1, 2, 3, \ldots, p - 1$ will $\equiv p - 1$ or $\equiv -1$. Q.E.D.

For example, for $p = 13$ the numbers $2, 3, 4, \ldots, 11$ will be associated as follows: 2 with 7; 3 with 9; 4 with 10; 5 with 8; 6 with 11. That is, $2 \cdot 7 \equiv 1$; $3 \cdot 9 \equiv 1$, etc. Thus $2 \cdot 3 \cdot 4 \ldots 11 \equiv 1$; and so $1 \cdot 2 \cdot 3 \ldots 12 \equiv -1$.

▶ 78. Wilson's theorem can be expressed more generally thus: *The product of all numbers less than a given number A and at the same time relatively prime to it, is congruent relative to A to plus or minus unity.* Minus unity is to be taken when A is of the form p^m or $2p^m$ where p is a prime number different from 2 and also when $A = 4$. Plus unity is taken in all other cases. The theorem as Wilson proposed it is contained in the former case. For example, for $A = 15$ the product of the numbers 1, 2, 4, 7, 8, 11, 13, 14 is $\equiv 1$ (mod. 15). For the sake of brevity we omit the proof and only observe that it can be done as in the preceding article except that the congruence $x^2 \equiv 1$ can have more than two roots because of certain peculiar considerations. We can also look for a proof from a consideration of indices as in article 75 if we include what we shall soon say about moduli that are not prime.

▶ 79. We return now to the enumeration of other propositions (art. 75).

The sum of all the terms of the period of any number is $\equiv 0$, just as in the example of article 75, $1 + 5 + 12 + 8 = 26 \equiv 0$ (mod. 13).

Demonstration. Let the number whose period we are considering $= a$, and the exponent to which it belongs $= t$, then the sum of all the terms of the period will be

$$\equiv 1 + a + a^2 + a^3 + \text{etc.} + a^{t-1} \equiv \frac{a^t - 1}{a - 1} \,(\text{mod. } p)$$

---

[14] "Miscellanea Analytica. Theorema a Cl. Waring sine demonstratione propositum."

But $a^t - 1 \equiv 0$; so this sum will always be $\equiv 0$ (art. 22), unless perhaps $a - 1$ is divisible by p, or $a \equiv 1$; and we can except this case if we are willing to consider only one term as a *period*.

▶ 80. *The product of all primitive roots will be* $\equiv 1$, except for the case when $p = 3$; for then there is only one primitive root, 2.

Demonstration. If any primitive root is taken as a base, the indices of all the primitive roots will be prime relative to $p - 1$ and at the same time less than $p - 1$. But the sum of these numbers, i.e. the index of the product of all primitive roots, is $\equiv 0$ (mod. $p - 1$) and so the product $\equiv 1$ (mod. p); for it is easy to see that if k is a number relatively prime to $p - 1$, $p - 1 - k$ will be prime relative to $p - 1$, and so the sum of numbers which are prime relative to $p - 1$ is composed of couples whose sum is divisible by $p - 1$ (k can never be equal to $p - 1 - k$ except for the case when $p - 1 = 2$ or $p = 3$; for clearly in all other cases $(p - 1)/2$ is not prime relative to $p - 1$).

▶ 81. *The sum of all primitive roots is either* $\equiv 0$ (when $p - 1$ is divisible by a *square*), *or* $\equiv \pm 1$ (mod. p) (when $p - 1$ is the product of unequal prime numbers; if the number of these is even the sign is positive but if the number is odd, the sign is negative).

Example. 1. For $p = 13$ the primitive roots are 2, 6, 7, 11, and the sum is $26 \equiv 0$ (mod. 13).

2. For $p = 11$ the primitive roots are 2, 6, 7, 8, and the sum is $23 \equiv +1$ (mod. 11).

3. For $p = 31$ the primitive roots are 3, 11, 12, 13, 17, 21, 22, 24, and the sum is $123 \equiv -1$ (mod. 31).

Demonstration. We showed above (art. 55.II) that if $p - 1 = a^\alpha b^\beta c^\gamma$ etc. (where a, b, c, etc. are unequal prime numbers) and A, B, C, etc. are any numbers belonging to the exponents a^α, b^β, c^γ, etc., respectively, all the products ABC etc. will be primitive roots. It is easy to show that any primitive root can be expressed as a product of this sort and indeed in a unique manner.[15]

It follows that these products can be taken in place of the primitive roots. But since in these products it is necessary to combine all values of A with all those of B etc., the sum of all these products will be equal to the product of the sum of all the values of A, multiplied by the sum of all the values of B, multiplied by all the values of C etc. as we know from the theory of combinations. Let all the values of A; B; etc.

[15]Let the numbers a, b, c, etc. be so determined that $a \equiv 1$ (mod. a^α) and $\equiv 0$ (mod. $b^\beta c^\gamma$, etc.); $b \equiv 1$ (mod. b^β) and $\equiv 0$ (mod. $a^\alpha c^\gamma$ etc.); etc. (see art. 32); and so $a + b + c$ etc. $\equiv 1$ (mod. $p - 1$) (art. 19). Now if any primitive root r is expressed by a product ABC etc. we will get $A \equiv r^a$, $B \equiv r^b$, $C \equiv r^c$, etc., and A will belong to the exponent a^α, B to the exponent b^β etc.; the product of all the A, B, C, etc. will be $\equiv r$ (mod. p); and it is easily seen that A, B, C, etc. cannot be determined in any other way.

be designated by A, A', A'', etc.; B, B', B'', etc.; etc., and the sum of all the primitive roots will be

$$\equiv (A + A' + \text{etc.})(B + B' + \text{etc.}) \text{ etc.}$$

I now claim that if the exponent $\alpha = 1$, the sum $A + A' + A''$ + etc. will be $\equiv -1$ (mod. p), but if $\alpha > 1$ the sum will be $\equiv 0$, and in like manner for the remaining β, γ, etc. If we prove these assertions the truth of our theorem will be manifest. For when $p - 1$ is divisible by a square, one of the exponents α, β, γ, etc. will be greater than unity, therefore one of the factors whose sum is congruent to the product of all the primitive roots will $\equiv 0$, and so will the product itself. But when $p - 1$ can be divided by no square, all the exponents α, β, γ, etc. will $= 1$ and so the sum of all the primitive roots will be congruent to the product of factors each of which is $\equiv -1$ and there will be as many of them as there are numbers a, b, c, etc. As a result the sum will be $\equiv \pm 1$ according as there is an even or an odd number of them. We prove this as follows.

1) When $\alpha = 1$ and A is a number belonging to the exponent a, the remaining numbers belonging to this exponent will be $A^2, A^3, \ldots A^{a-1}$. But

$$1 + A + A^2 + A^3 \ldots + A^{a-1}$$

is the sum of the complete period and thus $\equiv 0$ (art. 79) and

$$A + A^2 + A^3 \ldots + A^{a-1} \equiv -1$$

2) When however $\alpha > 1$ and A is a number belonging to the exponent a^α the remaining numbers belonging to this exponent will be found if from $A^2, A^3, A^4, \ldots A^{a^\alpha - 1}$ we delete A^a, A^{2a}, A^{3a}, etc. (see art. 53); and their sum will be

$$\equiv 1 + A + A^2 \ldots + A^{a^\alpha - 1} - (1 + A^a + A^{2a} \ldots + A^{a^\alpha - a})$$

i.e. congruent to the difference of two periods, and thus $\equiv 0$.　　Q.E.D.

▶ 82. All that we have said so far presupposes that the modulus is a prime number. It remains to consider the case when a composite number is used as modulus. But since here there are no such elegant properties as in the previous case and there is no need for subtle artifices in order to discover them (because almost everything can be deduced from an application of the preceding principles), it would be superfluous and tedious to treat all the details exhaustively. So we will explain only what this case has in common with the previous one and what is proper to itself.

▶ 83. The propositions in articles 45–48 have already been proven for the general case. But the proposition of article 49 must be changed as follows:

If f designates how many numbers there are relatively prime to m and less than m, i.e. if f = φm (art. 38); and if a is a given number relatively prime to m, then the exponent t of the lowest power of a that is congruent to unity relative to the modulus m will be = to f or a factor of f.

The demonstration of the proposition in article 49 has value in this case if we substitute m for p, f for $p - 1$ and if in place of the numbers $1, 2, 3, \ldots p - 1$ we substitute numbers which are relatively prime to m and less than m. Let the reader turn back and do this. But the other demonstrations we considered there (art. 50, 51) cannot be applied to this case without ambiguity. And with respect to the propositions of article 52 et seq. there is a great difference between moduli that are powers of prime numbers and those that are divisible by more than one prime number. We therefore consider moduli of the former type.

▶ 84. If p is a prime number and the modulus $m = p^n$, then we have $f = p^{n-1}(p - 1)$ [art. 38]. Now if we apply what we said in articles 53, 54 to this case, with the necessary changes as in the preceding article, we will discover that everything we said there will have place here provided we first prove that a congruence of the form $x^t - 1 \equiv 0$ (mod. p^n) cannot have more than t different roots. We showed that this was true relative to a prime modulus for a more general proposition in article 43; but this proposition is valid for prime moduli only and cannot be applied to this case. Nevertheless we will show by a special method that the proposition is true for this particular case. In Section VIII we will prove it still more easily.

▶ 85. We propose now to prove this theorem:

If the greatest common divisor of the numbers t and $p^{n-1}(p - 1)$ is e, the congruence $x^t \equiv 1$ (mod. p^n) will have e different roots.

Let $e = kp^v$ so that k does not involve the factor p. As a result it will divide the number $p - 1$. And the congruence $x^t \equiv 1$ relative to the modulus p will have k different roots. If we designate them A, B, C, etc. each root of the same congruence relative to the modulus p^n will be congruent relative to the modulus p to one of the numbers A, B, C, etc. Now we will show that the congruence $x^t \equiv 1$ (mod. p^n) has p^v roots congruent to A and as many congruent to B etc., all relative to the modulus p. From this we find that the number of all the roots will be kp^v or e as we said. We will do this by showing *first* that if α is a root congruent to A relative to the modulus p, then

$$\alpha + p^{n-v}, \quad \alpha + 2p^{n-v}, \quad \alpha + 3p^{n-v}, \ldots, \quad \alpha + (p^v - 1)p^{n-v}$$

will also be roots. *Second*, by showing that numbers congruent to A relative to modulus p cannot be roots unless they are in the form $\alpha + hp^{n-v}$ (h being an integer). As a result there will manifestly be p^v different roots and no more. The same will be true of the roots that are congruent to B, C, etc. *Third*, we will show how we can always find the root that is congruent to A relative to p.

▶ 86. THEOREM. *If as in the preceding article t is a number divisible by p^v, but not by p^{v+1}, we have*

$$(\alpha + hp^\mu)^t - \alpha^t \equiv 0 \text{ (mod. } p^{\mu+v}) \quad \text{and} \quad \equiv \alpha^{t-1}hp^\mu t \text{ (mod. } p^{\mu+v+1})$$

GOD CREATED THE INTEGERS

The second part of the theorem is not true when $p = 2$ and $\mu = 1$.

The demonstration of this theorem can be shown by expanding a binomial provided that we can show that all the terms after the second are divisible by $p^{\mu+\nu+1}$. But since a consideration of the denominators of the coefficients leads to many ambiguities, we prefer the following method.

Let us presume *first* that $\mu > 1$ and $\nu = 1$ and we will then have

$$x^t - y^t = (x - y)(x^{t-1} + x^{t-2}y + x^{t-3}y^2 + \text{etc.} + y^{t-1})$$
$$(\alpha + hp^\mu)^t - \alpha^t = hp^\mu[(\alpha + hp^\mu)^{t-1} + (\alpha + hp^\mu)^{t-2}\alpha + \text{etc.} + \alpha^{t-1}]$$

But

$$\alpha + hp^\mu \equiv \alpha \ (\text{mod. } p^2)$$

and so each term $(\alpha + hp^\mu)^{t-1}$, $(\alpha + hp^\mu)^{t-2}\alpha$, etc. will be $\equiv \alpha^{t-1}$ (mod. p^2), and the sum of all of them will be $\equiv t\alpha^{t-1}$ (mod. p^2); that is they will be of the form $t\alpha^{t-1} + Vp^2$ where V is any number whatsoever. Thus $(\alpha + hp^\mu)^t - \alpha^t$ will be of the form

$$\alpha^{t-1}hp^\mu t + Vhp^{\mu+2}, \quad \text{i.e.} \quad \equiv \alpha^{t-1}hp^\mu t \ (\text{mod. } p^{\mu+2})$$
$$\text{and} \quad \equiv 0 \ (\text{mod. } p^{\mu+1})$$

And thus the theorem is proven for this case.

Now keeping $\mu > 1$, if the theorem were not true for other values of ν, there would necessarily be a limit up to which the theorem would always be true and beyond which it would be false. Let the smallest value of ν for which it is false $= \phi$. It is easily seen that if t were divisible by $p^{\phi-1}$ but not by p^ϕ, the theorem would still be true, whereas if we substituted tp for t it would be false. So we have

$$(\alpha + hp^\mu)^t \equiv \alpha^t + \alpha^{t-1}hp^\mu t \ (\text{mod. } p^{\mu+\phi}),$$
$$\text{or} \quad \alpha^t + \alpha^{t-1}hp^\mu t + up^{\mu+\phi}$$

with u an integer. But since the theorem has already been proved for $\nu = 1$, we get

$$(\alpha^t + \alpha^{t-1}hp^\mu t + up^{\mu+\phi})^p \equiv \alpha^{tp} + \alpha^{tp-1}hp^{\mu+1}$$
$$+ \alpha^{t(p-1)}up^{\mu+\phi+1} \ (\text{mod. } p^{\mu+\phi+1})$$

and so also

$$(\alpha + hp^\mu)^{tp} \equiv \alpha^{tp} + \alpha^{tp-1}hp^\mu tp \ (\text{mod. } p^{\mu+\phi+1})$$

i.e. the theorem is true if tp is substituted for t, i.e. for $\nu = \phi$. But this is contrary to the hypothesis and so the theorem will be true for all values of ν.

▶ 87. There remains the case when $\mu = 1$. By a method very similar to the one we used in the preceding article we can show without the help of the binomial theorem that

$$(\alpha + hp)^{t-1} \equiv \alpha^{t-1} + \alpha^{t-2}(t-1)hp \ (\text{mod. } p^2)$$
$$\alpha(\alpha + hp)^{t-2} \equiv \alpha^{t-1} + \alpha^{t-2}(t-2)hp$$
$$\alpha^2(\alpha + hp)^{t-3} \equiv \alpha^{t-1} + \alpha^{t-2}(t-3)hp$$

etc.

and the sum (since the number of terms $= t$) will be

$$\equiv t\alpha^{t-1} + \frac{(t-1)t}{2}\alpha^{t-2}hp \,(\mathrm{mod.}\ p^2)$$

But since t is divisible by p, $(t-1)t/2$ will also be divisible by p in all cases except when $p = 2$, as we pointed out in the preceding article. In the remaining cases however we have $(t-1)t\alpha^{t-2}hp/2 \equiv 0 \,(\mathrm{mod.}\ p^2)$ and the sum will be $\equiv t\alpha^{t-1}$ (mod. p^2) as in the preceding article. The rest of the demonstration proceeds in the same way.

The general result except when $p = 2$ is

$$(\alpha + hp^{\mu})^t \equiv \alpha^t \,(\mathrm{mod.}\ p^{\mu+\nu})$$

and $(\alpha + hp^{\mu})^t$ is not $\equiv \alpha^t$ for any modulus which is a higher power of p than $p^{\mu+\nu}$, provided always that h is not divisible by p and that p^{ν} is the highest power of p that divides the number t.

From this we can immediately derive propositions 1 and 2 which we proposed to show in article 85:

First, if $\alpha^t \equiv 1$ we will also have $(\alpha + hp^{n-\nu})^t \equiv 1 \,(\mathrm{mod.}\ p^n)$.

Second, if some number α' is congruent relative to the modulus p to A and thus also to α but noncongruent to α relative to the modulus $p^{n-\nu}$, and if it satisfies the congruence $x^t \equiv 1 \,(\mathrm{mod.}\ p^n)$, we will let $\alpha' = \alpha + lp^{\lambda}$ in such a way that l is not divisible by p. It follows that $\lambda < n - \nu$ and so $(\alpha + lp^{\lambda})^t$ will be congruent to α^t relative to the modulus $p^{\lambda+\nu}$ but not relative to the modulus p^n which is a higher power. As a result α' cannot be a root of the congruence $x^t \equiv 1$.

▶ 88. *Third*, we proposed to find a root of the congruence $x^t \equiv 1 \,(\mathrm{mod.}\ p^n)$ which will be congruent to A. We can show here how this is done only if we already know a root of the same congruence relative to the modulus p^{n-1}. Manifestly this is sufficient, since we can go from the modulus p for which A is a root to the modulus p^2 and from there to all consecutive powers.

Therefore let α be a root of the congruence $x^t \equiv 1 \,(\mathrm{mod.}\ p^{n-1})$. We are looking for a root of the same congruence relative to the modulus p^n. Let us suppose it $= \alpha + hp^{n-\nu-1}$. From the preceding article it must have this form (we will consider separately $\nu = n - 1$, but note that ν can never be greater than $n - 1$). We will therefore have

$$(\alpha + hp^{n-\nu-1})^t \equiv 1 \,(\mathrm{mod.}\ p^{n-1})$$

But

$$(\alpha + hp^{n-\nu-1})^t \equiv \alpha^t + \alpha^{t-1}htp^{n-\nu-1} \,(\mathrm{mod.}\ p^n)$$

If, therefore, h is so chosen that $1 \equiv \alpha^t + \alpha^{t-1}htp^{n-\nu-1} \,(\mathrm{mod.}\ p^n)$ or [since by hypothesis $1 \equiv \alpha^t \,(\mathrm{mod.}\ p^{n-1})$ and t is divisible by p] so that $[(\alpha^t - 1)/p^{n-1}] + \alpha^{t-1}h(t/p^{\nu})$ is divisible by p, we will have the root we want. That this can be done is clear from the preceding section because we presuppose that t cannot be divided by a higher power of p than p^ν, and so $\alpha^{t-1}(t/p^\nu)$ will be prime relative to p.

But if $\nu = n - 1$, i.e. if t is divisible by p^{n-1} or by a higher power of p, any value A satisfying the congruence $x^t \equiv 1$ relative to the modulus p will satisfy it also relative to the modulus p^n. For if we let $t = p^{n-1}\tau$, we will have $t \equiv \tau \pmod{p-1}$; and so, since $A^t \equiv 1 \pmod{p}$, we will also have $A^\tau \equiv 1 \pmod{p}$. Let therefore $A^\tau = 1 + hp$ and we will have $A^t = (1 + hp)^{p^{n-1}} \equiv 1 \pmod{p^n}$ (art. 87).

▶ 89. Everything we have proven in articles 57 ff., with the aid of the theorem stating that the congruence $x^t \equiv 1$ can have no more than t different roots, is also true for a modulus that is the power of a prime number; and if we call *primitive roots* those numbers that belong to the exponent $p^{n-1}(p - 1)$, that is to say all numbers containing the numbers not divisible by p in their period, then we have primitive roots here to. Everything we said about indices and their use and about the solution of the congruence $x^t \equiv 1$ can be applied to this case also and, since the demonstrations offer no difficulty, it would be superfluous to repeat them. We showed earlier how to derive the roots of the congruence $x^t \equiv 1$ relative to the modulus p^n from the roots of the same congruence relative to the modulus p. Now we must add some observations for the case when the modulus is some power of the number 2, which we excepted above.

▶ 90. *If some power of the number 2 higher than the second, e.g. 2^n, is taken as modulus, the 2^{n-2}th power of any odd number is congruent to unity.*

For example, $3^8 = 6561 \equiv 1 \pmod{32}$.

For any odd number is of the form $1 + 4h$ or of the form $-1 + 4h$ and the proposition follows immediately (theorem in art. 86).

Since, therefore, the exponent belonging to any odd number relative to the modulus 2^n must be a divisor of 2^{n-2}, it is easy to judge to which of the numbers $1, 2, 4, 8, \ldots 2^{n-2}$ it belongs. Suppose the number proposed $= 4h \pm 1$ and the exponent of the highest power of 2 that divides h is $=m$ (m can $=0$ when h is odd); then the exponent to which the proposed number belongs will $= 2^{n-m-2}$ if $n > m + 2$; if however $n =$ or is $< m + 2$ the proposed number will $\equiv \pm 1$ and so will belong either to the exponent 1 or to the exponent 2. For it is easy to deduce from article 86 that a number of the form $\pm 1 + 2^{m+2}k$ (which is equivalent to $4h \pm 1$) is congruent to unity relative to the modulus 2^n if it is raised to the power 2^{n-m-2} and is noncongruent to unity if it is raised to a lower power of 2. Thus any number of the form $8k + 3$ or $8k + 5$ will belong to the exponent 2^{n-2}.

▶ 91. It follows that here we do not have *primitive roots* in the sense accepted above; that is, there are no numbers which include in their periods all numbers less than the modulus and relatively prime to it. But obviously we have an analogy. For we found that for a number of the form $8k + 3$ the power of an odd exponent is also of the form $8k + 3$ and the power of an even exponent is of the form $8k + 1$; no power then can be of the form $8k + 5$ or $8k + 7$. Since for a number of the form $8k + 3$ the period therefore consists of 2^{n-2} terms of the form $8k + 3$ or $8k + 1$, and since there are no more than 2^{n-2} of these less than the modulus, manifestly any number of the form $8k + 1$ or $8k + 3$ is congruent relative to the modulus 2^n to a power of some number of the form $8k + 3$. In a similar way we can show that the period of a number of the form $8k + 5$ includes all numbers of the form $8k + 1$ and $8k + 5$. Therefore, if we take a number of the form $8k + 5$ as base, we will find real indices for all numbers of the form $8k + 1$ and $8k + 5$ taken positively and for all numbers of the form $8k + 3$ and $8k + 7$ taken negatively. And we must still regard as equivalent indices which are congruent relative to 2^{n-2}. Table 1, in which 5 was always chosen as base for the moduli 16, 32, and 64, should be interpreted in this light (for modulus 8 no table is necessary). For example, the number 19 which is of the form $8n + 3$ and so must be taken *negatively* has the index 7 relative to the modulus 64. This means $5^7 \equiv -19$ (mod. 64). If we took numbers of the form $8n + 1$ and $8n + 5$ negatively and numbers of the form $8n + 3$ and $8n + 7$ positively, it would be necessary to give them imaginary indices, so to speak. If we did that, the calculation of indices could be reduced to a very simple algorithm. But since we would be led too far afield if we wished to treat this point with great rigor, we will save it for another occasion when we may be able to consider in more detail the theory of imaginary quantities. In our opinion no one has yet produced a clear treatment of this subject. Experienced mathematicians will find it easy to develop the algorithm. Those with less training can use our table provided only that they have a good grasp of the principles established above. They can do this in much the same way that people use logarithms even when they know nothing about modern studies of imaginary logarithms.

▶ 92. Almost all that pertains to residues of powers relative to a modulus composed of many prime numbers can be deduced from the general theory of congruences. We will show later at great length how to reduce congruences relating to moduli composed of several primes to congruences relating to moduli which are primes or powers of primes. For this reason we will not now dwell further on this subject. We only observe that the most elegant property that holds for other moduli is lacking here, namely the property which guarantees that we can always find numbers whose period includes all numbers relatively prime to the modulus. In one case, however, even here we can find such a number. It occurs when the modulus is twice a prime number or twice the power of a prime. For if the modulus m is reduced to the form $A^a B^b C^c$ etc. where A, B, C, etc. are different prime numbers and if we designate $A^{a-1}(A-1)$ by the letter α, $B^{b-1}(B-1)$ by the letter β, etc. and then choose a number z which is relatively prime to m, we get $z^\alpha \equiv 1$ (mod. A^a), $z^\beta \equiv 1$ (mod. B^b), etc. And if μ is the least common multiple of the numbers α, β, γ, etc. we have $z^\mu \equiv 1$ relative to all the moduli A^a, B^b, etc. and so also relative to m which is equal to their product. But except for the case where m is twice a prime or twice the power of a prime, the least common multiple of the numbers α, β, γ, etc. is less than their product (the numbers α, β, γ, etc. cannot be prime relative to each other since they have the common divisor 2). Thus no period can have as many terms as there are numbers which are relatively prime to the modulus and less than it because the number of these is equal to the product of α, β, γ, etc. For example, for $m = 1001$ the 60th power of any number relatively prime to m is congruent to unity because 60 is the least common multiple of the numbers 6, 10, 12. The case where the modulus is twice a prime number or twice the power of a prime number is completely similar to the one where it is a prime or the power of a prime.

▶ 93. We have already mentioned the writings of other geometers concerning the same subjects treated in this section. For those who want a more detailed discussion than brevity has permitted us, we recommend the following treatises of Euler because of the clarity and insight that has placed this great man far ahead of all other commentators: "Theoremata circa residua ex divisione potestatum relicta," *Novi comm. acad. Petrop.*, 7 [1758–59], 1761. "Demonstrationes circa residua ex divisione potestatum per numeros primos resultantia," *ibid.*, 18 [1773], 1774. To these we might add *Opuscula Analytica*, 1, Diss. 5; Diss. 8.[16]

SECTION IV

CONGRUENCES OF THE SECOND DEGREE

▶ 94. THEOREM. *If we take some number m as modulus, then of the numbers* 0, 1, 2, 3, ... $m - 1$ *no more than* $(m/2) + 1$ *can be congruent to a square when m is even and not more than* $(m/2) + (1/2)$ *of them when m is odd.*

Demonstration. Since the squares of congruent numbers are congruent to each other, any number that can be congruent to a square will also be congruent to some square whose root is $< m$. It is sufficient therefore to consider the least residues of the squares 0, 1, 4, 9, ... $(m - 1)^2$. But it is easy to see that $(m - 1)^2 \equiv \cdot 1$, $(m - 2)^2 \equiv 2^2$, $(m - 3)^2 \equiv 3^2$, etc. Thus also when m is even, the squares $[(m/2) - 1]^2$ and $[(m/2) + 1]^2$, $[(m/2) - 2]^2$ and $[(m/2) + 2]^2$, etc. will have the same least residues; and when m is odd, the squares $[(m/2) - (1/2)]^2$ and

---

[16]For title of Dissertation 5 see footnote 9. Dissertation 8: "De quibusdam eximiis proprietatibus circa divisores potestatum occurrentibus."

$[(m/2) + (1/2)]^2, [(m/2) - (3/2)]^2$ and $[(m/2) + (3/2)]^2$, etc. will be congruent. It follows that there are no other numbers congruent to a square than those which are congruent to one of the squares $0, 1, 4, 9, \ldots (m/2)^2$ when m is even; when it is odd, any number that is congruent to a square is necessarily congruent to one of the $0, 1, 4, 9, \ldots [(m/2) - (1/2)]^2$. In the former case therefore there will be at most $(m/2) + 1$ different least residues, in the latter case at most $(m/2) + (1/2)$. Q.E.D.

Example. Relative to the modulus 13 the least residues of the squares $0, 1, 2, 3, \ldots 6$ are found to be 0, 1, 4, 9, 3, 12, 10, and after this they occur in the reverse order 10, 12, 3, etc. Thus if a number is not congruent to one of these residues, that is, if it is congruent to 2, 5, 6, 7, 8, 11, it cannot be congruent to a square.

Relative to the modulus 15 we find the residues 0, 1, 4, 9, 1, 10, 6, 4, after which the same numbers occur in reverse order. Here therefore the number of residues which can be congruent to a square is less than $(m/2) + (1/2)$, since 0, 1, 4, 6, 9, 10 are the only ones that occur in the list. The numbers 2, 3, 5, 7, 8, 11, 12, 13, 14 and any number congruent to them cannot be congruent to a square relative to the modulus 15.

▶ 95. As a result, for any modulus all numbers can be divided into two classes, one which contains those numbers that can be congruent to a square, the other those that cannot. We will call the former *quadratic residues of the number which we take as modulus*[17] and the *latter quadratic nonresidues of this number.* When no ambiguity can arise we will call them simply *residues* and *nonresidues.* For the rest it is sufficient to classify all the numbers $0, 1, 2, \ldots m - 1$, since numbers that are congruent to these will be put in the same class.

We will begin in this investigation with prime moduli, and this is to be understood even if it is not said in so many words. But we must exclude the prime number 2 and so we will be considering *odd* primes only.

▲ 96. *If we take the prime number p as modulus, half the numbers $1, 2, 3, \ldots p - 1$ will be quadratic residues, the rest nonresidues; i.e. there will be $(p - 1)/2$ residues and the same number of nonresidues.*

It is easy to prove that all the squares $1, 4, 9, \ldots (p - 1)^2/4$ are not congruent. For if $r^2 \equiv (r')^2$ (mod. p) with r, r' unequal and not greater than $(p - 1)/2$ and $r > r'$, $(r - r')(r + r')$ will be positive and divisible by p. But each factor $(r - r')$ and $(r + r')$ is less than p, so the assumption is untenable (art. 13). We have therefore $(p - 1)/2$ quadratic residues among the numbers $1, 2, 3, \ldots p - 1$. There cannot be more because if we add the residue 0 we get $(p + 1)/2$ of them, and this is

[17]Actually in this case we give to these expressions a meaning different from that which we have used up to the present. When $r \equiv a^2$ (mod. m) we should say that r is a residue of the square a^2 relative to the modulus m; but for the sake of brevity in this section we will always call r the quadratic residue *of m itself,* and there is no danger of ambiguity. From now on we will not use the expression *residue* to signify the same thing as a congruent number unless perhaps we are treating of *least* residues, where there can be no doubt about what we mean.

larger than the number of all the residues together. Consequently the remaining numbers will be nonresidues and their number will $= (p - 1)/2$.

Since zero is always a residue we will exclude this and all numbers divisible by the modulus from our investigations. This case is clear and would contribute nothing to the elegance of our theorems. For the same reason we will also exclude 2 as modulus.

▶ 97. Since many of the things we will prove in this section can be derived from the principles of the preceding section, and since it is not out of place to discover the same truths by different methods, we will continue to point out the connection. It can be easily verified that all numbers congruent to a square have *even* indices and that numbers not congruent to a square have *odd* indices. And since $p - 1$ is an even number, there will be as many even indices as odd ones, namely $(p - 1)/2$ of each, and thus also there will be the same number of residues and nonresidues.

Examples.

Moduli	Residues
3	1
5	1, 4
7	1, 2, 4
11	1, 3, 4, 5, 9
13	1, 3, 4, 9, 10, 12
17	1, 2, 4, 8, 9, 13, 15, 16

etc.

and all other numbers less than the moduli are nonresidues.

▶ 98. THEOREM. *The product of two quadratic residues of a prime number p is a residue; the product of a residue and a nonresidue is a nonresidue; and finally the product of two nonresidues is a residue.*

Demonstration. I. Let A, B be residues of the squares a^2, b^2; that is $A \equiv a^2$, $B \equiv b^2$. The product AB will then be congruent to the square of the number ab; i.e. it will be a residue.

II. When A is a residue, for example $\equiv a^2$, and B a nonresidue, AB will be a nonresidue. For let us suppose that $AB \equiv k^2$, and let the value of the expression k/a (mod. p) $\equiv b$. We will have therefore $a^2 B \equiv a^2 b^2$ and $B \equiv b^2$; i.e. B is a residue contrary to the hypothesis.

Another way. Take all the numbers among $1, 2, 3, \ldots p - 1$ which are residues [there are $(p - 1)/2$ of them]. Multiply each by A and all the products will be quadratic residues and all noncongruent to each other. Now if the nonresidue B is multiplied by A, the product will not be congruent to any of the products we already have. And so if it were a residue, we would have $(p + 1)/2$ noncongruent residues, and we have not yet included the residue 0 among them. This contradicts article 96.

III. Let A, B be nonresidues. Multiply by A all the numbers among $1, 2, 3, \ldots p - 1$ which are residues. We still have $(p - 1)/2$ nonresidues noncongruent to one another (II); the product AB cannot be congruent to any of them; so if it were a nonresidue, we would have $(p + 1)/2$ nonresidues noncongruent to one another contrary to article 96. So the product etc.   Q.E.D.

These theorems could be derived more easily from the principles of the preceding section. For since the indices of residues are always even and the indices of nonresidues odd, the index of the product of two residues or nonresidues will be even, and the product itself will be a residue. On the other hand the index of the product of a residue and a nonresidue will be odd, and so the product itself will be a nonresidue.

Either method of proof can be used for the following theorems: *The value of the expression a/b (mod. p) will be a residue when the numbers a, b are both residues or both nonresidues; it will be a nonresidue when one of the numbers a, b is a residue and the other a nonresidue.* They can also be proved by converting the preceding theorems.

▶ 99. In general the product of any number of factors is a residue if all the factors are residues or if the number of nonresidues among them is even; if the number of nonresidues among the factors is odd, the product will be a nonresidue. It is easy therefore to judge whether or not a composite number is a residue as long as it is clear what the individual factors are. That is why in Table 2 we included only prime numbers. This is the arrangement of the table. Moduli[18] are listed in the margin, successive prime numbers along the tops when one of the latter is a residue of some modulus a dash has been placed in the space corresponding to each of them; when a prime number is a nonresidue of the modulus, the corresponding space has been left vacant.

▶ 100. Before we proceed to more difficult subjects, let us add a few remarks about moduli which are not prime.

If the modulus is p^n, a power of a prime number p (we assume p is not 2), half of all the numbers not divisible by p and less than p^n will be residues, half nonresidues; i.e. the number of each will $= [(p - 1)p^{n-1}]/2$.

For if r is a residue, it will be congruent to a square whose root is not greater than half the modulus (see art. 94). It is easy to see that there are $[(p - 1)p^{n-1}]/2$ numbers that are not divisible by p and less than half the modulus; it remains to show that the squares of all these numbers are noncongruent or that they produce different quadratic residues. Now if the squares of two numbers a, b not divisible by p and less than half the moduli were congruent, we would have $a^2 - b^2$ or

---

[18]We will soon see how to dispense with composite moduli.

$(a - b)(a + b)$ divisible by p^n (we suppose $a > b$). This cannot be unless *either* one of the numbers $(a - b), (a + b)$ is divisible by p^n, but this cannot be because each is $< p^n$, *or* one of them is divisible by p^m, the other by p^{n-m}, i.e. both by p. But this cannot be either. For this means that the sum and the difference $2a$ and $2b$ would be divisible by p and consequently a and b, contrary to the hypothesis. So finally among the numbers not divisible by p and less than the modulus, there are $[(p - 1)p^n]/2$ residues and the others, whose number is the same, will be nonresidues. Q.E.D. This theorem can be proven by considering indices just as in article 97.

▶ 101. *Any number not divisible by p, which is a residue of p, will also be a residue of p^n; if it is a nonresidue of p it will also be a nonresidue of p^n.*

The second part of this theorem is obvious. If the first part were false, among the numbers less than p^n and not divisible by p more would be residues of p than of p^n, i.e. more than $[p^{n-1}(p - 1)]/2$. But clearly the number of residues of p among these numbers is precisely $= [p^{n-1}(p - 1)]/2$.

It is just as easy to find a square congruent to a given residue relative to the modulus p^n if we have a square congruent to this residue relative to the modulus p.

If we have a square a^2 which is congruent to the given residue A relative to the modulus p^μ, we can find the square congruent to A relative to p^ν in the following way (we suppose here that $\nu > \mu$ and $=$ or $< 2\mu$). Suppose that the root of the square we seek $= \pm a + xp^\mu$. It is easy to see that it should have this form. And we should also have $a^2 \pm 2axp^\mu + x^2p^{2\mu} \equiv A$ (mod. p^ν) or if $2\mu > \nu, A - a^2 \equiv \pm 2axp^\mu$ (mod. p^ν). Let $A - a^2 = p^\mu d$ and x will be the value of the expression $\pm d/2a$ (mod. $p^{\nu\mu}$) which is equivalent to $\pm (A - a^2)/2ap^\mu$ (mod. p^ν).

Given therefore a square congruent to A relative to p, from it we derive a square congruent to A relative to the modulus p^2; from there we can go to the modulus p^4, then to p^8, etc.

Example. If we are given the residue 6, which is congruent to the square 1 relative to the modulus 5, we find the square 9^2 which is congruent to it relative to 25, 16^2 which is congruent to it relative to 125, etc.

▶ 102. Regarding numbers divisible by p, it is clear that their squares will be divisible by p^2, so that all numbers divisible by p but not by p^2 will be nonresidues of p^n. In general if we have a number p^kA with A not divisible by p we can distinguish the following cases:

1) When $k =$ or is $> n$, we have $p^kA \equiv 0$ (mod. p^n), i.e. a residue.

2) When $k < n$ and odd, p^kA will be a nonresidue.

For if we had $p^kA = p^{2x+1}A \equiv s^2$ (mod. p^n), s^2 would be divisible by p^{2x+1}. This is impossible unless s is divisible by p^{x+1}. But then p^{2x+2} would divide s^2 and also (since $2x + 2$ is certainly not greater than n) $p^k A$, i.e. $p^{2x+1} A$. And this means that p divides A, contrary to the hypothesis.

3) When $k < n$ and even. Then $p^k A$ will be a residue or nonresidue of p^n according as A is a residue or nonresidue of p. For when A is a residue of p, it is also a residue of p^{n-k}. But if we suppose that $A \equiv a^2$ (mod. p^{n-k}), we get $Ap^k \equiv a^2 p^k$ (mod. p^n) and $a^2 p^k$ is a square. When on the other hand A is a non-residue of p, $p^k A$ cannot be a residue of p^n. For if $p^k A \equiv a^2$ (mod. p^n), a^2 would necessarily be divisible by p^k. The quotient will be a square to which A is congruent relative to the modulus p^{n-k} and hence also relative to the modulus p; i.e. A will be a residue of p contrary to the hypothesis.

▶ 103. Since we have excluded the case where $p = 2$ above, we should say a few words about it. When the number 2 is the modulus, every number is a residue and there are no nonresidues. When 4 is the modulus, all odd numbers of the form $4k + 1$ will be residues, and all of the form $4k + 3$ nonresidues. When 8 or a higher power of the number 2 is the modulus, all odd numbers of the form $8k + 1$ are residues, all others of the forms $8k + 3$, $8k + 5$, $8k + 7$ are nonresidues. The last part of this proposition is clear from the fact that the square of any odd number, whether it be of the form $4k + 1$ or $4k - 1$, will be of the form $8k + 1$. We prove the first part as follows.

1) If either the sum or difference of two numbers is divisible by 2^{n-1} the squares of these numbers will be congruent relative to the modulus 2^n. For if one of the numbers $= a$, the other will be of the form $2^{n-1} h \pm a$, and the square of this is $\equiv a^2$ (mod. 2^n).

2) Any odd number which is a residue of 2^n will be congruent to a square whose root is odd and $< 2^{n-2}$. For let a^2 be a square to which the number is congruent and let $a \equiv \pm \alpha$ (mod. 2^{n-1}) so that α does not exceed half of the modulus (art. 4); $a^2 \equiv \alpha^2$ and so the given number will be $\equiv \alpha^2$. Manifestly both a and α will be odd and $\alpha < 2^{n-2}$.

3) The squares of all odd numbers less than 2^{n-2} will be noncongruent relative to 2^n. For let two such numbers be r and s. If their squares were congruent relative to 2^n, $(r - s)(r + s)$ would be divisible by 2^n (we assume $r > s$). It is easy to see that the numbers $r - s$, $r + s$ cannot both be divisible by 4. So if only one of them is divisible by 2, the other would be divisible by 2^{n-1} in order to have the product divisible by 2^n; Q.E.A., since each of them is $< 2^{n-2}$.

4) If these squares are reduced to their *least positive residues*, we will have 2^{n-3} different quadratic residues smaller than the modulus,[19] and each of them will be of

---

[19] Because the number of odd numbers less than 2^{n-2} is 2^{n-3}.

the form $8k + 1$. But since there are precisely 2^{n-3} numbers of the form $8k + 1$ less than the modulus, all these numbers must be residues.

To find a square which is congruent to a given number of the form $8k + 1$ relative to the modulus 2^n, a method can be used similar to the one in article 101; see also article 88. And finally with regard to even numbers, everything is valid that we said in general in article 102.

▶ 104. If A is a residue of p^n and if we are concerned with the number of different values (i.e. noncongruent relative to the modulus) that the expression $V = \sqrt{A}$ (mod. p^n) admits, we can easily draw the following conclusions from what has preceded. (We assume that the number p is prime as before, and for brevity's sake we include the case where $n = 1$.)

I. If A is not divisible by p, V will have one value for $p = 2$, $n = 1$, that is $V = 1$; *two* values when p is odd or when $p = 2$, $n = 2$, that is if one of them $\equiv v$, the other will $\equiv -v$; *four* values for $p = 2$, $n > 2$, that is if one of them $\equiv v$, the others will be $\equiv -v$, $2^{n-1} + v$, $2^{n-1} - v$.

II. If A is divisible by p but not by p^n, let the highest power of p that divides A be $p^{2\mu}$ (for manifestly this exponent has to be even) and we will have $A = ap^{2\mu}$. It is clear that all the values of V will be divisible by p^μ, and that all the quotients arising from this division will be values of the expression $V' = \sqrt{a}$ (mod. $p^{n-2\mu}$); we will then get all the values of V by multiplying all the values of V' that lie between 0 and $p^{n-\mu}$ by p^μ. Doing this we get
$$vp^\mu, \quad vp^\mu + p^{n-\mu}, \quad vp^\mu + 2p^{n-\mu}, \ldots vp^\mu + (p^\mu - 1)p^{n-\mu}$$
where v stands for all the *different* values of V' and so the number of them will be p^μ, $2p^\mu$, or $4p^\mu$ according as the number of values for V (by case I) is 1, 2, or 4.

III. If A is divisible by p^n it is easy to see that by letting $n = 2m$ or $n = 2m - 1$ according as it is even or odd, all numbers divisible by p^m will be values of V; but these will be the numbers 0, p^m, $2p^m$, $\ldots (p^{n-m} - 1)p^m$ and their number will be p^{n-m}.

▶ 105. It remains to consider the case where the modulus m is composed of several prime numbers. Let $m = abc \ldots$ where a, b, c, etc. are different prime numbers or powers of different primes. It is immediately clear that if n is a residue of m, it will also be a residue of each of the a, b, c, etc., and that if it is a nonresidue of any of the numbers a, b, c, etc. it will be a nonresidue of m. And, vice versa, if n is a residue of each of the a, b, c, etc. it will also be a residue of the product m. For suppose $n \equiv A^2$, B^2, C^2, etc. relative to the moduli a, b, c, etc. respectively. If we find the number N which is congruent to A, B, C, etc. relative to the moduli a, b, c, etc., respectively (art. 32), we will have $n \equiv N^2$ relative to all these moduli and so also relative to the product m. From the combination of *any* value of A or of the expression

$\sqrt{n}$ (mod. a), together with *any* value of B and *any* value of C etc., we get a value of N or of the expression $\sqrt{n}$ (mod. m), and from different combinations we get different values of N and from all combinations we get all values of N. As a result, the number of all the different values of N will be equal to the product of the various values of A, B, C, etc. which we showed how to determine in the preceding article. And clearly if one value of the expression $\sqrt{n}$ (mod. m) or of N were known, this would also be a value of all the A, B, C, etc.; and since we know from the preceding article how to deduce all the remaining values of these quantities from this one, it follows that from one value of N all the rest can be obtained.

Example. Let the modulus be 315. We want to know whether 46 is a residue or nonresidue. The prime divisors of 315 are 3, 5, 7, and the number 46 is a residue of each of them; therefore it is also a residue of 315. Further, since $46 \equiv 1$ and $\equiv 64$ (mod. 9); $\equiv 1$ and $\equiv 16$ (mod. 5); $\equiv 4$ and $\equiv 25$ (mod. 7), the roots of the squares to which 46 is congruent relative to the modulus 315 will be 19, 26, 44, 89, 226, 271, 289, 296.

▶ 106. From what we have just shown we see that if only we can determine whether a given *prime number* is a residue or a nonresidue of another *given prime*, all other cases can be reduced to this. We will therefore attempt to investigate certain criteria for this case. But before we do this we will show a criterion derived from the preceding section. Although it is of almost no practical use, it is worthy of mention because of its simplicity and generality.

Any number A not divisible by a prime number $2m + 1$ is a residue or nonresidue of this prime according as $A^m \equiv +1$ or $\equiv -1$ (mod. $2m + 1$).

For let a be the index of the number A for the modulus $2m + 1$ in any system whatsoever; a will be even when A is a residue of $2m + 1$, and odd when A is a nonresidue. But the index of the number A^m will be ma; i.e. it will $\equiv 0$ or $\equiv m$ (mod. $2m$), according as a is even or odd. In the former case A^m will be $\equiv +1$, and in the latter case it will $\equiv -1$ (mod. $2m + 1$) (see art. 57, 62).

Example. 3 is a residue of 13 because $3^6 \equiv 1$ (mod. 13), but 2 is a nonresidue of 13 because $2^6 \equiv -1$ (mod. 13).

But as soon as the numbers we are examining are even moderately large this criterion is practically useless because of the amount of calculation involved.

▶ 107. It is very easy, given a modulus, to characterize all the numbers that are residues or nonresidues. If the number $= m$, we determine the squares whose roots do not exceed half of m and also the numbers congruent to these squares relative to m (in practice there are still more expedient methods). All numbers congruent to any of these relative to m will be residues of m, all numbers congruent to none of them will be nonresidues. But the inverse question, *given a number, to assign all numbers of which it is a residue or a nonresidue*, is much more difficult. The statements in the preceding article depend on the solution to this problem. We will investigate it now, beginning with the simplest cases.

▶ 108. THEOREM. -1 *is a quadratic residue of all numbers of the form* $4n + 1$ *and a nonresidue of all numbers of the form* $4n + 3$.

Example. -1 is the residue of the numbers 5, 13, 17, 29, 37, 41, 53, 61, 73, 89, 97, etc. arising from the squares of the numbers 2, 5, 4, 12, 6, 9, 23, 11, 27, 34, 22, etc. respectively; on the other hand it is a nonresidue of the numbers 3, 7, 11, 19, 23, 31, 43, 47, 59, 67, 71, 79, 83, etc.

We mentioned this theorem in article 64, but it is best demonstrated as in article 106. For a number of the form $4n + 1$ we have $(-1)^{2n} \equiv 1$, for a number of the form $4n + 3$ we get $(-1)^{2n+1} \equiv -1$. This demonstration agrees with that of article 64, but because of the elegance and usefulness of the theorem we will show still another solution.

▶ 109. We will designate by the letter C the complex of all residues of a prime number p which are less than p and exclusive of the residue 0. Since the number of these residues is always $= (p - 1)/2$, C will obviously be even when p is of the form $4n + 1$, and odd when p is of the form $4n + 3$. By analogy with the language of article 77 where we spoke of numbers in general, we will call *associated residues* those whose product $\equiv 1$ (mod. p); for if r is a residue then $1/r$ (mod. p) is also a residue. And since the same residue cannot have more associates among the residues C, it is clear that all the residues C can be divided into classes each containing a pair of associated residues. Now if there is no residue that is its own associate, i.e. if each class contains a pair of *unequal* residues, the number of all residues will be double the number of classes; but if there are some residues which are their own associates, i.e. there are classes containing only one residue or, if you prefer, containing the same residue twice, the number of all the residues C will $= a + 2b$, where a is the number of classes of the second type and b the number of the first type. So when p is of the form $4n + 1$, a will be even, and when p is of the form $4n + 3$, a will be odd. But there are no numbers other than 1 and $p - 1$ which can be less than p and associates of themselves (see art. 77); but the first class certainly has 1 among its residues. So in the former case $p - 1$ (or -1, which comes to the same thing) must be a residue, in the latter case, a nonresidue. Otherwise in the first case we would have $a = 1$ and in the second $a = 2$, which is impossible.

▶ 110. This demonstration is due to Euler. He was also the first to discover the previous method (see *Opuscula Analytica, 1*). It is easy to see that it relies on principles very like those of our second demonstration of Wilson's theorem (art. 77). And if we suppose the truth of this theorem, the previous demonstration becomes much simplified. For among the numbers $1, 2, 3, \ldots p - 1$ there will be $(p - 1)/2$ quadratic residues of p and the same number of nonresidues; so the number of nonresidues will be even when p is of the form $4n + 1$, odd when p is of the form $4n + 3$. Thus the product of all the numbers $1, 2, 3, \ldots p - 1$ will be a residue in the former case, a nonresidue in the latter case (art. 99). But this product will always $\equiv -1 \pmod{p}$; and so in the former case -1 will also be a residue, in the latter case a nonresidue.

▶ 111. If therefore r is a residue of any prime number of the form $4n + 1$, $-r$ will also be a residue of this prime; and all nonresidues of such a number will remain nonresidues even if the sign is changed.[20] The contrary is true for prime numbers of the form $4n + 3$. Residues become nonresidues when the sign is changed and vice versa (see art. 98).

From what precedes we can derive the general rule: -1 *is a residue of all numbers not divisible by 4 or by any prime number of the form $4n + 3$; it is a nonresidue of all others* (see art. 103, 105).

▶ 112. Let us consider the residues $+2$ and -2.

If from Table 2 we collect all prime numbers whose residue is $+2$, we have 7, 17, 23, 31, 41, 47, 71, 73, 79, 89, 97. We notice that among these numbers none is of the form $8n + 3$ or $8n + 5$.

Let us see therefore whether this induction can be made a certitude.

First we observe that any composite number of the form $8n + 3$ or $8n + 5$ necessarily involves a prime factor of one or the other of the forms $8n + 3$ or $8n + 5$; for if only prime numbers of the form $8n + 1$, $8n + 7$ were involved, we would get numbers of the form $8n + 1$ or $8n + 7$. So if our induction is true in general, no number of the form $8n + 3$, $8n + 5$ can have $+2$ as a residue. Now certainly there is no number of this form less than 100 whose residue is $+2$. If however there are some such numbers greater than 100, let the least of all of them $= t$. This t will be either of the form $8n + 3$ or $8n + 5$; $+2$ will be a residue of t but a nonresidue of all similar numbers less than t. Let $2 \equiv a^2 \pmod{t}$. We can always find a such that it is odd and at the same time $< t$ (for a will have at least two positive values less than t whose sum $= t$, one of them being even, the other odd; see art. 104, 105). Let $a^2 = 2 + tu$ (that is $tu = a^2 - 2$); a^2 will be of the form $8n + 1$, tu of the form $8n - 1$; therefore u will be of the form $8n + 3$ or $8n + 5$ according as t is of the form $8n + 5$ or $8n + 3$. But from the equation $a^2 = 2 + tu$ we also have $2 \equiv a^2 \pmod{u}$; i.e. 2 is a residue of u. It is easy to see that $u < t$ and so t is not the smallest number in our induction, contrary to the hypothesis. And so what we have discovered by induction is proven true for the general case.

[20]Therefore when we speak of a number in so far as it is a residue or nonresidue of a number of the form $4n + 1$, we can ignore the sign completely or we can use the double sign $\pm$.

Combining this proposition with the propositions of article 111 we deduce the following theorems.

I. $+2$ *will be a nonresidue,* -2 *a residue of all prime numbers of the form* $8n + 3$.

II. $+2$ *and* -2 *will both be nonresidues of all prime numbers of the form* $8n + 5$.

▶ 113. By a similar induction from Table 2 we find the prime numbers whose residue is -2: 3, 11, 17, 19, 41, 43, 59, 67, 73, 83, 89, 97.[21] Since none of these is of the forms $8n + 5$, $8n + 7$ we will see whether this induction can lead us to a general theorem. We show as in the preceding article that no composite number of the form $8n + 5$ or $8n + 7$ can involve a prime factor of the form $8n + 5$ or $8n + 7$, so that if our induction is true in general, -2 can be the residue of no number of the form $8n + 5$ or $8n + 7$. If however such numbers did exist, let the least of them be $= t$, and we get $-2 = a^2 - tu$. If, as above, a is taken to be odd and less than t, u will be of the form $8n + 5$ or $8n + 7$, according as t is of the form $8n + 7$ or $8n + 5$. But from the fact that $a^2 + 2 = tu$ and $a < t$, it is easily proven that u also is less than t. Finally -2 will also be a residue of u; i.e. t will not be the smallest number of which -2 is a residue, contrary to the hypothesis. Thus -2 is necessarily a nonresidue of all numbers of the form $8n + 5$, $8n + 7$.

Combining this with the propositions of article 111, we get the following theorems.

I. *Both* -2 *and* $+2$ *are nonresidues of all prime numbers of the form* $8n + 5$, as we have already seen in the preceding article.

II. -2 *is a nonresidue,* $+2$ *a residue of all prime numbers of the form* $8n + 7$.

However in each demonstration we would have been able to take also an even value for a; but then we would have to distinguish the case where a was of the form $4n + 2$ from the one where a was of the form $4n$. The development would then proceed as above without any difficulty.

▶ 114. One case still remains; i.e. when the prime number is of the form $8n + 1$. The preceding method cannot be used here and a special device is needed.

Let $8n + 1$ be a prime modulus and a a primitive root. We will have (art. 62) $a^{4n} \equiv -1$ (mod. $8n + 1$). This congruence can be expressed in the form $(a^{2n} + 1)^2 \equiv 2a^{2n}$ (mod. $8n + 1$) or in the form $(a^{2n} - 1)^2 \equiv -2a^{2n}$. It follows that both $2a^{2n}$ and $-2a^{2n}$ are residues of $8n + 1$; but since a^{2n} is a square not divisible by the modulus, both $+2$ and -2 will also be residues (art. 98).

▶ 115. It will be of some use to add here another demonstration of this theorem. This is related to the preceding as the second demonstration (art. 109) of the theorem

---

[21]That is, by taking -2 as the product of $+2$ and -1 (see art. 111).

of article 108 is related to the first (art. 108). Skilled mathematicians can see that the two pairs of proofs are not so different as they seem at first.

I. For any prime modulus of the form $4m + 1$ among the numbers $1, 2, 3, \ldots 4m$ there are m numbers that can be congruent to a biquadratic whereas the other $3m$ cannot be.

This is easily derived from the principles of the preceding section but even without them it is not difficult to see. For we have shown that -1 is always a quadratic residue for such a modulus. Let therefore $f^2 \equiv -1$. Clearly if z is any number not divisible by the modulus, the biquadratics of the four numbers $+z, -z, +fz, -fz$ (two are obviously noncongruent) will be congruent among themselves. And the biquadratic of any number that is congruent to none of these four cannot be congruent to their biquadratics (otherwise the congruence $x^4 \equiv z^4$ which is of the fourth degree would have more than four roots, contrary to art. 43). Thus we deduce that all the numbers $1, 2, 3, \ldots 4m$ furnish only m noncongruent biquadratics and that among the same numbers there are m numbers congruent to these. The others can be congruent to no biquadratic.

II. Relative to a prime modulus of the form $8n + 1$, the number -1 can be made congruent to a biquadratic (-1 will be called the *biquadratic residue* of this prime number).

The number of all biquadratic residues less than $8n + 1$ (zero excluded) will be $= 2n$, i.e. even. It is easy to prove that if r is a biquadratic residue of $8n + 1$, the value of the expression $1/r$ (mod. $8n + 1$) will also be such a residue. So all biquadratic residues can be distributed into classes just as we distributed quadratic residues in article 109, and the rest of the demonstration proceeds in almost the same way as it did there.

III. Let $g^4 \equiv -1$, and h the value of the expression $1/g$ (mod. $8n + 1$). Then we will have
$$(g \pm h)^2 = g^2 + h^2 \pm 2gh \equiv g^2 + h^2 \pm 2$$
(because $gh \equiv 1$). But $g^4 \equiv -1$ and so $-h^2 \equiv g^4 h^2 \equiv g^2$. Thus $g^2 + h^2 \equiv 0$ and $(g \pm h)^2 \equiv \pm 2$; i.e. both $+2$ and -2 are quadratic residues of $8n + 1$.

Q.E.D.

▶ 116. From what precedes we can easily deduce the following general rule: *+2 is a residue of any number that cannot be divided by 4 or by any prime of the form $8n + 3$ or $8n + 5$, and a nonresidue of all others* (e.g. of all numbers of the forms $8n + 3$, $8n + 5$ whether they are prime or composite);

−2 is a residue of any number that cannot be divided by 4 or by any prime of the form $8n + 5$ or $8n + 7$, and a nonresidue of all others.

These elegant theorems were already known to the sagacious Fermat (*Opera Mathem.*[22]), but he never divulged the process which he claimed to have. Euler later searched in vain for it but it was Lagrange who published the first rigorous proof (*Nouv. mém. Acad. Berlin*, 1775). Euler seems still to have been unaware of it when he wrote his treatise in *Opuscula Analytica (1)*.[23]

▶ 117. We pass on to a consideration of the residues $+3$ and -3, and we begin with the second of them.

From Table 2 we find the residues of -3 to be 3, 7, 13, 19, 31, 37, 43, 61, 67, 73, 79, 97. Among them none is of the form $6n + 5$. We show as follows that outside of this table there are no primes of this form with -3 as a residue. First, any composite number of the form $6n + 5$ necessarily involves a prime factor of the same form. Thus when we show that no prime number of the form $6n + 5$ can have -3 as a residue, we also show that no such composite number can have this property. Suppose that there are such numbers outside our table. Let the smallest of them $= t$ and let $-3 = a^2 - tu$. Now if we take a as even and less than t, we will have $u < t$ and -3 a residue of u. But when a is of the form $6n + 2$, tu will be of the form $6n + 1$ and u of the form $6n + 5$. Q.E.A. because we supposed that t was the smallest number that contradicted our induction. Now if a is of the form $6n$, tu will be of the form $36n + 3$ and $tu/3$ of the form $12n + 1$. Thus $u/3$ will be of the form $6n + 5$. But clearly -3 is also a residue of $u/3$ and $u/3 < t$. Q.E.A. Manifestly therefore -3 can be the residue of no number of the form $6n + 5$.

Since every number of the form $6n + 5$ is necessarily contained among those of the form $12n + 5$ or $12n + 11$ and since the first of these is contained among numbers of the form $4n + 1$, the second among numbers of the form $4n + 3$, we have these theorems:

I. *Both -3 and $+3$ are nonresidues of all prime numbers of the form* $12n + 5$.

II. *-3 is a nonresidue, $+3$ a residue, of any prime number of the form* $12n + 11$.

▶ 118. From Table 2 we find that the numbers whose residue is $+3$ are the following: 3, 11, 13, 23, 37, 47, 59, 61, 71, 73, 83, 97. None of these is of the form $12n + 5$ or $12n + 7$. We can use the same method used in article 112, 113, 117 to show that there are no numbers at all of the forms $12n + 5$, $12n + 7$ which have $+3$ as a residue. We omit the details. Combining these results with those of article 111 we have the following theorems:

I. *Both $+3$ and -3 are nonresidues of all prime numbers of the form* $12n + 5$ (just as we found in the previous article).

II. *$+3$ is a nonresidue, -3 a residue of all prime numbers of the form* $12n + 7$.

[22] See footnote 1. The article in question here is a letter of Frénicle to Fermat, 2 August 1640.
[23] Diss. 8. See footnote 16.

▶ 119. We can learn nothing from this method about numbers of the form $12n + 1$, and so we must resort to special devices. By induction it is easy to see that $+3$ and -3 are residues of all prime numbers of this form. But we need only demonstrate this by showing that -3 is a residue because it follows necessarily that $+3$ will also be a residue (article 111). However we will show a more general result, i.e. that -3 is a residue of any prime number of the form $3n + 1$.

Let p be such a prime and a a number belonging to the exponent 3 for the modulus p (that there are such is clear from article 54, because 3 divides $p - 1$). We have therefore $a^3 \equiv 1$ (mod. p); i.e. $a^3 - 1$ or $(a^2 + a + 1)(a - 1)$ is divisible by p. But clearly a cannot be $\equiv 1$ (mod. p) because 1 belongs to the exponent 1; therefore $a - 1$ will not be divisible by p, but $a^2 + a + 1$ will be. Therefore $4a^2 + 4a + 4$ also will be; i.e. $(2a + 1)^2 \equiv -3$ (mod. p) and -3 is a residue of p.   Q.E.D.

It is clear that this demonstration (which is independent of the preceding ones) also includes prime numbers of the form $12n + 7$ which we treated in the preceding article.

We observe further that we could use the method of articles 109 and 115, but for brevity's sake we will not pursue it.

▶ 120. From these results we can easily deduce the following theorems (see art. 102, 103, 105).

I. -3 *is a residue of any number that cannot be divided by 8 or by 9 or by any prime number of the form $6n + 5$. It is a nonresidue of all others.*

II. $+3$ *is a residue of all numbers that cannot be divided by 4 or by 9 or by any prime of the form $12n + 5$ or $12n + 7$. It is a nonresidue of all others.*

Note this particular case:

-3 is a *residue* of all prime numbers of the form $3n + 1$, or what amounts to the same thing, *of all primes that are residues of* 3. It is a *nonresidue* of all prime numbers of the form $6n + 5$ or, with the exception of the number 2, of all primes of the form $3n + 2$; i.e. *of all primes that are nonresidues of* 3. All other cases follow naturally from this.

The propositions pertaining to the residues $+3$ and -3 were known to Fermat (*Opera Mathem. Wall.* [24]), but Euler was the first to give proofs[25] (*Novi comm. acad. Petrop.*, 8 [1760–61], 1763). This is why it is still more astonishing that proof of the propositions relative to the residues $+2$ and -2 kept eluding him, since they depend on similar devices. See also the commentary of Lagrange in *Nouv. mém. Acad. Berlin* (1775).

---

[24]The reference is to a letter: "Epistola XLVI, D. Fermatii ad D. Kenelmum Digby."
[25]"Supplementum quorundam theorematum arithmeticorum, quae in nonnullis demonstrationibus supponuntur."

▶ 121. Induction shows that $+5$ is a residue of no odd number of the form $5n + 2$ or $5n + 3$, i.e. of no odd number that is a nonresidue of 5. We show as follows that this rule has no exception. Let the smallest number that might be an exception to this rule $= t$. It is a nonresidue of the number 5, but 5 is a residue of t. Let $a^2 = 5 + tu$ so that a is even and less than t. Then u will be odd and less than t and $+5$ will be a residue of u. Now if a is not divisible by 5, neither will u be. But evidently tu is a residue of 5; but since t is a nonresidue of 5, u will also be a nonresidue. That is, there is a number less than t which is odd, a nonresidue of the number 5, and which has $+5$ as a residue, contrary to the hypothesis. If a is divisible by 5, let $a = 5b$ and $u = 5v$, then $tv \equiv -1 \equiv 4 \pmod{5}$; i.e. tv will be a residue of the number 5. From this point the demonstration proceeds as in the previous case.

▶ 122. Both $+5$ and -5 will be nonresidues therefore of all prime numbers which are at the same time nonresidues of 5 and of the form $4n + 1$, i.e. of all prime numbers of the form $20n + 13$ or $20n + 17$; and $+5$ will be a nonresidue, -5 a residue, of all prime numbers of the form $20n + 3$ or $20n + 7$.

In exactly the same way it can be shown that -5 is a nonresidue of all prime numbers of the forms $20n + 11$, $20n + 13$, $20n + 17$, $20n + 19$; and from this it easily follows that $+5$ is a residue of all prime numbers of the form $20n + 11$ or $20n + 19$, and a nonresidue of all of the form $20n + 13$ or $20n + 17$. And since any prime number, except 2 and 5 (which have ± 5 as residues), is contained in one of the forms $20n + 1, 3, 7, 9, 11, 13, 17, 19$, it is clear that we are able to judge concerning all primes except those of the form $20n + 1$ or of the form $20n + 9$.

▶ 123. By induction it is easy to establish that $+5$ and -5 are residues of all prime numbers of the form $20n + 1$ or $20n + 9$. And if this is true in general, we have the elegant law that $+5$ *is a residue of all prime numbers that are residues of* 5 (for these numbers are contained in the forms $5n + 1$ or $5n + 4$ or in one of $20n + 1$, 9, 11, 19; we have already considered the third and fourth of these) *and a nonresidue of all odd numbers that are nonresidues of* 5, as we have already demonstrated above. This theorem clearly suffices to judge whether $+5$ (or -5 by considering the product of $+5$ and -1) is a residue or nonresidue of any given number. And finally, observe the analogy between this theorem and the one of article 120 in which we treated the residue -3.

However the verification of this induction is not so easy. When a prime number of the form $20n + 1$, or more generally of the form $5n + 1$, is considered, the problem can be solved in a way similar to that of articles 114 and 119. Let a be some number belonging to the exponent 5 relative to the modulus $5n + 1$.

That there are such is clear from the preceding section. Thus we will have $a^5 \equiv 1$ or $(a - 1)(a^4 + a^3 + a^2 + a + 1) \equiv 0$ (mod. $5n + 1$). But since we cannot have $a \equiv 1$ and so not $a - 1 \equiv 0$, we must have $(a^4 + a^3 + a^2 + a + 1) \equiv 0$. Thus also $4(a^4 + a^3 + a^2 + a + 1) = (2a^2 + a + 2)^2 - 5a^2$ will $\equiv 0$; i.e. $5a^2$ will be a residue of $5n + 1$ and so also 5, because a^2 is a residue not divisible by $5n + 1$ (a is not divisible by $5n + 1$ because $a^5 \equiv 1$).   Q.E.D.

The case of a prime number of the form $5n + 4$ demands a more subtle device. But since the propositions we need here will be treated more generally later, we will touch on them only briefly.

I. If p is a prime number and b a given quadratic nonresidue of p, the value of the expression

$$(A) \ldots \frac{(x + \sqrt{b})^{p+1} - (x - \sqrt{b})^{p+1}}{\sqrt{b}}$$

(it is clear that the expansion contains no irrationals) will always be divisible by p, no matter what value is assumed for x. For it is clear from an inspection of the coefficients obtained in the expansion of A that all terms from the second to the penultimate (inclusive) will be divisible by p, and thus $A \equiv 2(p + 1)(x^p + xb^{(p-1)/2})$ (mod. p). But since b is a nonresidue of p we have $b^{(p-1)/2} \equiv -1$ (mod. p) (art. 106); but x^p is always $\equiv x$ (preceding section) and thus $A \equiv 0$.   Q.E.D.

II. In the congruence $A \equiv 0$ (mod. p), the indeterminate x has p dimensions and all the numbers $0, 1, 2, \ldots p - 1$ will be roots. Now let e be a divisor of $p + 1$. The expression

$$\frac{(x + \sqrt{b})^e - (x - \sqrt{b})^e}{\sqrt{b}}$$

(which we will designate by B) will be rational, the indeterminate x will have $e - 1$ dimensions, and it is clear from the fundamentals of analysis that A is divisible by B (indefinitely). Now I say that there are $e - 1$ values of x which, if substituted in B, make B divisible by p. For let $A \equiv BC$ and we find that x will have $p - e - 1$ dimensions in C and thus that the congruence $C \equiv 0$ (mod. p) will have no more than $p - e - 1$ roots. And it follows that the $e - 1$ numbers remaining among $0, 1, 2, 3, \ldots p - 1$ will be roots of the congruence $B \equiv 0$.

III. Now suppose p is of the form $5n + 4$, $e = 5$, b is a nonresidue of p, and the number a is so chosen that

$$\frac{(a + \sqrt{b})^5 - (a - \sqrt{b})^5}{\sqrt{b}}$$

is divisible by p. But this expression becomes

$$= 10a^4 + 20a^2b + 2b^2 = 2[(b + 5a^2)^2 - 20a^4]$$

Therefore we will also have $(b + 5a^2)^2 - 20a^4$ divisible by p; i.e. $20a^4$ is a residue of p, but since $4a^4$ is a residue not divisible by p (for it is easy to see that a cannot be divided by p), 5 will also be a residue of p.   Q.E.D.

Thus it is clear that the theorem enunciated in the beginning of this article is true generally.

We note that the demonstrations for both cases are due to Lagrange (*Nouv. mém. Acad. Berlin*, 1775).

▶ 124. By a similar method we can show:

   −7 is a nonresidue of any number that is a nonresidue of 7. By induction we can conclude,

   −7 is a residue of any prime number that is a residue of 7.

But no one thus far has proved this rigorously. The demonstration is easy for those residues of 7 that are of the form $4n - 1$; for by the method of the preceding articles it can be shown that $+7$ is always a nonresidue of such prime numbers, and so -7 is a residue. But we gain little by this, since the remaining cases cannot be handled in the same way. One case can be solved in the manner of articles 119 and 123. If p is a prime number of the form $7n + 1$, and a belongs to the exponent 7 relative to the modulus p, it is easy to see that

$$\frac{4(a^7 - 1)}{a - 1} = (2a^3 + a^2 - a - 2)^2 + 7(a^2 + a)^2$$

is divisible by p, and thus that $-7(a^2 + a)^2$ is a residue of p. But $(a^2 + a)^2$, as a square, is a residue of p and not divisible by p; for since we supposed that a belongs to the exponent 7, it can neither $\equiv 0$, nor $\equiv -1$ (mod. p); i.e. neither a nor $a + 1$ [nor therefore the square $(a + 1)^2 a^2$] will be divisible by p. Therefore 7 will also be a residue of p. Q.E.D. But prime numbers of the form $7n + 2$ or $7n + 4$ cannot be handled by any of the methods thus far considered. A proof however was discovered first by Lagrange in the same work. Below in Section VII we will show that the expression $4(x^p - 1)/(x - 1)$ can always be reduced to the form $X^2 \mp pY^2$ (where the upper sign should be taken when p is a prime number of the form $4n + 1$, the lower when it is of the form $4n + 3$). In this expression X, Y are rational expressions in x, free of fractions. Lagrange did not carry this analysis beyond the case $p = 7$.

▶ 125. Since the preceding methods are not sufficient to establish general demonstrations, we will now exhibit one that is. We begin with a theorem whose demonstration has long eluded us, although at first glance it seems so obvious that many authors believed there was no need for demonstration. It reads as follows: *Every number with the exception of squares taken positively is a nonresidue of some prime numbers.* But since we intend to use this theorem only as an aid in proving other considerations, we will explain here only those cases needed for this purpose. Other cases will be established later. We will show therefore that *any prime number of the form* $4n + 1$ *whether it be taken positively or negatively, is a nonresidue of some prime numbers*[26] and indeed (if it is >5) of some primes that are less than itself.

First when p a prime number of the form $4n + 1$ (>17 but $-13\,N\,3$, $-17\,N\,5$) is to be taken *negatively*,[27] let $2a$ be the even number immediately greater than $\sqrt{p}$; then $4a^2$ will always be $<2p$ or $4a^2 - p < p$. But $4a^2 - p$ is of the form $4n + 3$, and p a quadratic residue of $4a^2 - p$ (since $p \equiv 4a^2$ (mod. $4a^2 - p$)); and if $4a^2 - p$ is a prime number, $-p$ will be a nonresidue; if not, some factor of $4a^2 - p$ will necessarily be of the form $4n + 3$; and since $+p$ has to be a residue, $-p$ will be a nonresidue. Q.E.D.

For *positive* prime numbers we distinguish two cases. *First* let p be a prime number of the form $8n + 5$. Let a be any positive number $<\sqrt{(p/2)}$. Then $8n + 5 - 2a^2$ will be a positive number of the form $8n + 5$ or $8n + 3$ (according as a is even or odd) and so necessarily divisible by some prime number of the form $8n + 3$ or $8n + 5$, for a product of numbers of the form $8n + 1$ and $8n + 7$ cannot have the form $8n + 3$ or $8n + 5$. Let it here be denoted by q and we will have $8n + 5 \equiv 2a^2$ (mod. q). But 2 will be a nonresidue of q (art. 112) and so also $2a^2$ and $8n + 5$. Q.E.D.[28]

▶ 126. We have no such obvious devices to demonstrate that any prime number of the form $8n + 1$ taken positively is always a nonresidue of some prime number less than itself. But since this truth is of such great importance, we cannot omit a rigorous demonstration even though it is somewhat long. We begin as follows:

LEMMA. *Suppose we have two series of numbers*

A, B, C, etc. . . . . (I) A', B', C', etc. . . . . (II)

(whether the number of terms in each is the same or not, is a matter of indifference) *so arranged that if p is a prime number or the power of a prime number that divides one or more terms of the second series, there are at least as many terms of the first series divisible by p. I claim that the product of all numbers* (I) *is divisible by the product of all numbers* (II).

Example. Let (I) consist of the numbers 12, 18, 45; (II) of the numbers 3, 4, 5, 6, 9. Then if we take successively the numbers 2, 4, 3, 9, 5 we find that there are 2, 1, 3, 2, 1 terms in (I) and 2, 1, 3, 1, 1 terms in (II) which are, respectively, divisible by these numbers; and the product of all the terms (I) $=9720$ which is divisible by 3240, the product of all the terms (II).

---

[26] Obviously we must except $+1$.

[27] $-13\,N\,3$ means that -13 is a nonresidue of 3; $-17\,N\,5$ means that -17 is a nonresidue of 5.

[28] Article 98. It is clear that a^2 is a residue of q not divisible by q, for otherwise the prime number p would be divisible by q. Q.E.A.

Demonstration. Let the product of all the terms (I) $= Q$, the product of all the terms of the series (II) $= Q'$. It is clear that any prime number which is a divisor of Q' is also a divisor of Q. Now we will show that any prime factor of Q' has at least as many dimensions in Q as it has in Q'. Let such a divisor be p and let us suppose that in series (I) there are a terms divisible by p, b terms divisible by p^2, c terms divisible by p^3, etc. In a similar way determine the letters a', b', c', etc. for series (II). In Q, p will have $a + b + c +$ etc. dimensions and $a' + b' + c' +$ etc. dimensions in Q'. But a' is certainly not greater than a, b' not greater than b, etc. (by hypothesis); and therefore $a' + b' + c' +$ etc. will certainly not be $> a + b + c +$ etc. Since therefore no prime number can have higher dimensions in Q' than in Q, Q will be divisible by Q' (art. 17). Q.E.D.

▶ 127. LEMMA. *In the progression* $1, 2, 3, 4, \ldots n$ *there can not be more terms divisible by a number h than in the progression* $a, a + 1, a + 2, \ldots a + n - 1$, *which has the same number of terms.*

We see without any trouble that if n is a multiple of h, there are n/h terms in each progression divisible by h; if n is not a multiple of h, let $n = eh + f$ with $f < h$. In the first series there will be e terms divisible by h, in the second either e or $e + 1$.

As a corollary of this we have that well-known proposition from the theory of figurative numbers, namely that

$$\frac{a(a + 1)(a + 2) \ldots (a + n - 1)}{1 \cdot 2 \cdot 3 \ldots n}$$

is always an integer. But until now, as far as we know, no one has demonstrated it directly.

Finally, the lemma could have been expressed more generally as follows:

In the progression $a, a + 1, a + 2, \ldots, a + n - 1$ there are at least as many terms congruent relative to the modulus h to any given number r as there are terms in the progression $1, 2, 3, \ldots n$ divisible by h.

▶ 128. THEOREM. *Let a be any number of the form $8n + 1$; p some number which is prime relative to a having $+a$ as a residue; and let m be an arbitrary number. I claim that in the progression*

$$a, \tfrac{1}{2}(a - 1), \quad 2(a - 4), \quad \tfrac{1}{2}(a - 9), \quad 2(a - 16), \ldots \quad 2(a - m^2) \text{ or } \tfrac{1}{2}(a - m^2)$$

according as m is even or odd, there are at least as many terms divisible by p as there are in the progression

$$1, 2, 3, \ldots 2m + 1$$

We will designate the first of these by (I), the second by (II).

Demonstration. I. When $p = 2$, all the terms except the first, i.e. m terms, will be divisible in (I); the same number will be divisible in (II).

II. Let p be an odd number or double or four times an odd number and let $a \equiv r^2$ (mod. p). Then in the progression $-m$, $-(m-1)$, $-(m-2)$, $\ldots + m$ [it has as many terms as (II) and we will call it (III)] relative to the modulus p there will be as many terms congruent to r as there are in series (II) divisible by p (preceding article). But among them there can be no pairs which differ only in sign.[29] And each of them will have a corresponding value in series (I) which is divisible by p. This means that if $\pm b$ were a term of series (III) congruent to r relative to p, $a - b^2$ would be divisible by p. And so if b is even, the term $2(a - b^2)$ of series (I) will be divisible by p. But if b is odd, the term $(a - b^2)/2$ will be divisible by p; for manifestly $(a - b^2)/p$ is an *even* integer because $a - b^2$ is divisible by 8, whereas p is divisible at most by 4 (by hypothesis, a is of the form $8n + 1$, and b^2 which is the square of an odd number will be of the same form, and the difference will be of the form $8n$). Thus we conclude that in series (I) there are as many terms divisible by p as there are terms in (III) congruent to r relative to p, i.e. as many or more than there are in (II) divisible by p. Q.E.D.

III. Let p be of the form $8n$ and $a \equiv r^2$ (mod. $2p$). It is easy to see that a, which is by hypothesis a residue of p, is also a residue of $2p$. Then there will be at least as many terms in series (III) congruent to r relative to p as there are in (II) divisible by p, and they will all be unequal in value. But for each of them there will be a corresponding term in (I) which is divisible by p. For if $+b$ or $-b \equiv r$ (mod. p), we will have $b^2 \equiv r^2$ (mod. $2p$)[30] and the term $(a - b^2)/2$ will be divisible by p. Wherefore there will be at least as many terms in (I) divisible by p as there are in (II). Q.E.D.

▶ 129. THEOREM. *If a is a prime number of the form $8n + 1$ there is necessarily some prime number below $2\sqrt{a} + 1$ of which a is a nonresidue.*

Demonstration. If it is possible, let a be a residue of all primes less than $2\sqrt{a} + 1$. Then a would be a residue of all composite numbers less than $2\sqrt{a} + 1$ (cf. the rules for judging whether a given number is a residue of a composite number or not: art. 105). Let the number next smaller than $\sqrt{a}$ be $=m$. Then in the series

(I) $\qquad \ldots a, \frac{1}{2}(a - 1), 2(a - 4), \frac{1}{2}(a - 9), \ldots 2(a - m^2)$ or $\frac{1}{2}(a - m^2)$

there will be as many or more terms divisible by some number smaller than $2\sqrt{a} + 1$ as in

(II) $\qquad \ldots 1, 2, 3, 4, \ldots 2m + 1$ (preceding article)

And it follows that the product of all terms (I) will be divisible by the product of all terms (II) (art. 126). But the former product is either equal to

---

[29] For if $r \equiv -f \equiv +f$ (mod. p), $2f$ would be divisible by p and so also $2a$ [since $f^2 \equiv a$ (mod. p)]. This cannot be unless $p = 2$, since by hypothesis a and p are relatively prime. But we have already treated this case by itself.

[30] That is, we will have $b^2 - r^2 = (b - r)(b + r)$ composed of two factors, one of which is divisible by p (hypothesis), the other divisible by 2 (since both b and r are odd); and so $b^2 - r^2$ is divisible by $2p$.

$a(a - 1)(a - 4)\ldots(a - m^2)$ or to half of this product (according as m is even or odd). And so the product $a(a - 1)(a - 4)\ldots(a - m^2)$ will certainly be divisible by the product of all the terms (II) and since all these terms are prime relative to a, so will the product if we delete the factor a. But the product of all the terms (II) can be expressed as follows:

$$[m + 1][(m + 1)^2 - 1][(m + 1)^2 - 4]\ldots[(m + 1)^2 - m^2]$$

And therefore

$$\frac{1}{m + 1} \cdot \frac{a - 1}{(m + 1)^2 - 1} \cdot \frac{a - 4}{(m + 1)^2 - 4}\ldots\frac{a - m^2}{(m + 1)^2 - m^2}$$

will be an integer although it is the product of fractions which are less than unity; for since $\sqrt{a}$ is necessarily irrational, $m + 1 > \sqrt{a}$ and $(m + 1)^2 > a$. We conclude that our supposition cannot be true.

Q.E.D.

Now because a is certainly > 9, we will have $2\sqrt{a} + 1 < a$, and there exists a prime number $< a$, of which a is a nonresidue.

▶ 130. Now that we have rigorously established that any prime number of the form $4n + 1$, taken both positively and negatively, is a nonresidue of some prime less than itself, we proceed to a more exact and more general comparison of prime numbers, one of which is a residue or nonresidue of the other.

We have shown above that -3 and $+5$ are residues or nonresidues of all prime numbers that are residues or nonresidues of 3, 5 respectively.

By induction it is found that the numbers -7, -11, $+13$, $+17$, -19, -23, $+29$, -31, $+37$, $+41$, -43, -47, $+53$, -59, etc. are residues or nonresidues of all prime numbers which, taken positively, are residues or nonresidues of these primes respectively. This induction can be accomplished easily with the help of Table 2.

It can be seen that among these primes those of the form $4n + 1$ are positive, those of the form $4n + 3$ are negative.

▶ 131. We will soon show that what we have discovered by induction is true for the general case. But before we do this it will be necessary to discover all the consequences of this theorem, supposing it to be true:

If p is a prime number of the form $4n + 1$, $+p$ will be a residue or nonresidue of any prime number which taken positively is a residue or nonresidue of p. If p is of the form $4n + 3$, $-p$ will have the same property.

Since almost everything that can be said about quadratic residues depends on this theorem, the term *fundamental theorem* which we will use from now on should be acceptable.

To indicate our reasoning as briefly as possible, we will denote prime numbers of the form $4n + 1$ by the letters a, a', a'', etc. and prime numbers of the form

$4n + 3$ by the letters b, b', b'', etc.; any numbers of the form $4n + 1$ will be denoted by A, A', A'', etc., any numbers of the form $4n + 3$ by B, B', B'', etc.; finally the letter R placed between two quantities will indicate that the former is a residue of the latter, and the letter N will indicate the contrary. For example, $+5\,R\,11$, $\pm 2\,N\,5$ will indicate that $+5$ is a residue of 11, and $+2$ or -2 is a nonresidue of 5. Now with the help of the theorems in article 111 we can easily deduce the following propositions from the fundamental theorem.

	If	we have
1.	$\pm a\,R\,a'$	$\pm a'\,R\,a$
2.	$\pm a\,N\,a'$	$\pm a'\,N\,a$
3.	$+a\,R\,b$ $-a\,N\,b$	$\pm b\,R\,a$
4.	$+a\,N\,b$ $-a\,R\,b$	$\pm b\,N\,a$
5.	$\pm b\,R\,a$	$+a\,R\,b$ $-a\,N\,b$
6.	$\pm b\,N\,a$	$+a\,N\,b$ $-a\,R\,b$
7.	$+b\,R\,b'$ $-b\,N\,b'$	$+b'\,N\,b$ $-b'\,R\,b$
8.	$+b\,N\,b'$ $-b\,R\,b'$	$+b'\,R\,b$ $-b'\,N\,b$

▶ 132. This table includes all the cases when two prime numbers are compared; what follows pertains to any numbers, but the demonstrations of these is less obvious.

	If	we have
9.	$\pm a\,R\,A$	$\pm A\,R\,a$
10.	$\pm b\,R\,A$	$+A\,R\,b$ $-A\,N\,b$
11.	$+a\,R\,B$	$\pm B\,R\,a$
12.	$-a\,R\,B$	$\pm B\,N\,a$
13.	$+b\,R\,B$	$-B\,R\,b$ $+B\,N\,b$
14.	$-b\,R\,B$	$+B\,R\,b$ $-B\,N\,b$

Since the same principles lead to the demonstration of all these propositions, it will not be necessary to develop them all. A demonstration of proposition 9 which we will do as an example should be sufficient. First we observe that any number of the form $4n + 1$ has either no factor of the form $4n + 3$ or else two or four etc. of them; i.e. the number of such factors (including any that may be equal) will always be even; and that any number of the form $4n + 3$ involves an odd number of factors of the form $4n + 3$ (i.e. one or three or five etc.). The number of factors of the form $4n + 1$ remains undetermined.

Proposition 9 can be demonstrated as follows. Let A be the product of prime factors a', a'', a''', etc. b, b', b'', etc.; the number of factors b, b', b'', etc. will be even (or there may be none of them, which reduces to the same thing). Now if a is a residue of A, it will also be a residue of all the factors a', a'', a''', etc. b, b', b'', etc. By propositions 1, 3 of the preceding article, each of these factors will be residues of a and so will the product A as well as $-A$. And if $-a$ is a residue of A, by that very fact it is a residue of all the factors a', a'', etc. b, b', etc.; each of the a', a'', etc. will be residues of a, each of the b, b', etc. nonresidues. But since the latter are even in number, the product of all of them, i.e. A, will be a residue of a, and so also will $-A$.

▶ 133. We begin now a more general investigation. Consider any two odd, signed numbers P and Q which are relatively prime. Let us conceive of P without respect to its sign as resolved into its prime factors, and designate by p the number of these factors for which Q is a nonresidue. If any prime number of which Q is a nonresidue occurs many times among the factors of P, it is to be counted many times. Similarly let q be the number of prime factors of Q for which P is a nonresidue. The numbers p, q will have a certain mutual relation depending on the nature of the numbers P, Q. That is, if one of the numbers p, q is even or odd, the form of the numbers P, Q will indicate whether the other is even or odd. We show this relation in the following table.

The numbers p, q will both be even or both odd when the numbers P, Q have the forms:

1.	$+A$,	$+A'$
2.	$+A$,	$-A'$
3.	$+A$,	$+B$
4.	$+A$,	$-B$
5.	$-A$,	$-A'$
6.	$+B$,	$-B'$

On the other hand one of the numbers p, q will be even, the other odd, when the numbers P, Q have the forms:

7.	$-A$,	$+B$
8.	$-A$,	$-B$
9.	$+B$,	$+B'$
10.	$-B$,	$-B'$ [31]

[31] Let $l = 1$ if both P, $Q \equiv 3 \pmod 4$, otherwise $l = 0$. Let $m = 1$ if both P, Q are negative, otherwise $m = 0$. The relation then depends on $l + m$.

Example. Let the given numbers be -55 and $+1197$ which is the fourth case; 1197 is a nonresidue of one prime factor of 55, namely the number 5. But -55 is a nonresidue of three of the prime factors of 1197, namely 3, 3, 19.

If P and Q are prime numbers, these propositions reduce to those which we considered in article 131. Here p and q cannot be greater than 1, and so when p is even, it necessarily $= 0$; i.e. Q will be a residue of P. But when p is odd, Q will be a nonresidue of P, and vice versa. And so writing a, b in place of A, B it follows from 8 that if $-a$ is a residue or nonresidue of b, $-b$ will be a nonresidue or residue of a, which agrees with 3 and 4 of article 131.

In general Q cannot be a residue of P unless $p = 0$; if therefore p is odd, Q will certainly be a nonresidue of P.

The propositions of the preceding article can be derived from this fact without any difficulty.

It is apparent that this general representation is more than idle speculation because the demonstration of the fundamental theorem would be incomplete without it.

▶ 134. We attempt now to deduce these propositions.

I. Let us as before conceive of P as resolved into its prime factors neglecting all signs. Let Q be resolved into factors in any way whatsoever, but here we take into account the sign of Q. Now combine each of the former with each of the latter. Then if s designates the number of all combinations in which a factor of Q is a nonresidue of the factor of P, p and s will either both be even or both odd. Let the prime factors of P be f, f', f'', etc. And among the factors into which Q is resolved, let m of them be nonresidues of f, m' of them nonresidues of f', m'' of them nonresidues of f'', etc. Then obviously,

$$s = m + m' + m'' + \text{etc.}$$

and p will indicate how many numbers among m, m', m'', etc. are odd. Thus s will be even when p is even, odd when p is odd.

II. This is true in general no matter how Q is resolved into factors. Now to particular cases. For the first case, let one of the numbers, P, be positive and the other, Q, either of the form $+A$ or $-B$. Resolve P, Q into their prime factors, assigning to each of the factors of P a positive sign and to each of the factors of Q a positive or negative sign according as they are of the form a or b; manifestly Q will be either of the form $+A$ or $-B$, as required. Combine each of the factors of P with each of the factors of Q and designate as before s, the number of the combinations in which the factor of Q is a nonresidue of the factor of P. Similarly let t be the number of combinations in which the factor of P is a nonresidue of the factor of Q. But from the fundamental theorem it follows that these combinations must be identical and therefore $s = t$. Finally, from what we have just shown, $p \equiv s \pmod 2$, $q \equiv t \pmod 2$ and so $p \equiv q \pmod 2$.

Thus we have propositions 1, 3, 4, and 6 of article 133.

The other propositions can be demonstrated directly in a similar manner but they demand one new consideration; it is easier to derive them from the preceding in the following way.

III. Let us denote again P, Q any odd numbers which are relatively prime, p, q the number of prime factors of P, Q for which Q, P are respectively nonresidues. And let p' be the number of prime factors of P for which $-Q$ is a nonresidue (when Q is negative, manifestly $-Q$ will be positive). Now let all the prime factors of P be distributed into four classes:

1) factors of the form a of which Q is a residue
2) factors of the form b for which Q is a residue; let the number of these be χ
3) factors of the form a for which Q is a nonresidue; let the number of these be ψ
4) factors of the form b for which Q is a nonresidue; let the number of these be ω

It is easy to see that $p = \psi + \omega$, $p' = \chi + \psi$.

Now when P is of the form $+A$, $\chi + \omega$ and $\chi - \omega$ will be even; and thus $p' = p + \chi - \omega \equiv p$ (mod. 2). When P is of the form $\pm B$, we find by a similar computation that the numbers p, p' will be noncongruent relative to the modulus 2.

IV. Let us apply this to individual cases. Let both P and Q be of the form $+A$. From proposition 1 we have $p \equiv q$ (mod. 2); but $p' \equiv p$ (mod. 2) so also $p' \equiv q$ (mod. 2). This agrees with proposition 2. Similarly if P is of the form $-A$, Q of the form $+A$, we have $p \equiv q$ (mod. 2) from proposition 2 which we have just seen; then since $p' \equiv p$, we get $p' \equiv q$. Proposition 5 is thus demonstrated.

In the same way proposition 7 can be deduced from 3; proposition 8 from either 4 or 7; proposition 9 from 6; proposition 10 from 6.

▶ 135. We have not demonstrated the propositions of article 133 in the preceding article but we nevertheless showed that their truth depends on the truth of the fundamental theorem. By the method we followed it is clear that these propositions are true for the numbers P, Q if only the fundamental theorem is true for all the prime factors of these numbers compared among themselves, even if it were not generally true. Now we come to a demonstration of the fundamental theorem. We preface it with the following explanation:

We will say that the fundamental theorem is true up to some number M, if it is true for any two prime numbers neither of which is larger than M.

It should be understood in the same way if we say that the theorems of articles 131, 132, and 133 are true up to some term. If the fundamental theorem is true up to some term, it is clear that these propositions will be true up to the same term.

▶ 136. It is easy to confirm that the fundamental theorem is true for small numbers by induction and so the limit can be determined up to which it certainly applies. Let us suppose this induction established; how far we have carried it is a matter of indifference. Thus it would be sufficient to confirm it up to the number 5. This can be done by a single observation since $+5\ N\ 3$, $\pm3\ N\ 5$.

Now if the fundamental theorem were not generally true, there would be some limit T up to which it would be valid, but it would not be valid for the next greater number $T+1$. This is the same as saying that there are two prime numbers, the larger of which is $T+1$, such that they contradict the fundamental theorem when compared together. It would imply however that any other pair of prime numbers would satisfy the theorem if only both were less than $T+1$. From this it would follow that the propositions of articles 131, 132, and 133 would also be valid up to T. We will show now that this supposition is inconsistent. There are various cases we must distinguish according to the different forms that $T+1$ and the prime number less than $T+1$ can have. We will call that prime number p.

When both $T+1$ and p are of the form $4n+1$, the fundamental theorem can be false in two ways, i.e. if we had at the same time

either	$\pm p\ R(T+1)$	and	$\pm(T+1)N\,p$
or	$\pm p\ N(T+1)$	and	$\pm(T+1)R\,p$

When both $T+1$ and p are of the form $4n+3$, the fundamental theorem will be false if we had at the same time

either	$+p\ R(T+1)$	and	$-(T+1)N\,p$
[or what is the same thing	$-p\ N(T+1)$	and	$+(T+1)R\,p$]
or	$+p\ N(T+1)$	and	$-(Y+1)R\,p$
[or what is the same thing	$-p\ R(T+1)$	and	$+(T+1)N\,p$]

When $T+1$ is of the form $4n+1$, and p of the form $4n+3$, the fundamental theorem will be false if we had

either	$\pm p\ R(T+1)$	and	$+(T+1)N\,p$ [or	$-(T+1)R\,p$]
or	$\pm p\ N(T+1)$	and	$-(T+1)N\,p$ [or	$+(T+1)R\,p$]

When $T+1$ is of the form $4n+3$, and p of the form $4n+1$, the fundamental theorem will be false if we had

either	$+p\ R(T+1)$ [or $-p\ N(T+1)$]	and	$\pm(T+1)N\,p$
or	$+p\ N(T+1)$ [or $-p\ R(T+1)$]	and	$\pm(T+1)R\,p$

If it can be shown that none of these eight cases is valid, it will be certain likewise that the truth of the fundamental theorem is circumscribed by no limits. We proceed to this exposition now but, since some of the cases depend on others, we cannot preserve the same order we used to enumerate them.

▶ 137. *First case. When* $T + 1$ *is of the form* $4n + 1$ $(=a)$ *and* p *is of the same form, we cannot have* $\pm p\, R\, a$ *and* $\pm a\, N\, p$ *or we cannot have* $\pm a\, N\, b$ *if* $\pm p\, R\, a$.

Let $+p \equiv e^2$ (mod. a) where e is even and $< a$ (this is always possible). We can distinguish two cases.

I. When e is not divisible by p. Let $e^2 = p + af$. Here f will be positive, of the form $4n + 3$ (form B), $< a$ and not divisible by p. Further $e^2 \equiv p$ (mod. f); i.e. $p\, R\, f$ and so from proposition 11 of article 132 $\pm f\, R\, p$ (since $p, f < a$). But we have also $af\, R\, p$ and therefore $\pm a\, R\, p$.

II. When e is divisible by p, let $e = gp$ and $e^2 = p + aph$ or $pg^2 = 1 + ah$. Then h will be of the form $4n + 3$ (B) and relatively prime to p and g^2. Further we have $pg^2\, R\, h$ and so also $p\, R\, h$ and from this (prop. 11, art. 132) $\pm h\, R\, p$. But we also have $-ah\, R\, p$ because $-ah \equiv 1$ (mod. p); therefore $\mp a\, R\, p$.

▶ 138. *Second case. When* $T + 1$ *is of the form* $4n + 1$ $(=a)$, p *of the form* $4n + 3$, *and* $\pm p\, R(T + 1)$ *we cannot have* $+(T + 1)Np$ *or* $-(T + 1)Rp$. This was the fifth case above.

Let as above $e^2 = p + fa$ and e even and $< a$.

I. When e is not divisible by p, f will also not be divisible by p. Further, f will be positive, of the form $4n + 1$ (A), and $< a$; but $+p\, Rf$ and therefore (prop. 10, art. 132) $+f\, R\, p$. But also $+fa\, R\, p$; therefore $+a\, R\, p$ or $-a\, N\, p$.

II. When e is divisible by p, let $e = pg$ and $f = ph$. Therefore $g^2 p = 1 + ha$. Then h will be positive, of the form $4n + 3$ (B), and relatively prime to p and g^2. Further, $+g^2 p\, R\, h$ and so $+p\, R\, h$; as a result (prop. 13, art. 132) $-h\, R\, p$. But we have $-ha\, R\, p$ and so $+a\, R\, p$ and $-a\, N\, p$.

▶ 139. *Third case. When* $T + 1$ *is of the form* $4n + 1$ $(=a)$, p *of the same form and* $\pm p\, N\, a$, *we cannot have* $\pm a\, R\, p$ (second case above).

Take any prime number less than a of which $+a$ is a nonresidue. We have shown that there are such (art. 125, 129). But here we must consider two cases separately according as the prime number is of the form $4n + 1$ or $4n + 3$, for we did not show that there are such prime numbers of *each* form.

I. Let the prime number be of the form $4n + 1$ and $= a'$. Then we will have $\pm a'\, N\, a$ (art. 131) and therefore $\pm a' p\, R\, a$. Let now $e^2 \equiv a' p$ (mod. a) and e even and $< a$. Then again we have to distinguish four cases.

1) When e is not divisible by p or by a'. Let $e^2 = a' p \pm af$, taking whichever sign makes f positive. Then we have $f < a$, relatively prime to a' and p and, for the upper sign, of the form $4n + 3$, for the lower of the form $4n + 1$. For brevity's sake we will designate by $[x, y]$ the number of prime factors of the number y for which x is a nonresidue. Then we have $a' p\, Rf$ and so $[a' p, f] = 0$. Thus $[f, a' p]$ will be an even number (prop. 1, 3, art. 133), i.e. either $=0$ or $=2$. So f will be

a residue of each of the numbers a', p or of neither. The former case is impossible, since $\pm af$ is a residue of a' and $\pm a\,N\,a'$ (by hypothesis); therefore $\pm f\,N\,a'$. So f must be a nonresidue of each of the numbers a', p. But since $\pm af\,R\,p$, we will have $\pm a\,N\,p$. Q.E.D.

2) When e is divisible by p but not by a', let $e = gp$ and $g^2 = a' \pm ah$ with the sign so determined that h is positive. Then we will have $h < a$, relatively prime to a', g and p, and for the upper sign of the form $4n + 3$, for the lower of the form $4n + 1$. If we multiply the equation $g^2 p = a' \pm ah$ by p and a', we can deduce easily that $pa'\,R\,h \ldots (\alpha)$; $\pm ahp\,R\,a' \ldots (\beta)$; $aa'h\,R\,p \ldots (\gamma)$. It follows from (α) that $[pa', h] = 0$ and so (prop. 1, 3, art. 133) $[h, pa']$ is even; i.e. h is a nonresidue of both p, a' or of neither. *In the former case* it follows from (β) that $\pm ap\,N\,a'$ and, since by hypothesis $\pm a\,N\,a'$, we have $\pm p\,R\,a'$. Thus by the fundamental theorem which is valid for the numbers p, a', since they are less than $T + 1$, we get $\pm a'\,R\,p$. Now $h\,N\,p$, therefore by (γ) $\pm a\,N\,p$. Q.E.D. *In the latter case* it follows from (β) that $\pm ap\,R\,a'$, so $\pm p\,N\,a'$, $\pm a'\,N\,p$. But because $h\,R\,p$ we get from (γ) that $\pm a\,N\,p$. Q.E.D.

3) When e is divisible by a' but not by p. For this case the demonstration is almost the same as in the preceding and it will cause no difficulty to anyone who has understood it.

4) When e is divisible both by a' and by p and thus also by the product $a'p$ (we have supposed that a', p are *unequal* because otherwise the hypothesis $a\,N\,a'$ would contain $a\,N\,p$). Let $e = ga'p$ and $g^2 a'p = 1 \pm ah$. Then we will have $h < a$, relatively prime to a' and p, and of the form $4n + 3$ when the upper sign applies, of the form $4n + 1$ when the lower sign applies. From this equation we can easily deduce the following, $a'p\,R\,h \ldots (\alpha)$; $\pm ah\,R\,a' \ldots (\beta)$; $\pm ah\,R\,p \ldots (\gamma)$. From (α) which agrees with (α) in 2) the same result follows. That is we have either $h\,R\,p$, $h\,R\,a'$ or $h\,N\,p$, $h\,N\,a'$. But in the former case (β) would give us $a\,R\,a'$ contrary to the hypothesis. So therefore $h\,N\,p$ and by (γ) $a\,N\,p$ also.

II. When the prime number is of the form $4n + 3$, the demonstration is so like the preceding that it seems superfluous to include it. We only observe for those who wish to do it themselves (which we highly recommend), that it would be advantageous to consider each sign separately after developing the equation $e^2 = bp \pm af$ (b designates the prime number).

▶ 140. *Fourth case. When $T + 1$ is of the form $4n + 1$ $(= a)$, p of the form $4n + 3$ and $\pm p\,N\,a$, we cannot have $+a\,R\,p$ or $-a\,N\,p$* (sixth case above).

Since the demonstration of this is just like that of the third case, we omit it for the sake of brevity.

▶ 141. *Fifth case. When* $T + 1$ *is of the form* $4n + 3$ $(= b)$, p *of the same form and* $+p \, R \, b$ *or* $-p \, N \, b$, *we cannot have* $+b \, R \, p$ *or* $-b \, N \, p$ (third case above).

Let $p = e^2$ (mod. b) with e even and $< b$.

I. When e is not divisible by p. Let $e^2 = p + bf$ where f is positive, of the form $4n + 3$, $< b$, and relatively prime to p. Further $p \, R \, f$ and so (prop. 13, art. 132) $-f \, R \, p$. Thus also since $+bf \, R \, p$ we get $-b \, R \, p$ and $+b \, N \, p$. Q.E.D.

II. When e is divisible by p, let $e = pg$ and $g^2 p = 1 + bh$. Then h will be of the form $4n + 1$ and relatively prime to p; also $p \equiv g^2 p^2$ (mod. h), therefore $p \, R \, h$; from this we have $+h \, R \, p$ (prop. 10, art. 132) and then because $-bh \, R \, p$ it follows that $-b \, R \, p$ or $+b \, N \, p$. Q.E.D.

▶ 142. *Sixth case. When* $T + 1$ *is of the form* $4n + 3$ $(= b)$, p *of the form* $4n + 1$, *and* $p \, R \, b$, *we cannot have* $\pm b \, N \, p$ (seventh case above).

We omit the demonstration which is completely like the preceding.

▶ 143. *Seventh case. When* $T + 1$ *is of the form* $4n + 3 (= b)$, p *of the same form, and* $+p \, N \, b$ *or* $-p \, R \, b$, *we cannot have* $+b \, N \, p$ *or* $-b \, R \, p$ (fourth case above).

Let $-p \equiv e^2$ (mod. b) with e even and $< b$.

I. When e is not divisible by p. Let $-p = e^2 - bf$ where f is positive, of the form $4n + 1$, relatively prime to p, and less than b (for certainly e is not greater than $b - 1$, $p < b - 1$, and so $bf = e^2 + p < b^2 - b$; i.e. $f < b - 1$). Further we have $-p \, R \, f$ and from this (prop. 10, art. 132) $+f \, R \, p$. And since $+bf \, R \, p$ we get $+b \, R \, p$ or $-b \, N \, p$.

II. When e is divisible by p, let $e = pg$ and $g^2 p = -1 + bh$. Then h will be positive, of the form $4n + 3$, relatively prime to p, and $< b$. Further we get $-p \, R \, h$ and so (prop. 14, art. 132) $+h \, R \, p$. And since $bh \, R \, p$ it follows that $+b \, R \, p$ or $-b \, N \, p$. Q.E.D.

▶ 144. *Eighth case. When* $T + 1$ *is of the form* $4n + 3 (= b)$, p *of the form* $4n + 1$, *and* $+p \, N \, b$ *or* $-p \, R \, b$, *we cannot have* $\pm b \, R \, p$ (last case above).

The demonstration is the same as in the preceding case.

▶ 145. In the preceding demonstration we always took an even value for e (art. 137, 144); an odd value could have been used just as well but then many more distinctions would have to be introduced. (Those who enjoy these investigations will find it very useful to apply themselves to this task.) Furthermore the theorems pertaining to the residues $+2$ and -2 would have to be presupposed. But since our demonstration was accomplished without these theorems we will present them by a new method which should not be disdained, since it is more direct than the methods used above to show that ± 2 is a residue of any prime number of the form $8n + 1$. We will suppose the other cases (regarding prime numbers of the form $8n + 3$, $8n + 5$, $8n + 7$) have been established by the above methods and that the theorem has been established only by induction; in the following reflections this induction will be raised to a certainty.

If ± 2 is not a residue of all prime numbers of the form $8n + 1$, let the smallest prime for which ± 2 is a nonresidue $= a$, so that the theorem is valid for all primes less than a. Now take some prime number $< a/2$ for which a is a nonresidue (it is clear from article 129 that this can be done). Let this number be $= p$ and by the fundamental theorem $p \, N \, a$. And so $\pm 2p \, R \, a$. Let therefore $e^2 \equiv 2p$ (mod. a) so that e is odd and $< a$. Then two cases must be distinguished.

I. When e is not divisible by p, let $e^2 = 2p + aq$; q will be positive, of the form $8n + 7$ or $8n + 3$ (according as p is of the form $4n + 1$ or $4n + 3$), $< a$, and not divisible by p. Now all prime factors of q are divided into four classes: e of the form $8n + 1$, f of the form $8n + 3$, g of the form $8n + 5$, h of the form $8n + 7$; let the product of the factors of the first class be E, of the second, third, and fourth classes F, G, H respectively[32]. Let us consider *first* the case where p is of the form $4n + 1$ and q of the form $8n + 7$. Clearly we will have $2 \, R \, E$, $2 \, R \, H$ and therefore $p \, R \, E$, $p \, R \, H$, and finally $E \, R \, p$, $H \, R \, p$. Further, 2 will be a nonresidue of any factor of the form $8n + 3$ or $8n + 5$ and so also p; thus such a factor will be a nonresidue of p; and in conclusion FG will be a residue of p if $f + g$ is even, a nonresidue if $f + g$ is odd. But $f + g$ cannot be odd; for regardless of what e, f, g, h are individually, if $f + g$ is odd $EFGH$ or q will be of the form $8n + 3$ or $8n + 5$ in every case. And this is contrary to the hypothesis. We get therefore $F \, G \, R \, p$, $EFGH \, R \, p$, or $q \, R \, p$, but since $aq \, R \, p$ this implies that $a \, R \, p$, contrary to the hypothesis. *Second*, when p is of the form $4n + 3$ we can show in a similar way that $p \, R \, E$ and so $E \, R \, p$, $-p \, R \, F$ and consequently $F \, R \, p$; since $g + h$ is even we get $GH \, R \, p$ and it follows finally that $q \, R \, p$, $a \, R \, p$, contrary to the hypothesis.

II. When e is divisible by p, a demonstration can be developed in a similar way. The skillful mathematician (for whom this article is written) will be able to accomplish this with no difficulty. For brevity's sake we shall omit it.

▶ 146. By the fundamental theorem and the propositions pertaining to the residues -1 and ± 2 it can always be determined whether a given number is a residue or a nonresidue of any other given number. But it will be very useful to restate the conclusions arrived at above in order to bring together all that is necessary for the solution.

PROBLEM. *Given any two numbers P, Q, to discover whether Q is a residue or a nonresidue of P.*

Solution. I. Let $P = a^\alpha b^\beta c^\gamma$ etc. where a, b, c, etc. are unequal prime numbers taken positively (for obviously we must consider P absolutely). For brevity in this

---

[32]If there are no factors from any one of these classes, the number 1 should be written in place of the product.

article we shall speak simply of a *relation* of two numbers x, y meaning that the former x is a residue or nonresidue of the latter y. Thus the relation of Q, P depends on the relations of Q, a^α; Q, b^β; etc. (art. 105).

II. In order to determine the relation of Q, a^α (and of Q, b^β, etc. as well) it is necessary to distinguish two cases.

1. When Q is divisible by a. Set $Q = Q'a^e$ where Q' is not divisible by a. Then if $e = \alpha$ or $e > \alpha$ we have $Q \, R \, a^\alpha$; but if $e < \alpha$ and odd, we get $Q \, N \, a^\alpha$; and if $e < \alpha$ and even, Q will have the same relation to a^α as Q' has to $a^{\alpha-e}$. This reduces to case.

2. When Q is not divisible by a. Two cases must be again distinguished.

(A) When $a = 2$. Then always when $\alpha = 1$ we have $Q \, R \, a^\alpha$; when $\alpha = 2$ it is required that Q be of the form $4n + 1$; and when $\alpha = 3$ or > 3, Q will have to be of the form $8n + 1$. If this condition holds $Q \, R \, a^\alpha$.

(B) When a is any other prime number. Then Q will have the same relation to a^α as it has to a (see art. 101).

III. We will investigate the relation of any number Q to a prime number a (odd) as follows. When $Q > a$ substitute in place of Q its least positive residue relative to the modulus a.[33] This will have the same relation to a that Q has.

Resolve Q, or the number taken in its place, into its prime factors p, p', p'', etc. adjoining the factor -1 when Q is negative. Then clearly the relation of Q to a depends on the relations of the single numbers p, p', p'', etc. to a. That is, if among those factors there are $2m$ nonresidues of a we will have $Q \, R \, a$; if there are $2m + 1$ of them, we will have $Q \, N \, a$. For it is easy to see that if among the factors p, p', p'', etc. a pair or four or six or in general $2k$ are equal, these can be safely disregarded.

IV. If -1 and 2 appear among the factors p, p', p'', etc. their relation to a can be determined from articles 108, 112, 113, 114. The relation of the remaining factors to a depends on that of a to them (fundamental theorem and the propositions of article 131). Let p be one of them and we will find (by treating a, p as we treated Q and a which are respectively greater) that the relation between a and p can be determined by articles 108–114 [provided the least residue of a (mod. p) contains no odd prime factors] or that the relation depends on that of p to prime numbers that are less than p. The same holds for the other factors p', p'', etc. By continuing this operation we will come finally to numbers whose relations can be determined by the propositions of articles 108–114. This can be seen more clearly by example.

[33] *Residue* in accordance with the meaning in article 4. It will be especially useful to take the *absolutely* least residue.

Example. We want to find the relation of the number $+453$ to 1236: $1236 = 4 \cdot 3 \cdot 103$; $+453\,R\,4$ by II.2(A); $+453\,R\,3$ by II.1. Now to explore the relation of $+453$ to 103. It will be the same as that of $+41$ ($\equiv 453$, mod. 103) to 103; or the same as that of $+103$ to 41 (fundamental theorem) or of -20 to 41. But $-20\,R\,41$; for $-20 = -1 \cdot 2 \cdot 2 \cdot 5$; $-1\,R\,41$ (art. 108); and $+5\,R\,41$ and so $41 \equiv 1$ and is thus a residue of 5 (fundamental theorem). From this it follows that $+453\,R\,103$, and then $+453\,R\,1236$. It is true that $453 \equiv 297^2$ (mod. 1236).

▶ 147. If we are given a number A, certain *formulae* can be shown that contain all numbers relatively prime to A of which A is a residue, or all numbers that can be *divisors* of numbers of the form $x^2 - A$ (where x^2 is an undetermined square).[34] For brevity we will consider only these divisors which are odd and relatively prime to A, since all others can be easily reduced to these cases.

First let A be either a positive prime number of the form $4n + 1$ or negative of the form $4n - 1$. Then according to the fundamental theorem all prime numbers which taken positively are residues of A, will be divisors of $x^2 - A$; and all prime numbers (except 2 which is always a divisor) which are nonresidues of A, will be nondivisors of $x^2 - A$. Let all residues of A (excluding zero) which are less than A be denoted by r, r', r'', etc., all nonresidues by n, n', n'', etc. Then any prime number contained in one of the forms $Ak + r$, $Ak + r'$, $Ak + r''$, etc. will be a divisor of $x^2 - A$; but any prime contained in one of the forms $Ak + n$, $Ak + n'$, etc. will be a nondivisor. In these formulae k is an indeterminate integer. We will call the first set *forms of the divisors of $x^2 - A$*, the second set *forms of the nondivisors*. The number of members in each set will be $(A - 1)/2$. Further if C is an odd composite number and $A\,R\,B$, all the prime factors of B will be contained in one of the above forms and so also B. Therefore *any* odd number contained in a form of the nondivisors will be a nondivisor of the form $x^2 - A$. But this theorem cannot be inverted; for if B is an odd composite nondivisor of the form $x^2 - A$, some of the prime factors of B will be nondivisors and there will be an *even* number of them. But B itself will be found in a form of a divisor (see art. 99).

Example. For $A = -11$ the forms of the divisors of $x^2 + 11$ will be: $11k + 1$, 3, 4, 5, 9; the forms of the nondivisors will be $11k + 2$, 6, 7, 8, 10. Thus -11 will be a nonresidue of all odd numbers which are contained in the latter forms, a residue of all primes belonging to the former forms.

We will have similar forms for divisors and nondivisors of $x^2 - A$ no matter what number A is. Obviously we should consider only those values of A that are not

---

[34] We will call these numbers simply *divisors of $x^2 - A$*. It is obvious what we mean by *nondivisors*.

divisible by some square; for if $A = a^2A'$, all divisors[35] of $x^2 - A$ will also be divisors of $x^2 - A'$, and so for the nondivisors. But we must distinguish three cases: (1) when A is of the form $+(4n + 1)$ or $-(4n - 1)$; (2) when A is of the form $-(4n + 1)$ or $+(4n - 1)$; (3) when A is of the form $\pm(4n + 2)$, i.e. even.

▶ 148. The *first case*, when A is of the form $+(4n + 1)$ or $-(4n - 1)$. Resolve A into its prime factors, and to those of the form $4n + 1$ ascribe a positive sign, and to those of the form $4n - 1$ a negative sign (the product of all these will $=A$). Let these factors be a, b, c, d, etc. Now distribute all numbers which are less than A and relatively prime to A into two classes. In the first class will be all numbers that are nonresidues of none of the numbers a, b, c, d, etc., or of two of them, or of four of them, or in general of an even number of them; in the second class will be those numbers that are nonresidues of one of the numbers a, b, c, etc., or of three of them, or in general of an odd number of them. Designate the former by r, r', r'', etc., the latter by n, n', n'', etc. Then the forms $Ak + r$, $Ak + r'$, $Ak + r''$, etc. will be forms of the divisors of $x^2 - A$, the forms $Ak + n$, $Ak + n'$, etc. will be forms of the nondivisors of $x^2 - A$ (i.e. *every prime number except 2 will be a divisor or a nondivisor of $x^2 - A$ according as it is contained in one of the former or latter forms*). For if p is a positive prime number and a residue or nonresidue of one of the numbers a, b, c, etc., this number will be a residue or a nonresidue of p (fundamental theorem). So if among the numbers a, b, c, etc. there are m numbers of which p is a nonresidue, the same number will be nonresidues of p. Therefore if p is contained in one of the former forms, m will be even and $A \, R \, p$. But if it is contained in one of the latter forms, m will be odd and $A \, N \, p$.

Example. Let $A = +105 = (-3)(+5)(-7)$. Then the numbers r, r', r'', etc. will be these: 1, 4, 16, 46, 64, 79 (nonresidues of none of the numbers 3, 5, 7); 2, 8, 23, 32, 53, 92 (nonresidues of the numbers 3, 5); 26, 41, 59, 89, 101, 104 (nonresidues of the numbers 3, 7); 13, 52, 73, 82, 97, 103 (nonresidues of the numbers 5, 7). The numbers n, n', n'', etc. will be these: 11, 29, 44, 71, 74, 86; 22, 37, 43, 58, 67, 88; 19, 31, 34, 61, 76, 94; 17, 38, 47, 62, 68, 83. The first six are nonresidues of 3, the next six nonresidues of 5, then follow the nonresidues of 7, and finally those that are nonresidues of all three at the same time.

From the theory of combinations and articles 32 and 96 we can easily see that the number of numbers r, r', r'', etc. will

$$= t\left(1 + \frac{l(l - 1)}{1 \cdot 2} + \frac{l(l - 1)(l - 2)(l - 3)}{1 \cdot 2 \cdot 3 \cdot 4} + \cdots\right)$$

---

[35] That is, those relatively prime to A.

and of the numbers n, n', n'', etc.

$$= t\left(l + \frac{l(l-1)(l-2)}{1 \cdot 2 \cdot 3} + \frac{l(l-1)\ldots(l-4)}{1 \cdot 2 \ldots 5} + \ldots\right)$$

where l determines the number of numbers a, b, c, etc.;

$$t = 2^{-l}(a-1)(b-1)(c-1) \text{ etc.}$$

and each series is to be continued until it breaks off. (There will be t numbers which are residues of a, b, c, etc., $tl(l-1)/(1 \cdot 2)$ which are nonresidues of two of these etc., but brevity does not permit a fuller explanation.) The sum[36] of each of the series $= 2^{l-1}$. That is, the former is derived from

$$1 + (l-1) + \frac{(l-1)(l-2)}{1 \cdot 2} + \ldots$$

by adding the second and third term, the fourth and fifth etc., and the latter is derived from the same equation by adding the first and second, third and fourth, etc. There will therefore be as many forms for the divisors of $x^2 - A$ as there are forms for the nondivisors, namely $(a-1)(b-1)(c-1)$ etc./2.

▶ 149. We can consider the *second* and *third cases* together. We can express A as $=(-1)Q$, or $=(+2)Q$, or $=(-2)Q$ where Q is a number of the form $+(4n+1)$ or $-(4n-1)$, such as those we considered in the preceding article. In general let $A = \alpha Q$ where $\alpha = -1$ or ± 2. Then A will be a residue of all numbers for which both or neither α and Q are residues; and a nonresidue of all numbers for which only one of the numbers α, Q is a nonresidue. From this the forms of the divisors and nondivisors of $x^2 - A$ can be easily derived. If $\alpha = -1$ distribute all numbers less than $4A$ and relatively prime to it into two classes, putting into the first those numbers which are at the same time in some form of the divisors of $x^2 - Q$ and in the form $4n + 1$, and also those numbers which are at the same time in some form of the nondivisors of $x^2 - Q$ and in the form $4n + 3$; all other numbers will be put in the second class. Let the members of the former class be r, r', r'', etc. and those of the latter class, n, n', n'', etc. A will be a residue of all prime numbers contained in any of the forms $4Ak + r$, $4Ak + r'$, $4Ak + r''$, etc.; a nonresidue of all primes contained in any of the forms $4Ak + n$, $4Ak + n'$, etc.

If $\alpha = \pm 2$ distribute all numbers less than $8Q$ and relatively prime to it into two classes, putting into the first class those numbers which are at the same time contained in some form of the divisors of $x^2 - Q$ and in one of the forms $8n + 1$, $8n + 7$ when the upper sign holds, or in one of the forms $8n + 1$, $8n + 3$ when the lower sign holds; and also those numbers contained in some form of the nondivisors

---

[36]Disregarding the factor t.

of $x^2 - Q$ and at the same time in one of the forms $8n + 3$, $8n + 5$ when the upper sign holds, or in one of the forms $8n + 5$, $8n + 7$ when the lower sign holds. All other numbers will be put into the second class. Then if we designate numbers in the former class as r, r', r'', etc. and numbers of the latter class as n, n', n'', etc., $\pm 2Q$ will be a residue of all prime numbers contained in any of the forms $8Qk + r$, $8Qk + r'$, etc., and a nonresidue of all primes in any of the forms $8Qk + n$, $8Qk + n'$, $8Qk + n''$, etc. And it is easy to show here too that there are as many forms of the divisors of $x^2 - A$ as there are nondivisors.

Example. By this method we find that $+10$ is a residue of all prime numbers in any of the forms $40k + 1, 3, 9, 13, 27, 31, 37, 39$, a nonresidue of all primes in any of the forms $40k + 7, 11, 17, 19, 21, 23, 29, 33$.

▶ 150. These forms have many remarkable properties, but we indicate only one of them. If B is a composite number relatively prime to A, and if among its prime factors there are $2m$ of them contained in some form of the nondivisors of $x^2 - A$, B will be contained in some form of the divisors of $x^2 - A$; and if the number of prime factors of B contained in some form of the nondivisors of $x^2 - A$ is odd, B also will be contained in a form of the nondivisors. We omit the demonstration, which is not difficult. From all of this it follows that not only any prime number but also any odd composite number relatively prime to A contained in some form of the nondivisors will be itself a nondivisor; for some prime factor of such a number is necessarily a nondivisor.

▶ 151. The fundamental theorem must certainly be regarded as one of the most elegant of its type. No one has thus far presented it in as simple a form as we have done above. What is more surprising is that Euler already knew other propositions which depend on it and which should have led to its discovery. He was aware that certain forms existed which contain all the prime divisors of numbers of the form $x^2 - A$, that there were others containing all prime nondivisors of numbers of the same form, and that the two sets were mutually exclusive. And he knew further the method of finding these forms, but all his attempts at demonstration were in vain, and he succeeded only in giving a greater degree of verisimilitude to the truth that he had discovered by induction. In a memoir entitled "*Novae demonstrationes circa divisores numerorum formae xx + nyy*" which he read in the St. Petersburg Academy (Nov. 20, 1775) and which was published after his death,[37] he seems to believe that he had fulfilled his resolve. But an error did creep in, for he *tacitly* presupposed the existence of such forms of the divisors[38] and nondivisors and from this it was not difficult to discover *what* their *form* should be. But the method he used to prove

---

[37] *Nova acta acad. Petrop.*, 1 [1783], 1787.
[38] Namely that there do exist numbers r, r', r'', etc., n, n', n'', etc., all divisors and $<$ for A such that all prime divisors of $x^2 - A$ are contained in one of the forms $4Ak + r$, $4Ak + r'$ etc. (k being indescriminate), and all prime nondivisors in one of the forms $4Ak + n$, $4Ak + n'$, etc. (k being indeterminate).

this supposition does not seem to be suitable. In another paper, "De criteriis aequationis $fxx + gyy = hzz$ utrumque resolutionem admittat necne,"[39] *Opuscula Analytica, 1* (f, g, h are given, x, y, z are indeterminate) he finds by induction that if the equation is solvable for one value of $h = s$, it will also be solvable for any other value congruent to s relative to the modulus $4fg$ provided it is a prime number. From this proposition the supposition we spoke of can easily be demonstrated. But the demonstration of this theorem also eluded his efforts.[40] This is not remarkable because in our judgment it is necessary to start from the fundamental theorem. The truth of the proposition will flow automatically from what we will show in the following section.

After Euler, the renowned Legendre worked zealously on the same problem in his excellent tract, *"Recherches d'analyse indéterminée," Hist. Acad. Paris*, 1785. He arrived at a theorem basically the same as the fundamental theorem. He showed that if p, q are two positive prime numbers, the absolutely least residues of the powers $p^{(q-1)/2}$, $q^{(p-1)/2}$ relative to the moduli q, p respectively, are either both $+1$ or both -1 when either p or q is of the form $4n + 1$; but when both p and q are of the form $4n + 3$ one least residue will be $+1$, the other -1. From this, according to article 106, we derive the fact that the *relation* (taken according to the meaning in article 146) of p to q and of q to p will be the *same* when either p or q is of the form $4n + 1$, *opposite* when both p and q are of the form $4n + 3$. This proposition is contained among the propositions of article 131 and follows also from 1, 3, 9 of article 133; on the other hand the fundamental theorem can be derived from it. Legendre also attempted a demonstration and, since it is extremely ingenious, we will speak of it at some length in the following section. But since he presupposed many things without demonstration (as he himself confesses: *"Nous avons supposé seulement . . ."*), some of which have not been demonstrated by anyone up till now, and some of which cannot be demonstrated in our judgment without the help of the fundamental theorem itself, the road he has entered upon seems to lead to an impasse, and so our demonstration must be regarded as the first. Below we shall give *two other demonstrations* of this most important theorem, which are totally different from the preceding and from each other.

▶ 152. Thus far we have treated the pure congruence $x^2 \equiv A$ (mod. m) and we have learned to recognize whether or not it is solvable. By article 105 the investigation of the *roots themselves* was reduced to the case where m is either a prime or a power of a prime, and afterward by applying article 101 to the case where m is prime. For

---

[39] The original article reads "utrum ea" instead of "utrumque."

[40] As he himself confesses (*Opuscula Analytica, 1*): "A demonstration of this most elegant theorem is still sought even though it has been investigated in vain for so long and by so many. . . . And anyone who succeeds in finding such a demonstration must certainly be considered most outstanding." With what ardor this great man searched for a proof of this theorem and of others which are only special cases of the fundamental theorem can be seen in many other places, e.g. *Opuscula Analytica, 1* (Additamentum ad Diss. 8) and *2* (Diss. 13) and in many dissertations in *Comm. acad. Petrop.* which we have praised from time to time. [Dissertation 13 is entitled "De insigni promotione scientiae numerorum" *Ed. Note.*]

this case what we said in article 61 along with what we will show in sections V and VIII[41] embraces almost all that can be derived by direct methods. But in the cases where they are applicable they are infinitely more prolix than the indirect methods which we will show in section VI and so they are memorable not for their usefulness in practice but for their beauty. *Congruences of the second degree that are not pure can be reduced to pure congruences easily.* Suppose we are given the congruence

$$ax^2 + bx + c \equiv 0$$

which is to be solved relative to the modulus m. The following congruence is equivalent:

$$4a^2x^2 + 4abx + 4ac \equiv 0 \text{ (mod. } 4am)$$

i.e. any number that satisfies one will satisfy the other. The second congruence can be put in the form

$$(2ax + b)^2 \equiv b^2 - 4ac \text{ (mod. } 4am)$$

and from this all the values of $2ax + b$ less than $4m$ can be found, if any exists. If we designate them by r, r', r'', etc. all solutions of the proposed congruence can be deduced from the solutions of the congruences

$$2ax \equiv r - b, \qquad 2ax \equiv r' - b, \text{ etc. (mod. } 4am)$$

which we showed how to find in section II. But we observe that the solution can be shortened by various artifices; e.g. in place of the given congruence another can be found

$$a'x^2 + 2b'x + c' \equiv 0$$

which is equivalent to it and in which a' divides m. Brevity will not allow us to take up these considerations here, but see the last section.

---

[41]Section VIII was not published.

Augustin-Louis Cauchy

(1789–1857)

HIS LIFE AND WORK

E uclid earned an income teaching. Among his students was Ptolemy I, King of Egypt. King Ptolemy once asked his teacher for a shortcut to geometric knowledge, to which Euclid replied, "Sire, there is no royal road to geometry." Perhaps there is no royal road to mathematics in general and geometry particular. However, if any mathematician rode with the royals, it was Augustin-Louis Cauchy.

Augustin-Louis Cauchy was born during the infancy of the French Revolution, on August 21, 1789. His parents, Louis-François Cauchy and Marie-Madeleine Cauchy, neé Desestre, named him for the month of his birth and his father. Louis-François had been born to a Rouen metalsmith in 1760. Marie-Madeleine had been born to a family of Paris bureaucrats in 1767. They married in 1787.

Louis-François Cauchy had trained as a lawyer and was a commissioner of the Paris police under the royal regime. He lost that position as a result of the storming of the Bastille and had to take a position with the Bureau of Charities in order to make any kind of a living. As the Reign of Terror began to descend on Paris in 1792, Louis-François feared that his royalist ties might cost him his neck, and he fled to a country house in Arcueil with his family that by now included a second son.

This would be the first of Augustin-Louis Cauchy's flights from political upheaval.

The Cauchys' exile in Arcueil may have been hard at times but it had its benefits. Principal among them may have been the fact that their neighbors in Arcueil included the great mathematicians Pierre Simon Laplace and Joseph Louis Lagrange, the latter of whom is said to have predicted the young boy's scientific genius and warned his father about letting him see a mathematics text before the age of seventeen. Whether it was due to his own education in the classics or Lagrange's warning, Louis-François used his ample free time in Arcueil to begin the education of his eldest son, giving him a sound foundation in the study of Greek and Latin. Louis-François continued his teaching of his son when the family was able to move back to Paris in July 1794 after the fall of Robespierre and the end of the Reign of Terror.

Louis-François's fortunes rose with those of Napoleon, and when Napoleon became First Consul on January 1, 1800, Louis-François was elected Secretary-General of the newly created Senate whose membership included Laplace and Lagrange. At Lagrange's suggestion, Louis-François enrolled his son at the École Centrale du Panthéon just after his thirteenth birthday. At the École, young Cauchy spent two years studying ancient languages, drawing, and natural history. He excelled at ancient languages earning first-place prizes in Latin composition and Greek poetry competitions. Even though he had developed a distinguished record studying classical languages, Augustin-Louis Cauchy had his heart set on studying engineering at the École Polytechnique. Cauchy ranked second among all of the applicants to the École Polytechnique in 1805. When he enrolled in the fall, he had to choose the field of public service he would enter once he graduated. He chose civil engineering with a specific interest in highways and bridges. Cauchy soon rose to the top of his class, graduated with honors in 1807, and then enrolled for advanced study at the École des Ponts et Chausées (School of Highways and Bridges). At this school, students were in the classroom from December through March and spent the rest of the year on fieldwork.

In 1810, after once again graduating at the top of his class, Cauchy became a field engineer for the Department of Highways and Bridges in Cherbourg to work on a naval base from which Napoleon planned to invade England. In a letter to his father, Cauchy wrote that he had just four books in his possession when he left for Cherbourg: a collection of Virgil's poetry, Thomas à Kempis's *Imitation of Christ*, Laplace's *Mécanique céleste*, and Lagrange's *Traité des fonctiones analytiques*. We know little of Cauchy's time in Cherbourg other than the fact that he was introduced to Napoleon when the Emperor visited Cherbourg in May 1811.

In 1812, Cauchy returned to Paris to seek an academic appointment that would allow him the income and time to pursue his mathematical pursuits. He failed in this attempt in large part due to a protracted illness and had to be satisfied with an appointment as a member of the technical staff building the Oureq Canal. When his former mentor Lagrange died in 1813, Cauchy sought election as Lagrange's replacement on the mathematics faculty at the Insitut de France. Not only did he fail to win election to the post, but he was also eliminated on the first ballot. For the next two years, Cauchy sought every available position at the Insitut but never succeeded. Finally, at the very end of 1814, he received a bit of consolation with election to the Philomathic Society of Paris. Three months later, just as Cauchy was preparing to begin his teaching, France would once again be torn asunder following Napoleon's return to Paris from his exile in Elba.

Following the restoration of the Bourbon monarchy in 1814, many of France's leading institutions were purged of radical supporters of Napoleon. Cauchy tried and failed to win the chair in mechanics that had been held by Siméon-Denis Poisson. He finally won an appointment to a teaching post at the École Polytechnique, first as the "substitute" professor of analysis for the mathematician Louis Poinsot whose ill health had rendered him unable to teach for three years, and then a full appointment as professor of analysis and mechanics in his own right.

Having finally obtained a permanent position, Cauchy's parents urged him to marry. In fact, they did more than urge him to marry, his father chose the bride, Aloïse de Bure, daughter of the owner of a publishing company bearing the family's name. She came with a considerable dowry, and they had a wedding ceremony to suit their social position. Louis XVIII and the entire royal family signed the marriage contract.

All the evidence makes it seem as though Cauchy treated his wife and daughters as little more than adornments that could be left behind when necessary. He probably married just to satisfy the bourgeois demands of his parents. He and Aloïse had two daughters, Marie Françoise Alicia, born in 1819, and Marie Mathilde, born in 1823. They would each marry into the nobility.

Following his appointment, Cauchy set about to reorganize the mathematics curriculum at the École Polytechnique giving much more emphasis to pure mathematics. He initially focused on the calculus, the subject matter of the excerpts in this chapter. Newton and Leibniz had invented the calculus to solve particular mathematical problems. Newton, for example, needed to be able to solve problems in celestial dynamics involving ellipses and parabolas. As calculus evolved in the eighteenth century, it did so on uncertain foundations. The derivative of a function was considered to be a means to express the formula for the tangent to a curve at a particular point P.

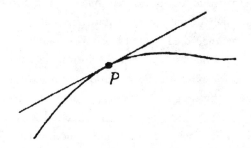

The integral of a function was considered to be an infinite sum of the values of an infinite number of infinitesimally thin rectangles whose heights were the value of the function and whose widths were the infinitesimal quantity expressed as **dx**. Indeed, Leibniz introduced the integral sign in

$$\int_a^b f(x)\, dx,$$

precisely because of its similarity to the letter S, the first letter of "summation." So long as particular, well-behaved functions were considered, for example, $f(x) = x^2$, the problem of determining the integral of a function was reduced to finding the limit of a particular series. For the function $f(x) = x^2$, the integral is considered to be the sum of the rectangles of height x^2 and width dx.

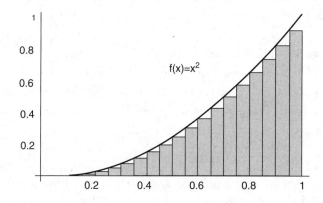

Consequently, the problem of finding the integral of the function $f(x) = x^2$ between 0 and 1 is nothing more than the problem of finding the limit of

$$\frac{(1/N)^2 + (2/N)^2 + (3/N)^2 + \ldots ((N-1)/N)^2}{N} \quad \text{as } N \to \infty,$$

which is

$$\frac{2N^3 - 3N^2 + N}{6N^3} \qquad \text{as } N \to \infty$$

$$= 1/3.$$

Early eighteenth-century mathematicians soon discovered that the integral and the derivative were inverses of each other. By the middle of the eighteenth century, Leonhard Euler and Johann Bernoulli considered the integral to be nothing more than the inverse of the derivative. In fact, Euler used Leibniz's concept of the integral as a summation only when making an approximate evaluation of an integral. Finding the formula for the integral of a function $f(x)$ entailed searching for another function $g(x)$ whose derivative was everywhere equal to $f(x)$. Thus, the formula for the integral of the function $f(x) = x^2$ was $g(x) = x^3/3$ simply because $f(x) = x^2$ is the derivative of $g(x) = x^3/3$.

These were the foundations on which the calculus stood when Cauchy began to teach at the École Polytechnique in 1818. There were methods of integrating well-behaved functions, but there was no theory of the integral. Perhaps spurred on by Fourier's demonstration that an *arbitrary* function could be represented by an infinite trigonometric series, Cauchy developed the first *theory of the integral*, independent of the particular function and independent of the differential calculus for which he also laid new foundations, independent of geometric intuitions. In the introduction to his *Cours d'Analyse* of 1821, Cauchy explained that he sought to bring the same degree of rigor found in Euclid's *Elements* to the field of mathematical analysis. He was particularly concerned to banish infinitesimals. A new definition of continuity was required. Cauchy gave the following as his definition of the continuity of a real function of one variable:

> *Let $f(x)$ be a function of the variable x and suppose that, for each value of x between two given bounds, this function constantly takes one finite value. If, from the value of x between those bounds, one attributes to the variable x an infinitely small increment α, α the function itself will receive as an increment the difference*
>
> $$f(x + \alpha) - f(x),$$
>
> *which will depend on the same time on the new variable and on the value of x. This being granted, the function $f(x)$ will be a continuous function of the variable x between the two assigned bounds if, for each value of x between those bounds, the numerical value of the difference $f(x + \alpha) - f(x)$ decreases indefinitely with α. In other words, the function $f(x)$ remains continuous with respect to x between the given bounds, if, between these bounds, an infinitely small increment in the variable always produces an infinitely small increment in the function itself.*

Cauchy devoted the first part of his *Résumé des Leçons sur le Calcul Infinitésimal* of 1823 to differential calculus focusing on the derivative of a function $y = f(x)$. He took the term derivative and the notation $f'(x)$ from his mentor Lagrange. However, Lagrange had continued to speak of the derivative in general in terms a tangent to a curve, finding formulae for particular derivatives as necessary. Cauchy went far beyond Lagrange and defined the derivative of f at x as the limit of the difference quotient

$$\Delta y/\Delta x = (f(x + i) - f(x)/i),$$

as i "tends" to 0, our modern, nongeometric definition of the derivative.

The second part of his *Calcul infinitésimal* focuses on the concept of the integral. Cauchy defined the integral

$$\int_a^b f(x)\,dx$$

of a function $f(x)$ continuous between a and b to be the limit of the sums

$$(x_1 - a)f(x_1) + (x_2 - x_1)f(x_2) + \ldots + (x_n - x_{n-1})f(x_n) + (b - x_n)f(b)$$

or

$$(x_1 - a)f(a) + (x_2 - x_1)f(x_1) + \ldots + (x_n - x_{n-1})f(x_{n-1}) + (b - x_n)f(x_n),$$

when the interval $[a, b]$ is divided at n points by $x_1, x_2, \ldots x_{n-1}, x_n$, such that $a < x_1 < x_2 < \ldots < x_{n-1} < x_n < b$.

Cauchy called the set of points $x_1, x_2, \ldots x_{n-1}, x_n$ a partition of the interval (a, b). He called the largest value of $(x_i - x_{i-1})$ the norm of the partition and called

$$S = (x_1 - a)f(a) + (x_2 - x_1)f(x_1) + \ldots + (x_n - x_{n-1})f(x_{n-1}) + (b - x_n)f(x_n)$$

the Cauchy sum of the partition.

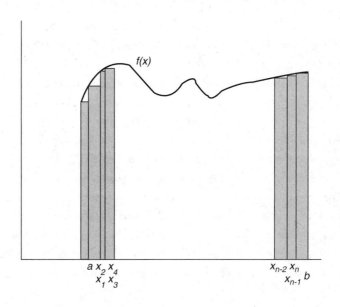

Using his definition of continuity, Cauchy showed that for any two partitions P and P' the corresponding sums S and S' could be made arbitrarily close to each other provided that the norms of P and P' are sufficiently small.

By defining a theory of the integral, Cauchy was then able to demonstrate the intimate relationship between the integral and the derivative as a theorem that has come to be known as the Fundamental Theorem of Calculus. It states that if f is an integrable function in the interval $[a, b]$ and if

$$G(x) = \text{integral of } f(t) \text{ from a to } x\,(\leq b),$$

then the derivative of $G(x)$ is identical to $f(x)$, that is, $G'(x) = f(x)$.

Although Cauchy was able to inspire his most gifted students, he never learned to adapt his courses to suit the level of the average student who came to the École seeking to augment their rudimentary knowledge of mathematics to suit the needs of, say, an engineering career. Because Cauchy's lectures were too ambitious to be presented in the allotted time, they often ran over. In April 1821, just before the end of a lecture that had run into overtime, several students booed and hissed Cauchy and then walked out on him.

Far from pleasing the faculty councils at the École, Cauchy's *Calcul Infinitésimal* incurred their wrath for being too theoretical and not sufficiently practical. In fact, in late 1823, the Minister of the Interior appointed a commission including Laplace and Poisson to ensure that the instruction in mathematics was attuned to the needs of engineering students. For the rest of the decade the École's administration continually monitored Cauchy's lectures to guarantee their suitability for engineering students.

France experienced a period of relative peace and stability during the reign of the restored monarch Louis XVIII. When Louis XVIII died in 1824, he was succeeded by his younger brother Charles X who attempted to restore the monarchy to the power it had prior to the revolution. As pockets of republican sympathy arose throughout France, the government sought to restrain them and enlisted the aid of organizations supporting the monarchy. An ardent royalist and Catholic, Cauchy was one of the founders of one of them, the Association for the Protection of the Catholic Religion.

After the July revolution of 1830 overthrew Charles X in favor of his cousin Louis-Phillipe, Duke of Orleans, academics, like all government officials, were required to take an oath of allegiance. Cauchy not only refused to do so, he went into exile of his own volition, perhaps concerned that the new regime would persecute devout Catholics like himself. Not surprisingly, Cauchy lost his appointments.

Leaving his wife and children in France, Cauchy first went to live with a group of Jesuits in the town of Fribourg, just over the Swiss border. The Jesuits soon recommended him to the King of Sardinia, one of their benefactors, and he offered Cauchy a chair at the University of Turin. However, Cauchy stayed there for little more than three years when he received an offer to join Charles X in Prague and teach the exiled King's grandson, the crown prince. Only then did Cauchy send for his family to join him after a four-year separation.

Charles X rewarded Cauchy by making him a Baron. By 1838, Charles X had died and the Crown Prince's education completed on his eighteenth birthday. With his work for the Court done, Cauchy returned to Paris, no doubt impelled by his mother's declining health. Having lost his academic positions after going into exile, the only position Cauchy could resume was his membership in the Académie des Sciences. Fortunately for him, his old teacher Prony died within a year of his return and Cauchy was elected to succeed his old mentor as Professor of Geometry at the Bureau of Longitudes.

Political conditions had eased and academics no longer had to swear an oath of allegiance. Although Cauchy's coolness to the new regime may not have jeopardized his life, it did hinder his career, costing Cauchy comfortable appointments at several government-sponsored institutions. In 1843, he was one of three candidates for a chair of mathematics that had become vacant at the Collège de France. Although clearly the most eminent of candidates, Cauchy's ultramonarchist and Jesuit sympathies swayed many of the electors away from him. In the end, he received only three votes from the twenty-four-member committee. Feeling humiliated by the result, Cauchy announced that he would no longer allow himself to be nominated for any position if there were even the slightest chance of his defeat.

He had no teaching post from 1838 to 1848. He only had his appointment to the Académie des Sciences to provide him a forum for presenting his scientific work and that he did in earnest, presenting 240 notes and studies to the Académie during this ten-year interval. When the Revolution of February 1848 overthrew Louis-Philippe, Cauchy hoped that his former pupil, the Duke of Bordeaux, now Count of Chambord, would take the thrown. That would not be the case. The Second Republic replaced Louis-Philippe. Although hardly a Republican, Cauchy benefited from the Republic's abolition of loyalty oaths. He soon won election to a chair at the Université de Paris. When Napoleon III became emperor in 1852, he exempted Cauchy from taking the oath.

Cauchy worked at a furious pace his entire life, publishing 789 mathematical and scientific articles. Publication of his *Oeuvres complètes* began in 1882 and only ended in 1970 with the publication of the twenty-seventh and final volume. His prodigious output only came to an end with his death on May 23, 1857, following a short bout with rheumatism.

SELECTION FROM *OEUVRES COMPLÈTES D'AUGUSTIN CAUCHY* DIFFERENTIAL CALCULUS*

THIRD LECTURE[1]

DERIVATIVES OF FUNCTIONS OF ONLY ONE VARIABLE

Whenever the function $y = f(x)$ remains continuous between two given limits of the variable x and a value contained between two given limits is assigned to this variable, an infinitely small increase made to the variable produces an infinitely small increase in the function itself. Consequently, if we set $\Delta x = i$ the two terms of the *ratio of the differences*,[2]

$$(1) \qquad \frac{\Delta y}{\Delta x} = \frac{f(x + i) - f(x)}{i},$$

will be infinitely small quantities. But, while the two terms, viz., the differences, simultaneously approach the limit zero, the ratio itself will be able to converge to another limit, be it positive or negative. This limit, if it exists, has a determinate value for each particular value of x; but it varies with x. For example, if we set $f(x) = x^m$, m designating a whole number, the ratio between the infinitely small differences will be[3]

$$\frac{(x + i)^m - x^m}{i} = mx^{m-1} + \frac{m(m - 1)}{1 \cdot 2}x^{m-2}i + \ldots + i^{m-1},$$

and, at its limit,[4] this ratio will be the quantity mx^{m-1}, i.e., a new function of the variable x. This will hold in general; only the form of the new function, which serves as the limit of the ratio $\frac{f(x + i) - f(x)}{i}$, will depend on the form of the proposed function $y = f(x)$. To indicate this dependence, we give the name *derivative*[5] to the new function and we designate this, with the aid of an accent, by the notation

$$y' \text{ or } f'(x).$$

In the study of the derivatives of functions of only one variable x, it is useful to distinguish the functions that are called *simple functions*, which are considered to be the result of only one operation carried out on this variable, from the functions that are construed by the aid of many operations, which are called *compound functions*.

*Translated by John Anders.
[1] *Leçon*. I chose "lecture" as opposed to "lesson" because it seems to me that calling these sections "lessons" might imply that Cauchy viewed these works as introductions into already well-known material. And despite the fact that the basic concepts of derivative and integral were well known before Cauchy and that Cauchy's organization and general style are pedagogical in many ways, Cauchy sees his approach to this already explored material as novel in that he thinks he has rigorously established this subject in a way that had not been done before.
[2] The "differences" are Δy and Δx.
[3] Cauchy arrives at this equation by expanding the binomial ("Pascal's expansion") then subtracting x^m and dividing by i.
[4] As the change in x, or i, approaches zero. In general, when Cauchy speaks of a limit he means a limit as a certain quantity or certain quantities tend to zero.
[5] *fonction dérivée*.

The simple functions that produce the operations of algebra and trigonometry (see Part 1 of *Cours d'Analyse*, Chap. 1) can be reduced to the following:

$$a + x, \quad a - x, \quad ax, \quad \frac{a}{x}, \quad x^a, \quad A^x, \quad Lx,$$

$$\sin x, \quad \cos x, \quad \arcsin x, \quad \arccos x,$$

A designates a constant number, $a = \pm A$ is a constant quantity,[6] and the letter L indicates a logarithm in the system of base A. In general, if we put a simple function in for y, it will be easy to obtain the derivative function y'. We find, for example, in the case of

$y = a + x$,

$$\frac{\Delta y}{\Delta x} = \frac{(a + x + i) - (a + x)}{i} = 1, \qquad y' = 1;$$

$y = a - x$,

$$\frac{\Delta y}{\Delta x} = \frac{(a - x - i) - (a - x)}{i} = -1, \qquad y' = -1;$$

$y = ax$,

$$\frac{\Delta y}{\Delta x} = \frac{a(x + i) - ax}{i} = a, \qquad y' = a;$$

$y = \dfrac{a}{x}$,

$$\frac{\Delta y}{\Delta x} = \frac{\dfrac{a}{x + i} - \dfrac{a}{x}}{i} = -\frac{a}{x(x + i)}, \qquad y' = -\frac{a}{x^2};$$

$y = \sin x$,

$$\frac{\Delta y}{\Delta x} = \frac{\sin \dfrac{1}{2} i}{\dfrac{1}{2} i} \cos\left(x + \frac{1}{2} i\right), \qquad y' = \cos x = \sin\left(x + \frac{\pi}{2}\right);$$

$y = \cos x$,

$$\frac{\Delta y}{\Delta x} = -\frac{\sin \dfrac{1}{2} i}{\dfrac{1}{2} i} \sin\left(x + \frac{1}{2} i\right), \qquad y' = -\sin x = \cos\left(x + \frac{\pi}{2}\right).$$

---

[6] Cauchy seems to use *nombre* to speak about what we would call today positive real numbers, and *quantité* to speak about all the real numbers (hence, quantities can be defined as + or − the numbers).

Furthermore, setting $i = \alpha x$, $A^i = 1 + \beta$, and $(1 + \alpha)^a = 1 + \gamma$, we will find for $y = Lx$,

$$\frac{\Delta y}{\Delta x} = \frac{L(x + i) - Lx}{i} = \frac{L(1 + \alpha)}{\alpha x} = \frac{L(1 + \alpha)^{\frac{1}{\alpha}}}{x}, \qquad y' = \frac{Le}{x};$$

for $y = A^x$,

$$\frac{\Delta y}{\Delta x} = \frac{A^{x+i} - A^x}{i} = \frac{A^i - 1}{i} A^x = \frac{A^x}{L(1 + \beta)^{\frac{1}{\beta}}}, \qquad y' = \frac{A^x}{Le};$$

for $y = x^a$,

$$\frac{\Delta y}{\Delta x} = \frac{(x + i)^a - x^a}{i} = \frac{(1 + \alpha)^a - 1}{\alpha} x^{a-1} = \frac{L(1 + \alpha)^{\frac{1}{\alpha}}}{L(1 + \gamma)^{\frac{1}{\gamma}}} a x^{a-1}, \qquad y' = a x^{a-1}.$$

In the above formulas, the letter e designates the number 2.718 . . . , which forms the limit[7] of the expression $(1 + \alpha)^{\frac{1}{\alpha}}$. If we set this number as the base of a system of logarithms, we will obtain the *Napierian* or *hyperbolic* logarithms,[8] which we always indicate with the aid of the letter l. That said, it will be evident that

$$le = 1, \qquad Le = \frac{Le}{LA} = \frac{le}{lA} = \frac{1}{lA}.$$

Moreover, we will find for $y = lx$,

$$y' = \frac{1}{x};$$

for $y = e^x$,

$$y' = e^x.$$

Because the different formulas above are determined only for the values of x that correspond to real values of y, we ought to suppose that x is positive in those formulas that contain the functions Lx, lx,[9] and even the function x^a, if a designates a fraction with an even denominator[10] or an irrational number.[11]

Now let z be a second function of x connected to the first function $y = f(x)$ by the formula

(2) $$z = F(y).$$

z or $F[f(x)]$ will be called a *function of a function* of the variable of x; and, if we designate the infinitesimal increases by Δx, Δy, and Δz, and the three variables by x, y, and z, we will find

$$\frac{\Delta z}{\Delta x} = \frac{F(y + \Delta y) - F(y)}{\Delta x} = \frac{F(y + \Delta y) - F(y)}{\Delta y} \frac{\Delta y}{\Delta x},$$

---

[7] Viz., as alpha goes to zero.

[8] Today this logarithm is called the "natural logarithm."

[9] Cauchy is worried about taking the logarithm of a negative value, which, in general, is not a real value.

[10] If x were negative and the fraction a had an even denominator then this fraction could be rewritten in a form in which a would be a rational number divided by 2, and this would mean that the entire expression would be the square root of a negative number all raised to a rational power. Thus for certain powers this expression would be imaginary.

[11] Here Cauchy is worried about cases like $(-3)^\pi$ and $(-5)^r$.

which, passing to the limits,

(3)
$$z' = y'F'(y) = f'(x)F'[f(x)].$$

For example, if we make $z = ay$ and $y = lx$, we will have

$$z' = ay' = \frac{a}{x}.$$

With the help of formula (3), we can easily determine the derivatives of the simple functions A^x, x^a, arcsin x, and arccos x, if the derivatives of the functions Lx, sin x, and cos x are known. We find, in conclusion,

for $y = A^x$,

$$Ly = x, \qquad y'\frac{Le}{y} = 1, \qquad y' = \frac{y}{Le} = A^x lA;$$

for $y = a^x$,

$$ly = alx, \qquad y'\frac{1}{y} = \frac{a}{x}, \qquad y' = a\frac{y}{x} = ax^{a-1};$$

for $y = $ arcsin x,

$$\sin y = x, \qquad y'\cos y = 1, \qquad y' = \frac{1}{\cos y} = \frac{1}{\sqrt{1-x^2}};$$

for $y = $ arccos x,

$$\cos y = x, \qquad y' \times (-\sin y) = 1, \qquad y' = \frac{-1}{\sin y} = \frac{1}{\sqrt{1-x^2}}.$$

Moreover, because the derivatives of the compound functions

$$A^y, \qquad e^y, \qquad \frac{1}{y}$$

are, by formula (3), respectively,

$$y'A^y lA, \qquad y'e^y, \qquad -\frac{y'}{y^2},$$

the derivatives of the following

$$A^{B^x}, \qquad e^{e^x}, \qquad \sec x = \frac{1}{\cos x}, \qquad \operatorname{cosec} x = \frac{1}{\sin x}$$

will be

$$A^{B^x}B^x lAlB, \qquad e^{e^x}e^x, \qquad \frac{\sin x}{\cos^2 x}, \qquad -\frac{\cos x}{\sin^2 x}.$$

Let us remark in closing that the derivatives of compound functions are often determined more easily than those of simple functions. Thus, for example, we find

for $y = \tan x = \dfrac{\sin x}{\cos x}$,

$$\frac{\Delta y}{\Delta x} = \frac{1}{i}\left[\frac{\sin(x+i)}{\cos(x+i)} - \frac{\sin x}{\cos x}\right] = \frac{\sin i}{i\cos x\cos(x+i)}, \qquad y' = \frac{1}{\cos^2 x};$$

for $y = \cot x = \dfrac{\cos x}{\sin x}$,

$$\frac{\Delta y}{\Delta x} = \frac{1}{i}\left[\frac{\cos(x+i)}{\sin(x+i)} - \frac{\cos x}{\sin x}\right] = \frac{\sin i}{i \sin x \sin(x+i)}, \qquad y' = \frac{1}{\sin^2 x};$$

and finally, for $y = \arctan x$,

$$\tan y = x, \qquad \frac{y'}{\cos^2 y} = 1, \qquad y' = \cos^2 y = \frac{1}{1+x^2};$$

for $y = \operatorname{arccot} x$,

$$\cot y = x, \qquad \frac{-y'}{\sin^2 y} = 1, \qquad y' = -\sin^2 y = -\frac{1}{1+x^2}.$$

FOURTH LECTURE

DIFFERENTIALS OF FUNCTIONS OF A SINGLE VARIABLE

Let $y = f(x)$ be a function of an independent variable x, i an infinitely small quantity, and h a finite quantity. If we set $i = \alpha h$, with α also being an infinitely small quantity, we will have

$$\frac{f(x+i) - f(x)}{i} = \frac{f(x+\alpha h) - f(x)}{\alpha h},$$

from which we conclude

(1) $$\frac{f(x+\alpha h) - f(x)}{\alpha} = \frac{f(x+i) - f(x)}{i}h.$$

The limit toward which the first term of equation (1) converges, while the variable α approaches zero and while the quantity h remains constant, we call the *differential*[12] of the function $y = f(x)$. We indicate this differential with a characteristic d, according to which we write

$$dy \quad \text{or} \quad df(x).$$

It is easy to obtain the value of the differential if we know the function y' or $f'(x)$. Consequently, taking the limits of both sides of equation (1), we will find in general

(2) $$df(x) = hf'(x).$$

In the particular case where $f(x) = x$, equation (2) reduces to[13]

(3) $$dx = h.$$

---

[12] *différentielle.*
[13] As Cauchy says, this dx is not Δx, nor a limit of it; it is the differential of the function $f(x) = x$.

Thus the differential of the independent variable x is nothing other than the finite constant h. That being said, equation (2) will become

(4) $$df(x) = f'(x)dx,$$

or, what comes to the same thing,

(5) $$dy = y'dx.$$

It follows from the last equation that the derivative $y' = f'(x)$ of a function $y = f(x)$ is precisely equal to $\frac{dy}{dx}$, which is to say that the derivative is equal to the ratio between the differential of the function and the differential of the variable or, if you like, that the derivative is equal to the coefficient by which we need to multiply the second differential in order to obtain the first. It is for this reason that we often call the derivative function the *differential coefficient*.

To differentiate a function is to find its *differential*. The operation by which one differentiates is called *differentiation*.

In fact from formula (4), we can immediately obtain the differentials of the functions whose derivatives we have calculated. If we first apply this formula to simple functions, we will find

$$d(a + x) = dx, \qquad d(a - x) = -dx, \qquad d(ax) = adx,$$

$$d\frac{a}{x} = d\frac{dx}{x^2}, \qquad dx^a = ax^{a-1}dx;$$

$$dA^x = A^x lAdx, \qquad de^x = e^x dx;$$

$$dLx = Le\frac{dx}{x}, \qquad dlx = \frac{dx}{x};$$

$$d\sin x = \cos x dx = \sin\left(x + \frac{\pi}{2}\right)dx,$$

$$d\cos x = -\sin x dx = \cos\left(x + \frac{\pi}{2}\right)dx;$$

$$d\arcsin x = \frac{dx}{\sqrt{1 - x^2}}, \qquad d\arccos x = -\frac{dx}{\sqrt{1 - x^2}}.$$

In the same manner, we can establish the equations

$$d\tan gx = \frac{dx}{\cos^2 x}, \qquad d\cot x = -\frac{dx}{\sin^2 x};$$

$$d\arctan x = \frac{dx}{1 + x^2}, \qquad d\text{arccot}\, x = -\frac{dx}{1 + x^2};$$

$$d\sec x = \frac{\sin x dx}{\cos^2 x}, \qquad d\text{cosec}\, x = -\frac{\cos x dx}{\sin^2 x}.$$

All these different equations, from those we have obtained in the preceding lectures all the way up through these very equations, should be considered as demonstrations only for the values of x that correspond to the real values of the functions whose derivatives we found.

Consequently, the simple functions whose differentials can be found for any real values of the variable x are

$$a + x, \; a - x, \; ax, \; \frac{a}{x}, \; A^x, \; e^x, \; \sin x, \; \cos x,$$

and the function x^a, if the numerical value of a reduces to a whole number or a fraction with an odd denominator.[14] But we must suppose that the variable x is constrained between the limits -1 and $+1$ in the differentials we found of the simple functions arcsin x and arccos x,[15] and that x is constrained between the limits 0 and ∞ in the differentials of the functions Lx and lx,[16] and also that x is constrained to being greater than or equal to zero in the case of the function x^a whenever the numeric value of a is a fraction with an even denominator or an irrational number.[17]

Again, it is essential to observe that, conforming to the conventions established in the Part 1 of *Cours d'Analyse*, we make use of the notations

arcsin x,   arccos x,   arctan x,   arccot x,   arcsec x,   arccosec x,

in order to represent, not merely any arc for which a certain trigonometric line is equal to x, but only those which have the smallest numerical value[18] or, if these arcs are pair wise equal and of opposite signs, those arcs which have the smallest positive value.[19] Consequently, arcsin x, arctan x, arccot x, arccosec x are arcs contained between the limits $-\frac{\pi}{2}$ and $+\frac{\pi}{2}$; and arccos x, arcsec x are arcs contained between the limits 0 and π.[20]

If we let $y = f(x)$ and $\Delta x = i = \alpha h$, equation (1) (the second term of which has dy as its limit) can be given in the form

$$\frac{\Delta y}{\alpha} = dy + \beta,$$

β designating an infinitely small quantity,[21] and we conclude from it that

(6) $$\Delta y = (dy + \beta)\alpha.$$

---

[14] See footnote 10.

[15] These functions are typically defined to take on values only from -1 to 1.

[16] See footnote 9.

[17] See footnotes 10 and 11.

[18] For example, the arcsin(1) could be thought of as answering the question: the sine of what arc is 1? However, because of the cyclic nature of the trigonometric functions, an infinite number of answers could be given to this question: $\frac{\pi}{2}, \frac{5\pi}{2}$, etc. If we allowed all of these values to stand as values of the function at 1, then our function would have multiple values for a given argument. Cauchy tells us to restrict our domain of answers to the smallest of these values in order to avoid this problem.

[19] For example, arccos $\left(\frac{\sqrt{2}}{2}\right)$ is equal to $\left(\frac{\pi}{4}\right) \pm n2\pi$, and $-\left(\frac{\pi}{4}\right) \pm n2\pi$. Even given Cauchy's last restriction, we would still have two values for this arccos. In general these values would come in pairs which were equal but opposite in sign. In order to ensure a unique result of the function for a given argument, Cauchy has us restrict our domain of function values to the positive portion of these pairs.

[20] The inverse trigonometric functions (i.e., "arc functions") take the results of the trigonometric functions as their arguments and return arcs for their function values. The arcs or function values of the "arc functions" are restricted as specified due to the restrictions Cauchy placed on the arguments above.

[21] As Cauchy hints, the second term of equation (1) can be thought of as the already found differential (which exists only at the limit) plus an infinitesimal quantity that separates the expression before we have taken the limit from the expression after we have taken the limit, viz., the differential itself.

Let z be a second function of the variable x. Similarly, we have

$$\Delta z = (dz + \gamma)\alpha,$$

γ again designating an infinitely small quantity. Immediately we find

$$\frac{\Delta z}{\Delta y} = \frac{dz + y}{dy + \beta},$$

which, passing to the limits,

(7) $$\lim\frac{\Delta z}{\Delta y} = \frac{dz}{dy} = \frac{z'}{y'}\frac{dx}{dx} = \frac{z'}{y'}.$$

Therefore, *the ratio between the different infinitely small values of two functions of the variable x has as its limit the ratio of their differentials or their derivatives.*

Now let us suppose that functions y and z are connected by the equation

(8) $$z = F(y).$$

We conclude from this that

$$\frac{\Delta z}{\Delta y} = \frac{F(y + \Delta y) - F(y)}{\Delta y},$$

which, in passing to the limit and taking note of formula (7), $\frac{\Delta z}{\Delta y} = \frac{z'}{y'} = F'(y)$,

(9) $$dz = F'(y)dy, \qquad z' = y'F'(y).$$

The second term of equation (9) coincides with equation (3) of the previous lecture.

Moreover, if we put $F(y)$ in place of z in the first expression, we will obtain

(10) $$dF(y) = F'(y)dy,$$

which is similar in form to equation (4) and serves to differentiate a function of y, even though y is not an independent variable.

Examples:

$$d(a + y) = dy, \qquad d(-y) = -dy, \qquad d(ay) = ady, \qquad de^y = e^y dy,$$

$$dly = \frac{dy}{y}, \qquad dly^2 = \frac{dy^2}{y^2} = \frac{2dy}{y}, \qquad d\frac{1}{2}ly^2 = \frac{dy}{y}, \quad \dots,$$

$$d(ax^m) = adx^m = max^{m-1}dx, \qquad de^{e^x} = e^{e^x}de^x = e^{e^x}e^x dx,$$

$$dl\sin x = \frac{d\sin x}{\sin x} = \frac{\cos x\, dx}{\sin x} = \frac{dx}{\tan x}, \qquad dl\tan x = \frac{dx}{\sin x \cos x}, \quad \dots.$$

The first of these formulas proves that *the addition of a constant to a function does not alter the differential and, consequently, does not alter the derivative.*

INTEGRAL CALCULUS
TWENTY-FIRST LECTURE
DEFINITE INTEGRALS

Let us suppose that the function $y = f(x)$ is continuous with respect to the variable x between two finite limits $x = x_0$ and $x = X$, that we designate new values of x interposed between these limits by $x_1, x_2, \ldots, x_{n-1}$, and that these values will always either increase or decrease from the first limit up until the second limit.[22] We will be able to use these values to divide the difference $X - x_0$ into elements

(1) $\qquad x_1 - x_0, \quad x_2 - x_1, \quad x_3 - x_2, \ldots, \quad X - x_{n-1},$

which will all have the same sign.[23] This being established, let us imagine that we multiply each element by the value of $f(x)$ corresponding to the *origin* of this very element, namely, the element $x_1 - x_0$ by $f(x_0)$, the element $x_2 - x_1$ by $f(x_1), \ldots,$ and finally the element $X - x_{n-1}$ by $f(x_{n-1})$; and let the sum of the products thus obtained be

(2) $\quad S = (x_1 - x_0)f(x_0) + (x_2 - x_1)f(x_1) + \ldots + (X - x_{n-1})f(x_{n-1}).$

Clearly, the quantity S will depend upon first the number of elements n into which we have divided the difference $X - x_{n-1}$, and second, the values themselves of these elements and, consequently, on the mode of division adopted. However, it is important to remark that if the numeric values of the elements become very small and the number n considerably large, the mode of division will thereafter have only an imperceptible influence[24] on the value of S. That is, in fact, what we are able to show as follows.

If we were to suppose that all the elements of the difference $X - x_0$ reduce to a single difference, which would be this self-same difference, we would simply have

(3) $\qquad\qquad S = (X - x_0)f(x_0).$

If, by contrast, we take the expressions in (1) for elements of the difference $X - x_0$, then the value of S, determined in this case by equation (2), is equal to the sum of the elements multiplied by an average between the coefficients

$$f(x_0), \quad f(x_1), \quad \ldots, \quad f(x_{n-1})$$

(see, in the introduction to the *Cours d'Analyse*, the corollary of theorem III).[25] Moreover, given that these coefficients are particular values of the expression

$$f[x_0 + \theta(X - x_0)]$$

that correspond to certain values of θ contained between zero and unity, we can prove

---

[22] Today we call this last condition "monotonicity."
[23] Because of the monotonicity requirement.
[24] *influence insensible.*
[25] Works of Cauchy, S. II, t. III.

(by reasoning similar to that which we used in the "Seventh Lecture"[26]) that the average we are interested in is another value of the same expression, which also corresponds to a value of θ contained between the same limits. Thus we can substitute for equation (2)

(4) $$S = (X - x_0)f[x_0 + \theta(X - x_0)],$$

in which θ is a number less than unity.[27]

To pass from the mode of division that we have considered to another in which the numeric values of the elements of $X - x_0$ will be smaller still, it will suffice to partition each of the expressions in (1) into new elements. Then, in the second term of equation (2) we will have to replace the product $(x_1 - x_0)f(x_0)$ by a sum of similar products for which we will be able to substitute an expression of the form

$$(x_1 - x_0)f[x_0 + \theta_0(x_1 - x_0)],$$

θ_0 being a number less than unity. We expect that there will be a relation between this sum and the product $(x_1 - x_0)f(x_0)$ similar to that which exists between the values of S furnished by equations (4) and (3). For the same reason, we will have to substitute for the product $(x_2 - x_1)f(x_1)$ a sum of terms which can be presented in the form

$$(x_2 - x_1)f[x_1 + \theta_1(x_2 - x_1)],$$

θ_1 again designating a number less than unity. Continuing in this way we will finish by concluding that, in the new mode of division, the value of S will be of the form

(5) $$S = (x_1 - x_0)f[x_0 + \theta_0(X - x_0)] + (x_2 - x_1)f[x_1 + \theta_1(x_2 - x_1)] + \ldots$$
$$+ (X - x_{n-1})f[x_{n-1} + \theta_{n-1}(X - x_{n-1})].$$

If in this last equation we set

$$f[x_0 + \theta_0(x_1 - x_0)] = f(x_0) \pm \varepsilon_0,$$
$$f[x_1 + \theta_1(x_2 - x_1)] = f(x_1) \pm \varepsilon_1,$$
$$\vdots$$
$$f[x_{n-1} + \theta_{n-1}(X - x_{n-1})] = f(x_{n-1}) \pm \varepsilon_{n-1},$$

we will conclude that

(6) $$S = (x_1 - x_0)f[x_0 \pm \varepsilon_0] + (x_2 - x_1)f[x_1 \pm \varepsilon_1] + \ldots$$
$$+ (X - x_{n-1})f[x_{n-1} \pm \varepsilon_{n-1}],$$

which, expanding the products, gives

(7) $$S = (x_1 - x_0)f(x_0) + (x_2 - x_1)f(x_1) + \ldots + (X - x_{n-1})f(x_{n-1})$$
$$\pm \varepsilon_0(x_1 - x_0) \pm \varepsilon_1(x_2 - x_1) \pm \ldots \pm \varepsilon_{n-1}(X - x_{n-1}).$$

Let us add that if the elements $x_1 - x_0, x_2 - x_1, x_3 - x_2, \ldots, X - x_{n-1}$ have very small numeric values, then each of the quantities $\pm\varepsilon_0, \pm\varepsilon_1, \ldots, \pm\varepsilon_{n-1}$ will differ very little from zero; and because of this, something similar can be said of the sum

[26] Not included. Though Cauchy does not prove this in our selection of his lectures, it is not too hard to see intuitively why the claim that follows should be true. Because $0 < \theta < 1$, and the function is monotonic, $f(x_0 + \theta(X - x_0))$ is between $f(x_0)$ and $f(X)$, so for the right θ, $f(x_0 + \theta(x - x_0))$ will be the average we want.

[27] As noted before, Cauchy reserves the term *nombre* for what we would call the positive real numbers. Thus this qualification means that theta is between zero and one (as Cauchy said explicitly above).

$$\pm\varepsilon_0(x_1 - x_0) \pm \varepsilon_1(x_2 - x_1) \pm \ldots \pm \varepsilon_{n-1}(X - x_{n-1}),$$

which is equivalent to the product of $X - x_0$ by an average between these different quantities. This being established, it results from comparing equations (2) and (7) that we will not perceptibly[28] alter the value of S calculated for a mode of division in which the elements of the difference $X - x_0$ have very small numeric values if we pass to a second mode of division in which each of these elements is subdivided into many others.

Let us now imagine that we consider at the same time two modes of division of the difference $X - x_0$ in each of which the elements have very small numeric values. We will be able to compare these two modes to a third so chosen that each element, be it of the first or the second mode, will be formed by the union of several elements of the third. Because this condition is fulfilled, all we need is that all the values of x, interposed between the limits x_0 and X in the first two modes be employed in the third, and it will be proven that the value of S is altered very little in passing from the first or the second mode to the third, and consequently the value of S will be altered very little in passing from the first to the second mode. Therefore, if the elements of the difference $X - x_0$ become infinitely small, the mode of division will have only an imperceptible influence[29] on the value of S; and if we decrease the numeric values of the elements indefinitely while augmenting their number, the value of S will result in being perceptibly[30] constant or, in other words, will result in attaining a certain limit that will depend uniquely[31] on the form of the function $f(x)$ and the limiting values x_0 and X attributed to the variable x. We call this limit a *definite integral*.

Now let us observe that if we designate by $\Delta x = h = dx$ a finite increase made to the variable x, the different terms of which constitute the value of S, such that the products

$$(x_1 - x_0)f(x_0), \qquad (x_2 - x_1)f(x_1), \qquad \ldots,$$

will all be contained in the general form

(8) $$hf(x) = f(x)\,dx,$$

from which the products are derived one after the other, while first setting

$$x = x_0 \qquad \text{and} \qquad h = x_1 - x_0,$$

then

$$x = x_1 \qquad \text{and} \qquad h = x_2 - x_1, \qquad \ldots.$$

Thus we can state that the quantity S is a sum of products resembling expression (8) that we always express with the aid of the characteristic symbol Σ, by writing

(9) $$S = \sum hf(x) = \sum f(x)\Delta x.$$

[28] *sensiblement.*
[29] *influence insensible.*
[30] *sensiblement.*
[31] I.e., for each function with given endpoints there is one and only one limit.

It is convenient to represent the definite integral, toward which the quantity S converges while the elements of the difference $X - x_0$ become infinitely small, by the notation $\int hf(x)$ or $\int f(x)dx$, where the letter $\int$ is substituted for the letter Σ, indicating no longer a sum of products similar to expression (8), but the limit of a sum of that sort. Moreover, as the value of the definite integral that we are considering depends on the limiting values x_0 and X attributed to the variable x, it is convenient to place the first of these two values underneath and the second above the symbol $\int$, both of which are written to the side of the integral, which are consequently designated by one of the following notations[32]

$$(10) \qquad \int_{x_0}^{X} f(x)\,dx, \qquad \int f(x)\,dx\begin{bmatrix} x_0 \\ X \end{bmatrix}, \qquad \int f(x)\,dx\begin{bmatrix} x=x_0 \\ x=X \end{bmatrix}.$$

The first of these notations, contrived by Fourier, is the most simple. In the particular case where the function $f(x)$ is replaced by a constant quantity, we find that whatever the mode of division of the difference $X - x_0$ may be,

$$S = a(X - x_0),$$

and we conclude from this that

$$(11) \qquad \int_{x_0}^{X} a\,dx = a(X - x_0).$$

If, in this last formula, we set $a = 1$ we would conclude that

$$(12) \qquad \int_{x_0}^{X} dx = (X - x_0).$$

---

[32] In the second and third expressions (which are no longer used) the upper and lower indices appear to be switched. Either this is a typographical error or Cauchy's description fits only the first notation.

TWENTY-SECOND LECTURE

FORMULAS FOR THE DETERMINATION OF EXACT VALUES OR VALUES APPROACHING DEFINITE INTEGRALS

According to what was said in the last lecture, if we divide $X - x_0$ into infinitely small elements $x_1 - x_0$, $x_2 - x_1$, $x_3 - x_2$, ..., $X - x_{n-1}$, the sum

$$(1) \qquad S = (x_1 - x_0) f(x_0) + (x_2 - x_1) f(x_1) + \ldots + (X - x_{n-1}) f(x_{n-1})$$

will converge toward a limit represented by the definite integral

$$(2) \qquad \int_{x_0}^{X} f(x)\, dx.$$

From the principles upon which we have founded this proposition, it follows that we would have reached the same limit if the value of S, instead of being determined by equation (1), were to be deduced from formulas similar to equations (5) and (6) ("Twenty-First Lecture"), that is to say, if we were to suppose

$$(3) \quad S = (x_1 - x_0) f[x_0 + \theta_0(x_1 - x_0)] + (x_2 - x_1) f[x_1 + \theta_1(x_2 - x_1)] + \ldots$$
$$+ (X - x_{n-1}) f[x_{n-1} + \theta_{n-1}(X - x_{n-1})],$$

with $\theta_0, \theta_1, \ldots, \theta_{n-1}$ designating any numbers less than unity, or

$$(4) \qquad S = (x_1 - x_0) f[x_0 + \varepsilon_0] + (x_2 - x_1) f[x_1 + \varepsilon_1] + \ldots$$
$$+ (X - x_{n-1}) f[x_{n-1} + \varepsilon_{n-1}],$$

with $\varepsilon_0, \varepsilon_1, \ldots, \varepsilon_{n-1}$ designating numbers that vanish with the elements of the difference $X - x_0$. The first of the two preceding formulas reduces to equation (1) if we set

$$\theta_0 = \theta_1 = \ldots = \theta_{n-1} = 0.$$

If, alternately, we set

$$\theta_0 = \theta_1 = \ldots = \theta_{n-1} = 1,$$

we find

$$(5) \qquad S = (x_1 - x_0) f(x_1) + (x_2 - x_1) f(x_2) + \ldots + (X - x_{n-1}) f(X).$$

In this last formula, if each of those values between the two quantities x_0 and X is switched such that all the terms are placed at equal distances from the two extremes in the series $x_0, x_1, \ldots, x_{n-1}, X$, we obtain a new value for S equal but opposite in sign to those furnished by equation (1). The limit toward which this new value of S will converge must, therefore, be equal but opposite in sign to integral (2), which we deduced by the mutual exchange of two quantities x_0, X. Thus in general we have

(6)
$$\int_{x_0}^{X} f(x)\, dx = -\int_{X}^{x_0} f(x)\, dx.$$

We frequently employ formulas (1) and (5) in the investigation of values approaching definite integrals. For the sake of simplicity, we ordinarily suppose that the quantities $x_0, x_1, \ldots, x_{n-1}, X$ contained in these formulas are in an arithmetic progression. Then the elements of the difference $X - x_0$ will all become equal to the fraction $\frac{X - x_0}{n}$; and designating this fraction by i, we find that equations (1) and (5) reduce to the two following equations:

(7) $S = i[f(x_0) + f(x_0 + i) + f(x_0 + 2i) + \ldots + f(X - 2i) + f(X - i)]$,

(8) $S = i[f(x_0 + i) + f(x_0 + 2i) + \ldots + f(X - 2i) + f(X - i) + f(X)]$.

Again we might suppose that the quantities $x_0, x_1, \ldots, x_{n-1}, X$ form a geometric progression whose rate of geometric growth or decay differs very little from unity. In adopting this hypothesis and setting $\left(\frac{X}{x_0}\right)^{\frac{1}{n}} = 1 + \alpha$, we draw two new values of S from formulas (1) and (5); the first is

(9) $S = \alpha \left\{ x_0 f(x_0) + x_0(1 + \alpha) f[x_0(1 + \alpha)] + \ldots + \dfrac{X}{1 + \alpha} f\left[\dfrac{X}{1 + \alpha}\right] \right\}.$

It is essential to observe that in many cases, we can deduce from equations (7) and (9) not only the values approaching the integral (2), but also its exact value or $\lim S$.[33] We find, for example,

(10)
$$\int_{x_0}^{X} x\, dx = \lim \frac{(X - x_0)(X + x_0 - i)}{2} = \frac{X^2 - x_0^2}{2},$$

(11)
$$\int_{x_0}^{X} A^x dx = \lim \frac{i(A^X - A^{x_0})}{A^i - 1} = \frac{A^X - A^{x_0}}{lA},$$

$$\int_{x_0}^{X} e^x dx = e^X - e^{x_0},$$

(12)
$$\int_{x_0}^{X} x^a dx = \lim \frac{\alpha(X^{a+1} - x_0^{a+1})}{(1 + \alpha)^{a+1} - 1} = \frac{X^{a+1} - x_0^{a+1}}{a + 1},$$

$$\int_{x_0}^{X} \frac{dx}{x} = \lim n\alpha = l\frac{X}{x_0}.$$

The last equation holds only in the case where the quantities x_0, X have the same sign.[34] Let us add that it is often easy to reduce the determination of a definite integral to that of another integral of the same sort. Thus, for example, from formula (1)

---

[33] Again, Cauchy means the limit as i and α approach zero.

[34] As in the "Third Lecture," Cauchy makes this qualification because he is worried about the possibility of taking logarithms of negative values.

we will have

(13) $\displaystyle\int_{x_0}^{X} a\varphi(x)\,dx = \lim a\left[(x_1 - x_0)\varphi(x_0) + \ldots + (X - x_{n-1})\varphi(x_{n-1})\right]$

$\displaystyle = a\int_{x_0}^{X}\varphi(x)\,dx,$

(14) $\displaystyle\int_{x_0}^{X} f(x + a)\,dx = \lim\left[(x_1 - x_0)f(x_0 + a) + \ldots + (X - x_{n-1})f(x_{n-1} + a)\right]$

$\displaystyle = \int_{x_0 + a}^{X + a} f(x)\,dx,$

(15) $\displaystyle\int_{x_0}^{X} f(x - a)\,dx = \int_{x_0 - a}^{X - a} f(x)\,dx, \qquad \int_{x_0}^{X}\frac{dx}{x - a} = \int_{x_0 - a}^{X - a}\frac{dx}{x} = l\,\frac{X - a}{x_0 - a},$

the last equation above holding only in the case where $x_0 - a$ and $X - a$ are quantities having the same sign.[35] Moreover, we conclude from formula (8), setting $x_0 = 0$ and replacing $f(x)$ by $f(X - x)$, that

(16) $\displaystyle\int_{0}^{X} f(X - x)\,dx = \lim i\left[f(X - i) + f(X - 2i) + \ldots + f(2i) + f(i) + f(0)\right]$

$\displaystyle = \int_{0}^{X} f(x)\,dx.$

Then we can conclude from formula (8), looking back to equation (14),

(17) $\displaystyle\int_{0}^{X - x_0} f(X - x)\,dx = \int_{0}^{X - x_0} f(x + x_0)\,dx = \int_{x_0}^{X} f(x)\,dx.$

Finally, if in formula (9) we set

$$f(x) = \frac{1}{xlx} \qquad \text{and} \qquad l(1 + \alpha) = \beta,$$

we will conclude from it

(18) $\displaystyle\int_{x_0}^{X}\frac{dx}{xlx} = \lim \beta\left(\frac{1}{lx_0} + \frac{1}{lx_0 + \beta} + \ldots + \frac{1}{lX - \beta}\right)\frac{e^{\beta} - 1}{\beta} = \int_{lx_0}^{lX}\frac{dx}{x} = l\,\frac{lX}{lx_0},$

where the above quantities x_0, X are positive and both are at the same time either greater than unity or less than unity.[36]

It is important to remark that the different forms under which the value of S presents itself in equations (4) and (5) of the preceding lecture converge to integral

[35] Again, Cauchy is worried about taking the logarithm of a negative value.
[36] The logarithm function changes sign about unity, so this condition, yet again, ensures that the arguments of the last term above (viz., the logarithm of the quotient of two logarithms) are positive.

(2) identically. In conclusion, if we substitute one of these equations for the other, while we subdivide either the difference $X - x_0$ or the quantities $x_1 - x_0, x_2 - x_1, x_3 - x_2, \ldots, X - x_{n-1}$ into infinitely small elements, they will still be true at the limit such that we will have

(19)
$$\int_{x_0}^{X} f(x)\,dx = (X - x_0)f[x_0 + \theta(X - x_0)]$$

and

(20)
$$\int_{x_0}^{X} f(x)\,dx = (x_1 - x_0)f[x_0 + \theta_0(x_1 - x_0)] + (x_2 - x_1)f[x_1 + \theta_1(x_2 - x_1)]$$
$$+ \ldots + (X - x_{n-1})f[x_{n-1} + \theta_{n-1}(X - x_{n-1})],$$

$\theta, \theta_0, \theta_1, \ldots, \theta_{n-1}$ designating unspecified numbers all less than unity. If, for the sake of simplicity, we suppose that the quantities $x_1 - x_0, x_2 - x_1, x_3 - x_2, \ldots, X - x_{n-1}$ are equal to each other, then setting $i = \frac{X - x_0}{n}$, we will find

(21)
$$\int_{x_0}^{X} f(x)\,dx = i[f(x_0 + \theta_0 i) + f(x_0 + i + \theta_1 i) + \ldots + f(X - i + \theta_{n-1}i)].$$

If the function $f(x)$ always increases or always decreases from $x = x_0$ up through $x = X$, the second member of formula (21) clearly remains constrained between the two values of S furnished by equations (7) and (8)—values whose difference is $\pm i[f(X) - f(x_0)]$. Consequently, on this hypothesis, substituting the half-sum of these two values or the expression

(22)
$$i\left[\frac{1}{2}f(x_0) + f(x_0 + i) + f(x_0 + 2i) + \ldots\right.$$
$$\left. + f(X - 2i) + f(X - i) + \frac{1}{2}f(X)\right]$$

for the value approached by integral (21), we are left with an error much smaller than the half-difference $\pm i[\frac{1}{2}f(X) - \frac{1}{2}f(x_0)]$.

Example:

If we suppose

$$f(x) = \frac{1}{1 + x^2}, \qquad x_0 = 0, \qquad X = 1, \qquad i = \frac{1}{4},$$

expression (22) becomes

$$\frac{1}{4}\left(\frac{1}{2} + 1\frac{6}{7} + \frac{4}{5} + \frac{1}{2}\frac{6}{5} + \frac{1}{4}\right) = 0.78 \ldots$$

Consequently the integral $\int_0^1 \frac{dx}{1+x^2}$ approaches the value 0.78. The error committed in this case cannot be greater than $\frac{1}{4}(\frac{1}{2} - \frac{1}{4}) = \frac{1}{16}$. In fact, it will be below $\frac{1}{100}$, as we will see much later.[37]

---

[37] The exact decimal values of the integral are 0.785398 . . . .

If the function $f(x)$ is sometimes increased and sometimes decreased between the limits $x = x_0$ and $x = X$, the error we commit in substituting one of the values of S furnished by equations (7) and (8) for the value approached by integral (2) is evidently less than the product of $ni = X - x_0$ by the largest numeric value that can be obtained from the difference

$$(23) \qquad f(x + \Delta x) - f(x) = \Delta x(f'(x + \theta \Delta x)),$$

where we suppose that y contains values of x between the limits x_0, X and values of Δx between the limits 0, i. Thus, if we call k the greatest numeric value that $f(x)$ receives while x varies from $x = x_0$ up through $x = X$, the error committed will certainly be constrained between the limits

$$-ki(X - x_0), \qquad +ki(X - x_0).$$

TWENTY-THIRD LECTURE

DECOMPOSITION OF A DEFINITE INTEGRAL INTO MANY OTHERS. IMAGINARY DEFINITE INTEGRALS. GEOMETRIC REPRESENTATION OF DEFINITE REAL INTEGRALS. DECOMPOSITION OF A FUNCTION UNDER THE SIGN ∫ INTO TWO FACTORS ONE OF WHICH ALWAYS CONSERVES THE SAME SIGN

To divide the definite integral

$$(1) \qquad \int_{x_0}^{X} f(x)\, dx$$

into many others of the same sort, it suffices to decompose into many parts either the function under the sign $\int$, or the difference $X - x_0$. Let us suppose initially that

$$f(x) = \varphi(x) + \chi(x) + \psi(x) + \ldots .$$

We will conclude from this that

$$(x_1 - x_0)f(x_0) + \ldots + (X - x_{n-1})f(x_{n-1})$$
$$= (x_1 - x_0)\varphi(x_0) + \ldots + (X - x_{n-1})\varphi(x_{n-1}) + (x_1 - x_0)\chi(x_0) + \ldots$$
$$+ (X - x_{n-1})\chi(x_{n-1}) + (x_1 - x_0)\psi(x_0) + \ldots$$
$$+ (X - x_{n-1})\psi(x_{n-1}) + \ldots ,$$

which, passing to the limits, becomes

$$\int_{x_0}^{X} f(x)\, dx = \int_{x_0}^{X} \varphi(x)\, dx + \int_{x_0}^{X} \chi(x)\, dx + \int_{x_0}^{X} \psi(x)\, dx + \ldots .$$

From this last formula, together with equation (13) (from the "Twenty-Second Lecture"), we will conclude, designating different functions of the variable x by u, v, w, and designating constant quantities by a, b, c, that

(2) $$\int_{x_0}^{X} (u + v + w + \ldots)\, dx = \int_{x_0}^{X} u\, dx + \int_{x_0}^{X} v\, dx + \int_{x_0}^{X} w\, dx + \ldots,$$

(3) $$\int_{x_0}^{X} (u + v)\, dx = \int_{x_0}^{X} u\, dx + \int_{x_0}^{X} v\, dx, \qquad \int_{x_0}^{X} (u - v)\, dx = \int_{x_0}^{X} u\, dx - \int_{x_0}^{X} v\, dx,$$

(4) $$\int_{x_0}^{X} (au + bv + cw + \ldots)\, dx = a\int_{x_0}^{X} u\, dx + b\int_{x_0}^{X} v\, dx + c\int_{x_0}^{X} w\, dx + \ldots.$$

If we stretch the definition that we gave for integral (1) to cases where the function $f(x)$ becomes imaginary, equation (4) remains true for the imaginary values of constants a, b, c, ... Consequently, we have

(5) $$\int_{x_0}^{X} (u + v\sqrt{-1})\, dx = \int_{x_0}^{X} u\, dx + \sqrt{-1} \int_{x_0}^{X} v\, dx.$$

Now let us suppose that before dividing the difference $X - x_0$ into a number of finite elements represented by $x_1 - x_0, x_2 - x_1, x_3 - x_2, \ldots, X - x_{n-1}$, we partition each of the elements into many others whose numeric values are infinitely small, and that, as a consequence, we modify the value of S furnished by equation (1) (from the "Twenty-Second Lecture"). The product $(x_1 - x_0)\, f(x_0)$ will be replaced by a sum of products resembling those having the definite integral $\int_{x_0}^{X} f(x)\, dx$ for their limit. Similarly, the products $(x_2 - x_1)f(x_1), \ldots,$ $(X - x_{n-1})f(x_{n-1})$ will be replaced by sums having the definite integrals $\int_{x_1}^{X_2} f(x)\, dx, \ldots, \int_{x_{n-1}}^{X} f(x)\, dx$ for their limits. Moreover, in reuniting these different sums, we will obtain as a result a total sum whose limit will be precisely integral (1). Thus, because the limit of a sum of many quantities is always equivalent to the sum of their limits, we have in general

(6) $$\int_{x_0}^{X} f(x)dx = \int_{x_0}^{x_1} f(x)dx + \int_{x_1}^{x_2} f(x)\, dx + \ldots + \int_{x_{n-1}}^{X} f(x)\, dx.$$

It is essential to recall that here we must attribute a finite value to the whole number n. If we interpose a single value of x represented by ξ between the limits x_0 and X, equation (6) reduces to

(7) $$\int_{x_0}^{X} f(x)\, dx = \int_{x_0}^{\xi} f(x)\, dx + \int_{\xi}^{X} f(x)\, dx.$$

It is easy to prove that equations (6) and (7) would remain true even in the case where some of the quantities $x_1, x_2, \ldots, x_{n-1}, \xi$ cease to be contained between the limits x_0, X, and in those cases where the differences $x_1 - x_0, x_2 - x_1, x_3 - x_2, \ldots, X - x_{n-1}$, $\xi - x_0, X - \xi$ are no longer quantities of the same sign. Let us say, for example, that the differences $\xi - x_0, X - \xi$ have opposite signs. Then, in addition to that, let us suppose that x_0 is contained between ξ and X, or rather, that X is contained between x_0 and ξ. We will find that

$$\int_{\xi}^{X} f(x)\,dx = \int_{\xi}^{x_0} f(x)\,dx + \int_{x_0}^{X} f(x)\,dx,$$

or rather

$$\int_{x_0}^{\xi} f(x)\,dx = \int_{x_0}^{X} f(x)\,dx + \int_{X}^{\xi} f(x)\,dx.$$

But formula (6) of the "Twenty-Second Lecture" suffices to show how the two equations that we are to obtain agree with equation (7). The latter being established in all hypothetical cases, we will be able to deduce equation (6) directly from it whatever $x_1, x_2, \ldots, x_{n-1}$ are.

In the preceding lecture, we saw how easy it is to find not only the values approaching integral (1) but also the limits of the errors committed if the function $f(x)$ is always increasing or always decreasing from $x = x_0$ up through $x = X$. When this latter condition fails to be satisfied, it is clear that with the aid of equation (6) we can decompose integral (1) into many other integrals, for each of which the same condition is fulfilled.

Let us now imagine that the limit X is greater than x_0, that the function $f(x)$ is positive from $x = x_0$ up through $x = X$, that x, y designate rectangular coordinates, and that A is the surface with one part contained between the x axis and the curve $y = f(x)$ and the other part contained between the ordinates $f(x_0), f(X)$. This surface, which has for its base the length $X - x_0$ measured along the x axis, will be an average between the areas of the two rectangles construed on the base $X - x_0$, with the heights respectively equal to the smallest and the greatest of the ordinates elevated from the different points of this base. Thus it will be equivalent to a rectangle construed on an average ordinate represented by an expression of the form $f[x_0 + \theta(X - x_0)]$; thus we have

(8) $A = (X - x_0)f[x_0 + \theta(X - x_0)],$

with θ designating a number less than unity. If we divide the base $X - x_0$ into very small elements $x_1 - x_0, x_2 - x_1, x_3 - x_2, \ldots, X - x_{n-1}$, the surface A will be divided into corresponding elements whose values will be given by an equation similar to formula (8).

Thus we have again

$$(9) \quad A = (x_1 - x_0)f[x_0 + \theta_0(x_1 - x_0)] + (x_2 - x_1)f[x_1 + \theta_1(x_2 - x_1)] + \ldots$$
$$+ (X - x_{n-1})f[x_{n-1} + \theta_{n-1}(X - x_{n-1})],$$

$\theta_0, \theta_1, \ldots, \theta_{n-1}$ designating numbers less than unity. If in this last equation we decrease the numeric values of the elements of $X - x_0$ indefinitely, we will conclude, passing to the limits, that

$$(10) \qquad\qquad A = \int_{x_0}^{X} f(x)\,dx.$$

Example:

Apply formula (10) to the curves $y = ax^2$, $xy = 1$, $y = e^x$, ...[38]

In concluding this lecture, let us reveal a remarkable property of definite, real integrals. If we suppose that $f(x) = \varphi(x)\chi(x)$, where $\varphi(x)$ and $\chi(x)$ are two new functions that remain continuous between the limits $x = x_0$ and $x = X$, and the second of which always keeps the same sign between these limits, the value of S given by equation (1) in the "Twenty-Second Lecture" will become

$$(11) \qquad S = (x_1 - x_0)\varphi(x_0)\chi(x_0) + (x_2 - x_1)\varphi(x_1)\chi(x_1) + \ldots$$
$$+ (X - x_{n-1})\varphi(x_{n-1})\chi(x_{n-1}),$$

and will be equivalent to the sum

$$(x_1 - x_0)\chi(x_0) + (x_2 - x_1)\chi(x_1) + \ldots + (X - x_{n-1})\chi(x_{n-1})$$

multiplied by an average between the coefficients $\varphi(x_0)$, $\varphi(x_1)$, ..., $\varphi(x_{n-1})$ or, what comes to the same thing, by a quantity of the form $\varphi(\xi)$, where ξ designates a value of x contained between x_0 and X. Thus we will have

$$(12) \qquad S = [(x_1 - x_0)\chi(x_0) + (x_2 - x_1)\chi(x_1) + \ldots + (X - x_{n-1})\chi(x_{n-1})]\varphi(\xi),$$

and we conclude from this that in looking at the limit of S,

$$(13) \qquad \int_{x_0}^{X} f(x)\,dx = \int_{x_0}^{X} \varphi(x)\chi(x)\,dx = \varphi(\xi)\int_{x_0}^{X} \chi(x)\,dx,$$

with ξ always designating a value of x contained between x_0 and X.

Examples:

If we successively set

$$\chi(x) = 1, \qquad \chi(x) = \frac{1}{x}, \qquad \chi(x) = \frac{1}{x-a},$$

we will obtain the formulas

$$(14) \qquad \int_{x_0}^{X} f(x)\,dx = f(\xi)\int_{x_0}^{X} dx = (X - x_0)f(\xi),$$

---

[38] Cauchy leaves this as an exercise to the reader.

(15)
$$\int_{x_0}^{X} f(x)\,dx = \xi f(\xi) \int_{x_0}^{X} \frac{dx}{x} = \xi f(\xi) l \frac{X}{x_0},$$

(16)
$$\int_{x_0}^{X} f(x)\,dx = (\xi - a)f(\xi - a) \int_{x_0}^{X} \frac{dx}{x - a} = (\xi - a) l \frac{X - a}{x_0 - a},$$

the first of which coincides with equation (19) of the "Twenty-Second Lecture." Let us add that the relation $\frac{X}{x_0}$ in the second formula and the relation $\frac{X-a}{x_0-a}$ in the third formula must always be assumed to be positive.[39]

[39] Again this ensures that the arguments of the logarithm function are positive.

TWENTY-FOURTH LECTURE

SOME DEFINITE INTEGRALS WHOSE VALUES ARE INFINITE OR INDETERMINATE. PRINCIPAL VALUES OF INDETERMINATE INTEGRALS

In the preceding lectures, we demonstrated many remarkable properties of the definite integral

(1)
$$\int_{x_0}^{X} f(x)\,dx,$$

but all the while supposing first that the limits x_0, X were finite quantities and second that the function $f(x)$ remained finite and continuous between these same limits. If these two kinds of conditions are fulfilled, then designating the new values of x interposed between the extremes x_0, X by $x_1, x_2, \ldots, x_{n-1}$, we have

(2)
$$\int_{x_0}^{X} f(x)\,dx = \int_{x_0}^{x_1} f(x)\,dx + \int_{x_1}^{x_2} f(x)\,dx + \ldots + \int_{x_{n-1}}^{X} f(x)\,dx.$$

When the intermediate values are reduced to two values, one differing very little from x_0 and represented by ξ_0, and the other differing very little from X and represented by ξ, equation (2) becomes

$$\int_{x_0}^{X} f(x)\,dx = \int_{x_0}^{\xi_0} f(x)\,dx + \int_{\xi_0}^{\xi} f(x)\,dx + \int_{\xi}^{X} f(x)\,dx,$$

which we can write as

$$\int_{x_0}^{X} f(x)\,dx = (\xi_0 - x_0)f[x_0 + \theta_0(\xi_0 - x_0)]$$

$$+ \int_{\xi_0}^{\xi} f(x)\,dx + (X - \xi)f[\xi + \theta(X - \xi)],$$

with θ_0, θ designating two numbers less than unity. If in the last formula, ξ_0 converges toward the limit x_0 and ξ toward the limit X, we will conclude, in passing to the limits, that

(3) $$\int_{x_0}^{X} f(x)\,dx = \lim \int_{\xi_0}^{\xi} f(x)\,dx.$$

If the extreme values x_0, X become infinite, or if the function $f(x)$ does not remain finite and continuous from $x = x_0$ up through $x = X$, we can no longer assert that the quantity designated by S in the preceding lectures will have a fixed limit, and consequently we can no longer see what meaning[40] we ought to attach to notation (1), which served to generally represent the limit of S. To remove all incertitude and render a clear and precise significance to notation (1) in every case, it suffices to stretch equations (2) and (3) by analogy[41] to cases where even they are not able to be rigorously demonstrated. This can be seen in some examples.

First let us consider the integral

(4) $$\int_{-\infty}^{\infty} e^x\,dx.$$

If we designate two variable quantities by ξ_0 and ξ, the first of which converges to the limit $-\infty$, and the second to the limit ∞, we will write for formula (3)

$$\int_{-\infty}^{\infty} e^x\,dx = \lim \int_{\xi_0}^{\xi} e^x\,dx = \lim (e^{\xi} - e^{\xi_0}) = e^{\infty} - e^{-\infty} = \infty.$$

Thus, integral (4) has a positive, infinite value.

Second, let us consider the integral

(5) $$\int_{0}^{\infty} \frac{dx}{x}$$

taken between two limits one of which is infinite while the other renders the function under the sign $\int$, viz., $\frac{1}{x}$, infinite. In designating two positive quantities by

[40] *sens.*
[41] *d'étendre par analogie.*

664

ξ_0 and ξ, the first of which converges toward the limit zero the second toward the limit ∞, we will write for formula (3)

$$\int_0^\infty \frac{dx}{x} = \lim \int_{\xi_0}^\xi \frac{dx}{x} = \lim l \frac{\xi}{\xi_0} = l \frac{\infty}{0} = \infty.$$

Thus integral (5) again has a positive, infinite value.

It is essential to observe that if both the variable x and the function $f(x)$ remain finite for one of the limits of integral (1), we will be able to reduce formula (3) to one of the following two:

(6) $$\int_{x_0}^X f(x)\,dx = \lim \int_{x_0}^\xi f(x)\,dx, \qquad \int_{x_0}^X f(x)\,dx = \lim \int_{\xi_0}^X f(x)\,dx.$$

We will draw the following results from these last equations:

(7) $$\int_{-\infty}^0 e^x dx = e^0 - e^{-\infty} = 1, \qquad \int_0^\infty e^x\,dx = e^\infty - e^0 = \infty,$$

$$\int_{-1}^0 \frac{dx}{x} = l0 = -\infty, \qquad \int_0^1 \frac{dx}{x} = l\frac{1}{0} = \infty.$$

Now let us consider the integral

(8) $$\int_{-1}^{+1} \frac{dx}{x},$$

in which the function under the $\int$ sign, viz., $\frac{1}{x}$, becomes infinite for the particular value $x = 0$, which is contained between the limits $x = -1$, $x = +1$. We will write for formula (2),

(9) $$\int_{-1}^{+1} \frac{dx}{x} = \int_{-1}^0 \frac{dx}{x} + \int_0^1 \frac{dx}{x} = -\infty + \infty.$$

The value of integral (8) thus appears to be indeterminate. To assure ourselves that it is in fact so, it suffices to observe that if we designate an infinitely small number by ε and two positive but arbitrary constants by μ, ν, we will have, in virtue of the formulas in (6),

(10) $$\int_{-1}^0 \frac{dx}{x} = \lim \int_{-1}^{-\varepsilon\mu} \frac{dx}{x}, \qquad \int_0^1 \frac{dx}{x} = \lim \int_{\varepsilon\nu}^1 \frac{dx}{x}.$$

Consequently, formula (9) will become

$$(11) \quad \int_{-1}^{+1} \frac{dx}{x} = \lim \left(\int_{-1}^{-\varepsilon\mu} \frac{dx}{x} + \int_{\varepsilon\nu}^{1} \frac{dx}{x} \right) = \lim \left(l\varepsilon\mu + l\frac{\mu}{\nu} \right) = 1\frac{\mu}{\nu}$$

and will furnish a completely indeterminate value for integral (8), because this value will be the Napierien logarithm[42] of the arbitrary constant $\frac{\mu}{\nu}$.

Now let us imagine that the function $f(x)$ becomes infinite between the limits $x = x_0$, $x = X$ for the particular values of x represented by $x_1, x_2, \ldots, x_m$. If we designate an infinitely small number by ε, and we designate positive but arbitrary constants by $\mu_1, \nu_1, \mu_2, \nu_2, \ldots, \mu_m, \nu_m$, we will write for formulas (2) and (3)

$$(12) \int_{x_0}^{X} f(x)\,dx = \int_{x_0}^{x_1} f(x)\,dx + \int_{x_1}^{x_2} f(x)\,dx + \ldots + \int_{x_m}^{X} f(x)\,dx$$

$$= \lim \left[\int_{x_0}^{x_1-\varepsilon\mu_1} f(x)\,dx + \int_{x_1+\varepsilon\nu_1}^{x_2-\varepsilon\mu_2} f(x)\,dx + \ldots + \int_{x_m+\varepsilon\nu_m}^{X} f(x)\,dx \right].$$

If the limits x_0, X are replaced by $-\infty$ and ∞, we will have

$$(13) \int_{-\infty}^{+\infty} f(x)\,dx = \lim \left[\int_{-\frac{1}{\varepsilon\mu}}^{x_1-\varepsilon\mu_1} f(x)\,dx + \int_{x_1+\varepsilon\nu_1}^{x_2-\varepsilon\mu_2} f(x)\,dx + \ldots + \int_{x_m+\varepsilon\nu_m}^{\frac{1}{\varepsilon\nu}} f(x)\,dx \right],$$

with μ, ν designating two new positive but arbitrary constants. Let us add that on the right-hand side of formula (13), we would have put X in the place of $\frac{1}{\varepsilon\nu}$ or x_0 in the place of $-\frac{1}{\varepsilon\nu}$ if one of the two quantities x_0, X had become infinite. In every case, the values of the integrals

$$(14) \qquad \int_{x_0}^{X} f(x)\,dx, \qquad \int_{-\infty}^{+\infty} f(x)\,dx$$

deduced from equations (12) and (13) can, depending on the nature of the function $f(x)$, be either infinite quantities or finite and determinate quantities or indeterminate quantities, which will depend on arbitrary values for the arbitrary constants μ, ν, $\mu_1, \nu_1, \ldots, \mu_m, \nu_m$.

If, in formulas (12) and (13), we reduce the arbitrary constants μ, ν, μ_1, $\nu_1, \ldots, \mu_m, \nu_m$ to unity, we will find

$$(15) \quad \int_{x_0}^{X} f(x)\,dx = \lim \left[\int_{x_0}^{x_1-\varepsilon} f(x)\,dx + \int_{x_1+\varepsilon}^{x_2-\varepsilon} f(x)\,dx + \ldots + \int_{x_m+\varepsilon}^{X} f(x)\,dx \right],$$

---

[42] Today this is called the "natural logarithm."

$$(16) \quad \int_{-\infty}^{+\infty} f(x)\,dx = \lim \left[\int_{-\frac{1}{\varepsilon}}^{x_1 - \varepsilon} f(x)\,dx + \int_{x_1 + \varepsilon}^{x_2 - \varepsilon} f(x)\,dx + \ldots + \int_{x_m + \varepsilon}^{\frac{1}{\varepsilon}} f(x)\,dx \right].$$

Whenever the integrals in (14) become indeterminate, equations (15) and (16) furnish for each of them a unique, particular value which we will call the *principal value*. If we take integral (8) as an example, we will recognize that its principal value reduces to zero though in general its value is indeterminate.

George Boole

(1815–1864)

HIS LIFE AND WORK

The class of people (drenched in rain) is identical to the class of people ([drenched in rain] and [drenched in rain]). This is an example of the principle of *idempotency* (literally "self-power"). George Boole discovered its full power. His wife's misunderstanding of it would lead to his premature death.

George Boole was born in Lincoln in the north of England, not far from the birthplace of the greatest English scientist of all time, Sir Isaac Newton. Boole's father, John Boole came from a family that had lived in Lincolnshire for many centuries but never flourished. Although trained as a shoemaker, John Boole developed a lively interest in scientific instruments, especially telescopes. In 1800 at the age of twenty-three, he went to London to seek his fortune.

London held no fortune for John, only a bride. In 1806, he met Mary Ann Joyce, a younger daughter from an old Berkshire family who was working in London as a lady's maid. They married after a short courtship but had to live apart at their places of employment for six months. Then, as now, London was by far the most expensive place to live in England. Realizing that he would never make his fortune in London, John Boole moved back to Lincoln in 1807 and set up a shoemaker's business.

The northern English air might not have suited Mary Ann Boole at first. For years she tried and failed to conceive a much hoped for child. She and her husband were overjoyed when at the age of thirty-four, after more than eight years of marriage, she became pregnant with their first child. On November 2, 1815, she gave birth to a son, George, who was named after his recently deceased paternal grandfather. Motherhood must have agreed with Mary Ann Boole. In short order she gave birth to a daughter and two more sons.

George Boole had a distinguished record at the local schools in spite of the shy and retiring character he would display his whole life. He supplemented his education in the classroom with much learning outside of it. As the first born, Boole was the apple of his parents' eyes. His father spent many evenings with him building cameras, kaleidoscopes, microscopes, telescopes, and sundials.

However, Boole first achieved notoriety thanks to his interest in classical languages. After learning Latin from a local bookseller, Boole taught himself classical Greek. At the age of fourteen, he made a metrical translation of an ode written by the Greek poet Meleager. The translation so impressed Boole's father that he persuaded the local *Lincoln Herald* to print it along with the translator's age. A local schoolmaster wrote to the editor that the translation far exceeded the powers of any fourteen-year-old. No doubt this schoolmaster would have been skeptical of Boole's mathematical genius.

Even in his teens, Boole had become a religious nonconformist. Although raised in the orthodox Church of England, his ability in languages allowed him to read a wide range of Christian theology. In some early notebooks, Boole compared the Christian trinity with the three dimensions of space. At an early age, Boole developed an attraction to the ancient Hebrew conception of God as an absolute unity. For a while, Boole even considered the idea of converting to Judaism. In the end he became a nonconformist Unitarian.

Had he come from a more well to do family, he certainly would have gone to Oxford or Cambridge. But that would never be his lot in life. As the son of a struggling shoemaker there was no way that Boole could even contemplate a university education, much less the issues that his religious unorthodoxy might raise.

So, rather than taking instruction, young George Boole decided to give instruction. At the age of seventeen, he became a schoolteacher in Doncaster. His nonconformist religious views made for a short stay there, and within a year Boole moved to a post in Liverpool. Convinced of his abilities as a teacher and an administrator, he left his Liverpool position after just six months to return to Lincoln where, in the Fall of 1834, he opened a Classical, Commercial, and Mathematical Academy, a day school for young boys and girls.

Boole's headmastership of the school must have been a success. In 1838, he took complete charge of a much larger academy in Waddington. With Waddington just four miles from Lincoln, Boole was able to maintain his ties with charitable and educational institutions in Lincoln. If Boole received his scientific tutelage from his father, he must have learned of the importance of good works from his mother. At the age of twenty, he became a cofounder and trustee of the Female Penitents Home in Lincoln, a center for the rehabilitation of wayward women. He was also an officer of the named Early Closing Association. This was not a temperance union but, instead, an organization seeking reasonable limits on working hours.

Boole's most important civic association in Lincoln was with the Mechanics Institution, whose founding coincided with his return in 1834. It was a combination trade school and lending library for the betterment of the working class. John Boole was its first curator and he made sure that the library was well stocked with Royal Society publications and great English and continental works such as Newton's *Principia*, Laplace's *Mécanique céleste*, and Lagrange's *Mécanique analytique*. George Boole devoured this material reading it not once, but again and again, if necessary, in order to master it.

Boole also benefited greatly from his acquaintance with one of the patrons of the Mechanics Institution, Sir Edward Bromhead, a Fellow of the Royal Society. Bromhead was a Cambridge graduate and a talented amateur mathematician who had an eye for mathematical genius. (He had recently been a sponsor of the Nottingham mathematician George Green.) Boole must have spent many hours visiting Bromhead at his family manor, Thurlby Hall, just outside Lincoln, and borrowing volumes from Bromhead's library.

The environment in and around Lincoln must have suited Boole. In 1838, he published his first paper *On Certain Theorems in the Calculus of Variations*. His early papers covered a wide range of topics: differential equations, integration, logic, probability, geometry, and linear transformations. These papers all appeared in the recently founded *Cambridge Mathematical Journal*. By 1843, Boole had the confidence to submit his paper *On a General Method in Analysis* to the Royal Society, the apex of British science. The Royal Society almost failed to consider Boole's paper because no one on the Society's council knew Boole. Fortunately, the head of the Society's mathematics committee suggested that two specialists review the paper. One of the specialists wanted to reject the paper. His opinion did not prevail. The other specialist not only recommended publication, but also nominated the paper for the gold medal that the Society was offering for the best paper submitted between 1841 and 1844. His recommendation carried the day, earning Boole the acclaim of the British mathematics community.

In spite of this award, Boole continued his habit of often refusing introductions to eminent mathematicians feeling that his work was inferior to theirs. In fact, as late as 1845 at almost the age of thirty, he considered taking an undergraduate degree at Cambridge. When he realized that he could not shirk his continuing financial responsibility for his parents, Boole reversed course and instead of seeking to enroll in a university he decided to seek a university professorship. Fortunately for Boole, the government of Sir Robert Peel passed a bill establishing three new Queen's Colleges in Ireland in 1846. Boole quickly applied for a professorship at these new colleges and hoped that he would be appointed within a year. That did not happen. The great Irish potato famine diverted the government's attention away from the colleges to more important matters. Boole finally received his appointment to Queen's College Cork in southwest Ireland in August 1849, shortly after the death of his father.

Boole's paper *On a General Method in Analysis* was a precursor to his groundbreaking work in logic. It focused on the structure and form of the objects of mathematical analysis, thereby laying the foundation for his attention to the structure and form of logic itself. In a postscript to his paper Boole wrote:

> *The position which I am most anxious to establish is that any great advance in the higher analysis must be sought for by an increased attention to the laws of the combination of symbols. The value of this principle can scarcely be overrated.*

At Queen's, Boole quickly earned appointment to be Dean of Faculty. In this position he gave an annual address to the College Faculty at the start of every college session. His 1851 address, titled *The Claims of Science, especially as founded in it Relations to Human Nature*, contains hints of what he would address three years later in *The Laws of Thought*. In this address he asked:

> *First, whether there exist, with reference to our mental faculties, such general laws as are necessary to constitute a science; for we have seen that it is essentially in recognition of general laws, not of particular facts, that science consists.*
>
> *Secondly, supposing that such general laws are discoverable, what is the nature of the relation which the mind sustains towards them? Is it, like that of external nature, a relation of necessary obedience, or is it a relation of some distinct kind having no example and no parallel in the material system? . . .*
>
> *If it is asked whether out of these common principles of the reason we are able to deduce the actual expression of its fundamental laws, I reply that this is possible, and that the results constitute the true basis*

of mathematics. I speak here not of the mathematics of number and quantity alone, but a mathematics in its larger, and I believe, truer sense, as universal reasoning expressed in symbolical forms, and conducted by laws, which have their ultimate abode in the human mind.

Aristotle and Leibniz, the greatest logicians of all time prior to Boole, believed that it should be possible to express the fundamental laws of logic in a mathematical form. Prior to Boole, many mathematicians had tried to do just that. Indeed, Leibniz had developed a scheme in which the expression "All X are Y" was expressed as "X/Y." By analogy to arithmetic,

$$X/Y * Y/Z = X/Z,$$

thereby representing that if all X are Y and all Y are Z then all X are Z. But this was about the limit to all attempts prior to Boole. All of the previous attempts to render logic in algebraic form had tried to force it into the algebra of real numbers. Boole realized that another algebra was needed. In *The Laws of Thought*, he progressed along the road to abstraction by performing algebraic operations on symbols representing entities not previously considered mathematical objects at all. The intersection of two classes was represented by a multiplication-like operation, and the union of two classes was represented by an addition-like operation.

Classical logic had concentrated on the "four forms" of statement — all X are Y, no X are Y, some X are Y, some X are not Y. Boole began by focusing on the first of these forms: All X are Y. He represented the assertion that all X are Y as $XY = Y$. Thus, if all X are Y and all Y are X, then $XY = YX$. In particular $XX = X$; that is, $X^2 = X$, the principle of idempotency, for all classes X, not just for two particular values in numeric algebra. With "$+$" representing class intersection, Boole demonstrated a distributive law of class intersection over class union, namely, $Z(X + Y) = ZX + ZY$. For example, the intersection of Z, the class of left-handed people with the union of X, the class of Englishmen, and Y the class of Irishmen is identical to the union of the class of left-handed Englishmen with the class of left-handed Irishmen.

Boole then demonstrated how purely symbolic manipulation of class representations could be used to represent classical schemes of implications without making reference to the schemes themselves in the argument. Here is how Boole proved that if all X are Y and all Y are Z, then all X are Z. By supposition, all X are Y and all Y are Z. Put symbolically, this is $X = XY$ and $Y = YZ$. Substituting $Y = YZ$ in $X = XY$ produces

$$X = XY = X(YZ) = (XY)Z = XZ$$

or $X = XZ$, which is the representation of all X are Z.

Boole achieved even more powerful results when he introduced the symbol 1 for the universal class, the symbol 0 for the empty class, and the operator "$-$" for complementation. (The class $P - Q$ is the set of elements in P not in Q.) By doing this, he was able to prove that for all X, $X(1 - X) = 0$, the principle of noncontradiction. (Nothing has a property and the negation of that property.)

$$X = X^2$$
$$X - X^2 = 0$$
$$X1 - X^2 = 0$$
$$X(1 - X) = 0$$

Aristotle had considered the principles of noncontradiction and idempotency to be of equal stature, both necessary postulates for logical reasoning. With his algebra of logic, George Boole had proven that the principle of noncontradiction could be deduced from the more intuitive principle of idempotency, that the class of objects with the property X and the property X (as many times as you like, in fact) is identical to the class of objects with the property X just once. Half a century after it was written, Bertrand Russell described *The Laws of Thought* as the work "in which pure mathematics was discovered."

If Cork was the scene of Boole's greatest mathematical accomplishment, it was also the site of the happiest time in his life. When he arrived at Cork, Boole soon became a close friend of John Ryall, Professor of Greek and a Vice President of the College. In 1850, Boole met Ryall's eighteen-year-old niece, Mary Everest. Mary Everest came from a family of country gentry in Gloucestershire. Her father Thomas was a Reverend in the Anglican Church. His older brother, Sir George Everest, had been Surveyor-General of India and was the first man to survey the world's tallest mountain, which he named after himself!

Unlike Boole, who had only known the harshness of the north of England as a child, Mary Everest had seen something of the world. Her father suffered from consumption and for several years took a leave in the healthier climate of France where he received homeopathic treatment. Homeopathy (literally "like the ailment") stressed treating an ailment by the administration of small doses of drugs that, if given to a healthy patient, would induce symptoms of the very ailment being treated. Reverend Everest also believed in long walks and ice cold baths before breakfast as invigorating morning constitutionals. The treatments worked well enough to allow the Everest family to return home to his parsonage.

In his diary, Boole confessed to being a hopeless romantic who had fallen in and out of love quite often. Yet he appeared to have little romantic interest in Mary Everest when they first met. He contented himself with teaching her mathematics during the summers between 1850 and 1852. Between 1852 and 1855, George and

Mary contented themselves with a correspondence on mathematical topics, never once meeting. That came to an end in June 1855 when Thomas Everest died suddenly, leaving Mary destitute. Ever the romantic, George Boole immediately proposed marriage. Mary accepted, and they married on September 11 of that year.

For the first two years after their marriage, they lived in a house called "College View," which was about a ten-minute walk to the college. It was convenient, but soon became too crowded for the Booles whose first daughter Mary Ellen, born in 1856, would soon be followed by another daughter named Margaret, born to them in 1858. Shortly before Margaret's birth, they moved to a rented house in the village of Blackrock, four miles from the College but only half a mile from the railway station. The Booles loved this house. It had a wonderful view of Cork's magnificent harbor.

But this house soon became too small for their ever-expanding family, now blessed with two more daughters, Alicia born in 1860 and Lucy born in 1862. In anticipation of yet another child, once again a daughter named Ethel, born in 1864, the Booles moved to a house named Litchfield Cottage in the village of Ballintemple, a mile closer to the college but much farther from the railway station. Mary Boole saw this as an opportunity for George to take long walks of the sort that her father had encouraged to help with the rheumatic affliction of which he often complained.

But the long walks did not cure Boole. One of them killed him. On November 24, 1864, Boole made the long walk from Ballintemple to the College in a torrential rain. After lecturing in his soaked clothes he made the long walk back to Ballintemple. When he got home, he collapsed into bed and became feverishly ill. Still convinced of the correctness of homeopathy, Mary Boole proceeded to pour buckets and buckets of cold water over her husband as a means of curing him. But, of course, the bronchial pneumonia Boole developed was not idempotent. The cold-water therapy only made him worse. Mary Boole finally agreed to call a medical doctor on December 5. But it was already too late. By then Boole was in a deep feverish coma. Three days later he died.

Boole achieved many honors in his life. Trinity College Dublin and Oxford Universities awarded him honorary doctorates. In 1857 he was made a Fellow of the Royal Society. But perhaps posterity has bestowed the greatest honor on Boole. Many computer languages have objects that take the values TRUE or FALSE. They are called "Booleans"!

AN INVESTIGATION OF THE LAWS OF THOUGHT

CHAPTER I

NATURE AND DESIGN OF THIS WORK

1. The design of the following treatise is to investigate the fundamental laws of those operations of the mind by which reasoning is performed; to give expression to them in the symbolical language of a Calculus, and upon this foundation to establish the science of Logic and construct its method; to make that method itself the basis of a general method for the application of the mathematical doctrine of Probabilities; and, finally, to collect from the various elements of truth brought to view in the course of these inquiries some probable intimations concerning the nature and constitution of the human mind.

2. That this design is not altogether a novel one it is almost needless to remark, and it is well known that to its two main practical divisions of Logic and Probabilities a very considerable share of the attention of philosophers has been directed. In its ancient and scholastic form, indeed, the subject of Logic stands almost exclusively associated with the great name of Aristotle. As it was presented to ancient Greece in the partly technical, partly metaphysical disquisitions of the Organon, such, with scarcely any essential change, it has continued to the present day. The stream of original inquiry has rather been directed towards questions of general philosophy, which, though they have arisen among the disputes of the logicians, have outgrown their origin, and given to successive ages of speculation their peculiar bent and character. The eras of Porphyry and Proclus, of Anselm and Abelard, of Ramus, and of Descartes, together with the final protests of Bacon and Locke, rise up before the mind as examples of the remoter influences of the study upon the course of human thought, partly in suggesting topics fertile of discussion, partly in provoking remonstrance against its own undue pretensions. The history of the theory of Probabilities, on the other hand, has presented far more of that character of steady growth which belongs to science. In its origin the early genius of Pascal,—in its maturer stages of development the most recondite of all the mathematical speculations of Laplace,—were directed to its improvement; to omit here the mention of other names scarcely less distinguished than these. As the study of Logic has been remarkable for the kindred questions of Metaphysics to which it has given occasion, so that of Probabilities also has been remarkable for the impulse which it has bestowed upon the higher

departments of mathematical science. Each of these subjects has, moreover, been justly regarded as having relation to a speculative as well as to a practical end. To enable us to deduce correct inferences from given premises is not the only object of Logic; nor is it the sole claim of the theory of Probabilities that it teaches us how to establish the business of life assurance on a secure basis; and how to condense whatever is valuable in the records of innumerable observations in astronomy, in physics, or in that field of social inquiry which is fast assuming a character of great importance. Both these studies have also an interest of another kind, derived from the light which they shed upon the intellectual powers. They instruct us concerning the mode in which language and number serve as instrumental aids to the processes of reasoning; they reveal to us in some degree the connexion between different powers of our common intellect; they set before us what, in the two domains of demonstrative and of probable knowledge, are the essential standards of truth and correctness,—standards not derived from without, but deeply founded in the constitution of the human faculties. These ends of speculation yield neither in interest nor in dignity, nor yet, it may be added, in importance, to the practical objects, with the pursuit of which they have been historically associated. To unfold the secret laws and relations of those high faculties of thought by which all beyond the merely perceptive knowledge of the world and of ourselves is attained or matured, is an object which does not stand in need of commendation to a rational mind.

3. But although certain parts of the design of this work have been entertained by others, its general conception, its method, and, to a considerable extent, its results, are believed to be original. For this reason I shall offer, in the present chapter, some preparatory statements and explanations, in order that the real aim of this treatise may be understood, and the treatment of its subject facilitated.

It is designed, in the first place, to investigate the fundamental laws of those operations of the mind by which reasoning is performed. It is unnecessary to enter here into any argument to prove that the operations of the mind are in a certain real sense subject to laws, and that a science of the mind is therefore *possible*. If these are questions which admit of doubt, that doubt is not to be met by an endeavour to settle the point of dispute *à priori*, but by directing the attention of the objector to the evidence of actual laws, by referring him to an actual science. And thus the solution of that doubt would belong not to the introduction to this treatise, but to the treatise itself. Let the assumption be granted, that a science of the intellectual powers is possible, and let us for a moment consider how the knowledge of it is to be obtained.

4. Like all other sciences, that of the intellectual operations must primarily rest upon observation,—the subject of such observation being the very operations and

processes of which we desire to determine the laws. But while the necessity of a foundation in experience is thus a condition common to all sciences, there are some special differences between the modes in which this principle becomes available for the
determination of general truths when the subject of inquiry is the mind, and when
the subject is external nature. To these it is necessary to direct attention.

The general laws of Nature are not, for the most part, immediate objects of perception. They are either inductive inferences from a large body of facts, the common
truth in which they express, or, in their origin at least, physical hypotheses of a causal
nature serving to explain phænomena with undeviating precision, and to enable us
to predict new combinations of them. They are in all cases, and in the strictest sense
of the term, *probable* conclusions, approaching, indeed, ever and ever nearer to certainty, as they receive more and more of the confirmation of experience. But of the
character of probability, in the strict and proper sense of that term, they are never
wholly divested. On the other hand, the knowledge of the laws of the mind does
not require as its basis any extensive collection of observations. The general truth is
seen in the particular instance, and it is not confirmed by the repetition of instances.
We may illustrate this position by an obvious example. It may be a question whether
that formula of reasoning, which is called the *dictum* of Aristotle, *de omni et nullo*,
expresses a primary law of human reasoning or not; but it is no question that it
expresses a general truth in Logic. Now that truth is made manifest in all its generality by reflection upon a single instance of its application. And this is both an evidence that the particular principle or formula in question is founded upon some
general law or laws of the mind, and an illustration of the doctrine that the perception
of such general truths is not derived from an induction from many instances, but is
involved in the clear apprehension of a single instance. In connexion with this truth
is seen the not less important one that our knowledge of the laws upon which the
science of the intellectual powers rests, whatever may be its extent or its deficiency,
is not probable knowledge. For we not only see in the particular example the general truth, but we see it also as a certain truth,—a truth, our confidence in which
will not continue to increase with increasing experience of its practical verifications.

5. But if the general truths of Logic are of such a nature that when presented
to the mind they at once command assent, wherein consists the difficulty of constructing the Science of Logic? Not, it may be answered, in collecting the materials
of knowledge, but in discriminating their nature, and determining their mutual place
and relation. All sciences consist of general truths, but of those truths some only are
primary and fundamental, others are secondary and derived. The laws of elliptic
motion, discovered by Kepler, are general truths in astronomy, but they are not its
fundamental truths. And it is so also in the purely mathematical sciences. An almost

boundless diversity of theorems, which are known, and an infinite possibility of others, as yet unknown, rest together upon the foundation of a few simple axioms; and yet these are all *general* truths. It may be added, that they are truths which to an intelligence sufficiently refined would shine forth in their own unborrowed light, without the need of those connecting links of thought, those steps of wearisome and often painful deduction, by which the knowledge of them is actually acquired. Let us define as fundamental those laws and principles from which all other general truths of science may be deduced, and into which they may all be again resolved. Shall we then err in regarding that as the true science of Logic which, laying down certain elementary laws, confirmed by the very testimony of the mind, permits us thence to deduce, by uniform processes, the entire chain of its secondary consequences, and furnishes, for its practical applications, methods of perfect generality? Let it be considered whether in any science, viewed either as a system of truth or as the foundation of a practical art, there can properly be any other test of the completeness and the fundamental character of its laws, than the completeness of its system of derived truths, and the generality of the methods which it serves to establish. Other questions may indeed present themselves. Convenience, prescription, individual preference, may urge their claims and deserve attention. But as respects the question of what constitutes science in its abstract integrity, I apprehend that no other considerations than the above are properly of any value.

6. It is designed, in the next place, to give expression in this treatise to the fundamental laws of reasoning in the symbolical language of a Calculus. Upon this head it will suffice to say, that those laws are such as to suggest this mode of expression, and to give to it a peculiar and exclusive fitness for the ends in view. There is not only a close analogy between the operations of the mind in general reasoning and its operations in the particular science of Algebra, but there is to a considerable extent an exact agreement in the laws by which the two classes of operations are conducted. Of course the laws must in both cases be determined independently; any formal agreement between them can only be established *a posteriori* by actual comparison. To borrow the notation of the science of Number, and then assume that in its new application the laws by which its use is governed will remain unchanged, would be mere hypothesis. There exist, indeed, certain general principles founded in the very nature of language, by which the use of symbols, which are but the elements of scientific language, is determined. To a certain extent these elements are arbitrary. Their interpretation is purely conventional: we are permitted to employ them in whatever sense we please. But this permission is limited by two indispensable conditions,— first, that from the sense once conventionally established we never, in the same process of reasoning, depart; secondly, that the laws by which the process is

conducted be founded exclusively upon the above fixed sense or meaning of the symbols employed. In accordance with these principles, any agreement which may be established between the laws of the symbols of Logic and those of Algebra can but issue in an agreement of processes. The two provinces of interpretation remain apart and independent, each subject to its own laws and conditions.

Now the actual investigations of the following pages exhibit Logic, in its practical aspect, as a system of processes carried on by the aid of symbols having a definite interpretation, and subject to laws founded upon that interpretation alone. But at the same time they exhibit those laws as identical in form with the laws of the general symbols of algebra, with this single addition, viz., that the symbols of Logic are further subject to a special law (Chap. II.), to which the symbols of quantity, as such, are not subject. Upon the nature and the evidence of this law it is not purposed here to dwell. These questions will be fully discussed in a future page. But as constituting the essential ground of difference between those forms of inference with which Logic is conversant, and those which present themselves in the particular science of Number, the law in question is deserving of more than a passing notice. It may be said that it lies at the very foundation of general reasoning,—that it governs those intellectual acts of conception or of imagination which are preliminary to the processes of logical deduction, and that it gives to the processes themselves much of their actual form and expression. It may hence be affirmed that this law constitutes the germ or seminal principle, of which every approximation to a general method in Logic is the more or less perfect development.

7. The principle has already been laid down (5) that the sufficiency and truly fundamental character of any assumed system of laws in the science of Logic must partly be seen in the perfection of the methods to which they conduct us. It remains, then, to consider what the requirements of a general method in Logic are, and how far they are fulfilled in the system of the present work.

Logic is conversant with two kinds of relations,—relations among things, and relations among facts. But as facts are expressed by propositions, the latter species of relation may, at least for the purposes of Logic, be resolved into a relation among propositions. The assertion that the fact or event A is an invariable consequent of the fact or event B may, to this extent at least, be regarded as equivalent to the assertion, that the truth of the proposition affirming the occurrence of the event B always implies the truth of the proposition affirming the occurrence of the event A. Instead, then, of saying that Logic is conversant with relations among things and relations among facts, we are permitted to say that it is concerned with relations among things and relations among propositions. Of the former kind of relations we have an example in the proposition—"All men are mortal;" of the latter kind in the proposition—"If the sun is totally eclipsed, the stars will become visible." The one expresses a relation between "men" and "mortal beings," the other between the elementary propositions—"The sun is totally eclipsed;" "The stars will become visible." Among such relations I suppose to be included those which affirm or deny existence with respect to things, and those which affirm or deny truth with respect to propositions. Now let those things or those propositions among which relation is expressed be termed the elements of the propositions by which such relation is expressed. Proceeding from this definition, we may then say that the *premises* of any logical argument express *given* relations among certain elements, and that the *conclusion* must express an *implied* relation among those elements, or among a part of them, i.e. a relation implied by or inferentially involved in the premises.

8. Now this being premised, the requirements of a general method in Logic seem to be the following:—

1st. As the conclusion must express a relation among the whole or among a part of the elements involved in the premises, it is requisite that we should possess the means of eliminating those elements which we desire not to appear in the conclusion, and of determining the whole amount of relation implied by the premises among the elements which we wish to retain. Those elements which do not present themselves in the conclusion are, in the language of the common Logic, called middle terms; and the species of elimination exemplified in treatises on Logic consists in deducing from two propositions, containing a common element or middle term, a conclusion connecting the two remaining terms. But the problem of elimination, as contemplated in this work, possesses a much wider scope. It proposes not merely the elimination of one middle term from two propositions, but the elimination generally of middle terms from propositions, without regard to the number of either of them, or to the nature of their connexion. To this object neither the processes of Logic nor those of Algebra, in their actual state, present any strict parallel. In the latter science the problem of elimination is known to be limited in the following manner:—From two equations we can eliminate one symbol of quantity; from three equations two symbols; and, generally, from n equations $n - 1$ symbols. But though this condition, necessary in Algebra, seems to prevail in the existing Logic also, it has no essential place in Logic as a science. There, no relation whatever can be proved to prevail between the number of terms to be eliminated and the number of propositions from which the elimination is to be effected. From the equation representing a single proposition, any number of symbols representing terms or elements in Logic may be eliminated; and from any number of equations representing propositions, one or any other number of symbols of this kind may be eliminated in a similar manner. For such elimination there exists one general process applicable to all cases.

This is one of the many remarkable consequences of that distinguishing law of the symbols of Logic, to which attention has been already directed.

2ndly. It should be within the province of a general method in Logic to express the final relation among the elements of the conclusion by any admissible *kind* of proposition, or in any selected *order* of terms. Among varieties of kind we may reckon those which logicians have designated by the terms categorical, hypothetical, disjunctive, &c. To a choice or selection in the order of the terms, we may refer whatsoever is dependent upon the appearance of particular elements in the subject or in the predicate, in the antecedent or in the consequent, of that proposition which forms the "conclusion." But waiving the language of the schools, let us consider what really distinct species of problems may present themselves to our notice. We have seen that the elements of the final or inferred relation may either be *things* or *propositions.* Suppose the former case; then it might be required to deduce from the premises a definition or description of some one thing, or class of things, constituting an element of the conclusion in terms of the other things involved in it. Or we might form the conception of some thing or class of things, involving more than one of the elements of the conclusion, and require its expression in terms of the other elements. Again, suppose the elements retained in the conclusion to be propositions, we might desire to ascertain such points as the following, viz., Whether, in virtue of the premises, any of those propositions, taken singly, are true or false?—Whether particular combinations of them are true or false?—Whether, assuming a particular proposition to be true, any consequences will follow, and if so, what consequences, with respect to the other propositions?—Whether any particular condition being assumed with reference to certain of the propositions, any consequences, and what consequences, will follow with respect to the others? and so on. I say that these are general questions, which it should fall within the scope or province of a general method in Logic to solve. Perhaps we might include them all under this one statement of the final problem of practical Logic. Given a set of premises expressing relations among certain elements, whether things or propositions: required explicitly the whole relation consequent among *any* of those elements under any proposed conditions, and in any proposed form. That this problem, under all its aspects, is resolvable, will hereafter appear. But it is not for the sake of noticing this fact, that the above inquiry into the nature and the functions of a general method in Logic has been introduced. It is necessary that the reader should apprehend what are the specific ends of the investigation upon which we are entering, as well as the principles which are to guide us to the attainment of them.

9. Possibly it may here be said that the Logic of Aristotle, in its rules of syllogism and conversion, sets forth the elementary processes of which all reasoning consists,

and that beyond these there is neither scope nor occasion for a general method. I have no desire to point out the defects of the common Logic, nor do I wish to refer to it any further than is necessary, in order to place in its true light the nature of the present treatise. With this end alone in view, I would remark:—1st. That syllogism, conversion, &c., are not the ultimate processes of Logic. It will be shown in this treatise that they are founded upon, and are resolvable into, ulterior and more simple processes which constitute the real elements of method in Logic. Nor is it true in fact that all inference is reducible to the particular forms of syllogism and conversion.—*Vide* Chap. XV. 2ndly. If all inference were reducible to these two processes (and it has been maintained that it is reducible to syllogism alone), there would still exist the same necessity for a general method. For it would still be requisite to determine in what order the processes should succeed each other, as well as their particular nature, in order that the desired relation should be obtained. By the desired relation I mean that full relation which, in virtue of the premises, connects any elements selected out of the premises at will, and which, moreover, expresses that relation in any desired form and order. If we may judge from the mathematical sciences, which are the most perfect examples of method known, this *directive* function of Method constitutes its chief office and distinction. The fundamental processes of arithmetic, for instance, are in themselves but the elements of a possible science. To assign their nature is the first business of its method, but to arrange their succession is its subsequent and higher function. In the more complex examples of logical deduction, and especially in those which form a basis for the solution of difficult questions in the theory of Probabilities, the aid of a directive method, such as a Calculus alone can supply, is indispensable.

10. Whence it is that the ultimate laws of Logic are mathematical in their form; why they are, except in a single point, identical with the general laws of Number; and why in that particular point they differ;—are questions upon which it might not be very remote from presumption to endeavour to pronounce a positive judgment. Probably they lie beyond the reach of our limited faculties. It may, perhaps, be permitted to the mind to attain a knowledge of the laws to which it is itself subject, without its being also given to it to understand their ground and origin, or even, except in a very limited degree, to comprehend their fitness for their end, as compared with other and conceivable systems of law. Such knowledge is, indeed, unnecessary for the ends of science, which properly concerns itself with what is, and seeks not for grounds of preference or reasons of appointment. These considerations furnish a sufficient answer to all protests against the exhibition of Logic in the form of a Calculus. It is not because we choose to assign to it such a mode of manifestation, but because the ultimate laws of thought render that mode possible, and prescribe

its character, and forbid, as it would seem, the perfect manifestation of the science in any other form, that such a mode demands adoption. It is to be remembered that it is the business of science not to create laws, but to discover them. We do not originate the constitution of our own minds, greatly as it may be in our power to modify their character. And as the laws of the human intellect do not depend upon our will, so the forms of the science, of which they constitute the basis, are in all essential regards independent of individual choice.

11. Beside the general statement of the principles of the above method, this treatise will exhibit its application to the analysis of a considerable variety of propositions, and of trains of propositions constituting the premises of demonstrative arguments. These examples have been selected from various writers, they differ greatly in complexity, and they embrace a wide range of subjects. Though in this particular respect it may appear to some that too great a latitude of choice has been exercised, I do not deem it necessary to offer any apology upon this account. That Logic, as a science, is susceptible of very wide applications is admitted; but it is equally certain that its ultimate forms and processes are mathematical. Any objection *à priori* which may therefore be supposed to lie against the adoption of such forms and processes in the discussion of a problem of morals or of general philosophy must be founded upon misapprehension or false analogy. It is not of the essence of mathematics to be conversant with the ideas of number and quantity. Whether as a general habit of mind it would be desirable to apply symbolical processes to moral argument, is another question. Possibly, as I have elsewhere observed,[1] the perfection of the method of Logic may be chiefly valuable as an evidence of the speculative truth of its principles. To supersede the employment of common reasoning, or to subject it to the rigour of technical forms, would be the last desire of one who knows the value of that intellectual toil and warfare which imparts to the mind an athletic vigour, and teaches it to contend with difficulties, and to rely upon itself in emergencies. Nevertheless, cases may arise in which the value of a scientific procedure, even in those things which fall confessedly under the ordinary dominion of the reason, may be felt and acknowledged. Some examples of this kind will be found in the present work.

12. The general doctrine and method of Logic above explained form also the basis of a theory and corresponding method of Probabilities. Accordingly, the development of such a theory and method, upon the above principles, will constitute a distinct object of the present treatise. Of the nature of this application it may be desirable to give here some account, more especially as regards the character of the

[1]Mathematical Analysis of Logic. London: G. Bell. 1847.

solutions to which it leads. In connexion with this object some further detail will be requisite concerning the forms in which the results of the logical analysis are presented.

The ground of this necessity of a prior method in Logic, as the basis of a theory of Probabilities, may be stated in a few words. Before we can determine the mode in which the expected frequency of occurrence of a particular event is dependent upon the known frequency of occurrence of any other events, we must be acquainted with the mutual dependence of the events themselves. Speaking technically, we must be able to express the event whose probability is sought, as a function of the events whose probabilities are given. Now this explicit determination belongs in all instances to the department of Logic. Probability, however, in its mathematical acceptation, admits of numerical measurement. Hence the subject of Probabilities belongs equally to the science of Number and to that of Logic. In recognising the co-ordinate existence of both these elements, the present treatise differs from all previous ones; and as this difference not only affects the question of the possibility of the solution of problems in a large number of instances, but also introduces new and important elements into the solutions obtained, I deem it necessary to state here, at some length, the peculiar consequences of the theory developed in the following pages.

13. The measure of the probability of an event is usually defined as a fraction, of which the numerator represents the number of cases favourable to the event, and the denominator the whole number of cases favourable and unfavourable; all cases being supposed equally likely to happen. That definition is adopted in the present work. At the same time it is shown that there is another aspect of the subject (shortly to be referred to) which might equally be regarded as fundamental, and which would actually lead to the same system of methods and conclusions. It may be added, that so far as the received conclusions of the theory of Probabilities extend, and so far as they are consequences of its fundamental definitions, they do not differ from the results (supposed to be equally correct in inference) of the method of this work.

Again, although questions in the theory of Probabilities present themselves under various aspects, and may be variously modified by algebraical and other conditions, there seems to be one general type to which all such questions, or so much of each of them as truly belongs to the theory of Probabilities, may be referred. Considered with reference to the *data* and the *quæsitum*, that type may be described as follows:— 1st. The data are the probabilities of one or more given events, each probability being either that of the absolute fulfilment of the event to which it relates, or the probability of its fulfilment under given supposed conditions. 2ndly. The *quæsitum*, or object sought, is the probability of the fulfilment, absolutely or conditionally, of some other event differing in expression from those in the data, but more or less involving the same elements. As concerns the data, they are either *causally given*,—as when the probability of a particular throw of a die is deduced from a knowledge of the constitution of the piece,—or they are derived from observation of repeated instances of the success or failure of events. In the latter case the probability of an event may be defined as the limit toward which the ratio of the favourable to the whole number of observed cases approaches (the uniformity of nature being presupposed) as the observations are indefinitely continued. Lastly, as concerns the nature or relation of the events in question, an important distinction remains. Those events are either *simple* or *compound.* By a compound event is meant one of which the expression in language, or the conception in thought, depends upon the expression or the conception of other events, which, in relation to it, may be regarded as *simple* events. To say "it rains," or to say "it thunders," is to express the occurrence of a simple event; but to say "it rains and thunders," or to say " it either rains or thunders," is to express that of a compound event. For the expression of that event depends upon the elementary expressions, "it rains," " it thunders." The criterion of simple events is not, therefore, any supposed simplicity in their nature. It is founded solely on the mode of their expression in language or conception in thought.

14. Now one general problem, which the existing theory of Probabilities enables us to solve, is the following, viz.:—Given the probabilities of any simple events: required the probability of a given compound event, i.e. of an event compounded in a given manner out of the given simple events. The problem can also be solved when the compound event, whose probability is required, is subjected to given conditions, i.e. to conditions dependent also in a given manner on the given simple events. Beside this general problem, there exist also particular problems of which the principle of solution is known. Various questions relating to *causes* and *effects* can be solved by known methods under the particular hypothesis that the causes are mutually exclusive, but apparently not otherwise. Beyond this it is not clear that any advance has been made toward the solution of what may be regarded as the general problem of the science, viz.: Given the probabilities of any events, simple or compound, conditioned or unconditioned: required the probability of any other event equally arbitrary in expression and conception. In the statement of this question it is not even postulated that the events whose probabilities are given, and the one whose probability is sought, should involve some common elements, because it is the office of a method to determine whether the data of a problem are sufficient for the end in view, and to indicate, when they are not so, wherein the deficiency consists.

This problem, in the most unrestricted form of its statement, is resolvable by the method of the present treatise; or, to speak more precisely, its theoretical solution is completely given, and its practical solution is brought to depend only upon processes purely mathematical, such as the resolution and analysis of equations. The order and character of the general solution may be thus described.

15. In the first place it is always possible, by the preliminary method of the Calculus of Logic, to express the event whose probability is sought as a logical function of the events whose probabilities are given. The result is of the following character: Suppose that X represents the event whose probability is sought, A, B, C, &c. the events whose probabilities are given, those events being either simple or compound. Then the *whole* relation of the event X to the events A, B, C, &c. is deduced in the form of what mathematicians term a *development*, consisting, in the most general case, of four distinct classes of terms. By the first class are expressed those combinations of the events A, B, C, which both necessarily accompany and necessarily indicate the occurrence of the event X; by the second class, those combinations which necessarily accompany, but do not necessarily imply, the occurrence of the event X; by the third class, those combinations whose occurrence in connexion with the event X is impossible, but not otherwise impossible; by the fourth class, those combinations whose occurrence is impossible under any circumstances. I shall not dwell upon this statement of the result of the logical analysis of the problem, further than to remark that the elements which it presents are precisely those by which the expectation of the event X, as dependent upon our knowledge of the events A, B, C, is, or alone can be, affected. General reasoning would verify this conclusion; but general reasoning would not usually avail to disentangle the complicated web of events and circumstances from which the solution above described must be evolved. The attainment of this object constitutes the first step towards the complete solution of the question proposed. It is to be noted that thus far the process of solution is logical, i.e. conducted by symbols of logical significance, and resulting in an equation interpretable into a *proposition*. Let this result be termed the *final logical equation.*

The second step of the process deserves attentive remark. From the final logical equation to which the previous step has conducted us, are deduced, by inspection, a series of algebraic equations implicitly involving the complete solution of the problem proposed. Of the mode in which this transition is effected let it suffice to say, that there exists a definite relation between the laws by which the probabilities of events are expressed as algebraic functions of the probabilities of other events upon which they depend, and the laws by which the logical connexion of the events is itself expressed. This relation, like the other coincidences of formal law which have been referred to, is not founded upon hypothesis, but is made known to us by observation (i. 4), and reflection. If, however, its reality were assumed *à priori* as the basis of the very definition of Probability, strict deduction would thence lead us to the received numerical definition as a necessary consequence. The Theory of Probabilities stands, as it has already been remarked (i. 12), in equally close relation to Logic and to Arithmetic; and it is indifferent, so far as results are concerned, whether we regard it as springing out of the latter of these sciences, or as founded in the mutual relations which connect the two together.

16. There are some circumstances, interesting perhaps to the mathematician, attending the general solutions deduced by the above method, which it may be desirable to notice.

1st. As the method is independent of the number and the nature of the data, it continues to be applicable when the latter are insufficient to render determinate the value sought. When such is the case, the final expression of the solution will contain terms with arbitrary constant coefficients. To such terms there will correspond terms in the final logical equation (i. 15), the interpretation of which will inform us what new data are requisite in order to determine the values of those constants, and thus render the numerical solution complete. If such data are not to be obtained, we can still, by giving to the constants their limiting values 0 and 1, determine the limits within which the probability sought must lie independently of all further experience. When the event whose probability is sought is *quite* independent of those whose probabilities are given, the limits thus obtained for its value will be 0 and 1, as it is evident that they ought to be, and the interpretation of the constants will only lead to a re-statement of the original problem.

2ndly. The expression of the final solution will in all cases involve a particular element of quantity, determinable by the solution of an algebraic equation. Now when that equation is of an elevated degree, a difficulty may seem to arise as to the selection of the proper root. There are, indeed, cases in which both the elements given and the element sought are so obviously restricted by necessary conditions that no choice remains. But in complex instances the discovery of such conditions, by unassisted force of reasoning, would be hopeless. A distinct method is requisite for this end,—a method which might not inappropriately be termed the Calculus of Statistical Conditions. Into the nature of this method I shall not here further enter than to say, that, like the previous method, it is based upon the employment of the "final logical equation," and that it definitely assigns, 1st, the conditions which must be fulfilled among the numerical elements of the data, in order that the problem may be real, i.e. derived from a *possible experience*; 2ndly, the numerical limits, within which the probability sought must have been confined, if, instead of being determined by theory, it had been deduced directly by observation from the same system of phænomena from which the data were derived. It is clear that these limits will be actual limits of the probability sought. Now, on supposing the data subject to the conditions above assigned to them, it appears in every instance which I have examined that there exists one root, and only one root, of the final algebraic equation which is subject to the required limitations. Every source of ambiguity is thus removed. It would even seem that new truths relating to the theory of algebraic equations are thus incidentally brought to light. It is remarkable that the special element of quantity, to which the previous discussion relates, depends only upon the *data*, and not at all upon the *quæsitum* of the problem proposed. Hence the solution of each particular problem unties the knot of difficulty for a system of problems, viz., for that system of problems which is marked by the possession of common data, independently of the nature of their *quæsita*. This circumstance is important whenever from a particular system of data it is required to deduce a series of connected conclusions. And it further gives to the solutions of particular problems that character of relationship, derived from their dependence upon a central and fundamental unity, which not unfrequently marks the application of general methods.

17. But though the above considerations, with others of a like nature, justify the assertion that the method of this treatise, for the solution of questions in the theory of Probabilities, is a general method, it does not thence follow that we are relieved in all cases from the necessity of recourse to hypothetical grounds. It has been observed that a solution may consist entirely of terms affected by arbitrary constant coefficients,—may, in fact, be wholly indefinite. The application of the method of this work to some of the most important questions within its range would—were the data of experience alone employed—present results of this character. To obtain a *definite* solution it is necessary, in such cases, to have recourse to hypotheses possessing more or less of independent probability, but incapable of exact verification. Generally speaking, such hypotheses will differ from the immediate results of experience in partaking of a logical rather than of a numerical character; in prescribing the conditions under which phænomena occur, rather than assigning the relative frequency of their occurrence. This circumstance is, however, unimportant. Whatever their nature may be, the hypotheses assumed must thenceforth be regarded as belonging to the actual data, although tending, as is obvious, to give to the solution itself somewhat of a hypothetical character. With this understanding as to the possible sources of the data actually employed, the method is perfectly general, but for the correctness of the hypothetical elements introduced it is of course no more responsible than for the correctness of the numerical data derived from experience.

In illustration of these remarks we may observe that the theory of the reduction of astronomical observations[2] rests, in part, upon hypothetical grounds. It assumes certain positions as to the nature of error, the equal probabilities of its occurrence in

[2] The author designs to treat this subject either in a separate work or in a future Appendix. In the present treatise he avoids the use of the integral calculus.

the form of excess or defect, &c., without which it would be impossible to obtain any *definite* conclusions from a system of conflicting observations. But granting such positions as the above, the residue of the investigation falls strictly within the province of the theory of Probabilities. Similar observations apply to the important problem which proposes to deduce from the records of the majorities of a deliberative assembly the mean probability of correct judgment in one of its members. If the method of this treatise be applied to the mere numerical data, the solution obtained is of that wholly indefinite kind above described. And to show in a more eminent degree the insufficiency of those data by themselves, the interpretation of the arbitrary constants (i. 16) which appear in the solution, merely produces a re-statement of the original problem. Admitting, however, the hypothesis of the independent formation of opinion in the individual mind, either absolutely, as in the speculations of Laplace and Poisson, or under limitations imposed by the actual data, as will be seen in this treatise, Chap. XXI., the problem assumes a far more definite character. It will be manifest that the ulterior value of the theory of Probabilities must depend very much upon the correct formation of such mediate hypotheses, where the purely experimental data are insufficient for *definite* solution, and where that further experience indicated by the interpretation of the final logical equation is unattainable. Upon the other hand, an undue readiness to form hypotheses in subjects which from their very nature are placed beyond human ken, must re-act upon the credit of the theory of Probabilities, and tend to throw doubt in the general mind over its most legitimate conclusions.

18. It would, perhaps, be premature to speculate here upon the question whether the methods of abstract science are likely at any future day to render service in the investigation of social problems at all commensurate with those which they have rendered in various departments of physical inquiry. An attempt to resolve this question upon pure *à priori* grounds of reasoning would be very likely to mislead us. For example, the consideration of human free-agency would seem at first sight to preclude the idea that the movements of the social system should ever manifest that character of orderly evolution which we are prepared to expect under the reign of a physical necessity. Yet already do the researches of the statist reveal to us facts at variance with such an anticipation. Thus the records of crime and pauperism present a degree of regularity unknown in regions in which the disturbing influence of human wants and passions is unfelt. On the other hand, the distemperature of seasons, the eruption of volcanoes, the spread of blight in the vegetable, or of epidemic maladies in the animal kingdom, things apparently or chiefly the product of natural causes, refuse to be submitted to regular and apprehensible laws. "Fickle as the wind," is a proverbial expression. Reflection upon these points teaches us in some degree to correct our earlier judgments. We learn that we are not to expect, under the dominion of necessity,

an order perceptible to human observation, unless the play of its producing causes is sufficiently simple; nor, on the other hand, to deem that free agency in the individual is inconsistent with regularity in the motions of the system of which he forms a component unit. Human freedom stands out as an apparent fact of our consciousness, while it is also, I conceive, a highly probable deduction of analogy (Chap. XXII.) from the nature of that portion of the mind whose scientific constitution we are able to investigate. But whether accepted as a fact reposing on consciousness, or as a conclusion sanctioned by the reason, it must be so interpreted as not to conflict with an established result of observation, viz.: that phænomena, in the production of which large masses of men are concerned, do actually exhibit a very remarkable degree of regularity, enabling us to collect in each succeeding age the elements upon which the estimate of its state and progress, so far as manifested in outward results, must depend. There is thus no sound objection à priori against the possibility of that species of data which is requisite for the experimental foundation of a science of social statistics. Again, whatever other object this treatise may accomplish, it is presumed that it will leave no doubt as to the existence of a system of abstract principles and of methods founded upon those principles, by which any collective body of social data may be made to yield, in an explicit form, whatever information they implicitly involve. There may, where the data are exceedingly complex, be very great difficulty in obtaining this information,—difficulty due not to any imperfection of the theory, but to the laborious character of the analytical processes to which it points. It is quite conceivable that in many instances that difficulty may be such as only united effort could overcome. But that we possess theoretically in all cases, and practically, so far as the requisite labour of calculation may be supplied, the means of evolving from statistical records the seeds of general truths which lie buried amid the mass of figures, is a position which may, I conceive, with perfect safety be affirmed.

19. But beyond these general positions I do not venture to speak in terms of assurance. Whether the results which might be expected from the application of scientific methods to statistical records, over and above those the discovery of which requires no such aid, would so far compensate for the labour involved as to render it worth while to institute such investigations upon a proper scale of magnitude, is a point which could, perhaps, only be determined by experience. It is to be desired, and it might without great presumption be expected, that in this, as in other instances, the abstract doctrines of science should minister to more than intellectual gratification. Nor, viewing the apparent order in which the sciences have been evolved, and have successively contributed their aid to the service of mankind, does it seem very improbable that a day may arrive in which similar aid may accrue from departments of the field of knowledge yet more intimately allied with the elements

of human welfare. Let the speculations of this treatise, however, rest at present simply upon their claim to be regarded as true.

20. I design, in the last place, to endeavour to educe from the scientific result of the previous inquiries some general intimations respecting the nature and constitution of the human mind. Into the grounds of the possibility of this species of inference it is not necessary to enter here. One or two general observations may serve to indicate the track which I shall endeavour to follow. It cannot but be admitted that our views of the science of Logic must materially influence, perhaps mainly determine, our opinions upon the nature of the intellectual faculties. For example, the question whether reasoning consists merely in the application of certain first or necessary truths, with which the mind has been originally imprinted, or whether the mind is itself a seat of law, whose operation is as manifest and as conclusive in the particular as in the general formula, or whether, as some not undistinguished writers seem to maintain, all reasoning is of particulars; this question, I say, is one which not merely affects the science of Logic, but also concerns the formation of just views of the constitution of the intellectual faculties. Again, if it is concluded that the mind is by original constitution a seat of law, the question of the nature of its subjection to this law,—whether, for instance, it is an obedience founded upon necessity, like that which sustains the revolutions of the heavens, and preserves the order of Nature,—or whether it is a subjection of some quite distinct kind, is also a matter of deep speculative interest. Further, if the mind is truly determined to be a subject of law, and if its laws also are truly assigned, the question of their probable or necessary influence upon the course of human thought in different ages is one invested with great importance, and well deserving a patient investigation, as matter both of philosophy and of history. These and other questions I propose, however imperfectly, to discuss in the concluding portion of the present work. They belong, perhaps, to the domain of probable or conjectural, rather than to that of positive, knowledge. But it may happen that where there is not sufficient warrant for the certainties of science, there may be grounds of analogy adequate for the suggestion of highly probable opinions. It has seemed to me better that this discussion should be entirely reserved for the sequel of the main business of this treatise,—which is the investigation of scientific truths and laws. Experience sufficiently instructs us that the proper order of advancement in all inquiries after truth is to proceed from the known to the unknown. There are parts, even of the philosophy and constitution of the human mind, which have been placed fully within the reach of our investigation. To make a due acquaintance with those portions of our nature the basis of all endeavours to penetrate amid the shadows and uncertainties of that conjectural realm which lies beyond and above them, is the course most accordant with the limitations of our present condition.

CHAPTER II

OF SIGNS IN GENERAL, AND OF THE SIGNS APPROPRIATE TO THE
SCIENCE OF LOGIC IN PARTICULAR; ALSO OF THE LAWS TO WHICH
THAT CLASS OF SIGNS ARE SUBJECT

1. That Language is an instrument of human reason, and not merely a medium for the expression of thought, is a truth generally admitted. It is proposed in this chapter to inquire what it is that renders Language thus subservient to the most important of our intellectual faculties. In the various steps of this inquiry we shall be led to consider the constitution of Language, considered as a system adapted to an end or purpose; to investigate its elements; to seek to determine their mutual relation and dependence; and to inquire in what manner they contribute to the attainment of the end to which, as co-ordinate parts of a system, they have respect.

In proceeding to these inquiries, it will not be necessary to enter into the discussion of that famous question of the schools, whether Language is to be regarded as an *essential* instrument of reasoning, or whether, on the other hand, it is possible for us to reason without its aid. I suppose this question to be beside the design of the present treatise, for the following reason, viz., that it is the business of Science to investigate laws; and that, whether we regard signs as the representatives of things and of their relations, or as the representatives of the conceptions and operations of the human intellect, in studying the laws of signs, we are in effect studying the manifested laws of reasoning. If there exists a difference between the two inquiries, it is one which does not affect the scientific expressions of formal law, which are the object of investigation in the present stage of this work, but relates only to the mode in which those results are presented to the mental regard. For though in investigating the laws of signs, *à posteriori*, the immediate subject of examination is Language, with the rules which govern its use; while in making the internal processes of thought the direct object of inquiry, we appeal in a more immediate way to our personal consciousness,—it will be found that in both cases the results obtained are formally equivalent. Nor could we easily conceive, that the unnumbered tongues and dialects of the earth should have preserved through a long succession of ages so much that is common and universal, were we not assured of the existence of some deep foundation of their agreement in the laws of the mind itself.

2. The elements of which all language consists are signs or symbols. Words are signs. Sometimes they are said to represent things; sometimes the operations by which the mind combines together the simple notions of things into complex conceptions; sometimes they express the relations of action, passion, or mere quality, which we perceive to exist among the objects of our experience; sometimes the emotions of the

perceiving mind. But words, although in this and in other ways they fulfil the office of signs, or representative symbols, are not the only signs which we are capable of employing. Arbitrary marks, which speak only to the eye, and arbitrary sounds or actions, which address themselves to some other sense, are equally of the nature of signs, provided that their representative office is defined and understood. In the mathematical sciences, letters, and the symbols $+$, $-$, $=$, &c., are used as signs, although the term "sign" is applied to the latter class of symbols, which represent operations or relations, rather than to the former, which represent the elements of number and quantity. As the real import of a sign does not in any way depend upon its particular form or expression, so neither do the laws which determine its use. In the present treatise, however, it is with written signs that we have to do, and it is with reference to these exclusively that the term "sign" will be employed. The essential properties of signs are enumerated in the following definition.

Definition.—A sign is an arbitrary mark, having a fixed interpretation, and susceptible of combination with other signs in subjection to fixed laws dependent upon their mutual interpretation.

3. Let us consider the particulars involved in the above definition separately.

(1.) In the first place, a sign is an *arbitrary* mark. It is clearly indifferent what particular word or token we associate with a given idea, provided that the association once made is permanent. The Romans expressed by the word "civitas" what we designate by the word "state." But both they and we might equally well have employed any other word to represent the same conception. Nothing, indeed, in the nature of Language would prevent us from using a mere letter in the same sense. Were this done, the laws according to which that letter would require to be used would be essentially the same with the laws which govern the use of "civitas" in the Latin, and of "state" in the English language, so far at least as the use of those words is regulated by any general principles common to all languages alike.

(2.) In the second place, it is necessary that each sign should possess, within the limits of the same discourse or process of reasoning, a fixed interpretation. The necessity of this condition is obvious, and seems to be founded in the very nature of the subject. There exists, however, a dispute as to the precise nature of the representative office of words or symbols used as names in the processes of reasoning. By some it is maintained, that they represent the conceptions of the mind alone; by others, that they represent things. The question is not of great importance here, as its decision cannot affect the laws according to which signs are employed. I apprehend, however, that the general answer to this and such like questions is, that in the processes of reasoning, signs stand in the place and fulfil the office of the conceptions and operations of the mind; but that as those conceptions and operations represent things, and the connexions and relations of things, so signs represent things with their connexions and relations; and lastly, that as signs stand in the place of the conceptions and operations of the mind, they are subject to the laws of those conceptions and operations. This view will be more fully elucidated in the next chapter; but it here serves to explain the third of those particulars involved in the definition of a sign, viz., its subjection to fixed laws of combination depending upon the nature of its interpretation.

4. The analysis and classification of those signs by which the operations of reasoning are conducted will be considered in the following Proposition:

Proposition I

All the operations of Language, as an instrument of reasoning, may be conducted by a system of signs composed of the following elements, viz.:

1st. *Literal symbols, as* x, y, *&c., representing things as subjects of our conceptions.*

2nd. *Signs of operation, as* +, −, ×, *standing for those operations of the mind by which the conceptions of things are combined or resolved so as to form new conceptions involving the same elements.*

3rd. *The sign of identity,* =.

And these symbols of Logic are in their use subject to definite laws, partly agreeing with and partly differing from the laws of the corresponding symbols in the science of Algebra.

Let it be assumed as a criterion of the true elements of rational discourse, that they should be susceptible of combination in the simplest forms and by the simplest laws, and thus combining should generate all other known and conceivable forms of language; and adopting this principle, let the following classification be considered.

Class I

5. *Appellative or descriptive signs, expressing either the name of a thing, or some quality or circumstance belonging to it.*

To this class we may obviously refer the substantive proper or common, and the adjective. These may indeed be regarded as differing only in this respect, that the former expresses the substantive existence of the individual thing or things to which it refers; the latter implies that existence. If we attach to the adjective the universally understood subject "being" or "thing," it becomes virtually a substantive, and may for all the essential purposes of reasoning be replaced by the substantive. Whether or not, in every particular of the mental regard, it is the same thing to say, "Water is a fluid thing," as to say, "Water is fluid;" it is at least equivalent in the expression of the processes of reasoning.

It is clear also, that to the above class we must refer any sign which may conventionally be used to express some circumstance or relation, the detailed exposition of which would involve the use of many signs. The epithets of poetic diction are very frequently of this kind. They are usually compounded adjectives, singly fulfilling the office of a many-worded description. Homer's "deep-eddying ocean" embodies a virtual description in the single word βαθυδίνης. And conventionally any other description addressed either to the imagination or to the intellect might equally be represented by a single sign, the use of which would in all essential points be subject to the same laws as the use of the adjective "good" or "great." Combined with the subject "thing," such a sign would virtually become a substantive; and by a single substantive the combined meaning both of thing and quality might be expressed.

6. Now, as it has been defined that a sign is an arbitrary mark, it is permissible to replace all signs of the species above described by letters. Let us then agree to represent the class of individuals to which a particular name or description is applicable, by a single letter, as x. If the name is "men," for instance, let x represent "all men," or the class "men." By a class is usually meant a collection of individuals, to each of which a particular name or description may be applied; but in this work the meaning of the term will be extended so as to include the case in which but a single individual exists, answering to the required name or description, as well as the cases denoted by the terms "nothing" and "universe," which as "classes" should be understood to comprise respectively "no beings," "all beings." Again, if an adjective, as "good," is employed as a term of description, let us represent by a letter, as y, all things to which the description "good" is applicable, i.e. "all good things," or the class "good things." Let it further be agreed, that by the combination xy shall be represented that class of things to which the names or descriptions represented by x and y are simultaneously applicable. Thus, if x alone stands for "white things," and y for "sheep," let xy stand for "white sheep;" and in like manner, if z stand for "horned things," and x and y retain their previous interpretations, let zxy represent "horned white sheep," i.e. that collection of things to which the name "sheep," and the descriptions "white" and "horned" are together applicable.

Let us now consider the laws to which the symbols x, y, &c., used in the above sense, are subject.

7. First, it is evident, that according to the above combinations, the order in which two symbols are written is indifferent. The expressions xy and yx equally represent that class of things to the several members of which the names or descriptions x and y are together applicable. Hence we have,

$$xy = yx. \tag{1}$$

In the case of x representing white things, and y sheep, either of the members of this equation will represent the class of "white sheep." There may be a difference as to the order in which the conception is formed, but there is none as to the individual things which are comprehended under it. In like manner, if x represent "estuaries," and y "rivers," the expressions xy and yx will indifferently represent "rivers that are estuaries," or "estuaries that are rivers," the combination in this case being in ordinary language that of two substantives, instead of that of a substantive and an adjective as in the previous instance. Let there be a third symbol, as z, representing that class of things to which the term "navigable" is applicable, and any one of the following expressions,

$$zxy, \quad zyx, \quad xyz, \quad \&c.,$$

will represent the class of "navigable rivers that are estuaries."

If one of the descriptive terms should have some implied reference to another, it is only necessary to include that reference expressly in its stated meaning, in order to render the above remarks still applicable. Thus, if x represent "wise" and y "counsellor," we shall have to define whether x implies wisdom in the absolute sense, or only the wisdom of counsel. With such definition the law $xy = yx$ continues to be valid.

We are permitted, therefore, to employ the symbols $x, y, z, \&c.$, in the place of the substantives, adjectives, and descriptive phrases subject to the rule of interpretation, that any expression in which several of these symbols are written together shall represent all the objects or individuals to which their several meanings are together applicable, and to the law that the order in which the symbols succeed each other is indifferent.

As the rule of interpretation has been sufficiently exemplified, I shall deem it unnecessary always to express the subject "things" in defining the interpretation of a symbol used for an adjective. When I say, let x represent "good," it will be understood that x only represents "good" when a subject for that quality is supplied by another symbol, and that, used alone, its interpretation will be "good things."

8. Concerning the law above determined, the following observations, which will also be more or less appropriate to certain other laws to be deduced hereafter, may be added.

First, I would remark, that this law is a law of thought, and not, properly speaking, a law of things. Difference in the order of the qualities or attributes of an object, apart from all questions of causation, is a difference in conception merely. The law (1) expresses as a general truth, that the same thing may be conceived in different ways, and states the nature of that difference; and it does no more than this.

Secondly, As a law of thought, it is actually developed in a law of Language, the product and the instrument of thought. Though the tendency of prose writing is toward uniformity, yet even there the order of sequence of adjectives absolute in their meaning, and applied to the same subject, is indifferent, but poetic diction borrows much of its rich diversity from the extension of the same lawful freedom to the substantive also. The language of Milton is peculiarly distinguished by this species of variety. Not only does the substantive often precede the adjectives by which it is qualified, but it is frequently placed in their midst. In the first few lines of the invocation to Light, we meet with such examples as the following:

"*Offspring of heaven first-born.*"

"The rising world of *waters dark and deep.*"

"Bright effluence of *bright essence increate.*"

Now these inverted forms are not simply the fruits of a poetic license. They are the natural expressions of a freedom sanctioned by the intimate laws of thought, but for reasons of convenience not exercised in the ordinary use of language.

Thirdly, The law expressed by (1) may be characterized by saying that the literal symbols x, y, z, are *commutative, like the symbols of Algebra.* In saying this, it is not affirmed that the process of multiplication in Algebra, of which the fundamental law is expressed by the equation

$$xy = yx,$$

possesses in itself any analogy with that process of logical combination which xy has been made to represent above; but only that if the arithmetical and the logical process are expressed in the same manner, their symbolical expressions will be subject to the same formal law. The evidence of that subjection is in the two cases quite distinct.

9. As the combination of two literal symbols in the form xy expresses the whole of that class of objects to which the names or qualities represented by x and y are together applicable, it follows that if the two symbols have exactly the same signification, their combination expresses no more than either of the symbols taken alone would do. In such case we should therefore have

$$xy = x.$$

As y is, however, supposed to have the same meaning as x, we may replace it in the above equation by x, and we thus get

$$xx = x.$$

Now in common Algebra the combination xx is more briefly represented by x^2. Let us adopt the same principle of notation here; for the mode of expressing a particular succession of mental operations is a thing in itself quite as arbitrary as the mode of expressing a single idea or operation (ii. 3). In accordance with this notation, then, the above equation assumes the form

$$x^2 = x, \tag{2}$$

and is, in fact, the expression of a second general law of those symbols by which names, qualities, or descriptions, are symbolically represented.

The reader must bear in mind that although the symbols x and y in the examples previously formed received significations distinct from each other, nothing prevents us from attributing to them precisely the same signification. It is evident that the more nearly their actual significations approach to each other, the more nearly does the class of things denoted by the combination xy approach to identity with the class denoted by x, as well as with that denoted by y. The case supposed in the demonstration of the equation (2) is that of *absolute* identity of meaning. The law which it expresses is practically exemplified in language. To say "good, good," in relation to any subject, though a cumbrous and useless pleonasm, is the same as to say "good." Thus "good, good" men, is equivalent to "good" men. Such repetitions of words are indeed sometimes employed to heighten a quality or strengthen an affirmation. But this effect is merely secondary and conventional; it is not founded in the intrinsic relations of language and thought. Most of the operations which we observe in nature, or perform ourselves, are of such a kind that their effect is augmented by repetition, and this circumstance prepares us to expect the same thing in language, and even to use repetition when we design to speak with emphasis. But neither in strict reasoning nor in exact discourse is there any just ground for such a practice.

10. We pass now to the consideration of another class of the signs of speech, and of the laws connected with their use.

Class II

11. *Signs of those mental operations whereby we collect parts into a whole, or separate a whole into its parts.*

We are not only capable of entertaining the conceptions of objects, as characterized by names, qualities, or circumstances, applicable to each individual of the group under consideration, but also of forming the aggregate conception of a group of objects consisting of partial groups, each of which is separately named or described. For this purpose we use the conjunctions "and," "or," &c. "Trees and minerals," "barren mountains, or fertile vales," are examples of this kind. In strictness, the words "and," "or," interposed between the terms descriptive of two or more classes of objects, imply that those classes are quite distinct, so that no member of one is found in another. In this and in all other respects the words "and" "or" are analogous with the sign $+$ in algebra, and their laws are identical. Thus the expression "men and women" is, conventional meanings set aside, equivalent with the expression "women and men." Let x represent "men," y, "women;" and let $+$ stand for "*and*" and "*or*," then we have

$$x + y = y + x, \tag{3}$$

an equation which would equally hold true if x and y represented *numbers*, and + were the sign of arithmetical addition.

Let the symbol z stand for the adjective "European," then since it is, in effect, the same thing to say "European men and women," as to say "European men and European women," we have

$$z(x + y) = zx + zy. \qquad (4)$$

And this equation also would be equally true were x, y, and z symbols of number, and were the juxtaposition of two literal symbols to represent their algebraic product, just as in the logical signification previously given, it represents the class of objects to which both the epithets conjoined belong.

The above are the laws which govern the use of the sign +, here used to denote the positive operation of aggregating parts into a whole. But the very idea of an operation effecting some positive change seems to suggest to us the idea of an opposite or negative operation, having the effect of undoing what the former one has done. Thus we cannot conceive it possible to collect parts into a whole, and not conceive it also possible to separate a part from a whole. This operation we express in common language by the sign *except*, as, "All men *except* Asiatics," "All states *except* those which are monarchical." Here it is implied that the things excepted form a part of the things from which they are excepted. As we have expressed the operation of aggregation by the sign +, so we may express the negative operation above described by − minus. Thus if x be taken to represent men, and y, Asiatics, i.e. Asiatic men, then the conception of "All men except Asiatics" will be expressed by $x − y$. And if we represent by x, "states," and by y the descriptive property "having a monarchical form," then the conception of "All states except those which are monarchical" will be expressed by $x − xy$.

As it is indifferent for all the *essential* purposes of reasoning whether we express excepted cases first or last in the order of speech, it is also indifferent in what order we write any series of terms, some of which are affected by the sign −. Thus we have, as in the common algebra,

$$x − y = −y + x. \qquad (5)$$

Still representing by x the class "men," and by y "Asiatics," let z represent the adjective "white." Now to apply the adjective "white" to the collection of men expressed by the phrase "Men except Asiatics," is the same as to say, "White men, except white Asiatics." Hence we have

$$z(x − y) = zx − zy. \qquad (6)$$

This is also in accordance with the laws of ordinary algebra.

The equations (4) and (6) may be considered as exemplification of a single general law, which may be stated by saying, *that the literal symbols, x, y, z, &c. are distributive in their operation.* The general fact which that law expresses is this, viz.:—If any quality or circumstance is ascribed to all the members of a group, formed either by aggregation or exclusion of partial groups, the resulting conception is the same as if the quality or circumstance were first ascribed to each member of the partial groups, and the aggregation or exclusion effected afterwards. That which is ascribed to the members of the whole is ascribed to the members of all its parts, howsoever those parts are connected together.

Class III

12. *Signs by which relation is expressed, and by which we form propositions.*

Though all verbs may with propriety be referred to this class, it is sufficient for the purposes of Logic to consider it as including only the substantive verb *is* or *are*, since every other verb may be resolved into this element, and one of the signs included under Class I. For as those signs are used to express quality or circumstance of every kind, they may be employed to express the active or passive relation of the subject of the verb, considered with reference either to past, to present, or to future time. Thus the Proposition, "Cæsar conquered the Gauls," may be resolved into "Cæsar is he who conquered the Gauls." The ground of this analysis I conceive to be the following:—Unless we understand what is meant by having conquered the Gauls, i.e. by the expression "One who conquered the Gauls," we cannot understand the sentence in question. It is, therefore, truly an element of that sentence; another element is "Cæsar," and there is yet another required, the copula *is*, to show the connexion of these two. I do not, however, affirm that there is no other mode than the above of contemplating the relation expressed by the proposition, "Cæsar conquered the Gauls;" but only that the analysis here given is a correct one for the particular point of view which has been taken, and that it suffices for the purposes of logical deduction. It may be remarked that the passive and future participles of the Greek language imply the existence of the principle which has been asserted, viz.: that the sign *is* or *are* may be regarded as an element of every personal verb.

13. The above sign, *is* or *are*, may be expressed by the symbol =. The laws, or as would usually be said, the axioms which the symbol introduces, are next to be considered.

Let us take the Proposition, "The stars are the suns and the planets," and let us represent stars by x, suns by y, and planets by z; we have then

$$x = y + z. \tag{7}$$

Now if it be true that the stars are the suns and the planets, it will follow that the stars, except the planets, are suns. This would give the equation

$$x - z = y, \tag{8}$$

which must therefore be a deduction from (7). Thus a term z has been removed from one side of an equation to the other by changing its sign. This is in accordance with the algebraic rule of transposition.

But instead of dwelling upon particular cases, we may at once affirm the general axioms:—

1st. If equal things are added to equal things, the wholes are equal.

2nd. If equal things are taken from equal things, the remainders are equal.

And it hence appears that we may add or subtract equations, and employ the rule of transposition above given just as in common algebra.

Again: If two classes of things, x and y, be identical, that is, if all the members of the one are members of the other, then those members of the one class which possess a given property z will be identical with those members of the other which possess the same property z. Hence if we have the equation

$$x = y;$$

then whatever class or property z may represent, we have also

$$zx = zy.$$

This is formally the same as the algebraic law:—If both members of an equation are multiplied by the same quantity, the products are equal.

In like manner it may be shown that if the corresponding members of two equations are multiplied together, the resulting equation is true.

14. Here, however, the analogy of the present system with that of algebra, as commonly stated, appears to stop. Suppose it true that those members of a class x which possess a certain property z are identical with those members of a class y which possess the same property z, it does not follow that the members of the class x universally are identical with the members of the class y. Hence it cannot be inferred from the equation

$$zx = zy,$$

that the equation

$$x = y$$

is also true. In other words, the axiom of algebraists, that both sides of an equation may be divided by the same quantity, has no formal equivalent here. I say no *formal equivalent*, because, in accordance with the general spirit of these inquiries, it is not even sought to determine whether the mental operation which is represented by removing a logical symbol, z, from a combination zx, is in itself analogous with the operation of division in Arithmetic. That mental operation is indeed identical with what is commonly termed Abstraction, and it will hereafter appear that its laws are dependent upon the laws already deduced in this chapter. What has now been shown is, that there does not exist among those laws anything analogous in *form* with a commonly received axiom of Algebra.

But a little consideration will show that even in common algebra that axiom does not possess the generality of those other axioms which have been considered. The deduction of the equation $x = y$ from the equation $zx = zy$ is only valid when it is known that z is not equal to 0. If then the value $z = 0$ is supposed to be admissible in the algebraic system, the axiom above stated ceases to be applicable, and the analogy before exemplified remains at least unbroken.

15. However, it is not with the symbols of quantity generally that it is of any importance, except as a matter of speculation, to trace such affinities. We have seen (ii. 9) that the symbols of Logic are subject to the special law,

$$x^2 = x.$$

Now of the symbols of Number there are but two, viz. 0 and 1, which are subject to the same formal law. We know that $0^2 = 0$, and that $1^2 = 1$; and the equation $x^2 = x$, considered as algebraic, has no other roots than 0 and 1. Hence, instead of determining the measure of formal agreement of the symbols of Logic with those of Number generally, it is more immediately suggested to us to compare them with symbols of quantity *admitting only of the values* 0 *and* 1. Let us conceive, then, of an Algebra in which the symbols x, y, z, &c. admit indifferently of the values 0 and 1, and of these values alone. The laws, the axioms, and the processes, of such an Algebra will be identical in their whole extent with the laws, the axioms, and the processes of an Algebra of Logic. Difference of interpretation will alone divide them. Upon this principle the method of the following work is established.

16. It now remains to show that those constituent parts of ordinary language which have not been considered in the previous sections of this chapter are either resolvable into the same elements as those which have been considered, or are subsidiary to those elements by contributing to their more precise definition.

The substantive, the adjective, and the verb, together with the particles *and, except,* we have already considered. The pronoun may be regarded as a particular form of the substantive or the adjective. The adverb modifies the meaning of the verb, but does not affect its nature. Prepositions contribute to the expression of circumstance or relation, and thus tend to give precision and detail to the meaning of the literal symbols. The conjunctions *if, either, or,* are used chiefly in the expression of relation among propositions, and it will hereafter be shown that the same relations can be completely expressed by elementary symbols analogous in interpretation, and identical in form and law with the symbols whose use and meaning have been explained in this Chapter. As to any remaining elements of speech, it will, upon examination, be found that they are used either to give a more definite significance to the terms of discourse, and thus enter into the interpretation of the literal symbols already considered, or to express some emotion or state of feeling accompanying the utterance of a proposition, and thus do not belong to the province of the understanding, with which alone our present concern lies. Experience of its use will testify to the sufficiency of the classification which has been adopted.

CHAPTER III

DERIVATION OF THE LAWS OF THE SYMBOLS OF LOGIC FROM THE LAWS OF THE OPERATIONS OF THE HUMAN MIND

1. The object of science, properly so called, is the knowledge of laws and relations. To be able to distinguish what is essential to this end, from what is only accidentally associated with it, is one of the most important conditions of scientific progress. I say, to *distinguish* between these elements, because a consistent devotion to science does not require that the attention should be altogether withdrawn from other speculations, often of a metaphysical nature, with which it is not unfrequently connected. Such questions, for instance, as the existence of a sustaining ground of phænomena, the reality of cause, the propriety of forms of speech implying that the successive states of things are connected by *operations*, and others of a like nature, may possess a deep interest and significance in relation to science, without being essentially scientific. It is indeed scarcely possible to express the conclusions of natural science without borrowing the language of these conceptions. Nor is there necessarily any practical inconvenience arising from this source. They who believe, and they who refuse to believe, that there is more in the relation of cause and effect than an invariable order of succession, agree in their interpretation of the conclusions of physical astronomy. But they only agree because they recognise a common element of scientific truth, which is independent of their particular views of the nature of causation.

2. If this distinction is important in physical science, much more does it deserve attention in connexion with the science of the intellectual powers. For the questions which this science presents become, in expression at least, almost necessarily mixed up with modes of thought and language, which betray a metaphysical origin. The idealist would give to the laws of reasoning one form of expression; the sceptic, if true to his principles, another. They who regard the phænomena with which we are concerned in this inquiry as the mere successive *states* of the thinking subject devoid of any causal connexion, and they who refer them to the *operations* of an active intelligence, would, if consistent, equally differ in their modes of statement. Like difference would also result from a difference of classification of the mental faculties. Now the principle which I would here assert, as affording us the only ground of confidence and stability amid so much of seeming and of real diversity, is the

following, viz., that if the laws in question are really deduced from observation, they have a real existence as laws of the human mind, independently of any metaphysical theory which may seem to be involved in the mode of their statement. They contain an element of truth which no ulterior criticism upon the nature, or even upon the reality, of the mind's operations, can essentially affect. Let it even be granted that the mind is but a succession of states of consciousness, a series of fleeting impressions uncaused from without or from within, emerging out of nothing, and returning into nothing again,—the last refinement of the sceptic intellect,—still, as laws of succession, or at least of a past succession, the results to which observation had led would remain true. They would require to be interpreted into a language from whose vocabulary all such terms as cause and effect, operation and subject, substance and attribute, had been banished; but they would still be valid as scientific truths.

Moreover, as any statement of the laws of thought, founded upon actual observation, must thus contain scientific elements which are independent of metaphysical theories of the nature of the mind, the practical application of such elements to the construction of a system or method of reasoning must also be independent of metaphysical distinctions. For it is upon the scientific elements involved in the statement of the laws, that any practical application will rest, just as the practical conclusions of physical astronomy are independent of any theory of the cause of gravitation, but rest only on the knowledge of its phænomenal effects. And, therefore, as respects both the determination of the laws of thought, and the practical use of them when discovered, we are, for all really scientific ends, unconcerned with the truth or falsehood of any metaphysical speculations whatever.

3. The course which it appears to me to be expedient, under these circumstances, to adopt, is to avail myself as far as possible of the language of common discourse, without regard to any theory of the nature and powers of the mind which it may be thought to embody. For instance, it is agreeable to common usage to say that we converse with each other by the communication of ideas, or conceptions, such communication being the office of words; and that with reference to any particular ideas or conceptions presented to it, the mind possesses certain powers or faculties by which the mental regard may be fixed upon some ideas, to the exclusion of others, or by which the given conceptions or ideas may, in various ways, be combined together. To those faculties or powers different names, as Attention, Simple Apprehension, Conception or Imagination, Abstraction, &c., have been given,—names which have not only furnished the titles of distinct divisions of the philosophy of the human mind, but passed into the common language of men. Whenever, then, occasion shall occur to use these terms, I shall do so without implying thereby that I accept the theory that the mind possesses such and such powers and faculties as

distinct elements of its activity. Nor is it indeed necessary to inquire whether such powers of the understanding have a distinct existence or not. We may merge these different titles under the one generic name of *Operations* of the human mind, define these operations so far as is necessary for the purposes of this work, and then seek to express their ultimate laws. Such will be the general order of the course which I shall pursue, though reference will occasionally be made to the names which common agreement has assigned to the particular states or operations of the mind which may fall under our notice.

It will be most convenient to distribute the more definite results of the following investigation into distinct Propositions.

Proposition I

4. *To deduce the laws of the symbols of Logic from a consideration of those operations of the mind which are implied in the strict use of language as an instrument of reasoning.*

In every discourse, whether of the mind conversing with its own thoughts, or of the individual in his intercourse with others, there is an assumed or expressed limit within which the subjects of its operation are confined. The most unfettered discourse is that in which the words we use are understood in the widest possible application, and for them the limits of discourse are co-extensive with those of the universe itself. But more usually we confine ourselves to a less spacious field. Sometimes, in discoursing of men we imply (without expressing the limitation) that it is of men only under certain circumstances and conditions that we speak, as of civilized men, or of men in the vigour of life, or of men under some other condition or relation. Now, whatever may be the extent of the field within which all the objects of our discourse are found, that field may properly be termed the universe of discourse.

5. Furthermore, this universe of discourse is in the strictest sense the ultimate *subject* of the discourse. The office of any name or descriptive term employed under the limitations supposed is not to raise in the mind the conception of all the beings or objects to which that name or description is applicable, but only of those which exist within the supposed universe of discourse. If that universe of discourse is the actual universe of things, which it always is when our words are taken in their real and literal sense, then by men we mean *all men that exist*; but if the universe of discourse is limited by any antecedent implied understanding, then it is of men under the limitation thus introduced that we speak. It is in both cases the business of the word *men* to direct a certain operation of the mind, by which, from the proper universe of discourse, we select or fix upon the individuals signified.

6. Exactly of the same kind is the mental operation implied by the use of an adjective. Let, for instance, the universe of discourse be the actual Universe. Then, as the word *men* directs us to select mentally from that Universe all the beings to which the term "men" is applicable; so the adjective "good," in the combination "good men," directs us still further to select mentally from the class of *men* all those who possess the further quality "good;" and if another adjective were prefixed to the combination "good men," it would direct a further operation of the same nature, having reference to that further quality which it might be chosen to express.

It is important to notice carefully the real nature of the operation here described, for it is conceivable, that it might have been different from what it is. Were the adjective simply *attributive* in its character, it would seem, that when a particular set of beings is designated by *men*, the prefixing of the adjective *good* would direct us to attach mentally to all those beings the quality of goodness. But this is not the real office of the adjective. The operation which we really perform is one of *selection according to a prescribed principle or idea.* To what faculties of the mind such an operation would be referred, according to the received classification of its powers, it is not important to inquire, but I suppose that it would be considered as dependent upon the two faculties of Conception or Imagination, and Attention. To the one of these faculties might be referred the formation of the general conception; to the other the fixing of the mental regard upon those individuals within the prescribed universe of discourse which answer to the conception. If, however, as seems not improbable, the power of Attention is nothing more than the power of continuing the exercise of any other faculty of the mind, we might properly regard the whole of the mental process above described as referrible to the mental faculty of Imagination or Conception, the first step of the process being the conception of the Universe itself, and each succeeding step limiting in a definite manner the conception thus formed. Adopting this view, I shall describe each such step, or any definite combination of such steps, as a *definite act of conception.* And the use of this term I shall extend so as to include in its meaning not only the conception of classes of objects represented by particular names or simple attributes of quality, but also the combination of such conceptions in any manner consistent with the powers and limitations of the human mind; indeed, any intellectual operation short of that which is involved in the structure of a sentence or proposition. The general laws to which such operations of the mind are subject are now to be considered.

7. Now it will be shown that the laws which in the preceding chapter have been determined *à posteriori* from the constitution of language, for the use of the literal symbols of Logic, are in reality the laws of that definite mental operation which has just been described. We commence our discourse with a certain understanding as to

the limits of its subject, i.e. as to the limits of its Universe. Every name, every term of description that we employ, directs him whom we address to the performance of a certain mental operation upon that subject. And thus is thought communicated. But as each name or descriptive term is in this view but the representative of an intellectual operation, that operation being also prior in the order of nature, it is clear that the laws of the name or symbol must be of a derivative character,—must, in fact, originate in those of the operation which they represent. That the laws of the symbol and of the mental process are identical in expression will now be shown.

8. Let us then suppose that the universe of our discourse is the actual universe, so that words are to be used in the full extent of their meaning, and let us consider the two mental operations implied by the words "white" and "men." The word "men" implies the operation of selecting in thought from its subject, the universe, all men; and the resulting conception, *men*, becomes the subject of the next operation. The operation implied by the word "white" is that of selecting from its subject, "men," all of that class which are white. The final resulting conception is that of "white men." Now it is perfectly apparent that if the operations above described had been performed in a converse order, the result would have been the same. Whether we begin by forming the conception of "*men*," and then by a second intellectual act limit that conception to "white men," or whether we begin by forming the conception of "white objects," and then limit it to such of that class as are "men," is perfectly indifferent so far as the result is concerned. It is obvious that the order of the mental processes would be equally indifferent if for the words "white" and "men" we substituted any other descriptive or appellative terms whatever, provided only that their meaning was fixed and absolute. And thus the indifference of the order of two successive acts of the faculty of Conception, the one of which furnishes the subject upon which the other is supposed to operate, is a general condition of the exercise of that faculty. It is a law of the mind, and it is the real origin of that law of the literal symbols of Logic which constitutes its formal expression (1) Chap. II.

9. It is equally clear that the mental operation above described is of such a nature that its effect is not altered by repetition. Suppose that by a definite act of conception the attention has been fixed upon men, and that by another exercise of the same faculty we limit it to those of the race who are white. Then any further repetition of the latter mental act, by which the attention is limited to white objects, does not in any way modify the conception arrived at, viz., that of white men. This is also an example of a general law of the mind, and it has its formal expression in the law ((2) Chap. II.) of the literal symbols.

10. Again, it is manifest that from the conceptions of two distinct classes of things we can form the conception of that collection of things which the two classes

taken together compose; and it is obviously indifferent in what order of position or of priority those classes are presented to the mental view. This is another general law of the mind, and its expression is found in (3) Chap. II.

11. It is not necessary to pursue this course of inquiry and comparison. Sufficient illustration has been given to render manifest the two following positions, viz.:

First, That the operations of the mind, by which, in the exercise of its power of imagination or conception, it combines and modifies the simple ideas of things or qualities, not less than those operations of the reason which are exercised upon truths and propositions, are subject to general laws.

Secondly, That those laws are mathematical in their form, and that they are actually developed in the essential laws of human language. Wherefore the laws of the symbols of Logic are deducible from a consideration of the operations of the mind in reasoning.

12. The remainder of this chapter will be occupied with questions relating to that law of thought whose expression is $x^2 = x$ (ii. 9), a law which, as has been implied (ii. 15), forms the characteristic distinction of the operations of the mind in its ordinary discourse and reasoning, as compared with its operations when occupied with the general algebra of quantity. An important part of the following inquiry will consist in proving that the symbols 0 and 1 occupy a place, and are susceptible of an interpretation, among the symbols of Logic; and it may first be necessary to show how particular symbols, such as the above, may with propriety and advantage be employed in the representation of distinct systems of thought.

The ground of this propriety cannot consist in any community of interpretation. For in systems of thought so truly distinct as those of Logic and Arithmetic (I use the latter term in its widest sense as the science of Number), there is, properly speaking, no community of subject. The one of them is conversant with the very conceptions of things, the other takes account solely of their numerical relations. But inasmuch as the forms and methods of any system of reasoning depend immediately upon the laws to which the symbols are subject, and only mediately, through the above link of connexion, upon their interpretation, there may be both propriety and advantage in employing the same symbols in different systems of thought, provided that such interpretations can be assigned to them as shall render their formal laws identical, and their use consistent. The ground of that employment will not then be community of interpretation, but the community of the formal laws, to which in their respective systems they are subject. Nor must that community of formal laws be established upon any other ground than that of a careful observation and comparison of those results which are seen to flow independently from the interpretations of the systems under consideration.

These observations will explain the process of inquiry adopted in the following Proposition. The literal symbols of Logic are universally subject to the law whose expression is $x^2 = x$. Of the symbols of Number there are two only, 0 and 1, which satisfy this law. But each of these symbols is also subject to a law peculiar to itself in the system of numerical magnitude, and this suggests the inquiry, what interpretations must be given to the literal symbols of Logic, in order that the same peculiar and formal laws may be realized in the logical system also.

Proposition II

13. *To determine the logical value and significance of the symbols* 0 *and* 1.

The symbol 0, as used in Algebra, satisfies the following formal law,

$$0 \times y = 0, \quad \text{or} \quad 0y = 0, \tag{1}$$

whatever *number* y may represent. That this formal law may be obeyed in the system of Logic, we must assign to the symbol 0 such an interpretation that the *class* represented by $0y$ may be identical with the class represented by 0, whatever the class y may be. A little consideration will show that this condition is satisfied if the symbol 0 represent Nothing. In accordance with a previous definition, we may term Nothing a class. In fact, Nothing and Universe are the two limits of class extension, for they are the limits of the possible interpretations of general names, none of which can relate to fewer individuals than are comprised in Nothing, or to more than are comprised in the Universe. Now whatever the class y may be, the individuals which are common to it and to the class "Nothing" are identical with those comprised in the class "Nothing," for they are none. And thus by assigning to 0 the interpretation Nothing, the law (1) is satisfied; and it is not otherwise satisfied consistently with the perfectly general character of the class y.

Secondly, The symbol 1 satisfies in the system of Number the following law, viz.,

$$1 \times y = y, \quad \text{or} \quad 1y = y,$$

whatever number y may represent. And all formal equation being assumed as equally valid in the system of this work, in which 1 and y represent classes, it appears that the symbol 1 must represent such a class that all the individuals which are found in *any* proposed class y are also all the individuals $1y$ that are common to that class y and the class represented by 1. A little consideration will here show that the class represented by 1 must be "the Universe," since this is the only class in which are found *all* the individuals that exist in *any* class. Hence the respective interpretations of the symbols 0 and 1 in the system of Logic are *Nothing* and *Universe*.

14. As with the idea of any class of objects as "men," there is suggested to the mind the idea of the contrary class of beings which are not men; and as the whole Universe is made up of these two classes together, since of every individual which it

comprehends we may affirm either that it is a man, or that it is not a man, it becomes important to inquire how such contrary names are to be expressed. Such is the object of the following Proposition.

Proposition III

If x represent any class of objects, then will $1 - x$ *represent the contrary or supplementary class of objects, i.e. the class including all objects which are not comprehended in the class x.*

For greater distinctness of conception let x represent the class *men*, and let us express, according to the last Proposition, the Universe by 1; now if from the conception of the Universe, as consisting of "men" and "not-men," we exclude the conception of "men," the resulting conception is that of the contrary class, "not-men." Hence the class "not-men" will be represented by $1 - x$. And, in general, whatever class of objects is represented by the symbol x, the contrary class will be expressed by $1 - x$.

15. Although the following Proposition belongs in strictness to a future chapter of this work, devoted to the subject of *maxims* or *necessary truths*, yet, on account of the great importance of that law of thought to which it relates, it has been thought proper to introduce it here.

Proposition IV

That axiom of metaphysicians which is termed the principle of contradiction, and which affirms that it is impossible for any being to possess a quality, and at the same time not to possess it, is a consequence of the fundamental law of thought, whose expression is $x^2 = x$.

Let us write this equation in the form

$$x - x^2 = 0,$$

whence we have

$$x(1 - x) = 0; \tag{1}$$

both these transformations being justified by the axiomatic laws of combination and transposition (ii. 13). Let us, for simplicity of conception, give to the symbol x the particular interpretation of *men*, then $1 - x$ will represent the class of "not-men" (Prop. III.) Now the formal product of the expressions of two classes represents that class of individuals which is common to them both (ii. 6). Hence $x(1 - x)$ will represent the class whose members are at once "men," and "not men," and the equation (1) thus express the principle, *that a class whose members are at the same time men and not men does not exist.* In other words, that *it is impossible for the same individual to be at the same time a man and not a man.* Now let the meaning of the symbol x be extended from the representing of "men," to that of any class of beings

711

characterized by the possession of any quality whatever; and the equation (1) will then express that it is impossible for a being to possess a quality and not to possess that quality at the same time. But this is identically that "principle of contradiction" which Aristotle has described as the fundamental axiom of all philosophy. "It is impossible that the same quality should both belong and not belong to the same thing. . . This is the most certain of all principles. . . Wherefore they who demonstrate refer to this as an ultimate opinion. For it is by nature the source of all the other axioms."[1]

The above interpretation has been introduced not on account of its immediate value in the present system, but as an illustration of a significant fact in the philosophy of the intellectual powers, viz., that what has been commonly regarded as the fundamental axiom of metaphysics is but the consequence of a law of thought, mathematical in its form. I desire to direct attention also to the circumstance that the equation (1) in which that fundamental law of thought is expressed is an equation of the second degree.[2] Without speculating at all in this chapter upon the question, whether that circumstance is necessary in its own nature, we may venture to assert that if it had not existed, the whole procedure of the understanding would have been different from what it is. Thus it is a consequence of the fact that the fundamental equation of thought is of the second degree, that we perform the operation of analysis and classification, by division into pairs of opposites, or, as it is technically said, by *dichotomy*. Now if the equation in question had been of the third degree, still admitting of interpretation as such, the mental division must have been threefold in character, and we must have proceeded by a species of *trichotomy*, the real nature of which it is impossible for us, with our existing faculties, adequately to conceive, but the laws of which we might still investigate as an object of intellectual speculation.

16. The law of thought expressed by the equation (1) will, for reasons which are made apparent by the above discussion, be occasionally referred to as the "law of duality."

---

[1]Τὸ γὰρ αὐτὸ ἅμα ὑπάρχειν τε καὶ μὴ ὑπάρχειν ἀδύνατον τῷ αὐτῷ καὶ κατὰ τὸ αὐτό . . . Αὕτη δὴ πασῶν ἐστὶ βεβαιοτάτη τῶν ἀρχῶν . . . Διὸ πάνες οἱ ἀποδεικνύντες εἰς ταύτην ἀνάγουσιν ἐσχάτην δόξαν· φύσει γὰρ ἀρχὴ καὶ τῶν ἄλλων ἀξιωμάτων αὕτη πάντων.—*Metaphysica*, III. 3.

[2] Should it here be said that the existence of the equation $x^2 = x$ necessitates also the existence of the equation $x^3 = x$, which is of the third degree, and then inquired whether that equation does not indicate a process of *trichotomy*; the answer is, that the equation $x^3 = x$ is not interpretable in the system of logic. For writing it in either of the forms

$$x(1 - x)(1 + x) = 0, \tag{2}$$
$$x(1 - x)(-1 - x) = 0, \tag{3}$$

we see that its interpretation, if possible at all, must involve that of the factor $1 + x$, or of the factor $-1 - x$. The former is not interpretable, because we cannot conceive of the addition of any class x to the universe 1; the latter is not interpretable, because the symbol -1 is not subject to the law $x(1 - x) = 0$, to which all class symbols are subject. Hence the equation $x^3 = x$ admits of no interpretation analogous to that of the equation $x^2 = x$. Were the former equation, however, true independently of the latter, i.e. were that act of the mind which is denoted by the symbol x, such that its second repetition should reproduce the result of a single operation, but not its first or mere repetition, it is presumable that we should be able to interpret one of the forms (2), (3), which under the actual conditions of thought we cannot do. There exist operations, known to the mathematician, the law of which may be adequately expressed by the equation $x^3 = x$. But they are of a nature altogether foreign to the province of general reasoning.

In saying that it is conceivable that the law of thought might have been different from what it is, I mean only that we can frame such an hypothesis, and study its consequences. The possibility of doing this involves no such doctrine as that the actual law of human reason is the product either of chance or of arbitrary will.

CHAPTER IV

OF THE DIVISION OF PROPOSITIONS INTO THE TWO CLASSES OF
"PRIMARY" AND "SECONDARY;" OF THE CHARACTERISTIC PROPERTIES
OF THOSE CLASSES, AND OF THE LAWS OF THE EXPRESSION OF
PRIMARY PROPOSITIONS

1. The laws of those mental operations which are concerned in the processes of Conception or Imagination having been investigated, and the corresponding laws of the symbols by which they are represented explained, we are led to consider the practical application of the results obtained: first, in the expression of the complex terms of propositions; secondly, in the expression of propositions; and lastly, in the construction of a general method of deductive analysis. In the present chapter we shall be chiefly concerned with the first of these objects, as an introduction to which it is necessary to establish the following Proposition:

Proposition I

All logical propositions may be considered as belonging to one or the other of two great classes, to which the respective names of "Primary" or "Concrete Propositions," and "Secondary" or "Abstract Propositions," may be given.

Every assertion that we make may be referred to one or the other of the two following kinds. Either it expresses a relation among *things*, or it expresses, or is equivalent to the expression of, a relation among *propositions*. An assertion respecting the properties of things, or the phænomena which they manifest, or the circumstances in which they are placed, is, properly speaking, the assertion of a relation among things. To say that "snow is white," is for the ends of logic equivalent to saying, that "snow is a white thing." An assertion respecting facts or events, their mutual connexion and dependence, is, for the same ends, generally equivalent to the assertion, that such and such propositions concerning those events have a certain relation to each other as respects their mutual truth or falsehood. The former class of propositions, relating to *things*, I call "Primary;" the latter class, relating to *propositions*, I call "Secondary." The distinction is in practice nearly but not quite co-extensive with the common logical distinction of propositions as categorical or hypothetical.

For instance, the propositions, "The sun shines," "The earth is warmed," are primary; the proposition, "If the sun shines the earth is warmed," is secondary. To say, "The sun shines," is to say, "The sun is that which shines," and it expresses a relation between two classes of things, viz., "the sun" and "things which shine." The secondary proposition, however, given above, expresses a relation of dependence between the two primary propositions, "The sun shines," and "The earth is warmed." I do not hereby affirm that the relation between these propositions is, like that which exists between the facts which they express, a relation of causality, but only that the relation among the propositions so implies, and is so implied by, the relation among the facts, that it may for the ends of logic be used as a fit representative of that relation.

2. If instead of the proposition, "The sun shines," we say, "It is true that the sun shines," we then speak not directly of things, but of a proposition concerning things, viz., of the proposition, "The sun shines." And, therefore, the proposition in which we thus speak is a secondary one. Every primary proposition may thus give rise to a secondary proposition, viz., to that secondary proposition which asserts its truth, or declares its falsehood.

It will usually happen, that the particles *if, either, or,* will indicate that a proposition is secondary; but they do not necessarily imply that such is the case. The proposition, "Animals are either rational or irrational," is primary. It cannot be resolved into "Either animals are rational or animals are irrational," and it does not therefore express a relation of dependence between the two propositions connected together in the latter disjunctive sentence. The particles, *either, or,* are in fact no *criterion* of the nature of propositions, although it happens that they are more frequently found in secondary propositions. Even the conjunction *if* may be found in primary propositions. "Men are, if wise, then temperate," is an example of the kind. It cannot be resolved into "If all men are wise, then all men are temperate."

3. As it is not my design to discuss the merits or defects of the ordinary division of propositions, I shall simply remark here, that the principle upon which the present classification is founded is clear and definite in its application, that it involves a real and fundamental distinction in propositions, and that it is of essential importance to the development of a general method of reasoning. Nor does the fact that a primary proposition may be put into a form in which it becomes secondary at all conflict with the views here maintained. For in the case thus supposed, it is not of the things connected together in the primary proposition that any direct account is taken, but only of the proposition itself considered as *true* or as *false.*

4. In the expression both of primary and of secondary propositions, the same symbols, subject, as it will appear, to the same laws, will be employed in this work. The difference between the two cases is a difference not of form but of interpretation. In both cases the actual relation which it is the object of the proposition to express will be denoted by the sign $=$. In the expression of primary propositions, the members thus connected will usually represent the "terms" of a proposition, or, as they are more particularly designated, its subject and predicate.

Proposition II

5. *To deduce a general method, founded upon the enumeration of possible varieties, for the expression of any class or collection of things, which may constitute a "term" of a Primary Proposition.*

First, If the class or collection of things to be expressed is defined only by names or qualities common to all the individuals of which it consists, its expression will consist of a single term, in which the symbols expressive of those names or qualities will be combined without any connecting sign, as if by the algebraic process of multiplication. Thus, if x represent opaque substances, y polished substances, z stones, we shall have,

xyz = opaque polished stones;

$xy(1 - z)$ = opaque polished substances which are not stones;

$x(1 - y)(1 - z)$ = opaque substances which are not polished, and are not stones;

and so on for any other combination. Let it be observed, that each of these expressions satisfies the same law of duality, as the individual symbols which it contains. Thus,

$$xyz \times xyz = xyz;$$
$$xy(1 - z) \times xy(1 - z) = xy(1 - z);$$

and so on. Any such term as the above we shall designate as a "class term," because it expresses a class of things by means of the common properties or names of the individual members of such class.

Secondly, If we speak of a collection of things, different portions of which are defined by different properties, names, or attributes, the expressions for those different portions must be separately formed, and then connected by the sign $+$. But if the collection of which we desire to speak has been formed by excluding from some wider collection a defined portion of its members, the sign $-$ must be prefixed to the symbolical expression of the excluded portion. Respecting the use of these symbols some further observations may be added.

6. Speaking generally, the symbol $+$ is the equivalent of the conjunctions "and," "or," and the symbol $-$, the equivalent of the preposition "except." Of the conjunctions "and" and "or," the former is usually employed when the collection to be described forms the subject, the latter when it forms the predicate, of a proposition. "The scholar *and* the man of the world desire happiness," may be taken as an illustration of one of these cases. "Things possessing utility are *either* productive of pleasure *or* preventive of pain," may exemplify the other. Now whenever an expression involving these particles presents itself in a primary proposition, it becomes very important to know whether the groups or classes separated in thought by them are intended to be quite distinct from each other and mutually exclusive, or not. Does the

715

expression, "Scholars and men of the world," include or exclude those who are both? Does the expression, "Either productive of pleasure or preventive of pain," include or exclude things which possess both these qualities? I apprehend that in strictness of meaning the conjunctions "and," "or," do possess the power of separation or exclusion here referred to; that the formula, "All x's are either y's or z's," rigorously interpreted, means, "All x's are either y's, but not z's," or, "z's but not y's." But it must at the same time be admitted, that the "jus et norma loquendi" seems rather to favour an opposite interpretation. The expression, "Either y's or z's," would generally be understood to include things that are y's and z's at the same time, together with things which come under the one, but not the other. Remembering, however, that the symbol $+$ does possess the separating power which has been the subject of discussion, we must resolve any disjunctive expression which may come before us into elements really separated in thought, and then connect their respective expressions by the symbol $+$.

And thus, according to the meaning implied, the expression, "Things which are either x's or y's," will have two different symbolical equivalents. If we mean, "Things which are x's, but not y's, or y's, but not x's," the expression will be

$$x(1 - y) + y(1 - x);$$

the symbol x standing for x's, y for y's. If, however, we mean, "Things which are either x's, or, if not x's, then y's," the expression will be

$$x + y(1 - x).$$

This expression supposes the admissibility of things which are both x's and y's at the same time. It might more fully be expressed in the form

$$xy + x(1 - y) + y(1 - x);$$

but this expression, on addition of the two first terms, only reproduces the former one.

Let it be observed that the expressions above given satisfy the fundamental law of duality (iii. 16). Thus we have

$$\{x(1 - y) + y(1 - x)\}^2 = x(1 - y) + y(1 - x),$$
$$\{x + y(1 - x)\}^2 = x + y(1 - x).$$

It will be seen hereafter, that this is but a particular manifestation of a general law of expressions representing "classes or collections of things."

7. The results of these investigations may be embodied in the following rule of expression.

RULE.—*Express simple names or qualities by the symbols x, y, z, &c., their contraries by $1 - x$, $1 - y$, $1 - z$, &c.; classes of things defined by common names or qualities, by connecting the corresponding symbols as in multiplication; collections of things, consisting of portions different from each other, by connecting the expressions of those portions by the sign $+$. In particular, let the expression, "Either x's or y's," be expressed* *by* $x(1-y)+y(1-x)$, *when the classes denoted by* x *and* y *are exclusive, by* $x+y(1-x)$ *when they are not exclusive. Similarly let the expression, "Either x's, or y's, or z's," be expressed by* $x(1-y)(1-z)+y(1-x)(1-z)+z(1-x)(1-y)$, *when the classes denoted by* x, y, *and* z, *are designed to be mutually exclusive, by* $x+y(1-x)+z(1-x)(1-y)$, *when they are not meant to be exclusive, and so on.*

8. On this rule of expression is founded the converse rule of interpretation. Both these will be exemplified with, perhaps, sufficient fulness in the following instances. Omitting for brevity the universal subject "things," or "beings," let us assume

$$x = \text{hard}, y = \text{elastic}, z = \text{metals};$$
and we shall have the following results:

"Non-elastic metals," will be expressed by $z(1-y)$;
"Elastic substances with non-elastic metals," by $y+z(1-y)$;
"Hard substances, except metals," by $x-z$;
"Metallic substances, except those which are neither hard nor elastic," by

$$z - z(1-x)(1-y), \text{ or by } z\{1-(1-x)(1-y)\}, \textit{vide} (6), \text{Chap. II.}$$

In the last example, what we had really to express was "Metals, except not hard, not elastic, metals." Conjunctions used between *adjectives* are usually superfluous, and, therefore, must not be expressed symbolically.

Thus, "Metals hard and elastic," is equivalent to "Hard elastic metals," and expressed by xyz.

Take next the expression, "Hard substances, except those which are metallic and non-elastic, and those which are elastic and non-metallic." Here the word *those* means hard substances, so that the expression really means, *Hard substances except hard substances, metallic, non-elastic, and hard substances non-metallic, elastic*; the word *except* extending to both the classes which follow it. The complete expression is

$$x - \{xz(1-y)+xy(1-z)\};$$
or,
$$x - xz(1-y) - xy(1-z).$$

9. The preceding Proposition, with the different illustrations which have been given of it, is a necessary preliminary to the following one, which will complete the design of the present chapter.

Proposition III

To deduce from an examination of their possible varieties a general method for the expression of Primary or Concrete Propositions.

A primary proposition, in the most general sense, consists of two terms, between which a relation is asserted to exist. These terms are not necessarily single-worded names, but may represent any collection of objects, such as we have been engaged in considering in the previous sections. The mode of expressing those terms is, therefore, comprehended in the general precepts above given, and it only remains to discover how the relations between the terms are to be expressed. This will evidently depend upon the nature of the relation, and more particularly upon the question whether, in that relation, the terms are understood to be universal or particular, i.e. whether we speak of the whole of that collection of objects to which a term refers, or indefinitely of the whole or of a part of it, the usual signification of the prefix, "some."

Suppose that we wish to express a relation of identity between the two classes, "Fixed Stars" and "Suns," i.e. to express that "All fixed stars are suns," and "All suns are fixed stars." Here, if x stand for fixed stars, and y for suns, we shall have

$$x = y$$

for the equation required.

In the proposition, "All fixed stars are suns," the term "all fixed stars" would be called the *subject*, and "suns" the *predicate*. Suppose that we extend the meaning of the terms *subject* and *predicate* in the following manner. By *subject* let us mean the first term of any affirmative proposition, i.e. the term which precedes the copula *is* or *are*; and by *predicate* let us agree to mean the second term, i.e. the one which follows the copula; and let us admit the assumption that either of these may be universal or particular, so that, in either case, the whole class may be implied, or only a part of it. Then we shall have the following Rule for cases such as the one in the last example:—

10. RULE.—*When both Subject and Predicate of a Proposition are universal, form the separate expressions for them, and connect them by the sign* =.

This case will usually present itself in the expression of the definitions of science, or of subjects treated after the manner of pure science. Mr. Senior's definition of wealth affords a good example of this kind, viz.:

"Wealth consists of things transferable, limited in supply, and either productive of pleasure or preventive of pain."

Before proceeding to express this definition symbolically, it must be remarked that the conjunction *and* is superfluous. Wealth is really defined by its possession of three properties or qualities, not by its composition out of three classes or collections of objects. Omitting then the conjunction *and*, let us make

w = wealth.

t = things transferable.

s = limited in supply.

p = productive of pleasure.

r = preventive of pain.

Now it is plain from the nature of the subject, that the expression, "Either productive of pleasure or preventive of pain," in the above definition, is meant to be equivalent to "Either productive of pleasure; or, if not productive of pleasure, preventive of pain." Thus the class of things which the above expression, taken alone, would define, would consist of all things productive of pleasure, together with all things not productive of pleasure, but preventive of pain, and its symbolical expression would be

$$p + (1 - p)r.$$

If then we attach to this expression placed in brackets to denote that both its terms are referred to, the symbols s and t limiting its application to things "transferable" and "limited in supply," we obtain the following symbolical equivalent for the original definition, viz.:

$$w = st\{p + r(1 - p)\}. \tag{1}$$

If the expression, "Either productive of pleasure or preventive of pain," were intended to point out merely those things which are productive of pleasure without being preventive of pain, $p(1 - r)$, or preventive of pain, without being productive of pleasure, $r(1 - p)$ (exclusion being made of those things which are both productive of pleasure and preventive of pain), the expression in symbols of the definition would be

$$w = st\{p(1 - r) + r(1 - p)\}. \tag{2}$$

All this agrees with what has before been more generally stated.

The reader may be curious to inquire what effect would be produced if we literally translated the expression, "Things productive of pleasure or preventive of pain," by $p + r$, making the symbolical equation of the definition to be

$$w = st(p + r). \tag{3}$$

The answer is, that this expression would be equivalent to (2), with the additional implication that the classes of things denoted by stp and str are quite distinct, so that of things transferable and limited in supply there exist none in the universe which are at the same time both productive of pleasure and preventive of pain. How the full import of any equation may be determined will be explained hereafter. What has been said may show that before attempting to translate our data into the rigorous language of symbols, it is above all things necessary to ascertain the *intended* import of the words we are using. But this necessity cannot be regarded as an evil by those who value correctness of thought, and regard the right employment of language as both its instrument and its safeguard.

11. Let us consider next the case in which the predicate of the proposition is particular, e.g. "All men are mortal."

In this case it is clear that our meaning is, "All men are some mortal beings," and we must seek the expression of the predicate, "some mortal beings." Represent then by v, a class indefinite in every respect but this, viz., that some of its members are mortal beings, and let x stand for "mortal beings," then will vx represent "some mortal beings." Hence if y represent men, the equation sought will be

$$y = vx.$$

From such considerations we derive the following Rule, for expressing an affirmative universal proposition whose predicate is particular:

RULE.—*Express as before the subject and the predicate, attach to the latter the indefinite symbol v, and equate the expressions.*

It is obvious that v is a symbol of the same kind as x, y, &c., and that it is subject to the general law,

$$v^2 = v, \quad \text{or} \quad v(1 - v) = 0.$$

Thus, to express the proposition, "The planets are either primary or secondary," we should, according to the rule, proceed thus:

Let x represent planets (the subject);

y = primary bodies;

z = secondary bodies;

then, assuming the conjunction "or" to separate absolutely the class of "primary" from that of "secondary" bodies, so far as they enter into our consideration in the proposition given, we find for the equation of the proposition

$$x = v\{y(1 - z) + z(1 - y)\}. \tag{4}$$

It may be worth while to notice, that in this case the *literal* translation of the premises into the form

$$x = v(y + z) \tag{5}$$

would be exactly equivalent, v being an indefinite class symbol. The form (4) is, however, the better, as the expression

$$y(1 - z) + z(1 - y)$$

consists of terms representing classes quite distinct from each other, and satisfies the fundamental law of duality.

If we take the proposition, "The heavenly bodies are either suns, or planets, or comets," representing these classes of things by w, x, y, z, respectively, its expression, on the supposition that none of the heavenly bodies belong at once to two of the divisions above mentioned, will be

$$w = v\{x(1 - y)(1 - z) + y(1 - x)(1 - z) + z(1 - x)(1 - y)\}.$$

If, however, it were meant to be implied that the heavenly bodies were either suns, or, if not suns, planets, or, if neither suns nor planets, fixed stars, a meaning which does not exclude the supposition of some of them belonging at once to two or to all three of the divisions of suns, planets, and fixed stars,—the expression required would be

$$w = v\{x + y(1 - x) + z(1 - x)(1 - y)\}. \tag{6}$$

The above examples belong to the class of descriptions, not definitions. Indeed the predicates of propositions are usually particular. When this is not the case, either the predicate is a singular term, or we employ, instead of the copula "is" or "are," some form of connexion, which implies that the predicate is to be taken universally.

12. Consider next the case of universal negative propositions, e.g. "No men are perfect beings."

Now it is manifest that in this case we do not speak of a class termed "no men," and assert of this class that all its members are "perfect beings." But we virtually make an assertion about "*all men*" to the effect that they are "*not perfect beings.*" Thus the true meaning of the proposition is this:

"All men (subject) are (copula) not perfect (predicate);" whence, if y represent "men," and x "perfect beings," we shall have

$$y = v(1 - x),$$

and similarly in any other case. Thus we have the following Rule:

RULE.—*To express any proposition of the form* "*No x's are y's,*" *convert it into the form* "*All x's are not y's,*" *and then proceed as in the previous case.*

13. Consider, lastly, the case in which the subject of the proposition is particular, e.g. "Some men are not wise." Here, as has been remarked, the negative *not* may properly be referred, certainly, at least, for the ends of Logic, to the predicate *wise*; for we do not mean to say that it is not true that "Some men are wise," but we intend to predicate of "some men" a want of wisdom. The requisite form of the given proposition is, therefore, "Some men are not-wise." Putting, then, y for "men," x for "wise," i.e. "wise beings," and introducing v as the symbol of a class indefinite in all respects but this, that it contains some individuals of the class to whose expression it is prefixed, we have

$$vy = v(1 - x).$$

14. We may comprise all that we have determined in the following general Rule:

General Rule for the Symbolical Expression of Primary Propositions

1st. *If the proposition is affirmative, form the expression of the subject and that of the predicate. Should either of them be particular, attach to it the indefinite symbol v, and then equate the resulting expressions.*

2ndly. *If the proposition is negative, express first its true meaning by attaching the negative particle to the predicate, then proceed as above.*

One or two additional examples may suffice for illustration.

Ex.—"No men are placed in exalted stations, and free from envious regards."

Let y represent "men," x, "placed in exalted stations," z, "free from envious regards."

Now the expression of the class described as "placed in exalted station," and "free from envious regards," is xz. Hence the contrary class, i.e. they to whom this description does not apply, will be represented by $1 - xz$, and to this class all men are referred. Hence we have

$$y = v(1 - xz).$$

If the proposition thus expressed had been placed in the equivalent form, "Men in exalted stations are not free from envious regards," its expression would have been

$$yx = v(1 - z).$$

It will hereafter appear that this expression is really equivalent to the previous one, on the particular hypothesis involved, viz., that v is an indefinite class symbol.

Ex.—"No men are heroes but those who unite self-denial to courage."

Let x = "men," y = "heroes," z = "those who practise self-denial," w, "those who possess courage."

The assertion really is, that "men who do not possess courage and practise self-denial are not heroes."

Hence we have

$$x(1 - zw) = v(1 - y)$$

for the equation required.

15. In closing this Chapter it may be interesting to compare together the great leading types of propositions symbolically expressed. If we agree to represent by X and Y the symbolical expressions of the "terms," or things related, those types will be

$$X = vY,$$
$$X = Y,$$
$$vX = vY.$$

In the first, the predicate only is particular; in the second, both terms are universal; in the third, both are particular. Some minor forms are really included under these. Thus, if $Y = 0$, the second form becomes

$$X = 0;$$

and if $Y = 1$ it becomes

$$X = 1;$$

both which forms admit of interpretation. It is further to be noticed, that the expressions X and Y, if founded upon a sufficiently careful analysis of the meaning of the "terms" of the proposition, will satisfy the fundamental law of duality which requires that we have

$$X^2 = X \quad \text{or} \quad X(1 - X) = 0,$$
$$Y^2 = Y \quad \text{or} \quad Y(1 - Y) = 0.$$

CHAPTER V

OF THE FUNDAMENTAL PRINCIPLES OF SYMBOLICAL REASONING, AND
OF THE EXPANSION OR DEVELOPMENT OF EXPRESSIONS
INVOLVING LOGICAL SYMBOLS

1. The previous chapters of this work have been devoted to the investigation of the fundamental laws of the operations of the mind in reasoning; of their development in the laws of the symbols of Logic; and of the principles of expression, by which that species of propositions called primary may be represented in the language of symbols. These inquiries have been in the strictest sense preliminary. They form an indispensable introduction to one of the chief objects of this treatise—the construction of a system or method of Logic upon the basis of an exact summary of the fundamental laws of thought. There are certain considerations touching the nature of this end, and the means of its attainment, to which I deem it necessary here to direct attention.

2. I would remark in the first place that the generality of a method in Logic must very much depend upon the generality of its elementary processes and laws. We have, for instance, in the previous sections of this work investigated, among other things, the laws of that logical process of *addition* which is symbolized by the sign +. Now those laws have been determined from the study of instances, in all of which it has been a necessary condition, that the classes or things added together in thought should be mutually exclusive. The expression $x + y$ seems indeed uninterpretable, unless it be assumed that the things represented by x and the things represented by y are entirely separate; that they embrace no individuals in common. And conditions analogous to this have been involved in those acts of conception from the study of which the laws of the other symbolical operations have been ascertained. The question then arises, whether it is necessary to restrict the application of these symbolical laws and processes by the same conditions of interpretability under which the knowledge of them was obtained. If such restriction is necessary, it is manifest that no such thing as a general method in Logic is possible. On the other hand, if such restriction is unnecessary, in what light are we to contemplate processes which appear to be uninterpretable in that sphere of thought which they are designed to aid? These questions do not belong to the science of Logic alone. They are equally pertinent to every developed form of human reasoning which is based upon the employment of a symbolical language.

3. I would observe in the second place, that this apparent failure of correspondency between process and interpretation does not manifest itself in the *ordinary* applications of human reason. For no operations are there performed of which the

meaning and the application are not seen; and to most minds it does not suffice that merely formal reasoning should connect their premises and their conclusions; but every step of the connecting train, every mediate result which is established in the course of demonstration, must be intelligible also. And without doubt, this is both an actual condition and an important safeguard, in the reasonings and discourses of common life.

There are perhaps many who would be disposed to extend the same principle to the general use of symbolical language as an instrument of reasoning. It might be argued, that as the laws or axioms which govern the use of symbols are established upon an investigation of those cases only in which interpretation is possible, we have no right to extend their application to other cases in which interpretation is impossible or doubtful, even though (as should be admitted) such application is employed in the intermediate steps of demonstration only. Were this objection conclusive, it must be acknowledged that slight advantage would accrue from the use of a symbolical method in Logic. Perhaps that advantage would be confined to the mechanical gain of employing short and convenient symbols in the place of more cumbrous ones. But the objection itself is fallacious. Whatever our *à priori* anticipations might be, it is an unquestionable fact that the validity of a conclusion arrived at by any symbolical process of reasoning, does not depend upon our ability to interpret the formal results which have presented themselves in the different stages of the investigation. There exist, in fact, certain general principles relating to the use of symbolical methods, which, as pertaining to the particular subject of Logic, I shall first state and I shall then offer some remarks upon the nature and upon the grounds of their claim to acceptance.

4. The conditions of valid reasoning, by the aid of symbols, are—

1st, That a fixed interpretation be assigned to the symbols employed in the expression of the data; and that the laws of the combination of those symbols be correctly determined from that interpretation.

2nd, That the formal processes of solution or demonstration be conducted throughout in obedience to all the laws determined as above, without regard to the question of the interpretability of the particular results obtained.

3rd, That the final result be interpretable in form, and that it be actually interpreted in accordance with that system of interpretation which has been employed in the expression of the data. Concerning these principles, the following observations may be made.

5. The necessity of a fixed interpretation of the symbols has already been sufficiently dwelt upon (ii. 3). The necessity that the fixed result should be in such a form as to admit of that interpretation being applied, is founded on the obvious principle, that

the use of symbols is a means towards an end, that end being the knowledge of some intelligible fact or truth. And that this end may be attained, the final result which expresses the symbolical conclusion must be in an interpretable form. It is, however, in connexion with the second of the above general principles or conditions (v. 4), that the greatest difficulty is likely to be felt, and upon this point a few additional words are necessary.

I would then remark, that the principle in question may be considered as resting upon a general law of the mind, the knowledge of which is not given to us *à priori*, i.e. antecedently to experience, but is derived, like the knowledge of the other laws of the mind, from the clear manifestation of the general principle in the particular instance. A single example of reasoning, in which symbols are employed in obedience to laws founded upon their interpretation, but without any sustained reference to that interpretation, the chain of demonstration conducting us through intermediate steps which are not interpretable, to a final result which is interpretable, seems not only to establish the validity of the particular application, but to make known to us the general law manifested therein. No accumulation of instances can properly add weight to such evidence. It may furnish us with clearer conceptions of that common element of truth upon which the application of the principle depends, and so prepare the way for its reception. It may, where the immediate force of the evidence is not felt, serve as a verification, *à posteriori*, of the practical validity of the principle in question. But this does not affect the position affirmed, viz., that the general principle must be seen in the particular instance,—seen to be general in application as well as true in the special example. The employment of the uninterpretable symbol $\sqrt{-1}$, in the intermediate processes of trigonometry, furnishes an illustration of what has been said. I apprehend that there is no mode of explaining that application which does not covertly assume the very principle in question. But that principle, though not, as I conceive, warranted by formal reasoning based upon other grounds, seems to deserve a place among those axiomatic truths which constitute, in some sense, the foundation of the possibility of general knowledge, and which may properly be regarded as expressions of the mind's own laws and constitution.

6. The following is the mode in which the principle above stated will be applied in the present work. It has been seen, that any system of propositions may be expressed by equations involving symbols x, y, z, which, whenever interpretation is possible, are subject to laws identical in form with the laws of a system of quantitative symbols, susceptible only of the values 0 and 1 (ii. 15). But as the formal processes of reasoning depend only upon the laws of the symbols, and not upon the nature of their interpretation, we are permitted to treat the above symbols, x, y, z, as if they were quantitative symbols of the kind above described. *We may in fact lay*

aside the logical interpretation of the symbols in the given equation; convert them into quantitative symbols, susceptible only of the values 0 and 1; perform upon them as such all the requisite processes of solution; and finally restore to them their logical interpretation. And this is the mode of procedure which will actually be adopted, though it will be deemed unnecessary to restate in every instance the nature of the transformation employed. The processes to which the symbols *x, y, z,* regarded as quantitative and of the species above described, are subject, are not limited by those conditions of thought to which they would, if performed upon purely logical symbols, be subject, and a freedom of operation is given to us in the use of them, without which, the inquiry after a general method in Logic would be a hopeless quest.

Now the above system of processes would conduct us to no intelligible result, unless the final equations resulting therefrom were in a form which should render their interpretation, after restoring to the symbols their logical significance, possible. There exists, however, a general method of reducing equations to such a form, and the remainder of this chapter will be devoted to its consideration. I shall say little concerning the way in which the method renders interpretation possible,—this point being reserved for the next chapter,—but shall chiefly confine myself here to the mere process employed, which may be characterized as a process of "development." As introductory to the nature of this process, it may be proper first to make a few observations.

7. Suppose that we are considering any class of things with reference to this question, viz., the relation in which its members stand as to the possession or the want of a certain property x. As every individual in the proposed class either possesses or does not possess the property in question, we may divide the class into two portions, the former consisting of those individuals which possess, the latter of those which do not possess, the property. This possibility of dividing in thought the whole class into two constituent portions, is antecedent to all knowledge of the constitution of the class derived from any other source; of which knowledge the effect can only be to inform us, more or less precisely, to what further conditions the portions of the class which possess and which do not possess the given property are subject. Suppose, then, such knowledge is to the following effect, viz., that the members of that portion which possess the property x, possess also a certain property u, and that these conditions united are a sufficient definition of them. We may then represent that portion of the original class by the expression ux (ii. 6). If, further, we obtain information that the members of the original class which do not possess the property x, are subject to a condition v, and are thus defined, it is clear, that those members will be represented by the expression $v(1 - x)$. Hence the class in its totality will be represented by

$$ux + v(1 - x);$$

which may be considered as a general developed form for the expression of any class of objects considered with reference to the possession or the want of a given property x.

The general form thus established upon purely logical grounds may also be deduced from distinct considerations of formal law, applicable to the symbols x, y, z, equally in their logical and in their quantitative interpretation already referred to (v. 6).

8. *Definition.*—Any algebraic expression involving a symbol x is termed a function of x, and may be represented under the abbreviated general form $f(x)$. Any expression involving two symbols, x and y, is similarly termed a function of x and y, and may be represented under the general form $f(x, y)$, and so on for any other case.

Thus the form $f(x)$ would indifferently represent any of the following functions, viz., x, $1 - x$, $\frac{1 + x}{1 - x}$, &c.; and $f(x, y)$ would equally represent any of the forms $x + y$, $x - 2y$, $\frac{x + y}{x - 2y}$, &c.

On the same principles of notation, if in any function $f(x)$, we change x into 1, the result will be expressed by the form $f(1)$; if in the same function we change x into 0, the result will be expressed by the form $f(0)$. Thus, if $f(x)$ represent the function $\frac{a + x}{a - 2x}$, $f(1)$ will represent $\frac{a + 1}{a - 2}$, and $f(0)$ will represent $\frac{a}{a}$.

9. *Definition.*—Any function $f(x)$, in which x is a logical symbol, or a symbol of quantity susceptible only of the values 0 and 1, is said to be developed, when it is reduced to the form $ax + b(1 - x)$, a and b being so determined as to make the result equivalent to the function from which it was derived.

This definition assumes, that it is possible to represent any function $f(x)$ in the form supposed. The assumption is vindicated in the following Proposition.

Proposition I

10. *To develop any function $f(x)$ in which x is a logical symbol.*

By the principle which has been asserted in this chapter, it is lawful to treat x as a quantitative symbol, susceptible only of the values 0 and 1.

Assume then,

$$f(x) = ax + b(1 - x),$$

and making $x = 1$, we have

$$f(1) = a.$$

Again, in the same equation making $x = 0$, we have

$$f(0) = b.$$

Hence the values of a and b are determined, and substituting them in the first equation, we have

$$f(x) = f(1)x + f(0)(1 - x); \tag{1}$$

as the development sought.[1] The second member of the equation adequately represents the function $f(x)$, whatever the form of that function may be. For x regarded as a quantitative symbol admits only of the values 0 and 1, and for each of these values the development

$$f(1) x + f(0)(1 - x),$$

assumes the same value as the function $f(x)$.

As an illustration, let it be required to develop the function $\frac{1 + x}{1 + 2x}$. Here, when $x = 1$, we find $f(1) = \frac{2}{3}$, and when $x = 0$, we find $f(0) = \frac{1}{1}$, or 1. Hence the expression required is

$$\frac{1 + x}{1 + 2x} = \frac{2}{3} x + 1 - x;$$

and this equation is satisfied for each of the values of which the symbol x is susceptible.

Proposition II

To expand or develop a function involving any number of logical symbols.

Let us begin with the case in which there are two symbols, x and y, and let us represent the function to be developed by $f(x, y)$.

First, considering $f(x, y)$ as a function of x alone, and expanding it by the general theorem (1), we have

$$f(x, y) = f(1, y) x + f(0, y)(1 - x); \qquad (2)$$

wherein $f(1, y)$ represents what the proposed function becomes, when in it for x we write 1, and $f(0, y)$ what the said function becomes, when in it for x we write 0.

Now, taking the coefficient $f(1, y)$, and regarding it as a function of y, and expanding it accordingly, we have

$$f(1, y) = f(1, 1) y + f(1, 0)(1 - y), \qquad (3)$$

wherein $f(1, 1)$ represents what $f(1, y)$ becomes when y is made equal to 1, and $f(1, 0)$ what $f(1, y)$ becomes when y is made equal to 0.

---

[1]To some it may be interesting to remark, that the development of $f(x)$ obtained in this chapter, strictly holds, in the logical system, the place o the expansion of $f(x)$ in ascending powers of x in the system of ordinary algebra. Thus it may be obtained by introducing into the expression of Taylor's well-known theorem, viz.:

$$f(x) = f(0) + f'(0)x + f''(0)\frac{x^2}{1 \cdot 2} + f'''(0)\frac{x^3}{1 \cdot 2 \cdot 3}, \&c. \qquad (1$$

the condition $x(1 - x) = 0$, whence we find $x^2 = x, x^3 = x$, &c., and

$$f(x) = f(0) + \left\{ f'(0) + \frac{f''(0)}{1 \cdot 2} + \frac{f'''(0)}{1 \cdot 2 \cdot 3} + \&c. \right\} x. \qquad (2$$

But making in (1), $x = 1$, we get

$$f(1) = f(0) + f'(0) + \frac{f''(0)}{1 \cdot 2} + \frac{f'''(0)}{1 \cdot 2 \cdot 3} + \&c.;$$

whence

$$f'(0) + \frac{f''(0)}{1 \cdot 2} + \&c. = f(1) - f(0),$$

and (2) becomes, on substitution,

$$f(x) = f(0) + \{f(1) - f(0)\}x,$$
$$= f(1)x + f(0)(1 - x),$$

the form in question. This demonstration in supposing $f(x)$ to be developable in a series of ascending powers of x is less general than the one in the text

In like manner, the coefficient $f(0, y)$ gives by expansion,

$$f(0, y) = f(0, 1)y + f(0, 0)(1 - y). \tag{4}$$

Substitute in (2) for $f(1, y)$, $f(0, y)$, their values given in (3) and (4), and we have

$$\begin{aligned} f(x, y) = {} & f(1, 1)xy + f(1, 0)x(1 - y) + f(0, 1)(1 - x)y \\ & + f(0, 0)(1 - x)(1 - y), \end{aligned} \tag{5}$$

for the expansion required. Here $f(1, 1)$ represents what $f(x, y)$ becomes when we make therein $x = 1$, $y = 1$; $f(1, 0)$ represents what $f(x, y)$ becomes when we make therein $x = 1$, $y = 0$, and so on for the rest.

Thus, if $f(x, y)$ represent the function $\frac{1-x}{1-y}$, we find

$$f(1, 1) = \frac{0}{0}, \quad f(1, 0) = \frac{0}{1} = 0, \quad f(0, 1) = \frac{1}{0}, \quad f(0, 0) = 1,$$

whence the expansion of the given function is

$$\frac{0}{0}xy + 0x(1 - y) + \frac{1}{0}(1 - x)y + (1 - x)(1 - y).$$

It will in the next chapter be seen that the forms $\frac{0}{0}$ and $\frac{1}{0}$, the former of which is known to mathematicians as the symbol of indeterminate quantity, admit, in such expressions as the above, of a very important logical interpretation.

Suppose, in the next place, that we have three symbols in the function to be expanded, which we may represent under the general form $f(x, y, z)$. Proceeding as before, we get

$$\begin{aligned} f(x, y, z) = {} & f(1, 1, 1)xyz + f(1, 1, 0)xy(1 - z) + f(1, 0, 1)x(1 - y)z \\ & + f(1, 0, 0)x(1 - y)(1 - z) + f(0, 1, 1)(1 - x)yz \\ & + f(0, 1, 0)(1 - x)y(1 - z) + f(0, 0, 1)(1 - x)(1 - y)z \\ & + f(0, 0, 0)(1 - x)(1 - y)(1 - z), \end{aligned}$$

in which $f(1, 1, 1)$ represents what the function $f(x, y, z)$ becomes when we make therein $x = 1$, $y = 1$, $z = 1$, and so on for the rest.

11. It is now easy to see the general law which determines the expansion of any proposed function, and to reduce the method of effecting the expansion to a rule. But before proceeding to the expression of such a rule, it will be convenient to premise the following observations:—

Each form of expansion that we have obtained consists of certain terms, into which the symbols x, y, &c. enter, multiplied by coefficients, into which those symbols do not enter. Thus the expansion of $f(x)$ consists of two terms, x and $1 - x$, multiplied by the coefficients $f(1)$ and $f(0)$ respectively. And the expansion of $f(x, y)$ consists of the four terms xy, $x(1 - y)$, $(1 - x)y$, and $(1 - x)$, $(1 - y)$, multiplied by the coefficients $f(1, 1)$, $f(1, 0)$, $f(0, 1)$, $f(0, 0)$, respectively. The terms x, $1 - x$, in the former case, and the terms xy, $x(1 - y)$, &c., in the latter, we shall call the *constituents* of the expansion. It is evident that they are in form independent of the form of the function to be expanded. Of the constituent xy, x and y are termed the *factors*.

The general rule of development will therefore consist of two parts, the first of which will relate to the formation of the *constituents* of the expansion, the second to the determination of their respective coefficients. It is as follows:

1st. *To expand any function of the symbols x, y, z.*—Form a series of constituents in the following manner: Let the first constituent be the product of the symbols change in this product any symbol z into $1 - z$, for the second constituent. Then in both these change any other symbol y into $1 - y$, for two more constituents Then in the four constituents thus obtained change any other symbol x into $1 - x$ for four new constituents, and so on until the number of possible changes is exhausted.

2ndly. *To find the coefficient of any constituent.*—If that constituent involves x as a factor, change in the original function x into 1; but if it involves $1 - x$ as a factor, change in the original function x into 0. Apply the same rule with reference to the symbols y, z, &c.: the final calculated value of the function thus transformed will be the coefficient sought.

The sum of the constituents, multiplied each by its respective coefficient, will be the expansion required.

12. It is worthy of observation, that a function may be developed with reference to symbols which it does not explicitly contain. Thus if, proceeding according to the rule, we seek to develop the function $1 - x$, with reference to the symbols x and y we have,

$$\text{When } x = 1 \text{ and } y = 1 \text{ the given function} = 0.$$
$$x = 1 \quad , \quad y = 0 \quad , \quad , \quad = 0.$$
$$x = 0 \quad , \quad y = 1 \quad , \quad , \quad = 1.$$
$$x = 0 \quad , \quad y = 0 \quad , \quad , \quad = 1.$$

Whence the development is

$$1 - x = 0 \, xy + 0 \, x(1 - y) + (1 - x) \, y + (1 - x)(1 - y);$$

and this is a true development. The addition of the terms $(1 - x)y$ and $(1 - x)(1 - y)$ produces the function $1 - x$.

The symbol 1 thus developed according to the rule, with respect to the symbol x, gives

$$x + 1 - x.$$

Developed with respect to x and y, it gives

$$xy + x(1 - y) + (1 - x) \, y + (1 - x)(1 - y).$$

Similarly developed with respect to any set of symbols, it produces a series consisting of all possible constituents of those symbols.

13. A few additional remarks concerning the nature of the general expansions may with propriety be added. Let us take, for illustration, the general theorem (5), which presents the type of development for functions of two logical symbols.

In the first place, that theorem is perfectly true and intelligible when x and y are quantitative symbols of the species considered in this chapter, whatever algebraic form may be assigned to the function $f(x, y)$, and it may therefore be intelligibly employed in any stage of the process of analysis intermediate between the change of interpretation of the symbols from the logical to the quantitative system above referred to, and the final restoration of the logical interpretation.

Secondly. The theorem is perfectly true and intelligible when x and y are logical symbols, provided that the form of the function $f(x, y)$ is such as to represent a *class or collection of things*, in which case the second member is always logically interpretable. For instance, if $f(x, y)$ represent the function $1 - x + xy$, we obtain on applying the theorem

$$1 - x + xy = xy + 0 \, x(1 - y) + (1 - x)y + (1 - x)(1 - y),$$
$$= xy + (1 - x)y + (1 - x)(1 - y),$$

and this result is intelligible and true.

Thus we may regard the theorem as true and intelligible for quantitative symbols of the species above described, *always*; for logical symbols, *always when interpretable*. Whensoever therefore it is employed in this work it must be understood that the symbols x, y are quantitative and of the particular species referred to, if the expansion obtained is not interpretable.

But though the expansion is not always immediately interpretable, it always conducts us at once to results which are interpretable. Thus the expression $x - y$ gives on development the form

$$x(1 - y) - y(1 - x),$$

which is not generally interpretable. We cannot take, in thought, from the class of things which are x's and not y's, the class of things which are y's and not x's, because the latter class is not contained in the former. But if the form $x - y$ presented itself as the first member of an equation, of which the second member was 0, we should have on development

$$x(1 - y) - y(1 - x) = 0.$$

Now it will be shown in the next chapter that the above equation, x and y being regarded as quantitative and of the species described, is resolvable at once into the two equations

$$x(1 - y) = 0, \quad y(1 - x) = 0,$$

and these equations are directly interpretable in Logic when logical interpretations are assigned to the symbols x and y. And it may be remarked, that though *functions*

do not necessarily become interpretable upon development, yet *equations* are alwa[
reducible by this process to interpretable forms.

14. The following Proposition establishes some important properties of co[
stituents. In its enunciation the symbol t is employed to represent indifferently a[
constituent of an expansion. Thus if the expansion is that of a function of two sy[
bols x and y, t represents any of the four forms xy, $x(1 - y)$, $(1 - x)y$, an[
$(1 - x)(1 - y)$. Where it is necessary to represent the constituents of an expansio[
by single symbols, and yet to distinguish them from each other, the distinction w[
be marked by suffixes. Thus t_1 might be employed to represent xy, t_2 to represe[
$x(1 - y)$, and so on.

Proposition III

Any single constituent t of an expansion satisfies the law of duality whose expression

$$t(1 - t) = 0.$$

The product of any two distinct constituents of an expansion is equal to 0, and the su[
of all the constituents is equal to 1.

1st. Consider the particular constituent xy. We have

$$xy \times xy = x^2y^2.$$

But $x^2 = x$, $y^2 = y$, by the fundamental law of class symbols; hence

$$xy \times xy = xy.$$

Or representing xy by t,

$$t \times t = t,$$

or

$$t(1 - t) = 0.$$

Similarly the constituent $x(1 - y)$ satisfies the same law. For we have

$$x^2 = x, \quad (1 - y)^2 = 1 - y,$$
$$\therefore \{x(1 - y)\}^2 = x(1 - y), \quad \text{or} \quad t(1 - t) = 0.$$

Now every factor of every constituent is either of the form x or of the form $1 - [$
Hence the square of each factor is equal to that factor, and therefore the square [
the product of the factors, i.e. of the constituent, is equal to the constituent; wher[
fore t representing any constituent, we have

$$t^2 = t, \quad \text{or} \quad t(1 - t) = 0.$$

2ndly. The product of any two constituents is 0. This is evident from the gener[
law of the symbols expressed by the equation $x(1 - x) = 0$; for whatev[
constituents in the same expansion we take, there will be at least one factor x in th[
one, to which will correspond a factor $1 - x$ in the other.

3rdly. The sum of all the constituents of an expansion is unity. This is evide[
from addition of the two constituents x and $1 - x$, or of the four constituents, $x[$
$x(1 - y)$, $(1 - x)y$, $(1 - x)(1 - y)$. But it is also, and more generally, proved [

expanding 1 in terms of any set of symbols (v. 12). The constituents in this case are formed as usual, and all the coefficients are unity.

15. With the above Proposition we may connect the following.

Proposition IV

If V represent the sum of any series of constituents, the separate coefficients of which are 1, then is the condition satisfied,

$$V(1 - V) = 0.$$

Let $t_1, t_2 \ldots t_n$ be the constituents in question, then

$$V = t_1 + t_2 \ldots + t_n.$$

Squaring both sides, and observing that $t_1^2 = t_1$, $t_1 t_2 = 0$, &c., we have

$$V^2 = t_1 + t_2 \ldots + t_n;$$

whence

$$V = V^2.$$

Therefore

$$V(1 - V) = 0.$$

CHAPTER VI

OF THE GENERAL INTERPRETATION OF LOGICAL EQUATIONS, AND THE RESULTING ANALYSIS OF PROPOSITIONS. ALSO, OF THE CONDITION OF INTERPRETABILITY OF LOGICAL FUNCTIONS

1. It has been observed that the complete expansion of any function by the general rule demonstrated in the last chapter, involves two distinct sets of elements, viz., the constituents of the expansion, and their coefficients. I propose in the present chapter to inquire, first, into the interpretation of constituents, and afterwards into the mode in which that interpretation is modified by the coefficients with which they are connected.

The terms "logical equation," "logical function," &c., will be employed generally to denote any equation or function involving the symbols x, y, &c., which may present itself either in the expression of a system of premises, or in the train of symbolical results which intervenes between the premises and the conclusion. If that function or equation is in a form not immediately interpretable in Logic, the symbols x, y, &c., must be regarded as quantitative symbols of the species described in previous chapters (ii. 15), (v. 6), as satisfying the law,

$$x(1 - x) = 0.$$

By the problem, then, of the interpretation of any such logical function or equation, is meant the reduction of it to a form in which, when logical values are assigned to the symbols x, y, &c., it shall become interpretable, together with the resulting interpretation. These conventional definitions are in accordance with the general principles for the conducting of the method of this treatise, laid down in the previous chapter.

Proposition I

2. *The constituents of the expansion of any function of the logical symbols x, y, &c., are interpretable, and represent the several exclusive divisions of the universe of discourse, formed by the predication and denial in every possible way of the qualities denoted by the symbols x, y, &c.*

For greater distinctness of conception, let it be supposed that the function expanded involves two symbols x and y, with reference to which the expansion has been effected. We have then the following constituents, viz.:

$$xy, \quad x(1-y), \quad (1-x)y, \quad (1-x)(1-y).$$

Of these it is evident, that the first xy represents that class of objects which at the same time possess both the elementary qualities expressed by x and y, and that the second $x(1-y)$ represents the class possessing the property x, but not the property y. In like manner the third constituent represents the class of objects which possess the property represented by y, but not that represented by x; and the fourth constituent $(1-x)(1-y)$, represents that class of objects, the members of which possess neither of the qualities in question.

Thus the constituents in the case just considered represent all the four classes of objects which can be described by affirmation and denial of the properties expressed by x and y. Those classes are distinct from each other. No member of one is a member of another, for each class possesses some property or quality contrary to a property or quality possessed by any other class. Again, these classes together make up the universe, for there is no object which may not be described by the presence or the absence of a proposed quality, and thus each individual thing in the universe may be referred to some one or other of the four classes made by the possible combination of the two given classes x and y, and their contraries.

The remarks which have here been made with reference to the constituents of $f(x, y)$ are perfectly general in character. The constituents of any expansion represent classes—those classes are mutually distinct, through the possession of contrary qualities, and they together make up the universe of discourse.

3. These properties of constituents have their expression in the theorems demonstrated in the conclusion of the last chapter, and might thence have been deduced. From the fact that every constituent satisfies the fundamental law of the individual symbols, it might have been conjectured that each constituent would represent a class.

734

From the fact that the product of any two constituents of an expansion vanishes, it might have been concluded that the classes they represent are mutually exclusive. Lastly, from the fact that the sum of the constituents of an expansion is unity, it might have been inferred, that the classes which they represent, together make up the universe.

4. Upon the laws of constituents and the mode of their interpretation above determined, are founded the analysis and the interpretation of logical equations. That all such equations admit of interpretation by the theorem of development has already been stated. I propose here to investigate the forms of possible solution which thus present themselves in the conclusion of a train of reasoning, and to show how those forms arise. Although, properly speaking, they are but manifestations of a single fundamental type or principle of expression, it will conduce to clearness of apprehension if the minor varieties which they exhibit are presented separately to the mind.

The forms, which are three in number, are as follows:

Form I

5. The form we shall first consider arises when any logical equation $V = 0$ is developed, and the result, after resolution into its component equations, is to be interpreted. The function is supposed to involve the logical symbols x, y, &c., in combinations which are not fractional. Fractional combinations indeed only arise in the class of problems which will be considered when we come to speak of the third of the forms of solution above referred to.

Proposition II

To interpret the logical equation $V = 0$.

For simplicity let us suppose that V involves but two symbols, x and y, and let us represent the development of the given equation by

$$axy + bx(1 - y) + c(1 - x)y + d(1 - x)(1 - y) = 0; \qquad (1)$$

a, b, c, and d being definite numerical constants.

Now, suppose that any coefficient, as a, does not vanish. Then multiplying each side of the equation by the constituent xy, to which that coefficient is attached, we have

$$axy = 0,$$

whence, as a does not vanish,

$$xy = 0,$$

and this result is quite independent of the nature of the other coefficients of the expansion. Its interpretation, on assigning to x and y their logical significance, is "No individuals belonging at once to the class represented by x, and the class represented by y, exist."

But if the coefficient *a does* vanish, the term *axy* does not appear in the development (1), and, therefore, the equation $xy = 0$ cannot thence be deduced.

In like manner, if the coefficient *b* does not vanish, we have

$$x(1 - y) = 0,$$

which admits of the interpretation, "There are no individuals which at the same time belong to the class *x*, and do not belong to the class *y*."

Either of the above interpretations may, however, as will subsequently be shown, be exhibited in a different form.

The sum of the distinct interpretations thus obtained from the several terms of the expansion whose coefficients do not vanish, will constitute the complete interpretation of the equation $V = 0$. The analysis is essentially independent of the number of logical symbols involved in the function *V*, and the object of the proposition will, therefore, in all instances, be attained by the following Rule:—

RULE.—*Develop the function V, and equate to 0 every constituent whose coefficient does not vanish. The interpretation of these results collectively will constitute the interpretation of the given equation.*

6. Let us take as an example the definition of "clean beasts," laid down in the Jewish law, viz., "Clean beasts are those which both divide the hoof and chew the cud," and let us assume

$$x = \text{clean beasts};$$
$$y = \text{beasts dividing the hoof};$$
$$z = \text{beasts chewing the cud}.$$

Then the given proposition will be represented by the equation

$$x = yz,$$

which we shall reduce to the form

$$x - yz = 0,$$

and seek that form of interpretation to which the present method leads. Fully developing the first member, we have

$$0 \, xyz + xy(1 - z) + x(1 - y)z + x(1 - y)(1 - z) - (1 - x)yz$$
$$+ \, 0(1 - x)y(1 - z) + 0(1 - x)(1 - y)z + 0(1 - x)(1 - y)(1 - z).$$

Whence the terms, whose coefficients do not vanish, give

$$xy(1 - z) = 0, \quad xz(1 - y) = 0, \quad x(1 - y)(1 - z) = 0, \quad (1 - x)yz = 0.$$

These equations express a denial of the existence of certain classes of objects, viz.:

1st. Of beasts which are clean, and divide the hoof, but do not chew the cud.

2nd. Of beasts which are clean, and chew the cud, but do not divide the hoof.

3rd. Of beasts which are clean, and neither divide the hoof nor chew the cud.

4th. Of beasts which divide the hoof, and chew the cud, and are not clean.

Now all these several denials are really involved in the original proposition. And conversely, if these denials be granted, the original proposition will follow as a necessary consequence. They are, in fact, the separate elements of that proposition. Every primary proposition can thus be resolved into a series of denials of the existence of certain defined classes of things, and may, from that system of denials, be itself reconstructed. It might here be asked, how it is possible to make an assertive proposition out of a series of denials or negations? From what source is the positive element derived? I answer, that the mind assumes the existence of a universe not *à priori* as a fact independent of experience, but either *à posteriori* as a deduction from experience, or *hypothetically* as a foundation of the possibility of assertive reasoning. Thus from the Proposition, "There are no men who are not fallible," which is a negation or denial of the existence of "infallible men," it may be inferred either hypothetically, "All men (if men exist) are fallible," or absolutely, (experience having assured us of the existence of the race), "All men are fallible."

The form in which conclusions are exhibited by the method of this Proposition may be termed the form of "Single or Conjoint Denial."

Form II

7. As the previous form was derived from the development and interpretation of an equation whose second member is 0, the present form, which is supplementary to it, will be derived from the development and interpretation of an equation whose second member is 1. It is, however, readily suggested by the analysis of the previous Proposition.

Thus in the example last discussed we deduced from the equation

$$x - yz = 0$$

the conjoint denial of the existence of the classes represented by the constituents

$$xy(1 - z), \quad xz(1 - y), \quad x(1 - y)(1 - z), \quad (1 - x)yz,$$

whose coefficients were not equal to 0. It follows hence that the remaining constituents represent classes which make up the universe. Hence we shall have

$$xyz + (1 - x)y(1 - z) + (1 - x)(1 - y)z + (1 - x)(1 - y)(1 - z) = 1.$$

This is equivalent to the affirmation that all existing things belong to some one or other of the following classes, viz.:

1st. Clean beasts both dividing the hoof and chewing the cud.

2nd. Unclean beasts dividing the hoof, but not chewing the cud.

3rd. Unclean beasts chewing the cud, but not dividing the hoof.

4th. Things which are neither clean beasts, nor chewers of the cud, nor dividers of the hoof.

This form of conclusion may be termed the form of "Single or Disjunctive Affirmation,"—single when but one constituent appears in the final equation; disjunctive when, as above, more constituents than one are there found.

Any equation, $V = 0$, wherein V satisfies the law of duality, may also be made to yield this form of interpretation by reducing it to the form $1 - V = 1$, and developing the first member. The case, however, is really included in the next general form. Both the previous forms are of slight importance compared with the following one.

Form III

8. In the two preceding cases the functions to be developed were equated to 0 and to 1 respectively. In the present case I shall suppose the corresponding function equated to any logical symbol w. We are then to endeavour to interpret the equation $V = w$, V being a function of the logical symbols x, y, z, &c. In the first place, however, I deem it necessary to show how the equation $V = w$, or, as it will usually present itself, $w = V$, arises.

Let us resume the definition of "clean beasts," employed in the previous examples, viz., "Clean beasts are those which both divide the hoof and chew the cud," and suppose it required to determine the relation in which "beasts chewing the cud" stand to "clean beasts" and "beasts dividing the hoof." The equation expressing the given proposition is

$$x = yz,$$

and our object will be accomplished if we can determine z as an interpretable function of x and y.

Now treating x, y, z as symbols of quantity subject to a peculiar law, we may deduce from the above equation, by solution,

$$z = \frac{x}{y}.$$

But this equation is not at present in an interpretable form. If we can reduce it to such a form it will furnish the relation required.

On developing the second member of the above equation, we have

$$z = xy + \frac{1}{0}x(1 - y) + 0(1 - x)y + \frac{0}{0}(1 - x)(1 - y),$$

and it will be shown hereafter (Prop. 3) that this admits of the following interpretation:

"Beasts which chew the cud consist of all clean beasts (which also divide the hoof), together with an indefinite remainder (some, none, or all) of unclean beasts which do not divide the hoof."

9. Now the above is a particular example of a problem of the utmost generality in Logic, and which may thus be stated:—"Given any logical equation connecting

the symbols x, y, z, w, required an interpretable expression for the relation of the class represented by w to the classes represented by the other symbols x, y, z, &c."

The solution of this problem consists in all cases in determining, from the equation *given*, the expression of the above symbol w, in terms of the other symbols, and rendering that expression interpretable by development. Now the equation given is always of the first degree with respect to each of the symbols involved. The required expression for w can therefore always be found. In fact, if we develop the given equation, whatever its form may be with respect to w, we obtain an equation of the form

$$Ew + E'(1 - w) = 0, \tag{1}$$

E and E' being functions of the remaining symbols. From the above we have

$$E' = (E' - E)\, w.$$

Therefore

$$w = \frac{E'}{E' - E} \tag{2}$$

and expanding the second member by the rule of development, it will only remain to interpret the result in logic by the next proposition.

If the fraction $\frac{E'}{E' - E}$ has common factors in its numerator and denominator, we are not permitted to reject them, unless they are mere numerical constants. For the symbols x, y, &c., regarded as quantitative, may admit of such values 0 and 1 as to cause the common factors to become equal to 0, in which case the algebraic rule of reduction fails. This is the case contemplated in our remarks on the failure of the algebraic axiom of division (ii. 14). To *express* the solution in the form (2), and without attempting to perform any unauthorized reductions, to interpret the result by the theorem of development, is a course strictly in accordance with the general principles of this treatise.

If the relation of the class expressed by $1 - w$ to the other classes, x, y, &c. is required, we deduce from (1), in like manner as above,

$$1 - w = \frac{E}{E - E'},$$

to the interpretation of which also the method of the following Proposition is applicable:

Proposition III

10. *To determine the interpretation of any logical equation of the form $w = V$, in which w is a class symbol, and V a function of other class symbols quite unlimited in its form.*

Let the second member of the above equation be fully expanded. Each coefficient of the result will belong to some one of the four classes, which, with their respective interpretations, we proceed to discuss.

1st. Let the coefficient be 1. As this is the symbol of the universe, and as the product of any two class symbols represents those individuals which are found in both classes, any constituent which has unity for its coefficient must be interpreted without limitation, i.e. the whole of the class which it represents is implied.

2nd. Let the coefficient be 0. As in Logic, equally with Arithmetic, this is the symbol of Nothing, no part of the class represented by the constituent to which it is prefixed must be taken.

3rd. Let the coefficient be of the form $\frac{0}{0}$. Now, as in Arithmetic, the symbol $\frac{0}{0}$ represents an *indefinite number*, except when otherwise determined by some special circumstance, analogy would suggest that in the system of this work the same symbol should represent an *indefinite class*. That this is its true meaning will be made clear from the following example:

Let us take the Proposition, "Men not mortal do not exist;" represent this Proposition by symbols; and seek, in obedience to the laws to which those symbols have been proved to be subject, a reverse definition of "mortal beings," in terms of "men."

Now if we represent "men" by y, and "mortal beings" by x, the Proposition, "Men who are not mortals do not exist," will be expressed by the equation

$$y(1 - x) = 0,$$

from which we are to seek the value of x. Now the above equation gives

$$y - yx = 0, \text{ or } yx = y.$$

Were this an ordinary algebraic equation, we should, in the next place, divide both sides of it by y. But it has been remarked in Chap. II. that the operation of division cannot be *performed* with the symbols with which we are now engaged. Our resource, then, is to *express* the operation, and develop the result by the method of the preceding chapter. We have, then, first,

$$x = \frac{y}{y},$$

and, expanding the second member as directed,

$$x = y + \frac{0}{0}(1 - y).$$

This implies that mortals (x) consist of all men (y), together with such a remainder of beings which are not men ($1 - y$), as will be indicated by the coefficient $\frac{0}{0}$. Now let us inquire what remainder of "not men" is implied by the premiss. It might happen that the remainder included all the beings who are not men, or it might include only some of them, and not others, or it might include none, and any one of these assumptions would be in perfect accordance with our premiss. In other words, whether those beings which are not men are *all*, or *some*, or *none*, of them *mortal*, the truth of the premiss which virtually asserts that all men are mortal, will be equally unaffected, and therefore the expression $\frac{0}{0}$ here indicates that *all*, *some*, or *none* of the class to whose expression it is affixed must be taken.

Although the above determination of the significance of the symbol $\frac{0}{0}$ is founded only upon the examination of a particular case, yet the principle involved in the demonstration is general, and there are no circumstances under which the symbol can present itself to which the same mode of analysis is inapplicable. We may properly term $\frac{0}{0}$ an *indefinite class symbol*, and may, if convenience should require, replace it by an uncompounded symbol v, subject to the fundamental law, $v(1 - v) = 0$.

4th. It may happen that the coefficient of a constituent in an expansion does not belong to any of the previous cases. To ascertain its true interpretation when this happens, it will be necessary to premise the following theorem:

11. THEOREM.—*If a function V, intended to represent any class or collection of objects, w, be expanded, and if the numerical coefficient, a, of any constituent in its development, do not satisfy the law.*

$$a(1 - a) = 0,$$

then the constituent in question must be made equal to 0.

To prove the theorem generally, let us represent the expansion given, under the form

$$w = a_1 t_1 + a_2 t_2 + a_3 t_3 + \&c., \tag{1}$$

in which t_1, t_2, t_3, &c. represent the constituents, and a_1, a_2, a_3, &c. the coefficients; let us also suppose that a_1 and a_2 do not satisfy the law

$$a_1(1 - a_1) = 0, \quad a_2(1 - a_2) = 0;$$

but that the other coefficients are subject to the law in question, so that we have

$$a_3^2 = a_3, \&c.$$

Now multiply each side of the equation (1) by itself. The result will be

$$w = a_1^2 t_1 + a_2^2 t_2 + \&c. \tag{2}$$

This is evident from the fact that it must represent the development of the equation

$$w = V^2,$$

but it may also be proved by actually squaring (1), and observing that we have

$$t_1^2 = t_1, \quad t_2^2 = t_2, \quad t_1 t_2 = 0, \&c.$$

by the properties of constituents. Now subtracting (2) from (1), we have

$$(a_1 - a_1^2) t_1 + (a_2 - a_2^2) t_2 = 0.$$

Or, $\qquad a_1(1 - a_1) t_1 + a_2(1 - a_2) t_2 = 0.$

Multiply the last equation by t_1; then since $t_1 t_2 = 0$, we have

$$a_1(1 - a_1) t_1 = 0, \quad \text{whence } t_1 = 0.$$

In like manner multiplying the same equation by t_2, we have

$$a_2(1 - a_2) t_2 = 0, \quad \text{whence } t_2 = 0.$$

741

Thus it may be shown generally that any constituent whose coefficient is not subject to the same fundamental law as the symbols themselves must be separately equated to 0. The usual form under which such coefficients occur is $\frac{1}{0}$. This is the algebraic symbol of infinity. Now the nearer any number approaches to infinity (allowing such an expression), the more does it depart from the condition of satisfying the fundamental law above referred to.

The symbol $\frac{0}{0}$, whose interpretation was previously discussed, does not necessarily disobey the law we are here considering, for it admits of the numerical values 0 and 1 indifferently. Its actual interpretation, however, as an indefinite class symbol, cannot, I conceive, except upon the ground of analogy, be deduced from its arithmetical properties, but must be established experimentally.

12. We may now collect the results to which we have been led, into the following summary:

1st. The symbol 1, as the coefficient of a term in a development, indicates that the whole of the class which that constituent represents, is to be taken.

2nd. The coefficient 0 indicates that none of the class are to be taken.

3rd. The symbol $\frac{0}{0}$ indicates that a perfectly *indefinite* portion of the class, i.e. *some*, *none*, or *all* of its members are to be taken.

4th. Any other symbol as a coefficient indicates that the constituent to which it is prefixed must be equated to 0.

It follows hence that if the solution of a problem, obtained by development, be of the form

$$w = A + 0B + \frac{0}{0}C + \frac{1}{0}D,$$

that solution may be resolved into the two following equations, viz.,

$$w = A + vC, \tag{3}$$
$$D = 0, \tag{4}$$

v being an indefinite class symbol. The interpretation of (3) shows what elements enter, or may enter, into the composition of w, the class of things whose definition is required; and the interpretation of (4) shows what relations exist among the elements of the original problem, in perfect independence of w.

Such are the canons of interpretation. It may be added, that they are universal in their application, and that their use is always unembarrassed by exception or failure.

13. *Corollary.*—If V be an independently interpretable logical function, it will satisfy the symbolical law, $V(1 - V) = 0$.

By an independently interpretable logical function, I mean one which is interpretable, without presupposing any relation among the things represented by the symbols which it involves. Thus $x(1 - y)$ is independently interpretable, but $x - y$ is not so. The latter function presupposes, as a condition of its interpretation, that the class represented by y is wholly contained in the class represented by x; the former function does not imply any such requirement.

Now if V be independently interpretable, and if w represent the collection of individuals which it contains, the equation $w = V$ will hold true without entailing as a consequence the vanishing of any of the constituents in the development of V; since such vanishing of constituents would imply relations among the classes of things denoted by the symbols in V. Hence the development of V will be of the form

$$a_1 t_1 + a_2 t_2 + \&c.$$

the coefficients a_1, a_2, &c. all satisfying the condition

$$a_1(1 - a_1) = 0, \quad a_2(1 - a_2) = 0, \&c.$$

Hence by the reasoning of Prop. 4, Chap. V. the function V will be subject to the law

$$V(1 - V) = 0.$$

This result, though evident *à priori* from the fact that V is supposed to represent a class or collection of things, is thus seen to follow also from the properties of the constituents of which it is composed. The condition $V(1 - V) = 0$ may be termed "the condition of interpretability of logical functions."

14. The general form of solutions, or logical conclusions developed in the last Proposition, may be designated as a "Relation between terms." I use, as before, the word "terms" to denote the parts of a proposition, whether simple or complex, which are connected by the copula "is" or "are." The classes of things represented by the individual symbols may be called the elements of the proposition.

15. Ex. 1.—Resuming the definition of "clean beasts," (vi. 6), required a description of "unclean beasts."

Here, as before, x standing for "clean beasts," y for "beasts dividing the hoof," z for "beasts chewing the cud," we have

$$x = yz; \tag{5}$$

whence

$$1 - x = 1 - yz;$$

and developing the second member,

$$1 - x = y(1 - z) + z(1 - y) + (1 - y)(1 - z);$$

which is interpretable into the following Proposition: *Unclean beasts are all which divide the hoof without chewing the cud, all which chew the cud without dividing the hoof, and all which neither divide the hoof nor chew the cud.*

Ex. 2.—The same definition being given, required a description of beasts which do not divide the hoof.

From the equation $x = yz$ we have

$$y = \frac{x}{z};$$

therefore,

$$1 - y = \frac{z - x}{z};$$

and developing the second member,

$$1 - y = 0\, xz + \frac{-1}{0}\, x(1 - z) + (1 - x)z + \frac{0}{0}(1 - x)(1 - z).$$

Here, according to the Rule, the term whose coefficients is $\frac{-1}{0}$, must be separately equated to 0, whence we have

$$1 - y = (1 - x)z + v(1 - x)(1 - z),$$
$$x(1 - z) = 0;$$

whereof the first equation gives by interpretation the Proposition: *Beasts which do not divide the hoof consist of all unclean beasts which chew the cud, and an indefinite remainder (some, none, or all) of unclean beasts which do not chew the cud.*

The second equation gives the Proposition: *There are no clean beasts which do not chew the cud.* This is one of the independent relations above referred to. We sought the direct relation of "Beasts not dividing the hoof," to "Clean beasts and beasts which chew the cud." It happens, however, that independently of any relation to beasts not dividing the hoof, there exists, in virtue of the premiss, a separate relation between clean beasts and beasts which chew the cud. This relation is also necessarily given by the process.

Ex. 3.—Let us take the following definition, viz.: "Responsible beings are all rational beings who are either free to act, or have voluntarily sacrificed their freedom," and apply to it the preceding analysis.

Let x stand for responsible beings.

y „   rational beings.

z „   those who are free to act,

w „   those who have voluntarily sacrificed their freedom of action.

In the expression of this definition I shall assume, that the two alternatives which it presents, viz.: "Rational beings free to act," and "Rational beings whose freedom of action has been voluntarily sacrificed," are mutually exclusive, so that no individuals are found at once in both these divisions. This will permit us to interpret the proposition literally into the language of symbols, as follows:

$$x = yz + yw. \tag{6}$$

Let us first determine hence the relation of "rational beings" to responsible beings, beings free to act, and beings whose freedom of action has been voluntarily abjured. Perhaps this object will be better stated by saying, that we desire to express the relation among the elements of the premiss in such a form as will enable us to determine how far rationality may be inferred from responsibility, freedom of action, a voluntary sacrifice of freedom, and their contraries.

From (6) we have

$$y = \frac{x}{z + x},$$

and developing the second member, but rejecting terms whose coefficients are 0,

$$y = \frac{1}{2}xzw + xz(1 - w) + x(1 - z)w + \frac{1}{0}x(1 - z)(1 - w)$$

$$+ \frac{0}{0}(1 - x)(1 - z)(1 - w),$$

whence, equating to 0 the terms whose coefficients are $\frac{1}{2}$ and $\frac{1}{0}$, we have

$$y = xz(1 - w) + xw(1 - z) + v(1 - x)(1 - z)(1 - w); \tag{7}$$
$$xzw = 0; \tag{8}$$
$$x(1 - z)(1 - w) = 0; \tag{9}$$

whence by interpretation—

DIRECT CONCLUSION.—*Rational beings are all responsible beings who are either free to act, not having voluntarily sacrificed their freedom, or not free to act, having voluntarily sacrificed their freedom, together with an indefinite remainder (some, none, or all) of beings not responsible, not free, and not having voluntarily sacrificed their freedom.*

FIRST INDEPENDENT RELATION.—*No responsible beings are at the same time free to act, and in the condition of having voluntarily sacrificed their freedom.*

SECOND.—*No responsible beings are not free to act, and at the same time in the condition of not having sacrificed their freedom.*

The independent relations above determined may, however, be put in another and more convenient form. Thus (8) gives

$$xw = \frac{0}{z} = 0\,z + \frac{0}{0}(1 - z), \text{ on development;}$$

or,
$$xw = v(1 - z); \tag{10}$$
and in like manner (9) gives

$$x(1 - w) = \frac{0}{1 - z} = \frac{0}{0}z + 0(1 - z);$$

or,
$$x(1 - w) = vz; \tag{11}$$
and (10) and (11) interpreted give the following Propositions:

1st. *Responsible beings who have voluntarily sacrificed their freedom are not free.*
2nd. *Responsible beings who have not voluntarily sacrificed their freedom are free.*

These, however, are merely different forms of the relations before determined.

16. In examining these results, the reader must bear in mind, that the sole province of a method of inference or analysis, is to determine those relations which are necessitated by the *connexion* of the terms in the original proposition. Accordingly, in estimating the completeness with which this object is effected, we have nothing whatever to do with those other relations which may be suggested to our minds by the *meaning* of the terms employed, as distinct from their expressed connexion. Thus it seems obvious to remark, that "They who have voluntarily sacrificed their freedom are not free," this being a relation implied in the very meaning of the terms. And hence it might appear, that the first of the two independent relations assigned by the method is on the one hand needlessly limited, and on the other hand superfluous. However, if regard be had merely to the connexion of the terms in the original premiss, it will be seen that the relation in question is not liable to either of these charges. The solution, as expressed in the direct conclusion and the independent relations, conjointly, is perfectly complete, without being in any way superfluous.

If we wish to take into account the implicit relation above referred to, viz., "They who have voluntarily sacrificed their freedom are not free," we can do so by making this a distinct proposition, the proper expression of which would be

$$w = v(1 - z).$$

This equation we should have to employ together with that expressive of the original premiss. The mode in which such an examination must be conducted will appear when we enter upon the theory of systems of propositions in a future chapter. The sole difference of result to which the analysis leads is, that the first of the independent relations deduced above is superseded.

17. Ex. 4.—Assuming the same definition as in Example 2, let it be required to obtain a description of irrational persons.

We have

$$1 - y = 1 - \frac{x}{z + w}$$

$$= \frac{z + w - x}{z + w}$$

$$= \frac{1}{2} xzw + 0\, xz(1 - w) + 0\, x(1 - z)w - \frac{1}{0} x(1 - z)(1 - w)$$

$$+ (1 - x)zw + (1 - x)z(1 - w) + (1 - x)(1 - z)w + \frac{0}{0}(1 - x)(1 - z)(1 - w)$$

$$= (1 - x)zw + (1 - x)z(1 - w) + (1 - x)(1 - z)w + v(1 - x)(1 - z)(1 - w)$$

$$= (1 - x)z + (1 - x)(1 - z)w + v(1 - x)(1 - z)(1 - w),$$

with $\quad xzw = 0, \qquad x(1 - z)(1 - w) = 0.$

The independent relations here given are the same as we before arrived at, as they evidently ought to be, since whatever relations prevail independently of the existence of a given class of objects y, prevail independently also of the existence of the contrary class $1 - y$.

The direct solution afforded by the first equation is:—*Irrational persons consist of all irresponsible beings who are either free to act, or have voluntarily sacrificed their liberty, and are not free to act; together with an indefinite remainder of irresponsible beings who have not sacrificed their liberty, and are not free to act.*

18. The propositions analyzed in this chapter have been of that species called definitions. I have discussed none of which the second or predicate term is particular, and of which the general type is $Y = vX$, Y and X being functions of the logical symbols x, y, z, &c., and v an indefinite class symbol. The analysis of such propositions is greatly facilitated (though the step is not an essential one) by the elimination of the symbol v, and this process depends upon the method of the next chapter. I postpone also the consideration of another important problem necessary to complete the theory of single propositions, but of which the analysis really depends upon the method of the reduction of systems of propositions to be developed in a future page of this work.

CHAPTER VII

ON ELIMINATION

1. In the examples discussed in the last chapter, all the elements of the original premiss re-appeared in the conclusion, only in a different order, and with a different connexion. But it more usually happens in common reasoning, and especially when we have more than one premiss, that some of the elements are required not to appear in the conclusion. Such elements, or, as they are commonly called, "middle terms," may be considered as introduced into the original propositions only for the sake of that connexion which they assist to establish among the other elements, which are alone designed to enter into the expression of the conclusion.

2. Respecting such intermediate elements, or middle terms, some erroneous notions prevail. It is a general opinion, to which, however, the examples contained in the last chapter furnish a contradiction, that inference consists peculiarly in the elimination of such terms, and that the elementary type of this process is exhibited in the elimination of one middle term from two premises, so as to produce a single resulting conclusion into which that term does not enter. Hence it is commonly held, that *syllogism* is the basis, or else the common type, of all inference, which may thus, however complex its form and structure, be resolved into a series of syllogisms. The propriety of this view will be considered in a subsequent chapter. At present I wish

to direct attention to an important, but hitherto unnoticed, point of difference between the system of Logic, as expressed by symbols, and that of common algebra, with reference to the subject of elimination. In the algebraic system we are able to eliminate one symbol from two equations, two symbols from three equations, and generally $n - 1$ symbols from n equations. There thus exists a definite connexion between the number of independent equations given, and the number of symbols of quantity which it is possible to eliminate from them. But it is otherwise with the system of Logic. No fixed connexion there prevails between the number of equations given representing propositions or premises, and the number of typical symbols of which the elimination can be effected. From a single equation an indefinite number of such symbols may be eliminated. On the other hand, from an indefinite number of equations, a single class symbol only may be eliminated. We may affirm, that in this peculiar system, the problem of elimination is resolvable under all circumstances alike. This is a consequence of that remarkable law of duality to which the symbols of Logic are subject. To the equations furnished by the premises given, there is added another equation or system of equations drawn from the fundamental laws of thought itself, and supplying the necessary means for the solution of the problem in question. Of the many consequences which flow from the law of duality, this is perhaps the most deserving of attention.

3. As in Algebra it often happens, that the elimination of symbols from a given system of equations conducts to a mere identity in the form $0 = 0$, no independent relations connecting the symbols which remain; so in the system of Logic, a like result, admitting of a similar interpretation, may present itself. Such a circumstance does not detract from the generality of the principle before stated. The object of the method upon which we are about to enter is to eliminate any number of symbols from any number of logical equations, and to exhibit in the result the actual relations which remain. Now it may be, that no such residual relations exist. In such a case the truth of the method is shown by its leading us to a merely identical proposition.

4. The notation adopted in the following Propositions is similar to that of the last chapter. By $f(x)$ is meant any expression involving the logical symbol x, with or without other logical symbols. By $f(1)$ is meant what $f(x)$ becomes when x is therein changed into 1; by $f(0)$ what the same function becomes when x is changed into 0.

Proposition I

5. *If $f(x) = 0$ be any logical equation involving the class symbol x, with or without other class symbols, then will the equation*

$$f(1)f(0) = 0$$

be true, independently of the interpretation of x; and it will be the complete result of the elimination of x from the above equation.

In other words, the elimination of x from any given equation, $f(x) = 0$, will be effected by successively changing in that equation x into 1, and x into 0, and multiplying the two resulting equations together.

Similarly the complete result of the elimination of any class symbols, x, y, &c., from any equation of the form $V = 0$, will be obtained by completely expanding the first member of that equation in constituents of the given symbols, and multiplying together all the coefficients of those constituents, and equating the product to 0.

Developing the first member of the equation $f(x) = 0$, we have (v. 10),

$$f(1)x + f(0)(1 - x) = 0;$$

or,
$$\{f(1) - f(0)\}x + f(0) = 0. \tag{1}$$

$$\therefore x = \frac{f(0)}{f(0) - f(1)};$$

and
$$1 - x = -\frac{f(1)}{f(0) - f(1)}.$$

Substitute these expressions for x and $1 - x$ in the fundamental equation

$$x(1 - x) = 0,$$

and there results

$$-\frac{f(0)f(1)}{\{f(0) - f(1)\}^2} = 0;$$

or,
$$f(1)f(0) = 0, \tag{2}$$

the form required.

6. It is seen in this process, that the elimination is really effected between the given equation $f(x) = 0$ and the universally true equation $x(1 - x) = 0$, expressing the fundamental law of logical symbols, *qua* logical. There exists, therefore, no need of more than one premiss or equation, in order to render possible the elimination of a term, the necessary law of thought virtually supplying the other premiss or equation. And though the demonstration of this conclusion may be exhibited in other forms, yet the same element furnished by the mind itself will still be virtually present. Thus we might proceed as follows:

Multiply (1) by x, and we have

$$f(1)x = 0, \tag{3}$$

and let us seek by the forms of ordinary algebra to eliminate x from this equation and (1).

Now if we have two algebraic equations of the form

$$ax + b = 0,$$
$$a'x + b' = 0;$$

it is well known that the result of the elimination of x is

$$ab' - a'b = 0. \tag{4}$$

But comparing the above pair of equations with (1) and (3) respectively, we find

$$a = f(1) - f(0), \quad b = f(0);$$
$$a' = f(1) \quad\quad b' = 0;$$

which, substituted in (4), give

$$f(1)f(0) = 0,$$

as before. In this form of the demonstration, the fundamental equation $x(1 - x) = 0$, makes its appearance in the derivation of (3) from (1).

7. I shall add yet another form of the demonstration, partaking of a half logical character, and which may set the demonstration of this important theorem in a clearer light.

We have as before

$$f(1)x + f(0)(1 - x) = 0.$$

Multiply this equation first by x, and secondly by $1 - x$, we get

$$f(1)x = 0, \quad f(0)(1 - x) = 0.$$

From these we have by solution and development,

$$f(1) = \frac{0}{x} = \frac{0}{0}(1 - x), \text{ on development,}$$

$$f(0) = \frac{0}{1 - x} = \frac{0}{0}x.$$

The direct interpretation of these equations is—

1st. Whatever individuals are included in the class represented by $f(1)$, are not x's.

2nd. Whatever individuals are included in the class represented by $f(0)$, are x's.

Whence by common logic, there are no individuals at once in the class $f(1)$ and in the class $f(0)$, i.e. there are no individuals in the class $f(1) f(0)$. Hence,

$$f(1) f(0) = 0. \tag{5}$$

Or it would suffice to multiply together the developed equations, whence the result would immediately follow.

8. The theorem (5) furnishes us with the following Rule:

TO ELIMINATE ANY SYMBOL FROM A PROPOSED EQUATION

RULE.—*The terms of the equation having been brought, by transposition if necessary, to the first side, give to the symbol successively the values 1 and 0, and multiply the resulting equations together.*

The first part of the Proposition is now proved.

9. Consider in the next place the general equation

$$f(x, y) = 0;$$

the first member of which represents any function of x, y, and other symbols.

By what has been shown, the result of the elimination of y from this equation will be

$$f(x, 1) f(x, 0) = 0;$$

for such is the form to which we are conducted by successively changing in the given equation y into 1, and y into 0, and multiplying the results together.

Again, if in the result obtained we change successively x into 1, and x into 0, and multiply the results together, we have

$$f(1, 1) f(1, 0) f(0, 1) f(0, 0) = 0; \tag{6}$$

as the final result of elimination.

But the four factors of the first member of this equation are the four coefficients of the complete expansion of $f(x, y)$, the first member of the original equation; whence the second part of the Proposition is manifest.

Examples

10. Ex. 1.—Having given the Proposition, "All men are mortal," and its symbolical expression, in the equation,

$$y = vx,$$

in which y represents "men," and x "mortals," it is required to eliminate the indefinite class symbol v, and to interpret the result.

Here bringing the terms to the first side, we have

$$y - vx = 0.$$

When $v = 1$ this becomes

$$y - x = 0;$$

and when $v = 0$ it becomes

$$y = 0;$$

and these two equations multiplied together, give

$$y - yx = 0,$$

or $\qquad\qquad y(1 - x) = 0,$

it being observed that $y^2 = y$.

The above equation is the required result of elimination, and its interpretation is, *Men who are not mortal do not exist,*—an obvious conclusion.

If from the equation last obtained we seek a description of beings who are not mortal, we have

$$x = \frac{y}{y},$$

$$\therefore 1 - x = \frac{0}{y}.$$

Whence, by expansion, $1 - x = \frac{0}{0}(1 - y)$, which interpreted gives, *They who are not mortal are not men.* This is an example of what in the common logic is called conversion by contraposition, or negative conversion.[1]

Ex. 2.—Taking the Proposition, "No men are perfect," as represented by the equation

$$y = v(1 - x),$$

wherein y represents "men," and x "perfect beings," it is required to eliminate v, and find from the result a description both of *perfect beings* and of *imperfect beings*. We have

$$y - v(1 - x) = 0.$$

Whence, by the rule of elimination,

$$\{y - (1 - x)\} \times y = 0,$$

or

$$y - y(1 - x) = 0,$$

or

$$yx = 0;$$

which is interpreted by the Proposition, *Perfect men do not exist.* From the above equation we have

$$x = \frac{0}{y} = \frac{0}{0}(1 - y) \text{ by development;}$$

whence, by interpretation, *No perfect beings are men.* Similarly,

$$1 - x = 1 - \frac{0}{y} = \frac{y}{y} = y + \frac{0}{0}(1 - y),$$

which, on interpretation, gives, *Imperfect beings are all men with an indefinite remainder of beings, which are not men.*

11. It will generally be the most convenient course, in the treatment of propositions, to eliminate first the indefinite class symbol v, wherever it occurs in the corresponding equations. This will only modify their form, without impairing their significance. Let us apply this process to one of the examples of Chap. IV. For the Proposition, "No men are placed in exalted stations and free from envious regards," we found the expression

$$y = v(1 - xz),$$

and for the equivalent Proposition, "Men in exalted stations are not free from envious regards," the expression

$$yx = v(1 - z);$$

and it was observed that these equations, v being an indefinite class symbol, were themselves equivalent. To prove this, it is only necessary to eliminate from each the symbol v. The first equation is

$$y - v(1 - xz) = 0,$$

[1] Whately's Logic, Book II. chap. II. sec. 4.

whence, first making $v = 1$, and then $v = 0$, and multiplying the results, we have
$$(y - 1 + xz)y = 0,$$
or
$$yxz = 0.$$
Now the second of the given equations becomes on transposition
$$yx - v(1 - z) = 0;$$
whence
$$(yx - 1 + z)yx = 0,$$
or
$$yxz = 0,$$
as before. The reader will easily interpret the result.

12. Ex. 3.—As a subject for the general method of this chapter, we will resume Mr. Senior's definition of wealth, viz.: "Wealth consists of things transferable, limited in supply, and either productive of pleasure or preventive of pain." We shall consider this definition, agreeably to a former remark, as including all things which possess at once both the qualities expressed in the last part of the definition, upon which assumption we have, as our representative equation,
$$w = st\{pr + p(1 - r) + r(1 - p)\},$$
or
$$w = st\{p + r(1 - p)\},$$
wherein

w stands for wealth.

s „   things limited in supply.

t „   things transferable.

p „   things productive of pleasure.

r „   things preventive of pain.

From the above equation we can eliminate any symbols that we do not desire to take into account, and express the result by solution and development, according to any proposed arrangement of subject and predicate.

Let us first consider what the expression for w, wealth, would be if the element r, referring to prevention of pain, were eliminated. Now bringing the terms of the equation to the first side, we get
$$w - st(p + r - rp) = 0.$$
Making $r = 1$, the first member becomes $w - st$, and making $r = 0$ it becomes $w - stp$; whence we have by the Rule,
$$(w - st)(w - stp) = 0, \qquad (7)$$
or
$$w - wstp - wst + stp = 0; \qquad (8)$$
whence
$$w = \frac{stp}{st + stp - 1};$$
the development of the second member of which equation gives
$$w = stp + \frac{0}{0} st(1 - p). \qquad (9)$$

Whence we have the conclusion,—*Wealth consists of all things limited in supply, transferable, and productive of pleasure, and an indefinite remainder of things limited in supply, transferable, and not productive of pleasure.* This is sufficiently obvious.

Let it be remarked that it is not necessary to perform the multiplication indicated in (7), and reduce that equation to the form (8), in order to determine the expression of w in terms of the other symbols. The process of development may in all cases be made to supersede that of multiplication. Thus if we develop (7) in terms of w, we find

$$(1 - st)(1 - stp)w + stp(1 - w) = 0,$$

whence

$$w = \frac{stp}{stp - (1 - st)(1 - stp)};$$

and this equation developed will give, as before,

$$w = stp + \frac{0}{0}st(1 - p).$$

13. Suppose next that we seek a description of things limited in supply, as dependent upon their relation to wealth, transferableness, and tendency to produce pleasure, omitting all reference to the prevention of pain.

From equation (8), which is the result of the elimination of r from the original equation, we have

$$w - s\left(wt + wtp - tp\right) = 0;$$

whence

$$s = \frac{w}{wt + wtp - tp}$$

$$= wtp + wt\left(1 - p\right) + \frac{1}{0}w\left(1 - t\right)p + \frac{1}{0}w\left(1 - t\right)\left(1 - p\right)$$

$$+ 0\left(1 - w\right)tp + \frac{0}{0}\left(1 - w\right)t\left(1 - p\right) + \frac{0}{0}\left(1 - w\right)\left(1 - t\right)p$$

$$+ \frac{0}{0}\left(1 - w\right)\left(1 - t\right)\left(1 - p\right).$$

We will first give the direct interpretation of the above solution, term by term; afterwards we shall offer some general remarks which it suggests; and, finally, show how the expression of the conclusion may be somewhat abbreviated.

First, then, the direct interpretation is, Things limited in supply consist of *All wealth transferable and productive of pleasure—all wealth transferable, and not productive of pleasure,—an indefinite amount of what is not wealth, but is either transferable, and not productive of pleasure, or intransferable and productive of pleasure, or neither transferable nor productive of pleasure.*

To which the terms whose coefficients are $\frac{1}{0}$ permit us to add the following independent relations, viz.:

1st. *Wealth that is intransferable, and productive of pleasure, does not exist.*

2ndly. *Wealth that is intransferable, and not productive of pleasure, does not exist.*

14. Respecting this solution I suppose the following remarks are likely to be made.

First, it may be said, that in the expression above obtained for "things limited in supply," the term "All wealth transferable," &c., is in part redundant; since all wealth is (as implied in the original proposition, and directly asserted in the *independent relations*) necessarily transferable.

I answer, that although in ordinary speech we should not deem it necessary to add to "wealth" the epithet "transferable," if another part of our reasoning had led us to express the conclusion, that there is no wealth which is not transferable, yet it pertains to the perfection of this method that it in all cases fully defines the objects represented by each term of the conclusion, by stating the relation they bear to each quality or element of distinction that we have chosen to employ. This is necessary in order to keep the different parts of the solution really distinct and independent, and actually prevents redundancy. Suppose that the pair of terms we have been considering had not contained the word "transferable," and had unitedly been "All wealth," we could then logically resolve the single term "All wealth" into the two terms "All wealth transferable," and "All wealth intransferable." But the latter term is shown to disappear by the "independent relations." Hence it forms no part of the description required, and is therefore redundant. The remaining term agrees with the conclusion actually obtained.

Solutions in which there cannot, by logical divisions, be produced any superfluous or redundant terms, may be termed *pure solutions*. Such are all the solutions obtained by the method of development and elimination above explained. It is proper to notice, that if the common algebraic method of elimination were adopted in the cases in which that method is possible in the present system, we should not be able to depend upon the purity of the solutions obtained. Its want of generality would not be its only defect.

15. In the second place, it will be remarked, that the conclusion contains two terms, the aggregate significance of which would be more conveniently expressed by a single term. Instead of "All wealth productive of pleasure, and transferable," and "All wealth not productive of pleasure, and transferable," we might simply say, "All wealth transferable." This remark is quite just. But it must be noticed that whenever any such simplifications are possible, they are immediately suggested by the form of the equation we have to interpret; and if that equation be reduced to its simplest

form, then the interpretation to which it conducts will be in its simplest form also. Thus in the original solution the terms wtp and $wt(1 - p)$, which have unity for their coefficient, give, on addition, wt; the terms $w(1 - t)p$ and $w(1 - t)(1 - p)$, which have $\frac{1}{0}$ for their coefficient give $w(1 - t)$; and the terms $(1 - w)(1 - t)p$ and $(1 - w)(1 - t)(1 - p)$, which have $\frac{0}{0}$ for their coefficient, give $(1 - w)(1 - t)$. Whence the complete solution is

$$s = wt + \frac{0}{0}(1 - w)(1 - t) + \frac{0}{0}(1 - w)t(1 - p),$$

with the independent relation,

$$w(1 - t) = 0, \text{ or } w = \frac{0}{0}t.$$

The interpretation would now stand thus:—

1st. *Things limited in supply consist of all wealth transferable, with an indefinite remainder of what is not wealth and not transferable, and of transferable articles which are not wealth, and are not productive of pleasure.*

2nd. *All wealth is transferable.*

This is the simplest form under which the general conclusion, with its attendant condition, can be put.

16. When it is required to eliminate two or more symbols from a proposed equation we can either employ (6) Prop. I., or eliminate them in succession, the order of the process being indifferent. From the equation

$$w = st(p + r - pr),$$

we have eliminated r, and found the result,

$$w - wst - wstp + stp = 0.$$

Suppose that it had been required to eliminate both r and t, then taking the above as the first step of the process, it remains to eliminate from the last equation t. Now when $t = 1$ the first member of that equation becomes

$$w - ws - wsp + sp,$$

and when $t = 0$ the same member becomes w. Whence we have

$$w(w - ws - wsp + sp) = 0,$$
$$w - ws = 0,$$

or

for the required result of elimination.

If from the last result we determine w, we have

$$w = \frac{0}{1 - s} = \frac{0}{0}s,$$

whence "*All wealth is limited in supply.*" As p does not enter into the equation, it is evident that the above is true, irrespectively of any relation which the elements of the conclusion bear to the quality "productive of pleasure."

Resuming the original equation, let it be required to eliminate s and t. We have

$$w = st(p + r - pr).$$

Instead, however, of separately eliminating s and t according to the Rule, it will suffice to treat st as a single symbol, seeing that it satisfies the fundamental law of the symbols by the equation

$$st(1 - st) = 0.$$

Placing, therefore, the given equation under the form

$$w - st(p + r - pr) = 0;$$

and making st successively equal to 1 and to 0, and taking the product of the results, we have

$$(w - p - r + pr)\, w = 0,$$

or

$$w - wp - wr + wpr = 0,$$

for the result sought.

As a particular illustration, let it be required to deduce an expression for "things productive of pleasure" (p), in terms of "wealth" (w), and "things preventive of pain" (r).

We have, on solving the equation,

$$p = \frac{w(1 - r)}{w(1 - r)}$$

$$= \frac{0}{0} wr + w(1 - r) + \frac{0}{0}(1 - w)r + \frac{0}{0}(1 - w)(1 - r)$$

$$= w(1 - r) + \frac{0}{0} wr + \frac{0}{0}(1 - w).$$

Whence the following conclusion:—*Things productive of pleasure are, all wealth not preventive of pain, an indefinite amount of wealth that is preventive of pain, and an indefinite amount of what is not wealth.*

From the same equation we get

$$1 - p = 1 - \frac{w(1 - r)}{w(1 - r)} = \frac{0}{w(1 - r)},$$

which developed, gives

$$w(1 - p) = \frac{0}{0} wr + \frac{0}{0}(1 - w) \cdot r + \frac{0}{0}(1 - w) \cdot (1 - r)$$

$$= \frac{0}{0} wr + \frac{0}{0}(1 - w).$$

Whence, *Things not productive of pleasure are either wealth, preventive of pain, or what is not wealth.*

Equally easy would be the discussion of any similar case.

17. In the last example of elimination, we have eliminated the compound symbol *st* from the given equation, by treating it as a single symbol. The same method is applicable to any combination of symbols which satisfies the fundamental law of individual symbols. Thus the expression $p + r - pr$ will, on being multiplied by itself, reproduce itself, so that if we represent $p + r - pr$ by a single symbol as y, we shall have the fundamental law obeyed, the equation

$$y = y^2, \text{ or } y(1 - y) = 0,$$

being satisfied. For the rule of elimination for symbols is founded upon the supposition that each individual symbol is subject to that law; and hence the elimination of any function or combination of such symbols from an equation, may be effected by a single operation, whenever that law is satisfied by the function.

Though the forms of interpretation adopted in this and the previous chapter show, perhaps better than any others, the direct significance of the symbols 1 and $\frac{0}{0}$, modes of expression more agreeable to those of common discourse may, with equal truth and propriety, be employed. Thus the equation (9) may be interpreted in the following manner: *Wealth is either limited in supply, transferable, and productive of pleasure, or limited in supply, transferable, and not productive of pleasure.* And reversely, *Whatever is limited in supply, transferable, and productive of pleasure, is wealth.* Reverse interpretations, similar to the above, are always furnished when the final development introduces terms having unity as a coefficient.

18. NOTE.—The fundamental equation $f(1)f(0) = 0$, expressing the result of the elimination of the symbol x from any equation $f(x) = 0$, admits of a remarkable interpretation.

It is to be remembered, that by the equation $f(x) = 0$ is implied some proposition in which the individuals represented by the class x, suppose "men," are referred to, together, it may be, with other individuals; and it is our object to ascertain whether there is implied in the proposition any relation among the other individuals, independently of those found in the class *men*. Now the equation $f(1) = 0$ expresses what the original proposition would become if *men* made up the universe, and the equation $f(0) = 0$ expresses what that original proposition would become if men ceased to exist, wherefore the equation $f(1)f(0) = 0$ expresses what in virtue of the original proposition would be equally true on either assumption, i.e. equally true whether "men" were "all things" or "nothing." Wherefore the theorem expresses that *what is equally true, whether a given class of objects embraces the whole universe or disappears from existence, is independent of that class altogether, and* vice versâ. Herein we see another example of the interpretation of formal results, immediately deduced from the mathematical laws of thought, into general axioms of philosophy.

CHAPTER VIII

On the Reduction of Systems of Propositions

1. In the preceding chapters we have determined sufficiently for the most essential purposes the theory of single primary propositions, or, to speak more accurately, of primary propositions expressed by a single equation. And we have established upon that theory an adequate method. We have shown how any element involved in the given system of equations may be eliminated, and the relation which connects the remaining elements deduced in any proposed form, whether of denial, of affirmation, or of the more usual relation of subject and predicate. It remains that we proceed to the consideration of systems of propositions, and institute with respect to them a similar series of investigations. We are to inquire whether it is possible from the equations by which a system of propositions is expressed to eliminate, *ad libitum*, any number of the symbols involved; to deduce by interpretation of the result the whole of the relations implied among the remaining symbols; and to determine in particular the expression of any single element, or of any interpretable combination of elements, in terms of the other elements, so as to present the conclusion in any admissible form that may be required. These questions will be answered by showing that it is possible to reduce any system of equations, or any of the equations involved in a system, to an equivalent single equation, to which the methods of the previous chapters may be immediately applied. It will be seen also, that in this reduction is involved an important extension of the theory of single propositions, which in the previous discussion of the subject we were compelled to forego. This circumstance is not peculiar in its nature. There are many special departments of science which cannot be completely surveyed from within, but require to be studied also from an external point of view, and to be regarded in connexion with other and kindred subjects, in order that their full proportions may be understood.

This chapter will exhibit two distinct modes of reducing systems of equations to equivalent single equations. The first of these rests upon the employment of arbitrary constant multipliers. It is a method sufficiently simple in theory, but it has the inconvenience of rendering the subsequent processes of elimination and development, when they occur, somewhat tedious. It was, however, the method of reduction first discovered, and partly on this account, and partly on account of its simplicity, it has been thought proper to retain it. The second method does not require the introduction of arbitrary constants, and is in nearly all respects preferable to the preceding one. It will, therefore, generally be adopted in the subsequent investigations of this work.

2. We proceed to the consideration of the first method.

Proposition I

Any system of logical equations may be reduced to a single equivalent equation, by multiplying each equation after the first by a distinct arbitrary constant quantity, and adding all the results, including the first equation, together.

By Prop. 2, Chap. VI., the interpretation of any single equation, $f(x, y..) = 0$ is obtained by equating to 0 those constituents of the development of the first member, whose coefficients do not vanish. And hence, if there be given two equations, $f(x, y..) = 0$, and $F(x, y..) = 0$, their united import will be contained in the system of results formed by equating to 0 all those constituents which thus present themselves in both, or in either, of the given equations developed according to the Rule of Chap. VI. Thus let it be supposed, that we have the two equations

$$xy - 2x = 0, \qquad (1)$$
$$x - y = 0; \qquad (2)$$

The development of the first gives

$$-xy - 2x(1 - y) = 0;$$

whence, $\qquad xy = 0, \quad x(1 - y) = 0. \qquad (3)$

The development of the second equation gives

$$x(1 - y) - y(1 - x) = 0;$$

whence, $\qquad x(1 - y) = 0, \quad y(1 - x) = 0. \qquad (4)$

The constituents whose coefficients do not vanish in both developments are xy, $x(1 - y)$, and $(1 - x)y$, and these would together give the system

$$xy = 0, \quad x(1 - y) = 0, \quad (1 - x)y = 0; \qquad (5)$$

which is equivalent to the two systems given by the developments separately, seeing that in those systems the equation $x(1 - y) = 0$ is repeated. Confining ourselves to the case of binary systems of equations, it remains then to determine a single equation, which on development shall yield the same constituents with coefficients which do not vanish, as the given equations produce.

Now if we represent by

$$V_1 = 0, \quad V_2 = 0,$$

the given equations, V_1 and V_2 being functions of the logical symbols x, y, z, &c.; then the single equation

$$V_1 + cV_2 = 0, \qquad (6)$$

c being an arbitrary constant quantity, will accomplish the required object. For let At represent any term in the full development V_1 wherein t is a constituent and A its numerical coefficient, and let Bt represent the corresponding term in the full development of V_2, then will the corresponding term in the development of (6) be

$$(A + cB)t.$$

The coefficient of t vanishes if A and B both vanish, but not otherwise. For if we assume that A and B do not both vanish, and at the same time make

$$A + cB = 0, \tag{7}$$

the following cases alone can present themselves.

1st. That A vanishes and B does not vanish. In this case the above equation becomes

$$cB = 0,$$

and requires that $c = 0$. But this contradicts the hypothesis that c is an *arbitrary* constant.

2nd. That B vanishes and A does not vanish. This assumption reduces (7) to

$$A = 0,$$

by which the assumption is itself violated.

3rd. That neither A nor B vanishes. The equation (7) then gives

$$c = \frac{-A}{B},$$

which is a definite value, and, therefore, conflicts with the hypothesis that c is arbitrary.

Hence the coefficient $A + cB$ vanishes when A and B both vanish, but not otherwise. Therefore, the same constituents will appear in the development of (6), with coefficients which do not vanish, as in the equations $V_1 = 0$, $V_2 = 0$, singly or together. And the equation $V_1 + cV_2 = 0$, will be equivalent to the system $V_1 = 0$, $V_2 = 0$.

By similar reasoning it appears, that the general system of equations

$$V_1 = 0, \quad V_2 = 0, \quad V_3 = 0, \&c.;$$

may be replaced by the single equation

$$V_1 + cV_2 + c'V_3 + \&c. = 0,$$

c, c', &c., being arbitrary constants. The equation thus formed may be treated in all respects as the ordinary logical equations of the previous chapters. The arbitrary constants c_1, c_2, &c., are not *logical* symbols. They do not satisfy the law,

$$c_1 (1 - c_1) = 0, \quad c_2 (1 - c_2) = 0.$$

But their introduction is justified by that general principle which has been stated in (ii. 15) and (v. 6), and exemplified in nearly all our subsequent investigations, viz., that equations involving the symbols of Logic may be treated in all respects as if those symbols were symbols of quantity, subject to the special law $x(1 - x) = 0$, until in the final stage of solution they assume a form interpretable in that system of thought with which Logic is conversant.

3. The following example will serve to illustrate the above method.

Ex. 1.—Suppose that an analysis of the properties of a particular class of substances has led to the following general conclusions, viz.:

1st. That wherever the properties A and B are combined, either the property C, or the property D, is present also; but they are not jointly present.

2nd. That wherever the properties B and C are combined, the properties A and D are either both present with them, or both absent.

3rd. That wherever the properties A and B are both absent, the properties C and D are both absent also; and *vice versâ*, where the properties C and D are both absent, A and B are both absent also.

Let it then be required from the above to determine what may be concluded in any particular instance from the presence of the property A with respect to the presence or absence of the properties B and C, paying no regard to the property D.

$$\text{Represent the property } A \text{ by } x;$$
$$\text{„ the property } B \text{ by } y;$$
$$\text{„ the property } C \text{ by } z;$$
$$\text{„ the property } D \text{ by } w.$$

Then the symbolical expression of the premises will be

$$xy = v\{w(1-z) + z(1-w)\};$$
$$yz = v\{xw + (1-x)(1-w)\};$$
$$(1-x)(1-y) = (1-z)(1-w).$$

From the first two of these equations, separately eliminating the indefinite class symbol v, we have

$$xy\{1 - w(1-z) - z(1-w)\} = 0;$$
$$yz\{1 - xw - (1-x)(1-w)\} = 0.$$

Now if we observe that by development

$$1 - w(1-z) - z(1-w) = wz + (1-w)(1-z),$$

and

$$1 - xw - (1-x)(1-w) = x(1-w) + w(1-x),$$

and in these expressions replace, for simplicity,

$$1 - x \text{ by } \bar{x}, \quad 1 - y \text{ by } \bar{y}, \text{ \&c.,}$$

we shall have from the three last equations,

$$xy(wz + \overline{w}\bar{z}) = 0; \tag{1}$$
$$yz(x\overline{w} + \bar{x}w) = 0; \tag{2}$$
$$\bar{x}\bar{y} = \overline{w}\bar{z}; \tag{3}$$

and from this system we must eliminate w.

Multiplying the second of the above equations by c, and the third by c', and adding the results to the first, we have

$$xy(wz + \overline{w}\bar{z}) + cyz(x\overline{w} + \bar{x}w) + c'(\bar{x}\bar{y} - \overline{w}\bar{z}) = 0.$$

When w is made equal to 1, and therefore $\bar{w}$ to 0, the first member of the above equation becomes

$$xyz + c\bar{x}yz + c'\overline{xy}.$$

And when in the same member w is made 0 and $\bar{w} = 1$, it becomes

$$xy\bar{z} + cxyz + c'\,\overline{xy} - c'\bar{z}.$$

Hence the result of the elimination of w may be expressed in the form

$$(xyz + c\bar{x}yz + c'\,\overline{xy})(xy\bar{z} + cxyz + c'\,\overline{xy} - c'\bar{z}) = 0; \qquad (4)$$

and from this equation x is to be determined.

Were we now to proceed as in former instances, we should multiply together the factors in the first member of the above equation; but it may be well to show that such a course is not at all necessary. Let us develop the first member of (4) with reference to x, the symbol whose expression is sought, we find

$$yz\,(y\bar{z} + cyz - c'\bar{z})x + (cyz + c'\bar{y})(c'\bar{y} - c'\bar{z})(1 - x) = 0;$$

or,

$$cyzx + (cyz + c'\bar{y})(c'\bar{y} - c'\bar{z})(1 - x) = 0;$$

whence we find,

$$x = \frac{(cyz + c'\bar{y})(c'\bar{y} - c'\bar{z})}{(cyz + c'\bar{y})(c'\bar{y} - c'\bar{z}) - cyz};$$

and developing the second member with respect to y and z,

$$x = 0yz + \frac{0}{0}\,y\bar{z} + \frac{c'^2}{c'^2}\,\bar{y}z + \frac{0}{0}\,\overline{yz};$$

or,

$$x = (1 - y)z + \frac{0}{0}\,y(1 - z) + \frac{0}{0}(1 - y)(1 - z);$$

or,

$$x = (1 - y)z + \frac{0}{0}(1 - z);$$

the interpretation of which is, *Wherever the property A is present, there either C is present and B absent, or C is absent.* And inversely, *Wherever the property C is present, and the property B absent, there the property A is present.*

These results may be much more readily obtained by the method next to be explained. It is, however, satisfactory to possess different modes, serving for mutual verification, of arriving at the same conclusion.

4. We proceed to the second method.

Proposition II

If any equations, $V_1 = 0$, $V_2 = 0$, &c., are such that the developments of their first members consist only of constituents with positive coefficients, those equations may be combined together into a single equivalent equation by addition.

For, as before, let At represent any term in the development of the function V_1, Bt the corresponding term in the development of V_2, and so on. Then will the corresponding term in the development of the equation

$$V_1 + V_2 + \&c. = 0, \tag{1}$$

formed by the addition of the several given equations, be

$$(A + B + \&c.)\, t.$$

But as by hypothesis the coefficients A, B, &c. are none of them negative, the aggregate coefficient $A + B$, &c. in the derived equation will only vanish when the separate coefficients A, B, &c. vanish together. Hence the same constituents will appear in the development of the equation (1) as in the several equations $V_1 = 0$, $V_2 = 0$, &c. of the original system taken collectively, and therefore the interpretation of the equation (1) will be equivalent to the collective interpretations of the several equations from which it is derived.

Proposition III

5. *If $V_1 = 0$, $V_2 = 0$, &c. represent any system of equations, the terms of which have by transposition been brought to the first side, then the combined interpretation of the system will be involved in the single equation,*

$$V_1{}^2 + V_2{}^2 + \&c. = 0,$$

formed by adding together the squares of the given equations.

For let any equation of the system, as $V_1 = 0$, produce on development an equation

$$a_1 t_1 + a_2 t_2 + \&c. = 0,$$

in which t_1, t_2, &c. are constituents, and a_1, a_2, &c. their corresponding coefficients. Then the equation $V_1{}^2 = 0$ will produce on development an equation

$$a_1{}^2 t_1 + a_2{}^2 t_2 + \&c. = 0,$$

as may be proved either from the law of the development or by squaring the function $a_1 t_1 + a_2 t_2$, &c. in subjection to the conditions

$$t_1{}^2 = t_1, \quad t_2{}^2 = t_2, \quad t_1 t_2 = 0,$$

assigned in Prop. 3, Chap. V. Hence the constituents which appear in the expansion of the equation $V_1{}^2 = 0$, are the same with those which appear in the expansion of the equation $V_1 = 0$, and they have positive coefficients. And the same remark applies to the equations $V_2 = 0$, &c. Whence, by the last Proposition, the equation

$$V_1{}^2 + V_2{}^2 + \&c. = 0$$

will be equivalent in interpretation to the system of equations

$$V_1 = 0, \quad V_2 = 0, \&c.$$

Corollary.—Any equation, $V = 0$, of which the first member already satisfies the condition

$$V^2 = V, \quad \text{or} \quad V(1 - V) = 0,$$

does not need (as it would remain unaffected by) the process of squaring. Such equations are, indeed, immediately developable into a series of constituents, with coefficients equal to 1, Chap. V. Prop. 4.

Proposition IV

6. *Whenever the equations of a system have by the above process of squaring, or by any other process, been reduced to a form such that all the constituents exhibited in their development have positive coefficients, any derived equations obtained by elimination will possess the same character, and may be combined with the other equations by addition.*

Suppose that we have to eliminate a symbol x from any equation $V = 0$, which is such that none of the constituents, in the full development of its first member, have negative coefficients. That expansion may be written in the form

$$V_1 x + V_0 (1 - x) = 0,$$

V_1 and V_0 being each of the form

$$a_1 t_1 + a_2 t_2 .. + a_n t_n,$$

in which $t_1 t_2 .. t_n$ are constituents of the other symbols, and $a_1 a_2 .. a_n$ in each case positive or vanishing quantities. The result of elimination is

$$V_1 V_2 = 0;$$

and as the coefficients in V_1 and V_2 are none of them negative, there can be no negative coefficients in the product $V_1 V_2$. Hence the equation $V_1 V_2 = 0$ may be added to any other equation, the coefficients of whose constituents are positive, and the resulting equation will combine the full significance of those from which it was obtained.

Proposition V

7. *To deduce from the previous Propositions a practical rule or method for the reduction of systems of equations expressing propositions in Logic.*

We have by the previous investigations established the following points, viz.:

1st. That any equations which are of the form $V = 0$, V satisfying the fundamental law of duality $V(1 - V) = 0$, may be combined together by simple addition.

2ndly. That any other equations of the form $V = 0$ may be reduced, by the process of squaring, to a form in which the same principle of combination by mere addition is applicable.

It remains then only to determine what equations in the actual expression of propositions belong to the former, and what to the latter, class.

Now the general types of propositions have been set forth in the conclusion of Chap. IV. The division of propositions which they represent is as follows:

1st. Propositions, of which the subject is universal, and the predicate particular.

The symbolical type (iv. 15) is

$$X = vY,$$

X and Y satisfying the law of duality. Eliminating v, we have

$$X(1 - Y) = 0, \tag{1}$$

and this will be found also to satisfy the same law. No further reduction by the process of squaring is needed.

2nd. Propositions of which both terms are universal, and of which the symbolical type is

$$X = Y,$$

X and Y separately satisfying the law of duality. Writing the equation in the form $X - Y = 0$, and squaring, we have

$$X - 2XY + Y = 0,$$

or $\qquad X(1 - Y) + Y(1 - X) = 0. \tag{2}$

The first member of this equation satisfies the law of duality, as is evident from its very form.

We may arrive at the same equation in a different manner. The equation

$$X = Y$$

is equivalent to the two equations

$$X = vY, \quad Y = vX,$$

(for to affirm that X's are identical with Y's is to affirm both that All X's are Y's, and that All Y's are X's). Now these equations give, on elimination of v,

$$X(1 - Y) = 0, \quad Y(1 - X) = 0,$$

which added, produce (2).

3rd. Propositions of which both terms are particular. The form of such propositions is

$$vX = vY,$$

but v is not quite arbitrary, and therefore must not be eliminated. For v is the representative of *some*, which, though it may include in its meaning *all*, does not include *none*. We must therefore transpose the second member to the first side, and square the resulting equation according to the rule.

The result will obviously be

$$vX(1 - Y) + vY(1 - X) = 0.$$

The above conclusions it may be convenient to embody in a Rule, which will serve for constant future direction.

8. RULE.—*The equations being so expressed as that the terms X and Y in the following typical forms obey the law of duality, change the equations*

$$X = vY \ into \ X(1 - Y) = 0,$$
$$X = Y \ into \ X(1 - Y) + Y(1 - X) = 0.$$
$$vX = vY \ into \ vX(1 - Y) + vY(1 - X) = 0.$$

Any equation which is given in the form $X = 0$ will not need transformation, and any equation which presents itself in the form $X = 1$ may be replaced by $1 - X = 0$, as appears from the second of the above transformations.

When the equations of the system have thus been reduced, any of them, as well as any equations derived from them by the process of elimination, may be combined by addition.

9. NOTE.—It has been seen in Chapter IV. that in literally translating the terms of a proposition, without attending to its real meaning, into the language of symbols, we may produce equations in which the terms X and Y do not obey the law of duality. The equation $w = st(p + r)$, given in (3) Prop. 3 of the chapter referred to, is of this kind. Such equations, however, as it has been seen, have a meaning. Should it, for curiosity, or for any other motive, be determined to employ them, it will be best to reduce them by the Rule (vi. 5).

10. Ex. 2.—Let us take the following Propositions of Elementary Geometry:

1st. Similar figures consist of all whose corresponding angles are equal, and whose corresponding sides are proportional.

2nd. Triangles whose corresponding angles are equal have their corresponding sides proportional, and *vice versâ*.

To represent these premises, let us make

s = similar.

t = triangles.

q = having corresponding angles equal.

r = having corresponding sides proportional.

Then the premises are expressed by the following equations:

$$s = qr, \tag{1}$$

$$tq = tr. \tag{2}$$

Reducing by the Rule, or, which amounts to the same thing, bringing the terms of these equations to the first side, squaring each equation, and then adding, we have

$$s + qr - 2qrs + tq + tr - 2tqr = 0. \tag{3}$$

Let it be required to deduce a description of dissimilar figures formed out of the elements expressed by the terms, *triangles*, having corresponding angles equal, having corresponding sides proportional.

We have from (3),

$$s = \frac{tq + qr + rt - 2tqr}{2qr - 1},$$

$$\therefore 1 - s = \frac{qr - tq - rt + 2tqr - 1}{2qr - 1}. \tag{4}$$

And fully developing the second member, we find

$$1 - s = 0tqr + 2tq(1 - r) + 2tr(1 - q) + t(1 - q)(1 - r)$$
$$+ 0(1 - t)qr + (1 - t)q(1 - r) + (1 - t)r(1 - q)$$
$$+ (1 - t)(1 - q)(1 - r). \tag{5}$$

In the above development two of the terms have the coefficient 2, these must be equated to 0 by the Rule, then those terms whose coefficients are 0 being rejected, we have

$$1 - s = t(1 - q)(1 - r) + (1 - t)q(1 - r) + (1 - t)r(1 - q)$$
$$+ (1 - t)(1 - q)(1 - r); \tag{6}$$
$$tq(1 - r) = 0; \tag{7}$$
$$tr(1 - q) = 0; \tag{8}$$

the direct interpretation of which is

1st. *Dissimilar figures consist of all triangles which have not their corresponding angles equal and sides proportional, and of all figures not being triangles which have either their angles equal, and sides not proportional, or their corresponding sides proportional, and angles not equal, or neither their corresponding angles equal nor corresponding sides proportional.*

2nd. *There are no triangles whose corresponding angles are equal, and sides not proportional.*

3rd. *There are no triangles whose corresponding sides are proportional and angles not equal.*

11. Such are the immediate interpretations of the final equation. It is seen, in accordance with the general theory, that in deducing a description of a particular class of objects, viz., dissimilar figures, in terms of certain other elements of the original premises, we obtain also the independent relations which exist among those elements in virtue of the same premises. And that this is not superfluous information, even as respects the immediate object of inquiry, may easily be shown. For example, the independent relations may always be made use of to reduce, if it be thought desirable, to a briefer form, the expression of that relation which is directly sought. Thus if we write (7) in the form

$$0 = tq(1 - r),$$

and add it to (6), we get, since

$$t(1 - q)(1 - r) + tq(1 - r) = t(1 - r),$$
$$1 - s = t(1 - r) + (1 - t)q(1 - r) + (1 - t)r(1 - q)$$
$$+ (1 - t)(1 - q)(1 - r),$$

which, on interpretation, would give for the first term of the description of dissimilar figures, "Triangles whose corresponding sides are not proportional," instead of the fuller description originally obtained. A regard to convenience must always determine the propriety of such reduction.

12. A reduction which is always advantageous (vii. 15) consists in collecting the terms of the immediate description sought, as of the second member of (5) or (6), into as few groups as possible. Thus the third and fourth terms of the second member of (6) produce by addition the single term $(1 - t)(1 - q)$. If this reduction be combined with the last, we have

$$1 - s = t(1 - r) + (1 - t) q (1 - r) + (1 - t)(1 - q),$$

the interpretation of which is

Dissimilar figures consist of all triangles whose corresponding sides are not proportional, and all figures not being triangles which have either their corresponding angles unequal, or their corresponding angles equal, but sides not proportional.

The fulness of the general solution is therefore not a superfluity. While it gives us all the information that we seek, it provides us also with the means of expressing that information in the mode that is most advantageous.

13. Another observation, illustrative of a principle which has already been stated, remains to be made. Two of the terms in the full development of $1 - s$ in (5) have 2 for their coefficients, instead of $\frac{1}{0}$. It will hereafter be shown that this circumstance indicates that the two premises were not independent. To verify this, let us resume the equations of the premises in their reduced forms, viz.,

$$s(1 - qr) + qr(1 - s) = 0,$$
$$tq(1 - r) + tr(1 - q) = 0.$$

Now if the first members of these equations have any common constituents, they will appear on multiplying the equations together. If we do this we obtain

$$stq(1 - r) + str(1 - q) = 0.$$

Whence there will result

$$stq(1 - r) = 0, \quad str(1 - q) = 0,$$

these being equations which are deducible from either of the primitive ones. Their interpretations are—

Similar triangles which have their corresponding angles equal have their corresponding sides proportional.

Similar triangles which have their corresponding sides proportional have their corresponding angles equal.

And these conclusions are equally deducible from either premiss *singly*. In this respect, according to the definitions laid down, the premises are not independent.

14. Let us, in conclusion, resume the problem discussed in illustration of the first method of this chapter, and endeavour to ascertain, by the present method, what may be concluded from the presence of the property C, with reference to the properties A and B.

We found on eliminating the symbols v the following equations, viz.:

$$xy(wz + \overline{w}\,\overline{z}) = 0, \tag{1}$$

$$yz(x\overline{w} + \overline{x}w) = 0, \tag{2}$$

$$\overline{x}\overline{y} = \overline{w}\,\overline{z}. \tag{3}$$

From these we are to eliminate w and determine z. Now (1) and (2) already satisfy the condition $V(1 - V) = 0$. The third equation gives, on bringing the terms to the first side, and squaring

$$\overline{x}\overline{y}(1 - \overline{w}\overline{z}) + \overline{w}\overline{z}(1 - \overline{x}\overline{y}) = 0. \tag{4}$$

Adding (1) (2) and (4) together, we have

$$xy(wz + \overline{w}\overline{z}) + yz(x\overline{w} + \overline{x}w) + \overline{x}\overline{y}(1 - \overline{w}\overline{z}) + \overline{w}\overline{z}(1 - \overline{x}\overline{y}) = 0.$$

Eliminating w, we get

$$(xyz + yz\overline{x} + \overline{x}\overline{y})\{xy\overline{z} + yzx + \overline{x}\overline{y}z + \overline{z}(1 - \overline{x}\overline{y})\} = 0.$$

Now, on multiplying the terms in the second factor by those in the first successively, observing that

$$x\overline{x} = 0, \quad y\overline{y} = 0, \quad z\overline{z} = 0,$$

nearly all disappear, and we have only left

$$xyz + \overline{x}\overline{y}z = 0; \tag{5}$$

whence

$$z = \frac{0}{xy + \overline{x}\overline{y}}$$

$$= 0xy + \frac{0}{0}x\overline{y} + \frac{0}{0}\overline{x}y + 0\overline{x}\overline{y}$$

$$= \frac{0}{0}x\overline{y} + \frac{0}{0}\overline{x}y,$$

furnishing the interpretation. *Wherever the property C is found, either the property A or the property B will be found with it, but not both of them together.*

From the equation (5) we may readily deduce the result arrived at in the previous investigation by the method of arbitrary constant multipliers, as well as any other proposed forms of the relation between x, y, and z; e.g. *If the property B is absent, either A and C will be jointly present, or C will be absent.* And conversely, *If A and C are jointly present, B will be absent.* The converse part of this conclusion is founded on the presence of a term xz with unity for its coefficient in the developed value of $\overline{y}$.

CHAPTER IX

On Certain Methods of Abbreviation

1. Though the three fundamental methods of development, elimination, and reduction, established and illustrated in the previous chapters, are sufficient for all the practical ends of Logic, yet there are certain cases in which they admit, and especially

the method of elimination, of being simplified in an important degree; and to these I wish to direct attention in the present chapter. I shall first demonstrate some propositions in which the principles of the above methods of abbreviation are contained, and I shall afterwards apply them to particular examples.

Let us designate as class terms any terms which satisfy the fundamental law $V(1 - V) = 0$. Such terms will individually be constituents; but, when occurring together, will not, as do the terms of a development, necessarily involve the same symbols in each. Thus $ax + bxy + cyz$ may be described as an expression consisting of three class terms, x, xy, and yz, multiplied by the coefficients a, b, c respectively. The principle applied in the two following Propositions, and which, in some instances, greatly abbreviates the process of elimination, is that of the *rejection of superfluous class terms*; those being regarded as superfluous which do not add to the constituents of the final result.

Proposition I

2. *From any equation, $V = 0$, in which V consists of a series of class terms having positive coefficients, we are permitted to reject any term which contains another term as a factor, and to change every positive coefficient to unity.*

For the significance of this series of positive terms depends only upon the number and nature of the constituents of its final expansion, i.e. of its expansion with reference to all the symbols which it involves, and not at all upon the actual values of the coefficients (vi. 5). Now let x be any term of the series, and xy any other term having x as a factor. The expansion of x with reference to the symbols x and y will be

$$xy + x(1 - y),$$

and the expansion of the sum of the terms x and xy will be

$$2xy + x(1 - y).$$

But by what has been said, these expressions occurring in the first member of an equation, of which the second member is 0, and of which all the coefficients of the first member are positive, are equivalent; since there must exist simply the two constituents xy and $x(1 - y)$ in the final expansion, whence will simply arise the resulting equations

$$xy = 0, \quad x(1 - y) = 0.$$

And, therefore, the aggregate of terms $x + xy$ may be replaced by the single term x.

The same reasoning applies to all the cases contemplated in the Proposition. Thus, if the term x is repeated, the aggregate $2x$ may be replaced by x, because under the circumstances the equation $x = 0$ must appear in the final reduction.

Proposition II

3. *Whenever in the process of elimination we have to multiply together two factors, each consisting solely of positive terms, satisfying the fundamental law of logical symbols, it is permitted to reject from both factors any common term, or from either factor any term which is divisible by a term in the other factor; provided always, that the rejected term be added to the product of the resulting factors.*

In the enunciation of this Proposition, the word "divisible" is a term of convenience, used in the algebraic sense, in which xy and $x(1 - y)$ are said to be divisible by x.

To render more clear the import of this Proposition, let it be supposed that the factors to be multiplied together are $x + y + z$ and $x + yw + t$. It is then asserted, that from these two factors we may reject the term x, and that from the second factor we may reject the term yw, provided that these terms be transferred to the final product. Thus, the resulting factors being $y + z$ and t, if to their product $yt + zt$ we add the terms x and yw, we have

$$x + yw + yt + zt,$$

as an expression equivalent to the product of the given factors $x + y + z$ and $x + yw + t$; equivalent namely in the process of elimination.

Let us consider, first, the case in which the two factors have a common term x, and let us represent the factors by the expressions $x + P$, $x + Q$, supposing P in the one case and Q in the other to be the sum of the positive terms additional to x.

Now,

$$(x + P)(x + Q) = x + xP + xQ + PQ. \tag{1}$$

But the process of elimination consists in multiplying certain factors together, and equating the result to 0. Either then the second member of the above equation is to be equated to 0, or it is a factor of some expression which is to be equated to 0.

If the former alternative be taken, then, by the last Proposition, we are permitted to reject the terms xP and xQ, inasmuch as they are positive terms having another term x as a factor. The resulting expression is

$$x + PQ,$$

which is what we should obtain by rejecting x from both factors, and adding it to the product of the factors which remain.

Taking the second alternative, the only mode in which the second member of (1) can affect the final result of elimination must depend upon the number and nature of its constituents, both which elements are unaffected by the rejection of the terms xP and xQ. For that development of x includes all possible constituents of which x is a factor.

Consider finally the case in which one of the factors contains a term, as xy, divisible by a term, x, in the other factor.

Let $x + P$ and $xy + Q$ be the factors. Now

$$(x + P)(xy + Q) = xy + xQ + xyP + PQ.$$

But by the reasoning of the last Proposition, the term xyP may be rejected as containing another positive term xy as a factor, whence we have

$$xy + xQ + PQ$$
$$= xy + (x + P)Q.$$

But this expresses the rejection of the term xy from the second factor, and its transference to the final product. Wherefore the Proposition is manifest.

Proposition III

4. *If t be any symbol which is retained in the final result of the elimination of any other symbols from any system of equations, the result of such elimination may be expressed in the form*

$$Et + E'(1 - t) = 0,$$

in which E is formed by making in the proposed system $t = 1$, and eliminating the same other symbols; and E' by making in the proposed system $t = 0$, and eliminating the same other symbols.

For let $\phi(t) = 0$ represent the final result of elimination. Expanding this equation, we have

$$\phi(1)\, t + \phi(0)\,(1 - t) = 0.$$

Now by whatever process we deduce the function $\phi(t)$ from the proposed system of equations, by the same process should we deduce $\phi(1)$, if in those equations t were changed into 1; and by the same process should we deduce $\phi(0)$, if in the same equations t were changed into 0. Whence the truth of the proposition is manifest.

5. Of the three propositions last proved, it may be remarked, that though quite unessential to the strict development or application of the general theory, they yet accomplish important ends of a practical nature. By Prop. 1 we can simplify the results of addition; by Prop. 2 we can simplify those of multiplication; and by Prop. 3 we can break up any tedious process of elimination into two distinct processes, which will in general be of a much less complex character. This method will be very frequently adopted, when the final object of inquiry is the determination of the value of t, in terms of the other symbols which remain after the elimination is performed.

6. Ex. 1.—Aristotle, in the Nicomachean Ethics, Book II. Cap. 3, having determined that actions are virtuous, not as possessing in themselves a certain character, but as implying a certain condition of mind in him who performs them, viz., that he perform them knowingly, and with deliberate preference, and for their own sakes, and upon fixed principles of conduct, proceeds in the two following chapters to consider

the question, whether virtue is to be referred to the genus of Passions, or Faculties, or Habits, together with some other connected points. He grounds his investigation upon the following premises, from which, also, he deduces the general doctrine and definition of moral virtue, of which the remainder of the treatise forms an exposition.

Premises

1. Virtue is either a passion (πάθος), or a faculty (δύναμις), or a habit (ἕξις).

2. Passions are not things according to which we are praised or blamed, or in which we exercise deliberate preference.

3. Faculties are not things according to which we are praised or blamed, and which are accompanied by deliberate preference.

4. Virtue is something according to which we are praised or blamed, and which is accompanied by deliberate preference.

5. Whatever art or science makes its work to be in a good state avoids extremes, and keeps the mean in view relative to human nature (τὸ μέσον . . . πρὸς ἡμᾶς).

6. Virtue is more exact and excellent than any art or science.

This is an argument *à fortiori*. If science and true art shun defect and extravagance alike, much more does virtue pursue the undeviating line of moderation. If *they* cause their work to be in a good state, much more reason have to we to say that Virtue causeth her peculiar work to be "in a good state." Let the final premiss be thus interpreted. Let us also pretermit all reference to praise or blame, since the mention of these in the premises accompanies only the mention of deliberate preference, and this is an element which we purpose to retain. We may then assume as our representative symbols—

v = virtue.

p = passions.

f = faculties.

h = habits.

d = things accompanied by deliberate preference.

g = things causing their work to be in a good state.

m = things keeping the mean in view relative to human nature.

Using, then, q as an indefinite class symbol, our premises will be expressed by the following equations:

$$v = q\{p(1 - f)(1 - h) + f(1 - p)(1 - h) + h(1 - p)(1 - f)\}.$$
$$p = q(1 - d).$$
$$f = q(1 - d).$$
$$v = qd.$$
$$g = qm.$$
$$v = qg.$$

And separately eliminating from these the symbols q,

$$v\{1 - p(1 - f)(1 - h) - f(1 - p)(1 - h) - h(1 - p)(1 - f)\} = 0. \quad (1)$$

$$pd = 0. \quad (2)$$

$$fd = 0. \quad (3)$$

$$v(1 - d) = 0. \quad (4)$$

$$g(1 - m) = 0. \quad (5)$$

$$v(1 - g) = 0. \quad (6)$$

We shall first eliminate from (2), (3), and (4) the symbol d, and then determine v in relation to p, f, and h. Now the addition of (2), (3), and (4) gives

$$(p + f)d + v(1 - d) = 0.$$

From which, eliminating d in the ordinary way, we find

$$(p + f)v = 0. \quad (7)$$

Adding this to (1), and determining v, we find

$$v = \frac{0}{p + f + 1 - p(1 - f)(1 - h) - f(1 - p)(1 - h) - h(1 - f)(1 - p)}.$$

Whence by development,

$$v = \frac{0}{0}h(1 - f)(1 - p).$$

The interpretation of this equation is: *Virtue is a habit, and not a faculty or a passion.*

Next, we will eliminate f, p, and g from the original system of equations, and then determine v in relation to h, d, and m. We will in this case eliminate p and f together. On addition of (1), (2), and (3), we get

$$v\{1 - p(1 - f)(1 - h) - f(1 - p)(1 - h) - h(1 - p)(1 - f)\} + pd + fd = 0.$$

Developing this with reference to p and f, we have

$$(v + 2d)pf + (vh + d)p(1 - f) + (vh + d)(1 - p)f$$
$$+ v(1 - h)(1 - p)(1 - f) = 0.$$

Whence the result of elimination will be

$$(v + 2d)(vh + d)(vh + d)v(1 - h) = 0.$$

Now $v + 2d = v + d + d$, which by Prop. I. is reducible to $v + d$. The product of this and the second factor is

$$(v + d)(vh + d),$$

which by Prop. II. reduces to

$$d + v(vh) \quad \text{or} \quad vh + d.$$

In like manner, this result, multiplied by the third factor, gives simply $vh + d$. Lastly, this multiplied by the fourth factor, $v(1 - h)$, gives, as the final equation,

$$vd(1 - h) = 0. \quad (8)$$

It remains to eliminate g from (5) and (6). The result is

$$v(1 - m) = 0. \quad (9)$$

Finally, the equations (4), (8), and (9) give on addition

$$v(1 - d) + vd(1 - h) + v(1 - m) = 0,$$

from which we have

$$v = \frac{0}{1 - d + d(1 - h) + 1 - m}.$$

And the development of this result gives

$$v = \frac{0}{0} hdm,$$

of which the interpretation is,—*Virtue is a habit accompanied by deliberate preference, and keeping in view the mean relative to human nature.*

Properly speaking, this is not a definition, but a description of virtue. It is *all,* however, that can be correctly inferred from the premises. Aristotle specially connects with it the necessity of prudence, to determine the safe and middle line of action; and there is no doubt that the ancient theories of virtue generally partook more of an intellectual character than those (the theory of utility excepted) which have most prevailed in modern days. Virtue was regarded as consisting in the right state and habit of the whole mind, rather than in the single supremacy of conscience or the moral faculty. And to some extent those theories were undoubtedly right. For though unqualified obedience to the dictates of conscience is an essential element of virtuous conduct, yet the conformity of those dictates with those unchanging principles of rectitude ($\alpha\dot{\iota}\acute{\omega}\nu\iota\alpha\ \delta\acute{\iota}\kappa\alpha\iota\alpha$) which are founded in, or which rather are themselves the foundation of the constitution of things, is another element. And generally this conformity, in any high degree at least, is inconsistent with a state of ignorance and mental hebetude. Reverting to the particular theory of Aristotle, it will probably appear to most that it is of too negative a character, and that the shunning of extremes does not afford a sufficient scope for the expenditure of the nobler energies of our being. Aristotle seems to have been imperfectly conscious of this defect of his system, when in the opening of his seventh book he spoke of an "heroic virtue"[1] rising above the measure of human nature.

7. I have already remarked (viii. 1) that the theory of single equations or propositions comprehends questions which cannot be fully answered, except in connexion with the theory of systems of equations. This remark is exemplified when it is proposed to determine from a given single equation the relation, not of some single elementary class, but of some compound class, involving in its expression more than one element, in terms of the remaining elements. The following particular example, and the succeeding general problem, are of this nature.

---

[1]$\tau\dot{\eta}\nu\ \dot{\upsilon}\pi\dot{\epsilon}\rho\ \dot{\eta}\mu\tilde{\alpha}\varsigma\ \dot{\alpha}\rho\epsilon\tau\dot{\eta}\nu\ \dot{\eta}\rho\omega\ddot{\iota}\kappa\dot{\eta}\nu\ \tau\iota\nu\alpha\ \kappa\alpha\dot{\iota}\ \theta\epsilon\dot{\iota}\alpha\nu.$—NIC. ETII. Book vii.

Ex. 2.—Let us resume the symbolical expression of the definition of wealth employed in Chap. VII., viz.,

$$w = st\{p + r(1 - p)\},$$

wherein, as before,

w = wealth,

s = things limited in supply,

t = things transferable,

p = things productive of pleasure,

r = things preventive of pain;

and suppose it required to determine hence the relation of things transferable and productive of pleasure, to the other elements of the definition, viz., wealth, things limited in supply, and things preventive of pain.

The expression for things transferable and productive of pleasure is tp. Let us represent this by a new symbol y. We have then the equations

$$w = st\{p + r(1 - p)\},$$
$$y = tp,$$

from which, if we eliminate t and p, we may determine y as a function of w, s, and r. The result interpreted will give the relation sought.

Bringing the terms of these equations to the first side, we have

$$w - stp - str(1 - p) = 0.$$
$$y - tp = 0. \tag{3}$$

And adding the squares of these equations together,

$$w + stp + str(1 - p) - 2wstp - 2wstr(1 - p) + y + tp - 2ytp = 0. \tag{4}$$

Developing the first member with respect to t and p, in order to eliminate those symbols, we have

$$(w + s - 2ws + 1 - y)\, tp + (w + sr - 2wsr + y)\, t(1 - p)$$
$$+ (w + y)(1 - t)p + (w + y)(1 - t)(1 - p); \tag{5}$$

and the result of the elimination of t and p will be obtained by equating to 0 the product of the four coefficients of

$$tp,\ t(1 - p),\ (1 - t)\, p,\ \text{and}\ (1 - t)(1 - p).$$

Or, by Prop. 3, the result of the elimination of t and p from the above equation will be of the form

$$Ey + E'(1 - y),$$

wherein E is the result obtained by changing in the given equation y into 1, and then eliminating t and p; and E' the result obtained by changing in the same equation y into 0, and then eliminating t and p. And the mode in each case of eliminating t and p is to multiply together the coefficients of the four constituents tp, $t(1 - p)$, &c.

If we make $y = 1$, the coefficients become—

1st. $w(1 - s) + s(1 - w)$.

2nd. $1 + w(1 - sr) + s(1 - w) r$, equivalent to 1 by Prop. I.

3rd and 4th. $1 + w$, equivalent to 1 by Prop. I.

Hence the value of E will be

$$w(1 - s) + s(1 - w).$$

Again, in (5) making $y = 0$, we have for the coefficients—

1st. $1 + w(1 - s) + s(1 - w)$, equivalent to 1.

2nd. $w(1 - sr) + sr(1 - w)$.

3rd and 4th. w.

The product of these coefficients gives

$$E' = w(1 - sr).$$

The equation from which y is to be determined, therefore, is

$$\{w(1 - s) + s(1 - w)\} y + w(1 - sr)(1 - y) = 0,$$

$$\therefore y = \frac{w(1 - sr)}{w(1 - sr) - w(1 - s) - s(1 - w)};$$

and expanding the second member,

$$y = \frac{0}{0} wsr + ws(1 - r) + \frac{1}{0} w(1 - s) r + \frac{1}{0} w(1 - s)(1 - r)$$

$$+ 0(1 - w)sr + 0(1 - w) s(1 - r) + \frac{0}{0}(1 - w)(1 - s) r$$

$$+ \frac{0}{0}(1 - w)(1 - s)(1 - r);$$

whence reducing

$$y = ws(1 - r) + \frac{0}{0} wsr + \frac{0}{0}(1 - w)(1 - s), \tag{6}$$

$$\text{with } w(1 - s) = 0. \tag{7}$$

The interpretation of which is—

1st. *Things transferable and productive of pleasure consist of all wealth (limited in supply and) not preventive of pain, an indefinite amount of wealth (limited in supply and) preventive of pain, and an indefinite amount of what is not wealth and not limited in supply.*

2nd. *All wealth is limited in supply.*

I have in the above solution written in parentheses that part of the full description which is implied by the accompanying independent relation (7).

8. The following problem is of a more general nature, and will furnish an easy practical rule for problems such as the last.

General Problem

Given any equation connecting the symbols x, y . . w, z . .

Required to determine the logical expression of any class expressed in any way by the symbols x, y . . in terms of the remaining symbols, w, z, &c.

Let us confine ourselves to the case in which there are but two symbols, x, y, and two symbols, w, z, a case sufficient to determine the general Rule.

Let $V = 0$ be the given equation, and let $\phi(x, y)$ represent the class whose expression is to be determined.

Assume $t = \phi(x, y)$, then, from the above two equations, x and y are to be eliminated.

Now the equation $V = 0$ may be expanded in the form

$$Axy + Bx(1 - y) + C(1 - x)y + D(1 - x)(1 - y) = 0, \qquad (1)$$

A, B, C, and D being functions of the symbols w and z.

Again, as $\phi(x, y)$ represents a class or collection of things, it must consist of a constituent, or series of constituents, whose coefficients are 1.

Wherefore if the *full* development of $\phi(x, y)$ be represented in the form

$$axy + bx(1 - y) + c(1 - x)y + d(1 - x)(1 - y),$$

the coefficients a, b, c, d must each be 1 or 0.

Now reducing the equation $t = \phi(x, y)$ by transposition and squaring, to the form

$$t\{1 - \phi(x, y)\} + \phi(x, y)(1 - t) = 0;$$

and expanding with reference to x and y, we get

$$\{t(1 - a) + a(1 - t)\}xy + \{t(1 - b) + b(1 - t)\}x(1 - y)$$
$$+ \{t(1 - c) + c(1 - t)\}(1 - x)y$$
$$+ \{t(1 - d) + d(1 - t)\}(1 - x)(1 - y) = 0;$$

whence, adding this to (1), we have

$$\{A + t(1 - a) + a(1 - t)\}xy$$
$$+ \{B + t(1 - b) + b(1 - t)\}x(1 - y) + \&c. = 0.$$

Let the result of the elimination of x and y be of the form

$$Et + E'(1 - t) = 0,$$

then E will, by what has been said, be the reduced product of what the coefficients of the above expansion become when $t = 1$, and E' the product of the same factors similarly reduced by the condition $t = 0$.

Hence E will be the reduced product

$$(A + 1 - a)(B + 1 - b)(C + 1 - c)(D + 1 - d).$$

Considering any factor of this expression, as $A + 1 - a$, we see that when $a = 1$ it becomes A, and when $a = 0$ it becomes $1 + A$, which reduces by Prop. I. to 1. Hence we may infer that E will be the product of the coefficients of those constituents in the development of V whose coefficients in the development of $\phi(x, y)$ are 1.

Moreover E' will be the reduced product
$$(A + a)(B + b)(C + c)(D + d).$$
Considering any one of these factors, as $A + a$, we see that this becomes A when $a = 0$, and reduces to 1 when $a = 1$; and so on for the others. Hence E' will be the product of the coefficients of those constituents in the development of y, whose coefficients in the development $\phi(x, y)$ are 0. Viewing these cases together, we may establish the following Rule:

9. *To deduce from a logical equation the relation of any class expressed by a given combination of the symbols x, y, &c, to the classes represented by any other symbols involved in the given equation.*

RULE.—*Expand the given equation with reference to the symbols x, y. Then form the equation*
$$Et + E'(1 - t) = 0,$$
in which E is the product of the coefficients of all those constituents in the above development, whose coefficients in the expression of the given class are 1, and E' the product of the coefficients of those constituents of the development whose coefficients in the expression of the given class are 0. The value of t deduced from the above equation by solution and interpretation will be the expression required.

NOTE.—*Although in the demonstration of this Rule V is supposed to consist solely of positive terms, it may easily be shown that this condition is unnecessary, and the Rule general, and that no preparation of the given equation is really required.*

10. Ex. 3.—The same definition of wealth being given as in Example 2, required an expression for *things transferable, but not productive of pleasure,* $t(1 - p)$, in terms of the other elements represented by w, s, and r.
The equation
$$w - stp - str(1 - p) = 0,$$
gives, when squared,
$$w + stp + str(1 - p) - 2wstp - 2wstr(1 - p) = 0;$$
and developing the first member with respect to t and p,
$$(w + s - 2ws)tp + (w + sr - 2wsr)t(1 - p) + w(1 - t)p$$
$$+ w(1 - t)(1 - p) = 0.$$
The coefficients of which it is best to exhibit as in the following equation;
$$\{w(1 - s) + s(1 - w)\}tp + \{w(1 - sr) + sr(1 - w)\}t(1 - p) + w(1 - t)p$$
$$+ w(1 - t)(1 - p) = 0.$$
Let the function $t(1 - p)$ to be determined, be represented by z; then the full development of z in respect of t and p is
$$z = 0\,tp + t(1 - p) + 0(1 - t)p + 0(1 - t)(1 - p).$$

Hence, by the last problem, we have

$$Ez + E'(1 - z) = 0;$$

where

$$E = w(1 - sr) + sr(1 - w);$$

$$E' = \{w(1 - s) + s(1 - w)\} \times w \times w = w(1 - s);$$

$$\therefore \{w(1 - sr) + sr(1 - w)\} z + w(1 - s)(1 - z) = 0.$$

Hence,

$$z = \frac{w(1 - s)}{2wsr - ws - sr}$$

$$= \frac{0}{0} wsr + 0 \, ws(1 - r) + \frac{1}{0} w(1 - s) r + \frac{1}{0} w(1 - s)(1 - r),$$

$$+ 0(1 - w) sr + \frac{0}{0}(1 - w) s(1 - r) + \frac{0}{0}(1 - w)(1 - s) r$$

$$+ \frac{0}{0}(1 - w)(1 - s)(1 - r).$$

Or,

$$z = \frac{0}{0} wsr + \frac{0}{0}(1 - w) s(1 - r) + \frac{0}{0}(1 - w)(1 - s),$$

with

$$w(1 - s) = 0.$$

Hence, *Things transferable and not productive of pleasure are either wealth (limited in supply and preventive of pain); or things which are not wealth, but limited in supply and not preventive of pain; or things which are not wealth, and are unlimited in supply.*

The following results, deduced in a similar manner, will be easily verified:

Things limited in supply and productive of pleasure which are not wealth,—are intransferable.

Wealth that is not productive of pleasure is transferable, limited in supply, and preventive of pain.

Things limited in supply which are either wealth, or are productive of pleasure, but not both,—are either transferable and preventive of pain, or intransferable.

11. From the domain of natural history a large number of curious examples might be selected. I do not, however, conceive that such applications would possess any independent value. They would, for instance, throw no light upon the true principles of classification in the science of zoology. For the discovery of these, some basis of positive knowledge is requisite,—some acquaintance with organic structure, with teleological adaptation; and this is a species of knowledge which can only be derived from the use of external means of observation and analysis. Taking, however, any collection of propositions in natural history, a great number of logical problems present themselves, without regard to the system of classification adopted. Perhaps in forming such examples, it is better to avoid, as superfluous, the mention of that property

of a class or species which is immediately suggested by its name, e.g. the ring-structure in the annelida, a class of animals including the earth-worm and the leech.

Ex. 4.—1. The annelida are soft-bodied, and either naked or enclosed in a tube.

2. The annelida consist of all invertebrate animals having red blood in a double system of circulating vessels.

Assume
$$a = \text{annelida};$$
$$s = \text{soft-bodied animals};$$
$$n = \text{naked};$$
$$t = \text{enclosed in a tube};$$
$$i = \text{invertebrate};$$
$$r = \text{having red blood, \&c.}$$

Then the propositions given will be expressed by the equations

$$a = vs\,\{n(1 - t) + t(1 - n)\};\qquad(1)$$
$$a = ir;\qquad(2)$$

to which we may add the implied condition,

$$nt = 0.\qquad(3)$$

On eliminating v, and reducing the system to a single equation, we have

$$a\{1 - sn(1 - t) - st(1 - n)\} + a(1 - ir) + ir(1 - a) + nt = 0.\qquad(4)$$

Suppose that we wish to obtain the relation in which soft-bodied animals enclosed in tubes are placed (by virtue of the premises) with respect to the following elements, viz., the possession of red blood, of an external covering, and of a vertebral column.

We must first eliminate a. The result is

$$ir\,\{1 - sn(1 - t) - st(1 - n)\} + nt = 0.$$

Then (ix. 9) developing with respect to s and t, and reducing the first coefficient by Prop. 1, we have

$$nst + ir(1 - n)\,s(1 - t) + (ir + n)(1 - s)\,t + ir(1 - s)(1 - t) = 0.\qquad(5)$$

Hence, if $st = w$, we find

$$nw + ir(1 - n) \times (ir + n) \times ir(1 - w) = 0;$$

or,
$$nw + ir(1 - n)(1 - w) = 0;$$

$$\therefore w = \frac{ir(1 - n)}{ir(1 - n) - n}$$

$$= 0\ irn + ir(1 - n) + 0i(1 - r)\,n + \frac{0}{0}i(1 - r)(1 - n)$$

$$+\ 0(1 - i)\,rn + \frac{0}{0}(1 - i)\,r(1 - n) + 0(1 - i)(1 - r)n$$

$$+\ \frac{0}{0}(1 - i)(1 - r)(1 - n);$$

or, $$w = ir(1 - n) + \frac{0}{0} i(1 - r)(1 - n) + \frac{0}{0}(1 - i)(1 - n).$$

Hence, *soft-bodied animals enclosed in tubes consist of all invertebrate animals having red blood and not naked, and an indefinite remainder of invertebrate animals not having red blood and not naked, and of vertebrate animals which are not naked.*

And in an exactly similar manner, the following reduced equations, the interpretation of which is left to the reader, have been deduced from the development (5).

$$s(1 - t) = irn + \frac{0}{0} i(1 - n) + \frac{0}{0}(1 - i)$$

$$(1 - s) t = \frac{0}{0}(1 - i) r(1 - n) + \frac{0}{0}(1 - r)(1 - n)$$

$$(1 - s)(1 - t) = \frac{0}{0} i(1 - r) + \frac{0}{0}(1 - i).$$

In none of the above examples has it been my object to exhibit in any special manner the power of the method. That, I conceive, can only be fully displayed in connexion with the mathematical theory of probabilities. I would, however, suggest to any who may be desirous of forming a correct opinion upon this point, that they examine by the rules of ordinary logic the following problem, *before* inspecting its solution; remembering at the same time, that whatever complexity it possesses might be multiplied indefinitely, with no other effect than to render its solution by the method of this work more operose, but not less certainly attainable.

Ex. 5. Let the observation of a class of natural productions be supposed to have led to the following general results.

1st, That in whichsoever of these productions the properties A and C are missing, the property E is found, together with one of the properties B and D, but not with both.

2nd, That wherever the properties A and D are found while E is missing, the properties B and C will either both be found, or both be missing.

3rd, That wherever the property A is found in conjunction with either B or E, or both of them, there either the property C or the property D will be found, but not both of them. And conversely, wherever the property C or D is found singly, there the property A will be found in conjunction with either B or E, or both of them.

Let it then be required to ascertain, first, what in any particular instance may be concluded from the ascertained presence of the property A, with reference to the properties B, C, and D; also whether any relations exist independently among the properties B, C, and D. Secondly, what may be concluded in like manner respecting the property B, and the properties A, C, and D.

It will be observed, that in each of the three data, the information conveyed respecting the properties A, B, C, and D, is complicated with another element, E, about which we desire to say nothing in our conclusion. It will hence be requisite to eliminate the symbol representing the property E from the system of equations, by which the given propositions will be expressed.

Let us represent the property A by x, B by y, C by z, D by w, E by v. The data are

$$\bar{x}\bar{z} = qv(y\bar{w} + w\bar{y}); \tag{1}$$
$$\bar{v}xw = q(yz + \bar{y}\bar{z}); \tag{2}$$
$$xy + xv\bar{y} = w\bar{z} + z\bar{w}; \tag{3}$$

$\bar{x}$ standing for $1 - x$, &c., and q being an indefinite class symbol. Eliminating q separately from the first and second equations, and adding the results to the third equation reduced by (5), Chap. VIII., we get

$$\bar{x}\bar{z}(1 - vy\bar{w} - vw\bar{y}) + \bar{v}xw(y\bar{z} + z\bar{y}) + (xy + xv\bar{y})(wz + \bar{w}\bar{z})$$
$$+ (w\bar{z} + z\bar{w})(1 - xy - xv\bar{y}) = 0. \tag{4}$$

From this equation v must be eliminated, and the value of x determined from the result. For effecting this object, it will be convenient to employ the method of Prop. 3 of the present chapter.

Let then the result of elimination be represented by the equation

$$Ex + E'(1 - x) = 0.$$

To find E make $x = 1$ in the first member of (4), we find

$$\bar{v}w(y\bar{z} + z\bar{y}) + (y + v\bar{y})(wz + \bar{w}\bar{z}) + (w\bar{z} + z\bar{w})\bar{v}\bar{y}.$$

Eliminating v, we have

$$(wz + \bar{w}\bar{z})\{w(y\bar{z} + z\bar{y}) + y(wz + \bar{w}\bar{z}) + \bar{y}(w\bar{z} + z\bar{w})\};$$

which, on actual multiplication, in accordance with the conditions $w\bar{w} = 0$, $z\bar{z} = 0$, &c., gives

$$E = wz + y\bar{w}\bar{z}.$$

Next, to find E' make $x = 0$ in (4), we have

$$z(1 - vy\bar{w} - v\bar{y}w) + w\bar{z} + z\bar{w}.$$

whence, eliminating v, and reducing the result by Propositions 1 and 2, we find

$$E' = w\bar{z} + z\bar{w} + \bar{y}\bar{w}\bar{z};$$

and, therefore, finally we have

$$(wz + y\bar{w}\bar{z})x + (w\bar{z} + z\bar{w} + \bar{y}\bar{w}\bar{z})\bar{x} = 0; \tag{5}$$

from which

$$x = \frac{w\bar{z} + z\bar{w} + \bar{y}\,\bar{w}\bar{z}}{w\bar{z} + z\bar{w} + \bar{y}\,\bar{w}\bar{z} - wz - y\bar{w}\bar{z}};$$

wherefore, by development,

$$x = 0yzw + yz\bar{w} + y\bar{z}w + 0y\bar{z}\bar{w}$$
$$+ 0\bar{y}zw + \bar{y}z\bar{w} + \bar{y}\bar{z}w + \bar{y}\bar{z}\bar{x};$$

or, collecting the terms in vertical columns,

$$x = z\overline{w} + \overline{z}w + \overline{y}\overline{z}\overline{w};\tag{6}$$

the interpretation of which is—

In whatever substances the property A is found, there will also be found either the property C or the property D, but not both, or else the properties B, C, and D, will all be wanting. And conversely, *where either the property C or the property D is found singly, or the properties B, C, and D, are together missing, there the property A will be found.*

It also appears that there is no independent relation among the properties B, C, and D.

Secondly, we are to find *y*. Now developing (5) with respect to that symbol,

$$(xwz + x\overline{w}\overline{z} + \overline{x}w\overline{z} + \overline{x}z\overline{w})y + (xwz + \overline{x}w\overline{z} + \overline{x}z\overline{w} + \overline{x}\overline{z}\overline{w})\overline{y} = 0;$$

whence, proceeding as before,

$$y = \overline{x}\,\overline{w}\,\overline{z} + \frac{0}{0}(\overline{x}wz + xw\overline{z} + xz\overline{w}),\tag{7}$$

$$xzw = 0;\tag{8}$$

$$\overline{x}\,\overline{z}w = 0;\tag{9}$$

$$\overline{x}z\overline{w} = 0;\tag{10}$$

From (10) reduced by solution to the form

$$\overline{x}z = \frac{0}{0}w;$$

we have the independent relation,—*If the property A is absent and C present, D is present.* Again, by addition and solution (8) and (9) give

$$xz + \overline{x}\,\overline{z} = \frac{0}{0}\overline{w}.$$

Whence we have for the general solution and the remaining independent relation:

1st. *If the property B be present in one of the productions, either the properties A, C, and D, are all absent, or some one alone of them is absent.* And conversely, *if they are all absent it may be concluded that the property A is present* (7).

2nd. *If A and C are both present or both absent, D will be absent, quite independently of the presence or absence of B* (8) and (9).

I have not attempted to verify these conclusions.

CHAPTER X

OF THE CONDITIONS OF A PERFECT METHOD

1. The subject of Primary Propositions has been discussed at length, and we are about to enter upon the consideration of Secondary Propositions. The interval of transition between these two great divisions of the science of Logic may afford a fit

occasion for us to pause, and while reviewing some of the past steps of our progress, to inquire what it is that in a subject like that with which we have been occupied constitutes perfection of method. I do not here speak of that perfection only which consists in power, but of that also which is founded in the conception of what is fit and beautiful. It is probable that a careful analysis of this question would conduct us to some such conclusion as the following, viz., that a perfect method should not only be an efficient one, as respects the accomplishment of the objects for which it is designed, but should in all its parts and processes manifest a certain unity and harmony. This conception would be most fully realized if even the very forms of the method were suggestive of the fundamental principles, and if possible of the one fundamental principle, upon which they are founded. In applying these considerations to the science of Reasoning, it may be well to extend our view beyond the mere analytical processes, and to inquire what is best as respects not only the mode or form of deduction, but also the system of data or premises from which the deduction is to be made.

2. As respects mere power, there is no doubt that the first of the methods developed in Chapter VIII. is, within its proper sphere, a perfect one. The introduction of arbitrary constants makes us independent of the forms of the premises, as well as of any conditions among the equations by which they are represented. But it seems to introduce a foreign element, and while it is a more laborious, it is also a less elegant form of solution than the second method of reduction demonstrated in the same chapter. There are, however, conditions under which the latter method assumes a more perfect form than it otherwise bears. To make the one fundamental condition expressed by the equation

$$x(1 - x) = 0,$$

the universal type of form, would give a unity of character to both processes and results, which would not else be attainable. Were brevity or convenience the only valuable quality of a method, no advantage would flow from the adoption of such a principle. For to impose upon every step of a solution the character above described, would involve in some instances no slight labour of preliminary reduction. But it is still interesting to know that this can be done, and it is even of some importance to be acquainted with the conditions under which such a form of solution would spontaneously present itself. Some of these points will be considered in the present chapter.

Proposition I

3. *To reduce any equation among logical symbols to the form* $V = 0$, *in which* V *satisfies the law of duality,*

$$V(1 - V) = 0.$$

It is shown in Chap. V. Prop. 4, that the above condition is satisfied whenever V is the sum of a series of constituents. And it is evident from Prop. 2, Chap. VI. that all equations are equivalent which, when reduced by transposition to the form $V = 0$, produce, by development of the first member, the same series of constituents with coefficients which do not vanish; the particular numerical values of those coefficients being immaterial.

Hence the object of this Proposition may always be accomplished by bringing all the terms of an equation to the first side, fully expanding that member, and changing in the result all the coefficients which do not vanish into unity, except such as have already that value.

But as the development of functions containing many symbols conducts us to expressions inconvenient from their great length, it is desirable to show how, in the only cases which do practically offer themselves to our notice, this source of complexity may be avoided.

The great primary forms of equations have already been discussed in Chapter VIII. They are—

$$X = vY,$$
$$X = Y,$$
$$vX = vY.$$

Whenever the conditions $X(1 - X) = 0$, $Y(1 - Y) = 0$, are satisfied, we have seen that the two first of the above equations conduct us to the forms

$$X(1 - Y) = 0, \tag{1}$$
$$X(1 - Y) + Y(1 - X) = 0; \tag{2}$$

and under the same circumstances it may be shown that the last of them gives

$$v\{X(1 - Y) + Y(1 - X)\} = 0; \tag{3}$$

all which results obviously satisfy, in their first members, the condition

$$V(1 - V) = 0.$$

Now as the above are the forms and conditions under which the equations of a logical system properly expressed do actually present themselves, it is always possible to reduce them by the above method into subjection to the law required. Though, however, the separate equations may thus satisfy the law, their equivalent sum (viii. 4) may not do so, and it remains to show how upon it also the requisite condition may be imposed.

Let us then represent the equation formed by adding the several reduced equations of the system together, in the form

$$v + v' + v'', \&c. = 0, \tag{4}$$

this equation being singly equivalent to the system from which it was obtained. We suppose v, v', v'', &c. to be class terms (ix. 1) satisfying the conditions

$$v(1 - v) = 0, \quad v'(1 - v') = 0, \&c.$$

Now the full interpretation of (4) would be found by developing the first member with respect to all the elementary symbols x, y, &c. which it contains, and equating to 0 all the constituents whose coefficients do not vanish; in other words, all the constituents which are found in either v, v', v'', &c. But those constituents consist of—1st, such as are found in v; 2nd, such as are not found in v, but are found in v', 3rd, such as are neither found in v nor v', but are found in v'', and so on. Hence they will be such as are found in the expression

$$v + (1 - v)\, v' + (1 - v)(1 - v')\, v'' + \&c.,\tag{5}$$

an expression in which no constituents are repeated, and which obviously satisfies the law $V(1 - V) = 0$.

Thus if we had the expression

$$(1 - t) + v + (1 - z) + tzw,$$

in which the terms $1 - t$, $1 - z$ are bracketed to indicate that they are to be taken as single class terms, we should, in accordance with (5), reduce it to an expression satisfying the condition $V(1 - V) = 0$, by multiplying all the terms after the first by t, then all after the second by $1 - v$; lastly, the term which remains after the third by z; the result being

$$1 - t + tv + t(1 - v)(1 - z) + t(1 - v)\, zw.\tag{6}$$

4. All logical equations then are reducible to the form $V = 0$, V satisfying the law of duality. But it would obviously be a higher degree of perfection if equations always presented themselves in such a form, without preparation of any kind, and not only exhibited this form in their original statement, but retained it unimpaired after those additions which are necessary in order to reduce systems of equations to single equivalent forms. That they do not spontaneously present this feature is not properly attributable to defect of method, but is a consequence of the fact that our premises are not always complete, and accurate, and independent. They are not complete when they involve material (as distinguished from formal) relations, which are not expressed. They are not accurate when they imply relations which are not intended. But setting aside these points, with which, in the present instance, we are less concerned, let it be considered in what sense they may fail of being independent.

5. A system of propositions may be termed independent, when it is not possible to deduce from any portion of the system a conclusion deducible from any other portion of it. Supposing the equations representing those propositions all reduced to the form

$$V = 0,$$

then the above condition implies that no constituent which can be made to appear in the development of a particular function V of the system, can be made to appear in the development of any other function V' of the same system. When this condition is not satisfied, the equations of the system are not independent. This may happen in various cases. Let all the equations satisfy in their first members the law of duality, then if there appears a positive term x in the expansion of one equation, and a term xy in that of another, the equations are not independent, for the term x is further developable into $xy + x(1 - y)$, and the equation

$$xy = 0$$

is thus involved in both the equations of the system. Again, let a term xy appear in one equation, and a term xz in another. Both these may be developed so as to give the common constituent xyz. And other cases may easily be imagined in which premises which appear at first sight to be quite independent are not really so. Whenever equations of the form $V = 0$ are thus not truly independent, though individually they may satisfy the law of duality,

$$V(1 - V) = 0,$$

the equivalent equation obtained by adding them together will not satisfy that condition, unless sufficient reductions by the method of the present chapter have been performed. When, on the other hand, the equations of a system both satisfy the above law, and are independent of each other, their sum will also satisfy the same law. I have dwelt upon these points at greater length than would otherwise have been necessary, because it appears to me to be important to endeavour to form to ourselves, and to keep before us in all our investigations, the pattern of an ideal perfection,—the object and the guide of future efforts. In the present class of inquiries the chief aim of improvement of method should be to facilitate, as far as is consistent with brevity, the transformation of equations, so as to make the fundamental condition above adverted to universal.

In connexion with this subject the following Propositions are deserving of attention.

Proposition II

If the first member of any equation $V = 0$ satisfy the condition $V(1 - V) = 0$, and if the expression of any symbol t of that equation be determined as a developed function of the other symbols, the coefficients of the expansion can only assume the forms $1, 0, \frac{0}{0}, \frac{1}{0}$.

For if the equation be expanded with reference to t, we obtain as the result,

$$Et + E'(1 - t), \tag{1}$$

E and E' being what V becomes when t is successively changed therein into 1 and 0. Hence E and E' will themselves satisfy the conditions

$$E(1 - E) = 0, \quad E'(1 - E') = 0. \tag{2}$$

Now (1) gives

$$t = \frac{E'}{E' - E''},$$

the second member of which is to be expanded as a function of the remaining symbols. It is evident that the only numerical values which E and E' can receive in the calculation of the coefficients will be 1 and 0. The following cases alone can therefore arise:

1st. $\quad E' = 1, \quad E = 1, \quad$ then $\dfrac{E'}{E' - E} = \dfrac{1}{0}.$

2nd. $\quad E' = 1, \quad E = 0, \quad$ then $\dfrac{E'}{E' - E} = 1.$

3rd. $\quad E' = 0, \quad E = 1, \quad$ then $\dfrac{E'}{E' - E} = 0.$

4th. $\quad E' = 0, \quad E = 0, \quad$ then $\dfrac{E'}{E' - E} = \dfrac{0}{0}.$

Whence the truth of the Proposition is manifest.

6. It may be remarked that the forms 1, 0, and $\frac{0}{0}$ appear in the solution of equations independently of any reference to the condition $V(1 - V) = 0$. But it is not so with the coefficient $\frac{1}{0}$. The terms to which this coefficient is attached when the above condition is satisfied may receive any other value except the three values 1, 0, and $\frac{0}{0}$, when that condition is not satisfied. It is permitted, and it would conduce to uniformity, to change any coefficient of a development not presenting itself in any of the four forms referred to in this Proposition into $\frac{1}{0}$, regarding this as the symbol proper to indicate that the coefficient to which it is attached should be equated to 0. This course I shall frequently adopt.

Proposition III

7. *The result of the elimination of any symbols x, y, &c. from an equation V = 0, of which the first member identically satisfies the law of duality,*

$$V(1 - V) = 0,$$

may be obtained by developing the given equation with reference to the other symbols, and equating to 0 the sum of those constituents whose coefficients in the expansion are equal to unity.

Suppose that the given equation $V = 0$ involves but three symbols, x, y, and t, of which x and y are to be eliminated. Let the development of the equation, with respect to t, be

$$At + B(1 - t) = 0, \tag{1}$$

A and B being free from the symbol t.

By Chap. IX. Prop. 3, the result of the elimination of x and y from the given equation will be of the form

$$Et + E'(1 - t) = 0, \tag{2}$$

in which E is the result obtained by eliminating the symbols x and y from the equation $A = 0$, E' the result obtained by eliminating from the equation $B = 0$.

Now A and B must satisfy the condition

$$A(1 - A) = 0, \quad B(1 - B) = 0.$$

Hence A (confining ourselves for the present to this coefficient) will either be 0 or 1, or a constituent, or the sum of a part of the constituents which involve the symbols x and y. If $A = 0$ it is evident that $E = 0$; if A is a single constituent, or the sum of a part of the constituents involving x and y, E will be 0. For the *full* development of A, with respect to x and y, will contain terms with vanishing coefficients, and E is the product of *all* the coefficients. Hence when $A = 1$, E is equal to A, but in other cases E is equal to 0. Similarly, when $B = 1$, E' is equal to B, but in other cases E' vanishes. Hence the expression (2) will consist of that part, if any there be, of (1) in which the coefficients A, B are unity. And this reasoning is general. Suppose, for instance, that V involved the symbols x, y, z, t, and that it were required to eliminate x and y. Then if the development of V, with reference to z and t, were

$$zt + xz(1 - t) + y(1 - z)t + (1 - z)(1 - t),$$

the result sought would be

$$zt + (1 - z)(1 - t) = 0,$$

this being that portion of the development of which the coefficients are unity.

Hence, if from any system of equations we deduce a single equivalent equation $V = 0$, V satisfying the condition

$$V(1 - V) = 0,$$

the ordinary processes of elimination may be entirely dispensed with, and the single process of development made to supply their place.

8. It may be that there is no practical advantage in the method thus pointed out, but it possesses a theoretical unity and completeness which render it deserving of regard, and I shall accordingly devote a future chapter (XIV.) to its illustration. The progress of applied mathematics has presented other and signal examples of the reduction of systems of problems or equations to the dominion of some central but pervading law.

9. It is seen from what precedes that there is one class of propositions to which all the special appliances of the above methods of preparation are unnecessary. It is that which is characterized by the following conditions:

First, That the propositions are of the ordinary kind, implied by the use of the copula *is* or *are*, the predicates being particular.

Secondly, That the terms of the proposition are intelligible without the supposition of any understood relation among the elements which enter into the expression of those terms.

Thirdly, That the propositions are independent.

We may, if such speculation is not altogether vain, permit ourselves to conjecture that these are the conditions which would be obeyed in the employment of language as an instrument of expression and of thought, by unerring beings, declaring simply what they mean, without suppression on the one hand, and without repetition on the other. Considered both in their relation to the idea of a perfect language, and in their relation to the processes of an exact method, these conditions are equally worthy of the attention of the student.

CHAPTER XI

OF SECONDARY PROPOSITIONS, AND OF THE PRINCIPLES OF THEIR SYMBOLICAL EXPRESSION

1. The doctrine has already been established in Chap. IV., that every logical proposition may be referred to one or the other of two great classes, viz., Primary Propositions and Secondary Propositions. The former of these classes has been discussed in the preceding chapters of this work, and we are now led to the consideration of Secondary Propositions, i.e. of Propositions concerning, or relating to, other propositions regarded as true or false. The investigation upon which we are entering will, in its general order and progress, resemble that which we have already conducted. The two inquiries differ as to the subjects of thought which they recognise, not as to the formal and scientific laws which they reveal, or the methods or processes which are founded upon those laws. Probability would in some measure favour the expectation of such a result. It consists with all that we know of the uniformity of Nature, and all that we believe of the immutable constancy of the Author of Nature, to suppose, that in the mind, which has been endowed with such high capabilities, not only for converse with surrounding scenes, but for the knowledge of itself, and for reflection upon the laws of its own constitution, there should exist a harmony and uniformity not less real than that which the study of the physical sciences makes known to us. Anticipations such as this are never to be made the primary rule of our inquiries, nor are they in any degree to divert us from those labours of patient research by which we ascertain what is the actual constitution of things within the particular province submitted to investigation. But when the grounds of resemblance have been properly and independently determined, it is not inconsistent, even with purely scientific ends, to make that resemblance a subject of meditation, to trace its

extent, and to receive the intimations of truth, yet undiscovered, which it may seem to us to convey. The necessity of a final appeal to fact is not thus set aside, nor is the use of analogy extended beyond its proper sphere,—the suggestion of relations which independent inquiry must either verify or cause to be rejected.

2. *Secondary Propositions are those which concern or relate to Propositions considered as true or false.* The relations of *things* we express by primary propositions. But we are able to make Propositions themselves also the subject of thought, and to express our judgments concerning them. The expression of any such judgment constitutes a secondary proposition. There exists no proposition whatever of which a competent degree of knowledge would not enable us to make one or the other of these two assertions, viz., either that the proposition is true, or that it is false; and each of these assertions is a secondary proposition. "It is true that the sun shines;" "It is not true that the planets shine by their own light;" are examples of this kind. In the former example the Proposition "The sun shines," is asserted to be true. In the latter, the Proposition, "The planets shine by their own light," is asserted to be false. Secondary propositions also include all judgments by which we express a relation or dependence among propositions. To this class or division we may refer conditional propositions, as, "If the sun shine the day will be fair." Also most disjunctive propositions, as, "Either the sun will shine, or the enterprise will be postponed." In the former example we express the dependence of the truth of the Proposition, "The day will be fair," upon the truth of the Proposition, "The sun will shine." In the latter we express a relation between the two Propositions, "The sun will shine," "The enterprise will be postponed," implying that the truth of the one excludes the truth of the other. To the same class of secondary propositions we must also refer all those propositions which assert the simultaneous truth or falsehood of propositions, as, "It is not true both that 'the sun will shine' and that 'the journey will be postponed.'" The elements of distinction which we have noticed may even be blended together in the same secondary proposition. It may involve both the disjunctive element expressed by either, *or*, and the conditional element expressed by *if*; in addition to which, the connected propositions may themselves be of a compound character. If "the sun shine," *and* "leisure permit," then *either* "the enterprise shall be commenced," *or* "some preliminary step shall be taken." In this example a number of propositions are connected together, not arbitrarily and unmeaningly, but in such a manner as to express a *definite* connexion between them,—a connexion having reference to their respective truth or falsehood. This combination, therefore, according to our definition, forms a Secondary Proposition.

The theory of Secondary Propositions is deserving of attentive study, as well on account of its varied applications, as for that close and harmonious analogy, already referred to, which it sustains with the theory of Primary Propositions. Upon each of these points I desire to offer a few further observations.

3. I would in the first place remark, that it is in the form of secondary propositions, at least as often as in that of primary propositions, that the reasonings of ordinary life are exhibited. The discourses, too, of the moralist and the metaphysician are perhaps less often concerning things and their qualities, than concerning principles and hypotheses, concerning truths and the mutual connexion and relation of truths. The conclusions which our narrow experience suggests in relation to the great questions of morals and society yet unsolved, manifest, in more ways than one, the limitations of their human origin; and though the existence of universal principles is not to be questioned, the partial formulæ which comprise our knowledge of their application are subject to conditions, and exceptions, and failure. Thus, in those departments of inquiry which, from the nature of their subject-matter, should be the most interesting of all, much of our actual knowledge is hypothetical. That there has been a strong tendency to the adoption of the same forms of thought in writers on speculative philosophy, will hereafter appear. Hence the introduction of a general method for the discussion of hypothetical and the other varieties of secondary propositions, will open to us a more interesting field of applications than we have before met with.

4. The discussion of the theory of Secondary Propositions is in the next place interesting, from the close and remarkable analogy which it bears with the theory of Primary Propositions. It will appear, that the formal laws to which the operations of the mind are subject, are identical in expression in both cases. The mathematical processes which are founded on those laws are, therefore, identical also. Thus the methods which have been investigated in the former portion of this work will continue to be available in the new applications to which we are about to proceed. But while the laws and processes of the method remain unchanged, the rule of interpretation must be adapted to new conditions. Instead of classes of things, we shall have to substitute propositions, and for the relations of classes and individuals, we shall have to consider the connexions of propositions or of events. Still, between the two systems, however differing in purport and interpretation, there will be seen to exist a pervading harmonious relation, an analogy which, while it serves to facilitate the conquest of every yet remaining difficulty, is of itself an interesting subject of study, and a conclusive proof of that unity of character which marks the constitution of the human faculties.

Proposition I

5. *To investigate the nature of the connexion of Secondary Propositions with the idea of Time.*

It is necessary, in entering upon this inquiry, to state clearly the nature of the analogy which connects Secondary with Primary Propositions.

Primary Propositions express relations among things, viewed as component parts of a universe within the limits of which, whether coextensive with the limits of the actual universe or not, the matter of our discourse is confined. The relations expressed are essentially *substantive*. Some, or all, or none, of the members of a given class, *are* also members of another class. The subjects to which primary propositions refer—the relations among those subjects which they express—are all of the above character.

But in treating of secondary propositions, we find ourselves concerned with another class both of subjects and relations. For the subjects with which we have to do are themselves propositions, so that the question may be asked,—Can we regard these subjects also as *things*, and refer them, by analogy with the previous case, to a universe of their own? Again, the relations among these subject propositions are relations of coexistent truth or falsehood, not of substantive equivalence. We do not say, when expressing the connexion of two distinct propositions, that the one *is* the other, but use some such forms of speech as the following, according to the meaning which we desire to convey: "*Either* the proposition X is true, *or* the proposition Y is true;" "If the proposition X is true, the proposition Y is true;" "The propositions X and Y are jointly true;" and so on.

Now, in considering any such relations as the above, we are not called upon to inquire into the whole extent of their possible meaning (for this might involve us in metaphysical questions of causation, which are beyond the proper limits of science); but it suffices to ascertain some meaning which they undoubtedly possess, and which is adequate for the purposes of logical deduction. Let us take, as an instance for examination, the conditional proposition, "If the proposition X is true, the proposition Y is true." An undoubted meaning of this proposition is, that the *time* in which the proposition X is true, is *time* in which the proposition Y is true. This indeed is only a relation of coexistence, and may or may not exhaust the meaning of the proposition, but it is a relation really involved in the statement of the proposition, and further, it suffices for all the purposes of logical inference.

The language of common life sanctions this view of the essential connexion of secondary propositions with the notion of time. Thus we limit the application of a primary proposition by the word "some," but that of a secondary proposition by the word "sometimes." To say, "Sometimes injustice triumphs," is equivalent to asserting that there are times in which the proposition "Injustice now triumphs," is a true proposition. There are indeed propositions, the truth of which is not thus limited to particular periods or conjunctures; propositions which are true throughout all time, and have received the appellation of "eternal truths." The distinction must be familiar to every reader of Plato and Aristotle, by the latter of whom, especially, it is employed to denote the contrast between the abstract verities of science, such as the propositions of geometry which are always true, and those contingent or phænomenal relations of things which are sometimes true and sometimes false. But the forms of language in which both kinds of propositions are expressed manifest a common dependence upon the idea of time; in the one case as limited to some finite duration, in the other as stretched out to eternity.

6. It may indeed be said, that in ordinary reasoning we are often quite unconscious of this notion of time involved in the very language we are using. But the remark, however just, only serves to show that we commonly reason by the aid of words and the forms of a well-constructed language, without attending to the ulterior grounds upon which those very forms have been established. The course of the present investigation will afford an illustration of the very same principle. I shall avail myself of the notion of time in order to determine the laws of the expression of secondary propositions, as well as the laws of combination of the symbols by which they are expressed. But when those laws and those forms are once determined, this notion of time (essential, as I believe it to be, to the above end) may practically be dispensed with. We may then pass from the forms of common language to the closely analogous forms of the symbolical instrument of thought here developed, and use its processes, and interpret its results, without any conscious recognition of the idea of time whatever.

Proposition II

7. *To establish a system of notation for the expression of Secondary Propositions, and to show that the symbols which it involves are subject to the same laws of combination as the corresponding symbols employed in the expression of Primary Propositions.*

Let us employ the capital letters X, Y, Z, to denote the elementary propositions concerning which we desire to make some assertion touching their truth or falsehood, or among which we seek to express some relation in the form of a secondary proposition. And let us employ the corresponding small letters x, y, z, considered as expressive of mental operations, in the following sense, viz.: Let x represent an act of the mind by which we fix our regard upon that portion of time for which the proposition X is true; and let this meaning be understood when it is asserted that x *denotes* the time for which the proposition X is true. Let us further employ the connecting signs $+$, $-$, $=$, &c., in the following sense, viz.: Let $x + y$ denote the aggregate of those portions of time for which the propositions X and Y are respectively true, those times being entirely separated from each other. Similarly let $x - y$ denote that remainder of time which is left when we take away from the portion of time for which X is true, that (by supposition) included portion for which Y is true. Also, let $x = y$ denote that the time for which the proposition X is true, is identical with the time for which the proposition Y is true. We shall term x the *representative symbol* of the proposition X, &c.

From the above definitions it will follow, that we shall always have

$$x + y = y + x,$$

for either member will denote the same aggregate of time.

Let us further represent by xy the performance in succession of the two operations represented by y and x, i.e. the whole mental operation which consists of the following elements, viz., 1st, The mental selection of that portion of time for which the proposition Y is true. 2ndly, The mental selection, out of that portion of time, of such portion as it contains of the time in which the proposition X is true,—the result of these successive processes being the fixing of the mental regard upon the whole of that portion of time for which the propositions X and Y are both true.

From this definition it will follow, that we shall always have

$$xy = yx. \tag{1}$$

For whether we select mentally, first that portion of time for which the proposition Y is true, then out of the result that contained portion for which X is true; or first, that portion of time for which the proposition X is true, then out of the result that contained portion of it for which the proposition Y is true; we shall arrive at the same final result, viz., that portion of time for which the propositions X and Y are both true.

By continuing this method of reasoning it may be established, that the laws of combination of the symbols x, y, z, &c., in the species of interpretation here assigned to them, are identical in expression with the laws of combination of the same symbols, in the interpretation assigned to them in the first part of this treatise. The reason of this final identity is apparent. For in both cases it is the same faculty, or the same combination of faculties, of which we study the operations; operations, the essential character of which is unaffected, whether we suppose them to be engaged upon that universe of things in which all existence is contained, or upon that whole of time in which all events are realized, and to some part, at least, of which all assertions, truths, and propositions, refer.

Thus, in addition to the laws above stated, we shall have by (4), Chap, II., the law whose expression is

$$x(y + z) = xy + xz; \tag{2}$$

and more particularly the fundamental law of duality (2) Chap, II., whose expression is

$$x^2 = x, \quad \text{or,} \quad x(1 - x) = 0; \tag{3}$$

a law, which while it serves to distinguish the system of thought in Logic from the system of thought in the science of quantity, gives to the processes of the former a completeness and a generality which they could not otherwise possess.

8. Again, as this law (3) (as well as the other laws) is satisfied by the symbols 0 and 1, we are led, as before, to inquire whether those symbols do not admit of interpretation in the present system of thought. The same course of reasoning which we before pursued shows that they do, and warrants us in the two following positions, viz.:

1st, That in the expression of secondary propositions, 0 represents *nothing* in reference to the element of time.

2nd, That in the same system 1 represents the universe, or whole of time, to which the discourse is supposed in any manner to relate.

As in primary propositions the universe of discourse is sometimes limited to a small portion of the actual universe of things, and is sometimes co-extensive with that universe; so in secondary propositions, the universe of discourse may be limited to a single day or to the passing moment, or it may comprise the whole duration of time. It may, in the most literal sense, be "eternal." Indeed, unless there is some limitation expressed or implied in the nature of the discourse, the proper interpretation of the symbol 1 in secondary propositions is "eternity;" even as its proper interpretation in the primary system is the actually existent universe.

9. Instead of appropriating the symbols x, y, z, to the representation of the truths of propositions, we might with equal propriety apply them to represent the occurrence of events. In fact, the occurrence of an event both implies, and is implied by, the truth of a proposition, viz., of the proposition which asserts the occurrence of the event. The one signification of the symbol x necessarily involves the other. It will greatly conduce to convenience to be able to employ our symbols in either of these really equivalent interpretations which the circumstances of a problem may suggest to us as most desirable; and of this liberty I shall avail myself whenever occasion requires. In problems of pure Logic I shall consider the symbols x, y, &c. as representing elementary propositions, among which relation is expressed in the premises. In the mathematical theory of probabilities, which, as before intimated (i. 12), rests upon a basis of Logic, and which it is designed to treat in a subsequent portion of this work, I shall employ the same symbols to denote the simple events, whose implied or required frequency of occurrence it counts among its elements.

Proposition III

10. *To deduce general Rules for the expression of Secondary Propositions.*

In the various inquiries arising out of this Proposition, fulness of demonstration will be the less necessary, because of the exact analogy which they bear with similar inquiries already completed with reference to primary propositions. We shall first consider the expression of terms; secondly, that of the propositions by which they are connected.

As 1 denotes the whole duration of time, and x that portion of it for which the proposition X is true, $1 - x$ will denote that portion of time for which the proposition X is false.

Again, as xy denotes that portion of time for which the propositions X and Y are both true, we shall, by combining this and the previous observation, be led to the following interpretations, viz.:

The expression $x(1 - y)$ will represent the time during which the proposition X is true, and the proposition Y false. The expression $(1 - x)(1 - y)$ will represent the time during which the propositions X and Y are simultaneously false.

The expression $x(1 - y) + y(1 - x)$ will express the time during which either X is true or Y true, but not both; for that time is the sum of the times in which they are singly and exclusively true. The expression $xy + (1 - x)(1 - y)$ will express the time during which X and Y are either both true or both false.

If another symbol z presents itself, the same principles remain applicable. Thus xyz denotes the time in which the propositions X, Y, and Z are simultaneously true; $(1 - x)(1 - y)(1 - z)$ the time in which they are simultaneously false; and the sum of these expressions would denote the time in which they are either true or false together.

The general principles of interpretation involved in the above examples do not need any further illustrations or more explicit statement.

11. The laws of the expression of propositions may now be exhibited and studied in the distinct cases in which they present themselves. There is, however, one principle of fundamental importance to which I wish in the first place to direct attention. Although the principles of expression which have been laid down are perfectly general, and enable us to limit our assertions of the truth or falsehood of propositions to any particular portions of that whole of time (whether it be an unlimited eternity, or a period whose beginning and whose end are definitely fixed, or the passing moment) which constitutes the universe of our discourse, yet, in the actual procedure of human reasoning, such limitation is not commonly employed. When we assert that a proposition is true, we generally mean that it is true throughout the whole duration of the time to which our discourse refers; and when different assertions of the unconditional truth or falsehood of propositions are jointly made as the premises of a logical demonstration, it is to the same universe of time that those assertions are referred, and not to particular and limited parts of it. In that necessary matter which is the object or field of the exact sciences every assertion of a truth may be the assertion of an "eternal truth." In reasoning upon transient phænomena (as of some social conjuncture)

each assertion may be qualified by an immediate reference to the present time, "Now." But in both cases, unless there is a distinct expression to the contrary, it is to the same period of duration that each separate proposition relates. The cases which then arise for our consideration are the following:

1st. *To express the Proposition, "The proposition X is true."*

We are here required to express that within those limits of time to which the matter of our discourse is confined the proposition X is true. Now the time for which the proposition X is true is denoted by x, and the extent of time to which our discourse refers is represented by 1. Hence we have

$$x = 1 \qquad (4)$$

as the expression required.

2nd. *To express the Proposition, "The proposition X is false."*

We are here to express that within the limits of time to which our discourse relates, the proposition X is false; or that within those limits there is no portion of time for which it is true. Now the portion of time for which it is true is x. Hence the required equation will be

$$x = 0. \qquad (5)$$

This result might also be obtained by equating to the whole duration of time 1, the expression for the time during which the proposition X is false, viz., $1 - x$. This gives

$$1 - x = 1,$$

whence

$$x = 0.$$

3rd. *To express the disjunctive Proposition, "Either the proposition X is true or the proposition Y is true;" it being thereby implied that the said propositions are mutually exclusive, that is to say, that one only of them is true.*

The time for which either the proposition X is true or the proposition Y is true, but not both, is represented by the expression $x(1 - y) + y(1 - x)$. Hence we have

$$x(1 - y) + y(1 - x) = 1, \qquad (6)$$

for the equation required.

If in the above Proposition the particles *either*, *or*, are supposed not to possess an absolutely disjunctive power, so that the possibility of the simultaneous truth of the propositions X and Y is not excluded, we must add to the first member of the above equations the term xy. We shall thus have

$$xy + x(1 - y) + (1 - x)y = 1,$$

or

$$x + (1 - x)y = 1. \qquad (7)$$

4th. *To express the conditional Proposition, "If the proposition Y is true, the proposition X is true."*

Since whenever the proposition Y is true, the proposition X is true, it is necessary and sufficient here to express, that the time in which the proposition Y is true is time in which the proposition X is true; that is to say, that it is some indefinite portion of the whole time in which the proposition X is true. Now the time in which the proposition Y is true is y, and the whole time in which the proposition X is true is x. Let v be a symbol of time indefinite, then will vx represent an indefinite portion of the whole time x. Accordingly, we shall have

$$y = vx$$

as the expression of the proposition given.

12. When v is thus regarded as a symbol of time indefinite, vx may be understood to represent the whole, or an indefinite part, or no part, of the whole time x; for any one of these meanings may be realized by a particular determination of the arbitrary symbol v. Thus, if v be determined to represent a time in which the whole time x is included, vx will represent the whole time x. If v be determined to represent a time, some part of which is included in the time x, but which does not fill up the measure of that time, vx will represent a part of the time x. If, lastly, v is determined to represent a time, of which no part is common with any part of the time x, vx will assume the value 0, and will be equivalent to "no time," or "never."

Now it is to be observed that the proposition, "If Y is true, X is true," contains no assertion of the truth of either of the propositions X and Y. It may equally consist with the supposition that the truth of the proposition Y is a condition indispensable to the truth of the proposition X, in which case we shall have $v = 1$; or with the supposition that although Y expresses a condition which, when realized, assures us of the truth of X, yet X may be true without implying the fulfilment of that condition, in which case v denotes a time, some part of which is contained in the whole time x; or, lastly, with the supposition that the proposition Y is not true at all, in which case v represents some time, no part of which is common with any part of the time x. All these cases are involved in the general supposition that v is a symbol of time indefinite.

5th. *To express a proposition in which the conditional and the disjunctive characters both exist.*

The general form of a conditional proposition is, "If Y is true, X is true," and its expression is, by the last section, $y = vx$. We may properly, in analogy with the usage which has been established in primary propositions, designate Y and X as the *terms* of the conditional proposition into which they enter; and we may further adopt the language of the ordinary Logic, which designates the term Y, to which the particle *if* is attached, the "antecedent" of the proposition, and the term X the "consequent."

Now instead of the terms, as in the above case, being simple propositions, let each or either of them be a disjunctive proposition involving different terms connected by the particles *either*, *or*, as in the following illustrative examples, in which X, Y, Z, &c. denote simple propositions.

1st. If either X is true or Y is true, then Z is true.

2nd. If X is true, then either Y is true or Z true.

3rd. If either X is true or Y is true, then either Z and W are both true, or they are both false.

It is evident that in the above cases the relation of the antecedent to the consequent is not affected by the circumstance that one of those terms or both are of a disjunctive character. Accordingly it is only necessary to obtain, in conformity with the principles already established, the proper expressions for the antecedent and the consequent, to affect the latter with the indefinite symbol v, and to equate the results. Thus for the propositions above stated we shall have the respective equations,

1st. $\quad\quad\quad\quad x(1 - y) + (1 - x)y = vz.$

2nd. $\quad\quad\quad\quad x = v\{y(1 - z) + z(1 - y)\}.$

3rd. $\quad\quad x(1 - y) + y(1 - x) = v\{zw + (1 - z)(1 - w)\}.$

The rule here exemplified is of general application.

Cases in which the disjunctive and the conditional elements enter in a manner different from the above into the expression of a compound proposition, are conceivable, but I am not aware that they are ever presented to us by the natural exigencies of human reason, and I shall therefore refrain from any discussion of them. No serious difficulty will arise from this omission, as the general principles which have formed the basis of the above applications are perfectly general, and a slight effort of thought will adapt them to any imaginable case.

13. In the laws of expression above stated those of interpretation are implicitly involved. The equation

$$x = 1$$

must be understood to express that the proposition X is true; the equation

$$x = 0,$$

that the proposition X is false. The equation

$$xy = 1$$

will express that the propositions X and Y are both true together; and the equation

$$xy = 0$$

that they are not both together true.

In like manner the equations

$$x(1 - y) + y(1 - x) = 1,$$
$$x(1 - y) + y(1 - x) = 0,$$

will respectively assert the truth and the falsehood of the disjunctive Proposition, "Either X is true or Y is true." The equations

$$y = vx$$
$$y = v(1 - x)$$

will respectively express the Propositions, "If the proposition Y is true, the proposition X is true." "If the proposition Y is true, the proposition X is false."

Examples will frequently present themselves, in the succeeding chapters of this work, of a case in which some terms of a particular member of an equation are affected by the indefinite symbol v, and others not so affected. The following instance will serve for illustration. Suppose that we have

$$y = xz + vx(1 - z).$$

Here it is implied that the time for which the proposition Y is true consists of all the time for which X and Z are together true, together with an indefinite portion of the time for which X is true and Z false. From this it may be seen, 1st, That if Y is true, either X and Z are together true, or X is true and Z false; 2ndly, If X and Z are together true, Y is true. The latter of these may be called the reverse interpretation, and it consists in taking the antecedent out of the second member, and the consequent from the first member of the equation. The existence of a term in the second member, whose coefficient is unity, renders this latter mode of interpretation possible. The general principle which it involves may be thus stated:

14. PRINCIPLE.—*Any constituent term or terms in a particular member of an equation which have for their coefficient unity, may be taken as the antecedent of a proposition, of which all the terms in the other member form the consequent.*

Thus the equation

$$y = xz + vx(1 - z) + (1 - x)(1 - z)$$

would have the following interpretations:

DIRECT INTERPRETATION.—*If the proposition Y is true, then either X and Z are true, or X is true and Z false, or X and Z are both false.*

REVERSE INTERPRETATION.—*If either X and Z are true, or X and Z are false, Y is true.*

The aggregate of these partial interpretations will express the whole significance of the equation given.

15. We may here call attention again to the remark, that although the idea of time appears to be an essential element in the theory of the interpretation of secondary propositions, it may practically be neglected as soon as the laws of expression and of interpretation are definitely established. The forms to which those laws give rise seem, indeed, to correspond with the forms of a perfect language. Let us imagine any known or existing language freed from idioms and divested of superfluity, and let us express in that language any given proposition in a manner the most simple and literal,—the most in accordance with those principles of pure and universal thought upon which all languages are founded, of which all bear the manifestation, but from which all have more or less departed. The transition from such a language to the notation of analysis would consist

of no more than the substitution of one set of signs for another, without essential change either of form or character. For the elements, whether things or propositions, among which relation is expressed, we should substitute letters; for the disjunctive conjunction we should write +; for the connecting copula or sign of relation, we should write =. This analogy I need not pursue. Its reality and completeness will be made more apparent from the study of those forms of expression which will present themselves in subsequent applications of the present theory, viewed in more immediate comparison with that imperfect yet noble instrument of thought—the English language.

16. Upon the general analogy between the theory of Primary and that of Secondary Propositions, I am desirous of adding a few remarks before dismissing the subject of the present chapter.

We might undoubtedly have established the theory of Primary Propositions upon the simple notion of space, in the same way as that of secondary propositions has been established upon the notion of time. Perhaps, had this been done, the analogy which we are contemplating would have been in somewhat closer accordance with the view of those who regard space and time as merely "forms of the human understanding," conditions of knowledge imposed by the very constitution of the mind upon all that is submitted to its apprehension. But this view, while on the one hand it is incapable of demonstration, on the other hand ties us down to the recognition of "place," τὸ ποῦ as an essential category of existence. The question, indeed, whether it is so or not, lies, I apprehend, beyond the reach of our faculties; but it may be, and I conceive has been, established, that the formal processes of reasoning in primary propositions do not require, as an essential condition, the manifestation in space of the things about which we reason; that they would remain applicable, with equal strictness of demonstration, to forms of existence, if such there be, which lie beyond the realm of sensible extension. It is a fact, perhaps, in some degree analogous to this, that we are able in many known examples in geometry and dynamics, to exhibit the formal analysis of problems founded upon some intellectual conception of space different from that which is presented to us by the senses, or which can be realized by the imagination.[*] I conceive, therefore, that the idea of space is not essential to the development of a theory of primary propositions, but am

---

*Space is presented to us in perception, as possessing the three dimensions of length, breadth, and depth. But in a large class of problems relating to the properties of curved surfaces, the rotations of solid bodies around axes, the vibrations of elastic media, &c., this limitation appears in the analytical investigation to be of an arbitrary character, and if attention were paid to the processes of solution alone, no reason could be discovered why space should not exist in four or in any greater number of dimensions. The intellectual procedure in the imaginary world thus suggested can be apprehended by the clearest light of analogy.

The existence of space in three dimensions, and the views thereupon of the religious and philosophical mind of antiquity, are thus set forth by Aristotle:—Μέγεθος δὲ τὸ μὲν ἐφ᾽ ἕν, γραμμὴ τὸ δ᾽ ἐπὶ δύο ἐπίπεδον, τὸ δ᾽ ἐπὶ τρία σῶμα· Καὶ παρὰ ταῦτα οὐκ ἔστιν ἄλλο μέγεθος, διὰ τὸ τριά πάντα εἶναι καὶ τὸ τρὶς πάντῃ. Καθάπερ γάρ φασι καὶ οἱ Πυθαγόρειοι, τὸ πᾶν καὶ τὰ πάντα τοῖς τρισὶν ὥρισται. Τελευτὴ γὰρ καὶ μέσον καὶ ᾽ ᾽ χὴ τὸν ἀριθμὸν ἔχει τὸν τοῦ παντός· ταῦτα δὲ τὸν τῆς τριάδος. Διὸ παρὰ τῆς φύσεως εἰληφότες ὥσπερ νόμους ἐκείνης, καὶ πρὸς τὰς ἁγιστείας χρώμεθα τῶν θεῶν τῷ ἀριθμῷ τούτῳ.—De Cælo, 1.

disposed, though desiring to speak with diffidence upon a question of such extreme difficulty, to think that the idea of time is essential to the establishment of a theory of secondary propositions. There seem to be grounds for thinking, that without any change in those faculties which are concerned in *reasoning*, the manifestation of space to the human mind might have been different from what it is, but not (at least the same) grounds for supposing that the manifestation of time could have been otherwise than we perceive it to be. Dismissing, however, these speculations as possibly not altogether free from presumption, let it be affirmed that the real ground upon which the symbol 1 represents in primary propositions the universe of things, and not the space they occupy, is, that the sign of identity = connecting the members of the corresponding equations, implies that the things which they represent are identical, not simply that they are found in the same portion of space. Let it in like manner be affirmed, that the reason why the symbol 1 in secondary propositions represents, not the universe of events, but the eternity in whose successive moments and periods they are evolved, is, that the same sign of identity connecting the logical members of the corresponding equations implies, not that the events which those members represent are identical, but that the times of their occurrence are the same. These reasons appear to me to be decisive of the immediate question of interpretation. In a former treatise on this subject (Mathematical Analysis of Logic), following the theory of Wallis respecting the Reduction of Hypothetical Propositions, I was led to interpret the symbol 1 in secondary propositions as the universe of "cases" or "conjunctures of circumstances;" but this view involves the necessity of a definition of what is meant by a "case," or "conjuncture of circumstances;" and it is certain, that whatever is involved in the term beyond the notion of time is alien to the objects, and restrictive of the processes, of formal Logic.

CHAPTER XII

OF THE METHODS AND PROCESSES TO BE ADOPTED IN THE TREATMENT OF SECONDARY PROPOSITIONS

1. It has appeared from previous researches (xi. 7) that the laws of combination of the literal symbols of Logic are the same, whether those symbols are employed in the expression of primary or in that of secondary propositions, the sole existing difference between the two cases being a difference of interpretation. It has also been established (v. 6), that whenever distinct systems of thought and interpretation are connected with the same system of formal laws, i. e., of laws relating to the combination and use of symbols, the attendant processes, intermediate between the expression of the primary conditions of a problem and the interpretation of its symbolical solution, are the same

in both. Hence, as between the systems of thought manifested in the two forms of primary and of secondary propositions, this community of formal law exists, the processes which have been established and illustrated in our discussion of the former class of propositions will, without any modification, be applicable to the latter.

2. Thus the laws of the two fundamental processes of elimination and development are the same in the system of secondary as in the system of primary propositions. Again, it has been seen (Chap. VI. Prop. 2) how, in primary propositions, the interpretation of any proposed equation devoid of fractional forms may be effected by developing it into a series of constituents, and equating to 0 every constituent whose coefficient does not vanish. To the equations of secondary propositions the same method is applicable, and the interpreted result to which it finally conducts us is, as in the former case (vi. 6), a system of co-existent denials. But while in the former case the force of those denials is expended upon the existence of certain classes of things, in the latter it relates to the truth of certain combinations of the elementary propositions involved in the *terms* of the given premises. And as in primary propositions it was seen that the system of denials admitted of conversion into various other forms of propositions (vi. 7), &c., such conversion will be found to be possible here also, the sole difference consisting not in the forms of the equations, but in the nature of their interpretation.

3. Moreover, as in primary propositions, we can find the expression of any element entering into a system of equations, in terms of the remaining elements (vi. 10), or of any selected number of the remaining elements, and interpret that expression into a logical inference, the same object can be accomplished by the same means, difference of interpretation alone excepted, in the system of secondary propositions. The elimination of those elements which we desire to banish from the final solution, the reduction of the system to a single equation, the algebraic solution and the mode of its development into an interpretable form, differ in no respect from the corresponding steps in the discussion of primary propositions.

To remove, however, any possible difficulty, it may be desirable to collect under a general Rule the different cases which present themselves in the treatment of secondary propositions.

RULE.—*Express symbolically the given propositions* (xi. 11).

Eliminate separately from each equation in which it is found the indefinite symbol v (vii. 5).

Eliminate the remaining symbols which it is desired to banish from the final solution: always before elimination reducing to a single equation those equations in which the symbol or symbols to be eliminated are found (viii. 7). *Collect the resulting equations into a single equation* $V = 0$.

Then proceed according to the particular form in which it is desired to express the final relation, as—

1st. *If in the form of a denial, or system of denials, develop the function V, and equate to 0 all those constituents whose coefficients do not vanish.*

2ndly. *If in the form of a disjunctive proposition, equate to 1 the sum of those constituents whose coefficients vanish.*

3rdly. *If in the form of a conditional proposition having a simple element, as x or $1 - x$, for its antecedent, determine the algebraic expression of that element, and develop that expression.*

4thly. *If in the form of a conditional proposition having a compound expression, as xy, $xy + (1 - x)(1 - y)$, &c., for its antecedent, equate that expression to a new symbol t, and determine t as a developed function of the symbols which are to appear in the consequent, either by ordinary methods or by the special method* (ix. 9).

5thly. *Interpret the results by* (xi. 13, 14).

If it only be desired to ascertain whether a particular elementary proposition x is true or false, we must eliminate all the symbols but x; then the equation $x = 1$ will indicate that the proposition is true, $x = 0$ that it is false, $0 = 0$ that the premises are insufficient to determine whether it is true or false.

4. Ex.1.—The following prediction is made the subject of a curious discussion in Cicero's fragmentary treatise, De Fato:—"Si quis (Fabius) natus est oriente Canicula, is in mari non morietur." I shall apply to it the method of this chapter. Let y represent the proposition, "Fabius was born at the rising of the dogstar;" x the proposition, "Fabius will die in the sea." In saying that x *represents* the proposition, "Fabius, &c.," it is only meant that x is a symbol so appropriated (xi. 7) to the above proposition, that the equation $x = 1$ declares, and the equation $x = 0$ denies, the truth of that proposition. The equation we have to discuss will be

$$y = v(1 - x). \tag{1}$$

And, first, let it be required to reduce the given proposition to a negation or system of negations (xii. 3). We have, on transposition,

$$y - v(1 - x) = 0.$$

Eliminating v,

$$y\{y - (1 - x)\} = 0,$$

or,
$$y - y(1 - x) = 0,$$
or,
$$yx = 0. \tag{2}$$

The interpretation of this result is:—"It is not true that Fabius was born at the rising of the dogstar, and will die in the sea."

Cicero terms this form of proposition, "Conjunctio ex repugnantibus;" and he remarks that Chrysippus thought in this way to evade the difficulty which he imagined to exist in contingent assertions respecting the future: "Hoc loco Chrysippus æstuans falli sperat Chaldæos cæterosque divinos, neque eos usuros esse conjunctionibus ut ita sua percepta pronuntient: Si quis natus est oriente Caniculâ is in mari non morietur; sed potius ita dicant: Non et natus est quis oriente Caniculâ, et in mari morietur. O licentiam jocularem! . . . . Multa genera sunt enuntiandi, nec ullum distortius quam hoc quo Chrysippus sperat Chaldæos contentos Stoicorum causâ fore."—*Cic. De Fato*, 7, 8.

5. To reduce the given proposition to a disjunctive form.

The constituents not entering into the first member of (2) are

$$x(1 - y), (1 - x)y, (1 - x)(1 - y).$$

Whence we have

$$y(1 - x) + x(1 - y) + (1 - x)(1 - y) = 1. \tag{3}$$

The interpretation of which is:—*Either Fabius was born at the rising of the dogstar, and will not perish in the sea; or he was not born at the rising of the dogstar, and will perish in the sea; or he was not born at the rising of the dogstar, and will not perish in the sea.*

In cases like the above, however, in which there exist constituents differing from each other only by a single factor, it is, as we have seen (vii. 15), most convenient to collect such constituents into a single term. If we thus connect the first and third terms of (3), we have

$$(1 - y)x + 1 - x = 1;$$

and if we similarly connect the second and third, we have

$$y(1 - x) + 1 - y = 1.$$

These forms of the equation severally give the interpretations—

Either Fabius was not born under the dogstar, and will die in the sea, or he will not die in the sea.

Either Fabius was born under the dogstar, and will not die in the sea, or he was not born under the dogstar.

It is evident that these interpretations are strictly equivalent to the former one.

Let us ascertain, in the form of a conditional proposition, the consequences which flow from the hypothesis, that "Fabius will perish in the sea."

In the equation (2), which expresses the result of the elimination of v from the original equation, we must seek to determine x as a function of y.

We have

$$x = \frac{0}{y} = 0 \quad y + \frac{0}{0}(1 - y) \text{ on expansion,}$$

or,

$$x = \frac{0}{0}(1 - y);$$

the interpretation of which is,—*If Fabius shall die in the sea, he was not born at the rising of the dogstar.*

These examples serve in some measure to illustrate the connexion which has been established in the previous sections between primary and secondary propositions, a connexion of which the two distinguishing features are identity of process and analogy of interpretation.

6. Ex. 2.—There is a remarkable argument in the second book of the Republic of Plato, the design of which is to prove the immutability of the Divine Nature. It is a very fine example both of the careful induction from familiar instances by which Plato arrives at general principles, and of the clear and connected logic by which he deduces from them the particular inferences which it is his object to establish. The argument is contained in the following dialogue:

"Must not that which departs from its proper form be changed either by itself or by another thing? Necessarily so. Are not things which are in the best state least changed and disturbed, as the body by meats and drinks, and labours, and every species of plant by heats and winds, and such like affections? Is not the healthiest and strongest the least changed? Assuredly. And does not any trouble from without least disturb and change that soul which is strongest and wisest? And as to all made vessels, and furnitures, and garments, according to the same principle, are not those which are well wrought, and in a good condition, least changed by time and other accidents? Even so. And whatever is in a right state, either by nature or by art, or by both these, admits of the smallest change from any other thing. So it seems. But God and things divine are in every sense in the best state. Assuredly. In this way, then, God should least of all bear many forms? Least, indeed, of all. Again, should He transform and change Himself? Manifestly He must do so, if He is changed at all. Changes He then Himself to that which is more good and fair, or to that which is worse and baser? Necessarily to the worse, if he be changed. For never shall we say that God is indigent of beauty or of virtue. You speak most rightly, said I, and the matter being so, seems it to you, O Adimantus, that God or man *willingly* makes himself in any sense worse? Impossible, said he. Impossible, then, it is, said I, that a god should wish to change himself; but ever being fairest and best, each of them ever remains absolutely in the same form."

The premises of the above argument are the following:

1st. If the Deity suffers change, He is changed either by Himself or by another.

2nd. If He is in the best state, He is not changed by another.

3rd. The Deity is in the best state.

4th. If the Deity is changed by Himself, He is changed to a worse state.

5th. If He acts willingly, He is not changed to a worse state.

6th. The Deity acts willingly.

Let us express the elements of these premises as follows:

Let x represent the proposition, "The Deity suffers change."

    y, He is changed by Himself.

    z, He is changed by another.

    s, He is in the best state.

    t, He is changed to a worse state.

    w, He acts willingly.

Then the premises expressed in symbolical language yield, after elimination of the indefinite class symbols v, the following equations:

$$xyz + x(1 - y)(1 - z) = 0, \tag{1}$$
$$sz = 0, \tag{2}$$
$$s = 1, \tag{3}$$
$$y(1 - t) = 0, \tag{4}$$
$$wt = 0, \tag{5}$$
$$w = 1. \tag{6}$$

Retaining x, I shall eliminate in succession z, s, y, t, and w (this being the order in which those symbols occur in the above system), and interpret the successive results.

Eliminating z from (1) and (2), we get

$$xs(1 - y) = 0. \tag{7}$$

Eliminating s from (3) and (7),

$$x(1 - y) = 0. \tag{8}$$

Eliminating y from (4) and (8),

$$x(1 - t) = 0. \tag{9}$$

Eliminating t from (5) and (9),

$$xw = 0. \tag{10}$$

Eliminating w from (6) and (10),

$$x = 0. \tag{11}$$

These equations, beginning with (8), give the following results:

From (8) we have $x = \frac{0}{0}y$, therefore, *If the Deity suffers change, He is changed by Himself.*

From (9), $x = \frac{0}{0}t$, *If the Deity suffers change, He is changed to a worse state.*

From (10), $x = \frac{0}{0}(1 - w)$. *If the Deity suffers change, He does not act willingly.*

From (11), *The Deity does not suffer change.* This is Plato's result.

Now I have before remarked, that the order of elimination is indifferent. Let us in the present case seek to verify this fact by eliminating the same symbols in a reverse order, beginning with w. The resulting equations are,

$$t = 0, \quad y = 0, \quad x(1 - z) = 0, \quad z = 0, \quad x = 0;$$

yielding the following interpretations:

God is not changed to a worse state.
He is not changed by Himself.
If He suffers change, He is changed by another.
He is not changed by another.
He is not changed.

We thus reach by a different route the same conclusion.

Though as an exhibition of the *power* of the method, the above examples are of slight value, they serve as well as more complicated instances would do, to illustrate its nature and character.

7. It may be remarked, as a final instance of analogy between the system of primary and that of secondary propositions, that in the latter system also the fundamental equation,

$$x(1 - x) = 0,$$

admits of interpretation. It expresses the axiom, *A proposition cannot at the same time be true and false.* Let this be compared with the corresponding interpretation (iii. 15). Solved under the form

$$x = \frac{0}{1 - x} = \frac{0}{0}x,$$

by development, it furnishes the respective axioms: "A thing is what it is:" "If a proposition is true, it is true:" forms of what has been termed "The principle of identity." Upon the nature and the value of these axioms the most opposite opinions have been entertained. Some have regarded them as the very pith and marrow of philosophy. Locke devoted to them a chapter, headed, "On Trifling Propositions."[1] In both these views there seems to have been a mixture of truth and error. Regarded as supplanting experience, or as furnishing materials for the vain and wordy janglings of the schools, such propositions are worse than trifling. Viewed, on the other hand, as intimately allied with the very laws and conditions of thought, they rise into at least a speculative importance.

[1] Essay on the Human Understanding, Book IV. Chap. VIII.

Georg Friedrich Bernhard Riemann

(1826–1866)

HIS LIFE AND WORK

Boston and Chicago are two great seats of mathematical research located in major American cities. Until they won in 2004, if you asked a baseball fan in Boston what they most hoped to see in their lifetime, they would have answered a World Series win for the Boston Red Sox. Chicago Cub fans are still waiting. Ask a mathematician in either of these cities or anywhere else in the world what they would most hope to see in their lifetime, and they would most likely answer: "A proof of the Riemann hypothesis!" Perhaps mathematicians, like Red Sox fans, will have their prayers answered in our lifetimes, or at least before the Cubs win the World Series.

Bernhard Riemann's life spanned just under forty years, a life span that produced only a single volume of collected works. Some of these works were not published until after Riemann's early death in 1866. In spite of, or perhaps because of, the slimness of his collected works, everything he touched revolutionized mathematics.

Georg Friedrich Bernhard Riemann was born on September 17, 1826, in the town of Breselenz in northern Germany, the second of six children born to

Friedrich Bernhard Riemann and his wife, the former Charlotte Ebell. A year later, the family moved to the nearby town of Quickborn. Nothing in his family foreshadowed Riemann's mathematical genius. His father came from a long line of Lutheran ministers. His maternal grandfather was a counselor at the Hanoverian court. Nevertheless, Riemann's genius soon became readily apparent to all, in spite of a lifelong shyness that made it hard for him to engage with people. Riemann so excelled at the local school in Quickborn that the school's administration provided him with an individual tutor to teach him advanced arithmetic and geometry. Before long, the teacher realized that he was learning more from Riemann's sophisticated solutions than he was teaching Riemann.

Mindful of his son's talents, Riemann's father insisted that young Bernhard enroll at the prestigious Gymnasium in Hanover when he turned fourteen. Being away from his parents for the first time was a great hardship for Riemann, who soon wrote home of his overwhelming loneliness. Only the fact that he was able to live with his maternal grandmother made it possible for the painfully shy boy to bear the separation from his parents. When Riemann's grandmother died two years later, his parents reluctantly allowed him transfer to the Gymnasium in nearby Löneberg to complete his secondary education.

The Gymnasium at Lüneberg proved to be a most fortunate choice for Riemann. Recognizing the new student's gifts, Lüneberg's director allowed Riemann access to his private library, which contained many advanced mathematical works. When Riemann asked him to recommend a book that was not too easy, the director suggested Legendre's massive *Theory of Numbers*. A week later when Riemann returned the book, the director asked him if it had been too much of a challenge to which Riemann had responded that he was glad to have been given a book that took him a whole week to master. All 859 pages of it! Legendre's *Theory of Numbers* undoubtedly provided Riemann his first introduction to the study of the distribution of prime numbers, the subject of the third selection included here. Two years later, Riemann asked that he be examined on Legendre's book as part of his graduation examination. Not surprisingly, he answered every question perfectly, in spite of not having looked at the work in two years!

Ever dutiful to his parents, Riemann enrolled at the University of Göttingen planning to take a degree in theology and philosophy so that he could follow his father into the ministry. Much as he tried to satisfy his family's wishes, Riemann could not resist the attraction of Göttingen's one and only star mathematician: Carl Friedrich Gauss. Riemann was enthralled by Gauss's lectures on the method of least squares and decided to make mathematics his career. By this time, Gauss was past seventy and ailing. He suggested that Riemann transfer to the University of Berlin where he could study under a new generation of German mathematicians: Steiner, Jacobin, Eisenstein, and especially Dirichlet. Riemann's concern about Berlin's distance from Quickborn was partially relieved when his parents provided their blessing that he follow Gauss's advice.

After receiving his baccalaureate degree in 1849, Riemann returned to Göttingen to pursue a doctorate under Gauss. In November 1851, Riemann submitted to Gauss his doctoral thesis, entitled *Foundations of a General Theory of Functions of a Complex Variable.* Near to the end of his career at Göttingen, Gauss realized that he had finally found a worthy successor and lavished praise on Riemann's dissertation:

> *The dissertation submitted by Herr Riemann offers convincing evidence of the author's thorough and penetrating investigations in those parts of the subject treated in the dissertation, of a creative, active, truly mathematical mind, and of a gloriously fertile originality. The presentation is perspicuous and concise and, in places, beautiful. The majority of readers would have preferred a greater clarity of arrangement. The whole is a substantial, valuable work, which not only satisfies the standards demanded for doctoral dissertations, but far exceeds them.*

A month later, Riemann passed his dissertation defense with flying colors.

Because of the few teaching jobs available in German universities, freshly minted PhD's had to take unpaid positions and then write a second dissertation, called an *Habilitation* paper, and then give an *Habilitation* lecture to demonstrate their ability to hold a paid teaching post. For his *Habilitation* paper, Riemann chose to write *On the representability of a function by means of a trigonometric series*, to respond to his teacher Dirichlet's query about the meaning of the integrals in Fourier's formulae. Riemann wrote this paper, the first selection presented here, while working as an unpaid assistant to Wilhelm Weber's seminar in mathematical physics.

For Leibniz, the definite integral $\int y(x)dx$ was the sum of infinitely many small summands, $\Sigma y_i(x)\Delta_i(x)$.

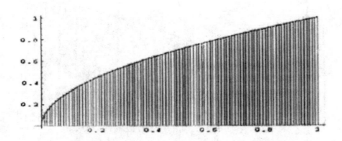

However, when the integral was discovered to be the inverse of the tangent, mathematicians shunned the computation of sums in favor of taking the difference of the indefinite integral. Fourier's work on the representation of functions by trigonometric series forced mathematicians to deal with integrals of functions without obvious indefinite integrals. It forced them to find a method of integration when indefinite integral primitive functions could not be found in general. Following Cauchy's approach to integration, mathematicians could easily establish the integrability of piecewise continuous functions which Dirichlet did in 1829. Dirichlet left the task of devising a method to integrate a wider class of functions to his successors, and Riemann was the one who found a solution. Cauchy had taken the value of the function at one of the two endpoints of each subinterval when evaluating Cauchy sums for convergence to the definite integral.

Cauchy Integral (evaluated at left endpoints)

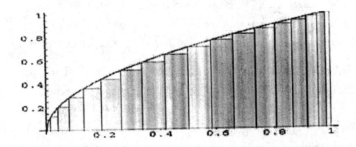

In contrast, Riemann required that *any* value in each subinterval could be chosen to evaluate the Riemann sums that must converge to a limit as the norm of the partition decreases to 0.

Riemann Integral (evaluated at arbitrary points)

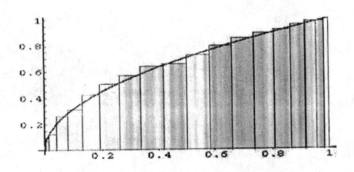

The existence of the Cauchy integral for a function $f(\text{x})$ in the interval $[a, b]$ had depended on the **continuity** of $f(x)$ in the interval $[a, b]$. The existence of the Riemann integral for a function $f(x)$ depended on a related but radically different concept: **variation**. Riemann, like Cauchy, begins his theory of the integral by partitioning the interval $[a, b]$ with the set of points $x_1, x_2, x_3, \ldots x_{n-1}, x_n$, such that $a = x_0 < x_1 < x_2 < x_3 < \ldots < x_{n-1} < x_n = b$. He then considers the variation of the function $f(x)$ in each subinterval $[x_{k-1}, x_k]$, that is, the maximum value of the difference of the value of the function in the subinterval $[x_{k-1}, x_k]$

$$D_k = \max|f(s) - f(t)|$$

for any s and t, such that $x_{k-1} \leq s, t \leq x_k$.

The variation of the function $f(x)$ in the interval $[a, b]$ is the sum of the products of each of the D_k times the corresponding width of the subinterval $[x_k, x_{k-1}]$; that is,

$$\Sigma D_k\, (x_k - x_{k-1})$$

Riemann recognized that this sum had to approach 0 as the norm of the partition; that is, the greatest value of $(x_k - x_{k-1})$, approached 0. Informally, this can be described as being continuous, except in intervals of arbitrarily small size. Thus, for example, the Dirichlet function, $D(x)$, which takes the following values:

$$D(x) = \tfrac{1}{b} \text{ if } x \text{ is rational and can be expressed in lowest terms}$$
$$\text{as the fraction } \tfrac{a}{b}$$
$$= 0 \text{ if } x \text{ is irrational}$$

can be integrated, because the values of $D(x)$ at rational values of x are all less than 1, and the rational values of x can be enumerated and the nth such value enclosed in an interval of size $\tfrac{\varepsilon}{2^n}$ so that they sum to ε, a value chosen to be arbitrarily small.

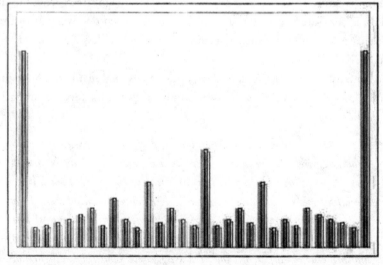

Small intervals about rational values of $x = \tfrac{a}{b}$ with $D(x) = \tfrac{1}{b}$.

Riemann never bothered to publish this paper during his lifetime. His friend and colleague Richard Dedekind published it in 1868, two years after Riemann's death. In these few pages, Riemann provided the impetus for the vigorous development of theories of measure and the integral for fifty years culminating with Lebesgue's theory of measure and the integral in 1904, presented later in this volume.

With his *Habilitation* paper done, Riemann had one final hurdle to clear before he could take up a paid teaching post, his *Habilitation* lecture. Gauss asked Riemann to suggest three topics for this lecture. Riemann responded with the following topics:

- *The history of the question of the representability of a function as a trigonometric series*
- *Solutions of two quadratic equations in two unknowns*
- *On the hypotheses that underlie the foundations of geometry*

expecting that Gauss would choose the first topic and allowing him to present the results of his *Habilitation* paper. To Riemann's great surprise, Gauss chose the third topic. Dedekind thought that it was because Gauss was curious to hear how such a young person would handle so difficult a topic. After overcoming his initial surprise, Riemann realized that he now had the opportunity to write an essay that could be intelligible to a wider audience at the university and strove to make it so.

Riemann begins his analysis by distinguishing discrete manifolds from continuous manifolds. He notes that for discrete manifolds, such as the natural numbers, comparison is done by counting. In contrast, for continuous manifolds such as space, comparison is done by measuring. Measuring is done by superposition using one chosen magnitude as a standard. Until a standard magnitude is chosen, the most we can do is compare two magnitudes to each other when one is a part of the other in which case we can tell which is greater and which is less qualitatively but not quantitatively.

Riemann asks the reader to consider a one-dimensional manifold as one in which continuous progress from a point is possible only on two sides, forwards or backwards. Riemann then suggests

> If one now supposes that this manifoldness in its turn passes over into another entirely different, and again in a definite way, namely so that each point passes over into a definite point of the other, then all the specialisations so obtained form a doubly extended manifoldness. In a similar manner one obtains a triply extended manifoldness, if one imagines a doubly extended one passing over in a definite way to another entirely different; and it is easy to see how this construction

may be continued. If one regards the variable object instead of the determinable notion of it, this construction may be described as a composition of a variability of **n + 1** *dimensions out of a variability of n dimensions and a variability of one dimension.*

Riemann explains that in a similar fashion fixing the value of one dimension of an *n*-dimensional manifold results in a manifold of **n − 1** dimensions. And then Riemann makes his critical move.

Having constructed the notion of a manifoldness of n dimensions, and found that its true character consists in the property that the determination of position in it may be reduced to n determinations of magnitude, we come to the second of the problems proposed above, viz. the study of the measure-relations of which such a manifoldness is capable, and of the conditions which suffice to determine them. These measure-relations can only be studied in abstract notions of quantity, and their dependence on one another can only be represented by formulae.

The measure of interest to Riemann is, of course, the distance between two points, which is, equivalently, the length of the line between them.

Position-fixing being reduced to quantity-fixings, and the position of a point in the n-dimensioned manifoldness being consequently expressed by means of n variables $x_1, x_2, x_3, \ldots, x_n$, the determination of a line comes to the giving of these quantities as functions of one variable. The problem consists then in establishing a mathematical expression for the length of a line, and to this end we must consider the quantities x as expressible in terms of certain units.

Riemann considers a very general class of metrics for which the distance, *ds*, between points close together, say $<x_1, x_2, x_3, \ldots, x_n>$ and $<x_1 + \Delta x_1, x_2 + \Delta x_2, x_3 + \Delta x_3, \ldots, x_n + \Delta x_n>$ can be expressed as

$$\left(\sum_{i,j=1}^{n} G_{ij} \Delta x_i \Delta x_j \right)^{\frac{1}{2}}$$

and the sum (*i.e.*, the integral) of the *ds* gives the length of the line between two arbitrary points. Riemann notes that the Euclidean distance from the origin to the point defined by the *n* variables $(x_1, x_2, x_3, \ldots, x_n)$, namely, the square root of the sum of the squares of the variables.

$$\left((x_1)^2 + (x_2)^2 + (x_3)^2 + \ldots + (x_n)^2 \right)^{\frac{1}{2}}$$

is just one of many particular forms of manifoldness ($G_{ij} = 1$ if $i = j$, $G_{ij} = 0$ otherwise), the one for which space is *flat*!

Riemann recognized that in spaces on nonconstant curvature bodies may move about without stretching. He also demonstrated that if the sum of the angles of a triangle is always greater than two right angles, then the space must be a space of positive curvature, if the sum is always less than two right angles then the space must be a space of negative curvature, and if the sum is always identical to two right angles then the space must be everywhere flat. Of course, space need not be flat, it need not even be of constant curvature as it must be for the sum of the angles of a triangle to be invariant.

Aware of Gauss's futile attempts to survey triangles whose angles added up to a value different than two right angles, Riemann wrote:

> If we suppose that bodies exist independently of position, the curvature is everywhere constant, and it then results from astronomical measurements that it cannot be different from zero; or at any rate its reciprocal must be an area in comparison with which the range of our telescopes may be neglected. But if this independence of bodies from position does not exist, we cannot draw conclusions from metric relations of the great, to those of the infinitely small; in that case the curvature at each point may have an arbitrary value in three directions, provided that the total curvature of every measurable portion of space does not differ sensibly from zero. Still more complicated relations may exist if we no longer suppose the linear element expressible as the square root of a quadric differential. Now it seems that the empirical notions on which the metrical determinations of space are founded, the notion of a solid body and of a ray of light, cease to be valid for the infinitely small. We are therefore quite at liberty to suppose that the metric relations of space in the infinitely small do not conform to the hypotheses of geometry; and we ought in fact to suppose it, if we can thereby obtain a simpler explanation of phenomena.
>
> The question of the validity of the hypotheses of geometry in the infinitely small is bound up with the question of the ground of the metric relations of space.

Riemann had begun his lecture by noting that the physical space of our experience is one of several forms of three-dimensional space. The metrical properties of our three-dimensional space, he says, must be derived from experience, they cannot be derived solely from the axioms and postulates of geometry. Riemann closed his lecture by stating

> *Either therefore the reality which underlies space must form a discrete manifoldness, or we must seek the ground of its metric relations outside it, in binding forces which act upon it.*
>
> *The answer to these questions can only be got by starting from the conception of phenomena which has hitherto been justified by experience, and which Newton assumed as a foundation, and by making in this conception the successive changes required by facts which it cannot explain. Researches starting from general notions, like the investigation we have just made, can only be useful in preventing this work from being hampered by too narrow views, and progress in knowledge of the interdependence of things from being checked by traditional prejudices. This leads us into the domain of another science, of physic, into which the object of this work does not allow us to go to day.*

Little did Riemann know that fifty years after his death, this work would provide the mathematical foundations on which Einstein constructed the General Theory of Relativity.

Riemann's lecture, the second selection presented in this volume, must have met with success because Gauss and Weber each responded with great praise. Having passed the final hurdle, Riemann finally began his teaching career in the fall of 1854 giving a lecture series on partial differential equations to a grand total of eight students, more than double the number he had expected. However, this was hardly a way to make a living. When Riemann's father died in 1855, his three surviving sisters had to seek financial support from their other brother, a postal clerk, who earned far more than the economically destitute mathematician. When Gauss died in the same year, Riemann's friends knew that Gauss's chair would go to Riemann's former Berlin professor P. G. L. Dirichlet. They tried to persuade the university to appoint Riemann to an assistant professorship but that effort came to naught.

The loss of his father and the mounting financial strain proved to be too much for Riemann's weak disposition and led to a nervous breakdown in late 1855. To relieve the strain, Riemann retreated to the mountains where he refreshed himself by taking long hikes with his colleague Richard Dedekind. Good fortune seemed to strike Riemann in 1857 when the University offered him an assistant professorship. In 1858, the Italian mathematicians Enrico Betti, Felice Casorati, and Francesco Brioschi visited Riemann in Göttingen to discuss his mathematical works.

As soon as Riemann began to feel financially secure, his brother died leaving him to support his three unmarried sisters. Whether coincidence or not, Riemann first began to suffer from consumption during this most troubling of financial circumstances. Within a year, Riemann's burdens eased somewhat with the premature death of his sister Marie and then a year later in 1859, Riemann finally achieved the professional success of which he was so worthy when he succeeded Dirichlet to the chair of mathematics at Göttingen, following his former teacher's untimely death.

The year 1859 also marked Riemann's election to the Berlin Academy of Science. To commemorate his appointment, Riemann wrote his only paper on number theory, entitled *On the Number of Prime Numbers Less Than a Given Magnitude*. Undoubtedly his interest in the subject grew out of his reading Legendre's *Theory of Numbers* while a student at the Gymnasium in Lüneberg fifteen years earlier.

Legendre and Gauss had each conjectured that $\pi(x)$, the function counting all the primes less than x, asymptotically approached $Li(x)$, (*i.e.*, the ratio $\pi(x)/Li(x) \rightarrow 1$), where

$$Li(n) = \int_2^n \frac{dx}{ln\ x}$$

as x grows large. In this epochal paper, which I consider mathematical virtuosity of the highest order, Riemann introduced radically new methods to attack this problem. He begins by introducing the infinite series of inverse s powers

$$\left(\frac{1}{1^s}\right) + \left(\frac{1}{2^s}\right) + \left(\frac{1}{3^s}\right) + \left(\frac{1}{4^s}\right) + \ldots + \left(\frac{1}{n^s}\right) + \ldots$$

a series first studied by Euler in the mid-eighteenth century who proved that for $s = 2$ the series sums to the limit $\frac{\pi^2}{6}$. Riemann names this the zeta function after the Greek letter ζ. More formally, the zeta function is presented as

$$\zeta(s) = \sum_{n=1}^{\infty} \frac{1}{n^s}$$

At first glance, it appears far from obvious how an infinite series involving all of the positive integers can relate to the primes. However, Euler had provided the connection by first noticing that thanks to the unique factorization of the integers, the infinite sum of inverse s powers of the positive integers

$$\left(\frac{1}{1^s}\right) + \left(\frac{1}{2^s}\right) + \left(\frac{1}{3^s}\right) + \left(\frac{1}{4^s}\right) + \ldots + \left(\frac{1}{n^s}\right) + \ldots$$

can be rewritten as the infinite products of the infinite geometric series for each prime number p

$$\left\{\left(1 + \left(\frac{1}{2^1}\right) + \left(\frac{1}{2^2}\right) + \left(\frac{1}{2^3}\right) + \ldots + \left(\frac{1}{2^k}\right) + \ldots\right\}\right.$$

$*$

$$\left\{\left(1 + \left(\frac{1}{3^1}\right) + \left(\frac{1}{3^2}\right) + \left(\frac{1}{3^3}\right) + \ldots + \left(\frac{1}{3^k}\right) + \ldots\right\}\right.$$

$*$

$$\left\{\left(1 + \left(\frac{1}{5^1}\right) + \left(\frac{1}{5^2}\right) + \left(\frac{1}{5^3}\right) + \ldots + \left(\frac{1}{5^k}\right) + \ldots\right\}\right.$$

$*$

$\ldots$

$*$

$$\left\{\left(1 + \left(\frac{1}{p^1}\right) + \left(\frac{1}{p^2}\right) + \left(\frac{1}{p^3}\right) + \ldots + \left(\frac{1}{p^k}\right) + \ldots\right\}\right.$$

$*$

$\ldots$

and then applying a basic property of infinite geometric series that

$$1 + \left(\frac{1}{p^s}\right) + \left(\frac{1}{p^{2s}}\right) + \left(\frac{1}{p^{3s}}\right) + \ldots + \left(\frac{1}{p^{ks}}\right) + \ldots$$

sums to $\frac{1}{(1 - p^{-s})}$ to restate the zeta function as the infinite product

$$\zeta(s) = \prod_p \frac{1}{1 - p^{-s}}$$

These are just the initial steps in this mathematical *tour de force*. As defined by these infinite sum and product representations, the zeta function only converges for complex numbers with real parts greater than 1. Drawing on breakthrough results from his doctoral dissertation, he extends the zeta function to all complex numbers having an infinte singularity only for $s = 1$.

During the course of the paper, Riemann makes the casual remark that

> It is very probable that all roots [*of a related function* $\xi(x)$] *are real. Certainly one would wish for a stricter proof here; Meanwhile, I have temporarily put aside the search for this after some fleeting futile attempts.*

(A value x is a root of the function $f(x)$ if $f(x) = 0$.) A root of the function $\xi(x)$ is real if and only if a root of the zeta function is a complex number with real part equal to $\frac{1}{2}$.) Making this assumption about the roots of the zeta function, Riemann concludes:

> *The well-known approximation formula* $F(x) = Li(x)$ [$F(x)$ *is also called* $\pi(x)$, *is the function that counts the primes less than* x] *is thus correct only up to values of the order of magnitudes* $x^{\frac{1}{2}}$.

The values of the roots of the zeta function determine the magnitude of the difference between $\pi(x)$ and $Li(x)$. The hypothesis that all the roots of the zeta function have real part equal to $\frac{1}{2}$ has come to be known as the *Riemann hypothesis*.

After a few years, Riemann's prospects brightened briefly, and in June 1862 he married Elise Koch, a friend of his sisters. However, within a month of his marriage, Riemann once again fell ill with pleurisy followed by consumption. Influential friends persuaded the university authorities to allow Riemann the funds to travel to Italy for the winter to convalesce. Thinking himself recovered, Riemann left Italy in the spring and foolishly walked in deep snow while traveling through the alpine passes in Switzerland. Following a relapse, he returned to Italy once again in August 1863 stopping first in Pisa where his wife gave birth to their one and only child, Ida. While in Pisa, Riemann occasionally attended lectures at its University which soon offered him a professorship at the urging of Betti and Casorati, members of its mathematics faculty, and Brisochi, Minister of Education. Although the terms of his contract at Göttingen prevented him from accepting the offer, the German university allowed him to make an extended stay in Pisa in the hope that he would regain his health. Unfortunately, that was not to be. Riemann died on July 20, 1866, in the village of Selasco on the shores of Lake Maggiore, two months short of his fortieth birthday.

For several decades after the publication of Riemann's paper on the distribution of primes, mathematicians wondered how he arrived at his conjecture that all of the zeroes of the zeta function have a real part of $\frac{1}{2}$. The paper not only gives no clue how Riemann arrived at this conjecture, but also doesn't even provide grounds for knowing that any zero of the zeta function has a real part of $\frac{1}{2}$. Some mathematicians simply supposed that Riemann possessed some wonderful insights.

In fact, Riemann gained his knowledge of the behavior of the zeta function with great computational effort. When he died, his wife rescued most of his private papers from an overzealous housemaid who had begun to burn them in the furnace. Having saved them from destruction, Elise Riemann kept them locked away for the rest of her life. They only became public in the 1920s, and when they did, they provided a treasure trove to their editor, the mathematician C. L. Siegel. Siegel saw that Riemann had discovered several high-powered computational techniques that other mathematicians had rediscovered over the intervening sixty years. Ever the perfectionist, Riemann left unpublished these methods because he lacked proofs to demonstrate their effectiveness. We can only imagine what Riemann might have done had he been fortunate enough to enjoy two or three more decades of life.

For thirty years following Riemann's death, virtually no progress was made on the problem of Prime Number distribution. Then in the 1890s, Hadamard, von

Mangoldt, and de la Vallée-Poussin exploited ideas in Riemann's paper proving his main formula for $\pi(x)$ and the Prime Number Theorem that

$$\pi(x) \sim Li(x),$$

where

$$Li(n) = \int_2^n \frac{dx}{ln\ x}$$

originally conjectured by Gauss and Legendre a century earlier.

David Hilbert made the Riemann hypothesis the eighth of his twenty-three problems presented to the International Congress of Mathematicians in 1900. Hilbert confidently expected the Riemann hypothesis to be proven in a decade or so. Hilbert revised this estimate as the Riemann hypothesis resisted all attempts to crack it. Shortly before his death in 1943, someone asked Hilbert what his first question would be if he were revived in 500 years. The great mathematician immediately responded, I would ask: "Has someone proved the Riemann Hypothesis?"

A century later it remains unproven, and there is now a $1,000,000 prize for whoever proves or disproves the hypothesis. No one expects it to be disproved. Mathematicians have proven that at least 40% of the solutions must have a real part equal to $\frac{1}{2}$; indeed, the first *one and a half billion* known solutions all have real part equal to $\frac{1}{2}$. Yet it remains unproven.

Riemann died at far too early an age. We can only ask whether he would have rigorously proved his eponymous hypothesis had he been granted the biblical life span of three score and ten years. Perhaps we will live long enough to see Riemann's hypothesis proved. If we are fortunate for that to happen, we shouldn't be surprised if the roots of that proof lie in Riemann's epochal paper itself.

ON THE REPRESENTABILITY OF A FUNCTION BY MEANS OF A TRIGONOMETRIC SERIES*

(from: Essays of the Royal Society of Sciences (Göttingen), Volume Thirteen.)[1]

XII

The following essay on trigonometric series consists of two essentially different parts. The first part contains a history of research and opinions on arbitrary functions (graphically displayed) and the ability to represent them through trigonometric series. In composing it, I was able to make use of a few hints of the famous mathematician to whom we owe the first in-depth work on this subject. In the second part, I present an investigation of whether functions can be represented through trigonometric series, which includes the cases thus far unsolved. It was necessary to preface it with a short essay on the concept of a definite integral and the scope of its validity.

Whether a Given Arbitrary Function can be Represented by a Trigonometric Series: A History of the Question

1.

Trigonometric series, so called by Fourier, *i.e.*, series of the form

$$a_1 \sin x + a_2 \sin 2x + a_3 \sin 3x + \ldots$$
$$+ 1/2\, b_0 + b_1 \cos x + b_2 \cos 2x + b_3 \cos 3x \ldots$$

play a significant role in that branch of Mathematics where completely arbitrary functions occur; indeed, there is reason to assert that the most significant advances in this branch of Mathematics, so important to Physics, have depended on obtaining clearer insight into the nature of these series. Even in the first mathematical investigations that led to arbitrary functions being considered, there was discussion of the question of whether such an entirely arbitrary function could be expressed through a series of the above sort.

This happened in the middle of the preceding century in the course of investigations of vibrating strings, a problem occupying the most famous mathematicians of the day. Their views on our current topic, naturally, cannot be presented without going into this problem.

---

*Translated by Michael Ansaldi, with technical assistance by John Anders.
[1]This essay was submitted by the author in 1854 as his *Habilitationsschrift* for the faculty of Philosophy at the University of Göttingen. Although the author apparently did not intend for it to be published, even so, its publication here, unchanged, will doubtless seem sufficiently justified by the great interest of the subject itself as well as by the method it sets out for handling the most important principles of infinitesimal calculus.
Braunschweig, July 1867.   — R. Dedekind

On certain assumptions, which in fact are true in an approximate sort of way, it is well known that the form of a vibrating string stretched along a surface, where x represents the distance of any given point on it from its point of origin, and y the point's distance from the resting place at time t, is defined by the partial differential equation

$$\frac{\partial^2 y}{\partial t^2} = \alpha\alpha \frac{\partial^2 y}{\partial x^2}$$

where α is independent of t and, with a string of even thickness throughout, of x.

The first to give a general solution of this differential equation was d'Alembert.

He showed[2] that every function y of x and t that solves the above equation must be contained in the form

$$f(x + \alpha t) + \varphi(x - \alpha t)$$

which results from introducing the independent variable $x + \alpha t$ and $x - \alpha t$ instead of x and t, whereby

$$\frac{\partial^2 y}{\partial x^2} - \frac{1}{\alpha\alpha}\frac{\partial^2 y}{\partial t^2} \quad \text{is converted into} \quad 4\frac{\partial \dfrac{\partial y}{\partial(x + \alpha t)}}{\partial(x - \alpha t)}.$$

Apart from this partial differential equation, which results from general laws of movement, y must now still fulfill the condition of always being $=0$ at the string's points of attachment. Thus, if $x = 0$ at one of these points, and $x = l$ at the other, we get

$$f(\alpha t) = -\varphi(-\alpha t), f(l + \alpha t) = -\varphi(l - \alpha t)$$

and consequently

$$f(z) = -\varphi(-z) = -\varphi(l - (l + z)) = f(2l + z),$$
$$y = f(\alpha t + x) - f(\alpha t - x).$$

After d'Alembert had made this contribution to the general solution of the problem, in a continuation[3] of his essay he dealt with the equation $f(z) = f(2l + z)$; *i.e.*, he sought analytic expressions that remain unchanged if z grows by $2l$.

It was a substantial accomplishment of Euler (who in the following year's *Berliner Abhandlungen*[4] made a new presentation of this work of d'Alembert's) that he more correctly recognized the nature of the conditions that the function $f(z)$ must satisfy. He noted that by the nature of the problem, the movement of the string is completely defined at any point in time when the string's shape and the speed of any point (thus y and $\frac{\partial y}{\partial t}$) are given, and showed that, if one thought of both these functions as being defined by arbitrarily drawn curves, d'Alembert's curve $f(z)$ could be found by simple geometric construction. In fact, if one assumes that for

$$t = 0, y = g(x) \text{ and } \partial y/\partial t = h(x),$$

[2]Mémoires de l'académie de Berlin (1747), p. 214.
[3]Ibid., p. 220.
[4]Mémoires de l'académie de Berlin (1748), p. 69.

from that one gets, for values of x between 0 and l,

$$f(x) - f(-x) = g(x), \quad f(x) + f(-x) = \frac{1}{\alpha} \int h(x) \, dx$$

and consequently the function $f(z)$ between $-l$ and l. But from this we can obtain its value for every other value of z by means of the equation

$$f(z) = f(2l + z).$$

This, presented in abstract but nowadays generally familiar terms, is Euler's definition of the function $f(z)$.

D'Alembert, however, immediately[5] took issue with this expansion of his method by Euler, because his method would necessarily assume that y can be analytically expressed in terms of t and x.

Before Euler could give an answer to this, a third treatment of this subject, quite different from these two, appeared, by Daniel Bernoulli.[6] Even before d'Alembert, Taylor[7] had already seen that $\frac{\partial^2 y}{\partial t^2} = \frac{\alpha \alpha \partial^2 y}{\partial x^2}$ and also that, for $x = 0$ and for $x = l$, y was always equal to 0, if $y = \sin\frac{n\pi x}{l} \cos\frac{n\pi \alpha t}{l}$ and n is a whole number. From this he explained the physical fact that, in addition to its own tonic, a string could also produce the tonic of a string that was $\frac{1}{2}, \frac{1}{3}, \frac{1}{4} \ldots$ as long (made that way at any rate), and considered his particular solution to be general, i.e., he believed that the string's vibration, if the whole number n were defined according to the tone's pitch, would always come very close, at least, to being expressed by the equation. The observation that a string could at the same time produce the various tones it could now led Bernoulli to note that (in theory) the string could also vibrate according to the equation

$$y = \sum a_n \sin\frac{n\pi x}{l} \cos\frac{n\pi \alpha}{l}(t - \beta_n)$$

and because all observed modifications of the phenomenon could be explained from this equation, he believed it the most general.[8] In support of this opinion, he studied the vibrations of a stretched filament of no mass that, at individual points, was weighted down with finite masses, and showed that its vibrations can always be broken down into a number of vibrations (equal to the number of finite-mass points), every one of which lasts the same time for all masses.

This work of Bernoulli's elicited a new essay from Euler, printed directly after it in the *Abhandlungen der Berliner Akademie*.[9] There, in response to d'Alembert he

[5] Mémoires de l'académie de Berlin (1750), p. 358. "In fact, it seems to me, one may not express y analytically in a more general way except by supposing it a function of t and of x. But in supposing this, one only finds a solution to the problem for cases where the different figures of the vibrating string can be contained in one and the same equation." (Quoted in French.)

[6] Mémoires de l'académie de Berlin (1753), p. 147.

[7] Taylor, De methodo incrementorum.

[8] Loc. cit., p. 157, art. XIII.

[9] Mémoires de l'académie de Berlin (1753), p. 196.

maintains[10] that the function $f(z)$ can be completely arbitrary between the limits $-l$ and l, and notes[11] that Bernoulli's solution (which he had earlier already established as a particular solution) would be general if and only if the series

$$a_1 \sin\frac{x\pi}{l} + a_2 \sin\frac{2x\pi}{l} + \ldots$$

$$+ \frac{1}{2}b_0 + b_1 \cos\frac{x\pi}{l} + b_2\cos\frac{2x\pi}{l} + \ldots$$

could represent, for abscissa x, the coordinates of a totally arbitrary curve between 0 and l. Now, at that time, no one doubted that all the transformations one could perform on an analytic expression—whether finite or infinite—were valid for any values of undetermined magnitude or, if they were not, that they could not be applied only in quite special cases. Thus it seemed impossible to use the above expression to represent any algebraic curve or, more generally, an analytically represented non-periodic curve, and thus Euler believed he needed to decide the question against Bernoulli.

The dispute between Euler and d'Alembert, however, was still not settled. This caused a young, at that time still little-known mathematician, Lagrange, to try to solve the problem in an entirely new way, by which he arrived at Euler's results. He undertook[12] to define the vibrations of a mass-free thread weighted down with a finite, indeterminate number of equal masses at equal intervals, and then investigated how these vibrations change as the number of masses grows towards infinity. However much skill, however much analytical artfulness he may have expended to carry out the first part of this study, the transition from finite to infinite nonetheless left much to be desired, such that d'Alembert, in a work he placed at the very top of his *Opuscules mathématiques*, could continue to proclaim the renown of his solution to the widest public. Thus, the views of the famous mathematicians of the day were, and continued to be, divided on this subject; for even in their later works each basically held to his position.

In conclusion, then, to sum up the viewpoints that this problem gave them the occasion to develop regarding arbitrary functions and whether the latter could be represented by a trigonometric series, Euler was the first to introduce these functions into the Calculus and, supported by a geometric perspective, he applied infinitesimal calculus to them. Lagrange[13] held Euler's results (his geometric construction of the vibration process) to be correct; but Euler's geometric treatment of these functions did not satisfy him. D'Alembert, by contrast, took Euler's differential equation approach and limited himself to championing the correctness of his results, since

---

[10]Loc. cit., p. 214.
[11]Loc. cit., art. III–X.
[12]Miscellanea Taurinensia. Tom. I Recherches sur la nature et la propagation du son.
[13]Miscellanea Taurinensia. Tom. II. Pars math., p. 18.

with entirely arbitrary functions it was impossible to know whether their differential quotients were continuous. As for Bernoulli's solution, all three agreed that it could not be considered general; but whereas d'Alembert,[14] so as to declare Bernoulli's solution less general than his own, had to maintain that even an analytically represented periodic function could not always be represented by a trigonometric series, Lagrange[15] believed he could prove this possibility.

2.

Nearly fifty years went by without any significant progress being made on the question of whether arbitrary functions could be represented analytically. Then, a comment of Fourier's cast new light on this subject. A new era began in the development of this branch of mathematics, which quickly manifested itself, outside Mathematics proper, in spectacular enhancements to mathematical physics. Fourier noted that in the trigonometric series

$$f(x) = \begin{cases} a_1 \sin x + a_2 \sin 2x + \ldots \\ + \dfrac{1}{2} b_0 + b_1 \cos x + b_2 \cos 2x + \ldots \end{cases}$$

the coefficients could be defined by the formulæ

$$a_n = \frac{1}{\pi} \int_{-\pi}^{\pi} f(x) \sin nx\, dx, \quad b_n = \frac{1}{\pi} \int_{-\pi}^{\pi} f(x) \cos nx\, dx.$$

He saw that this way of defining them continued to be applicable even if the function $f(x)$ was completely arbitrary. For $f(x)$ he posited a so-called discontinuous function (the coordinates of a broken line for the abscissa x) and thus obtained a series that in fact gave the value of the function continuously.

When Fourier, in one of his first works on heat presented to the French Academy[16] (December 21, 1807), first set out the theorem that a given arbitrary function, drawn in a graph, could be expressed through a trigonometric series, this assertion so took the elderly Lagrange by surprise that he was most decisive in opposing it. There is supposedly[17] a writing hereon still to be found in the archives of the Paris academy. That notwithstanding, Poisson, wherever he uses trigonometric series to represent arbitrary functions, refers[18] to a passage in Lagrange's works on vibrating strings where this method of representation is allegedly found. To refute this assertion, which can be explained merely in terms of the known rivalry between Fourier and

[14]Opuscules mathématiques. Tome I, p. 42. art. XXIV.
[15]Misc. Taur. Tom. III. Pars math., p. 221. art. XXV.
[16]Bulletin des sciences p. la soc. Philomatique. Tome I, p. 112.
[17]By oral report of Professor Dirichlet.
[18]In his expanded Traité de mécanique, no. 323, p. 638, *inter alia*.

Poisson,[19] we are compelled to go back once again to Lagrange's essay, for there is nothing in print about that proceeding in the Academy.

In the passage[20] cited by Poisson, we in fact find the formula:

$$„y = 2\int Y \sin X\pi \, dX \times \sin x\pi + 2\int Y \sin 2X\pi \, dX \times \sin 2x\pi$$
$$+ 2\int Y \sin 3X\pi \, dX \times \sin 3x\pi + \text{etc.} + 2\int Y \sin n \, X\pi \, dX \times \sin nx\pi,$$

so that, when $x = X$, we will get $y = Y$, Y being the ordinate corresponding to the abscissa X."

To be sure, this formula looks quite like the Fourier series, so that it is easy to be confused at a fleeting glance. But this appearance derives merely from Lagrange's use of the symbol $\int dX$, where nowadays he would have used the symbol $\sum \Delta X$. It provides the solution to the problem, *viz.* how to define the finite series of sine functions

$$a_1 \sin x\pi + a_2 \sin 2x\pi + \ldots + a_n \sin nx\pi$$

in such a way that it should contain given values for the values

$$\frac{1}{n + 1}, \frac{2}{n + 1}, \ldots, \frac{n}{n + 1}$$

of x, which Lagrange indeterminately designates as X. If in this formula Lagrange had let n become infinitely large, he would certainly have come to Fourier's result. But if one reads through his essay, one sees he is a long way from believing that a totally arbitrary function could really be represented by an infinite series of sines. Rather, he had undertaken the entire project because he believed that these arbitrary functions could not be represented through a formula, and he believed that trigonometric series could represent any analytically represented periodic function. Of course it now seems to us almost inconceivable that Lagrange should not have arrived at the Fourier series from his summation formula. But this can be explained by the fact that, through the Euler-d'Alembert dispute, he had formed a definite predisposition about which road to go down. He thought he first had to completely solve the vibration problem for an indefinite finite number of masses before applying his ideas on limits. These require a fairly extensive investigation,[21] which was unnecessary if he was familiar with the Fourier series. The nature of trigonometric series was, indeed, recognized by Fourier[22] altogether correctly. Since then, they have been much used in mathematical physics to represent arbitrary functions, and in every individual case one became easily convinced that the Fourier series in fact converged toward the function's value. But it was a long time before this important theorem could be given a general proof.

[19] The report in the *Bulletin des sciences* on the essay Fourier presented to the Academy is by Poisson.
[20] Misc. Taur. Tom. III. Pars math., p. 261.
[21] Misc. Taur. Tom. III. Pars math. p. 251.
[22] Bulletin d. sc. Tom I., p. 115. Coefficients *a, a', a",* . . . having thus been determined etc. (Quoted in French).

The proof that Cauchy gave,[23] in an essay read to the Paris academy on February 27, 1826, was inadequate, as Dirichlet showed.[24] Cauchy assumes that, if in an arbitrarily given periodic function $f(x)$ one supplied for x a complex argument $x + yi$, this function would be finite for every value of y. But this happens only if the function is equal to a constant quantity. However, it is easy to see that this assumption is not needed to make more wide-ranging conclusions. It is sufficient that there be a function $\varphi(x + yi)$ that is finite for all positive values of y and that the real part of it, for $y = 0$, be equal to the given periodic function $f(x)$. By assuming that theorem, which in fact is correct,[25] the path on which Cauchy set out does indeed lead where we wished to go (just as, conversely, the theorem can be derived from the Fourier series).

3.

Not until January 1829, in Crelle's Journal,[26] did Dirichlet's essay appear in which, for completely integrable functions without an infinite number of maxima and minima, he gave an entirely rigorous answer to the question of whether arbitrary functions could be represented by trigonometric series.

He recognized which path to take to solve this problem from the insight that infinite series can be divided into two essentially different categories, depending on whether they remain convergent or not when one makes all their members positive. For the former, the members can be rearranged at will, whereas the value of the latter is dependent on the order of the members. In fact, if in a series of the second class one calls the positive members of the series, in order,

$$a_1, a_2, a_3, \ldots,$$

and the negative members

$$-b_1, -b_2, -b_3, \ldots,$$

it is clear that both $\sum a$ as well as $\sum b$ must be infinite; for if both were finite, the series would converge even after the signs were made equal; but if one were infinite, the series would diverge. Obviously by a suitable arrangement of the members the series can come to contain any arbitrarily given value C. For if one alternates and takes positive members of the series up to the point where its value becomes greater than C and then takes negative members up to the point where its value becomes less than C, the deviation from C will never amount to more than the value of the member preceding the last sign change. Now since both the a and b values eventually become infinitely small as the index increases, so too will the deviations from C

---

[23] Mémoires de l'ac. d. sc. de Paris. Tom. VI, p. 603.
[24] Crelle Journal für die Mathematik. Bd. IV, p. 157 & 158.
[25] The proof is found in the author's inaugural dissertation.
[26] Bd. IV, p. 157.

become arbitrarily small if we go far enough in the series, *i.e.*, the series will converge to C. Only to series of the first class are the rules for finite sums applicable; only they can really be regarded as the embodiment of their members, while series of the second class cannot; a circumstance overlooked by mathematicians of the last century, no doubt mainly because series which progress by raising the powers of a variable quantity belong, generally speaking (*i.e.*, excepting individual values of this quantity), to the first class.

Obviously, the Fourier series does not necessarily belong to the first class; its convergence could certainly not be derived, as Cauchy in vain[27] sought to do, from the rule by which members were decreased. What needed to be shown, rather, was that the finite series

$$\frac{1}{\pi}\int_{-\pi}^{\pi} f(\alpha)\sin\alpha\, d\alpha\,\sin x + \frac{1}{\pi}\int_{-\pi}^{\pi} f(\alpha)\sin 2\alpha\, d\alpha\,\sin 2x + \ldots$$

$$+ \frac{1}{\pi}\int_{-\pi}^{\pi} f(\alpha)\sin n\alpha\, d\alpha\,\sin nx$$

$$+ \frac{1}{2\pi}\int_{-\pi}^{\pi} f(\alpha)\, d\alpha + \frac{1}{\pi}\int_{-\pi}^{\pi} f(\alpha)\cos\alpha\, d\alpha\,\cos x + \frac{1}{\pi}\int_{-\pi}^{\pi} f(\alpha)\cos 2\alpha\, d\alpha\,\cos 2x + \ldots$$

$$+ \frac{1}{\pi}\int_{-\pi}^{\pi} f(\alpha)\cos n\alpha\, d\alpha\,\cos nx,$$

or, what amounts to the same thing, the integral

$$\frac{1}{2\pi}\int_{-\pi}^{\pi} f(\alpha)\frac{\sin\frac{2n+1}{2}(x-\alpha)}{\sin\frac{x-\alpha}{2}}\, d\alpha$$

approaches infinitely close to the value $f(x)$ if n grows to infinity. Dirichlet supports this proof with two theorems:

1) If $0 < c \leq \frac{\pi}{2}$, eventually $\int_0^c \varphi(\beta)\frac{\sin(2n+1)\beta}{\sin\beta}\, d\beta$ approaches infinitely close to the value $\frac{\pi}{2}\varphi(0)$ as n increases;

2) If $0 < b < c \leq \frac{\pi}{2}$, eventually $\int_b^c \varphi(\beta)\frac{\sin(2n+1)\beta}{\sin\beta}\, d\beta$ infinitely approaches the value 0 as n increases,

---

[27] Dirichlet in Crelle's Journal. Bd. IV, p. 158. Be that as it may with this first observation . . . insofar as n is increasing. (Quoted in French.)

assuming that the function $\varphi(\beta)$ is either always decreasing or always increasing within the limits of this integral.

Aided by these two theorems, if function f does not shift from increasing to decreasing, or decreasing to increasing, infinitely often, the integral

$$\frac{1}{2\pi}\int\limits_{-\pi}^{\pi} f(\alpha)\,\frac{\sin\frac{2n+1}{2}(x-\alpha)}{\sin\frac{x-\alpha}{2}}\,d\alpha$$

can obviously be broken down into a **finite** number of members, one[28] of which will converge to $\frac{1}{2}\,f(x+0)$, another to $\frac{1}{2}\,f(x-0)$, but the rest to 0, if x grows towards infinity. It follows from this that a trigonometric series can represent **every** function periodically repeating itself in 2π intervals that

1. is completely integrable,

2. does not have an infinite number of minima and maxima, and

3. where its value changes in sudden jumps, it adopts the average of the limits on either side.

A function that has the first two properties but not the third can obviously not be represented by a trigonometric series. For a trigonometric series that represented it apart from the discontinuities would diverge from it at the discontinuous points themselves. But whether and, if so, under what circumstances a function that does not fulfill the first two conditions can be represented by a trigonometric series, is left unanswered by this study.

Dirichlet's work provided a firm foundation for a whole host of important analytical investigations. Inasmuch as he completely illumined the point on which Euler erred, he succeeded in settling a question that had occupied so many eminent mathematicians for more than seventy years (since 1753). In fact, it was completely settled for all cases in Nature with which we are concerned, for however great our ignorance of temporo-spatial changes in material forces and states in the infinitely small, even so we can safely assume that the functions to which Dirichlet's investigation does not extend do not occur in Nature.

Even so, the cases not solved by Dirichlet seem to merit attention for two reasons. First, as Dirichlet himself notes at the end of his essay, this subject is very closely tied to the principles of infinitesimal calculus and can thus serve to bring greater clarity and determinacy to these principles. In this respect, treatment of the subject possesses an immediate interest. Second, however, the applicability of the Fourier

---

[28]It is not hard to prove that the value of a function f without an infinite number of maxima and minima will always, both when the decreasing and the increasing argument's value is equal to x, either approach the fixed limit values $f(x+0)$ and $f(x-0)$ (following Dirichlet's notation in Dove's Repertorium der Physik, Bd. 1, p. 170), or must become infinitely large. (^{1})

series is not limited to physical investigations. It is now also successfully being applied in an area of pure mathematics, number theory, and it is precisely here that those functions not investigated by Dirichlet for their representability by a trigonometric series are of importance.

At the conclusion of his essay Dirichlet indeed promises to return to these cases later, but thus far this promise has gone unfulfilled. Nor do the works of Dirksen and Bessel on cosine and sine series fill this gap. Rather, they are very much Dirichlet's inferior in rigor and generality. Dirksen's essay,[29] almost entirely contemporaneous with the latter's, but obviously written without knowledge of it, goes down the right road in general, but in its particulars contains a number of inaccuracies. For leaving aside the fact that, in a special case,[30] he comes up with an incorrect result for the sum of the series, in a side remark he relies on a series whose expansion is possible only in particular cases,[31] so his proof is complete only for functions with first differential quotients that are always finite. Bessel[32] seeks to simplify Dirichlet's proof. But the changes to this proof do not vouch-safe any fundamental simplification in its conclusions, at most rather serving to clothe it in more familiar concepts, thereby significantly impairing its rigor and generality.

Thus, the question of whether a function can be represented by a trigono-metric series has so far been answered only on the basis of two assumptions: that the function be completely integrable and that it not have an infinite number of minima and maxima. If the latter assumption is not made, both of Dirichlet's integral theorems are inadequate to answer the question. But if the former falls by the wayside, then Fourier's coefficient definition is no longer applicable. As shall be seen, this determines the path we set out on in what follows, where this question is to be investigated without any particular assumptions about the nature of the function. As things stand, a path as direct as Dirichlet's is not possible.

THE CONCEPT OF A DEFINITE INTEGRAL AND THE SCOPE OF ITS VALIDITY

4.

The indeterminacy which still prevails on a number of fundamental points of the theory of definite integrals compels us to make some prefatory remarks about the concept of a definite integral and the scope of its validity.

[29]Crelle's Journal, Bd. IV, p. 170.
[30]Loc. cit. Formula 22.
[31]Loc. cit. Art. 3
[32]Schumacher, Astronomische Nachrichten, no. 374 (Bd. 16, p. 229).

First, then: what do we understand by $\int_a^b f(x)\, dx$?

To determine this, let us assume a series of values $x_1, x_2, \ldots, x_{n-1}$ increasing in size between a and b, and for the sake of brevity let us use δ_1 to indicate $x_1 - a$, δ_2 to indicate $x_2 - x_1, \ldots , \delta_n$ to indicate $b - x_{n-1}$, and ε to indicate a positive real fraction. Then, the value of the sum

$$S = \delta_1 f(a + \varepsilon_1 \delta_1) + \delta_2 f(x_1 + \varepsilon_2 \delta_2) + \delta_3 f(x_2 + \varepsilon_3 \delta_3) + \ldots$$
$$+ \delta_n f(x_{n-1} + \varepsilon_n \delta_n)$$

will depend on the choice of intervals δ and quantities ε. Now, however δ and ε may be chosen, if it has the property of approaching infinitely close to a fixed limit A once all δ's become infinitely small, this value we then would call $\int_a^b f(x)\, dx$.

If it does not have this property, then $\int_a^b f(x)\, dx$ has no meaning.

However even in that case, attempts have several times been made to attribute a meaning to this symbol, and among these expansions of the concept of a definite integral there is **one** that is accepted by all mathematicians. To wit, if the function $f(x)$ becomes infinitely large as the argument approaches a particular value c in the interval (a, b), then obviously the sum S, however small we make the δ's, can obtain any value whatsoever. Thus it has no limiting value, and $\int_a^b f(x)\, dx$, as per above, would have no meaning. But, should a_1 and a_2 become infinitely small, if

$$\int_a^{c-\alpha_1} f(x)\, dx + \int_{c+\alpha_2}^b f(x)\, dx$$

at that point approaches a fixed limit, $\int_a^b f(x)\, dx$ is understood to be this limit value.

Other statements by Cauchy regarding the concept of the definite integral in cases where, under the basic concept, one does not exist may be useful for individual investigations. However, they were not introduced at a general level and, given their great arbitrariness, are certainly scarcely suited therefor.

5.

Second, let us now look at the scope of validity of this concept, or at the question: in which cases is a function integrable and in which not?

We shall first look at the concept of integral in the narrow sense, *i.e.*, we assume that the sum S will converge as all δ's become infinitely small. So if we call the greatest variation of the function between a and x_1, *i.e.*, the difference between its largest and smallest values in this interval, D_1, that between x_1 and x_2, D_2 ..., that between x_{n-1} and b, D_n, then

$$\delta_1 D_1 + \delta_2 D_2 + \ldots + \delta_n D_n$$

must become infinitely small along with the values of δ. We further assume that, as long as all δ's remain smaller than d, the greatest value that this sum can have is Δ. Δ will at that point be a function of d that is always decreasing along with d and becomes infinitely small together with this value. Now if the total size of the intervals in which the variations are greater than σ is $= s$, the contribution these intervals make to the sum $\delta_1 D_1 + \delta_2 D_2 + \ldots + \delta_n D_n$ is plainly $\geq \sigma s$. Thus one gets

$$\sigma s \leq \delta_1 D_1 + \delta_2 D_2 + \ldots + \delta_n D_n \leq \Delta, \text{ consequently } s \leq \frac{\Delta}{\sigma}.$$

Now if σ is given, $\frac{\Delta}{\sigma}$ can always be made infinitely small by an suitable choice of d; hence the same is true of s, and it results in the following:

For the sum S to converge as all δ's become infinitely small, not only does the function $f(x)$ need to be finite, but also the total size of the intervals in which the variations are $> \sigma$ (whatever σ may be) must be able to be made arbitrarily small by a suitable choice of d.

This theorem can also be turned around:

If the function $f(x)$ is always finite and, for intervals where the variations of function $f(x)$ are larger than a given value δ, if the total size s of those intervals ends up continually becoming infinitely small as all values of σ infinitely decrease, the sum S will converge if all δ's become infinitely small.

For those intervals where the variations are $> \sigma$ contribute an amount smaller than s to the sum $\delta_1 D_1 + \delta_2 D_2 + \ldots + \delta_n D_n$, multiplied by the greatest change in value of the function between a and b, which (*ex hypothesi*) is finite. The remaining intervals contribute an amount $< \sigma$ $(b - a)$. Obviously one can first assume a σ to be arbitrarily small and then always (*ex hypothesi*) define the size of the intervals so that s too becomes arbitrarily small, whereby the sum $\delta_1 D_1 + \delta_2 D_2 + \ldots + \delta_n D_n$ can be given any miniscule size one wishes, and consequently the sum S can be contained within arbitrarily narrow limits.

Thus we have found conditions that are necessary and sufficient for the sum S to converge to a finite value as the values of δ infinitely decrease and in this case we can speak in the narrow sense of an integral of the function $f(x)$ between a and b.[2]

If the concept of the integral is now expanded as above, the latter of the two conditions we found is obviously still needed even in that case in order for a complete integration to be possible. But in place of the condition that the function always be finite comes the condition that the function become infinite only as the argument approaches particular values, and that a determinate limit-value emerge if the integration limits are brought infinitely close to this value.

6.

After we have investigated the conditions for a definite integral to be possible in general, *i.e.*, without particular assumptions about the nature of the function to be integrated, this investigation should now partly be applied, partly be further elaborated in particular cases, to wit, in functions first of all that, between any two limits however close, have infinitely many discontinuities.

Since these functions have not yet been looked at by anyone, it will be good to proceed on the basis of a specific example. For the sake of brevity, let us use (x) to indicate the excess of x over the next whole number, or, if x is midway between two values and this definition becomes ambiguous, (x) indicates the average of both values $\frac{1}{2}$ and $-\frac{1}{2}$, *i.e.*, zero. Furthermore let us use n to indicate a whole number and p an odd number, and after that let us construct the series

$$f(x) = \frac{(x)}{1} + \frac{(2x)}{4} + \frac{(3x)}{9} + \ldots = \sum_{1,\infty} \frac{(nx)}{nn};$$

as can easily be seen, this series converges for every value of x. If the value supplied by the argument (both when steadily decreasing and increasing) becomes equal to x, the value of the series continually approaches a fixed limit, and in fact, if $x = \frac{p}{2n}$ (where p and n are relatively prime numbers)

$$f(x + 0) = f(x) - \frac{1}{2nn}\left(1 + \frac{1}{9} + \frac{1}{25} + \ldots\right) = f(x) - \frac{\pi\pi}{16nn},$$

$$f(x - 0) = f(x) + \frac{1}{2nn}\left(1 + \frac{1}{9} + \frac{1}{25} + \ldots\right) = f(x) + \frac{\pi\pi}{16nn},$$

but otherwise everywhere $f(x + 0) = f(x), f(x - 0) = f(x)$.

This function is thus discontinuous for every rational value of x, which, reduced to its lowest terms, is a fraction with an even denominator. Thus the function has infinitely many discontinuities between any two limits however close, though in such a way that the number of jumps larger than a given value is always finite. It is completely integrable. Besides its finiteness, two properties are in fact sufficient for this purpose: that for every value of x it have a limit $f(x + 0)$ and $f(x - 0)$ on either side, and that the number of jumps greater than or equal to a given value σ always be finite. For if we apply our above investigation, an obvious consequence of these two circumstances is that d can always be assumed to be small enough that, in all intervals not containing these jumps, the variations are smaller than σ and the total size of the intervals which contain these jumps becomes arbitrarily small.

It is worth noting that functions which do not have an infinite number of maxima and minima—the function just examined, incidentally is not one of the latter—where they do not become infinite, always possess both properties, and thus are integrable wherever they do not become infinite, as can also be shown directly. (^{3})

To take a closer look now at the case where the function to be integrated, $f(x)$, becomes infinitely large for a particular value, let us assume that this happens for $x = 0$, so that for a decreasing positive x its value eventually grows beyond every given limit.

It can then easily be shown that, as x decreases starting from a finite limit a, x $f(x)$ cannot perpetually remain larger than a finite value c. For that would mean that

$$\int_x^a f(x)\,dx > c\int_x^a \frac{dx}{x},$$

would thus be larger than $c(\log\frac{1}{x} - \log\frac{1}{a})$, which value would eventually grow to infinity as x decreased. Thus $x\,f(x)$, if this function does not have an infinite number of maxima and minima in the vicinity of $x = 0$, must necessarily become infinitely small along with x in order for $f(x)$ to be integrable. On the other hand, if

$$f(x)x^\alpha = \frac{f(x)\,dx(1-\alpha)}{d(x^{1-\alpha})}$$

becomes infinitely small along with x with a value of $\alpha < 1$, it is clear that the integral converges as the lower limit becomes infinitely small.

One likewise finds that, where the integrals converge, the functions

$$f(x)x\log\frac{1}{x} = \frac{f(x)dx}{-d\log\log\frac{1}{x}}, \quad f(x)x\log\frac{1}{x}\log\log\frac{1}{x} = \frac{f(x)dx}{-d\log\log\log\frac{1}{x}} \cdots,$$

$$f(x)x\log\frac{1}{x}\log\log\frac{1}{x}\cdots\log^{n-1}\frac{1}{x}\log^n\frac{1}{x} = \frac{f(x)dx}{-d\log^{1+n}\frac{1}{x}}$$

cannot, as x decreases starting from a finite limit, perpetually remain greater than a finite value and thus, if they do not have an infinite number of maxima and minima, must become infinitely small along with x. On the other hand, we find that the integral $\int f(x)\,dx$ converges as the lower limit becomes infinitely small, once

$$f(x)x\log\frac{1}{x}\cdots\log^{n-1}\frac{1}{x}\left(\log^n\frac{1}{x}\right)^\alpha = \frac{f(x)\,dx(1-\alpha)}{-d\left(\log^n\frac{1}{x}\right)^{1-\alpha}}$$

becomes infinitely small along with x, for $\alpha > 1$.

But if function $f(x)$ has an infinite number of maxima and minima, nothing can be determined about the order in which it becomes infinite. In fact, if we assume that the function is given by its absolute value, on which alone the order in which it becomes infinite depends, then by appropriately defining the sign one can always cause the integral $\int f(x)\, dx$ to converge as the lower limit becomes infinitely smaller. The function

$$\frac{d\left(x \cos e^{\frac{1}{x}}\right)}{dx} = \cos e^{\frac{1}{x}} + \frac{1}{x}\, e^{\frac{1}{x}} \sin e^{\frac{1}{x}}$$

is an example of a function that becomes infinite in such a way that its order is infinitely large (the order of $\frac{1}{x}$ taken as a unit).

Let that suffice for this topic, which fundamentally belongs in another area. Let us move now to our actual task, namely a general investigation of whether a function can be represented by a trigonometric series.

INVESTIGATION OF WHETHER A FUNCTION CAN BE REPRESENTED BY A TRIGONOMETRIC SERIES WITH NO PARTICULAR ASSUMPTIONS ABOUT THE NATURE OF THE FUNCTION

7.

Works on this subject up till now had the goal of proving Fourier series for cases occurring in nature. Thus the proof for an entirely arbitrary function could be begun, and later the workings of the function could be made subject to arbitrary limitations to assist the proof as long as they did not impair that goal. For our purpose, we can only subject the proof to those conditions necessary for a function to be representable. Thus the conditions necessary for representability must first be sought out and then, from these, those sufficient for representability must be selected. Hence, whereas prior works showed: if a function has these and these properties, it can then be represented by a Fourier series; we must proceed from the reverse question: if a function is representable by a trigonometric series, what is the consequence of that for how the function works, for its changes of value as the argument is continually changed?

Accordingly, let us look at the series

$$a_1 \sin x + a_2 \sin 2x + \ldots$$

$$+ \frac{1}{2} b_0 + b_1 \cos x + b_2 \cos 2x + \ldots$$

or, if for brevity's sake we set

$$\frac{1}{2} b_0 = A_0,\ a_1 \sin x + b_1 \cos x = A_1,\ a_2 \sin 2x + b_2 \cos 2x = A_2,$$

we can regard the series as given by

$$A_0 + A_1 + A_2 + \ldots.$$

We call this expression Ω and its value $f(x)$, so that this function is present only for those values of x where the series converges.

For a series to converge, it is eventually necessary for its members to become infinitely small. If the coefficients a_n and b_n decrease as n increases towards infinity, the members of the series Ω eventually become infinitely small for every value of x; otherwise this can only happen for particular values of x. Each of these two cases needs to be treated separately.

8.

Thus, let us first assume that the members of the series Ω eventually become infinitely small for every value of x.

On this assumption, the series

$$C + C'x + A_0\frac{xx}{2} - A_1 - \frac{A_2}{4} - \frac{A_3}{9}\ldots = F(x),$$

which series we get from Ω by integrating every member with respect to x twice, converges for every value of x. Its value $f(x)$ continually changes with x, and this function F of x consequently is everywhere integrable.

To get an understanding of both of these—the convergence of the series and the continuity of the function $f(x)$—, let us use N to indicate the sum of the members up to $-\frac{A_n}{nn}$, R to indicate the rest of the series, i.e., the series

$$-\frac{A_{n+1}}{(n+1)^2} - \frac{A_{n+2}}{(n+2)^2} - \ldots$$

and ε to indicate the greatest value of A_m for $m > n$. At that point, however far along one continues the series, the value of R obviously continues to be excluded from the expression

$$< \varepsilon\left(\frac{1}{(n+1)^2} + \frac{1}{(n+2)^2} + \ldots\right) < \frac{\varepsilon}{n}$$

and can remain contained within arbitrarily small limits, as long as one assumes a sufficiently large n; hence the series converges.

Furthermore, the function $f(x)$ is continuous; i.e., the change in it can be given any degree of smallness if one prescribes a sufficient degree of smallness for the corresponding change in x. For the change of $f(x)$ is made up of the change in R and in N; now obviously one can begin by assuming n to be large enough—whatever x is—for R, and hence the change in R, to be arbitrarily small for every change of x, and then assume the change in x to be small enough that the change in N also becomes arbitrarily small.

It will be a good idea to begin by setting out a few theorems about this function $f(x)$, the proof of which will interrupt the thread of the investigation.

Theorem 1. If series Ω converges,

$$\frac{F(x + \alpha + \beta) - F(x + \alpha - \beta) - F(x - \alpha + \beta) + F(x - \alpha - \beta)}{4\alpha\beta},$$

converges with the same value as the series if α and β become infinitely small in such a way that their ratio remains finite.

In fact,

$$\frac{F(x + \alpha + \beta) - F(x + \alpha - \beta) - F(x - \alpha + \beta) + F(x - \alpha - \beta)}{4\alpha\beta}$$

$$= A_0 + A_1 \frac{\sin\alpha}{\alpha} \frac{\sin\beta}{\beta} + A_2 \frac{\sin 2\alpha}{2\alpha} \frac{\sin 2\beta}{2\beta} + A_3 \frac{\sin 3\alpha}{3\alpha} \frac{\sin 3\beta}{3\beta} + \ldots$$

or, to solve the simpler case where $\beta = \alpha$ first,

$$\frac{F(x + 2\alpha) - 2F(x) + F(x - 2\alpha)}{4\alpha\alpha} = A_0 + A_1\left(\frac{\sin\alpha}{\alpha}\right)^2 + A_2\left(\frac{\sin 2\alpha}{2\alpha}\right)^2 + \ldots$$

If the infinite series

$$A_0 + A_1 + A_2 + \ldots = f(x),$$

is the series

$$A_0 + A_1 + \ldots A_{n-1} = f(x) + \varepsilon_n,$$

then, for any given value of σ whatsoever, it must be possible to specify a value m of n such that, if $n > m$, then $\varepsilon_n < \delta$. If we now assume α to be so small that $m\alpha < \pi$, and, substituting

$$A_n = \varepsilon_{n+1} - \varepsilon_n,$$

if we further assume $\sum_{0,\infty} \left(\frac{\sin n\alpha}{n\alpha}\right)^2 A_n$ in the form

$$f(x) + \sum_{1,\infty} \varepsilon_n \left\{ \left(\frac{\sin(n-1)\alpha}{(n-1)\alpha}\right)^2 - \left(\frac{\sin n\alpha}{n\alpha}\right)^2 \right\},$$

and if we divide the latter infinite series into three parts, by summing up

1. the members of the series from 1 to m,

2. the members of the series from $m + 1$ to the greatest whole number lying below $\frac{\pi}{\alpha}$, which is s,

3. the members of the series from $s + 1$ to infinity

the first part then consists of a finite number of continuously changing members and can thus be brought arbitrarily close to its limit value 0 if one lets α become sufficiently small. The second part, since the factor of ε_n is consistently positive, is obviously excluded from the expression

$$< \delta \left\{ \left(\frac{\sin m\alpha}{m\alpha}\right)^2 - \left(\frac{\sin s\alpha}{s\alpha}\right)^2 \right\};$$

Finally, in order for the third part to be contained within limits, one should analyze the general member into

$$\varepsilon_n\left\{\left(\frac{\sin(n-1)\alpha}{(n-1)\alpha}\right)-\left(\frac{\sin(n-1)\alpha}{n\alpha}\right)^2\right\}$$

and

$$\varepsilon_n\left\{\left(\frac{\sin(n-1)\alpha}{n\alpha}\right)^2-\left(\frac{\sin n\alpha}{n\alpha}\right)^2\right\}=-\varepsilon_n\frac{\sin(2n-1)\alpha\sin\alpha}{(n\alpha)^2};$$

it is then obvious that it is

$$<\delta\left\{\frac{1}{(n-1)^2\alpha\alpha}-\frac{1}{nn\alpha\alpha}\right\}+\delta\frac{1}{nn\alpha}$$

and consequently the sum from $n=s+1$ up to $n=\infty$ is

$$<\delta\left\{\frac{1}{(s\alpha)^2}+\frac{1}{s\alpha}\right\},$$

which value, for an infinitely small α, becomes

$$\delta\left\{\frac{1}{\pi\pi}+\frac{1}{\pi}\right\}.$$

As α decreases, the series

$$\sum\varepsilon_n\left\{\left(\frac{\sin(n-1)\alpha}{(n-1)\alpha}\right)^2-\left(\frac{\sin n\alpha}{n\alpha}\right)^2\right\}$$

thus approaches a limiting value that cannot be greater than

$$\delta\left\{1+\frac{1}{\pi}+\frac{1}{\pi\pi}\right\}$$

thus it must be zero, and consequently

$$\frac{F(x+2\alpha)-2F(x)+F(x-2\alpha)}{4\alpha\alpha},$$

which

$$=f(x)+\sum\varepsilon_n\left\{\left(\frac{\sin(n-1)\alpha}{(n-1)\alpha}\right)^2-\left(\frac{\sin n\alpha}{n\alpha}\right)^2\right\},$$

converges along with $f(x)$ as α becomes infinitely small, which proves our theorem for the case where $\beta=\alpha$.

To prove it generally, let

$$F(x+\alpha+\beta)-2F(x)+F(x-\alpha-\beta)=(\alpha+\beta)^2(f(x)+\delta_1)$$
$$F(x+\alpha-\beta)-2F(x)+F(x-\alpha+\beta)=(\alpha-\beta)^2(f(x)+\delta_2),$$

from which

$$F(x+\alpha+\beta)-F(x+\alpha-\beta)-F(x-\alpha+\beta)+F(x-\alpha-\beta)$$
$$=4\alpha\beta f(x)+(\alpha+\beta)^2\delta_1-(\alpha-\beta)^2\delta_2.$$

As a result of what was just proven, δ_1 and δ_2 now become infinitely small once α and β become infinitely small;

$$\frac{(\alpha + \beta)^2}{4\alpha\beta} \delta_1 - \frac{(\alpha - \beta)^2}{4\alpha\beta} \delta_2$$

also becomes infinitely small, if the coefficients of δ_1 and δ_2 do not at the same time become infinitely large, which does not happen if $\frac{\beta}{\alpha}$ at the same time remains finite; and consequently

$$\frac{F(x + \alpha + \beta) - F(x + \alpha - \beta) - F(x - \alpha + \beta) + F(x - \alpha - \beta)}{4\alpha\beta}$$

thereupon converges along with $f(x)$, $\qquad$ Q.E.D.

Theorem 2.

$$\frac{F(x + 2\alpha) + F(x - 2\alpha) - 2F(x)}{2\alpha}$$

always becomes infinitely small along with α.

To prove this, let us divide the series

$$\sum A_n \left(\frac{\sin n\alpha}{n\alpha}\right)^2$$

into three groups, the first of which contains all members up to a fixed index m, after which A_n always stays smaller than ε, the second includes all the successive members for which $n\alpha$ is $\leq$ a fixed value c, and the third includes the rest of the series. It then is easy to see that, if α becomes infinitely small, the sum of the first finite group remains finite, i.e., less than a fixed value Q; the sum of the second is $< \varepsilon \frac{c}{\alpha}$, that for the third

$$< \varepsilon \sum_{c < n\alpha} \frac{1}{nn\alpha\alpha} < \frac{\varepsilon}{\alpha c}.$$

Accordingly, there remains

$\frac{F(x + 2\alpha) + F(x - 2\alpha) - 2F(x)}{2\alpha}$, which is $= 2\alpha \sum A_n \left(\frac{\sin n\alpha}{n\alpha}\right)^2$, $< 2\left(Q\alpha + \varepsilon\left(c + \frac{1}{c}\right)\right)$,

from which the proposition to be proven follows.

Theorem 3. If we use b and c to indicate any two constants, of which c is the larger, and $\lambda(x)$ to indicate a function which, along with its first differential quotient, is always continuous between b and c and becomes equal to zero at the limits, and whose second differential quotient does not have an infinite number of maxima and minima, the integral

$$\mu\mu \int_b^c F(x) \cos \mu(x - a)\lambda(x)\,dx,$$

eventually becomes smaller than any given value if μ grows to infinity.

If one replaces $F(x)$ with its expansion through the series, then for

$$\mu\mu \int_b^c F(x) \cos \mu(x - a)\lambda(x)\,dx$$

one gets the series Φ

$$\mu\mu \int_b^c \left(C + C'x + A_0 \frac{xx}{2} \right) \cos \mu(x - a)\lambda(x)dx$$

$$- \sum_{1,\infty} \frac{\mu\mu}{nn} \int_b^c A_n \cos \mu(x - a)\lambda(x)dx.$$

Now since $A_n \cos\mu(x - a)$ can obviously be expressed as an aggregate of $\cos(\mu + n)(x - a)$, $\sin(\mu + n)(x + a)$, $\cos(\mu - n)(x - a)$, $\sin(\mu - n)(x - a)$, and if in it one uses $B_{\mu+n}$ to indicate the sum of the first two members and $B_{\mu-n}$ to designate the sum of the last two members, then we get $\cos \mu(x - a) A_n = B_{\mu+n} + B_{\mu-n}$,

$$\frac{d^2 B_{\mu+n}}{dx^2} = -(\mu + n)^2 B_{\mu+n}, \quad \frac{d^2 B_{\mu-n}}{dx^2} = -(\mu - n)^2 B_{\mu-n},$$

and whatever x may be, $B_{\mu+n}$ and $B_{\mu-n}$ eventually become infinitely small as n increases.

The general expression for a member of the series (ϕ)

$$- \frac{\mu\mu}{nn} \int_b^c A_n \cos \mu(x - a)\lambda(x)dx$$

thus becomes

$$= \frac{\mu^2}{n^2(\mu + n)^2} \int_b^c \frac{d^2 B_{\mu+n}}{dx^2} \lambda(x)dx + \frac{\mu^2}{n^2(\mu - n)^2} \int_b^c \frac{d^2 B_{\mu-n}}{dx^2} \lambda(x)dx$$

or, after integrating by parts twice, in which first $\lambda(x)$ and then $\lambda'(x)$ is treated as a constant,

$$= \frac{\mu^2}{n^2(\mu + n)^2} \int_b^c B_{\mu+n}\lambda''(x)\,dx + \frac{\mu^2}{n^2(\mu - n)^2} \int_b^c B_{\mu-n}\lambda''(x)dx,$$

since $\lambda(x)$ and $\lambda'(x)$ and thus the members of the series under the integral sign become $= 0$ at the limits.

One is easily convinced that $\int_b^c B_{\mu \pm n} \lambda''(x)\, dx$, if μ grows to infinity, becomes infinitely small, whatever n may be; for this expression is equal to an aggregate of the integrals

$$\int_b^c \cos(\mu \pm n)(x - a)\lambda''(x)dx, \quad \int_b^c \sin(\mu \pm n)(x - \alpha)\lambda''(x)dx,$$

and if $\mu \pm n$ becomes infinitely large, so do these integrals, but if not, their coefficients in this expression become infinitely small because then n becomes infinitely large.

Thus, for the sum

$$\sum \frac{\mu^2}{(\mu - n)^2 n^2}$$

expanded to all whole values of n satisfying the conditions $n < -c'$, $c'' < n < \mu - c'''$ and $\mu + c^{IV} < n$ (for any values of the c variables whatsoever), it is obviously enough to prove our theorem if it can be shown that the sum remains finite if μ becomes infinitely large. For apart from the members for which $-c' < n < c''$ and $\mu - c''' < n < \mu + c^{IV}$, which obviously become infinitely small and are finite in number, the series (Φ) obviously remains smaller than this sum, multiplied by the greatest value of $\int_b^c B_{\mu \pm n} \lambda''(x)\, dx$, which becomes infinitely small.

But now, if the values of $c > 1$, the sum

$$\sum \frac{\mu^2}{(\mu - n)^2 n^2} = \frac{1}{\mu} \sum \frac{\frac{1}{\mu}}{\left(1 - \frac{n}{\mu}\right)^2 \left(\frac{n}{\mu}\right)^2},$$

is, within the above limits, smaller than

$$\frac{1}{\mu} \int \frac{dx}{(1 - x)^2 x^2},$$

expanded from

$$-\infty \text{ to } -\frac{c' - 1}{\mu}, \frac{c'' - 1}{\mu} \text{ to } 1 - \frac{c''' - 1}{\mu}, 1 + \frac{c^{IV} - 1}{\mu} \text{ to } \infty;$$

for if we break down the whole interval from $-\infty$ to $+\infty$, starting from zero, into intervals of size $\frac{1}{\mu}$, and if we everywhere replace the function under the integral sign by the smallest value in every interval, since this function does not have a maximum anywhere between the integration limits, we get all the members of the series. If we carry out the integration, we get

$$\frac{1}{\mu} \int \frac{dx}{x^2(1 - x)^2} = \frac{1}{\mu}\left(-\frac{1}{x} + \frac{1}{1 - x} + 2\log x - 2\log(1 - x)\right) + \text{const.}$$

and thus a value between the above limits that does not become infinitely small along with μ.([4])

9.

With the assistance of these theorems, the following can be determined about the ability to represent a function through a trigonometric series the members of which eventually become infinitely small for every argument value:

I. For a function $f(x)$, a 2π periodic function, to be representable by a trigonometric series whose members eventually become infinitely small for every value of x, there must be a continuous function $f(x)$, on which $f(x)$ depends in such a way that

$$\frac{F(x + \alpha + \beta) - F(x + \alpha - \beta) - F(x - \alpha + \beta) + F(x - \alpha - \beta)}{4\alpha\beta},$$

converges along with $f(x)$ if α and β remain infinitely small and at the same time the ratio between them remains finite.

Furthermore,

$$\mu\mu \int_b^c F(x) \cos \mu(x - a)\lambda(x)dx,$$

must eventually become infinitely small as μ increases, if $\lambda(x)$ and $\lambda'(x) = 0$ at the integral's limits and $\lambda(x)$ and $\lambda'(x)$ are always continuous between them, and if $\lambda''(x)$ does not have an infinite number of maxima and minima.

II. Conversely, if both these conditions are fulfilled, there is then a trigonometric series in which the coefficients eventually become infinitely small and which represents the function, where it converges.

For if we define the values C' and A_0 in such a way that

$$F(x) - C'x - A_0\frac{(xx)}{2}$$

is a function that periodically repeats in a 2π interval, and if we develop the latter by **Fourier**'s method into the trigonometric series

$$C - \frac{A_1}{1} - \frac{A_2}{4} - \frac{A_3}{9} - \dots,$$

while setting the following equal to each other:

$$\frac{1}{2\pi} \int_{-\pi}^{\pi} \left(F(t) - C't - A_0\frac{tt}{2} \right)dt = C,$$

$$\frac{1}{\pi} \int_{-\pi}^{\pi} \left(F(t) - C't - A_0\frac{tt}{2} \right) \cos n(x - t)\,dt = -\frac{A_n}{nn}$$

847

then

$$A_n = -\frac{nn}{\pi}\int_{-\pi}^{\pi}\left(F(t) - C't - A_0\frac{tt}{2}\right)\cos n(x - t)\,dt$$

must (*ex hypothesi*) eventually become infinitely small as n increases, the result of which, under Theorem 1 of the previous Article, is that the series

$$A_0 + A_1 + A_2 + \ldots$$

converges with $f(x)$ everywhere it converges.([5])

III. Let $b < x < c$, and $\rho(t)$ be a function such that $\rho(t)$ and $\rho'(t)$ have the value 0 for $t = b$ and $t = c$ and continuously change between these two values, and let $\rho''(t)$ not have an infinite number of maxima and minima, and furthermore, for $t = x$, let $\rho(t) = 1$, $\rho'(t) = 0$ and $\rho''(t) = 0$, but, finally, let $\rho'''(t)$ and $\rho^{IV}(t)$ be finite and continuous; thus the difference between the series

$$A_0 + A_1 + \ldots + A_n$$

and the integral

$$\frac{1}{2\pi}\int_b^c F(t)\frac{dd\dfrac{\sin\dfrac{2n+1}{2}(x-t)}{\sin\dfrac{(x-t)}{2}}}{dt^2}\rho(t)dt$$

eventually becomes infinitely small as n increases. The series

$$A_0 + A_1 + A_2 + \ldots$$

will then either converge or not converge, depending on whether

$$\frac{1}{2\pi}\int_b^c F(t)\frac{dd\dfrac{\sin\dfrac{2n+1}{2}(x-t)}{\sin\dfrac{(x-t)}{2}}}{dt^2}\rho(t)dt$$

eventually does or does not approach a fixed limit as n increases.

In fact,

$$A_1 + A_2 + \ldots A_n = \frac{1}{\pi}\int_{-\pi}^{\pi}\left(F(t) - C't - A_0\frac{tt}{2}\right)\sum_{1,n} - nn\,\cos n(x - t)\,dt,$$

or, since

$$2\sum_{1,n} - nn \cos n(x - t) = 2\sum_{1,n} \frac{d^2 \cos n(x - t)}{dt^2} = \frac{dd\,\dfrac{\sin\dfrac{2n + 1}{2}(x - t)}{\sin\dfrac{x - t}{2}}}{dt^2}$$

then

$$= \frac{1}{2\pi}\int_{-\pi}^{\pi}\left(F(t) - C't - A_0\frac{tt}{2}\right)\frac{dd\,\dfrac{\sin\dfrac{2n + 1}{2}(x - t)}{\sin\dfrac{x - t}{2}}}{dt^2}\,dt.$$

But now, under Theorem 3 of the previous Article,

$$= \frac{1}{2\pi}\int_{-\pi}^{\pi}\left(F(t) - C't - A_0\frac{tt}{2}\right)\frac{dd\,\dfrac{\sin\dfrac{2n + 1}{2}(x - t)}{\sin\dfrac{x - t}{2}}}{dt^2}\lambda(t)\,dt$$

becomes infinitely small as n infinitely increases, if $\lambda(t)$ along with its first differential quotient is continuous, $\lambda''(t)$ does not have an infinite number of maxima and minima, and, for $t = x$, $\lambda(t) = 0$, $\lambda''(t) = 0$, but $\lambda'''(t)$ and $\lambda^{IV}(t)$ are finite and continuous.(6)

If one here sets $\lambda(t)$ equal to 1 outside the limits b, c, and between these limits $= 1 - \rho(t)$, which obviously is possible to do, it follows that the difference between the series $A_1 + \ldots + A_n$ and the integral

$$= \frac{1}{2\pi}\int_{b}^{c}\left(F(t) - C't - A_0\frac{tt}{2}\right)\frac{dd\,\dfrac{\sin\dfrac{2n + 1}{2}(x - t)}{\sin\dfrac{x - t}{2}}}{dt^2}\rho(t)\,dt$$

eventually becomes infinitely small as n increases. But one is easily convinced by integrating by parts that

$$= \frac{1}{2\pi}\int_{b}^{c}\left(C't - A_0\frac{tt}{2}\right)\frac{dd\,\dfrac{\sin\dfrac{2n + 1}{2}(x - t)}{\sin\dfrac{x - t}{2}}}{dt^2}\rho(t)\,dt,$$

converges with A_0 if n becomes infinitely large, from which one gets the above theorem.

10.

Thus, the result of this investigation has been that, if the coefficients of the series Ω eventually become infinitely small, then the convergence of the series for a defined value of x depends on the behavior of the function $f(x)$ only in the immediate vicinity of this value.

Now whether the coefficients of the series eventually become infinitely small must in many cases be decided not from their expansion into definite integrals but in some other way. Even so, it is worth highlighting a case in which this can immediately be decided from the nature of the function, to wit if the function $f(x)$ remains finite throughout and is integrable.

In this case, if we break down the series over the entire interval from $-\pi$ to π piece by piece into parts of size

$$\delta_1, \delta_2, \delta_3, \ldots$$

and use D_1 to designate the function's greatest variation in the first, D_2 in the second, etc., then

$$\delta_1 D_1 + \delta_2 D_2 + \delta_3 D_3 + \ldots$$

must become infinitely small, once all δ's become infinitely small.

But if one breaks down the integral $\int\limits_{-\pi}^{\pi} f(x) \sin n\,(x - a)\,dx$, (the form in which the series' coefficients, apart from the factor $\frac{1}{\pi}$, are contained), or, what amounts to the same thing, $\int\limits_{a}^{a+2\pi} f(x) \sin n\,(x - a)\,dx$ von $x = a$ (starting with $x = a$) into intervals of the size $\frac{(2\pi)}{n}$, each of these contributes to the sum an amount smaller than $\frac{2}{n}$, multiplied by the greatest variation in this interval, and its sum is thus smaller than a value that *ex hypothesi* must become infinitely small along with $\frac{(2\pi)}{n}$.

In fact: these integrals have the form

$$\int\limits_{a+\frac{i}{n}2\pi}^{a+\frac{i+1}{n}2\pi} f(x) \sin n\,(x - a)\,dx.$$

The sine is positive in the first half, negative in the second. If one uses M to indicate the greatest value of $f(x)$ in the integral's interval, and m to indicate the smallest, it is clear that the integral is increased if we replace $f(x)$ with M in the first half and with m in the second, but that the integral is decreased if we replace $f(x)$ with m in the first half and with M in the second. But in the first half we get the value $\frac{2}{n}(M - m)$, and in the second $\frac{2}{n}(m - M)$. Thus, this integral, apart from the sign, is smaller than $\frac{2}{n}(M - m)$, and the integral

$$\int\limits_{a}^{a+2\pi} f(x) \sin n(x - a)\,dx$$

is smaller than

$$\frac{2}{n}(M_1 - m_1) + \frac{2}{n}(M_2 - m_2) + \frac{2}{n}(M_3 - m_3) + \ldots,$$

if we use M_3 to indicate the greatest, and m_3 the smallest value of $f(x)$ in the s-th interval; but, if $f(x)$ is integrable, this sum must become infinitely small as n becomes infinitely large and thus the size of the $\frac{(2\pi)}{n}$ intervals becomes infinitely small.

Thus, in the case set out above, the coefficients of the series become infinitely small.

11.

Now we still must investigate the case where the members of the series Ω eventually become infinitely small for argument value x, without this happening for every argument value. This case can be reduced to the previous one.

To wit, if in the series, for argument value $x + t$ and $x - t$, one adds in the members of equal range, one gets the series

$$2A_O + 2A_1 \cos t + 2A_2 \cos 2t + \ldots,$$

the members of which eventually become infinitely small for every value of t and to which the previous investigation can thus be applied.

To this end, if we use $G(x)$ to indicate the value of the infinite series

$$C + C'x + A_0\frac{xx}{2} + A_0\frac{tt}{2} - A_1\frac{\cos t}{1} - A_2\frac{\cos 2t}{4} - A_3\frac{\cos 3t}{9} - \ldots$$

so that $\frac{F(x+t) + F(x-t)}{2}$ is $= 0$ everywhere that the series for $F(x + t)$ and $F(x + t)$ converge, that results in the following:

I. If the members of the series Ω eventually become infinitely small for argument value x, then

$$\mu\mu \int_b^c G(t) \cos \mu(t - a)\lambda(t)dt,$$

must eventually become infinitely small as λ increases, if μ is a function as above in Article 9.

The value of this integral is made up of the following components

$$\mu\mu \int_b^c \frac{F(x+t)}{2} \cos \mu(t - a)\lambda(t)\,dt \text{ and } \mu\mu \int_b^c \frac{F(x-t)}{2} \cos \mu(t - a)\lambda(t)\,dt,$$

provided that these expressions have a value. Its becoming infinitely small is thus caused by the behavior of the function F at two points symmetrically located on either side of x. But it must be noted that there must be points here where each component does not become infinitely small on its own; for otherwise the members of the series would eventually become infinitely small for every argument value. Thus the contributions of the two points symmetrically located on either side of x must cancel each other out, so that their sum becomes infinitely small for an infinite μ.

From this it follows that the series Ω can only converge for those values of the variable x where the points are located symmetrically, where

$$\mu\mu \int_b^c F(x) \cos \mu(x - a)\lambda(x)dx$$

does not become infinitely small for an infinite μ. Thus only if the number of these points is infinitely great can a trigonometric series with coefficients not decreasing to infinity converge for an infinite number of argument values.

Conversely,

$$A_n = -nn\frac{2}{\pi} \int_0^\pi \left(G(t) - A_0 \frac{tt}{2} \right) \cos nt\, dt$$

and thus eventually becomes infinitely small as n increases, if

$$\mu\mu \int_b^c G(t) \cos \mu(t - a)\lambda(t)dt$$

always becomes infinitely small for an infinite μ.

II. If the members of the series Ω eventually become infinitely small for certain values of x, then it depends on how the function $G(t)$ works for an infinitely small t whether or not the series converges, and in fact the difference between

$$A_0 + A_1 + \ldots + A_n$$

and the integral

$$\frac{1}{\pi} \int_0^b G(t) \frac{dd\, \dfrac{\sin \dfrac{2n+1}{2}t}{\sin \dfrac{t}{2}}}{dt^2} \rho(t)dt$$

eventually becomes infinitely small as n increases, if b is a constant, however small, contained between 0 and π and $\rho(t)$ is a function such that $\rho(t)$ and $\rho'(t)$ are always continuous and are equal to zero for $t = b$, $\rho''(t)$ does not have an infinite number of maxima and minima, and $\rho(t) = 1$, $\rho'(t) = 0$ and $\rho''(t) = 0$ for $t = 0$, but $\rho'''(t)$ and $\rho''''(t)$ are finite and continuous.

12.

The conditions for a function to be represented by a trigonometric series, naturally, can be narrowed somewhat further, thereby somewhat advancing our investigations without making particular assumptions about the nature of the function.

852

Thus, for example, in the last theorem we obtained, the condition that $\rho''(0) = 0$ can be left out if, in the integral

$$dd\frac{\sin\dfrac{2n+1}{2}t}{\sin\dfrac{t}{2}}$$

$$\frac{1}{\pi}\int_0^b G(t)\frac{}{dt^2}\rho(t)dt$$

we replace $G(t)$ with $G(t) - G(0)$. But that does not accomplish anything of significance. So as we turn to look at particular cases, we first want to try to give to the investigation of functions not having an infinite number of maxima and minima that complete treatment which, after Dirichlet's work, it is still open to receiving.

It was noted above that such a function can be integrated wherever it does not become infinite, and it is apparent that this can occur only for a finite number of argument values. Also, if we leave out the unnecessary assumption that the function is continuous, there is in fact nothing left to be desired in Dirichlet's proof that, in the integral expression for the nth member of the series and for the sum of its first n members, the contribution of every section, as n increases, eventually becomes infinitely small (except for sections where the function becomes infinite and those lying infinitely close to the value of the series' argument), and that

$$\int_x^{x+b} f(t)\frac{\sin\dfrac{2n+1}{2}(x-t)}{\sin\dfrac{x-t}{2}}dt,$$

converges with $\pi f(x + 0)$, if $0 < b < \pi$ and $f(t)$ does not become infinite within the integral's limits. Thus all we still have to investigate is in which cases in these integral expressions the contribution of those points where the function becomes infinite eventually becomes infinitely small as n increases. This investigation has not yet been carried out. Rather, on occasion Dirichlet merely showed that this happens if we assume that the function to be represented is integrable, which is not necessary.

We saw above that, if the members of the series Ω eventually become infinitely small for every value of x, the function $F(x)$, whose second differential quotient is $f(x)$, must be finite and continuous, and that

$$\frac{F(x + \alpha) - 2F(x) + F(x - \alpha)}{\alpha}$$

must continually become infinitely small along with α. Now if $F'(x + t) - F'(x - t)$ does not have an infinite number of maxima and minima, it must, if t becomes zero, converge to a fixed limit value L or become infinitely large, and it is then obvious that

$$\frac{1}{\alpha}\int_0^\alpha (F'(x + t) - F'(x - t))dt = \frac{F(x + \alpha) - 2F(x) + F(x - \alpha)}{\alpha}$$

must likewise converge to L or to ∞ and thus can only become infinitely small if $F'(x + t) - F'(x - t)$ converges to zero. Thus, if $f(x)$ becomes infinitely large for $x = a$, $f(a + t) + f(a - t)$ must indeed always be integrable up to $t = 0$. This is sufficient for

$$\left(\int_b^{a-\varepsilon} + \int_{a+\varepsilon}^c \right) dx\, (f(x)\cos n(x - a))$$

to converge as ε decreases and to become infinitely small as n increases. Furthermore, because the function $F(x)$ is finite and continuous, then $F'(x)$ must be integrable up to $x = a$ and $(x - a)\,F'(x)$ must become infinitely small along with $(x - a)$, if this function does not have an infinite number of maxima and minima; from which it follows that

$$\frac{d(x - a)F'(x)}{dx} = (x - a)f(x) + F'(x)$$

and thus also that $(x - a)\,f(x)$ is integrable up to $x = a$. Thus, $\int f(x) \sin n\,(x - a)\,dx$ can also be integrated up to $x = a$, and for the coefficients of the series eventually to become infinitely small, obviously the only thing still necessary is that

$$\int_b^c f(x) \sin n(x - a)\,dx, \text{ where } b < a < c,$$

eventually become infinitely small as n increases. If we set

$$f(x)(x - a) = \varphi(x),$$

then, as Dirichlet showed, if this function does not have an infinite number of maxima and minima,

$$\int_b^c f(x) \sin n(x - a)dx = \int_b^c \frac{\varphi(x)}{x - a} \sin n(x - a)dx = \pi \frac{\varphi(a + 0) + \varphi(a - 0)}{2},$$

for an infinite n. Thus,

$$\varphi(a + t) + \varphi(a - t) = f(a + t)t - f(a - t)t$$

must become infinitely small along with t, and since

$$f(a + t) + f(a - t)$$

can be integrated up to $t = 0$ and thus

$$f(a + t)t + f(a - t)t$$

also becomes infinitely small together with t, both $f(a + t)t$ and $f(a - t)t$ must eventually become infinitely small as t decreases. Apart from functions having an infinite number of maxima and minima, for the function $f(x)$ to be representable by a trigonometric series with coefficients decreasing to infinity, it is necessary and sufficient (if it becomes infinite for $x = a$) for $f(a + t)t$ and $f(a - t)t$ to become infinitely small along with t, and for $f(a + t)t$ and $f(a - t)t$ to be integrable up to $t = 0$.

A function $f(x)$ not having an infinite number of maxima and minima (since only for a finite number of points does

$$\mu\mu \int_b^c F(x) \cos \mu(x - a)\lambda(x)\,dx$$

not become infinitely small for an infinite μ) can be represented—also for only a finite number of argument values—by a trigonometric series whose coefficients do not eventually become infinitely small, and on this no more time need be spent.

13.

As for functions that do have an infinite number of maxima and minima, it is probably not excessive to note that a function $f(x)$ having an infinite number of maxima and minima can be completely integrable, without being representable by a Fourier series. ([7])

This happens, for example, when $f(x)$ between 0 and 2π is equal to

$$\frac{d\left(x^v \cos \dfrac{1}{x}\right)}{dx}, \text{ and } 0 < v < \frac{1}{2}$$

For in the integral $\int_0^{2\pi} f(x) \cos n(x - a)\,dx$ with an increasing n, generally speaking, the contribution of those points where x is nearly $= \sqrt{\frac{1}{n}}$ eventually becomes infinitely large, so that the ratio of this integral to

$$\frac{1}{2}\sin\left(2\sqrt{n} - na + \frac{\pi}{4}\right)\sqrt{\pi}\,n^{\frac{1-2v}{4}}$$

converges with 1, as we shall see shortly. To provide a more general example of this, so that the essentials of the matter should better stand out, let us set

$$\int f(x)\,dx = \varphi(x) \cos \psi(x)$$

and let us assume that $\varphi(x)$ becomes infinitely small and $\psi(x)$ infinitely large for an infinitely small x, but besides that, let these functions along with their differential quotients be continuous and not have an infinite number of maxima and minima. Then

$$f(x) = \varphi'(x) \cos \psi(x) - \varphi(x)\psi'(x) \sin \psi(x)$$

and

$$\int f(x) \cos n(x - a)\, dx$$

become equal to the sum of the four integrals

$$\frac{1}{2} \int \varphi'(x) \cos\left(\psi(x) \pm n(x - a)\right) dx,$$

$$-\frac{1}{2} \int \varphi(x)\psi'(x) \sin\left(\psi(x) \pm n(x - a)\right) dx.$$

Taking $\psi(x)$ to be positive, let us now look at the member

$$-\frac{1}{2} \int \varphi(x)\psi'(x) \sin\left(\psi(x) + n(x - a)\right) dx$$

and look for the point in this integral where the sine function changes sign most slowly. If we set

$$\psi(x) + n(x - a) = y,$$

this happens where $dy/dx = 0$, and thus, assuming

$$\psi'(\alpha) + n = 0,$$

this happens for $x = \alpha$. So let us look at the behavior of the integral

$$-\frac{1}{2} \int_{\alpha-\varepsilon}^{\alpha+\varepsilon} \varphi(x)\psi'(x) \sin y\, dx$$

for the case where s becomes infinitely small for an infinite n, and for this purpose let us introduce the variable y. If we set

$$\psi(\alpha) + n(\alpha - a) = \beta,$$

we get, for a sufficiently small ε,

$$y = \beta + \psi''(\alpha)\frac{(x - \alpha)^2}{2} + \dots$$

and of course $\psi''(\alpha)$ is positive, since $\psi(x)$ becomes infinite in the positive direction for an infinitely small x; we further get

$$\frac{dy}{dx} = \psi''(\alpha)(x - \alpha) = \pm\sqrt{2\psi''(\alpha)(y - \beta)},$$

depending on whether $x - a > 0$ or $x - a < 0$, and

$$-\frac{1}{2} \int_{\alpha-\varepsilon}^{\alpha+\varepsilon} \varphi(x)\psi'(x) \sin y\, dx$$

$$= \frac{1}{2}\left(\int_{\beta+\psi''(\alpha)\frac{\varepsilon\varepsilon}{2}}^{\beta} - \int_{\beta}^{\beta+\psi''(\alpha)\frac{\varepsilon\varepsilon}{2}} \right)\left(\sin y \, \frac{dy}{\sqrt{y-\beta}} \right)\frac{\varphi(\alpha)\psi'(\alpha)}{\sqrt{2\psi''(\alpha)}}$$

$$= -\int_{0}^{\psi''(\alpha)\frac{\varepsilon\varepsilon}{2}} \sin(y+\beta)\frac{dy}{\sqrt{y}}\,\frac{\varphi(\alpha)\psi'(\alpha)}{\sqrt{2\psi''(\alpha)}}\,.$$

If, as n increases, we let the value ε decrease so that $\psi''(\alpha)\varepsilon\varepsilon$ becomes infinitely large, then if

$$\int_{0}^{\infty} \sin(y+\beta)\frac{dy}{\sqrt{y}},$$

—which as we know is equal to $\sin(\beta + \frac{\pi}{4})\sqrt{\pi}$—is not zero, apart from lower order values we get

$$-\frac{1}{2}\int_{\alpha-\varepsilon}^{\alpha+\varepsilon} \varphi(x)\psi'(x)\sin(\psi(x)+n(x-a))\,dx = -\sin\left(\beta + \frac{\pi}{4}\right)\frac{\sqrt{\pi}\varphi(\alpha)\psi'(\alpha)}{\sqrt{2\psi''(\alpha)}}$$

Thus, if this value does not become infinitely small, the ratio of

$$\int_{0}^{2\pi} f(x)\cos n(x-a)\,dx$$

to this value will converge to 1 as n infinitely increases, since the value's remaining components become infinitely small.

If one assumes that, for an infinitely small x, $\varphi(x)$ and $\psi'(x)$ have powers of x of the same order, specifically $\varphi(x)$ of x^- and $\psi'(x)$ as $x^{-\mu-1}$, so that necessarily $\nu > 0$ and $\mu \geq 0$, then for an infinite n we get

$$\frac{\varphi(\alpha)\psi'(\alpha)}{\sqrt{2\psi''(\alpha)}}$$

of the same order as $\alpha^{\nu-\mu/2}$, and thus it is not infinitely small, if $\mu \geq 2\nu$. But in general, if $x\psi'(x)$—or what amounts to the same thing, $\frac{\psi(x)}{\log x}$—is infinitely large for an infinitely small x, $\varphi(x)$ can always be chosen in such a way as to be infinitely small for an infinitely small x, but

$$\varphi(x)\frac{\psi'(x)}{\sqrt{2\psi''(x)}} = \frac{\varphi(x)}{\sqrt{-2\dfrac{d}{dx}\dfrac{1}{\psi'(x)}}} = \frac{\varphi(x)}{\sqrt{-2\lim\dfrac{1}{x\psi'(x)}}}$$

becomes infinitely large, and accordingly $\int_x f(x)\, dx$ can be extended up to $x = 0$, while

$$\int_0^{2\pi} f(x)\, \cos\, n(x - a)\, dx$$

will not become infinitely small for an infinite n. As we can see, in the integral $\int_x f(x)\, dx$, the increments of the integral cancel each other out as x gets infinitely small, even though the ratio they have to the changes in x grows very rapidly, on account of the function $f(x)$'s rapid change of signs; but then when the factor $\cos n(x - a)$ is brought in, that causes these increments to be summed together here.

But accordingly just as, even though a function can be completely integrable, its Fourier series does not converge (and even a member of the series can eventually become infinitely large)—so too, despite $f(x)$'s being completely unintegrable, between any two values, however close, there can lie an infinite number of x-values for which the series Ω converges.

Taking (nx) to mean what it does above (Article 6), an example is supplied by the function given by the series

$$\sum_{1,\infty} \frac{(nx)}{n},$$

a function which exists for all rational values of x and which can be represented by the trigonometric series

$$\sum_{1,\infty}^{n} \frac{\sum^\theta - (-1)^\theta}{n\pi} \sin 2nx\pi, \; (^8)$$

(with θ replaced by all parts of n), a series, though, which is not contained between finite limits in any size interval, however small, and consequently not anywhere integrable.

We can get another example if, for $c_0, c_1, c_2, \ldots$ in the series

$$\sum_{0,\infty} c_n \cos nnx, \quad \sum_{0,\infty} c_n \sin nnx$$

we supply positive values which are always decreasing and which eventually become infinitely small, while $\sum_{1,n}^{s} c_s$ becomes infinitely large together with n. For if the ratio of x to 2π is rational and, when reduced to its lowest terms, a fraction with denominator m, obviously these series will converge or grow to infinity, depending on whether

$$\sum_{0,m-1} \cos nnx, \quad \sum_{0,m-1} \sin nnx$$

858

are or are not equal to zero. But according to a well-known theorem of circle division,[33] for an infinite number of x-values, both cases occur between any two limits, however close they may be.

The series Ω can converge just as easily, without the value of the series

$$C' + A_0 x - \sum \frac{1}{nn} \frac{dA_n}{dx},$$

—which we get from integrating every member of Ω—being integrable by an interval of however small a size.

For example, if we take the expression

$$\sum_{1,\infty} \frac{1}{n^3} (1 - q^n) \log \left(\frac{-\log (1 - q^n)}{q^n} \right),$$

where the logarithms should be taken so as to disappear for $q = 0$, and expand it by increasing the powers of q, setting $q = e^{xi}$, the imaginary part of it constitutes a trigonometric series, which when differentiated twice with respect to x in every size interval converges infinitely often, while its first differential quotient becomes infinite at infinitely many points.

To the same extent, i.e., infinitely often between any two argument values however close, the trigonometric series too can itself converge if its coefficients do not eventually become infinitely small. A simple example of such a series is the series

$\sum_{1,\infty} \sin (n! \, x\pi)$, where $n!$, as usual

$$= 1 \cdot 2 \cdot 3 \ldots n,$$

a series which does not converge merely for every rational value of x, by changing into a finite series, but rather also converges for an infinite number of irrational values, the simplest of which are $\sin 1$, $\cos 1$, $2/e$ and their multiples, odd multiples of

$$c, \frac{e - \frac{1}{e}}{4}, \text{ etc.}(^9)$$

NOTES

(1) If we assume that function $f(x)$ does not increase in interval Δ between x and $x_1 > x$, and if we use g to indicate the upper limit of its values, which assumes $f(x + \xi)$ for $0 < \xi < \Delta$, i.e., a value that is not exceeded by any of these function values but reached with any degree of approximation one pleases, then $g - f(x + \xi)$ will never decrease as ξ increases, whereas it will, though, become arbitrarily small, i.e., it is $\lim_{\xi = 0} (g - f(x + \xi)) = 0$, $g = f(x + 0)$. The theorem that a system of real numbers **S**, whose individual instances s, finite or infinite in number,

[33]Disquis. ar. p. 636 art. 356 (Gauss, Werke, Bd. I, p. 442).

never exceed a finite numerical value, has an upper limit probably was first precisely articulated and proven by Weierstrass (*cf.* O. Biermann, Theorie der analytischen Funktionen, § 16. Leipzig: Teubner, 1884.) The proof is quite easy based on Dedekind's views on irrational numbers (Stetigkeit und irrationale Zahlen. Braunschweig: Vieweg, 1872). To wit, if one divides the real number series into two parts, A and B, such that every number a in A is exceeded by numbers of the **S** system and every number b in B not exceeded, these two parts, A and B, are divided by an existing number g, which has the properties of the upper limit of **S**.

(2) On this point, there is a fragmentary manuscript note of Riemann's which we attempt to produce in what follows, since it is necessary to complete the proof that the disappearance of Δ together with d is also a sufficient condition for the convergence of S. It might seem as though, albeit for two different subdivisions, in which the δ', δ'' intervals are smaller than d and, consequently, the difference between the greatest and least value (upper and lower limit) of the sum S—which for the two divisions is indicated by S' and S''—is smaller than a given value δ, the sums S' and S'' themselves could, even so, lie around a finite portion. To see the impossibility of this, let us make a third subdivision δ, corresponding to sum S, while simultaneously carrying out δ' and δ''. Now since every δ' element consists of a whole number of δ elements, if we look at any value of S, the sum of the members of S corresponding to these δ elements will then lie between the greatest and least value of the member of S'' corresponding to the δ' element, and consequently the entire sum S will also lie between the greatest and least value of S' and likewise also between the greatest and least value of S''; consequently S, S' and S'' cannot differ from each other by more than ε.

(3) Here is how we prove that every finite function $f(x)$ which does not increase between limits a and b, and accordingly every function which does not have an infinite number of maxima and minima, is integrable.

Let

$$a = x_1 < x_2 < x_3 \ldots < x_n = b,$$
$$\delta_1 = x_2 - x_1, \delta_2 = x_3 - x_2, \ldots \delta_{n-1} = x_n - x_{n-1},$$
$$D_1 = f(x_1) - f(x_2), D_2 = f(x_2) - f(x_3) \ldots D_{n-1} = f(x_{n-1}) - f(x_n),$$
$$D_1 + D_2 + \ldots + D_{n-1} = f(a) - f(b).$$

Since by hypothesis $f(x)$ does not increase, the values $D_1, D_2, \ldots D_{n-1}$ are the greatest variations in the intervals $\delta_1, \delta_2 \ldots, \delta_{n-1}$ and all are positive, or at least none of them is negative. If m is the number of those intervals in which $D > \sigma$, then $m\sigma < f(a) - f(b)$, or

$$m < \frac{f(a) - f(b)}{\sigma}.$$

If all intervals δ are smaller than d, then the total size of the intervals whose greatest variation is greater than σ, $< \frac{f(a) - f(b)}{\sigma} d$, and thus becomes infinitely small along with d,   Q.E.D.

(4) The $B_{\mu \pm n}$ values used here are expressed as

$$B_{\mu+n} = \tfrac{1}{2} \cos(\mu + n)(x - a)(a_n \sin na + b_n \cos na)$$
$$+ \tfrac{1}{2} \sin(\mu + n)(x - a)(a_n \cos na + b_n \sin na),$$
$$B_{\mu-n} = \tfrac{1}{2} \cos(\mu - n)(x - a)(a_n \sin na + b_n \cos na)$$
$$- \tfrac{1}{2} \sin(\mu - n)(x - a)(a_n \cos na + b_n \sin na).$$

To complete this, it must still be proved that

$$\mu\mu \int_b^c \left(C + C'x + A_0 \frac{xx}{2} \right) \cos \mu(x - a) \lambda(x)\, dx$$

also has the limit value 0. The easiest way to get that is if one sets the following values

$$\left(C + C'x + A_0 \frac{xx}{2} \right) \cos \mu(x - a) = -\frac{1}{\mu\mu} \frac{d^2 B}{dx^2},$$

$$B = \left(C - \frac{3A_0}{\mu\mu} + C'x + A_0 \frac{xx}{2} \right) \cos \mu(x - a) - 2(C' + A_0 x) \frac{\sin \mu(x - \alpha)}{\mu}$$

and twice performs a partial integration. That integrals like

$$\int_b^c \cos \mu(x - a) \lambda''(x)\, dx, \quad \int_b^c \sin \mu(x - a) \lambda''(x)\, dx$$

disappear along with an infinitely increasing μ can be proved either by the Dirichlet method, or more simply by the du Bois-Reymond Average theorem, according to which, if $\varphi(x)$ is a function that does not increase or decrease between limits b and c and ξ is a value between b and c,

$$\int_b^c f(x)\varphi(x)dx = \varphi(b) \int_b^\xi f(x)\, dx + \varphi(c) \int_\xi^c f(x)\, dx.$$

(5) The theorems set out under II require discussion. Since function $f(x)$ is assumed to have periods of 2π, then

$$F(x + 2\pi) - F(x) = \varphi(x)$$

must have the property that

$$\frac{\varphi(x + \alpha + \beta) - \varphi(x + \alpha - \beta) - \varphi(x - \alpha + \beta) + \varphi(x - \alpha - \beta)}{4\alpha\beta}$$

approaches 0 along with α and β under the assumption made in the text. Therefore, $\varphi(x)$ is a linear function of x, and thus the constants C' and A_0 may be defined in such a way that

$$\Phi(x) = F(x) - C'x - A_0 \frac{xx}{2}$$

is a function of x with 2π periods.

Now about function $F(x)$ it is further assumed that for any limits b and c,

$$\mu\mu \int_b^c F(x) \cos \mu(x - a) \lambda(x) \, dx$$

approaches the limit 0 along with μ as it grows infinitely large, if $\lambda(x)$ satisfies the conditions in the text, from which it follows that, on the same assumptions,

$$\mu\mu \int_b^c \Phi(x) \cos \mu(x - a) \lambda(x) \, dx$$

approaches the limit 0.

Now let $b < -\pi$ and $c > \pi$ and, as is permissible, let us take $\lambda(x) = 1$ in the interval from $-\pi$ to $+\pi$, then it also follows that:

$$\mu\mu \int_b^{-\pi} \Phi(x) \cos \mu(x - a) \lambda(x) \, dx + \mu\mu \int_\pi^c \Phi(x) \cos \mu(x - a) \lambda(x) \, dx$$

$$+ \mu\mu \int_{-\pi}^{+\pi} \Phi(x) \cos \mu(x - a) \, dx$$

has zero as a limit. Now in view of $\Phi(x)$'s periodicity, if μ is a whole number n, we can set for this sum:

$$nn \int_{b+2\pi}^c \Phi(x) \cos n(x - a) \lambda_1(x) \, dx + nn \int_{-\pi}^{+\pi} \Phi(x) \cos n(x - a) \, dx,$$

if $\lambda_1(x) = \lambda(x - 2\pi)$ in the interval from $b + 2\pi$ to π and $\lambda_1(x) = \lambda(x)$ in the interval from π to c, so that between the limits $b + 2\pi$ and c $\lambda_1(x)$ satisfies the conditions placed on $\lambda(x)$.

Accordingly, the first member of the above sum in itself has 0 as a limit value, and consequently the limit value of

$$\mu\mu \int_{-\pi}^{+\pi} \Phi(x) \cos \mu(x - a) \, dx$$

is also equal to zero.

(6) Here, for function $\lambda(x)$, it appears the condition must be added that it repeat in 2π intervals (which is consistent with the assumption afterwards made). In fact, the integral being discussed, for example, would not approach the limit 0 if we were to set $F(t) - C't - A_0 tt/2 = $ constant, and $\lambda(t) = (x - t)^3$. On the other hand, assuming $\lambda(x)$'s periodicity, it is easy to demonstrate the vanishing of this integral by carrying out the differentiation

$$dd \frac{\sin \dfrac{2n + 1}{2}(x - t)}{\sin \dfrac{x - t}{2}} \Big/ dt^2$$

through application of Theorem 3, Art. 8 and a procedure similar to that in note (5).

Ascoli, in an essay on trigonometric series (Accademia dei Lincei, 1880), raised a number of doubts about Notes (5) and (6), which in the first edition of these Collected Works were numbered (1) and (2). Nonetheless, they shall remain unchanged here; the following may merely be added:

The proof of the theorem whereby function $\varphi(x)$ in Note (5) must be linear (for which I refer to an essay by G. Cantor in Crelle's Journal, Bd. 72, p. 141) does indeed presuppose that function $f\Phi(x)$ exists for every value of x (thus, also is finite). But I appear to make complete sense of numbers I and II of Art. 9 at all only if this existence is assumed, indeed if, as Ascoli would have it, one also assumes as a condition for function $f(x)$ a requirement of being transformed into a periodic function through the addition of an expression $- C'x - A_0 xx/2$. If we leave out the assumption that $f(x)$ exists everywhere, there can be an infinite number of different $F(x)$ functions, which do not differ from one another merely by linear expressions. Of course, Art. 9, III still retains its meaning even if we do not assume that $f(x)$ exists everywhere, but rather if $F(x)$ is defined by the series $C - A_1/1 - A_2/4 - A_3/9 \ldots$, as in Art. 8.

As for Note (6), it must be conceded that in the text's formulæ, since function $\lambda(t)$ occurs only in the interval $-\pi$ to $+\pi$, it is sufficient to assume not the periodicity of $\lambda(t)$ and $\lambda'(t)$ but only the formulæ $\lambda(\pi) = \lambda(-\pi)$ and $\lambda'(\pi) = \lambda'(-\pi)$, thus not periodicity properly speaking but the possibility of continuous periodic progression. But since the function $F(t) - C't - A_0/2\, t^2$ is defined by the series $C - A_1/1 - A_2/4 - A_3/9 \ldots$ and not $-A_1/1 - A_2/4 - A_3/9 \ldots$, as Ascoli assumes, the example's assumption, that $F(t) - c't - A_0/2\, t^2$ is a constant different from zero, is very much allowable.

The procedure I applied, by analogy to Note (5), to prove the disappearance of the integral

$$\frac{1}{2\pi}\int_{-\pi}^{+\pi}\Phi(t)\,\frac{dd\,\dfrac{\sin\dfrac{2n+1}{2}(x-t)}{\sin\dfrac{x-t}{2}}}{dt^2}\,\lambda(t)\,dt$$

also needs to be laid out somewhat more precisely.

If one carries out the differentiation under the integral, one gets an expression in several parts, the first term of which (if to abbreviate we set $\frac{2n+1}{2}=\mu$) is

$$-\mu^2\int_{-\pi}^{+\pi}\Phi(t)\,\frac{\lambda(t)}{\sin\dfrac{x-t}{2}}\sin\mu(x-t)\,dt,$$

or if we set $\lambda(t)=\lambda_1(t)\sin\frac{x-t}{2}$ and $x=a+\pi$,

$$(-1)^n\mu^2\int_{-\mu}^{+\pi}\Phi(t)\lambda_1(t)\,\cos\,\mu(a-t)\,dt.$$

Now select b such that the interval from b to c includes the interval from $-\pi$ to $+\pi$, and in the first interval define a function $\lambda(t)$ such that, between $-\pi$ and $+\pi$, $\lambda(t)=\lambda_1(t)$ but $\lambda(t)$ and $\lambda'(t)$ disappear at the limits, and further define a function $\lambda_2(t)$ in the interval from $b+2\pi$ to c such that, between $b+2\pi$ and π, $\lambda_2(t)=-\lambda(t-2\pi)$ and, between π and c, $\lambda_2(t)=\lambda(t)$, which thus assumes that $\lambda_2(\pi)=-\lambda(-\pi)$ and $\lambda'_2(\pi)=\lambda'_1(-\pi)$. Then, as in Note (5), this gives

$$\mu^2\int_{-\pi}^{+\pi}\Phi(t)\lambda_1(t)\,\cos\,\mu(a-t)\,dt=\mu^2\int_{b}^{c}\Phi(t)\lambda(t)\,\cos\,\mu(a-t)\,dt$$

$$-\mu^2\int_{b+2\pi}^{c}\Phi(t)\lambda_2(t)\,\cos\,\mu(a-t)\,dt$$

and both components on the right disappear under Theorem 3, Art. 8 as μ increases to infinity. One proceeds in the same manner with the remaining components of the integral we have before us.

(7) Reference may be made here to the works of P. du Bois-Reymond, which made further significant advances in the theory of trigonometric series after Riemann. In them, examples are given that show that there even are completely finite and

continuous functions with an infinite number of maxima and minima which may not be represented through trigonometric series.

(8) The symbol $\Sigma^\theta - (-1)^\theta$ should be understood as a sum of positive and negative units such that, to every even factor of n, there corresponds a negative member, to every odd one a positive. We find this development (albeit in a not totally unobjectionable way) if we express the function (x) through the well-known formula

$$- \sum_{1,\infty}^{m} (-1)^m \frac{\sin 2m\pi x}{m\pi}$$

insert this into the sum $\Sigma \frac{(nx)}{n}$ and transpose the order of the summations.

(9) The value $x = \frac{1}{4}(e - \frac{1}{e})$ does not belong, as Genochi notes in an essay dealing with these examples (Intorno ad alcune serie, Turin, 1875), to the values of x for which the series $\sum_{1,\infty} \sin(n! x\pi)$ converges. But even for $x = \frac{1}{2}(e - \frac{1}{e})$, the series is not convergent, as Genochi claims.

———— ◆ ————

ON THE HYPOTHESES WHICH LIE AT THE BASES OF GEOMETRY*

[*Nature*, Vol. VIII. Nos. 183, 184, pp. 14–17, 36, 37.]

PLAN OF THE INVESTIGATION

It is known that geometry assumes, as things given, both the notion of space and the first principles of constructions in space. She gives definitions of them which are merely nominal, while the true determinations appear in the form of axioms. The relation of these assumptions remains consequently in darkness; we neither perceive whether and how far their connection is necessary, nor *a priori*, whether it is possible.

From Euclid to Legendre (to name the most famous of modern reforming geometers) this darkness was cleared up neither by mathematicians nor by such philosophers as concerned themselves with it. The reason of this is doubtless that the general notion of multiply extended magnitudes (in which space-magnitudes are included) remained entirely unworked. I have in the first place, therefore, set myself the task of constructing the notion of a multiply extended magnitude out of general notions of magnitude. It will follow from this that a multiply extended magnitude is capable of different measure-relations, and consequently that space is only a

*Translated by William Kingdon Clifford.

particular case of a triply extended magnitude. But hence flows as a necessary consequence that the propositions of geometry cannot be derived from general notions of magnitude, but that the properties which distinguish space from other conceivable triply extended magnitudes are only to be deduced from experience. Thus arises the problem, to discover the simplest matters of fact from which the measure-relations of space may be determined; a problem which from the nature of the case is not completely determinate, since there may be several systems of matters of fact which suffice to determine the measure-relations of space—the most important system for our present purpose being that which Euclid has laid down as a foundation. These matters of fact are—like all matters of fact—not necessary, but only of empirical certainty; they are hypotheses. We may therefore investigate their probability, which within the limits of observation is of course very great, and inquire about the justice of their extension beyond the limits of observation, on the side both of the infinitely great and of the infinitely small.

I. NOTION OF AN N-PLY EXTENDED MAGNITUDE

In proceeding to attempt the solution of the first of these problems, the development of the notion of a multiply extended magnitude, I think I may the more claim indulgent criticism in that I am not practised in such undertakings of a philosophical nature where the difficulty lies more in the notions themselves than in the construction; and that besides some very short hints on the matter given by Privy Councillor Gauss in his second memoir on Biquadratic Residues, in the *Göttingen Gelehrte Anzeige*, and in his Jubileebook, and some philosophical researches of Herbart, I could make use of no previous labours.

§ 1. Magnitude-notions are only possible where there is an antecedent general notion which admits of different specialisations. According as there exists among these specialisations a continuous path from one to another or not, they form a *continuous* or *discrete* manifoldness; the individual specialisations are called in the first case points, in the second case elements, of the manifoldness. Notions whose specialisations form a *discrete* manifoldness are so common that at least in the cultivated languages any things being given it is always possible to find a notion in which they are included. (Hence mathematicians might unhesitatingly found the theory of discrete magnitudes upon the postulate that certain given things are to be regarded as equivalent.) On the other hand, so few and far between are the occasions for forming notions whose specialisations make up a *continuous* manifoldness, that the only simple notions whose specialisations form a multiply extended manifoldness are the positions of perceived objects and colours. More frequent occasions for the creation and development of these notions occur first in the higher mathematic.

Definite portions of a manifoldness, distinguished by a mark or by a boundary, are called Quanta. Their comparison with regard to quantity is accomplished in the case of discrete magnitudes by counting, in the case of continuous magnitudes by measuring. Measure consists in the superposition of the magnitudes to be compared; it therefore requires a means of using one magnitude as the standard for another. In the absence of this, two magnitudes can only be compared when one is a part of the other; in which case also we can only determine the more or less and not the how much. The researches which can in this case be instituted about them form a general division of the science of magnitude in which magnitudes are regarded not as existing independently of position and not as expressible in terms of a unit, but as regions in a manifoldness. Such researches have become a necessity for many parts of mathematics, *e.g.*, for the treatment of many-valued analytical functions; and the want of them is no doubt a chief cause why the celebrated theorem of Abel and the achievements of Lagrange, Pfaff, Jacobi for the general theory of differential equations, have so long remained unfruitful. Out of this general part of the science of extended magnitude in which nothing is assumed but what is contained in the notion of it, it will suffice for the present purpose to bring into prominence two points; the first of which relates to the construction of the notion of a multiply extended manifoldness, the second relates to the reduction of determinations of place in a given manifoldness to determinations of quantity, and will make clear the true character of an n-fold extent.

§ 2. If in the case of a notion whose specialisations form a continuous manifoldness, one passes from a certain specialisation in a definite way to another, the specialisations passed over form a simply extended manifoldness, whose true character is that in it a continuous progress from a point is possible only on two sides, forwards or backwards. If one now supposes that this manifoldness in its turn passes over into another entirely different, and again in a definite way, namely so that each point passes over into a definite point of the other, then all the specialisations so obtained form a doubly extended manifoldness. In a similar manner one obtains a triply extended manifoldness, if one imagines a doubly extended one passing over in a definite way to another entirely different; and it is easy to see how this construction may be continued. If one regards the variable object instead of the determinable notion of it, this construction may be described as a composition of a variability of $n + 1$ dimensions out of a variability of n dimensions and a variability of one dimension.

§ 3. I shall show how conversely one may resolve a variability whose region is given into a variability of one dimension and a variability of fewer dimensions. To this end let us suppose a variable piece of a manifoldness of one dimension—reckoned from a fixed origin, that the values of it may be comparable with one another—which has for every point of the given manifoldness a definite value, varying continuously with the point; or, in other words, let us take a continuous function of position within the given manifoldness, which, moreover, is not constant throughout any part of that manifoldness. Every system of points where the function has a constant value, forms then a continuous manifoldness of fewer dimensions than the given one. These manifoldnesses pass over continuously into one another as the function changes; we may therefore assume that out of one of them the others proceed, and speaking generally this may occur in such a way that each point passes over into a definite point of the other; the cases of exception (the study of which is important) may here be left unconsidered. Hereby the determination of position in the given manifoldness is reduced to a determination of quantity and to a determination of position in a manifoldness of less dimensions. It is now easy to show that this manifoldness has $n - 1$ dimensions when the given manifold is n-ply extended. By repeating then this operation n times, the determination of position in an n-ply extended manifoldness is reduced to n determinations of quantity, and therefore the determination of position in a given manifoldness is reduced to a finite number of determinations of quantity *when this is possible.* There are manifoldnesses in which the determination of position requires not a finite number, but either an endless series or a continuous manifoldness of determinations of quantity. Such manifoldnesses are, for example, the possible determinations of a function for a given region, the possible shapes of a solid figure, &c.

II. MEASURE-RELATIONS OF WHICH A MANIFOLDNESS OF *N* DIMENSIONS IS CAPABLE ON THE ASSUMPTION THAT LINES HAVE A LENGTH INDEPENDENT OF POSITION, AND CONSEQUENTLY THAT EVERY LINE MAY BE MEASURED BY EVERY OTHER

Having constructed the notion of a manifoldness of n dimensions, and found that its true character consists in the property that the determination of position in it may be reduced to n determinations of magnitude, we come to the second of the problems proposed above, viz. the study of the measure-relations of which such a manifoldness is capable, and of the conditions which suffice to determine them. These measure-relations can only be studied in abstract notions of quantity, and their dependence on one another can only be represented by formulæ. On certain assumptions, however, they are decomposable into relations which, taken separately, are capable of geometric representation; and thus it becomes possible to express geometrically the calculated results. In this way, to come to solid ground, we cannot, it is true, avoid abstract considerations in our formulæ, but at least the results of calculation may subsequently be presented in a geometric form. The foundations of these two parts of the question are established in the celebrated memoir of Gauss, *Disqusitiones generales circa superficies curvas.*

§ 1. Measure-determinations require that quantity should be independent of position, which may happen in various ways. The hypothesis which first presents itself, and which I shall here develop, is that according to which the length of lines is independent of their position, and consequently every line is measurable by means of every other. Position-fixing being reduced to quantity-fixings, and the position of a point in the n-dimensioned manifoldness being consequently expressed by means of n variables x_1, x_2, x_3, . . . , x_n, the determination of a line comes to the giving of these quantities as functions of one variable. The problem consists then in establishing a mathematical expression for the length of a line, and to this end we must consider the quantities x as expressible in terms of certain units. I shall treat this problem only under certain restrictions, and I shall confine myself in the first place to lines in which the ratios of the increments dx of the respective variables vary continuously. We may then conceive these lines broken up into elements, within which the ratios of the quantities dx may be regarded as constant; and the problem is then reduced to establishing for each point a general expression for the linear element ds starting from that point, an expression which will thus contain the quantities x and the quantities dx. I shall suppose, secondly, that the length of the linear element, to the first order, is unaltered when all the points of this element undergo the same infinitesimal displacement, which implies at the same time that if all the quantities dx are increased in the same ratio, the linear element will vary also in the same ratio. On these suppositions, the linear element may be any homogeneous function of the first degree of the quantities dx, which is unchanged when we change the signs of all the dx, and in which the arbitrary constants are continuous functions of the quantities x. To find the simplest cases, I shall seek first an expression for manifoldnesses of $n - 1$ dimensions which are everywhere equidistant from the origin of the linear element; that is, I shall seek a continuous function of position whose values distinguish them from one another. In going outwards from the origin, this must either increase in all directions or decrease in all directions; I assume that it increases in all directions, and therefore has a minimum at that point. If, then, the first and second differential coefficients of this function are finite, its first differential must vanish, and the second differential cannot become negative; I assume that it is always positive. This differential expression, of the second order remains constant when ds remains constant, and increases in the duplicate ratio when the dx, and therefore also ds, increase in the same ratio; it must therefore be ds^2 multiplied by a constant, and consequently

ds is the square root of an always positive integral homogeneous function of the second order of the quantities dx, in which the coefficients are continuous functions of the quantities x. For Space, when the position of points is expressed by rectilinear co-ordinates, $ds = \sqrt{\Sigma (dx)^2}$; Space is therefore included in this simplest case. The next case in simplicity includes those manifoldnesses in which the line-element may be expressed as the fourth root of a quartic differential expression. The investigation of this more general kind would require no really different principles, but would take considerable time and throw little new light on the theory of space, especially as the results cannot be geometrically expressed; I restrict myself, therefore, to those man-ifoldnesses in which the line element is expressed as the square root of a quadric dif-ferential expression. Such an expression we can transform into another similar one if we substitute for the n independent variables functions of n new independent vari-ables. In this way, however, we cannot transform any expression into any other; since the expression contains $\frac{1}{2}n(n + 1)$ coefficients which are arbitrary functions of the independent variables; now by the introduction of new variables we can only satisfy n conditions, and therefore make no more than n of the coefficients equal to given quantities. The remaining $\frac{1}{2}n(n - 1)$ are then entirely determined by the nature of the continuum to be represented, and consequently $\frac{1}{2}n\ (n - 1)$ functions of positions are required for the determination of its measure-relations. Manifoldnesses in which, as in the Plane and in Space, the line-element may be reduced to the form $\sqrt{\Sigma\,dx^2}$, are therefore only a particular case of the manifoldnesses to be here investigated; they require a special name, and therefore these manifoldnesses in which the square of the line-element may be expressed as the sum of the squares of complete differentials I will call *flat*. In order now to review the true varieties of all the continua which may be represented in the assumed form, it is necessary to get rid of difficulties arising from the mode of representation, which is accomplished by choosing the variables in accor-dance with a certain principle.

§ 2. For this purpose let us imagine that from any given point the system of shortest limes going out from it is constructed; the position of an arbitrary point may then be determined by the initial direction of the geodesic in which it lies, and by its distance measured along that line from the origin. It can therefore be expressed in terms of the ratios dx_0 of the quantities dx in this geodesic, and of the length s of this line. Let us introduce now instead of the dx_0 linear functions dx of them, such that the initial value of the square of the line-element shall equal the sum of the squares of these expressions, so that the independent varaibles are now the length s and the ratios of the quantities dx. Lastly, take instead of the dx quanti-ties $x_1, x_2, x_3, \ldots, x_n$ proportional to them, but such that the sum of their squares $= s^2$. When we introduce these quantities, the square of the line-element is $\sum dx^2$ for infinitesimal values of the x, but the term of next order in it is equal to a homogeneous function of the second order of the $\frac{1}{2} n(n-1)$ quantities $(x_1 \, dx_2 - x_2 \, dx_1), (x_1 \, dx_3 - x_3 \, dx_1) \ldots$ an infinitesimal, therefore, of the fourth order; so that we obtain a finite quantity on dividing this by the square of the infinitesimal triangle, whose vertices are $(0, 0, 0, \ldots), (x_1, x_2, x_3, \ldots), (dx_1, dx_2, dx_3, \ldots)$. This quantity retains the same value so long as the x and the dx are included in the same binary linear form, or so long as the two geodesics from 0 to x and from 0 to dx remain in the same surface-element; it depends therefore only on place and direction. It is obviously zero when the manifold represented is flat, *i.e.*, when the squared line-element is reducible to $\sum dx^2$, and may therefore be regarded as the measure of the deviation of the manifoldness from flatness at the given point in the given surface-direction. Multiplied by $-\frac{3}{4}$ it becomes equal to the quantity which Privy Councillor Gauss has called the total curvature of a surface. For the determination of the measure-relations of a manifoldness capable of representation in the assumed form we found that $\frac{1}{2} n(n-1)$ place-functions were necessary; if, therefore, the curvature at each point in $\frac{1}{2} n(n-1)$ surface-directions is given, the measure-relations of the continuum may be determined from them—provided there be no identical relations among these values, which in fact, to speak generally, is not the case. In this way the measure-relations of a manifoldness in which the line-element is the square root of a quadric differential may be expressed in a manner wholly independent of the choice of independent variables. A method entirely similar may for this purpose be applied also to the manifoldness in which the line-element has a less simple expression, *e.g.*, the fourth root of a quartic differential. In this case the line-element, generally speaking, is no longer reducible to the form of the square root of a sum of squares, and therefore the deviation from flatness in the squared line-element is an infinitesimal of the second order, while in those manifoldnesses it was of the fourth order. This property of the last-named continua may thus be called flatness of the smallest parts. The most important property of these continua for our present purpose, for whose sake alone they are here investigated, is that the relations of the twofold ones may be geometrically represented by surfaces, and of the morefold ones may be reduced to those of the surfaces included in them; which now requires a short further discussion.

§ 3. In the idea of surfaces, together with the intrinsic measure-relations in which only the length of lines on the surfaces is considered, there is always mixed up the position of points lying out of the surface. We may, however, abstract from external relations if we consider such deformations as leave unaltered the length of lines—*i.e.*, if we regard the surface as bent in any way without stretching, and treat all surfaces so related to each other as equivalent. Thus, for example, any cylindrical or conical surface counts as equivalent to a plane, since it may be made out of one by mere bending, in which the intrinsic measure-relations remain, and all theorems about a plane—therefore the whole of planimetry—retain their validity. On the other hand they count as essentially different from the sphere, which cannot be changed into a plane without stretching. According to our previous investigation the intrinsic measure-relations of a twofold extent in which the line-element may be expressed as the square root of a quadric differential, which is the case with surfaces, are characterised by the total curvature. Now this quantity in the case of surfaces is capable of a visible interpretation, viz., it is the product of the two curvatures of the surface, or multiplied by the area of a small geodesic triangle, it is equal to the spherical excess of the same. The first definition assumes the proposition that the product of the two radii of curvature is unaltered by mere bending; the second, that in the same place the area of a small triangle is proportional to its spherical excess. To give an intelligible meaning to the curvature of an n-fold extent at a given point and in a given surface-direction through it, we must start from the fact that a geodesic proceeding from a point is entirely determined when its initial direction is given. According to this we obtain a determinate surface if we prolong all the geodesics proceeding from the given point and lying initially in the given surface-direction; this surface has at the given point a definite curvature, which is also the curvature of the n-fold continuum at the given point in the given surface-direction.

§ 4. Before we make the application to space, some considerations about flat manifoldness in general are necessary; *i.e.*, about those in which the square of the line-element is expressible as a sum of squares of complete differentials.

In a flat n-fold extent the total curvature is zero at all points in every direction; it is sufficient, however (according to the preceding investigation), for the determination of measure-relations, to know that at each point the curvature is zero in $\frac{1}{2}n(n-1)$ independent surface directions. Manifoldnesses whose curvature is constantly zero may be treated as a special case of those whose curvature is constant. The common character of those continua whose curvature is constant may be also expressed thus, that figures may be viewed in them without stretching. For clearly figures could not be arbitrarily shifted and turned round in them if the curvature at each point were not the same in all directions. On the other hand, however, the measure-relations of the manifoldness are entirely determined by the curvature; they are therefore exactly the same in all directions at one point as at another, and consequently the same constructions can be made from it: whence it follows that in aggregates with constant curvature figures may have any arbitrary position given them. The measure-relations of these manifoldnesses depend only on the value of the curvature, and in relation to the analytic expression it may be remarked that if this value is denoted by α, the expression for the line-element may be written

$$\frac{1}{1 + \frac{1}{4}\alpha \sum x^2} \sqrt{\sum dx^2}.$$

§ 5. The theory of *surfaces* of constant curvature will serve for a geometric illustration. It is easy to see that surface whose curvature is positive may always be rolled on a sphere whose radius is unity divided by the square root of the curvature; but to review the entire manifoldness of these surfaces, let one of them have the form of a sphere and the rest the form of surfaces of revolution touching it at the equator. The surfaces with greater curvature than this sphere will then touch the sphere internally, and take a form like the outer portion (from the axis) of the surface of a ring; they may be rolled upon zones of spheres having new radii, but will go round more than once. The surfaces with less positive curvature are obtained from spheres of larger radii, by cutting out the lune bounded by two great half-circles and bringing the section-lines together. The surface with curvature zero will be a cylinder standing on the equator; the surfaces with negative curvature will touch the cylinder externally and be formed like the inner portion (towards the axis) of the surface of a ring. If we regard these surfaces as *locus in quo* for surface-regions moving in them, as Space is *locus in quo* for bodies, the surfaceregions can be moved in all these surfaces without stretching. The surfaces with positive curvature can always be so formed that surface-regions may also be moved arbitrarily about upon them without *bending*, namely (they may be formed) into sphere-surfaces; but not those with negative-curvature. Besides this independence of surface-regions from position there is in surfaces of zero curvature also an independence of *direction* from position, which in the former surfaces does not exist.

III. APPLICATION TO SPACE

§ 1. By means of these inquiries into the determination of the measure-relations of an *n*-fold extent the conditions may be declared which are necessary and sufficient to determine the metric properties of space, if we assume the independence of line-length from position and expressibility of the line-element as the square root of a quadric differential, that is to say, flatness in the smallest parts.

First, they may be expressed thus: that the curvature at each point is zero in three surface-directions; and thence the metric properties of space are determined if the sum of the angles of a triangle is always equal to two right angles.

Secondly, if we assume with Euclid not merely an existence of lines independent of position, but of bodies also, it follows that the curvature is everywhere constant; and then the sum of the angles is determined in all triangles when it is known in one.

Thirdly, one might, instead of taking the length of lines to be independent of position and direction, assume also an independence of their length and direction from position. According to this conception changes or differences of position are complex magnitudes expressible in three independent units.

§ 2. In the course of our previous inquiries, we first distinguished between the relations of extension or partition and the relations of measure, and found that with the same extensive properties, different measure-relations were conceivable; we then investigated the system of simple size-fixings by which the measure-relations of space are completely determined, and of which all propositions about them are a necessary consequence; it remains to discuss the question how, in what degree, and to what extent these assumptions are borne out by experience. In this respect there is a real distinction between mere extensive relations, and measure-relations; in so far as in the former, where the possible cases form a discrete manifoldness, the declarations of experience are indeed not quite certain, but still not inaccurate; while in the latter, where the possible cases form a continuous manifoldness, every determination from experience remains always inaccurate: be the probability ever so great that it is nearly exact. This consideration becomes important in the extensions of these empirical determinations beyond the limits of observation to the infinitely great and infinitely small; since the latter may clearly become more inaccurate beyond the limits of observation, but not the former.

In the extension of space-construction to the infinitely great, we must distinguish between *unboundedness* and *infinite extent*, the former belongs to the extent relations, the latter to the measure-relations. That space is an unbounded three-fold manifoldness, is an assumption which is developed by every conception of the outer world; according to which every instant the region of real perception is completed and the possible positions of a sought object are constructed, and which by these applications is for ever confirming itself. The unboundedness of space possesses in this way a greater empirical certainty than any external experience. But its infinite extent by no means follows from this; on the other hand if we assume independence of bodies from position, and therefore ascribe to space constant curvature, it must necessarily be finite provided this curvature has ever so small a positive value. If we prolong all the geodesics starting in a given surface-element, we should obtain an unbounded surface of constant curvature, *i.e.*, a surface which in a *flat* manifoldness of three dimensions would take the form of a sphere, and consequently be finite.

§ 3. The questions about the infinitely great are for the interpretation of nature useless questions. But this is not the case with the questions about the infinitely small. It is upon the exactness with which we follow phenomena into the infinitely small that our knowledge of their causal relations essentially depends. The progress of recent centuries in the knowledge of mechanics depends almost entirely on the exactness of the construction which has become possible through the invention of the infinitesimal calculus, and through the simple principles discovered by Archimedes, Galileo, and Newton, and used by modern physic. But in the natural sciences which are still in want of simple principles for such constructions, we seek to discover the causal relations by following the phenomena into great minuteness, so far as the microscope permits. Questions about the measure-relations of space in the infinitely small are not therefore superfluous questions.

If we suppose that bodies exist independently of position, the curvature is everywhere constant, and it then results from astronomical measurements that it cannot be different from zero; or at any rate its reciprocal must be an area in comparison with which the range of our telescopes may be neglected. But if this independence of bodies from position does not exist, we cannot draw conclusions from metric relations of the great, to those of the infinitely small; in that case the curvature at each point may have an arbitrary value in three directions, provided that the total curvature of every measurable portion of space does not differ sensibly from zero. Still more complicated relations may exist if we no longer suppose the linear element expressible as the square root of a quadric differential. Now it seems that the empirical notions on which the metrical determinations of space are founded, the notion of a solid body and of a ray of light, cease to be valid for the infinitely small. We are therefore quite at liberty to suppose that the metric relations of space in the infinitely small do not conform to the hypotheses of geometry; and we ought in fact to suppose it, if we can thereby obtain a simpler explanation of phenomena.

The question of the validity of the hypotheses of geometry in the infinitely small is bound up with the question of the ground of the metric relations of space. In this last question, which we may still regard as belonging to the doctrine of space, is found the application of the remark made above; that in a discrete manifoldness, the ground of its metric relations is given in the notion of it, while in a continuous manifoldness, this ground must come from outside. Either therefore the reality which underlies space must form a discrete manifoldness, or we must seek the gound of its metric relations outside it, in binding forces which act upon it.

The answer to these questions can only be got by starting from the conception of phenomena which has hitherto been justified by experience, and which Newton assumed as a foundation, and by making in this conception the successive changes required by facts which it cannot explain. Researches starting from general notions, like the investigation we have just made, can only be useful in preventing this work from being hampered by too narrow views, and progress in knowledge of the interdependence of things from being checked by traditional prejudices.

This leads us into the domain of another science, of physic, into which the object of this work does not allow us to go to-day.

ON THE NUMBER OF PRIME NUMBERS LESS THAN A GIVEN QUANTITY*

(Monthly Reports of the Berlin Academy, November 1859)

VII

I believe I can best express my thanks for the honor the Academy has done me in including me amongst its Corresponding Members by making the speediest use of the privilege thereby accorded me, to report on a study of the frequency with which prime numbers occur, a topic that, through the interest Gauss and Dirichlet have long devoted to it, will perhaps not appear entirely unworthy of such reporting.

Serving me as the starting point for this study was Euler's observation that the product

$$\Pi \frac{1}{1 - \dfrac{1}{p^s}} = \Sigma \frac{1}{n^s},$$

if for p we substitute each prime number, and for n every whole numbers. The function of the complex variable s, which is represented by these two functions as long as they converge, I shall call $\zeta(s)$. Both of them only converge as long as the real part of s is greater than 1; at the same time, it is easy to find an expression of the function that always remains valid. By applying the equation

$$\int_0^\infty e^{-nx} x^{s-1} \, dx = \frac{\Pi(s-1)}{n^s}$$

*Translated by Michael Ansaldi, with technical assistance by John Anders.

first of all we get

$$\Pi(s - 1)\zeta(s) = \int\limits_{0}^{\infty} \frac{x^{s-1}dx}{e^x - 1}.$$

Now if we look at the integral

$$\int \frac{(-x)^{s-1}dx}{e^x - 1}$$

extended positively from $+\infty$ to $+\infty$ around a domain of values containing the value 0 but no other discontinuous value of the function under the integral sign, it is just as easy to write

$$(e^{-\pi si} - e^{\pi si}) \int\limits_{0}^{\infty} \frac{x^{s-1}dx}{e^x - 1},$$

provided that in the multivalued function $(-x)^{s-1} = e^{(s-1)\log(-x)}$ the logarithm of $-x$ is defined in such a way that it becomes real for negative x. Thus we have

$$2\sin \pi s\, \Pi(s - 1)\zeta(s) = i\int\limits_{\infty}^{\infty} \frac{(-x)^{s-1}dx}{e^x - 1},$$

where the integral means what was just stated.

Now this equation gives the value of the function $\zeta(s)$ for every arbitrary complex s value, showing that it is a unique and finite function for all finite values of s except 1, and also that it vanishes if s is equal to a negative even number. (^{1})

If the real part of s is negative, the integral, instead of being extended positively around the given domain of values, can also be extended negatively around the domain of values that contains all the remaining complex values, since the integral then is infinitely small because it has values that have an infinitely large modulus. But within this domain of values, the function under the integral symbol becomes discontinuous only if x is equal to a whole number multiple of $\pm 2\pi i$ and the integral is thus equal to the sum of the integrals taken negatively around these values. But the integral around the value $n2\pi i$ is $= (-n2\pi i)^{s-1}(-2\pi i)$, and thus we get

$$2\sin \pi s\, \Pi(s - 1)\, \zeta(s) = (2\pi)^{s} \Sigma n^{s-1}((-i)^{s-1} + i^{s-1}),$$

hence a relationship between $\zeta(s)$ and $\zeta(1 - s)$ that, by using well-known properties of the function Π, can be expressed as follows:

$$\Pi\left(\frac{s}{2} - 1\right)\pi^{-\frac{s}{2}}\zeta(s)$$

This remains unchanged if s becomes $1 - s$.

This property of the function allowed me to substitute, instead of $\Pi(s-1)$, the integral $\Pi(\frac{s}{2}-1)$ into the general member of the series $\Sigma\frac{1}{n^s}$, from which we get a very convenient expression for the function $\zeta(s)$. In fact we have

$$\frac{1}{n^s}\Pi\left(\frac{s}{2}-1\right)\pi^{-\frac{s}{2}} = \int_0^\infty e^{-nn\pi x}x^{\frac{s}{2}-1}dx,$$

and so, if we set

$$\sum_1^\infty e^{-nn\pi x} = \psi(x)$$

we get

$$\Pi\left(\frac{s}{2}-1\right)\pi^{-\frac{s}{2}}\zeta(s) = \int_0^\infty \psi(x)x^{\frac{s}{2}-1}\,dx,$$

or since

$$2\psi(x)+1 = x^{-\frac{1}{2}}\left(2\psi\left(\frac{1}{x}\right)+1\right), \text{(Jacobi, Fund. p. 184)[1]}$$

$$\Pi\left(\frac{s}{2}-1\right)\pi^{-\frac{s}{2}}\zeta(s) = \int_1^\infty \psi(x)x^{\frac{s}{2}-1}dx + \int_0^1 \psi\left(\frac{1}{x}\right)x^{\frac{s-3}{2}}\,dx$$

$$+\frac{1}{2}\int_0^1(x^{\frac{s-3}{2}}-x^{\frac{s}{2}-1})\,dx$$

$$=\frac{1}{s(s-1)} + \int_1^\infty \psi(x)(x^{\frac{s}{2}-1}+x^{-\frac{1+s}{2}})\,dx.$$

I now set $s = \frac{1}{2}+ti$ and

$$\Pi\left(\frac{s}{2}\right)(s-1)\pi^{-\frac{s}{2}}\zeta(s) = \xi(t),$$

so that

$$\xi(t) = \frac{1}{2}-\left(tt+\frac{1}{4}\right)\int_1^\infty \psi(x)x^{-\frac{3}{4}}\cos\left(\frac{1}{2}t\log x\right)dx$$

or also

$$\xi(t) = 4\int_1^\infty \frac{d(x^{\frac{3}{2}}\psi'(x))}{dx}x^{-\frac{1}{4}}\cos\left(\frac{1}{2}t\log x\right)dx.$$

[1] Jacobi's Collected Works, Vol. I.

This function is finite for all finite values of t, and can be expanded into a very rapidly converging series in powers of tt. Since, for a value of s whose real component is greater than 1, $\log \zeta(s) = -\Sigma \log(1 - p^{-s})$ remains finite, and the same holds true for the logarithms of the remaining factors of $\xi(t)$, then the function $\xi(t)$ will vanish only if the imaginary part of t lies between $\frac{1}{2}i$ and $-\frac{1}{2}i$. The number of roots of $\xi(t) = 0$ whose real part lies between 0 and T is around

$$= \frac{T}{2\pi} \log \frac{T}{2\pi} - \frac{T}{2\pi};$$

for the integral $\int d \log \xi(t)$, positively extended around the peak of the t values (the imaginary part of which lies between $\frac{1}{2}i$ and $-\frac{1}{2}i$, and the real part between 0 and T is (up to a fraction of the order of magnitude of $\frac{1}{T}$) equal to $(T \log \frac{T}{(2\pi)} - T)i$; but this integral is equal to the number of roots of $\xi(t) = 0$ lying within this domain, multiplied by $2\pi i$. Now one does in fact find about that many real roots within these limits, and it is quite likely that all the roots are real. Certainly a rigorous proof of this would be desirable; however, after several fleeting attempts to no avail, I have temporarily set aside the search for this proof, since for the immediate purpose of my investigation it appeared to be dispensable.

If we use α to indicate every root of the equation $\xi(\alpha) = 0$, then we can express $\log \xi(t)$ through

$$\sum \log\left(1 - \frac{tt}{\alpha\alpha}\right) + \log \xi(0)$$

for since the frequency of the roots of the variable t increases together with t only as long as $\log \frac{t}{(2\pi)}$ is increasing, this expression thus converges and, where t is infinite, becomes infinite only as $t \log t$ does; thus it differs from $\log \xi(t)$ by a function of tt which remains continuous and finite where t is finite, and when divided by tt becomes infinitely small where t is infinite. This difference is thus a constant, the value of which can be defined by setting $t = 0$.

Aided by the above, we can now define the number of prime numbers smaller than x.

Let $F(x)$ be equal to that number if x is not already a prime number, but if x is a prime number, then let it be larger than that number by $\frac{1}{2}$, such that for an x at which $F(x)$ has discontinuous changes,

$$F(x) = \frac{F(x + 0) + F(x - 0)}{2}.$$

Now if in

$$\log \xi(s) = -\sum \log(1 - p^{-3}) = \sum p^{-3} + \frac{1}{2}\sum p^{-2s} + \frac{1}{2}\sum p^{-3s} + \dots$$

we replace

$$p^{-s} \text{ by } s\int_p^\infty x^{-s-1}dx, \quad p^{-2s} \text{ by } s\int_{p^2}^\infty x^{-s-1}dx, \ldots,$$

we then get

$$\frac{\log \zeta(s)}{s} = \int_1^\infty f(x)x^{-s-1}dx,$$

if we use $f(x)$ to indicate

$$F(x) + \frac{1}{2}F(x^{\frac{1}{2}}) + \frac{1}{3}F(x^{\frac{1}{3}}) + \ldots.$$

This equation is valid for every complex value $a + bi$ of s_1, if $a > 1$. But if to that extent the equation

$$g(s) = \int_0^\infty h(x)x^{-s}d\log x$$

is valid, then with the help of the Fourier theorem we can express function h through function g. If $h(x)$ is real and

$$g(a + bi) = g_1(b) + ig_2(b),$$

then the equation breaks down into both the following:

$$g_1(b) = \int_0^\infty h(x)x^{-a}\cos(b\log x)\,d\log x,$$

$$ig_2(b) = -i\int_0^\infty h(x)x^{-a}\sin(b\log x)\,d\log x.$$

If both equations are multiplied by $(\cos(b\log y) + i\sin(b\log y))\,db$ and are integrated from $-\infty$ to $+\infty$, then by Fourier's theorem in both of them we get $\pi h(y)y^{-a}$ on the right side, hence if both equations are added and multiplied by iy^a,

$$2\pi ih(y) = \int_{a-\infty i}^{a+\infty i} g(s)y^s ds,$$

where the integration is to be carried out in such a way that the real part of s remains constant.([2])

For a value of y for which function $h(y)$ has a discontinuous change, the integral represents the average of the values of function h on either side of the discontinuity. The way function $f(x)$ is defined here, it has the same property, and hence we get, with complete generality,

$$f(y) = \frac{1}{2\pi i} \int_{a-\infty i}^{a+\infty i} \frac{\log \zeta(s)}{s} y^s ds.$$

For $\log \zeta$ we can now substitute the expression we found before:

$$\frac{s}{2} \log \pi - \log(s-1) - \log\Pi\left(\frac{s}{2}\right) + \sum^{\alpha} \log\left(1 + \frac{(s - \frac{1}{2})^2}{\alpha\alpha}\right) + \log \xi(0)$$

but then the integrals of the individual members of this expression, extended to infinity, do not converge, which is why it is useful to transform the equation by integrating by parts into

$$f(x) = - \frac{1}{2\pi i} \frac{1}{\log x} \int_{a-\infty i}^{a+\infty i} \frac{d\frac{\log \zeta(s)}{s}}{ds} x^s ds$$

Since

$$-\log\Pi\left(\frac{s}{2}\right) = \lim\left(\sum_{n=1}^{n=m} \log\left(1 + \frac{s}{2n}\right) - \frac{s}{2} \log m\right),$$

when $m = \infty$, we have

$$-\frac{d\frac{1}{s}\log \Pi\left(\frac{s}{2}\right)}{ds} = \sum_1^\infty \frac{d\frac{1}{s}\log\left(1 + \frac{s}{2n}\right)}{ds},$$

then all members of the expression for $f(x)$, except for

$$\frac{1}{2\pi i} \frac{1}{\log x} \int_{a-\infty i}^{a+\infty i} \frac{1}{ss} \log \xi(0) x^s ds = \log \xi(0)$$

take the form

$$\pm \frac{1}{2\pi i} \frac{1}{\log x} \int_{a-\infty i}^{a+\infty i} \frac{d\left(\frac{1}{s} \log\left(1 - \frac{s}{\beta}\right)\right)}{ds} x^s ds.$$

But now,

$$\frac{d\left(\frac{1}{s}\log\left(1 - \frac{s}{\beta}\right)\right)}{d\beta} = \frac{1}{(\beta - s)\beta},$$

and, if the real part of s is greater than the real part of β,

$$-\frac{1}{2\pi i} \int_{a-\infty i}^{a+\infty i} \frac{x^s ds}{(\beta - s)\beta} = \frac{x^\beta}{\beta} = \int_\infty^\infty x^{\beta-1} dx,$$

or

$$= \int_0^x x^{\beta-1}dx,$$

depending on whether the real part of β is negative or positive. Thus we get

$$\frac{1}{2\pi i}\frac{1}{\log x}\int_{a-\infty i}^{a+\infty i}\frac{d\left(\frac{1}{s}\log\left(1-\frac{s}{\beta}\right)\right)}{ds}x^s\,ds$$

$$= -\frac{1}{2\pi i}\int_{a-\infty i}^{a+\infty i}\frac{1}{s}\log\left(1-\frac{s}{\beta}\right)x^s ds$$

$$= \int_\infty^x \frac{x^{\beta-1}}{\log x}dx + \text{Constant of integration in the first case}$$

and

$$= \int_0^x \frac{x^{\beta-1}}{\log x}dx + \text{Constant of integration in the second.}$$

In the first case, the constant of integration is determined if we let the real part of β go to negative infinity; in the second case, the integral takes on values between 0 and x that differ by $2\pi i$, depending on whether the integration is done using complex values with a positive or a negative arc. On the former scenario, it becomes infinitely small if the coefficient of i with value β goes to positive infinity, but on the latter, if such coefficient goes to negative infinity. From this we get a way to define $\log(1-\frac{s}{\beta})$ on the left side so that the integration constant is eliminated.

By substituting these values in the expression for $f(x)$, we get

$$f(x) = Li(x) - \Sigma^\alpha(Li(x^{\frac{1}{2}+\alpha i}) + Li(x^{\frac{1}{2}-\alpha i}))$$

$$+ \int_x^\infty \frac{1}{x^2-1}\frac{dx}{x\log x} + \log\xi(0),\,(^8)$$

if for α in $\sum^\alpha$ we substitute all positive roots (or roots containing a positive real part) of the equation $\xi(\alpha) = 0$ arranged according to their size. With the help of a more precise discussion of the function ξ, it can easily be shown that, ordered thus, the value of the series

$$\sum(Li(x^{\frac{1}{2}+\alpha i}) + Li(x^{\frac{1}{2}-\alpha i}))\log x$$

coincides with the limit value that

$$\frac{1}{2\pi i} \int_{a-bi}^{a+bi} \frac{d\frac{1}{s} \Sigma \log\left(1 + \frac{(s-\frac{1}{2})^2}{\alpha\alpha}\right)}{ds} x^s \, ds$$

converges to as the value b keeps on growing; but if the order is changed it could have any arbitrary real value.

From $f(x)$ we can find $F(x)$, after inverting the relation

$$f(x) = \sum \frac{1}{n} F\left(x^{\frac{1}{n}}\right),$$

by means of the resultant equation

$$F(x) = \sum (-1)^\alpha \frac{1}{m} f\left(x^{\frac{1}{m}}\right),$$

where the series of those numbers not divisible by any square except 1 is sequentially substituted for m and where μ indicates the number of prime factors of m.

If we limit $\sum^\alpha$ to a finite number of members, the derivative of the expression for $f(x)$, or

$$\frac{1}{\log x} - 2 \sum^\alpha \frac{\cos(\alpha \log x) x^{-\frac{1}{2}}}{\log x}$$

(up to a part that very rapidly decreases as x increases), gives an approximate expression for the frequency of prime numbers + half the frequency of prime number squares $+\frac{1}{3}$ the frequency of prime number cubes, etc., of the value x.

The well-known approximation formula $F(x) = Li(x)$ is thus correct only up to values of the order of magnitude $x^{\frac{1}{2}}$, and gives a value that is somewhat too large; for apart from values that do not grow to infinity along with x, the non-periodic members in the expression for $F(x)$ are

$$Li(x) - \frac{1}{2} Li\left(x^{\frac{1}{2}}\right) - \frac{1}{3} Li\left(x^{\frac{1}{3}}\right) - \frac{1}{5} Li\left(x^{\frac{1}{5}}\right) + \frac{1}{6} Li\left(x^{\frac{1}{6}}\right) - \frac{1}{7} Li\left(x^{\frac{1}{7}}\right) + \dots$$

In fact, when Gauss and Goldschmidt compared $Li(x)$ with the number of prime numbers under x (up to $x =$ three million), the number of prime numbers became smaller than $Li(x)$ as early as $x =$ one hundred thousand. And in fact the difference, with many fluctuations, slowly grows along with x. But when these enumerations were done, attention has also been drawn to how, depending on the periodic members, prime numbers sporadically show up with great frequency or only rarely. (No regular pattern had been noted, however.) Should any new enumeration be done, it would be interesting to trace the influence exerted by individual periodic members on the expression for prime number frequency. More regular in its workings than $F(x)$ would be function $f(x)$, which right in the first hundred already very clearly shows itself as coinciding for the most part with $Li(x) + \log \xi(0)$.

Notes

In a letter a draft of which is in the estate, we find the following observation, after the results of the work had been reported:

"I still have not completely carried out the proof, and in that regard I would still like . . . to add the observation that both theorems, which I there merely cited,

- that between 0 and T, there are around $\frac{T}{(2\pi)} \log \frac{T}{(2\pi)} - \frac{T}{(2\pi)}$ real roots of the equation $\xi(\alpha)$, and
- that the series $\sum^{\alpha} (Li(x^{\frac{1}{2}+\alpha i}) + Li(x^{\frac{1}{2}-\alpha i}))$, if members are ordered by an increasing α, converges with the same limit value as

$$\frac{1}{2\pi i \log x} \int_{a-bi}^{a+bi} d\frac{1}{s} \log \frac{\xi((s-\frac{1}{2})i)}{\xi(0)} \, ds \, x^s, \text{ as the value } b \text{ continues to grow}$$

follow from a new development of function ξ but one which I had still not simplified enough to be able to report on it."

Despite many later studies (Scheibner, Pilz, Stieltjes), the obscurities of this work have still not been fully illuminated, however.

(1) The behavior of function $\zeta(s)$ results from using the second form of this function

$$2\zeta(s) = \pi i \, \Pi(-s) \int_{\infty}^{\infty} \frac{(-x)^{s-1} dx}{e^x - 1}$$

and in light of the fact that $\frac{1}{e^x - 1} + \frac{1}{2}$ expanded in increasing powers of x contains only non-even powers.

(2) This theorem is not quite accurately expressed. If both integration limits 0 and ∞ are applied to $\log x$, both equations, treated individually in the manner indicated, yield $\pi y^{-\alpha} (h(y) \pm h(\frac{1}{y}))$, and thus only in their summation do they yield the formula in the text.

(3) For real values of x greater than 1, the function $Li(x)$ is to be defined by the integral $\int_0^x \frac{dx}{\log x} \pm \pi i$, where the upper or lower sign should be used depending on whether the integration is done by complex numbers with a positive or a negative

arc. From that we can easily derive the development given by Scheibner (Schlömilch's Zeitschrift, Vol. V)

$$Li(x) = \log\log x - \Gamma'(1) + \sum_{1,\,\infty}^{x} \frac{(\log x)^n}{n \cdot nl},$$

which is valid for all values of x, and for negative real values results in a discontinuity. (*Cf.* Gauss – Bessel correspondence.)

If one follows the calculation hinted at by Riemann, in the formula one finds $\log\frac{1}{2}$ instead of $\log\xi(0)$. Possibly this is only a slip of the pen or a typographical error, $\log\xi(0)$ instead of $\log\zeta(0)$, since $\zeta(0) = \frac{1}{2}$.

Karl Weierstrass

(1815–1897)

HIS LIFE AND WORK

Imagine what it would be like to be a mathematician whose name would be destined to live on forever. Not bad as fantasies go. Now, imagine what it would be like if you had to spend the first twelve years of your career teaching in a high school, not just teaching mathematics, but calligraphy and even physical education. That is the part of Karl Weierstrass's autobiography that even Weierstrass could only remember as a time of "unending dreariness and boredom." For every epsilon, there is a delta!

Karl Weierstrass was born in Ostenfelde, Westphalia, now a part of Germany, on October 31, 1815, four and a half months after Wellington's defeat of Napoleon at the battle of Waterloo. There is little in Weierstrass's forebears to suggest intellectual achievement, much less mathematical greatness. He was the eldest child of Wilhelm and Theodora (neé Vonderforst). We know nothing of his mother's family except that they, like the Weierstrass family, were moderate Catholics. We know that his father was reasonably well educated in arts and sciences. Wilhelm Weierstrass had a career as a petty bureaucrat in the Westphalian and then the Prussian government. At age eight, he entered the tax service and began to transfer frequently. Karl's already tumultuous home life took a more serious turn for the worse when his mother died suddenly in 1827 after she had given Wilhelm two daughters and another son. The elder Weierstrass remarried within two years, adding yet one more upset into young Karl's life.

In 1828, Karl entered the Catholic Gymnasium (high school) in the Westphalian town of Paderborn, yet another new hometown necessitated by his father's move to yet another new job, assistant tax commissioner of Paderborn. Home finances must have been very difficult for the Weierstrass family. At the age of fourteen, young Karl went to work as a bookkeeper for a local merchant's wife. His father thought that it would give Karl a chance to use his prowess at arithmetic! Little could father Weierstrass imagine that it would give Karl the spare time to read the Gymnasium's copies of the *Journal of Pure and Applied Mathematics,* the leading mathematics journal in all of Europe! Devouring these copies, Karl Weierstrass made up his mind that he was going to be a mathematician.

But Wilhelm Weierstrass had other ideas. In spite of the many mathematics prizes that his son won on his Gymnasium graduation in 1834, Wilhelm insisted that Karl enroll in a public finance and administration program at the nearby University of Bonn. Young Karl very grudgingly agreed. Great physical and mental strain filled the four years he spent at Bonn. Weierstrass often indulged himself in bouts of drinking and fencing as a means of rebellion. Several times he barely avoided serious injury in a fencing duel.

After four long years, young Weierstrass left Bonn without sitting for the degree examinations. His father was distraught. Four years at University and still Karl was no closer to a position as a petty bureaucrat. Within a year, a family friend suggested a reasonable compromise: Karl could enroll at the nearby Theological and Philosophical Academy in nearby Münster and sit an examination for a teaching degree in just a year.

Weierstrass did sit the examination just a year later in 1840 and asked for a particularly challenging problem on the mathematics exam. Little did Weierstrass know how well he had answered the problem. The school authorities merely told him that he had passed the mathematics exam along with the other portions of the exam. They didn't bother to tell him that the external examiner for the mathematics exam was the mathematician Christoph Gudermann, who had been a student of the great Gauss. Gudermann recognized that Weierstrass's examination answer had made important advances in a branch of mathematics called elliptical function theory. But Weierstrass wasn't to learn of this praise for almost a decade.

Thus, Weierstrass embarked on a fifteen-year career of teaching mathematics to disinterested students and lacking colleagues possessing even a fraction of his gift. Amazingly, Weierstrass published his first research papers in the Gymnasium's prospectus, designed to interest fathers looking to place their sons in school. We can only imagine what the fathers' reactions must have been to reading mathematics that was completely incomprehensible to them because it was groundbreaking!

To maintain his sanity in the midst of the dreary monotony of the Gymnasium years, Weierstrass immersed himself in whatever contemporary mathematics he could find. Years of such stress took its toll. By 1850, Weierstrass began to suffer persistent attacks of vertigo that could last as long as an hour and would only come to a halt with a violent bout of vomiting.

But success lay just around the corner. In 1854, Weierstrass finally submitted his paper "On the Theory of Abelian Functions" to the *Journal of Pure and Applied Mathematics*. It was the watershed event of his life. Weierstrass finally began to receive the attention he deserved. The University of Königsberg (now Danzig) offered him an honorary doctorate. Hoping to keep him, the Gymnasium at Braunsberg promoted him to Senior Lecturer and offered a yearlong sabbatical in 1855. But Weierstrass hoped for something greater, a true University appointment. In 1855, at nearly forty years of age, Weierstrass sought his first University appointment, as Professor of Mathematics at the University of Breslau in East Prussia (now Wroclaw in Poland). He failed to secure that post, but then the offers began to pour in.

Shortly after he published a second paper on Abellian functions in 1856, he attended a conference in Vienna and was offered a chair at any Austrian University of his choice. At the time, Austria was something of mathematical backwater so Weierstrass remained unsure of what to do. Then the Industrial Institute of Berlin offered him a full professorship. Weierstrass knew that it wasn't the most prestigious of institutions, but was better than Austria!

However, it was not better than the prestigious University of Berlin. As fate would have it, the University of Berlin offered Weierstrass an Associate Professorship just after he accepted the offer from the Industrial Institute. Much as he wanted to accept the offer from the University of Berlin, Weierstrass knew that he had to honor his seven-year commitment to the Industrial Institute. Although not among the most prestigious institutions of higher learning in Germany, the Industrial Institute finally provided Weierstrass the opportunity to pursue his research and present it to students eager to learn them.

Weierstrass is best known for his development of what is now known as the *"epsilon* method" (after the Greek letter epsilon ε), the contribution presented in this chapter. Weierstrass began developing the epsilon method in his class titled *Introduction to Analysis* during the academic year 1859 to 1860. The epsilon method provided a rigorous way for mathematicians to work with the notion of an infinite sequence or series reaching a limit. A few examples will help to illustrate the power of Weierstrass's epsilon method.

Consider the geometric series that begins with 1 and continues with successive powers of $\frac{1}{2}$.

$$1 + \frac{1}{2} + \frac{1}{4} + \frac{1}{8} + \frac{1}{16} + \frac{1}{32} \ldots$$

Continued on "to infinity." Many readers will remember the formula for the sum of a *finite* number of terms of a geometric progression:

$$a \sum_{k=0}^{n-1} r^k = a \frac{r^n - 1}{r - 1}$$

This formula can be applied to any finite number n of terms in the series given above by setting $a = 1$ and $r = \frac{1}{2}$, yielding

$$\frac{1 - \frac{1}{2}^n}{1 - \frac{1}{2}}.$$

Mathematicians argued, intuitively, that the numerator of this expression reached 1 in the limit as n reached infinity since $\frac{1}{2}^n$ reached 0 as n reached infinity. But what did it mean for n "to reach infinity"? That had puzzled mathematicians and philosophers since the days of the ancient Greek mathematician and philosopher Zeno of Elea (died about 425 B.C.).

Weierstrass's epsilon method provided a solution: n no longer had to reach infinity. Instead the limit was defined for n being in the process of reaching infinity. Weierstrass defined an infinite sequence as having a limit if for any epsilon ε, you could find an integer n so that for all integers $m \geq n$, the mth term of the sequence was always within ε of the limit. Consider how easily the epsilon method can be applied to the problem of proving that the limit of $\frac{1}{2}^n$ reaches 0 as n reaches infinity. For any ε, just find the power of $\frac{1}{2}$ that is less than ε and it is obvious that all greater powers will also be less than ε.

The epsilon method had an even greater impact in the theory of functions. Prior to the middle of the nineteenth century, mathematicians had concentrated on functions that, like the sine wave, could be drawn, at least in parts, without lifting a pencil from the paper, or perhaps, lifting it, at most, occasionally. This was a very geometric notion of continuity.

The epsilon method freed mathematicians from thinking about continuity in such a geometric manner. Weierstrass defined continuity of a function f at a point x_0 as follows.

If for every ε, a δ (the Greek letter delta) be found so that for all values of x such that

$$x_0 - \delta < x < x_0 - \varepsilon,$$

then

$$|f(x) - f(x_0)| < \varepsilon$$

(i.e., the difference between $f(x)$ and $f(x_0)$ is less than ε).
This definition certainly fit the intuitive notion that as you drew a continuous function near a point, you had to be getting close to the value of the function at that point. But it could do much more.

Earlier in the nineteenth century, the German mathematician Gustav Dirichlet had defined two rather strange functions. The first, $\chi(x)$ is called the characteristic function of the rationals in the reals. It takes the following values:

$\chi(x) = 1$ if x is rational
$\quad\quad = 0$ if x is irrational

The second function, not surprisingly called the Dirichlet function, $D(x)$, is even stranger. It takes the following values:

$D(x) = \frac{1}{b}$ if x is rational and can be expressed in lowest terms
$\quad\quad$ as the fraction $\frac{a}{b}$
$\quad\quad = 0$ if x is irrational

Neither of these functions can be "drawn" the way a sine curve can be drawn. That's not surprising for the first of them, since it is discontinuous at every point. (Just take an ε less than 1. There will always be another point nearby whose value differs by 1 since the rationals and the irrationals are intermixed with each other everywhere.)

Not all mathematicians initially found much value in monstrous functions such as these. Very early in the twentieth century the great French mathematician Henri Poincaré complained of this and other functions.

> *Logic sometimes makes monsters. Since half a century we have seen arise a crown of bizarre functions which seem to try to resemble as little as possible the honest functions which serve some purpose.*

In 1861, within two years of first formulating the epsilon method, Weierstrass proved that the Dirichlet function $D(x)$ was continuous at any irrational value of x. Suppose that we want to apply the epsilon-delta method for $x = \sqrt{2}$ with $\varepsilon = 0.1$; that is, $\varepsilon = \frac{1}{10}$. We know that $D(\sqrt{2}) = 0$. What value can δ take so that all values of $D(x)$ for x within δ of $\sqrt{2}$ are not more than 0.1, our choice for ε, away from 0.0, the value of $D(x)$ for $x = \sqrt{2}$?

The demonstration is remarkably straightforward once you realize that the only rational numbers for which $D(x)$ is greater than 0.1 (our ε) must have a denominator no greater than 10. All of these rational numbers near $\sqrt{2}$ can be listed out and the one closest to $\sqrt{2}$ determined. We have the benefit of using a spreadsheet!

Here is a chart of the values of the values for $D(x)$ for all the rational numbers between 1.0 and 2.0 having a denominator no greater than 10.

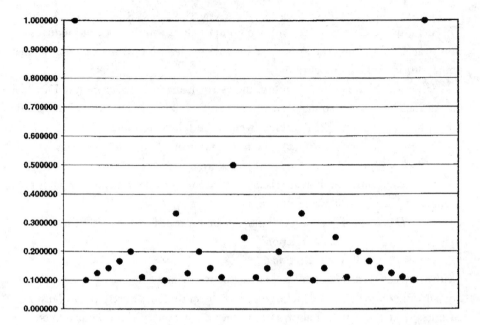

It turns out that the rational number with a value of $D(x)$ closest to 0 is $x = 1.30$, or $\frac{13}{10}$.

The initial spate of mathematical activity in Berlin greatly wearied Weierstrass. Shortly after the groundbreaking researches of 1859 to 1861, he suffered a complete collapse from what was termed "brain spasms." Ultimately, he had to take a year's sabbatical not just from teaching, but from mathematical research. When he returned to teaching in 1863, he always lectured from a seating position. An advanced student dealt with the physical labor of writing the mathematics on a blackboard.

Finally, after eight long years he was able to move to a position at the esteemed University of Berlin. Ever the perfectionist and always careful of his physical exertions, Weierstrass refrained from publishing his epsilon method until the middle of the 1880s. Perhaps not surprisingly the best of a generation of German mathematics students matriculated at Weierstrass's footsteps at the University of Berlin, the only source for learning the epsilon method firsthand.

Georg Cantor, whose work appears later in this book, was perhaps the greatest of all of Weierstrass's students. Cantor's work also became the cause of friction between Weierstrass and one of his oldest and best friends, Leopold Kronecker, who left us with the quotation, "God created the integers, all else is the work of man," used as the title of this book. Kronecker could abide all sorts of numbers: rational numbers, real numbers, even imaginary numbers. However, Kronecker could not abide the infinite, the focus of Cantor's work. Kronecker simply thought that Cantor's infinite mathematical objects could not exist and that Cantor's work was meaningless and useless.

Aside from his mathematics, Weierstrass's later years were not very happy. Like many mathematicians, Weierstrass never married and neither did his brother nor either of his two sisters. Weierstrass spent the last three years of his life confined to a wheelchair, completely dependent on others and immobile. He died in Berlin in early 1897 several months after his eighty-first birthday.

The great French mathematician Henri Poincaré, who had earlier complained of the surge of monstrous mathematical objects such as the Dirichlet function, had high praise for Weierstrass. Poincaré considered Weierstrass to be Germany's third greatest mathematician of the nineteenth century, surpassed by only the incomparable Gauss and the inestimable Riemann.

A THEORY OF FUNCTIONS*

7. UNIFORM CONTINUITY

We define a function of n variables $u_1, u_2, \ldots, u_n$ for an open[1] continuum where we allow that the whole n dimensional manifold could itself be this continuum. (Yet we exclude the boundaries of the continuum in the same way that we did earlier for the case of a function of one variable. Earlier we defined this function either for all values from $-\infty$ to ∞ or between a and b, but left it undetermined whether the function even made sense for these choices of boundaries.[2]

Now if one and only one value of the function x corresponds to each system of values $(u_1, u_2, \ldots, u_n)$, then we call this a *unique function* of the variables $u_1, u_2, \ldots, u_n$.[3] However, with this definition alone little is accomplished; it will be fruitful for our analysis only after introducing the concept of *continuity*. We define the concept of continuity in the usual manner as follows: the function x of the variables $(u_1, u_2, \ldots, u_n)$ is continuous in the neighborhood[4] of the point[5] $a = (a_1, \ldots, a_n)$, if, given any arbitrarily small, positive magnitude ε, it is possible to determine a δ such that

$$\text{if } |u_\lambda - a_\lambda| < \delta \qquad (\lambda = 1, 2, \ldots, n),$$
$$\text{then } |x - b| < \varepsilon,$$

where b refers to the value the function x takes for $(a_1, \ldots, a_n)$.

We can also characterize continuity thus: the function x is continuous in the neighborhood of a if for any given ε, it is possible to determine a region around a, with ρ, such that

$$\text{if } \sum_1^n (u_\lambda - a_\lambda)^2 < \rho^2,$$
$$\text{then } |x - b| < \varepsilon.$$

That both of these definitions completely agree with each other clearly needs no further explanation.

Finally, having correctly grasped the *concept of the infinitely small*,[6] we can thereby define the concept of the continuity of a function in the neighborhood of

---

*Translated by John Anders.

[1] "open" translates *unabgeschlossen*, which literally means "un-closed." In a previous section (not published here), Weierstrass defines an "open continuum" in the following discussion: ". . . if the boundary points [of a manifold] are entirely contained within a closed set of points then every point not contained in [this] set of points is contained in the continuum (or, if the manifold is partitioned into several continua through [certain] underlying sets of points, then [every point not contained in a given set of points] is contained in [one of] these continua). We call such a continuum . . . an *open continuum*."

[2] This earlier section is not included.

[3] $(u_1, u_2, \ldots, u_n)$ are given values which are appropriately inserted into the n-place function of the variables $u_1, u_2, \ldots, u_n$; Weierstrass calls this function x and this is the same function he defined in the opening sentence. Counter to the current convention, Weierstrass uses "x" to denote the dependent variable (what we usually call $f(x)$ or y) and a given "u" to denote a given independent variable (what we call x today).

[4] Here and below "neighborhood" translates "*Naehe*."

[5] Here and below "point" translates both "*Stelle*" and "*Punkt*." Here Weierstrass uses *Stelle*.

[6] Here Weierstrass makes a small joke by punning on "endlich" (finally, at last) and "unendlich" (infinite, without end).

a: to an infinitely small change of the argument, there corresponds an infinitely small change of the value of the function in the neighborhood of *a*. Also we can easily extend this definition to cases where the function is not defined at certain points of the continuum, if we always apply the definition only to points $(u_1, u_2, \ldots, u_n)$ for which the function is defined in the first place.[7]

From the concept of the continuity of a function we move on to an additional concept, which is a consequence of it, but has been frequently overlooked in the past–namely, the concept of *uniform continuity*.[8] We call a function uniformly continuous within some determinate intervals[9] $[a \ldots b]$ if, given any arbitrarily small, positive magnitude ε, it is possible to determine a magnitude δ such that the following condition is fulfilled for arbitrarily many value pairs *u* and *u'* within the intervals $[a \ldots b]$:

$$\text{if } |u' - u| < \delta,$$
$$\text{then } |f(u') - f(u)| < \varepsilon.$$

As we mentioned above, uniform continuity is a consequence of continuity as such,[10] and the proof of this which we are about to give for a function of one variable[11] can be applied to a function of several variables with no essential changes.[12] (This proof rests upon the fundamental propositions from the doctrine of manifolds we comprehensively established above.[13])

Proof:

We partition [*teilen*] the singular[14] manifold (which arises in the first place when we assign [*zuerteilen*][15] all values from $-\infty$ to ∞ to a variable magnitude *u*) into intervals in such a way that, after choosing some positive number *n*, we allow there to be a magnitude μ which ranges through all positive and negative whole numbers. And thereby we form the intervals

$$\left[\frac{\mu}{n} \ldots \frac{\mu + 1}{n} \right] \qquad (\mu = 0, \pm 1, \pm 2, \ldots, \pm \infty).[16]$$

---

[7] "in the first place" translates "*ueberhaupt.*"

[8] "*gleichmaessigen Stetigkeit*" which is literally "equal-measure continuity" or "equally-measured continuity." "Uniform continuity" is the modern English parlance. However, Weierstrass's term is more descriptive than our own; "uniform" continuity is uniform precisely with respect to how it is determined or measured by epsilons and deltas. See Weierstrass's own definition below.

[9] *Intervals.* For reasons that will be specified later, I have introduced square brackets where Weierstrass had open brackets in order to make it clear that he is dealing with closed intervals. Throughout these lectures Weierstrass used only open brackets.

[10] "as such" translates "*ueberhaupt.*" Alternately, one could say "in general." Weierstrass does not stress that fact that for a given interval of a given function, though its being uniformly continuous always implies its being continuous, in general, its being continuous *does not* imply its being uniformly continuous. (The classic example of an interval of a function which is continuous but not uniformly continuous is the $(0,1)$ interval of $f(x) = x^{-1}$.) In fact here and elsewhere his phrasing is misleading in that it seems to suggest that the converse holds *in general*, which is false. Of course, in many special cases continuity does imply uniform continuity, and Weierstrass does indeed deal with such a special case. (See footnote below.)

[11] Note that just above, Weierstrass moved from speaking about what it means for a function of *n variables* to be continuous to what it means for a function of *one variable* to be uniformly continuous. Here Weierstrass alludes to his later formulation of what it means for a function of *n variables* to be uniformly continuous. In the proof given just below, Weierstrass deals with a function of only one variable.

[12] This Weierstrass does in the second section below (which was delivered in 1866).

[13] Again these earlier sections are not included. It seems that Weierstrass is here referring to what is today called the Bolzano–Weierstrass accumulation point theorem.

[14] *einfache.*

[15] The connection between "*teilen*" and "*zuteilen*" is impossible to bring out in English.

[16] As mentioned in footnote 9 above, I am changing Weierstrass's notation: where he uses round brackets throughout I am using square brackets in the appropriate places to indicate that the intervals thus formed are closed, not open intervals. It is crucial that Weierstrass work with closed intervals. This is one condition that must hold if uniform continuity is to be derived from continuity.

All the values of u are contained in one of these intervals[17], and as long as $\frac{\mu}{n} \leq u < \frac{\mu + 1}{n}$, u must also be contained in the interval $[\frac{\mu}{n} \ldots \frac{\mu + 1}{n}]$. Now we give n and μ values such that the interval so defined by n and μ contains values of u for which the function is defined. For the values of $f(u)$ so contained in the interval, there is surely an upper and lower limit g and g' such that $g - g'$ represents the *largest variation* of the values of the function within this interval.[18] Thus

$$|f(u') - f(u'')| < |g - g'|,$$

where u' and u'' are arbitrary values of the arguments in these given intervals, and $|f(u') - f(u'')|$ is a positive, variable magnitude, the upper limit of which is $g - g'$ when u' and u'' lie in $[\frac{\mu}{n} \ldots \frac{\mu + 1}{n}]$.

We now consider the totality[19] of the intervals $[\frac{\mu}{n} \ldots \frac{\mu + 1}{n}]$ ($\mu = 0, \pm 1, \pm 2, \ldots, \pm\infty$), and we keep only those intervals for which the function as such[20] is defined. Now we imagine that for each of these intervals we have determined the greatest change in value within each interval, and then we select that change in value which is the greatest, namely, that which is $= g_n$. Clearly g_n depends on n.[21] (It is not important that there may perhaps be many μ for which the variation in value reaches this maximum value g_n in the interval $[\frac{\mu}{n} \ldots \frac{\mu + 1}{n}]$.)

Now there are infinitely many such intervals $[\frac{\mu}{n} \ldots \frac{\mu + 1}{n}]$, because infinitely many values can be posited for n. In order to recognize this we can use n to represent, for example, the values of each of the prime numbers. Then each interval occurs only once. Therefore, because the number of prime numbers is infinite, there are in this case actually infinitely many of them in the intervals defined above. Now one such interval is completely determined by means of the point $\frac{\mu}{n}$, and according to the fundamental lemma from our magnitude doctrine[22] there must be at least one point in whose neighborhood the thus defined points $\frac{\mu}{n}$ arbitrarily accumulate.[23] If there are many such points, then we select one of the them and call it u_0. Then we select an arbitrarily small magnitude ε, and because the function we chose constantly changes in the neighborhood of u_0,[24] we are able to determine a magnitude δ such that if $|u - u_0| < \delta$, then $|f(u) - f(u_0)| < \varepsilon$.

---

[17] This is true because these intervals span all the real numbers.

[18] The fact that Weierstrass here assumes that there is an upper and lower limit for the function on the given interval is another condition that must hold for Weierstrass to prove that uniform continuity follows from continuity. Without this assumption (i.e., in the general case) it is not true that continuity necessarily implies uniform continuity. In general, uniform continuity is the stronger condition.

[19] Here and below "totality" translates "*Gesamtheit.*"

[20] "as such" translates "*ueberhaupt.*"

[21] Because the interval that will contain the greatest change in value, viz., g_n, will, in general, depend on how finely the intervals are divided, which, in turn, is dictated by the choice n.

[22] Again, this is not included.

[23] Today this is called the Bolzano–Weierstrass theorem. In modern terms it states that every bounded infinite set of points on the real line has an accumulation point.

[24] I.e., the function is continuous by supposition. See the definition of continuity above.

Now if we consider the interval $[u_0 - \delta \ldots u_0 + \delta]$, then in every neighborhood of u_0 there lie infinitely many points $\frac{\mu}{n}$. It follows from this that we can find $\frac{\mu}{n}$ and $\frac{\mu + 1}{n}$ such that $[\frac{\mu}{n} \ldots \frac{\mu + 1}{n}]$ lies entirely within $[u_0 - \delta \ldots u_0 + \delta]$, and it is clear that to accomplish this, all we have to do is select a large enough n. But then in the interval $[\frac{\mu}{n} \ldots \frac{\mu + 1}{n}]$, the difference between two argument values, one of which is $=u_0$, is surely $< \delta$; therefore, because the condition above[25] is fulfilled, $|f(u) - f(u_0)| < \varepsilon$. Consequently, if u' and u'' are the argument values that yield the smallest and the largest values of $f(u)$ in $[\frac{\mu}{n} \ldots \frac{\mu + 1}{n}]$, then $|f(u') - f(u'')| < 2\varepsilon$, thus $g_n < 2\varepsilon$. Now because in every given interval $\frac{\mu}{n} \ldots \frac{\mu + 1}{n}$ the greatest change in value is less than or equal to g_n, it follows that the difference between two arbitrary values whose arguments lie in this interval is smaller than the magnitude 2ε, which itself can be made arbitrarily small.

Thereby the proposition is proven; it has been shown that we can partition the domain[26] of the variable u into intervals such that in each of them the difference between two arbitrary function values is smaller than any arbitrarily small, prescribed magnitude.

Thus, given that $|u' - u''| < \delta$, in order for it to be the case that $|f(u') - f(u'')| < \varepsilon$, all we have to do is select an n such that $g_n < \frac{1}{n}\varepsilon$, and then $\delta < \frac{1}{n}$ fulfills our conditions. For even if u' and u'' were to lie in two successive intervals (which is the only remaining case if u' and u'' were not to lie in the same intervals – I say this because it must always be the case that $|u' - u''| < \frac{1}{n}$), it would still be the case that $|f(u') - f(u)| < 2\varepsilon$. And, of course, 2ε can be made arbitrarily small along with ε.

Therefore the proposition is completely proven; *uniform continuity* can be derived as a *consequence of continuity as such*.[27] The importance of this proposition consists precisely in that from now on, for any particular function for which continuity has been proven, the proof of uniform continuity is already accomplished[28]–a proof that will probably be very laborious in many special cases.

[25] Namely, if $|u' - u| < \delta$, then $|f(u') - f(u)| < \varepsilon$.
[26] Here and below "domain" translates "*Bereich.*"
[27] "as such" translates "*ueberhaupt*".
[28] Literally "overcome" ["*ueberhoben*"]. Of course, as we stated above, continuity implies uniform continuity only for certain intervals of certain functions, i.e., it does not hold in general.

Saturday, *June* 26, 1886

Now we progress from proving the proposition for functions of one variable to expanding the proof to functions of arbitrarily many variables. Let $f(u_1, \ldots, u_n)$ be a unique function[29] of n variables $(u_1, u_2, \ldots, u_n)$, which is defined for either a closed or an open continuum[30] and is continuous in the *region*[31] in which it is defined. Thus it is our task to prove uniform continuity from the continuity of a function in this region.

We call one such function $f(u_1, \ldots, u_n)$ uniformly continuous if, given any arbitrarily small magnitude ε, it is possible to determine another magnitude δ such that

$$\text{if} \quad |u'_\lambda - u''_\lambda| < \delta \quad (\lambda = 1, 2, \ldots, n),$$
$$\text{then} \quad |f(u'_1, \ldots, u'_n) - f(u''_1, \ldots, u''_n)| < \varepsilon,$$

where this holds for two points u (the above inequalities holding, of course, only in this region in which the function is defined). Now our proof of the proposition for functions of n variables is completely analogous to the proof for functions of one variable.

Proof:

We call the totality of all points $(u_1, u_2, \ldots, u_n)$ for which $a_\lambda \leq u_\lambda \leq a_\lambda + d_\lambda$ $(\lambda = 1, \ldots, n)$ (where d_λ is a prescribed, positive magnitude) a known *domain* of the variables $(u_1, u_2, \ldots, u_n)$, and we designate this domain by

$$[a_1 \ldots a_1 + d_1, a_2 \ldots a_2 + d_2, \ldots, a_n \ldots a_n + d_n].$$

Now we consider the domain

$$\left[\frac{\mu_1}{m} \ldots \frac{\mu_1 + 1}{m}, \ldots, \frac{\mu_n}{m} \ldots \frac{\mu_n + 1}{m} \right],$$

where μ_λ can take on all possible positive and negative whole number values and m is a fixed, positive number. Thus every point $(u_1, u_2, \ldots, u_n)$ will belong to one and only one of these domains, and the totality of them forms our n dimensional manifold. Because d_λ has simply the same value, viz., $\frac{1}{m}$, for all λ, we are already able to characterize the domain through a specification of numbers μ_λ $(\lambda = 1, \ldots, n)$. If we imagine two points, $(u'_1, u'_2, \ldots, u'_n)$ and $(u''_1, u''_2, \ldots, u''_n)$, in one of these intervals, then for each $|f(u'_1, \ldots, u'_n) - f(u''_1, \ldots, u''_n)|$ there is a variable magnitude with an upper limit g that has a determinate value in all of the considered intervals.[32]

Now we imagine those intervals for which the function is defined in the first place;[33] the number of them is itself infinite, because the function is defined for an

---

[29] See first page of this section where this term is introduced.
[30] For the terms "closed continuum" and "open continuum," see footnote 1 for Weierstrass's definition.
[31] Here and below "region" translates "*Gebiete.*"
[32] This is the crucial assumption that allows Weierstrass to prove uniform continuity from continuity. Compare footnote 18 above.
[33] "in the first place" translates "*ueberhaupt.*"

infinite continuum. For each of these intervals, there is an upper limit of the absolute value of the difference given above,[34] and again among them there is either one that is the largest or there are several that are the largest.[35] If $\left[\frac{\mu_1}{m}, \frac{\mu_2}{m}, \ldots, \frac{\mu_n}{m}\right]$[36] is one of the domains just mentioned, in which g reaches its highest value, then it is clear that if we now allow it to vary as far as a given, fixed m, we will obtain infinitely many such intervals. Thus there must be at least one point $(a_1, a_2, \ldots, a_n)$ at which infinitely many of the defined points $\left(\frac{\mu_1}{m}, \ldots, \frac{\mu_n}{m}\right)$ are present in each neighborhood of it.[37]

If we focus on one such point $(a_1, a_2, \ldots, a_n)$, then, because of the presupposition of continuity, we are able to determine a magnitude δ such that if $(u'_1, u'_2, \ldots, u'_n)$ and $(u''_1, u''_2, \ldots, u''_n)$ are contained within the interval $[a_1 - \delta \ldots a_1 + \delta, \ldots, a_n - \delta \ldots a_n + \delta]$, then it is the case that $|f(u'_1, \ldots, u'_n) - f(u''_1, \ldots, u''_n)| < \varepsilon$.

Furthermore, following some selection of m, we can determine the meaning of $(a_1, a_2, \ldots, a_n)$ such that the domain $\left[\frac{\mu_1}{m}, \ldots, \frac{\mu_n}{m}\right]$ lies entirely within $[a_1 - \delta \ldots a_1 + \delta, \ldots, a_n - \delta \ldots a_n + \delta]$, because all we have to do first is to determine the quotients $\frac{\mu_\lambda}{m}$ such that the a_λ come as close to them as we wish. From this we can make $\frac{1}{m}$ as small as δ by suitably enlarging m, while both $\frac{\mu_\lambda}{m}$ and $\frac{\mu_\lambda + 1}{m}$ lie within $[a_\lambda - \delta \ldots a_\lambda + \delta]$. Then the above inequality[38] applies to all points $(u'_1, u'_2, \ldots, u'_n)$ and $(u''_1, u''_2, \ldots, u''_n)$, which are contained in the domain $\left[\frac{\mu_1}{m}, \ldots, \frac{\mu_n}{m}\right]$; thus $g < \varepsilon$ is the greatest difference between the values in this domain. In each remaining interval, the difference is $\leq g$, thus *a fortiori*, $< \varepsilon$.

Therefore, we have the entire continuum for which the function is defined divided into a number of parts of the domain in such a way that in each of them the greatest change in value of the function is smaller than an arbitrarily small, given magnitude ε.

Now, after choosing an arbitrary point $(a'_1, \ldots, a'_n)$ at which we know that the function is defined, we can demarcate a domain around it $[a'_1 - d'_1 \ldots a'_n + d''_1, \ldots, a'_n - d'_n \ldots a'_n + d''_n]$ such that for all points within this domain the differences between the two function values is $< \varepsilon$; to do this, first we need to select an m large enough that for all parts of the domain $\left[\frac{\mu_1}{m}, \ldots, \frac{\mu_n}{m}\right]$, the greatest variation in value is $< \frac{1}{2}\varepsilon$. According to the above, we can always do this. For if we select the part of the domain in which $(a'_1, \ldots, a'_n)$ lies, and choose δ'_λ

---

[34] Viz., $|f(u'_1, \ldots, u'_n) - f(u''_1, \ldots, u''_n)|$ at every pair of points.

[35] That is to say, that either the largest g occurs within some one interval (all other intervals having a smaller g) or the largest g is to be found in several intervals (which happen to share the same g).

[36] Author's note: Which from now on is how we will more briefly refer to our partition of the domain from the basis given above.

[37] This follows from Weierstrass's accumulation-point theorem. See footnote 23 above.

[38] viz. $|f(u'_1, \ldots, u'_n) - f(u''_1, \ldots, u''_n)| < \varepsilon$.

and δ''_λ to be $< \frac{1}{2}m$, then either all of $[a'_1 - d'_1 \ldots a'_n + d''_1, \ldots, a'_n - d'_n \ldots$ $a'_n + d''_n]$ belongs to the part of the domain that is $[\frac{\mu_1}{m}, \ldots, \frac{\mu_n}{m}]$ or it simultaneously belongs to it and to $2^n - 1$ of the $3^n - 1$ neighboring parts of the domain.[39] In the former case, the second difference between any two functions contained in the interval $[a'_1 - d'_1 \ldots a'_1 + d''_1, \ldots]$ is not only $< \varepsilon$ but also $< \frac{1}{2}\varepsilon$; in the latter case, where both argument values clearly belong to two different sections of the 2^n parts under consideration, it is $< \varepsilon$.

Accordingly, around *every* point in the continuum for which the function is defined, we can define a domain such that in it the greatest variation in value of the function will be arbitrarily small. Therefore, the proposition put forward at the beginning of the section is proven, and indeed, to emphasize it again, *uniform* continuity has been deduced as a consequence of continuity as such.[40]

Example: The case $n = 2$ (even), $3^n - 1 = 8$, $2^n - 1 = 3$.

---

[39] See Weierstrass's diagram.
[40] *ueberhaupt.*

Richard Julius Wilhelm Dedekind

(1831–1916)

HIS LIFE AND WORK

Richard Julius Wilhelm Dedekind was born on October 6, 1831, in Brunswick, Germany, which was also the birthplace of Carl Friedrich Gauss. Dedekind was the youngest of the four children of Julius Levin Ulrich Dedekind and Caroline Marie Henriette Dedekind (neé Emperius). He grew up in the intellectually stimulating home one would expect of an academically accomplished family. His father was a Professor of Law at the Collegium Carolinum in Brunswick whose own father had been a distinguished chemist and physician. His maternal grandfather had also been a professor at the Collegium. The other Dedekind son became a court judge, and his sister Julie published several novels.

Dedekind followed in his brother's footsteps and began his formal education at the Gymnasium Martino-Catharineum in Brunswick. At first, chemistry and physics were his favorite subjects, with mathematics just a useful tool in their service. However, by the time he enrolled at the local Collegium Carolinum in 1848, Dedekind had turned away from chemistry and physics because of what seemed to him to be their chaotic logical structure in favor of mathematics. During his two years at the Collegium, Dedekind studied nothing but mathematics and the most abstract of the physical sciences. In short order, he mastered analytic geometry, differential and integral calculus, and the foundations of analysis.

When Dedekind enrolled at the nearby University of Göttingen in 1850, he was by far the best prepared student of his entering class. Dedekind sped through Göttingen's mathematics offerings. He attended Gauss's lectures on the theory of least squares. Fifty years later he would remember them as the most beautiful and logical exposition of mathematics he ever encountered. While at Göttingen he attended a seminar in number theory, the most advanced offering of the mathematics department. At this seminar he met the slightly older Bernhard Riemann, whose champion he would become. In the spring of 1852, Dedekind wrote his doctoral dissertation on the theory of Eulerian integrals under Gauss's supervision. He would be the last of the great Gauss's six doctoral students.

In spite of having received his doctorate, Dedekind considered his mathematical education to be incomplete. He spent two years attending classes to fill in what he considered to be the gaps in his knowledge before he took the examination that would allow him to be a university lecturer. Of course, he passed with flying colors enabling him to take a lowly teaching post at Göttingen as a *Privatdozent*, lecturing undergraduates on geometry and probability.

Dedekind felt a particularly great loss when Gauss died in 1855. However, Dedekind's loss was soon followed by the great satisfaction of working with Gauss's successor, Peter Gustave Lejeune Dirichlet. Although a member of the faculty himself, Dedekind had no compunction about attending Dirichlet's courses on a wide variety of topics including number theory, potential theory, definite integrals, and partial differential equations. The two men were so close that many members of the Göttingen mathematics community joked that they were inseparable twins. But if anyone were Dedekind's twin, it was Riemann. They were truly inseparable spending long hours together walking in the woods near Göttingen discussing mathematical topics. When Dedekind received an offer of a better post at the Zurich Polytechnic, he agonized over leaving Göttingen and Riemann. In the end, Dedekind accepted the offer in far off Zurich.

Although physically in Switzerland, Dedekind remained spiritually in Germany. When Riemann was elected to the Berlin Academy in 1859, Dedekind leapt at the chance to accompany his friend on his visit to Berlin. It turned out to be wise decision. While in Berlin, he met the leading figures of the German mathematical community: Kummer, Brochardt, Kronecker, and especially Weierstrass. These contacts soon bore fruit. In 1862, they arranged for Dedekind to receive an offer at the Brunswick Polytechnic, the successor of the Carolinum Collegium. Dedekind returned to his alma mater, where his father had taught. Although more prestigious appointments would be offered in later years, Dedekind felt very much at home in his birthplace and never took another position.

A course on differential calculus that Dedekind taught his first year in Zurich kindled his interest in the foundations of mathematics. As the course unfolded, he realized that he was unable to present his students with a sound theoretical foundation for the continuum of real numbers comprising the real line. He set to work to remedy this situation. Fourteen years later, he published his classic work *Continuity and Irrational Numbers* (1872) that gave the first rigorous foundation for the continuum of real numbers. It is included in this volume along with his essay *The Nature and Meaning of Numbers* (1887).

The nature of the real line had bedeviled mathematics ever since the Pythagoreans discovered the irrationality of $\sqrt{2}$ in the fifth century B.C. Surveying the development of mathematics, Dedekind recognized the successive extension of one class of numbers into a larger class of which they were a part. First, the whole numbers 1, 2, 3, . . . were extended to the entire class of integers: positive, zero, and negative. Mathematical operations on negative integers could be expressed in terms of whole numbers. Then the integers were extended to the rational numbers of which they were a part. Most recently, Dedekind's mentor Gauss had extended the real numbers to the complex numbers having real and imaginary parts.

Dedekind employed the concept of a *cut* to ground the whole of the real line, rational and irrational numbers, in the rational numbers. He began by noticing that every rational number a divides the set of rational numbers into two classes:

- A_1, the set of rational numbers less than a
- A_2, the set of rational numbers greater than a

Moreover, every rational number belonging to A_1 is less than every rational number belonging to A_2. Thus the rational number a *cuts* the rational numbers into two distinct classes of numbers separated by the order relationship *less than/greater than*.

Dedekind realized that many other cuts could be created among the rational numbers. For example, $\sqrt{2}$ cuts the rational numbers into two classes:

- A_1, the set of positive rational numbers whose square is less than $\sqrt{2}$ (plus 0 and all the negative rational numbers)
- A_2, the set of positive rational numbers whose square greater than $\sqrt{2}$

Once again, every rational number belonging to A_1 is less than every rational number belonging to A_2. Similarly, π can be defined as cutting the rational numbers into two classes:

• A_1, the set of positive rational numbers of the form a/b whose ratio is less than ration of a circle's circumference to its diameter (plus 0 and all the negative rational numbers)

• A_2, the set of positive rational numbers of the form a/b whose ratio is greater than ration of a circle's circumference to its diameter

Here too, every rational number belonging to A_1 is less than every rational number belonging to A_2.

Having given examples of *cuts*, Dedekind generalized the concept so that any exhaustive division of the rational numbers into two sets A_1 and A_2 such that every member of A_1 is less than every member of A_2 defines a *cut* that is identified with a real number. If the real number identified is not a rational number, as in the case of the cut that defines $\sqrt{2}$, then it must be an irrational number.

After defining the real numbers in terms of cuts, Dedekind went on to show that arithmetic on the real numbers is just as we would expect it to be precisely because the basic arithmetic operations of addition and subtraction preserve the *less than/greater than* relationship among positive rational numbers. Consider, for example, the cuts created by $\sqrt{2}$ and $\sqrt{3}$ into the sets A_1 and A_2 and B_1 and B_2 respectively. Take any rational number in A_1 and any rational number in B_1. Their product will be less than the product of any rational number in A_2 and any rational number in B_2, and the cut corresponds to $\sqrt{6}$, just as one would expect.

Dedekind frequently spent his summers in Switzerland or the Tyrolean Alps. During his 1874 holiday, he met Georg Cantor for the first time at the city of Interlaken. Dedekind's groundbreaking analysis of the continuum served as one of Cantor's major sources of inspiration for his own revolutionary work on transfinite numbers presented later in this volume. It is through Cantor's correspondence with Dedekind in the 1870s and 1880s that historians have come to understand the development of Cantor's ideas.

As Göttingen rose to the prominence of Berlin in mathematics, Brunswick remained a backwater. Dedekind never had any doctoral students at Brunswick. The light demands at Brunswick afforded Dedekind the opportunity to oversee the publication of two mathematical masterpieces. In 1863, Dedekind published Dirichlet's *Lectures on Number Theory*, mostly from notes he had taken. By 1894, three more editions had appeared. When Riemann died prematurely in 1866, Dedekind took it on himself to publish his friend's *Collected Works*. That classic volume appeared ten years after his friend's early death. He also supervised the publication of the collected works of his mentor Carl Friedrich Gauss.

The light demands at Brunswick also allowed Dedekind to become the Polytechnic's Director from 1872 to 1875. Although he disliked administrative duties, Dedekind did enjoy the fact that he had taken on a position that had previously been his father's and his maternal grandfather's. Dedekind retired from his university post in 1894. Even though retired, he continued to lecture until well into his 70s. Unlike many mathematicians he enjoyed lecturing.

Dedekind received many honors over the course of his career. The first came from Göttingen which elected him a corresponding member of its Academy in 1862. In 1880, he received a similar honor from the Academy of Science in Berlin, in 1900 from the Académie des Sciences in Paris and the Academy in Rome. In 1902, Dedekind received many honorary doctorates on the fiftieth anniversary of his doctorate at Göttingen.

Content with his life in Brunswick, Dedekind never married and lived with his sister Julie until her death in 1914. He died after a short illness on February 12, 1916. Even though Europe was in the midst of the Great War, mathematicians in France and England mourned his passing, knowing that the last direct link to the great Gauss had been lost.

ESSAYS ON THE
THEORY OF NUMBERS

PREFACE TO THE FIRST EDITION

In science nothing capable of proof ought to be accepted without proof. Though this demand seems so reasonable yet I cannot regard it as having been met even in the most recent methods of laying the foundations of the simplest science; viz., that part of logic which deals with the theory of numbers.[1] In speaking of arithmetic (algebra, analysis) as a part of logic I mean to imply that I consider the number-concept entirely independent of the notions or intuitions of space and time, that I consider it an immediate result from the laws of thought. My answer to the problems propounded in the title of this paper is, then, briefly this: numbers are free creations of the human mind; they serve as a means of apprehending more easily and more sharply the difference of things. It is only through the purely logical process of building up the science of numbers and by thus acquiring the continuous number-domain that we are prepared accurately to investigate our notions of space and time by bringing them into relation with this number-domain created in our mind.[2] If we scrutinise closely what is done in counting an aggregate or number of things, we are led to consider the ability of the mind to relate things to things, to let a thing correspond to a thing, or to represent a thing by a thing, an ability without which no thinking is possible. Upon this unique and therefore absolutely indispensable foundation, as I have already affirmed in an announcement of this paper,[3] must, in my judgment, the whole science of numbers be established. The design of such a presentation I had formed before the publication of my paper on *Continuity*, but only after its appearance and with many interruptions occasioned by increased official duties and other necessary labors, was I able in the years 1872 to 1878 to commit to paper a first rough draft which several mathematicians examined and partially discussed with me. It bears the same title and contains, though not arranged in the best order, all the essential fundamental ideas of my present paper, in which they are more carefully elaborated. As such main points I mention here the sharp distinction between finite and infinite (64), the notion of the number

---

[1] Of the works which have come under my observation I mention the valuable *Lehrbuch der Arithmetik und Algebra* of E. Schröder (Leipzig, 1873), which contains a bibliography of the subject, and in addition the memoirs of Kronecker and von Helmholtz upon the Number-Concept and upon Counting and Measuring (in the collection of philosophical essays published in honor of E. Zeller, Leipzig, 1887). The appearance of these memoirs has induced me to publish my own views in many respects similar but in foundation essentially different, which I formulated many years ago in absolute independence of the works of others.

[2] See Section III. of my memoir, *Continuity and Irrational Numbers* (Braunschweig, 1872).

[3] Dirichlet's *Vorlesungen über Zahlentheorie*, third edition, 1879, § 163.

[*Anzahl*] of things (161), the proof that the form of argument known as complete induction (or the inference from n to $n + 1$) is really conclusive (59), (60), (80), and that therefore the definition by induction (or recursion) is determinate and consistent (126).

This memoir can be understood by any one possessing what is usually called good common sense; no technical philosophic, or mathematical, knowledge is in the least degree required. But I feel conscious that many a reader will scarcely recognise in the shadowy forms which I bring before him his numbers which all his life long have accompanied him as faithful and familiar friends; he will be frightened by the long series of simple inferences corresponding to our step-by-step understanding, by the matter-of-fact dissection of the chains of reasoning on which the laws of numbers depend, and will become impatient at being compelled to follow out proofs for truths which to his supposed inner consciousness seem at once evident and certain. On the contrary in just this possibility of reducing such truths to others more simple, no matter how long and apparently artificial the series of inferences, I recognise a convincing proof that their possession or belief in them is never given by inner consciousness but is always gained only by a more or less complete repetition of the individual inferences. I like to compare this action of thought, so difficult to trace on account of the rapidity of its per-formance, with the action which an accomplished reader performs in reading; this reading always remains a more or less complete repetition of the individual steps which the beginner has to take in his wearisome spelling-out; a very small part of the same, and therefore a very small effort or exertion of the mind, is suffi-cient for the practised reader to recognise the correct, true word, only with very great probability, to be sure; for, as is well known, it occasionally happens that even the most practised proof-reader allows a typographical error to escape him, i.e., reads falsely, a thing which would be impossible if the chain of thoughts associated with spelling were fully repeated. So from the time of birth, continu-ally and in increasing measure we are led to relate things to things and thus to use that faculty of the mind on which the creation of numbers depends; by this practice continually occurring, though without definite purpose, in our earliest years and by the attending formation of judgments and chains of reasoning we acquire a store of real arithmetic truths to which our first teachers later refer as to something simple, self-evident, given in the inner consciousness; and so it hap-pens that many very complicated notions (as for example that of the number [*Anzahl*] of things) are erroneously regarded as simple. In this sense which I wish to express by the word formed after a well-known saying ἀεὶ ὁ ἄνθρωπος ἀριθμητίζει, I hope that the following pages, as an attempt to establish the science of numbers upon a uniform foundation will find a generous welcome and that other mathematicians will be led to reduce the long series of inferences to more moderate and attractive proportions.

In accordance with the purpose of this memoir I restrict myself to the consideration of the series of so-called natural numbers. In what way the gradual extension of the number-concept, the creation of zero, negative, fractional, irrational and complex numbers are to be accomplished by reduction to the earlier notions and that without any introduction of foreign conceptions (such as that of measurable magnitudes, which according to my view can attain perfect clearness only through the science of numbers), this I have shown at least for irrational numbers in my former memoir on *Continuity* (1872); in a way wholly similar, as I have already shown in Section III. of that memoir, may the other extensions be treated, and I propose sometime to present this whole subject in systematic form. From just this point of view it appears as something self-evident and not new that every theorem of algebra and higher analysis, no matter how remote, can be expressed as a theorem about natural numbers,—a declaration I have heard repeatedly from the lips of Dirichlet. But I see nothing meritorious—and this was just as far from Dirichlet's thought—in actually performing this wearisome circumlocution and insisting on the use and recognition of no other than rational numbers. On the contrary, the greatest and most fruitful advances in mathematics and other sciences have invariably been made by the creation and introduction of new concepts, rendered necessary by the frequent recurrence of complex phenomena which could be controlled by the old notions only with difficulty. On this subject I gave a lecture before the philosophic faculty in the summer of 1854 on the occasion of my admission as privat-docent in Göttingen. The scope of this lecture met with the approval of Gauss; but this is not the place to go into further detail.

Instead of this I will use the opportunity to make some remarks relating to my earlier work, mentioned above, on *Continuity and Irrational Numbers*. The theory of irrational numbers there presented, wrought out in the fall of 1853, is based on the phenomenon (Section IV.) occurring in the domain of rational numbers which I designate by the term cut [*Schnitt*] and which I was the first to investigate carefully; it culminates in the proof of the continuity of the new domain of real numbers (Section V., iv.). It appears to me to be somewhat simpler, I might say easier, than the two theories, different from it and from each other, which have been proposed by Weierstrass and G. Cantor, and which likewise are perfectly rigorous. It has since been adopted without essential modification by U. Dini in his *Fondamenti per la teorica delle funzioni di variabili reali* (Pisa, 1878); but the fact that in the course of this exposition my name happens to be mentioned, not in the description of the purely arithmetic phenomenon of the cut but when the author discusses the existence of a measurable quantity corresponding to the cut, might easily lead to the supposition that my theory rests upon the consideration of such quantities. Nothing could be further from the truth; rather have I in Section III. of my paper advanced several reasons why I wholly reject the introduction of measurable quantities; indeed, at the end of the paper I have pointed out with respect to their existence that for a great part of the science of space the continuity of its configurations is not even a necessary condition, quite aside from the fact that in works on geometry arithmetic is only casually mentioned by name but is never clearly defined and therefore cannot be employed in demonstrations. To explain this matter more clearly I note the following example: If we select three non-collinear points A, B, C at pleasure, with the single limitation that the ratios of the distances AB, AC, BC are algebraic numbers,[4] and regard as existing in space only those points M, for which the ratios of AM, BM, CM to AB are likewise algebraic numbers, then is the space made up of the points M, as is easy to see, everywhere discontinuous; but in spite of this discontinuity, and despite the existence of gaps in this space, all constructions that occur in Euclid's *Elements*, can, so far as I can see, be just as accurately effected as in perfectly continuous space; the discontinuity of this space would not be noticed in Euclid's science, would not be felt at all. If any one should say that we cannot conceive of space as anything else than continuous, I should venture to doubt it and to call attention to the fact that a far advanced, refined scientific training is demanded in order to perceive clearly the essence of continuity and to comprehend that besides rational quantitative relations, also irrational, and besides algebraic, also transcendental quantitative relations are conceivable. All the more beautiful it appears to me that without any notion of measurable quantities and simply by a finite system of simple thought-steps man can advance to the creation of the pure continuous number-domain; and only by this means in my view is it possible for him to render the notion of continuous space clear and definite.

The same theory of irrational numbers founded upon the phenomenon of the cut is set forth in the *Introduction à la théorie des fonctions d'une variable* by J. Tannery (Paris, 1886). If I rightly understand a passage in the preface to this work, the author

---

[4] Dirichlet's *Vorlesungen über Zahlentheorie*, § 159 of the second edition, § 160 of the third.

has thought out his theory independently, that is, at a time when not only my paper, but Dini's *Fondamenti* mentioned in the same preface, was unknown to him. This agreement seems to me a gratifying proof that my conception conforms to the nature of the case, a fact recognised by other mathematicians, e.g., by Pasch in his *Einleitung in die Differential- und Integral-rechnung* (Leipzig, 1883). But I cannot quite agree with Tannery when he calls this theory the development of an idea due to J. Bertrand and contained in his *Traité d'arithmétique*, consisting in this that an irrational number is defined by the specification of all rational numbers that are less and all those that are greater than the number to be defined. As regards this statement which is repeated by Stolz—apparently without careful investigation—in the preface to the second part of his *Vorlesungen über allgemeine Arithmetik* (Leipzig, 1886), I venture to remark the following: That an irrational number is to be considered as fully defined by the specification just described, this conviction certainly long before the time of Bertrand was the common property of all mathematicians who concerned themselves with the notion of the irrational. Just this manner of determining it is in the mind of every computer who calculates the irrational root of an equation by approximation, and if, as Bertrand does exclusively in his book, (the eighth edition, of the year 1885, lies before me,) one regards the irrational number as the ratio of two measurable quantities, then is this manner of determining it already set forth in the clearest possible way in the celebrated definition which Euclid gives of the equality of two ratios (*Elements*, V., 5). This same most ancient conviction has been the source of my theory as well as that of Bertrand and many other more or less complete attempts to lay the foundations for the introduction of irrational numbers into arithmetic. But though one is so far in perfect agreement with Tannery, yet in an actual examination he cannot fail to observe that Bertrand's presentation, in which the phenomenon of the cut in its logical purity is not even mentioned, has no similarity whatever to mine, inasmuch as it resorts at once to the existence of a measurable quantity, a notion which for reasons mentioned above I wholly reject. Aside from this fact this method of presentation seems also in the succeeding definitions and proofs, which are based on the postulate of this existence, to present gaps so essential that I still regard the statement made in my paper (Section VI.), that the theorem $\sqrt{2} \cdot \sqrt{3} = \sqrt{6}$ has nowhere yet been strictly demonstrated, as justified with respect to this work also, so excellent in many other regards and with which I was unacquainted at that time.

R. DEDEKIND.

HARZBURG, October 5, 1887.

PREFACE TO THE SECOND EDITION

The present memoir soon after its appearance met with both favorable and unfavorable criticisms; indeed serious faults were charged against it. I have been unable to convince myself of the justice of these charges, and I now issue a new edition of the memoir, which for some time has been out of print, without change, adding only the following notes to the first preface.

The property which I have employed as the definition of the infinite system had been pointed out before the appearance of my paper by G. Cantor (*Ein Beitrag zur Mannigfaltigkeitslehre, Crelle's Journal,* Vol. 84, 1878), as also by Bolzano (*Paradoxien des Unendlichen,* § 20, 1851). But neither of these authors made the attempt to use this property for the definition of the infinite and upon this foundation to establish with rigorous logic the science of numbers, and just in this consists the content of my wearisome labor which in all its essentials I had completed several years before the appearance of Cantor's memoir and at a time when the work of Bolzano was unknown to me even by name. For the benefit of those who are interested in and understand the difficulties of such an investigation, I add the following remark. We can lay down an entirely different definition of the finite and infinite, which appears still simpler since the notion of similarity of transformation is not even assumed, viz.:

"A system S is said to be finite when it may be so transformed in itself (36) that no proper part (6) of S is transformed in itself; in the contrary case S is called an infinite system."

Now let us attempt to erect our edifice upon this new foundation! We shall soon meet with serious difficulties, and I believe myself warranted in saying that the proof of the perfect agreement of this definition with the former can be obtained only (and then easily) when we are permitted to assume the series of natural numbers as already developed and to make use of the final considerations in (131); and yet nothing is said of all these things in either the one definition or the other! From this we can see how very great is the number of steps in thought needed for such a remodeling of a definition.

About a year after the publication of my memoir I became acquainted with G. Frege's *Grundlagen der Arithmetik,* which had already appeared in the year 1884. However different the view of the essence of number adopted in that work is from my own, yet it contains, particularly from § 79 on, points of very close contact with my paper, especially with my definition (44). The agreement, to be sure, is not easy to discover on account of the different form of expression; but the positiveness with which the author speaks of the logical inference from n to $n + 1$ (page 93, below) shows plainly that here he stands upon the same ground with me. In the meantime E. Schröder's *Vorlesungen über die Algebra der Logik* has been almost completed (1890–1891). Upon the importance of this extremely suggestive work, to which I pay my highest tribute, it is impossible here to enter further; I will simply confess that in spite of the remark made on p. 253 of Part I., I have retained my somewhat clumsy symbols (8) and (17); they make no claim to be adopted generally but are intended simply to serve the purpose of this arithmetic paper to which in my view they are better adapted than sum and product symbols.

R. DEDEKIND.

HARZBURG, August 24, 1893.

CONTINUITY AND IRRATIONAL NUMBERS

My attention was first directed toward the considerations which form the subject of this pamphlet in the autumn of 1858. As professor in the Polytechnic School in Zürich I found myself for the first time obliged to lecture upon the elements of the differential calculus and felt more keenly than ever before the lack of a really scientific foundation for arithmetic. In discussing the notion of the approach of a variable magnitude to a fixed limiting value, and especially in proving the theorem that every magnitude which grows continually, but not beyond all limits, must certainly approach a limiting value, I had recourse to geometric evidences. Even now such resort to geometric intuition in a first presentation of the differential calculus, I regard as exceedingly useful, from the didactic standpoint, and indeed indispensable, if one does not wish to lose too much time. But that this form of introduction into the differential calculus can make no claim to being scientific, no one will deny. For myself this feeling of dissatisfaction was so overpowering that I made the fixed resolve to keep meditating on the question till I should find a purely arithmetic and perfectly rigorous foundation for the principles of infinitesimal analysis. The statement is so frequently made that the differential calculus deals with continuous magnitude, and yet an explanation of this continuity is nowhere given; even the most rigorous expositions of the differential calculus do not base their proofs upon continuity but, with more or less consciousness of the fact, they either appeal to geometric notions or those suggested by geometry, or depend upon theorems which are never established in a purely arithmetic manner. Among these, for example, belongs the above-mentioned theorem, and a more careful investigation convinced me that this theorem, or any one equivalent to it, can be regarded in some way as a sufficient basis for infinitesimal analysis. It then only remained to discover its true origin in the elements of arithmetic and thus at the same time to secure a real definition of the essence of continuity. I succeeded Nov. 24, 1858, and a few days afterward I communicated the results of my meditations to my dear friend Durège with whom I had a long and lively discussion. Later I explained these views of a scientific basis of arithmetic to a few of my pupils, and here in Braunschweig read a paper upon the subject before the scientific club of professors, but I could not make up my mind to its publication, because in the first place, the presentation did not seem altogether simple, and further, the theory itself had little promise. Nevertheless I had already half determined to select this theme as subject for this occasion, when a few days ago, March 14, by the kindness of the author, the paper *Die Elemente der Funktionenlehre* by E. Heine (*Crelle's Journal*, Vol. 74) came into my hands and confirmed me in my decision. In the main I fully agree with the substance of this memoir, and indeed I could hardly do otherwise, but I will frankly acknowledge that my own presentation seems to me to be simpler in form and to bring out the vital point more clearly. While writing this preface (March 20, 1872), I am just in receipt of the interesting paper *Ueber die Ausdehnung eines Satzes aus der Theorie der trigonometrischen Reihen*, by G. Cantor (*Math. Annalen*, Vol. 5), for which I owe the ingenious author my hearty thanks. As I find on a hasty perusal, the axiom given in Section II. of that paper, aside from the form of presentation, agrees with what I designate in Section III. as the essence of continuity. But what advantage will be gained by even a purely abstract definition of real numbers of a higher type, I am as yet unable to see, conceiving as I do of the domain of real numbers as complete in itself.

I

PROPERTIES OF RATIONAL NUMBERS

The development of the arithmetic of rational numbers is here presupposed, but still I think it worth while to call attention to certain important matters without discussion, so as to show at the outset the standpoint assumed in what follows. I regard the whole of arithmetic as a necessary, or at least natural, consequence of the simplest arithmetic act, that of counting, and counting itself as nothing else than the successive creation of the infinite series of positive integers in which each individual is defined by the one immediately preceding; the simplest act is the passing from an already-formed individual to the consecutive new one to be formed. The chain of these numbers forms in itself an exceedingly useful instrument for the human mind;

it presents an inexhaustible wealth of remarkable laws obtained by the introduction of the four fundamental operations of arithmetic. Addition is the combination of any arbitrary repetitions of the above-mentioned simplest act into a single act; from it in a similar way arises multiplication. While the performance of these two operations is always possible, that of the inverse operations, subtraction and division, proves to be limited. Whatever the immediate occasion may have been, whatever comparisons or analogies with experience, or intuition, may have led thereto; it is certainly true that just this limitation in performing the indirect operations has in each case been the real motive for a new creative act; thus negative and fractional numbers have been created by the human mind; and in the system of all rational numbers there has been gained an instrument of infinitely greater perfection. This system, which I shall denote by R, possesses first of all a completeness and self-containedness which I have designated in another place[1] as characteristic of a *body of numbers* [Zahlkörper] and which consists in this that the four fundamental operations are always performable with any two individuals in R, i.e., the result is always an individual of R, the single case of division by the number zero being excepted.

For our immediate purpose, however, another property of the system R is still more important; it may be expressed by saying that the system R forms a well-arranged domain of one dimension extending to infinity on two opposite sides. What is meant by this is sufficiently indicated by my use of expressions borrowed from geometric ideas; but just for this reason it will be necessary to bring out clearly the corresponding purely arithmetic properties in order to avoid even the appearance as if arithmetic were in need of ideas foreign to it.

To express that the symbols a and b represent one and the same rational number we put $a = b$ as well as $b = a$. The fact that two rational numbers a, b are different appears in this that the difference $a - b$ has either a positive or negative value. In the former case a is said to be *greater* than b, b *less* than a; this is also indicated by the symbols $a > b$, $b < a$.[2] As in the latter case $b - a$ has a positive value it follows that $b > a$, $a < b$. In regard to these two ways in which two numbers may differ the following laws will hold:

I. If $a > b$, and $b > c$, then $a > c$. Whenever a, c are two different (or unequal) numbers, and b is greater than the one and less than the other, we shall, without hesitation because of the suggestion of geometric ideas, express this briefly by saying: b lies between the two numbers a, c.

II. If a, c are two different numbers, there are infinitely many different numbers lying between a, c.

[1] *Vorlesungen über Zahlentheorie*, by P. G. Lejeune Dirichlet. 2d ed. § 159.
[2] Hence in what follows the so-called "algebraic" greater and less are understood unless the word "absolute" is added.

III. If a is any definite number, then all numbers of the system R fall into two classes, A_1 and A_2, each of which contains infinitely many individuals; the first class A_1 comprises all numbers a_1 that are $<a$, the second class A_2 comprises all numbers a_2 that are $>a$; the number a itself may be assigned at pleasure to the first or second class, being respectively the greatest number of the first class or the least of the second. In every case the separation of the system R into the two classes A_1, A_2 is such that every number of the first class A_1 is less than every number of the second class A_2.

II

COMPARISON OF THE RATIONAL NUMBERS WITH THE POINTS OF A STRAIGHT LINE

The above-mentioned properties of rational numbers recall the corresponding relations of position of the points of a straight line L. If the two opposite directions existing upon it are distinguished by "right" and "left," and p, q are two different points, then either p lies to the right of q, and at the same time q to the left of p, or conversely q lies to the right of p and at the same time p to the left of q. A third case is impossible, if p, q are actually different points. In regard to this difference in position the following laws hold:

I. If p lies to the right of q, and q to the right of r, then p lies to the right of r; and we say that q lies between the points p and r.

II. If p, r are two different points, then there always exist infinitely many points that lie between p and r.

III. If p is a definite point in L, then all points in L fall into two classes, P_1, P_2, each of which contains infinitely many individuals; the first class P_1 contains all the points p_1, that lie to the left of p, and the second class P_2 contains all the points p_2 that lie to the right of p; the point p itself may be assigned at pleasure to the first or second class. In every case the separation of the straight line L into the two classes or portions P_1, P_2, is of such a character that every point of the first class P_1 lies to the left of every point of the second class P_2.

This analogy between rational numbers and the points of a straight line, as is well known, becomes a real correspondence when we select upon the straight line a definite origin or zero-point o and a definite unit of length for the measurement of segments. With the aid of the latter to every rational number a a corresponding length can be constructed and if we lay this off upon the straight line to the right or left of o according as a is positive or negative, we obtain a definite end-point p, which may be regarded as the point corresponding to the number a; to the rational number zero corresponds the point o. In this way to every rational number a, i.e., to every individual in R, corresponds one and only one point p, i.e., an individual in L. To the two numbers a, b respectively correspond the two points p, q, and if $a > b$, then p lies to the right of q. To the laws I, II, III of the previous Section correspond completely the laws I, II, III of the present.

III

Continuity of the Straight Line

Of the greatest importance, however, is the fact that in the straight line L there are infinitely many points which correspond to no rational number. If the point p corresponds to the rational number a, then, as is well known, the length op is commensurable with the invariable unit of measure used in the construction, i.e., there exists a third length, a so-called common measure, of which these two lengths are integral multiples. But the ancient Greeks already knew and had demonstrated that there are lengths incommensurable with a given unit of length, e.g., the diagonal of the square whose side is the unit of length. If we lay off such a length from the point o upon the line we obtain an end-point which corresponds to no rational number. Since further it can be easily shown that there are infinitely many lengths which are incommensurable with the unit of length, we may affirm: The straight line L is infinitely richer in point-individuals than the domain R of rational numbers in number-individuals.

If now, as is our desire, we try to follow up arithmetically all phenomena in the straight line, the domain of rational numbers is insufficient and it becomes absolutely necessary that the instrument R constructed by the creation of the rational numbers be essentially improved by the creation of new numbers such that the domain of numbers shall gain the same completeness, or as we may say at once, the same *continuity*, as the straight line.

The previous considerations are so familiar and well known to all that many will regard their repetition quite superfluous. Still I regarded this recapitulation as necessary to prepare properly for the main question. For, the way in which the irrational numbers are usually introduced is based directly upon the conception of extensive magnitudes—which itself is nowhere carefully defined—and explains number as the result of measuring such a magnitude by another of the same kind.[3] Instead of this I demand that arithmetic shall be developed out of itself.

That such comparisons with non-arithmetic notions have furnished the immediate occasion for the extension of the number-concept may, in a general way, be

[3] The apparent advantage of the generality of this definition of number disappears as soon as we consider complex numbers. According to my view, on the other hand, the notion of the ratio between two numbers of the same kind can be clearly developed only after the introduction of irrational numbers.

granted (though this was certainly not the case in the introduction of complex numbers); but this surely is no sufficient ground for introducing these foreign notions into arithmetic, the science of numbers. Just as negative and fractional rational numbers are formed by a new creation, and as the laws of operating with these numbers must and can be reduced to the laws of operating with positive integers, so we must endeavor completely to define irrational numbers by means of the rational numbers alone. The question only remains how to do this.

The above comparison of the domain R of rational numbers with a straight line has led to the recognition of the existence of gaps, of a certain incompleteness or discontinuity of the former, while we ascribe to the straight line completeness, absence of gaps, or continuity. In what then does this continuity consist? Everything must depend on the answer to this question, and only through it shall we obtain a scientific basis for the investigation of *all* continuous domains. By vague remarks upon the unbroken connection in the smallest parts obviously nothing is gained; the problem is to indicate a precise characteristic of continuity that can serve as the basis for valid deductions. For a long time I pondered over this in vain, but finally I found what I was seeking. This discovery will, perhaps, be differently estimated by different people; the majority may find its substance very commonplace. It consists of the following. In the preceding section attention was called to the fact that every point p of the straight line produces a separation of the same into two portions such that every point of one portion lies to the left of every point of the other. I find the essence of continuity in the converse, i.e., in the following principle:

"If all points of the straight line fall into two classes such that every point of the first class lies to the left of every point of the second class, then there exists one and only one point which produces this division of all points into two classes, this severing of the straight line into two portions."

As already said I think I shall not err in assuming that every one will at once grant the truth of this statement; the majority of my readers will be very much disappointed in learning that by this commonplace remark the secret of continuity is to be revealed. To this I may say that I am glad if every one finds the above principle so obvious and so in harmony with his own ideas of a line; for I am utterly unable to adduce any proof of its correctness, nor has any one the power. The assumption of this property of the line is nothing else than an axiom by which we attribute to the line its continuity, by which we find continuity in the line. If space has at all a real existence it is *not* necessary for it to be continuous; many of its properties would remain the same even were it discontinuous. And if we knew for certain that space was discontinuous there would be nothing to prevent us, in case we so desired, from filling up its gaps, in thought, and thus making it continuous; this filling up would consist in a creation of new point-individuals and would have to be effected in accordance with the above principle.

IV

CREATION OF IRRATIONAL NUMBERS

From the last remarks it is sufficiently obvious how the discontinuous domain R of rational numbers may be rendered complete so as to form a continuous domain. In Section I it was pointed out that every rational number a effects a separation of the system R into two classes such that every number a_1 of the first class A_1 is less than every number a_2 of the second class A_2; the number a is either the greatest number of the class A_1 or the least number of the class A_2. If now any separation of the system R into two classes A_1, A_2, is given which possesses only *this* characteristic property that every number a_1 in A_1 is less than every number a_2 in A_2, then for brevity we shall call such a separation a *cut* [Schnitt] and designate it by (A_1, A_2). We can then say that every rational number a produces one cut or, strictly speaking, two cuts, which, however, we shall not look upon as essentially different; this cut possesses, *besides*, the property that either among the numbers of the first class there exists a greatest or among the numbers of the second class a least number. And conversely, if a cut possesses this property, then it is produced by this greatest or least rational number.

But it is easy to show that there exist infinitely many cuts not produced by rational numbers. The following example suggests itself most readily.

Let D be a positive integer but not the square of an integer, then there exists a positive integer λ such that

$$\lambda^2 < D < (\lambda + 1)^2.$$

If we assign to the second class A_2, every positive rational number a_2 whose square is $>D$, to the first class A_1 all other rational numbers a_1, this separation forms a cut (A_1, A_2), i.e., every number a_1 is less than every number a_2. For if $a_1 = 0$, or is negative, then on that ground a_1 is less than any number a_2, because, by definition, this last is positive; if a_1 is positive, then is its square $\leqq D$, and hence a_1 is less than any positive number a_2 whose square is $>D$.

But this cut is produced by no rational number. To demonstrate this it must be shown first of all that there exists no rational number whose square $=D$. Although this is known from the first elements of the theory of numbers, still the following indirect proof may find place here. If there exist a rational number whose square $=D$, then there exist two positive integers t, u, that satisfy the equation

$$t^2 - Du^2 = 0,$$

and we may assume that u is the *least* positive integer possessing the property that its square, by multiplication by D, may be converted into the square of an integer t. Since evidently

$$\lambda u < t < (\lambda + 1)u,$$

the number $u' = t - \lambda u$ is a positive integer certainly *less* than u. If further we put

$$t' = Du - \lambda t,$$

t' is likewise a positive integer, and we have

$$t'^2 - Du'^2 = (\lambda^2 - D)(t^2 - Du^2) = 0,$$

which is contrary to the assumption respecting u.

Hence the square of every rational number x is either $< D$ or $> D$. From this it easily follows that there is neither in the class A_1 a greatest, nor in the class A_2 a least number. For if we put

$$y = \frac{x(x^2 + 3D)}{3x^2 + D},$$

we have

$$y - x = \frac{2x(D - x^2)}{3x^2 + D}$$

and

$$y^2 - D = \frac{(x^2 - D)^2}{(3x^2 + D)^2}.$$

If in this we assume x to be a positive number from the class A_1, then $x^2 < D$, and hence $y > x$ and $y^2 < D$. Therefore y likewise belongs to the class A_1. But if we assume x to be a number from the class A_2, then $x^2 > D$, and hence $y < x, y > 0$, and $y^2 > D$. Therefore y likewise belongs to the class A_2. This cut is therefore produced by no rational number.

In this property that not all cuts are produced by rational numbers consists the incompleteness or discontinuity of the domain R of all rational numbers.

Whenever, then, we have to do with a cut (A_1, A_2) produced by no rational number, we create a new, an *irrational* number α, which we regard as completely defined by this cut (A_1, A_2); we shall say that the number α corresponds to this cut, or that it produces this cut. From now on, therefore, to every definite cut there corresponds a definite rational or irrational number, and we regard two numbers as *different* or *unequal* always and only when they correspond to essentially different cuts.

In order to obtain a basis for the orderly arrangement of all *real*, i.e., of all rational and irrational numbers we must investigate the relation between any two cuts (A_1, A_2) and (B_1, B_2) produced by any two numbers α and β. Obviously a cut (A_1, A_2) is given completely when one of the two classes, e.g., the first A_1 is known, because the second A_2 consists of all rational numbers not contained in A_1, and the characteristic property of such a first class lies in this that if the number a_1 is contained in it, it also contains all numbers less than a_1. If now we compare two such first classes A_1, B_1 with each other, it may happen

1. That they are perfectly identical, i.e., that every number contained in A_1 is also contained in B_1, and that every number contained in B_1 is also contained in A_1. In this case A_2 is necessarily identical with B_2, and the two cuts are perfectly identical, which we denote in symbols by $\alpha = \beta$ or $\beta = \alpha$.

But if the two classes A_1, B_1 are not identical, then there exists in the one, e.g., in A_1, a number $a'_1 = b'_2$ not contained in the other B_1 and consequently found in B_2; hence all numbers b_1 contained in B_1 are certainly less than this number $a'_1 = b'_2$ and therefore all numbers b_1 are contained in A_1.

2. If now this number a'_1 is the only one in A_1 that is not contained in B_1, then is every other number a_1 contained in A_1 also contained in B_1 and is consequently $<a'_1$, i.e., a'_1 is the greatest among all the numbers a_1, hence the cut (A_1, A_2) is produced by the rational number $\alpha = a'_1 = b'_2$. Concerning the other cut (B_1, B_2) we know already that all numbers b_1 in B_1 are also contained in A_1 and are less than the number $a'_1 = b'_2$ which is contained in B_2; every other number b_2 contained in B_2 must, however, be greater than b'_2, for otherwise it would be less than a'_1, therefore contained in A_1 and hence in B_1; hence b'_2 is the least among all numbers contained in B_2, and consequently the cut (B_1, B_2) is produced by the same rational number $\beta = b'_2 = a'_1 = \alpha$. The two cuts are then only unessentially different.

3. If, however, there exist in A_1 at least two different numbers $a'_1 = b'_2$ and $a''_1 = b''_2$, which are not contained in B_1, then there exist infinitely many of them, because all the infinitely many numbers lying between a'_1 and a''_1 are obviously contained in A_1 (Section I, II) but not in B_1. In this case we say that the numbers α and β corresponding to these two essentially different cuts (A_1, A_2) and (B_1, B_2) are *different*, and further that α is *greater* than β, that β is *less* than α, which we express in symbols by $\alpha > \beta$ as well as $\beta < \alpha$. It is to be noticed that this definition coincides completely with the one given earlier, when α, β are rational.

The remaining possible cases are these:

4. If there exists in B_1 one and only one number $b'_1 = a'_2$, that is not contained in A_1 then the two cuts (A_1, A_2) and (B_1, B_2) are only unessentially different and they are produced by one and the same rational number $\alpha = a'_2 = b'_1 = \beta$.

5. But if there are in B_1 at least two numbers which are not contained in A_1, then $\beta > \alpha$, $\alpha < \beta$.

As this exhausts the possible cases, it follows that of two different numbers one is necessarily the greater, the other the less, which gives two possibilities. A third case is impossible. This was indeed involved in the use of the *comparative* (greater, less) to designate the relation between α, β; but this use has only now been justified. In just such investigations one needs to exercise the greatest care so that even with the best intention to be honest he shall not, through a hasty choice of expressions borrowed from other notions already developed, allow himself to be led into the use of inadmissible transfers from one domain to the other.

If now we consider again somewhat carefully the case $\alpha > \beta$ it is obvious that the less number β, if rational, certainly belongs to the class A_1; for since there is in A_1 a number $a'_1 = b'_2$ which belongs to the class B_2, it follows that the number β, whether the greatest number in B_1 or the least in B_2 is certainly $\leqq a'_1$ and hence contained in A_1. Likewise it is obvious from $\alpha > \beta$ that the greater number α, if rational, certainly belongs to the class B_2, because $\alpha \geqq a'_1$. Combining these two considerations we get the following result: If a cut is produced by the number α then any rational number belongs to the class A_1 or to the class A_2 according as it is less or greater than α; if the number α is itself rational it may belong to either class.

From this we obtain finally the following: If $\alpha > \beta$, i.e., if there are infinitely many numbers in A_1 not contained in B_1 then there are infinitely many such numbers that at the same time are different from α and from β; every such rational number c is $< \alpha$, because it is contained in A_1 and at the same time it is $> \beta$ because contained in B_2.

V

CONTINUITY OF THE DOMAIN OF REAL NUMBERS

In consequence of the distinctions just established the system $\mathfrak{R}$ of all real numbers forms a well-arranged domain of one dimension; this is to mean merely that the following laws prevail:

I. If $\alpha > \beta$, and $\beta > \gamma$, then is also $\alpha > \gamma$. We shall say that the number β lies between α and γ.

II. If α, γ are any two different numbers, then there exist infinitely many different numbers β lying between α, γ.

III. If α is any definite number then all numbers of the system $\mathfrak{R}$ fall into two classes $\mathfrak{A}_1$ and $\mathfrak{A}_2$ each of which contains infinitely many individuals; the first class $\mathfrak{A}_1$ comprises all the numbers α_1 that are less than α, the second $\mathfrak{A}_2$ comprises all the numbers α_2 that are greater than α; the number α itself may be assigned at pleasure to the first class or to the second, and it is respectively the greatest of the first or the least of the second class. In each case the separation of the system $\mathfrak{R}$ into the two classes $\mathfrak{A}_1$, $\mathfrak{A}_2$ is such that every number of the first class $\mathfrak{A}_1$ is smaller than every number of the second class $\mathfrak{A}_2$ and we say that this separation is produced by the number α.

For brevity and in order not to weary the reader I suppress the proofs of these theorems which follow immediately from the definitions of the previous section.

Beside these properties, however, the domain $\mathfrak{R}$ possesses also *continuity*; i.e., the following theorem is true:

IV.   If the system $\mathfrak{R}$ of all real numbers breaks up into two classes $\mathfrak{A}_1$, $\mathfrak{A}_2$ such that every number α_1 of the class $\mathfrak{A}_1$ is less than every number α_2 of the class $\mathfrak{A}_2$ then there exists one and only one number α by which this separation is produced.

Proof.   By the separation or the cut of $\mathfrak{R}$ into $\mathfrak{A}_1$ and $\mathfrak{A}_2$ we obtain at the same time a cut (A_1, A_2) of the system R of all rational numbers which is defined by this that A_1 contains all rational numbers of the class $\mathfrak{A}_1$ and A_2 all other rational numbers, i.e., all rational numbers of the class $\mathfrak{A}_2$. Let α be the perfectly definite number which produces this cut (A_1, A_2). If β is any number different from α, there are always infinitely many rational numbers c lying between α and β. If $\beta < \alpha$, then $c < \alpha$; hence c belongs to the class A_1 and consequently also to the class $\mathfrak{A}_1$, and since at the same time $\beta < c$ then β also belongs to the same class $\mathfrak{A}_1$, because every number in $\mathfrak{A}_2$ is greater than every number c in $\mathfrak{A}_1$. But if $\beta > \alpha$, then is $c > \alpha$; hence c belongs to the class A_2 and consequently also to the class $\mathfrak{A}_2$, and since at the same time $\beta > c$, then β also belongs to the same class $\mathfrak{A}_2$, because every number in $\mathfrak{A}_1$ is less than every number c in $\mathfrak{A}_2$. Hence every number β different from α belongs to the class $\mathfrak{A}_1$ or to the class $\mathfrak{A}_2$ according as $\beta < \alpha$ or $\beta > \alpha$; consequently α itself is either the greatest number in $\mathfrak{A}_1$ or the least number in $\mathfrak{A}_2$, i.e., α is one and obviously the only number by which the separation of R into the classes $\mathfrak{A}_1$, $\mathfrak{A}_2$ is produced. Which was to be proved.

VI

OPERATIONS WITH REAL NUMBERS

To reduce any operation with two real numbers α, β to operations with rational numbers, it is only necessary from the cuts (A_1, A_2), (B_1, B_2) produced by the numbers α and β in the system R to define the cut (C_1, C_2) which is to correspond to the result of the operation, γ. I confine myself here to the discussion of the simplest case, that of addition.

If c is any rational number, we put it into the class C_1, provided there are two numbers one a_1 in A_1 and one b_1 in B_1 such that their sum $a_1 + b_1 \geqq c$; all other rational numbers shall be put into the class C_2. This separation of all rational numbers into the two classes C_1, C_2 evidently forms a cut, since every number c_1 in

C_1 is less than every number c_2 in C_2. If both α and β are rational, then every number c_1 contained in C_1 is $\leqq \alpha + \beta$, because $a_1 \leqq \alpha$, $b_1 \leqq \beta$, and therefore $a_1 + b_1 \leqq \alpha + \beta$; further, if there were contained in C_2 a number $c_2 < \alpha + \beta$, hence $\alpha + \beta = c_2 + p$, where p is a positive rational number, then we should have

$$c_2 = (\alpha - \tfrac{1}{2}p) + (\beta - \tfrac{1}{2}p),$$

which contradicts the definition of the number c_2, because $\alpha - \tfrac{1}{2}p$ is a number in A_1, and $\beta - \tfrac{1}{2}p$ a number in B_1; consequently every number c_2 contained in C_2 is $\geqq \alpha + \beta$. Therefore in this case the cut (C_1, C_2) is produced by the sum $\alpha + \beta$. Thus we shall not violate the definition which holds in the arithmetic of rational numbers if in all cases we understand by the sum $\alpha + \beta$ of any two real numbers α, β that number γ by which the cut (C_1, C_2) is produced. Further, if only one of the two numbers α, β is rational, e.g., α, it is easy to see that it makes no difference with the sum $\gamma = \alpha + \beta$ whether the number α is put into the class A_1 or into the class A_2.

Just as addition is defined, so can the other operations of the so-called elementary arithmetic be defined, viz., the formation of differences, products, quotients, powers, roots, logarithms, and in this way we arrive at real proofs of theorems (as, e. g., $\sqrt{2} \cdot \sqrt{3} = \sqrt{6}$), which to the best of my knowledge have never been established before. The excessive length that is to be feared in the definitions of the more complicated operations is partly inherent in the nature of the subject but can for the most part be avoided. Very useful in this connection is the notion of an *interval*, i.e., a system A of rational numbers possessing the following characteristic property; if a and a' are numbers of the system A, then are all rational numbers lying between a and a' contained in A. The system R of all rational numbers, and also the two classes of any cut are intervals. If there exist a rational number a_1 which is less and a rational number a_2 which is greater than every number of the interval A, then A is called a finite interval; there then exist infinitely many numbers in the same condition as a_1 and infinitely many in the same condition as a_2; the whole domain R breaks up into three parts A_1, A, A_2 and there enter two perfectly definite rational or irrational numbers α_1, α_2 which may be called respectively the lower and upper (or the less and greater) *limits* of the interval; the lower limit α_1 is determined by the cut for which the system A_1 forms the first class and the upper α_2 by the cut for which the system A_2 forms the second class. Of every rational or irrational number α lying between α_1 and α_2 it may be said that it lies *within* the interval A. If all numbers of an interval A are also numbers of an interval B, then A is called a portion of B.

Still lengthier considerations seem to loom up when we attempt to adapt the numerous theorems of the arithmetic of rational numbers (as, e.g., the theorem $(a + b)c = ac + bc$) to any real numbers. This, however, is not the case. It is easy to see that it all reduces to showing that the arithmetic operations possess a certain continuity. What I mean by this statement may be expressed in the form of a general theorem:

"If the number λ is the result of an operation performed on the numbers α, β, γ, . . . and λ lies within the interval L, then intervals A, B, C, . . . can be taken within which lie the numbers α, β, γ, . . . such that the result of the same operation in which the numbers α, β, γ, . . . are replaced by arbitrary numbers of the intervals A, B, C, . . . is always a number lying within the interval L." The forbidding clumsiness, however, which marks the statement of such a theorem convinces us that something must be brought in as an aid to expression; this is, in fact, attained in the most satisfactory way by introducing the ideas of *variable magnitudes, functions, limiting values*, and it would be best to base the definitions of even the simplest arithmetic operations upon these ideas, a matter which, however, cannot be carried further here.

VII

INFINITESIMAL ANALYSIS

Here at the close we ought to explain the connection between the preceding investigations and certain fundamental theorems of infinitesimal analysis.

We say that a variable magnitude x which passes through successive definite numerical values approaches a fixed limiting value α when in the course of the process x lies finally between two numbers between which α itself lies, or, what amounts to the same, when the difference $x - \alpha$ taken absolutely becomes finally less than any given value different from zero.

One of the most important theorems may be stated in the following manner: "If a magnitude x grows continually but not beyond all limits it approaches a limiting value."

I prove it in the following way. By hypothesis there exists one and hence there exist infinitely many numbers α_2 such that x remains continually $< \alpha_2$; I designate by $\mathfrak{A}_2$ the system of all these numbers α_2, by $\mathfrak{A}_1$ the system of all other numbers α_1; each of the latter possesses the property that in the course of the process x becomes finally $\geqq \alpha_1$, hence every number α_1 is less than every number α_2 and consequently there exists a number α which is either the greatest in $\mathfrak{A}_1$ or the least in $\mathfrak{A}_2$ (V, IV). The former cannot be the case since x never ceases to grow, hence α is the least number in $\mathfrak{A}_2$. Whatever number α_1 be taken we shall have finally $\alpha_1 < x < \alpha$, i.e., x approaches the limiting value α.

This theorem is equivalent to the principle of continuity, i.e., it loses its validity as soon as we assume a single real number not to be contained in the domain $\mathfrak{R}$; or otherwise expressed: if this theorem is correct, then is also theorem IV. in V. correct.

Another theorem of infinitesimal analysis, likewise equivalent to this, which is still oftener employed, may be stated as follows: "If in the variation of a magnitude x we can for every given positive magnitude δ assign a corresponding position from and after which x changes by less than δ then x approaches a limiting value."

This converse of the easily demonstrated theorem that every variable magnitude which approaches a limiting value finally changes by less than any given positive magnitude can be derived as well from the preceding theorem as directly from the principle of continuity. I take the latter course. Let δ be any positive magnitude (i.e., $\delta > 0$), then by hypothesis a time will come after which x will change by less than δ, i.e., if at this time x has the value a, then afterwards we shall continually have $x > a - \delta$ and $x < a + \delta$. I now for a moment lay aside the original hypothesis and make use only of the theorem just demonstrated that all later values of the variable x lie between two assignable finite values. Upon this I base a double separation of all real numbers. To the system $\mathfrak{A}_2$ I assign a number α_2 (e.g., $a + \delta$) when in the course of the process x becomes finally $\leqq \alpha_2$; to the system $\mathfrak{A}_1$ I assign every number not contained in $\mathfrak{A}_2$; if α_1 is such a number, then, however far the process may have advanced, it will still happen infinitely many times that $x > \alpha_2$. Since every number α_1 is less than every number α_2 there exists a perfectly definite number α which produces this cut $(\mathfrak{A}_1, \mathfrak{A}_2)$ of the system $\mathfrak{R}$ and which I will call the upper limit of the variable x which always remains finite. Likewise as a result of the behavior of the variable x a second cut $(\mathfrak{B}_1, \mathfrak{B}_2)$ of the system $\mathfrak{R}$ is produced; a number β_2 (e.g., $a - \delta$) is assigned to $\mathfrak{B}_2$ when in the course of the process x becomes finally $\geqq \beta$; every other number β_2, to be assigned to $\mathfrak{B}_2$, has the property that x is never finally $\geqq \beta_2$; therefore infinitely many times x becomes $< \beta_2$; the number β by which this cut is produced I call the lower limiting value of the variable x. The two numbers α, β are obviously characterised by the following property: if ϵ is an arbitrarily small positive magnitude then we have always finally $x < \alpha + \epsilon$ and $x > \beta - \epsilon$, but never finally $x < \alpha - \epsilon$ and never finally $x > \beta + \epsilon$. Now two cases are possible. If α and β are different from each other, then necessarily $\alpha > \beta$, since continually $\alpha_2 \geqq \beta_2$; the variable x oscillates, and, however far the process advances, always undergoes changes whose amount surpasses the value $(\alpha - \beta) - 2\epsilon$ where ϵ is an arbitrarily small positive magnitude. The original hypothesis to which I now return contradicts this consequence; there remains only the second case $\alpha = \beta$ and since it has already been shown that, however small be the positive magnitude ϵ, we always have finally $x < \alpha + \epsilon$ and $x > \beta - \epsilon$, x approaches the limiting value α, which was to be proved.

These examples may suffice to bring out the connection between the principle of continuity and infinitesimal analysis.

THE NATURE AND MEANING
OF NUMBERS

I

SYSTEMS OF ELEMENTS

1. In what follows I understand by *thing* every object of our thought. In order to be able easily to speak of things, we designate them by symbols, e.g., by letters, and we venture to speak briefly of the thing a or of a simply, when we mean the thing denoted by a and not at all the letter a itself. A thing is completely determined by all that can be affirmed or thought concerning it. A thing a is the same as b (identical with b), and b the same as a, when all that can be thought concerning a can also be thought concerning b, and when all that is true of b can also be thought of a. That a and b are only symbols or names for one and the same thing is indicated by the notation $a = b$, and also by $b = a$. If further $b = c$, that is, if c as well as a is a symbol for the thing denoted by b, then is also $a = c$. If the above coincidence of the thing denoted by a with the thing denoted by b does not exist, then are the things a, b said to be different, a is another thing than b, b another thing than a; there is some property belonging to the one that does not belong to the other.

2. It very frequently happens that different things, a, b, c, . . . for some reason can be considered from a common point of view, can be associated in the mind, and we say that they form a *system S*; we call the things a, b, c, . . . *elements* of the system S, they are *contained* in S; conversely, S *consists* of these elements. Such a system S (an aggregate, a manifold, a totality) as an object of our thought is likewise a thing (1); it is completely determined when with respect to every thing it is determined whether it is an element of S or not.[1] The system S is hence the same as the system T, in symbols $S = T$, when every element of S is also element of T, and every element of T is also element of S. For uniformity of expression it is advantageous to include also the special case where a system S consists of a *single* (one and only one) element a, i.e., the thing a is element of S, but every thing different from a is not an element of S. On the other hand, we intend here for certain reasons wholly to

---

[1] In what manner this determination is brought about, and whether we know a way of deciding upon it, is a matter of indifference for all that follows; the general laws to be developed in no way depend upon it; they hold under all circumstances. I mention this expressly because Kronecker not long ago (*Crelle's Journal*, Vol. 99, pp. 334–336) has endeavored to impose certain limitations upon the free formation of concepts in mathematics which I do not believe to be justified; but there seems to be no call to enter upon this matter with more detail until the distinguished mathematician shall have published his reasons for the necessity or merely the expediency of these limitations.

exclude the empty system which contains no element at all, although for other investigations it may be appropriate to imagine such a system.

3. Definition.   A system A is said to be *part* of a system S when every element of A is also element of S. Since this relation between a system A and a system S will occur continually in what follows, we shall express it briefly by the symbol $A3S$. The inverse symbol $S\mathcal{E}A$, by which the same fact might be expressed, for simplicity and clearness I shall wholly avoid, but for lack of a better word I shall sometimes say that S is *whole* of A, by which I mean to express that among the elements of S are found all the elements of A. Since further every element s of a system S by (2) can be itself regarded as a system, we can hereafter employ the notation $s3S$.

4. Theorem. $A3A$, by reason of (3).

5. Theorem.   If $A3B$ and $B3A$, then $A = B$.

The proof follows from (3), (2).

6. Definition.   A system A is said to be a *proper* [*echter*] part of S, when A is part of S, but different from S. According to (5) then S is not a part of A, i.e., there is in S an element which is not an element of A.

7. Theorem.   If $A3B$ and $B3C$, which may be denoted briefly by $A3B3C$, then is $A3C$, and A is certainly a proper part of C, if A is a proper part of B or if B is a proper part of C.

The proof follows from (3), (6).

8. Definition.   By the system *compounded* out of any systems $A, B, C, \ldots$ to be denoted by $\mathfrak{M}\ (A, B, C, \ldots)$ we mean that system whose elements are determined by the following prescription: a thing is considered as element of $\mathfrak{M}\ (A, B, C, \ldots)$ when and only when it is element of some one of the systems $A, B, C, \ldots$, i.e., when it is element of A, *or* B, *or* C, $\ldots$ We include also the case where only a single system A exists; then obviously $\mathfrak{M}\ (A) = A$. We observe further that the system $\mathfrak{M}\ (A, B, C, \ldots)$ compounded out of $A, B, C, \ldots$ is carefully to be distinguished from the system whose elements are the systems $A, B, C, \ldots$ themselves.

9. Theorem.   The systems $A, B, C, \ldots$ are parts of $\mathfrak{M}\ (A, B, C, \ldots)$.

The proof follows from (8), (3).

10. Theorem.   If $A, B, C, \ldots$ are parts of a system S, then is $\mathfrak{M}\ (A, B, C, \ldots)3S$.

The proof follows from (8), (3).

11. Theorem.   If P is part of one of the systems $A, B, C, \ldots$ then is $P3\mathfrak{M}(A, B, C, \ldots)$.

The proof follows from (9), (7).

12. Theorem.   If each of the systems $P, Q, \ldots$ is part of one of the systems $A, B, C, \ldots$ then is $\mathfrak{M}\ (P, Q, \ldots)3\mathfrak{M}(A, B, C, \ldots)$.

The proof follows from (11), (10).

13. Theorem. If A is compounded out of any of the systems P, Q, . . . then is $A\, 3\,\mathfrak{M}$ $(P,\ Q,\ \ldots)$.

Proof. For every element of A is by (8) element of one of the systems P, Q, . . . , consequently by (8) also element of $\mathfrak{M}$ $(P,\ Q,\ \ldots)$, whence the theorem follows by (3).

14. Theorem. If each of the systems A, B, C, . . . is compounded out of any of the systems P, Q, . . . then is

$$\mathfrak{M}\ (A,\ B,\ C,\ \ldots)\, 3\,\mathfrak{M}(P,\ Q,\ \ldots)$$

The proof follows from (13), (10).

15. Theorem. If each of the systems P, Q, . . . is part of one of the systems A, B, C, . . . , and if each of the latter is compounded out of any of the former, then is

$$\mathfrak{M}\ (P,\ Q,\ \ldots) = \mathfrak{M}\ (A,\ B,\ C,\ \ldots).$$

The proof follows from (12), (14), (5).

16. Theorem. If

$$A = \mathfrak{M}\ (P,\ Q)\ \text{and}\ B = \mathfrak{M}\ (Q,\ R)$$
$$\text{then is}\ \mathfrak{M}\ (A,\ R) = \mathfrak{M}\ (P,\ B).$$

Proof. For by the preceding theorem (15)

$$\mathfrak{M}\ (A,\ R)\ \text{as well as}\ \mathfrak{M}\ (P,\ B) = \mathfrak{M}\ (P,\ Q,\ R).$$

17. Definition. A thing g is said to be *common* element of the systems A, B, C, . . . , if it is contained in each of these systems (that is in A *and* in B *and* in C . . . .). Likewise a system T is said to be a *common part* of A, B, C, . . . when T is part of each of these systems; and by the *community* [*Gemeinheit*] of the systems A, B, C, . . . we understand the perfectly determinate system $\mathfrak{G}$ $(A$, B, C, . . .) which consists of all the common elements g of A, B, C, . . . and hence is likewise a common part of those systems. We again include the case where only a single system A occurs; then $\mathfrak{G}$ (A) (is to be put) $=A$. But the case may also occur that the systems A, B, C, . . . possess no common element at all, therefore no common part, no community; they are then called systems *without* common part, and the symbol $\mathfrak{G}$ $(A$, B, C, . . .) is meaningless (compare the end of (2)). We shall however almost always in theorems concerning communities leave it to the reader to add in thought the condition of their existence and to discover the proper interpretation of these theorems for the case of non-existence.

18. Theorem. Every common part of A, B, C, . . . is part of $\mathfrak{G}$ $(A$, B, C, . . .).
The proof follows from (17).

19. Theorem. Every part of $\mathfrak{G}$ $(A$, B, C, . . .) is common part of A, B, C, . . .
The proof follows from (17), (7).

20. Theorem.   If each of the systems A, B, C, . . . is whole (3) of one of the systems P, Q, . . . then is

$$\mathfrak{G}(P, Q, \ldots) \mathbf{3} \, \mathfrak{G}(A, B, C, \ldots)$$

Proof.   For every element of $\mathfrak{G}$ $(P, Q, \ldots)$ is common element of P, Q, . . . , therefore also common element of A, B, C, . . . , which was to be proved.

II

TRANSFORMATION OF A SYSTEM

21. Definition.[2]   By a *transformation* [*Abbildung*] ϕ of a system S we under-stand a law according to which to every determinate element s of S there *belongs* a determinate thing which is called the *transform* of s and denoted by $\phi(s)$; we say also that $\phi(s)$ *corresponds* to the element s, that $\phi(s)$ *results* or is *produced* from s by the transformation ϕ, that s is *transformed* into $\phi(s)$ by the transformation ϕ. If now T is any part of S, then in the transformation ϕ of S is likewise con-tained a determinate transformation of T, which for the sake of simplicity may be denoted by the same symbol ϕ and consists in this that to every element t of the system T there corresponds the same transform $\phi(t)$, which t possesses as ele-ment of S; at the same time the system consisting of all transforms $\phi(t)$ shall be called the transform of T and be denoted by $\phi(T)$, by which also the significance of $\phi(S)$ is defined. As an example of a transformation of a system we may regard the mere assignment of determinate symbols or names to its elements. The sim-plest transformation of a system is that by which each of its elements is trans-formed into itself; it will be called the *identical* transformation of the system. For convenience, in the following theorems (22), (23), (24), which deal with an arbi-trary transformation ϕ of an arbitrary system S, we shall denote the transforms of elements s and parts T respectively by s' and T'; in addition we agree that small and capital italics without accent shall always signify elements and parts of this system S.

22. Theorem.[3]   If $A\mathbf{3}B$, then $A'\mathbf{3}B'$.

Proof.   For every element of A' is the transform of an element contained in A, and therefore also in B, and is therefore element of B', which was to be proved.

23. Theorem.   The transform of $\mathfrak{M}$ $(A, B, C, \ldots)$ is $\mathfrak{M}$ $(A', B', C', \ldots)$.

Proof.   If we denote the system $\mathfrak{M}$ $(A, B, C, \ldots)$ which by (10) is likewise part of S by M, then is every element of its transform M' the transform m' of an element m of M; since therefore by (8) m is also element of one of the systems

[2]See Dirichlet's *Vorlesungen über Zahlentheorie,* 3d edition, 1879, § 163.
[3]See theorem 27.

A, B, C, . . . and consequently m' element of one of the systems A', B', C', . . . , and hence by (8) also element of $\mathfrak{M}(A', B', C', \ldots)$, we have by (3)

$$M' \mathbf{3} \mathfrak{M}(A', B', C', \ldots).$$

On the other hand, since A, B, C, . . . are by (9) parts of M, and hence A', B', C', . . . by (22) parts of M', we have by (10)

$$\mathfrak{M}(A', B', C', \ldots) \mathbf{3} M'.$$

By combination with the above we have by (5) the theorem to be proved

$$M' = \mathfrak{M}(A', B', C', \ldots).$$

24. Theorem.[4]   The transform of every common part of A, B, C, . . . , and therefore that of the community $\mathfrak{G}$ $(A, B, C, \ldots)$ is part of $\mathfrak{G}$ $(A', B', C', \ldots)$.

Proof.   For by (22) it is common part of A', B', C', . . . , whence the theorem follows by (18).

25. Definition and theorem.   If ϕ is a transformation of a system S, and ψ a transformation of the transform $S' = \phi(S)$, there always results a transformation θ of S, *compounded* [5] out of ϕ and ψ, which consists of this that to every element s of S there corresponds the transform

$$\theta(s) = \psi(s') = \psi(\phi(s)),$$

where again we have put $\phi(s) = s'$. This transformation θ can be denoted briefly by the symbol $\psi.\phi$ or $\psi\phi$, the transform $\theta(s)$ by $\psi\phi(s)$ where the order of the symbols ϕ, ψ is to be considered, since in general the symbol $\phi\psi$ has no interpretation and actually has meaning only when $\psi(s') \mathbf{3} s$. If now χ signifies a transformation of the system $\psi(s') = \psi\phi(s)$ and η the transformation $\chi\psi$ of the system S' compounded out of ψ and χ, then is $\chi\theta(s) = \chi\psi(s') = \eta(s') = \eta\phi(s)$; therefore the compound transformations $\chi\theta$ and $\eta\phi$ coincide for every element s of S, i.e., $\chi\theta = \eta\phi$. In accordance with the meaning of θ and η this theorem can finally be expressed in the form

$$\chi \cdot \psi\phi = \chi\psi \cdot \phi,$$

and this transformation compounded out of ϕ, ψ, χ can be denoted briefly by $\chi\psi\phi$.

III

SIMILARITY OF A TRANSFORMATION. SIMILAR SYSTEMS

26. Definition.   A transformation ϕ of a system S is said to be *similar* [*ähnlich*] or *distinct*, when to different elements a, b of the system S there always correspond different transforms $a' = \phi(a)$, $b' = \phi(b)$. Since in this case conversely from $s' = t'$ we always have $s = t$, then is every element of the system $S' = \phi(S)$ the transform s'

---

[4]See theorem 29.
[5]A confusion of this compounding of transformations with that of systems of elements is hardly to be feared.

of a single, perfectly determinate element s of the system S, and we can therefore set over against the transformation ϕ of S an *inverse* transformation of the system S', to be denoted by $\bar{\phi}$, which consists in this that to every element s' of S' there corresponds the transform $\bar{\phi}(s') = s$, and obviously this transformation is also similar. It is clear that $\bar{\phi}(S') = S$, that further ϕ is the inverse transformation belonging to $\bar{\phi}$ and that the transformation $\bar{\phi}\phi$ compounded out of ϕ and $\bar{\phi}$ by (25) is the identical transformation of S (21). At once we have the following additions to II., retaining the notation there given.

27. Theorem.[6]   If $A' \, 3 \, B'$, then $A \, 3 \, B$.

Proof.   For if a is an element of A then is a' an element of A', therefore also of B', hence $= b'$, where b is an element of B; but since from $a' = b'$ we always have $a = b$, then is every element of A also element of B, which was to be proved.

28. Theorem.   If $A' = B'$, then $A = B$.

The proof follows from (27), (4), (5).

29. Theorem.[7]   If $G = \mathfrak{G}\,(A, B, C, \ldots)$, then
$$G' = \mathfrak{G}\,(A', B', C', \ldots).$$

Proof.   Every element of $\mathfrak{G}\,(A', B', C', \ldots)$ is certainly contained in S', and is therefore the transform g' of an element g contained in S; but since g' is common element of $A', B', C', \ldots$ then by (27) must g be common element of $A, B, C, \ldots$ therefore also element of G; hence every element of $\mathfrak{G}\,(A', B', C', \ldots)$ is transform of an element g of G, therefore element of G', i.e., $\mathfrak{G}\,(A', B', C', \ldots) \, 3 \, G'$, and accordingly our theorem follows from (24), (5).

30. Theorem.   The identical transformation of a system is always a similar transformation.

31. Theorem.   If ϕ is a similar transformation of S and ψ a similar transformation of $\phi(S)$, then is the transformation $\psi\phi$ of S, compounded of ϕ and ψ, a similar transformation, and the associated inverse transformation $\overline{\psi\phi} = \bar{\phi}\bar{\psi}$.

Proof.   For to different elements a, b of S correspond different transforms $a' = \phi(a)$, $b' = \phi(b)$, and to these again different transforms $\psi(a') = \psi\phi(a)$, $\psi(b') = \psi\phi(b)$ and therefore $\psi\phi$ is a similar transformation. Besides every element $\psi\phi(s) = \psi(s')$ of the system $\psi\phi(S)$ is transformed by $\bar{\psi}$ into $s' = \phi(s)$ and this by $\bar{\phi}$ into s, therefore $\psi\phi(s)$ is transformed by $\bar{\phi}\bar{\psi}$ into s, which was to be proved.

32. Definition.   The systems R, S are said to be *similar* when there exists such a similar transformation ϕ of S that $\phi(S) = R$, and therefore $\bar{\phi}(R) = S$. Obviously by (30) every system is similar to itself.

---

[6]See theorem 22.
[7]See theorem 24.

33. Theorem. If R, S are similar systems, then every system Q similar to R is also similar to S.

Proof. For if ϕ, ψ are similar transformations of S, R such that $\phi(S) = R$, $\psi(R) = Q$, then by (31) $\psi\phi$ is a similar transformation of S such that $\psi\phi(S) = Q$, which was to be proved.

34. Definition. We can therefore separate all systems into *classes* by putting into a determinate class all systems Q, R, S, . . . , and only those, that are similar to a determinate system R, the *representative* of the class; according to (33) the class is not changed by taking as representative any other system belonging to it.

35. Theorem. If R, S are similar systems, then is every part of S also similar to a part of R, every proper part of S also similar to a proper part of R.

Proof. For if ϕ is a similar transformation of S, $\phi(S) = R$, and $T3S$, then by (22) is the system similar to T $\phi(T)3R$; if further T is proper part of S, and s an element of S not contained in T, then by (27) the element $\phi(s)$ contained in R cannot be contained in $\phi(T)$; hence $\phi(T)$ is proper part of R, which was to be proved.

IV

TRANSFORMATION OF A SYSTEM IN ITSELF

36. Definition. If ϕ is a similar or dissimilar transformation of a system S, and $\phi(S)$ part of a system Z, then ϕ is said to be a transformation of S *in* Z, and we say S is transformed by ϕ in Z. Hence we call ϕ a transformation of the system S *in itself*, when $\phi(S)3S$, and we propose in this paragraph to investigate the general laws of such a transformation ϕ. In doing this we shall use the same notations as in II. and again put $\phi(s) = s'$, $\phi(T) = T'$. These transforms s', T' are by (22), (7) themselves again elements or parts of S, like all things designated by italic letters.

37. Definition. K is called a *chain* [*Kette*], when $K'3K$. We remark expressly that this name does not in itself belong to the part K of the system S, but is given only with respect to the particular transformation ϕ; with reference to another transformation of the system S in itself K can very well not be a chain.

38. Theorem. S is a chain.

39. Theorem. The transform K' of a chain K is a chain.

Proof. For from $K'3K'$ it follows by (22) that $(K')'3K'$, which was to be proved.

40. Theorem. If A is part of a chain K, then is also $A'3K$.

Proof. For from $A3K$ it follows by (22) that $A'3K'$, and since by (37) $K'3K$, therefore by (7) $A'3K$, which was to be proved.

41. Theorem. If the transform A' is part of a chain L, then is there a chain K, which satisfies the conditions $A3K$, $K'3L$; and $\mathfrak{M}(A, L)$ is just such a chain K.

Proof. If we actually put $K = \mathfrak{M}(A, L)$, then by (9) the one condition $A3K$ is fulfilled. Since further by (23) $K' = \mathfrak{M}(A', L')$ and by hypothesis $A'3L$, $L'3L$, then by (10) is the other condition $K'3L$ also fulfilled and hence it follows because by (9) $L3K$, that also $K'3K$, i.e., K is a chain, which was to be proved.

42. Theorem. A system M compounded simply out of chains A, B, C, . . . is a chain.

Proof. Since by (23) $M' = \mathfrak{M}(A', B', C', . . .)$ and by hypothesis $A'3B$, $B'3B$, $C'3C$, . . . therefore by (12) $M'3M$, which was to be proved.

43. Theorem. The community G of chains A, B, C, . . . is a chain.

Proof. Since by (17) G is common part of A, B, C, . . . , therefore by (22) G' common part of A', B', C', . . ., and by hypothesis $A'3A$, $B'3B$, $C'3C$, . . . , then by (7) G' is also common part of A, B, C, . . . and hence by (18) also part of G, which was to be proved.

44. Definition. If A is any part of S, we will denote by A_o the community of all those chains (e.g., S) of which A is part; this community A_o exists (17) because A is itself common part of all these chains. Since further by (43) A_o is a chain, we will call A_o the *chain of the system A*, or briefly the chain of A. This definition too is strictly related to the fundamental determinate transformation ϕ of the system S in itself, and if later, for the sake of clearness, it is necessary we shall at pleasure use the symbol $\phi_o(A)$ instead of A_o, and likewise designate the chain of A corresponding to another transformation ω by $\omega_o(A)$. For this very important notion the following theorems hold true.

45. Theorem. $A3A_o$.

Proof. For A is common part of all those chains whose community is A_o, whence the theorem follows by (18).

46. Theorem. $(A_o)'3A_o$.

Proof. For by (44) A_o is a chain (37).

47. Theorem. If A is part of a chain K, then is also A_o3K.

Proof. For A_o is the community and hence also a common part of all the chains K, of which A is part.

48. Remark. One can easily convince himself that the notion of the chain A_o defined in (44) is completely characterised by the preceding theorems, (45), (46), (47).

49. Theorem. $A'3(A_o)'$.

The proof follows from (45), (22).

50. Theorem. $A'3A_o$.

The proof follows from (49), (46), (7).

51. Theorem. if A is a chain, then $A_o = A$.

Proof. Since A is part of the chain A, then by (47) $A_o 3A$, whence the theorem follows by (45), (5).

52. Theorem. If $B3A$, then $B3A_o$.

The proof follows from (45), (7).

53. Theorem. If $B3A_o$, then $B_o 3A_o$, and conversely.

Proof. Because A_o is a chain, then by (47) from $B3A_o$, we also get $B_o 3A_o$; conversely, if $B_o 3A_o$, then by (7) we also get $B3A_o$, because by (45) $B3B_o$.

54. Theorem. If $B3A$, then is $B_o 3A_o$.

The proof follows from (52), (53).

55. Theorem. If $B3A_o$, then is also $B'3A_o$.

Proof. For by (53) $B_o 3A_o$, and since by (50) $B'3B_o$, the theorem to be proved follows by (7). The same result, as is easily seen, can be obtained from (22), (46), (7), or also from (40).

56. Theorem. If $B3A_o$, then is $(B_o)'3(A_o)'$.

The proof follows from (53), (22).

57. Theorem and definition. $(A_o)' = (A')_o$, i.e., the transform of the chain of A is at the same time the chain of the transform of A. Hence we can designate this system in short by A'_o and at pleasure call it the *chain-transform* or *transform-chain* of A. With the clearer notation given in (44) the theorem might be expressed by $\phi(\phi_o(A)) = \phi_o(\phi(A))$.

Proof. If for brevity we put $(A')_o = L$, L is a chain (44) and by (45) $A'3L$; hence by (41) there exists a chain K satisfying the conditions $A3K$, $K'3L$; hence from (47) we have $A_o 3K$, therefore $(A_o)'3K'$, and hence by (7) also $(A_o)'3L$, i.e.,

$$(A_o)'3(A')_o.$$

Since further by (49) $A'3(A_o)'$, and by (44), (39) $(A_o)'$ is a chain, then by (47) also

$$(A')_o 3(A_o)',$$

whence the theorem follows by combining with the preceding result (5).

58. Theorem. $A_o = \mathfrak{M}(A, A'_o)$, i.e., the chain of A is compounded out of A and the transform-chain of A.

Proof. If for brevity we again put

$$L = A'_o = (A_o)' = (A')_o \text{ and } K = \mathfrak{M}(A, L),$$

then by (45) $A'3L$, and since L is a chain, by (41) the same thing is true of K; since further $A3K$ (9), therefore by (47)

$$A_o 3K.$$

On the other hand, since by (45) $A3A_o$, and by (46) also $L3A_o$, then by (10) also

$$K3A_o,$$

whence the theorem to be proved $A_o = K$ follows by combining with the preceding result (5).

59. Theorem of complete induction. In order to show that the chain A_o is part of any system Σ— be this latter part of S or not—it is sufficient to show,

ρ. that $A \mathbf{3} \Sigma$, and

σ. that the transform of every common element of A_o and Σ is likewise element of Σ.

Proof. For if ρ is true, then by (45) the community $G = \mathfrak{G}(A_o, \Sigma)$ certainly exists, and by (18) $A \mathbf{3} G$; since besides by (17)

$$G \mathbf{3} A_o,$$

then is G also part of our system S, which by ϕ is transformed in itself and at once by (55) we have also $G' \mathbf{3} A_o$. If then σ is likewise true, i.e., if $G' \mathbf{3} \Sigma$, then must G' as common part of the systems A_o, Σ by (18) be part of their community G, i.e., G is a chain (37), and since, as above noted, $A \mathbf{3} G$, then by (47) is also

$$A_o \mathbf{3} G,$$

and therefore by combination with the preceding result $G = A_o$, hence by (17) also $A_o \mathbf{3} \Sigma$, which was to be proved.

60. The preceding theorem, as will be shown later, forms the scientific basis for the form of demonstration known by the name of complete induction (the inference from n to $n + 1$); it can also be stated in the following manner: In order to show that all elements of the chain A_o possess a certain property $\mathfrak{E}$ (or that a theorem $\mathfrak{S}$ dealing with an undetermined thing n actually holds good for all elements n of the chain A_o) it is sufficient to show

ρ. that all elements a of the system A possess the property $\mathfrak{E}$ (or that $\mathfrak{S}$ holds for all a's) and

σ. that to the transform n' of every such element n of A_o possessing the property $\mathfrak{E}$, belongs the same property $\mathfrak{E}$ (or that the theorem $\mathfrak{S}$, as soon as it holds for an element n of A_o, certainly must also hold for its transform n').

Indeed, if we denote by Σ the system of all things possessing the property $\mathfrak{E}$ (or for which the theorem $\mathfrak{S}$ holds) the complete agreement of the present manner of stating the theorem with that employed in (59) is immediately obvious.

61. Theorem. The chain of $\mathfrak{M}(A, B, C, \ldots)$ is $\mathfrak{M}(A_o, B_o, C_o, \ldots)$.

Proof. If we designate by M the former, by K the latter system, then by (42) K is a chain. Since then by (45) each of the systems $A, B, C, \ldots$ is part of one of the systems $A_o, B_o, C_o, \ldots$, and therefore by (12) $M \mathbf{3} K$, then by (47) we also have

$$M_o \mathbf{3} K.$$

On the other hand, since by (9) each of the systems $A, B, C, \ldots$ is part of M, and hence by (45), (7) also part of the chain M_o, then by (47) must also each of the systems $A_o, B_o, C_o, \ldots$ be part of M_o, therefore by (10)

$$K \mathbf{3} M_o$$

whence by combination with the preceding result follows the theorem to be proved $M_o = K$ (5).

62. Theorem. The chain of $\mathfrak{G}\,(A, B, C, \ldots)$ is part of $\mathfrak{G}\,(A_o, B_o, C_o, \ldots)$.

Proof. If we designate by G the former, by K the latter system, then by (43) K is a chain. Since then each of the systems $A_o, B_o, C_o, \ldots$ by (45) is whole of one of the systems $A, B, C, \ldots$, and hence by (20) $G\,3\,K$, therefore by (47) we obtain the theorem to be proved $G_o\,3\,K$.

63. Theorem. If $K'\,3L3\,K$, and therefore K is a chain, L is also a chain. If the same is proper part of K, and U the system of all those elements of K which are not contained in L, and if further the chain U_o is proper part of K, and V the system of all those elements of K which are not contained in U_o, then is $K = \mathfrak{M}(U_o, V)$ and $L = \mathfrak{M}(U'_o, V)$. If finally $L = K'$ then $V\,3\,V'$.

The proof of this theorem of which (as of the two preceding) we shall make no use may be left for the reader.

V

THE FINITE AND INFINITE

64. Definition.[8]   A system S is said to be *infinite* when it is similar to a proper part of itself (32); in the contrary case S is said to be a *finite* system.

65. Theorem.   Every system consisting of a single element is finite.

Proof.   For such a system possesses no proper part (2), (6).

66. Theorem.   There exist infinite systems.

Proof.[9]   My own realm of thoughts, i.e., the totality S of all things, which can be objects of my thought, is infinite. For if s signifies an element of S, then is the thought s', that s can be object of my thought, itself an element of S. If we regard this as transform $\phi(s)$ of the element s then has the transformation ϕ of S, thus determined, the property that the transform S' is part of S; and S' is certainly proper part of S, because there are elements in S (e.g., my own ego) which are different from such thought s' and therefore are not contained in S'. Finally it is clear that if a, b are different elements of S, their transforms a', b' are also different, that therefore the transformation ϕ is a distinct (similar) transformation (26). Hence S is infinite, which was to be proved.

67. Theorem.   If R, S are similar systems, then is R finite or infinite according as S is finite or infinite.

---

[8] If one does not care to employ the notion of similar systems (32) he must say: S is said to be infinite, when there is a proper part of S (6) in which S can be distinctly (similarly) transformed (26), (36). In this form I submitted the definition of the infinite which forms the core of my whole investigation in September, 1882, to G. Cantor and several years earlier to Schwarz and Weber. All other attempts that have come to my knowledge to distinguish the infinite from the finite seem to me to have met with so little success that I think I may be permitted to forego any criticism of them.

[9] A similar consideration is found in § 13 of the *Paradoxien des Unendlichen* by Bolzano (Leipzig, 1851).

Proof. If S is infinite, therefore similar to a proper part S' of itself, then if R and S are similar, S' by (33) must be similar to R and by (35) likewise similar to a proper part of R, which therefore by (33) is itself similar to R; therefore R is infinite, which was to be proved.

68. Theorem. Every system S, which possesses an infinite part is likewise infinite; or, in other words, every part of a finite system is finite.

Proof. If T is infinite and there is hence such a similar transformation ψ of T, that $\psi(T)$ is a proper part of T, then, if T is part of S, we can extend this transformation ψ to a transformation ϕ of S in which, if s denotes any element of S, we put $\phi(s) = \psi(s)$ or $\phi(s) = s$ according as s is element of T or not. This transformation ϕ is a similar one; for, if a, b denote different elements of S, then if both are contained in T, the transform $\phi(a) = \psi(a)$ is different from the transform $\phi(b) = \psi(b)$, because ψ is a similar transformation; if further a is contained in T, but b not, then is $\phi(a) = \psi(a)$ different from $\phi(b) = b$, because $\psi(a)$ is contained in T; if finally neither a nor b is contained in T then also is $\phi(a) = a$ different from $\phi(b) = b$, which was to be shown. Since further $\psi(T)$ is part of T, because by (7) also part of S, it is clear that also $\phi(S)3S$. Since finally $\psi(T)$ is proper part of T there exists in T and therefore also in S, an element t, not contained in $\psi(T) = \phi(T)$; since then the transform $\phi(s)$ of every element s not contained in T is equal to s, and hence is different from t, t cannot be contained in $\phi(S)$; hence $\phi(S)$ is proper part of S and consequently S is infinite, which was to be proved.

69. Theorem. Every system which is similar to a part of a finite system, is itself finite.

The proof follows from (67), (68).

70. Theorem. If a is an element of S, and if the aggregate T of all the elements of S different from a is finite, then is also S finite.

Proof. We have by (64) to show that if ϕ denotes any similar transformation of S in itself, the transform $\phi(S)$ or S' is never a proper part of S but always $=S$. Obviously $S = \mathfrak{M}(a, T)$ and hence by (23), if the transforms are again denoted by accents, $S' = \mathfrak{M}(a', T')$, and, on account of the similarity of the transformation ϕ, a' is not contained in T' (26). Since further by hypothesis $S'3S$, then must a' and likewise every element of T' either $=a$, or be element of T. If then—a case which we will treat first—a is not contained in T', then must $T'3T$ and hence $T' = T$, because ϕ is a similar transformation and because T is a finite system; and since a', as remarked, is not contained in T', i.e., not in T', then must $a' = a$, and hence in this case we actually have $S' = S$ as was stated. In the opposite case when a is contained in T' and hence is the transform b' of an element b contained in T, we will denote by U the aggregate of all those elements u of T, which are different from b;

then $T = \mathfrak{M}(b, U)$ and by (15) $S = \mathfrak{M}(a, b, U)$, hence $S' = \mathfrak{M}(a', a, U')$. We now determine a new transformation ψ of T in which we put $\psi(b) = a'$, and generally $\psi(u) = u'$, whence by (23) $\psi(T) = \mathfrak{M}(a', U')$. Obviously ψ is a similar transformation, because ϕ was such, and because a is not contained in U and therefore also a' not in U'. Since further a and every element u is different from b then (on account of the similarity of ϕ) must also a' and every element u' be different from a and consequently contained in T; hence $\psi(T)\mathfrak{Z}T$ and since T is finite, therefore must $\psi(T) = T$, and $\mathfrak{M}(a', U') = T$. From this by (15) we obtain

$$\mathfrak{M}(a', a, U') = \mathfrak{M}(a, T)$$

i.e., according to the preceding $S' = S$. Therefore in this case also the proof demanded has been secured.

VI

SIMPLY INFINITE SYSTEMS. SERIES OF NATURAL NUMBERS

71. Definition.   A system N is said to be *simply infinite* when there exists a similar transformation ϕ of N in itself such that N appears as chain (44) of an element not contained in $\phi(N)$. We call this element, which we shall denote in what follows by the symbol 1, the *base-element* of N and say the simply infinite system N is *set in order* [*geordnet*] by this transformation ϕ. If we retain the earlier convenient symbols for transforms and chains (IV) then the essence of a simply infinite system N consists in the existence of a transformation ϕ of N and an element 1 which satisfy the following conditions α, β, γ, δ:

α. $N'\mathfrak{Z}N$.

β. $N = 1_o$.

γ.   The element 1 is not contained in N'.

δ.   The transformation ϕ is similar.

Obviously it follows from α, γ, δ that every simply infinite system N is actually an infinite system (64) because it is similar to a proper part N' of itself.

72. Theorem.   In every infinite system S a simply infinite system N is contained as a part.

Proof.   By (64) there exists a similar transformation ϕ of S such that $\phi(S)$ or S' is a proper part of S; hence there exists an element 1 in S which is not contained in S'. The chain $N = 1_o$, which corresponds to this transformation ϕ of the system S in itself (44), is a simply infinite system set in order by ϕ; for the characteristic conditions α, β, γ, δ in (71) are obviously all fulfilled.

73. Definition.   If in the consideration of a simply infinite system N set in order by a transformation ϕ we entirely neglect the special character of the elements;

simply retaining their distinguishability and taking into account only the relations to one another in which they are placed by the order-setting transformation ϕ, then are these elements called *natural numbers* or *ordinal numbers* or simply *numbers*, and the base-element 1 is called the *base-number* of the *number-series* N. With reference to this freeing the elements from every other content (abstraction) we are justified in calling numbers a free creation of the human mind. The relations or laws which are derived entirely from the conditions α, β, γ, δ in (71) and therefore are always the same in all ordered simply infinite systems, whatever names may happen to be given to the individual elements (compare 134), form the first object of the *science of numbers* or *arithmetic*. From the general notions and theorems of IV. about the transformation of a system in itself we obtain immediately the following fundamental laws where a, b, . . . m, n, . . . always denote elements of N, therefore numbers, A, B, C, . . . parts of N, a', b', . . . m', n', . . . A', B', C' . . . the corresponding transforms, which are produced by the order-setting transformation ϕ and are always elements or parts of N; the transform n' of a number n is also called the number *following* n.

74. Theorem. Every number n by (45) is contained in its chain n_o and by (53) the condition $n 3 m_o$ is equivalent to $n_o 3 m_o$.

75. Theorem. By (57) $n'_o = (n_o)' = (n')_o$.

76. Theorem. By (46) $n'_o 3 n_o$.

77. Theorem. By (58) $n_o = \mathfrak{M}(n, n'_o)$.

78. Theorem. $N = \mathfrak{M}(1, N')$, hence every number different from the base-number 1 is element of N', i.e., transform of a number.

The proof follows from (77) and (71).

79. Theorem. N is the only number-chain containing the base-number 1.

Proof. For if 1 is element of a number-chain K, then by (47) the associated chain $N 3 K$, hence $N = K$, because it is self-evident that $K 3 N$.

80. Theorem of complete induction (inference from n to n'). In order to show that a theorem holds for all numbers n of a chain m_o, it is sufficient to show,

ρ. that it holds for $n = m$, and

σ. that from the validity of the theorem for a number n of the chain m_o its validity for the following number n' always follows.

This results immediately from the more general theorem (59) or (60). The most frequently occurring case is where $m = 1$ and therefore m_o is the complete number-series N.

VII

Greater and Less Numbers

81. Theorem. Every number n is different from the following number n'.

Proof by complete induction (80):

ρ. The theorem is true for the number $n = 1$, because it is not contained in N' (71), while the following number $1'$ as transform of the number 1 contained in N is element of N'.

σ. If the theorem is true for a number n and we put the following number $n' = p$, then is n different from p, whence by (26) on account of the similarity (71) of the order-setting transformation ϕ it follows that n', and therefore p, is different from p'. Hence the theorem holds also for the number p following n, which was to be proved.

82. Theorem. In the transform-chain n'_o of a number n by (74), (75) is contained its transform n', but not the number n itself.

Proof by complete induction (80):

ρ. The theorem is true for $n = 1$, because $1'_o = N'$, and because by (71) the base-number 1 is not contained in N'.

σ. If the theorem is true for a number n, and we again put $n' = p$, then is n not contained in p_o, therefore is it different from every number q contained in p_o, whence by reason of the similarity of ϕ it follows that n', and therefore p, is different from every number q' contained in p'_o, and is hence not contained in p'_o. Therefore the theorem holds also for the number p following n, which was to be proved.

83. Theorem. The transform-chain n'_o is proper part of the chain n_o.

The proof follows from (76), (74), (82).

84. Theorem. From $m_o = n_o$ it follows that $m = n$.

Proof. Since by (74) m is contained in m_o, and

$$m_o = n_o = \mathfrak{M}(n, n'_o)$$

by (77), then if the theorem were false and hence m different from n, m would be contained in the chain n'_o, hence by (74) also $m_o 3 n'_o$, i.e., $n_o 3 n'_o$; but this contradicts theorem (83). Hence our theorem is established.

85. Theorem. If the number n is not contained in the number-chain K, then is $K 3 n'_o$.

Proof by complete induction (80):

ρ. By (78) the theorem is true for $n = 1$.

σ. If the theorem is true for a number n, then is it also true for the following number $p = n'$; for if p is not contained in the number-chain K, then by (40) n also

940

cannot be contained in K and hence by our hypothesis $K3n'_o$; now since by (77) $n'_o = p_o = \mathfrak{M}(p, p'_o)$, hence $K3\mathfrak{M}(p, p'_o)$ and p is not contained in K, then must $K3p'_o$, which was to be proved.

86. Theorem.   If the number n is not contained in the number-chain K, but its transform n' is, then $K = n'_o$.

Proof.   Since n is not contained in K, then by (85) $K3n'_o$, and since $n'3K$, then by (47) is also n'_o3K, and hence $K = n'_o$, which was to be proved.

87. Theorem.   In every number-chain K there exists one, and by (84) only one, number k, whose chain $k_o = K$.

Proof.   If the base-number 1 is contained in K, then by (79) $K = N = 1_o$. In the opposite case let Z be the system of all numbers not contained in K; since the base-number 1 is contained in Z, but Z is only a proper part of the number-series N, then by (79) Z cannot be a chain, i.e., Z' cannot be part of Z; hence there exists in Z a number n, whose transform n' is not contained in Z, and is therefore certainly contained in K; since further n is contained in Z, and therefore not in K, then by (86) $K = n'_o$, and hence $k = n'$, which was to be proved.

88. Theorem.   If m, n are different numbers then by (83), (84) one and only one of the chains m_o, n_o is proper part of the other and either $n_o3m'_o$ or $m_o3n'_o$.

Proof.   If n is contained in m_o, and hence by (74) also n_o3m_o, then m can not be contained in the chain n_o (because otherwise by (74) we should have m_o3n_o, therefore $m_o = n_o$, and hence by (84) also $m = n$) and thence it follows by (85) that $n_o3m'_o$. In the contrary case, when n is not contained in the chain m_o, we must have by (85) $m_o3n'_o$, which was to be proved.

89. Definition.   The number m is said to be *less* than the number n and at the same time n *greater* than m, in symbols

$$m < n, \quad n > m,$$

when the condition

$$n_o3m'_o$$

is fulfilled, which by (74) may also be expressed

$$n3m'_o.$$

90. Theorem.   If m, n are any numbers, then always one and only one of the following cases λ, μ, ν occurs:

$$\lambda. \quad m = n, \quad n = m, \quad \text{i.e., } m_o = n_o$$
$$\mu. \quad m < n, \quad n > m, \quad \text{i.e., } n_o3m'_o$$
$$\nu. \quad m > n, \quad n < m, \quad \text{i.e., } m_o3n'_o.$$

Proof.   For if λ occurs (84) then can neither μ nor ν occur because by (83) we never have $n_o3n'_o$. But if λ does not occur then by (88) one and only one of the cases μ, ν occurs, which was to be proved.

91. Theorem. $n < n'$.

Proof. For the condition for the case ν in (90) is fulfilled by $m = n'$.

92. Definition. To express that m is either $= n$ or $< n$, hence not $> n$ (90) we use the symbols

$$m \leqq n \text{ or also } n \geqq m$$

and we say m is *at most equal* to n, and n is *at least equal* to m.

93. Theorem. Each of the conditions

$$m \leqq n, \quad m < n', \quad n_o 3 m_o$$

is equivalent to each of the others.

Proof. For if $m \leqq n$, then from λ, μ in (90) we always have $n_o 3 m_o$, because by (76) $m'_o 3 m$. Conversely, if $n_o 3 m_o$, and therefore by (74) also $n 3 m_o$, it follows from $m_o = \mathfrak{M}(m, m'_o)$ that either $n = m$, or $n 3 m'_o$, i.e., $n > m$. Hence the condition $m \leqq n$ is equivalent to $n_o 3 m_o$. Besides it follows from (22), (27), (75) that this condition $n_o 3 m_o$ is again equivalent to $n'_o 3 m'_o$, i.e., by μ in (90) to $m < n'$, which was to be proved.

94. Theorem. Each of the conditions

$$m' \leqq n, \quad m' < n', \quad m < n$$

is equivalent to each of the others.

The proof follows immediately from (93), if we replace in it m by m', and from μ in (90).

95. Theorem. If $l < m$ and $m \leqq n$ or if $l \leqq m$, and $m < n$, then is $l < n$. But if $l \leqq m$ and $m \leqq n$, then is $l \leqq n$.

Proof. For from the corresponding conditions (89), (93) $m_o 3 l'_o$ and $n_o 3 m_o$, we have by (7) $n_o 3 l'_o$ and the same thing comes also from the conditions $m_o 3 l_o$ and $n_o 3 m'_o$, because in consequence of the former we have also $m'_o 3 l'_o$. Finally from $m_o 3 l_o$ and $n_o 3 m_o$ we have also $n_o 3 l_o$, which was to be proved.

96. Theorem. In every part T of N there exists one and only one *least* number k, i.e., a number k which is less than every other number contained in T. If T consists of a single number, then is it also the least number in T.

Proof. Since T_o is a chain (44), then by (87) there exists one number k whose chain $k_o = T_o$. Since from this it follows by (45), (77) that $T 3 \mathfrak{M}(k, k'_o)$, then first must k itself be contained in T (because otherwise $T 3 k'_o$, hence by (47) also $T_o 3 k'_o$, i.e., $k 3 k'_o$, which by (83) is impossible), and besides every number of the system T, different from k, must be contained in k'_o, i.e., be $> k$ (89), whence at once from (90) it follows that there exists in T one and only one least number, which was to be proved.

97. Theorem. The least number of the chain n_o is n, and the base-number 1 is the least of all numbers.

Proof. For by (74), (93) the condition $m 3 n_o$ is equivalent to $m \geqq n$. Or our theorem also follows immediately from the proof of the preceding theorem, because if in that we assume $T = n_o$, evidently $k = n$ (51).

98. Definition. If n is any number, then will we denote by Z_n the system of all numbers that are *not greater* than n, and hence *not* contained in n'_o. The condition

$$m 3 Z_n$$

by (92), (93) is obviously equivalent to each of the following conditions:

$$m \leqq n, \quad m < n', \quad n_o 3 m_o.$$

99. Theorem. $1 3 Z_n$ and $n 3 Z_n$.

The proof follows from (98) or from (71) and (82).

100. Theorem. Each of the conditions equivalent by (98)

$$m 3 Z_n, \quad m \leqq n, \quad m < n', \quad n_o 3 m_o$$

is also equivalent to the condition

$$Z_m 3 Z_n.$$

Proof. For if $m 3 Z_n$, and hence $m \leqq n$, and if $l 3 Z_m$, and hence $l \leqq m$, then by (95) also $l \leqq n$, i.e., $l 3 Z_n$; if therefore $m 3 Z_n$, then is every element l of the system Z_m also element of Z_n, i.e., $Z_m 3 Z_n$. Conversely, if $Z_m 3 Z_n$, then by (7) must also $m 3 Z_n$, because by (99) $m 3 Z_m$, which was to be proved.

101. Theorem. The conditions for the cases λ, μ, ν in (90) may also be put in the following form:

$$\lambda. \quad m = n, \quad n = m, \quad Z_m = Z_n$$
$$\mu. \quad m < n, \quad n > m, \quad Z_{m'} 3 Z_n$$
$$\nu. \quad m > n, \quad n < m, \quad Z_{n'} 3 Z_m.$$

The proof follows immediately from (90) if we observe that by (100) the conditions $n_o 3 m_o$ and $Z_m 3 Z_n$ are equivalent.

102. Theorem. $Z_1 = 1$.

Proof. For by (99) the base-number 1 is contained in Z_1, while by (78) every number different from 1 is contained in $1'_o$, hence by (98) not in Z_1, which was to be proved.

103. Theorem. By (98) $N = \mathfrak{M}(Z_n, n'_o)$.

104. Theorem. $n = \mathfrak{G}(Z_n, n_o)$, i.e., n is the only common element of the system Z_n and n_o.

Proof. From (99) and (74) it follows that n is contained in Z_n and n_o; but every element of the chain n_o different from n by (77) is contained in n'_o, and hence by (98) not in Z_n, which was to be proved.

105. Theorem. By (91), (98) the number n' is not contained in Z_n.

106. Theorem. If $m < n$, then is Z_m proper part of Z_n and conversely.

Proof. If $m < n$, then by (100) $Z_m 3 Z_n$, and since the number n, by (99) contained in Z_n, can by (98) not be contained in Z_m because $n > m$, therefore Z_m is proper part of Z_n. Conversely if Z_m is proper part of Z_n then by (100) $m \leq n$, and since m cannot be $= n$, because otherwise $Z_m = Z_n$, we must have $m < n$, which was to be proved.

107. Theorem. Z_n is proper part of $Z_{n'}$.

The proof follows from (106), because by (91) $n < n'$.

108. Theorem. $Z_{n'} = \mathfrak{M}(Z_n, n')$.

Proof. For every number contained in $Z_{n'}$ by (98) is $\leq n'$, hence either $= n'$ or $< n'$, and therefore by (98) element of Z_n. Therefore certainly $Z_{n'} 3 \mathfrak{M}(Z_n, n')$. Since conversely by (107) $Z_n 3 Z_{n'}$ and by (99) $n' 3 Z_{n'}$, then by (10) we have

$$\mathfrak{M}(Z_n, n') 3 Z_{n'},$$

whence our theorem follows by (5).

109. Theorem. The transform Z'_n of the system Z_n is proper part of the system $Z_{n'}$.

Proof. For every number contained in Z'_n is the transform m' of a number m contained in Z_n, and since $m \leq n$, and hence by (94) $m' \leq n'$, we have by (98) $Z'_n 3 Z_{n'}$. Since further the number 1 by (99) is contained in $Z_{n'}$, but by (71) is not contained in the transform Z'_n, then is Z'_n proper part of $Z_{n'}$, which was to be proved.

110. Theorem. $Z_{n'} = \mathfrak{M}(1, Z'_n)$.

Proof. Every number of the system $Z_{n'}$ different from 1 by (78) is the transform m' of a number m and this must be $\leq n$, and hence by (98) contained in Z_n (because otherwise $m > n$, hence by (94) also $m' > n'$ and consequently by (98) m' would not be contained in $Z_{n'}$); but from $m 3 Z_n$ we have $m' 3 Z'_n$, and hence certainly

$$Z_{n'} 3 \mathfrak{M}(1, Z'_n).$$

Since conversely by (99) $1 3 Z_n$, and by (109) $Z'_n 3 Z_{n'}$, then by (10) we have $\mathfrak{M}(1, Z'_n) 3 Z_{n'}$ and hence our theorem follows by (5).

111. Definition. If in a system E of numbers there exists an element g, which is greater than every other number contained in E, then g is said to be the *greatest* number of the system E, and by (90) there can evidently be only one such greatest number in E. If a system consists of a single number, then is this number itself the greatest number of the system.

112. Theorem. By (98) n is the greatest number of the system Z_n.

113. Theorem. If there exists in E a greatest number g, then is $E 3 Z_g$.

Proof. For every number contained in E is $\leq g$, and hence by (98) contained in Z_g, which was to be proved.

114. Theorem.   If E is part of a system Z_n, or what amounts to the same thing, there exists a number n such that all numbers contained in E are $\leqq n$, then E possesses a greatest number g.

Proof.   The system of all numbers p satisfying the condition $E\mathbf{3}Z_p$—and by our hypothesis such numbers exist—is a chain (37), because by (107), (7) it follows also that $E\mathbf{3}Z_{p'}$, and hence by (87) $=g_o$, where g signifies the least of these numbers (96), (97). Hence also $E\mathbf{3}Z_g$, therefore by (98) every number contained in E is $\leqq g$, and we have only to show that the number g is itself contained in E. This is immediately obvious if $g = 1$, for then by (102) Z_g, and consequently also E consists of the single number 1. But if g is different from 1 and consequently by (78) the transform f' of a number f, then by (108) is $E\mathbf{3}\mathfrak{M}(Z_f, g)$; if therefore g were not contained in E, then would $E\mathbf{3}Z_f$, and there would consequently be among the numbers p a number f by (91) $<g$, which is contrary to what precedes; hence g is contained in E, which was to be proved.

115. Definition.   If $l < m$ and $m < n$ we say the number m *lies between* l and n (also between n and l).

116. Theorem.   There exists no number lying between n and n'.

Proof.   For as soon as $m < n'$, and hence by (93) $m \leqq n$, then by (90) we cannot have $n < m$, which was to be proved.

117. Theorem.   If t is a number in T, but not the least (96), then there exists in T one and only one *next less* number s, i.e., a number s such that $s < t$, and that there exists in T no number lying between s and t. Similarly, if t is not the greatest number in T (111) there always exists in T one and only one *next greater* number u, i.e., a number u such that $t < u$, and that there exists in T no number lying between t and u. At the same time in T t is next greater than s and next less than u.

Proof.   If t is not the least number in T, then let E be the system of all those numbers of T that are $<t$; then by (98) $E\mathbf{3}Z_t$, and hence by (114) there exists in E a greatest number s obviously possessing the properties stated in the theorem, and also it is the only such number. If further t is not the greatest number in T, then by (96) there certainly exists among all the numbers of T, that are $>t$, a least number u, which and which alone possesses the properties stated in the theorem. In like manner the correctness of the last part of the theorem is obvious.

118. Theorem.   In N the number n' is next greater than n, and n next less than n'. The proof follows from (116), (117).

VIII

FINITE AND INFINITE PARTS OF THE NUMBER-SERIES

119. Theorem. Every system Z_n in (98) is finite.

Proof by complete induction (80).

ρ. By (65), (102) the theorem is true for $n = 1$.

σ. If Z_n is finite, then from (108) and (70) it follows that $Z_{n'}$ is also finite, which was to be proved.

120. Theorem. If m, n are different numbers, then are Z_m, Z_n dissimilar systems.

Proof. By reason of the symmetry we may by (90) assume that $m < n$; then by (106) Z_m is proper part of Z_n, and since by (119) Z_n is finite, then by (64) Z_m and Z_n cannot be similar, which was to be proved.

121. Theorem. Every part E of the number-series N, which possesses a greatest number (111), is finite.

The proof follows from (113), (119), (68).

122. Theorem. Every part U of the number-series N, which possesses no greatest number, is simply infinite (71).

Proof. If u is any number in U, there exists in U by (117) one and only one next greater number than u, which we will denote by $\psi(u)$ and regard as transform of u. The thus perfectly determined transformation ψ of the system U has obviously the property

$$\alpha. \quad \psi(U) \mathbf{3} U,$$

i.e., U is transformed in itself by ψ. If further u, v are different numbers in U, then by symmetry we may by (90) assume that $u < v$; thus by (117) it follows from the definition of ψ that $\psi(u) \leqq v$ and $v < \psi(v)$, and hence by (95) $\psi(u) < \psi(v)$; therefore by (90) the transforms $\psi(u)$, $\psi(v)$ are different, i.e.,

$$\delta. \quad \text{the transformation } \psi \text{ is similar.}$$

Further, if u_1 denotes the least number (96) of the system U, then every number u contained in U is $\geqq u_1$, and since generally $u < \psi(u)$, then by (95) $u_1 < \psi(u)$, and therefore by (90) u_1 is different from $\psi(u)$, i.e.,

$$\gamma. \quad \text{the element } u_1 \text{ of } U \text{ is not contained in } \psi(U).$$

Therefore $\psi(U)$ is proper part of U and hence by (64) U is an infinite system. If then in agreement with (44) we denote by $\psi_o(V)$, when V is any part of U, the chain of V corresponding to the transformation ψ, we wish to show finally that

$$\beta. \quad U = \psi_o(u_1).$$

In fact, since every such chain $\psi_o(V)$ by reason of its definition (44) is a part of the system U transformed in itself by ψ, then evidently is $\psi_o(u_1) \mathbf{3} U$; conversely it is first

946

of all obvious from (45) that the element u_1 contained in U is certainly contained in $\psi_o(u_1)$; but if we assume that there exist elements of U, that are not contained in $\psi_o(u_1)$, then must there be among them by (96) a least number w, and since by what precedes this is different from the least number u_1 of the system U, then by (117) must there exist in U also a number v which is next less than w, whence it follows at once that $w = \phi(v)$; since therefore $v < w$, then must v by reason of the definition of w certainly be contained in $\psi_o(u_1)$; but from this by (55) it follows that also $\psi(v)$, and hence w must be contained in $\psi_o(u_1)$, and since this is contrary to the definition of w, our foregoing hypothesis is inadmissible; therefore $U 3 \psi_o(u_1)$ and hence also $U = \psi_o(u_1)$, as stated. From α, β, γ, δ it then follows by (71) that U is a simply infinite system set in order by ψ, which was to be proved.

123. Theorem.   In consequence of (121), (122) any part T of the number-series N is finite or simply infinite, according as a greatest number exists or does not exist in T.

IX

DEFINITION OF A TRANSFORMATION OF THE NUMBER-SERIES BY INDUCTION

124. In what follows we denote numbers by small Italics and retain throughout all symbols of the previous sections VI. to VIII., while Ω designates an arbitrary system whose elements are not necessarily contained in N.

125. Theorem.   If there is given an arbitrary (similar or dissimilar) transformation θ of a system Ω in itself, and besides a determinate element ω in Ω, then to every number n corresponds one transformation ψ_n and one only of the associated number-system Z_n explained in (98), which satisfies the conditions:[10]

 I. $\psi_n(Z_n) 3 \Omega$

 II. $\psi_n(1) = \omega$

 III. $\psi_n(t') = \theta\psi_n(t)$, if $t < n$, where the symbol $\theta\psi_n$ has the meaning given in (25).

Proof by complete induction (80).

ρ. The theorem is true for $n = 1$. In this case indeed by (102) the system Z_n consists of the single number 1, and the transformation ψ, is therefore completely defined by II alone so that I is fulfilled while III drops out entirely.

σ. If the theorem is true for a number n then we show that it is also true for the following number $p = n'$, and we begin by proving that there can be only a

---

[10]For clearness here and in the following theorem (126) I have especially mentioned condition I., although properly it is a consequence of II. and III.

single corresponding transformation ψ_p of the system Z_p. In fact, if a transformation ψ_p satisfies the conditions

I'. $\psi_p(Z_p)3\Omega$

II'. $\psi_p(1) = \omega$

III'. $\psi_p(m') = \theta\psi_p(m)$, when $m < p$, then there is also contained in it by (21), because $Z_n 3 Z_p$ (107) a transformation of Z_n which obviously satisfies the same conditions I, II, III as ψ_n, and therefore coincides throughout with ψ_n; for all numbers contained in Z_n, and hence (98) for all numbers m which are $<p$, i.e., $\leqq n$, must therefore

$$\psi_p(m) = \psi_n(m) \qquad (m)$$

whence there follows, as a special case,

$$\psi_p(n) = \psi_n(n); \qquad (n)$$

since further by (105), (108) p is the only number of the system Z_p not contained in Z_n, and since by III' and (n) we must also have

$$\psi_p(p) = \theta\psi_n(n) \qquad (p)$$

there follows the correctness of our foregoing statement that there can be only one transformation ψ_p of the system Z_p satisfying the conditions I', II', III', because by the conditions (m) and (p) just derived ψ_p is completely reduced to ψ_n. We have next to show conversely that this transformation ψ_p of the system Z_p completely determined by (m) and (p) actually satisfies the conditions I', II', III'. Obviously I' follows from (m) and (p) with reference to I, and because $\theta(\Omega)3\Omega$. Similarly II' follows from (m) and II, since by (99) the number 1 is contained in Z_n. The correctness of III' follows first for those numbers m which are $<n$ from (m) and III, and for the single number $m = n$ yet remaining it results from (p) and (n). Thus it is completely established that from the validity of our theorem for the number n always follows its validity for the following number p, which was to be proved.

126. Theorem of the definition by induction. If there is given an arbitrary (similar or dissimilar) transformation θ of a system Ω in itself, and besides a determinate element ω in Ω, then there exists one and only one transformation ψ of the number-series N, which satisfies the conditions

I. $\psi(N)3\Omega$

II. $\psi(1) = \omega$

III. $\psi(n') = \theta\psi(n)$, where n represents every number.

Proof. Since, if there actually exists such a transformation ψ, there is contained in it by (21) a transformation ψ_n of the system Z_n, which satisfies the conditions I, II, III stated in (125), then because there exists one and only one such transformation ψ_n must necessarily

$$\psi(n) = \psi_n(n). \qquad (n)$$

Since thus ψ is completely determined it follows also that there can exist only one such transformation ψ (see the closing remark in (130)). That conversely the transformation ψ determined by (n) also satisfies our conditions I, II, III, follows easily from (n) with reference to the properties I, II and (p) shown in (125), which was to be proved.

127. Theorem. Under the hypotheses made in the foregoing theorem,
$$\psi(T') = \theta\psi(T),$$
where T denotes any part of the number-series N.

Proof. For if t denotes every number of the system T, then $\psi(T')$ consists of all elements $\psi(t')$, and $\theta\psi(T)$ of all elements $\theta\psi(t)$; hence our theorem follows because by III in (126) $\psi(t') = \theta\psi(t)$.

128. Theorem. If we maintain the same hypotheses and denote by θ_o the chains (44) which correspond to the transformation θ of the system Ω in itself, then is
$$\psi(N) = \theta_o(\omega).$$
Proof. We show first by complete induction (80) that
$$\psi(N)3\theta_o(\omega),$$
i.e., that every transform $\psi(n)$ is also element of $\theta_o(\omega)$. In fact,

ρ. this theorem is true for $n = 1$, because by (126, II) $\psi(1) = \omega$, and because by (45) $\omega3\theta_o(\omega)$.

σ. If the theorem is true for a number n, and hence $\psi(n)3\theta_o(\omega)$, then by (55) also $\theta(\psi(n))3\theta_o(\omega)$, i.e., by (126, III) $\psi(n')3\theta_o(\omega)$, hence the theorem is true for the following number n', which was to be proved.

In order further to show that every element ν of the chain $\theta_o(\omega)$ is contained in $\psi(N)$, therefore that
$$\theta_o(\omega)3\psi(N)$$
we likewise apply complete induction, i.e., theorem (59) transferred to Ω and the transformation θ. In fact,

ρ. the element $\omega = \psi(1)$, and hence is contained in $\psi(N)$.

σ. If ν is a common element of the chain $\theta_o(\omega)$ and the system $\psi(N)$, then $\nu = \psi(n)$, where n denotes a number, and by (126, III) we get $\theta(\nu) = \theta\psi(n) = \psi(n')$, and therefore $\theta(\nu)$ is contained in $\psi(N)$, which was to be proved.

From the theorems just established, $\psi(N)3\theta_o(\omega)$ and $\theta_o(\omega)3\psi(N)$, we get by (5) $\psi(N) = \theta_o(\omega)$, which was to be proved.

129. Theorem. Under the same hypotheses we have generally:
$$\psi(n_o) = \theta_o(\psi(n)).$$
Proof by complete induction (80). For

ρ. By (128) the theorem holds for $n = 1$, since $1_o = N$ and $\psi(1) = \omega$.

σ. If the theorem is true for a number n, then
$$\theta(\psi(n_o)) = \theta(\theta_o(\psi(n)));$$

since by (127), (75)

$$\theta(\psi(n_o)) = \psi(n'_o),$$

and by (57), (126, III)

$$\theta(\theta_o(\psi(n))) = \theta_o(\theta(\psi(n))) = \theta_o(\psi(n')),$$

we get

$$\psi(n'_o) = \theta_o(\psi(n')),$$

i.e., the theorem is true for the number n' following n, which was to be proved.

130. Remark. Before we pass to the most important applications of the theorem of definition by induction proved in (126), (sections X–XIV), it is worth while to call attention to a circumstance by which it is essentially distinguished from the theorem of demonstration by induction proved in (80) or rather in (59), (60), however close may seem the relation between the former and the latter. For while the theorem (59) is true quite generally for every chain A_o where A is any part of a system S transformed in itself by any transformation ϕ (IV), the case is quite different with the theorem (126), which declares only the existence of a consistent (or one-to-one) transformation ψ of the simply infinite system 1_o. If in the latter theorem (still maintaining the hypotheses regarding Ω and θ) we replace the number-series 1_o by an arbitrary chain A_o out of such a system S, and define a transformation ψ of A_o in Ω in a manner analogous to that in (126, II, III) by assuming that

ρ. to every element a of A there is to correspond a determinate element $\psi(a)$ selected from Ω, and

σ. for every element n contained in A_o and its transform $n' = \phi(n)$, the condition $\psi(n') = \theta\psi(n)$ is to hold, then would the case very frequently occur that such a transformation ψ does not exist, since these conditions ρ, σ may prove incompatible, even though the freedom of choice contained in ρ be restricted at the outset to conform to the condition σ. An example will be sufficient to convince one of this. If the system S consisting of the different elements a and b is so transformed in itself by ϕ that $a' = b$, $b' = a$, then obviously $a_o = b_o = S$; suppose further the system Ω consisting of the different elements α, β and γ be so transformed in itself by θ that $\theta(\alpha) = \beta, \theta(\beta) = \gamma, \theta(\gamma) = \alpha$; if we now demand a transformation ψ of a_o in Ω such that $\psi(a) = \alpha$, and that besides for every element n contained in a_o always $\psi(n') = \theta\psi(n)$, we meet a contradiction; since for $n = a$, we get $\psi(b) = \theta(\alpha) = \beta$, and hence for $n = b$, we must have $\psi(a) = \theta(\beta) = \gamma$, while we had assumed $\psi(a) = \alpha$.

But if there exists a transformation ψ of A_o in Ω, which satisfies the foregoing conditions ρ, σ without contradiction, then from (60) it follows easily that it is completely determined; for if the transformation χ satisfies the same conditions, then we have, generally, $\chi(n) = \psi(n)$, since by ρ this theorem is true for all elements

$n = a$ contained in A, and since if it is true for an element n of A_o it must by σ be true also for its transform n'.

131. In order to bring out clearly the import of our theorem (126), we will here insert a consideration which is useful for other investigations also, e.g., for the so-called group-theory.

We consider a system Ω, whose elements allow a certain combination such that from an element ν by the effect of an element ω, there always results again a determinate element of the same system Ω, which may be denoted by $\omega \cdot \nu$ or $\omega\nu$, and in general is to be distinguished from $\nu\omega$. We can also consider this in such a way that to every determinate element ω, there corresponds a determinate transformation of the system Ω in itself (to be denoted by $\dot{\omega}$), in so far as every element ν furnishes the determinate transform $\dot{\omega}(\nu) = \omega\nu$. If to this system Ω and its element ω we apply theorem (126), designating by $\dot{\omega}$ the transformation there denoted by θ, then there corresponds to every number n a determinate element $\psi(n)$ contained in Ω, which may now be denoted by the symbol ω^n and sometimes called the nth power of ω; this notion is completely defined by the conditions imposed upon it

$$\text{II.} \quad \omega^1 = \omega$$
$$\text{III.} \quad \omega^{n'} = \omega\omega^n,$$

and its existence is established by the proof of theorem (126).

If the foregoing combination of the elements is further so qualified that for arbitrary elements μ, ν, ω, we always have $\omega(\nu\mu) = \omega\nu(\mu)$, then are true also the theorems

$$\omega^{n'} = \omega^n\omega, \quad \omega^m\omega^n = \omega^n\omega^m,$$

whose proofs can easily be effected by complete induction and may be left to the reader.

The foregoing general consideration may be immediately applied to the following example. If S is a system of arbitrary elements, and Ω the associated system whose elements are all the transformations ν of S in itself (36), then by (25) can these elements be continually compounded, since $\nu(S)3S$, and the transformation $\omega\nu$ compounded out of such transformations ν and ω is itself again an element of Ω. Then are also all elements ω^n transformations of S in itself, and we say they arise by repetition of the transformation ω. We will now call attention to a simple connection existing between this notion and the notion of the chain $\omega_o(A)$ defined in (44), where A again denotes any part of S. If for brevity we denote by A_n the transform $\omega^n(A)$ produced by the transformation ω^n, then from III and (25) it follows that $\omega(A_n) = A_{n'}$. Hence it is easily shown by complete induction (80) that all these systems A_n are parts of the chain $\omega_o(A)$; for

ρ. by (50) this statement is true for $n = 1$, and

σ. if it is true for a number n, then from (55) and from $A_{n'} = \omega(A_n)$ it follows that it is also true for the following number n', which was to be proved. Since further by (45) $A\mathbf{3}\omega_o(A)$, then from (10) it results that the system K compounded out of A and all transforms A_n is part of $\omega_o(A)$. Conversely, since by (23) $\omega(K)$ is compounded out of $\omega(A) = A_1$ and all systems $\omega(A_n) = A_{n'}$, therefore by (78) out of all systems A_n, which by (9) are parts of K, then by (10) is $\omega(K)\mathbf{3}K$, i.e., K is a chain (37), and since by (9) $A\mathbf{3}K$, then by (47) it follows also that that $\omega_o(A)\mathbf{3}K$. Therefore $\omega_o(A) = K$, i.e., the following theorem holds: If ω is a transformation of a system S in itself, and A any part of S, then is the chain of A corresponding to the transformation ω compounded out of A and all the transforms $\omega^n(A)$ resulting from repetitions of ω. We advise the reader with this conception of a chain to return to the earlier theorems (57), (58).

X

THE CLASS OF SIMPLY INFINITE SYSTEMS

132. Theorem.   All simply infinite systems are similar to the number-series N and consequently by (33) also to one another.

Proof.   Let the simply infinite system Ω be set in order (71) by the transformation θ, and let ω be the base-element of Ω thus resulting; if we again denote by θ_o the chains corresponding to the transformation θ (44), then by (71) is the following true:

α. $\theta(\Omega)\mathbf{3}\Omega$.

β. $\Omega = \theta_o(\omega)$.

γ. ω is not contained in $\theta(\Omega)$.

δ.   The transformation θ is similar.

If then ψ denotes the transformation of the number-series N defined in (126), then from β and (128) we get first

$$\psi(N) = \Omega,$$

and hence we have only yet to show that ψ is a similar transformation, i.e., (26) that to different numbers m, n correspond different transforms $\psi(m)$, $\psi(n)$. On account of the symmetry we may by (90) assume that $m > n$, hence $m\mathbf{3}n'_o$, and the theorem to prove comes to this that $\psi(n)$ is not contained in $\psi(n'_o)$, and hence by (127) is not contained in $\theta\psi(n_o)$. This we establish for every number n by complete induction (80). In fact,

ρ. this theorem is true by γ for $n = 1$, since $\psi(1) = \omega$ and $\psi(1_o) = \psi(N) = \Omega$.

σ. If the theorem is true for a number n, then is it also true for the following number n'; for if $\psi(n')$, i.e., $\theta\psi(n)$, were contained in $\theta\psi(n'_o)$, then by δ and (27),

$\psi(n)$ would also be contained in $\psi(n'_o)$ while our hypothesis states just the opposite; which was to be proved.

133. Theorem. Every system which is similar to a simply infinite system and therefore by (132), (33) to the number-series N is simply infinite.

Proof. If Ω is a system similar to the number-series N, then by (32) there exists a similar transformation ψ of N such that

$$\text{I.} \quad \psi(N) = \Omega;$$

then we put

$$\text{II.} \quad \psi(1) = \omega.$$

If we denote, as in (26), by $\overline{\psi}$ the inverse, likewise similar transformation of Ω, then to every element ν of Ω there corresponds a determinate number $\overline{\psi}(\nu) = n$, viz., that number whose transform $\psi(n) = \nu$. Since to this number n there corresponds a determinate following number $\phi(n) = n'$, and to this again a determinate element $\psi(n')$ in Ω there belongs to every element ν of the system Ω a determinate element $\psi(n')$ of that system which as transform of ν we shall designate by $\theta(\nu)$. Thus a transformation θ of Ω in itself is completely determined,[11] and in order to prove our theorem we will show that by θ Ω is set in order (71) as a simply infinite system, i.e., that the conditions α, β, γ, δ stated in the proof of (132) are all fulfilled. First α is immediately obvious from the definition of θ. Since further to every number n corresponds an element $\nu = \phi(n)$, for which $\theta(\nu) = \psi(n')$, we have generally,

$$\text{III.} \quad \psi(n') = \theta\psi(n),$$

and thence in connection with I, II, α it results that the transformations θ, ψ fulfill all the conditions of theorem (126); therefore β follows from (128) and I. Further by (127) and I

$$\psi(N') = \theta\psi(N) = \theta(\Omega),$$

and thence in combination with II and the similarity of the transformation ψ follows γ, because otherwise $\psi(1)$ must be contained in $\psi(N')$, hence by (27) the number 1 in N', which by (71, γ) is not the case. If finally μ, ν denote elements of Ω and m, n the corresponding numbers whose transforms are $\psi(m) = \mu, \psi(n) = \nu$, then from the hypothesis $\theta(\mu) = \theta(\nu)$ it follows by the foregoing that $\psi(m') = \psi(n')$, thence on account of the similarity of ψ, ϕ that $m' = n'$, $m = n$, therefore also $\mu = \nu$; hence also δ is true, which was to be proved.

134. Remark. By the two preceding theorems (132), (133) all simply infinite systems form a class in the sense of (34). At the same time, with reference to (71), (73) it is clear that every theorem regarding numbers, i.e., regarding the elements n of the simply infinite system N set in order by the transformation ϕ' and indeed

---

[11]Evidently θ is the transformation $\psi\phi\overline{\psi}$ compounded by (25) out of $\overline{\psi}$, ϕ, ψ.

every theorem in which we leave entirely out of consideration the special character of the elements n and discuss only such notions as arise from the arrangement ϕ, possesses perfectly general validity for every other simply infinite system Ω set in order by a transformation θ and its elements ν, and that the passage from N to Ω (e.g., also the translation of an arithmetic theorem from one language into another) is effected by the transformation ψ considered in (132), (133), which changes every element n of N into an element ν of Ω, i.e., into $\psi(n)$. This element ν can be called the nth element of Ω and accordingly the number n is itself the nth number of the number series N. The same significance which the transformation ϕ possesses for the laws in the domain N, in so far as every element n is followed by a determinate element $\phi(n) = n'$, is found, after the change effected by ψ, to belong to the transformation θ for the same laws in the domain Ω, in so far as the element $\nu = \psi(n)$ arising from the change of n is followed by the element $\theta(\nu) = \psi(n')$ arising from the change of n'; we are therefore justified in saying that by $\psi\phi$ is changed into θ, which is symbolically expressed by $\theta = \psi\phi\overline{\psi}, \phi = \overline{\psi}\theta\psi$. By these remarks, as I believe, the definition of the notion of numbers given in (73) is fully justified. We now proceed to further applications of theorem (126).

XI

ADDITION OF NUMBERS

135. Definition.   It is natural to apply the definition set forth in theorem (126) of a transformation ψ of the number-series N, or of the *function* $\psi(n)$ determined by it to the case, where the system there denoted by Ω in which the transform $\psi(N)$ is to be contained, is the number-series N itself, because for this system Ω a transformation θ of Ω in itself already exists, viz., that transformation ϕ by which N is set in order as a simply infinite system (71), (73). Then is also $\Omega = N, \theta(n) = \phi(n) = n'$, hence

$$\text{I.} \quad \psi(N)3N,$$

and it remains in order to determine ψ completely only to select the element ω from Ω, i.e., from N, at pleasure. If we take $\omega = 1$, then evidently ψ becomes the identical transformation (21) of N, because the conditions

$$\psi(1) = 1, \quad \psi(n') = (\psi(n))'$$

are generally satisfied by $\psi(n) = n$. If then we are to produce another transformation ψ of N, then for ω we must select a number m' different from 1, by (78) contained in N, where m itself denotes any number; since the transformation ψ is obviously dependent upon the choice of this number m, we denote the corresponding transform $\psi(n)$ of an arbitrary number n by the symbol $m + n$, and call this number the *sum* which arises from the number m by the *addition* of the number n, or in short the sum of the numbers m, n. Therefore by (126) this sum is completely determined by the conditions[12]

$$\text{II.} \quad m + 1 = m',$$
$$\text{III.} \quad m + n' = (m + n)'.$$

136. Theorem. $m' + n = m + n'$.

Proof by complete induction (80). For

ρ. the theorem is true for $n = 1$, since by (135, II)

$$m' + 1 = (m')' = (m + 1)',$$

and by (135, III) $\quad (m + 1)' = m + 1'.$

σ. If the theorem is true for a number n, and we put the following number $n' = p$, then is $m' + n = m + p$, hence also $(m' + n)' = (m + p)'$, whence by (135, III) $m' + p = m + p'$; therefore the theorem is true also for the following number p, which was to be proved.

137. Theorem. $m' + n = (m + n)'$.

The proof follows from (136) and (135, III).

138. Theorem. $1 + n = n'$.

Proof by complete induction (80). For

ρ. by (135, II) the theorem is true for $n = 1$.

σ. If the theorem is true for a number n and we put $n' = p$, then $1 + n = p$, therefore also $(1 + n)' = p'$, whence by (135, III) $1 + p = p'$, i.e., the theorem is true also for the following number p, which was to be proved.

139. Theorem. $1 + n = n + 1$.

The proof follows from (138) and (135, II).

140. Theorem. $m + n = n + m$.

Proof by complete induction (80). For

ρ. by (139) the theorem is true for $n = 1$.

σ. If the theorem is true for a number n, then there follows also $(m + n)' = (n + m)'$, i.e., by (135, III) $m + n' = n + m'$, hence by (136) $(m + n)' = (n + m)'$, therefore the theorem is also true for the following number n', which was to be proved.

141. Theorem. $(l + m) + n = l + (m + n)$.

Proof by complete induction (80). For

ρ. the theorem is true for $n = 1$, because by (135, II, III, II) $(l + m) + 1 = (l + m)' = l + m' = l + (m + 1)$.

---

[12] The above definition of addition based immediately upon theorem (126) seems to me to be the simplest. By the aid of the notion developed in (131) we can, however, define the sum $m + n$ by $\phi^n(m)$ or also by $\phi^m(n)$, where ϕ has again the foregoing meaning. In order to show the complete agreement of these definitions with the foregoing, we need by (126) only to show that if $\phi^n(m)$ or $\phi^m(n)$ is denoted by $\psi(n)$, the conditions $\psi(1) = m'$, $\psi(n') = \phi\psi(n)$ are fulfilled which is easily done with the aid of complete induction (80) by the help of (131).

σ. If the theorem is true for a number n, then there follows also $((l + m) + n)' = (l + (m + n))'$, i.e., by (135, III)

$$(l + m) + n' = l + (m + n)' = l + (m + n'),$$

therefore the theorem is also true for the following number n', which was to be proved.

142. Theorem. $m + n > m$.

Proof by complete induction (80). For

ρ. by (135, II) and (91) the theorem is true for $n = 1$.

σ. If the theorem is true for a number n, then by (95) it is also true for the following number n', because by (135, III) and (91)

$$m + n' = (m + n)' > m + n,$$

which was to be proved.

143. Theorem. The conditions $m > a$ and $m + n > a + n$ are equivalent.

Proof by complete induction (80). For

ρ. by (135, II) and (94) the theorem is true for $n = 1$.

σ. If the theorem is true for a number n, then is it also true for the following number n', since by (94) the condition $m + n > a + n$ is equivalent to $(m + n)' > (a + n)'$, hence by (135, III) also equivalent to

$$m + n' > a + n',$$

which was to be proved.

144. Theorem. If $m > a$ and $n > b$, then is also

$$m + n > a + b.$$

Proof. For from our hypotheses we have by (143) $m + n > a + n$ and $n + a > b + a$ or, what by (140) is the same, $a + n > a + b$, whence the theorem follows by (95).

145. Theorem. If $m + n = a + n$, then $m = a$.

Proof. For if m does not $= a$, hence by (90) either $m > a$ or $m < a$, then by (143) respectively $m + n > a + n$ or $m + n < a + n$, therefore by (90) we surely cannot have $m + n = a + n$, which was to be proved.

146. Theorem. If $l > n$, then there exists one and by (157) only one number m which satisfies the condition $m + n = l$.

Proof by complete induction (80). For

ρ. the theorem is true for $n = 1$. In fact, if $l > 1$, i.e., (89) if l is contained in N', and hence is the transform m' of a number m, then by (135, II) it follows that $l = m + 1$, which was to be proved.

σ. If the theorem is true for a number n, then we show that it is also true for the following number n'. In fact, if $l > n'$, then by (91), (95) also $l > n$, and hence there exists a number k which satisfies the condition $l = k + n$; since by (138) this is different from 1 (otherwise l would be $= n'$) then by (78) is it the transform m'

of a number m, consequently $l = m' + n$, therefore also by (136) $l = m + n'$, which was to be proved.

XII

MULTIPLICATION OF NUMBERS

147. Definition.   After having found in XI an infinite system of new transformations of the number-series N in itself, we can by (126) use each of these in order to produce new transformations ψ of N. When we take $\Omega = N$, and $\theta(n) = m + n = n + m$, where m is a determinate number, we certainly again have

$$\text{I.} \quad \psi(N) \mathbf{3} N,$$

and it remains, to determine ψ completely only to select the element ω from N at pleasure. The simplest case occurs when we bring this choice into a certain agreement with the choice of θ, by putting $\omega = m$. Since the thus perfectly determinate ψ depends upon this number m, we designate the corresponding transform $\psi(n)$ of any number n by the symbol $m \times n$ or $m \cdot n$ or mn, and call this number the *product* arising from the number m by *multiplication* by the number n, or in short the product of the numbers m, n. This therefore by (126) is completely determined by the conditions

$$\text{II.} \quad m \cdot 1 = m$$
$$\text{III.} \quad mn' = mn + m,$$

148. Theorem. $m'n = mn + n$.

Proof by complete induction (80). For

ρ. by (147, II) and (135, II) the theorem is true for $n = 1$.

σ. If the theorem is true for a number n, we have

$$m'n + m' = (mn + n) + m'$$

and consequently by (147, III), (141), (140), (136), (141), (147, III)

$$m'n' = mn + (n + m') = mn + (m' + n) = mn + (m + n')$$
$$= (mn + m) + n' = mn' + n';$$

therefore the theorem is true for the following number n', which was to be proved.

149. Theorem. $1 \cdot n = n$.

Proof by complete induction (80). For

ρ. by (147, II) the theorem is true for $n = 1$.

σ. If the theorem is true for a number n, then we have $1 \cdot n + 1 = n + 1$, i.e., by (147, III), (135, II) $1 \cdot n' = n'$, therefore the theorem also holds for the following number n', which was to be proved.

150. Theorem. $mn = nm$.

Proof by complete induction (80). For

ρ. by (147, II), (149) the theorem is true for $n = 1$.

σ. If the theorem is true for a number n, then we have

$$mn + m = nm + m,$$

i.e., by (147, III), (148) $mn' = n'm$, therefore the theorem is also true for the following number n', which was to be proved.

151. Theorem. $l(m + n) = lm + ln$.

Proof by complete induction (80). For

ρ. by (135, II), (147, III), (147, II) the theorem is true for $n = 1$.

σ. If the theorem is true for a number n, we have

$$l(m + n) + l = (lm + ln) + l;$$

but by (147, III), (135, III) we have

$$l(m + n) + l = l(m + n)' = l(m + n'),$$

and by (141), (147, III)

$$(lm + ln) + l = lm + (ln + l) = lm + ln',$$

consequently $l(m + n') = lm + ln'$, i.e., the theorem is true also for the following number n', which was to be proved.

152. Theorem. $(m + n)l = ml + nl$.

The proof follows from (151), (150).

153. Theorem. $(lm)n = l(mn)$.

Proof by complete induction (80). For

ρ. by (147, II) the theorem is true for $n = 1$.

σ. If the theorem is true for a number n, then we have

$$(lm)n + lm = l(mn) + lm,$$

i.e., by (147, III), (151), (147, III)

$$(lm)n' = l(mn + m) = l(mn'),$$

hence the theorem is also true for the following number n', which was to be proved.

154. Remark. If in (147) we had assumed no relation between ω and θ, but had put $\omega = k$, $\theta(n) = m + n$, then by (126) we should have had a less simple transformation ψ of the number-series N; for the number 1 would $\psi(1) = k$ and for every other number (therefore contained in the form n') would $\psi(n') = mn + k$; since thus would be fulfilled, as one could easily convince himself by the aid of the foregoing theorems, the condition $\psi(n') = \theta\psi(n)$, i.e., $\psi(n') = m + \psi(n)$ for all numbers n.

XIII

INVOLUTION OF NUMBERS

155. Definition. If in theorem (126) we again put $\Omega = N$, and further $\omega = a$, $\theta(n) = an = na$, we get a transformation ψ of N which still satisfies the condition

I. $\psi(N)3N$;

the corresponding transform $\psi(n)$ of any number n we denote by the symbol a^n, and call this number a *power of the base a*, while n is called the *exponent* of this power of a. Hence this notion is completely determined by the conditions

$$\text{II.} \quad a^1 = a$$

$$\text{III.} \quad a^{n'} = a \cdot a^n = a^n \cdot a.$$

156. Theorem. $a^{m+n} = a^m \cdot a^n$.

Proof by complete induction (80). For

ρ. by (135, II), (155, III), (155, II) the theorem is true for $n = 1$.

σ. If the theorem is true for a number n, we have

$$a^{m+n} \cdot a = \left(a^m \cdot a^n\right)a;$$

but by (155, III), (135, III) $a^{m+n} \cdot a = a^{(m+n)'} = a^{m+n'}$, and by (153), (155, III) $\left(a^m \cdot a^n\right)a = a^m(a^n \cdot a) = a^m \cdot a^{n'}$; hence $a^{m+n'} = a^m \cdot a^{n'}$, i.e., the theorem is also true for the following number n', which was to be proved.

157. Theorem. $(a^m)^n = a^{mn}$.

Proof by complete induction (80). For

ρ. by (155, II), (147, II) the theorem is true for $n = 1$.

σ. If the theorem is true for a number n, we have

$$(a^m)^n \cdot a^m = a^{mn} \cdot a^m$$

but by (155, III) $(a^m)^n \cdot a^m = (a^m)^{n'}$, and by (156), (147, III) $a^{mn} \cdot a^m = a^{mn+m} = a^{mn'}$; hence $(a^m)^{n'} = a^{mn'}$, i.e., the theorem is also true for the following number n', which was to be proved.

158. Theorem. $(ab)^n = a^n \cdot b^n$.

Proof by complete induction (80). For

ρ. by (155, II) the theorem is true for $n = 1$.

σ. If the theorem is true for a number n, then by (150), (153), (155, III) we have also $(ab)^n \cdot a = a(a^n \cdot b^n) = (a \cdot a^n)b^n = a^{n'} \cdot b^n$, and thus $((ab)^n \cdot a)b = (a^{n'} \cdot b^n)b$; but by (153), (155, III) $((ab)^n \cdot a)b = (ab)^n \cdot (ab) = (ab)^{n'}$, and likewise

$$(a^{n'} \cdot b^n)b = a^{n'} \cdot (b^n \cdot b) = a^{n'} \cdot b^{n'};$$

therefore $(ab)^{n'} = a^{n'} \cdot b^{n'}$, i.e., the theorem is also true for the following number n', which was to be proved.

XIV

NUMBER OF THE ELEMENTS OF A FINITE SYSTEM

159. Theorem.   If Σ is an infinite system, then is every one of the number-systems Z_n defined in (98) similarly transformable in Σ (i.e., similar to a part of Σ), and conversely.

Proof.   If Σ is infinite, then by (72) there certainly exists a part T of Σ, which is simply infinite, therefore by (132) similar to the number-series N, and consequently by (35) every system Z_n as part of N is similar to a part of T, therefore also to a part of Σ, which was to be proved.

The proof of the converse—however obvious it may appear—is more complicated. If every system Z_n is similarly transformable in Σ, then to every number n corresponds such a similar transformation α_n of Z_n that $\alpha_n(Z_n)3\Sigma$. From the existence of such a series of transformations α_n, regarded as given, but respecting which nothing further is assumed, we derive first by the aid of theorem (126) the existence of a new series of such transformations ψ_n possessing the special property that whenever $m \leqq n$, hence by (100) $Z_m 3 Z_n$, the transformation ψ_m of the part Z_m is contained in the transformation ψ_n of Z_n (21), i.e., the transformations ψ_m and ψ_n completely coincide with each other for all numbers contained in Z_m, hence always

$$\psi_m(m) = \psi_n(m).$$

In order to apply the theorem stated to gain this end we understand by Ω that system whose elements are all possible similar transformations of all systems Z_n in Σ, and by aid of the given elements α_n, likewise contained in Ω, we define in the following manner a transformation θ of Ω in itself. If β is any element of Ω, thus, e.g., a similar transformation of the determinate system Z_n in Σ, then the system $\alpha_{n'}(Z_{n'})$ cannot be part of $\beta(Z_n)$, for otherwise $Z_{n'}$ would be similar by (35) to a part of Z_n, hence by (107) to a proper part of itself, and consequently infinite, which would contradict theorem (119); therefore there certainly exists in $Z_{n'}$ one number or several numbers p such that $\alpha_{n'}(p)$ is not contained in $\beta(Z_n)$; from these numbers p we select—simply to lay down something determinate—always the least k (96) and, since $Z_{n'}$ by (108) is compounded out of Z_n and n', define a transformation γ of $Z_{n'}$ such that for all numbers m contained in Z_n the transform $\gamma(m) = \beta(m)$ and besides $\gamma(n') = \alpha_{n'}(k)$; this obviously similar transformation γ of $Z_{n'}$ in Σ we consider then as a transform $\theta(\beta)$ of the transformation β, and thus a transformation θ of the system Ω in itself is completely defined. After the things named Ω and θ in (126) are determined we select finally for the element of Ω denoted by ω the given transformation α_1; thus by (126) there is determined a transformation ψ of the number-series N in Ω, which, if we denote the transform belonging to an arbitrary number n, not by $\psi(n)$ but by ψ_n, satisfies the conditions

II. $\quad \psi_1 = \alpha_1$

III. $\quad \psi_{n'} = \theta(\psi_n)$

By complete induction (80) it results first that ψ_n is a similar transformation of Z_n in Σ; for

ρ. by II this is true for $n = 1$.

σ. if this statement is true for a number n, it follows from III and from the character of the above described transition θ from β to γ, that the statement is also

960

true for the following number n', which was to be proved. Afterward we show likewise by complete induction (80) that if m is any number the above stated property

$$\psi_n(m) = \psi_m(m)$$

actually belongs to all numbers n, which are $\geqq m$, and therefore by (93), (74) belong to the chain m_o; in fact,

ρ. this is immediately evident for $n = m$, and

σ. if this property belongs to a number n it follows again from III and the nature of θ, that it also belongs to the number n', which was to be proved. After this special property of our new series of transformations ψ_n has been established, we can easily prove our theorem. We define a transformation χ of the number-series N, in which to every number n we let the transform $\chi(n) = \psi_n(n)$ correspond; obviously by (21) all transformations ψ_n are contained in this one transformation χ. Since ψ_n was a transformation of Z_n in Σ, it follows first that the number series N is likewise transformed by χ in Σ, hence $\chi(N)3\Sigma$. If further m, n are different numbers we may by reason of symmetry according to (90) suppose $m < n$; then by the foregoing $\chi(m) = \psi_m(m) = \psi_n(m)$, and $\chi(n) = \psi_n(n)$; but since ψ_n was a similar transformation of Z_n in Σ, and m, n are different elements ot Z_n, then is $\psi_n(m)$ different from $\psi_n(n)$, hence also $\chi(m)$ different from $\chi(n)$, i.e., χ is a similar transformation of N. Since further N is an infinite system (71), the same thing is true by (67) of the system $\chi(N)$ similar to it and by (68), because $\chi(N)$ is part of Σ, also of Σ, which was to be proved.

160. Theorem.   A system Σ is finite or infinite, according as there does or does not exist a system Z_n similar to it.

Proof.   If Σ is finite, then by (159) there exist systems Z_n which are not similarly transformable in Σ; since by (102) the system Z_1 consists of the single number 1, and hence is similarly transformable in every system, then must the least number k (96) to which a system Z_k not similarly transformable in Σ corresponds be different from 1 and hence by (78) $=n'$, and since $n < n'$ (91) there exists a similar transformation ψ of Z_n in Σ; if then $\psi(Z_n)$ were only a proper part of Σ, i.e., if there existed an element α in Σ not contained in $\psi(Z_n)$, then since $Z_{n'} = \mathfrak{M}(Z_n, n')$ (108) we could extend this transformation ψ to a similar transformation ψ of $Z_{n'}$ in Σ by putting $\psi(n') = \alpha$, while by our hypothesis $Z_{n'}$ is not similarly transformable in Σ. Hence $\psi(Z_n) = \Sigma$, i.e., Z_n and Σ are similar systems. Conversely, if a system Σ is similar to a system Z_n, then by (119), (67) Σ is finite, which was to be proved.

161. Definition.   If Σ is a finite system, then by (160) there exists one and by (120), (33) only one single number n to which a system Z_n similar to the system Σ corresponds; this number n is called the *number [Anzahl]* of the elements contained in Σ (or also the *degree* of the system Σ) and we say Σ consists of or is a system of

n elements, or the number n shows *how many* elements are contained in Σ.[13] If numbers are used to express accurately this determinate property of finite systems they are called *cardinal numbers*. As soon as a determinate similar transformation ψ of the system Z_n is chosen by reason of which $\psi(Z_n) = Z$, then to every number m contained in Z_n (i.e., every number m which is $\leqq n$) there corresponds a determinate element $\psi(m)$ of the system Σ, and conversely by (26) to every element of Σ by the inverse transformation $\overline{\psi}$ there corresponds a determinate number m in Z_n. Very often we denote all elements of Σ by a single letter, e.g., α, to which we append the distinguishing numbers m as indices so that $\psi(m)$ is denoted by α_m. We say also that these elements are *counted and set in order* by ψ in determinate manner, and call α_m the mth element of Σ; if $m < n$ then $\alpha_{m'}$ is called the element *following* α_m, and α_n is called the *last* element. In this counting of the elements therefore the numbers m appear again as ordinal numbers (73).

162. Theorem. All systems similar to a finite system possess the same number of elements.

The proof follows immediately from (33), (161).

163. Theorem. The number of numbers contained in Z_n, i.e., of those numbers which are $\leqq n$, is n.

Proof. For by (32) Z_n is similar to itself.

164. Theorem. If a system consists of a single element, then is the number of its elements $=1$, and conversely.

The proof follows immediately from (2), (26), (32), (102), (161).

165. Theorem. If T is proper part of a finite system Σ, then is the number of the elements of T less than that of the elements of Σ.

Proof. By (68) T is a finite system, therefore similar to a system Z_m, where m denotes the number of the elements of T; if further n is the number of elements of Σ, therefore Σ similar to Z_n, then by (35) T is similar to a proper part E of Z_n and by (33) also Z_m and E are similar to each other; if then we were to have $n \leqq m$, hence $Z_n 3 Z_m$, by (7) E would also be proper part of Z_m, and consequently Z_m an infinite system, which contradicts theorem (119); hence by (90), $m < n$, which was to be proved.

166. Theorem. If $\Gamma = \mathfrak{M}(B, \gamma)$, where B denotes a system of n elements, and γ an element of Γ not contained in B, then Γ consists of n' elements.

Proof. For if $B = \psi(Z_n)$, where ψ denotes a similar transformation of Z_n, then by (105), (108) it may be extended to a similar transformation ψ of $Z_{n'}$, by putting $\psi(n') = \gamma$, and we get $\psi(Z_{n'}) = \Gamma$, which was to be proved.

---

[13]For clearness and simplicity in what follows we restrict the notion of the number throughout to finite systems; if then we speak of a number of certain things, it is always understood that the system whose elements these things are is a finite system.

167. Theorem. If γ is an element of a system Γ consisting of n' elements, then is n the number of all other elements of Γ.

Proof. For if B denotes the aggregate of all elements in Γ different from γ, then is $\Gamma = \mathfrak{M}(B, \gamma)$; if then b is the number of elements of the finite system B, by the foregoing theorem b' is the number of elements of Γ, therefore $=n'$, whence by (26) we get $b = n$, which was to be proved.

168. Theorem. If A consists of m elements, and B of n elements, and A and B have no common element, then $\mathfrak{M}(A, B)$ consists of $m + n$ elements.

Proof by complete induction (80). For

ρ. by (166), (164), (135, II) the theorem is true for $n = 1$.

σ. If the theorem is true for a number n, then is it also true for the following number n'. In fact, if Γ is a system of n' elements, then by (167) we can put $\Gamma = \mathfrak{M}(B, \gamma)$ where γ denotes an element and B the system of the n other elements of Γ. If then A is a system of m elements each of which is not contained in Γ, therefore also not contained in B, and we put $\mathfrak{M}(A, B) = \Sigma$, by our hypothesis $m + n$ is the number of elements of Σ, and since γ is not contained in Σ, then by (166) the number of elements contained in $\mathfrak{M}(\Sigma, \gamma) = (m + n')$, therefore by (135, III) $= m + n'$; but since by (15) obviously $\mathfrak{M}(\Sigma, \gamma) = \mathfrak{M}(A, B, \gamma) = \mathfrak{M}(A, \Gamma)$, then is $m + n'$ the number of the elements of $\mathfrak{M}(A, \Gamma)$, which was to be proved.

169. Theorem. If A, B are finite systems of m, n elements respectively, then is $\mathfrak{M}(A, B)$ a finite system and the number of its elements is $\leqq m + n$.

Proof. If $B \, 3 \, A$, then $\mathfrak{M}(A, B) = A$, and the number m of the elements of this system is by (142) $< m + n$, as was stated. But if B is not part of A, and T is the system of all those elements of B that are not contained in A, then by (165) is their number $p \leqq n$, and since obviously

$$\mathfrak{M}(A, B) = \mathfrak{M}(A, T),$$

then by (143) is the number $m + p$ of the elements of this system $\leqq m + n$, which was to be proved.

170. Theorem. Every system compounded out of a number n of finite systems is finite.

Proof by complete induction (80). For

ρ. by (8) the theorem is self-evident for $n = 1$.

σ. If the theorem is true for a number n, and if Σ is compounded out of n' finite systems, then let A be one of these systems and B the system compounded out of all the rest; since their number by (167) $= n$, then by our hypothesis B is a finite system. Since obviously $\Sigma = \mathfrak{M}(A, B)$, it follows from this and from (169) that Σ is also a finite system, which was to be proved.

171. Theorem.   If ψ is a dissimilar transformation of a finite system Σ of n elements, then is the number of elements of the transform $\psi(\Sigma)$ less than n.

Proof.   If we select from all those elements of Σ that possess one and the same transform, always one and only one at pleasure, then is the system T of all these selected elements obviously a proper part of Σ, because ψ is a dissimilar transformation of Σ (26). At the same time it is clear that the transformation by (21) contained in ψ of this part T is a similar transformation, and that $\psi(T) = \psi(\Sigma)$; hence the system $\psi(\Sigma)$ is similar to the proper part T of Σ, and consequently our theorem follows by (162), (165).

172. Final remark.   Although it has just been shown that the number m of the elements of $\psi(\Sigma)$ is less than the number n of the elements of Σ, yet in many cases we like to say that the number of elements of $\psi(\Sigma) = n$. The word number is then, of course, used in a different sense from that used hitherto (161); for if α is an element of Σ and a the number of all those elements of Σ, that possess one and the same transform $\psi(\alpha)$ then is the latter as element of $\psi(\Sigma)$ frequently regarded still as representative of a elements, which at least from their derivation may be considered as different from one another, and accordingly counted as a-fold element of $\psi(\Sigma)$. In this way we reach the notion, very useful in many cases, of systems in which every element is endowed with a certain frequency-number which indicates how often it is to be reckoned as element of the system. In the foregoing case, e.g., we would say that n is the number of the elements of $\psi(\Sigma)$ counted in this sense, while the number m of the actually different elements of this system coincides with the number of the elements of T. Similar deviations from the original meaning of a technical term which are simply extensions of the original notion, occur very frequently in mathematics; but it does not lie in the line of this memoir to go further into their discussion.

Georg Cantor

(1845–1918)

HIS LIFE AND WORK

Georg Cantor scaled the peaks of infinity and then plunged into the deepest abysses of the mind: mental depression.

Georg Ferdinand Ludwig Phillipp Cantor was the firstborn son and namesake of a Protestant father and a Catholic mother. His father, Georg Waldemar Cantor, was a German-born Protestant who moved to St Petersburg, the capital of Tsarist Russia, to become a stockbroker. Cantor ultimately became famous because of the mathematical talent on his father's side. However, he first achieved notoriety for his fine violin playing, no doubt a talent derived from his mother, Marie Böhm, a native Russian who came from a musical family renowned for its violin virtuosi.

Any records of Cantor's early schooling in St Petersburg must have been lost when the family moved back to Germany and settled in Frankfurt so that his father would no longer have to endure the harsh Russian winters. Cantor had a distinguished record at Gymnasia in Frankfurt and nearby Wiesbaden. Cantor's father thought that young Georg's love of mathematics could enable him to be "a shining star in the engineer firmament." Cantor acquiesced to his father's firm suggestion that the Polytechnic school in Zurich would be a good school for studying engineering. After a semester, young Georg finally summoned up the courage to ask his father's permission to transfer to the University of Berlin, where he could study pure mathematics.

965

Much to Cantor's surprise, his father agreed to the change. In June 1863, as he was finishing his year in Zurich, Cantor learned of his father's sudden death. The elder Cantor would never know of even his son's first accomplishments as a shining star in the mathematical firmament! Cantor sped through the mathematics curriculum at Berlin where he was a pupil of the great Weierstrass, newly resident in the chair of mathematics. Within four years he had both his undergraduate and doctoral degrees.

After teaching at a local girls school for two years, Cantor received his first university teaching appointment at the University in Halle, the birthplace of the composer George Friedrich Handel, about 100 miles south of Berlin. Ten years later, he received a full professorship. He would remain there for the rest of his career.

When Cantor arrived at Halle, his new colleague, the mathematician Heinrich Heine, challenged him to prove the uniqueness of a functions representation by a trigonometric series, a generalization of a Fourier series. This was the research that led Cantor to his study of the infinite in the 1870s.

Cantor was not the first mathematician to formalize the concept of the infinite. Prior to Cantor, Richard Dedekind (1831–1916) made the first giant step by deciding how to **recognize** the infinite, rather than **construct** it, thereby avoiding objections such as the following one made by the great Gauss:

> I protest against the use of infinite magnitude as something completed, which in mathematics is never permissible. Infinity is merely a façon de parler, the real meaning being a limit which certain ratios approach indefinitely near, while others are permitted to increase without restriction.

Dedekind took the natural numbers, 0, 1, 2, 3, 4, …, as the paradigm example of an infinite set and defined a set as infinite if the natural numbers could be put into a one-to-one correspondence with that set, or a subset of it. Thus, the natural numbers are infinite by definition and so are the integers, the rational numbers, and the real numbers because every natural number is also an integer, a rational number, and a real number.

With Dedekind having accomplished that, Cantor asked two interesting questions: First, can infinity be recognized without making reference to the natural numbers? Second, are there different degrees of infinity?

Cantor answered his first question by defining a set as being infinite if it could be put into a one-to-one correspondence with a proper subset of itself, that is, a subset other than itself. It should be immediately obvious that the natural numbers satisfy this condition. Just consider the mapping that takes $0 \to 1$, $1 \to 2$, $2 \to 3$, $3 \to 4$, … Indeed, this mapping shows that any set that satisfies Dedekind's definition of the infinite automatically satisfies Cantor's definition. To show the opposite requires a piece of twentieth-century mathematics called "the Axiom of Choice."

The bulk of Cantor's groundbreaking work concerned his second question, and it built on his answer to the first question. By making the infinite more general than the natural numbers, Cantor opened the possibility of there being different degrees of the infinite. Cantor defined two sets as being *equinumerous* if they could be put into a one-to-one correspondence with each other. Thus, the positive integers are equinumerous with the negative integers by the mapping $n \leftrightarrow -n$ (for a positive integer n).

Similarly, the positive real numbers are equinumerous with the negative real numbers by a similar mapping. (Try proving to yourself that the positive integers are equinumerous with all of the integers.)

As he made these proofs, Cantor introduced the distinction between a set's *cardinality*, that is, how many members it has, and its *order-type*. He noted that although the positive and the negative integers have the same cardinality, they have a different order-type. There is a first positive integer but no last one using the standard greater-than/less-than ordering. In contrast, there is a last negative integer but no first one using the standard greater-than/less-than ordering. Cantor used Hebrew letters to denote cardinal numbers. $\aleph_0$ (The Hebrew letter Aleph with a subscript of 0 denotes the first infinite cardinal.) Greek letters denote order-types. The Greek letter ω denotes the order-type of a set like the positive integers with a first element but no last element and ω^* (ω with an asterisk as a superscript) denotes a set like the negative integers with a last element but no first element. (The order-type $\omega + 1$ describes a set like the positive integers plus a single element greater than all of the positive integers. The integers as a whole have order-type $\omega^* + \omega$. Can you guess what kind of set the order-type $\omega + \omega^*$ describes? It is not hard. It is just not one you're used to dealing with!)

Then Cantor demonstrated the power of his definition based on equinumerosity. First, he demonstrated that the positive rational numbers are equinumerous with the positive integers. Cantor realized that the positive rational numbers couldn't be put into greater-than/less-than, so he rearranged!

1/1	2/1	3/1	4/1	5/1	6/1	7/1	8/1	9/1	10/1
1/2	2/2	3/2	4/2	5/2	6/2	7/2	8/2	9/2	10/2
1/3	2/3	3/3	4/3	5/3	6/3	7/3	8/3	9/3	10/3
1/4	2/4	3/4	4/4	5/4	6/4	7/4	8/4	9/4	10/4
1/5	2/5	3/5	4/5	5/5	6/5	7/5	8/5	9/5	10/5
1/6	2/6	3/6	4/6	5/6	6/6	7/6	8/6	9/6	10/6
1/7	2/7	3/7	4/7	5/7	6/7	7/7	8/7	9/7	10/7
1/8	2/8	3/8	4/8	5/8	6/8	7/8	8/8	9/8	10/8
1/9	2/9	3/9	4/9	5/9	6/9	7/9	8/9	9/9	10/9
1/10	2/10	3/10	4/10	5/10	6/10	7/10	8/10	9/10	10/10

He rearranged them by the sum of their numerator and denominator and then started enumerating them as follows: 1/1, 2/1, 1/2, 1/3, 2/2, 3/1, 4/1, 3/2, ... thereby demonstrating that the positive rational numbers are equinumerous with the positive integers.

Cantor then tackled the real numbers and proved that they are not equinumerous with the integers. There are strictly more real numbers than integers. Here is the gist of Cantor's proof (for real numbers greater than zero and less than one): If the positive real numbers greater than zero and less than one are equinumerous with the positive integers, then they can all be listed out in a sequence such as

1st	0.**2**092119644443...
2nd	0.3**1**08131969619...
3rd	0.24**2**5129315441...
4th	0.348**0**075650872...
5th	0.0415**8**10010525...
6th	0.47027**4**2494171...
7th	0.659837**1**022485...
8th	0.4153943**6**69555...
9th	0.88325973**6**2598...
10th	0.247564657**6**200...
11th	0.7400378254**5**61...
12th	0.65230954343**7**1...
13th	0.351396247085**1**...

Now comes Cantor's great insight. He considers the highlighted digits on the diagonal and he changes all of their values adding 1 if the value is 0 through 8 and changing a 9 to a 0. This constructs the real number **0.3231952777682 ...** that cannot be in the table because it differs from every real number listed in the table! Cantor used a particular form of a *reduction ad absurdum* proof called a *diagonalization argument* to demonstrate that there are strictly more real numbers than there are integers. We will see diagonalization arguments reappear with gusto in the works of Kurt Gödel and Alan Turing.

Because the word *infinity* had a long history with much baggage attached, Cantor introduced the term *transfinite numbers* to denote all of his infinite numbers, both the cardinal numbers and the ordinal numbers (order-types).

Then Cantor asked a question that remains open to this day: Given that there are strictly more real numbers than integers, how many different types of infinite subsets of the real numbers are there? Are there only two types of infinite subsets of the real numbers, those equinumerous to the real numbers and those equinumerous to the integers? Or are there more types of subsets in between these two types? Cantor believed that there are only two types of infinite subsets of the real numbers, but he could not prove it. This conjecture, now known as Cantor's Continuum Hypothesis, has not yet been proven or disproven despite the best efforts of mathematicians, even when they have added axioms to Cantorian set theory.

Cantor's early years in Halle must have been filled with joy. In 1874, he married Vally Guttmann, a friend of his sister. They honeymooned in Switzerland and returned to a house Cantor had built with the inheritance from his father. It would be the birthplace of their five children over the next twelve years.

Cantor's growing family must have made financial demands on a professor paid at relatively meager provincial standards. Hoping to alleviate these problems, Cantor sought a professorship at his alma mater in Berlin. Cantor's work on transfinite numbers had drawn wide praise in the world of mathematics, praise that he hoped would win him an appointment at a prestigious university such as Berlin. Weierstrass, his old mentor, had been particularly full of praise for Cantor's work. But there were pockets of those opposed to any talk of actual infinities. Among them was Leopold Kronecker, a high-ranking professor in Berlin.

Mathematics, according to Kronecker, dealt with constructions, precisely what Cantor and Dedekind had avoided in their treatment of the infinite. In spite of Weierstrass's efforts, Kronecker was able to block all attempts to get Cantor a mathematics professorship at Berlin.

Just before turning forty, the combination of personal and professional stresses became too much for Cantor to bear. He had his first bout of deep mental depression, spending a few weeks in a sanitarium. After his release, Cantor wrote to a fellow mathematician:

> *I don't know when I shall return to the continuation of my scientific work. At the moment I can do absolutely nothing with it, and limit myself to the most necessary duty of my lectures; how much happier I would be to be scientifically active, if only I had the necessary mental freshness.*

Indeed, Cantor's best years as a mathematician had ended. Perhaps recognizing this, Cantor devoted significant energy to building an association of mathematicians across the newly unified Germany. He served as its first president from its founding in 1890 until 1893 when he had his next round of depression.

Cantor would be in and out of mental hospitals for the last twenty-five years of his life, fighting depression. In 1894, he published an exceptionally strange paper for a mathematician of his renown. While in the mental hospital, he had become fascinated by Goldbach's famous conjecture that every even number could be expressed as the sum of two primes. In 1894, he published a paper demonstrating all of the ways the even numbers up to 1,000 could be written as the sum of two primes, forty years after an obscure mathematician had done the same for all of the even numbers up to 10,000.

As the years went on, Cantor's mental state got worse and worse. In these later years, he devoted himself to the study of Shakespeare. He even attempted to prove that the Bard and the philosopher Francis Bacon were one and the same person!

The German mathematical community had planned to have a major celebration in 1915 in honor of Cantor's seventieth birthday. However, the privations of World War I made that impossible. Cantor entered a mental hospital for the last time in June 1917. On January 6, 1918, he died, unaware that Imperial Germany would also perish by the end of the same year.

SELECTIONS FROM *CONTRIBUTIONS TO THE FOUNDING OF THE THEORY OF TRANSFINITE NUMBERS*

(FIRST ARTICLE)

"Hypotheses non fingo."
    "Neque enim leges intellectui aut rebus damus ad arbitrium nostrum, sed tanquam scribæ fideles ab ipsius naturæ voce latas et prolatas excipimus et describimus."
    "Veniet tempus, quo ista quæ nunc latent, in lucem dies extrahat et longioris ævi diligentia."

§ 1
THE CONCEPTION OF POWER
OR CARDINAL NUMBER

By an "aggregate" (*Menge*) we are to understand any collection into a whole (*Zusammenfassung zu einem Ganzen*) M of definite and separate objects *m* of our intuition or our thought. These objects are called the "elements" of M.

In signs we express this thus:

$$M = \{m\}. \qquad (1)$$

We denote the uniting of many aggregates M, N, P, ..., which have no common elements, into a single aggregate by

$$(M, N, P, \ldots). \qquad (2)$$

The elements of this aggregate are, therefore, the elements of M, of N, of P, ..., taken together.

We will call by the name "part" or "partial aggregate" of an aggregate M any other aggregate M_1 whose elements are also elements of M.

If M_2 is a part of M_1 and M_1 is a part of M, then M_2 is a part of M.

Every aggregate M has a definite "power," which we will also call its "cardinal number."

We will call by the name "power" or "cardinal number" of M the general concept which, by means of our active faculty of thought, arises from the aggregate M when we make abstraction of the nature of its various elements *m* and of the order in which they are given.

[482]

We denote the result of this double act of abstraction, the cardinal number c power of M, by

$$\overline{\overline{M}}. \qquad (3$$

Since every single element m, if we abstract from its nature, becomes a "unit," th cardinal number $\overline{\overline{M}}$ is a definite aggregate composed of units, and this number ha existence in our mind as an intellectual image or projection of the given aggregate M

We say that two aggregates M and N are "equivalent," in signs

$$M \sim N \quad \text{or} \quad N \sim M, \qquad (4$$

if it is possible to put them, by some law, in such a relation to one another that t every element of each one of them corresponds one and only one element of th other. To every part M_1 of M there corresponds, then, a definite equivalent part N of N, and inversely.

If we have such a law of co-ordination of two equivalent aggregates, then, apart fror the case when each of them consists only of one element, we can modify this law i many ways. We can, for instance, always take care that to a special element m_0 of M special element n_0 of N corresponds. For if, according to the original law, the element m_0 and n_0 do not correspond to one another, but to the element m_0 of M the elemen n_1 of N corresponds, and to the element n_0 of N the element m_1 of M corresponds, w take the modified law according to which m_0 corresponds to n_0 and m_1 to n_1 and fo the other elements the original law remains unaltered. By this means the end is attained

Every aggregate is equivalent to itself:

$$M \sim M. \qquad (5$$

If two aggregates are equivalent to a third, they are equivalent to one another; tha is to say:

$$\text{from } M \sim P \text{ and } N \sim P \text{ follows } M \sim N. \qquad (6$$

Of fundamental importance is the theorem that two aggregates M and N hav the same cardinal number if, and only if, they are equivalent: thus,

$$\text{from } M \sim N \quad \text{we get} \quad \overline{\overline{M}} = \overline{\overline{N}}, \qquad (7$$

and

$$\text{from } \overline{\overline{M}} = \overline{\overline{N}} \quad \text{we get} \quad M \sim N. \qquad (8$$

Thus the equivalence of aggregates forms the necessary and sufficient condition fo the equality of their cardinal numbers.

[483]

In fact, according to the above definition of power, the cardinal number $\overline{\overline{M}}$ remain unaltered if in the place of each of one or many or even all elements m of M othe things are substituted. If, now, M ~ N, there is a law of co-ordination by means of which M and N are uniquely and reciprocally referred to one another; and by it to the element m of M corresponds the element n of N. Then we can imagine, in the place of every element m of M, the corresponding element n of N substituted, and, in this way, M transforms into N without alteration of cardinal number. Consequently

$$\overline{\overline{M}} = \overline{\overline{N}}.$$

The converse of the theorem results from the remark that between the elements of M and the different units of its cardinal number M a reciprocally univocal (or bi-univocal) relation of correspondence subsists. For, as we saw, $\overline{\overline{M}}$ grows, so to speak, out of M in such a way that from every element m of M a special unit of M arises. Thus we can say that

$$M \sim \overline{\overline{M}}. \tag{9}$$

In the same way N ~ $\overline{\overline{N}}$. If then $\overline{\overline{M}} = \overline{\overline{N}}$, we have, by (6), M ~ N.

We will mention the following theorem, which results immediately from the conception of equivalence. If M, N, P, ... are aggregates which have no common elements, M′, N′, P′, ... are also aggregates with the same property, and if

$$M \sim M', \quad N \sim N', \quad P \sim P', \ldots ,$$

then we always have

$$(M, N, P, \ldots) \sim (M', N', P', \ldots).$$

§ 2
"GREATER" AND "LESS" WITH POWERS

If for two aggregates M and N with the cardinal numbers a = $\overline{\overline{M}}$ and b = $\overline{\overline{N}}$, both the conditions:

(*a*) There is no part of M which is equivalent to N,

(*b*) There is a part N_1 of N, such that $N_1 \sim M$,

are fulfilled, it is obvious that these conditions still hold if in them M and N are replaced by two equivalent aggregates M′ and N′. Thus they express a definite relation of the cardinal numbers a and b to one another.

[484]

Further, the equivalence of M and N, and thus the equality of a and b, is excluded; for if we had M ~ N, we would have, because $N_1 \sim M$, the equivalence $N_1 \sim N$, and then, because M ~ N, there would exist a part M_1 of M such that $M_1 \sim M$, and therefore we should have $M_1 \sim N$; and this contradicts the condition (*a*).

Thirdly, the relation of a to b is such that it makes impossible the same relation of b to a; for if in (*a*) and (*b*) the parts played by M and N are interchanged, two conditions arise which are contradictory to the former ones.

We express the relation of a to b characterized by (*a*) and (*b*) by saying: a is "less" than b or b is "greater" than a; in signs

$$a < b \quad \text{or} \quad b > a. \tag{1}$$

We can easily prove that,

$$\text{if } a < b \text{ and } b < c, \text{ then we always have } a < c. \tag{2}$$

Similarly, from the definition, it follows at once that, if P_1 is part of an aggregate P, from $a < \overline{\overline{P}}_1$ follows $a < \overline{\overline{P}}$ and from $\overline{\overline{P}} < b$ follows $\overline{\overline{P}}_1 < b$.

We have seen that, of the three relations

$$a = b, \quad a < b, \quad b < a,$$

each one excludes the two others. On the other hand, the theorem that, with any two cardinal numbers a and b, one of those three relations must necessarily be realized, is by no means self-evident and can hardly be proved at this stage.

Not until later, when we shall have gained a survey over the ascending sequence of the transfinite cardinal numbers and an insight into their connexion, will result the truth of the theorem:

A. If a and b are any two cardinal numbers, then either a = b or a < b or a > b.

From this theorem the following theorems, of which, however, we will here make no use, can be very simply derived:

B. If two aggregates M and N are such that M is equivalent to a part N_1 of N and N to a part M_1 of M, then M and N are equivalent;

C. If M_1 is a part of an aggregate M, M_2 is a part of the aggregate M_1, and if the aggregates M and M_2 are equivalent, then M_1, is equivalent to both M and M_2;

D. If, with two aggregates M and N, N is equivalent neither to M nor to a part of M, there is a part N_1 of N that is equivalent to M;

E. If two aggregates M and N are not equivalent, and there is a part N_1 of N that is equivalent to M, then no part of M is equivalent to N.

[485]

§ 3
THE ADDITION AND MULTIPLICATION OF POWERS

The union of two aggregates M and N which have no common elements was denoted in § 1, (2), by (M, N). We call it the "union-aggregate (*Vereinigungsmenge*) of M and N."

If M' and N' are two other aggregates without common elements, and if M ~ M' and N ~ N', we saw that we have
$$(M, N) \sim (M', N').$$
Hence the cardinal number of (M, N) only depends upon the cardinal numbers $\overline{\overline{M}} = a$ and $\overline{\overline{N}} = b$.

This leads to the definition of the sum of a and b. We put
$$a + b = (\overline{\overline{M, N}}). \tag{1}$$
Since in the conception of power, we abstract from the order of the elements, we conclude at once that
$$a + b = b + a; \tag{2}$$
and, for any three cardinal numbers a, b, c, we have
$$a + (b + c) = (a + b) + c. \tag{3}$$

We now come to multiplication. Any element m of an aggregate M can be thought to be bound up with any element n of another aggregate N so as to form a new element (m, n); we denote by $(M \cdot N)$ the aggregate of all these bindings (m, n), and call it the "aggregate of bindings (*Verbindungsmenge*) of M and N." Thus
$$(M \cdot N) = \{(m, n)\}. \tag{4}$$
We see that the power of $(M \cdot N)$ only depends on the powers $\overline{\overline{M}} = a$ and $\overline{\overline{N}} = b$; for, if we replace the aggregates M and N by the aggregates
$$M' = \{m'\} \quad \text{and} \quad N' = \{n'\}$$
respectively equivalent to them, and consider m, m' and n, n' as corresponding elements, then the aggregate
$$(M' \cdot N') = \{(m', n')\}$$
is brought into a reciprocal and univocal correspondence with $(M \cdot N)$ by regarding (m, n) and (m', n') as corresponding elements. Thus
$$(M' \cdot N') \sim (M \cdot N). \tag{5}$$
We now define the product $a \cdot b$ by the equation
$$a \cdot b = (\overline{\overline{M \cdot N}}). \tag{6}$$

[486]

An aggregate with the cardinal number $a \cdot b$ may also be made up out of two aggregates M and N with the cardinal numbers a and b according to the following rule: We start from the aggregate N and replace in it every element n by an aggregate $M_n \sim M$; if, then, we collect the elements of all these aggregates M_n to a whole S, we see that
$$S \sim (M \cdot N), \tag{7}$$
and consequently
$$\overline{\overline{S}} = a \cdot b.$$

For, if, with any given law of correspondence of the two equivalent aggregates M and M_n, we denote by m the element of M which corresponds to the element m_n of M_n, we have

$$S = \{m_n\}; \tag{8}$$

and thus the aggregates S and $(M \cdot N)$ can be referred reciprocally and univocally to one another by regarding m_n and (m, n) as corresponding elements.

From our definitions result readily the theorems:

$$a \cdot b = b \cdot a, \tag{9}$$
$$a \cdot (b \cdot c) = (a \cdot b) \cdot c, \tag{10}$$
$$a(b + c) = ab + ac; \tag{11}$$

because:

$$(M \cdot N) \sim (N \cdot M),$$
$$(M \cdot (N \cdot P)) \sim ((M \cdot N) \cdot P),$$
$$(M \cdot (N, P)) \sim ((M \cdot N), (M \cdot P)).$$

Addition and multiplication of powers are subject, therefore, to the commutative, associative, and distributive laws.

§ 4
THE EXPONENTIATION OF POWERS

By a "covering of the aggregate N with elements of the aggregate M," or, more simply, by a "covering of N with M," we understand a law by which with every element n of N a definite element of M is bound up, where one and the same element of M can come repeatedly into application. The element of M bound up with n is, in a way, a one-valued function of n, and may be denoted by $f(n)$; it is called a "covering function of n." The corresponding covering of N will be called $f(N)$.

[487]

Two coverings $f_1(N)$ and $f_2(N)$ are said to be equal if, and only if, for all elements n of N the equation

$$f_1(n) = f_2(n) \tag{1}$$

is fulfilled, so that if this equation does not subsist for even a single element $n = n_0$, $f_1(N)$ and $f_2(N)$ are characterized as different coverings of N. For example, if m_0 is a particular element of M, we may fix that, for all n's

$$f(n) = m_0;$$

this law constitutes a particular covering of N with M. Another kind of covering results if m_0 and m_1 are two different particular elements of M and n_0 a particular element of N, from fixing that

$$f(n_0) = m_0$$
$$f(n) = m_1,$$

for all n's which are different from n_0.

The totality of different coverings of N with M forms a definite aggregate with the elements $f(N)$; we call it the "covering-aggregate (*Belegungsmenge*) of N with M" and denote it by $(N \mid M)$. Thus:

$$(N \mid M) = \{f(N)\}. \tag{2}$$

If $M \sim M'$ and $N \sim N'$, we easily find that

$$(N \mid M) \sim (N' \mid M'). \tag{3}$$

Thus the cardinal number of $(N \mid M)$ depends only on the cardinal numbers $\overline{\overline{M}} = a$ and $\overline{\overline{N}} = b$; it serves us for the definition of a^b:

$$a^b = \overline{\overline{(N \mid M)}}. \tag{4}$$

For any three aggregates, M, N, P, we easily prove the theorems:

$$((N \mid M) \cdot (P \mid M)) \sim ((N, P) \mid M), \tag{5}$$
$$((P \mid M) \cdot (P \mid N)) \sim (P \mid (M \cdot N)), \tag{6}$$
$$(P \mid (N \mid M)) \sim ((P \cdot N) \mid M), \tag{7}$$

from which, if we put $\overline{\overline{P}} = c$, we have, by (4) and by paying attention to § 3, the theorems for any three cardinal numbers, a, b, and c:

$$a^b \cdot a^c = a^{b+c}, \tag{8}$$
$$a^c \cdot b^c = (a \cdot b)^c, \tag{9}$$
$$(a^b)^c = a^{b \cdot c}. \tag{10}$$

[488]

We see how pregnant and far-reaching these simple formulæ extended to powers are by the following example. If we denote the power of the linear continuum X (that is, the totality X of real numbers x such that $x \geqq$ and $\leqq 1$) by 0, we easily see that it may be represented by, amongst others, the formula:

$$0 = 2^{\aleph_0}, \tag{11}$$

where § 6 gives the meaning of $\aleph_0$. In fact, by (4), $2^{\aleph}$ is the power of all representations

$$x = \frac{f(1)}{2} + \frac{f(2)}{2^2} + \cdots + \frac{f(v)}{2^v} + \cdots \qquad \text{(where } f(v) = 0 \text{ or } 1) \tag{12}$$

of the numbers x in the binary system. If we pay attention to the fact that every number x is only represented once, with the exception of the numbers $x = \frac{2v + 1}{2^\mu} < 1$, which are represented twice over, we have, if we denote the "enumerable" totality of the latter by $\{s_\nu\}$,

$$2^{\aleph_0} = \overline{\overline{(\{s_\nu\}, X)}}.$$

If we take away from X any "enumerable" aggregate $\{t_\nu\}$ and denote the remainder by X_1, we have:

$$X = (\{t_\nu\}, X_1) = (\{t_{2\nu-1}\}, \{t_{2\nu}\}, X_1),$$
$$(\{s_\nu\}, X) = (\{s_\nu\}, \{t_\nu\}, X_1),$$
$$\{t_{2\nu-1}\} \sim \{s_\nu\}, \quad \{t_{2\nu}\} \sim \{t_\nu\}, \quad X_1 \sim X_1;$$
$$X \sim (\{s_\nu\}, X),$$

and thus (§ 1)

$$2^{\aleph_0} = \overline{\overline{X}} = 0.$$

From (11) follows by squaring (by § 6, (6))

$$0 \cdot 0 = 2^{\aleph_0} \cdot 2^{\aleph_0} = 2^{\aleph_0 + \aleph_0} = 2^{\aleph_0} = 0,$$

and hence, by continued multiplication by 0,

$$0^\nu = 0, \tag{13}$$

where ν is any finite cardinal number.

If we raise both sides of (11) to the power[1] $\aleph_0$ we get

$$0^{\aleph_0} = (2^{\aleph_0})^{\aleph_0} = 2^{\aleph_0 \cdot \aleph_0}.$$

But since, by § 6, (8), $\aleph_0 \cdot \aleph_0 = \aleph_0$, we have

$$0^{\aleph_0} = 0. \tag{14}$$

The formulæ (13) and (14) mean that both the ν-dimensional and the $\aleph_0$-dimensional continuum have the power of the one-dimensional continuum. Thus the whole contents of my paper in Crelle's *Journal*, vol. lxxxiv, 1878, are derived purely algebraically with these few strokes of the pen from the fundamental formulæ of the calculation with cardinal numbers.

[489]

§ 5
THE FINITE CARDINAL NUMBERS

We will next show how the principles which we have laid down, and on which later on the theory of the actually infinite or transfinite cardinal numbers will be built, afford also the most natural, shortest, and most rigorous foundation for the theory of finite numbers.

To a single thing e_0, if we subsume it under the concept of an aggregate $E_0 = (e_0)$, corresponds as cardinal number what we call "one" and denote by 1; we have

$$1 = \overline{\overline{E}}_0. \tag{1}$$

[1] In English there is an ambiguity.

Let us now unite with E_0 another thing e_1, and call the union-aggregate E_1, so that

$$E_1 = (E_0, e_1) = (e_0, e_1). \tag{2}$$

The cardinal number of E_1 is called "two" and is denoted by 2:

$$2 = \overline{\overline{E}}_1. \tag{3}$$

By addition of new elements we get the series of aggregates

$$E_2 = (E_1, e_2), \quad E_3 = (E_2, e_3), \ldots,$$

which give us successively, in unlimited sequence, the other so-called "finite cardinal numbers" denoted by 3, 4, 5, ... The use which we here make of these numbers as suffixes is justified by the fact that a number is only used as a suffix when it has been defined as a cardinal number. We have, if by $v - 1$ is understood the number immediately preceding v in the above series,

$$v = \overline{\overline{E}}_{v-1}, \tag{4}$$

$$E_v = (E_{v-1}, e_v) = (e_0, e_1, \ldots e_v). \tag{5}$$

From the definition of a sum in § 3 follows:

$$\overline{\overline{E}}_v = \overline{\overline{E}}_{v-1} + 1; \tag{6}$$

that is to say, every cardinal number, except 1, is the sum of the immediately preceding one and 1.

Now, the following three theorems come into the foreground:

A. The terms of the unlimited series of finite cardinal numbers

$$1, 2, 3, \ldots, \quad v, \ldots$$

are all different from one another (that is to say, the condition of equivalence established in § 1 is not fulfilled for the corresponding aggregates).

[490]

B. Every one of these numbers v is greater than the preceding ones and less than the following ones (§ 2).

C. There are no cardinal numbers which, in magnitude, lie between two consecutive numbers v and $v + 1$ (§ 2).

We make the proofs of these theorems rest on the two following ones, D and E. We shall, then, in the next place, give the latter theorems rigid proofs.

D. If M, is an aggregate such that it is of equal power with none of its parts, then the aggregate (M, e), which arises from M by the addition of a single new element e, has the same property of being of equal power with none of its parts.

E. If N is an aggregate with the finite cardinal number v, and N_1 is any part of N, the cardinal number of N_1 is equal to one of the preceding numbers $1, 2, 3, \ldots, v - 1$.

Proof of D.—Suppose that the aggregate (M, *e*) is equivalent to one of its part which we will call N. Then two cases, both of which lead to a contradiction, are to be distinguished:

(*a*) The aggregate N contains *e* as element; let N = (M$_1$, *e*); then M$_1$ is a par of M because N is a part of (M, *e*). As we saw in § 1, the law of correspondence o the two equivalent aggregates (M, *e*) and (M$_1$, *e*) can be so modified that the ele ment *e* of the one corresponds to the same element *e* of the other; by that, then, N and M$_1$ are referred reciprocally and univocally to one another. But this contradict the supposition that M is not equivalent to its part M$_1$.

(*b*) The part N of (M, *e*) does not contain *e* as element, so that N is either N or a part of M. In the law of correspondence between (M, *e*) and N, which lies a the basis of our supposition, to the element *e* of the former let the element *f* of the latter correspond. Let N = (M$_1$, *f*); then the aggregate M is put in a reciprocally univocal relation with M$_1$. But M$_1$ is a part of N and hence of M. So here too N would be equivalent to one of its parts, and this is contrary to the supposition.

Proof of E.—We will suppose the correctness of the theorem up to a certain *v* and then conclude its validity for the number *v* + 1 which immediately follows, in the following manner:—We start from the aggregate E$_v$ = (*e*$_0$, *e*$_1$, ..., *e*$_v$) as an aggre gate with the cardinal number *v* + 1. If the theorem is true for this aggregate, it truth for any other aggregate with the same cardinal number *v* + 1 follows at once by § 1. Let E′ be any part of E$_v$; we distinguish the following cases:

(*a*) E′ does not contain *e*$_v$ as element, then E is either E$_{v-1}$

[491]

or a part of E$_{v-1}$, and so has as cardinal number either *v* or one of the number 1, 2, 3, ..., *v* − 1, because we supposed our theorem true for the aggregate E$_{v-1}$ with the cardinal number *v*.

(*b*) E′ consists of the single element *e*$_v$, then $\overline{\overline{E}}'$ = 1.

(*c*) E′ consists of *e*$_v$ and an aggregate E″, so that E′ = (E″, *e*$_v$). E″ is a par of E$_{v-1}$ and has therefore by supposition as cardinal number one of the number 1, 2, 3, ..., *v* − 1. But now $\overline{\overline{E}}'$ = $\overline{\overline{E}}''$ + 1, and thus the cardinal number of E′ i one of the numbers 2, 3, ..., *v*.

Proof of A.—Every one of the aggregates which we have denoted by E$_v$ has the property of not being equivalent to any of its parts. For if we suppose that this is so as far as a certain *v*, it follows from the theorem D that it is so for the imme diately following number *v* + 1. For *v* = 1, we recognize at once that the aggre gate E$_1$ = (*e*$_0$, *e*$_1$) is not equivalent to any of its parts, which are here (*e*$_0$) and (*e*$_1$).

980

Consider, now, any two numbers μ and ν of the series 1, 2, 3, ...; then, if μ is the earlier and ν the later, $E_{\mu-1}$ is a part of $E_{\nu-1}$. Thus $E_{\mu-1}$ and $E_{\nu-1}$ are not equivalent, and accordingly their cardinal numbers $\mu = \overline{\overline{E}}_{\mu-1}$ and $\nu = \overline{\overline{E}}_{\nu-1}$ are not equal.

Proof of B.—If of the two finite cardinal numbers μ and ν the first is the earlier and the second the later, then $\mu < \nu$. For consider the two aggregates $M = E_{\mu-1}$ and $N = E_{\nu-1}$; for them each of the two conditions in § 2 for $\overline{\overline{M}} < \overline{\overline{N}}$ is fulfilled. The condition (*a*) is fulfilled because, by theorem E, a part of $M = E_{\mu-1}$ can only have one of the cardinal numbers 1, 2, 3, ..., $\mu - 1$, and therefore, by theorem A, cannot be equivalent to the aggregate $N = E_{\nu-1}$. The condition (*b*) is fulfilled because M itself is a part of N.

Proof of C.—Let a be a cardinal number which is less than $\nu + 1$. Because of the condition (*b*) of § 2, there is a part of E_ν with the cardinal number a. By theorem E, a part of E_ν can only have one of the cardinal numbers 1, 2, 3, ..., ν. Thus a is equal to one of the cardinal numbers 1, 2, 3, ..., ν. By theorem B, none of these is greater than ν. Consequently there is no cardinal number a which is less than $\nu + 1$ and greater than ν.

Of importance for what follows is the following theorem:

F. If K is any aggregate of different finite cardinal numbers, there is one, κ_1, amongst them which is smaller than the rest, and therefore the smallest of all.

[492]

Proof.—The aggregate K either contains the number 1, in which case it is the least, $\kappa_1 = 1$, or it does not. In the latter case, let J be the aggregate of all those cardinal numbers of our series, 1, 2, 3, ..., which are smaller than those occurring in K. If a number ν belongs to J, all numbers less than ν belong to J. But J must have one element ν_1 such that $\nu_1 + 1$, and consequently all greater numbers, do not belong to J, because otherwise J would contain all finite numbers, whereas the numbers belonging to K are not contained in J. Thus J is the segment (*Abschnitt*) $(1, 2, 3, ..., \nu_1)$. The number $\nu_1 + 1 = \kappa_1$ is necessarily an element of K and smaller than the rest.

From F we conclude :

G. Every aggregate $K = \{\kappa\}$ of different finite cardinal numbers can be brought into the form of a series

$$K = (\kappa_1, \kappa_2, \kappa_3, ...)$$

such that

$$\kappa_1 < \kappa_2 < \kappa_3, ...$$

§ 6
THE SMALLEST TRANSFINITE CARDINAL
NUMBER ALEPH-ZERO

Aggregates with finite cardinal numbers are called "finite aggregates," all others we will call "transfinite aggregates" and their cardinal numbers "transfinite cardinal numbers."

The first example of a transfinite aggregate is given by the totality of finite cardinal numbers v; we call its cardinal number (§ 1) "Aleph-zero" and denote it by $\aleph_0$; thus we define

$$\aleph_0 = \{\overline{\overline{v}}\}. \tag{1}$$

That $\aleph_0$ is a *transfinite* number, that is to say, is not equal to any finite number μ, follows from the simple fact that, if to the aggregate $\{v\}$ is added a new element e_0, the union-aggregate $(\{v\}, e_0)$ is equivalent to the original aggregate $\{v\}$. For we can think of this reciprocally univocal correspondence between them: to the element e_0 of the first corresponds the element 1 of the second, and to the element v of the first corresponds the element $v + 1$ of the other. By § 3 we thus have

$$\aleph_0 + 1 = \aleph_0. \tag{2}$$

But we showed in § 5 that $\mu + 1$ is always different from μ, and therefore $\aleph_0$ is not equal to any finite number μ.

The number $\aleph_0$ is greater than any finite number μ:

$$\aleph_0 > \mu. \tag{3}$$

[493]

This follows, if we pay attention to § 3, from the three facts that $\mu = (1, 2, 3, \ldots, \mu)$, that no part of the aggregate $(1, 2, 3, \ldots, \mu)$ is equivalent to the aggregate $\{v\}$, and that $(1, 2, 3, \ldots, \mu)$ is itself a part of $\{v\}$.

On the other hand, $\aleph_0$ is the least transfinite cardinal number. If a is any transfinite cardinal number different from $\aleph_0$, then

$$\aleph_0 < a. \tag{4}$$

This rests on the following theorems:

A. Every transfinite aggregate T has parts with the cardinal number $\aleph_0$.

Proof.—If, by any rule, we have taken away a finite number of elements t_1, t_2, $\ldots$, t_{v-1}, there always remains the possibility of taking away a further element t_v. The aggregate $\{t_v\}$, where v denotes any finite cardinal number, is a part of T with the cardinal number $\aleph_0$, because $\{t_v\} \sim \{v\}$ (§ 1).

982

B. If S is a transfinite aggregate with the cardinal number $\aleph_0$, and S_1 is any transfinite part of S, then $\overline{\overline{S_1}} = \aleph_0$.

Proof.—We have supposed that $S \sim \{v\}$. Choose a definite law of correspondence between these two aggregates, and, with this law, denote by s_v that element of S which corresponds to the element v of $\{v\}$, so that

$$S = \{s_v\}.$$

The part S_1 of S consists of certain elements s_κ of S, and the totality of numbers κ forms a transfinite part K of the aggregate $\{v\}$. By theorem G of § 5 the aggregate K can be brought into the form of a series

$$K = \{\kappa_v\},$$

where

$$\kappa_v < \kappa_{v+1};$$

consequently we have

$$S_1 = \{s_{\kappa_v}\}.$$

Hence follows that $S_1 \sim S$, and therefore $\overline{\overline{S_1}} = \aleph_0$.

From A and B the formula (4) results, if we have regard to § 2.

From (2) we conclude, by adding 1 to both sides,

$$\aleph_0 + 2 = \aleph_0 + 1 = \aleph_0,$$

and, by repeating this

$$\aleph_0 + v = \aleph_0. \tag{5}$$

We have also

$$\aleph_0 + \aleph_0 = \aleph_0. \tag{6}$$

[494]

For, by (1) of § 3, $\aleph_0 + \aleph_0$ is the cardinal number $(\overline{\overline{\{a_v\}, \{b_v\}}})$ because

$$\overline{\overline{\{a_v\}}} = \overline{\overline{\{b_v\}}} = \aleph_0.$$

Now, obviously

$$\{v\} = (\{2v - 1\}, \{2v\}),$$
$$(\{2v - 1\}, \{2v\}) \sim (\{a_v\}, \{b_v\}),$$

and therefore

$$(\overline{\overline{\{a_v\}, \{b_v\}}}) = \overline{\overline{\{v\}}} = \aleph_0.$$

The equation (6) can also be written

$$\aleph_0 \cdot 2 = \aleph_0;$$

and, by adding $\aleph_0$ repeatedly to both sides, we find that

$$\aleph_0 \cdot v = v \cdot \aleph_0 = \aleph_0. \tag{7}$$

We also have

$$\aleph_0 \cdot \aleph_0 = \aleph_0. \tag{8}$$

Proof.—By (6) of § 3, $\aleph_0 \cdot \aleph_0$ is the cardinal number of the aggregate of bindings
$$\{(\mu, v)\},$$
where μ and v are any finite cardinal numbers which are independent of one another. If also λ represents any finite cardinal number, so that $\{\lambda\}$, $\{\mu\}$, and $\{v\}$ are only different notations for the same aggregate of all finite numbers, we have to show that
$$\{(\mu, v)\} \sim \{\lambda\}.$$
Let us denote $\mu + v$ by ρ; then ρ takes all the numerical values 2, 3, 4, ..., and there are in all $\rho - 1$ elements (μ, v) for which $\mu + v = \rho$, namely:
$$(1, \rho - 1), (2, \rho - 2), \ldots, (\rho - 1, 1).$$
In this sequence imagine first the element $(1, 1)$, for which $\rho = 2$, put, then the two elements for which $\rho = 3$, then the three elements for which $\rho = 4$, and so on. Thus we get all the elements (μ, v) in a simple series:
$$(1, 1); (1, 2), (2, 1); (1, 3), (2, 2), (3, 1); (1, 4), (2, 3), \ldots,$$
and here, as we easily see, the element (μ, v) comes at the λth place, where
$$\lambda = \mu + \frac{(\mu + v - 1)(\mu + v - 2)}{2}. \tag{9}$$
The variable λ takes every numerical value 1, 2, 3; ..., once. Consequently, by means of (9), a reciprocally univocal relation subsists between the aggregates $\{v\}$ and $\{(\mu, v)\}$.

[495]

If both sides of the equation (8) are multiplied by $\aleph_0$, we get $\aleph_0^3 = \aleph_0^2 = \aleph_0$, and, by repeated multiplications by $\aleph_0$, we get the equation, valid for every finite cardinal number v:
$$\aleph_0^v = \aleph_0. \tag{10}$$
The theorems E and A of § 5 lead to this theorem on finite aggregates:

C. Every finite aggregate E is such that it is equivalent to none of its parts.

This theorem stands sharply opposed to the following one for transfinite aggregates:

D. Every transfinite aggregate T is such that it has parts T_1 which are equivalent to it.

Proof.—By theorem A of this paragraph there is a part $S = \{t_v\}$ of T with the cardinal number $\aleph_0$. Let $T = (S, U)$, so that U is composed of those elements of T which are different from the elements t_v. Let us put $[S_1 = \{t_{v+1}\}$, $T_1 = (S_1, U)$; then T_1 is a part of T, and, in fact, that part which arises out of T if we leave out the single element t_1. Since $S \sim S_1$, by theorem B of this paragraph, and $U \sim U$, we have, by § 1, $T \sim T_1$.

984

In these theorems C and D the essential difference between finite and transfinite aggregates, to which I referred in the year 1877, in volume lxxxiv [1878] of Crelle's *Journal*, appears in the clearest way.

After we have introduced the least transfinite cardinal number $\aleph_0$ and derived its properties that lie the most readily to hand, the question arises as to the higher cardinal numbers and how they proceed from $\aleph_0$. We shall show that the transfinite cardinal numbers can be arranged according to their magnitude, and, in this order, form, like the finite numbers, a "well-ordered aggregate" in an extended sense of the words. Out of $\aleph_0$ proceeds, by a definite law, the next greater cardinal number $\aleph_1$, out of this by the same law the next greater $\aleph_2$, and so on. But even the unlimited sequence of cardinal numbers

$$\aleph_0, \aleph_1, \aleph_2, \ldots, \aleph_\nu, \ldots$$

does not exhaust the conception of transfinite cardinal number. We will prove the existence of a cardinal number which we denote by $\aleph_\omega$ and which shows itself to be the next greater to all the numbers $\aleph_\nu$; out of it proceeds in the same way as $\aleph_1$ out of $\aleph$ a next greater $\aleph_{\omega+1}$, and so on, without end.

[496]

To every transfinite cardinal number a there is a next greater proceeding out of it according to a unitary law, and also to every unlimitedly ascending well-ordered aggregate of transfinite cardinal numbers, {a}, there is a next greater proceeding out of that aggregate in a unitary way.

For the rigorous foundation of this matter, discovered in 1882 and exposed in the pamphlet *Grundlagen einer allgemeinen Mannichfaltigkeitslehre* (Leipzig, 1883) and in volume xxi of the *Mathematische Annalen*, we make use of the so-called "ordinal types" whose theory we have to set forth in the following paragraphs.

§ 7
THE ORDINAL TYPES OF SIMPLY ORDERED AGGREGATES

We call an aggregate M "simply ordered" if a definite "order of precedence" (*Rangordnung*) rules over its elements m, so that, of every two elements m_1 and m_2, one takes the "lower" and the other the "higher" rank, and so that, if of three elements m_1, m_2, and m_3, m_1, say, is of lower rank than m_2, and m_2 is of lower rank than m_3, then m_1 is of lower rank than m_3.

The relation of two elements m_1 and m_2, in which m_1 has the lower rank in the given order of precedence and m_2 the higher, is expressed by the formulæ:

$$m_1 < m_2, \quad m_2 > m_1. \tag{1}$$

Thus, for example, every aggregate P of points defined on a straight line is a simply ordered aggregate if, of every two points p_1 and p_2 belonging to it, that one whose co-ordinate (an origin and a positive direction having been fixed upon) is the lesser is given the lower rank.

It is evident that one and the same aggregate can be "simply ordered" according to the most different laws. Thus, for example, with the aggregate R of all positive rational numbers p/q (where p and q are relatively prime integers) which are greater than 0 and less than 1, there is, firstly, their "natural" order according to magnitude; then they can be arranged (and in this order we will denote the aggregate by R_0) so that, of two numbers p_1/q_1 and p_2/q_2 for which the sums $p_1 + q_1$ and $p_2 + q_2$ have different values, that number for which the corresponding sum is less takes the lower rank, and, if $p_1 + q_1 = p_2 + q_2$, then the smaller of the two rational numbers is the lower.

[497]

In this order of precedence, our aggregate, since to one and the same value of $p + q$ only a finite number of rational numbers p/q belongs, evidently has the form

$$R_0 = (r_1, r_2, \ldots, r_v, \ldots) = (\tfrac{1}{2}, \tfrac{1}{3}, \tfrac{1}{4}, \tfrac{2}{3}, \tfrac{1}{5}, \tfrac{1}{6}, \tfrac{2}{5}, \tfrac{3}{4}, \ldots),$$

where

$$r_v < r_{v+1}.$$

Always, then, when we speak of a "simply ordered" aggregate M, we imagine laid down a definite order or precedence of its elements, in the sense explained above.

There are doubly, triply, v-ply and a-ply ordered aggregates, but for the present we will not consider them. So in what follows we will use the shorter expression "ordered aggregate" when we mean "simply ordered aggregate."

Every ordered aggregate M has a definite "ordinal type," or more shortly a definite "type," which we will denote by

$$\overline{M}. \tag{2}$$

By this we understand the general concept which results from M if we only abstract from the nature of the elements m, and retain the order of precedence among them. Thus the ordinal type $\overline{M}$ is itself an ordered aggregate whose elements are units which have the same order of precedence amongst one another as the corresponding elements of M, from which they are derived by abstraction.

We call two ordered aggregates M and N "similar"(*ähnlich*) if they can be put into a biunivocal correspondence with one another in such a manner that, if m_1 and

m_2 are any two elements of M and n_1 and n_2 the corresponding elements of N, then the relation of rank of m_1 to m_2 in M is the same as that of n_1 to n_2 in N. Such a correspondence of similar aggregates we call an "imaging" (*Abbildung*) of these aggregates on one another. In such an imaging, to every part—which obviously also appears as an ordered aggregate—M_1 of M corresponds a similar part N_1 of N.

We express the similarity of two ordered aggregates M and N by the formula:

$$M \simeq N. \tag{3}$$

Every ordered aggregate is similar to itself.

If two ordered aggregates are similar to a third, they are similar to one another.

[498]

A simple consideration shows that two ordered aggregates have the same ordinal type if, and only if, they are similar, so that, of the two formulæ

$$\overline{M} = \overline{N}, \quad M \simeq N, \tag{4}$$

one is always a consequence of the other.

If, with an ordinal type $\overline{M}$ we also abstract from the order of precedence of the elements, we get (§ 1) the cardinal number $\overline{\overline{M}}$ of the ordered aggregate M, which is, at the same time, the cardinal number of the ordinal type $\overline{M}$. From $\overline{M} = \overline{N}$ always follows $\overline{\overline{M}} = \overline{\overline{N}}$, that is to say, ordered aggregates of equal types always have the same power or cardinal number; from the similarity of ordered aggregates follows their equivalence. On the other hand, two aggregates may be equivalent without being similar.

We will use the small letters of the Greek alphabet to denote ordinal types. If α is an ordinal type, we understand by

$$\overline{\alpha} \tag{5}$$

its corresponding cardinal number.

The ordinal types of finite ordered aggregates offer no special interest. For we easily convince ourselves that, for one and the same finite cardinal number v, all simply ordered aggregates are similar to one another, and thus have one and the same type. Thus the finite simple ordinal types are subject to the same laws as the finite cardinal numbers, and it is allowable to use the same signs 1, 2, 3, ..., v, ... for them, although they are conceptually different from the cardinal numbers. The case is quite different with the transfinite ordinal types; for to one and the same cardinal number belong innumerably many different types of simply ordered aggregates, which, in their totality, constitute a particular "class of types" (*Typenclasse*). Every one of these classes of types is, therefore, determined by the transfinite cardinal number a which is common to all the types belonging to the class. Thus we call it for short the class of types [a]. That class which naturally presents itself first to us, and whose complete investigation must, accordingly, be the next special aim of the theory of transfinite aggregates, is the class of types $[\aleph_0]$ which embraces all the types with the least transfinite cardinal number $\aleph_0$. From the cardinal number which *determines* the class of types [a] we have to distinguish that cardinal number a′ which for its part

[499]

is determined by the class of types [a]. The latter is the cardinal number which (§ 1) the class [a] has, in so far as it represents a well-defined aggregate whose elements are all the types α with the cardinal number a. We will see that a′ is different from a, and indeed always greater than a.

If in an ordered aggregate M all the relations of precedence of its elements are inverted, so that "lower" becomes "higher" and "higher" becomes "lower" everywhere, we again get an ordered aggregate, which we will denote by

$$*M \tag{6}$$

and call the "inverse" of M. We denote the ordinal type of *M, if $\alpha = \overline{\overline{M}}$, by

$$*\alpha. \tag{7}$$

It may happen that $*\alpha = \alpha$, as, for example, in the case of finite types or in that of the type of the aggregate of all rational numbers which are greater than 0 and less than 1 in their natural order of precedence. This type we will investigate under the notation η.

We remark further that two similarly ordered aggregates can be imaged on one another either in one manner or in many manners; in the first case the type in question is similar to itself in only one way, in the second case in many ways. Not only all finite types, but the types of transfinite "well-ordered aggregates," which will occupy us later and which we call transfinite "ordinal numbers," are such that they allow only a single imaging on themselves. On the other hand, the type η is similar to itself in an infinity of ways.

We will make this difference clear by two simple examples. By ω we understand the type of a well-ordered aggregate

$$(e_1, e_2, \ldots, e_\nu, \ldots),$$

in which

$$e_\nu < e_{\nu+1},$$

and where ν represents all finite cardinal numbers in turn. Another well-ordered aggregate

$$(f_1, f_2, \ldots, f_\nu, \ldots),$$

with the condition

$$f_\nu < f_{\nu+1},$$

of the same type ω can obviously only be imaged on the former in such a way that e_ν and f_ν are corresponding elements. For e_1, the lowest element in rank of the first, must, in the process of imaging, be correlated to the lowest element f_1 of the second, the next after e_1 in rank (e_2) to f_2, the next after f_1, and so on.

[500]

Every other bi-univocal correspondence of the two equivalent aggregates $\{e_\nu\}$ and $\{f_\nu\}$ is not an "imaging" in the sense which we have fixed above for the theory of types.

On the other hand, let us take an ordered aggregate of the form

$$\{e_\nu\},$$

where ν represents all positive and negative finite integers, including 0, and where likewise

$$e_\nu < e_{\nu+1}.$$

This aggregate has no lowest and no highest element in rank. Its type is, by the definition of a sum given in § 8,

$$^*\omega + \omega.$$

It is similar to itself in an infinity of ways. For let us consider an aggregate of the same type

$$\{f_{\nu'}\},$$

where

$$f_{\nu'} < f_{\nu'+1}.$$

Then the two ordered aggregates can be so imaged on one another that, if we understand by ν_0' a definite one of the numbers ν', to the element $e_{\nu'}$ of the first the element $f_{\nu_0'+\nu'}$ of the second corresponds. Since ν_0' is arbitrary, we have here an infinity of imagings.

The concept of "ordinal type" developed here, when it is transferred in like manner to "multiply ordered aggregates," embraces, in conjunction with the concept of "cardinal number" or "power" introduced in § 1, everything capable of being numbered (*Anzahlmässige*) that is thinkable, and in this sense cannot be further generalized. It contains nothing arbitrary, but is the natural extension of the concept of number. It deserves to be especially emphasized that the criterion of equality (4) follows with absolute necessity from the concept of ordinal type and consequently permits of no alteration. The chief cause of the grave errors in G. Veronese's *Grundzüge der Geometrie* (German by A. Schepp, Leipzig, 1894) is the non-recognition of this point.

989

The "number (*Anzahl oder Zahl*) of an ordered group" is defined in exactly the same way as what we have called the "ordinal type of a simply ordered aggregate" (*Zur Lehre vom Transfiniten*, Halle, 1890; reprinted from the *Zeitschr. für Philos. und philos. Kritik* for 1887).

[501]

But Veronese thinks that he must make an addition to the criterion of equality. He says: "Numbers whose units correspond to one another uniquely and in the same order and of which the one is neither a part of the other nor equal to a part of the other are equal."[2] This definition of equality contains a circle and thus is meaningless. For what is the meaning of "not equal to a part of the other" in this addition? To answer this question, we must first know when two numbers are equal or unequal. Thus, apart from the arbitrariness of his definition of equality, it presupposes a definition of equality, and this again presupposes a definition of equality, in which we must know again what equal and unequal are, and so on *ad infinitum*. After Veronese has, so to speak, given up of his own free will the indispensable foundation for the comparison of numbers, we ought not to be surprised at the lawlessness with which, later on, he operates with his pseudo-transfinite numbers, and ascribes properties to them which they cannot possess simply because they themselves, in the form imagined by him, have no existence except on paper. Thus, too, the striking similarity of his "numbers" to the very absurd "infinite numbers" in Fontenelle's *Géométrie de l'Infini* (Paris, 1727) becomes comprehensible. Recently, W. Killing has given welcome expression to his doubts concerning the foundation of Veronese's book in the *Index lectionum* of the Münster Academy for 1895–1896.[3]

§ 8
ADDITION AND MULTIPLICATION
OF ORDINAL TYPES

The union-aggregate (M, N) of two aggregates M and N can, if M and N are ordered, be conceived as an ordered aggregate in which the relations of precedence of the elements of M among themselves as well as the relations of precedence of the elements of N among themselves remain the same as in M or N respectively, and all elements of M have a lower rank than all the elements of N. If M' and N' are two other ordered aggregates, M ≃ M' and N ≃ N',

---

[2] In the original Italian edition this passage runs: "Numeri le unità dei quali si corrispondono univocamente e nel medesimo ordine, e di cui l' uno non è parte o uguale ad una parte dell' altro, sono uguali."

[3] Veronese replied to this in *Math. Ann.*, vol. xlvii, 1897. *Cf.* Killing, *ibid.*, vol. xlviii, 1897.

[502]

then $(M, N) \simeq (M', N')$; so the ordinal type of (M, N) depends only on the ordinal types $\overline{M} = \alpha$ and $\overline{N} = \beta$. Thus, we define:

$$\alpha + \beta = (\overline{M, N}). \tag{1}$$

In the sum $\alpha + \beta$ we call α the "augend" and β the "addend."

For any three types we easily prove the associative law:

$$\alpha + (\beta + \gamma) = (\alpha + \beta) + \gamma. \tag{2}$$

On the other hand, the commutative law is not valid, in general, for the addition of types. We see this by the following simple example.

If ω is the type, already mentioned in § 7, of the well-ordered aggregate

$$E = (e_1, e_2, \ldots, e_v \ldots), \quad e_v \prec e_{v+1},$$

then $1 + \omega$ is not equal to $\omega + 1$. For, if f is a new element, we have by (1):

$$1 + \omega = (\overline{f, E}),$$
$$\omega + 1 = (\overline{E, f}).$$

But the aggregate

$$(f, E) = (f, e_1, e_2, \ldots, e_v, \ldots)$$

is similar to the aggregate E, and consequently

$$1 + \omega = \omega.$$

On the contrary, the aggregates E and (E, f) are not similar, because the first has no term which is highest in rank, but the second has the highest term f. Thus $\omega + 1$ is different from $\omega = 1 + \omega$.

Out of two ordered aggregates M and N with the types α and β we can set up an ordered aggregate S by substituting for every element n of N an ordered aggregate M_n which has the same type α as M, so that

$$\overline{M}_n = \alpha; \tag{3}$$

and, for the order of precedence in

$$S = \{M_n\} \tag{4}$$

we make the two rules:

(1) Every two elements of S which belong to one and the same aggregate M_n are to retain in S the same order of precedence as in M_n;

(2) Every two elements of S which belong to two different aggregates M_{n_1} and M_{n_2} have the same relation of precedence as n_1 and n_2 have in N.

The ordinal type of S depends, as we easily see, only on the types α and β; we define

$$\alpha \cdot \beta = \overline{S}. \tag{5}$$

[503]

In this product α is called the "multiplicand" and β the "multiplier."

In any definite imaging of M on M_n let m_n be the element of M_n that corresponds to the element m of M; we can then also write

$$S = \{m_n\}. \tag{6}$$

Consider a third ordered aggregate $P = \{p\}$ with the ordinal type $\overline{P} = \gamma$, then, by (5),

$$\alpha \cdot \beta = \{\overline{m_n}\}, \quad \beta \cdot \gamma = \{\overline{n_p}\}, \quad (\alpha \cdot \beta) \cdot \gamma = \{\overline{(m_n)_p}\},$$
$$\alpha \cdot (\beta \cdot \gamma) = \{\overline{m_{(n_p)}}\}.$$

But the two ordered aggregates $\{(m_n)_p\}$ and $\{m_{(n_p)}\}$ are similar, and are imaged on one another if we regard the elements $(m_n)_p$ and $m_{(n_p)}$ as corresponding. Consequently, for three types α, β, and γ the associative law

$$(\alpha \cdot \beta) \cdot \gamma = \alpha \cdot (\beta \cdot \gamma) \tag{7}$$

subsists. From (1) and (5) follows easily the distributive law

$$\alpha \cdot (\beta + \gamma) = \alpha \cdot \beta + \alpha \cdot \gamma; \tag{8}$$

but only in this form, where the factor with two terms is the multiplier.

On the contrary, in the multiplication of types as in their addition, the commutative law is not generally valid. For example, $2 \cdot \omega$ and $\omega \cdot 2$ are different types; for, by (5),

$$2 \cdot \omega = \overline{(e_1, f_1; e_2, f_2; \ldots; e_\nu, f_\nu; \ldots)} = \omega;$$

while

$$\omega \cdot 2 = \overline{(e_1, e_2, \ldots, e_\nu, \ldots; f_1, f_2, \ldots, f_\nu, \ldots)}$$

is obviously different from ω.

If we compare the definitions of the elementary operations for cardinal numbers, given in § 3, with those established here for ordinal types, we easily see that the cardinal number of the sum of two types is equal to the sum of the cardinal numbers of the single types, and that the cardinal number of the product of two types is equal to the product of the cardinal numbers of the single types. Every equation between ordinal types which proceeds from the two elementary operations remains correct, therefore, if we replace in it all the types by their cardinal numbers.

[504]

§ 9

THE ORDINAL TYPE η OF THE AGGREGATE R OF ALL RATIONAL NUMBERS WHICH ARE GREATER THAN 0 AND SMALLER THAN 1, IN THEIR NATURAL ORDER OF PRECEDENCE

By R we understand, as in § 7, the system of all rational numbers p/q (p and q being relatively prime) which > 0 and < 1, in their natural order of precedence, where the magnitude of a number determines its rank. We denote the ordinal type of R by η:

$$\eta = \overline{R}. \tag{1}$$

But we have put the same aggregate in another order of precedence in which we call it R_0. This order is determined, in the first place, by the magnitude of $p + q$, and in the second place—for rational numbers for which $p + q$ has the same value—by the magnitude of p/q, itself. The aggregate R_0 is a well-ordered aggregate of type ω:

$$R_0 = (r_1, r_2, \ldots, r_v, \ldots), \text{ where } r_v < r_{v+1}, \tag{2}$$
$$\overline{R}_0 = \omega. \tag{3}$$

Both R and R_0 have the same cardinal number since they only differ in the order of precedence of their elements, and, since we obviously have $\overline{\overline{R}}_0 = \aleph_0$, we also have

$$\overline{\overline{R}} = \overline{\eta} = \aleph_0. \tag{4}$$

Thus the type η belongs to the class of types $[\aleph_0]$.

Secondly, we remark that in R there is neither an element which is lowest in rank nor one which is highest in rank. Thirdly, R has the property that between every two of its elements others lie. This property we express by the words: R is "everywhere dense" (*überalldicht*).

We will now show that these three properties characterize the type η of R, so that we have the following theorem:

If we have a simply ordered aggregate M such that

(*a*) $\overline{\overline{M}} = \aleph_0$;

(*b*) M has no element which is lowest in rank, and no highest;

(*c*) M is everywhere dense;

then the ordinal type of M is η:

$$\overline{M} = \eta.$$

Proof.—Because of the condition (*a*), M can be brought into the form

[505]

of a well-ordered aggregate of type ω; having fixed upon such a form, we denote it by M_0 and put

$$M_0 = (m_1, m_2, \ldots, m_v, \ldots). \tag{5}$$

We have now to show that

$$M \simeq R; \tag{6}$$

that is to say, we must prove that M can be imaged on R in such a way that the relation of precedence of any and every two elements in M is the same as that of the two corresponding elements in R.

Let the element r_1 in R be correlated to the element m_1 in M. The element r_2 has a definite relation of precedence to r_1 in R. Because of the condition (*b*), there

993

are infinitely many elements m_ν of M which have the same relation of precedence in M to m_1 as r_2 to r_1 in R; of them we choose that one which has the smallest index in M_0, let it be m_{l_2} and correlate it to r_2. The element r_3 has in R definite relations of precedence to r_1 and r_2; because of the conditions (*b*) and (*c*) there is an infinity of elements m_ν of M which have the same relation of precedence to m_1 and m_{l_2} in M as r_3 to r_1 and r_2 to R; of them we choose that—let it be m_{l_2}—which has the smallest index in M_0, and correlate it to r_3. According to this law we imagine the process of correlation continued. If to the ν elements

$$r_1, r_2, r_3, \ldots, r_\nu$$

of R are correlated, as images, definite elements

$$m_1, m_{l_2}, m_{l_3}, \ldots, m_{l_\nu},$$

which have the same relations of precedence amongst one another in M as the corresponding elements in R, then to the element $r_{\nu+1}$ of R is to be correlated that element $m_{l_{\nu+1}}$ of M which has the smallest index in M_0 of those which have the same relations of precedence to

$$m_1, m_{l_2}, m_{l_3}, \ldots, m_{l_\nu}$$

in M as $r_{\nu+1}$ to $r_1, r_2, \ldots, r_\nu$, in R.

In this manner we have correlated definite elements m_{l_ν} of M to all the elements r_ν of R, and the elements m_{l_ν} have in M the same order of precedence as the corresponding elements r_ν in R. But we have still to show that the elements m_{l_ν} include *all* the elements m_ν of M, or, what is the same thing, that the series

$$1, l_2, l_3, \ldots, l_\nu, \ldots$$

[506]

is only a permutation of the series

$$1, 2, 3, \ldots \nu, \ldots$$

We prove this by a complete induction: we will show that, if the elements m_1, $m_2, \ldots, m_\nu$ appear in the imaging, that is also the case with the following element $m_{\nu+1}$.

Let λ be so great that, among the elements

$$m_1, m_{l_2}, m_{l_3}, \ldots, m_{l_\lambda},$$

the elements

$$m_1, m_2, \ldots, m_\nu,$$

which, by supposition, appear in the imaging, are contained. It may be that also $m_{\nu+1}$ is found among them; then $m_{\nu+1}$ appears in the imaging. But if $m_{\nu+1}$ is not among the elements

$$m_1, m_{l_2}, m_{l_3}, \ldots, m_{l_\lambda},$$

994

then $m_{\nu+1}$ has with respect to these elements a definite ordinal position in M; infinitely many elements in R have the same ordinal position in R with respect to $r_1, r_2, \ldots, r_\lambda$, amongst which let $r_{\lambda+\sigma}$ be that with the least index in R_0. Then $m_{\nu+1}$ has, as we can easily make sure, the same ordinal position with respect to

$$m_1, m_{l_2}, m_{l_3}, \ldots, m_{l_{\lambda+\sigma-1}}$$

in M as $r_{\lambda+\sigma}$ has with respect to

$$r_1, r_2, \ldots, r_{\lambda+\sigma-1}$$

in R. Since $m_1, m_2, \ldots, m_\nu$ have already appeared in the imaging, $m_{\nu+1}$ is that - element with the smallest index in M which has this ordinal position with respect to

$$m_1, m_{l_2}, \ldots, m_{l_{\lambda+\sigma-1}}.$$

Consequently, according to our law of correlation,

$$m_{l_{\lambda+\sigma}} = m_{\nu+1}.$$

Thus, in this case too, the element $m_{\nu+1}$ appears in the imaging, and $r_{\lambda+\sigma}$ is the element of R which is correlated to it.

We see, then, that by our manner of correlation, the *whole aggregate* M is imaged on the *whole aggregate* R; M and R are similar aggregates, which was to be proved.

From the theorem which we have just proved result, for example, the following theorems:

[507]

The ordinal type of the aggregate of all negative and positive rational numbers, including zero, in their natural order of precedence, is η.

The ordinal type of the aggregate of all rational numbers which are greater than a and less than b, in their natural order of precedence, where a and b are any real numbers, and $a < b$, is η.

The ordinal type of the aggregate of all real algebraic numbers in their natural order of precedence is η.

The ordinal type of the aggregate of all real algebraic numbers which are greater than a and less than b, in their natural order of precedence, where a and b are any real numbers and $a < b$, is η.

For all these ordered aggregates satisfy the three conditions required in our theorem for M (see Crelle's *Journal*, vol. lxxvii).

If we consider, further, aggregates with the types—according to the definitions given in § 8—written $\eta + \eta$, $\eta\eta$, $(1 + \eta)\eta$, $(\eta + 1)\eta$, $(1 + \eta + 1)\eta$, we find that those three conditions are also fulfilled with them. Thus we have the theorems:

$$\eta + \eta = \eta, \tag{7}$$
$$\eta\eta = \eta, \tag{8}$$
$$(1 + \eta)\eta = \eta, \tag{9}$$
$$(\eta + 1)\eta = \eta, \tag{10}$$
$$(1 + \eta + 1)\eta = \eta. \tag{11}$$

The repeated application of (7) and (8) gives for every finite number v:

$$\eta \cdot v = \eta, \tag{12}$$
$$\eta^v = \eta. \tag{13}$$

On the other hand we easily see that, for $v > 1$, the types $1 + \eta, \eta + 1, v \cdot \eta$, $1 + \eta + 1$ are different both from one another and from η. We have

$$\eta + 1 + \eta = \eta, \tag{14}$$

but $\eta + v + \eta$, for $v > 1$, is different from η.

Finally, it deserves to be emphasized that

$$*\eta = \eta. \tag{15}$$

[508]

§ 10

THE FUNDAMENTAL SERIES CONTAINED IN A TRANSFINITE ORDERED AGGREGATE

Let us consider any simply ordered transfinite aggregate M. Every part of M is itself an ordered aggregate. For the study of the type $\overline{M}$, those parts of M which have the types ω and $*\omega$ appear to be especially valuable; we call them "fundamental series of the first order contained in M," and the former—of type ω—we call an "ascending" series, the latter—of type $*\omega$—a "descending" one. Since we limit ourselves to the consideration of fundamental series of the first order (in later investigations fundamental series of higher order will also occupy us), we will here simply call them "fundamental series." Thus an "ascending fundamental series" is of the form

$$\{a_v\}, \quad \text{where} \quad a_v < a_{v+1}; \tag{1}$$

a "descending fundamental series" is of the form

$$\{b_v\}, \quad \text{where} \quad b_v > b_{v+1}. \tag{2}$$

The letter v, as well as κ, λ, and μ, has everywhere in our considerations the signification of an arbitrary finite cardinal number or of a finite type (a finite ordinal number).

We call two ascending fundamental series $\{a_\nu\}$ and $\{a_\nu'\}$ in M "coherent" (*zusammengehörig*), in signs

$$\{a_\nu\} \parallel \{a_\nu'\}, \tag{3}$$

if, for every element a_ν there are elements a_λ' such that

$$a_\nu < a_\lambda',$$

and also for every element a_ν' there are elements a_μ such that

$$a_\nu' < a_\mu.$$

Two descending fundamental series $\{b_\nu\}$ and $\{b_\nu'\}$ in M are said to be "coherent," in signs

$$\{b_\nu\} \parallel \{b_\nu'\}, \tag{4}$$

if for every element b_ν there are elements b_λ' such that

$$b_\nu > b_\lambda',$$

and for every element b_ν' there are elements b_μ such that

$$b_\nu' > b_\mu.$$

An ascending fundamental series $\{a_\nu\}$ and a descending one $\{b_\nu\}$ are said to be "coherent," in signs

[509]

$$\{a_\nu\} \parallel \{b_\nu\}, \tag{5}$$

if (*a*) for all values of ν and μ

$$a_\nu < b_\mu,$$

and (*b*) in M exists at most one (thus either only one or none at all) element m_0 such that, for all ν's,

$$a_\nu < m_0 < b_\nu.$$

Then we have the theorems:

A. If two fundamental series are coherent to a third, they are also coherent to one another.

B. Two fundamental series proceeding in the same direction of which one is part of the other are coherent.

If there exists in M an element m_0 which has such a position with respect to the ascending fundamental series $\{a_\nu\}$ that:

(*a*) for every ν

$$a_\nu < m_0,$$

(*b*) for every element m of M that precedes m_0 there exists a certain number ν_0 such that

$$a_\nu > m, \quad \text{for} \quad \nu \gtreqless \nu_0,$$

then we will call m_0 a "limiting element (*Grenzelement*) of $\{a_\nu\}$ in M" and also a "principal element (*Hauptelement*) of M." In the same way we call m_0 a "principal

997

element of M" and also "limiting element of $\{b_\nu\}$ in M" if these conditions are satisfied:

(a) for every ν

$$b_\nu > m_0,$$

(b) for every element m of M that follows m_0 exists a certain number ν_0 such that

$$b_\nu > m, \quad \text{for} \quad \nu \gtreqless \nu_0.$$

A fundamental series can never have more than one limiting element in M; but M has, in general, many principal elements.

We perceive the truth of the following theorems:

C. If a fundamental series has a limiting element in M, all fundamental series coherent to it have the same limiting element in M.

D. If two fundamental series (whether proceeding in the same or in opposite directions) have one and the same limiting element in M, they are coherent.

If M and M′ are two similarly ordered aggregates, so that

$$\overline{M} = \overline{M}', \tag{6}$$

and we fix upon any imaging of the two aggregates, then we easily see that the following theorems hold:

[510]

E. To every fundamental series in M corresponds as image a fundamental series in M′, and inversely; to every ascending series an ascending one, and to every descending series a descending one; to coherent fundamental series in M correspond as images coherent fundamental series in M′, and inversely.

F. If to a fundamental series in M belongs a limiting element in M, then to the corresponding fundamental series in M′ belongs a limiting element in M′, and inversely; and these two limiting elements are images of one another in the imaging.

G. To the principal elements of M correspond as images principal elements of M′, and inversely.

If an aggregate M consists of principal elements, so that every one of its elements is a principal element, we call it an "aggregate which is dense in itself (*insichdichte Menge*)." If to every fundamental series in M there is a limiting element in M, we call M a "closed (*abgeschlossene*) aggregate." An aggregate which is both "dense in itself" and "closed" is called a "perfect aggregate." If an aggregate has one of these three predicates, every similar aggregate has the same predicate; thus these predicates can also be ascribed to the corresponding ordinal types, and so there are "types which are dense in themselves," "closed types," "perfect types," and also "everywhere-dense types" (§ 9).

For example, η is a type which is "dense in itself," and, as we showed in § 9, it is also "everywhere-dense," but it is not "closed." The types ω and $*\omega$ have no principal elements, but $\omega + \nu$ and $\nu + *\omega$ each have a principal element, and are "closed" types. The type $\omega \cdot 3$ has two principal elements, but is not "closed"; the type $\omega \cdot 3 + \nu$ has three principal elements, and is "closed."

§ 11
THE ORDINAL TYPE θ OF THE LINEAR CONTINUUM X

We turn to the investigation of the ordinal type of the aggregate $X = \{x\}$ of all real numbers x, such that $x \geqq 0$ and $\leqq 1$, in their natural order of precedence, so that, with any two of its elements x and x',

$$x \prec x', \quad \text{if} \quad x < x'.$$
Let the notation for this type be

$$\overline{X} = \theta. \tag{1}$$

[511]

From the elements of the theory of rational and irrational numbers we know that every fundamental series $\{x_\nu\}$ in X has a limiting element x_0 in X, and that also, inversely, every element x of X is a limiting element of coherent fundamental series in X. Consequently X is a "perfect aggregate" and θ is a "perfect type."

But θ is not sufficiently characterized by that; besides that we must fix our attention on the following property of X. The aggregate X contains as part the aggregate R of ordinal type η investigated in § 9, and in such a way that, between any two elements x_0 and x_1 of X, elements of R lie.

We will now show that these properties, taken together, characterize the ordinal type θ of the linear continuum X in an exhaustive manner, so that we have the theorem:

If an ordered aggregate M is such that (a) it is "perfect," and (b) in it is contained an aggregate S with the cardinal number $\overline{\overline{S}} = \aleph_0$ and which bears such a relation to M that, between any two elements m_0 and m_1 of M elements of S lie, then $\overline{M} = \theta$.

Proof.—If S had a lowest or a highest element, these elements, by (b), would bear the same character as elements of M; we could remove them from S without S losing thereby the relation to M expressed in (b). Thus, we suppose that S is without lowest or highest element, so that, by § 9, it has the ordinal type η. For since S is a part of M, between any two elements s_0 and s_1 of S other elements of S must, by (b), lie. Besides, by (b) we have $\overline{\overline{S}} = \aleph_0$. Thus the aggregates S and R are "similar" to one another.

$$S \backsimeq R.$$

We fix on any "imaging" of R on S, and assert that it gives a definite "imaging" of X on M in the following manner:

Let all elements of X which, at the same time, belong to the aggregate R correspond as images to those elements of M which are, at the same time, elements of S and, in the supposed imaging of R on S, correspond to the said elements of R. But if x_0 is an element of X which does not belong to R, x_0 may be regarded as a limiting element of a fundamental series $\{x_\nu\}$ contained in X, and this series can be replaced by a coherent fundamental series $\{r_{\kappa_\nu}\}$ contained in R. To this

[512]

corresponds as image a fundamental series $\{s_{\lambda_\nu}\}$ in S and M, which, because of (a), is limited by an element m_0 of M that does not belong to S (F, § 10). Let this element m_0 of M (which remains the same, by E, C, and D of § 10, if the fundamental series $\{x_\nu\}$ and $\{r_{\kappa_\nu}\}$ are replaced by others limited by the same element x_0 in X) be the image of x_0 in X. Inversely, to every element m_0 of M which does not occur in S belongs a quite definite element x_0 of X which does not belong to R and of which m_0 is the image.

In this manner a bi-univocal correspondence between X and M is set up, and we have now to show that it gives an "imaging" of these aggregates.

This is, of course, the case for those elements of X which belong to R, and for those elements of M which belong to S. Let us compare an element r of R with an element x_0 of X which does not belong to R; let the corresponding elements of M be s and m_0. If $r < x_0$, there is an ascending fundamental series $\{r_{\kappa_\nu}\}$, which is limited by x_0 and, from a certain ν_0 on,

$$r < r_{\kappa_\nu} \quad \text{for} \quad \nu \gtreqless \nu_0.$$

The image of $\{r_{\kappa_\nu}\}$ in M is an ascending fundamental series $\{s_{\lambda_\nu}\}$, which will be limited by an m_0 of M, and we have (§ 10) $s_{\lambda_\nu} < m_0$ for every ν, and $s < s_{\lambda_\nu}$ for $\nu \gtreqless \nu_0$. Thus (§ 7) $s < m_0$.

If $r > x_0$, we conclude similarly that $s > m_0$.

Let us consider, finally, two elements x_0 and x_0' not belonging to R and the elements m_0 and m_0' corresponding to them in M; then we show, by an analogous consideration, that, if $x_0 < x_0'$, then $m_0 < m_0'$.

The proof of the similarity of X and M is now finished, and we thus have

$$\overline{M} = \theta.$$

HALLE, *March* 1895.

CONTRIBUTIONS TO THE FOUNDING OF THE THEORY OF TRANSFINITE NUMBERS

(SECOND ARTICLE)

§ 12
WELL-ORDERED AGGREGATES

Among simply ordered aggregates "well-ordered aggregates" deserve a special place; their ordinal types, which we call "ordinal numbers," form the natural material for an exact definition of the higher transfinite cardinal numbers or powers,—a definition which is throughout conformable to that which was given us for the least transfinite cardinal number Aleph-zero by the system of all finite numbers v (§ 6).

We call a simply ordered aggregate F (§ 7) "well-ordered" if its elements f ascend in a definite succession from a lowest f_1 in such a way that:

I.   There is in F an element f_1 which is lowest in rank.

II.   If F' is any part of F and if F has one or many elements of higher rank than all elements of F', then there is an element f' of F which follows immediately after the totality F', so that no elements in rank between f' and F' occur in F.[1]

In particular, to every single element f of F, if it is not the highest, follows in rank as next higher another definite element f'; this results from the condition II if for F' we put the single element f. Further, if, for example, an infinite series of consecutive elements

$$e' < e'' < e''' < \cdots < e^{(v)} < e^{(v+1)} \ldots$$

is contained in F in such a way, however, that there are also in F elements of

[208]

higher rank than all elements $e^{(v)}$, then, by the second condition, putting for F' the totality $\{e^{(v)}\}$, there must exist an element f' such that not only

$$f' > e^{(v)}$$

---

[1] This definition of "well-ordered aggregates," apart from the wording, is identical with that which was introduced in vol. xxi of the *Math. Ann.* (*Grundlagen einer allgemeinen Mannichfaltigkeitslehre*).

for all values of v, but that also there is no element g in F which satisfies the two conditions

$$g < f',$$
$$g > e^{(v)}$$

for all values of v.

Thus, for example, the three aggregates

$$(a_1, a_2, \ldots, a_v, \ldots),$$
$$(a_1, a_2, \ldots, a_v, \ldots, b_1, b_2 \ldots, b_\mu, \ldots),$$
$$(a_1, a_2, \ldots, a_v, \ldots b_1, b_2, \ldots, b_\mu, \ldots c_1, c_2, c_3),$$

where

$$a_v < a_{v+1} < b_\mu < b_{\mu+1} < c_1 < c_2 < c_3,$$

are well-ordered. The two first have no highest element, the third has the highest element c_3; in the second and third b_1 immediately follows all the elements a_v, in the third c_1 immediately follows all the elements a_v and b_μ.

In the following we will extend the use of the signs $<$ and $>$, explained in § 7, and there used to express the ordinal relation of two elements, to groups of elements, so that the formulæ

$$M < N,$$
$$M > N$$

are the expression for the fact that in a given order all the elements of the aggregate M have a lower, or higher, respectively, rank than all elements of the aggregate N.

A. Every part F_1 of a well-ordered aggregate F has a lowest element.

Proof.—If the lowest element f_1 of F belongs to F_1, then it is also the lowest element of F_1. In the other case, let F′ be the totality of all elements of F′ which have a lower rank than all elements F_1, then, for this reason, no element of F lies between F′ and F_1. Thus, if f' follows (II) next after F′, then it belongs necessarily to F and here takes the lowest rank.

B. If a simply ordered aggregate F is such that both F and every one of its parts have a lowest element, then F is a well-ordered aggregate.

[209]

Proof.—Since F has a lowest element, the condition I is satisfied. Let F′ be a part of F such that there are in F one or more elements which follow F′; let F_1 be the totality of all these elements and f' the lowest element of F_1, then obviously f' is the element of F which follows next to F′. Consequently, the condition II is also satisfied, and therefore F is a well-ordered aggregate.

C. Every part F′ of a well-ordered aggregate F is also a well-ordered aggregate.

Proof.—By theorem A, the aggregate F′ as well as every part F″ of F′ (since it is also a part of F) has a lowest element; thus by theorem B, the aggregate F′ is well-ordered.

D. Every aggregate G which is similar to a well-ordered aggregate F is also a well-ordered aggregate.

Proof.—If M is an aggregate which has a lowest element, then, as immediately follows from the concept of similarity (§ 7), every aggregate N similar to it has a lowest element. Since, now, we are to have G ≃ F, and F has, since it is a well-ordered aggregate, a lowest element, the same holds of G. Thus also every part G′ of G has a lowest element; for in an imaging of G on F, to the aggregate G′ corresponds a part F′ of F as image, so that

$$G' \simeq F'.$$

But, by theorem A, F′ has a lowest element, and therefore also G′ has. Thus, both G and every part of G have lowest elements. By theorem B, consequently, G is a well-ordered aggregate.

E. If in a well-ordered aggregate G, in place of its elements g well-ordered aggregates are substituted in such a way that, if F_g and $F_{g'}$ are the well-ordered aggregates which occupy the places of the elements g and g' and $g < g'$, then also $F_g < F_{g'}$, then the aggregate H, arising by combination in this manner of the elements of all the aggregates F_g, is well-ordered.

Proof.—Both H and every part H_1 of H have lowest elements, and by theorem B this characterizes H as a well-ordered aggregate. For, if g_1 is the lowest element of G, the lowest element of F_{g_1} is at the same time the lowest element of H. If, further, we have a part H_1 of H, its elements belong to definite aggregates F_g which form, when taken together, a part of the well-ordered aggregate $\{F_g\}$, which consists of the elements F_g and is similar to the aggregate G. If, say, F_{g_0} is the lowest element of this part, then the lowest element of the part of H_1 contained in F_{g_0} is at the same time the lowest element of H.

[210]

§ 13
THE SEGMENTS OF WELL-ORDERED AGGREGATES

If f is any element of the well-ordered aggregate F which is different from the initial element f_1, then we will call the aggregate A of all elements of F which precede f a "segment (*Abschnitt*) of F," or, more fully, "the segment of F which is defined by the element f." On the other hand, the aggregate R of all the other elements of F, including f, is a "remainder of F," and, more fully, "the remainder which is

1003

determined by the element f." The aggregates A and R are, by theorem C of § 12, well-ordered, and we may, by § 8 and § 12, write:

$$F = (A, \ R), \tag{1}$$
$$R = (f, R'), \tag{2}$$
$$A < R. \tag{3}$$

R′ is the part of R which follows the initial element f and reduces to 0 if R has, besides f, no other element.

For example, in the well-ordered aggregate

$$F = (a_1, a_2, ..., a_v, ... b_1, b_2, ..., b_\mu, ... c_1, c_2, c_3),$$

the segment

$$(a_1, a_2)$$

and the corresponding remainder

$$(a_3, a_4, ... a_{v+2}, ... b_1, b_2, ... b_\mu, ... c_1, c_2, c_3)$$

are determined by the element a_3; the segment

$$(a_1, a_2, ..., a_v, ...)$$

and the corresponding remainder

$$(b_1, b_2, ..., b_\mu, ... c_1, c_2, c_3)$$

are determined by the element b_1; and the segment

$$(a_1, a_2, ..., a_v, ... b_1, b_2, ..., b_\mu, ... c_1)$$

and the corresponding segment

$$(c_2, c_3)$$

by the element c_2.

If A and A′ are two segments of F, f and f' their determining elements, and

$$f' < f, \tag{4}$$

then A′ is a segment of A. We call A′ the "less," and A the "greater" segment of F:

$$A' < A. \tag{5}$$

Correspondingly we may say of every A of F that it is "less" than F itself:

$$A < F.$$

[211]

A. If two similar well-ordered aggregates F and G are imaged on one another, then to every segment A of F corresponds a similar segment B of G, and to every segment B of G corresponds a similar segment A of F, and the elements f and g of F and G by which the corresponding segments A and B are determined also correspond to one another in the imaging.

Proof.—If we have two similar simply ordered aggregates M and N imaged on one another, m and n are two corresponding elements, and M′ is the aggregate of all elements of M which precede m and N′ is the aggregate of all elements of N which precede n, then in the imaging M′ and N′ correspond to one another. For, to every element $m′$ of M that precedes m must correspond, by § 7, an element $n′$ of N that precedes n, and inversely. If we apply this general theorem to the well-ordered aggregates F and G we get what is to be proved.

B. A well-ordered aggregate F is not similar to any of its segments A.

Proof.—Let us suppose that F ≃ A, then we will imagine an imaging of F on A set up. By theorem A the segment A′ of A corresponds to the segment A of F, so that A′ ≃ A. Thus also we would have A′ ≃ F and A′ < A. From A′ would result, in the same manner, a smaller segment A″ of F, such that A″ ≃ F and A″ < A′; and so on. Thus we would obtain an infinite series

$$A > A′ > A″ \ldots A^{(v)} > A^{(v+1)} \ldots$$

of segments of F, which continually become smaller and all similar to the aggregate F. We will denote by $f, f′, f″, \ldots, f^{(v)}, \ldots$ the elements of F which determine these segments; then we would have

$$f > f′ > f″ > \cdots > f^{(v)} > f^{(v+1)} \ldots$$

We would therefore have an infinite part

$$(f, f′, f″, \ldots, f^{(v)}, \ldots)$$

of F in which no element takes the lowest rank. But by theorem A of § 12 such parts of F are not possible. Thus the supposition of an imaging F on one of its segments leads to a contradiction, and consequently the aggregate F is not similar to any of its segments.

Though by theorem B a well-ordered aggregate F is not similar to any of its segments, yet, if F is infinite, there are always

[212]

other parts of F to which F is similar. Thus, for example, the aggregate

$$F = (a_1, a_2, \ldots, a_v, \ldots)$$

is similar to every one of its remainders

$$(a_{\kappa+1}, a_{\kappa+2}, \ldots, a_{\kappa+v}, \ldots).$$

Consequently, it is important that we can put by the side of theorem B the following:

C. A well-ordered aggregate F is similar to no part of any one of its segments A.

Proof.—Let us suppose that F′ is a part of a segment A of F and F′ ≃ F. We imagine an imaging of F on F′; then, by theorem A, to a segment A of the well-ordered aggregate F corresponds as image the segment F″ of F′; let this segment be determined by the element $f′$ of F′. The element $f′$ is also an element of A, and determines a segment A′ of A of which F″ is a part. The supposition of a part F′ of a segment A of F such that F′ ≃ F leads us consequently to a part F″ of a segment

A' of A such that $F'' \simeq A$. The same manner of conclusion gives us a part F''' of a segment A'' of A' such that $F''' \simeq A'$. Proceeding thus, we get, as in the proof of theorem B, an infinite series of segments of F which continually become smaller:

$$A > A' > A'' \dots A^{(v)} > A^{(v+1)} \dots,$$

and thus an infinite series of elements determining these segments:

$$f > f' > f'' \dots f^{(v)} > f^{(v+1)} \dots,$$

in which is no lowest element, and this is impossible by theorem A of § 12. Thus there is no part F' of a segment A of F such that $F' \simeq F$.

D. Two different segments A and A' of a well-ordered aggregate F are not similar to one another.

Proof.—If $A' < A$, then A' is a segment of the well-ordered aggregate A, and thus, by theorem B, cannot be similar to A.

E. Two similar well-ordered aggregates F and G can be imaged on one another only in a single manner.

Proof.—Let us suppose that there are two different imagings of F on G, and let f be an element of F to which in the two imagings different images g and g' in G correspond. Let A be the segment of F that is determined by f, and B and B' the segments of G that are determined by g and g'. By theorem A, both $A \simeq B$

[213]

and $A \simeq B'$, and consequently $B \simeq B'$, contrary to theorem D.

F. If F and G are two well-ordered aggregates, a segment A of F can have at most one segment B in G which is similar to it.

Proof.—If the segment A of F could have two segments B and B' in G which were similar to it, B and B' would be similar to one another, which is impossible by theorem D.

G. If A and B are similar segments of two well-ordered aggregates F and G, for every smaller segment $A' < A$ of F there is a similar segment $B' < B$ of G and for every smaller segment $B' < B$ of G a similar segment $A' < A$ of F.

The proof follows from theorem A applied to the similar aggregates A and B.

H. If A and A' are two segments of a well-ordered aggregate F, B and B' are two segments similar to those of a well-ordered aggregate G, and $A' < A$, then $B' < B$.

The proof follows from the theorems F and G.

I. If a segment B of a well-ordered aggregate G is similar to no segment of a well-ordered aggregate F, then both every segment $B' > B$ of D and G itself are similar neither to a segment of F nor F itself.

1006

The proof follows from theorem G.

K. If for any segment A of a well-ordered aggregate F there is a similar segment B of another well-ordered aggregate G, and also inversely, for every segment B of G a similar segment A of F, then F ≃ G.

Proof.—We can image F and G on one another according to the following law: Let the lowest element f_1 of F correspond to the lowest element g_1 of G. If $f > f_1$ is any other element of F, it determines a segment A of F. To this segment belongs by supposition a definite similar segment B of G, and let the element g of G which determines the segment B be the image of F. And if g is any element of G that follows g_1, it determines a segment B of G, to which by supposition a similar segment A of F belongs. Let the element f which determines this segment A be the image of g. It easily follows that the bi-univocal correspondence of F and G defined in this manner is an imaging in the sense of § 7. For if f and f' are any two elements of F, g and g'

[214]

the corresponding elements of G, A and A′ the segments determined by f and f', B and B′ those determined by g and g', and if, say,

$$f' < f,$$

then

$$A' < A.$$

By theorem H, then, we have

$$B' < B,$$

and consequently

$$g' < g.$$

L. If for every segment A of a well-ordered aggregate F there is a similar segment B of another well-ordered aggregate G, but if, on the other hand, there is at least one segment of G for which there is no similar segment of F, then there exists a definite segment B_1 of G such that $B_1 ≃ F$.

Proof.—Consider the totality of segments of G for which there are no similar segments in F. Amongst them there must be a least segment which we will call B_1. This follows from the fact that, by theorem A of § 12, the aggregate of all the elements determining these segments has a lowest element; the segment B_1 of G determined by that element is the least of that totality. By theorem I, every segment of G which is greater than B_1 is such that no segment similar to it is present in F. Thus the segments B of G which correspond to similar segments of F must all be less than B_1, and to every segment $B < B_1$ belongs a similar segment A of F, because B_1 is the least segment of G among those to which no similar segments in F correspond. Thus, for every segment A of F there is a similar segment B of B_1, and for every segment B of B_1 there is a similar segment A of F. By theorem K, we thus have

$$F \simeq B_1.$$

M. If the well-ordered aggregate G has at least one segment for which there is no similar segment in the well-ordered aggregate F, then every segment A of F must have a segment B similar to it in G.

Proof.—Let B_1 be the least of all those segments of G for which there are no similar segments in F.[2] If there were segments in F for which there were no corresponding segments in G, amongst these, one, which we will call A_1, would be the least. For every segment of A_1 would then exist a similar segment of B_1, and also for every segment of B_1 a similar segment of A_1. Thus, by theorem K, we would have

$$B_1 \simeq A_1.$$

[215]

But this contradicts the datum that for B_1 there is no similar segment of F. Consequently, there cannot be in F a segment to which a similar segment in G does not correspond.

N. If F and G are any two well-ordered aggregates, then either:

(*a*) F and G are similar to one another, or

(*b*) there is a definite segment B_1 of G to which F is similar, or

(*c*) there is a definite segment A_1 of F to which G is similar;

and each of these three cases excludes the two others.

Proof.—The relation of F to G can be any one of the three:

(*a*) To every segment A of F there belongs a similar segment B of G, and inversely, to every segment B of G belongs a similar one A of F;

(*b*) To every segment A of F belongs a similar segment B of G, but there is at least one segment of G to which no similar segment in F corresponds;

(*c*) To every segment B of G belongs a similar segment A of F, but there is at least one segment of F to which no similar segment in G corresponds.

The case that there is both a segment of F to which no similar segment in G corresponds and a segment of G to which no similar segment in F corresponds is not possible; it is excluded by theorem M.

By theorem K, in the first case we have

$$F \simeq G.$$

[2] See the above proof of L.

In the second case there is, by theorem L, a definite segment B_1 of B such that

$$B_1 \simeq F;$$

and in the third case there is a definite segment A_1 of F such that

$$A_1 \simeq G.$$

We cannot have $F \simeq G$ and $F \simeq B_1$ simultaneously, for then we would have $G \simeq B_1$, contrary to theorem B; and, for the same reason, we cannot have both $F \simeq G$ and $G \simeq A_1$. Also it is impossible that both $F \simeq B_1$ and $G \simeq A_1$, for, by theorem A, from $F \simeq B_1$ would follow the existence of a segment B_1' of B_1 such that $A_1 \simeq B_1'$. Thus we would have $G \simeq B_1'$, contrary to theorem B.

O. If a part F′ of a well-ordered aggregate F is not similar to any segment of F, it is similar to F itself.

Proof.—By theorem C of § 12, F′ is a well-ordered aggregate. If F′ were similar neither to a segment of F nor to F itself, there would be, by theorem N, a segment F_1' of F′ which is similar to F. But F_1' is a part of that segment A of F which

[216]

is determined by the same element as the segment F_1' of F′. Thus the aggregate F would have to be similar to a part of one of its segments, and this contradicts the theorem C.

§ 14
THE ORDINAL NUMBERS OF WELL-ORDERED AGGREGATES

By § 7, every simply ordered aggregate M has a definite ordinal type $\overline{M}$; this type is the general concept which results from M if we abstract from the nature of its elements while retaining their order of precedence, so that out of them proceed units (*Einsen*) which stand in a definite relation of precedence to one another. All aggregates which are similar to one another, and only such, have one and the same ordinal type. We call the ordinal type of a well-ordered aggregate F its "ordinal number."

If α and β are any two ordinal numbers, one can stand to the other in one of three possible relations. For if F and G are two well-ordered aggregates such that

$$\overline{F} = \alpha, \quad \overline{G} = \beta,$$

then, by theorem N of § 13, three mutually exclusive cases are possible:

(*a*) $\qquad\qquad\qquad F \simeq G;$

(*b*) There is a definite segment B_1 of G such that

$$F \simeq B_1;$$

(c) There is a definite segment A_1 of F such that
$$G \simeq A_1.$$

As we easily see, each of these cases still subsists if F and G are replaced by aggregates respectively similar to them. Accordingly, we have to do with three mutually exclusive relations of the types α and β to one another. In the first case $\alpha = \beta$; in the second we say that $\alpha < \beta$; in the third we say that $\alpha > \beta$. Thus we have the theorem:

A. If α and β are any two ordinal numbers, we have either $\alpha = \beta$ or $\alpha < \beta$ or $\alpha > \beta$.

From the definition of minority and majority follows easily:

B. If we have three ordinal numbers α, β, γ, and if $\alpha < \beta$ and $\beta < \gamma$, then $\alpha < \gamma$.

Thus the ordinal numbers form, when arranged in order of magnitude, a simply ordered aggregate; it will appear later that it is a well-ordered aggregate.

[217]

The operations of addition and multiplication of the ordinal types of any simply ordered aggregates, defined in § 8, are, of course, applicable to the ordinal numbers. If $\alpha = \overline{F}$ and $\beta = \overline{G}$, where F and G are two well-ordered aggregates, then
$$\alpha + \beta = \overline{(F, G)}. \tag{1}$$

The aggregate of union (F, G) is obviously a well-ordered aggregate too; thus we have the theorem:

C. The sum of two ordinal numbers is also an ordinal number.

In the sum $\alpha + \beta$, α is called the "augend" and β the "addend."

Since F is a segment of (F, G), we have always
$$\alpha < \alpha + \beta. \tag{2}$$
On the other hand, G is not a segment but a remainder of (F, G), and may thus, as we saw in § 13, be similar to the aggregate (F, G). If this is not the case, G is, by theorem O of § 13, similar to a segment of (F, G). Thus
$$\beta \leqq \alpha + \beta. \tag{3}$$

Consequently we have:

D. The sum of the two ordinal numbers is always greater than the augend, but greater than or equal to the addend. If we have $\alpha + \beta = \alpha + \gamma$, we always have $\beta = \gamma$.

In general $\alpha + \beta$ and $\beta + \alpha$ are not equal. On the other hand, we have, if γ is a third ordinal number,
$$(\alpha + \beta) + \gamma = \alpha + (\beta + \gamma). \tag{4}$$

That is to say:

E. In the addition of ordinal numbers the associative law always holds.

If we substitute for every element g of the aggregate G of type β an aggregate F_g of type α, we get, by theorem E of § 12, a well-ordered aggregate H whose type is completely determined by the types α and β and will be called the product $\alpha \cdot \beta$:

$$\overline{F}_g = \alpha, \tag{5}$$

$$\alpha \cdot \beta = \overline{H}. \tag{6}$$

F. The product of two ordinal numbers is also an ordinal number.

In the product $\alpha \cdot \beta$, α is called the "multiplicand" and β the "multiplier."

In general $\alpha \cdot \beta$ and $\beta \cdot \alpha$ are not equal. But we have (§ 8)

$$(\alpha \cdot \beta) \cdot \gamma = \alpha \cdot (\beta \cdot \gamma). \tag{7}$$

That is to say:

[218]

G. In the multiplication of ordinal numbers the associative law holds.

The distributive law is valid, in general (§ 8), only in the following form:

$$\alpha \cdot (\beta + \gamma) = \alpha \cdot \beta + \alpha \cdot \gamma. \tag{8}$$

With reference to the magnitude of the product, the following theorem, as we easily see, holds:

H. If the multiplier is greater than 1, the product of two ordinal numbers is always greater than the multiplicand, but greater than or equal to the multiplier. If we have $\alpha \cdot \beta = \alpha \cdot \gamma$, then it always follows that $\beta = \gamma$.

On the other hand, we evidently have

$$\alpha \cdot 1 = 1 \cdot \alpha = \alpha. \tag{9}$$

We have now to consider the operation of subtraction. If α and β are two ordinal numbers, and α is less than β, there always exists a definite ordinal number which we will call $\beta - \alpha$, which satisfies the equation

$$\alpha + (\beta - \alpha) = \beta. \tag{10}$$

For if $\overline{G} = \beta$, G has a segment B such that $\overline{B} = \alpha$; we call the corresponding remainder S, and have

$$G = (B, S),$$

$$\beta = \alpha + \overline{S};$$

and therefore

$$\beta - \alpha = \overline{S}. \tag{11}$$

The determinateness of $\beta - \alpha$ appears clearly from the fact that the segment B of G is a completely definite one (theorem D of § 13), and consequently also S is uniquely given.

We emphasize the following formulæ, which follow from (4), (8), and (10):

$$(\gamma + \beta) - (\gamma + \alpha) = \beta - \alpha, \tag{12}$$

$$\gamma(\beta - \alpha) = \gamma\beta - \gamma\alpha. \tag{13}$$

It is important to reflect that an infinity of ordinal numbers can be summed so that their sum is a definite ordinal number which depends on the sequence of the summands. If

$$\beta_1, \cdot \beta_2, \ldots, \beta_\nu, \ldots$$

is any simply infinite series of ordinal numbers, and we have

$$\beta_\nu = \overline{G}_\nu, \tag{14}$$

[219]

then, by theorem E of § 12,

$$G = (G_1, G_2, \ldots, G_\nu, \ldots) \tag{15}$$

is also a well-ordered aggregate whose ordinal number represents the sum of the numbers β_ν. We have, then,

$$\beta_1 + \beta_2 + \cdots + \beta_\nu + \cdots = \overline{G} = \beta, \tag{16}$$

and, as we easily see from the definition of a product, we always have

$$\gamma \cdot (\beta_1 + \beta_2 + \cdots + \beta_\nu + \cdots) = \gamma \cdot \beta_1 + \gamma \cdot \beta_2 + \cdots + \gamma \cdot \beta_\nu + \cdots \tag{17}$$

If we put

$$\alpha_\nu = \beta_1 + \beta_2 + \cdots + \beta_\nu, \tag{18}$$

then

$$\alpha_\nu = \overline{(G_1, G_2, \ldots G_\nu)}. \tag{19}$$

We have

$$\alpha_{\nu+1} > \alpha_\nu, \tag{20}$$

and, by (10), we can express the numbers β_ν by the numbers α_ν as follows:

$$\beta_1 = \alpha_1; \quad \beta_{\nu+1} = \alpha_{\nu+1} - \alpha_\nu. \tag{21}$$

The series

$$\alpha_1, \alpha_2, \ldots, \alpha_\nu, \ldots$$

thus represents *any* infinite series of ordinal numbers which satisfy the condition (20); we will call it a "fundamental series" of ordinal numbers (§ 10). Between it and β subsists a relation which can be expressed in the following manner:

(*a*) The number β is greater than α_ν for every ν, because the aggregate $(G_1, G_2, \ldots, G_\nu)$, whose ordinal number is α_ν, is a segment of the aggregate G which has the ordinal number β;

(*b*) If β' is any ordinal number less than β, then, from a certain ν onwards, we always have

$$\alpha_\nu > \beta'.$$

For, since $\beta' < \beta$, there is a segment B$'$ of the aggregate G which is of type β'. The element of G which determines this segment must belong to one of the parts G_ν; we will call this part G_{ν_0}. But then B$'$ is also a segment of $(G_1, G_2, \ldots, G_{\nu_0})$, and consequently $\beta' < \alpha_{\nu_0}$. Thus

$$\alpha_\nu > \beta'$$

for $\nu \geqq \nu_0$.

Thus β is the ordinal number which follows next in order of magnitude after all the numbers α_ν; accordingly we will call it the "limit" (*Grenze*) of the numbers α_ν for increasing ν and denote it by $\underset{\nu}{\text{Lim}}\ \alpha_\nu$, so that, by (16) and (21):

$$\underset{\nu}{\text{Lim}}\ \alpha_\nu = \alpha_1 + (\alpha_2 - \alpha_1) + \cdots + (\alpha_{\nu+1} - \alpha_\nu) + \cdots \tag{22}$$

[220]

We may express what precedes in the following theorem:

1. To every fundamental series $\{\alpha_\nu\}$ of ordinal numbers belongs an ordinal number $\underset{\nu}{\text{Lim}}\ \alpha_\nu$ which follows next, in order of magnitude, after all the numbers α_ν; it is represented by the formula (22).

If by γ we understand any constant ordinal number, we easily prove, by the aid of the formulæ (12), (13), and (17), the theorems contained in the formulæ:

$$\underset{\nu}{\text{Lim}}\ (\gamma + \alpha_\nu) = \gamma + \underset{\nu}{\text{Lim}}\ \alpha_\nu; \tag{23}$$

$$\underset{\nu}{\text{Lim}}\ \gamma \cdot \alpha_\nu = \gamma \cdot \underset{\nu}{\text{Lim}}\ \alpha_\nu. \tag{24}$$

We have already mentioned in § 7 that all simply ordered aggregates of given finite cardinal number ν have one and the same ordinal type. This may be proved here as follows. Every simply ordered aggregate of finite cardinal number is a well-ordered aggregate; for it, and every one of its parts, must have a lowest element,— and this, by theorem B of § 12, characterizes it as a well-ordered aggregate. The types of finite simply ordered aggregates are thus none other than finite ordinal numbers. But two different ordinal numbers α and β cannot belong to the same finite cardinal number ν. For if, say, $\alpha < \beta$ and $\overline{G} = \beta$, then, as we know, there exists a segment B of G such that $\overline{B} = \alpha$. Thus the aggregate G and its part B would have the same finite cardinal number ν. But this, by theorem C of § 6, is impossible. Thus the finite ordinal numbers coincide in their properties with the finite cardinal numbers.

The case is quite different with the transfinite ordinal numbers; to one and the same transfinite cardinal number a belong an infinity of ordinal numbers which form a unitary and connected system. We will call this system the "number-class Z(a)," and it is a part of the class of types [a] of § 7. The next object of our consideration is the number-class $Z(\aleph_0)$, which we will call "the second number-class." For in this connexion we understand by "the first number-class" the totality $\{v\}$ of finite ordinal numbers.

[221]

§ 15

THE NUMBERS OF THE SECOND NUMBER-CLASS $Z(\aleph_0)$

The second number-class $Z(\aleph_0)$ is the totality $\{\alpha\}$ of ordinal types α of well-ordered aggregates of the cardinal number $\aleph_0$ (§ 6).

A. The second number-class has a least number $\omega = \underset{v}{\text{Lim}}\, v$.

Proof.—By ω we understand the type of the well-ordered aggregate

$$F_0 = (f_1, f_2, \ldots, f_v, \ldots), \tag{1}$$

where v runs through all finite ordinal numbers and

$$f_v < f_{v+1}. \tag{2}$$

Therefore (§ 7)

$$\omega = \overline{F}_0, \tag{3}$$

and (§ 6)

$$\overline{\omega} = \aleph_0. \tag{4}$$

Thus ω is a number of the second number-class, and indeed the least. For if γ is any ordinal number less than ω, it must (§ 14) be the type of a segment of F_0. But F_0 has only segments

$$A = (f_1, f_2, \ldots, f_v),$$

with *finite* ordinal number v. Thus $\gamma = v$. Therefore there are no transfinite ordinal numbers which are less than ω, and thus ω is the least of them. By the definition of $\text{Lim}\,\alpha_v$ given in § 14, we obviously have $\omega = \underset{v}{\text{Lim}}\, v$.

B. If α is any number of the second number-class, the number $\alpha + 1$ follows it as the next greater number of the same number-class.

Proof.—Let F be a well-ordered aggregate of the type α and of the cardinal number $\aleph_0$:

$$\overline{F} = \alpha, \tag{5}$$
$$\overline{\alpha} = \aleph_0. \tag{6}$$

We have, where by g is understood a new element,

$$\alpha + 1 = (\overline{F, g}). \tag{7}$$

Since F is a segment of (F, g), we have

$$\alpha + 1 > \alpha. \tag{8}$$

We also have
$$\overline{\alpha + 1} = \overline{\alpha} + 1 = \aleph_0 + 1 = \aleph_0 \ (\S \ 6).$$
Therefore the number $\alpha + 1$ belongs to the second number-class. Between α and $\alpha + 1$ there are no ordinal numbers; for every number γ

[222]

which is less than $\alpha + 1$ corresponds, as type, to a segment of (F, g), and such a segment can only be either F or a segment of F. Therefore γ is either equal to or less than α.

C. If air $\alpha_1, \alpha_2, \ldots, \alpha_\nu, \ldots$ is any fundamental series of numbers of the first or second number-class, then the number $\underset{\nu}{\text{Lim}} \ \alpha_\nu$ ($\S$ 14) following them next in order of magnitude belongs to the second number-class.

Proof.—By $\S$ 14 there results from the fundamental series $\{\alpha_\nu\}$ the number $\underset{\nu}{\text{Lim}} \ \alpha_\nu$ if we set up another series $\beta_1, \beta_2, \ldots, \beta_\nu, \ldots$, where
$$\beta_1 = \alpha_1, \quad \beta_2 = \alpha_2 - \alpha_1, \ldots, \quad \beta_{\nu+1} = \alpha_{\nu+1} - \alpha_\nu, \ldots$$
If, then, $G_1, G_2, \ldots, G_\nu, \ldots$ are well-ordered aggregates such that
$$\overline{G}_\nu = \beta_\nu,$$
then also
$$G = (G_1, G_2, \ldots, G_\nu, \ldots)$$
is a well-ordered aggregate and
$$\underset{\nu}{\text{Lim}} \ \alpha_\nu = \overline{G}.$$
It only remains to prove that
$$\overline{\overline{G}} = \aleph_0.$$
Since the numbers $\beta_1, \beta_2, \ldots, \beta_\nu, \ldots$ belong to the first or second number-class, we have
$$\overline{\overline{G}}_\nu \leqq \aleph_0,$$
and thus
$$\overline{\overline{G}} \leqq \aleph_0 \cdot \aleph_0 = \aleph_0.$$
But, in any case, G is a transfinite aggregate, and so the case $\overline{\overline{G}} < \aleph_0$ is excluded.

We will call two fundamental series $\{\alpha_\nu\}$ and $\{\alpha_\nu'\}$ of numbers of the first or second number-class ($\S$ 10) "coherent," in signs:
$$\{\alpha_\nu\} \ \| \ \{\alpha_\nu'\}, \tag{9}$$
if for every ν there are finite numbers λ_0 and μ_0 such that
$$\alpha_\lambda' > \alpha_\nu, \quad \lambda \geqq \lambda_0, \tag{10}$$
and
$$\alpha_\mu > \alpha_\nu', \quad \mu \geqq \mu_0. \tag{11}$$

D. The limiting numbers $\operatorname{Lim}_{\nu} \alpha_\nu$ and $\operatorname{Lim}_{\nu} \alpha_\nu'$ belonging respectively to two fundamental series $\{\alpha_\nu\}$ and $\{\alpha_\nu'\}$ are equal when, and only when, $\{\alpha_\nu\} \parallel \{\alpha_\nu'\}$.

Proof.—For the sake of shortness we put $\operatorname{Lim}_{\nu} \alpha_\nu = \beta$, $\operatorname{Lim}_{\nu} \alpha_\nu' = \gamma$. We will first suppose that $\{\alpha_\nu\} \parallel \{\alpha_\nu'\}$; then we assert that $\beta = \gamma$. For if β were not equal to γ, one of these two numbers would have to be the smaller. Suppose that $\beta < \gamma$. From a certain ν onwards we would have $\alpha_\nu' > \beta$ (§ 14), and consequently, by (11), from a certain μ onwards we would have $\alpha_\mu > \beta$. But this is impossible because $\beta = \operatorname{Lim}_{\nu} \alpha_\nu$. Thus for all μ's we have $\alpha_\mu < \beta$.

If, inversely, we suppose that $\beta = \gamma$, then, because $\alpha_\nu < \gamma$, we must conclude that, from a certain λ onwards, $\alpha_\lambda' > \alpha_\nu$, and, because $\alpha_\nu' < \beta$, we must conclude that, from a certain μ onwards, $\alpha_\mu > \alpha_\nu'$. That is to say, $\{\alpha_\nu\} \parallel \{\alpha_\nu'\}$.

E. If α is any number of the second number-class and ν_0 any finite ordinal number, we have $\nu_0 + \alpha = \alpha$, and consequently also $\alpha - \nu_0 = \alpha$.

Proof.—We will first of all convince ourselves of the correctness of the theorem when $\alpha = \omega$. We have

$$\omega = \overline{(f_1, f_2, \ldots, f_\nu, \ldots)},$$
$$\nu_0 = \overline{(g_1, g_2, \ldots g_{\nu_0})},$$

and consequently

$$\nu_0 + \omega = \overline{(g_1, g_2, \ldots g_{\nu_0}, f_1, f_2, \ldots, f_\nu, \ldots)} = \omega.$$

But if $\alpha > \omega$, we have

$$\alpha = \omega + (\alpha - \omega),$$
$$\nu_0 + \alpha = (\nu_0 + \omega) + (\alpha - \omega) = \omega + (\alpha - \omega) = \alpha.$$

F. If ν_0 is any finite ordinal number, we have $\nu_0 \cdot \omega = \omega$.

Proof.—In order to obtain an aggregate of the type $\nu_0 \cdot \omega$ we have to substitute for the single elements f_ν of the aggregate $(f_1, f_2, \ldots, f_\nu, \ldots)$ aggregates $(g_{\nu,1}, g_{\nu,2}, \ldots, g_{\nu,\nu_0})$ of the type ν_0. We thus obtain the aggregate

$$(g_{1,1}, g_{1,2}, \ldots, g_{1,\nu_0}, g_{2,1}, \ldots, g_{2,\nu_0}, \ldots, g_{\nu,1}, g_{\nu,2}, \ldots, g_{\nu,\nu_0}, \ldots),$$

which is obviously similar to the aggregate $\{f_\nu\}$. Consequently

$$\nu_0 \cdot \omega = \omega.$$

The same result is obtained more shortly as follows. By (24) of § 14 we have, since $\omega = \operatorname{Lim}_{\nu} \nu$,

$$\nu_0 \omega = \operatorname{Lim}_{\nu} \nu_0 \nu.$$

On the other hand,

$$\{\nu_0 \, \nu\} \parallel \{\nu\},$$

and consequently

$$\mathrm{Lim}_v \, v_0 \, v = \mathrm{Lim}_v \, v = \omega;$$

so that

$$v_0 \, \omega = \omega.$$

[224]

G. We have always

$$(\alpha + v_0)\omega = \alpha\omega,$$

where α is a number of the second number-class and v_0 a number of the first number-class.

Proof.—We have

$$\mathrm{Lim}_v \, v = \omega.$$

By (24) of § 14 we have, consequently,

$$(\alpha + v_0)\omega = \mathrm{Lim}_v \, (\alpha + v_0)v.$$

But

$$(\alpha + v_0)v = \overset{1}{(\alpha + v_0)} + \overset{2}{(\alpha + v_0)} + \cdots + \overset{v}{(\alpha + v_0)}$$

$$= \alpha + \overset{1}{(v_0 + \alpha)} + \overset{2}{(v_0 + \alpha)} \cdots \overset{v-1}{(v_0 + \alpha)} + v_0$$

$$= \overset{1}{\alpha} + \overset{2}{\alpha} + \cdots + \overset{v}{\alpha} + v_0$$

$$= \alpha v + v_0.$$

Now we have, as is easy to see,

$$\{\alpha v + v_0\} \parallel \{\alpha v\},$$

and consequently

$$\mathrm{Lim}_v \, (\alpha + v_0)v = \mathrm{Lim}_v \, (\alpha v + v_0) = \mathrm{Lim}_v \, \alpha v = \alpha\omega.$$

H. If α is any number of the second number-class, then the totality $\{\alpha'\}$ of numbers α' of the first and second number-classes which are less than α form, in their order of magnitude, a well-ordered aggregate of type α.

Proof.—Let F be a well-ordered aggregate such that $\overline{\overline{F}} = \alpha$, and let f_1 be the lowest element of F. If α' is any ordinal number which is less than α, then, by § 14, there is a definite segment A′ of F such that

$$\overline{A}' = \alpha',$$

and inversely every segment A′ determines by its type $\overline{A}' = \alpha'$ a number $\alpha' < \alpha$ of the first or second number-class. For, since $\overline{\overline{F}} = \aleph_0$, $\overline{A}'$ can only be either a finite cardinal number or $\aleph_0$. The segment A′ is determined by an element $f' > f_1$ of F, and inversely every element $f' > f_1$ of F determines a segment A′ of F. If f' and f'' are two elements of F which follow f_1 in rank, A′ and A″ are the segments of F determined by them, α' and α'' are their ordinal types, and, say $f' < f''$, then, by § 13, $A' < A''$ and consequently $\alpha' < \alpha''$.

If, then, we put $F = (f_1, F')$, we obtain, when we make the element f' of F' correspond to the element α' of $\{\alpha'\}$, an imaging of these two aggregates. Thus we have

$$\overline{\{\alpha'\}} = \overline{F'}.$$

But $\overline{F'} = \alpha - 1$, and, by theorem E, $\alpha - 1 = \alpha$. Consequently

$$\overline{\{\alpha'\}} = \alpha.$$

Since $\overline{\overline{\alpha}} = \aleph_0$, we also have $\overline{\overline{\{\alpha'\}}} = \aleph_0$; thus we have the theorems:

I. The aggregate $\{\alpha'\}$ of numbers α' of the first and second number-classes which are smaller than a number α of the second number-class has the cardinal number $\aleph_0$.

K. Every number α of the second number-class is either such that (a) it arises out of the next smaller number α_{-1} by the addition of 1:

$$\alpha = \alpha_{-1} + 1,$$

or (b) there is a fundamental series $\{\alpha_\nu\}$ of numbers of the first or second number-class such that

$$\alpha = \operatorname{Lim}_\nu \alpha_\nu.$$

Proof.—Let $\alpha = \overline{F}$. If F has an element g which is highest in rank, we have $F = (A, g)$, where A is the segment of F which is determined by g. We have then the first case, namely,

$$\alpha = \overline{A} + 1 = \alpha_{-1} + 1.$$

There exists, therefore, a next smaller number which is that called α_1.

But if F has no highest element, consider the totality $\{\alpha'\}$ of numbers of the first and second number-classes which are smaller than α. By theorem H, the aggregate $\{\alpha'\}$, arranged in order of magnitude, is similar to the aggregate F; among the numbers α', consequently, none is greatest. By theorem 1, the aggregate $\{\alpha'\}$ can be brought into the form $\{\alpha_\nu'\}$ of a simply infinite series. If we set out from α_1', the next following elements α_2', α_3', ... in this order, which is different from the order of magnitude, will, in general, be smaller than α_1'; but in every case, in the further course of the process, terms will occur which are greater than α_1'; for α_1' cannot be greater than all other terms, because among the numbers $\{\alpha_\nu'\}$ there is no greatest. Let that number α_ν' which has the least index of those greater than α_1' be α_{ρ_2}'. Similarly, let α_{ρ_3}' be that number of the series $\{\alpha_\nu'\}$ which has the least index of those which are greater than α_{ρ_2}'. By proceeding in this way, we get an infinite series of increasing numbers, a fundamental series in fact,

$$\alpha_1', \alpha_{\rho_2}', \alpha_{\rho_3}', \ldots, \alpha_{\rho_\nu}', \ldots$$

[226]

We have

$$1 < \rho_2 < \rho_3 < \cdots < \rho_v < \rho_{v+1} \cdots,$$
$$\alpha_1' < \alpha_{\rho_2}' < \alpha_{\rho_3}' < \cdots < \alpha_{\rho_v}' < \alpha_{\rho_{v+1}}' \ldots,$$
$$\alpha_\mu' < \alpha_{\rho_v}' \quad \text{always if} \quad \mu < \rho_v';$$

and since obviously $v \gtreqless \rho_v$, we always have

$$\alpha_v' \leqq \alpha_{\rho_v}'.$$

Hence we see that every number α_v', and consequently every number $\alpha' < \alpha$, is exceeded by numbers α_{ρ_v}' for sufficiently great values of v. But α is the number which, in respect of magnitude, immediately follows all the numbers α', and consequently is also the next greater number with respect to all α_{ρ_v}'. If, therefore, we put $\alpha_1' = \alpha_1$, $\alpha_{\rho_{v+1}} = \alpha_{v+1}$, we have

$$\alpha = \operatorname*{Lim}_v \alpha_v.$$

From the theorems B, C, . . ., K it is evident that the numbers of the second number-class result from smaller numbers in two ways. Some numbers, which we call "numbers of the first kind (*Art*)," are got from a next smaller number α_{-1} by addition of 1 according to the formula

$$\alpha = \alpha_{-1} + 1;$$

The other numbers, which we call "numbers of the second kind," are such that for any one of them there is not a next smaller number α_{-1}, but they arise from fundamental series $\{\alpha_v\}$ as limiting numbers according to the formula

$$\alpha = \operatorname*{Lim}_v \alpha_v.$$

Here α is the number which follows next in order of magnitude to all the numbers α_v.

We call these two ways in which greater numbers proceed out of smaller ones "the first and the second principle of generation of numbers of the second number-class."

§ 16

THE POWER OF THE SECOND NUMBER-CLASS IS EQUAL TO THE SECOND GREATEST TRANSFINITE CARDINAL NUMBER ALEPH-ONE

Before we turn to the more detailed considerations in the following paragraphs of the numbers of the second number-class and of the laws which rule them, we will answer the question as to the cardinal number which is possessed by the aggregate $Z(\aleph_0) = \{\alpha\}$ of all these numbers.

[227]

A. The totality $\{\alpha\}$ of all numbers α of the second number-class forms, when arranged in order of magnitude, a well-ordered aggregate.

Proof.—If we denote by A_α the totality of numbers of the second number-class which are smaller than a given number α, arranged in order of magnitude, then A_α is a well-ordered aggregate of type $\alpha - \omega$. This results from theorem H of § 14. The aggregate of numbers α' of the first and second number-class which was there denoted by $\{\alpha'\}$, is compounded out of $\{v\}$ and A_α, so that

$$\{\alpha'\} = (\{v\}, A_\alpha).$$

Thus

$$\{\overline{\alpha'}\} = \{\overline{v}\} + \overline{A}_\alpha;$$

and since

$$\{\overline{\alpha'}\} = \alpha, \quad \{\overline{v}\} = v,$$

we have

$$\overline{A}_\alpha = \alpha - v.$$

Let J be any part of $\{\alpha\}$ such that there are numbers in $\{\alpha\}$ which are greater than all the numbers of J. Let, say, α be one of these numbers. Then J is also a part of A_{α_0+1}, and indeed such a part that at least the number α_0 of A_{α_0+1} is greater than all the numbers of J. Since A_{α_0+1} is a well-ordered aggregate, by § 12 a number α' of A_{α_0+1}, and therefore also of $\{\alpha\}$, must follow next to all the numbers of J. Thus the condition II of § 12 is fulfilled in the case of $\{\alpha\}$; the condition I of § 12 is also fulfilled because $\{\alpha\}$ has the least number ω.

Now, if we apply to the well-ordered aggregate $\{\alpha\}$ the theorems A and C of § 12, we get the following theorems:

B. Every totality of different numbers of the first and second number-classes has a least number.

C. Every totality of different numbers of the first and second number-classes arranged in their order of magnitude forms a well-ordered aggregate.

We will now show that the power of the second number-class is different from that of the first, which is $\aleph_0$.

D. The power of the totality $\{\alpha\}$ of all numbers α of the second number-class is not equal to $\aleph_0$.

Proof.—If $\{\overline{\overline{\alpha}}\}$ were equal to $\aleph_0$, we could bring the totality $\{\alpha\}$ into the form of a simply infinite series

$$\gamma_1, \gamma_2, \ldots, \gamma_v, \ldots$$

such that $\{\gamma_v\}$ would represent the totality of numbers of the second

[228]

number-class in an order which is different from the order of magnitude, and $\{\gamma_\nu\}$ would contain, like $\{\alpha\}$, no greatest number.

Starting from γ_1, let γ_{ρ_2} be the term of the series which has the least index of those greater than γ_1, γ_{ρ_3} the term which has the least index of those greater than γ_{ρ_2}, and so on. We get an infinite series of increasing numbers,

$$\gamma_1, \gamma_{\rho_2}, \ldots, \gamma_{\rho_\nu}, \ldots,$$

such that

$$1 < \rho_2 < \rho_3 \ldots < \rho_\nu < \rho_{\nu+1} < \ldots,$$
$$\gamma_1 < \gamma_{\rho_2} < \gamma_{\rho_3} \ldots < \gamma_{\rho_\nu} < \gamma_{\rho_\nu+1} < \ldots,$$
$$\gamma_\nu \leqq \gamma_{\rho_\nu}.$$

By theorem C of § 15, there would be a definite number δ of the second number-class, namely,

$$\delta = \operatorname*{Lim}_\nu \gamma_{\rho_\nu},$$

which is greater than all numbers γ_{ρ_ν}. Consequently we would have

$$\delta > \gamma_\nu$$

for every ν. But $\{\gamma_\nu\}$ contains *all* numbers of the second number-class, and consequently also the number δ; thus we would have, for a definite ν_0,

$$\delta = \gamma_{\nu_0},$$

which equation is inconsistent with the relation $\delta > \gamma_{\nu_0}$. The supposition $\overline{\overline{\{\alpha\}}} = \aleph_0$ consequently leads to a contradiction.

E. Any totality $\{\beta\}$ of different numbers β of the second number-class has, if it is infinite, either the cardinal number $\aleph_0$ or the cardinal number $\overline{\overline{\{\alpha\}}}$ of the second number-class.

Proof.—The aggregate $\{\beta\}$, when arranged in its order of magnitude, is, since it is a part of the well-ordered aggregate $\{\alpha\}$, by theorem 0 of § 13, similar either to a segment A_{α_0}, which is the totality of all numbers of the same number-class which are less than α_0, arranged in their order of magnitude, or to the totality $\{\alpha\}$ itself. As was shown in the proof of theorem A, we have

$$\overline{A}_{\alpha_0} = \alpha_0 - \omega.$$

Thus we have either $\{\overline{\beta}\} = \alpha_0 - \omega$ or $\{\overline{\beta}\} = \{\overline{\alpha}\}$, and consequently $\{\overline{\overline{\beta}}\}$ is either equal to $\overline{\overline{\alpha_0 - \omega}}$ or $\overline{\overline{\{\alpha\}}}$ But $\overline{\overline{\alpha_0 - \omega}}$ is either a finite cardinal number or is equal to $\aleph_0$ (theorem 1 of § 15). The first case is here excluded because $\{\beta\}$ is supposed to be an infinite aggregate. Thus the cardinal number $\{\overline{\overline{\beta}}\}$ is either equal to $\aleph_0$ or $\{\overline{\overline{\alpha}}\}$.

F. The power of the second number-class $\{\alpha\}$ is the second greatest transfinite cardinal number Aleph-one.

Proof.—There is no cardinal number a which is greater than $\aleph_0$ and less than $\{\overline{\overline{\alpha}}\}$. For if not, there would have to be, by § 2, an infinite part $\{\beta\}$ of $\{\alpha\}$ such that $\{\overline{\overline{\beta}}\}$ = a. But by the theorem E just proved, the part $\{\beta\}$ has either the cardinal number $\aleph_0$ or the cardinal number $\{\overline{\overline{\alpha}}\}$. Thus the cardinal number $\{\overline{\overline{\alpha}}\}$ is necessarily the cardinal number which immediately follows $\aleph_0$ in magnitude; we call this new cardinal number $\aleph_1$.

In the second number-class $Z(\aleph_0)$ we possess, consequently, the natural representative for the second greatest transfinite cardinal number Aleph-one.

§ 17
THE NUMBERS OF THE FORM
$$\omega^\mu v_0 + \omega^{\mu-1}v_1 + \cdots + v_\mu.$$

It is convenient to make ourselves familiar, in the first place, with those numbers of $Z(\aleph_0)$ which are whole algebraic functions of finite degree of ω. Every such number can be brought—and brought in only one way—into the form

$$\phi = \omega^\mu v_0 + \omega^{\mu-1}v_1 + \cdots + v_\mu, \tag{1}$$

where μ, v_0 are finite and different from zero, but $v_1, v_2, \ldots, v_\mu$ may be zero. This rests on the fact that

$$\omega^{\mu'}v' + \omega^\mu v = \omega^\mu v, \tag{2}$$

if $\mu' < \mu$ and $v > 0$, $v' > 0$. For, by (8) of § 14, we have

$$\omega^{\mu'}v' + \omega^\mu v = \omega^{\mu'}(v' + \omega^{\mu-\mu'}v),$$

and, by theorem E of § 15,

$$v' + \omega^{\mu-\mu'}v = \omega^{\mu-\mu'}v.$$

Thus, in an aggregate of the form

$$\cdots + \omega^{\mu'}v' + \omega^\mu v + \cdots,$$

all those terms which are followed towards the right by terms of higher degree in ω may be omitted. This method may be continued until the form given in (1) is reached. We will also emphasize that

$$\omega^\mu v + \omega^\mu v' = \omega^\mu(v + v'). \tag{3}$$

Compare, now, the number ϕ with a number ψ of the same kind:

$$\psi = \omega^\lambda \rho_0 + \omega^{\lambda-1}\rho_1 + \cdots + \rho_\lambda. \tag{4}$$

If μ and λ are different and, say, $\mu < \lambda$, we have by (2) $\phi + \psi = \psi$, and therefore $\phi < \psi$.

[230]

If $\mu = \lambda$, v_0, and ρ_0 are different, and, say, $v_0 < \rho_0$, we have by (2)
$$\phi + (\omega^\lambda (\rho_0 - v_0) + \omega^{\lambda-1}\rho_1 + \cdots + \rho_\mu) = \psi,$$
and therefore
$$\phi < \psi.$$
If, finally,
$$\mu = \lambda, \quad v_0 = \rho_0, \quad v_1 = \rho_1, \ldots v_{\sigma-1} = \rho_{\sigma-1}, \quad \sigma \gtreqless \mu,$$
but v_σ is different from ρ_σ and, say, $v_\sigma < \rho_\sigma$, we have by (2)
$$\phi + (\omega^{\lambda-\sigma}(\rho_\sigma - v_\sigma) + \omega^{\lambda-\sigma-1}\rho_{\sigma+1} + \cdots + \rho_\mu) = \psi,$$
and therefore again
$$\phi < \psi.$$

Thus, we see that only in the case of complete identity of the expressions ϕ and ψ can the numbers represented by them be equal.

The *addition* of ϕ and ψ leads to the following result:

(a) If $\mu < \lambda$, then, as we have remarked above,
$$\phi + \psi = \psi;$$
(b) If $\mu = \lambda$, then we have
$$\phi + \psi = \omega^\lambda(v_0 + \rho_0) + \omega^{\lambda-1}\rho_1 + \cdots + \rho_\lambda;$$
(c) If $\mu > \lambda$, we have
$$\phi + \psi = \omega^\mu v_0 + \omega^{\mu-1}v_1 + \cdots + \omega^{\lambda+1}v_{\mu-\lambda-1} + \omega^\lambda(v_{\mu-\lambda} + \rho_0)$$
$$+ \omega^{\lambda-1}\rho_1 + \cdots + \rho_\lambda.$$

In order to carry out the *multiplication* of ϕ and ψ, we remark that, if ρ is a finite number which is different from zero, we have the formula:
$$\phi\rho = \omega^\mu v_0 \rho + \omega^{\mu-1}v_1 + \cdots + v_\mu. \tag{5}$$
It easily results from the carrying out of the sum consisting of ρ terms $\phi + \phi + \cdots + \phi$. By means of the repeated application of the theorem G of § 15 we get, further, remembering the theorem F of § 15,
$$\phi\omega = \omega^{\mu+1}, \tag{6}$$
and consequently also
$$\phi\omega^\lambda = \omega^{\mu+\lambda}. \tag{7}$$
By the distributive law, numbered (8) of § 14, we have
$$\phi\psi = \phi\omega^\lambda\rho_0 + \phi\omega^{\lambda-1}\rho_1 + \cdots + \phi\omega\rho_{\lambda-1} + \phi\rho_\lambda.$$
Thus the formulæ (4), (5), and (7) give the following result:

(a) If $\rho_\lambda = 0$, we have
$$\phi\psi = \omega^{\mu+\lambda}\rho_0 + \omega^{\mu+\lambda-1}\rho_1 + \cdots + \omega^{\mu+1}\rho_{\lambda-1} = \omega^\mu\psi;$$

1023

(b) If ρ_λ is not equal to zero, we have

$$\phi\psi = \omega^{\mu+\lambda}\rho_0 + \omega^{\mu+\lambda-1}\rho_1 + \cdots + \omega^{\mu+1}\rho_{\lambda-1} + \omega^\mu r_0\rho_\lambda + \omega^{\mu-1}v_1 + \cdots + v_\mu.$$

[231]

We arrive at a remarkable resolution of the numbers ϕ in the following manner. Let

$$\phi = \omega^\mu\kappa_0 + \omega^{\mu_1}\kappa_1 + \cdots + \omega^{\mu_\tau}\kappa_\tau, \qquad (8)$$

where

$$\mu > \mu_1 > \mu_2 > \cdots > \mu_\tau \geqq 0$$

and $\kappa_0, \kappa_1, \ldots, \kappa_\tau$ are finite numbers which are different from zero. Then we have

$$\phi = (\omega^{\mu_1}\kappa_1 + \omega^{\mu_2}\kappa_2 + \cdots + \omega^{\mu_\tau}\kappa_\tau)(\omega^{\mu-\mu_1}\kappa+1).$$

By the repeated application of this formula we get

$$\phi = \omega^{\mu_\tau}\kappa_\tau(\omega^{\mu_{\tau-1}-\mu_\tau}\kappa_{\tau-1} + 1)(\omega^{\mu_{\tau-2}-\mu_{\tau-1}}\kappa_{\tau-2} + 1)\ldots(\omega^{\mu-\mu_1}\kappa_0 + 1).$$

But, now,

$$\omega^\lambda\kappa + 1 = (\omega^\lambda + 1)\kappa,$$

if κ is a finite number which is different from zero; so that:

$$\phi = \omega^{\mu_\tau}\kappa_\tau(\omega^{\mu_{\tau-1}-\mu_\tau} + 1)\kappa_{\tau-1}(\omega^{\mu_{\tau-2}-\mu_{\tau-1}} + 1)\kappa_{\tau-2}\ldots(\omega^{\mu-\mu_1} + 1)\kappa_0. \quad (9)$$

The factors $\omega^\lambda + 1$ which occur here are all irresoluble, and a number ϕ can be represented in this product-form in only one way. If $\mu_\tau = 0$, then ϕ is of the first kind, in all other cases it is of the second kind.

The apparent deviation of the formulæ of this paragraph from those which were given in *Math. Ann.*, vol. xxi (or *Grundlagen*), is merely a consequence of the different writing of the product of two numbers: we now put the multiplicand on the left and the multiplicator on the right, but then we put them in the contrary way.

§ 18
THE POWER[3] γ^α IN THE DOMAIN OF THE SECOND NUMBER-CLASS

Let ξ be a variable whose domain consists of the numbers of the first and second number-classes including zero. Let γ and δ be two constants belonging to the same domain, and let

$$\delta > 0, \quad \gamma > 1.$$

We can then assert the following theorem:

---

[3] Here obviously it is *Potenz* and not *Mächtigkeit*.

A. There is one wholly determined one-valued function $f(\xi)$ of the variable ξ such that:

(a)
$$f(0) = \delta.$$

(b) If ξ' and ξ'' are any two values of ξ, and if
$$\xi' < \xi'',$$
then
$$f(\xi') < f(\xi'').$$

[232]

(c) For every value of ξ we have
$$f(\xi + 1) = f(\xi)\gamma.$$

(d) If $\{\xi_\nu\}$ is any fundamental series, then $\{f(\xi_\nu)\}$ is one also, and if we have
$$\xi = \operatorname*{Lim}_\nu \xi_\nu,$$
then
$$f(\xi) = \operatorname*{Lim}_\nu f(\xi_\nu).$$

Proof.—By (a) and (c), we have
$$f(1) = \delta\gamma, \quad f(2) = \delta\gamma\gamma, \quad f(3) = \delta\gamma\gamma\gamma, \ldots,$$
and, because $\delta > 0$ and $\gamma > 1$, we have
$$f(1) < f(2) < f(3) < \cdots < f(\nu) < f(\nu + 1) < \cdots$$
Thus the function $f(\xi)$ is wholly determined for the domain $\xi < \omega$. Let us now suppose that the theorem is valid for all values of ξ which are less than α, where α is any number of the second number-class, then it is also valid for $\xi \leqq \alpha$. For if α is of the first kind, we have from (c):
$$f(\alpha) = f(\alpha_{-1})\gamma > f(\alpha_{-1});$$
so that the conditions (b), (c), and (d) are satisfied for $\xi \leqq \alpha$. But if α is of the second kind and $\{\alpha_\nu\}$ is a fundamental series such that $\operatorname*{Lim}_\nu \alpha_\nu = \alpha$, then it follows from (b) that also $\{f(\alpha_\nu)\}$ is a fundamental series, and from (d) that $f(\alpha) = \operatorname*{Lim}_\nu f(\alpha_\nu)$. If we take another fundamental series $\{\alpha_\nu'\}$ such that $\operatorname*{Lim}_\nu \alpha_\nu' = \alpha$, then, because of (b), the two fundamental series $\{f(\alpha_\nu)\}$ and $\{f(\alpha_\nu')\}$ are coherent, and thus also $f(\alpha) = \operatorname*{Lim}_\nu f(\alpha_\nu')$. The value of $f(\alpha)$ is, consequently, uniquely determined in this case also.

If α' is any number less than α, we easily convince ourselves that $f(\alpha') < f(\alpha)$. The conditions (b), (c), and (d) are also satisfied for $\xi \gtreqless \alpha$. Hence follows the validity of the theorem *for all values* of ξ. For if there were exceptional values of ξ for which it did not hold, then, by theorem B of § 16, one of them, which we will call α, would have to be the least. Then the theorem would be valid for $\xi < \alpha$, but not for $\xi \leqq \alpha$, and this would be in contradiction with what we have proved. Thus there is for the whole domain of ξ one and only one function $f(\xi)$ which satisfies the conditions (a) to (d).

If we attribute to the constant δ the value 1 and then denote the function $f(\xi)$ by

$$\gamma^\xi,$$

we can formulate the following theorem:

B. If γ is any constant greater than 1 which belongs to the first or second number-class, there is a wholly definite function γ^ξ of ξ such that:

(a) $\gamma^0 = 1$;

(b) If $\xi' < \xi''$ then $\gamma^{\xi'} < \gamma^{\xi''}$;

(c) For every value of ξ we have $\gamma^{\xi+1} = \gamma^\xi\gamma$;

(d) If $\{\xi_v\}$ is a fundamental series, then $\{\gamma^{\xi_v}\}$ is such a series, and we have, if $\xi = \underset{v}{\text{Lim}}\ \xi_v$, the equation

$$\gamma^\xi = \underset{v}{\text{Lim}}\ \gamma^{\xi_v}.$$

We can also assert the theorem:

C. If $f(\xi)$ is the function of ξ which is characterized in theorem A, we have

$$f(\xi) = \delta\gamma^\xi.$$

Proof.—If we pay attention to (24) of § 14, we easily convince ourselves that the function $\delta\gamma^\xi$ satisfies, not only the conditions (a), (b), and (c) of theorem A, but also the condition (d) of this theorem. On account of the uniqueness of the function $f(\xi)$, it must therefore be identical with $\delta\gamma^\xi$.

D. If α and β are any two numbers of the first or second number-class, including zero, we have

$$\gamma^{\alpha+\beta} = \gamma^\alpha\gamma^\beta.$$

Proof.—We consider the function $\phi(\xi) = \gamma^{\alpha+\xi}$. Paying attention to the fact that, by formula (23) of § 14,

$$\underset{v}{\text{Lim}}\ (\alpha + \xi_v) = \alpha + \underset{v}{\text{Lim}}\ \xi_v,$$

we recognize that $\phi(\xi)$ satisfies the following four conditions:

(a) $\phi(0) = \gamma^\alpha$;

(b) If $\xi' < \xi''$, then $\phi(\xi') < \phi(\xi'')$;

(c) For every value of ξ we have $\phi(\xi + 1) = \phi(\xi)\gamma$;

(d) If $\{\xi_v\}$ is a fundamental series such that $\underset{v}{\text{Lim}}\ \xi_v = \xi$, we have

$$\phi(\xi) = \underset{v}{\text{Lim}}\ \phi(\xi_v).$$

By theorem C we have, when we put $\delta = \gamma^\alpha$,

$$\phi(\xi) = \gamma^\alpha\gamma^\xi.$$

If we put $\xi = \beta$ in this, we have
$$\gamma^{\alpha+\beta} = \gamma^{\alpha}\gamma^{\beta}.$$

E. If α and β are any two numbers of the first or second number-class, including zero, we have
$$\gamma^{\alpha\beta} = (\gamma^{\alpha})^{\beta}.$$

[234]

Proof.—Let us consider the function $\psi(\xi) = \gamma^{\alpha\xi}$ and remark that, by (24) of § 14, we always have $\operatorname*{Lim}_{v} \alpha\xi_{v} = \alpha \operatorname*{Lim}_{v} \xi_{v}$, then we can, by theorem D, assert the following:

(*a*) $\psi(0) = 1$;

(*b*) If $\xi' < \xi''$, then $\psi(\xi') < \psi(\xi'')$;

(*c*) For every values of ξ we have $\psi(\xi + 1) = \psi(\xi)\gamma^{\alpha}$;

(*d*) If $\{\xi_{v}\}$ is a fundamental series, then $\{\psi(\xi_{v})\}$ is also such a series, and we have, if $\xi = \operatorname*{Lim}_{v} \xi_{v}$, the equation $\psi(\xi) = \operatorname*{Lim}_{v} \psi(\xi_{v})$.

Thus, by theorem C, if we substitute in it 1 for δ and γ^{α} for γ,
$$\psi(\xi) = (\gamma^{\alpha})^{\xi}.$$

On the *magnitude* of γ^{ξ} in comparison with ξ we can assert the following theorem:

F. If $\gamma > 1$, we have, for every value of ξ,
$$\gamma^{\xi} \geqq \xi.$$

Proof.—In the cases $\xi = 0$ and $\xi = 1$ the theorem is immediately evident. We now show that, if it holds for all values of ξ which are smaller than a given number $\alpha > 1$, it also holds for $\xi = \alpha$.

If α is of the first kind, we have, by supposition,
$$\alpha_{-1} \leqq \gamma^{\alpha-1},$$

and consequently
$$\alpha_{-1}\gamma \gtreqless \gamma^{\alpha-1}\gamma = \gamma^{\alpha}.$$

Hence
$$\gamma^{\alpha} \geqq \alpha_{-1} + \alpha_{-1}(\gamma - 1).$$

Since both α_{-1} and $\gamma - 1$ are at least equal to 1, and $\alpha_{-1} + 1 = \alpha$, we have
$$\gamma^{\alpha} \geqq \alpha.$$

If, on the other hand, α is of the second kind and
$$\alpha = \operatorname*{Lim}_{v} \alpha_{v},$$

then, because $\alpha_{v} < \alpha$, we have by supposition
$$\alpha_{v} \leqq \gamma^{\alpha_{v}}.$$

Consequently
$$\operatorname*{Lim}_{v} \alpha_{v} \leqq \operatorname*{Lim}_{v} \gamma^{\alpha_{v}},$$

that is to say,

$$\alpha \leqq \gamma^{\alpha}.$$

If, now, there were values of ξ for which

$$\xi > \gamma^{\xi},$$

one of them, by theorem B of § 16, would have to be the least. If this number is denoted by α, we would have, for $\xi < \alpha$,

[235]

$$\xi \leqq \gamma^{\xi};$$

but

$$\alpha > \gamma^{\alpha},$$

which contradicts what we have proved above. Thus we have for all values of ξ

$$\gamma^{\xi} \geqq \xi.$$

§ 19
THE NORMAL FORM OF THE NUMBERS OF THE SECOND NUMBER-CLASS

Let α be any number of the second number-class. The power ω^{ξ} will be, for sufficiently great values of ξ, greater than α. By theorem F of § 18, this is always the case for $\xi > \alpha$; but in general it will also happen for smaller values of ξ.

By theorem B of § 16, there must be, among the values of ξ for which

$$\omega^{\xi} > \alpha,$$

one which is the least. We will denote it by β, and we easily convince ourselves that it cannot be a number of the second kind. If, indeed, we had

$$\beta = \operatorname*{Lim}_{\nu} \beta_{\nu},$$

we would have, since $\beta_{\nu} < \beta$,

$$\omega^{\beta_{\nu}} \lessgtr \alpha,$$

and consequently

$$\operatorname*{Lim}_{\nu} \omega^{\beta_{\nu}} \lessgtr \alpha.$$

Thus we would have

$$\omega^{\beta} \lessgtr \alpha,$$

whereas we have

$$\omega^{\beta} > \alpha.$$

Therefore β is of the first kind. We denote β_{-1} by α_0, so that $\beta = \alpha_0 + 1$, and consequently can assert that there is a wholly determined number α_0 of the first or second number-class which satisfies the two conditions:

$$\omega^0 \leqq \alpha, \quad \omega^{\alpha_0}\omega > \alpha. \tag{1}$$

From the second condition we conclude that

$$\omega^{\alpha_0}v \leqq \alpha$$

does not hold for all finite values of v, for if it did we would have

$$\operatorname*{Lim}_{v} \omega^{\alpha_0}v = \omega^{\alpha_0}\omega \leqq \alpha.$$

The least finite number v for which

$$\omega^{\alpha_0}v > \alpha$$

will be denoted by $\kappa_0 + 1$. Because of (1), we have $\kappa_0 > 0$.

[236]

There is, therefore, a wholly determined number κ_0 of the first number-class such that

$$\omega^{\alpha_0}\kappa_0 \leqq \alpha, \quad \omega^{\alpha_0}(\kappa_0 + 1) > \alpha. \tag{2}$$

If we put $\alpha - \omega^{\alpha_0}\kappa_0 = \alpha'$, we have

$$\alpha = \omega^{\alpha_0}\kappa_0 + \alpha' \tag{3}$$

and

$$0 \leqq \alpha' < \omega^{\alpha_0}, \quad 0 < \kappa_0 < \omega. \tag{4}$$

But α can be represented in the form (3) under the conditions (4) in only a single way. For from (3) and (4) follow inversely the conditions (2) and thence the conditions (1). But only the number $\alpha_0 = \beta_{-1}$ satisfies the conditions (1), and by the conditions (2) the finite number κ_0 is uniquely determined. From (1) and (4) follows, by paying attention to theorem F of § 18, that

$$\alpha' < \alpha, \quad \alpha_0 \leqq \alpha.$$

Thus we can assert the following theorem:

A. Every number α of the second number-class can be brought, and brought in only one way, into the form

$$\alpha = \omega^{\alpha_0}\kappa_0 + \alpha',$$

where

$$0 \leqq \alpha' < \omega^{\alpha_0}, \quad 0 < \kappa_0 < \omega,$$

and α' is always smaller than α, but α_0 is smaller than or equal to α.

If α' is a number of the second number-class, we can apply theorem A to it, and we have

$$\alpha' = \omega^{\alpha_1}\kappa_1 + \alpha'' \tag{5}$$

where

$$0 \leqq \alpha'' < \omega^{\alpha_1}, \quad 0 < \kappa_1 < \omega,$$

and

$$\alpha_1 < \alpha_0, \quad \alpha'' < \alpha'.$$

In general we get a further sequence of analogous equations:

$$\alpha'' = \omega^{\alpha_2}\kappa_2 + \alpha''', \qquad (6)$$

$$\alpha''' = \omega^{\alpha_3}\kappa_3 + \alpha^{iv}. \qquad (7)$$

. . . .

But this sequence cannot be infinite, but must necessarily break off. For the numbers α, α', α'', ... decrease in magnitude:

$$\alpha > \alpha' > \alpha'' > \alpha''' \ldots$$

If a series of decreasing transfinite numbers were infinite, then no term would be the least; and this is impossible by theorem B of § 16. Consequently we must have, for a certain finite numerical value τ,

$$\alpha^{(\tau+1)} = 0.$$

[237]

If we now connect the equations (3), (5), (6), and (7) with one another, we get the theorem:

B. Every number α of the second number-class can be represented, and represented in only one way, in the form

$$\alpha = \omega^{\alpha_0}\kappa_0 + \omega^{\alpha_1}\kappa_1 + \cdots + \omega^{\alpha_\tau}\kappa_\tau,$$

where α_0, α_1, ... α_τ are numbers of the first or second number-class, such that:

$$\alpha_0 > \alpha_1 > \alpha_2 > \cdots > \alpha_\tau \geqq 0$$

while κ_0, κ_1, ... κ_τ, $\tau + 1$ are numbers of the first number-class which are different from zero.

The form of numbers of the second number-class which is here shown will be called their "normal form"; α_0 is called the "degree" and α_τ the "exponent" of α. For $\tau = 0$, degree and exponent are equal to one another.

According as the exponent α_τ is equal to or greater than zero, α is a number of the first or second kind.

Let us take another number β in the normal form:

$$\beta = \omega^{\beta_0}\lambda_0 + \omega^{\beta_1}\lambda_1 + \cdots + \omega^{\beta_\sigma}\lambda_\sigma. \qquad (8)$$

The formulæ:

$$\omega^{\alpha'}\kappa' + \omega^{\alpha'}\kappa = \omega^{\alpha'}(\kappa' + \kappa), \qquad (9)$$

$$\omega^{\alpha'}\kappa' + \omega^{\alpha''}\kappa'' = \omega^{\alpha''}\kappa'', \quad \alpha' < \alpha'', \qquad (10)$$

where κ, κ', κ'' here denote finite numbers, serve both for the comparison of α with β and for the carrying out of their sum and difference. These are generalizations of the formulæ (2) and (3) of § 17.

For the formation of the product $\alpha\beta$, the following formulæ come into consideration:

$$\alpha\lambda = \omega^{\alpha_0}\kappa_0\lambda + \omega^{\alpha_1}\kappa_1 + \cdots + \omega^{\alpha_\tau}\kappa_\tau, \quad 0 < \lambda < \omega; \tag{11}$$

$$\alpha\omega = \omega^{\alpha_0+1}; \tag{12}$$

$$\alpha\omega^{\beta'} = \omega^{\alpha_0+\beta'}, \quad \beta' > 0. \tag{13}$$

The exponentiation α^β can be easily carried out on the basis of the following formulæ:

$$\alpha^\lambda = \omega^{\alpha_0\lambda}\kappa_0 + \cdots, \quad 0 < \lambda < \omega. \tag{14}$$

The terms not written on the right have a lower degree than the first. Hence follows readily that the fundamental series $\{\alpha^\lambda\}$ and $\{\omega^{\alpha_0\lambda}\}$ are coherent, so that

$$\alpha^\omega = \omega^{\alpha_0\omega}, \quad \alpha_0 > 0. \tag{15}$$

Thus, in consequence of theorem E of § 18, we have:

$$\alpha^{\omega^{\beta'}} = \omega^{\alpha_0\omega^{\beta'}}, \quad \alpha_0 > 0, \quad \beta' > 0. \tag{16}$$

By the help of these formulæ we can prove the following theorems:

[238]

C. If the first terms $\omega^{\alpha_0}\kappa_0$, $\omega^{\beta_0}\lambda_0$ of the normal forms of the two numbers α and β are not equal, then α is less or greater than β according as $\omega^{\alpha_0}\kappa_0$ is less or greater than $\omega^{\beta_0}\lambda_0$. But if we have

$$\omega^{\alpha_0}\kappa_0 = \omega^{\beta_0}\lambda_0, \omega^{\alpha_1}\kappa_1 = \omega^{\beta_1}\lambda_1, \ldots, \omega^{\alpha_\rho}\kappa_\rho = \omega^{\beta_\rho}\lambda_\rho,$$

and if $\omega^{\alpha_{\rho+1}}\kappa_{\rho+1}$ is less or greater than $\omega^{\beta_{\rho+1}}\lambda_{\rho+1}$, then α is correspondingly less or greater than β.

D. If the degree α_0 of α is less than the degree β_0, of β, we have

$$\alpha + \beta = \beta.$$

If $\alpha_0 = \beta_0$, then

$$\alpha + \beta = \omega^{\beta_0}(\kappa_0 + \lambda_0) + \omega^{\beta_1}\lambda_1 + \cdots + \omega^{\beta_\sigma}\lambda_\sigma.$$

But if

$$\alpha_0 > \beta_0, \alpha_1 > \beta_0, \ldots, \alpha_\rho \geqq \beta_0, \alpha_{\rho+1} < \beta_0,$$

then

$$\alpha + \beta = \omega^{\alpha_0}\kappa_0 + \cdots + \omega^{\alpha_\rho}\kappa_\rho + \omega^{\beta_0}\lambda_0 + \omega^{\beta_1}\lambda_1 + \cdots + \omega^{\beta_\sigma}\lambda_\sigma.$$

E. If β is of the second kind ($\beta_\sigma > 0$), then

$$\alpha\beta = \omega^{\alpha_0+\beta_0}\lambda_0 + \omega^{\alpha_0+\beta_1}\lambda_1 + \cdots + \omega^{\alpha_0+\beta_\sigma}\lambda_\sigma = \omega^{\alpha_0}\beta;$$

But if β is of the first kind ($\beta_\sigma = 0$), then

$$\alpha\beta = \omega^{\alpha_0+\beta_0}\lambda_0 + \omega^{\alpha_0+\beta_1}\lambda_1 + \cdots + \omega^{\alpha_0+\beta_{\sigma-1}}\lambda_{\sigma-1} + \omega^{\alpha_0}\kappa_0\lambda_\sigma$$
$$+ \omega^{\alpha_1}\kappa_1 + \cdots + \omega^{\alpha_\tau}\kappa_\tau.$$

F. If β is of the second kind ($\beta_\sigma > 0$), then

$$\alpha^\beta = \omega^{\alpha_0}\beta.$$

But if β is of the first kind ($\beta_\sigma = 0$), and indeed $\beta = \beta' + \lambda_\sigma$, where β' is of the second kind, we have:

$$\alpha^\beta = \omega^{\alpha_0 \beta'} \alpha^{\lambda_\sigma}.$$

G. Every number α of the second number-class can be represented, in only one way, in the product-form:

$$\alpha = \omega^{\gamma_0} \kappa_\tau (\omega^{\gamma_1} + 1) \kappa_{\tau-1} (\omega^{\gamma_2} + 1) \kappa_{\tau-2} \ldots (\omega^{\gamma_\tau} + 1) \kappa_0,$$

and we have

$$\gamma_0 = \alpha_\tau, \quad \gamma_1 = \alpha_{\tau-1} - \alpha_\tau, \quad \gamma_2 = \alpha_{\tau-2} - \alpha_{\tau-1}, \ldots, \quad \gamma_\tau = \alpha_0 - \alpha_1,$$

whilst $\kappa_0, \kappa_1, \ldots \kappa_\tau$ have the same denotation as in the normal form. The factors $\omega^\gamma + 1$ are all irresoluble.

H. Every number α of the second kind which belongs to the second number-class can be represented, and represented in only one way, in the form

$$\alpha = \omega^{\gamma_0} \alpha',$$

where $\gamma_0 > 0$ and α' is a number of the first kind which belongs to the first or second number-class.

[239]

I. In order that two numbers α and β of the second number-class should satisfy the relation

$$\alpha + \beta = \beta + \alpha,$$

it is necessary and sufficient that they should have the form

$$\alpha = \gamma\mu, \quad \beta = \gamma\upsilon,$$

where μ and υ are numbers of the first number-class.

K. In order that two numbers α and β of the second number-class, which are both of the first kind, should satisfy the relation

$$\alpha\beta = \beta\alpha,$$

it is necessary and sufficient that they should have the form

$$\alpha = \gamma^\mu, \quad \beta = \gamma^\upsilon,$$

where μ and υ are numbers of the first number-class.

In order to exemplify the extent of the *normal form* dealt with and the *product-form* immediately connected with it, of the numbers of the second number-class, the proofs, which are founded on them, of-the two last theorems, I and K, may here follow.

From the supposition

$$\alpha + \beta = \beta + \alpha$$

we first conclude that the degree α_0 of α must be equal to the degree β_0 of β. For if, say, $\alpha_0 < \beta_0$, we would have, by theorem D,

$$\alpha + \beta = \beta,$$

and consequently

$$\beta + \alpha = \beta,$$

which is not possible, since, by (2) of § 14,

$$\beta + \alpha > \beta.$$

Thus we may put

$$\alpha = \omega^{\alpha_0}\mu + \alpha', \quad \beta = \omega^{\alpha_0}\nu + \beta',$$

where the degrees of the numbers α' and β' are less than α_0, and μ and ν are infinite numbers which are different from zero. Now, by theorem D we have

$$\alpha + \beta = \omega^{\alpha_0}(\mu + \nu) + \beta', \quad \beta + \alpha = \omega^{\alpha_0}(\mu + \nu) + \alpha',$$

and consequently

$$\omega^{\alpha_0}(\mu + \nu) + \beta' = \omega^{\alpha_0}(\mu + \nu) + \alpha'.$$

By theorem D of § 14 we have consequently

$$\beta' = \alpha'.$$

Thus we have

$$\alpha = \omega^{\alpha_0}\mu + \alpha', \quad \beta = \omega^{\alpha_0}\nu + \alpha',$$

[240]

and if we put

$$\omega^{\alpha_0} + \alpha' = \gamma$$

we have, by (11):

$$\alpha = \gamma\mu, \quad \beta = \gamma\nu.$$

Let us suppose, on the other hand, that α and β are two numbers which belong to the second number-class, are of the first kind, and satisfy the condition

$$\alpha\beta = \beta\alpha,$$

and we suppose that

$$\alpha > \beta.$$

We will imagine both numbers, by theorem G, in their product-form, and let

$$\alpha = \delta\alpha', \quad \beta = \delta\beta',$$

where α' and β' are without a common factor (besides 1) at the left end. We have then

$$\alpha' > \beta',$$

and

$$\alpha'\delta\beta' = \beta'\delta\alpha'.$$

All the numbers which occur here and farther on are of the first kind, because this was supposed of α and β.

The last equation, when we refer to theorem G, shows that α' and β' cannot be both transfinite, because, in this case, there would be a common factor at the left end. Neither can they be both finite; for then δ would be transfinite, and, if κ is the finite factor at the left end of δ, we would have

$$\alpha'\kappa = \beta'\kappa,$$

and thus

$$\alpha' = \beta'.$$

Thus there remains only the possibility that

$$\alpha' > \omega, \quad \beta' < \omega.$$

But the finite number β' must be 1:

$$\beta' = 1,$$

because otherwise it would be contained as part in the finite factor at the left end of α'.

We arrive at the result that $\beta = \delta$, consequently

$$\alpha = \beta\alpha',$$

where α' is a number belonging to the second number-class, which is of the first kind, and must be less than α:

$$\alpha' < \alpha.$$

Between α' and β the relation

$$\alpha'\beta = \beta\alpha'$$

subsists.

[241]

Consequently if also $\alpha' > \beta$, we conclude in the same way the existence of a transfinite number of the first kind α'' which is less than α' and such that

$$\alpha' = \beta\alpha'', \quad \alpha''\beta = \beta\alpha''.$$

If also α'' is greater than β, there is such a number α''' less than α'', such that

$$\alpha'' = \beta\alpha''', \quad \alpha'''\beta = \beta\alpha''',$$

and so on. The series of decreasing numbers, $\alpha, \alpha', \alpha'', \alpha''', \ldots$, must, by theorem B of § 16, break off. Thus, for a definite finite index ρ_0, we must have

$$\alpha^{(\rho_0)} \leqq \beta.$$

If

$$\alpha^{(\rho_0)} = \beta,$$

we have

$$\alpha = \beta^{\rho_0+1}, \quad \beta = \beta;$$

the theorem K would then be proved, and we would have

$$\gamma = \beta, \quad \mu = \rho_0 + 1, \quad \nu = 1.$$

But if

$$\alpha^{(\rho_0)} < \beta,$$

then we put

$$\alpha^{(\rho_0)} = \beta_1,$$

and have
$$\alpha = \beta^{\rho_0}\beta_1, \quad \beta\beta_1 = \beta_1\beta, \quad \beta_1 < \beta.$$
Thus there is also a finite number ρ_1 such that
$$\beta = \beta_1^{\rho_1}\beta_2, \quad \beta_1\beta_2 = \beta_2\beta_1, \quad \beta_2 < \beta_1.$$
In general, we have analogously:
$$\beta_1 = \beta_2^{\rho_2}\beta_3, \quad \beta_2\beta_3 = \beta_3\beta_2, \quad \beta_3 < \beta_2,$$
and so on. The series of decreasing numbers $\beta_1, \beta_2, \beta_3, \ldots$ also must, by theorem B of § 16, break off. Thus there exists a finite number κ such that
$$\beta_{\kappa-1} = \beta_\kappa^{\rho_\kappa}.$$
If we put
$$\beta_\kappa = \gamma,$$
then
$$\alpha = \gamma^\mu, \quad \beta = \gamma^\nu,$$
where μ and ν are numerator and denominator of the continued fraction.
$$\frac{\mu}{\nu} = \rho_0 + \cfrac{1}{\rho_1 + \cfrac{1}{\ddots + \cfrac{1}{\rho_\kappa}}}$$

[242]

§ 20
THE ε-NUMBERS OF THE SECOND NUMBER-CLASS

The degree α_0 of a number α is, as is immediately evident from the normal form:
$$\alpha = \omega^{\alpha_0}\kappa_0 + \omega^{\alpha_1}\kappa_1 + \cdots, \quad \alpha_0 > \alpha_1 > \cdots, \quad 0 < \kappa_\nu < \omega, \tag{1}$$
when we pay attention to theorem F of § 18, never greater than α; but it is a question whether there are not numbers for which $\alpha_0 = \alpha$. In such a case the normal form of α would reduce to the first term, and this term would be equal to ω^α, that is to say, α would be a root of the equation
$$\omega^\xi = \xi. \tag{2}$$
On the other hand, every root α of this equation would have the normal form ω^α; its degree would be equal to itself.

The numbers of the second number-class which are equal to their degree coincide, therefore, with the roots of the equation (2). It is our problem to determine these roots in their totality. To distinguish them from all other numbers we will call them the "ε-numbers of the second number-class." That there *are* such ε-numbers results from the following theorem:

A. If γ is any number of the first or second number-class which does not satisfy the equation (2), it determines a fundamental series $\{\gamma\}$ by means of the equations
$$\gamma_1 = \omega^\gamma, \quad \gamma_2 = \omega^{\gamma_1}, \ldots, \quad \gamma_\nu = \omega^{\gamma_{\nu-1}}, \ldots$$

1035

The limit $\operatorname*{Lim}_{v} \gamma_{v} = E(\gamma)$ of this fundamental series is always an ε-number.

Proof.—Since γ is not an ε-number, we have $\omega^{\gamma} > \gamma$, that is to say, $\gamma_{1} > \gamma$. Thus, by theorem B of § 18, we have also $\omega^{\gamma_1} > \omega^{\gamma}$, that is to say, $\gamma_{2} > \gamma_{1}$; and in the same way follows that $\gamma_{3} > \gamma_{2}$, and so on. The series $\{\gamma_{v}\}$ is thus a fundamental series. We denote its limit, which is a function of γ, by $E(\gamma)$ and have:

$$\omega^{E}(\gamma) = \operatorname*{Lim}_{v} \omega^{\gamma} = \operatorname*{Lim}_{v} \gamma_{v+1} = E(\gamma).$$

Consequently $E(\gamma)$ is an ε-number.

B. The number $\varepsilon_{0} = E(1) = \operatorname*{Lim}_{v} \omega_{v}$, where

$$\omega_{1} = \omega, \quad \omega_{2} = \omega^{\omega_1}, \quad \omega_{3} = \omega^{\omega_2}, \ldots, \quad \omega_{v} = \omega^{\omega_{v-1}}, \ldots,$$

is the least of all the ε-numbers.

[243]

Proof.—Let ε' be any ε-number, so that

$$\omega^{\varepsilon'} = \varepsilon'.$$

Since $\varepsilon' > \omega$, we have $\omega^{\varepsilon'} > \omega^{\omega}$, that is to say, $\varepsilon' > \omega_{1}$. Similarly $\omega^{\varepsilon'} > \omega^{\omega_1}$, that is to say, $\varepsilon' > \omega_{2}$, and so on. We have in general

$$\varepsilon' > \omega_{v},$$

and consequently

$$\varepsilon' \geqq \operatorname*{Lim}_{v} \omega_{v},$$

that is to say,

$$\varepsilon' \geqq \varepsilon_{0}.$$

Thus $\varepsilon_{0} = E(1)$ is the least of all ε-numbers.

C. If ε' is any ε-number, ε'' is the next greater ε-number, and γ is any number which lies between them:

$$\varepsilon' < \gamma < \varepsilon'',$$

then $E(\gamma) = \varepsilon''$.

Proof.—From

$$\varepsilon' < \gamma < \varepsilon''$$

follows

$$\omega^{\varepsilon'} < \omega^{\gamma} < \omega^{\varepsilon''},$$

that is to say,

$$\varepsilon' < \gamma_{1} < \varepsilon''.$$

Similarly we conclude

$$\varepsilon' < \gamma_{2} < \varepsilon'',$$

and so on. We have, in general,

$$\varepsilon' < \gamma_{v} < \varepsilon'',$$

and thus
$$\varepsilon' < E(\gamma) \leqq \varepsilon''.$$
By theorem A, $E(\gamma)$ is an ε-number. Since ε'' is the ε-number which follows ε' next in order of magnitude, $E(\gamma)$ cannot be less than ε'', and thus we must have
$$E(\gamma) = \varepsilon''.$$
Since $\varepsilon' + 1$ is not an ε-number, simply because all ε-numbers, as follows from the equation of definition $\xi = \omega^\xi$, are of the second kind, $\varepsilon' + 1$ is certainly less than ε'', and thus we have the following theorem:

D. If ε' is any ε-number, then $E(\varepsilon' + 1)$ is the next greater ε-number.

To the least ε-number, ε_0, follows, then, the next greater one:
$$\varepsilon_1 = E(\varepsilon_0 + 1),$$

[244]

to this the next greater number:
$$\varepsilon_2 = E(\varepsilon_1 + 1),$$
and so on. Quite generally, we have for the $(v + 1)$th ε-number in order of magnitude the formula of recursion
$$\varepsilon_v = E(\varepsilon_{v-1} + 1). \tag{3}$$
But that the infinite series
$$\varepsilon_0, \varepsilon_1, \ldots \varepsilon_v, \ldots$$
by no means embraces the totality of ε-numbers results from the following theorem:

E. If $\varepsilon, \varepsilon', \varepsilon'', \ldots$ is any infinite series of e-numbers such that
$$\varepsilon < \varepsilon' < \varepsilon'' \ldots \varepsilon^{(v)} < \varepsilon^{(v+1)} < \ldots,$$
then $\operatorname{Lim}_v \operatorname{Lim} \varepsilon^{(v)}$ is an ε-number, and, in fact, the ε-number which follows next in order of magnitude to all the numbers $\varepsilon^{(v)}$.

Proof.—
$$\omega^{\operatorname{Lim}_v \omega(v)} = \operatorname{Lim}_v \omega^{\varepsilon^{(v)}} = \operatorname{Lim}_v \varepsilon^{(v)}.$$

That $\operatorname{Lim}_v \varepsilon^{(v)}$ is the ε-number which follows next in order of magnitude to all the numbers $\varepsilon^{(v)}$ results from the fact that $\operatorname{Lim}_v \varepsilon^{(v)}$ is the number of the second number-class which follows next in order of magnitude to all the numbers $\varepsilon^{(v)}$.

F. The totality of ε-numbers of the second number-class forms, when arranged in order of magnitude, a well-ordered aggregate of the type Ω of the second number-class in its order of magnitude, and has thus the power Aleph-one.

Proof.—The totality of ε-numbers of the second number-class, when arranged in their order of magnitude, forms, by theorem C of § 16, a well-ordered aggregate:

$$\varepsilon_0, \varepsilon_1, \ldots, \varepsilon_\nu, \ldots \varepsilon_{\omega+1}, \ldots \varepsilon_{\alpha'} \ldots, \tag{4}$$

whose law of formation is expressed in the theorems D and E. Now, if the index α' did not successively take all the numerical values of the second number-class, there would be a least number α which it did not reach. But this would contradict the theorem D, if α were of the first kind, and theorem E, if α were of the second kind. Thus α' takes all numerical values of the second number-class.

If we denote the type of the second number-class by Ω, the type of (4) is

$$\omega + \Omega = \omega + \omega^2 + (\Omega - \omega^2).$$

[245]

But since $\omega + \omega^2 = \omega^2$, we have

$$\omega + \Omega = \Omega;$$

and consequently

$$\overline{\omega + \Omega} = \overline{\Omega} = \aleph_1.$$

G. If ε is any ε-number and α is any number of the first or second number-class which is less than ε:

$$\alpha < \varepsilon,$$

then ε satisfies the three equations:

$$\alpha + \varepsilon = \varepsilon, \quad \alpha\varepsilon = \varepsilon \quad \alpha^\varepsilon = \varepsilon.$$

Proof.—If α_0 is the degree of α, we have $\alpha_0 \leqq \alpha$, and consequently, because of $\alpha < \varepsilon$, we also have $\alpha_0 < \varepsilon$. But the degree of $\varepsilon = \omega^\varepsilon$ is ε; thus α has a less degree than ε. Consequently, by theorem D of § 19,

$$\alpha + \varepsilon = \varepsilon,$$

and thus

$$\alpha_0 + \varepsilon = \varepsilon.$$

On the other hand, we have, by formula (13) of § 19,

$$\alpha\varepsilon = \alpha\omega^\varepsilon = \omega^{\alpha_0+\varepsilon} = \omega^\varepsilon = \varepsilon,$$

and thus

$$\alpha_0\varepsilon = \varepsilon.$$

Finally, paying attention, to the formula (16) of § 19,

$$\alpha^\varepsilon = \alpha\omega^\varepsilon = \omega^{\alpha_0\omega^\varepsilon} = \omega^{\alpha_0\varepsilon} = \omega^\varepsilon = \varepsilon.$$

H. If α is any number of the second number-class, the equation

$$\alpha^\xi = \xi$$

has no other roots than the ε-numbers which are greater than α.

Proof.—Let β be a root of the equation

$$\alpha^\xi = \xi,$$

so that
$$\alpha^\beta = \beta.$$
Then, in the first place, from this formula follows that
$$\beta > \alpha.$$
On the other hand, β must be of the second kind, since, if not, we would have
$$\alpha^\beta > \beta.$$
Thus we have, by theorem F of § 19,
$$\alpha^\beta = {}^{\alpha_0\beta}.$$
and consequently
$$\omega^{\alpha_0\beta} = \beta.$$

[246]

By theorem F of § 19, we have
$$\omega^{\alpha_0\beta} \geqq \alpha_0\beta,$$
and thus
$$\beta \geqq \alpha_0\beta.$$
But β cannot be greater than $\alpha_0\beta$; consequently
$$\alpha_0\beta = \beta,$$
and thus
$$\omega^\beta = \beta.$$
Therefore β is an ε-number which is greater than α.

HALLE, *March* 1897.

Henri Lebesgue

(1875–1941)

HIS LIFE AND WORK

I remember once hearing a story about an historian of Babylonian astronomy whose colleagues asked him for help interpreting an ancient astronomical manuscript. He stroked his beard for a while and pondered the odd symbols on the manuscript his colleagues had placed in front of him. Then suddenly, he began to explain the manuscript to his colleagues as he rotated it 90° from portrait mode to landscape mode. Henri Lebesgue made just such a change in perspective for the theory of the integral.

On June 28, 1875, Henri Lebesgue was born into a middle-class family in the town of Beauvais some fifty miles north of Pairs. Henri grew up in a home that encouraged intellectual pursuit. Although his father was a typesetter and his mother an elementary school teacher, they had a remarkable library that allowed young Henri to read widely. The local schools soon recognized his brilliance and arranged for local philanthropies to finance his education when his father died of tuberculosis.

Lebesgue had an outstanding academic record, first at the local Collège de Beauvais, and then in Paris at the Lycée Saint Louis, and then at the Lycée Louis-le-Grand. In 1894 at the age of nineteen, he entered the prestigious École Normal Supérieure in Paris. While at the École, Lebesgue often indulged himself in his mathematical studies, neglecting other subjects that interested him less. His diary revealed that he only managed to pass a chemistry examination by speaking in hushed tones to his deaf examiner and not writing anything on the blackboard.

Lebesgue graduated from the École Normal Supérieure in 1897. For the two years following his graduation, Lebesgue worked in the École's library and began his attack on a new theory of measure and the integral as he pursued his graduate studies at the Sorbonne. Lebesgue secured a teaching position at the Lycée Central in Nancy near the German border. While in Nancy, Lebesgue finished the development of the ideas that blossomed into the dissertation for which he received a doctorate from the Sorbonne in 1902. It was immediately published by the *Annali di Matematica* of Milan and hailed as a revolutionary breakthrough.

Prior to Lebesgue, mathematicians believed that Riemann had developed the theory of the integral as completely as possible. In previous selections, we have seen that Cauchy and then Riemann had based a theory of the integral on the process of partitioning the x axis and taking a value of the function $f(x)$ somewhere in each of the segments of the partition.

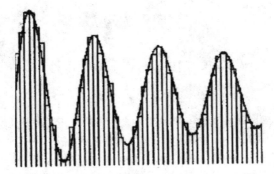

Lebesgue's great insight was to partition the y axis, rather than the x axis, in much the same way that Cauchy and Riemann had partitioned the x axis.

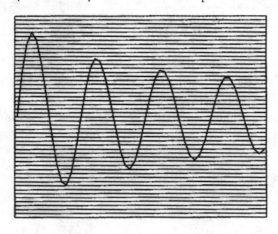

Having partitioned the y axis, Lebesgue then considered each element of the partition and proceeded to take **_the measure_** of that of the portion of the x axis that the function being integrated mapped into that element of the partition.

Here is an example. Consider the interval (y_n, y_{n+1}), one element of a partition of the y axis.

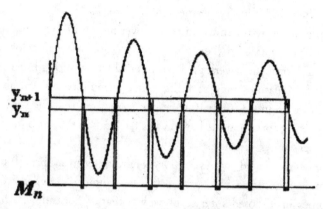

In this example, the set of points on the x axis that take a value in the interval (y_n, y_{n+1}) forms the set M_n. In order to determine the contribution of the interval (y_n, y_{n+1}) to the Lebesgue integral being evaluated, Lebesgue multiplies the measure of the set M_n and any value in the interval (y_n, y_{n+1}). Using Weierstrass's rigorous epsilon-delta method, Lebesgue says that if a limit exists as the size of the "widest" element of the partition of the y axis approaches 0, then that limit is the value of the Lebesgue interval being evaluated.

By defining the integral in this manner, Lebesgue had reduced the theory of the integral to the theory of measure. If we consider the problem from a geometrical perspective, we see that Lebesgue had reduced the problem of determining the area of a two-dimensional object to the problem of determining the measure of a set of points embedded in the one-dimensional real line. In Lebesgue's seminal paper, a large portion of which is presented here, he first addresses the topic of measuring a set of points embedded in the one-dimensional real line and then progresses to his theory of the integral.

Lebesgue begins by enunciating properties a measure must have to satisfy our intuitions.

• The measure of the interval $[a, b]$ (i.e., the set of real numbers x such that $a \leq x \leq b$) is simply the value $b - a$.

• If the sets of real numbers E_n are pairwise disjoint (i.e., no two sets have any points in common), then the measure of the union of the E_n is equal to the sum of the measures of the individual E_n.

• If a set of real numbers E has measure m, then the set of points formed by adding the constant value a to every real number in E results in a set (denoted by $E + a$) that also has measure m.

Before defining a set's measure, Lebesgue first defined its outer measure and its inner measure. If a set has an outer measure and an inner measure and if these values are identical, then that value is the measure of the set.

A set's outer measure is defined in terms coverings of intervals that form a superset of the set being measured (i.e., the set in question is a subset of the union of the intervals that form the covering.) Each interval in a covering has a measure, and the sum of these measures may be finite. Since a bounded set can always be covered by the entire bounded interval, Lebesgue recognized that the sum of the measures of the coverings of a bounded set always have a finite lower bound. This lower bound is the set's outer measure.

Lebesgue demonstrated that the rationals have outer measure of 0 by employing Cantor's demonstration that the rationals can be enumerated and enclosing the nth rational number in a closed interval of width, that is, measure ε^n. Summing the measures of all of these intervals results in ε which can be made as small as desired. Thus, the rationals have an outer measure of 0, which makes them a measurable set of measure 0.

Lebesgue defined the inner measure of a set E bounded by the interval $[a, b]$ in terms of the outer measure of the set of points contained in the interval $[a, b]$ less the set of points contained in the set E. To wit, the inner measure of a set E bounded by the interval $[a, b]$ is the measure of the interval $[a, b]$ minus the outer measure of set of points contained in the interval $[a, b]$ less the set of points contained in the set E. This is written as:

$$meas_i(E) = b - a - meas_e([a, b] - E),$$

where

$$meas_i \text{ is inner measure}$$

and

$$meas_e \text{ is outer measure.}$$

Using Lebesgue's definition, it becomes readily apparent that the set of irrational numbers in the unit interval $[0, 1]$ has inner measure 1 as a result of the set of rational numbers in the unit interval $[0, 1]$ having outer measure 0. An immediate consequence is that the set of irrational numbers in the unit interval $[0, 1]$ has measure 1, and the set of rational numbers in the unit interval $[0, 1]$ has measure 0.

With this radically different approach to the theory of the integral, a wide range of functions became integrable and fell within the embrace of mathematical analysis. The most obvious example of a function that is Lebesgue integrable but not Riemann integrable is $\chi(x)$, the characteristic function of the rationals in the reals. It takes the following values:

$$\chi(x) = 1 \text{ if } x \text{ is rational}$$
$$= 0 \text{ if } x \text{ is irrational}$$

This function is not Riemann integrable, because every element of a partition of the x axis, no matter how fine grained, will contain infinitely many rational numbers and infinitely many irrational numbers. Thus, the Riemann sums can never converge. On the other hand, the only elements of a partition of that matter for determining a Lebesgue integral are the elements including the values 0 and 1. As we have seen, the set of points on the x axis that $\chi(x)$ maps to 1 has measure 0. The measure of the irrationals in any interval considered is the width of the interval. Therefore, the Lebesgue integral of $\chi(x)$ over any interval $[a, b]$ is simply 0.

With the publication of his dissertation, Lebesgue quickly had his choice of academic appointments. In 1902 to 1903, he taught at the university at Rennes. A year later, he moved to the Collège de France in Paris where he remained for three years until he accepted a position at the University of Poitiers. In 1910, the mathematics faculty at the Sorbonne appointed him *maître de conferences* (lecture master). When France entered World War I, Lebesgue immediately offered his services to the Ministry of Defense which put him to work determining how frequently troop trains should run to the front lines. Following the war, the Sorbonne promoted Lebesgue to a full professorship, which he held for only three years until he accepted a chaired professorship at the Collège de France. Lebesgue remained in this position for the rest of his career.

Although his ideas revolutionized mathematical analysis, Lebesgue himself had very little personal influence in the mathematical community. Uninterested in politics inside or outside of the academy, Lebesgue seemed content to give his lectures, write his articles, and lead a solitary life. In spite of the prominence of his work, he supervised only one doctoral dissertation during his career. The French mathematical community had the good sense to recognize the importance of his contributions. Lebesgue received the Prix Houllevige in 1912, the Prix Poncelet in 1914, the Prix Santour in 1917, and the Prix Petit d'Ormoy in 1919. His 1922 election to the Académie des Sciences topped the list of honors he received.

Lebesgue lived such a modest life that very little is known of his family life. While at Rennes in 1903, he married Louise-Marguerite Valle, the sister of a fellow student at the École Normal Supérieure. They had a daughter named Suzanne and a son named Jacques before they divorced in 1916 after little more than a dozen years of marriage. Lebesgue lived alone for the rest of his life often suffering from ill health. Dispirited by the German occupation, he died on July 26, 1941, at the age of 66 following a short illness.

SELECTIONS FROM *INTEGRAL, LENGTH AND AREA**

INTRODUCTION

In this work, I attempt to give definitions as general and precise as possible of some of the numbers one looks at in Calculus: a definite integral, the length of a curve, and the area of a surface.

Mr. Jordan, in the second edition of his *Cours d'Analyse*, made an in-depth study of these numbers. Nonetheless, it seemed useful to me to take up that study again, and here is why. We know that there are derived functions that are not integrable when one adopts, as Jordan does, the definition of an integral given by Riemann; with the result that integration, as defined by Riemann, does not in all cases let us solve the fundamental problem of integral calculus:

How to find a function if we know its derivative.

It may then seem natural to look for another definition of an integral such that, even in the most general case, integration would be the inverse of differentiation.

Moreover, as Jordan notes, there is no definition for the area of a surface whose tangent planes do not vary continuously; and the statements one might be tempted to accept because they are analogous to the definition of the length of a curve cannot be accepted.[1] Thus, there is good reason to look for a definition of area, and also perhaps to modify the definition of length such that both definitions should be as closely analogous as possible.

In studying questions related to the theory of functions with real variables, it was frequently recognized that it would be useful if, to certain sets of points, we could attach numbers bearing certain properties of their segment lengths or polygonal areas. Different definitions were proposed for these numbers, known as set measures;[2] the one that was most often adopted is set out and studied in Mr. Jordan's book.

In the first chapter, along with Mr. Borel, I define the measure of a set by its basic properties. After supplementing and further refining the somewhat cursory indications given by Mr. Borel,[3] I show how this definition of measure relates to that of Mr. Jordan. The definition I adopt applies to spaces of many dimensions; from the idea of the measure of a set whose elements are points on a plane, one can deduce the idea of the area of a plane domain; if the elements are points in ordinary space, from that one can deduce the idea of volume, etc.

---

*Translated by Michael Ansaldi, with technical assistance by John Anders.

[1] See Schwarz, letter to Genocchi. This letter is reproduced in the lithographic edition of the *Cours professé à la Faculté des sciences* (Course given to the Faculty of Sciences) by Ch. Hermite, during the second semester of 1882. (Second printing, page 25.)—See also, Peano, *Atti della Accademia dei Lincei*, 1890.

[2] On these definitions, see Schœnflies, *Jahresbericht der deutschen Mathematiker-Vereinigung*, 1900.

[3] *Leçons sur la théorie des Fonctions.*

Once these preliminaries have been laid out, nothing stops us from defining the integral of a continuous function as the area of a plane domain; and this method in fact has the advantage of leading us to define the integral of a bounded discontinuous function as the measure of a certain set of points. This is the geometric definition I adopt in chapter II. On a side note, this definition can be replaced by an analytic definition, under which the integral then appears as the limit of a series of sums pretty much analogous to those Riemann considers in his definition. The functions to which this geometric definition applies I call *summable.*

I know of no function that is not summable, not being aware of whether any exist. Any function that can be defined using arithmetic operations and progression to a limit is summable. All Riemann-integrable functions are summable, and both definitions of integral lead to the same number. Any bounded derived function is summable.

The integral of a bounded derivative, viewed as a function of the upper limit of integration, is a primitive function of the given derivative; thus the fundamental problem of integral calculus is theoretically solved whenever the given derived function is bounded.

To obtain more general results, a definition of an integral must be given that applies to unbounded functions. Such a definition can easily be found, but the one that appeared simplest and most natural to me does not apply to all unbounded derived functions; such that, for unbounded functions, the problem in studying primitive functions is not solved for all cases. As defined by me, I find that *the necessary and sufficient condition for a derived function to have an integral is for its primitive function to have a bounded variation. Whenever the integral exists, a primitive function is revealed.*

The actual calculation of an integral depends largely on the way the function to be integrated is given. Where the function is defined using series, one can make use of the following property, a particular case of which was obtained by Mr. Osgood:[4] *a series whose terms have integrals, and the remainders of which have an absolute value less than a fixed number, is integrable term for term.*

The definition of integral is at that point extended to functions with several variables.

In the first chapter, I have developed a generalized concept of the length of a segment; a different sort of generalization gives the concept of the length of a curve. In the third chapter, where I deal with this concept, I adopt the following definition: *the length of a curve C is the lowest limit of the lengths of the polygonal lines*

[4] *American Journal,* 1897.

uniformly tending towards C. This definition is exactly equivalent to the classical definition.[5] A curve with a finite limit is said to be rectifiable. I promptly reobtain Mr. Jordan's main results relative to these curves.

The search for an expression for the length of a curve that has tangents leads to a new application for integrals as defined in the preceding chapter. *If f', φ', and ψ' exist, the necessary and sufficient condition that allows the curve*

$$x = f(t), \qquad y = \varphi(t), \qquad z = \psi(t)$$

to be rectifiable is that the integral of $\sqrt{f'^2 + \varphi'^2 + \psi'^2}$ exist. Whenever this integral exists, it represents the length of the curve. The definition adopted by P. Du Bois Reymond[6] is thus a particular case of the classical definition, even when the meaning of integral is extended as I have done.

In the fourth chapter, *I define the area of a surface L as the lowest limit of the areas of the polyhedral surfaces uniformly tending towards L.* From that one can deduce a definition of area analogous to that for the length of a curve, defined as the limit of the lengths of the inscribed polygons.

The subject of representing area using a double integral is broached only in the very special case where a surface allows tangent planes varying continuously; I reobtain the classic integral $\int\int \sqrt{EG - F^2}\, du\, dv$.

The last two chapters are devoted to rather different investigations. Examples are looked at to see whether the extended meaning given to the words "length" and "area" may not entail corresponding modifications to propositions made or arguments adopted in surface geometry. These arguments generally suppose surfaces and curves which are analytic, or at least defined using functions having a certain number of derivatives.

The first problem I set for myself is how to find a surface that can be applied to a plane: to look for surfaces that correspond point for point to a plane, so that lengths are preserved. I find on the one hand that *there are surfaces that can be applied to a plane and that have no straight-line segments,* and on the other that *there are skew curves having at each point an osculating plane and whose tangents make up a surface that cannot be applied to the plane.* The elementary procedures I employed did not show me how all surfaces can be applied to a plane, but they made me aware of the necessary and sufficient conditions for a cylindrical surface, a conical surface, a surface formed by the tangents of a skew curve, and a surface of revolution to be applicable to the plane; finally, they show that, when they are applied, their areas are preserved.

[5] Scheeffer, *Acta Mathematica*, 5; Jordan, *Cours d'Analyse.*
[6] *Mathematische Annalen*, Bd. 15, p. 287 and *Acta Mathematica*, 6.

The second problem is that of Lagrange or Plateau: given a fixed contour, find a surface limited to this shape the area of which is the smallest possible. I show that this problem is always possible and it allows an infinite number of solutions.

It would be very interesting to know whether, among all these surfaces which are solutions, there may not be an analytic surface. The method that immediately comes to mind and that consists of successively demonstrating the existence of each derivative needed to formulate the expression in terms of partial derivatives for the smallest possible surfaces seems exceedingly difficult to apply. However, some very elementary arguments allowed me, for a particular case, to demonstrate the existence of tangent planes touching one of the solution surfaces.

The methods in this last chapter are analogous to those that allowed Mr. Hilbert[7] to resume the study of Dirichlet's problem by the Riemann procedure. The results Mr. Hilbert obtained and those I just referred to, however incomplete they may be, appear to show that it would be worthwhile, at least for the moment, to set aside the partial-differential equations given by the ordinary methods of the calculus of variations, and to argue straight from the integral that is being minimized.

I have indicated the principal results of this work in various notes in *Comptes Rendus de l'Académie des Sciences*. (June 19 and November 27, 1899, November 26 and December 3, 1900, April 29, 1901.)

<hr>

CHAPTER I
MEASURE OF A SET

§ 1. A set of points is said to be bounded if the distance between any two of its points has an upper limit. Two sets are said to be equal if, by rearranging one of them, we can get them to coincide. Given sets $E_1, E_2, \ldots$, the sum E is made up of the points belonging to at least one of the E_i. We will only ever have to consider a finite number or a countable infinity of sets E_i, and we shall set

$$E = E_1 + E_2 + \ldots.$$

If every point in E_2 is a point of E_1, we say that E_1 contains E_2, and the set of points in E_1 that do not belong to E_2 we call the difference between E_1 and E_2 ($E_1 - E_2$). It should be noted that if E_2 contains E_3, the sets

$$E_1 + (E_2 - E_3) \quad \text{and} \quad (E_1 + E_2) - E_3$$

differ if there are points at once common to E_1, E_2, and E_3.

[7] *Nouvelles Annales de Mathématiques*, August 1900.

These definitions having been given,

Let us assign to each bounded set either a positive number or zero, which we shall call its measure, which satisfies the following conditions:

1. *There are sets whose measure is non-zero.*
2. *Two equal sets have the same measure.*
3. *The measure of the sum of a finite number of sets, or a countable infinity of sets with no common points between them, is the sum of the measures of these sets.*

We shall solve this *problem of measure* only for sets that we call measurable. Moreover, this problem allows different solutions depending on whether we limit ourselves to sets whose points are all on a straight line, or to sets whose points are all on a plane, etc. To make this distinction, we shall say, as need be: linear measure, surface measure, etc.

Let us note that if the problem of measure has a solution, then we get another system of measures by multiplying all the measures obtained by the same number. We shall not consider these new measures to be different from such solutions, with the result that, without any impairment of generality, we can assign the measure 1 to any set with a non-zero measure.

I. THE ELEMENTS OF A SET ARE POINTS ON A STRAIGHT LINE

§ 2. Let us suppose that the problem of measure is possible. A set made up of one point has a measure of zero since a bounded set containing an infinite number of points must have a finite measure. The set of points of segment MN thus has the same measure whether or not M and N are part of the set; furthermore, MN cannot have a measure of zero without the same being true for every bounded set.

Let us choose a segment MN and assign it a measure of 1. We know that if we take MN as a unit of measure, one may assign to each segment PQ a number, its length; this number is also the measure of the set of PQ's points. To convince one-self of this, it is sufficient to recall that, if the length l of PQ is measurable and equal to $\frac{\alpha}{\beta}$, there is a segment RS contained α times in PQ and β times in MN; that if l is not measurable, to every number λ less than l there corresponds a segment contained in PQ of length λ; and to every number λ greater than l there corresponds a segment containing PQ of length λ.

To satisfy the third condition of the problem of measure, the length of a segment that is the sum of a finite or infinite number of other segments, none of which overlap, must be the sum of the lengths of these segments.

From the properties of lengths, it follows that this is indeed the case if the component segments are finite in number; this continues to be so even if they are infinite in number. (See *Leçons sur la Théorie des fonctions*, by Mr. Borel.)

§ 3. Given a set E, there are an infinite number of ways to enclose its points within a finite number or a countable infinity of intervals. The set E_1 of the points of these intervals contains E; thus the measure of E, $m(E)$, is at most equal to the measure of E_1, $m(E_1)$, i.e., at most equal to the sum of the lengths of the intervals in question. The lower limit of this sum is an upper limit of $m(E)$, which we shall call the exterior measure of E, $m_e(E)$.

Let us suppose that all points of E belong to a segment AB. We shall call the set $AB - E$ the complement of E relative to AB, $C_{AB}(E)$. Since the measure of $C_{AB}(E)$ is at most $m_e[C_{AB}(E)]$, the measure of E is at least $m(AB) - m_e[C_{AB}(E)]$. This number does not depend on the number of segments AB we chose to cover E; we shall call the interior measure of E, $m_i(E)$. Two equal sets have equal interior and equal exterior measures. Furthermore, since we have

$$m_e(E) + m_e[C_{AB}(E)] \geq m(AB),$$

the exterior measure is never less than the interior measure. If the problem of measure is possible, the measure of a set E is contained between the two numbers $m_e(E)$ and $m_i(E)$ that we have just defined.

§ 4. We shall call *measurable sets*[8] *those whose exterior and interior measures are equal*; the value these two numbers have in common will be the measure of the set, if the problem of measure is possible. From the properties that follow, the result will be that the number $m(E)$, thus defined, does indeed satisfy the conditions of the problem of measure *if one confines oneself to considering only measurable sets*.

The definition of measurable sets is equivalent to this: A set E is said to be measurable if it is possible to enclose its points within α intervals, and its complement's points within β intervals, in such a way that the sum of the lengths of the parts common to α and β is as small as one likes. Whether there be a finite or a countable infinity of measurable sets $E_1, E_2, \ldots$, let us show that their sum E is measurable.

We assume that all sets E_i are made up of the points of a segment AB whose complements we shall take. Let us enclose the points of E_1 in α_1 intervals, none of which overlap, and the points of $C(E_1)$ in β_1 intervals; with the parts common to α_1 and β_1 being of an arbitrarily chosen total length ε_1. Let us enclose E_2 in α_2 intervals and $C(E_2)$ in β_2 intervals, having in common a total length ε_2. Let α'_2 and β'_2 be the parts of α_2 and β_2 common to β_1. To E_3 there correspond the intervals α_3 and β_3 and a number ε_3; let α'_3 and β'_3 be the parts of α_3 and β_3 common to β'_2, and so forth.

[8]In adopting this definition, we are modifying the language adopted by Mr. Borel.

E's points can be enclosed in intervals $\alpha_1, \alpha'_2, \alpha'_3 \ldots$. Those of $C(E)$ can be enclosed in β'_i intervals, whatever i may be. Now these two series of intervals have parts in common of a total length at most equal to

$$l_i = \varepsilon_1 + \varepsilon_2 + \ldots + \varepsilon_i + m(\alpha'_{i+1}) + m(\alpha'_{i+2}) + \ldots.$$

The series $\sum m(\alpha'_i)$ is convergent. Thus if one has chosen the ε'_i such that the series $\sum \varepsilon_i$ is convergent and has ε as its sum, i can be made large enough for l_i to be less than 2ε. Thus E is summable.—Since the sum of a finite number or a countable infinity of summable sets is a summable set, it makes sense to pose the problem of measure only for measurable sets. If $E_1, E_2, \ldots$ have no point in common between them, the points of E_i are inside the α'_i intervals such that $m(\alpha'_i) - m(E_i)$ is at most equal to ε_i. Now $m(E)$ differs from

$$m(\alpha_1) + m(\alpha'_2) + m(\alpha'_3) + \ldots$$

by less than

$$\varepsilon_1 + \varepsilon_2 + \varepsilon_3 + \ldots,$$

thus we get

$$m(E) = m(E_1) + m(E_2) + \ldots,$$

and the 3rd condition of the problem of measure is satisfied.

§ 5. The problem of measure is thus possible for measurable sets; and it allows for only one solution, since the arguments that served us when defining the two numbers m_e and m_i, when applied to a measurable set, bring only measurable sets into play.

In no sense has it been shown that the problem of measure is impossible for sets (if there are any) whose interior and exterior measures are unequal. But in what follows, we shall encounter only measurable sets. In fact, the procedures we shall employ to define a set can always be reduced to the following two:

1. Get the sum of a finite number or a countable infinity of previously defined sets;

2. Consider the set of points common to a finite number or a countable infinity of given sets;

and these two procedures, when applied to measurable sets, yield measurable sets. The first of these we saw; let us now demonstrate the second.

Let $E_1, E_2, \ldots$ be given sets; the set we are seeking e_i can be defined to have as its complement the sum of the complements of $E_1, E_2, \ldots$, which demonstrates the proposition.

Let e_i be a set analogous to e_1, relative to the sequence $E_i, E_{i+1}, \ldots$; the sum of e_i is formed by the points common to all the E_i, at least beginning with a certain value of i, and furthermore variable from one point to another; as the sum of measurable sets, it is measurable.

Here is another application of the 2nd procedure. Where E_1 contains E_2, if $E_1 - E_2$ is the set of points common to E_1 and $C(E_2)$, then, if E_1 and E_2 are measurable, $E_1 - E_2$ is as well. Moreover, since we have

$$E_1 = (E_1 - E_2) + E_2$$
$$m(E_1 - E_2) = m(E_1) - m(E_2).$$

§ 6. Since we are familiar with measurable sets (those formed from all the points of an interval) the two prior procedures, applied a finite number of times, let us define new ones. The ones we can obtain by this method, and their complements, are those that Mr. Borel calls measurable,[9] and that we shall call *(B) measurable sets*. They are defined by a countable infinity of conditions; the set containing them has the power of the continuum. Among these sets, we must point out those that are sums of intervals and that are closed sets, i.e., those containing their derivative,[10] the complements of which are sums of intervals.

The set E, made up of points on the abscissa

$$x = \frac{a_1}{3} + \frac{a_2}{3^2} + \frac{a_3}{3^3} + \dots,$$

where the a_i are equal to 0 or 2, and being perfect, is (B) measurable. Its complement is formed by one interval $(\frac{1}{3}, \frac{2}{3})$ of length $\frac{1}{3}$, two intervals $(\frac{1}{9}, \frac{2}{9}), (\frac{2}{3} + \frac{1}{9}, \frac{2}{3} + \frac{2}{9})$ of length $\frac{1}{3^2}$, four intervals of length $\frac{1}{3^3}$, etc., and thus has as its measure

$$\frac{1}{3} + 2\frac{1}{3^2} + 2^2\frac{1}{3^3} + \dots = 1,$$

and consequently E has a measure of zero. E has the power of the continuum, and thus with the points of E one can form an infinite number of sets all of which, because they have an exterior measure of zero, are measurable. The power of the set of these sets is that of the set of the sets of points; thus there are measurable sets that are not (B) measurable, and the power of the set of measurable sets is that of the set of the sets of points.

§ 7. Let E be a measurable set. Let us choose some numbers $\varepsilon_1, \varepsilon_2 \dots$ decreasing to zero. One can enclose E in a countable infinity of intervals α_i, of measure $m(E) + \varepsilon_i$. The set E_1 of the points that belong at the same time to sets $\alpha_1, \alpha_2 \dots$ is (B) measurable; it has as its measure $m(E)$ and contains E. The set $E_1 - E$ has a measure of zero. It can be enclosed in intervals β_i, contained within the α_i, and having a measure of ε_i. The set e of points common to all the β_i is (B) measurable, with a measure of zero. The set $E_2 = E_1 - e$ is thus (B) measurable, with a measure of $m(E)$; such that *any measurable set is contained in a set E_1 and contains a set E_2, with E_1 and E_2 being (B) measurable and having the same measure.* Thus, the sets that we

---

[9] *Leçons sur la théorie des fonctions*, pp. 46–50.
[10] These are the sets that Mr. Jordan calls "perfect" and Mr. Borel "relatively perfect." This sort of set contains its boundary (which shall be defined later).

call measurable are those Mr. Borel's procedures allow us to measure, provided we take account of the remarks made at the end of page 48[11] (*loc. cit.*).

By analogy, it can be shown that the exterior measure of a set E is the lower limit of the measures of the measurable sets containing E, and that there does in fact exist a (B) measurable set containing E and with a measure of $m_e(E)$. Likewise, $m_i(E)$ is the upper limit of the measures of the measurable sets contained within E, and there does in fact exist a (B) measurable set contained within E and having $m_i(E)$ as its measure.

§ 8. In his treatise on the Calculus, Mr. Jordan gives the following definitions. A point M is an interior point of a set E if it is inside a segment all the points of which are points of E. The boundary of E is the set of points which are not inside either E or $C(E)$.

Let us divide the segment AB containing E into partial intervals. Let l be the sum of the lengths of the specific intervals all of whose points are inside E, and let L be the sum of the lengths of those intervals that contain the points of E or those that contain the points at its boundary. It can be shown that, when one varies the division of AB in any way, such that the maximum length of the partial intervals tends towards zero, the two numbers l and L tend towards determinate limits, the *interior and exterior extensions of E*. From this definition, it results that the exterior extension is at least equal to the exterior measure and that the interior extension is at most equal to the interior measure. Mr. Jordan calls those sets whose exterior and interior extensions are both equal measurable; these sets that we shall call (*J*) *measurable*, thus, are measurable in the sense that we have adopted, and both definitions of measure agree whenever both are applicable. It may additionally be stated that the interior extension of E is the measure of the set of its internal points, and since the said set is open,[12] i.e., it does not contain any point of its boundary, the set has as its complement a closed set, and thus is (B) measurable. The exterior extension of E is the measure of the sum of E and its boundary, and since the union is closed, it is (B) measurable. Thus for a set to be (J) measurable, it is necessary and sufficient for its boundary to have a measure of zero.

A closed set, having its exterior extension as its measure, when it has a measure of zero we can say that it is (J) measurable. In particular, the "perfect set" defined in § 6 is (J) measurable; the same is true for all those sets that can be formed with their points. Thus the set of (J) measurable sets has the same power as the set of the sets of points, and there are (J) measurable sets that are not (B) measurable.

---

[11]"However, if a set E contains all the elements of a measurable set E_1, of measure α_1, we may say that the measure of E is greater than α_1, without being troubled by whether E is measurable or not. Inversely, . . . The words 'greater than' and 'less than', incidentally, do not exclude being equal to."

[12]All points of such a set are internal to the set.

§ 9. We have just assigned a measure to certain sets, and it now remains for us to investigate how to calculate this number. This obviously depends on the way the set is given.

Suppose that, for a given interval (a, b), we know how to recognize whether in (a, b) there are points of the given set E or of its boundary, and whether there are points of $C(E)$. Then, by a finite number of operations, we can calculate any number of terms of the two sequences (one of which we can assume to be decreasing, the other increasing), the limits of which are the exterior and interior extensions of E. The usual way we have of handling series leads us to regard the extensions as being well defined.

Thus we can calculate the measure of a (J) measurable set, and simultaneously considering both sequences allows us to have an upper limit for errors caused by stopping at any particular term.

It is much more difficult to calculate $m_e(E)$ and $m_i(E)$ for any set. These numbers are in fact defined by considering an uncountable infinity of numbers; to find a sequence of numbers tending towards $m_e(E)$, we would have to consider dividing the segment AB containing E into partial intervals that would depend on the set E.

If a set is (B) measurable and is defined on the basis of a series of intervals, by using the two operations we have indicated it is easy to calculate the set's measure by relying on the 3rd condition for the problem of measure and on this property: the set E of points common to all measurable sets $E_1, E_2, \ldots$, which are sets such that each one contains all those that follow, is the lower limit of the series $m(E_1), m(E_2), \ldots$.

In fact, $C(E)$ is the sum of the sets with no points in common $C(E_1)$,
$$[C(E_2) - C(E_1)], [C(E_3) - C(E_2)], \ldots.$$
Thus
$$m[C(E)] = m[C(E_1)] + m[C(E_2) - C(E_1)] + \ldots$$
and
$$m(E) = m(E_1) + [m(E_2) - m(E_1)] + \ldots.$$

II. THE ELEMENTS OF A SET ARE POINTS ON A PLANE

§ 10. The foregoing considerations can with no problem be extended to sets whose elements are points in a space of many dimensions; we shall confine ourselves to the case of a plane. Arguing as in § 2, we can see that any bounded set of points on a straight line has a surface measure of zero, and that the set of points of a square cannot have a measure of 0. Let us, then, arbitrarily assign a measure of 1 to a square $MNPQ$.

The arguments used in basic geometry to find the area of a triangle prove that the measure of the set of points of a triangle cannot be other than half the product of the numbers that measure its side and its height, MN being the unit of length.

Having thus defined the measure of a triangle, it must be shown that *the measure of a triangle which is the sum of nonoverlapping triangles is the sum of the measures of these triangles.* The arguments set out by Mr. Hadamard in note D of his *Géometrie élémentaire* prove that this is so if the component triangles are finite in number. The case where they are infinite in number is handled by an argument similar to the one we found useful for the case of a set of points on a straight line (Borel, *Théorie des fonctions*, page 42).

§ 11. We can now give definitions analogous to those in § 3.

The exterior measure of a set E, $m_e(E)$, is the lower limit of the sum of the measures of the triangles (whether finite or infinite in number) within which the points of E can be enclosed. Since E is internal to a triangle ABC, by definition,

$$C_{ABC}(E) = (ABC) - E.$$

The interior measure of E will, by definition, be

$$m_i(E) = m(ABC) - m_e[C_{ABC}(E)].$$

A set for which both numbers, as thus defined, are equal we shall call measurable, and the value these numbers have in common will be its measure.

As in § 4, we can show that the problem of measure is possible and only has a unique solution when we limit ourselves to measurable sets; and that the two procedures of § 5, when applied to measurable sets, yield measurable sets. The two procedures, when applied a finite number of times to sets each of which is formed by the points of a triangle, yield plane sets that we shall call, along with their complements, (B) measurable sets.

Let E be an open set, each of whose points M is inside E. Corresponding to M, then, we can have a square with M as its center, with sides parallel to given rectilinear directions, and defined as being the largest square for which all its interior points are inside E. Since E is the sum of those squares corresponding to points both of whose coordinates are rational, it is (B) measurable.

The complement of a closed set is an open set, and thus every closed set is (B) measurable. We shall define the exterior and interior extensions of a set as we did for a straight line, replacing the division of the set-containing portion of the line into a finite number of segments with the division into a finite number of squares in the set-containing portion of the plane. Hence we get the concept of a (J) measurable set.

All these sets and all these numbers have the same relations between them as the sets and numbers with the same names that we encountered previously.

III. THE PROBLEM OF AREA[13]

§ 12. We know that "plane curve" is the name given to the set of the two equations

$$x = f(t), \qquad y = \varphi(t), \tag{1}$$

f and φ being continuous in the *finite* interval (a, b) in which they are defined. Corresponding to each value of t we may have a point whose coordinates are the corresponding values of x and y. A curve thus defines a set of points, and this set is perfect.[14] A point is said to be multiple if it corresponds to several values of t. For a curve without multiple points, knowing the set of points of the curve is enough to define it, since we do not consider curve (1) to be different from those that are deduced from it by replacing t by a function $\theta(t)$ that is always increasing or always decreasing.

A curve is said to be closed with no multiple points if it has no multiple point other than a double point corresponding to $t = a$ and $t = b$. We regard this curve as defined by the set of its points. Given such a curve, we know that it divides the plane into two regions, one inside and the other outside.[15]

We shall give the name *domain* to the set of points inside a closed curve C with no multiple points. C is the boundary of the domain, which is an open set. We shall say that a domain D is the sum of the domains $D_1, D_2, \ldots$, finite or infinite in number, if every point of D belongs to one and only one of the D_i, or to at least one of the boundaries of the D_i.

§ 13. *We propose assigning to each domain a positive number that we shall call its area, while satisfying the following conditions:*

1. *Two equal domains have the same area.*
2. *The area of a domain that is the sum of a finite or infinite number of other domains is the sum of the areas of these domains.*

This is the *problem of area*.

If this problem is possible, it is so in an infinite number of ways, and one may arbitrarily attribute to a square $MNPQ$ the area 1. The well-known arguments of basic geometry prove that the area of a rectangle cannot be anything other than the product of the lengths of its sides, MN being the unit of length, i.e., of the surface measure of the rectangle.

Being an open set, a domain, as we have seen, is the sum of a countable infinity of rectangles, which we may assume do not overlap with one another. Thus its area cannot be anything other than the sum of the areas of these rectangles, i.e., of the surface measure of the domain regarded as a set of points.

---

[13] Jordan, Volume I.—J. Hadamard, *Géométrie Elémentaire*.
[14] This would not be the case if, as frequently happens in mechanics, the interval (a, b) was infinite.
[15] Jordan, *Cours d'Analyse*, 2nd Edition, Volume I.

Let D_1 and D_2 be domains with no point in common, having in common one and only one boundary arc $\alpha\beta$. Domain D – the sum of the set of points of D_1, the set of points of D_2, the set of points of $\alpha\beta$ (other than α and β) – has as its measure the sum of the measures of these three sets, all three of which are measurable since the first two are open and the third is perfect, apart from points α and β. To satisfy the 2nd condition of the problem of area, it is thus necessary for the surface measure of arc $\alpha\beta$ to be zero.

The problem of area is thus possible only if one looks merely at domains whose boundary has a surface measure of zero.

We shall call these domains *squarable domains*, and curves whose surface measure is zero *squarable curves*.

Let D be a squarable domain which is the sum of squarable domains $D_1, D_2, \ldots$. The set of D points contains the sum of the sets without a common point between them, $D_1, D_2, \ldots$, and since none of the D_i has a zero measure, they constitute at most a countable infinity. The boundaries have a surface measure of zero, and they can be ignored when calculating the measure or area of D.

The problem of areas is thus possible for squarable domains and has a unique solution, if one sets the unit of area.

§ 14. Now we shall assume the 2nd condition of the problem of area to be modified as follows:

The area of a domain which is the sum of two others is the sum of the areas of the two others.[16] Resuming the arguments already employed, we shall see that the area of a domain D is contained between the interior and exterior extensions of this domain. The result is that the area of a squarable domain is also well defined.

The area of a nonsquarable domain D, limited by a nonsquarable curve C, is contained between the numbers $m(D)$ and $m(D) + m(C)$.[17]

Let us show that the problem of area, as thus formulated, is indeterminate for nonsquarable domains. We shall base ourselves on this property: when two domains D_1 and D_2 have a common boundary arc $\alpha\beta$, we can find a domain D containing $\alpha\beta$ (perhaps without $\alpha\beta$) such that either every point of D inside D_1 is inside D_2 and vice versa, or that any point of D inside D_1 does not belong to D_2 and vice versa.[18] In the first case, we shall say that D_1 and D_2 are on the same side of $\alpha\beta$, and in the second case, that they are on different sides.

[16]This is how Mr. Hadamard poses the question for the areas of polygons. (*Géométrie Elémentaire*, Note D.)

[17]There are nonsquarable curves because there are curves that pass through all the points of a square. To form a nonsquarable curve without multiple points, we need only slightly modify the method employed by Mr. Hilbert to define a curve passing through all the points of a square (*Mathematische Annalen*, Vol. 38 or Plcard, *Traité d'Analyse*, 2nd Edition, Volume I). We shall replace each of the squares in Mr. Hilbert's definition with a polygon inside this square, rather large in area, chosen such that the boundaries of two of these polygons have in common only the vertex, if any, by which the curve passes from one into the other.

[18]This can easily be demonstrated.

Let there be an arc of curve $\alpha\beta$, with no multiple points and not squarable, and let us assume that it is part of the boundary of a domain Δ. Let there be any domain D, bounded by curve C. C and $\alpha\beta$ may have arcs in common (we ignore those common points, if any, not on such arcs). Let E be the set of the specific arcs along which D and Δ are on the same side of $\alpha\beta$, and let E_1 be the set of arcs for which this is not true. Let us arbitrarily choose, once and for all, a number θ contained between 0 and 1. To D we shall assign the area

$$m(D) + \frac{1}{2}m(C - E - E_1) + \theta m(E) + (1 - \theta)m(E_1).$$

It will be very easy to show that the area, as thus defined, satisfies the second condition of the problem of area, as formulated at the beginning of this section.[19]

In summary, the problem of area is simultaneously possible and well defined only for squarable domains. In what follows, we shall speak of area only for the case of a squarable domain. Arguments analogous to the foregoing can be made on the topic of the *volumes of domains* in ordinary space, and more generally of the *extension of a domain* in a space of any number of dimensions.

CHAPTER II

INTEGRALS

I. DEFINITE INTEGRAL OF FUNCTIONS WITH ONE VARIABLE

§ 15. From a geometric perspective, the problem of integration can be posed as follows:

For a curve C given by the equation $y = f(x)$ (f being a continuous positive function, and the axes rectilinear), find the area of the domain bounded by an arc of C, a segment of $0x$ and two lines parallel to the y axis for given abscissa values a and b, where $a < b$.

This area is called the definite integral of f between the limits a and b, and is represented by $\int_a^b f(x)\,dx$.

In squaring a segment of a parabola, Archimedes solved a particular case of this problem. The classical method applicable to the general case essentially consists in evaluating the interior and exterior extensions of the domain by dividing the plane into rectangles whose sides are parallel to $0x$ and $0y$. To get these rectangles,

---

[19] For that, we will have to rely on the following property: if a domain is the sum of two domains D_1 and D_2, D_1 and D_2 have one and only one common boundary arc, and no point in common outside of this arc.

let us first trace lines parallel to $0y$, then let us divide the bands we get by segments parallel to $0x$, with y-coordinates varying from one band to the next. If R_1, one of the rectangles R thereby formed, must be considered in order to calculate one of the extensions, all rectangles R located in the same band and contained between R_1 and $0x$ must also be considered for calculating the same extension. The extensions are thus limits of the sums of the areas of rectangles having their base on $0x$.

Let $\delta_1, \delta_2, \ldots$ be the lengths of these bases; let $m_1, m_2, \ldots, M_1, M_2$ be the lower and upper values of f in the corresponding intervals. If we assume the δ as given, i.e., the lines traced parallel to $0y$, and if we choose the segments parallel to $0x$ so as to obtain values which approximate the extensions as closely as possible, for these approximate values we get

$$s = \sum \delta_i\, m_i, \qquad S = \sum \delta_i\, M_i.$$

Thus, we know how to calculate the two extensions of the domain; we can show that they are equal, that the problem that we set ourselves is thus meaningful, and that we know how to solve it.

Relative to bounded f functions of any kind, Mr. Darboux has demonstrated[20] that the two sums s and S tend towards perfectly determinate limits; these are called *deficient integrals* and *excess integrals*. When the two integrals are equal, and this happens for functions other than continuous functions, the function is said to be integrable and, since Riemann,[21] the limit common to s and S has been called *the definite integral of f between a and b*.

§ 16. To interpret these numbers geometrically, let us assign to each positive function f defined in (a, b) the set E of the points whose coordinates simultaneously satisfy the two inequalities

$$a \leq x \leq b, \qquad 0 \leq y \leq f(y).$$

The two sums s and S are clearly values that approximate the interior and exterior extensions of E, and consequently s and S have well-defined limits: these extensions. Thus, from a geometric perspective, the existence of deficient integrals and excess integrals is a consequence of the existence of the interior and exterior extensions of a bounded set. For the function f to be integrable, it is necessary and sufficient that E be (J) measurable; the measure of E is the integral.

Of whatever sign the function f is, corresponding to it we shall have the set E of points whose coordinates satisfy the following three inequalities:

$$a \leq x \leq b, \qquad xf(x) \geq 0, \qquad 0 \leq y^2 \leq \overline{f(x)}^2.$$

---

[20]Darboux, *Mémoire sur les fonctions discontinues*. Annales de l'Ecole Normale, 1875.
[21]Riemann, *The Representation of Functions through Trigonometric Series*.

The set E is the sum of two sets E_1 and E_2 formed by the points with positive y-axis values for E_1 and negative values for E_2.[22] The deficient integral is the interior extension of E_1 less the exterior extension of E_2; the excess integral is the exterior extension of E_1 less the interior extension of E_2. If E is (J) measurable (in which case E_1 and E_2 are as well), the function is integrable, the integral being $m(E_1) - m(E_2)$.

§ 17. These results immediately suggest the following generalization: *if the set E is measurable (in which case E_1 and E_2 are as well), we shall call the quantity*

$$m(E_1) - m(E_2)$$

the definite integral of f between a and b. The corresponding functions f will be said to be summable.

Relative to nonsummable functions, if there are any, we shall define the lower and upper integrals as being equal to

$$m_i(E_1) - m_e(E_2), \qquad m_e(E_1) - m_i(E_2).$$

These two numbers are contained between the deficient integrals and the excess integrals.

§ 18. We shall now define summable functions analytically.

Since E is measurable, it is contained in set E' and it contains set E'', with E' and E'' being (B) measurable and having a measure of $m(E)$ (§ 7). Moreover, the arguments that gave us this result prove that one may assume E' and E'' to be made up of segments parallel to $0y$ and having their base on $0x$, i.e., corresponding to two functions f_1 and f_2 (where $f_1 \geq f_2$).

Let e, e', and e'' be sets formed by points from E, E', and E'' whose coordinates on the y-axis are greater than a given number $m > 0$; e, e', and e'' are measurable and have the same measure. Let s, s', and s'' be the sections of these sets sectioned off by the straight line $y = m + h$; s and s' are linearly (B) measurable, with $m(s)$ and $m(s')$ not decreasing as h tends towards zero. Let S' and S'' be their limits. Let us show that they are equal.—In fact, were it otherwise, for an h that was small enough we would always get

$$m(s') \geq m(s'') + \varepsilon.$$

And we can find an h_1 and h_2, where $h_1 < h_2$, small enough that

$$m[s'(h_1)] \geq m[s'(h_2)] \geq m[s''(h_1)] + \frac{\varepsilon}{2}.$$

If e'_1 and e''_1 are points of e' and e'' contained between $y = h_1$ and $y = h_2$, we get

$$m(e'_1) \geq (h_2 - h_1)\, m[s'(h_2)]$$
$$m(e''_1) \leq (h_2 - h_1)\, m[s''(h_1)].$$

---

[22] It scarcely matters whether the points on the x-axis be considered as part of E_1 or E_2.

Thus

$$m(e'_1) \geq m(e''_1) + (b_2 - b_1)\frac{\varepsilon}{2},$$

which is impossible, because e'_1 and e''_1 must have the same measure.

Thus s' and s'' have the same linear measure and thus s is measurable; i.e., *the set of values of x for which $f(x)$ is greater than $m > 0$ is measurable*. Likewise, *the set of values of x for which $f(x)$ is less than $m < 0$ is measurable*.

The result of this is that the set of values of x for which $f(x)$ is less than or equal to $m > 0$ (greater than or equal to $m < 0$) is measurable; thus the set of points for which we have either $a \geq f(x) > b > 0$, or $0 > c > f(x) \geq d$, or $e \geq f(x) \geq g$ (where $eg < 0$) is measurable; and by getting b to tend towards a, or d towards c, or e and g towards 0, we see that the set of points for which y has a given value is measurable. In summary, without needing to bother over the signs of a and b, *if f is summable, the set of values of x for which we have*

$$a > f(x) > b$$

is measurable.

§ 19. Reciprocally: *for any a and b, if the set of values of x for which $a > f(x) > b$ is measurable, and if the function $f(x)$ is bounded, it is summable.*

In fact, let us divide the interval of variation of $f(x)$: let $a_0, a_1, a_2, \ldots, a_n$ be the points of division.

Let e_i (with $i = 0, 1, \ldots, n$) be the set of values of x for which $f(x) = a_i$.

Let e'_i (with $i = 0, 1, \ldots, n - 1$) be the set of values of x for which $a_i < f(x) < a_{i+1}$.

The points of set E assigned to $f(x)$, corresponding to the values of x that belong to e_i, form a measurable set in the plane and of measure $|a_i| \cdot m_l(e_i)$, where $m_l(e_i)$ indicates a linear measure. The points of E that correspond to those of e'_i form a set containing a measurable set of measure $|a_i| \cdot m_l(e'_i)$, and are contained within a measurable set of measure $|a_{i+1}| \cdot m_l(e'_i)$.

Thus, E contains a set of measure

$$\sum_0^n |a_i| m_l(e_i) + \sum_1^n |a_{i-1}| m_l(e'_i),$$

and is contained in a set of measure

$$\sum_0^n |a_i| m_l(e_i) + \sum_1^n |a_i| m_l(e'_i).$$

These two measures differ by less than $(a_n - a_0)\alpha$, where α is the name we give to the maximum of $a_i - a_{i-1}$; thus one may make them as close as one likes. And E is measurable and thus f is summable.

Furthermore, we know how to calculate the measure of E; thus, *if f is positive the integral is the common limit of the two sums*

$$\sigma = \sum_0^n a_i m_l(e_i) + \sum_1^n a_{i-1} m_l(e_i'),$$

$$\Sigma = \sum_0^n a_i m_l(e_i) + \sum_1^n a_i m_l(e_i'),$$

when $a_{i-1} - a_i$ tends towards zero.

Now if f is not always positive, the limit of the sum of those terms of σ or Σ that are positive gives the measure of the set that we have called E_1 (§ 16), and the limit of the sum of the negative terms gives $-m(E_2)$; thus *in all cases σ and Σ define the integral.*

§ 20. It is perhaps of some use to show that analytic arguments could have led us to consider summable functions and what we have just called their integrals.

Let $f(x)$ be a constantly increasing continuous function defined between α and β (with $\alpha < \beta$), and varying between a and b (where $a < b$). For x, let us arbitrarily take the values

$$x_0 = \alpha < x_1 < x_2 \ldots < x_n = \beta,$$

to which, for $f(x)$, there correspond the values

$$a_0 = a < a_1 < a_2 \ldots < a_n = b.$$

The definite integral, in the ordinary sense of the word, is the limit common to the two sums

$$\sum_1^n (x_i - x_{i-1})a_{i-1}$$

$$\sum_1^n (x_i - x_{i-1})a_i$$

when the maximum of $x_i - x_{i-1}$ tends towards zero.

But x_i is given if a_i is too, and $x_i - x_{i-1}$ tends towards zero if $a_i - a_{i-1}$ tends towards zero. Thus, *to define the integral of a continually increasing function $f(x)$, we can take a_i, i.e., the division of the interval of variation of $f(x)$, instead of x_i, i.e., the division of the interval of variation of x.*

Looking to proceed this way, first of all in the simple case of continuous functions varying in any interval, with only a finite number of minima and maxima, and then in the case of any continuous function, we are easily led to this property. Let $f(x)$ be a continuous function defined in (α, β) and varying between a and b (where $a < b$). Let us arbitrarily choose

$$a = a_0 < a_1 < a_2 \ldots < a_n = b;$$

$f(x) = a_i$ for the points of a closed set e_i (where $i = 0, 1 \ldots n$). Let $a_i < f(x) < a_{i+1}$ for the points of a set that is the sum of the intervals e'_i (where $i = 0, 1, 2 \ldots n - 1$). Let sets e_i and e'_i be measurable.

The two quantities

$$\sigma = \sum_0^n a_i m(e_i) + \sum_1^n a_i m(e'_i),$$

$$\Sigma = \sum_0^n a_i m(e_i) + \sum_1^n a_{i+1} m(e'_i),$$

tend towards $\int_a^b f(x)dx$ when the number of a_i increases such that the maximum of $a_i - a_{i-1}$ tends towards zero.

Once this property has been obtained, we can take it as the definition of the integral of $f(x)$. But both quantities σ and Σ have a meaning for functions other than continuous functions, that is, for summable functions. We shall demonstrate that for these functions, σ and Σ have the same limit independent of the choice of a_i. This limit will be, by definition, the integral of $f(x)$ between α and β.

When one introduces new points of division among the a_i, σ does not decrease and Σ does not increase, and thus ρ and Σ have limits. They are equal, since $\Sigma - \sigma$ is at most equal to $(\beta - \alpha)$ multiplied by the maximum of $(a_i - a_{i-1})$.

Let there now be another method of division of the variation of $f(x)$, using points b_i, and let σ' and Σ' be the values corresponding to σ and Σ. Let σ'' and Σ'' be the values corresponding to the method of division in which one uses both a_i and the b_i at the same time. The two series of inequalities

$$\sigma \leq \sigma'' \leq \Sigma'' \leq \Sigma',$$
$$\sigma' \leq \sigma'' \leq \Sigma'' \leq \Sigma,$$

prove that the six sums σ, σ', σ'', Σ, Σ', and Σ'' have the same limit.

This demonstrates the existence of the integral. If one adopts this method of exposition, which incidentally is little different from the preceding one, it is not self-evident that Riemann's definition of an integral will never be at odds with the preceding one. To demonstrate it, we shall rely on this fact: the points of discontinuity of an integrable function form a set having a measure of zero.[23] — Let $f(x)$ be an integrable function, and let E be the set of points for which

$$a \leq f(x) \leq b$$

where a and b are any two numbers. The points that are E's limits and not part of E are points of discontinuity; thus they form a set e having a measure of zero. Since

---

[23] Riemann states this property as follows: For a function to be integrable, it is necessary that "the sum total of intervals for which the variations are greater than σ (whatever its value) can be made infinitely small."

it is closed, $E + e$ is measurable, and thus E is as well. That is sufficient to let us conclude that f is summable.

If in an interval of length l, the maximum of f is M and minimum is m, the integral (in the sense in which we are using the word) is contained between lM and lm. Furthermore, if $a_1, a_2, \ldots a_n$ are increasing numbers, we get

$$\int\limits_{a_1}^{a_2} + \int\limits_{a_2}^{a_3} + \ldots + \int\limits_{a_{n-1}}^{a_n} = \int\limits_{a_1}^{a_n}$$

(where the integrals have the meaning we have given them).

The result of this is that the integral of a summable function is contained between the deficient integral and the excess integral, and in particular that the two definitions of an integral are in agreement wherever both are applicable.

§ 21. The integral between the limits a and b has only been defined if a is less than b. We shall complete the definition by the identity

$$\int\limits_{a}^{b} f(x)dx + \int\limits_{b}^{a} f(x)dx = 0.$$

As a result of that we get

$$\int\limits_{a}^{b} f(x)dx + \int\limits_{b}^{c} f(x)dx + \ldots + \int\limits_{k}^{l} f(x)dx = \int\limits_{a}^{l} f(x)dx,$$

which we get whatever $a, b, \ldots, l$ may be.

Thus, we shall also need the concept of an integral of a function f that has been defined only for the points of set E.[24] Where AB is a segment containing E, let us define a function φ as equal to f for the points of E and equal to 0 for the points of $C_{AB}(E)$. The integral of f within E is, by definition, the integral of φ within AB. It is obvious that the integral of f as thus defined does not depend on the choice of the segment AB containing E.

If E is the sum of $E_1, E_2, \ldots$, where all these sets are measurable and with no point in common between them, and if the function f is summable within E, then we have

$$\int\limits_{E} f(x)\,dx = \sum \int\limits_{E_i} f(x)\,dx.$$

Let us again note that one may define the lower integral of a function f as the upper limit of the integrals of the summable functions φ not greater than f. There is one of these φ functions the integral of which is equal to the lower integral of f. An analogous property could be stated for the upper integral.

---

[24] We might first have defined summable functions within E, and then their integrals using the same definitions as before, provided we extracted out all points not belonging to E.

§ 22. Now we shall show that elementary arithmetic operations applied to summable functions yield summable functions.

Let f and φ be two summable functions that stay contained within m and M. Let us divide the interval (m, M), and let the points of division be

$$m_0 = m < m_1 < m_2 \ldots < m_n = M.$$

Where $i = 1, 2, \ldots n$, let e_i be the set of values of x for which: $m_{i-1} < f \le m_i$. Let e'_1 be the corresponding set for φ.

Let e_{ij} be the set of points common to e_i and e'_j. For the points of this set we have:

$$m_{i-1} + m_{j-1} < f + \varphi \le m_i + m_j;$$

e_i, e'_j, and e_{ij} are measurable.

Let a and b be any two numbers. Let E be the sum of those e_{ij} whose points are contained *between* a and b. E is measurable.

Let us infinitely increase the number of m_i such that the maximum of $m_i - m_{i-1}$ tends towards zero. We will get an infinite series of sets E the sum of which—which is measurable—is the set of values of x for which we have

$$a < f + \varphi < b,$$

and thus $f + \varphi$ is summable.

The integral of f is the sum of the integrals of f within sets e_{ij}, and thus we have

$$\sum m(e_{ij}) m_{i-1} < \int f(x)dx < \sum m(e_{ij}) m_i.$$

And likewise,

$$\sum m(e_{ij}) m_{j-1} < \int \varphi(x) dx < \sum m(e_{ij}) m_j.$$

And arguing in the same way for the function $f + \varphi$,

$$\sum m(e_{ij})(m_{i-1} + m_{j-1}) < \int (f + \varphi) dx < \sum m(e_{ij})(m_i + m_j).$$

From this, it follows that

$$\left| \int (f + \varphi)dx - \int f dx - \int \varphi dx \right| < \Sigma m(e_{ij})(m_i + m_j - m_{i-1} - m_{j-1}),$$

therefore,

$$\int (f + \varphi)dx - \int f\, dx + \int \varphi dx. [25]$$

This property can immediately be generalized such that: *the sum of any number of swmnable functions is a summable function and the integral is the sum of the integrals.*

---

[25] Had we introduced the concepts of lower and exterior integrals, we could have confined ourselves to the 2nd part of the preceding argument.

We shall likewise show that *the product of two summable functions is a summable function*; and that *the inverse of a summable function f that satisfies the inequality*

$$0 < m < |f| < M$$

is a summable function; that *the mth arithmetic root of a summable function f, for which this root exists, is summable*; that, *if f and φ are two summable functions such that f (φ) is meaningful, the function f (φ) is summable if f and φ are*, etc.

A more important proposition is the following: *If a bounded function f is the limit of a series of summable functions f_i, f is summable.*

In fact, let e_i be the set of values for which f_i is contained *between a* and *b*. The set *e* of points common to all e_i, at least starting with a certain value of *i*, is the set of values of *x* for which *f* is contained *between a* and *b*. Now since the e_i are measurable, *e* is as well, and thus *f* is summable.

§ 23. The propositions we have just obtained allow us to define an important class of summable functions.

We shall base our reasoning on the fact that $y = h$ and $y = x$ are summable functions; thus, $k x^m$ is summable and every polynomial is summable.

Since Weierstrass, we have known that every continuous function is the limit of a series of polynomials, and thus that continuous functions are summable. But there are functions besides continuous functions that are limits of polynomials, the functions studied by Mr. Baire, which he called "first-class functions." (*Sur les fonctions des variables réelles*, Annali di Matematica, 1899.) First-class functions are thus summable. The limits of first-class functions or of second-class functions are summable, etc. All functions of the set that Mr. Baire calls *E* (*loc. cit.*, page 70) are summable.

These results provide us with numerous examples of summable functions that are discontinuous and not Riemann-integrable. Furthermore, we can obtain examples of that sort as follows.—The argument that allowed us, in § 20, to demonstrate that every integrable function is summable proves that: If, when we extract a set that has a measure of zero, there is still a set at each point where a function is continuous, this function is summable. Thus, if *f* and φ are two continuous functions, the function *F*—defined as being equal to *f*, except at the points of a set *E* with a measure of zero, for which

$$F = f + \varphi,$$

—is summable. Now if φ is never zero and if *E* is dense in every interval, all the points are points of discontinuity for *F*, which is thus not Riemann-integrable.

This procedure lets us construct summable functions that form a set whose power is equal to that of the set of functions.

§ 24. The geometric method that we made use of at the beginning of this chapter, being based on the idea of measure of a bounded set, was applied only to bounded functions.[26] By contrast, the analytic method referenced in § 20 applies virtually unchanged to functions whose absolute value has no upper limit.

A function shall be said to be summable if, for any a and b, the set of values of x for which

$$a < f(x) < b$$

is measurable. We shall distinguish bounded summable functions, those with which we have thus far concerned ourselves, from unbounded summable functions.

Let $f(x)$ be a summable function. Let us choose numbers

$$\ldots < m_{-2} < m_{-1} < m_0 < m_1 < m_2 < \ldots$$

varying between $-\infty$ and $+\infty$, such that $m_i - m_{i-1}$ has an upper limit on its absolute value. Let e_i always be the set of values of x for which $f(x)$ equals m_i, and let e'_i be the set of values of x for which

$$m_i < f(x) < m_{i+1}.$$

Let us consider the two sums

$$\sigma = \sum m_i \cdot m(e_i) + \sum m_i \cdot m(e'_i), \qquad \sum = \sum m_i \cdot m(e_{i+1}) + \sum m_{i+1} \cdot m(e'_i)$$

in which the Σ signs represent the sums of two series, one with positive terms, the other with negative terms. These series may be convergent or divergent. If those that figure in σ are convergent, i.e., if σ is meaningful, Σ is meaningful, and vice versa; and this is true whatever the m_i are selected to be.

Arguing as in § 20, we shall see that the two sums σ and Σ tend towards the same limit, independent of the m_i selected, when we increase the number of m_i such that the maximum of $m_i - m_{i-1}$ tends towards zero.—This limit is the integral.

Even after thus extending the meaning of the words *summable function* and *integral*, all statements made up to this point still remain correct.[27] But it must be recalled that an unbounded summable function does not necessarily have an integral.[28]

§ 25. Calculating the integral of a given function presents the same problems as calculating the measure of a given set.

As most discontinuous functions heretofore considered in the Calculus have been defined using series, knowing of the following theorem is of some interest:

If a series of summable functions, having integrals $f_1, f_2, f_3, \ldots$ has a limit f and if $|f - f_n|$ (whatever the value of n) remains less than a fixed number M, f has an integral that is the limit of the integrals of functions f_n.[29]

---

[26] As an aside, there would be no difficulty about raising the problem of the measure of sets of points for all sets, bounded or not.

[27] The first theorem in § 22, however, requires some explanations. If f and φ are summable, $f + \varphi$ is as well, and in fact we get $\int (f + \varphi) = \int f + \int \varphi$ if $\int f$ and $\int \varphi$ are meaningful; but $\int (f + \varphi)$ may be meaningful without the same being true for $\int f$ and $\int \varphi$.

[28] Still to be examined would be the case where f becomes infinite for certain values of x. If these values are finite in number, the integral only need be defined by extracting the values that render f infinite, in order for the ordinary theorems to be unchanged.

[29] The most interesting particular case of this theorem, where f and the f_i are continuous functions, has already been obtained, using very different reasoning, by Mr. Osgood, in his Memorandum on nonuniform convergence (*American Journal*, 1894).

In fact we have
$$f = f_n + (f - f_n).$$
Since both functions on the right-hand side of the equation have integrals, the same is true for f and the integral of f is the sum of the integrals of f_n and $f - f_n$. Let us look for an upper limit for this second integral.

Let us arbitrarily choose a positive number Σ. If e_n is the set of values of x for which, for any positive value of p (or zero), we do not get
$$|f - f_{n+p}| < \varepsilon,$$
e_n is measurable.

If E is the measurable set within which the integrals are taken, we get
$$\left| \int (f - f_n) dx \right| \leq M \cdot m(e_n) + \varepsilon \left[m(E) - m(e_n) \right].$$

Now every set e_n contains all sets with larger indices, and there is no point which all e_n have in common at one time. Thus, $m(e_n)$ tends towards zero along with $\frac{1}{n}$, and thus the same is true for
$$\left| \int (f - f_n) dx \right|.$$

When f is bounded, the proposition can be stated as follows: When a series of summable functions $f_1, f_2 \ldots$, with an upper limit on the absolute value of their set, has a limit of f, the integral of f is the limit of the integrals of the functions f_n.

Here, in another form, is the theorem for the general case:

When the set of the remainders of a convergent series of functions that have integrals has an upper limit on its absolute value, the series is integrable term by term.

As a very special case, we have the theorem on the integration of uniformly convergent series.

II. INDEFINITE INTEGRALS AND PRIMITIVE FUNCTIONS
OF FUNCTIONS WITH ONE VARIABLE

§ 26. We shall say that an indefinite integral of a function $f(x)$ having a definite integral in an interval (α, β) is a definite function $F(x)$ defined within (α, β) such that, whatever a and b (contained within α and β) may be, we have
$$\int_a^b f(x)dx = F(b) - F(a).$$

From this equality can be drawn
$$F(x) = \int_a^x f(x)dx + F(a).$$

Thus any function that has a definite integral allows for an infinite number of indefinite integrals differing only by a constant $F(a)$.

The *indefinite integral is a continuous function*;[30] this is obvious if the function $f(x)$ is bounded. To demonstrate this for the general case, let us return to the notation in paragraph 24. With a arbitrarily selected, we must show that, once h has an absolute value less than a certain quantity, we obtain

$$\left| F(a+h) - F(a) \right| = \left| \int_{a}^{a+h} f(x)\,dx \right| < \varepsilon.$$

To simplify, we are going to assume that h is positive.

At most, there is a countable infinity of numbers m_i such that the corresponding sets e_i have a nonzero measure. Thus we can assume that the m_i were not included among these exceptional values, i.e., we shall make $m(e_i) = 0$, which simplifies the sums σ and Σ.

So if we assume that $m_0 = 0$,[31] we can write

$$\int_{a}^{a+h} f(x)\,dx = \lim \left\{ \sum_{0}^{\infty} m_{i+1}\, m\left[e'_i(h) \right] + \sum_{-1}^{-\infty} m_i\, m\left[e'_i(h) \right] \right\},$$

indicating by $e'_i(h)$ that portion of e'_i contained between a and $a + h$. Let us consider a fixed system of numbers m_i. The two series of the second member will only vary if h varies, and we can assume h to be small enough that the absolute value of any finite number of terms of the second member is however small one likes. Thus one may assume h to be small enough that both series of the second member, one of which has positive terms and the other negative, will have an absolute value as small as one wishes.

But to progress to the limit, new members have to be introduced among the m_i that were chosen; this operation makes the two series of the second member decrease in absolute value. Thus, it has been well demonstrated that $\int_{a}^{a+h} f(x)\,dx$ can be made as small as one likes. The indefinite integral is thus indeed a continuous function.

§ 27. If M and m are the maximum and minimum of $f(x)$ within $(a, a + h)$, we have

$$m h < \int_{a}^{a+h} f(x)\,dx < M h,$$

[30]One might also add that its variation is bounded; this variation being at most equal to $\int | (f - f_0)| \, dx$. The demonstration of this is the same as that employed by Mr. Jordan, *Cours d'Analyse*, § 81. This explains the results obtained further on (§§ 30, 31, 32).
[31]Then e_0 would perhaps not have a measure of zero, but that is not important for what follows.

and hence

$$m < \frac{F(a + h) - F(a)}{h} < M;$$

thus for $x = a$, if $f(x)$ is continuous at this point, $F(x)$ has a derivative equal to $f(a)$.

If $f(x)$ is continuous for any value of x, $F(x)$ is any one of its functions that has $f(x)$ as a derivative, i.e., one of the *primitive functions* of $f(x)$.

For continuous functions, then, looking for the primitive functions is identical to looking for the indefinite integrals of a given function. This well-known result remains true when dealing with a Riemann-integrable derived function.[32] But there are derived functions that are not Riemann-integrable;[33] given one of these functions, one cannot calculate its primitive functions using Riemann integration.

We shall see that any bounded derived function has an indefinite integral that is one of its primitive functions. We shall thus be able to calculate the primitive function (if there is one) of a given bounded function.

As regards unbounded derived functions, we shall demonstrate that if they have integrals, their primitive functions are identical to their indefinite integrals.

§ 28. The derivative of a function $f(x)$, as h tends towards zero, is the limit of the expression

$$\frac{f(x + h) - f(x)}{h} = \varphi(x),$$

which, since h is fixed, represents a continuous function. Thus the derivative is a limit of continuous functions and is summable.

Let us assume that the derivative f always has an absolute value less than M. By virtue of the finite increments theorem $\varphi(x) = f'(x + \theta h)$, the functions $f(x)$ are bounded in their set, and consequently we have (§ 25)

$$\int_a^b f'x \, dx = \lim \int_a^b \varphi(x) \, dx = \left[\lim_{h=0} \frac{1}{h} \int_x^{x+h} f(x) \, dx \right]_a^b,$$

and so

$$\int_a^b f'x \, dx = f(b) - f(a).$$

Any bounded derived function has its primitive functions as indefinite integrals. This result is still true if we are dealing with a derivative bounded on the right or on the

---

[32] See Darboux, *Mémoire sur les fonctions discontinues.*
[33] Mr. Volterra was the first to give an actual example of these functions (*Giornale de Battaglini*, vol. XIX, 1881). This example is reproduced further on.

left; also if it involves a limit that $\varphi(x)$ tends towards for certain values of h tending towards zero.

§ 29. In application of the foregoing, we shall look to see if there are primitive functions for the function defined as follows:

Let E be a nondense, closed set in every portion of $(0, 1)$, with a nonzero measure. Let (a, b) be an interval contiguous to E,[34] and c the midpoint of this interval. The function

$$\varphi(x - a) = 2(x - a) \sin \frac{1}{x - a} - \cos \frac{1}{x - a}$$

zeroes out an infinite number of times between a and c. Let $a + d$ be the point closest to c between a and c for which the function is zero.

The function $f(x)$ with which we shall be concerned is zero for all points of E. In every interval (a, b) contiguous to E, it is equal to $\varphi(x - a)$ between a and $a + d$, zero between $a + d$ and $b - d$, and equal to $-\varphi(b - x)$ between $b - d$ and b.

This function is continuous in every interval contiguous to E, discontinuous for all points of E that are points of discontinuity of the second sort.

Furthermore, $f(x)$ is always contained between -3 and $+3$. In order for $f(x)$ to have a primitive function, it first of all has to have a definite integral in the interval $(0, 1)$. This integral, if it exists, is equal to the integral within E plus the integral within $C(E)$, assuming it exists. Now the integral in E exists and is zero; the integral in $C(E)$ also exists, since it is the sum of the integrals in the intervals contiguous to E, which are zero.

In consequence of this, the function $F(x)$, being zero for all points of E and defined in every interval (a, b) contiguous to E through the equalities

$$F(x) = (x - a)^2 \sin \frac{1}{(x - a)} \qquad \text{between } a \text{ and } a + d,$$

$$F(x) = d^2 \sin \frac{1}{d} \qquad \text{between } a + d \text{ and } b - d,$$

$$F(x) = (b - x)^2 \sin \frac{1}{(b - x)} \qquad \text{between } b - d \text{ and } b,$$

is equal to $\int f(x)\, dx$.

Thus, if $f(x)$ has primitive functions, $F(x)$ is one of them.

For all points where $f(x)$ is continuous, i.e., for all points of $C(E)$, we obviously get

$$F'(x) = f(x).$$

---

[34] That is, an interval containing no points of E and the endpoints of which are points of E.—This expression is Mr. Baire's.

Let a be a point of E. If a is an endpoint of an interval contiguous to E, located to the right of a, $F(x)$ obviously has a derivative whose right side is zero. Let us assume that, to the right of a, there are an infinite number of points of E that have a as their limiting point. If α_i is one of these points, for x greater than α_i, the absolute value of the ratio

$$r(x) = \frac{F(x) - F(a)}{x - a}$$

is less than

$$\frac{(x - \alpha_i)^2}{x - a} < x - a,$$

and thus tends towards zero when x tends towards a.

At all points of E, $F(x)$ thus has a derivative whose right side is zero, and one would also see that it has a derivative whose left side is zero, and thus for any value of x contained between 0 and 1 we get

$$F'(x) = f(x).$$

The function $f(x)$ is thus a derived function; it is not Riemann-integrable, since the set of its points of discontinuity has a nonzero measure.

This example of a non-Riemann-integrable function we owe to Mr. Volterra, *Giornale de Battaglini,* volume XIX.[35]

§ 30. The primitive functions we have just found have a bounded variation.[36] We shall show that: *the necessary and sufficient condition for the existence of an integral of the derivative (bounded or not) of a differentiable function is that this function have a bounded variation. If it does, the function is one of the indefinite integrals of its derivative.*

Since $f'(x)$ is summable, in investigating its integral we shall proceed as in paragraph 24. We shall assume all e_i have a measure of zero and furthermore that $m_0 = 0$, which is possible if, instead of arguing on the basis of the given function $f(x)$, we argue on the basis of $f(x) + Kx$, where K has been conveniently chosen.

Corresponding to every point x_0 of e'_i we can have an interval (α, β) such that if we have

$$\alpha < a \le x_0 \le \beta,$$

we would also have

$$m_i < r(a, b) = \frac{f(b) - f(a)}{b - a} < m_{i+1}.$$

---

[35]Certain uniformly convergent series, the terms of which are functions analogous to the one we have just been looking at, allow Mr. Volterra to give examples of derived functions that are not integrable in any interval. Term for term integration of these series will give us the primitive functions.

[36]Here we shall make use of certain properties of these functions (see Jordan, *Comptes Rendus de l'Académie des Sciences* 1881 and *Cours d'Analyse* 2nd Edition, volume I). Most of these properties are taken up again in the following chapter, such that paragraphs 30 to 35 could have been placed in that chapter. The order followed by the text allows us to put in one place everything related to the investigation of primitive functions.

We shall define (α, β) as the largest possible interval of a length at most equal to a given number σ and that has x_0 as its midpoint.

If $m_i - m_{i-1}$ is always less than η, $(b - a) r (a, b)$ is equal, except for an amount *of* $\eta(b - a)$, to $f'(x_0)(b - a)$.

Let $E_i(\sigma)$ be the sum of the intervals corresponding to the points of e'_i. $E_i(\sigma)$ can be considered the sum of nonoverlapping intervals. If (a, b) is one of these intervals and if we have

$$a < \alpha < \beta < b,$$

then we also have

$$m_i < r(\alpha, \beta) < m_{i+1},$$

provided that, between α and β, there is at least one point of e'_i.

$E_i(\sigma)$ contains e'_i. Let us make σ tend towards zero, and let x_0 be a point belonging to an infinite number of sets $E_i(\sigma)$. $f'(x)$ is the limit of the values of $r(\alpha, \beta)$ relative to the intervals of $E_i(\sigma)$ that contain x_0. Thus x_0 is a point of e_i, of e'_i, or of e_{i+1}. Consequently, the series E_i, made up of the points common to an infinite number of $E_i(\sigma)$ relative to the values of σ that tend towards zero, contains e'_i and points of $e_i + e_{i+1}$. Thus it has the same measure as e'_i. Furthermore, since each $E_i(\sigma)$ contains the sets relative to the smallest values of σ, $m(E_i)$ is the limit of $m[Ei(\sigma)]$.

Thus one can choose the numbers

$$\ldots \sigma_{-2}, \sigma_{-1}, \sigma_0, \sigma_1, \sigma_2 \ldots$$

such that the sum D is as small as one pleases,

$$D = \sum_{-\infty}^{+\infty} |m_i| \cdot \Big(m[E_i(\sigma_i)] - m(E_i) \Big).$$

That having been said, let us note that $\int f' dx$ and $\int |f'| dx$ exist at the same time, such that $\int f' dx$ exists if the series

$$\sum_{-\infty}^{+\infty} |m_i| \cdot m(e'_i) = \sum_{-\infty}^{+\infty} |m_i| \cdot m(E_i)$$

is convergent, i.e., if the same is true for

$$V = \sum_{-\infty}^{+\infty} |m_i| \cdot m[E_i(\sigma_i)].$$

From among the intervals making up $E_i(\sigma_i)$, one can always choose a finite number such that the contributions of these intervals A within V is as large as one likes if V is divergent, and as close as one likes to the value of V if this series is convergent. Let us eliminate enough of these intervals A without changing the sum of the intervals so that none of the intervals retained is within other retained intervals. The contribution within V of the intervals eliminated is less than D.

Let us consider two overlapping intervals (a_i, b_i) and (a_j, b_j) relative to e'_i and e'_j. Let us assume we have

$$a_i < a_j < b_i < b_j.$$

Between a_j and b_i there cannot at one time be points of e'_i and e'_j without $r(a_j + \varepsilon, b_i - \varepsilon)$ being simultaneously contained between m_i and m_{i+1} and between m_j and m_{j+1}. Thus between a_j and b_i we can find a point c such that between a_i and c there is a point of e'_i, and, between c and b_j, there is a point of e'_j. Thus we have

$$|f(c) - f(a_i)| + |f(b_j) - f(c)| = (c - a_i)|r(a_i, c)| + (b_j - c)|r(c, b_j)|.$$

Thus, the first member, i.e., the variation of $f(x)$ between a_i and b_j when we consider the division

$$a_i \qquad c \qquad b_j,$$

is equal to the contribution within V of the two intervals (a_i, b_i) and (a_j, b_j), apart from an amount less than $(b_i - a_j)|m_j - m_i| + \eta(b_j - a_i)$. The quantity $(b_i - a_j)|m_j - m_i|$ is less than the contribution within D of the interval (a_j, b_i).

Along the same lines, we are led to consider a series of increasing values $x_0, x_1, x_2, \ldots$, that is finite in number. The sum $\Sigma |f(x_i) - f(x_{i+1})|$ differs from the contributions within V of the intervals A by less than $D + \eta\, m(A)$. Now this sum is less than the total variation of $f(x)$. Thus, the limit of V, i.e., $\int |f|\, dx$, is less than, or at most equal to, the total variation of $f(x)$. That is, if $f(x)$ has a bounded variation, $\int |f'|\, dx$ exists and is less than the variation of $f(x)$.

§ 31. Let us assume that the integral $\int |f'|\, dx$ exists.

Let us enclose the points of e_i within a countable infinity of intervals A_i, and one may opt to have $m(A_i)$ be as small as one likes. Corresponding to each point x_0 of e_i, let us have the largest interval (α, β) of a length less than σ'_i, with x_0 as its midpoint, and the whole thing within A_i, such that

$$\alpha < a \leq x_0 \leq b < \beta$$

entails

$$m_i - \varepsilon_i < r(a, b) < m_i + \varepsilon_i.$$

Let $e_i(\sigma'_i)$ be the sum of these intervals. As long as we choose suitable σ'_i and ε'_i, the sum D' will be however small one likes,

$$D' = \sum_{-\infty}^{+\infty} |m_i| \cdot m[e_i(\sigma'_i)].$$

Each $E_i(\sigma_i)$ or $e_i(\sigma'_i)$ is the sum of a countable infinity of nonoverlapping intervals. If $f(x)$ is defined within (a, b), each point *inside* (a, b) is *inside* at least one of these intervals and furthermore a and b are endpoints of such intervals. Thus according to a set theorem we can choose, from among the intervals that make up the $E_i(\sigma_i)$ and the $e_i(\sigma_i)$, a finite number of intervals B such that any point within (a, b) is within one of the B.

We shall assume this choice to be made in such a way that no retained interval is within other retained intervals. The contribution within V of the unused intervals of $E_i(\sigma)$ is at most equal to $D + D_i$, D_i being the integral of $|f'|$ in the set of $e_i(\sigma'_i)$.

In arguing based on the B as well as on the A, we are led to consider the numbers

$$x_0 = a < x_1 < x_2 \ldots < x_n = b.$$

The sum $\Sigma \, |f(x_i) - f(x_{i-1})|$ is equal to the contribution within V of the retained intervals coming from the $E_i(\sigma_i)$, apart from $D + D' + \eta(b - a)$.

Now if two consecutive numbers x_i come from the same interval belonging to one of the $E_i(\sigma_i)$ or the $e_i(\sigma'_i)$, which interval can be broken down into intervals of lengths at most equal to $2\sigma_i$ or $2\sigma'_i$, the sum of the variations corresponding to such a division always differs from V by less than $2D + D_1 + D' + \eta(b - a)$. Now by introducing these new points of division while making the maximum σ of the σ_i and the σ'_i small enough, we render the sum $\Sigma \, |f(x_i) - f(x_{i-1})|$ as close as desired to the total variation of $f(x)$ within (a, b).

From this it follows that, if $\int f'(x)\,dx$ exists, the function $f(x)$ has a bounded variation, this variation being equal to the value of the integral $\int |f'|\,dx$.

§ 32. Thus we have found the necessary and sufficient condition for the integral of $|f'|$ to exist, and we are aware of what it means.

But the preceding argument supplies other results. Let us in fact resume the argument and turn our attention to the sets $e_i, E_i, E_i(\sigma), E_i$, etc., with positive indices.

We see that the total positive variation of $f(x)$ between a and x is equal to the integral of $f'(x)$ extended over the set of points for which f' is positive, i.e., over the integral

$$p(x) = \frac{1}{2} \int_a^x (f' + |f'|)\,dx.$$

Likewise, for the negative variation we have

$$-n(x) = \frac{1}{2} \int_a^x (f' + |f'|)\,dx.$$

And since we have

$$f(x) - f(a) = p(x) - n(x),$$

$$f(x) = f(a) + \int_a^x f'(x)\,dx.$$

Thus, for a given function $f(x)$ we know how to recognize whether it is the derivative of a function with a bounded variation and, if so, we know how to find its primitive functions.

If $f(x)$ is bounded, its primitive functions, if any, have a bounded variation, and we can find them.

But integration, as defined by us, does not allow us to know whether a given function has primitive functions with unbounded variation.[37]

§ 33. The function $f(x)$ given by $f(x) = x^2 \sin\frac{1}{x^2}$, for $x \neq 0$ and $f(0) = 0$, is continuous and has an unbounded variation.

In fact

$$f\left(\frac{1}{\sqrt{k\pi}}\right) = 0,$$

$$f\left(\frac{1}{\sqrt{k\pi + \frac{\pi}{2}}}\right) = (-1)^k \frac{1}{k\pi + \frac{\pi}{2}},$$

and thus the sum of the variations is $\sum \dfrac{1}{k\pi + \dfrac{\pi}{2}}$, a divergent series.

This function has a derivative $f'(x)$,

$$f'(x) = 2x \sin \frac{1}{x^2} - \frac{2}{x} \cos \frac{1}{x^2}, \qquad \text{where } x \neq 0$$

and

$$f'(0) = 0.$$

$f'(x)$ offers us an example of an unbounded summable function that has no integral. When $f'(x)$ is given, the foregoing methods do not allow us to find $f(x)$.

It is interesting to note that the classical definition of the integral of a function that becomes infinite in the neighborhood of a point lets us find $f(x)$ if we know $f'(x)$. This is because, for the case where the function to be integrated is not bounded, the definition that we have adopted is not a generalization of the classical definition, but agrees with it whenever both apply. Moreover, it would be very easy to generalize the notion of the definite integral such that the classical definition and the one we have adopted become particular cases of a more general definition. To simplify the following theorems, however, we shall preserve the previously adopted sense of the words *definite integral*, but we shall extend the meaning of the words *indefinite integral*.

[37] The latter result was obvious since (§ 26, Note) any indefinite integral has a bounded variation. The foregoing demonstration shows that to obtain a function $f(x)$ with an unbounded variation when we know its derivative, using a method analogous to the one we used when f has a bounded variation, the terms of series such as $\Sigma\, m_i m\big[E_i(\sigma_i)\big]$ and $\Sigma\, m_i m(e_i')$ would have to be put in a certain order.

Generalizing the notion of indefinite integral given in the following paragraph allows us to obtain this result in certain cases; it will also give us primitive functions with unbounded variation.

We saw that every indefinite integral was continuous. If we now consider this property as one part of the definition of indefinite integrals, we are led to say

A function $f(x)$, defined in (α, β), has in that interval an indefinite integral $F(x)$, if there exists one and only one continuous function $F(x)$, unique except for an additive constant, such that

$$F(b) - F(a) = \int_a^b f(x)\,dx$$

for all systems of numbers a and b (between α and β) chosen in such a way that the second member is meaningful.[38]

§ 34. The indefinite integral of a derived function is always one of its primitive functions since a primitive function is continuous and, based on what we just said, properly satisfies the equality

$$F(b) - F(a) = \int_a^b f(x)\,dx$$

whenever the second member is meaningful.

We shall thus be able to find the primitive function of § 30's function $f'(x)$.

But it is easy to form derived functions that do not have indefinite integrals.

Let $\varphi(x)$ be a differentiable function defined between 0 and 1 which zeroes out for 0 and 1, such that its derivative has a bounded variation in every interval *inside* (0, 1) and an unbounded variation in every interval one of whose endpoints is 0 or 1.

We can find $\varphi(x)$ when $\varphi'(x)$ is known since $\varphi(x)$ is the indefinite integral of $\varphi'(x)$ which zeroes out when $x = 0$.

Let us look at a closed set E, which is nondense in every part of (0, 1) and has a nonzero measure. [For example, the one we get by extracting from (0, 1) an indefinite series of intervals whose midpoints are points with rational abscissa values and the sum of whose lengths is less than 1.]

Let us define a function $f(x)$ to be continuous, through the condition that it be equal to $(b - a)^2\, \varphi(\frac{x - a}{b - a})$ in every interval (a, b) contiguous to set E. Thus, $f(x)$ is zero at all points of E. This function is differentiable, its derivative is zero for all points of E, and equal to $(b - a)\, \varphi'(\frac{x - a}{b - a})$ at every point of an interval (a, b) contiguous to E.

This derivative $f'(x)$ does not allow for an indefinite integral. In fact, if it did, its indefinite integrals would be $f(x) +$ *constant*. But let $\psi(x)$ be the function that

---

[38]Compare this definition with that given by Mr. Jordan for the definite integral of an unbounded function. *Cours d'Analyse*, 2nd Edition, Volume II.

represents the measure of the set of those points of E which are in the interval $(0, x)$. Let $\psi(x)$ be a continuous function, constant in any interval contiguous to E. Thus $f(x) + \psi(x)$ satisfies the equality

$$\left[f(x) + \psi(x) \right]_{\alpha}^{\beta} = \int_{\alpha}^{\beta} f'(x)\, dx$$

for all systems α, β for which the second member is meaningful.

The definitions we have given are thus insufficient for us to be able to speak of indefinite integrals of $f'(x)$.

The problem of the search for primitive functions is not completely solved.[39]

§ 35. Let $f(x)$ be a continuous function; h can be given a series of values tending towards zero such that

$$\frac{f(x_0 + h) - f(x_0)}{h}$$

has a limit. The set of numbers thereby defined, corresponding to the positive values of h, has an upper limit Λ_d and a lower limit λ_d, which are the extreme oscillations for the point x_0 which occur to the right of the function $f(x)$. We define Λ_g and λ_g the same way. These four numbers are the derivate numbers; for certain problems, they provide the same services that the derivative does.[40]

The following problem: *finding a function when one of its derived numbers is known,*[41] is thus a generalization of the problem we have just been dealing with.

Certain particular cases of the problem can be solved by Riemann-integration (Dini, *loc. cit.*). Integration, the way we have defined it, would allow it to be solved in more extended cases. We shall confine ourselves to the following remarks.

First and foremost, if one of the four derived numbers is always finite, Λ_d for example, the function is summable. Let us in fact look at the set E of values of x for which Λ_d is greater than a given number M. Let us give h all rational positive values less than ε_1; corresponding to each of them is a function $\frac{f(x + h) - f(x)}{h} = \varphi(x, h)$. Corresponding to $\varphi(x, h)$ is a measurable set $E(h)$ made up of all points for which

$$\varphi(x, h) > M.$$

Let $E(\varepsilon_1)$ be the sum of all $E(h)$; it is measurable.

[39] It could be said that we can solve the problem when x interval of variation can be considered to be the sum of a nondense set E and the set of intervals (α, β) contiguous to E, where set E is reducible and the proposed function has an indefinite integral $F(x)$ in every interval (α, β).

Even if E is not reducible, the problem can be solved provided that the function is integrable in E and the series of quantities $[F(\beta) - F(\alpha)]$ is absolutely convergent. This is true in the preceding example, but it is not the case in general.

[40] See Dini: *Fondamenti per la teorica delle funzioni di variabili reali.*

[41] This problem is meaningful; i.e., all functions that have a given derivate number in common differ only by a constant (Volterra, *Sui principii del Calcolo Integrale.* Giornale de Battaglini, XIX).

Corresponding to $\varepsilon_2, \varepsilon_3, \ldots$ are $E(\varepsilon_2), E(\varepsilon_3) \ldots$

If the ε tend towards zero, the set common to all $E(\varepsilon_i)$ (which is measurable) contains the set we are looking for, plus points for which $\Lambda_d = M$. That is enough to allow us to conclude that the function Λ_d is summable.

Let us now suppose that one of the four derivate numbers is bounded, in which case all the others are as well.[42] Λ_d will then have an integral.

Let us consider a series of positive numbers decreasing to zero, $h_1, h_2, \ldots$ and the functions

$$\varphi(x, h_i) = \frac{f(x + h_i) - f(x)}{h_i}.$$

To every value of x there corresponds a value n such that, for $i \geq n$,

$$\varphi(x, h_i) < \Lambda_d(x) + \varepsilon,$$

Let E_k be the set of values of x for which $n \leq k$. The complement $C(E_k)$, taken in relation to the interval being considered, has a measure that tends towards zero along with $\frac{1}{k}$, and for the points of this set we have

$$|\varphi(x, h_k) - \Lambda_d(x)| \leq M$$

if M is the upper limit of the absolute value of Λ_d. Thus[43]

$$\int_a^b \varphi(x, h_k)\, dx < \int_a^b \Lambda_d(x)\, dx + \varepsilon\, m(E_k) + M\, m\left[C(E_k)\right].$$

Let us evaluate the first member

$$\int_a^b \varphi(x, h_k)\, dx = \int_a^b \frac{f(x + h_k) - f(x)}{h_k} = f(b + \theta\, h_k) - f(a + \theta'\, h_k).$$

When k indefinitely increases, this quantity tends towards $f(b) - f(a)$, and thus we get

$$f(b) - f(a) < \int_a^b \Lambda_d(x)\, dx.$$

In the same way, we would find

$$\int_a^b \lambda_d(x)\, dx < f(b) - f(a).$$

---

[42] For if Λ_d is always contained between A and B, the ratio $\frac{f(b) - f(a)}{b - a}$ is always contained between A and B. (See DINI, *Fondamenti*, etc.)

[43] In order for $\int_a^b \varphi(x, h_k)\, dx$ to be meaningful, it is necessary that $f(x)$ be defined in $a, b + h_k$. It is sufficient to define $f(x)$ as a constant equal to $f(b)$ for an x greater than b.

Thus, *if both derivate numbers to the right (or to the left) of a function f(x) are bounded and have the same integral, their indefinite integrals are equal to f(x) except for an additive constant.*

III. DEFINITE INTEGRALS OF FUNCTIONS WITH SEVERAL VARIABLES

§ 36. There is no difficulty in extending the results we obtained to functions with several variables.

A function f will be called summable if the set of points for which

$$a < f < b$$

is measurable, whatever the values of a and b.

Functions that are continuous with respect to the set of their variables are summable. The sum, the product of two summable functions, the limit of a series of summable functions, are summable functions. Thus the discontinuous functions that Mr. Baire calls first-class functions, second-class functions, etc., are summable.

Functions with n variables, continuous with respect to each of them, are functions of the $n - 1$ class at most,[44] and thus they are summable.

Let f be a summable function. Let us consider the numbers

$$\ldots < m_{-2} < m_{-1} < m_0 < m_1 < m_2 < \ldots$$

such that $m_i - m_{i-1}$ has a maximum of η.

$f = m_i$ for the points of a measurable set e_i; $m_i < f < m_{i+1}$ for the points of a measurable set e'_i.

The two sums

$$\sigma = \sum m_i m(e_i) + \sum m_i m(e'_i), \quad \Sigma = \sum m_i m(e_i) + \sum m_{i+1} m(e'_i)$$

are at the same time either both meaningful or both not.

If they are meaningful, this is so for whatever values are chosen for m_i, and these two sums tend towards one and the same limit when η tends towards zero.

This limit is the integral of f. σ and Σ are meaningful when f is bounded, such that *every bounded summable function has an integral.*

The preceding definitions apply whether the function is defined in a domain or for the points of a set, which would necessarily have to be measurable for the function to be summable. Let f be a bounded function defined on a measurable set E. If f is not summable, there is an infinite number of bounded summable functions φ such that we always have

$$f(x) > \varphi(x).$$

Let $\varphi_1(x), \varphi_2(x), \ldots$ be a series of those functions whose integrals tend towards the upper limit of the integrals of the φ functions. Let $\psi(x)$ be a function that, for every

---

[44] See Lebesgue, *Sur l'approximation des fonctions.* (Bulletin des Sciences Mathématiques, 1898.)

value of x_0, is equal to the upper limit of the numbers $\varphi_1(x_0)$, $\varphi_2(x_0)$, ... It can easily be shown that $\psi(x)$ is summable. Its integral is not less than the integrals of the functions $\varphi_i(x)$, and since $\psi(x)$ is a function $\varphi(x)$, the integral of $\psi(x)$ is exactly equal to the upper limit of the integrals of the $\varphi(x)$ functions. Thus, given a bounded function f, there is a summable function ψ not greater than f and the integral of which is the upper limit of the integrals of the summable functions not greater than f. *This is the lower integral of f.*

The upper integral would be defined in the same way.[45]

§ 37. We are going to look at whether the calculation of a multiple integral can be reduced to the calculations of simple integrals. Confining ourselves to the case of two variables, we shall attempt to generalize the classical formula

$$\int\int f\,dx\,dy = \int\left(\int f\,dy\right)dx.$$

The simplest case we have to examine corresponds to $f = 1$. Then we have to evaluate the surface measure of a set as a function of the linear measures of its sections. The considerations developed in §§ 18 and 19 solve a particular case of this problem.

Let E be a measurable plane set. By $E(x_0)$ we shall indicate the set of points of E whose abscissa is x_0, i.e., the section of E where $x = x_0$. $E(x_0)$ is not necessarily measurable if there are sets of points on a straight line that are not linearly measurable, since every bounded set of points on a straight line is measurable along its surface. But $E(x_0)$ will be linearly (B) measurable if E is (B) measurable along its surface. Now we know that E contains a (B) measurable set, E_1, with a measure of $m(E)$. Thus the measure of $E_1(x_0)$ will at most be equal to the internal measure of $E(x_0)$; i.e., it will be at least equal to the lower integral, where $x = x_0$, of the function φ equal to 1 for the points of E, equal to zero for the other points. Thus,

$$m_l[E_1(x_0)] \le \int_{\text{inf.}} \varphi(x_0, y)\,dy.$$

Reverting to § 7, where the existence of E_1 was demonstrated, we see that E_1 was defined as made up of the points common to all sets in the series $A_1, A_2, \ldots$. And set A_i was defined as the sum of a countable infinity of nonoverlapping rectangles whose sides are parallel to $0x$ and $0y$. A_i contains $A_{i+1}, A_{i+2}, \ldots$.

Now $m_l[A_i(x)]$ is the sum of the measures of the sections to the right of abscissa x of the rectangles C_{ij} that make up A_i. Thus we have

$$m_l[A_i(x)] = \sum_j m_l[C_{ij}(x)].$$

---

[45]All these conditions are interpreted geometrically, as for the case of one variable. If there are n variables, one should think in terms of a space with $n + 1$ dimensions.

In this series, the absolute values of the remainders have an upper limit, since, because E_1 is bounded, all A_i are located in one and the same bounded domain, and consequently the set (relative to i and to x) of numbers $m_l[A_i(x)]$ is bounded. This series is thus integrable term for term (§ 25).

The integral of $m_l[C_{ij}(x)]$ is the area of C_{ij}, thus

$$m_s(A_i) = \int m_l[A_i(x)]\,dx.$$

Now the limits for an infinite i of the numbers $m_l[A_i(x)]$ and $m_s(A_i)$ are $m_l[E_i(x)]$ and $m_s(E_1)$, and since the set of these numbers is limited, we have

$$\int m_l[E_1(x)]\,dx = \lim_{i=\infty} \int m_l[A_i(x)]\,dx = \lim_{i=\infty} m_s(A_i) = m_s(E_1)(*).[46]$$

Thus we can conclude that

$$m_s(E) \le \int_{\text{inf.}} m_{l,\text{int.}}[E(x)]\,dx$$

and also that

$$m_s(E) \le \int_{\text{inf.}} \left(\int_{\text{inf.}} \varphi(x,y)\,dy \right) dx.$$

In the same way, it will be shown that

$$m_s(E) \ge \int_{\text{sup.}} \left(\int_{\text{sup.}} \varphi(x,y)\,dy \right) dx.$$

From that, we conclude

$$m_s(E) = \int_{\text{inf.}} \left(\int_{\text{inf.}} \varphi(x,y)\,dy \right) dx = \int_{\text{sup.}} \left(\int_{\text{sup.}} \varphi(x,y)\,dy \right) dx(**).[47]$$

For the function $f = 1$, defined only in the summable set E, whose sections are perhaps not all measurable,

$$\int\int^{E} f\,dx\,dy = \int_{\text{inf.int.}}^{e} \left(\int_{\text{inf.int}}^{E(x)} f\,dy \right) dx = \int_{\text{sup.ext.}}^{e} \left(\int_{\text{sup.ext.}}^{E(x)} f\,dy \right) dx,$$

with the integral with respect to x being extended to the set e (a projection of E onto $0x$), and the second integral being extended to the set $E(x)$. But these two sets are perhaps not linearly measurable. The sign $\int_{\text{inf.int.}}^{A}$ indicates the upper limit of the lower integrals of f extended to the measurable sets contained within A.

---

[46] This argument may be interpreted as follows: $m_l[E_1(x)]$ is a function of at least the second class.
[47] Up to this point the integrals are extended to certain segments of the x or y axes.

The foregoing equality can also be written

$$m_s(E) = \int_{\text{inf.int.}}^{e} m_{l,\text{int.}}[E(x)]dx = \int_{\text{sup.ext.}}^{e} m_{l,\text{ext.}}[E(x)]dx.$$

§ 38. The formula found to express $\iint_E f\,dx\,dy$ is general, and it applies to all bounded summable functions.

Let us use $\varphi(x, y)$ to indicate a function that equals f for the points of E, and equals zero for the other points. If E is entirely within rectangle $OACB$, whose sides OA and OB are held by $0x$ and $0y$, the formula to be demonstrated is equivalent to the following

$$\iint_{OACB} \varphi(x, y)dx\,dy = \int_{O,\text{inf.}}^{A} \left(\int_{O,\text{inf.}}^{B} \varphi(x, y)dy \right)dx = \int_{O,\text{sup.}}^{A} \left(\int_{O,\text{sup.}}^{B} \varphi(x, y)dy \right)dx.$$

This is the formula that we shall demonstrate.

Let $m_0, m_1, \ldots, m_n$ be the divisions of the $\varphi(x, y)$ interval of variation. Let us use $\varphi_p(x, y)$ to indicate the function equal to φ the points of e_p (using § 19 notation), and equal to zero for the other points, and use $\varphi'_p(x, y)$ to indicate the function equal to φ for the points of e'_p, and equal to zero for the other points.

We have

$$\iint \varphi(x, y)dx\,dy = \sum_p \iint \varphi_p(x, y)dx\,dy + \sum_p \iint \varphi'_p(x, y)dx\,dy,$$

$\iint \varphi_p(x, y)\,dx\,dy$ is equal to $m_p \cdot m(e_p)$ and, following the preceding section, we get

$$\iint \varphi_p(x, y)dx\,dy = \int_{\text{inf.}} \left(\int_{\text{inf.}} \varphi_p(x, y)dy \right)dx,$$

$\iint \varphi'_p(x, y)\,dx\,dy$ is contained between $m_p\,m(e'_p)$ and $m_{p+1}m(e'_p)$. Furthermore if we replace φ'_p by a function ψ, equal to m_p or m_{p+1} at all points where φ_p is different from zero, and equal to zero when φ'_p is zero, the interval is changed by less than $(m_{p+1} - m_p)\,m(e'_p)$. Moreover, the two expressions

$$\int_{\text{inf.}} \left(\int_{\text{inf,}} \psi\,dy \right)dx \qquad \int_{\text{inf.}} \left(\int_{\text{inf,}} \varphi'\,dy \right)dx$$

are also different by less than $(m_{p+1} - m_p)\,m(e'_p)$. Thus, except for an amount less than $2(m_{p+1} - m_p)\,m(e'_p)$, we have

$$\iint \varphi'_p(x, y)dx\,dy = \int_{\text{inf.}} \left(\int_{\text{inf.}} \varphi'_p(x, y)dy \right)dx.$$

Let η be the maximum of $m_{p+1} - m_p$; except for an amount less than $2\eta m$ $(OACB)$ we will get

$$\iint \varphi(x,y)\,dx\,dy = \sum_p \int_{\text{inf.}} \left(\int_{\text{inf.}} \varphi(x,y)\,dy \right) dx + \sum_p \int_{\text{inf.}} \left(\int_{\text{inf.}} \varphi'_p(x,y)\,dy \right) dx.$$

Now we have

$$\int_{\text{inf.}} \left(\int_{\text{inf.}} \varphi(x,y)\,dy \right) dx \leq \sum_p \left\{ \int_{\text{inf.}} \left(\int_{\text{inf.}} \varphi_p(x,y)\,dy \right) dx + \int_{\text{inf.}} \left(\int_{\text{inf.}} \varphi'_p(x,y)\,dy \right) dx \right\}.$$

Thus we get (whatever η is)

$$\iint \varphi(x,y)\,dx\,dy \leq \int_{\text{inf.}} \left(\int_{\text{inf.}} \varphi(x,y)\,dy \right) dx + 2\eta \cdot m(OABC).$$

The result of this inequality, and the analogous one related to upper integrals, is the formula stated.

§ 39. If the given function is such that all sets e'_p are (B) measurable, in which case we can say that the function is (B) summable, the formula is simplified and becomes

$$\iint_{OACB} \varphi(x,y)\,dx\,dy = \int_O^A \left(\int_O^B \varphi(x,y)\,dy \right) dx.$$

This is the classical formula. We know that this formula must be replaced by a more complicated formula, analogous to the one we obtained, when dealing with Riemann-integration, applied in all its generality.[48]

Among (B) summable functions, we can single out continuous functions, the limits of continuous functions or first-class functions, the limits of first-class functions or second-class functions, and in a general way all functions of class n, where n is finite.

In particular, the simple classical formula is applicable to functions of n variables continuous relative to each of them.

This formula is also applicable to the functions f''_{xy} and f''_{yx}[49] if they exist and are bounded. Thus we have

$$f(x,y) - f(0,y) - f(x,0) + f(0,0) = \int_0^x \left(\int_0^y f''_{xy}\,dy \right) dx = \iint f''_{xy}\,dx\,dy.$$

This formula solves the problem that, for the case of two variables, is analogous to that concerning the search for primitive functions.

---

[48] See Jordan (*loc. cit.*) §§ 56, 57, 58.
[49] Since they are at most of the second class.

§ 40. Some of the foregoing results may be extended to unbounded functions.

An unbounded nonsummable function can have a lower integral and an upper integral. Without having to resume the previous arguments, we see that we have

$$\int\int \varphi(x, y)\, dx\, dy = \int_{\text{inf.}}\left(\int_{\text{inf.}} \varphi(x, y)\, dy\right) dx = \int_{\text{sup.}}\left(\int_{\text{sup.}} \varphi(x, y)\, dy\right) dx$$

whenever the integrals that appear in this formula are meaningful. The same is true with the classical formula

$$\int\int \varphi(x, y)\, dx\, dy = \int\left(\int \varphi(x, y)\, dy\right) dx.$$

The arguments employed in this chapter have led to a generalization of the concept of definite integral.

For such a generalization to be serviceable, it must satisfy certain conditions that can easily be discerned and that may be imposed *a priori*.

Here are some of those conditions. We must have

$$\int_a^c = \int_a^b + \int_b^c, \qquad \int f + \varphi = \int f + \int \varphi.$$

The definition adopted must contain Riemann's definition as a particular case.

There must not be significant differences between the case with one variable and that with several variables.

Finally, if we want integration to allow us to solve the fundamental problem of integral calculus: how to find a function when its derivative is known, the definite integral of a derived function, considered as a function of its upper limit, must be a primitive function of f.

The definition I have adopted, at least for the case where the function to be integrated is bounded, does indeed fulfill all these conditions. But these conditions are insufficient to define the integral of a bounded function (except where the function is an algebraic sum of Riemann-integrable functions and derived functions), such that chapter one's methods[50] could not be used. Though unable to demonstrate that the proposed definition is the only one that satisfies the conditions imposed, I have sought to show that it was natural and, from a geometric perspective, appeared almost as a matter of necessity.

---

[50] These methods are analogous to those of Mr. Drach. (Essai sur une théorie générale de l'Intégration—Introduction à l'Etude de la Théorie des nombres et de l'Algèbre supérieure.)

On this topic, see note 1 on page 48 of Mr. Borel's work.

See also Hadamard, *Géométrie Elémentaire*, Part 1, Note D.

I have also tried to show that it was useful: in fact it allows us to solve the fundamental problem of differential calculus in all cases where the derived function is bounded, and consequently permits integration of differential functions that can be reduced to quadratures. For example, if $f(x)$ is any bounded function, we are able to tell whether the equation

$$y' + ay = f(x)$$

has any solutions, and if so how to find them.[51]

In the following chapters, where we are dealing with the concepts of length and area, we shall find geometric applications for integration.

---

[51]This remark leads to some interesting problems. For example, if $f(x)$ and $\varphi(x)$ are bounded, are all solutions of the equation

$$y' + f(x)y = \varphi(x)$$

included in the classical formula $y = \int \varphi(x) e^{\int f(x)dx} dx \cdot e^{-\int f(x)dx}$?

Kurt Gödel

(1906–1978)

HIS LIFE AND WORK

Mathematicians are well known for their eccentricities. The film and book *A Beautiful Mind* portrays the eccentricities of the mathematician John Forbes Nash during his time at MIT and Princeton. Although Nash might have been the most eccentric mathematician during his days at MIT, he certainly was not the most eccentric mathematician at Princeton. That honor undoubtedly belongs to Kurt Gödel, the greatest mathematical logician of all time and certainly the maddest. The very same mind able to prove some of the deepest results of modern mathematics was also a lifelong hypochondriac who saw conspiracies behind every door.

Kurt Gödel was born on April 28, 1906, in Brno, Moravia, then a province of Austria-Hungary, now part of the Czech Republic. He was the second of two sons born to Rudolf and Marianne Gödel, Protestant ethnic Germans. No one on either side of Gödel's family had ever shown a significant aptitude for math or science. Kurt Gödel did.

Gödel sped through the early years of his education taking only four years to complete the equivalent of the first eight years of schooling. He entered the

Realgymnasium (the equivalent of high school) in 1918, just as Czechoslovakia became an independent nation following the end of World War I. Gödel almost always received top marks throughout his schooling. Ironically, the only time he received less than a top mark was in a math class!

Nothing survives from these early years to indicate the greatness Gödel would achieve. Given the Gödels' German ethnicity, it is not surprising that in the fall of 1924 Kurt followed his older brother Rudolf to the University of Vienna, about seventy-five miles from Brno, where Rudolf was studying medicine. Gödel originally intended to pursue a career in physics. From the collection of library slips he saved (he was an inveterate pack rat), we know that Gödel read many physics texts during his early years at Vienna.

Vienna hosted much intellectual ferment during the 1920s. A group of scientists and philosophers who called themselves the Vienna Circle met on a weekly basis attempting to set both mathematics and the physical sciences on a firm philosophical foundation. In 1926, Gödel's mentor, the mathematician Hans Hahn, invited him to attend the Circle's meetings. Gödel did and that undoubtedly turned his attention from physics to logic.

In 1900, David Hilbert, the world's most eminent mathematician, delivered an address to the second International Congress of Mathematics in Paris at which he presented twenty-three problems "from the discussion of which an advancement of science may be expected." Hilbert's first problem was to determine the cardinal number of the continuum of real numbers, that is, to solve Cantor's Continuum problem (presented in Chapter 14). Hilbert's second problem was to demonstrate the consistency of the axioms of arithmetic that form the foundations of mathematics. Gödel completely solved Hilbert's second problem with an entirely unexpected answer. He also achieved the first substantial result in response to Hilbert's first problem.

Gödel's library slips show that his interests definitely turned from physics to logic between the summer and fall of 1928 as he was seeking a topic for his doctoral dissertation. Gödel chose the completeness of the predicate calculus as his topic. The predicate calculus introduces *Universal Quantifiers* ("FOR ALL x") and *Existential Quantifiers* ("THERE EXISTS AN x") into logic. This allows formalization of Aristotle's logic for which the assertion "All men are mortal" is a typical example. In the predicate calculus this is expressed as

FOR ALL x (IF x is a man THEN x is mortal)

A typical theorem of the predicate calculus is

THERE EXISTS AN x such that (property A holds for x AND property B holds for x)
IMPLIES

THERE EXISTS AN x such that (property A holds for x)
AND
THERE EXISTS AN x such that (property B holds for x)
Gödel's doctoral dissertation demonstrated a twofold result.

• The *completeness* of the predicate calculus, namely, that **all** true propositions can be proven.

• The *consistency* of the predicate calculus, namely, that **only** true propositions can be proven. (Otherwise it would be inconsistent.)

Gödel worked on his doctoral dissertation in the first half of 1929, even as his mother and Aunt Anna moved in with his brother Rudolf and him following the sudden death of Gödel's father in February. Gödel's advisors approved his dissertation on July 6, 1929, and he received his Ph.D. seven months later in February 1930.

Gödel's completeness theorem won him some recognition from the mathematical community. It solved a problem that had been posed by a prominent mathematician in the expected manner. In the Austro-German University environment of that era, a doctoral dissertation and Ph.D. degree were not sufficient to take up a University position. An additional piece of research called an *Habilitationsschrift* was needed. Gödel sought a more challenging problem for his *Habilitationsschrift*: to demonstrate the consistency of the axioms of arithmetic, the foundations of mathematics. Gödel set out on the journey leading to his *Habilitationsschrift* in early 1930 and finished by early fall of that year. Like Columbus, his journey led him far from his expected destination!

Following Hilbert, Gödel took the axioms of the Italian mathematician Giuseppe Peano as his foundation for the formal arithmetic of the Natural Numbers.

• **Axiom 1** 0 is a natural number.

• **Axiom 2** Every natural number n has a successor (informally $n + 1$) that is also a natural number.

• **Axiom 3** No natural number has 0 as its successor.

• **Axiom 4** Different natural numbers have different successors.

• **Axiom 5** If for some property P, P holds for 0, and if P holds for a natural number n then P also holds for n's successor, then P holds for all natural numbers.

In his *Habilitationsschrift* Gödel demonstrated

• The *incompleteness* of formal arithmetic. He showed that in any such formal system there always exists a statement that cannot be proven within the system even though its truth is apparent.

• If formal arithmetic is ***consistent*** then that ***consistency*** cannot be proven *within* formal arithmetic itself as Hilbert had sought.

In the process, Gödel clearly distinguished ***provability*** from ***truth***.

This is the work of Gödel's presented in this volume. Gödel's remarkable result rests on two cornerstones. The first is Gödel numbering, the assignment of natural number values to each sequence of terms in the formal system. Consider the simple statement:

$$0 = 0' \quad (0 \text{ equals the successor of } 0, \text{ i.e., } 0 + 1)$$

Suppose that the following odd numbers have been assigned to the three terms appearing in this statement:

term	odd number
0	23
=	15
'	21

Then the number

$$2^{23} * 3^{15} * 5^{23} * 7^{21}$$

uniquely expresses the sequence of terms $0 = 0'$. (Literally, it says that the first and third terms are "0," the second term is "=," and the fourth term is "'".) Gödel then shows how to recognize the Gödel numbers of valid proofs inside the formal system.

The second cornerstone of Gödel's incompleteness theorem is a formalized version of the statement

"This statement is unprovable"

that asserts its own unprovability. This is a subtle variant of a form of the Liar's paradox ("This sentence is false") that goes back to ancient Greece. The sentence in the Liar's paradox can neither be true nor false without leading to absurdity.

By focusing on provability rather than on truth, Gödel's sentence avoids the paradox. If formal arithmetic is consistent, meaning that only true statements can be proven, then Gödel's statement *must be* true. If it were false then it *could be* proven, contrary to the consistency! Furthermore, it cannot be proven, because that would demonstrate just the opposite of what it asserts, its unprovability!

Moreover, Gödel showed that if the consistency of the formal system could be demonstrated inside the system itself, then the informal argument just given could be formalized and the formalized version of the statement, "This statement is unprovable," would itself be proven, thereby contradicting itself and demonstrating the *inconsistency* of the system!

Gödel's two incompleteness theorems were published in March 1931. They made an immediate impact throughout the mathematical community. Hilbert was at first angry to learn that his grand scheme for securing the foundations of mathematics had been shown to be impossible. However, Hilbert's assistant Paul Bernays convinced him that Gödel had achieved a truly important advance. They also had an impact on Gödel's career.

Gödel submitted this work as his *Habilitationsschrift* a year later allowing him to take up a paid position at the University. More importantly, this work drew attention in the American mathematical community that led to an invitation for Gödel to be a visiting scholar at the newly founded Institute for Advanced Studies in Princeton, New Jersey.

Gödel was one of twenty-four visiting scholars at the Institute during 1933 to 1934, its first year of operation. Albert Einstein was one of the Institute's eight original permanent faculty members. Gödel and Einstein met during Gödel's first year at the Institute but did not become close friends for nearly another decade.

Gödel returned to a Vienna in tumult after an uneventful year in the peace and quiet of Princeton. Just a month after his return, a band of Nazis stormed the Chancellery and assassinated the Austrian President. After his return, Gödel's physical health declined in parallel to Austria's political well-being. He suffered from a toothache all summer long. By October, he complained of mental exhaustion and needed to spend a week convalescing at a sanitarium outside of Vienna.

Gödel regained some measure of health and in September 1935, he returned to the Institute in Princeton for another year's visit. However, within two months time, he once again fell into the throes of depression and in November decided to return to Vienna to recuperate. This time, he needed several months in a sanitarium to recover.

After his recovery, Gödel began to work on the third great achievement mentioned earlier: a partial solution to Hilbert's first problem of determining the cardinal number of the continuum of real numbers. Cantor had conjectured that the power of the continuum is $\aleph_1$, the second transfinite cardinal. By 1938, Gödel proved that if set theory is consistent, then adding the axiom that $\aleph_1$ is the power of the continuum does not make it inconsistent.

Gödel would visit America one more time before the start of World War II. He had a good reason for his inability to leave for America in September 1938, in time for the start of the academic year. In that month, Gödel finally married the only woman with whom he ever formed a romantic attachment: Adele Nimsburger, neé Porkert.

Kurt first met Adele in late 1928 or early 1929, while working on his dissertation. Both of his parents immediately objected to his relationship with Adele. Kurt was only twenty-two years old. She was twenty-nine about to be divorced, a lower class Catholic, and worst of all in their eyes, she worked as a nightclub dancer! Little is known of their courtship. In contrast to the voluminous store of the most insignificant minutiae he saved during his life, not a single letter or card has been found from the Gödel's nearly decade-long courtship. The marriage came as a complete surprise to Gödel's closest friends; they were completely unaware of Kurt and Adele's relationship.

Adele was Kurt's protectress, putting up with his ways and taking care of as much of his personal life as she could. Unknown to Gödel's family, Adele had frequently visited him during his stay at the sanitarium in 1936, often tasting his food. Unfortunately for Kurt, Adele could only take care of domestic matters.

Gödel's last visit to America before World War II reveals the manner in which Gödel often ignored his own affairs. He should have informed his Dean of Faculty at Vienna that he was taking a leave of absence before he left. Characteristically for the eccentric Gödel, he waited until the very end of October to do so. Gödel also let his authorization to teach at the University of Vienna lapse while he was in America.

In June 1939, Gödel returned to a thoroughly Nazified Austria. Following Germany's annexation of Austria in the *Anschluss* of March 1938, Gödel and his wife had automatically become German citizens. Shortly after his return from America, the Reich served him with a notice to report to his draft board. The German Army classified Gödel as unfit for combat but fit for garrison duty behind the lines. Fortunately for Gödel, the Nazis never learned of his hospitalizations for depression. If they had, they might have sent him to a concentration camp rather than classify him as fit for garrison duty.

Gödel, of course, no longer wanted any part of Austria or Germany and sought to leave permanently to reside in the United States. Throughout the fall of 1939, as the German army raced through Poland, the Institute for Advanced Study and Gödel attempted to negotiate American entry visas and German exit visas for both Gödel and his wife. The Institute director coyly wrote a letter to the German embassy in Washington urging that Gödel was an Aryan and one of the greatest mathematicians in the world and that surely the German government would consider it more important for Gödel to continue his mathematical work than to serve in the army.

The Gödels finally secured all of the necessary travel documents in January 1940. They feared taking the direct route to America across the North Atlantic. The British might attack and sink their ship. If the British captured them, they would be interned because they were German citizens. Consequently, the Gödels came to America the long way around. They took the Trans-Siberian railway to Vladivostok, one ship to Yokohama, Japan, and finally an American flagship to San Francisco. All told, the journey took seven weeks. When an American immigration official asked Gödel if he had ever been a patient in a mental institution Gödel wisely and falsely replied "No." Ironically, all of Gödel's great mathematical achievements had been accomplished by the time he took up permanent residence at the Institute in 1940. One great scientific achievement remained. In the late 1940s, Gödel turned his attention to cosmology. In anticipation of a paper in a volume honoring Einstein, Gödel constructed a *rotating* model of the universe that satisfied Einstein's equations. Gödel showed that in such a rotating universe, there can be no privileged notion of universal time that can be regarded as absolute throughout the cosmos. Indeed, in such a rotating universe, closed, timelike lines, that is, travel into the (distant) past, is theoretically possible. Like his incompleteness theorems, this result upset previously held expectations.

Gödel and Einstein became close friends when Gödel returned to Princeton. In many ways, they were polar opposites; Gödel was cold and aloof, very serious, and suspicious of common sense. Einstein was warm and affable. Perhaps Einstein recognized Gödel's need for someone to take care of him. Einstein may also have valued Gödel's willingness to engage in intellectual disagreements with him. For example, Gödel doubted that a unified theory of quantum mechanics and gravitation could ever be achieved. This was the focus of Einstein's intellectual efforts in his later years.

Einstein played a role in one of the strangest and most amusing incidents in Gödel's life. Gödel became an American citizen in December 1947. He studied diligently for his hearing, much more diligently than necessary. The economist Oskar Morgenstern, one of his closest friends, noticed Gödel becoming more and more upset as the hearing date approached. Morgenstern simply thought that Gödel was nervous in anticipation of the hearing. A few days before the hearing Gödel confided to Morgenstern that he had found a serious flaw in the American Constitution. The President could fill vacancies without Senate approval while the Senate was in recess. This, Gödel reasoned, could lead to a dictatorship!

Morgenstern realized that he and Einstein needed to persuade Gödel that pursuing this point during his citizenship hearing might well jeopardize his chances of being granted citizenship. Einstein joined Morgenstern who drove Gödel to Trenton for the hearing. During the drive Gödel's two friends tried to persuade him to stay away from this topic, but they failed. When Judge Phillip Forman asked Gödel, "Whether a dictatorship similar to the one in Germany could arise in the US?" Gödel, following his inclination to focus on the logical rather than the practical, answered in the affirmative and began to expound on the supposed flaw in the constitution. Fortunately for Gödel, Judge Forman, who had also presided at Einstein's citizenship hearing, found Einstein's presence to be a sufficient reference for Gödel, and quickly steered him to other topics.

Gödel's mental and physical well-being deteriorated significantly over the last thirty years of his life. He became even more reclusive and more of a hypochondriac. He began to eat even less. At times, he would refuse to leave his home when distinguished European mathematicians visited Princeton fearing that they might be intent on killing him.

In his later years, Gödel would sometimes see visitors in his house during the afternoon. He always made sure to ask his wife to interrupt after thirty minutes to remind him that it was time to take his nap. I once heard a British philosopher repeat this story with a surprising twist. When Adele came in to tell Kurt that it was time to take his nap Gödel replied, "But I like this man!" Gödel gave his last public lecture in 1951. After 1952, he published nothing except revisions of earlier work. His note-books show that he attempted to continue his work during the 1950s and 1960s but failed to accomplish any finished results. He kept abreast of developments in logic and set theory through the 1960s. Most notable among these was Paul J. Cohen's 1963 proof that the axioms of set theory could be consistently extended by suppos-ing that the power of the continuum takes a power other than $\aleph_1$; in other words, the independence of Cantor's continuum hypothesis from the axioms of set theory.

Gödel withdrew more and more from the life of the Institute during the last decade of his life. Most of the members who were there when he arrived had long since died. His notebooks reveal that he expended almost all of his intellectual energies on two problems. The first of these problems attempted to continue his earlier accomplish-ment on the continuum hypothesis. Gödel believed that the true power of the con-tinuum was $\aleph_2$ the next cardinal number after $\aleph_1$. He tried in vain to find axioms for set theory from which this power of the continuum could be derived. The second prob-lem consuming Gödel was completely different from anything else he ever worked on. He attempted to formalize St Anselm's ontological proof of the existence of God! Many mathematicians experience a decline in their mathematical output as they age as Gödel did. Few experience deterioration in their mental and physical health as he did.

Gödel was always obsessed with his diet and health, especially with his bowel movements. Starting in 1946 he recorded his daily consumption of laxatives. In 1951, Gödel needed to be hospitalized to treat bleeding ulcers. After his discharge, he put himself on a very strict diet that included one quarter pound of butter, three whole eggs, and two egg whites a day. He hardly ever ate any meat. Late in 1954, Gödel suffered another serious bout of depression. Undoubtedly, his recovery from these episodes must be ascribed to Adele's tender loving care.

Throughout the 1960s, Gödel's physical and mental health remained precarious but not life threatening. All that changed again in early 1970 when he once again became paranoid about his health and ate even less than before. According to Morgenstern, Gödel "looked like a living corpse." Unfortunately for Gödel, his wife Adele who was six years older than him began to suffer from her own serious health problems including a series of small strokes. She was no longer able to take care of him.

Gödel entered the hospital for the final time just after Christmas 1977. He died on January 11, 1978, weighing only sixty-five pounds. The death certificate testified to the fact that he had starved himself to death.

Gödel took on only the biggest problems, made decisive breakthroughs, and then left further developments to others. He never directed a graduate student during his career and he never had a teaching appointment once he settled in America. Never-theless, Kurt Gödel left his unmistakable imprint on the foundations of mathematics.

ON FORMALLY UNDECIDABLE PROPOSITIONS OF PRINCIPIA MATHEMATICA AND RELATED SYSTEMS [1]

1

The development of mathematics in the direction of greater exactness has—as is well known—led to large tracts of it becoming formalized, so that proofs can be carried out according to a few mechanical rules. The most comprehensive formal systems yet set up are, on the one hand, the system of Principia Mathematica (PM)[2] and, on the other, the axiom system for set theory of Zermelo-Fraenkel (later extended by J. v. Neumann).[3] These two systems are so extensive that all methods of proof used in mathematics today have been formalized in them, i.e. reduced to a few axioms and rules of inference. It may therefore be surmised that these axioms and rules of inference are also sufficient to decide *all* mathematical questions which can in any way at all be expressed formally in the systems concerned. It is shown below that this is not the case, and that in both the systems mentioned there are in fact relatively simple problems in the theory of ordinary whole numbers[4] which cannot be decided from the axioms. This situation is not due in some way to the special [174] nature of the systems set up, but holds for a very extensive class of formal systems, including, in particular, all those arising from the addition of a finite number of axioms to the two systems mentioned,[5] provided that thereby no false propositions of the kind described in footnote 4 become provable.

Before going into details, we shall first indicate the main lines of the proof, naturally without laying claim to exactness. The formulae of a formal system—we restrict

---

[1] Cf. the summary of the results of this work, published in *Anzeiger der Akad. d. Wiss. in Wien* (math.-naturw. Kl.) 1930, No. 19.

[2] A. Whitehead and B. Russell, *Principia Mathematica*, 2nd edition, Cambridge 1925. In particular, we also reckon among the axioms of PM the axiom of infinity (in the form: there exist denumerably many individuals), and the axioms of reducibility and of choice (for all types).

[3] Cf. A. Fraenkel, 'Zehn Vorlesungen über die Grundlegung der Mengenlehre', *Wissensch. u. Hyp.*, Vol. XXXI; J. v. Neumann, 'Die Axiomatisierung der Mengenlehre', *Math. Zeitschr.* 27, 1928, *Journ. f. reine u. angew. Math.* 154 (1925), 160 (1929). We may note that in order to complete the formalization, the axioms and rules of inference of the logical calculus must be added to the axioms of set-theory given in the above-mentioned papers. The remarks that follow also apply to the formal systems presented in recent years by D. Hilbert and his colleagues (so far as these have yet been published). Cf. D. Hilbert, *Math. Ann.* 88, *Abh. aus d. math. Sem. der Univ. Hamburg* I (1922), VI (1928); P. Bernays, *Math. Ann.* 90; J. v. Neumann, *Math. Zeitschr.* 26 (1927); W. Ackermann, *Math. Ann.* 93.

[4] I.e., more precisely, there are undecidable propositions in which, besides the logical constants $-$ (not), $\vee$ (or), (x) (for all) and $=$ (identical with), there are no other concepts beyond $+$ (addition) and $\cdot$ (multiplication), both referred to natural numbers, and where the prefixes (x) can also refer only to natural numbers.

[5] In this connection, only such axioms in PM are counted as distinct as do not arise from each other purely by change of type.

ourselves here to the system PM—are, looked at from outside, finite series of basic signs (variables, logical constants and brackets or separation points), and it is easy to state precisely just *which* series of basic signs are meaningful formulae and which are not.[6] Proofs, from the formal standpoint, are likewise nothing but finite series of formulae (with certain specifiable characteristics). For metamathematical purposes it is naturally immaterial what objects are taken as basic signs, and we propose to use natural numbers[7] for them. Accordingly, then, a formula is a finite series of natural numbers,[8] and a particular proof-schema is a finite series of finite series of natural numbers. Metamathematical concepts and propositions thereby become concepts and propositions concerning natural numbers, or series of them,[9] and therefore at least partially expressible in the symbols of the system PM itself. In particular, it can be shown that the concepts, "formula", "proof-schema", "provable formula" are definable in the system PM, i.e. one can give[10] a formula $F(v)$ of PM—for example—with one free variable v (of the type of a series of numbers), such that $F(v)$—interpreted as to content— states: v is a provable formula. We now obtain an undecidable proposition of the system PM, i.e. a proposition A, for which neither A nor *not-A* are provable, in the following manner:

A formula of PM with just one free variable, and that of the type of the natural numbers (class of classes), we shall designate a **class-sign**. We think of the class-signs as being somehow arranged in a series,[11] and denote the n-th one by $R(n)$; and we note that the concept "class-sign" as well as the ordering relation R are definable in the system PM. Let α be any class-sign; by $[\alpha; \, n]$ we designate that formula which is derived on replacing the free variable in the class-sign α by the sign for the natural number n. The three-term relation $x = [y; z]$ also proves to be definable in PM. We now define a class K of natural numbers, as follows:

$$n \, \varepsilon \, K \equiv \overline{Bew} \, [R(n); \, n][11a] \tag{1}$$

(where *Bew x* means: x is a provable formula). Since the concepts which appear in the definiens are all definable in PM, so too is the concept K which is constituted from them, i.e. there is a class-sign S,[12] such that the formula $[S; \, n]$—interpreted as to its content—states that the natural number n belongs to K. S, being a class-sign, is identical with some determinate $R(q)$, i.e.

$$S = R(q)$$

---

[6] Here and in what follows, we shall always understand the term "formula of PM" to mean a formula written without abbreviations (i.e. without use of definitions). Definitions serve only to abridge the written text and are therefore in principle superfluous.

[7] I.e. we map the basic signs in one-to-one fashion on the natural numbers.

[8] I.e. a covering of a section of the number series by natural numbers. (Numbers cannot in fact be put into a spatial order.)

[9] In other words, the above-described procedure provides an isomorphic image of the system PM in the domain of arithmetic, and all metamathematical arguments can equally well be conducted in this isomorphic image. This occurs in the following outline proof, i.e. "formula", "proposition", "variable", etc. **are always to be understood as the corresponding objects of the isomorphic image.**

[10] It would be very simple (though rather laborious) actually to write out this formula.

[11] Perhaps according to the increasing sums of their terms and, for equal sums, in alphabetical order.

[11a] The bar-sign indicates negation.

[12] Again there is not the slightest difficulty in actually writing out the formula S.

holds for some determinate natural number q. We now show that the proposition $[R(q); q][13]$ is undecidable in PM. For supposing the proposition $[R(q); q]$ were provable, it would also be correct; but that means, as has been said, that q would belong to K, i.e. according to (1), $\overline{Bew}\,[R(q); q]$ would hold good, in contradiction to our initial assumption. If, on the contrary, the negation of $[R(q); q]$ were provable, then $\overline{n\ \varepsilon\ K}$, i.e. $Bew\ [R(q); q]$ would hold good. $[R(q); q]$ would thus be provable at the same time as its negation, which again is impossible.

The analogy between this result and Richard's antinomy leaps to the eye; there is also a close relationship with the "liar" antinomy,[14] since the undecidable proposition $[R(q); q]$ states precisely that q belongs to K, i.e. according to (1), that $[R(q); q]$ is not provable. We are therefore confronted with a proposition which asserts its own unprovability.[15] The method of proof just exhibited can clearly be applied to every formal system having the following features: firstly, interpreted as to content, it disposes of sufficient means of expression to define the concepts occurring in the above argument (in particular the concept "provable formula"); secondly, every provable formula in it is also correct as regards content. The exact statement of the above proof, which now follows, will have among others the task of substituting for the second of these assumptions a purely formal and much weaker one.

From the remark that $[R(q); q]$ asserts its own unprovability, it follows at once that $[R(q); q]$ is correct, since $[R(q); q]$ is certainly unprovable (because undecidable). So the proposition which is undecidable *in the system* PM yet turns out to be decided by metamathematical considerations. The close analysis of this remarkable circumstance leads to surprising results concerning proofs of consistency of formal systems, which are dealt with in more detail in Section 4 (Proposition XI).

2

We proceed now to the rigorous development of the proof sketched above, and begin by giving an exact description of the formal system P, for which we seek to demonstrate the existence of undecidable propositions. P is essentially the system obtained by superimposing on the Peano axioms the logic of PM[16] (numbers as individuals, relation of successor as undefined basic concept).

[13]Note that "$[R(q); q]$" (or—what comes to the same thing—"$[S; q]$") is merely a **metamathematical description** of the undecidable proposition. But as soon as one has ascertained the formula S, one can naturally also determine the number q, and thereby effectively write out the undecidable proposition itself.
[14]Every epistemological antinomy can likewise be used for a similar undecidability proof.
[15]In spite of appearances, there is nothing circular about such a proposition, since it begins by asserting the unprovability of a wholly determinate formula (namely the q-th in the alphabetical arrangement with a definite substitution), and only subsequently (and in some way by accident) does it emerge that this formula is precisely that by which the proposition was itself expressed.
[16]The addition of the Peano axioms, like all the other changes made in the system PM, serves only to simplify the proof and can in principle be dispensed with.

The basic signs of the system P are the following:

I. Constants: "$\sim$" (not), "$\vee$" (or), "$\prod$" (for all), "0" (nought), "f" (the successor of), "$($", "$)$" (brackets).

II. Variables of first type (for individuals, i.e. natural numbers including 0): "x_1", "y_1", "z_1", ...

Variables of second type (for classes of individuals): "x_2", "y_2", "z_2", ...

Variables of third type (for classes of classes of individuals): "x_3", "y_3", "z_3", ...

and so on for every natural number as type.[17]

Note: Variables for two-termed and many-termed functions (relations) are superfluous as basic signs, since relations can be defined as classes of ordered pairs and ordered pairs again as classes of classes, e.g. the ordered pair a, b by $((a), (a, b))$, where (x, y) means the class whose only elements are x and y, and (x) the class whose only element is x.[18]

By a **sign of first type** we understand a combination of signs of the form:

$$a, fa, ffa, fffa \ldots \text{etc.}$$

where a is either 0 or a variable of first type. In the former case we call such a sign a **number-sign**. For $n > 1$ we understand by a **sign of n-th type** the same as **variable of n-th type**. Combinations of signs of the form $a(b)$, where b is a sign of n-th and a a sign of $(n + 1)$-th type, we call **elementary formulae**. The class of **formulae** we define as the smallest class[19] containing all elementary formulae and, also, along with any a and b the following: $\sim (a)$, $(a) \vee (b)$, $x \prod (a)$ (where x is any given variable).[18a] We term $(a) \vee (b)$ the **disjunction** of a and b, $\sim (a)$ the **negation** and $x \prod (a)$ a **generalization** of a. A formula in which there is no free variable is called a **propositional formula** (**free variable** being defined in the usual way). A formula with just n free individual variables (and otherwise no free variables) we call an **n-place relation-sign** and for $n = 1$ also a **class-sign**.

By Subst $a\binom{v}{b}$ (where a stands for a formula, v a variable and b a sign of the same type as v) we understand the formula derived from a, when we replace v in it, wherever it is free, by b.[20] We say that a formula a is a **type-lift** of another one b, if a derives from b, when we increase by the same amount the type of all variables appearing in b.

[17] It is presupposed that for every variable type denumerably many signs are available.

[18] Unhomogeneous relations could also be defined in this manner, e.g. a relation between individuals and classes as a class of elements of the form: $((x_2), (x_1), (x_2))$. As a simple consideration shows, all the provable propositions about relations in PM are also provable in this fashion.

[18a] Thus $x \prod (a)$ is also a formula if x does not occur, or does not occur free, in a. In that case $x \prod (a)$ naturally means the same as a.

[19] With regard to this definition (and others like it occurring later), cf. J. Lukasiewicz and A. Tarski, 'Untersuchungen über den Aussagenkalkül', *Comptes Rendus des séances de la Société des Sciences et des Lettres de Varsovie* XXIII, 1930, Cl. III.

[20] Where v does not occur in a as a free variable, we must put Subst $a\binom{v}{b} = a$. Note that "Subst" is a sign belonging to metamathematics.

The following formulae (I-V) are called **axioms** (they are set out with the help of the customarily defined abbreviations: ., $\supset$, $\equiv$, (Ex), $=$,[21] and subject to the usual conventions about omission of brackets)[22]:

I. 1. $\sim (fx_1 = 0)$

2. $fx_1 = fy_1 \supset x_1 = y_1$

3. $x_2(0) \cdot x_1 \prod (x_2(x_1) \supset x_2(fx_1)) \supset x_1 \prod (x_2(x_1))$

II. Every formula derived from the following schemata by substitution of any formulae for p, q and r.

1. $p \lor p \supset p$ 3. $p \lor q \supset q \lor p$

2. $p \supset p \lor q$ 4. $(p \supset q) \supset (r \lor p \supset r \lor q)$

III. Every formula derived from the two schemata

1. $v \prod (a) \supset \text{Subst } a \binom{v}{c}$

2. $v \prod (b \lor a) \supset b \lor v \prod (a)$

by making the following substitutions for a, v, b, c (and carrying out in 1 the operation denoted by "Subst"): for a any given formula, for v any variable, for b any formula in which v does not appear free, for c a sign of the same type as v, provided that c contains no variable which is bound in a at a place where v is free.[23]

IV. Every formula derived from the schema

1. $(Eu)(v \prod (u(v) \equiv a))$

on substituting for v or u any variables of types n or $n + 1$ respectively, and for a a formula which does not contain u free. This axiom represents the axiom of reducibility (the axiom of comprehension of set theory).

V. Every formula derived from the following by type-lift (and this formula itself):

1. $x_1 \prod (x_2(x_1) \equiv y_2(x_1)) \supset x_2 = y_2$.

This axiom states that a class is completely determined by its elements.

A formula c is called an **immediate consequence** of a and b, if a is the formula $(\sim(b)) \lor (c)$, and an **immediate consequence** of a, if c is the formula $v \prod (a)$, where v denotes any given variable. The class of **provable formulae** is defined as the smallest class of formulae which contains the axioms and is closed with respect to the relation "immediate consequence of".[24]

---

[21]As in PM I, *13, $x_1 = y_1$ is to be thought of as defined by $x_2 \prod (x_2(x_1) \supset x_2(y_1))$ (and similarly for higher types.)

[22]To obtain the axioms from the schemata presented (and in the cases of II, III and IV, after carrying out the permitted substitutions), one must therefore still

1. eliminate the abbreviations,
2. add the suppressed brackets.

Note that the resultant expressions must be "formulae" in the above sense.

[23]c is therefore either a variable or 0 or a sign of the form $f \ldots fu$ where u is either 0 or a variable of type 1. With regard to the concept "free (bound) at a place in a" cf. section I A5 of the work cited in footnote 24.

[24]The rule of substitution becomes superfluous, since we have already dealt with all possible substitutions in the axioms themselves (as is also done in J. v. Neumann, 'Zur Hilbertschen Beweistheorie', *Math. Zeitschr.* 26, 1927).

The basic signs of the system P are now ordered in one-to-one correspondence with natural numbers, as follows:

"0" ... 1   "∨" ... 7   "(" ... 11
"f" ... 3   "Π" ... 9   ")" ... 13
"~" ... 5

Furthermore, variables of type n are given numbers of the form p^n (where p is a prime number >13). Hence, to every finite series of basic signs (and so also to every formula) there corresponds, one-to-one, a finite series of natural numbers. These finite series of natural numbers we now map (again in one-to-one correspondence) on to natural numbers, by letting the number $2^{n_1} \cdot 3^{n_2} \ldots p_k^{n_k}$ correspond to the series $n_1, n_2, \ldots n_k$, where p_k denotes the k-th prime number in order of magnitude. A natural number is thereby assigned in one-to-one correspondence, not only to every basic sign, but also to every finite series of such signs. We denote by $\Phi(a)$ the number corresponding to the basic sign or series of basic signs a. Suppose now one is given a class or relation $R(a_1, a_2, \ldots a_n)$ of basic signs or series of such. We assign to it that class (or relation) $R'(x_1, x_2, \ldots x_n)$ of natural numbers, which holds for $x_1, x_2, \ldots x_n$ when and only when there exist $a_1, a_2, \ldots a_n$ such that $x_i = \Phi(a_i)$ $(i = 1, 2, \ldots n)$ and $R(a_1, a_2, \ldots a_n)$ holds. We represent by the same words in italics those classes and relations of natural numbers which have been assigned in this fashion to such previously defined metamathematical concepts as "variable", "formula", "propositional formula", "axiom", "provable formula", etc. The proposition that there are undecidable problems in the system P would therefore read, for example, as follows: There exist *propositional formulae a* such that neither *a* nor the *negation* of *a* are *provable formulae.*

We now introduce a parenthetic consideration having no immediate connection with the formal system P, and first put forward the following definition: A number-theoretic function[25] $\phi(x_1, x_2, \ldots x_n)$ is said to be **recursively defined** by the number-theoretic functions $\psi(x_1, x_2, \ldots x_{n-1})$ and $\mu(x_1, x_2, \ldots x_{n+1})$, if for all $x_2, \ldots x_n, k$[26] the following hold:

$$\phi(0, x_2, \ldots x_n) = \psi(x_2 \ldots x_n)$$
$$\phi(k+1, x_2, \ldots x_n) = \mu(k, \phi(k, x_2, \ldots x_n), x_2, \ldots x_n). \qquad (2)$$

A number-theoretic function ϕ is called **recursive**, if there exists a finite series of number-theoretic functions $\phi_1, \phi_2, \ldots \phi_n$, which ends in ϕ and has the property that every function ϕ_k of the series is either recursively defined by two of the earlier ones,

[25] I.e. its field of definition is the class of non-negative whole numbers (or n-tuples of such), respectively, and its values are non-negative whole numbers.
[26] In what follows, small italic letters (with or without indices) are always variables for non-negative whole numbers (failing an express statement to the contrary).

or is derived from any of the earlier ones by substitution,[27] or, finally, is a constant or the successor function $x + 1$. The length of the shortest series of ϕ_i, which belongs to a recursive function ϕ, is termed its **degree**. A relation $R(x_1 \ldots x_n)$ among natural numbers is called recursive,[28] if there exists a recursive function $\phi(x_1 \ldots x_n)$ such that for all $x_1, x_2, \ldots x_n$

$$R(x_1 \ldots x_n) \sim [\phi(x_1 \ldots x_n) = 0].[29]$$

The following propositions hold:

I. *Every function (or relation) derived from recursive functions (or relations) by the substitution of recursive functions in place of variables is recursive; so also is every function derived from recursive functions by recursive definition according to schema* (2).

II. *If R and S are recursive relations, then so also are* $\bar{R}, R \vee S$ *(and therefore also R & S).*

III. *If the functions* $\phi(\mathfrak{x})$ *and* $\psi(\mathfrak{y})$ *are recursive, so also is the relation:* $\phi(\mathfrak{x}) = \psi(\mathfrak{y})$.[30]

IV. *If the function* $\phi(\mathfrak{x})$ *and the relation* $R(x, \mathfrak{y})$ *are recursive, so also then are the relations S, T*

$$S(\mathfrak{x}, \mathfrak{y}) \sim (Ex)[x \leqq \phi(\mathfrak{x}) \& R(x, \mathfrak{y})]$$
$$T(\mathfrak{x}, \mathfrak{y}) \sim (x)[x \leqq \phi(\mathfrak{x}) \rightarrow R(x, \mathfrak{y})]$$

and likewise the function ψ

$$\psi(\mathfrak{x}\,\mathfrak{y}) = \varepsilon x [x \leqq \phi(\mathfrak{x}) \& R(x, \mathfrak{y})],$$

where $\varepsilon x F(x)$ means: the smallest number x for which $F(x)$ holds and 0 if there is no such number.

Proposition I follows immediately from the definition of "recursive". Propositions II and III are based on the readily ascertainable fact that the number-theoretic functions corresponding to the logical concepts $\bar{}$, $\vee$, $=$

$$\alpha(x), \beta(x, y), \gamma(x, y)$$

namely

$$\alpha(0) = 1; \alpha(x) = 0 \quad \text{for} \quad x \neq 0$$
$$\beta(0, x) = \beta(x, 0) = 0; \beta(x, y) = 1, \text{If } x, y \text{ both} \neq 0$$
$$\gamma(x, y) = 0, \text{ if } x = y; \gamma(x, y) = 1, \text{ if } x \neq y$$

are recursive. The proof of Proposition IV is briefly as follows: According to the assumption there exists a recursive $\rho(x, \mathfrak{y})$ such that

$$R(x, \mathfrak{y}) \sim [\rho(x, \mathfrak{y}) = 0].$$

---

[27]More precisely, by substitution of certain of the foregoing functions in the empty places of the preceding, e.g. $\phi_k(x_1, x_2) = \phi_p[\phi_q(x_1, x_2), \phi_r(x_2)]$ ($p, q, r < k$). Not all the variables on the left-hand side must also occur on the right (and similarly in the recursion schema (2)).

[28]We include classes among relations (one-place relations). Recursive relations R naturally have the property that for every specific n-tuple of numbers it can be decided whether $R(x_1 \ldots x_n)$ holds or not.

[29]For all considerations as to content (more especially also of a metamathematical kind) the Hilbertian symbolism is used, cf. Hilbert-Ackermann, *Grundzüge der theoretischen Logik*, Berlin 1928.

[30]We use gothic letters $\mathfrak{x}$, $\mathfrak{y}$, as abbreviations for given n-tuple sets of variables, e.g. $x_1, x_2 \ldots x_n$.

We now define, according to the recursion schema (2), a function $\chi(x, \mathfrak{y})$ in the following manner:

$$\chi(0, \mathfrak{y}) = 0$$
$$\chi(n + 1, \mathfrak{y}) = (n + 1) \cdot a + \chi(n, \mathfrak{y}) \cdot \alpha(a)[31]$$

where

$$a = \alpha[\alpha(\rho(0, \mathfrak{y}))] \cdot \alpha[\rho(n + 1, \mathfrak{y})] \cdot \alpha[\chi(n, \mathfrak{y})].$$

$\chi(n + 1, \mathfrak{y})$ is therefore either $= n + 1$ (if $a = 1$) or $= \chi(n, \mathfrak{y})$ (if $a = 0$).[32] The first case clearly arises if and only if all the constituent factors of a are 1, i.e. if

$$\overline{R}(0, \mathfrak{y}) \ \& \ R(n + 1, \mathfrak{y}) \ \& \ [\chi(n, \mathfrak{y}) = 0].$$

From this it follows that the function $\chi(n, \mathfrak{y})$ (considered as a function of n) remains 0 up to the smallest value of n for which $R(n, \mathfrak{y})$ holds, and from then on is equal to this value (if $R(0, \mathfrak{y})$ is already the case, the corresponding $\chi(x, \mathfrak{y})$ is constant and $=0$). Therefore:

$$\psi(\mathfrak{x}, \mathfrak{y}) = \chi(\phi(\mathfrak{x}), \mathfrak{y})$$
$$S(\mathfrak{x}, \mathfrak{y}) \sim R[\psi(\mathfrak{x}, \mathfrak{y}), \mathfrak{y}].$$

The relation T can be reduced by negation to a case analogous to S, so that Proposition IV is proved.

The functions $x + y$, $x \cdot y$, x^y, and also the relations $x < y$, $x = y$ are readily found to be recursive; starting from these concepts, we now define a series of functions (and relations) 1–45, of which each is defined from the earlier ones by means of the operations named in Propositions I to IV. This procedure, generally speaking, puts together many of the definition steps permitted by Propositions I to IV. Each of the functions (relations) 1–45, containing, for example, the concepts "*formula*", "*axiom*", and "*immediate consequence*", is therefore recursive.

1. $x/y \equiv (Ez)[z \leqq x \ \& \ x = y \cdot z][33]$
   x is divisible by y.[34]
2. Prim $(x) \equiv (\overline{Ez})[z \leqq x \ \& \ z \neq 1 \ \& \ z \neq x \ \& \ x/z] \ \& \ x > 1$
   x is a prime number.
3. $0 \ Pr \ x \equiv 0$
   $(n + 1) \ Pr \ x \equiv \varepsilon y[y \leqq x \ \& \ \text{Prim}(y) \ \& \ x/y \ \& \ y > n \ Pr \ x]$
   $n \ Pr \ x$ is the n-th (in order of magnitude) prime number contained in x.[34a]

---

[31] We take it to be recognized that the functions $x + y$ (addition) and $x \cdot y$ (multiplication) are recursive.

[32] a cannot take values other than 0 and 1, as is evident from the definition of α.

[33] The sign $\equiv$ is used to mean "equivalence by definition", and therefore does duty in definitions either for $=$ or for $\sim$ (otherwise the symbolism is Hilbertian).

[34] Wherever in the following definitions one of the signs (x), (Ex), εx occurs, it is followed by a limitation on the value of x. This limitation merely serves to ensure the recursive nature of the concept defined. (Cf. Proposition IV.) On the other hand, the **range** of the defined concept would almost always remain unaffected by its omission.

[34a] For $0 < n \leqq z$, where z is the number of distinct prime numbers dividing into x. Note that for $n = z + 1$, $n \ Pr \ x = 0$.

4. $0! \equiv 1$

$(n + 1)! \equiv (n + 1) \cdot n!$

5. $Pr(0) \equiv 0$

$Pr(n + 1) \equiv \varepsilon y \left[y \leq \{Pr(n)\}! + 1 \,\&\, \mathrm{Prim}(y) \,\&\, y > Pr(n) \right]$

$Pr(n)$ is the n-th prime number (in order of magnitude).

6. $n \, Gl \, x \equiv \varepsilon y \left[y \leq x \,\&\, x / (n \, Pr \, x)^y \,\&\, \overline{x / (n \, Pr \, x)^{y+1}} \right]$

$n \, Gl \, x$ is the n-th term of the series of numbers assigned to the number x (for $n > 0$ and n not greater than the length of this series).

7. $l(x) \equiv \varepsilon y \left[y \leq x \,\&\, y \, Pr \, x > 0 \,\&\, (y + 1) \, Pr \, x = 0 \right]$

$l(x)$ is the length of the series of numbers assigned to x.

8. $x * y \equiv \varepsilon z \left[z \leq \left[Pr\{l(x) + l(y)\} \right]^{x+y} \right.$

$\&\, (n) \left[n \leq l(x) \to n \, Gl \, z = n \, Gl \, x \right]$

$\left. \&\, (n) \left[0 < n \leq l(y) \to \{n + l(x)\} \, Gl \, z = n \, Gl \, y \right] \right]$

$x * y$ corresponds to the operation of "joining together" two finite series of numbers.

9. $R(x) \equiv 2^x$

$R(x)$ corresponds to the number-series consisting only of the number x (for $x > 0$).

10. $E(x) \equiv R(11) * x * R(13)$

$E(x)$ corresponds to the operation of "bracketing" [11 and 13 are assigned to the basic signs "(" and ")"].

11. $n \, \mathrm{Var} \, x \equiv (Ez) \left[13 < z \leq x \,\&\, \mathrm{Prim}(z) \,\&\, x = z^n \right] \,\&\, n \neq 0$

x is a *variable of n-th type*.

12. $\mathrm{Var}(x) \equiv (En) \left[n \leq x \,\&\, n \, \mathrm{Var} \, x \right]$

x is a *variable*.

13. $\mathrm{Neg}(x) \equiv R(5) * E(x)$

$\mathrm{Neg}(x)$ is the *negation* of x.

14. $x \, \mathrm{Dis} \, y \equiv E(x) * R(7) * E(y)$

$x \, \mathrm{Dis} \, y$ is the *disjunction* of x and y.

15. $x \, \mathrm{Gen} \, y \equiv R(x) * R(9) * E(y)$

$x \, \mathrm{Gen} \, y$ is the *generalization* of y by means of the *variable* x (assuming x is a *variable*).

16. $0 \, N \, x \equiv x$

$(n + 1) \, N \, x \equiv R(3) * n \, N \, x$

$n \, N \, x$ corresponds to the operation: "n-fold prefixing of the sign 'f' before x."

17. $Z(n) \equiv n \, N \left[R(1) \right]$

$Z(n)$ is the *number-sign* for the number n.

18. $\mathrm{Typ}_1'(x) \equiv (Em, n)\{m, n \leq x \,\&\, [m = 1 \lor 1\ \mathrm{Var}\ m]$

    $\&\ x = n\,N\,[R(m)]\}[34\mathrm{b}]$

x is a *sign of first type*.

19. $\mathrm{Typ}_n(x) \equiv [n = 1 \,\&\, \mathrm{Typ}_1'(x)] \lor [n > 1 \,\&\, (Ev)\{v \leq x$

    $\&\ n\ \mathrm{Var}\ v \,\&\, x = R(v)\}]$

x is a *sign of n-th type*.

20. $\mathrm{Elf}(x) \equiv (Ey, z, n)\, [\, y, z, n \leq x \,\&\, \mathrm{Typ}_n(y)$

    $\&\ \mathrm{Typ}_{n+1}(z) \,\&\, x = z * E(y)]$

x is an *elementary formula*.

21. $Op(x, y, z) \equiv x = \mathrm{Neg}(y) \lor x = y\ \mathrm{Dis}\ z \lor (Ev)\, [\, v \leq x$

    $\&\ \mathrm{Var}(v) \,\&\, x = v\ \mathrm{Gen}\ y]$

22. $FR(x) \equiv (n)\, \{0 < n \leq l(x) \to \mathrm{Elf}(n\ Gl\ x) \lor (Ep, q)$

    $[\,0 < p, q < n \,\&\, Op(n\ Gl\ x, p\ Gl\ x, q\ Gl\ x)]\}$

    $\&\ l(x) > 0$

x is a series of *formulae* of which each is either an *elementary formula* or arises from those preceding by the operations of *negation*, *disjunction* and *generalization*.

23. $\mathrm{Form}\,(x) \equiv (En)\, \{n \leq (Pr\,[l(x)^2])^{x\,\cdot\,[l(x)]^2}$

    $\&\ FR(n) \,\&\, x = [l(n)]\ Gl\ n\}[35]$

x is a *formula* (i.e. last term of a *series of formulae n*).

24. $v\ \mathrm{Geb}\ n, x \equiv \mathrm{Var}\,(v) \,\&\, \mathrm{Form}\,(x) \,\&\, (E\,a, b, c)\,[\,a, b, c \leq x$

    $\&\ x = a * (v\ \mathrm{Gen}\ b) * c \,\&\, \mathrm{Form}\,(b)$

    $\&\ l(a) + 1 \leq n \leq l(a) + l(v\ \mathrm{Gen}\ b)]$

The *variable v* is *bound* at the *n*-th place in *x*.

25. $v\ \mathrm{Fr}\ n, x \equiv \mathrm{Var}\,(v) \,\&\, \mathrm{Form}\,(x) \,\&\, v = n\ Gl\ x$

    $\&\ n \leq l(x) \,\&\, v\ \mathrm{Geb}\ n, x$

The *variable v* is *free* at the *n*-th place in *x*.

26. $v\ \mathrm{Fr}\ x \equiv (En)\, [n \leq l(x) \,\&\, v\ \mathrm{Fr}\ n, x]$

v occurs in x as a *free variable*.

27. $\mathrm{Su}\ x\binom{n}{y} \equiv \varepsilon z\, \{z \leq [Pr\,(l(x) + l(y))]^{x+y}$

    $\&\ [(Eu, v)\, u, v \leq x \,\&\, x = u * R(n\ Gl\ x) * v$

    $\&\ z = u * y * v \,\&\, n = l(u) + 1]\}$

$\mathrm{Su}\ x\binom{n}{y}$ derives from x on substituting y in place of the *n*-th term of x (it being assumed that $0 < n \leq l(x)$).

[34b] $m, n \leq x$ stands for: $m \leq x \,\&\, n \leq x$ (and similarly for more than two variables).

[35] The limitation $n \leq (Pr\,[l(x)]^2)^{x\,\cdot\,[l(x)]^2}$ means roughly this: The length of the shortest series of formulae belonging to x can at most be equal to the number of constituent formulae of x. There are however at most $l(x)$ constituent formulae of length 1, at most $l(x) - 1$ of length 2, etc. and in all, therefore, at most $\frac{1}{2}[l(x)\{l(x) + 1\}] \leq [l(x)]^2$. The prime numbers in n can therefore all be assumed smaller than $Pr\,\{[l(x)]^2\}$, their number $\leq [l(x)]^2$ and their exponents (which are constituent formulae of x) $\leq x$.

28. $O\,St\,v, x \equiv \varepsilon\,n\,\{n \leq l(x) \;\&\; v\,Fr\,n, x \;\&\; \overline{(Ep)}\,[n < p \leq l(x)$
$\;\&\; v\,Fr\,p, x]\}$
$(k + 1)\,St\,v, x \equiv \varepsilon\,n\,\{n < k\,St\,v, x$
$\;\&\; v\,Fr\,n, x \;\&\; (Ep)\,[n < p < k\,St\,v, x \;\&\; v\,Fr\,p, x]\}$

$k\,St\,v, x$ is the $(k + 1)$th place in x (numbering from the end of the *formula* x) at which v is free in x (and 0, if there is no such place).

29. $A\,(v, x) \equiv \varepsilon\,n\,\{n \leq l(x) \;\&\; n\,St\,v, x = 0\}$

$A\,(v, x)$ is the number of places at which v is *free* in x.

30. $Sb_0(x_y^v) \equiv x$

$$Sb_{k+1}(x_y^v) \equiv Su\,[Sb_k(x_y^v)]\binom{k\,St\,v, x}{y}$$

31. $Sb\,(x_y^v) \equiv Sb_{A(v, x)}(x_y^v)[36]$

$Sb(x_y^v)$ is the concept *Subst* $a(_b^v)$, defined above.[37]

32. $x\,Imp\,y \equiv [Neg\,(x)]\,Dis\,y$
$x\,Con\,y \equiv Neg\,\{[Neg\,(x)]Dis\,[Neg\,(y)]\}$
$x\,Aeq\,y \equiv (x\,Imp\,y)\,Con\,(y\,Imp\,x)$
$v\,Ex\,y \equiv Neg\,\{v\,Gen\,[Neg\,(y)]\}$

33. $nTh\,x \equiv \varepsilon\,y\,\{y \leq x^{(x^n)} \;\&\; (k)\,[k \leq l(x) \rightarrow (k\,Gl\,x \leq 13$
$\;\&\; k\,Gl\,y = k\,Gl\,x) \vee (k\,Gl\,x > 13$
$\;\&\; k\,Gl\,y = k\,Gl\,x\,.\,[1\,Pr\,(k\,Gl\,x)]^n)]\}$

$n\,Th\,x$ is the n-th *type-lift* of x (in the case when x and $n\,Th\,x$ are *formulae*).

To the axioms I, 1 to 3, there correspond three determinate numbers, which we denote by z_1, z_2, z_3, and we define:

34. $Z - Ax\,(x) \equiv (x = z_1 \;\vee\; x = z_2 \;\vee\; x = z_3)$

35. $A_1 - Ax\,(x) \equiv (Ey)\,[y \leq x \;\&\; Form\,(y)$
$\;\&\; x = (y\,Dis\,y)\,Imp\,y]$

x is a *formula* derived by substitution in the axiom-schema II, 1. Similarly $A_2 - Ax, A_3 - Ax, A_4 - Ax$ are defined in accordance with the axioms II, 2 to 4.

36. $A - Ax(x) \equiv A_1 - Ax(x) \;\vee\; A_2 - Ax(x) \;\vee$
$A_3 - Ax(x) \;\vee\; A_4 - Ax(x)$

x is a *formula* derived by substitution in an axiom of the sentential calculus.

37. $Q(z, y, v) \equiv (En, m, w)\,[n \leq l(y) \;\&\; m \leq l(z) \;\&\; w \leq z$
$\;\&\; w = m\,Gl\,x \;\&\; w\,Geb\,n, y \;\&\; v\,Fr\,n, y]$

z contains no *variable bound* in y at a position where v is *free*.

---

[36] Where v is not a *variable* or x not a *formula*, then $Sb\,(x_y^v) = x$.
[37] Instead of $Sb\,[Sb\,(x_y^v)_w^z]$ we write: $Sb\,(x_{\;y\;z}^{\;v\;w})$ (and similarly for more than two *variables*).

38. $L_1 - Ax(x) \equiv (Ev, y, z, n) \{v, y, z, n \leqq x \ \& \ n \operatorname{Var} v$
$\& \ \operatorname{Typ}_n(z) \ \& \ \operatorname{Form}(y) \ \& \ Q(z, y, v)$
$\& \ x = (v \operatorname{Gen} y) \operatorname{Imp} [Sb\binom{v}{z})]\}$

x is a *formula* derived from the axiom-schema III, 1 by substitution.

39. $L_2 - Ax(x) \equiv (Ev, q, p) \{v, q, p \leqq x \ \& \ \operatorname{Var}(v)$
$\& \ \operatorname{Form}(p) \ \& \ v \operatorname{Fr} p \ \& \ \operatorname{Form}(q)$
$\& \ x = [v \operatorname{Gen}(p \operatorname{Dis} q)] \operatorname{Imp} [p \operatorname{Dis}(v \operatorname{Gen} q)]\}$

x is a *formula* derived from the axiom-schema III, 2 by substitution.

40. $R - Ax(x) \equiv (E u, v, y, n) [u, v, y, n \leqq x \ \& \ n \operatorname{Var} v$
$\& \ (n + 1) \operatorname{Var} u \ \& \ u \operatorname{Fr} y \ \& \ \operatorname{Form}(y)$
$\& \ x = u \operatorname{Ex} \{v \operatorname{Gen} [[R(u) * E(R(v))] \operatorname{Aeq} y]\}]$

x is a *formula* derived from the axiom-schema IV, 1 by substitution.

To the axiom V, 1 there corresponds a determinate number z_4 and we define:

41. $M - Ax(x) \equiv (En) [n \leqq x \ \& \ x = nTh z_4]$

42. $Ax(x) \equiv Z - Ax(x) \lor A - Ax(x) \lor L_1 - Ax(x) \lor$
$L_2 - Ax(x) \lor R - Ax(x) \lor M - Ax(x)$

x is an *axiom*.

43. $Fl(x y z) \equiv y = z \operatorname{Imp} x \ \lor \ (E v) [v \leqq x \ \& \ \operatorname{Var}(v)$
$\& \ x = v \operatorname{Gen} y]$

x is an *immediate consequence* of y and z.

44. $Bw(x) \equiv (n) \{0 < n \leqq l(x) \rightarrow Ax(n \operatorname{Gl} x)$
$\lor (Ep, q) [0 < p, q < n \ \& \ Fl(n \operatorname{Gl} x, p \operatorname{Gl} x, q \operatorname{Gl} x)]\}$
$\& \ l(x) > 0$

x is a *proof-schema* (a finite series of *formulae*, of which each is either an axiom or an *immediate consequence* of two previous ones).

45. $x B y \equiv Bw(x) \ \& \ [l(x)] \operatorname{Gl} x = y$

x is a *proof* of the *formula* y.

46. $\operatorname{Bew}(x) = (Ey) y B x$

x is a *provable formula*. [Bew (x) is the only one of the concepts 1–46 of which it cannot be asserted that it is recursive.]

The following proposition is an exact expression of a fact which can be vaguely formulated in this way: every recursive relation is definable in the system P (interpreted as to content), regardless of what interpretation is given to the formulae of P:

Proposition V: To every recursive relation $R(x_1 \ldots x_n)$ there corresponds an n-place *relation-sign* r (with the *free variables*[38] $u_1, u_2, \ldots u_n$) such that for every

---

[38] The *variables* $u_1 \ldots u_n$ could be arbitrarily allotted. There is always, e.g., an r with the *free variables* 17, 19, 23 . . . etc., for which (3) and (4) hold.

n-tuple of numbers $(x_1 \ldots x_n)$ the following hold:

$$R(x_1 \ldots x_n) \rightarrow \text{Bew}\left[Sb\left(r \begin{array}{ccc} u_1 & \ldots & u_n \\ Z(x_n) & \ldots & Z(x_n) \end{array}\right)\right] \tag{3}$$

$$R(x_1 \ldots x_n) \rightarrow \text{Bew}\left[\text{Neg } Sb\left(r \begin{array}{ccc} u_1 & \ldots & u_n \\ Z(x_1) & \ldots & Z(x_n) \end{array}\right)\right] \tag{4}$$

We content ourselves here with indicating the proof of this proposition in out-line, since it offers no difficulties of principle and is somewhat involved.[39] We prove the proposition for all relations $R(x_1 \ldots x_n)$ of the form: $x_1 = \phi(x_2 \ldots x_n)$[40] (where ϕ is a recursive function) and apply mathematical induction on the degree of ϕ. For functions of the first degree (i.e. constants and the function $x + 1$) the proposition is trivial. Let ϕ then be of degree m. It derives from functions of lower degree $\phi_1 \ldots \phi_k$ by the operations of substitution or recursive definition. Since, by the inductive assumption, everything is already proved for $\phi_1 \ldots \phi_k$, there exist corresponding *relation-signs* $r_1 \ldots r_k$ such that (3) and (4) hold. The processes of definition whereby ϕ is derived from $\phi_1 \ldots \phi_k$ (substitution and recursive definition) can all be formally mapped in the system P. If this is done, we obtain from $r_1 \ldots r_k$ a new *relation-sign* r[41], for which we can readily prove the validity of (3) and (4) by use of the inductive assumption. A *relation-sign* r, assigned in this fashion to a recursive relation,[42] will be called recursive.

We now come to the object of our exercises:

Let c be any class of *formulae*. We denote by Flg (c) (set of consequences of c) the smallest set of *formulae* which contains all the *formulae* of c and all *axioms*, and which is closed with respect to the relation *"immediate consequence of"*. c is termed ω-consistent, if there is no *class-sign* a such that:

$$(n)\left[Sb\left(a \begin{array}{c} v \\ Z(n) \end{array}\right) \varepsilon \text{ Flg}(c)\right] \& \left[\text{Neg}(v \text{ Gen } a)\right] \varepsilon \text{ Flg}(c)$$

where v is the *free variable* of the *class-sign* a.

Every ω-consistent system is naturally also consistent. The converse, however, is not the case, as will be shown later.

The general result as to the existence of undecidable propositions reads:

Proposition VI: To every ω-consistent recursive class c of *formulae* there correspond recursive *class-signs* r, such that neither v Gen r nor Neg $(v$ Gen $r)$ belongs to Flg (c) (where v is the *free variable* of r).

---

[39] Proposition V naturally is based on the fact that for any recursive relation R, it is decidable, for every n-tuple of numbers, **from the axioms of the system P,** whether the relation R holds or not.

[40] From this there follows immediately its validity for every recursive relation, since any such relation is equivalent to $0 = \phi(x_1 \ldots x_n)$, where ϕ is recursive.

[41] In the precise development of this proof, r is naturally defined, not by the roundabout route of indicating its content, but by its purely formal constitution.

[42] Which thus, as regards content, expresses the existence of this relation.

Proof: Let c be any given recursive ω-consistent class of *formulae*. We define:

$$Bw_c(x) \equiv (n)\,[n \leq l(x) \rightarrow A\,x\,(n\,Gl\,x) \ \lor \ (n\,Gl\,x)\,\varepsilon\,c \ \lor$$
$$(Ep,\,q)\,\{0 < p,\,q < n \ \& \ Fl\,(n\,Gl\,x,\,p\,Gl\,x,\,q\,Gl\,x)\}]$$
$$\& \ l(x) > 0 \tag{5}$$

(cf. the analogous concept 44)

$$x\,B_c y \equiv Bw_c(x) \ \& \ [l(x)]\,Gl\,x = y \tag{6}$$
$$Bew_c(x) \equiv (Ey)\,y\,B_c x \tag{6.1}$$

(cf. the analogous concepts 45, 46)

The following clearly hold:

$$(x)\,[Bew_c(x) \sim x\,\varepsilon\,Flg\,(c)] \tag{7}$$
$$(x)\,[Bew(x) \rightarrow Bew_c(x)] \tag{8}$$

We now define the relation:

$$Q(x,\,y) \equiv x\,B_c\left[\overline{Sb\!\left(y\,\genfrac{}{}{0pt}{}{19}{Z(y)}\right)}\right]. \tag{8.1}$$

Since $x\,B_c\,y$ [according to (6), (5)] and $Sb(y_{Z(y)}^{19})$ (according to definitions 17, 31) are recursive, so also is $Q(x,\,y)$. According to Proposition V and (VIII) there is therefore a *relation-sign* q (with the *free variables* 17, 19) such that

$$x\,B_c\left[\overline{Sb\!\left(y\,\genfrac{}{}{0pt}{}{19}{Z(y)}\right)}\right] \rightarrow Bew_c\left[Sb\!\left(q\,\genfrac{}{}{0pt}{}{17\ \ \ 19}{Z(x)\ Z(y)}\right)\right] \tag{9}$$

$$x\,B_c\left[Sb\!\left(y\,\genfrac{}{}{0pt}{}{19}{Z(y)}\right)\right] \rightarrow Bew_c\left[Neg\,Sb\!\left(q\,\genfrac{}{}{0pt}{}{17\ \ \ 19}{Z(x)\ Z(y)}\right)\right] \tag{10}$$

We put

$$p = 17\,Gen\,q \tag{11}$$

(p is a *class-sign* with the *free variable* 19)

and

$$r = Sb\!\left(q\,\genfrac{}{}{0pt}{}{19}{Z(p)}\right) \tag{12}$$

(r is a recursive *class-sign* with the *free variable* 17).[43] Then

$$Sb\!\left(p\,\genfrac{}{}{0pt}{}{19}{Z(p)}\right) = Sb\!\left([17\,Gen\,q]\,\genfrac{}{}{0pt}{}{19}{Z(p)}\right)$$

$$= 17\,Gen\,Sb\!\left(q\,\genfrac{}{}{0pt}{}{19}{Z(p)}\right) \tag{13}$$

$$= 17\,Gen\,r[44]$$

[because of (11) and (12)] and furthermore:

$$Sb\!\left(q\,\genfrac{}{}{0pt}{}{17\ \ \ 19}{Z(x)\ Z(p)}\right) = Sb\!\left(r\,\genfrac{}{}{0pt}{}{17}{Z(x)}\right) \tag{14}$$

---

[43] r is derived, in fact, from the recursive *relation-sign* q on replacement of a *variable* by a determinate number (p).
[44] The operations Gen and Sb are naturally always commutative, wherever they refer to different *variables*.

[according to (12)]. If now in (9) and (10) we substitute p for y, we find, in virtue of (13) and (14):

$$x \, B_c \, (17 \text{ Gen } r) \rightarrow \text{Bew}_c \left[Sb\left(r \, \frac{17}{Z(x)} \right) \right] \tag{15}$$

$$x \, B_c \, (17 \text{ Gen } r) \rightarrow \text{Bew}_c \left[\text{Neg } Sb\left(r \, \frac{17}{Z(x)} \right) \right] \tag{16}$$

Hence:

1. 17 Gen r is not *c-provable*.[45] For if that were so, there would (according to 6.1) be an n such that $n \, B_c$ (17 Gen r). By (16) it would therefore be the case that:

$$\text{Bew}_c \left[\text{Neg } Sb\left(r \, \frac{17}{Z(n)} \right) \right]$$

while—on the other hand—from the *c-provability* of 17 Gen r there follows also that of $Sb(r\frac{17}{Z(n)})$. c would therefore be inconsistent (and, *a fortiori*, ω-inconsistent.

2. Neg (17 Gen r) is not *c-provable*. Proof: As shown above, 17 Gen r is not *c-provable*, i.e. (according to 6.1) the following holds: $(n) \, n \, B_c$ (17 Gen r). Whence it follows, by (15), that $(n) \, \text{Bew}_c [Sb(r\frac{17}{Z(n)})]$, which together with $\text{Bew}_c [\text{Neg } (17 \text{ Gen } r)]$ would conflict with the ω-consistency of c.

17 Gen r is therefore undecidable in c, so that Proposition VI is proved.

One can easily convince oneself that the above proof is constructive,[45a] i.e. that the following is demonstrated in an intuitionistically unobjectionable way: Given any recursively defined class c of *formulae*: If then a formal decision (in c) be given for the (effectively demonstrable) *propositional formula* 17 Gen r, we can effectively state:

1. A *proof* for Neg (17 Gen r).

2. For any given n, a *proof* for $[Sb(r\frac{17}{Z(n)})]$, i.e. a formal decision of 17 Gen r would lead to the effective demonstrability of an ω-inconsistency.

We shall call a relation (class) of natural numbers $R(x_1 \dots x_n)$ **calculable** [*entscheidungsdefinit*], if there is an n-place *relation-sign* r such that (3) and (4) hold (cf Proposition V). In particular, therefore, by Proposition V, every recursive relation is calculable. Similarly, a *relation-sign* will be called **calculable**, if it be assigned in this manner to a calculable relation. It is, then, sufficient for the existence of undecidable propositions, to assume of the class c that it is ω-consistent and calculable. For the property of being calculable carries over from c to $x \, B_c y$ (cf. (5), (6)) and to $Q(x, y)$ (cf. 8.1), and only these are applied in the above proof. The undecidable proposition has in this case the form v Gen r, where r is a calculable *class-sign* (it is in fact enough that c should be calculable in the system extended by adding c).

[45] "x is *c-provable*" signifies: $x \varepsilon$ Flg (c), which, by (7), states the same as Bew_c (x).

[45a] Since all existential assertions occurring in the proof are based on Proposition V, which, as can easily be seen, is intuitionistically unobjectionable.

KURT GÖDEL

If, instead of ω-consistency, mere consistency as such is assumed for c, then there follows, indeed, not the existence of an undecidable proposition, but rather the existence of a property (r) for which it is possible neither to provide a counter-example nor to prove that it holds for all numbers. For, in proving that 17 Gen r is not c-provable, only the consistency of c is employed and from $\overline{Bew}_c$ (17 Gen r) it follows, according to (15), that for every number x, $sb\left(r_{Z(x)}^{17}\right)$ is c-provable, and hence that Neg $Sb\left(r_{Z(x)}^{17}\right)$ is not c-provable for any number.

By adding Neg (17 Gen r) to c, we obtain a consistent but not ω-consistent *class of formulae c' · c'* is consistent, since otherwise 17 Gen r would be c-provable. c' is not however ω-consistent, since in virtue of $\overline{Bew}_c$ (17 Gen r) and (15) we have: $(x)\, Sb\left(r_{Z(x)}^{17}\right)$, and so *a fortiori*:

$(x)\, Bew_{c'}\, Sb\left(r_{Z(x)}^{17}\right)$, and on the other hand, naturally: $Bew_{c'}$ [Neg (17 Gen r)].[46]

A special case of Proposition VI is that in which the class c consists of a finite number of *formulae* (with or without those derived therefrom by *type-lift*). Every finite class α is naturally recursive. Let a be the largest number contained in α. Then in this case the following holds for c:

$$x \,\varepsilon\, c \sim (Em,\, n)\,[m \leqq x \,\&\, n \leqq a \,\&\, n\, \varepsilon\, \alpha \,\&\, x = m\ Thn]$$

c is therefore recursive. This allows one, for example, to conclude that even with the help of the axiom of choice (for all types), or the generalized continuum hypothesis, not all propositions are decidable, it being assumed that these hypotheses are ω-consistency.

In the proof of Proposition VI the only properties of the system P employed were the following:

1. The class of axioms and the rules of inference (i.e. the relation "immediate consequence of") are recursively definable (as soon as the basic signs are replaced in any fashion by natural numbers).

2. Every recursive relation is definable in the system P (in the sense of Proposition V).

Hence in every formal system that satisfies assumptions 1 and 2 and is ω-consistent, undecidable propositions exist of the form $(x)\, F(x)$, where F is a recursively defined property of natural numbers, and so too in every extension of such a system made by adding a recursively definable ω-consistent class of axioms. As can be easily confirmed, the systems which satisfy assumptions 1 and 2 include the Zermelo-Fraenkel and the v. Neumann axiom systems of set theory,[47] and also the axiom system of number theory which consists of the Peano axioms, the operation

---

[46] Thus the existence of consistent and not ω-consistent c's can naturally be proved only on the assumption that, in general, consistent c's do exist (i.e. that P is consistent).

[47] The proof of assumption 1 is here even simpler than that for the system P, since there is only one kind of basic variable (or two for J. v. Neumann).

of recursive definition [according to schema (2)] and the logical rules.[48] Assumption 1 is in general satisfied by every system whose rules of inference are the usual ones and whose axioms (like those of P) are derived by substitution from a finite number of schemata.[48a]

3

From Proposition VI we now obtain further consequences and for this purpose give the following definition:

A relation (class) is called **arithmetical**, if it can be defined solely by means of the concepts $+$, $\cdot$ [addition and multiplication, applied to natural numbers][49] and the logical constants $\vee$, $\overline{}$, (x), $=$, where (x) and $=$ are to relate only to natural numbers.[50] The concept of "arithmetical proposition" is defined in a corresponding way. In particular the relations "greater" and "congruent to a modulus" are arithmetical, since

$$x > y \sim \overline{(Ez)}\left[y = x + z\right]$$
$$x \equiv y \pmod n \sim (Ez)[x = y + z \cdot n \vee y = x + z \cdot n]$$

We now have:

Proposition VII: Every recursive relation is arithmetical.

We prove this proposition in the form: Every relation of the form $x_0 = \phi(x_1 \ldots x_n)$, where ϕ is recursive, is arithmetical, and apply mathematical induction on the degree of ϕ. Let ϕ be of degree s $(s > 1)$. Then either

1. $\phi(x_1 \ldots x_n) = \rho[\chi_1(x_1 \ldots x_n),$
   $\chi_2(x_1 \ldots x_n) \ldots \chi_m(x_1 \ldots x_n)][51]$
   (where ρ and all the χ's have degrees smaller than s) or

2. $\phi(0, x_2 \ldots x_n) = \psi(x_2 \ldots x_n)$
   $\phi(k + 1, x_2 \ldots x_n) = \mu[k, \phi(k, x_2 \ldots x_n), x_2 \ldots x_n]$
   (where ψ, μ are of lower degree than s).

In the first case we have:

$$x_0 = \phi(x_1 \ldots x_n) \sim (Ey_1 \ldots y_m)\left[R(x_0\, y_1 \ldots y_m)\right.$$
$$\left. \&\ S_1(y_1, x_1 \ldots x_n)\ \&\ \ldots\ \&\ S_m(y_m, x_1 \ldots x_n)\right],$$

where R and S_i are respectively the arithmetical relations which by the inductive assumption exist, equivalent to $x_0 = \rho(y_1 \ldots y_m)$ and $y = \chi_i(x_1 \ldots x_n)$. In this case, therefore, $x_0 = \phi(x_1 \ldots x_n)$ is arithmetical.

[48] Cf. Problem III in D. Hilbert's lecture: 'Probleme der Grundlegung der Mathematik', *Math. Ann.* 102.

[48a] The true source of the incompleteness attaching to all formal systems of mathematics, is to be found—as will be shown in Part II of this essay— in the fact that the formation of ever higher types can be continued into the transfinite (cf. D. Hilbert, 'Über das Unendliche', *Math. Ann.* 95), whereas in every formal system at most denumerably many types occur. It can be shown, that is, that the undecidable propositions here presented always become decidable by the adjunction of suitable higher types (e.g. of type ω for the system P). A similar result also holds for the axiom system of set theory.

[49] Here, and in what follows, zero is always included among the natural numbers.

[50] The definiens of such a concept must therefore be constructed solely by means of the signs stated, variables for natural numbers $x, y \ldots$ and the signs 0 and 1 (function and set variables must not occur). (Any other number-variable may naturally occur in the prefixes in place of x.)

[51] It is not of course necessary that all $x_1 \ldots x_n$ should actually occur in x_i [cf. the example in footnote 27].

In the second case we apply the following procedure: The relation $x_0 = \phi(x_1 \ldots x_n)$ can be expressed with the help of the concept "series of numbers" (f)[52] as follows:

$$x_0 = \phi(x_1 \ldots x_n) \sim (Ef)\{f_0 = \psi(x_2 \ldots x_n)$$
$$\& (k)[k < x_1 \to f_{k+1} = \mu(k, f_k, x_2 \ldots x_n)]$$
$$\& x_0 = f_{xi}\}$$

If $S(y, x_2 \ldots x_n)$ and $T(z, x_1 \ldots x_{n+1})$ are respectively the arithmetical relations—which by the inductive assumption exist—equivalent to

$$y = \psi(x_2 \ldots x_n) \quad \text{and} \quad z = \mu(x_1 \ldots x_{n+1}),$$

the following then holds:

$$x_0 = \phi(x_1 \ldots x_n) \sim (Ef)\{S(f_0, x_2 \ldots x_n)$$
$$\& (k)[k < x_1 \to T(f_{k+1}, k, f_k, x_2 \ldots x_n)]$$
$$\& x_0 = f_{xi}\} \tag{17}$$

We now replace the concept "series of numbers" by "pair of numbers", by assigning to the number pair n, d the number series $f^{(n,d)}$ ($f_k^{(n,d)} = [n]_{1+(k+1)d}$), where $[n]_p$ denotes the smallest non-negative residue of n modulo p.

We then have the following:

Lemma 1: If f is any series of natural numbers and k any natural number, then there exists a pair of natural numbers n, d, such that $f^{(n,d)}$ and f agree in the first k terms.

Proof: Let l be the largest of the numbers $k, f_0, f_1 \ldots f_{k-1}$. Let n be so determined that

$$n = f_i (\text{mod } (1 + (i + 1)\, l!)] \text{ for } i = 0, 1 \ldots k - 1$$

which is possible, since every two of the numbers $1 + (i + 1)l!$ ($i = 0, 1 \ldots k - 1$) are relatively prime. For a prime number contained in two of these numbers would also be contained in the difference $(i_1 - i_2)l!$ and therefore, because $|i_1 - i_2| < l$, in $l!$, which is impossible. The number pair $n, l!$ thus accomplishes what is required.

Since the relation $x = [n]_p$ is defined by $x \equiv n \pmod{p}$ & $x < p$ and is therefore arithmetical, so also is the relation $P(x_0, x_1 \ldots x_n)$ defined as follows:

$$P(x_0 \ldots x_n) \equiv (En, d)\{S([n]_{d+1}, x_2 \ldots x_n)$$
$$\& (k)[k < x_1 \to T([n]_{1+d(k+2)}, k, [n]_{1+d(k+1)},$$
$$x_2 \ldots x_n)] \& x_0 = [n]_{1+d(x_1+1)}\}$$

which, according to (17) and Lemma 1, is equivalent to $x_0 = \phi(x_1 \ldots x_n)$ (we are concerned with the series f in (17) only in its course up to the $x_1 + 1$-th term). Thereby Proposition VII is proved.

According to Proposition VII there corresponds to every problem of the form $(x) F(x)$ (F recursive) an equivalent arithmetical problem, and since the whole proof of Proposition VII can be formalized (for every specific F) within the system P, this equivalence is provable in P. Hence:

---

[52] f signifies here a variable, whose domain of values consists of series of natural numbers. f_k denotes the $k + 1$-th term of a series f (f_0 being the first).

Proposition VIII: In every one of the formal systems[53] referred to in Proposition VI there are undecidable arithmetical propositions.

The same holds (in virtue of the remarks at the end of Section 3) for the axiom system of set theory and its extensions by ω-consistent recursive classes of axioms.

We shall finally demonstrate the following result also:

Proposition IX: In all the formal systems referred to in Proposition VI[53] there are undecidable problems of the restricted predicate calculus[54] (i.e. formulae of the restricted predicate calculus for which neither universal validity nor the existence of a counter-example is provable).[55]

This is based on

Proposition X: Every problem of the form $(x) \, F(x)$ (F recursive) can be reduced to the question of the satisfiability of a formula of the restricted predicate calculus (i.e. for every recursive F one can give a formula of the restricted predicate calculus, the satisfiability of which is equivalent to the validity of $(x) \, F(x)$).

We regard the restricted predicate calculus (r.p.c.) as consisting of those formulae which are constructed out of the basic signs: $\overline{}$, $\vee$, (x), $=$; $x, y \ldots$ (individual variables) and $F(x)$, $G(x, y)$, $H(x, y, z) \ldots$ (property and relation variables)[56] where (x) and $=$ may relate only to individuals. To these signs we add yet a third kind of variables $\phi(x)$, $\psi(x\, y)$, $\chi(x\, y\, z)$ etc. which represent object functions; i.e. $\phi(x)$, $\psi(x\, y)$, etc. denote one-valued functions whose arguments and values are individuals.[57] A formula which, besides the first mentioned signs of the r.p.c., also contains variables of the third kind, will be called a formula in the wider sense (i.w.s.).[58] The concepts of "satisfiable" and "universally valid" transfer immediately to formulae i.w.s. and we have the proposition that for every formula i.w.s. A we can give an ordinary formula of the r.p.c. B such that the satisfiability of A is equivalent to that of B. We obtain B from A, by replacing the variables of the third kind $\phi(x)$, $\psi(x\, y) \ldots$ appearing in A by expressions of the form $(\imath z) \, F(z\, x)$, $(\imath z) \, G(z, x\, y) \ldots$, by eliminating the "descriptive" functions on the lines of PM I $*$ 14, and by logically multiplying[59] the resultant formula by an expression, which states that all the $F, G \ldots$ substituted for the $\phi, \psi \ldots$ are strictly one-valued with respect to the first empty place.

---

[53] These are the ω-consistent systems derived from P by addition of a recursively definable class of axioms.

[54] Cf. Hilbert-Ackermann, *Grundzüge der theoretischen Logik*. In the system P, formulae of the restricted predicate calculus are to be understood as those derived from the formulae of the restricted predicate calculus of PM on replacement of relations by classes of higher type.

[55] In my article 'Die Vollständigkeit der Axiome des logischen Funktionenkalküls', *Monatsh. f. Math. u. Phys.* XXXVII, 2, I have shown of every formula of the restricted predicate calculus that it is either demonstrable as universally valid or else that a counter-example exists; but in virtue of Proposition IX the existence of this counter-example is not always demonstrable (in the formal systems in question).

[56] D. Hilbert and W. Ackermann, in the work already cited, do not include the sign $=$ in the restricted predicate calculus. But for every formula in which the sign $=$ occurs, there exists a formula without this sign, which is satisfiable simultaneously with the original one (cf. the article cited in footnote 55).

[57] And of course the domain of the definition must always be the **whole** domain of individuals.

[58] Variables of the third kind may therefore occur at all empty places instead of individual variables, e.g. $y = \phi(x), F(x, \phi\,(y)), G\,[\psi\,(x, \phi\,(y)), x]$ etc.

[59] I.e. forming the conjunction.

We now show, that for every problem of the form (x) $F(x)$ (F recursive) there is an equivalent concerning the satisfiability of a formula i.w.s., from which Proposition X follows in accordance with what has just been said.

Since F is recursive, there is a recursive function $\Phi(x)$ such that $F(x) \sim [\Phi(x) = 0]$, and for Φ there is a series of functions $\Phi_1, \Phi_2 \ldots \Phi_n$, such that $\Phi_n = \Phi$, $\Phi_1(x) = x + 1$ and for every Φ_k ($1 < k \leqq n$) either

1. $(x_2 \ldots x_m) \left[\Phi_k(0, x_2 \ldots x_m) = \Phi_p(x_2 \ldots x_m) \right]$

$(x, x_2 \ldots x_m) \left\{ \Phi_k \left[\Phi_1(x), x_2 \ldots x_m \right] \right.$

$$= \Phi_q \left[x, \Phi_k(x, x_2 \ldots x_m), x_2 \ldots x_m \right] \right\} \tag{18}$$

$$p, q < k$$

or

2. $(x_1 \ldots x_m) \left[\Phi_k(x_1 \ldots x_m) = \Phi_r(\Phi_{i_1}(\mathfrak{X}_1) \ldots \phi_{i_s}(\mathfrak{X}_s)) \right][60] \tag{19}$

$$r < k, i_v < k \,(\text{for } v = 1, 2 \ldots s)$$

or

3. $(x_1 \ldots x_m) \left[\Phi_k(x_1 \ldots x_m) = \Phi_1(\Phi_1 \ldots \Phi_1(0)) \right] \tag{20}$

In addition, we form the propositions:

$$(x)\, \Phi_1(x) = 0 \,\&\, (x\,y) \left[\Phi_1(x) = \Phi_1(y) \rightarrow x = y \right] \tag{21}$$

$$(x) \left[\Phi_n(x) = 0 \right] \tag{22}$$

In all the formulae (18), (19), (20) (for $k = 2, 3, \ldots n$) and in (21), (22), we now replace the functions Φ_i by the function variable ϕ_i, the number 0 by an otherwise absent individual variable x_0 and form the conjunction C of all the formulae so obtained.

The formula $(E\,x_0)\, C$ then has the required property, i.e.

1. If $(x) \left[\Phi(x) = 0 \right]$ is the case, then $(E\,x_0)\, C$ is satisfiable, since when the functions $\Phi_1, \Phi_2, \ldots \Phi_n$ are substituted for $\phi_1, \phi_2, \ldots \phi_n$ in $(E\,x_0)\, C$ they obviously yield a correct proposition.

2. If $(E\,x_0)\, C$ is satisfiable, then $(x) \left[\Phi(x) = 0 \right]$ is the case.

Proof: Let $\Psi_1, \Psi_2 \ldots \Psi_n$ be the functions presumed to exist, which yield a correct proposition when substituted for $\phi_1, \phi_2 \ldots \phi_n$ in $(E\,x_0)\, C$. Let its domain of individuals be I. In view of the correctness of $(E\,x_0)\, C$ for all functions Ψ_i, there is an individual a (in I) such that all the formulae (18) to (22) transform into correct propositions (18′) to (22′) on replacement of the Φ_i by Ψ_i and of 0 by a. We now form the smallest sub-class of I, which contains a and is closed with respect to the operation $\Psi_1(x)$. This sub-class (I') has the property that every one of the functions Ψ_i, when applied to elements of I', again yields elements of I'. For this holds of Ψ_1 in virtue of the definition of I'; and by reason of (18′), (19′), (20′) this property carries over from Ψ_i of lower index to those of higher. The functions derived from

---

[60]$\mathfrak{X}_i(i = 1 \ldots s)$ represents any complex of the variables $x_1, x_2 \ldots x_m$, e.g. $x_1\, x_3\, x_2$.

Ψ_i by restriction to the domain of individuals I', we shall call Ψ_i'. For these functions also the formulae (18) to (22) all hold (on replacement of 0 by a and Φ_i by Ψ_i').

Owing to the correctness of (21) for Ψ_1' and a, we can map the individuals of I' in one-to-one correspondence on the natural numbers, and this in such a manner that a transforms into 0 and the function Ψ_1' into the successor function Φ_1. But, by this mapping, all the functions Ψ_i' transform into the functions Φ_i, and owing to the correctness of (22) for Ψ'_n and a, we get $(x) \left[\Phi_n(x) = 0 \right]$ or $(x) \left[\Phi(x) = 0 \right]$, which was to be proved.[61]

Since the considerations leading to Proposition X (for every specific F) can also be restated within the system P, the equivalence between a proposition of the form $(x)\, F(x)$ (F recursive) and the satisfiability of the corresponding formula of the r.p.c. is therefore provable in P, and hence the undecidability of the one follows from that of the other, whereby Proposition IX is proved.[62]

4

From the conclusions of Section 2 there follows a remarkable result with regard to a consistency proof of the system P (and its extensions), which is expressed in the following proposition:

Proposition XI: If c be a given recursive, consistent class[63] of *formulae*, then the *propositional formula* which states that c is consistent is not *c-provable*; in particular, the consistency of P is unprovable in P,[64] it being assumed that P is consistent (if not, of course, every statement is provable).

The proof (sketched in outline) is as follows: Let c be any given recursive class of *formulae*, selected once and for all for purposes of the following argument (in the simplest case it may be the null class). For proof of the fact that 17 Gen r is not *c-provable*,[65] only the consistency of c was made use of, as appears from 1, page 1111; i.e.

$$\text{Wid}\,(c) \rightarrow \overline{\text{Bew}_c}\,(17\ \text{Gen}\ r) \tag{23}$$

i.e. by (6.1):

$$\text{Wid}\,(c) \rightarrow \overline{(x)\ x\ B_c\,(17\ \text{Gen}\ r)}$$

By (13), 17 Gen $r = Sb\left(p\ \dfrac{19}{Z(p)} \right)$ and hence:

$$\text{Wid}\,(c) \rightarrow \overline{(x)\ x\ B_c\ Sb\left(p\ \dfrac{19}{Z\,(p)} \right)}$$

---

[61] From Proposition X it follows, for example, that the Fermat and Goldbach problems would be soluble, if one had solved the decision problem for the r.p.c.

[62] Proposition IX naturally holds also for the axiom system of set theory and its extensions by recursively definable ω-consistent classes of axioms, since in these systems also there certainly exist undecidable theorems of the form $(x)\, F(x)$ (F recursive).

[63] c is consistent (abbreviated as Wid (c)) is defined as follows: Wid $(c) = (E\,x)$ [Form (x) & $\overline{\text{Bew}_c}(x)$].

[64] This follows if c is replaced by the null class of *formulae*.

[65] r naturally depends on c (just as p does).

1117

i.e. by (8.1):

$$\text{Wid } (c) \rightarrow (x) \, Q \, (x, p) \tag{24}$$

We now establish the following: All the concepts defined (or assertions proved) in Sections 2[66] and 4 are also expressible (or provable) in P. For we have employed throughout only the normal methods of definition and proof accepted in classical mathematics, as formalized in the system P. In particular c (like any recursive class) is definable in P. Let w be the *propositional formula* expressing Wid (c) in P. The relation $Q \, (x, y)$ is expressed, in accordance with (8.1), (9) and (10), by the *relation-sign* q, and $Q \, (x, p)$, therefore, by r [since by (12) $r = Sb \, (q \, {}^{19}_{Z(p)})$] and the proposition (x) $Q \, (x, p)$ by 17 Gen r.

In virtue of (24) w Imp (17 Gen r) is therefore *provable* in P[67] (and *a fortiori c-provable*). Now if w were *c-provable*, 17 Gen r would also be *c-provable* and hence it would follow, by (23), that c is not consistent.

It may be noted that this proof is also constructive, i.e. it permits, if a *proof* from c is produced for w, the effective derivation from c of a contradiction. The whole proof of Proposition XI can also be carried over word for word to the axiom-system of set theory M, and to that of classical mathematics A,[68] and here too it yields the result that there is no consistency proof for M or for A which could be formalized in M or A respectively, it being assumed that M and A are consistent. It must be expressly noted that Proposition XI (and the corresponding results for M and A) represent no contradiction of the formalistic standpoint of Hilbert. For this standpoint presupposes only the existence of a consistency proof effected by finite means, and there might conceivably be finite proofs which **cannot** be stated in P (or in M or in A).

Since, for every consistent class c, w is not *c-provable*, there will always be propositions which are undecidable (from c), namely w, so long as Neg (w) is not *c-provable*; in other words, one can replace the assumption of ω-consistency in Proposition VI by the following: The statement "c is inconsistent" is not *c*-provable. (Note that there are consistent c's for which this statement is *c*-provable.)

Throughout this work we have virtually confined ourselves to the system P, and have merely indicated the applications to other systems. The results will be stated and proved in fuller generality in a forthcoming sequel. There too, the mere outline proof we have given of Proposition XI will be presented in detail.

(Received: 17. xi. 1930.)

---

[66]From the definition of "recursive" up to the proof of Proposition VI inclusive.

[67]That the correctness of w Imp (17 Gen r) can be concluded from (23), is simply based on the fact that—as was remarked at the outset—the undecidable proposition 17 Gen r asserts its own unprovability.

[68]Cf. J. v. Neumann, 'Zur Hilbertschen Beweistheorie', *Math. Zeitschr.* 26, 1927.

Alan Mathison Turing

(1912–1954)

HIS LIFE AND WORK

As the American physicist J. Robert Oppenheimer waited to testify in front of a Congressional committee about to strip him of his top-secret clearance, Oppenheimer's friend and colleague Isadore Rabi tried to console him. Rabi told him, "Oppy, if this were Britain, you would have been given a knighthood for your work on the atomic bomb."

If Oppenheimer deserves the leading credit for bringing the war in the Pacific to a close thanks to his work leading the Manhattan project, then Alan Mathison Turing deserves similar leading credit for bringing the war in Europe to a close earlier rather than later thanks to his work in decoding Enigma, the Nazi's encryption machine.

Turing received an honor below a knighthood following World War II. A knighthood would have required public acknowledgment of Turing's cracking of the enigma code, something the British kept hidden from the public for well after twenty-five years, long after Turing's death. Ironically, this honor provided Turing no help when he was arrested for acts of "gross indecency." Instead, it accentuated his violation of public morals. Not only had he committed acts of gross indecency but he had done it with someone from the working class.

Alan Turing came from a family of modest distinction. The Turings could trace their ancestry back to a fourteenth-century family from Aberdeenshire in northern Scotland. They must have been supporters of the Stuart dynasty who came south to England when James I ascended to the throne in 1603. Charles II awarded a baronetcy to Sir John Turing in 1638. Sir John nearly paid for the honor with his head during the English Civil War.

Turing's mother, Ethel Stoney, came from a Protestant Irish family that received its land holdings from William of Orange in 1688. A distant cousin, George Johnstone Stoney coined the term *electron* in 1894, just as the atomic era dawned. Ethel's father, Edward Stoney, had a distinguished career building railways and bridges for the British Raj in India. She met Alan's father, Julius Mathison Turing, on a sea voyage from India in the spring of 1907. He had made a career in the Indian Civil Service after receiving his degree at Oxford. They married that October and returned to India in the new year where their first child, John, was born in September. Alan, their second child, was born in London on June 23, 1912, while the whole family was on leave in England.

Britons in the employ of the Indian Civil Service lived a life with benefits and costs. While in India, they lived in near luxury, much better than they could back home. But India was not Britain, especially when it came to children who often spent long stretches back home without one or both parents. Such was the life of young Alan whose father returned to India when he was nine months old. His mother returned to India six months later, leaving her two sons with the Wards, an unrelated retired army couple, in a small town near Hastings.

The Wards had little time for the Turing boys. They had four daughters of their own. Colonel Ward was stiff and remote, and Alan disappointed Mrs. Ward when he failed to show an interest in military toys. And so it was for Turing growing up with his parents in faraway India during most of his formative years. As luck would have it, Julius Turing resigned from the Indian Civil Service in 1926, just as Alan entered Sherborne, one of the older public boarding schools in England.

Alan never quite fit in at Sherborne. At the end of his first year, Alan's headmaster wrote to his parents, "He should do very well when he finds his *métier*; but meanwhile he would do much better if he would try to do his best as a member of this school — he should have more *esprit de corps*." Alan complemented his poor showing on the playing fields with equally poor showings in the classrooms, ranking at the bottom of his nonscience classes for his first year and a half. But mathematics and science were another matter. Alan soon began to excel in these classes, and a fortunate case of the mumps during the second half of his second year gave him the opportunity to pursue these subjects while paying much less attention to the others.

While in the school sanatorium, he read Einstein's popularized accounts of his theories of special and general relativity.

While at Sherborne, Turing developed a deep intimate bond with Christopher Morcom, a student a year ahead of him. Morcom was everything that Turing was not. He excelled in all of classes and won many school prizes. They had a shared passion for mathematics which they each hoped to pursue at Trinity College Cambridge, the college of Isaac Newton and Bertrand Russell, the very center of English mathematics.

In December 1929, they took the Trinity scholarship exams. Christopher passed with flying colors. Alan did not. He was distraught at the thought of being at another Cambridge college, unable to share rooms with Christopher, but there would be no time together at Cambridge. In February 1930, Christopher suddenly became ill and died from complications of tuberculosis, a disease he had contracted as a young boy.

Christopher Morcom's death left Alan Turing utterly alone. While his parents only abandoned him for a year or two at a time, Christopher had abandoned him forever. Turing resolved that he would have to go to Cambridge and by himself achieve enough for the two of them. He did! Turing finally won a scholarship to King's College Cambridge where he matriculated in the autumn of 1931. In many ways, King's was a much better match for Turing than Trinity would have been. The college of John Maynard Keynes and E. M. Forster, King's was not as hidebound to tradition as Trinity. Its dons even went out of their way to mix with the students! Doing mathematics at Cambridge was a joy to Turing. He was finally in an environment where the qualities stressed by Sherborne class, sports talent, or charm did not matter. The only thing that mattered was mathematical ability, and Alan Turing had that in trumps!

For his undergraduate dissertation, Turing developed a new proof of the central limit theorem of probability, and he passed his baccalaureate examinations with marks of distinction. These accomplishments earned him a stipend to do postgraduate work at King's and this led to his greatest mathematical achievement: his 1936 paper *On Computable Numbers, With an Application to the Entscheidungsproblem*, reproduced here.

In 1928, the German mathematician David Hilbert, the greatest mathematician alive, repeated three challenges to the world of mathematics that he had first made at the great mathematical Congress held at Paris in 1900:

1. To prove that all true mathematical statements could be proven, that is, the *completeness* of mathematics.

2. To prove that only true mathematical statements could be proven, that is, the consistency of mathematics.

3. To prove the decidability of mathematics, that is, the existence of a *decision procedure* to decide the truth or falsity of any given mathematical proposition.

Late in 1934, Turing learned that Kurt Gödel had disproven the first two of these challenges. The third challenge remained. Turing set out to answer it, working on it from the spring of 1935 to the spring of 1936. In order to do so, he needed to make precise the notion of a decision procedure. He needed to formalize it. Perhaps inspired by a childhood interest in typewriters, Turing did this by expressing the concept of a decision procedure in terms of machines, machines now known as Turing machines.

Turing recognized that typewriters can only write onto a sheet of paper. They have no ability to interpret the sheet of paper. That is left to the human typist. Turing realized that to eliminate the human component, the machine needed to be able to read input as well as write output. Abstracting and simplifying as much as possible, Turing supposed that his machines operated on a tape composed of a square that could either contain a mark or be blank. The tape extended indefinitely to the left and to the right. Turing gave the machine the ability to read the tape, to write a mark on a square, and to erase a mark on a square. In order to avoid human intervention and be purely mechanical, it was necessary that for each combination of the machine's configuration and symbol scanned, the machine's construction would determine

• whether to write a symbol in a blank square, erase a symbol found in a square, or neither write nor erase the square;

• whether to remain in the same configuration (i.e., set of rules) or to change to another configuration;

• whether to move to the left, move to the right, or stay in place.

Writing out all of this information defining an automated machine would create a "table of behavior" of a finite size that would completely define the machine's behavior. From the abstract perspective, the table was the machine.

There could be many such tables, each defining a machine with a different behavior. A fairly simple machine could mechanically emulate the process of adding two numbers. Suppose each number is represented as consecutively marked squares on the tape and the two numbers are separated by an empty square. The machine starts on the leftmost marked square of the two numbers. It moves right leaving symbols in place as they are found, and then writes a symbol when it encounters the first empty square, which represents the end of the first number and the start of the second. Now change the configuration to the following set of rules. Keep moving right leaving symbols in place as they are found in marked squares until coming to an empty square. This marks the end of the second number. Move left and erase the symbol found. A Turing machine with a slightly more complicated table of behavior can be defined to multiply two numbers. But these were simple processes.

Turing realized that, in contrast, a decision-making machine of the sort required by Hilbert's decision problem could not be directly constructed. At most, its existence could only be inferred. And if no such machine existed, that could only be inferred, perhaps by taking an inventory of all such machines, which is precisely what Turing did.

He supposed that all of the machines were put in some definite order. He then realized that a square with a symbol written on it could be interpreted as a "1," and an empty square could be interpreted as a "0." Having made these interpretations, the machines could be interpreted as generating tapes of 0s and 1s. Supposing these sequences of 0s and 1s to be preceded by a decimal point, Turing was able to interpret each of the machines as a real number (in binary form) between 0 and 1. Turing called these numbers computable numbers, and he turned to Cantor's diagonalization procedure to produce a real number that was not a computable number, a real number different from every number in the table. Such a number could not be the output of any of the machines that had been listed. It was an uncomputable number.

But Turing was not done. Perhaps the very process of Cantor diagonalization was itself a mechanical procedure that was nothing more than computing the first square using the table for the first machine (and then changing its value), then changing to the table for the second machine to compute the second square (and then changing its value), and so on, *ad infinitum* using the table for the 2,241,985th machine to compute the 2,241,985th square (and changing its value), *ad inifintum*. But, Turing realized, this supposed that the nth machine actually got as far as the nth square, and there was no guarantee that the nth machine wouldn't get into a loop and never get to the nth square. In the absence of a decision procedure, whose very existence was in question, there was no way to examine the table for the machine and tell whether it would get into a loop or not. The only way to tell if a given machine avoided such a loop would be to let the machine execute until it got to the nth square. However, that would need to be done for *all* of the machines that had been listed originally in order to show that the Cantor diagonalization procedure was itself mechanical. Doing this for *all* of the machines would require an infinite number of steps, not the finite number of steps required for the process to be mechanical in the sense defined by Turing. Therefore, the Cantor diagonalization procedure was not mechanical in the precise sense defined by Turing.

Turing had shown that there was no decision procedure. In the space of six years, Gödel and Turing had shattered Hilbert's dream.

Typically for Turing, his work on computable numbers was completely original rather than a derivative of someone else's groundbreaking result. In fact, it was so original that Turing didn't even know that the American logician Alonzo Church had arrived at the same result for Hilbert's decision problem by radically different means. Turing's proof was so revolutionary that the London Mathematical Society decided that only Church himself could referee the paper. Church not only refereed the paper but also invited Turing to come to Princeton University to study with him. Turing's Cambridge mentor Max Newman approved.

Princeton's mock gothic architecture reminded Turing very much of his days at Cambridge. However, if Cambridge embodied class, then Princeton bespoke wealth, a wealth totally unfamiliar to Turing. The clock tower of its graduate college was an imitation of the clock tower at Magdalen College Oxford. Students jokingly called it the "Ivory Tower," not because of the often ethereal intellectual activity it overlooked, but because it overlooked "Procter Hall," the graduate college's chief public room, built with a donation from William Cooper Procter, a founder of the American company Procter & Gamble, the manufacturers of Ivory Soap!

Within the confines of the mathematics department, Turing could pursue whatever he pleased. There was no need to conform. He felt at home in the department. Yet, in spite of John von Neumann's offer to stay on at the Institute for Advanced Study as a research assistant, Turing decided to return to England after receiving his Ph.D. in 1938. He realized that America, in general, and Presbyterian Princeton, in particular, would not easily tolerate a nonconformist such as himself.

In the summer of 1938, he returned to a Europe on the brink of war. Before heading up to Cambridge, Turing stopped at the Government Code and Cypher School (GCCS) run by the Foreign Office to let them know that he had begun thinking about encryption and especially decryption techniques. When war broke out in September 1939, Turing left his Cambridge fellowship and immediately reported to the facility the GCCS had established in the small town of Bletchley Park, the town where the rail line from Oxford to Cambridge intersected the main rail line from London to the north!

While at Bletchley Park, Turing had his only serious relationship with a woman. Joan Clarke had been preparing for her final examinations for a Cambridge mathematics degree when she reported to Bletchley Park in June 1940. Her older brother had been at King's College Cambridge, and she had once met Turing before the war. Turing began dating Joan in the spring of 1941. Working the same shift allowed them the opportunity to play chess or go to the movies together, even go on daylong cycling trips when they shared a day off.

After a few months of dating, Turing proposed marriage and Joan immediately accepted. Within a few days Turing, already having second thoughts, told Joan of his homosexual tendencies expecting her to break off the engagement. To Turing's surprise, she did nothing of the sort and took him to London to meet her parents.

Turing and Joan agreed to keep their engagement a secret from all but their closest confidants. She never wore her engagement ring at work where they continued to address each other as "Mr. Turing" and "Miss Clarke." After several months, Turing realized that he could not go through with the marriage. He loved Joan, but it was in much the same way that he imagined he would love a sister, or perhaps in the same way that he loved Christopher Morcom's mother. There was deep affection but not the same passion that he had felt toward Christopher. After breaking his engagement with Joan, Turing made sure that they never worked the same shifts.

Turing could have resumed his Cambridge fellowship following the war and pursued a career in mathematics. However, the work on Enigma had given him a passion to "build a brain." Cambridge was not the place to do this. The National Physical Laboratory in London was! After the war, von Neumann (at Princeton) and Turing each worked on projects to develop the first real computer. They each envisioned a computer containing stored instructions. But that is where their similarity ended. von Neumann had intended that the instructions stored in the computer could not be modified. They could only be called as stored in a predetermined sequence. Turing went much further. He suggested that computers store a set of instruction tables that could be called as desired with the desired inputs. Turing had devised the art of *programming an automatic electronic digital computer with internal program storage*!

While at The National Physical Lab, the British government awarded Turing the Order of the British Empire (OBE) for his war work. The OBE award had little impact on Turing. He was greatly relieved when the awards ceremony was cancelled due to the King's illness. Having the letter painted on his office door only meant that he had to respond whenever someone asked him what he had done to win the award. In the end, he simply stashed the medal in a toolbox.

Within two years of starting the computer project, Turing could see that it was running into bureaucratic quicksand. Realizing this was not for him, he fled London on an early sabbatical, retreating to the refuge of Cambridge where he resumed his fellowship and taught mathematics for the year. While back at Cambridge, Turing began an erotic relationship with a third-year math undergraduate named Neville Johnson.

At the end of the year, Neville graduated from Cambridge and moved away. So did Turing. Instead of going back to the NPL at London he accepted a position at Manchester University in the British midlands where his old mentor Max Newman was leading a team that had been making significant progress building a computer. Manchester's sheer shabbiness appealed to Turing. It lacked the airs of either Cambridge or Princeton. It also lacked Cambridge's tolerance for nonconformity.

In 1952, Turing went to the Manchester police to report that he had been bur-
gled in his home. The culprit was someone named "Harry," an acquaintance of
Turing's newly found homosexual partner. Being wiser in the ways of the world than
Turing, Harry had assumed that Turing was an easy mark for robbery. As a homo-
sexual, Turing had forfeited the protection of the law in mid-twentieth-century
Britain. But Harry had miscalculated. Turing did go to the legal authorities. With
Turing's naïve help, the authorities identified Harry as the culprit and prosecuted
him for the burglary.

Harry may have miscalculated, but Turing made a much more significant
miscalculation. He fared far worse than Harry. The authorities charged him with
the crime of "Gross Indecency." Far worse, perhaps, was the fact that Turing had
crossed class lines and become intimate with someone from the working classes.
While he kept his sexual liaisons within the university, no one noticed outside
its cloistered halls.

Following his conviction, Turing chose to submit himself to estrogen therapy
rather than a two-year prison sentence. He preferred to give up his emotional desires
rather than an environment conducive to intellectual pursuits. But he chose poorly.
The estrogen therapy led to impotence (as intended) and severe depression. At one
point, Turing remarked to a friend, "I'm growing breasts!" Writing to another, he
reformulated the Liar's paradox as it applied to himself:

Turing believes that machines think.

Turing lies with men.

Therefore machines do not think.

Turing committed suicide in 1954.

We can only imagine what great works might have been in his future. And we can
only imagine what the consequences for Britain and the free world would have been
had Turing's ordeal occurred fifteen years earlier, before he led the effort that cracked
the Nazi's enigma code in World War II. However, Turing had proven that no machine
would ever be able to decide this question.

ON COMPUTABLE NUMBERS, WITH AN APPLICATION TO THE ENTSCHEIDUNGSPROBLEM

[RECEIVED 28 MAY, 1936.—READ 12 NOVEMBER, 1936.]

The "computable" numbers may be described briefly as the real numbers whose expressions as a decimal are calculable by finite means. Although the subject of this paper is ostensibly the computable *numbers*, it is almost equally easy to define and investigate computable functions of an integral variable or a real or computable variable, computable predicates, and so forth. The fundamental problems involved are, however, the same in each case, and I have chosen the computable numbers for explicit treatment as involving the least cumbrous technique. I hope shortly to give an account of the relations of the computable numbers, functions, and so forth to one another. This will include a development of the theory of functions of a real variable expressed in terms of computable numbers. According to my definition, a number is computable if its decimal can be written down by a machine.

In §§ 9, 10 I give some arguments with the intention of showing that the computable numbers include all numbers which could naturally be regarded as computable. In particular, I show that certain large classes of numbers are computable. They include, for instance, the real parts of all algebraic numbers, the real parts of the zeros of the Bessel functions, the numbers π, e, etc. The computable numbers do not, however, include all definable numbers, and an example is given of a definable number which is not computable.

Although the class of computable numbers is so great, and in many ways similar to the class of real numbers, it is nevertheless enumerable. In § 8 I examine certain arguments which would seem to prove the contrary. By the correct application of one of these arguments, conclusions are reached which are superficially similar to those of Gödel[1]. These results have valuable applications. In particular, it is shown (§ 11) that the Hilbertian Entscheidungsproblem can have no solution.

In a recent paper Alonzo Church[2] has introduced an idea of "effective calculability", which is equivalent to my "computability", but is very differently defined.

---

[1]Gödel, "Über formal unentscheidbare Sätze der Principia Mathematica und verwandter Systeme, I", *Monatshefte Math. Phys.*, 38 (1931).
[2]Alonzo Church, "An unsolvable problem of elementary number theory", *American J. of Math.*, 58 (1936).

Church also reaches similar conclusions about the Entscheidungsproblem.[3] The proof of equivalence between "computability" and "effective calculability" is outlined in an appendix to the present paper.

1. COMPUTING MACHINES

We have said that the computable numbers are those whose decimals are calculable by finite means. This requires rather more explicit definition. No real attempt will be made to justify the definitions given until we reach § 9. For the present I shall only say that the justification lies in the fact that the human memory is necessarily limited.

We may compare a man in the process of computing a real number to a machine which is only capable of a finite number of conditions $q_1, q_2, \ldots, q_R$ which will be called "m-configurations". The machine is supplied with a "tape" (the analogue of paper) running through it, and divided into sections (called "squares") each capable of bearing a "symbol". At any moment there is just one square, say the r-th, bearing the symbol $\mathfrak{S}(r)$ which is "in the machine". We may call this square the "scanned square". The symbol on the scanned square may be called the "scanned symbol". The "scanned symbol" is the only one of which the machine is, so to speak, "directly aware". However, by altering its m-configuration the machine can effectively remember some of the symbols which it has "seen" (scanned) previously. The possible behaviour of the machine at any moment is determined by the m-configuration q_n and the scanned symbol $\mathfrak{S}(r)$. This pair q_n, $\mathfrak{S}(r)$ will be called the "configuration": thus the configuration determines the *possible* behaviour of the machine. In some of the configurations in which the scanned square is blank (*i.e.* bears no symbol) the machine writes down a new symbol on the scanned square: in other configurations it erases the scanned symbol. The machine may also change the square which is being scanned, but only by shifting it one place to right or left. In addition to any of these operations the m-configuration may be changed. Some of the symbols written down will form the sequence of figures which is the decimal of the real number which is being computed. The others are just rough notes to "assist the memory". It will only be these rough notes which will be liable to erasure.

It is my contention that these operations include all those which are used in the computation of a number. The defence of this contention will be easier when the theory of the machines is familiar to the reader. In the next section I therefore

[3]Alonzo Church, "A note on the Entscheidungsproblem", *J. of Symbolic Logic*, 1 (1936).

proceed with the development of the theory and assume that it is understood what is meant by "machine", "tape", "scanned", etc.

2. DEFINITIONS

Automatic Machines

If at each stage the motion of a machine (in the sense of § 1) is *completely* determined by the configuration, we shall call the machine an "automatic machine" (or *a*-machine).

For some purposes we might use machines (choice machines or *c*-machines) whose motion is only partially determined by the configuration (hence the use of the word "possible" in § 1). When such a machine reaches one of these ambiguous configurations, it cannot go on until some arbitrary choice has been made by an external operator. This would be the case if we were using machines to deal with axiomatic systems. In this paper I deal only with automatic machines, and will therefore often omit the prefix *a*-.

Computing Machines

If an *a*-machine prints two kinds of symbols, of which the first kind (called figures) consists entirely of 0 and 1 (the others being called symbols of the second kind), then the machine will be called a computing machine. If the machine is supplied with a blank tape and set in motion, starting from the correct initial *m*-configuration, the subsequence of the symbols printed by it which are of the first kind will be called the *sequence computed by the machine*. The real number whose expression as a binary decimal is obtained by prefacing this sequence by a decimal point is called the *number computed by the machine*.

At any stage of the motion of the machine, the number of the scanned square, the complete sequence of all symbols on the tape, and the *m*-configuration will be said to describe the *complete configuration* at that stage. The changes of the machine and tape between successive complete configurations will be called the *moves* of the machine.

Circular and Circle-Free Machines

If a computing machine never writes down more than a finite number of symbols of the first kind, it will be called *circular*. Otherwise it is said to be *circle-free*.

A machine will be circular if it reaches a configuration from which there is no possible move, or if it goes on moving, and possibly printing symbols of the second kind, but cannot print any more symbols of the first kind. The significance of the term "circular" will be explained in § 8.

Computable Sequences and Numbers

A sequence is said to be computable if it can be computed by a circle-free machine. A number is computable if it differs by an integer from the number computed by a circle-free machine.

We shall avoid confusion by speaking more often of computable sequences than of computable numbers.

3. EXAMPLES OF COMPUTING MACHINES

I. A machine can be constructed to compute the sequence 010101 . . . . The machine is to have the four *m*-configurations "ƀ", " c ", "f", "e" and is capable of printing "0" and "1". The behaviour of the machine is described in the following table in which "*R*" means "the machine moves so that it scans the square immediately on the right of the one it was scanning previously". Similarly for "*L*". "*E*" means "the scanned symbol is erased" and "*P*" stands for "prints". This table (and all succeeding tables of the same kind) is to be understood to mean that for a configuration described in the first two columns the operations in the third column are carried out successively, and the machine then goes over into the *m*-configuration described in the last column. When the second column is left blank, it is understood that the behaviour of the third and fourth columns applies for any symbol and for no symbol. The machine starts in the *m*-configuration ƀ with a blank tape.

Configuration		Behaviour	
m-config.	*symbol*	*operations*	*final m-config.*
ƀ	None	P0, R	c
c	None	R	e
e	None	P1, R	f
f	None	R	ƀ

If (contrary to the description in § 1) we allow the letters *L*, *R* to appear more than once in the operations column we can simplify the table considerably.

m-config.	*symbol*	*operations*	*final m-config.*
ƀ	None	P0	ƀ
	0	R, R, P1	ƀ
	1	R, R, P0	ƀ

II. As a slightly more difficult example we can construct a machine to compute the sequence 001011011101111011111 . . . . The machine is to be capable of five *m*-configurations, viz. "ɔ", "q", "p", "f", "ƀ" and of printing "ə", "x", "0", "1". The first three symbols on the tape will be "ə ə0"; the other figures follow on alternate squares. On the intermediate squares we never print anything but "*x*". These letters serve to "keep the place" for us and are erased when we have finished with them. We also arrange that in the sequence of figures on alternate squares there shall be no blanks.

Configuration		*Behaviour*	
m-config.	*symbol*	*operations*	*final m-config.*
b		$P\partial, R, P\partial, R, P0, R, R, P0, L, L$	o
o	1	R, Px, L, L, L	o
	0		q
q	Any (0 or 1)	R, R	q
	None	$P1, L$	p
p	x	E, R	q
	∂	R	f
	None	L, L	p
f	Any	R, R	f
	None	$P0, L, L$	o

To illustrate the working of this machine a table is given below of the first few complete configurations. These complete configurations are described by writing down the sequence of symbols which are on the tape, with the *m*-configuration written below the scanned symbol. The successive complete configurations are separated by colons.

```
  : ə ə 0        0 : ə ə 0     0 : ə ə 0     0 : ə ə 0   0      : ə ə 0   0 1 :
 b           o             q              q               q                p
ə ə 0   0   1 : ə ə 0      0 1 : ə ə 0    0 1 : ə ə 0    0 1 :
       p              p                 f                   f
ə ə 0   0   1 : ə ə 0      0 1    : ə ə 0   0 1   0 :
       f              f                  o
ə ə 0   0   1 x 0 : . . . .
       o
```

This table could also be written in the form

$$ b : ə ə 0 0 \quad 0 : ə ə q 0 \quad 0 : \ldots , \qquad\qquad (C) $$

in which a space has been made on the left of the scanned symbol and the *m*-configuration written in this space. This form is less easy to follow, but we shall make use of it later for theoretical purposes.

The *convention* of writing the figures only on alternate squares is very useful: I shall always make use of it. I shall call the one sequence of alternate squares *F*-squares and the other sequence *E*-squares. The symbols on *E*-squares will be liable to erasure. The symbols on *F*-squares form a continuous sequence. There are no blanks until the end is reached. There is no need to have more than one E-square between each pair of F-squares: an apparent need of more E-squares can be satisfied by having a sufficiently rich variety of symbols capable of being printed on E-squares. If a symbol β is on an F-square S and a symbol α is on the E-square next on the right of S, then S and β will be said to be *marked* with α. The process of printing this α will be called marking β (or S) with α.

4. ABBREVIATED TABLES

There are certain types of process used by nearly all machines, and these, in some machines, are used in many connections. These processes include copying down sequences of symbols, comparing sequences, erasing all symbols of a given form, etc. Where such processes are concerned we can abbreviate the tables for the m-configurations considerably by the use of "skeleton tables". In skeleton tables there appear capital German letters and small Greek letters. These are of the nature of "variables". By replacing each capital German letter throughout by an m-configuration and each small Greek letter by a symbol, we obtain the table for an m-configuration.

The skeleton tables are to be regarded as nothing but abbreviations: they are not essential. So long as the reader understands how to obtain the complete tables from the skeleton tables, there is no need to give any exact definitions in this connection.

Let us consider an example:

m-config.	Symbol	Behaviour	Final m-config.	
$\mathfrak{f}(\mathfrak{C}, \mathfrak{B}, \alpha)$	$\ni$	L	$\mathfrak{f}_1(\mathfrak{C}, \mathfrak{B}, \alpha)$	From the m-configuration
	not $\ni$	L	$\mathfrak{f}(\mathfrak{C}, \mathfrak{B}, \alpha)$	$\mathfrak{f}(\mathfrak{C}, \mathfrak{B}, \alpha)$ the machine finds the
$\mathfrak{f}_1(\mathfrak{C}, \mathfrak{B}, \alpha)$	α		$\mathfrak{C}$	symbol of form α which is
	not α	R	$\mathfrak{f}_1(\mathfrak{C}, \mathfrak{B}, \alpha)$	farthest to the left (the "first α")
	None	R	$\mathfrak{f}_2(\mathfrak{C}, \mathfrak{B}, \alpha)$	and the m-configuration then
$\mathfrak{f}_2(\mathfrak{C}, \mathfrak{B}, \alpha)$	α		$\mathfrak{C}$	becomes $\mathfrak{C}$. If there is no α then
	not α	R	$\mathfrak{f}_1(\mathfrak{C}, \mathfrak{B}, \alpha)$	the m-configuration becomes $\mathfrak{B}$.
	None	R	$\mathfrak{B}$	

If we were to replace $\mathfrak{C}$ throughout by $\mathfrak{q}$ (say), $\mathfrak{B}$ by $\mathfrak{r}$, and α by x, we should have a complete table for the m-configuration $\mathfrak{f}(\mathfrak{q}, \mathfrak{r}, x)$. $\mathfrak{f}$ is called an "m-configuration function" or "m-function".

The only expressions which are admissible for substitution in an m-function are the m-configurations and symbols of the machine. These have to be enumerated more or less explicity: they may include expressions such as $\mathfrak{p}(\mathfrak{e}, x)$; indeed they must if there are any m-functions used at all. If we did not insist on this explicit enumeration, but simply stated that the machine had certain m-configurations (enumerated) and all m-configurations obtainable by substitution of m-configurations in certain m-functions, we should usually get an infinity of m-configurations; *e.g.*, we might say that the machine was to have the m-configuration $\mathfrak{q}$ and all m-configurations obtainable by substituting an m-configuration for $\mathfrak{C}$ in $\mathfrak{p}(\mathfrak{C})$. Then it would have $\mathfrak{q}$, $\mathfrak{p}(\mathfrak{q})$, $\mathfrak{p}\big(\mathfrak{p}(\mathfrak{q})\big)$, $\mathfrak{p}\big(\mathfrak{p}(\mathfrak{p}(\mathfrak{q}))\big)$, . . . as m-configurations.

Our interpretation rule then is this. We are given the names of the m-configurations of the machine, mostly expressed in terms of m-functions. We are also given skeleton tables. All we want is the complete table for the m-configurations of the machine. This is obtained by repeated substitution in the skeleton tables.

Further examples.

(In the explanations the symbol "$\rightarrow$" is used to signify "the machine goes into the m-configuration. . . .")

$e\,(\mathfrak{C}, \mathfrak{B}, \alpha)$	$\mathfrak{f}\big(e_1(\mathfrak{C}, \mathfrak{B}, \alpha), \mathfrak{B}, \alpha)\big)$	From $e(\mathfrak{C}, \mathfrak{B}, \alpha)$ the first α is erased and $\rightarrow \mathfrak{C}$. If there is no $\alpha \rightarrow \mathfrak{B}$.
$e_1(\mathfrak{C}, \mathfrak{B}, \alpha)$ $\quad E$	$\mathfrak{C}$	
$e\,(\mathfrak{B}, \alpha)$	$e\big(e\,(\mathfrak{B}, \alpha), \mathfrak{B}, \alpha\big)$	From $e(\mathfrak{B}, \alpha)$ all letters α are erased and $\rightarrow \mathfrak{B}$.

The last example seems somewhat more difficult to interpret than most. Let us suppose that in the list of m-configurations of some machine there appears $e(\mathfrak{b}, x)$ ($=\mathfrak{q}$, say). The table is

$e(\mathfrak{b}, x)$	$e\big(e(\mathfrak{b}, x), \mathfrak{b}, x\big)$
or $\qquad \mathfrak{q}$	$e(\mathfrak{q}, \mathfrak{b}, x)$.

Or, in greater detail:

$\mathfrak{q}$	$e(\mathfrak{q}, \mathfrak{b}, x)$
$e(\mathfrak{q}, \mathfrak{b}, x)$	$\mathfrak{f}\big(e_1(\mathfrak{q}, \mathfrak{b}, x), \mathfrak{b}, x\big)$
$e_1\,(\mathfrak{q}, \mathfrak{b}, x) \qquad E$	$\mathfrak{q}$.

In this we could replace $e_1(\mathfrak{q}, \mathfrak{b}, x)$ by $\mathfrak{q}'$ and then give the table for $\mathfrak{f}$ (with the right substitutions) and eventually reach a table in which no m-functions appeared.

$\mathfrak{p}e(\mathfrak{C}, \beta)$		$\mathfrak{f}\big(\mathfrak{p}e_1(\mathfrak{C}, \beta), \mathfrak{C}, \mathfrak{d}\big)$	From $\mathfrak{p}e\,(\mathfrak{C}, \beta)$ the machine prints β at the end of the sequence of symbols and $\rightarrow \mathfrak{C}$.
$\mathfrak{p}e_1(\mathfrak{C}, \beta)$	$\begin{cases} \text{Any} & R, R \\ \text{None} & P\beta \end{cases}$	$\mathfrak{p}e_1(\mathfrak{C}, \beta)$ \quad $\mathfrak{C}$	
$\mathfrak{l}(\mathfrak{C})$	L	$\mathfrak{C}$	From $\mathfrak{f}'(\mathfrak{C}, \mathfrak{B}, \alpha)$ it does
$\mathfrak{r}(\mathfrak{C})$	R	$\mathfrak{C}$	the same as for $\mathfrak{f}(\mathfrak{C}, \mathfrak{B}, \alpha)$ but moves to the left before $\rightarrow \mathfrak{C}$.

$\mathfrak{f}'(\mathfrak{C}, \mathfrak{B}, \alpha)$	$\mathfrak{f}(\mathfrak{l}(\mathfrak{C}), \mathfrak{B}, \alpha)$	
$\mathfrak{f}''(\mathfrak{C}, \mathfrak{B}, \alpha)$	$\mathfrak{f}(\mathfrak{r}(\mathfrak{C}), \mathfrak{B}, \alpha)$	
$\mathfrak{c}(\mathfrak{C}, \mathfrak{B}, \alpha)$	$\mathfrak{f}'(\mathfrak{c}_1(\mathfrak{C}), \mathfrak{B}, \alpha)$	$\mathfrak{c}(\mathfrak{C}, \mathfrak{B}, \alpha)$. The machine
$\mathfrak{c}_1(\mathfrak{C})$ $\quad\quad\beta$	$\mathfrak{pe}(\mathfrak{C}, \beta)$	writes at the end the first symbol marked α and $\rightarrow \mathfrak{C}$.

The last line stands for the totality of lines obtainable from it by replacing β by any symbol which may occur on the tape of the machine concerned.

$\mathfrak{ce}(\mathfrak{C}, \mathfrak{B}, \alpha)$	$\mathfrak{c}\big(\mathfrak{e}(\mathfrak{C}, \mathfrak{B}, \alpha), \mathfrak{B}, \alpha\big)$	$\mathfrak{ce}(\mathfrak{B}, \alpha)$. The machine copies down in order at the end
$\mathfrak{ce}(\mathfrak{B}, \alpha)$	$\mathfrak{ce}\big(\mathfrak{ce}(\mathfrak{B}, \alpha), \mathfrak{B}, \alpha\big)$	all symbols marked α and erases the letters α; $\rightarrow \mathfrak{B}$.

$\mathfrak{re}(\mathfrak{C}, \mathfrak{B}, \alpha, \beta)$	$\mathfrak{f}\big(\mathfrak{re}_1(\mathfrak{C}, \mathfrak{B}, \alpha, \beta), \mathfrak{B}, \alpha\big)$	$\mathfrak{re}(\mathfrak{C}, \mathfrak{B}, \alpha, \beta)$. The machine replaces the first α by
$\mathfrak{re}_1(\mathfrak{C}, \mathfrak{B}, \alpha, \beta)$ $E, P\beta$	$\mathfrak{C}$	β and $\rightarrow \mathfrak{C} \rightarrow \mathfrak{B}$ if there is no α. $\mathfrak{re}(\mathfrak{B}, \alpha, \beta)$.) The machine
$\mathfrak{re}(\mathfrak{B}, \alpha, \beta)$	$\mathfrak{re}\big(\mathfrak{re}(\mathfrak{B}, \alpha, \beta), \mathfrak{B}, \alpha, \beta\big)$	replaces all letters α by β; $\rightarrow \mathfrak{B}$.

$\mathfrak{cr}(\mathfrak{C}, \mathfrak{B}, \alpha)$	$\mathfrak{c}\big(\mathfrak{re}(\mathfrak{C}, \mathfrak{B}, \alpha, a), \mathfrak{B}, \alpha\big)$	$\mathfrak{cr}(\mathfrak{B}, \alpha)$ differs from $\mathfrak{ce}(\mathfrak{B}, \alpha)$ only in that the letters α are
$\mathfrak{cr}(\mathfrak{B}, \alpha)$	$\mathfrak{cr}\big(\mathfrak{cr}(\mathfrak{B}, \alpha), \mathfrak{re}(\mathfrak{B}, a, \alpha), \alpha\big)$	not erased. The m-configuration $\mathfrak{cr}(\mathfrak{B}, \alpha)$ is taken up when no letters "a" are on the tape.

$\mathfrak{cp}(\mathfrak{C}, \mathfrak{A}, \mathfrak{E}, \alpha, \beta)$	$\mathfrak{f}'\big(\mathfrak{cp}_1(\mathfrak{E}_1\mathfrak{A}, \beta), \mathfrak{f}(\mathfrak{A}, \mathfrak{E}, \beta), \alpha\big)$	
$\mathfrak{cp}_1(\mathfrak{C}, \mathfrak{A}, \beta)$ $\quad\quad\gamma$	$\mathfrak{f}'\big(\mathfrak{cp}_2(\mathfrak{C}, \mathfrak{A}, \gamma), \mathfrak{A}, \beta\big)$	
$\mathfrak{cp}_2(\mathfrak{C}, \mathfrak{A}, \gamma)$ $\left\{ \begin{array}{l} \gamma \\ \text{not } \gamma \end{array}\right.$	$\begin{array}{l} \mathfrak{C} \\ \mathfrak{A}. \end{array}$	

The first symbol marked α and the first marked β are compared. If there is neither α nor β, $\rightarrow \mathfrak{E}$. If there are both and the symbols are alike, $\rightarrow \mathfrak{C}$. Otherwise $\rightarrow \mathfrak{A}$.

$$\mathfrak{cpe}(\mathfrak{C}, \mathfrak{A}, \mathfrak{E}, \alpha, \beta) \qquad \mathfrak{cp}\big(\mathfrak{e}(\mathfrak{e}(\mathfrak{C}, \mathfrak{C}, \beta), \mathfrak{C}, \alpha), \mathfrak{A}, \mathfrak{E}, \alpha, \beta\big)$$

$\mathfrak{cpe}(\mathfrak{C}, \mathfrak{A}, \mathfrak{E}, \alpha, \beta)$ differs from $\mathfrak{cp}(\mathfrak{C}, \mathfrak{A}, \mathfrak{E}, \alpha, \beta)$ in that in the case when there is similarity the first α and β are erased.

$$\mathfrak{cpe}(\mathfrak{A}, \mathfrak{E}, \alpha, \beta) \qquad \mathfrak{cpe}\big(\mathfrak{cpe}(\mathfrak{A}, \mathfrak{E}, \alpha, \beta), \mathfrak{A}, \mathfrak{E}, \alpha, \beta\big).$$

$\mathfrak{cpe}(\mathfrak{A}, \mathfrak{E}, \alpha, \beta)$. The sequence of symbols marked α is compared with the sequence marked β. $\rightarrow \mathfrak{E}$ if they are similar. Otherwise $\rightarrow \mathfrak{A}$. Some of the symbols α and β are erased.

m-config.	Symbol	Operations	Final m-config.	Description
q(ℭ)	Any / None	R / R	q(ℭ) / q₁(ℭ)	q(ℭ, α). The machine finds the last symbol of form α. → ℭ.
q₁(ℭ)	Any / None	R	q(ℭ) / ℭ	
q(ℭ, α)			q(q₁(ℭ, α))	
q₁(ℭ, α)	α / not α	/ L	ℭ / q₁(ℭ, α)	
pe₂(ℭ, α, β)			pe(pe(ℭ, β), α)	pe₂(ℭ, α, β). The machine prints α β at the end.
ce₂(𝔅, α, β)			ce(ce(𝔅, β), α)	ce₃(𝔅, α, β, γ). The machine copies down at the end first the symbols marked α, then those marked β, and finally those marked γ; it erases the symbols α, β, γ.
ce₃(𝔅, α, β, γ)			ce(ce₂(𝔅, β, γ), α)	
e(ℭ)	ə / Not ə	R / L	e₁(ℭ) / e(ℭ)	From e(ℭ) the marks are erased from all marked symbols.
e₁(ℭ)	Any / None	R, E, R	e₁(ℭ) / ℭ	→ ℭ.

5. ENUMERATION OF COMPUTABLE SEQUENCES

A computable sequence γ is determined by a description of a machine which computes γ. Thus the sequence $001011011101111 \ldots$ is determined by the table on p. 1131, and, in fact, any computable sequence is capable of being described in terms of such a table.

It will be useful to put these tables into a kind of standard form. In the first place let us suppose that the table is given in the same form as the first table, for example, I on p. 1130. That is to say, that the entry in the operations column is always of one of the forms $E : E, R : E, L : P\alpha : P\alpha, R : P\alpha, L : R : L :$ or no entry at all. The table can always be put into this form by introducing more m-configurations. Now let us give numbers to the m-configurations, calling them $q_1, \ldots, q_R$ as in § 1. The initial m-configuration is always to be called q_1. We also give numbers to the symbols $S_1, \ldots, S_m$ and, in particular, blank $= S_0$, $0 = S_1$, $1 = S_2$. The lines of the table are now of form

m-config.	Symbol	Operations	Final m-config.	
q_i	S_j	PS_k, L	q_m	(N_1)
q_i	S_j	PS_k, R	q_m	(N_2)
q_i	S_j	PS_k	q_m	(N_3)

Lines such as

$$q_i \qquad S_j \qquad E, R \qquad q_m$$

are to be written as

$$q_i \qquad S_j \qquad PS_0, R \qquad q_m$$

and lines such as

$$q_i \qquad S_j \qquad R \qquad q_m$$

to be written as

$$q_i \qquad S_j \qquad PS_j, R \qquad q_m$$

In this way we reduce each line of the table to a line of one of the forms (N_1), (N_2), (N_3).

From each line of form (N_1) let us form an expression $q_i\ S_j\ S_k\ Lq_m$; from each line of form (N_2) we form an expression $q_i\ S_j\ S_k\ Rq_m$; and from each line of form (N_3) we form an expression $q_i\ S_j\ S_k\ Nq_m$.

Let us write down all expressions so formed from the table for the machine and separate them by semi-colons. In this way we obtain a complete description of the machine. In this description we shall replace q_i by the letter "D" followed by the letter "A" repeated i times, and S_j by "D" followed by "C" repeated j times. This new description of the machine may be called the *standard description* (S.D). It is made up entirely from the letters "A", "C", "D", "L", "R", "N", and from ";".

If finally we replace "A" by "1", "C" by "2", "D" by "3", "L" by "4", "R" by "5", "N" by "6", and ";" by "7" we shall have a description of the machine in the form of an arabic numeral. The integer represented by this numeral may be called a *description number* (D.N) of the machine. The D.N determine the S.D and the structure of the machine uniquely. The machine whose D.N is n may be described as $\mathcal{M}(n)$.

To each computable sequence there corresponds at least one description number, while to no description number does there correspond more than one computable sequence. The computable sequences and numbers are therefore enumerable.

Let us find a description number for the machine I of § 3. When we rename the m-configurations its table becomes:

q_1	S_0	PS_1, R	q_2
q_2	S_0	PS_0, R	q_3
q_3	S_0	PS_2, R	q_4
q_4	S_0	PS_0, R	q_1

Other tables could be obtained by adding irrelevant lines such as

| q_1 | S_1 | PS_1, R | q_2 |

Our first standard form would be

$$q_1 S_0 S_1 R q_2;\quad q_2 S_0 S_0 R q_3;\quad q_3 S_0 S_2 R q_4;\quad q_4 S_0 S_0 R q_1;.$$

The standard description is

DADDCRDAA ;DAADDRDAAA ;DAAADDCCRDAAAA ;DAAAADDRDA ;

A description number is

31332531173113353111731113322531111731111335317

and so is

31332531173113353111731113322531111731111335317311331323253117

A number which is a description number of a circle-free machine will be called a *satisfactory* number. In § 8 it is shown that there can be no general process for determining whether a given number is satisfactory or not.

6. THE UNIVERSAL COMPUTING MACHINE

It is possible to invent a single machine which can be used to compute any computable sequence. If this machine $\mathcal{U}$ is supplied with a tape on the beginning of which is written the S.D of some computing machine $\mathcal{M}$, then $\mathcal{U}$ will compute the same sequence as $\mathcal{M}$. In this section I explain in outline the behaviour of the machine. The next section is devoted to giving the complete table for $\mathcal{U}$.

Let us first suppose that we have a machine $\mathcal{M}'$ which will write down on the *F*-squares the successive complete configurations of $\mathcal{M}$. These might be expressed in the same form as on p. 1131, using the second description, (C), with all symbols on one line. Or, better, we could transform this description (as in § 5) by replacing each *m*-configuration by "*D*" followed by "*A*" repeated the appropriate number of times, and by replacing each symbol by "*D*" followed by "*C*" repeated the appropriate number of times. The numbers of letters "*A*" and "*C*" are to agree with the numbers chosen in § 5, so that, in particular, "0" is replaced by "*DC*", "1" by "*DCC*", and the blanks by "*D*". These substitutions are to be made after the complete configurations have been put together, as in (C). Difficulties arise if we do the substitution first. In each complete configuration the blanks would all have to be replaced by "*D*", so that the complete configuration would not be expressed as a finite sequence of symbols.

If in the description of the machine II of § 3 we replace "ə" by "*DAA*", "ə" by "*DCCC*", "q" by "*DAAA*", then the sequence (C) becomes:

DA : DCCCDCCCDAADCDDC : DCCCDCCCDAAADCDDC: . . . . (C₁)

(This is the sequence of symbols on *F*-squares.)

It is not difficult to see that if $\mathcal{M}$ can be constructed, then so can $\mathcal{M}'$. The manner of operation of $\mathcal{M}'$ could be made to depend on having the rules of operation (*i.e.*, the S.D) of $\mathcal{M}$ written somewhere within itself (*i.e.* within $\mathcal{M}'$); each step could be carried out by referring to these rules. We have only to regard the rules as being capable of being taken out and exchanged for others and we have something very akin to the universal machine.

One thing is lacking: at present the machine $\mathcal{M}'$ prints no figures. We may correct this by printing between each successive pair of complete configurations the figures which appear in the new configuration but not in the old. Then (C_1) becomes

$$DDA : 0 : 0 : DCCCDCCCDAADCDDC : DCCC. \ldots \qquad (C_2)$$

It is not altogether obvious that the E-squares leave enough room for the necessary "rough work", but this is, in fact, the case.

The sequences of letters between the colons in expressions such as (C_1) may be used as standard descriptions of the complete configurations. When the letters are replaced by figures, as in § 5, we shall have a numerical description of the complete configuration, which may be called its description number.

7. DETAILED DESCRIPTION OF THE UNIVERSAL MACHINE

A table is given below of the behaviour of this universal machine. The m-configurations of which the machine is capable are all those occurring in the first and last columns of the table, together with all those which occur when we write out the unabbreviated tables of those which appear in the table in the form of m-functions. $E.g.$, $e(\mathfrak{anf})$ appears in the table and is an m-function. Its unabbreviated table is (see p. 1135)

$e(\mathfrak{anf})$	$\begin{cases} \ni \\ \text{not } \ni \end{cases}$	$\begin{matrix} R \\ L \end{matrix}$	$\begin{matrix} e_1(\mathfrak{anf}) \\ e(\mathfrak{anf}) \end{matrix}$
$e_1(\mathfrak{anf})$	$\begin{cases} \text{Any} \\ \text{None} \end{cases}$	$\begin{matrix} R, E, R \\ \ \end{matrix}$	$\begin{matrix} e_1(\mathfrak{anf}) \\ \mathfrak{anf} \end{matrix}$

Consequently $e_1(\mathfrak{anf})$ is an m-configuration of $\mathcal{U}$.

When $\mathcal{U}$ is ready to start work the tape running through it bears on it the symbol $\ni$ on an F-square and again $\ni$ on the next E-square; after this, on F-squares only, comes the S.D of the machine followed by a double colon "::" (a single symbol, on an F-square). The S.D consists of a number of instructions, separated by semi-colons.

Each instruction consists of five consecutive parts

(i) "D" followed by a sequence of letters "A". This describes the relevant m-configuration.

(ii) "D" followed by a sequence of letters "C". This describes the scanned symbol.

(iii) "D" followed by another sequence of letters "C". This describes the symbol into which the scanned symbol is to be changed.

(iv) "L", "R", or "N", describing whether the machine is to move to left, right, or not at all.

(v) "D" followed by a sequence of letters "A". This describes the final m-configuration.

The machine $\mathcal{U}$ is to be capable of printing "A", "C", "D", "0", "1", "u", "v", "w", "x", "y", "z". The S.D is formed from "$;$", "A", "C", "D", "L", "R", "N".

Subsidiary skeleton table.

$\mathfrak{con}(\mathfrak{C}, \alpha)$	$\begin{cases} \text{Not } A \\ A \end{cases}$	$\begin{matrix} R, R \\ L, P\alpha, R \end{matrix}$	$\begin{matrix} \mathfrak{con}(\mathfrak{C}, \alpha) \\ \mathfrak{con}_1(\mathfrak{C}, \alpha) \end{matrix}$	$\mathfrak{con}(\mathfrak{C}, \alpha)$. Starting from an F-square, S say, the sequence C of symbols describing a configuration closest on the right of S is marked out with letters α. $\rightarrow \mathfrak{C}$.
$\mathfrak{con}_1(\mathfrak{C}, \alpha)$	$\begin{cases} A \\ D \end{cases}$	$\begin{matrix} R, P\alpha, R \\ R, P\alpha, R \end{matrix}$	$\begin{matrix} \mathfrak{con}_1(\mathfrak{C}, \alpha) \\ \mathfrak{con}_2(\mathfrak{C}, \alpha) \end{matrix}$	
$\mathfrak{con}_2(\mathfrak{C}, \alpha)$	$\begin{cases} C \\ \text{Not } C \end{cases}$	$\begin{matrix} R, P\alpha, R \\ R, R \end{matrix}$	$\begin{matrix} \mathfrak{con}_2(\mathfrak{C}, \alpha) \\ \mathfrak{C} \end{matrix}$	$\mathfrak{con}(\mathfrak{C},)$. In the final configuration the machine is scanning the square which is four squares to the right of the last square of C. C is left unmarked.

The table for $\mathcal{U}$.

$\mathfrak{b}$			$\mathfrak{f}(\mathfrak{b}_1, \mathfrak{b}_1, ::)$	$\mathfrak{b}$. The machine prints.
$\mathfrak{b}_1$	$R, R, P :, R, R, PD, R, R, PA$		$\mathfrak{anf}$	$: DA$ on the F-squares after $::$ $\rightarrow \mathfrak{anf}$.
$\mathfrak{anf}$			$\mathfrak{q}(\mathfrak{anf}_1, :)$	$\mathfrak{anf}$. The machine marks the configuration in the last complete configuration with $y.$ $\rightarrow \mathfrak{fom}$.
$\mathfrak{anf}_1$			$\mathfrak{con}(\mathfrak{fom}, y)$	
$\mathfrak{fom}$	$\begin{cases} ; \\ z \\ \text{not } z \text{ nor } ; \end{cases}$	$\begin{matrix} R, Pz, L \\ L, L \\ L \end{matrix}$	$\begin{matrix} \mathfrak{con}(\mathfrak{fmp}, x) \\ \mathfrak{fom} \\ \mathfrak{fom} \end{matrix}$	$\mathfrak{fom}$. The machine finds the last semi-colon not marked with z. It marks this semi-colon with z and the configuration following it with x. $\rightarrow$kmp
$\mathfrak{fmp}$			$\mathfrak{cpe}\big(\mathfrak{e}(\mathfrak{fom}, x, y), \mathfrak{sim}, x, y\big)$	$\mathfrak{fmp}$. The machine compares the sequences marked x and y. It erases all letters x and $y.$ $\rightarrow \mathfrak{sim}$ if they are alike. Otherwise $\rightarrow \mathfrak{fom}$.

$\mathfrak{anf}$. Taking the long view, the last instruction relevant to the last configuration is found. It can be recognized afterwards as the instruction following the last semi-colon marked z. $\rightarrow \mathfrak{sim}$.

m-config		symbol	operations	final config
$\mathfrak{sim}$				$\mathfrak{f}'(\mathfrak{sim}_1, \mathfrak{sim}_1, z)$
$\mathfrak{sim}_1$				$\mathfrak{con}(\mathfrak{sim}_2,)$
$\mathfrak{sim}_2$	$\Big\{$	A		$\mathfrak{sim}_3$
		not A	L, Pu, R, R, R	$\mathfrak{sim}_2$
$\mathfrak{sim}_3$	$\Big\{$	not A	L, Py	$\mathfrak{e}(\mathfrak{mf}, z)$
		A	L, Py, R, R, R	$\mathfrak{sim}_3$
$\mathfrak{mf}$				$\mathfrak{q}(\mathfrak{mf}_1, :)$
$\mathfrak{mf}_1$	$\Big\{$	not A	R, R	$\mathfrak{mf}_1$
		A	L, L, L, L	$\mathfrak{mf}_2$
$\mathfrak{mf}_2$	$\Bigg\{$	C	R, Px, L, L, L	$\mathfrak{mf}_2$
		$:$		$\mathfrak{mf}_4$
		D	R, Px, L, L, L	$\mathfrak{mf}_3$
$\mathfrak{mf}_3$	$\Big\{$	not $:$	R, Pv, L, L, L	$\mathfrak{mf}_3$
		$:$		$\mathfrak{mf}_4$
$\mathfrak{mf}_4$				$\mathfrak{con}\big(\mathfrak{l}(\mathfrak{l}(\mathfrak{mf}_5)),\big)$
$\mathfrak{mf}_5$	$\Big\{$	Any	R, Pw, R	$\mathfrak{mf}_5$
		None	$P :$	$\mathfrak{sh}$
$\mathfrak{sh}$				$\mathfrak{f}(\mathfrak{sh}_1, \mathfrak{inst}, u)$
$\mathfrak{sh}_1$			L, L, L	$\mathfrak{sh}_2$
$\mathfrak{sh}_2$	$\Big\{$	D	R, R, R, R	$\mathfrak{sh}_2\ \mathfrak{sh}_3$
		not D		$\mathfrak{inst}$
$\mathfrak{sh}_3$	$\Big\{$	C	R, R	$\mathfrak{sh}_4$
		not C		$\mathfrak{inst}$
$\mathfrak{sh}_4$	$\Big\{$	C	R, R	$\mathfrak{sh}_5$
		not C		$\mathfrak{pe}_2(\mathfrak{inst}, 0, :)$
$\mathfrak{sh}_5$	$\Big\{$	C		$\mathfrak{inst}$
		not C		$\mathfrak{pe}_2(\mathfrak{inst}, 1, :)$

$\mathfrak{sim}$. The machine marks out the instructions. That part of the instructions which refers to operations to be carried out is marked with u, and the final m-configuration with y. The letters z are erased.

$\mathfrak{mf}$. The last complete configuration is marked out into four sections. The configuration is left unmarked. The symbol directly preceding it is marked with x. The remainder of the complete configuration is divided into two parts, of which the first is marked with v and the last with w. A colon is printed after the whole. $\rightarrow \mathfrak{sh}$.

$\mathfrak{sh}$. The instructions (marked u) are examined. If it is found that they involve "Print 0" or "Print 1", then 0 : or 1 : is printed at the end.

1140

inst			$q\left(l(\text{inst}_1),\, u\right)$	inst. The next complete
inst₁	α	R, E	inst₁(α)	configuration is written
inst₁(L)			ce₅(ob, v, y, x, u, w)	down, carrying out the
inst₁(R)			ce₅(ob, v, x, u, y, w)	marked instructions. The let-
inst₁(N)			ce₅(ob, v, x, y, u, w)	ters u, v, w, x, y are erased.
ob			e(anf)	→anf.

8. APPLICATION OF THE DIAGONAL PROCESS

It may be thought that arguments which prove that the real numbers are not enumerable would also prove that the computable numbers and sequences cannot be enumerable.[4] It might, for instance, be thought that the limit of a sequence of computable numbers must be computable. This is clearly only true if the sequence of computable numbers is defined by some rule.

Or we might apply the diagonal process. "If the computable sequences are enumerable, let α_n be the n-th computable sequence, and let $\phi_n(m)$ be the m-th figure in α_n. Let β be the sequence with $1 - \phi_n(n)$ as its n-th figure. Since β is computable, there exists a number K such that $1 - \phi_n(n) = \phi_K(n)$ all n. Putting $n = K$, we have $1 = 2\phi_K(K)$, i.e. 1 is even. This is impossible. The computable sequences are therefore not enumerable".

The fallacy in this argument lies in the assumption that β is computable. It would be true if we could enumerate the computable sequences by finite means, but the problem of enumerating computable sequences is equivalent to the problem of finding out whether a given number is the D.N of a circle-free machine, and we have no general process for doing this in a finite number of steps. In fact, by applying the diagonal process argument correctly, we can show that there cannot be any such general process.

The simplest and most direct proof of this is by showing that, if this general process exists, then there is a machine which computes β. This proof, although perfectly sound, has the disadvantage that it may leave the reader with a feeling that "there must be something wrong". The proof which I shall give has not this disadvantage, and gives a certain insight into the significance of the idea "circle-free". It depends not on constructing β, but on constructing β', whose n-th figure is $\phi_n(n)$.

Let us suppose that there is such a process; that is to say, that we can invent a machine $\mathfrak{D}$ which, when supplied with the S.D of any computing machine $\mathcal{M}$ will test this S.D and if $\mathcal{M}$ is circular will mark the S.D with the symbol "u" and

[4] Cf. Hobson, *Theory of functions of a real variable* (2nd ed., 1921).

if it is circle-free will mark it with "s". By combining the machines $\mathcal{D}$ and $\mathcal{U}$ we could construct a machine $\mathcal{H}$ to compute the sequence β'. The machine $\mathcal{D}$ may require a tape. We may suppose that it uses the E-squares beyond all symbols on F-squares, and that when it has reached its verdict all the rough work done by $\mathcal{D}$ is erased.

The machine $\mathcal{H}$ has its motion divided into sections. In the first $N - 1$ sections, among other things, the integers $1, 2, \ldots, N - 1$ have been written down and tested by the machine $\mathcal{D}$. A certain number, say $R(N - 1)$, of them have been found to be the D.N's of circle-free machines. In the N-th section the machine $\mathcal{D}$ tests the number N. If N is satisfactory, *i.e.*, if it is the D.N of a circle-free machine, then $R(N) = 1 + R(N - 1)$ and the first $R(N)$ figures of the sequence of which a D.N is N are calculated. The $R(N)$-th figure of this sequence is written down as one of the figures of the sequence β' computed by $\mathcal{H}$. If N is not satisfactory, then $R(N) = R(N - 1)$ and the machine goes on to the $(N + 1)$-th section of its motion.

From the construction of $\mathcal{H}$ we can see that $\mathcal{H}$ is circle-free. Each section of the motion of $\mathcal{H}$ comes to an end after a finite number of steps. For, by our assumption about $\mathcal{D}$, the decision as to whether N is satisfactory is reached in a finite number of steps. If N is not satisfactory, then the N-th section is finished. If N is satisfactory, this means that the machine $\mathcal{M}(N)$ whose D.N is N is circle-free, and therefore its $R(N)$-th figure can be calculated in a finite number of steps. When this figure has been calculated and written down as the $R(N)$-th figure of β', the N-th section is finished. Hence $\mathcal{H}$ is circle-free.

Now let K be the D.N of $\mathcal{H}$. What does $\mathcal{H}$ do in the K-th section of its motion? It must test whether K is satisfactory, giving a verdict "s" or "u". Since K is the D.N of $\mathcal{H}$ and since $\mathcal{H}$ is circle-free, the verdict cannot be "u". On the other hand the verdict cannot be "s". For if it were, then in the K-th section of its motion $\mathcal{H}$ would be bound to compute the first $R(K - 1) + 1 = R(K)$ figures of the sequence computed by the machine with K as its D.N and to write down the $R(K)$-th as a figure of the sequence computed by $\mathcal{H}$. The computation of the first $R(K) - 1$ figures would be carried out all right, but the instructions for calculating the $R(K)$-th would amount to "calculate the first $R(K)$ figures computed by H and write down the $R(K)$-th". This $R(K)$-th figure would never be found. *I.e.*, $\mathcal{H}$ is circular, contrary both to what we have found in the last paragraph and to the verdict "s". Thus both verdicts are impossible and we conclude that there can be no machine $\mathcal{D}$.

We can show further that *there can be no machine $\mathcal{E}$ which, when supplied with the S.D of an arbitrary machine $\mathcal{M}$, will determine whether $\mathcal{M}$ ever prints a given symbol (0 say)*.

We will first show that, if there is a machine $\mathscr{E}$, then there is a general process for determining whether a given machine $\mathscr{M}$ prints 0 infinitely often. Let $\mathscr{M}_1$ be a machine which prints the same sequence as $\mathscr{M}$, except that in the position where the first 0 printed by $\mathscr{M}$ stands, $\mathscr{M}_1$ prints $\bar{0}$. $\mathscr{M}_2$ is to have the first two symbols 0 replaced by $\bar{0}$, and so on. Thus, if $\mathscr{M}$ were to print

$$A\ B\ A\ 0\ 1\ A\ A\ B\ 0\ 0\ 1\ 0\ A\ B\ \ldots ,$$

then $\mathscr{M}_1$ would print

$$A\ B\ A\ \bar{0}\ 1\ A\ A\ B\ 0\ 0\ 1\ 0\ A\ B\ \ldots$$

and $\mathscr{M}_2$ would print

$$A\ B\ A\ \bar{0}\ 1\ A\ A\ B\ \bar{0}\ 0\ 1\ 0\ A\ B\ \ldots .$$

Now let $\mathscr{F}$ be a machine which, when supplied with the S.D of $\mathscr{M}$, will write down successively the S.D of $\mathscr{M}$, of $\mathscr{M}_1$, of $\mathscr{M}_2$, . . . (there is such a machine). We combine $\mathscr{F}$ with $\mathscr{E}$ and obtain a new machine, $\mathscr{G}$. In the motion of $\mathscr{G}$ first $\mathscr{F}$ is used to write down the S.D of $\mathscr{M}$, and then $\mathscr{E}$ tests it, : 0 : is written if it is found that $\mathscr{M}$ never prints 0; then $\mathscr{F}$ writes the S.D of $\mathscr{M}_1$, and this is tested, : 0 : being printed if and only if $\mathscr{M}_1$ never prints 0, and so on. Now let us test $\mathscr{G}$ with $\mathscr{E}$. If it is found that $\mathscr{G}$ never prints 0, then $\mathscr{M}$ prints 0 infinitely often; if $\mathscr{G}$ prints 0 sometimes, then $\mathscr{M}$ does not print 0 infinitely often.

Similarly there is a general process for determining whether $\mathscr{M}$ prints 1 infinitely often. By a combination of these processes we have a process for determining whether $\mathscr{M}$ prints an infinity of figures, *i.e.* we have a process for determining whether $\mathscr{M}$ is circle-free. There can therefore be no machine $\mathscr{E}$.

The expression "there is a general process for determining . . . ." has been used throughout this section as equivalent to "there is a machine which will determine . . . ". This usage can be justified if and only if we can justify our definition of "computable". For each of these "general process" problems can be expressed as a problem concerning a general process for determining whether a given integer n has a property $G(n)$ [*e.g.* $G(n)$ might mean "n is satisfactory" or "n is the Gödel representation of a provable formula"], and this is equivalent to computing a number whose n-th figure is 1 if $G(n)$ is true and 0 if it is false.

9. THE EXTENT OF THE COMPUTABLE NUMBERS

No attempt has yet been made to show that the "computable" numbers include all numbers which would naturally be regarded as computable. All arguments which can be given are bound to be, fundamentally, appeals to intuition, and for this reason rather unsatisfactory mathematically. The real question at issue is "What are the possible processes which can be carried out in computing a number?"

The arguments which I shall use are of three kinds.

(*a*) A direct appeal to intuition.

(*b*) A proof of the equivalence of two definitions (in case the new definition has a greater intuitive appeal).

(*c*) Giving examples of large classes of numbers which are computable.

Once it is granted that computable numbers are all "computable", several other propositions of the same character follow. In particular, it follows that, if there is a general process for determining whether a formula of the Hilbert function calculus is provable, then the determination can be carried out by a machine.

I. [Type (*a*)]. This argument is only an elaboration of the ideas of § 1.

Computing is normally done by writing certain symbols on paper. We may suppose this paper is divided into squares like a child's arithmetic book. In elementary arithmetic the two-dimensional character of the paper is sometimes used. But such a use is always avoidable, and I think that it will be agreed that the two-dimensional character of paper is no essential of computation. I assume then that the computation is carried out on one-dimensional paper, *i.e.* on a tape divided into squares. I shall also suppose that the number of symbols which may be printed is finite. If we were to allow an infinity of symbols, then there would be symbols differing to an arbitrarily small extent.[5] The effect of this restriction of the number of symbols is not very serious. It is always possible to use sequences of symbols in the place of single symbols. Thus an Arabic numeral such as 17 or 999999999999999 is normally treated as a single symbol. Similarly in any European language words are treated as single symbols (Chinese, however, attempts to have an enumerable infinity of symbols). The differences from our point of view between the single and compound symbols is that the compound symbols, if they are too lengthy, cannot be observed at one glance. This is in accordance with experience. We cannot tell at a glance whether 9999999999999999 and 999999999999999 are the same.

The behaviour of the computer at any moment is determined by the symbols which he is observing, and his "state of mind" at that moment. We may suppose that there is a bound B to the number of symbols or squares which the computer can observe at one moment. If he wishes to observe more, he must use successive observations. We will also suppose that the number of states of mind which need be taken into account is finite. The reasons for this are of the same character as those which restrict the number of symbols. If we admitted an infinity of states of mind, some of them will be "arbitrarily close" and will be confused. Again, the restriction

---

[5] If we regard a symbol as literally printed on a square we may suppose that the square is $0 \leqslant x \leqslant 1$, $0 \leqslant y \leqslant 1$. The symbol is defined as a set of points in this square, viz. the set occupied by printer's ink. If these sets are restricted to be measurable, we can define the "distance" between two symbols as the cost of transforming one symbol into the other if the cost of moving unit area of printer's ink unit distance is unity, and there is an infinite supply of ink at $x = 2$, $y = 0$. With this topology the symbols form a conditionally compact space.

is not one which seriously affects computation, since the use of more complicated states of mind can be avoided by writing more symbols on the tape.

Let us imagine the operations performed by the computer to be split up into "simple operations" which are so elementary that it is not easy to imagine them further divided. Every such operation consists of some change of the physical system consisting of the computer and his tape. We know the state of the system if we know the sequence of symbols on the tape, which of these are observed by the computer (possibly with a special order), and the state of mind of the computer. We may suppose that in a simple operation not more than one symbol is altered. Any other changes can be split up into simple changes of this kind. The situation in regard to the squares whose symbols may be altered in this way is the same as in regard to the observed squares. We may, therefore, without loss of generality, assume that the squares whose symbols are changed are always "observed" squares.

Besides these changes of symbols, the simple operations must include changes of distribution of observed squares. The new observed squares must be immediately recognisable by the computer. I think it is reasonable to suppose that they can only be squares whose distance from the closest of the immediately previously observed squares does not exceed a certain fixed amount. Let us say that each of the new observed squares is within L squares of an immediately previously observed square.

In connection with "immediate recognisability", it may be thought that there are other kinds of square which are immediately recognisable. In particular, squares marked by special symbols might be taken as immediately recognisable. Now if these squares are marked only by single symbols there can be only a finite number of them, and we should not upset our theory by adjoining these marked squares to the observed squares. If, on the other hand, they are marked by a sequence of symbols, we cannot regard the process of recognition as a simple process. This is a fundamental point and should be illustrated. In most mathematical papers the equations and theorems are numbered. Normally the numbers do not go beyond (say) 1000. It is, therefore, possible to recognise a theorem at a glance by its number. But if the paper was very long, we might reach Theorem 157767733443477; then, further on in the paper, we might find ". . . hence (applying Theorem 157767733443477) we have . . .". In order to make sure which was the relevant theorem we should have to compare the two numbers figure by figure, possibly ticking the figures off in pencil to make sure of their not being counted twice. If in spite of this it is still thought that there are other "immediately recognisable" squares, it does not upset my contention so long as these squares can be found by some process of which my type of machine is capable. This idea is developed in III below.

The simple operations must therefore include:

(*a*) Changes of the symbol on one of the observed squares.

(*b*) Changes of one of the squares observed to another square within L squares of one of the previously observed squares.

It may be that some of these changes necessarily involve a change of state of mind. The most general single operation must therefore be taken to be one of the following:

(*A*) A possible change (*a*) of symbol together with a possible change of state of mind.

(*B*) A possible change (*b*) of observed squares, together with a possible change of state of mind.

The operation actually performed is determined by the state of mind of the computer and the observed symbols. In particular, they determine the state of mind of the computer after the operation is carried out.

We may now construct a machine to do the work of this computer. To each state of mind of the computer corresponds an "*m*-configuration" of the machine. The machine scans B squares corresponding to the B squares observed by the computer. In any move the machine can change a symbol on a scanned square or can change any one of the scanned squares to another square distant not more than L squares from one of the other scanned squares. The move which is done, and the succeeding configuration, are determined by the scanned symbol and the *m*-configuration. The machines just described do not differ very essentially from computing machines as defined in § 2, and corresponding to any machine of this type a computing machine can be constructed to compute the same sequence, that is to say the sequence computed by the computer.

II. [Type (*b*)].

If the notation of the Hilbert functional calculus[6] is modified so as to be systematic, and so as to involve only a finite number of symbols, it becomes possible to construct an automatic[7] machine $\mathcal{K}$, which will find all the provable formulae of the calculus.[8]

Now let α be a sequence, and let us denote by $G_\alpha(x)$ the proposition "The x-th figure of α is 1", so that[9] $-G_\alpha(x)$ means "The x-th figure of α is 0". Suppose further that we can find a set of properties which define the sequence α and which can be

---

[6] The expression "the functional calculus" is used throughout to mean the *restricted* Hilbert functional calculus.

[7] It is most natural to construct first a choice machine (§ 2) to do this. But it is then easy to construct the required automatic machine. We can suppose that the choices are always choices between two possibilities 0 and 1. Each proof will then be determined by a sequence of choices $i_1, i_2, \ldots, i_n$ ($i_1 = 0$ or 1, $i_2 = 0$ or 1, ... $i_n = 0$ or 1), and hence the number $2^n + i_1 2^{n-1} + i_2 2^{n-2} + \ldots + i_n$ completely determines the proof. The automatic machine carries out successively proof 1, proof 2, proof 3, ...

[8] The author has found a description of such a machine.

[9] The negation sign is written before an expression and not over it.

expressed in terms of $G_\alpha(x)$ and of the propositional functions $N(x)$ meaning "x is a non-negative integer" and $F(x, y)$ meaning "$y = x + 1$". When we join all these formulae together conjunctively, we shall have a formula, $\mathfrak{A}$ say, which defines α. The terms of $\mathfrak{A}$ must include the necessary parts of the Peano axioms, viz.,

$$(\exists u)N(u)\&(x)\big(N(x) \to (\exists y)\, F(x, y)\big) \,\&\, \big(F(x, y) \to N(y)\big),$$

which we will abbreviate to P.

When we say "$\mathfrak{A}$ defines α", we mean that $-\mathfrak{A}$ is not a provable formula, and also that, for each n, one of the following formulae (A_n) or (B_n) is provable.

$$\mathfrak{A} \,\&\, F^{(n)} \to G_\alpha(u^{(n)}), \qquad\qquad (A_n)[10]$$

$$\mathfrak{A} \,\&\, F^{(n)} \to \big(-G_\alpha(u^{(n)})\big), \qquad\qquad (B_n),$$

where $F^{(n)}$ stands for $F(u, u') \,\&\, F(u', u'') \,\&\, \ldots\, F(u^{(n-1)}, u^{(n)})$.

I say that α is then a computable sequence: a machine $\mathcal{K}_\alpha$ to compute α can be obtained by a fairly simple modification of $\mathcal{K}$.

We divide the motion of $\mathcal{K}_\alpha$ into sections. The n-th section is devoted to finding the n-th figure of α. After the $(n-1)$-th section is finished a double colon : : is printed after all the symbols, and the succeeding work is done wholly on the squares to the right of this double colon. The first step is to write the letter "A" followed by the formula (A_n) and then "B" followed by (B_n). The machine $\mathcal{K}_\alpha$ then starts to do the work of $\mathcal{K}$, but whenever a provable formula is found, this formula is compared with (A_n) and with (B_n). If it is the same formula as (A_n), then the figure "1" is printed, and the n-th section is finished. If it is (B_n), then "0" is printed and the section is finished. If it is different from both, then the work of $\mathcal{K}$ is continued from the point at which it had been abandoned. Sooner or later one of the formulae (A_n) or (B_n) is reached; this follows from our hypotheses about α and $\mathfrak{A}$, and the known nature of $\mathcal{K}$. Hence the n-th section will eventually be finished. $\mathcal{K}_\alpha$ is circle-free; α is computable.

It can also be shown that the numbers α definable in this way by the use of axioms include all the computable numbers. This is done by describing computing machines in terms of the function calculus.

It must be remembered that we have attached rather a special meaning to the phrase "$\mathfrak{A}$ defines α". The computable numbers do not include all (in the ordinary sense) definable numbers. Let δ be a sequence whose n-th figure is 1 or 0 according as n is or is not satisfactory. It is an immediate consequence of the theorem of § 8 that δ is not computable. It is (so far as we know at present) possible that any assigned number of figures of δ can be calculated, but not by a uniform process. When sufficiently many figures of δ have been calculated, an essentially new method is necessary in order to obtain more figures.

[10]A sequence of r primes is denoted by $^{(r)}$.

III. This may be regarded as a modification of I or as a corollary of II.

We suppose, as in I, that the computation is carried out on a tape; but we avoid introducing the "state of mind" by considering a more physical and definite counterpart of it. It is always possible for the computer to break off from his work, to go away and forget all about it, and later to come back and go on with it. If he does this he must leave a note of instructions (written in some standard form) explaining how the work is to be continued. This note is the counterpart of the "state of mind". We will suppose that the computer works in such a desultory manner that he never does more than one step at a sitting. The note of instructions must enable him to carry out one step and write the next note. Thus the state of progress of the computation at any stage is completely determined by the note of instructions and the symbols on the tape. That is, the state of the system may be described by a single expression (sequence of symbols), consisting of the symbols on the tape followed by Δ (which we suppose not to appear elsewhere) and then by the note of instructions. This expression may be called the "state formula". We know that the state formula at any given stage is determined by the state formula before the last step was made, and we assume that the relation of these two formulae is expressible in the functional calculus. In other words, we assume that there is an axiom $\mathfrak{A}$ which expresses the rules governing the behaviour of the computer, in terms of the relation of the state formula at any stage to the state formula at the preceding stage. If this is so, we can construct a machine to write down the successive state formulae, and hence to compute the required number.

10. EXAMPLES OF LARGE CLASSES OF NUMBERS WHICH ARE COMPUTABLE

It will be useful to begin with definitions of a computable function of an integral variable and of a computable variable, etc. There are many equivalent ways of defining a computable function of an integral variable. The simplest is, possibly, as follows. If γ is a computable sequence in which 0 appears infinitely[11] often, and n is an integer, then let us define $\xi(\gamma, n)$ to be the number of figures 1 between the n-th and the $(n + 1)$-th figure 0 in γ. Then $\phi(n)$ is computable if, for all n and some γ, $\phi(n) = \xi(\gamma, n)$. An equivalent definition is this. Let $H(x, y)$ mean $\phi(x) = y$. Then, if we can find a contradiction-free axiom $\mathfrak{A}_\phi$, such that $\mathfrak{A}_\phi \rightarrow P$, and if for each integer n there exists an integer N, such that

$$\mathfrak{A}_\phi \ \& \ F^{(N)} \rightarrow H(u^{(n)}, u^{(\phi(n))}),$$

[11] If $\mathcal{M}$ computes γ, then the problem whether $\mathcal{M}$ prints 0 infinitely often is of the same character as the problem whether $\mathcal{M}$ is circle-free.

and such that, if $m \neq \phi(n)$, then, for some N',
$$\mathfrak{A}_\phi \& F^{(N')} \to \left(-H(u^{(n)}, u^{(m)})\right),$$
then ϕ may be said to be a computable function.

We cannot define general computable functions of a real variable, since there is no general method of describing a real number, but we can define a computable function of a computable variable. If n is satisfactory, let γ_n be the number computed by $\mathcal{M}(n)$, and let
$$\alpha_n = \tan\left(\pi(\gamma_n - \tfrac{1}{2})\right),$$
unless $\gamma_n = 0$ or $\gamma_n = 1$, in either of which cases $\alpha_n = 0$. Then, as n runs through the satisfactory numbers, α_n runs through the computable numbers.[12] Now let $\phi(n)$ be a computable function which can be shown to be such that for any satisfactory argument its value is satisfactory.[13] Then the function f, defined by $f(\alpha_n) = \alpha_{\phi(n)}$, is a computable function and all computable functions of a computable variable are expressible in this form.

Similar definitions may be given of computable functions of several variables, computable-valued functions of an integral variable, etc.

I shall enunciate a number of theorems about computability, but I shall prove only (ii) and a theorem similar to (iii).

(i) A computable function of a computable function of an integral or computable variable is computable.

(ii) Any function of an integral variable defined recursively in terms of computable functions is computable. *I.e.* if $\phi(m, n)$ is computable, and r is some integer, then $\eta(n)$ is computable, where
$$\eta(0) = r,$$
$$\eta(n) = \phi\left(n, \eta(n - 1)\right).$$

(iii) If $\phi(m, n)$ is a computable function of two integral variables, then $\phi(n, n)$ is a computable function of n.

(iv) If $\phi(n)$ is a computable function whose value is always 0 or 1, then the sequence whose n-th figure is $\phi(n)$ is computable.

Dedekind's theorem does not hold in the ordinary form if we replace "real" throughout by "computable". But it holds in the following form:

(v) If $G(\alpha)$ is a propositional function of the computable numbers and
$$(a) \quad (\exists\alpha)(\exists\beta)\{G(\alpha) \& (-G(\beta))\},$$
$$(b) \quad G(\alpha) \& (-G(\beta)) \to (\alpha < \beta),$$

[12]A function α_n may be defined in many other ways so as to run through the computable numbers.
[13]Although it is not possible to find a general process for determining whether a given number is satisfactory, it is often possible to show that certain classes of numbers are satisfactory.

and there is a general process for determining the truth value of $G(\alpha)$, then there is a computable number ξ such that

$$G(\alpha) \to \alpha \leqslant \xi,$$
$$-G(\alpha) \to \alpha \geqslant \xi.$$

In other words, the theorem holds for any section of the computables such that there is a general process for determining to which class a given number belongs.

Owing to this restriction of Dedekind's theorem, we cannot say that a computable bounded increasing sequence of computable numbers has a computable limit. This may possibly be understood by considering a sequence such as

$$-1, -\tfrac{1}{2}, -\tfrac{1}{4}, -\tfrac{1}{8}, -\tfrac{1}{16}, \tfrac{1}{2}, \ldots .$$

On the other hand, (v) enables us to prove

(vi) If α and β are computable and $\alpha < \beta$ and $\phi(\alpha) < 0 < \phi(\beta)$, where $\phi(\alpha)$ is a computable increasing continuous function, then there is a unique computable number γ, satisfying $\alpha < \gamma < \beta$ and $\phi(\gamma) = 0$.

Computable Convergence

We shall say that a sequence β_n of computable numbers *converges computably* if there is a computable integral valued function $N(\epsilon)$ of the computable variable ϵ, such that we can show that, if $\epsilon > 0$ and $n > N(\epsilon)$ and $m > N(\epsilon)$, then $|\beta_n - \beta_m| < \epsilon$.

We can then show that

(vii) A power series whose coefficients form a computable sequence of computable numbers is computably convergent at all computable points in the interior of its interval of convergence.

(viii) The limit of a computably convergent sequence is computable.

And with the obvious definition of "uniformly computably convergent":

(ix) The limit of a uniformly computably convergent computable sequence of computable functions is a computable function. Hence

(x) The sum of a power series whose coefficients form a computable sequence is a computable function in the interior of its interval of convergence.

From (viii) and $\pi = 4(1 - \tfrac{1}{3} + \tfrac{1}{5} - \ldots)$ we deduce that p is computable.

From $e = 1 + 1 + \tfrac{1}{2!} + \tfrac{1}{3!} + \ldots$ we deduce that e is computable.

From (vi) we deduce that all real algebraic numbers are computable.

From (vi) and (x) we deduce that the real zeros of the Bessel functions are computable.

Proof of (ii).

Let $H(x, y)$ mean "$\eta(x) = y$", and let $K(x, y, z)$ mean "$\phi(x, y) = z$". $\mathfrak{A}_\phi$ is the axiom for $\phi(x, y)$. We take $\mathfrak{A}_\eta$ to be

$$\mathfrak{A}_\phi \ \& \ P \ \& \ \big(F(x, y) \to G(x, y)\big) \ \& \ \big(G(x, y) \ \& \ G(y, z) \to G(x, z)\big)$$
$$\& \ \big(F^{(r)} \to H(u, u^{(r)})\big) \ \& \ \big(F(v, w) \ \& \ H(v, x) \ \& \ K(w, x, z) \to H(w, z)\big)$$
$$\& \ \big[H(w, z) \ \& \ G(z, t) \ \mathrm{v} \ G(t, z) \to \big(-H(w, t)\big)\big].$$

I shall not give the proof of consistency of $\mathfrak{A}_\eta$. Such a proof may be constructed by the methods used in Hilbert and Bernays, *Grundlagen der Mathematik* (Berlin, 1934). The consistency is also clear from the meaning.

Suppose that, for some n, N, we have shown

$$\mathfrak{A}_\eta \ \& \ F^{(N)} \to H(u^{(n-1)}, u^{(\eta(n-1))}),$$

then, for some M,

$$\mathfrak{A}_\phi \ \& \ F^{(M)} \to K(u^{(n)}, u^{(\eta(n-1))}, u^{(\eta(n))}),$$
$$\mathfrak{A}_\eta \ \& \ F^{(M)} \to F(u^{(n-1)}, u^{(n)}) \ \& \ H(u^{(n-1)}, u^{(\eta(n-1))})$$
$$\& \ K(u^{(n)}, u^{(\eta(n-1))}, u^{(\eta(n))}),$$

and

$$\mathfrak{A}_\eta \ \& \ F^{(M)} \to \big[F(u^{(n-1)}, u^{(n)}) \ \& \ H(u^{(n-1)}, u^{(\eta(n-1))})$$
$$\& \ K(u^{(n)}, u^{(\eta(n-1))}, u^{(\eta(n))}) \to H(u^{(n)}, u^{(\eta(n))})\big].$$

Hence
$$\mathfrak{A}_\eta \ \& \ F^{(M)} \to H(u^{(n)}, u^{(\eta(n))}).$$

Also
$$\mathfrak{A}_\eta \ \& \ F^{(r)} \to H(u, u^{(\eta(0))}).$$

Hence for each n some formula of the form

$$\mathfrak{A}_\eta \ \& \ F^{(M)} \to H(u^{(n)}, u^{(\eta(n))})$$

is provable. Also, if $M' \geqslant M$ and $M' \geqslant m$ and $m \neq \eta(u)$, then

$$\mathfrak{A}_\eta \ \& \ F^{(M')} \to G(u^{\eta((n))}, u^{(m)}) \ \mathrm{v} \ G(u^{(m)}, u^{(\eta(n))})$$

and

$$\mathfrak{A}_\eta \ \& \ F^{(M')} \to \big[\{G\big(u^{(\eta(n))}, u^{(m)}\big) \ \mathrm{v} \ G\big(u^{(m)}, u^{(\eta(n))}\big)$$
$$\& \ H\big(u^{(n)}, u^{(\eta(n))}\big)\} \to \big(-H\big(u^{(n)}, u^{(m)}\big)\big)\big].$$

Hence
$$\mathfrak{A}_\eta \ \& \ F^{(M')} \to \big(-H(u^{(n)}, u^{(m)})\big).$$

The conditions of our second definition of a computable function are therefore satisfied. Consequently η is a computable function.

Proof of a modified form of (iii).

Suppose that we are given a machine $\mathfrak{N}$, which, starting with a tape bearing on it ǝ ǝ followed by a sequence of any number of letters "F" on F-squares and in the m-configuration b, will compute a sequence γ_n depending on the number n of letters "F". If $\phi_n(m)$ is the m-th figure of γ_n, then the sequence β whose n-th figure is $\phi_n(n)$ is computable.

We suppose that the table for $\mathfrak{N}$ has been written out in such a way that in each line only one operation appears in the operations column. We also suppose that

Ξ, Θ, $\bar{0}$, and $\bar{1}$ do not occur in the table, and we replace $\ni$ throughout by Θ, 0 by $\bar{0}$, and 1 by $\bar{1}$. Further substitutions are then made. Any line of form

$\mathfrak{A}$	α	$P\bar{0}$	$\mathfrak{B}$

we replace by

$\mathfrak{A}$	α	$P\bar{0}$	$\mathfrak{re}(\mathfrak{B}, \mathfrak{u}, h, k)$

and any line of the form

$\mathfrak{A}$	α	$P\bar{1}$	$\mathfrak{B}$

by

$\mathfrak{A}$	α	$P\bar{1}$	$\mathfrak{re}(\mathfrak{B}, \mathfrak{v}, h, k)$

and we add to the table the following lines:

$\mathfrak{u}$		$\mathfrak{pe}(\mathfrak{u}_1, 0)$
$\mathfrak{u}_1$	$R, Pk, R, P\Theta, R, P\Theta$	$\mathfrak{u}_2$
$\mathfrak{u}_2$		$\mathfrak{re}(\mathfrak{u}_3, \mathfrak{u}_3, k, h)$
$\mathfrak{u}_3$		$\mathfrak{pe}(\mathfrak{u}_2, F)$

and similar lines with $\mathfrak{v}$ for $\mathfrak{u}$ and 1 for 0 together with the following line

$\mathfrak{c}$	$R, P\Xi, R, Ph$	$\mathfrak{b}.$

We then have the table for the machine $\mathcal{N}'$ which computes β. The initial m-configuration is $\mathfrak{c}$, and the initial scanned symbol is the second $\ni$.

11. APPLICATION TO THE ENTSCHEIDUNGSPROBLEM

The results of § 8 have some important applications. In particular, they can be used to show that the Hilbert Entscheidungsproblem can have no solution. For the present I shall confine myself to proving this particular theorem. For the formulation of this problem I must refer the reader to Hilbert and Ackermann's *Grundzüge der Theoretischen Logik* (Berlin, 1931), chapter 3.

I propose, therefore, to show that there can be no general process for determining whether a given formula $\mathfrak{A}$ of the functional calculus K is provable, *i.e.* that there can be no machine which, supplied with any one $\mathfrak{A}$ of these formulae, will eventually say whether $\mathfrak{A}$ is provable.

It should perhaps be remarked that what I shall prove is quite different from the well-known results of Gödel.[14] Gödel has shown that (in the formalism of Principia Mathematica) there are propositions $\mathfrak{A}$ such that neither $\mathfrak{A}$ nor $-\mathfrak{A}$ is provable. As a consequence of this, it is shown that no proof of consistency of Principia Mathematica (or of K) can be given within that formalism. On the other hand, I shall show that there is no general method which tells whether a given formula $\mathfrak{A}$ is provable in K, or, what comes to the same, whether the system consisting of K with $-\mathfrak{A}$ adjoined as an extra axiom is consistent.

[14]*Loc. cit.*

If the negation of what Gödel has shown had been proved, *i.e.* if, for each $\mathfrak{A}$, either $\mathfrak{A}$ or $-\mathfrak{A}$ is provable, then we should have an immediate solution of the Entscheidungsproblem. For we can invent a machine $\mathcal{K}$ which will prove consecutively all provable formulae. Sooner or later $\mathcal{K}$ will reach either $\mathfrak{A}$ or $-\mathfrak{A}$. If it reaches $\mathfrak{A}$, then we know that $\mathfrak{A}$ is provable. If it reaches $-\mathfrak{A}$, then, since K is consistent (Hilbert and Ackermann), we know that $\mathfrak{A}$ is not provable.

Owing to the absence of integers in K the proofs appear somewhat lengthy. The underlying ideas are quite straightforward.

Corresponding to each computing machine $\mathcal{M}$ we construct a formula Un ($\mathcal{M}$) and we show that, if there is a general method for determining whether Un ($\mathcal{M}$) is provable, then there is a general method for determining whether $\mathcal{M}$ ever prints 0.

The interpretations of the propositional functions involved are as follows:

$R_{Si}(x, y)$ is to be interpreted as "in the complete configuration x (of $\mathcal{M}$) the symbol on the square y is S".

$I(x, y)$ is to be interpreted as "in the complete configuration x the square y is scanned".

$K_{qm}(x)$ is to be interpreted as "in the complete configuration x the m-configuration is q_m.

$F(x, y)$ is to be interpreted as "y is the immediate successor of x".

Inst $\{q_i\, S_j\, S_k\, Lq_l\}$ is to be an abbreviation for

$$(x, y, x', y')\, \{(R_{s_j}(x, y)\, \&\, I(x, y)\, \&\, K_{q_i}(x)\, \&\, F(x, x')\, \&\, F(y', y))$$
$$\rightarrow (I(x', y')\, \&\, R_{s_k}(x', y)\, \&\, K_{q_l}(x')$$
$$\&\, (z)\left[F(y', z)\, v\, (R_{Sj}(x, z)\rightarrow R_{S_k}(x', z))\right])]\}.$$

Inst $\{q_iS_jS_kRq_l\}$ and   Inst $\{q_iS_jS_kNq_l\}$

are to be abbreviations for other similarly constructed expressions.

Let us put the description of $\mathcal{M}$ into the first standard form of § 6. This description consists of a number of expressions such as "$q_iS_jS_kLq_l$" (or with R or N substituted for L). Let us form all the corresponding expressions such as Inst $\{q_i\, S_j\, S_k\, Lq_l\}$ and take their logical sum. This we call Des ($\mathcal{M}$).

The formula Un ($\mathcal{M}$) is to be

$$(\exists u)\left[N(u)\, \&\, (x)\left(N(x)\rightarrow(\exists x')\, F(x, x')\right)\right.$$
$$\&\, (y, z)\left(F(y, z)\rightarrow N(y)\, \&\, N(z)\right)\, \&\, (y)\, R_{S_0}(u, y)$$
$$\&\, I(u, u)\, \&\, K_{q_1}(u)\, \&\, \text{Des}(\mathcal{M})]$$
$$\rightarrow(\exists s)(\exists t)\left[N(s)\, \&\, N(t)\, \&\, R_{S_1}(s, t)\right].$$

$[N(u)\, \&\, \ldots\, \&\, \text{Des}(\mathcal{M})]$ may be abbreviated to $A(\mathcal{M})$.

When we substitute the meanings suggested above, we find that Un ($\mathcal{M}$) has the interpretation "in some complete configuration of $\mathcal{M}$, S_1 (*i.e.* 0) appears on the tape". Corresponding to this I prove that

(a) If S_1 appears on the tape in some complete configuration of $\mathcal{M}$, then $\mathrm{Un}(\mathcal{M})$ is provable.

(b) If $\mathrm{Un}(\mathcal{M})$ is provable, then S_1 appears on the tape in some complete configuration of $\mathcal{M}$.

When this has been done, the remainder of the theorem is trivial.

Lemma 1. If S_1 Appears on the Tape in Some Complete Configuration of $\mathcal{M}$, then Un $(\mathcal{M})$ is Provable.

We have to show how to prove Un $(\mathcal{M})$. Let us suppose that in the n-th complete configuration the sequence of symbols on the tape is $S_{r(n,\,0)}$, $S_{r(n,\,1)}$, . . . , $S_{r(n,\,n)}$, followed by nothing but blanks, and that the scanned symbol is the $i\,(n)$-th, and that the m-configuration is $q_{k(n)}$. Then we may form the proposition

$$R_{S_{r(n,0)}}(u^{(n)},\,u)\ \&\ R_{S_{r(n,1)}}(u^{(n)},\,u')\ \&\ .\,.\,.\ \&\ R_{S_{r(n,ni)}}(u^{(n)},\,u^{(n)})$$
$$\&\ I\,(u^{(n)},\,u^{(i(n))})\ \&\ K_{qk(n)}(u^{(n)})$$
$$\&\ (y)\,F\big((y,\,u')\ \mathrm{v}\ F(u,\,y)\ \mathrm{v}\ F(u',\,y)\ \mathrm{v}\ .\,.\,.\ \mathrm{v}\ F(u^{(n-1)},\,y)\,\mathrm{v}\,R_{S_0}(u^{(n)},\,y)\big),$$

which we may abbreviate to CC_n.

As before, $F(u,\,u')\ \&\ F(u',\,u'')\ \&\ .\,.\,.\ F(u^{(r-1)},\,u^{(r)})$ is abbreviated to $F^{(r)}$.

I shall show that all formulae of the form $A(\mathcal{M})\ \&\ F^{(n)} \rightarrow CCn$ (abbreviated to CF_n) are provable. The meaning of CF_n is "The n-th complete configuration of $\mathcal{M}$ is so and so", where "so and so" stands for the actual n-th complete configuration of $\mathcal{M}$. That CF_n should be provable is therefore to be expected.

CF_0 is certainly provable, for in the complete configuration the symbols are all blanks, the m-configuration is q_1, and the scanned square is u, i.e. CC_0 is

$$(y)\,R_{S_0}(u,\,y)\ \&\ I\,(u,\,u)\ \&\ K_{q1}(u).$$

$A(\mathcal{M}) \rightarrow CC_0$ is then trivial.

We next show that $CF_n \rightarrow CF_{n+1}$ is provable for each n. There are three cases to consider, according as in the move from the n-th to the $(n+1)$-th configuration the machine moves to left or to right or remains stationary. We suppose that the first case applies, i.e. the machine moves to the left. A similar argument applies in the other cases. If $r\big(n,\,i(n)\big) = a$, $r\big(n+1,\,i(n+1)\big) = c$, $k\big(i(n)\big) = b$, and $k\big(i(n+1)\big) = d$, then Des $(\mathcal{M})$ must include-Inst $\{q_a\,S_b\,S_d\,Lq_c\}$ as one of its terms, i.e.

$$\mathrm{Des}(\mathcal{M})\ \rightarrow\ \mathrm{Inst}\ \{q_a\,S_b\,S_d\,Lq_c\}.$$

Hence $\qquad A(\mathcal{M})\ \&\ F^{(n+1)} \rightarrow \mathrm{Inst}\ \{q_a S_b S_d Lq_c\}\ \&\ F^{(n+1)}.$

But $\qquad \mathrm{Inst}\ \{q_a S_b S_d Lq_c\}\ \&\ F^{(n+1)} \rightarrow (CC_n \rightarrow CC_{n+1})$

is provable, and so therefore is

$$A(\mathcal{M})\ \&\ F^{(n+1)} \rightarrow (CC_n \rightarrow CC_{n+1})$$

and
$$\left(A(\mathcal{M}) \ \& \ F^{(n)} \to CC_n\right) \to \left(A \ (\mathcal{M}) \ \& \ F^{(n+1)} \to CC_{n+1}\right),$$

i.e.
$$CF_n \to CF_{n+1}.$$

CF_n is provable for each n. Now it is the assumption of this lemma that S_1 appears somewhere, in some complete configuration, in the sequence of symbols printed by $\mathcal{M}$; that is, for some integers N, K, CC_N has $R_{S_1}(u^{(N)}, u^{(K)})$ as one of its terms, and therefore $CC_N \to R_{S_1}(u^{(N)}, u^{(K)})$ is provable. We have then
$$CC_N \to R_{S_1}(u^{(N)}, u^{(K)})$$

and
$$A(\mathcal{M}) \ \& \ F^{(N)} \to CC^N.$$

We also have
$$(\exists u)A \ (\mathcal{M}) \to (\exists u)(\exists u') \ldots (\exists u^{(N')})\left(A(\mathcal{M}) \ \& \ F^{(N)}\right),$$

where $N' = \max(N, K)$. And so
$$(\exists u)A \ (\mathcal{M}) \to (\exists u)(\exists u') \ldots (\exists u^{(N)})R_{S_1}(u^{(N)}, u^{(K)}),$$
$$(\exists u)A \ (\mathcal{M}) \to (\exists u^{(N)})(\exists u^{(K)})R_{S_1}(u^{(N)}, u^{(K)}),$$
$$(\exists u)A \ (\mathcal{M}) \to (\exists s)(\exists t)R_{S_1}(s, t),$$

i.e. $\mathrm{Un}(\mathcal{M})$ is provable.

This completes the proof of Lemma 1.

Lemma 2. If $\mathrm{Un}(\mathcal{M})$ is Provable, then S_1 Appears on the Tape in some Complete Configuration of $\mathcal{M}$.

If we substitute any propositional functions for function variables in a provable formula, we obtain a true proposition. In particular, if we substitute the meanings tabulated on p. 1153 in $\mathrm{Un}(\mathcal{M})$, we obtain a true proposition with the meaning "S_1 appears somewhere on the tape in some complete configuration of $\mathcal{M}$".

We are now in a position to show that the Entscheidungsproblem cannot be solved. Let us suppose the contrary. Then there is a general (mechanical) process for determining whether $\mathrm{Un}(\mathcal{M})$ is provable. By Lemmas 1 and 2, this implies that there is a process for determining whether $\mathcal{M}$ ever prints 0, and this is impossible, by § 8. Hence the Entscheidungsproblem cannot be solved.

In view of the large number of particular cases of solutions of the Entscheidungsproblem for formulae with restricted systems of quantors, it is interesting to express $\mathrm{Un}(\mathcal{M})$ in a form in which all quantors are at the beginning. $\mathrm{Un}(\mathcal{M})$ is, in fact, expressible in the form
$$(u)(\exists x)(w)(\exists u_1) \ldots (\exists u_n)\mathfrak{B}, \tag{I}$$

where $\mathfrak{B}$ contains no quantors, and $n = 6$. By unimportant modifications we can obtain a formula, with all essential properties of $\mathrm{Un}(\mathcal{M})$, which is of form (I) with $n = 5$.

Added 28 *August*, 1936.

APPENDIX

COMPUTABILITY AND EFFECTIVE CALCULABILITY

The theorem that all effectively calculable (λ-definable) sequences are computable and its converse are proved below in outline. It is assumed that the terms "well-formed formula" (W.F.F.) and "conversion" as used by Church and Kleene are understood. In the second of these proofs the existence of several formulae is assumed without proof; these formulae may be constructed straightforwardly with the help of, *e.g.*, the results of Kleene in "A theory of positive integers in formal logic", *American Journal of Math.*, 57 (1935).

The W.F.F. representing an integer n will be denoted by N_n. We shall say that a sequence γ whose n-th figure is $\phi_\gamma(n)$ is λ-definable or effectively calculable if $1 + \phi_\gamma(u)$ is a λ-definable function of n, *i.e.* if there is a W.F.F. M_γ such that, for all integers n,

$$\{M_\gamma\}\,(N_n)\ \text{conv}\ N_{\phi_\gamma(n)+1},$$

i.e. $\{M_\gamma\}\,(N_n)$ is convertible into $\lambda xy' \cdot x\big(x(y)\big)$ or into $\lambda xy \cdot x(y)$ according as the n-th figure of λ is 1 or 0.

To show that every λ-definable sequence γ is computable, we have to show how to construct a machine to compute γ. For use with machines it is convenient to make a trivial modification in the calculus of conversion. This alteration consists in using $x, x', x'', \ldots$ as variables instead of $a, b, c, \ldots$. We now construct a machine $\mathscr{L}$ which, when supplied with the formula M_γ, writes down the sequence γ. The construction of $\mathscr{L}$ is somewhat similar to that of the machine $\mathscr{K}$ which proves all provable formulae of the functional calculus. We first construct a choice machine $\mathscr{L}_1$, which, if supplied with a W.F.F., M say, and suitably manipulated, obtains any formula into which M is convertible. $\mathscr{L}_1$ can then be modified so as to yield an automatic machine $\mathscr{L}_2$ which obtains successively all the formulae into which M is convertible. The machine $\mathscr{L}$ includes $\mathscr{L}_2$ as a part. The motion of the machine $\mathscr{L}$ when supplied with the formula M_γ is divided into sections of which the n-th is devoted to finding the n-th figure of γ. The first stage in this n-th section is the formation of $\{M_\gamma\}\,(N_n)$. This formula is then supplied to the machine $\mathscr{L}_2$, which converts it successively into various other formulae. Each formula into which it is convertible eventually appears, and each, as it is found, is compared with

$$\lambda x\big[\lambda x'\,[\{x\}(\{x\}(x'))]\big],\ \textit{i.e.}\ N_2,$$

and with

$$\lambda x\big[\lambda x'[\{x\}(x')]\big],\ \textit{i.e.}\ N_1.$$

If it is identical with the first of these, then the machine prints the figure 1 and the n-th section is finished. If it is identical with the second, then 0 is printed and the section is finished. If it is different from both, then the work of $\mathcal{L}_2$ is resumed. By hypothesis, $\{M_\gamma\}(N_n)$ is convertible into one of the formulae N_2 or N_1; consequently the n-th section will eventually be finished, *i.e.* the n-th figure of γ will eventually be written down.

To prove that every computable sequence γ is λ-definable, we must show how to find a formula M_γ such that, for all integers n,
$$\{M_\gamma\}(N_n) \text{ conv } N_{1+\phi_\gamma(n)}.$$

Let $\mathcal{M}$ be a machine which computes γ and let us take some description of the complete configurations of $\mathcal{M}$ by means of numbers, *e.g.* we may take the D.N of the complete configuration as described in § 6. Let $\xi(n)$ be the D.N of the n-th complete configuration of $\mathcal{M}$. The table for the machine $\mathcal{M}$ gives us a relation between $\xi(n+1)$ and $\xi(n)$ of the form
$$\xi(n+1) = \rho_\gamma(\xi(n)),$$

where ρ_γ is a function of very restricted, although not usually very simple, form: it is determined by the table for $\mathcal{M}$. ρ_γ is λ-definable (I omit the proof of this), *i.e.* there is a W.F.F. A_γ such that, for all integers n,
$$\{A_\gamma\}(N_{\xi(n)}) \text{ conv } N_{\xi(n+1)}.$$

Let U stand for
$$\lambda u\left[\{\{u\}(A_\gamma)\}(N_r)\right],$$

where $r = \xi(0)$; then, for all integers n,
$$\{U_\gamma\}(N_n) \text{ conv } N_{\xi(n)}.$$

It may be proved that there is a formula V such that

$$\{\{V\}(N_{\xi(n+1)})\}(N_{\xi(n)}) \begin{cases} \text{conv } N_1 & \text{if, in going from the } n\text{-th to the } (n+1)\text{-th} \\ & \text{complete configuration, the figure } 0 \text{ is} \\ & \text{printed.} \\ \text{conv } N_2 & \text{if the figure } 1 \text{ is printed.} \\ \text{conv } N_3 & \text{otherwise.} \end{cases}$$

Let W_γ stand for
$$\lambda u\left[\left\{\{V\}\left(\{A_\gamma\}(\{U_\gamma\}(u))\right)\right\}(\{U_\gamma\}(u))\right],$$

so that, for each integer n,
$$\{\{V\}(N_{\xi(n+1)})\}(N_{\xi(n)}) \text{ conv } \{W_\gamma\}(N_n),$$

and let Q be a formula such that
$$\{\{Q\}(W_\gamma)\}(N_s) \text{ conv } N_{r(z)},$$

where $r(s)$ is the s-th integer q for which $\{W_\gamma\}(N_q)$ is convertible into either N_1 or N_2. Then, if M_γ stands for
$$\lambda w\left[\{W_\gamma\}(\{\{Q\}(W_\gamma)\}(w))\right],$$

it will have the required property.[1]

The Graduate College,

Princeton University,

New Jersey, U.S.A.

[1]In a complete proof of the λ-definability of computable sequences it would be best to modify this method by replacing the numerical description of the complete configurations by a description which can be handled more easily with our apparatus. Let us choose certain integers to represent the symbols and the m-configurations of the machine. Suppose that in a certain complete configuration the numbers representing the successive symbols on the tape are $s_1 s_2 \ldots s_n$, that the m-th symbol is scanned, and that the m-configuration has the number t; then we may represent this complete configuration by the formula

$$[[N_{h_1}, N_{h_2}, \ldots, N_{h_{m-1}}], [N_t, N_{rm}], [N_{h_{m+1}}, \ldots, N_{h_n}]],$$

where
$$[a, b] \text{ stands for } \lambda u \left[\{\{u\}(a)\}(b)\right],$$
$$[a, b, c] \text{ stands for } \lambda u \left[\{\{\{u\}(a)\}(b)\}(c)\right],$$

etc.

ON COMPUTABLE NUMBERS, WITH AN APPLICATION TO THE ENTSCHEIDUNGSPROBLEM. A CORRECTION

In a paper entitled "On computable numbers, with an application to the Entscheidungsproblem"[1] the author gave a proof of the insolubility of the Entscheidungsproblem of the "engere Funktionenkalkül". This proof contained some formal errors[2] which will be corrected here: there are also some other statements in the same paper which should be modified, although they are not actually false as they stand.

The expression for Inst $\{q_i \, S_j \, S_k \, Lq_l\}$ on p. 1154 of the paper quoted should read

$$(x, y, x', y') \Big\{ \big(R_{S_j}(x, y) \,\&\, I(x, y) \,\&\, K_{qi}(x) \,\&\, F(x, x') \,\&\, F(y', y)\big)$$

$$\rightarrow \Big(I(x', y') \,\&\, R_{Sk}(x', y) \,\&\, K_{ql}(x') \,\&\, F(y', z) \,\mathrm{v}\, \big[\big(R_{S_0}(x, z) \rightarrow R_{S_0}(x', z)\big)$$

$$\,\&\, \big(R_{S_1}(x, z) \rightarrow R_{S_1}(x', z)\big) \,\&\, \ldots \,\&\, \big(R_{S_u}(x, z) \rightarrow R_{S_u}(x', z)\big)\big]\Big)\Big\},$$

[1] *Proc. London Math. Soc.* (2), 42 (1936–7).
[2] The author is indebted to P. Bernays for pointing out these errors.

$S_0, S_1, \ldots, S_M$ being the symbols which $\mathcal{M}$ can print. The statement on p. 1155, line 33, viz.

$$\text{``Inst}\{q_a\, S_b\, S_d\, Lq_c\}\ \&\ F^{(n+1)} \rightarrow (CC_n \rightarrow CC_{n+1})$$

is provable" is false (even with the new expression for Inst $\{q_a\, S_b\, S_d\, Lq_c\}$) : we are unable for example to deduce $F^{(n+1)} \rightarrow (-F(u, u''))$ and therefore can never use the term

$$F(y', z)\ \text{v}\ \big[(R_{S_0}(x, z) \rightarrow R_{S_0}(x', z))\ \&\ \ldots\ \&\ (R_{S_M}(x, z) \rightarrow R_{S_M}(x', z))\big]$$

in Inst $\{q_a\, S_b\, S_d\, Lq_c\}$. To correct this we introduce a new functional variable $G\,[\,G(x, y)$ to have the interpretation "x precedes y"]. Then, if Q is an abbreviation for

$$(x)(\exists w)(y, z)\{F(x, w)\ \&\ (F(x, y) \rightarrow G(x, y))\ \&\ (F(x, z)\ \&\ G(z, y) \rightarrow G(x, y))$$
$$\&\ \big[G(z, x)\ \text{v}\ (G(x, y)\ \&\ F(y, z))\ \text{v}\ (F(x, y)\ \&\ F(z, y)) \rightarrow (-F(x, z))\big]$$

the corrected formula Un $(\mathcal{M})$ is to be

$$(\exists u) A(\mathcal{M}) \rightarrow (\exists s)(\exists t)\, R_{S_1}(s, t),$$

where $A(\mathcal{M})$ is an abbreviation for

$$Q\ \&\ (y)\ R_{S_0}(u, y)\ \&\ I(u, u)\ \&\ K_{q_1}(u)\ \&\ \text{Des}(\mathcal{M}).$$

The statement on page 1155 (line 33) must then read

$$\text{Inst}\{q_a\, S_b\, S_d\, Lq_c\}\ \&\ Q\ \&\ F^{(n+1)} \rightarrow (CC_n \rightarrow CC_{n+1}),$$

and line 29 should read

$$r(n, i(n)) = b, \quad r(n + 1, i(n)) = d, \quad k(n) = a, \quad k(n + 1) = c.$$

For the words "logical sum" on p. 1153, line 29, read "conjunction". With these modifications the proof is correct. Un $(\mathcal{M})$ may be put in the form (I) (p. 1155) with $n = 4$.

Some difficulty arises from the particular manner in which "computable number" was defined (p. 1130). If the computable numbers are to satisfy intuitive requirements we should have:

If we can give a rule which associates with each positive integer n two rationals a_n, b_n satisfying $a_n \leqslant a_{n+1} < b_{n+1} \leqslant b_n$, $b_n - a_n < 2^{-n}$, then there is a computable number a for which $a_n \leqslant a \leqslant b_n$ each n. (A)

A proof of this may be given, valid by ordinary mathematical standards, but involving an application of the principle of excluded middle. On the other hand the following is false:

There is a rule whereby, given the rule of formation of the sequences a_n, b_n in (A) we can obtain a D.N. for a machine to compute α. (B)

That (B) is false, at least if we adopt the convention that the decimals of numbers of the form $m/2^n$ shall always terminate with zeros, can be seen in this way. Let $\mathfrak{N}$ be some machine, and define c_n as follows: $c_n = \frac{1}{2}$ if $\mathfrak{N}$ has not printed a figure 0 by the time the n-th complete configuration is reached $c_n = \frac{1}{2} - 2^{-m-3}$ if 0 had first been printed at the m-th complete configuration ($m \leqslant n$). Put

$a_n = c_n - 2^{-n-2}$, $b_n = c_n + 2^{-n-2}$. Then the inequalities of (A) are satisfied, and the first figure of α is 0 if $\mathfrak{N}$ ever prints 0 and 1 otherwise. If (B) were true we should have a means of finding the first figure of α given the D.N. of $\mathfrak{N}$: *i.e.* we should be able to determine whether $\mathfrak{N}$ ever prints 0, contrary to the results of § 8 of the paper quoted. Thus although (A) shows that there must be machines which compute the Euler constant (for example) we cannot at present describe any such machine, for we do not yet know whether the Euler constant is of the form $m/2^n$.

This disagreeable situation can be avoided by modifying the manner in which computable numbers are associated with computable sequences, the totality of computable numbers being left unaltered. It may be done in many ways[3] of which this is an example. Suppose that the first figure of a computable sequence γ is i and that this is followed by 1 repeated n times, then by 0 and finally by the sequence whose r-th figure is c_r; then the sequence γ is to correspond to the real number

$$(2i - 1)n + \sum_{r=1}^{\infty} (2c_r - 1)\left(\frac{2}{3}\right)^r.$$

If the machine which computes γ is regarded as computing also this real number then (B) holds. The uniqueness of representation of real numbers by sequences of figures is now lost, but this is of little theoretical importance, since the D.N.'s are not unique in any case.

The Graduate College,
Princeton, N.J., U.S.A.

---

[3] This use of overlapping intervals for the definition of real numbers is due originally to Brouwer.

Acknowledgments

This book would not have been possible without the help of a number of talented people who made different contributions at various stages of the book's development. Among those deserving special thanks are Michael Rosin, a consultant to Running Press, Leonard Mlodinow, Mrs. Karen Sime, assistant to Professor Stephen Hawking, John Anders and Michael Ansaldi for their fantastic translation skills, and John Probst and his fantastic team at Techbooks.

Thanks are also due to current and past employees of Running Press: Deborah Grandinetti, Diana von Glahn, Kathleen Greczylo, Julia Ludwig, and Nicole Smith.

STEPHEN HAWKING

Stephen Hawking is considered the most brilliant theoretical physicist since Einstein. He has also done much to popularize science. His book, *A Brief History of Time*, has sold more than 10 million copies in 40 languages achieving the kind of success almost unheard of in the history of science writing. His subsequent books, *The Universe in A Nutshell*, and *The Future of Spacetime*, have also been well-received.

He was born in Oxford, England on January 8, 1942 (300 years after the death of Galileo). He studied physics at University College, Oxford, received his Ph.D. in Cosmology at Cambridge and since 1979, has held the post of Lucasian Professor of mathematics. The chair was founded in 1663 by Sir Isaac Newton, and is reserved for those individuals considered the most brilliant thinkers of their time.

Professor Hawking has worked on the basic laws that govern the universe. With Roger Penrose, he showed that Einstein's General Theory of Relativity implied space and time would have a beginning in the Big Bang and an end in black holes. The results indicated it was necessary to unify General Relativity with Quantum, Theory the other great scientific development of the first half of the twentieth century. One consequence of such a unification that he discovered was that black holes should not be completely black but should emit radiation and eventually disappear. Another conjecture is that the universe has no edge or boundary in imaginary time.

Stephen Hawking has twelve honorary degrees, and is the recipient of many awards, medals and prizes. He is a Fellow of the Royal Society and a Member of the US National Academy of Sciences. He continues to combine family life (he has three children and one grandchild) and his research into theoretical physics together with an extensive program of travel and public lectures.